Energy Principles and Variational Methods in Applied Mechanics

Energy Principles and Variational Methods in Applied Mechanics

Third Edition

J. N. Reddy
Texas A&M University, USA

WILEY

Registered Offices
John Wiley & Sons Inc., 111 River Street, Hoboken, NJ07030, USA
John Wiley & Sons Ltd, The Atrium, Southern Gate, Chichester, West Sussex, PO19 8SQ, UK

Editorial Office
The Atrium, Southern Gate, Chichester, West Sussex, PO19 8SQ, UK

For details of our global editorial offices, customer services, and more information about Wiley products visit us at www.wiley.com.

Wiley also publishes its books in a variety of electronic formats and by print-on-demand. Some content that appears in standard print versions of this book may not be available in other formats.

Library of Congress Cataloging-in-Publication Data applied for

Hardback: 9781119087373

Cover design by Wiley

Cover images: (Top) © inakiantonana/Gettyimages; (Bottom) © VII-photo/Gettyimages

Set in 11/13.5pt, Computer Modern by SPi Global, Chennai, India

To

The Red, White, and Blue
(The United States of America)
The Land of Opportunities and Freedom
with sincere gratitude and appreciation

Contents

About the Author

J. N. Reddy is a university distinguished professor, regents' professor, and the holder of Oscar S. Wyatt Endowed Chair in the Department of Mechanical Engineering at Texas A&M University. He is known worldwide for his significant contributions to the field of applied mechanics through the authorship of widely used textbooks on the linear and nonlinear finite element analysis, variational methods, composite materials and structures, applied functional analysis, and continuum mechanics.

Professor Reddy's earlier research focused primarily on mathematics of finite elements, variational principles of mechanics, shear deformation and layerwise theories of laminated composite plates and shells, analysis of bimodular materials, modeling of geological and geophysical phenomena, penalty finite elements for flows of viscous incompressible fluids, and least-squares finite element models of fluid flows and solid continua. His pioneering works on the development of shear deformation theories (that bear his name in the literature as the Reddy third-order plate theory and the Reddy layerwise theory) have had a major impact and have led to new research developments and applications. The finite element formulations and models he developed have been implemented in commercial software like Abaqus, NISA, and HyperXtrude. In recent years, his research involved the development of 7- and 12-parameter shell theories, non-local and non-classical continuum mechanics problems, and problems involving couple stresses, surface stress effects, discrete fracture and flow, micropolar cohesive damage, and continuum plasticity of metals from considerations of non-equilibrium thermodynamics.

His most significant awards and honors are: the *Worcester Reed Warner Medal* and *Charles Russ Richards Memorial Award* from the American Society of Mechanical Engineers; *Archie Higdon Distinguished Educator Award* from the Mechanics Division of the American Society for Engineering Education; *Nathan M. Newmark Medal* and *Raymond D. Mindlin Medal* from the American Society of Civil Engineers; *Excellence in the Field of Composites* and *Distinguished Research Award* from the American Society for Composites; *Computational Solid Mechanics* award from the US Association for Computational Mechanics; *The IACM O.C. Zienkiewicz Award* from the International Association of Computational Mechanics; *William Prager Medal* from the Society of Engineering Science; and *ASME Medal* from the American Society of Mechanical Engineers. He is a member of the US National Academy of Engineering and a foreign fellow of the Indian National Academy of Engineering. He is a fellow of all professional societies of his research (e.g., AIAA, ASC, ASCE, ASME, AAM, USACM, IACM).

Professor Reddy is one of the original top 100 ISI Highly Cited Researchers in Engineering around the world, with over 21,000 citations and an h-index of over 68 per Web of Science; the number of citations is over 52,200 with an h-index of 92 according to Google Scholar. He is the founding editor in chief of *Mechanics of Advanced Materials and Structures* and *International Journal for Computational Methods in Engineering Science and Mechanics* and coeditor of *International Journal of Structural Stability and Dynamics*; he also serves on the editorial boards of more than two dozen journals. A more complete resume with links to journal papers can be found at **http://mechanics.tamu.edu**.

"What we have done for ourselves alone dies with us; what we have done for others and the world, remains and is immortal." – *Albert Pike* (American attorney, soldier, and writer)

Note: Quotes by various people included in this book were found at different web sites. For example, visit:

http://naturalscience.com/dsqhome.html,
http://thinkexist.com/quotes/david_hilbert/,
http://www.yalescientific.org/2010/10/from-the-editor-imagination-in-science/,
https://www.brainyquote.com/quotes/topics/topic_science.html. https://www.wikipedia.org/.

This author is motivated to include the quotes at various places in his book for their wit and wisdom; The author cannot vouch for their accuracy.

About the Companion Website

Don't forget to visit the companion website for this book:

www.wiley.com/go/reddy/applied_mechanics_EPVM

There you will find valuable material designed to enhance your learning, including:

1. Solutions manual

Scan this QR code to visit the companion website

Preface to the Third Edition

The development of new material systems paved the way for the development of new devices and improved components of existing structures. Development of mathematical models and their numerical simulations are central to design. While a select number of courses on elasticity and structural mechanics provide engineers with background to formulate a suitable mathematical model using the laws of physics (such as the second law of Newton), refined mathematical models require the tools of variational calculus to formulate and evaluate them. The latest commercial codes like *COMSOL Multiphysics*® and *FEniCS*® require the user to just give the weak forms of the governing equations that they like to solve using the finite element method. It is in this connection that a course on energy principles and variational methods provides the needed background.

Most books on energy principles and variational methods traditionally treat the topics, not incorporating the latest developments and generalizing the classical methods to a broader class of problems. This textbook is unique (i.e., there is no parallel to this book in its class) because most available books in the market are very old (some are even out of print) and do not treat topics that are of current interest. In particular, this book provides a systematic coverage of old and new topics (e.g., finite elements, functionally graded structures, relationships between classical and refined theories) and it contains illustrative examples and problem sets that enable readers to test their understanding of the subject matter.

The third edition of *Energy Principles and Variational Methods* has the same objective as its previous editions, namely, to facilitate an easy and thorough understanding of the concepts and tools necessary to develop variational formulations and associated numerical evaluations. *The book offers easy-to-understand treatment of the subject of energy principles and variational methods.* The new edition is extensively reorganized and contains a substantially large amount of new material. In particular, Chapters 1 and 2 are combined into a single chapter (Chapter 1 in the new edition); Chapter 2 on the review of equations of solid mechanics is an extensively revised version of Chapter 3 from the second edition; Chapters 3 to 9 in the new edition correspond to Chapters 4 to 10 of the second edition but with additional explanations, examples, and exercise problems; Chapter 10 in this new edition is entirely new and devoted to functionally graded beams and plates. In general, all of the chapters of the third edition contain additional explanations, detailed example problems, and fresh exercise problems. Thus the new edition more than replaces the previous editions, and it is hoped that it is acquired by the library of every institution of higher learning by serious structural analysts.

Since the publication of the previous editions, many users of the book communicated their comments and compliments as well as errors they found, for which the author thanks them. The author is grateful to the following professional colleagues for their friendship, encouragement, and support over the years:

Marcilio Alves, University of São Paulo, Brazil
Marco Amabili, McGill University, Canada
Erasmo Carrera, University of Torino, Italy
Antonio Ferreira, University of Porto, Portugal
Somnath Ghosh, Johns Hopkins University
Antonio Grimaldi, University of Rome II, Italy
Yonggang Huang, Northwestern University
S. Kitipornchai, University of Queensland, Australia
K. M. Liew, City University of Hong Kong
C. W. Lim, City University of Hong Kong
Franco Maceri, University of Rome II, Italy
Cristovão Mota Soares, Technical University of Lisbon, Portugal
Antonio Miravete, Zaragoza University, Spain
Glaucio Paulino, Georgia Institute of Technology
Amirtham Rajagopal, Indian Institute of Technology, Hyderabad, India
Jani Romanoff, Aalto University, Finland
Jose Roesset, Texas A&M University
Debasish Roy, Indian Institute of Science, Bangalore, India
Elio Sacco, University of Cassino and Southern Lazio, Italy
E. C. N. Silva, University of São Paulo, Brazil
Arun Srinivasa, Texas A&M University
Karan Surana, University of Kansas
Liqun Tang, South China University of Technology
C. M. Wang, National University of Singapore
Y. B. Yang, National University of Taiwan

Drafts of the manuscript of this book prior to its publication were read by the author's doctoral students, who have made suggestions for improvements. In particular, the author wishes to thank the following former and current students (listed in alphabetical order): Ronald Averill, K. Chandrashekhara, Paul Heyliger, Filis Kokkinos, Filipa Moleiro, Asghar Nosier, Felix Palmerio, Gregory Payette, Grama Praveen, Donald Robbins, Jr., Samit Roy, Ginu Unnikrishnan, Vinu Unnikrishnan, Archana Arbind, Parisa Khodabakhshi, Jinseok Kim, and Namhee Kim. The author requests readers to send their comments and corrections to *jnreddy@tamu.edu*.

J. N. Reddy
College Station, TX
March 2017

Preface to the Second Edition

The increasing use of numerical and computational methods in engineering and applied sciences has shed new light on the importance of energy principles and variational methods. The number of engineering courses that make use of energy principles and variational formulations and methods has also grown very rapidly in recent years. In view of the increase in the use of variational formulations and methods (including the finite element method), there is a need to introduce the concepts of energy principles and variational methods and their use in the formulation and solution of problems of mechanics to both undergraduate and beginning graduate students. This book, an extensively revised version of the author's earlier book *Energy and Variational Methods in Applied Mechanics*, is intended for senior undergraduate students and beginning graduate students of aerospace, civil, and mechanical engineering and applied mechanics who have had a course in fundamental engineering and ordinary and partial differential equations.

The book is organized into 10 chapters and is self-contained as far as the subject matter is concerned. Chapter 1 presents a general introduction to the subject of variational principles. Chapter 2 contains a brief review of the algebra and calculus of vectors and Cartesian tensors, whereas Chapter 3 reviews of the basic equations of linear solid continuum mechanics, which will be frequently referred to in subsequent chapters. Much of the material presented in Chapters 1 to 3 can be assigned as a reading material, especially in a graduate class.

Chapter 4 deals with the concepts of work and energy and basic topics of variational calculus, including the Euler equations, fundamental lemma of calculus of variations, essential and natural boundary conditions, and minimization of functionals with and without equality constraints. Principles of virtual work and energy and energy methods of solid and structural mechanics are presented in Chapter 5. Chapter 6 is devoted to a discussion of Hamilton's principle for dynamical systems. Classical variational methods of approximation (e.g., the methods of Ritz, Galerkin, and Kantorovich) are presented in Chapter 7. All of the concepts and methods presented in Chapters 4 to 7 are illustrated using bars and beams although the methods discussed in Chapter 7 are readily applicable to field problems whose differential equations resemble those of bars and beams. Chapter 8 is dedicated to applications of the energy principles and variational methods developed in earlier chapters to circular and rectangular plates. In the interest of completeness and for use as a reference for approximate solutions, exact solutions are also included. The finite element method is introduced in Chapter 9, with applications to beams and plates. Displacement finite element models of Euler–Bernoulli and Timoshenko beam theories and classical and first-order shear deformation plate theories are presented. A unified approach, more general than that found in most solid mechanics books, is used to introduce the finite element method. As a result, the student can readily extend the method to other subject areas of solid mechanics and other branches of engineering. Lastly, the mixed variational principles of Hellinger and Reissner for elasticity are derived in Chapter 10. Mixed variational formulations, including mixed finite element models of beams and plates, are discussed.

Each chapter of the book contains many example problems and exercises that illustrate, test, and broaden the understanding of the topics covered. A list of references, by no means complete or up to date, is also provided at the end of each chapter. Answers to selective problems are included at the end of the book.

The book is suitable as a textbook for a senior undergraduate course or a first-year graduate course on energy principles and variational methods taught in aerospace, civil, and mechanical engineering and applied mechanics departments. To gain the most from the text, the student should have a senior undergraduate or first-year graduate standing in engineering. Some familiarity with basic courses in differential equations, mechanics of materials, and dynamics would also be helpful.

The author has professionally benefited from the works, encouragement, and support of many colleagues and students who have taught him how to explain complicated concepts in simple terms. While it is not possible to name all of them, without their help and support, it would not have been possible for the author to modestly contribute to the field of mechanics through his teaching, research, and writing. Special thanks are due to his teacher Professor J. T. Oden (University of Texas at Austin) and Professor C. W. Bert (University of Oklahoma, Norman) for their mentorship, advice, and support.

J. N. Reddy
College Station, TX
August 2002

Preface to the First Edition

The increasing use of finite element methods in engineering and applied science has shed new light on the importance of energy and variational methods. The number of engineering courses and research papers that make use of variational and energy methods has also grown very rapidly in recent years. In view of the increased use of variational methods (including the finite element method), there is a need to introduce the concepts of energy and variational methods and their use in the formulation and solution of equations of mechanics to both undergraduate and beginning graduate students. This book is intended for senior undergraduate students and beginning graduate students of aerospace, civil, and mechanical engineering and applied mechanics, who have had a course in ordinary and partial differential equations. The text is organized into four chapters. Chapter 1 is essentially a review, especially for graduate students, of the equations of applied mechanics. Much of the chapter can be assigned as reading material to the student. The equations of bars, beams, torsion, and plane elasticity presented in Section 1.7 are used to illustrate concepts from energy and variational methods. Chapter 2 deals with the study of the basic topics from variational calculus, virtual work and energy principles, and energy methods of mechanics. The instructor can omit Section 2.4 on stationary principles and Section 2.5 on Hamilton's principle if he or she wants to cover all of Chapter 4. Classical variational methods of approximation (e.g., the methods of Ritz, Galerkin, and Kantorovich) and the finite element method are introduced and illustrated in Chapter 3 via linear problems of science and engineering, especially solid mechanics. A unified approach, more general than that found in most solid mechanics books, is used to introduce the variational methods. As a result, the student can readily extend the methods to other subject areas of solid mechanics and other branches of engineering.

The classical variational methods and the finite element method are put to work in Chapter 4 in the derivation and approximate solution of the governing equations of elastic plates and shells. In the interest of completeness, and for use as a reference for approximate solutions, exact solutions of plates and shells are also included. Keeping in mind the current developments in composite material structures, a brief but reasonably complete discussion of laminated plates and shells is included in Sections 4.3 and 4.4. The book contains many example and exercise problems that illustrate, test, and broaden the understanding of the topics covered. A long list of references, by no means complete or up to date, is provided in Bibliography at the end of the book.

The author wishes to acknowledge, with great pleasure and appreciation, the encouragement and support by Professor Daniel Frederick (Head, ESM Department at Virginia Tech) during the course of this writing and the skillful typing of the manuscript by Vanessa McCoy. The author is also thankful to the many students who, through their comments, contributed to the improvement of this book. Special thanks to K. Chandrashekhara, Glenn Creamer, C. F. Liu, and Paul Heyliger for their help in proofreading the galleys and pages and to Dr. Ozden Ochoa for constructive comments on the preliminary draft of the manuscript. It is a pleasure to acknowledge, with many thanks, the cooperation of the technical staff at Wiley, New York (Frank Cerra, Christina Mikulak, and Lisa Morano).

J. N. Reddy
Blacksburg, VA
June 1984

Introduction and Mathematical Preliminaries

1.1 Introduction

1.1.1 Preliminary Comments

The phrase "energy principles" or "energy methods" in the present study refers to methods that make use of the total potential energy (i.e., strain energy and potential energy due to applied loads) of a system to obtain values of an unknown displacement or force, at a specific point of the system. These include Castigliano's theorems, unit dummy load and unit dummy displacement methods, and Betti's and Maxwell's theorems. These methods are often limited to the (exact) determination of generalized displacements or forces at fixed points in the structure; in most cases, they cannot be used to determine the complete solution (i.e., displacements and/or forces) as a function of position in the structure. The phrase "variational methods," on the other hand, refers to methods that make use of the variational principles, such as the principles of virtual work and the principle of minimum total potential energy, to determine approximate solutions as continuous functions of position in a body. In the classical sense, a *variational principle* has to do with the minimization or finding stationary values of a functional with respect to a set of undetermined parameters introduced in the assumed solution. The functional represents the total energy of the system in solid and structural mechanics problems, and in other problems it is simply an integral representation of the governing equations. In all cases, the functional includes all the intrinsic features of the problem, such as the governing equations, boundary and/or initial conditions, and constraint conditions.

1.1.2 The Role of Energy Methods and Variational Principles

Variational principles have always played an important role in mechanics. Variational formulations can be useful in three related ways. First, many problems of mechanics are posed in terms of finding the extremum (i.e., minima or maxima) and thus, by their nature, can be formulated in terms of variational state-

Energy Principles and Variational Methods in Applied Mechanics, Third Edition. J.N. Reddy.
©2017 John Wiley & Sons Ltd. Published 2017 by John Wiley & Sons Ltd.
Companion Website: www.wiley.com/go/reddy/applied_mechanics_EPVM

ments. Second, there are problems that can be formulated by other means, such as by vector mechanics (e.g., Newton's laws), but these can also be formulated by means of variational principles. Third, variational formulations form a powerful basis for obtaining approximate solutions to practical problems, many of which are intractable otherwise. The principle of minimum total potential energy, for example, can be regarded as a substitute to the equations of equilibrium of an elastic body, as well as a basis for the development of displacement finite element models that can be used to determine approximate displacement and stress fields in the body. Variational formulations can also serve to unify diverse fields, suggest new theories, and provide a powerful means for studying the existence and uniqueness of solutions to problems. In many cases they can also be used to establish upper and/or lower bounds on approximate solutions.

1.1.3 A Brief Review of Historical Developments

In modern times, the term "variational formulation" applies to a wide spectrum of concepts having to do with weak, generalized, or direct variational formulations of boundary- and initial-value problems. Still, many of the essential features of variational methods remain the same as they were over 200 years ago when the first notions of variational calculus began to be formulated.[1]

Although Archimedes (287–212 B.C.) is generally credited as the first to use work arguments in his study of levers, the most primitive ideas of variational theory (the minimum hypothesis) are present in the writings of the Greek philosopher Aristotle (384–322 B.C.), to be revived again by the Italian mathematician/engineer Galileo (1564–1642), and finally formulated into a principle of least time by the French mathematician Fermat (1601–1665). The phrase *virtual velocities* was used by Jean Bernoulli in 1717 in his letter to Varignon (1654–1722). The development of early variational calculus, by which we mean the classical problems associated with minimizing certain functionals, had to await the works of Newton (1642–1727) and Leibniz (1646–1716). The earliest applications of such variational ideas included the classical *isoperimetric problem* of finding among closed curves of given length the one that encloses the greatest area, and Newton's problem of determining the solid of revolution of "minimum resistance." In 1696, Jean Bernoulli proposed the problem of the *brachistochrone*: among all curves connecting two points, find the curve traversed in the shortest time by a particle under the influence of gravity. It stood as a challenge to the mathematicians of their day to solve the problem using the rudimentary tools of analysis then available to them or whatever new ones they were capable of developing. Solutions to this problem were presented by some of the greatest mathematicians of the time: Leibniz, Jean Bernoulli's older brother Jacques Bernoulli, L'Hopital, and Newton.

---

[1]Many of the developments came from European scientists, whose works appeared in their native language and were not accessible to the whole scientific community.

The first step toward developing a general method for solving variational problems was given by the Swiss genius Leonhard Euler (1707–1783) in 1732 when he presented a "general solution of the isoperimetric problem," although Maupertuis is credited to have put forward a law of minimal property of potential energy for stable equilibrium in his *Mémoires de lÁcadémie des Sciences* in 1740. It was in Euler's 1732 work and subsequent publication of the principle of least action (in his book *Methodus inveniendi lineas curvas ...*) in 1744 that variational concepts found a welcome and permanent home in mechanics. He developed all ideas surrounding the principle of minimum potential energy in his work on the *elastica*, and he demonstrated the relationship between his variational equations and those governing the flexure and buckling of thin rods.

A great impetus to the development of variational mechanics began in the writings of Lagrange (1736–1813), first in his correspondence with Euler. Euler worked intensely in developing Lagrange's method but delayed publishing his results until Lagrange's works were published in 1760 and 1761. Lagrange used D'Alembert's principle to convert dynamics to statics and then used the principle of virtual displacements to derive his famous equations governing the laws of dynamics in terms of kinetic and potential energy. Euler's work, together with Lagrange's *Mécanique analytique* of 1788, laid down the basis for the variational theory of dynamical systems. Further generalizations appeared in the fundamental work of Hamilton in 1834. Collectively, all these works have had a monumental impact on virtually every branch of mechanics.

A more solid mathematical basis for variational theory began to be developed in the eighteenth and early nineteenth century. Necessary conditions for the existence of "minimizing curves" of certain functionals were studied during this period, and we find among contributors of that era the familiar names of Legendre, Jacobi, and Weierstrass. Legendre gave criteria for distinguishing between maxima and minima in 1786, and Jacobi gave sufficient conditions for existence of extrema in 1837. A more rigorous theory of existence of extrema was put together by Weierstrass, who established in 1865 the conditions on extrema for variational problems.

During the last half of the nineteenth century, the use of variational ideas was widespread among leaders in theoretical mechanics. We mention the works of Kirchhoff on plate theory; Lamé, Green, and Kelvin on elasticity; and the works of Betti, Maxwell, Castigliano, Menabrea, and Engesser on discrete structural systems. Lamé was the first in 1852 to prove a work equation, named after his colleague Claperon, for deformable bodies. Lamé's equation was used by Maxwell [1][2] to the solution of redundant frame-works using the unit dummy load technique. In 1875 Castigliano published an extremum version of this technique but attributed the idea to Menabrea. A generalization of Castigliano's work is due to Engesser [2].

---

[2] The references are listed at the end of the book.

Among the prominent contributors to the subject near the end of the nine-teenth century and in the early years of the twentieth century, particularly in the area of variational methods of approximation and their applications, were Rayleigh [3], Ritz [4], and Galerkin [5]. Modern variational principles began in the works of Hellinger [6], Hu [7], and Reissner [8–10] on mixed variational principles for elasticity problems. A short historical account of early varia-tional methods in mechanics can be found in the book of Lanczos [11] and Truesdell and Toupin [12]; additional information can be found in Dugas [13] and Timoshenko [14], and historical development of energetical principles in elastomechanics can be found in the paper by Oravas and McLean [15, 16]. Reference to much of the relevant contemporary literature can be found in the books by Washizu [17] and Oden and Reddy [18]. Additional historical papers and textbooks on variational principles and methods can be found in [19-60].

1.1.4   Preview

The objective of the present book is to introduce energy methods and variational principles of solid and structural mechanics and to illustrate their use in the derivation and solution of the equations of applied mechanics, including plane elasticity, beams, frames, and plates. Of course, variational formulations and methods presented in this book are also applicable to problems outside solid mechanics. To keep the scope of the book within reasonable limits, mostly linear problems of solid and structural mechanics are considered.

In the remaining part of the chapter, we review the algebra and calculus of vectors and tensors. In Chapter 2, a brief review of the equations of solid me-chanics is presented, and the concepts of work and energy and elements from calculus of variations are discussed in Chapter 3. Principles of virtual work and their special cases are presented in Chapter 4. The chapter also includes energy theorems of structural mechanics, namely, Castigliano's theorems I and II, dummy displacement and dummy force methods, and Betti's and Maxwell's reciprocity theorems of elasticity. Chapter 5 is dedicated to Hamilton's princi-ple for dynamical systems of solid mechanics. In Chapter 6 we introduce the Ritz, Galerkin, and weighted-residual methods. Chapter 7 contains the appli-cations of variational methods to the formulation of plate theories and their solution by variational methods. For the sake of completeness and comparison, analytical solutions of bending, vibration, and buckling of circular and rectan-gular plates are also presented. An introduction to the finite element method and its application to displacement finite element models of beams and plates are discussed in Chapter 8. Chapter 9 is devoted to the discussion of mixed variational principles and mixed finite element models of beams and plates. Fi-nally, theories and analytical as well as finite element solutions of functionally graded beams and plates are presented in Chapter 10.

1.2 Vectors

1.2.1 Introduction

Our approach in this book is evolutionary, that is, we wish to begin with concepts that are simple and intuitive and then generalize these concepts to a broader and more abstract body of analysis. This is a natural inductive approach, more or less in accord with the development of the subject of variational methods.

In analyzing physical phenomena, we set up, with the help of physical principles, relations between various quantities that represent the phenomena. As a means of expressing a natural law, a coordinate system in a chosen frame of reference can be introduced, and the various physical quantities involved can be expressed in terms of measurements made in that system. The mathematical form of the law thus depends upon the chosen coordinate system and may appear different in another type of coordinate system. The laws of nature, however, should be independent of the artificial choice of a coordinate system, and we may seek to represent the law in a manner independent of a particular coordinate system. A way of doing this is provided by vector and tensor analysis. When vector notation is used, a particular coordinate system need not be introduced. Consequently, the use of vector notation in formulating natural laws leaves them *invariant* to coordinate transformations. A study of physical phenomena by means of vector equations often leads to a deeper understanding of the problem in addition to bringing simplicity and versatility into the analysis.

The term *vector* is used often to imply a *physical* vector that has "magnitude and direction" and obeys certain rules of vector addition and scalar multiplication. In the sequel we consider more general, abstract objects than physical vectors, which are also called vectors. It transpires that the physical vector is a special case of what is known as a "vector from a linear vector space." Then the notion of vectors in modern mathematical analysis is an abstraction of the elementary notion of a physical vector. While the definition of a vector in abstract analysis does not require the vector to have a magnitude, in nearly all cases of practical interest, the vector is endowed with a magnitude, in which case the vector is said to belong to a normed vector space.

Like physical vectors, which have direction and magnitude and satisfy the parallelogram law of addition, *tensors* are more general objects that are endowed with a magnitude and multiple direction(s) and satisfy rules of tensor addition and scalar multiplication. In fact, vectors are often termed the first-order tensors. As will be shown shortly, the stress (i.e., force per unit area) requires a magnitude and two directions – one normal to the plane on which the stress is measured and the other is the direction of the force – to specify it uniquely. For additional details, References [61–88] listed at the end of the book may be consulted.

1.2.2   Definition of a Vector

In the analysis of physical phenomena, we are concerned with quantities that may be classified according to the information needed to specify them completely. Consider the following two groups:

Scalars	**Nonscalars**
Mass	Force
Temperature	Moment
Density	Stress
Volume	Acceleration
Time	Displacement

After units have been selected, the scalars are given by a single number. Nonscalars need not only a magnitude specified but also additional information, such as direction. Nonscalars that obey certain rules (such as the parallelogram law of addition) are called *vectors*. Not all nonscalar quantities are vectors. The specification of a stress requires not only a force, which is a vector, but also an area upon which the force acts. A stress is a second-order tensor, as will be shown shortly.

In written or typed material, it is customary to place an arrow or a bar over the letter denoting the vector, such as $\vec{A}$. Sometimes the typesetter's mark of a tilde under the letter is used. In printed material the vector letter is denoted by a boldface letter, $\mathbf{A}$, such as used in this book. The magnitude of the vector $\mathbf{A}$ is denoted by $|\mathbf{A}|$ or just A. The computation of the magnitude of a vector will be defined in the sequel, after the concept of scalar product of vectors is discussed.[3]

Two vectors $\mathbf{A}$ and $\mathbf{B}$ are equal if their magnitudes are equal, $|\mathbf{A}| = |\mathbf{B}|$, and if their directions and sense are equal. Consequently a vector is not changed if it is moved parallel to itself. This means that the position of a vector in space may be chosen arbitrarily. In certain applications, however, the actual point of location of a vector may be important (for instance, a moment or a force acting on a body). A vector associated with a given point is known as a localized or bound vector.

Let $\mathbf{A}$ and $\mathbf{B}$ be any two vectors. Then we can add them as shown in Fig. 1.2.1(a). The combination of the two diagrams in Fig. 1.2.1(a) gives the parallelogram shown in Fig. 1.2.1(b). Thus we say the vectors add according to the *parallelogram law* of addition so that

$$\mathbf{C} = \mathbf{A} + \mathbf{B} = \mathbf{B} + \mathbf{A}. \tag{1.2.1}$$

We thus see that vector addition is *commutative*.

---

[3]Mathematically, the length of a vector can be computed only when its components with respect to a basis are known.

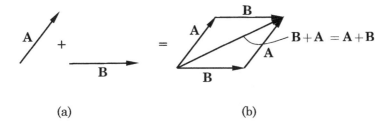

Fig. 1.2.1 (a) Addition of vectors. (b) Parallelogram law of addition.

Subtraction of vectors is carried out along the same lines. To form the difference $\mathbf{A} - \mathbf{B}$, we write

$$\mathbf{A} - \mathbf{B} = \mathbf{A} + (-\mathbf{B}) \tag{1.2.2}$$

and subtraction reduces to the operation of addition. The negative vector $-\mathbf{B}$ has the same magnitude as $\mathbf{B}$ but has the opposite *sense*.

With the rules of addition in place, we can define a (geometric) vector. *A vector is a quantity that possesses both magnitude and direction and obeys the parallelogram law of addition.* Obeying the law is important because there are quantities having both magnitude and direction that do not obey this law. A finite rotation of a rigid body is not a vector although infinitesimal rotations are. The definition given above is a *geometrical* definition. That vectors can be represented graphically is an *incidental* rather than a fundamental feature of the vector concept.

A vector of unit length is called a *unit vector*. The unit vector may be defined as follows:

$$\hat{\mathbf{e}}_A = \frac{\mathbf{A}}{A}. \tag{1.2.3}$$

We may now write

$$\mathbf{A} = A\hat{\mathbf{e}}_A. \tag{1.2.4}$$

Thus *any vector may be represented as a product of its magnitude and a unit vector.* A unit vector is used to designate direction. It does not have any physical dimensions. We denote a unit vector by a "hat" (caret) above the boldface letter.

A vector of zero magnitude is called a *zero vector* or a *null vector*. All null vectors are considered equal to each other without consideration as to direction:

$$\mathbf{A} + \mathbf{0} = \mathbf{A} \quad \text{and} \quad 0\mathbf{A} = \mathbf{0}. \tag{1.2.5}$$

The laws that govern addition, subtraction, and scalar multiplication of vectors are identical with those governing the operations of scalar algebra.

1.2.3   Scalar and Vector Products

Besides addition, subtraction, and multiplication by a scalar, we must consider the multiplication of two vectors. There are several ways the product of two vectors can be defined. We consider first the so-called scalar product. Let us recall the concept of work. When a force $\mathbf{F}$ acts on a mass point and moves through an infinitesimal displacement vector $\mathbf{ds}$, the work done by the force vector is defined by the *projection* of the force in the direction of the displacement times the magnitude of the displacement (see Fig. 1.2.2). Such an operation may be defined for any two vectors. Since the result of the product is a scalar, it is called the *scalar product*. We denote this product as follows:

$$\mathbf{F} \cdot \mathbf{ds} = F \, ds \cos\theta, \qquad 0 \leq \theta \leq \pi. \tag{1.2.6}$$

The scalar product is also known as the *dot product* or *inner product*.

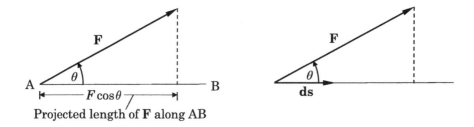

Fig. 1.2.2 Representation of work.

To understand the vector product, consider the concept of the *moment* due to a force. Let us describe the moment about a point O of a force $\mathbf{F}$ acting at a point P, as shown in Fig. 1.2.3(a). By definition, the magnitude of the moment is given by

$$M = F\,l, \quad F = |\mathbf{F}| = \sqrt{\mathbf{F} \cdot \mathbf{F}}, \tag{1.2.7}$$

where l is the lever arm for the force about the point O. If $\mathbf{r}$ denotes the vector $\mathbf{OP}$ and θ the angle between $\mathbf{r}$ and $\mathbf{F}$ as shown, such that $0 \leq \theta \leq \pi$, we have

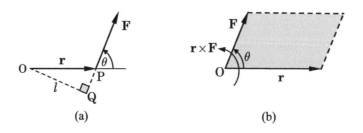

Fig. 1.2.3 (a) Representation of a moment. (b) Direction of rotation.

$l = r \sin \theta$, and thus

$$M = Fr \sin \theta. \tag{1.2.8}$$

A direction can now be assigned to the moment. Drawing the vectors **F** and **r** from the common origin O, we note that the rotation due to **F** tends to bring **r** into **F** [see Fig. 1.2.3(b)]. We now set up an axis of rotation perpendicular to the plane formed by **F** and **r**. Along this axis of rotation we set up a preferred direction as that in which a right-handed screw would advance when turned in the direction of rotation due to the moment [see Fig. 1.2.4(a)]. Along this axis of rotation, we draw a unit vector $\hat{e}_M$ and agree that it represents the direction of the moment **M**. Thus we have

$$\mathbf{M} = Fr \sin \theta \, \hat{e}_M = \mathbf{r} \times \mathbf{F}. \tag{1.2.9}$$

According to this expression, **M** may be looked upon as resulting from a special operation between the two vectors **F** and **r**. It is thus the basis for defining a product between any two vectors. Since the result of such a product is a vector, it may be called the *vector product*.

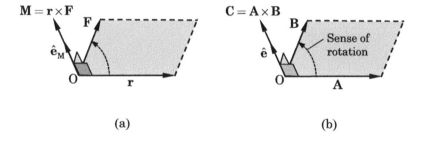

Fig. 1.2.4 (a) Axis of rotation. (b) Representation of the vector.

The vector product of two vectors **A** and **B** is a vector **C** whose magnitude is equal to the product of the magnitude of **A** and **B** times the sine of the angle measured from **A** to **B** such that $0 \leq \theta \leq \pi$, and whose direction is specified by the condition that **C** be perpendicular to the plane of the vectors **A** and **B** and points to the direction where a right-handed screw advances when turned so as to bring **A** into **B**.

The vector product is usually denoted by

$$\mathbf{C} = \mathbf{A} \times \mathbf{B} = AB \, \sin(\mathbf{A}, \mathbf{B}) \, \hat{e}, \tag{1.2.10}$$

where $\sin(\mathbf{A}, \mathbf{B})$ denotes the sine of the angle between vectors **A** and **B**. This product is called the *cross product, skew product*, and also *outer product*, as well as the vector product [see Fig. 1.2.4(b)].

Now consider the various products of three vectors:

$$\mathbf{A}(\mathbf{B} \cdot \mathbf{C}), \quad \mathbf{A} \cdot (\mathbf{B} \times \mathbf{C}), \quad \mathbf{A} \times (\mathbf{B} \times \mathbf{C}). \tag{1.2.11}$$

The product $\mathbf{A}(\mathbf{B} \cdot \mathbf{C})$ is merely a multiplication of the vector $\mathbf{A}$ by the scalar $\mathbf{B} \cdot \mathbf{C}$. The product $\mathbf{A} \cdot (\mathbf{B} \times \mathbf{C})$ is a scalar. It can be seen that the product $\mathbf{A} \cdot (\mathbf{B} \times \mathbf{C})$, except for the algebraic sign, is the volume of the parallelepiped formed by the vectors $\mathbf{A}$, $\mathbf{B}$, and $\mathbf{C}$, as shown in Fig. 1.2.5.

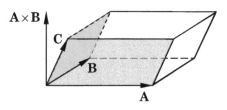

Volume of the parallelepiped is
$$\mathbf{A} \times \mathbf{B} \cdot \mathbf{C} \equiv (\mathbf{A} \times \mathbf{B}) \cdot \mathbf{C} = -\mathbf{A} \times \mathbf{C} \cdot \mathbf{B}$$

Fig. 1.2.5 Scalar triple product as the volume of a parallelepiped.

We also note the following properties:

1. The dot and cross can be interchanged without changing the value:

$$\mathbf{A} \cdot \mathbf{B} \times \mathbf{C} = \mathbf{A} \times \mathbf{B} \cdot \mathbf{C} \equiv [\mathbf{ABC}]. \tag{1.2.12}$$

2. A cyclical permutation of the order of the vectors leaves the result unchanged:

$$\mathbf{A} \cdot \mathbf{B} \times \mathbf{C} = \mathbf{C} \cdot \mathbf{A} \times \mathbf{B} = \mathbf{B} \cdot \mathbf{C} \times \mathbf{A} \equiv [\mathbf{ABC}]. \tag{1.2.13}$$

3. If the cyclic order is changed, the sign changes:

$$\mathbf{A} \cdot \mathbf{B} \times \mathbf{C} = -\mathbf{A} \cdot \mathbf{C} \times \mathbf{B} = -\mathbf{C} \cdot \mathbf{B} \times \mathbf{A} = -\mathbf{B} \cdot \mathbf{A} \times \mathbf{C}. \tag{1.2.14}$$

4. A necessary and sufficient condition for any three vectors, $\mathbf{A}$, $\mathbf{B}$, and $\mathbf{C}$ to be coplanar is that $\mathbf{A} \cdot (\mathbf{B} \times \mathbf{C}) = 0$. Note also that the scalar triple product is zero when any two vectors are the same.

The product $\mathbf{A} \times (\mathbf{B} \times \mathbf{C})$ is a vector normal to the plane formed by $\mathbf{A}$ and $(\mathbf{B} \times \mathbf{C})$. The vector $(\mathbf{B} \times \mathbf{C})$, however, is perpendicular to the plane formed by $\mathbf{B}$ and $\mathbf{C}$. This means that $\mathbf{A} \times (\mathbf{B} \times \mathbf{C})$ lies in the plane formed by $\mathbf{B}$ and $\mathbf{C}$

and is perpendicular to $\mathbf{A}$ (see Fig. 1.2.6). Thus $\mathbf{A} \times (\mathbf{B} \times \mathbf{C})$ can be expressed as a linear combination of $\mathbf{B}$ and $\mathbf{C}$:

$$\mathbf{A} \times (\mathbf{B} \times \mathbf{C}) = m_1 \mathbf{B} + n_1 \mathbf{C}. \tag{1.2.15}$$

Likewise, we would find that

$$(\mathbf{A} \times \mathbf{B}) \times \mathbf{C} = m_2 \mathbf{A} + n_2 \mathbf{B}. \tag{1.2.16}$$

Thus the parentheses *cannot* be interchanged or removed. It can be shown that

$$m_1 = \mathbf{A} \cdot \mathbf{C}, \qquad n_1 = -\mathbf{A} \cdot \mathbf{B},$$

and hence that

$$\mathbf{A} \times (\mathbf{B} \times \mathbf{C}) = (\mathbf{A} \cdot \mathbf{C})\mathbf{B} - (\mathbf{A} \cdot \mathbf{B})\mathbf{C}. \tag{1.2.17}$$

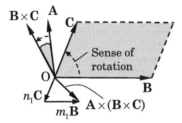

$$\mathbf{A} \times (\mathbf{B} \times \mathbf{C}) = m_1 \mathbf{B} + n_1 \mathbf{C} = (\mathbf{A} \cdot \mathbf{C})\mathbf{B} - (\mathbf{A} \cdot \mathbf{B})\mathbf{C}$$

Fig. 1.2.6 The vector triple product.

Example 1.2.1

Find the equation of a plane perpendicular to a vector $\mathbf{A}$ and passing through the terminal point of vector $\mathbf{B}$ without the use of any coordinate system (see Fig. 1.2.7).

Solution: Let O be the origin and B the terminal point of vector $\mathbf{B}$. Draw a directed line segment from O to Q, such that $\mathbf{OQ}$ is parallel to $\mathbf{A}$ and Q is in the plane. Then $\mathbf{OQ} = \alpha \mathbf{A}$, where α is a scalar. Let P be an arbitrary point on the line BQ. If the position vector of the point P is $\mathbf{r}$, then

$$\mathbf{BP} = \mathbf{OP} - \mathbf{OB} = \mathbf{r} - \mathbf{B}.$$

Since $\mathbf{BP}$ is perpendicular to $\mathbf{OQ} = \alpha \mathbf{A}$, we must have

$$\mathbf{BP} \cdot \mathbf{OQ} = 0 \quad \text{or} \quad (\mathbf{r} - \mathbf{B}) \cdot \mathbf{A} = 0,$$

which is the equation of the plane in question.

The perpendicular distance from point O to the plane is the magnitude of $\mathbf{OQ}$. However, we do not know its magnitude (or α is not known). The distance is also given by the projection

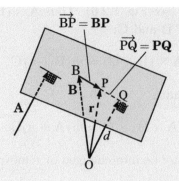

Fig. 1.2.7 Plane perpendicular to **A** and passing through the terminal point of **B**.

of vector **B** along **OQ**:

$$d = \mathbf{B} \cdot \frac{\mathbf{OQ}}{|\mathbf{OQ}|} = \mathbf{B} \cdot \hat{\mathbf{e}}_A,$$

where $\hat{\mathbf{e}}_A$ is the unit vector along **A**, $\hat{\mathbf{e}}_A = \mathbf{A}/A$.

Example 1.2.2

Let **A** and **B** be any two vectors in space. Then express the vector **A** in terms of components along (i.e., parallel) and perpendicular to **B**.

Solution: The component of **A** along **B** is given by $(\mathbf{A} \cdot \hat{\mathbf{e}}_B)$, where $\hat{\mathbf{e}}_B = \mathbf{B}/B$. The component of **A** perpendicular to **B** and in the plane of **A** and **B** is given by the vector triple product $\hat{\mathbf{e}}_B \times (\mathbf{A} \times \hat{\mathbf{e}}_B)$. Thus,

$$\mathbf{A} = (\mathbf{A} \cdot \hat{\mathbf{e}}_B)\hat{\mathbf{e}}_B + \hat{\mathbf{e}}_B \times (\mathbf{A} \times \hat{\mathbf{e}}_B). \tag{1}$$

Alternately, using Eq. (1.2.17) with $\mathbf{A} = \mathbf{C} = \hat{\mathbf{e}}_B$ and $\mathbf{B} = \mathbf{A}$, we obtain

$$\hat{\mathbf{e}}_B \times (\mathbf{A} \times \hat{\mathbf{e}}_B) = \mathbf{A} - (\hat{\mathbf{e}}_B \cdot \mathbf{A})\hat{\mathbf{e}}_B. \tag{2}$$

1.2.4   Components of a Vector

So far we have proceeded on a geometrical description of a vector as a directed line segment. We now embark on an analytical description of a vector and some of the operations associated with this description. Such a description yields a connection between vectors and ordinary numbers and relates operation on vectors with those on numbers. The analytical description is based on the notion of components of a vector.

In what follows, we shall consider a three-dimensional space, and the extensions to n dimensions will be evident (except for a few exceptions). A set of n vectors is said to be linearly dependent if a set of n numbers $\beta_1, \beta_2, \cdots, \beta_n$ can be found such that

$$\beta_1 \mathbf{A}_1 + \beta_2 \mathbf{A}_2 + \cdots + \beta_n \mathbf{A}_n = \mathbf{0}, \tag{1.2.18}$$

where $\beta_1, \beta_2, \cdots, \beta_n$ cannot all be zero. If this expression cannot be satisfied, the vectors are said to be *linearly independent*.

In a three-dimensional space, a set of no more than three linearly independent vectors can be found. Let us choose any set and denote it as follows:

$$\mathbf{e}_1, \mathbf{e}_2, \mathbf{e}_3. \tag{1.2.19}$$

This set is called a *basis* (or a base system).

It is clear from the concept of linear dependence that we can represent any vector in three-dimensional space as a linear combination of the basis vectors (see Fig. 1.2.8):

$$\mathbf{A} = A_1 \mathbf{e}_1 + A_2 \mathbf{e}_2 + A_3 \mathbf{e}_3. \tag{1.2.20}$$

The vectors $A_1 \mathbf{e}_1$, $A_2 \mathbf{e}_2$, and $A_3 \mathbf{e}_3$ are called the *vector components* of $\mathbf{A}$, and A_1, A_2, and A_3 are called *scalar components* of $\mathbf{A}$ associated with the basis $(\mathbf{e}_1, \mathbf{e}_2, \mathbf{e}_3)$. Also, we use the notation $\mathbf{A} = (A_1, A_2, A_3)$ to denote a vector by its components.

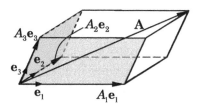

Fig. 1.2.8 Components of a vector.

1.2.5 Summation Convention

It is useful to abbreviate a summation of terms by understanding that a repeated index means summation over all values of that index. Thus the summation

$$\mathbf{A} = \sum_{i=1}^{3} A_i \mathbf{e}_i \tag{1.2.21}$$

can be shortened to

$$\mathbf{A} = A_i \mathbf{e}_i. \tag{1.2.22}$$

The repeated index is a *dummy index* and thus can be replaced by *any other symbol that has not already been used*. Thus we can also write

$$\mathbf{A} = A_i \mathbf{e}_i = A_m \mathbf{e}_m, \text{ and so on.}$$

When a basis is unit and orthogonal, that is, orthonormal, we have

$$[\mathbf{e}_1 \mathbf{e}_2 \mathbf{e}_3] = 1. \tag{1.2.23}$$

In many situations an *orthonormal basis* simplifies the calculations.

For an orthonormal basis, the vectors $\mathbf{A}$ and $\mathbf{B}$ can be written as

$$\mathbf{A} = A_1\hat{\mathbf{e}}_1 + A_2\hat{\mathbf{e}}_2 + A_3\hat{\mathbf{e}}_3 = A_i\hat{\mathbf{e}}_i$$
$$\mathbf{B} = B_1\hat{\mathbf{e}}_1 + B_2\hat{\mathbf{e}}_2 + B_3\hat{\mathbf{e}}_3 = B_i\hat{\mathbf{e}}_i,$$

where $\mathbf{e}_i \equiv \hat{\mathbf{e}}_i$ ($i = 1, 2, 3$) is the orthonormal basis and A_i and B_i are the corresponding *physical components* (i.e., the components have the same physical dimensions as the vector).

It is convenient at this time to introduce the Kronecker delta δ_{ij} and alternating symbol ε_{ijk} for representing the dot product and cross product of two orthonormal vectors in a right-handed basis system. We define the dot product $\hat{\mathbf{e}}_i \cdot \hat{\mathbf{e}}_j$ between the orthonormal basis vectors of a right-handed system as

$$\hat{\mathbf{e}}_i \cdot \hat{\mathbf{e}}_j \equiv \delta_{ij} = \begin{cases} 1, \text{ if } i = j, \text{ for any fixed value of } i, j \\ 0, \text{ if } i \neq j, \text{ for any fixed value of } i, j, \end{cases} \quad (1.2.24)$$

where δ_{ij} is called the *Kronecker delta symbol*. Similarly, we define the cross product $\hat{\mathbf{e}}_i \times \hat{\mathbf{e}}_j$ for a right-handed system as

$$\hat{\mathbf{e}}_i \times \hat{\mathbf{e}}_j \equiv \varepsilon_{ijk}\hat{\mathbf{e}}_k, \quad (1.2.25)$$

where

$$\varepsilon_{ijk} = \begin{cases} 1, \text{ if } i, j, k \text{ are in cyclic order} \\ \quad \text{and not repeated } (i \neq j \neq k), \\ -1, \text{ if } i, j, k \text{ are not in cyclic order} \\ \quad \text{and not repeated } (i \neq j \neq k), \\ 0, \text{ if any of } i, j, k \text{ are repeated.} \end{cases} \quad (1.2.26)$$

The symbol ε_{ijk} is called the *alternating symbol* or *permutation symbol*.

In an orthonormal basis, the scalar and vector products can be expressed in the index form using the Kronecker delta and alternating symbols as

$$\mathbf{A} \cdot \mathbf{B} = (A_i\hat{\mathbf{e}}_i) \cdot (B_j\hat{\mathbf{e}}_j) = A_iB_j\delta_{ij} = A_iB_i,$$
$$\mathbf{A} \times \mathbf{B} = (A_i\hat{\mathbf{e}}_i) \times (B_j\hat{\mathbf{e}}_j) = A_iB_j\varepsilon_{ijk}\hat{\mathbf{e}}_k. \quad (1.2.27)$$

Thus, the length of a vector in an orthonormal basis can be expressed as $A = \sqrt{\mathbf{A} \cdot \mathbf{A}} = \sqrt{A_iA_i}$. The Kronecker delta and the permutation symbol are related by the identity, known as the ε-δ *identity*:

$$\varepsilon_{ijk}\varepsilon_{imn} = \delta_{jm}\delta_{kn} - \delta_{jn}\delta_{km}. \quad (1.2.28)$$

The permutation symbol and the Kronecker delta prove to be very useful in proving vector identities. Since a vector form of any identity is invariant (i.e., valid in any coordinate system), it suffices to prove it in one coordinate system. In particular, an orthonormal system is very convenient because of the permutation symbol and the Kronecker delta. The following example illustrates some of the uses of δ_{ij} and ε_{ijk}.

Example 1.2.3 ————————————————————————

Express the vector operation $(\mathbf{A} \times \mathbf{B}) \cdot (\mathbf{C} \times \mathbf{D})$ in an alternate vector form.

Solution: We have

$$
\begin{aligned}
(\mathbf{A} \times \mathbf{B}) \cdot (\mathbf{C} \times \mathbf{D}) &= (A_i B_j \varepsilon_{ijk} \hat{\mathbf{e}}_k) \cdot (C_m D_n \varepsilon_{mnp} \hat{\mathbf{e}}_p) \\
&= A_i B_j C_m D_n \varepsilon_{ijk} \varepsilon_{mnp} \delta_{kp} \\
&= A_i B_j C_m D_n \varepsilon_{ijk} \varepsilon_{mnk} \\
&= A_i B_j C_m D_n (\delta_{im} \delta_{jn} - \delta_{in} \delta_{jm}) \\
&= A_i B_j C_m D_n \delta_{im} \delta_{jn} - A_i B_j C_m D_n \delta_{in} \delta_{jm},
\end{aligned}
$$

where we have used the ε–δ identity in Eq. (1.2.28). Since $C_m \delta_{im} = C_i$ (or $A_i \delta_{im} = A_m$, etc.), we have

$$
\begin{aligned}
(\mathbf{A} \times \mathbf{B}) \cdot (\mathbf{C} \times \mathbf{D}) &= A_i B_j C_i D_j - A_i B_j C_j D_i \\
&= A_i C_i B_j D_j - A_i D_i B_j C_j \\
&= (\mathbf{A} \cdot \mathbf{C})(\mathbf{B} \cdot \mathbf{D}) - (\mathbf{A} \cdot \mathbf{D})(\mathbf{B} \cdot \mathbf{C}).
\end{aligned}
$$

Although the above vector identity is established in an orthonormal coordinate system, it holds in a general coordinate system. That is, the vector identity is invariant.

——

We can establish the relationship between the components of two different orthonormal coordinate systems, say, unbarred and barred. Consider the unbarred coordinate basis $(\hat{\mathbf{e}}_1, \hat{\mathbf{e}}_2, \hat{\mathbf{e}}_3)$ and the barred coordinate basis $(\hat{\bar{\mathbf{e}}}_1, \hat{\bar{\mathbf{e}}}_2, \hat{\bar{\mathbf{e}}}_3)$. Then, we can express the same vector in the two coordinate systems as

$$
\begin{aligned}
\mathbf{A} &= A_j \hat{\mathbf{e}}_j \quad \text{in unbarred basis,} \\
&= \bar{A}_j \hat{\bar{\mathbf{e}}}_j \quad \text{in barred basis.}
\end{aligned}
$$

Now taking the dot product of the both sides with the vector $\hat{\bar{\mathbf{e}}}_i$ (from the left), we obtain the following relation between the components of a vector in two different coordinate systems:

$$
\bar{A}_i = \beta_{ij} A_j, \quad \beta_{ij} = \hat{\bar{\mathbf{e}}}_i \cdot \hat{\mathbf{e}}_j. \tag{1.2.29}
$$

Thus, the relationship between the components $(\bar{A}_1, \bar{A}_2, \bar{A}_3)$ and (A_1, A_2, A_3) is called the *transformation rule* between the barred and unbarred components in the two orthogonal coordinate systems. The coefficients β_{ij} are the *direction cosines* of the barred coordinate system with respect to the unbarred coordinate system:

$$
\beta_{ij} = \text{cosine of the angle between } \hat{\bar{\mathbf{e}}}_i \text{ and } \hat{\mathbf{e}}_j. \tag{1.2.30}
$$

Note that the first subscript of β_{ij} comes from the barred coordinate system and the second subscript from the unbarred system. Obviously, β_{ij} is not symmetric (i.e., $\beta_{ij} \neq \beta_{ji}$). The direction cosines allow us to relate components of a vector (or a tensor) in the unbarred coordinate system to components of the same vector (or tensor) in the barred coordinate system. **Example 1.2.4** illustrates the computation of direction cosines.

Example 1.2.4 ———————————————————————————

Let $\hat{\mathbf{e}}_i$ $(i = 1, 2, 3)$ be a set of orthonormal base vectors, and define new right-handed coordinate base vectors by $(\hat{\bar{\mathbf{e}}}_1.\hat{\bar{\mathbf{e}}}_2 = 0)$:

$$\hat{\bar{\mathbf{e}}}_1 = \frac{2\hat{\mathbf{e}}_1 + 2\hat{\mathbf{e}}_2 + \hat{\mathbf{e}}_3}{3}, \qquad \hat{\bar{\mathbf{e}}}_2 = \frac{\hat{\mathbf{e}}_1 - \hat{\mathbf{e}}_2}{\sqrt{2}}.$$

Determine the direction cosines of the transformation between the two coordinate systems.

Solution: First we compute the third base vector in the barred coordinate system by

$$\hat{\bar{\mathbf{e}}}_3 = \hat{\bar{\mathbf{e}}}_1 \times \hat{\bar{\mathbf{e}}}_2 = \frac{\hat{\mathbf{e}}_1 + \hat{\mathbf{e}}_2 - 4\hat{\mathbf{e}}_3}{3\sqrt{2}}.$$

An arbitrary vector **A** can be represented in either coordinate system:

$$\mathbf{A} = A_i \hat{\mathbf{e}}_i = \bar{A}_i \hat{\bar{\mathbf{e}}}_i.$$

The components of the vector in the two different coordinate systems are related by

$$\{\bar{A}\} = [\beta]\{A\}, \qquad \beta_{ij} = \hat{\bar{\mathbf{e}}}_i \cdot \hat{\mathbf{e}}_j.$$

For the case at hand, we have

$$\beta_{11} = \hat{\bar{\mathbf{e}}}_1 \cdot \hat{\mathbf{e}}_1 = \frac{2}{3}, \qquad \beta_{12} = \hat{\bar{\mathbf{e}}}_1 \cdot \hat{\mathbf{e}}_2 = \frac{2}{3}, \qquad \beta_{13} = \hat{\bar{\mathbf{e}}}_1 \cdot \hat{\mathbf{e}}_3 = \frac{1}{3},$$

$$\beta_{21} = \hat{\bar{\mathbf{e}}}_2 \cdot \hat{\mathbf{e}}_1 = \frac{1}{\sqrt{2}}, \qquad \beta_{22} = \hat{\bar{\mathbf{e}}}_2 \cdot \hat{\mathbf{e}}_2 = -\frac{1}{\sqrt{2}}, \qquad \beta_{23} = \hat{\bar{\mathbf{e}}}_2 \cdot \hat{\mathbf{e}}_3 = 0,$$

$$\beta_{31} = \hat{\bar{\mathbf{e}}}_3 \cdot \hat{\mathbf{e}}_1 = \frac{1}{3\sqrt{2}}, \qquad \beta_{32} = \hat{\bar{\mathbf{e}}}_3 \cdot \hat{\mathbf{e}}_2 = \frac{1}{3\sqrt{2}}, \qquad \beta_{33} = \hat{\bar{\mathbf{e}}}_3 \cdot \hat{\mathbf{e}}_3 = -\frac{4}{3\sqrt{2}},$$

or

$$[\beta] = \frac{1}{3\sqrt{2}} \begin{bmatrix} 2\sqrt{2} & 2\sqrt{2} & \sqrt{2} \\ 3 & -3 & 0 \\ 1 & 1 & -4 \end{bmatrix}.$$

When the basis vectors are constant, that is, with fixed lengths (with the same units) and directions, the basis is called *Cartesian*. The general Cartesian system is oblique. When the basis vectors are unit and orthogonal (orthonormal), the basis system is called *rectangular Cartesian*, or simply *Cartesian*. In much of our study, we shall deal with Cartesian bases.

Let us denote an orthonormal Cartesian basis by

$$\{\hat{\mathbf{e}}_x, \hat{\mathbf{e}}_y, \hat{\mathbf{e}}_z\} \qquad \text{or} \qquad \{\hat{\mathbf{e}}_1, \hat{\mathbf{e}}_2, \hat{\mathbf{e}}_3\}.$$

The Cartesian coordinates are denoted by (x, y, z) or (x^1, x^2, x^3). The familiar rectangular Cartesian coordinate system is shown in Fig. 1.2.9. We shall always use right-handed coordinate systems.

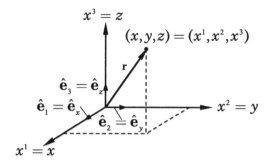

Fig. 1.2.9 Rectangular Cartesian coordinates.

A position vector to an arbitrary point (x, y, z) or (x^1, x^2, x^3), measured from the origin, is given by

$$\mathbf{r} = x\hat{\mathbf{e}}_x + y\hat{\mathbf{e}}_y + z\hat{\mathbf{e}}_z$$
$$= x^1\hat{\mathbf{e}}_1 + x^2\hat{\mathbf{e}}_2 + x^3\hat{\mathbf{e}}_3,$$

or, in summation notation, by

$$\mathbf{r} = x^j\hat{\mathbf{e}}_j. \tag{1.2.31}$$

The distance between two infinitesimally removed points is given by

$$d\mathbf{r} \cdot d\mathbf{r} = (ds)^2 = dx^j dx^j$$
$$= (dx)^2 + (dy)^2 + (dz)^2. \tag{1.2.32}$$

1.2.6   Vector Calculus

The basic notions of vector and scalar calculus, especially with regard to physical applications, are closely related to the rate of change of a scalar field with distance. Let us denote a scalar field by $\phi = \phi(\mathbf{r})$. In general coordinates we can write $\phi = \phi(q^1, q^2, q^3)$. The coordinate system (q^1, q^2, q^3) is referred to as the *unitary system*.

We now define the unitary basis $(\mathbf{e}_1, \mathbf{e}_2, \mathbf{e}_3)$ as follows:

$$\mathbf{e}_1 \equiv \frac{\partial \mathbf{r}}{\partial q^1}, \qquad \mathbf{e}_2 \equiv \frac{\partial \mathbf{r}}{\partial q^2}, \qquad \mathbf{e}_3 \equiv \frac{\partial \mathbf{r}}{\partial q^3}. \tag{1.2.33}$$

Hence, an arbitrary vector $\mathbf{A}$ is expressed as

$$\mathbf{A} = A^1\mathbf{e}_1 + A^2\mathbf{e}_2 + A^3\mathbf{e}_3, \tag{1.2.34}$$

and a differential distance is denoted by

$$d\mathbf{r} = dq^1\mathbf{e}_1 + dq^2\mathbf{e}_2 + dq^3\mathbf{e}_3 = dq^i\mathbf{e}_i. \tag{1.2.35}$$

Observe that the A's and dq's have superscripts, whereas the unitary basis $(\mathbf{e}_1, \mathbf{e}_2, \mathbf{e}_3)$ has subscripts. The dq^i are referred to as the *contravariant components* of the differential vector $d\mathbf{r}$, and A^i are the contravariant components of vector $\mathbf{A}$. The unitary basis can be described in terms of the rectangular Cartesian basis $(\hat{\mathbf{e}}_x, \hat{\mathbf{e}}_y, \hat{\mathbf{e}}_z) = (\hat{\mathbf{e}}_1, \hat{\mathbf{e}}_2, \hat{\mathbf{e}}_3)$ as follows:

$$\mathbf{e}_1 = \frac{\partial \mathbf{r}}{\partial q^1} = \frac{\partial x}{\partial q^1}\hat{\mathbf{e}}_x + \frac{\partial y}{\partial q^1}\hat{\mathbf{e}}_y + \frac{\partial z}{\partial q^1}\hat{\mathbf{e}}_z,$$

$$\mathbf{e}_2 = \frac{\partial \mathbf{r}}{\partial q^2} = \frac{\partial x}{\partial q^2}\hat{\mathbf{e}}_x + \frac{\partial y}{\partial q^2}\hat{\mathbf{e}}_y + \frac{\partial z}{\partial q^2}\hat{\mathbf{e}}_z,$$

$$\mathbf{e}_3 = \frac{\partial \mathbf{r}}{\partial q^3} = \frac{\partial x}{\partial q^3}\hat{\mathbf{e}}_x + \frac{\partial y}{\partial q^3}\hat{\mathbf{e}}_y + \frac{\partial z}{\partial q^3}\hat{\mathbf{e}}_z.$$

In the summation convention, we have

$$\mathbf{e}_i \equiv \frac{\partial \mathbf{r}}{\partial q^i} = \frac{\partial x^j}{\partial q^i}\hat{\mathbf{e}}_j, \qquad i = 1, 2, 3. \tag{1.2.36}$$

Associated with any arbitrary basis is another basis that can be derived from it. We can construct this basis in the following way: Taking the scalar product of the vector $\mathbf{A}$ in Eq. (1.2.34) with the cross product $\mathbf{e}_1 \times \mathbf{e}_2$, we obtain

$$\mathbf{A} \cdot (\mathbf{e}_1 \times \mathbf{e}_2) = A^3 \mathbf{e}_3 \cdot (\mathbf{e}_1 \times \mathbf{e}_2)$$

since $\mathbf{e}_1 \times \mathbf{e}_2$ is perpendicular to both $\mathbf{e}_1$ and $\mathbf{e}_2$. Solving for A^3 gives

$$A^3 = \mathbf{A} \cdot \frac{\mathbf{e}_1 \times \mathbf{e}_2}{\mathbf{e}_3 \cdot (\mathbf{e}_1 \times \mathbf{e}_2)} = \mathbf{A} \cdot \frac{\mathbf{e}_1 \times \mathbf{e}_2}{[\mathbf{e}_1 \mathbf{e}_2 \mathbf{e}_3]}.$$

In similar fashion, we can obtain expressions for A^1 and A^2. Thus, we have

$$A^1 = \mathbf{A} \cdot \frac{\mathbf{e}_2 \times \mathbf{e}_3}{[\mathbf{e}_1 \mathbf{e}_2 \mathbf{e}_3]}, \quad A^2 = \mathbf{A} \cdot \frac{\mathbf{e}_3 \times \mathbf{e}_1}{[\mathbf{e}_1 \mathbf{e}_2 \mathbf{e}_3]}, \quad A^3 = \mathbf{A} \cdot \frac{\mathbf{e}_1 \times \mathbf{e}_2}{[\mathbf{e}_1 \mathbf{e}_2 \mathbf{e}_3]}. \tag{1.2.37}$$

We thus observe that we can obtain the components A^1, A^2, and A^3 by taking the scalar product of the vector $\mathbf{A}$ with special vectors, which we denote as follows:

$$\mathbf{e}^1 = \frac{\mathbf{e}_2 \times \mathbf{e}_3}{[\mathbf{e}_1 \mathbf{e}_2 \mathbf{e}_3]}, \quad \mathbf{e}^2 = \frac{\mathbf{e}_3 \times \mathbf{e}_1}{[\mathbf{e}_1 \mathbf{e}_2 \mathbf{e}_3]}, \quad \mathbf{e}^3 = \frac{\mathbf{e}_1 \times \mathbf{e}_2}{[\mathbf{e}_1 \mathbf{e}_2 \mathbf{e}_3]}. \tag{1.2.38}$$

The set of vectors $(\mathbf{e}^1, \mathbf{e}^2, \mathbf{e}^3)$ is called the *dual* or *reciprocal* basis. Notice from the basic definitions that we have the following relations:

$$\mathbf{e}^i \cdot \mathbf{e}_j = \delta_j^i = \begin{cases} 1, & i = j \\ 0, & i \neq j \end{cases} \tag{1.2.39}$$

It is possible, since the dual basis is linearly independent (the reader should verify this), to express a vector $\mathbf{A}$ in terms of the dual basis:

$$\mathbf{A} = A_1\mathbf{e}^1 + A_2\mathbf{e}^2 + A_3\mathbf{e}^3. \tag{1.2.40}$$

Notice now that the components associated with the dual basis have subscripts, and A_i are the *covariant components* of $\mathbf{A}$.

By an analogous process as that above, we can show that the original basis can be expressed in terms of the dual basis in the following way:

$$\mathbf{e}_1 = \frac{\mathbf{e}^2 \times \mathbf{e}^3}{[\mathbf{e}^1\mathbf{e}^2\mathbf{e}^3]}, \qquad \mathbf{e}_2 = \frac{\mathbf{e}^3 \times \mathbf{e}^1}{[\mathbf{e}^1\mathbf{e}^2\mathbf{e}^3]}, \qquad \mathbf{e}_3 = \frac{\mathbf{e}^1 \times \mathbf{e}^2}{[\mathbf{e}^1\mathbf{e}^2\mathbf{e}^3]}. \tag{1.2.41}$$

Of course in the evaluation of the cross products, we shall always use the right-hand rule. It follows from the above expressions that

$$\begin{aligned} A^1 = \mathbf{A} \cdot \mathbf{e}^1, \quad A^2 = \mathbf{A} \cdot \mathbf{e}^2, \quad A^3 = \mathbf{A} \cdot \mathbf{e}^3, \quad &\text{or} \quad A^i = \mathbf{A} \cdot \mathbf{e}^i, \\ A_1 = \mathbf{A} \cdot \mathbf{e}_1, \quad A_2 = \mathbf{A} \cdot \mathbf{e}_2, \quad A_3 = \mathbf{A} \cdot \mathbf{e}_3, \quad &\text{or} \quad A_i = \mathbf{A} \cdot \mathbf{e}_i. \end{aligned} \tag{1.2.42}$$

Returning to the scalar field ϕ, the differential change is given by

$$d\phi = \frac{\partial \phi}{\partial q^1} dq^1 + \frac{\partial \phi}{\partial q^2} dq^2 + \frac{\partial \phi}{\partial q^3} dq^3. \tag{1.2.43}$$

The differentials dq^1, dq^2, and dq^3 are components of $d\mathbf{r}$ (see Eq. (1.2.35)). We would now like to write $d\phi$ in such a way that we elucidate *the direction* as well as the magnitude of $d\mathbf{r}$. Since $\mathbf{e}^1 \cdot \mathbf{e}_1 = 1$, $\mathbf{e}^2 \cdot \mathbf{e}_2 = 1$, and $\mathbf{e}^3 \cdot \mathbf{e}_3 = 1$, we can write

$$\begin{aligned} d\phi &= \mathbf{e}^1 \frac{\partial \phi}{\partial q^1} \cdot \mathbf{e}_1 dq^1 + \mathbf{e}^2 \frac{\partial \phi}{\partial q^2} \cdot \mathbf{e}_2 dq^2 + \mathbf{e}^3 \frac{\partial \phi}{\partial q^3} \cdot \mathbf{e}_3 dq^3 \\ &= (dq^1\mathbf{e}_1 + dq^2\mathbf{e}_2 + dq^3\mathbf{e}_3) \cdot \left(\mathbf{e}^1 \frac{\partial \phi}{\partial q^1} + \mathbf{e}^2 \frac{\partial \phi}{\partial q^2} + \mathbf{e}^3 \frac{\partial \phi}{\partial q^3} \right) \\ &= d\mathbf{r} \cdot \left(\mathbf{e}^1 \frac{\partial \phi}{\partial q^1} + \mathbf{e}^2 \frac{\partial \phi}{\partial q^2} + \mathbf{e}^3 \frac{\partial \phi}{\partial q^3} \right). \end{aligned} \tag{1.2.44}$$

Let us now denote the magnitude of $d\mathbf{r}$ by $ds \equiv |d\mathbf{r}|$. Then $\hat{\mathbf{e}} = d\mathbf{r}/ds$ is a unit vector in the direction of $d\mathbf{r}$, and we have

$$\left(\frac{d\phi}{ds} \right)_{\hat{\mathbf{e}}} = \hat{\mathbf{e}} \cdot \left(\mathbf{e}^1 \frac{\partial \phi}{\partial q^1} + \mathbf{e}^2 \frac{\partial \phi}{\partial q^2} + \mathbf{e}^3 \frac{\partial \phi}{\partial q^3} \right). \tag{1.2.45}$$

The derivative $(d\phi/ds)_{\hat{\mathbf{e}}}$ is called the *directional derivative* of ϕ. We see that it is the *rate of change* of ϕ with respect to distance and that it depends on the direction $\hat{\mathbf{e}}$ in which the distance is taken.

The vector that is scalar multiplied by $\hat{\mathbf{e}}$ can be obtained immediately whenever the scalar field is given. Because the magnitude of this vector is equal to the maximum value of the directional derivative, it is called the *gradient vector* and is denoted by grad ϕ:

$$\text{grad } \phi \equiv \mathbf{e}^1 \frac{\partial \phi}{\partial q^1} + \mathbf{e}^2 \frac{\partial \phi}{\partial q^2} + \mathbf{e}^3 \frac{\partial \phi}{\partial q^3}. \tag{1.2.46}$$

From this representation it can be seen that

$$\frac{\partial \phi}{\partial q^1}, \qquad \frac{\partial \phi}{\partial q^2}, \qquad \frac{\partial \phi}{\partial q^3}$$

are the *covariant components* of the gradient vector.

When the scalar function $\phi(\mathbf{r})$ is set equal to a constant, $\phi(\mathbf{r}) = \text{constant}$, a family of surfaces is generated. A different surface is designated by different values of the constant, and each surface is called a *level surface* (see Fig. 1.2.10). If the direction in which the directional derivative is taken lies within a level surface, then $d\phi/ds$ is zero, since ϕ is a constant on a level surface. In this case the unit vector $\hat{\mathbf{e}}$ is tangent to a level surface. It follows, therefore, that if $d\phi/ds$ is zero, then grad ϕ must be perpendicular to $\hat{\mathbf{e}}$ and thus *perpendicular to a level surface*. Thus if any surface is given by $\phi(\mathbf{r}) = \text{constant}$, the unit normal to the surface is determined by

$$\hat{\mathbf{n}} = \pm \frac{\text{grad } \phi}{|\text{grad } \phi|}. \tag{1.2.47}$$

The plus or minus sign appears because the direction of $\hat{\mathbf{n}}$ may point in either direction away from the surface. If the surface is closed, the usual convention is to take $\hat{\mathbf{n}}$ pointing outward.

It is convenient to write the gradient vector as

$$\text{grad } \phi \equiv \left(\mathbf{e}^1 \frac{\partial}{\partial q^1} + \mathbf{e}^2 \frac{\partial}{\partial q^2} + \mathbf{e}^3 \frac{\partial}{\partial q^3} \right) \phi \tag{1.2.48}$$

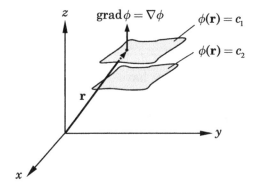

Fig. 1.2.10 Level surfaces and gradient to a surface.

and interpret grad ϕ as some operator operating on ϕ, that is, grad $\phi \equiv \boldsymbol{\nabla}\phi$. This operator is denoted by

$$\boldsymbol{\nabla} \equiv \mathbf{e}^1 \frac{\partial}{\partial q^1} + \mathbf{e}^2 \frac{\partial}{\partial q^2} + \mathbf{e}^3 \frac{\partial}{\partial q^3} \qquad (1.2.49)$$

and is called the "del operator." The del operator is a *vector differential* operator, and the "components" $\partial/\partial q^1$, $\partial/\partial q^2$, and $\partial/\partial q^3$ appear as covariant components.

It is important to note that whereas the del operator has some of the properties of a vector, it does not have them all, because it is an operator. For instance, $\boldsymbol{\nabla} \cdot \mathbf{A}$ is a scalar (called the divergence of $\mathbf{A}$), whereas $\mathbf{A} \cdot \boldsymbol{\nabla}$ is a scalar *differential operator*. Thus the del operator does not commute in this sense.

In the rectangular Cartesian system, we have the simple form

$$\boldsymbol{\nabla} \equiv \hat{\mathbf{e}}_x \frac{\partial}{\partial x} + \hat{\mathbf{e}}_y \frac{\partial}{\partial y} + \hat{\mathbf{e}}_z \frac{\partial}{\partial z},$$

or, in the summation convention, we have

$$\boldsymbol{\nabla} \equiv \hat{\mathbf{e}}_i \frac{\partial}{\partial x_i}. \qquad (1.2.50)$$

The dot product of del operator with a vector is called the *divergence of a vector* and denoted by

$$\boldsymbol{\nabla} \cdot \mathbf{A} \equiv \operatorname{div}\mathbf{A} = \frac{\partial A_i}{\partial x_i}. \qquad (1.2.51)$$

If we take the divergence of the gradient vector, we have

$$\operatorname{div}(\operatorname{grad}\,\phi) \equiv \boldsymbol{\nabla} \cdot \boldsymbol{\nabla}\phi = (\boldsymbol{\nabla} \cdot \boldsymbol{\nabla})\phi = \nabla^2\phi. \qquad (1.2.52)$$

The notation $\nabla^2 = \boldsymbol{\nabla}\cdot\boldsymbol{\nabla}$ is called the *Laplacian operator*. In Cartesian systems this reduces to the simple form

$$\nabla^2\phi = \frac{\partial^2\phi}{\partial x^2} + \frac{\partial^2\phi}{\partial y^2} + \frac{\partial^2\phi}{\partial z^2} = \frac{\partial^2\phi}{\partial x_i \partial x_i}. \qquad (1.2.53)$$

The Laplacian of a scalar appears frequently in the partial differential equations governing physical phenomena.

The curl of a vector is defined as the del operator operating on a vector by means of the cross product:

$$\operatorname{curl}\mathbf{A} = \boldsymbol{\nabla} \times \mathbf{A} = \hat{\mathbf{e}}_j \frac{\partial}{\partial x_j} \times \hat{\mathbf{e}}_k A_k = \frac{\partial A_k}{\partial x_j}\,(\hat{\mathbf{e}}_j \times \hat{\mathbf{e}}_k) = \frac{\partial A_k}{\partial x_j}\,\varepsilon_{jki}\,\hat{\mathbf{e}}_i. \qquad (1.2.54)$$

Thus the ith component of $(\boldsymbol{\nabla} \times \mathbf{A})$ is $\frac{\partial A_k}{\partial x_j}\,\varepsilon_{jki}$.

Example 1.2.5 ————————————————————————————————————

Using the index-summation notation, prove the following vector identity:

$$\nabla \times (\nabla \times \mathbf{v}) \equiv \nabla(\nabla \cdot \mathbf{v}) - \nabla^2 \mathbf{v},$$

where $\mathbf{v}$ is a vector function of the coordinates, x_i.

Solution: Observe that

$$\nabla \times (\nabla \times \mathbf{v}) = \hat{\mathbf{e}}_i \frac{\partial}{\partial x_i} \times \left(\hat{\mathbf{e}}_j \frac{\partial}{\partial x_j} \times v_k \hat{\mathbf{e}}_k \right)$$

$$= \hat{\mathbf{e}}_i \frac{\partial}{\partial x_i} \times \left(\varepsilon_{jk\ell} \frac{\partial v_k}{\partial x_j} \hat{\mathbf{e}}_\ell \right)$$

$$= \varepsilon_{i\ell m} \, \varepsilon_{jk\ell} \frac{\partial^2 v_k}{\partial x_i \partial x_j} \hat{\mathbf{e}}_m.$$

Using the ε-δ identity, we obtain

$$\nabla \times (\nabla \times \mathbf{v}) \equiv (\delta_{mj}\delta_{ik} - \delta_{mk}\delta_{ij}) \frac{\partial^2 v_k}{\partial x_i \partial x_j} \hat{\mathbf{e}}_m$$

$$= \frac{\partial^2 v_i}{\partial x_i \partial x_j} \hat{\mathbf{e}}_j - \frac{\partial^2 v_k}{\partial x_i \partial x_i} \hat{\mathbf{e}}_k = \hat{\mathbf{e}}_j \frac{\partial}{\partial x_j} \left(\frac{\partial v_i}{\partial x_i} \right) - \frac{\partial^2}{\partial x_i \partial x_i} (v_k \hat{\mathbf{e}}_k)$$

$$= \nabla(\nabla \cdot \mathbf{v}) - \nabla^2 \mathbf{v}.$$

This result is sometimes used as the definition of the Laplacian of a vector, that is,

$$\nabla^2 \mathbf{v} = \nabla(\nabla \cdot \mathbf{v}) - \nabla \times (\nabla \times \mathbf{v}).$$

A summary of vector operations in both general vector notation and in Cartesian component form is given in Table 1.2.1, and some useful vector operations for cylindrical and spherical coordinate systems (see Fig. 1.2.11) are presented in Table 1.2.2.

1.2.7 Gradient, Divergence, and Curl Theorems

Useful expressions for the integrals of the gradient, divergence, and curl of a vector can be established between volume integrals and surface integrals. Let Ω denote a region in space surrounded by the closed surface Γ. Let $d\Gamma$ be a differential element of surface and $\hat{\mathbf{n}}$ the unit outward normal, and let $d\Omega$ be a differential volume element. The following integral relations are proven to be useful in the coming chapters.

Gradient theorem:

$$\int_\Omega \operatorname{grad} \phi \, d\Omega = \oint_\Gamma \hat{\mathbf{n}} \phi \, d\Gamma \quad \left[\int_\Omega \hat{\mathbf{e}}_i \frac{\partial \phi}{\partial x_i} \, d\Omega = \oint_\Gamma \hat{\mathbf{e}}_i n_i \phi \, d\Gamma \right]. \tag{1.2.55}$$

Curl theorem:

$$\int_\Omega \text{curl } \mathbf{A} \, d\Omega = \oint_\Gamma \hat{\mathbf{n}} \times \mathbf{A} \, d\Gamma \quad \left[\int_\Omega \varepsilon_{ijk} \hat{\mathbf{e}}_k \frac{\partial A_j}{\partial x_i} \, d\Omega = \oint_\Gamma \varepsilon_{ijk} \hat{\mathbf{e}}_k n_i A_j \, d\Gamma \right].$$

$$(1.2.56)$$

Divergence theorem:

$$\int_\Omega \text{div } \mathbf{A} \, d\Omega = \oint_\Gamma \hat{\mathbf{n}} \cdot \mathbf{A} \, d\Gamma \quad \left[\int_\Omega \frac{\partial A_i}{\partial x_i} \, d\Omega = \oint_\Gamma n_i A_i \, d\Gamma \right]. \qquad (1.2.57)$$

Table 1.2.1 Vector expressions and their Cartesian component forms ($\mathbf{A}$, $\mathbf{B}$, and $\mathbf{C}$ are vector functions, U is a scalar function, $\mathbf{x}$ is the position vector, and ($\hat{\mathbf{e}}_1, \hat{\mathbf{e}}_2, \hat{\mathbf{e}}_3$) are the Cartesian unit vectors in a rectangular Cartesian coordinate system; see Fig. 1.2.9).

No.	Vector form and its equivalence	Component form
1.	$\mathbf{A} \cdot \mathbf{B}$	$A_i B_i$
2.	$\mathbf{A} \times \mathbf{B}$	$\varepsilon_{ijk} A_i B_j \hat{\mathbf{e}}_k$
3.	$\mathbf{A} \cdot (\mathbf{B} \times \mathbf{C})$	$\varepsilon_{ijk} A_i B_j C_k$
4.	$\mathbf{A} \times (\mathbf{B} \times \mathbf{C}) = \mathbf{B}(\mathbf{A} \cdot \mathbf{C}) - \mathbf{C}(\mathbf{A} \cdot \mathbf{B})$	$\varepsilon_{ijk} e_{klm} A_j B_l C_m \hat{\mathbf{e}}_i$
5.	$\nabla \mathbf{A}$	$\frac{\partial A_j}{\partial x_i} \hat{\mathbf{e}}_i \hat{\mathbf{e}}_j$
6.	$\nabla \cdot \mathbf{A}$	$\frac{\partial A_i}{\partial x_i}$
7.	$\nabla \times \mathbf{A}$	$\varepsilon_{ijk} \frac{\partial A_j}{\partial x_i} \hat{\mathbf{e}}_k$
8.	$\nabla \cdot (\nabla \times \mathbf{A}) = 0$	$\varepsilon_{ijk} \frac{\partial^2 A_j}{\partial x_i \partial x_k}$
9.	$\nabla \times (\nabla U) = 0$	$\varepsilon_{ijk} \hat{\mathbf{e}}_k \frac{\partial^2 U}{\partial x_i \partial x_j}$
10.	$\nabla \cdot (\mathbf{A} \times \mathbf{B}) = \mathbf{B} \cdot (\nabla \times \mathbf{A}) - \mathbf{A} \cdot (\nabla \times \mathbf{B})$	$\varepsilon_{ijk} \frac{\partial}{\partial x_i} (A_j B_k)$
11.	$(\nabla \times \mathbf{A}) \times \mathbf{B} = \mathbf{B} \cdot [\nabla \mathbf{A} - (\nabla \mathbf{A})^\mathrm{T}]$	$\varepsilon_{ijk} \varepsilon_{klm} B_l \frac{\partial A_j}{\partial x_i} \hat{\mathbf{e}}_m$
12.	$\mathbf{A} \times (\nabla \times \mathbf{A}) = \frac{1}{2} \nabla (\mathbf{A} \cdot \mathbf{A}) - (\mathbf{A} \cdot \nabla) \mathbf{A}$	$\varepsilon_{nim} \varepsilon_{jkm} A_i \frac{\partial A_k}{\partial x_j} \hat{\mathbf{e}}_n$
13.	$\nabla \cdot (\nabla \mathbf{A}) = \nabla^2 \mathbf{A}$	$\frac{\partial^2 A_j}{\partial x_i \partial x_i} \hat{\mathbf{e}}_j$
14.	$\nabla \times (\nabla \times \mathbf{A}) = \nabla (\nabla \cdot \mathbf{A}) - (\nabla \cdot \nabla) \mathbf{A}$	$\varepsilon_{mil} \varepsilon_{jkl} \frac{\partial^2 A_k}{\partial x_i \partial x_j} \hat{\mathbf{e}}_m$

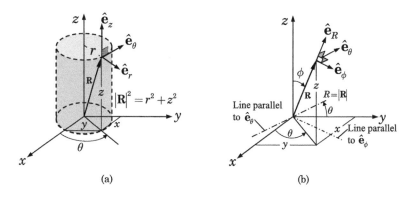

Fig. 1.2.11 (a) Cylindrical coordinate system. (b) Spherical coordinate system.

Table 1.2.2 Base vectors and operations with the del operator in cylindrical and spherical coordinate systems; see Fig. 1.2.11.

- *Cylindrical coordinate system* (r, θ, z)

$x = r \cos \theta, \ y = r \sin \theta, \ z = z, \quad \mathbf{R} = r\,\hat{\mathbf{e}}_r + z\,\hat{\mathbf{e}}_z, \quad \mathbf{A} = A_r\hat{\mathbf{e}}_r + A_\theta\hat{\mathbf{e}}_\theta + A_z\hat{\mathbf{e}}_z$ (a vector)

$\hat{\mathbf{e}}_r = \cos\theta\,\hat{\mathbf{e}}_x + \sin\theta\,\hat{\mathbf{e}}_y, \ \hat{\mathbf{e}}_\theta = -\sin\theta\,\hat{\mathbf{e}}_x + \cos\theta\,\hat{\mathbf{e}}_y, \ \hat{\mathbf{e}}_z = \hat{\mathbf{e}}_z$

$\frac{\partial\hat{\mathbf{e}}_r}{\partial\theta} = -\sin\theta\,\hat{\mathbf{e}}_x + \cos\theta\,\hat{\mathbf{e}}_y = \hat{\mathbf{e}}_\theta, \ \frac{\partial\hat{\mathbf{e}}_\theta}{\partial\theta} = -\cos\theta\,\hat{\mathbf{e}}_x - \sin\theta\,\hat{\mathbf{e}}_y = -\hat{\mathbf{e}}_r$

All other derivatives of the base vectors are zero.

$\nabla = \hat{\mathbf{e}}_r\frac{\partial}{\partial r} + \frac{1}{r}\hat{\mathbf{e}}_\theta\frac{\partial}{\partial\theta} + \hat{\mathbf{e}}_z\frac{\partial}{\partial z}, \quad \nabla^2 = \frac{1}{r}\left[\frac{\partial}{\partial r}\left(r\frac{\partial}{\partial r}\right) + \frac{1}{r}\frac{\partial^2}{\partial\theta^2} + r\frac{\partial^2}{\partial z^2}\right]$

$\nabla \cdot \mathbf{A} = \frac{1}{r}\left[\frac{\partial(rA_r)}{\partial r} + \frac{\partial A_\theta}{\partial\theta} + r\frac{\partial A_z}{\partial z}\right]$

$\nabla \times \mathbf{A} = \left(\frac{1}{r}\frac{\partial A_z}{\partial\theta} - \frac{\partial A_\theta}{\partial z}\right)\hat{\mathbf{e}}_r + \left(\frac{\partial A_r}{\partial z} - \frac{\partial A_z}{\partial r}\right)\hat{\mathbf{e}}_\theta + \frac{1}{r}\left[\frac{\partial(rA_\theta)}{\partial r} - \frac{\partial A_r}{\partial\theta}\right]\hat{\mathbf{e}}_z$

$\nabla\mathbf{A} = \frac{\partial A_r}{\partial r}\hat{\mathbf{e}}_r\hat{\mathbf{e}}_r + \frac{\partial A_\theta}{\partial r}\hat{\mathbf{e}}_r\hat{\mathbf{e}}_\theta + \frac{1}{r}\left(\frac{\partial A_r}{\partial\theta} - A_\theta\right)\hat{\mathbf{e}}_\theta\hat{\mathbf{e}}_r + \frac{\partial A_z}{\partial r}\hat{\mathbf{e}}_r\hat{\mathbf{e}}_z + \frac{\partial A_r}{\partial z}\hat{\mathbf{e}}_z\hat{\mathbf{e}}_r$

$\qquad + \frac{1}{r}\left(A_r + \frac{\partial A_\theta}{\partial\theta}\right)\hat{\mathbf{e}}_\theta\hat{\mathbf{e}}_\theta + \frac{1}{r}\frac{\partial A_z}{\partial\theta}\hat{\mathbf{e}}_\theta\hat{\mathbf{e}}_z + \frac{\partial A_\theta}{\partial z}\hat{\mathbf{e}}_z\hat{\mathbf{e}}_\theta + \frac{\partial A_z}{\partial z}\hat{\mathbf{e}}_z\hat{\mathbf{e}}_z$

- *Spherical coordinate system* (R, ϕ, θ)

$x = R\sin\phi\cos\theta, \ y = R\sin\phi\sin\theta, \ z = R\cos\phi, \quad \mathbf{R} = R\hat{\mathbf{e}}_R, \quad \mathbf{A} = A_R\hat{\mathbf{e}}_R + A_\phi\hat{\mathbf{e}}_\phi + A_\theta\hat{\mathbf{e}}_\theta$

$\hat{\mathbf{e}}_R = \sin\phi\,(\cos\theta\,\hat{\mathbf{e}}_x + \sin\theta\,\hat{\mathbf{e}}_y) + \cos\phi\,\hat{\mathbf{e}}_z, \ \hat{\mathbf{e}}_\phi = \cos\phi\,(\cos\theta\,\hat{\mathbf{e}}_x + \sin\theta\,\hat{\mathbf{e}}_y) - \sin\phi\,\hat{\mathbf{e}}_z$

$\hat{\mathbf{e}}_\theta = -\sin\theta\,\hat{\mathbf{e}}_x + \cos\theta\,\hat{\mathbf{e}}_y$

$\hat{\mathbf{e}}_x = \cos\theta\,(\sin\phi\,\hat{\mathbf{e}}_R + \cos\phi\,\hat{\mathbf{e}}_\phi) - \sin\theta\,\hat{\mathbf{e}}_\theta, \ \hat{\mathbf{e}}_y = \sin\theta\,(\sin\phi\,\hat{\mathbf{e}}_R + \cos\phi\,\hat{\mathbf{e}}_\phi) + \cos\theta\,\hat{\mathbf{e}}_\theta$

$\hat{\mathbf{e}}_z = \cos\phi\,\hat{\mathbf{e}}_R - \sin\phi\,\hat{\mathbf{e}}_\phi$

$\frac{\partial\hat{\mathbf{e}}_R}{\partial\phi} = \hat{\mathbf{e}}_\phi, \ \frac{\partial\hat{\mathbf{e}}_R}{\partial\theta} = \sin\phi\,\hat{\mathbf{e}}_\theta, \ \frac{\partial\hat{\mathbf{e}}_\phi}{\partial\phi} = -\hat{\mathbf{e}}_R, \ \frac{\partial\hat{\mathbf{e}}_\phi}{\partial\theta} = \cos\phi\,\hat{\mathbf{e}}_\theta, \ \frac{\partial\hat{\mathbf{e}}_\theta}{\partial\theta} = -\sin\phi\,\hat{\mathbf{e}}_R - \cos\phi\,\hat{\mathbf{e}}_\phi$

All other derivatives of the base vectors are zero.

$\nabla = \hat{\mathbf{e}}_R\frac{\partial}{\partial R} + \frac{\hat{\mathbf{e}}_\phi}{R}\frac{\partial}{\partial\phi} + \frac{\hat{\mathbf{e}}_\theta}{R\sin\phi}\frac{\partial}{\partial\theta}, \ \nabla^2 = \frac{1}{R^2}\left[\frac{\partial}{\partial R}\left(R^2\frac{\partial}{\partial R}\right) + \frac{1}{\sin\phi}\frac{\partial}{\partial\phi}\left(\sin\phi\frac{\partial}{\partial\phi}\right) + \frac{1}{\sin^2\phi}\frac{\partial^2}{\partial\theta^2}\right]$

$\nabla \cdot \mathbf{A} = \frac{2A_R}{R} + \frac{\partial A_R}{\partial R} + \frac{1}{R\sin\phi}\frac{\partial(A_\phi\sin\phi)}{\partial\phi} + \frac{1}{R\sin\phi}\frac{\partial A_\theta}{\partial\theta}$

$\nabla \times \mathbf{A} = \frac{1}{R\sin\phi}\left[\frac{\partial(\sin\phi A_\theta)}{\partial\phi} - \frac{\partial A_\phi}{\partial\theta}\right]\hat{\mathbf{e}}_R + \left[\frac{1}{R\sin\phi}\frac{\partial A_R}{\partial\theta} - \frac{1}{R}\frac{\partial(RA_\theta)}{\partial R}\right]\hat{\mathbf{e}}_\phi$

$\qquad + \frac{1}{R}\left[\frac{\partial(RA_\phi)}{\partial R} - \frac{\partial A_R}{\partial\phi}\right]\hat{\mathbf{e}}_\theta$

$\nabla\mathbf{A} = \frac{\partial A_R}{\partial R}\hat{\mathbf{e}}_R\hat{\mathbf{e}}_R + \frac{\partial A_\phi}{\partial R}\hat{\mathbf{e}}_R\hat{\mathbf{e}}_\phi + \frac{1}{R}\left(\frac{\partial A_R}{\partial\phi} - A_\phi\right)\hat{\mathbf{e}}_\phi\hat{\mathbf{e}}_R + \frac{\partial A_\theta}{\partial R}\hat{\mathbf{e}}_R\hat{\mathbf{e}}_\theta$

$\qquad + \frac{1}{R\sin\phi}\left(\frac{\partial A_R}{\partial\theta} - A_\theta\sin\phi\right)\hat{\mathbf{e}}_\theta\hat{\mathbf{e}}_R + \frac{1}{R}\left(A_R + \frac{\partial A_\phi}{\partial\phi}\right)\hat{\mathbf{e}}_\phi\hat{\mathbf{e}}_\phi + \frac{1}{R}\frac{\partial A_\theta}{\partial\phi}\hat{\mathbf{e}}_\phi\hat{\mathbf{e}}_\theta$

$\qquad + \frac{1}{R\sin\phi}\left(\frac{\partial A_\phi}{\partial\theta} - A_\theta\cos\phi\right)\hat{\mathbf{e}}_\theta\hat{\mathbf{e}}_\phi + \frac{1}{R\sin\phi}\left(A_R\sin\phi + A_\phi\cos\phi + \frac{\partial A_\theta}{\partial\theta}\right)\hat{\mathbf{e}}_\theta\hat{\mathbf{e}}_\theta$

Let $\mathbf{A} = \operatorname{grad}\phi$ in Eq. (1.2.57). Then the divergence theorem gives

$$\int_\Omega \operatorname{div}(\operatorname{grad}\phi)\,dv \equiv \int_\Omega \nabla^2\phi\,dv = \oint_\Gamma \hat{\mathbf{n}}\cdot\operatorname{grad}\phi\,ds. \qquad (1.2.58)$$

The quantity $\hat{\mathbf{n}}\cdot\operatorname{grad}\phi$ is called the *normal derivative* of ϕ on the surface s and is denoted by (n is the coordinate along the unit normal vector $\hat{\mathbf{n}}$)

$$\frac{\partial\phi}{\partial n} \equiv \hat{\mathbf{n}}\cdot\operatorname{grad}\phi = \hat{\mathbf{n}}\cdot\nabla\phi. \qquad (1.2.59)$$

In a Cartesian system, this becomes

$$\frac{\partial \phi}{\partial n} = \frac{\partial \phi}{\partial x}n_x + \frac{\partial \phi}{\partial y}n_y + \frac{\partial \phi}{\partial z}n_z,$$

where n_x, n_y and n_z are the direction cosines of the unit normal,

$$\hat{\mathbf{n}} = n_x\hat{\mathbf{e}}_x + n_y\hat{\mathbf{e}}_y + n_z\hat{\mathbf{e}}_z. \tag{1.2.60}$$

The next example illustrates the relation between the integral relations Eqs. (1.2.55) to (1.2.57) and the so-called integration by parts.

Example 1.2.6 ─────────────────────────────────

Consider a rectangular region $R = \{(x,y) : 0 < x < a, 0 < y < b\}$ with boundary C, which is the union of line segments C_1, C_2, C_3, and C_4 (see Fig. 1.2.12). Evaluate the integral $\int_R \nabla^2 \phi\, dx dy$ over the rectangular region.

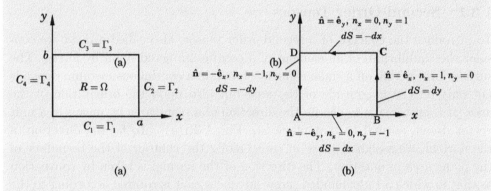

Fig. 1.2.12 Integration over rectangular regions.

Solution: From Eq. (1.2.58) we have

$$\int_R \nabla^2 \phi\, dx dy = \int_R \nabla \cdot (\nabla \phi)\, dx dy = \oint_C \frac{\partial \phi}{\partial n}\, ds.$$

The line integral can be simplified for the region under consideration as follows (note that in two dimensions, the volume integral becomes an area integral):

$$\oint_C \frac{\partial \phi}{\partial n}\, ds = \int_{C_1} \frac{\partial \phi}{\partial n}\, ds + \int_{C_2} \frac{\partial \phi}{\partial n}\, ds + \int_{C_3} \frac{\partial \phi}{\partial n}\, ds + \int_{C_4} \frac{\partial \phi}{\partial n}\, ds$$

$$= \int_0^a \left(-\frac{\partial \phi}{\partial y}\right)\Bigg|_{y=0} dx + \int_0^b \left(\frac{\partial \phi}{\partial x}\right)\Bigg|_{x=a} dy$$

$$+ \int_a^0 \left(\frac{\partial \phi}{\partial y}\right)\Bigg|_{y=b} (-dx) + \int_b^0 \left(-\frac{\partial \phi}{\partial x}\right)\Bigg|_{x=0} (-dy)$$

$$= \int_0^a \left[\left(\frac{\partial \phi}{\partial y}\right)_{y=b} - \left(\frac{\partial \phi}{\partial y}\right)_{y=0}\right] dx + \int_0^b \left[\left(\frac{\partial \phi}{\partial x}\right)_{x=a} - \left(\frac{\partial \phi}{\partial x}\right)_{x=0}\right] dy.$$

The same result can be obtained by means of integration by parts:

$$\int_R \nabla^2 \phi \, dx dy = \int_0^b \int_0^a \left(\frac{\partial^2 \phi}{\partial x^2} + \frac{\partial^2 \phi}{\partial y^2} \right) dx dy$$

$$= \int_0^b \int_0^a \frac{\partial}{\partial x} \left(\frac{\partial \phi}{\partial x} \right) dx dy + \int_0^a \int_0^b \frac{\partial}{\partial y} \left(\frac{\partial \phi}{\partial y} \right) dy dx$$

$$= \int_0^b \left(\frac{\partial \phi}{\partial x} \right) \Big|_{x=0}^{x=a} dy + \int_0^a \left(\frac{\partial \phi}{\partial y} \right) \Big|_{y=0}^{y=b} dx$$

$$= \int_0^b \left[\left(\frac{\partial \phi}{\partial x} \right)_{x=a} - \left(\frac{\partial \phi}{\partial x} \right)_{x=0} \right] dy + \int_0^a \left[\left(\frac{\partial \phi}{\partial y} \right)_{y=b} - \left(\frac{\partial \phi}{\partial y} \right)_{y=0} \right] dx.$$

Thus integration by parts is a special case of the gradient or the divergence theorem.

1.3 Tensors

1.3.1 Second-Order Tensors

To introduce the concept of a second-order tensor, also called a *dyad*, we consider the equilibrium of an element of a continuum acted upon by forces. The surface force acting on a small element of area in a continuous medium depends not only on the magnitude of the area but also upon the orientation of the area. It is customary to denote the direction of a plane area by means of a unit vector drawn normal to that plane [see Fig. 1.3.1(a)]. To fix the direction of the normal, we assign a *sense of travel* along the contour of the boundary of the plane area in question. The direction of the normal is taken by convention as that in which a right-handed screw advances as it is rotated according to the sense of travel along the boundary curve or contour [see Fig. 1.3.1(b)]. Let the unit normal vector be given by $\hat{\mathbf{n}}$. Then the area can be denoted by $\mathbf{s} = s\hat{\mathbf{n}}$.

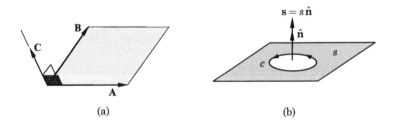

(a)   (b)

Fig. 1.3.1 (a) Plane area as a vector. (b) Unit normal vector and sense of travel.

If we denote by $\Delta \mathbf{F}(\hat{\mathbf{n}})$ the force on a small area $\hat{\mathbf{n}} \Delta s = \Delta \mathbf{s}$ located at the position $\mathbf{r}$ (see Fig. 1.3.2), the *stress vector* can be defined as follows:

$$\mathbf{t}(\hat{\mathbf{n}}) = \lim_{\Delta s \to 0} \frac{\Delta \mathbf{F}(\hat{\mathbf{n}})}{\Delta s}. \tag{1.3.1}$$

We see that the stress vector is a point function of the unit normal $\hat{n}$, which denotes the orientation of the surface Δs. The component of $\mathbf{t}$ that is in the direction of $\hat{n}$ is called the *normal stress*. The component of $\mathbf{t}$ that is normal to $\hat{n}$ is called a *shear* stress. Because of Newton's third law for action and reaction, we see that $\mathbf{t}(-\hat{n}) = -\mathbf{t}(\hat{n})$.

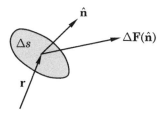

Fig. 1.3.2 Force on an area element.

At a fixed point $\mathbf{r}$ for each given unit vector $\hat{n}$, there is a stress vector $\mathbf{t}(\hat{n})$ acting on the plane normal to $\hat{n}$. Note that $\mathbf{t}(\hat{n})$ is, in general, not in the direction of $\hat{n}$. It is fruitful to establish a relationship between $\mathbf{t}$ and $\hat{n}$. To do this we now set up an infinitesimal tetrahedron in Cartesian coordinates, as shown in Fig. 1.3.3.

If $-\mathbf{t}_1, -\mathbf{t}_2, -\mathbf{t}_3$, and $\mathbf{t}$ denote the stress vectors in the outward directions on the faces of the infinitesimal tetrahedron whose areas are $\Delta s_1, \Delta s_2, \Delta s_3$, and Δs, respectively, we have by Newton's second law for the mass inside the tetrahedron:

$$\mathbf{t}\Delta s - \mathbf{t}_1 \Delta s_1 - \mathbf{t}_2 \Delta s_2 - \mathbf{t}_3 \Delta s_3 + \rho \Delta v \mathbf{f} = \rho \Delta v \mathbf{a}, \qquad (1.3.2)$$

where Δv is the volume of the tetrahedron, ρ is the density, $\mathbf{f}$ is the body force per unit mass, and $\mathbf{a}$ is the acceleration. Since the total vector area of a closed surface is zero (see the gradient theorem; set $\phi = 1$ in Eq. (1.2.55)), we have

$$\Delta s \hat{n} - \Delta s_1 \hat{e}_1 - \Delta s_2 \hat{e}_2 - \Delta s_3 \hat{e}_3 = \mathbf{0}.$$

It follows that

$$\Delta s_1 = (\hat{n} \cdot \hat{e}_1)\Delta s, \quad \Delta s_2 = (\hat{n} \cdot \hat{e}_2)\Delta s, \quad \Delta s_3 = (\hat{n} \cdot \hat{e}_3)\Delta s. \qquad (1.3.3)$$

The volume of the element Δv can be expressed as

$$\Delta v = \frac{\Delta h}{3}\Delta s, \qquad (1.3.4)$$

where Δh is the perpendicular distance from the origin to the slant face. The result in Eq. (1.3.4) can also be obtained from the divergence theorem in Eq. (1.2.57) by setting $\mathbf{A} = \mathbf{r}$.

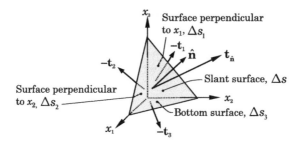

Fig. 1.3.3 Tetrahedral element in Cartesian coordinates.

Substitution of Eqs (1.3.3) and (1.3.4) in Eq. (1.3.2) and dividing throughout by Δs reduces it to

$$\mathbf{t} = (\hat{\mathbf{n}} \cdot \hat{\mathbf{e}}_1)\mathbf{t}_1 + (\hat{\mathbf{n}} \cdot \hat{\mathbf{e}}_2)\mathbf{t}_2 + (\hat{\mathbf{n}} \cdot \hat{\mathbf{e}}_3)\mathbf{t}_3 + \rho\frac{\Delta h}{3}(\mathbf{a} - \mathbf{f}).$$

In the limit when the tetrahedron shrinks to a point, $\Delta h \to 0$, we are left with

$$\begin{aligned}
\mathbf{t} &= (\hat{\mathbf{n}} \cdot \hat{\mathbf{e}}_1)\mathbf{t}_1 + (\hat{\mathbf{n}} \cdot \hat{\mathbf{e}}_2)\mathbf{t}_2 + (\hat{\mathbf{n}} \cdot \hat{\mathbf{e}}_3)\mathbf{t}_3 \\
&= (\hat{\mathbf{n}} \cdot \hat{\mathbf{e}}_i)\mathbf{t}_i.
\end{aligned} \tag{1.3.5}$$

It is now convenient to display the above equation as

$$\mathbf{t} = \hat{\mathbf{n}} \cdot (\hat{\mathbf{e}}_1\mathbf{t}_1 + \hat{\mathbf{e}}_2\mathbf{t}_2 + \hat{\mathbf{e}}_3\mathbf{t}_3). \tag{1.3.6}$$

The terms in the parenthesis are to be treated as a dyad, called *stress dyad* or *stress tensor* $\boldsymbol{\sigma}$:

$$\boldsymbol{\sigma} \equiv \hat{\mathbf{e}}_1\mathbf{t}_1 + \hat{\mathbf{e}}_2\mathbf{t}_2 + \hat{\mathbf{e}}_3\mathbf{t}_3. \tag{1.3.7}$$

The stress tensor is a property of the medium that is independent of the $\hat{\mathbf{n}}$. Thus, we have

$$\mathbf{t}(\hat{\mathbf{n}}) = \hat{\mathbf{n}} \cdot \boldsymbol{\sigma} \quad (t_i = n_j\sigma_{ji}) \tag{1.3.8}$$

and the dependence of $\mathbf{t}$ on $\hat{\mathbf{n}}$ has been explicitly displayed. Equation (1.3.8) is known as *Cauchy's formula*.

It is useful to resolve the stress vectors $\mathbf{t}_1, \mathbf{t}_2$, and $\mathbf{t}_3$ into their orthogonal components. We have

$$\begin{aligned}
\mathbf{t}_i &= \sigma_{i1}\hat{\mathbf{e}}_1 + \sigma_{i2}\hat{\mathbf{e}}_2 + \sigma_{i3}\hat{\mathbf{e}}_3 \\
&= \sigma_{ij}\hat{\mathbf{e}}_j
\end{aligned} \tag{1.3.9}$$

for $i = 1, 2, 3$. Hence, the stress dyad can be expressed in summation notation as

$$\boldsymbol{\sigma} = \hat{\mathbf{e}}_i\mathbf{t}_i = \sigma_{ij}\hat{\mathbf{e}}_i\hat{\mathbf{e}}_j. \tag{1.3.10}$$

The component σ_{ij} represents the stress (force per unit area) on an area perpendicular to the ith coordinate and in the jth coordinate direction (see Fig. 1.3.4). The stress vector **t** represents the vectorial stress on an area perpendicular to the direction $\hat{n}$. Equation (1.3.8) is known as the *Cauchy stress formula* and σ is termed the *Cauchy stress tensor*.

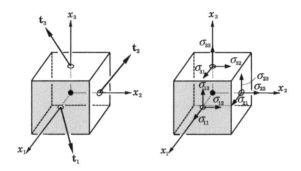

Fig. 1.3.4 Definition of stress components in Cartesian rectangular coordinates.

1.3.2 General Properties of a Dyadic

Because of its utilization in physical applications, a dyad is defined as two vectors standing side by side and acting as a unit. A linear combination of dyads is called a *dyadic*. Let $\mathbf{A}_1, \mathbf{A}_2, \ldots, \mathbf{A}_n$ and $\mathbf{B}_1, \mathbf{B}_2, \cdots, \mathbf{B}_n$ be arbitrary vectors. Then we can represent a dyadic as

$$\mathbf{\Phi} = \mathbf{A}_1\mathbf{B}_1 + \mathbf{A}_2\mathbf{B}_2 + \cdots + \mathbf{A}_n\mathbf{B}_n. \tag{1.3.11}$$

Here, we limit our discussion to Cartesian tensors. For a Cartesian tensor, the basis vectors are constants and thus do not take roles as variables in differentiation and integration.

One of the properties of a dyadic is defined by the dot product with a vector, say **V**:

$$\begin{aligned}
\mathbf{\Phi} \cdot \mathbf{V} &= \mathbf{A}_1(\mathbf{B}_1 \cdot \mathbf{V}) + \mathbf{A}_2(\mathbf{B}_2 \cdot \mathbf{V}) + \cdots + \mathbf{A}_n(\mathbf{B}_n \cdot \mathbf{V}), \\
\mathbf{V} \cdot \mathbf{\Phi} &= (\mathbf{V} \cdot \mathbf{A}_1)\mathbf{B}_1 + (\mathbf{V} \cdot \mathbf{A}_2)\mathbf{B}_2 + \cdots + (\mathbf{V} \cdot \mathbf{A}_n)\mathbf{B}_n.
\end{aligned} \tag{1.3.12}$$

The dot operation with a vector produces another vector. In the first case, the dyadic acts as a *prefactor* and in the second case as a *postfactor*. The two operations in general produce different vectors.

The conjugate, or transpose, of a dyadic is defined as the result obtained by the interchange of the two vectors in each of the dyads:

$$\mathbf{\Phi}^{\mathrm{T}} = \mathbf{B}_1\mathbf{A}_1 + \mathbf{B}_2\mathbf{A}_2 + \cdots + \mathbf{B}_n\mathbf{A}_n. \tag{1.3.13}$$

It is clear that we have

$$\mathbf{V} \cdot \mathbf{\Phi} = \mathbf{\Phi}^{\mathrm{T}} \cdot \mathbf{V},$$
$$\mathbf{\Phi} \cdot \mathbf{V} = \mathbf{V} \cdot \mathbf{\Phi}^{\mathrm{T}}. \tag{1.3.14}$$

1.3.3 Nonion Form and Matrix Representation of a Dyad

We can display all of the components of a dyad $\mathbf{\Phi} = \phi_{ij}\, \hat{\mathbf{e}}_i \hat{\mathbf{e}}_j$ by letting the j index run to the right and the i index run downward:

$$\begin{aligned} \mathbf{\Phi} = \; & \phi_{11}\hat{\mathbf{e}}_1\hat{\mathbf{e}}_1 + \phi_{12}\hat{\mathbf{e}}_1\hat{\mathbf{e}}_2 + \phi_{13}\hat{\mathbf{e}}_1\hat{\mathbf{e}}_3 \\ & + \phi_{21}\hat{\mathbf{e}}_2\hat{\mathbf{e}}_1 + \phi_{22}\hat{\mathbf{e}}_2\hat{\mathbf{e}}_2 + \phi_{23}\hat{\mathbf{e}}_2\hat{\mathbf{e}}_3 \\ & + \phi_{31}\hat{\mathbf{e}}_3\hat{\mathbf{e}}_1 + \phi_{32}\hat{\mathbf{e}}_3\hat{\mathbf{e}}_2 + \phi_{33}\hat{\mathbf{e}}_3\hat{\mathbf{e}}_3. \end{aligned} \tag{1.3.15}$$

This form is called the *nonion* form. Equation (1.3.15) illustrates that a dyad in three-dimensional space, or what we shall call a second-order tensor, has nine independent components in general, each component associated with a certain dyadic pair. The components are thus said to be ordered. When the ordering is understood, such as suggested by the nonion form in Eq. (1.3.15), the explicit writing of the dyads can be suppressed and the dyadic written as an array:

$$[\mathbf{\Phi}] = \begin{bmatrix} \phi_{11} & \phi_{12} & \phi_{13} \\ \phi_{21} & \phi_{22} & \phi_{23} \\ \phi_{31} & \phi_{32} & \phi_{33} \end{bmatrix} \quad \text{and} \quad \mathbf{\Phi} = \begin{Bmatrix} \hat{\mathbf{e}}_1 \\ \hat{\mathbf{e}}_2 \\ \hat{\mathbf{e}}_3 \end{Bmatrix}^{\mathrm{T}} [\mathbf{\Phi}] \begin{Bmatrix} \hat{\mathbf{e}}_1 \\ \hat{\mathbf{e}}_2 \\ \hat{\mathbf{e}}_3 \end{Bmatrix}. \tag{1.3.16}$$

This representation is simpler than Eq. (1.3.16), but it is taken to mean the same.

This rectangular array $[\mathbf{\Phi}]$ of scalars ϕ_{ij} is called a *matrix*, and the quantities ϕ_{ij} are called the *elements* of $[\mathbf{\Phi}]$.[4] If a matrix has m rows and n columns, we say that is m by n ($m \times n$), the number of rows is always being listed first. The element in the ith row and jth column of a matrix $[A]$ is generally denoted by a_{ij}, and we will sometimes designate a matrix by $[A] = [a_{ij}]$. A square matrix is one that has the same number of rows as columns. An $n \times n$ matrix is said to be of *order n*. The elements of a square matrix for which the row number and the column number are the same (i.e., a_{ij} for $i = j$) are called *diagonal elements* or simply the *diagonal*. A square matrix is said to be a *diagonal matrix* if all of the off-diagonal elements are zero. An *identity matrix*, denoted by $[I]$ (i.e., matrix representation of the second-order identity tensor $\mathbf{I}$), is a diagonal matrix whose elements are all 1's.

---

[4]The word "matrix" was first used in 1850 by James Sylvester (1814–1897), an English algebraist. However, Arthur Caley (1821–1895), professor of mathematics at Cambridge, was the first one to explore properties of matrices. Significant contributions in the early years were made by Charles Hermite, Georg Frobenius, and Camille Jordan, among others.

If the matrix has only one row or one column, we normally use only a single subscript to designate its elements. For example,

$$\{X\} = \begin{Bmatrix} x_1 \\ x_2 \\ x_3 \end{Bmatrix} \quad \text{and} \quad \{Y\} = \{y_1 \ y_2 \ y_3\}$$

denote a column matrix and a row matrix, respectively. Row and column matrices can be used to represent the components of a vector.

The reader is expected to have a working knowledge of matrix theory, that is, addition of matrices, multiplication of a matrix by a scalar, and product of two matrices, determinant of a matrix, inverse of a matrix, and so on. Readers who wish to refresh their background on this topic may consult the textbooks [73, 74].

In the general scheme that is thus developed, vectors are called *first-order tensors* and dyads are called *second-order tensors*. Scalars are called *zeroth-order tensors*. The generalization to *third-order tensors* thus leads, or is derived from, *triads*, or three vectors standing side by side. It follows that higher-order tensors are developed from *polyads*.

Example 1.3.1

With reference to a rectangular Cartesian system (x_1, x_2, x_3), the components of the stress dyadic at a certain point of a continuous medium are given by

$$[\sigma] = \begin{bmatrix} 200 & 400 & 300 \\ 400 & 0 & 0 \\ 300 & 0 & -100 \end{bmatrix} \text{psi.}$$

Determine the stress vector **t** at the point and normal to the plane, $p(x_1, x_2, x_3) = x_1 + 2x_2 + 2x_3 - 6 = 0$, and then compute the normal and tangential components of the stress vector at the point.

Solution: First we should find the unit normal to the plane on which we are required to find the stress vector. The unit normal is given by (see Eq. (1.2.47))

$$\hat{\mathbf{n}} = \frac{\nabla p}{|\nabla p|}, \qquad p(x_1, x_2, x_3) = x_1 + 2x_2 + 2x_3 - 6,$$

$$\hat{\mathbf{n}} = \frac{1}{3}(\hat{\mathbf{e}}_1 + 2\hat{\mathbf{e}}_2 + 2\hat{\mathbf{e}}_3).$$

The components of the stress vector are displayed in an array

$$\begin{Bmatrix} t_1 \\ t_2 \\ t_3 \end{Bmatrix} = \begin{bmatrix} 200 & 400 & 300 \\ 400 & 0 & 0 \\ 300 & 0 & -100 \end{bmatrix} \frac{1}{3} \begin{Bmatrix} 1 \\ 2 \\ 2 \end{Bmatrix} = \frac{1}{3} \begin{Bmatrix} 1600 \\ 400 \\ 100 \end{Bmatrix} \text{psi,}$$

or

$$\mathbf{t}(\hat{\mathbf{n}}) = \frac{1}{3}(1600\hat{\mathbf{e}}_1 + 400\hat{\mathbf{e}}_2 + 100\hat{\mathbf{e}}_3) \text{ psi.}$$

The normal component t_n of the stress vector **t** on the plane is given by

$$t_n = \mathbf{t}(\hat{\mathbf{n}}) \cdot \hat{\mathbf{n}} = \frac{2600}{9} \text{ psi},$$

and the tangential component is given by (the Pythagorean theorem)

$$t_s = \sqrt{|\mathbf{t}|^2 - t_n^2} = \frac{10^2}{9}\sqrt{(256 + 16 + 1)9 - 26 \times 26} \text{ psi}$$

$$= 100\frac{\sqrt{1781}}{9} = 468.9 \text{ psi}.$$

A second-order Cartesian tensor $\boldsymbol{\Phi}$ may be represented in unbarred and barred coordinate systems as

$$\boldsymbol{\Phi} = \phi_{ij}\hat{\mathbf{e}}_i\hat{\mathbf{e}}_j$$
$$= \bar{\phi}_{kl}\hat{\bar{\mathbf{e}}}_k\hat{\bar{\mathbf{e}}}_l. \tag{1.3.17}$$

The unit base vectors in the unbarred and barred systems are related by

$$\hat{\mathbf{e}}_i = \frac{\partial \bar{x}_j}{\partial x_i}\hat{\bar{\mathbf{e}}}_j \equiv \beta_{ji}\hat{\bar{\mathbf{e}}}_j, \quad \beta_{ij} = \hat{\mathbf{e}}_i \cdot \hat{\bar{\mathbf{e}}}_j, \tag{1.3.18}$$

where β_{ij} denote the direction cosines between unbarred and barred systems (see Eq. (1.2.29)). Thus the components of a second-order tensor transform according to

$$\bar{\phi}_{kl} = \phi_{ij}\beta_{ki}\beta_{lj} \quad \text{or} \quad [\bar{\phi}] = [\beta][\phi][\beta]^\mathrm{T}. \tag{1.3.19}$$

Equation (1.3.19) is used to define a second-order tensor, that is, $\boldsymbol{\Phi}$ is a second-order tensor if and only if its components ϕ_{ij} transform according to Eq. (1.3.19). In a right-handed orthogonal system, the determinant of the transformation matrix is unity, and we have

$$[\beta]^{-1} = [\beta]^\mathrm{T}. \tag{1.3.20}$$

The unit tensor is defined as

$$\mathbf{I} = \hat{\mathbf{e}}_i\hat{\mathbf{e}}_i. \tag{1.3.21}$$

With the help of the Kronecker delta symbol, this can be written alternatively as

$$\mathbf{I} = \delta_{ij}\hat{\mathbf{e}}_i\hat{\mathbf{e}}_j. \tag{1.3.22}$$

Clearly the unit tensor is symmetric.

The sum of the diagonal terms of a Cartesian tensor is called the *trace of the tensor*:

$$\text{trace } \boldsymbol{\Phi} = \phi_{ii}. \tag{1.3.23}$$

The trace of a tensor is *invariant*, called the first invariant, and it is denoted by I_1, that is, it is invariant with coordinate transformations ($\phi_{ii} = \bar{\phi}_{ii}$). The three invariants of a Cartesian tensor are given by

$$I_1 = \phi_{ii}, \quad I_2 = \frac{1}{2}\left(\phi_{ij}\phi_{ij} - \phi_{ii}\phi_{jj}\right), \quad I_3 = \det[\phi] = |\phi|. \qquad (1.3.24)$$

The double-dot product between two dyadics is very useful in many problems. The double-dot product between a dyad ($\mathbf{AB}$) and another ($\mathbf{CD}$) is defined as the scalar:

$$(\mathbf{AB}) : (\mathbf{CD}) \equiv (\mathbf{B} \cdot \mathbf{C})(\mathbf{A} \cdot \mathbf{D}). \qquad (1.3.25)$$

The double-dot product, by this definition, is commutative. The double-dot product between two dyads is given by

$$\begin{aligned}
\mathbf{\Phi} : \mathbf{\Psi} &= (\phi_{ij}\hat{\mathbf{e}}_i\hat{\mathbf{e}}_j) : (\psi_{mn}\hat{\mathbf{e}}_m\hat{\mathbf{e}}_n) \\
&= \phi_{ij}\psi_{mn}(\hat{\mathbf{e}}_i \cdot \hat{\mathbf{e}}_n)(\hat{\mathbf{e}}_j \cdot \hat{\mathbf{e}}_m) \\
&= \phi_{ij}\psi_{mn}\delta_{in}\delta_{jm} \\
&= \phi_{ij}\psi_{ji}.
\end{aligned} \qquad (1.3.26)$$

Note that the double-dot product of a Cartesian tensor $\mathbf{\Phi}$ with the unit tensor $\mathbf{I}$ produces its trace $I_1 = \phi_{ii}$.

We note that the gradient of a vector is a second-order tensor:

$$\begin{aligned}
\nabla\mathbf{A} &= \hat{\mathbf{e}}_i\frac{\partial}{\partial x_i}(A_j\hat{\mathbf{e}}_j) \\
&= \frac{\partial A_j}{\partial x_i}\hat{\mathbf{e}}_i\hat{\mathbf{e}}_j.
\end{aligned} \qquad (1.3.27)$$

It can be expressed as the sum of

$$\nabla\mathbf{A} = \frac{1}{2}\left(\frac{\partial A_j}{\partial x_i} + \frac{\partial A_i}{\partial x_j}\right)\hat{\mathbf{e}}_i\hat{\mathbf{e}}_j + \frac{1}{2}\left(\frac{\partial A_j}{\partial x_i} - \frac{\partial A_i}{\partial x_j}\right)\hat{\mathbf{e}}_i\hat{\mathbf{e}}_j. \qquad (1.3.28)$$

Analogously to the divergence of a vector, the divergence of a (second-order) Cartesian tensor is defined as

$$\begin{aligned}
\mathrm{div}\mathbf{\Phi} &= \nabla \cdot \mathbf{\Phi} \\
&= \hat{\mathbf{e}}_i\frac{\partial}{\partial x_i} \cdot (\phi_{mn}\hat{\mathbf{e}}_m\hat{\mathbf{e}}_n) \\
&= \frac{\partial\phi_{mn}}{\partial x_i}(\hat{\mathbf{e}}_i \cdot \hat{\mathbf{e}}_m)\hat{\mathbf{e}}_n \\
&= \frac{\partial\phi_{in}}{\partial x_i}\hat{\mathbf{e}}_n.
\end{aligned}$$

Thus the divergence of a second-order tensor is a vector.

The integral theorems of vectors presented in Section 1.2.7 are also valid for tensors (second-order and higher):

$$\int_\Omega \text{grad}\mathbf{A}\, dv = \oint_\Gamma \hat{\mathbf{n}}\mathbf{A}\, ds,$$

$$\int_\Omega \text{div}\boldsymbol{\Phi}\, dv = \oint_\Gamma \hat{\mathbf{n}} \cdot \boldsymbol{\Phi}\, ds, \qquad (1.3.29)$$

$$\int_\Omega \text{curl}\boldsymbol{\Phi}\, dv = \oint_\Gamma \hat{\mathbf{n}} \times \boldsymbol{\Phi}\, ds.$$

It is important that the order of the operations be observed in the above expressions.

1.3.4   Eigenvectors Associated with Dyads

It is conceptually useful to regard a dyadic as an operator that changes a vector into another vector (by means of the dot product). In this regard it is of interest to inquire whether there are certain vectors that have only their lengths, and not their orientation, changed when operated upon by a given dyadic or tensor. If such vectors exist, they must satisfy the equation

$$\boldsymbol{\Phi} \cdot \mathbf{A} = \lambda \mathbf{A}. \qquad (1.3.30)$$

The vectors $\mathbf{A}$ are called *characteristic vectors, or eigenvectors*, associated with $\boldsymbol{\Phi}$. The parameter λ is called an *eigenvalue*, and it characterizes the change in length (and possibly sense) of the eigenvector $\mathbf{A}$ after it has been operated upon by $\boldsymbol{\Phi}$. The eigenvalues of a stress tensor are known as the *principal stresses* and the eigenvectors are called the *principal planes*.

Since $\mathbf{A}$ can be expressed as $\mathbf{A} = \mathbf{I} \cdot \mathbf{A}$, Eq. (1.3.30) can also be written as

$$(\boldsymbol{\Phi} - \lambda \mathbf{I}) \cdot \mathbf{A} = \mathbf{0}. \qquad (1.3.31)$$

When written in matrix for Cartesian components, this equation becomes

$$\begin{bmatrix} \phi_{11} - \lambda & \phi_{12} & \phi_{13} \\ \phi_{21} & \phi_{22} - \lambda & \phi_{23} \\ \phi_{31} & \phi_{32} & \phi_{33} - \lambda \end{bmatrix} \begin{Bmatrix} A_1 \\ A_2 \\ A_3 \end{Bmatrix} = \begin{Bmatrix} 0 \\ 0 \\ 0 \end{Bmatrix}. \qquad (1.3.32)$$

Because this is a homogeneous set of equations for A_1, A_2, and A_3, a nontrivial solution will not exist unless the determinant of the matrix $[\boldsymbol{\Phi} - \lambda\mathbf{I}]$ vanishes. The vanishing of this determinant yields a cubic equation for λ, called the *characteristic equation*, the solution of which yields three values of λ, that is, three eigenvalues $\lambda_1, \lambda_2,$ and λ_3. The character of these eigenvalues depends on the character of the dyadic $\boldsymbol{\Phi}$. At least one of the eigenvalues must be real. The other two may be real and distinct, real and repeated, or complex conjugates.

In the preponderance of practical problems, the dyadic $\mathbf{\Phi}$ is symmetric, that is, $\mathbf{\Phi} = \mathbf{\Phi}^{\mathrm{T}}$ (e.g., Cauchy stress tensor). Of course, $\mathbf{\Phi}$ is always real in our considerations. For example, the moment-of-inertia dyadic is symmetric, and the stress tensor $\boldsymbol{\sigma}$ is usually but not always symmetric. We limit our discussion to symmetric dyadics.

The vanishing of the determinant assures that three eigenvectors are not unique to within a multiplicative constant; however, an infinite number of solutions exist having at least three different orientations. Since only orientation is important, it is thus useful to represent the three eigenvectors by three unit eigenvectors $\hat{\mathbf{e}}_1^*$, $\hat{\mathbf{e}}_2^*$, and $\hat{\mathbf{e}}_3^*$, denoting three different orientations, each associated with a particular eigenvalue.

Suppose now that λ_1 and λ_2 are two distinct eigenvalues and $\mathbf{A}_1$ and $\mathbf{A}_2$ are their corresponding eigenvectors:

$$\mathbf{\Phi} \cdot \mathbf{A}_1 = \lambda_1 \mathbf{A}_1,$$
$$\mathbf{\Phi} \cdot \mathbf{A}_2 = \lambda_2 \mathbf{A}_2. \tag{1.3.33}$$

Scalar product of the first equation by $\mathbf{A}_2$ and the second by $\mathbf{A}_1$, and then subtraction, yields

$$\mathbf{A}_2 \cdot \mathbf{\Phi} \cdot \mathbf{A}_1 - \mathbf{A}_1 \cdot \mathbf{\Phi} \cdot \mathbf{A}_2 = (\lambda_1 - \lambda_2)\mathbf{A}_1 \cdot \mathbf{A}_2. \tag{1.3.34}$$

Since $\mathbf{\Phi}$ is symmetric, one can establish that the left-hand side of this equation vanishes. Thus

$$0 = (\lambda_1 - \lambda_2)\mathbf{A}_1 \cdot \mathbf{A}_2. \tag{1.3.35}$$

Now suppose that λ_1 and λ_2 are complex conjugates such that $\lambda_1 - \lambda_2 = 2i\lambda_{1i}$, where $i = \sqrt{-1}$ and λ_{1i} is the imaginary part of λ_1. Then $\mathbf{A}_1 \cdot \mathbf{A}_2$ is always positive since $\mathbf{A}_1$ and $\mathbf{A}_2$ are complex conjugate vectors associated with λ_1 and λ_2. It then follows from Eq. (1.3.35) that $\lambda_{1i} = 0$ and hence that the *three eigenvalues associated with a symmetric dyadic are all real.*

Now assume that λ_1 and λ_2 are real and distinct such that $\lambda_1 - \lambda_2$ is not zero. It then follows from Eq. (1.3.35) that $\mathbf{A}_1 \cdot \mathbf{A}_2 = 0$. Thus the *eigenvectors associated with distinct eigenvalues of a symmetric dyadic are orthogonal.* If the three eigenvalues are all distinct, then the three eigenvectors are mutually orthogonal.

If λ_1 and λ_2 are distinct, but λ_3 is repeated, say $\lambda_3 = \lambda_2$, then $\mathbf{A}_3$ must also be perpendicular to $\mathbf{A}_1$ as deducted by an argument similar to that for $\mathbf{A}_2$ stemming from Eq. (1.3.35). Neither $\mathbf{A}_2$ nor $\mathbf{A}_3$ is preferred, and they are both arbitrary, except insofar as they are both perpendicular to $\mathbf{A}_1$. It is useful, however, to select $\mathbf{A}_3$ such that it is perpendicular to both $\mathbf{A}_1$ and $\mathbf{A}_2$. We do this by choosing $\mathbf{A}_3 = \mathbf{A}_1 \times \mathbf{A}_2$ and thus establishing a mutually orthogonal set of eigenvectors. This sort of behavior arises when there is an axis of symmetry present in a problem.

In a Cartesian system the characteristic equation associated with a dyadic can be expressed in the form

$$\lambda^3 - I_1\lambda^2 - I_2\lambda - I_3 = 0, \tag{1.3.36}$$

where I_1, I_2, and I_3 are the invariants associated with the matrix of $\boldsymbol{\Phi}$. The invariants can also be expressed in terms of the eigenvalues:

$$I_1 = \lambda_1 + \lambda_2 + \lambda_3, \quad I_2 = -(\lambda_1\lambda_2 + \lambda_2\lambda_3 + \lambda_3\lambda_1), \quad I_3 = \lambda_1\lambda_2\lambda_3. \tag{1.3.37}$$

Finding the roots of the cubic Eq.(1.3.36) is not always easy. However, when the matrix under consideration is of the form

$$\begin{bmatrix} \phi_{11} & 0 & 0 \\ 0 & \phi_{22} & \phi_{23} \\ 0 & \phi_{32} & \phi_{33} \end{bmatrix},$$

one of the roots is $\lambda_1 = \phi_{11}$, and the remaining two roots can be found from the quadratic equation

$$(\phi_{22} - \lambda)(\phi_{33} - \lambda) - \phi_{23}\phi_{32} = 0.$$

That is,

$$\lambda_{2,3} = \frac{\phi_{22} + \phi_{33}}{2} \pm \tfrac{1}{2}\sqrt{(\phi_{22} + \phi_{33})^2 - 4(\phi_{22}\phi_{33} - \phi_{23}\phi_{32})}. \tag{1.3.38}$$

In cases where one of the roots is not obvious, an alternative procedure given below proves to be useful.

In the alternative method we seek the eigenvalues of the so-called *deviatoric tensor* associated with $\boldsymbol{\Phi}$:

$$\phi'_{ij} \equiv \phi_{ij} - \tfrac{1}{3}\phi_{kk}\delta_{ij}. \tag{1.3.39}$$

Note that

$$\phi'_{ii} = \phi_{ii} - \phi_{kk} = 0. \tag{1.3.40}$$

That is, the first invariant I'_1 of the deviatoric tensor is zero. As a result the characteristic equation associated with the deviatoric tensor is of the form,

$$(\lambda')^3 - I'_2\lambda' - I'_3 = 0, \tag{1.3.41}$$

where λ' is the eigenvalue of the deviatoric tensor. The eigenvalues associated with ϕ_{ij} itself can be computed from

$$\lambda = \lambda' + \tfrac{1}{3}\phi_{kk}. \tag{1.3.42}$$

The cubic equation in Eq. (1.3.41) is of a special form that allows a direct computation of its roots. Equation (1.3.41) can be solved explicitly by introducing the transformation

$$\lambda' = 2(\tfrac{1}{3}I_2')^{1/2}\cos\alpha, \tag{1.3.43}$$

which transforms Eq. (1.3.41) into

$$2(\tfrac{1}{3}I_2')^{3/2}[4\cos^3\alpha - 3\cos\alpha] = I_3'. \tag{1.3.44}$$

The expression in square brackets is equal to $\cos 3\alpha$. Hence

$$\cos 3\alpha = \frac{I_3'}{2}\left(\frac{3}{I_2'}\right)^{3/2}. \tag{1.3.45}$$

If α_1 is the angle satisfying $0 \le 3\alpha_1 \le \pi$ whose cosine is given by Eq. (1.3.45), then $3\alpha_1$, $3\alpha_1 + 2\pi$, and $3\alpha_1 - 2\pi$ all have the same cosine, and furnish three independent roots of Eq. (1.3.41):

$$\lambda_i' = 2\left(\tfrac{1}{3}I_2'\right)^{1/2}\cos\alpha_i, \quad i = 1, 2, 3, \tag{1.3.46}$$

where

$$\alpha_1 = \tfrac{1}{3}\left\{\cos^{-1}\left[\frac{I_3'}{2}\left(\frac{3}{I_2'}\right)^{3/2}\right]\right\}, \quad \alpha_2 = \alpha_1 + \tfrac{2}{3}\pi, \quad \alpha_3 = \alpha_1 - \tfrac{2}{3}\pi. \tag{1.3.47}$$

Finally we can compute λ_i from Eq. (1.3.42).

Example 1.3.2 ────────────────────────────────────

Determine the eigenvalues and eigenvectors of the matrix:

$$[\phi] = \begin{bmatrix} 2 & 1 & 0 \\ 1 & 4 & 1 \\ 0 & 1 & 2 \end{bmatrix}.$$

Solution: The characteristic equation is obtained by setting $\det(\phi_{ij} - \lambda\,\delta_{ij})$ to zero:

$$\begin{vmatrix} 2-\lambda & 1 & 0 \\ 1 & 4-\lambda & 1 \\ 0 & 1 & 2-\lambda \end{vmatrix} = (2-\lambda)[(4-\lambda)(2-\lambda) - 1] - 1\cdot(2-\lambda) = 0,$$

or

$$(2-\lambda)[(4-\lambda)(2-\lambda) - 2] = 0.$$

Hence

$$\lambda_1 = 3 + \sqrt{3} = 4.7321, \quad \lambda_2 = 3 - \sqrt{3} = 1.2679, \quad \lambda_3 = 2.$$

Alternatively,

$$[\phi'] = \begin{bmatrix} 2 - \frac{8}{3} & 1 & 0 \\ 1 & 4 - \frac{8}{3} & 1 \\ 0 & 1 & 2 - \frac{8}{3} \end{bmatrix}$$

$$I_2' = \frac{1}{2}(\phi_{ij}'\phi_{ij}' - \phi_{ii}'\phi_{jj}') = \frac{1}{2}\phi_{ij}'\phi_{ij}'$$

$$= \frac{1}{2}\left[\left(-\frac{2}{3}\right)^2 + \left(-\frac{2}{3}\right)^2 + \left(\frac{4}{3}\right)^2 + 2 + 2\right] = \frac{10}{3}$$

$$I_3' = \det(\phi_{ij}') = \frac{52}{27}.$$

From Eq. (1.3.47),

$$\alpha_1 = \frac{1}{3}\left\{\cos^{-1}\left[\frac{52}{54}\left(\frac{9}{10}\right)^{3/2}\right]\right\} = 11.565°$$

$$\alpha_2 = 131.565°, \quad \alpha_3 = -108.435°,$$

and from Eq. (1.3.46),

$$\lambda_1' = 2.065384, \quad \lambda_2' = -1.3987, \quad \lambda_3' = -0.66667.$$

Finally, using Eq. (1.3.42), we obtain the eigenvalues

$$\lambda_1 = 4.7321, \quad \lambda_2 = 1.2679, \quad \lambda_3 = 2.00.$$

The eigenvector corresponding to $\lambda_3 = 2$, for example, is calculated as follows. From $(\phi_{ij} - \lambda_3\delta_{ij})A_j = 0$, we have

$$\begin{bmatrix} 2 - 2 & 1 & 0 \\ 1 & 4 - 2 & 1 \\ 0 & 1 & 2 - 2 \end{bmatrix} \begin{Bmatrix} A_1 \\ A_2 \\ A_3 \end{Bmatrix} = \begin{Bmatrix} 0 \\ 0 \\ 0 \end{Bmatrix}.$$

This gives

$$A_2 = 0, \quad A_1 = -A_3.$$

Using $A_1^2 + A_2^2 + A_3^2 = 1$ (called the normalization of the eigenvectors; the normalization of eigenvectors is not necessary as we are only interested in the planes represented by the vectors), we obtain

$$\hat{\mathbf{A}}_3 = \pm\frac{1}{\sqrt{2}}(1, 0, -1), \quad \text{for } \lambda_3 = 2.$$

Similarly, the eigenvectors corresponding to $\lambda_{1,2} = 3 \pm \sqrt{3}$ are calculated as

$$\hat{\mathbf{A}}_1 = \pm\frac{(3 - \sqrt{3})}{12}\left(1, \left(1 + \sqrt{3}\right), 1\right), \quad \text{for } \lambda_1 = 3 + \sqrt{3},$$

$$\hat{\mathbf{A}}_2 = \pm\frac{(3 + \sqrt{3})}{12}\left(1, \left(1 - \sqrt{3}\right), 1\right), \quad \text{for } \lambda_2 = 3 - \sqrt{3}.$$

When matrix $[\phi]$ represents the matrix associated with the stress tensor $[\sigma]$, the eigenvalues are called the *principal stresses* (i.e., maximum and minimum values of the stress at a point) and eigenvectors are called the *principal planes* (or directions).

1.4   Summary

In this chapter a brief review of vectors and tensors is presented. Operations with vectors and tensors, such as the scalar product (dot product) and vector product (cross product), and calculus of vectors and tensors are discussed. The index notation and summation convention are also introduced. The stress vector and Cauchy stress tensor are introduced and Cauchy's formula is derived. The determination of eigenvalues and eigenvectors of a second-order tensor is detailed, which provides a procedure for determining the principal values and principal planes of stress and strain tensors in solid and structural mechanics problems. The ideas presented in this chapter will be used in the coming chapters.

The main results of this chapter are summarized here using the rectangular Cartesian system.

Kronecker delta [Eq. (1.2.24)]:

$$\delta_{ij} = \begin{cases} 1, \text{ if } i = j, \text{ for any fixed value of } i, j \\ 0, \text{ if } i \neq j, \text{ for any fixed value of } i, j. \end{cases} \tag{1.4.1}$$

Permutation symbol [Eq. (1.2.26)]:

$$\varepsilon_{ijk} = \begin{cases} 1, \text{ if } i, j, k \text{ are in cyclic order} \\ \quad \text{and not repeated } (i \neq j \neq k), \\ -1, \text{ if } i, j, k \text{ are not in cyclic order} \\ \quad \text{and not repeated } (i \neq j \neq k), \\ 0, \text{ if any of } i, j, k \text{ are repeated.} \end{cases} \tag{1.4.2}$$

ε-δ identity [Eq. (1.2.28)]:

$$\varepsilon_{ijk}\varepsilon_{imn} = \delta_{jm}\delta_{kn} - \delta_{jn}\delta_{km}. \tag{1.4.3}$$

Scalar and vector products of vectors [Eq. (1.2.27)]:

$$\mathbf{A} \cdot \mathbf{B} = A_i B_i, \quad \mathbf{A} \times \mathbf{B} = A_i B_j \varepsilon_{ijk} \hat{\mathbf{e}}_k. \tag{1.4.4}$$

Transformation of the rectangular Cartesian components of vectors [Eq. (1.2.29)]:

$$\bar{A}_i = \beta_{ij} A_j, \quad \beta_{ij} = \hat{\bar{\mathbf{e}}}_i \cdot \hat{\mathbf{e}}_j. \tag{1.4.5}$$

The "nabla" operator in the rectangular Cartesian coordinate system [(Eq. (1.2.50)]:

$$\boldsymbol{\nabla} = \hat{\mathbf{e}}_i \frac{\partial}{\partial x_i} = \hat{\mathbf{e}}_x \frac{\partial}{\partial x} + \hat{\mathbf{e}}_y \frac{\partial}{\partial y} + \hat{\mathbf{e}}_z \frac{\partial}{\partial z}. \tag{1.4.6}$$

The gradient, curl, and divergence operations in the rectangular Cartesian coordinate system [Eqs (1.2.46), (1.2.54), and (1.2.51)]:

$$\nabla \phi \equiv \operatorname{grad}\phi = \hat{\mathbf{e}}_i \frac{\partial \phi}{\partial x_i},$$

$$\nabla \times \mathbf{A} \equiv \operatorname{curl}\mathbf{A} = \varepsilon_{ijk}\frac{\partial A_j}{\partial x_i}\,\hat{\mathbf{e}}_k, \tag{1.4.7}$$

$$\nabla \cdot \mathbf{A} \equiv \operatorname{div}\mathbf{A} = \frac{\partial A_i}{\partial x_i}.$$

The gradient, curl, and divergence theorems in the rectangular Cartesian coordinate system [Eqs (1.2.55), (1.2.56), and (1.2.57)]:

$$\int_\Omega \operatorname{grad}\phi\, d\Omega = \oint_\Gamma \hat{\mathbf{n}}\phi\, d\Gamma \quad \left[\int_\Omega \hat{\mathbf{e}}_i \frac{\partial \phi}{\partial x_i}\, d\Omega = \oint_\Gamma \hat{\mathbf{e}}_i n_i \phi\, d\Gamma\right]. \tag{1.4.8}$$

$$\int_\Omega \operatorname{curl}\mathbf{A}\, d\Omega = \oint_\Gamma \hat{\mathbf{n}} \times \mathbf{A}\, d\Gamma \quad \left[\int_\Omega \varepsilon_{ijk}\hat{\mathbf{e}}_k \frac{\partial A_j}{\partial x_i}\, d\Omega = \oint_\Gamma \varepsilon_{ijk}\hat{\mathbf{e}}_k n_i A_j\, d\Gamma\right]. \tag{1.4.9}$$

$$\int_\Omega \operatorname{div}\mathbf{A}\, d\Omega = \oint_\Gamma \hat{\mathbf{n}} \cdot \mathbf{A}\, d\Gamma \quad \left[\int_\Omega \frac{\partial A_i}{\partial x_i}\, d\Omega = \oint_\Gamma n_i A_i\, d\Gamma\right]. \tag{1.4.10}$$

Cauchy's formula and stress tensor [Eqs (1.3.8) and (1.3.10)]:

$$\mathbf{t} = \hat{\mathbf{n}} \cdot \boldsymbol{\sigma} \quad (t_i = n_j \sigma_{ji}); \quad \boldsymbol{\sigma} = \sigma_{ij}\,\hat{\mathbf{e}}_i\,\hat{\mathbf{e}}_j. \tag{1.4.11}$$

Transformation of the rectangular Cartesian components of second-order tensors [Eq. (1.3.19)]:

$$[\bar{\phi}] = [\beta][\phi][\beta]^{\mathrm{T}}; \quad \bar{\phi}_{ij} = \beta_{im}\beta_{jn}\phi_{mn}. \tag{1.4.12}$$

Eigenvalues of a second-order tensor [Eqs (1.3.31) and (1.3.36)]:

$$|\mathbf{S} - \lambda\mathbf{I}| = 0 \;\Rightarrow\; \lambda^3 - I_1\lambda^2 - I_2\lambda - I_3 = 0, \tag{1.4.13}$$

where

$$I_1 = s_{kk}, \quad I_2 = -\frac{1}{2}(s_{ii}s_{jj} - s_{ij}s_{ji}), \quad I_3 = \det\mathbf{S} = |\mathbf{S}|. \tag{1.4.14}$$

are the three invariants of the tensor $\mathbf{S}$.

Problems

1.1 Find the equation of a line (or a set of lines) passing through the terminal point of a vector $\mathbf{A}$ and in the direction of vector $\mathbf{B}$.

1.2 Find the equation of a plane connecting the terminal points of vectors $\mathbf{A}$, $\mathbf{B}$, and $\mathbf{C}$. Assume that all three vectors are referred to a common origin.

1.3 Prove with the help of vectors that the diagonals of a parallelogram bisect each other.

1.4 Prove the following vector identity without the use of a coordinate system:

$$\mathbf{A} \times (\mathbf{B} \times \mathbf{C}) = (\mathbf{A} \cdot \mathbf{C})\mathbf{B} - (\mathbf{A} \cdot \mathbf{B})\mathbf{C}.$$

1.5 If $\hat{\mathbf{e}}$ is any unit vector and $\mathbf{A}$ an arbitrary vector, show that

$$\mathbf{A} = (\mathbf{A} \cdot \hat{\mathbf{e}})\hat{\mathbf{e}} + \hat{\mathbf{e}} \times (\mathbf{A} \times \hat{\mathbf{e}}).$$

This identity shows that a vector can resolved into a component parallel to and one perpendicular to an arbitrary direction $\hat{\mathbf{e}}$.

1.6 Verify the following identities:

(a) $\delta_{ii} = 3$.

(b) $\delta_{ij}\delta_{ij} = \delta_{ii}$.

(c) $\delta_{ij}\delta_{jk} = \delta_{ik}$.

(d) $\varepsilon_{mjk}\varepsilon_{njk} = 2\delta_{mn}$.

(e) $\varepsilon_{ijk}\varepsilon_{ijk} = 6$.

(f) $A_i A_j \varepsilon_{ijk} = 0$.

1.7 Using the index notation, prove the identity

$$(\mathbf{A} \times \mathbf{B}) \cdot (\mathbf{B} \times \mathbf{C}) \times (\mathbf{C} \times \mathbf{A}) = (\mathbf{A} \cdot (\mathbf{B} \times \mathbf{C}))^2.$$

1.8 Prove the following vector identity in an orthonormal system using index-summation notation:

$$(\mathbf{A} \times \mathbf{B}) \times (\mathbf{C} \times \mathbf{D}) = [\mathbf{A} \cdot (\mathbf{C} \times \mathbf{D})]\mathbf{B} - [\mathbf{B} \cdot (\mathbf{C} \times \mathbf{D})]\mathbf{A}.$$

1.9 Determine whether the following set of vectors is linearly independent:

$$\mathbf{A} = \hat{\mathbf{e}}_1 + \hat{\mathbf{e}}_2, \quad \mathbf{B} = \hat{\mathbf{e}}_2 + \hat{\mathbf{e}}_4, \quad \mathbf{C} = \hat{\mathbf{e}}_3 + \hat{\mathbf{e}}_4, \quad \mathbf{D} = \hat{\mathbf{e}}_1 + \hat{\mathbf{e}}_2 + \hat{\mathbf{e}}_3 + \hat{\mathbf{e}}_4.$$

Here $\hat{\mathbf{e}}_i$ are orthonormal unit base vectors in a four-dimensional space.

1.10 Determine whether the following set of vectors is linearly independent:

$$\mathbf{A} = 2\hat{\mathbf{e}}_1 - \hat{\mathbf{e}}_2 + \hat{\mathbf{e}}_3, \quad \mathbf{B} = \hat{\mathbf{e}}_2 - \hat{\mathbf{e}}_3, \quad \mathbf{C} = -\hat{\mathbf{e}}_1 + \hat{\mathbf{e}}_2.$$

Here $\hat{\mathbf{e}}_i$ are orthonormal unit base vectors in $\Re^3$.

1.11 Determine which of the following sets of vectors span $\Re^3$:

(a) $\mathbf{A} = \hat{\mathbf{e}}_1 + 3\hat{\mathbf{e}}_2 - \hat{\mathbf{e}}_3, \quad \mathbf{B} = -4\hat{\mathbf{e}}_1 + 3\hat{\mathbf{e}}_2 - 5\hat{\mathbf{e}}_3, \quad \mathbf{C} = 2\hat{\mathbf{e}}_1 + \hat{\mathbf{e}}_2 + \hat{\mathbf{e}}_3$.

(b) $\mathbf{A} = \hat{\mathbf{e}}_1 + \hat{\mathbf{e}}_2, \quad \mathbf{B} = \hat{\mathbf{e}}_1 + \hat{\mathbf{e}}_2 - 2\hat{\mathbf{e}}_3, \quad \mathbf{C} = \hat{\mathbf{e}}_1 - \hat{\mathbf{e}}_3$.

Here $\hat{\mathbf{e}}_i$ are orthonormal unit base vectors in $\Re^3$.

1.12 Consider two rectangular Cartesian coordinate systems that are rotated with respect to each other and have a common origin. Let one system be denoted as a barred system, so that a position vector can be written in each of the systems as

$$\mathbf{r} = x_i \hat{\mathbf{e}}_i,$$
$$= \bar{x}_j \hat{\bar{\mathbf{e}}}_j,$$

where $\{\hat{e}_j\}$ and $\{\hat{\bar{e}}_j\}$ are the respective orthonormal Cartesian bases in the unbarred and barred systems. By requiring that the position vector $\mathbf{r}$ be invariant under a rotation of the coordinate systems, deduce that the transformation between the coordinates is given by

$$\bar{x}_1 = a_{11}x_1 + a_{12}x_2 + a_{13}x_3$$
$$\bar{x}_2 = a_{21}x_1 + a_{22}x_2 + a_{23}x_3$$
$$\bar{x}_3 = a_{31}x_1 + a_{32}x_2 + a_{33}x_3,$$

or more compactly,

$$\bar{x}_i = a_{ij}x_j, \quad i, j = 1, 2, 3,$$

where the terms a_{ij} can be identified as the direction cosines

$$a_{ij} \equiv \hat{\bar{e}}_i \cdot \hat{e}_j = \cos\left(\hat{\bar{e}}_i, \hat{e}_j\right).$$

Deduce further that the basis vectors obey the same transformation

$$\hat{\bar{e}}_i = a_{ij}\hat{e}_j,$$

and that the following orthogonality conditions hold:

$$a_{ij}a_{kj} = \delta_{ik}.$$

1.13 Determine the transformation matrix relating the orthonormal basis vectors $(\hat{e}_1, \hat{e}_2, \hat{e}_3)$ and the orthonormal basis vectors $(\hat{e}'_1, \hat{e}'_2, \hat{e}'_3)$, when $\hat{e}'_i$ are given by

(a) $\hat{e}'_1$ along the vector $\hat{e}_1 - \hat{e}_2 + \hat{e}_3$ and $\hat{e}'_2$ is perpendicular to the plane $2x_1 + 3x_2 + x_3 - 5 = 0$.

(b) $\hat{e}'_1$ along the line segment connecting point $(1, -1, 3)$ to $(2, -2, 4)$ and $\hat{e}'_3 = (-\hat{e}_1 + \hat{e}_2 + 2\hat{e}_3)/\sqrt{6}$.

(c) $\hat{e}'_3 = \hat{e}_3$, and the angle between x'_1-axis and x_1-axis is $30°$.

1.14 The angles between the barred and unbarred coordinate lines are given as follows:

	$\hat{e}_1$	$\hat{e}_2$	$\hat{e}_3$
$\hat{\bar{e}}_1$	$60°$	$30°$	$90°$
$\hat{\bar{e}}_2$	$150°$	$60°$	$90°$
$\hat{\bar{e}}_3$	$90°$	$90°$	$0°$

Determine the direction cosines of the transformation.

1.15 The angles between the barred and unbarred coordinate lines are given as follows:

	x_1	x_2	x_3
$\bar{x}_1$	$45°$	$90°$	$45°$
$\bar{x}_2$	$60°$	$45°$	$120°$
$\bar{x}_3$	$120°$	$45°$	$60°$

Determine the transformation matrix.

1.16 In a rectangular Cartesian coordinate system, find the length and direction cosines of a vector $\mathbf{A}$ that extends from the point $(1, -1, 3)$ to the midpoint of the line segment from the origin to the point $(6, -6, 4)$.

1.17 The vectors $\mathbf{A}$ and $\mathbf{B}$ are defined as follows:

$$\mathbf{A} = 3\hat{\mathbf{i}} - 4\hat{\mathbf{k}},$$
$$\mathbf{B} = 2\hat{\mathbf{i}} - 2\hat{\mathbf{j}} + \hat{\mathbf{k}},$$

where $\hat{\mathbf{i}}, \hat{\mathbf{j}}$, and $\hat{\mathbf{k}}$ are an orthonormal basis.

(a) Find the orthogonal projection of $\mathbf{A}$ in the direction of $\mathbf{B}$.

(b) Find the angle between the positive directions of the vectors.

1.18 Prove the following identities (see Eq. (1.2.12) for the definition of $[\mathbf{ABC}]$):

(a) $\frac{d}{dt}[\mathbf{ABC}] = \left[\frac{d\mathbf{A}}{dt}\mathbf{BC}\right] + \left[\mathbf{A}\frac{d\mathbf{B}}{dt}\mathbf{C}\right] + \left[\mathbf{AB}\frac{d\mathbf{C}}{dt}\right]$.

(b) $\frac{d}{dt}\left[\mathbf{A}\frac{d\mathbf{A}}{dt}\frac{d^2\mathbf{A}}{dt^2}\right] = \left[\mathbf{A}\frac{d\mathbf{A}}{dt}\frac{d^3\mathbf{A}}{dt^3}\right]$.

1.19 Let $\mathbf{r}$ denote a position vector $\mathbf{r} = x_i\hat{\mathbf{e}}_i$ ($r^2 = x_ix_i$) and $\mathbf{A}$ an arbitrary constant vector. Show that (div $= \nabla\cdot$; grad $= \nabla$; curl $= \nabla\times$):

(a) $\mathrm{grad}(r) = \frac{\mathbf{r}}{r}$.

(b) $\mathrm{grad}(r^n) = nr^{n-2}\mathbf{r}$.

(c) $\nabla^2(r^n) = n(n+1)r^{n-2}$.

(d) $\mathrm{grad}\,(\mathbf{r}\cdot\mathbf{A}) = \mathbf{A}$.

(e) $\mathrm{div}(\mathbf{r}\times\mathbf{A}) = 0$.

(f) $\mathrm{curl}(\mathbf{r}\times\mathbf{A}) = -2\mathbf{A}$.

(g) $\mathrm{div}(r\mathbf{A}) = \frac{1}{r}(\mathbf{r}\cdot\mathbf{A})$.

(h) $\mathrm{curl}\,(r\mathbf{A}) = \frac{1}{r}(\mathbf{r}\times\mathbf{A})$.

1.20 Let $\mathbf{A}$ and $\mathbf{B}$ be continuous vector functions of the position vector $\mathbf{r}$ with continuous first derivatives, and let F and G be continuous scalar functions of $\mathbf{r}$ with continuous first and second derivatives. Show that (div $= \nabla\cdot$; grad $= \nabla$; curl $= \nabla\times$):

(a) $\mathrm{curl}\,(\mathrm{grad}\,F) = 0$.

(b) $\mathrm{div}\,(\mathrm{curl}\,\mathbf{A}) = 0$.

(c) $\mathrm{div}\,(\mathrm{grad}\,F\times\mathrm{grad}\,G) = 0$.

(d) $\mathrm{grad}\,(FG) = F\,\mathrm{grad}\,G + G\,\mathrm{grad}\,F$.

(e) $\mathrm{div}\,(F\mathbf{A}) = \mathbf{A}\cdot\mathrm{grad}\,F + F\,\mathrm{div}\,\mathbf{A}$.

(f) $\mathrm{curl}\,(F\mathbf{A}) = F\,\mathrm{curl}\,\mathbf{A} - \mathbf{A}\times\mathrm{grad}\,F$.

(g) $\mathrm{grad}\,(\mathbf{A}\cdot\mathbf{B}) = \mathbf{A}\cdot\mathrm{grad}\mathbf{B} + \mathbf{B}\cdot\mathrm{grad}\mathbf{A} + \mathbf{A}\times\mathrm{curl}\mathbf{B} + \mathbf{B}\times\mathrm{curl}\mathbf{A}$.

(h) $\mathrm{div}(\mathbf{A}\times\mathbf{B}) = \mathbf{B}\cdot\mathrm{curl}\mathbf{A} - \mathbf{A}\cdot\mathrm{curl}\mathbf{B}$.

(i) $\mathrm{curl}(\mathbf{A}\times\mathbf{B}) = \mathbf{B}\cdot\nabla\mathbf{A} - \mathbf{A}\cdot\nabla\mathbf{B} + \mathbf{A}\,\mathrm{div}\mathbf{B} - \mathbf{B}\,\mathrm{div}\mathbf{A}$.

(j) $\nabla^2(FG) = F\nabla^2G + 2\nabla F\cdot\nabla G + G\nabla^2F$.

1.21 Find the gradient of a vector $\mathbf{A}$ in the (a) cylindrical and (b) spherical coordinate systems.

1.22 Show that the vector area of a closed surface is zero, that is,

$$\oint_\Gamma \hat{\mathbf{n}}\,ds = \mathbf{0}.$$

1.23 Show that the volume enclosed by a surface Γ is

$$\text{volume} = \frac{1}{6}\oint_\Gamma \mathrm{grad}(r^2)\cdot\hat{\mathbf{n}}\,ds,$$

or

$$\text{volume} = \frac{1}{3}\oint_\Gamma \mathbf{r}\cdot\hat{\mathbf{n}}\,ds.$$

1.24 Let $\phi(\mathbf{r})$ be a scalar field. Show that

$$\int_\Omega \nabla^2\phi \, dv = \oint_\Gamma \frac{\partial\phi}{\partial n} \, ds,$$

where $\partial\phi/\partial n \equiv \hat{\mathbf{n}} \cdot \operatorname{grad}\phi$ is the derivative of ϕ in the outward direction normal to the boundary Γ of the domain Ω.

1.25 In the divergence theorems, set $\mathbf{A} = \phi\operatorname{grad}\psi$ and $\mathbf{A} = \psi\operatorname{grad}\phi$ successively and obtain the integral forms

$$\int_\Omega \left[\phi\nabla^2\psi + \nabla\phi \cdot \nabla\psi\right] dr = \oint_\Gamma \phi\frac{\partial\psi}{\partial n} \, ds, \tag{1}$$

$$\int_\Omega \left[\phi\nabla^2\psi - \psi\nabla^2\phi\right] dr = \oint_\Gamma \left[\phi\frac{\partial\psi}{\partial n} - \psi\frac{\partial\phi}{\partial n}\right] ds, \tag{2}$$

$$\int_\Omega \left[\phi\nabla^4\psi - \nabla^2\phi\nabla^2\psi\right] dr = \oint_\Gamma \left[\phi\frac{\partial}{\partial n}(\nabla^2\psi) - \nabla^2\psi\frac{\partial\phi}{\partial n}\right] ds, \tag{3}$$

where Ω denotes a (2D or 3D) region with boundary Γ. The first two identities are sometimes called Green's first and second theorems.

1.26 Determine the rotation transformation matrix such that the new base vector $\hat{\mathbf{e}}_1$ is along $\hat{\mathbf{e}}_1 - \hat{\mathbf{e}}_2 + \hat{\mathbf{e}}_3$, and $\hat{\mathbf{e}}_2$ is along the normal to the plane $2x_1 + 3x_2 + x_3 = 5$. If $\mathbf{T}$ is the dyadic whose components in the unbarred system are given by $T_{11} = 1$, $T_{12} = 0$, $T_{13} = -1$, $T_{22} = 3$, $T_{23} = -2$, and $T_{33} = 0$, find the components in the barred coordinates.

1.27 Show that the characteristic equation for a second-order tensor σ_{ij} can be expressed as

$$\lambda^3 - I_1\lambda^2 - I_2\lambda - I_3 = 0,$$

where

$$I_1 = \sigma_{kk},$$

$$I_2 = -\frac{1}{2}(\sigma_{ii}\sigma_{jj} - \sigma_{ij}\sigma_{ji}),$$

$$I_3 = \frac{1}{6}(2\sigma_{ij}\sigma_{jk}\sigma_{ki} - 3\sigma_{ij}\sigma_{ji}\sigma_{kk} + \sigma_{ii}\sigma_{jj}\sigma_{kk}) = \det(\sigma_{ij})$$

are the three invariants of the tensor.

1.28 Find the eigenvalues and eigenvectors of the following matrices:

$$\text{(a)} \begin{bmatrix} 4 & -4 & 0 \\ -4 & 0 & 0 \\ 0 & 0 & 3 \end{bmatrix}, \qquad \text{(b)} \begin{bmatrix} 2 & -\sqrt{3} & 0 \\ -\sqrt{3} & 4 & 0 \\ 0 & 0 & 4 \end{bmatrix},$$

$$\text{(c)} \begin{bmatrix} 1 & 0 & 0 \\ 0 & 3 & -1 \\ 0 & -1 & 3 \end{bmatrix}, \qquad \text{(d)} \begin{bmatrix} 2 & -1 & 1 \\ -1 & 0 & 1 \\ 1 & 1 & 2 \end{bmatrix},$$

$$\text{(e)} \begin{bmatrix} 3 & 5 & 8 \\ 5 & 1 & 0 \\ 8 & 0 & 2 \end{bmatrix}, \qquad \text{(f)} \begin{bmatrix} 1 & -1 & 0 \\ -1 & 2 & -1 \\ 0 & -1 & 2 \end{bmatrix}.$$

1.29 Evaluate the three invariants of the matrices in **Problem 1.28** and check them against the invariants obtained by using the eigenvalues.

1.30 The components of a stress dyadic at a point, referred to the (x_1, x_2, x_3) system, are (in ksi $= 1000$ psi):

$$\text{(a)} \begin{bmatrix} 12 & 9 & 0 \\ 9 & -12 & 0 \\ 0 & 0 & 6 \end{bmatrix}, \quad \text{(b)} \begin{bmatrix} 9 & 0 & 12 \\ 0 & -25 & 0 \\ 12 & 0 & 16 \end{bmatrix}, \quad \text{(c)} \begin{bmatrix} 1 & -3 & \sqrt{2} \\ -3 & 1 & -\sqrt{2} \\ \sqrt{2} & -\sqrt{2} & 4 \end{bmatrix}.$$

Find the following:

(a) The stress vector acting on a plane perpendicular to the vector $2\hat{e}_1 - 2\hat{e}_2 + \hat{e}_3$ passing through the point

(b) The magnitude of the stress vector and the angle between the stress vector and the normal to the plane

(c) The magnitudes of the normal and tangential components of the stress vector

1.31 If $\mathbf{A}$ is an arbitrary vector, $\boldsymbol{\Phi}$ is an arbitrary dyad, and $\mathbf{I}$ is the identity tensor, verify that

(a) $\mathbf{I} \cdot \boldsymbol{\Phi} = \boldsymbol{\Phi} \cdot \mathbf{I} = \boldsymbol{\Phi}$.

(b) $(\mathbf{I} \times \mathbf{A}) \cdot \boldsymbol{\Phi} = \mathbf{A} \times \boldsymbol{\Phi}$.

(c) $(\mathbf{A} \times \mathbf{I}) \cdot \boldsymbol{\Phi} = \mathbf{A} \times \boldsymbol{\Phi}$.

(d) $(\boldsymbol{\Phi} \times \mathbf{A})^{\mathrm{T}} = -\mathbf{A} \times \boldsymbol{\Phi}^{\mathrm{T}}$.

1.32 If $p(x) = a_0 + a_1 x + a_2 x^2 + \cdots + a_n x^n$, and $[A]$ is any square matrix, we define the polynomial in $[A]$ by

$$p([A]) = a_0[I] + a_1[A] + a_2[A]^2 + \cdots + a_n[A]^n.$$

If

$$[A] = \begin{bmatrix} 1 & -1 \\ -1 & 1 \end{bmatrix},$$

and $p(x) = 1 - 2x + x^2$, compute $p(A)$.

1.33 *Cayley–Hamilton theorem* Consider a square matrix $[S]$ of order n. Denote by $p(\lambda)$ the determinant of $[S] - \lambda[I]$ (i.e., $p(\lambda) \equiv |(S - \lambda I)|$), called the *characteristic polynomial*. Then the Cayley–Hamilton theorem states that $p([S]) = 0$ (i.e., every matrix satisfies its own characteristic equation). Here $p([S])$ is as defined in **Problem 1.32**. Use matrix computation to verify the Cayley–Hamilton theorem for each of the following matrices:

(a) $\begin{bmatrix} 1 & -1 \\ 2 & 1 \end{bmatrix}$,   (b) $\begin{bmatrix} 2 & -1 & 1 \\ 0 & 1 & 0 \\ 1 & -2 & 1 \end{bmatrix}$.

1.34 Consider the matrix

$$[S] = \begin{bmatrix} 2 & 1 & 0 \\ 1 & 4 & 1 \\ 0 & 1 & 2 \end{bmatrix}.$$

Verify the Cayley–Hamilton theorem and use it to compute the inverse of $[S]$.

"If a man is in too big a hurry to give up an error, he is liable to give up some truth with it."
— *Wilbur Wright* (developed the world's first successful airplane)

Review of Equations of Solid Mechanics

2.1 Introduction

2.1.1 Classification of Equations

The objective of this chapter is to record the governing equations of a solid continuum in a form suitable for use later in this book. The word *continuum* needs to be explained first. Microscopically a medium occupied by a matter, solid or fluid, is made of discrete particles of protons, neutrons, and electrons. Macroscopically, the medium is assumed to contain no gaps or voids between material points of the medium so that it can be divided indefinitely into smaller and smaller parts without encountering a void. This concept allows us to shrink an arbitrarily small region to a point so that all spatial derivatives of various quantities associated with the medium can be defined. For example, if mass density $\rho(\mathbf{x}, t)$ of the medium is defined to be the mass per unit volume, we assume that the limit

$$\rho(\mathbf{x}, t) = \lim_{\Delta V \to 0} \frac{\Delta m}{\Delta V} \tag{2.1.1}$$

exists (i.e., ρ is finite). Here Δm denotes the mass of the matter occupying an infinitesimal volume ΔV, $\mathbf{x}$ is the position vector, and t denotes time. Of course, the continuum assumption may be violated when one considers matter at nanometer (10^{-9}m) or atomic scale, and the continuum equations reviewed in this chapter may not be applicable to problems at such scales.

The governing equations of a continuum are derived using the following conservation principles (or laws of physics):

1. Principle of conservation of mass
2. Principle of balance of linear momentum
3. Principle of balance of angular momentum
4. Principle of conservation of energy

The above principles do not explicitly account for geometric changes or mechanical response of the continuum. Without these, the equations derived from the

Energy Principles and Variational Methods in Applied Mechanics, Third Edition. J.N. Reddy.
©2017 John Wiley & Sons Ltd. Published 2017 by John Wiley & Sons Ltd.
Companion Website: www.wiley.com/go/reddy/applied_mechanics_EPVM

conservation and balance laws are insufficient to determine the total response of a continuum. Therefore, we must consider the following additional two sets of equations:

5. Kinematics (strain–displacement equations)
6. Constitutive equations (e.g., stress–strain relations)

Kinematics is a study of the geometry of motion and deformation without consideration of the forces causing the motion. The constitutive equations describe the mechanical behavior of the continuum and relate the dependent variables introduced in the kinetic description (i.e., balance of momenta and conservation of energy) to those in the kinematic description. Since the resulting equations involve derivatives with respect to spatial (i.e., position) coordinates and time, appropriate boundary and initial conditions are also needed to determine the solution. For a detailed derivation of the equations, the reader may consult the books on continuum mechanics, elasticity, and mechanics of materials [81, 84, 89–98].

2.1.2   Descriptions of Motion

There are two alternative descriptions used to study the motion of a continuum. In the first, one considers the motion of all matter passing through a *fixed spatial location*, as shown in Fig. 2.1.1(a). Here one is interested in various properties (e.g., velocity, pressure, temperature, density, and so on) of the matter that instantly occupies the fixed spatial location. This description is called the *spatial description* or the *Eulerian description*. In the second, one focuses attention on a *fixed set of material particles*, irrespective of their spatial locations [see Fig. 2.1.1(b)]. The relative displacements of these particles and the stress caused by external forces and temperature are of interest in this case. This description is known as the *material description* or the *Lagrangian description*. The Eulerian description is most commonly used to study fluid flows and coupled heat transfer and fluid flow, while the Lagrangian description is generally used to study heat transfer, stress, and deformation of solid bodies.

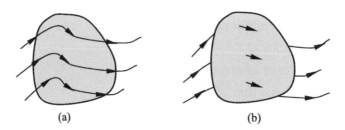

(a)   (b)

Fig. 2.1.1 (a) Fixed region in space and mass moving through it. (b) Motion of a fixed collection of material particles.

In order to understand the difference between the material and spatial descriptions, consider a continuum and identify a region of the continuum for our study. Let **X** denote the position of an arbitrarily fixed point in the region (in both descriptions) at time $t = 0$, and let us label the material particle that occupies position **X** with X (a name given to the particle). For time $t > 0$, the point **X** in a spatial description remains the same, but denoted by **x**. Although the current position **x** and initial position **X** occupied by the particle X are the same in the spatial description, the particles occupying the position **x** at two different times are not the same (i.e., particle X no longer occupies the position **x** at time $t > 0$), unless there is no motion. Thus, in the spatial description particles come and go, while we focus on the spatial location **x** (and not on the particle that happens to instantaneously occupies the location). On the other hand, in a material description, where attention is focused on a material particle, the particle X occupying the position **X** at time $t = 0$ moves to a new position **x** at time $t > 0$ (in other words, we keep track of the motion of the material particle X).

To gain further understanding, consider representing a scalar quantity such as the mass density ρ in the two descriptions. In a material description, the functional variation of ρ is described with respect to the coordinate **X** occupied by the material particle X in the region as

$$\rho = \rho(\mathbf{X}, t). \tag{2.1.2}$$

In spatial description ρ is expressed in terms of position in space, **x**, currently occupied by a material particle in the continuum at time t as

$$\rho = \rho(\mathbf{x}, t). \tag{2.1.3}$$

In Eq. (2.1.2) a change in time t implies that the same material particle X has a different density. The particle's current position **x** has been expressed in terms of its position **X** at time $t = 0$. Thus, in material description, attention is focused on the material particle. In Eq. (2.1.3), however, a change in time t implies that a different density is observed at the same spatial location **x**, now probably occupied by a different material particle. Therefore, in spatial description, attention is focused on spatial location.

Since much of the present study is primarily concerned with deformable solids without large strains, the Lagrangian (material) description is adopted here. In this case, the principle of conservation of mass is trivially satisfied as no mass is added or removed from any portion of the body under consideration. Also, we will not consider thermodynamic aspects that require the use of the energy equation. Thus, the review of equations presented here is limited to the balance of linear and angular momenta and kinematics of deformation. For additional reading on this subject, the readers may consult the textbooks on continuum mechanics by Malvern [81] and Reddy [84].

2.2   Balance of Linear and Angular Momenta

2.2.1   Equations of Motion

The principle of conservation of linear momentum, or Newton's second law of motion, applied to a continuous medium can be stated as *the total time rate of change of (linear) momentum equals the net force exerted on the continuum.* Written in vector form for a material particle or a collection of material particles with mass m and velocity $\mathbf{v}$, symbolically the principle can be stated as

$$\frac{d}{dt}(m\mathbf{v}) = \mathbf{F}, \qquad (2.2.1)$$

where $\mathbf{F}$ is the resultant force on the collection of particles. For constant mass, Eq. (2.2.1) becomes

$$\mathbf{F} = m\frac{d\mathbf{v}}{dt} = m\mathbf{a}, \qquad (2.2.2)$$

which is the familiar form of Newton's second law (i.e., force = mass × acceleration). To derive the equation of motion applied to a solid continuum (i.e., a fixed material body that is in motion), we must identify the forces acting on it.

Forces acting on a control volume can be classified as *internal* and *external*. The internal forces resist the tendency of one part of the body to be separated from another part. The internal force per unit area is termed stress, as defined in Section 1.3 [see Eq. (1.3.1)]. The external forces are those transmitted by the body. The external forces can be further classified as *body (or volume) forces* and *surface forces*.

Body forces act on the distribution of mass inside the body. Examples of body forces are provided by the gravitational and electromagnetic forces. Body forces are usually measured per unit mass or unit volume of the body. Let $\mathbf{f}$ denote the body force per unit mass in the deformed body. Consider an elemental volume dv inside the deformed body. The body force of the elemental volume is equal to $\rho dv\mathbf{f}$. Hence, the total body force is

$$\int_\Omega \rho\mathbf{f}\, dv, \qquad (2.2.3)$$

where Ω denotes the volume occupied by the body at time $t > 0$.

Surface forces are contact forces acting on the boundary surface of the body. Examples of surface forces are provided by applied forces on the surface of the body. Surface forces are reckoned per unit area. Let $\mathbf{t}$ denote the surface force per unit area (or surface stress vector). The surface force on an elemental surface ds of the volume is $\mathbf{t}ds$. The total surface force acting on the closed surface of the body, denoted by Γ, is

$$\oint_\Gamma \mathbf{t}\, ds. \qquad (2.2.4)$$

Since the stress vector **t** on the surface is related to the (internal) stress tensor $\boldsymbol{\sigma}$ by Cauchy's formula [see Eq. (1.3.8)]

$$\mathbf{t} = \hat{\mathbf{n}} \cdot \boldsymbol{\sigma}, \tag{2.2.5}$$

where $\hat{\mathbf{n}}$ denotes the unit normal to the surface, we can express the surface force as

$$\oint_\Gamma \hat{\mathbf{n}} \cdot \boldsymbol{\sigma} \, ds.$$

Using the divergence theorem in Eq. (1.2.57), we can write

$$\oint_\Gamma \hat{\mathbf{n}} \cdot \boldsymbol{\sigma} \, ds = \int_\Omega \boldsymbol{\nabla} \cdot \boldsymbol{\sigma} \, dv, \tag{2.2.6}$$

where $\boldsymbol{\nabla}$ is the vector differential operator with respect to **x**. Recall from the discussion of Section 1.3 that $\boldsymbol{\sigma}$ is the Cauchy stress tensor, which is measured as the force per unit area of the deformed body.

The principle of conservation of linear momentum as applied to a deformed solid continuum yields the result

$$\boldsymbol{\nabla} \cdot \boldsymbol{\sigma} + \rho \mathbf{f} = \rho \frac{\partial^2 \mathbf{u}}{\partial t^2}, \tag{2.2.7}$$

where **u** is the displacement vector. In Cartesian rectangular system, we have

$$\frac{\partial \sigma_{ji}}{\partial x_j} + \rho f_i = \rho \frac{\partial^2 u_i}{\partial t^2} \quad (i = 1, 2, 3). \tag{2.2.8}$$

For static equilibrium, we set the right-hand side of Eqs. (2.2.7) and (2.2.8) to zero and obtain the equations of equilibrium:

$$\boldsymbol{\nabla} \cdot \boldsymbol{\sigma} + \rho \mathbf{f} = \mathbf{0} \quad \left(\frac{\partial \sigma_{ji}}{\partial x_j} + \rho f_i = 0, \quad i = 1, 2, 3 \right). \tag{2.2.9}$$

The Cauchy stress tensor is the most natural and physical measure of the state of stress at a point in the deformed body and measured per unit area of the deformed body. Since the geometry of the deformed body is not known (and to be determined), the governing equations must be written in terms of the known reference configuration,[1] say, configuration at $t = 0$. This need gives rise to other measures of stress. These measures are purely mathematical but facilitate analysis. One such measure is the second Piola–Kirchhoff stress tensor, and its meaning is described next.

Consider an undeformed body Ω_0 subjected to a set of forces that results in the deformed configuration Ω. Let the force vector on an elemental area da with normal $\hat{\mathbf{n}}$ in the deformed body be $d\mathbf{f}$. Suppose that the area element in the undeformed configuration that corresponds to da is dA with normal $\hat{\mathbf{N}}$.

---

[1] We shall use the term *configuration* to mean the simultaneous position of all material points of a body.

The force $d\mathbf{f}$ can be expressed in terms of a stress vector $\mathbf{t}$ times the deformed area da as

$$d\mathbf{f} = \mathbf{t}(\hat{\mathbf{n}})\,da.$$

The *second Piola–Kirchhoff stress tensor* $\mathbf{S}$, which is used in the study of large deformation analysis, is introduced as the stress tensor associated with the force $d\mathcal{F}$ on the undeformed elemental area dA that corresponds to the force $d\mathbf{f}$ on the deformed elemental area da:

$$d\mathcal{F} = \mathbf{F}^{-1} \cdot d\mathbf{f} = \hat{\mathbf{N}} \cdot \mathbf{S}dA, \tag{2.2.10}$$

where $\mathbf{F}$ and $\mathbf{F}^{-1}$ are the deformation gradient and its inverse, respectively:

$$\mathbf{F} = (\boldsymbol{\nabla}_0 \mathbf{x})^{\mathrm{T}}, \quad \mathbf{F}^{-1} = (\boldsymbol{\nabla}\mathbf{X})^{\mathrm{T}}. \tag{2.2.11}$$

Here $\boldsymbol{\nabla}_0$ denotes the gradient operator with respect to the material coordinates $\mathbf{X}$. Thus, the second Piola–Kirchhoff stress tensor gives the *transformed current force per undeformed area*. The area and volume elements in the two configurations are related by (see Reddy [84] for details)

$$\hat{\mathbf{n}}\,da = J\,\mathbf{F}^{-\mathrm{T}} \cdot \hat{\mathbf{N}}\,dA, \quad dv = J\,dV, \tag{2.2.12}$$

where J denotes the determinant of the deformation gradient $\mathbf{F}$, $J = |\mathbf{F}|$. The second Piola–Kirchhoff stress tensor $\mathbf{S}$ and the Cauchy stress tensor $\boldsymbol{\sigma}$ can be shown to be related according to

$$\mathbf{S}^{\mathrm{T}} = J\mathbf{F}^{-1} \cdot \boldsymbol{\sigma}^{\mathrm{T}} \cdot \mathbf{F}^{-\mathrm{T}}, \quad \boldsymbol{\sigma}^{\mathrm{T}} = \frac{1}{J}\mathbf{F} \cdot \mathbf{S}^{\mathrm{T}} \cdot \mathbf{F}^{\mathrm{T}}. \tag{2.2.13}$$

Clearly, $\mathbf{S}$ is symmetric whenever $\boldsymbol{\sigma}$ is symmetric. Cartesian component representation of $\mathbf{S}$ is

$$\mathbf{S} = S_{IJ}\hat{\mathbf{E}}_I\hat{\mathbf{E}}_J, \tag{2.2.14}$$

where $(\hat{\mathbf{E}}_1, \hat{\mathbf{E}}_2, \hat{\mathbf{E}}_3)$ is the orthonormal basis in the material coordinate system (X_1, X_2, X_3).

We can now express the equation of motion in terms of the second Piola–Kirchhoff stress tensor $\mathbf{S}$ as

$$\boldsymbol{\nabla}_0 \cdot (\mathbf{S}^{\mathrm{T}} \cdot \mathbf{F}^{\mathrm{T}}) + \rho_0 \mathbf{f} = \rho_0 \frac{\partial^2 \mathbf{u}}{\partial t^2}, \tag{2.2.15}$$

where ρ_0 is the mass density measured in the undeformed body

$$\int_{\Omega} \rho\mathbf{f}(\mathbf{x})\,dv = \int_{\Omega_0} \rho_0\mathbf{f}(\mathbf{X})\,dV. \tag{2.2.16}$$

Clearly, the equations of motion expressed in terms of the second Piola–Kirchhoff stress tensor are nonlinear, and this nonlinearity is in addition to any nonlinearity that may come from the strain–displacement relations and constitutive relations.

The equations of equilibrium in Eq. (2.2.9) can also be derived by applying Newton's second law of motion to an infinitesimal volume element in the deformed body (without distorting the sides of the cube). Consider the stresses and body forces on an infinitesimal parallelepiped element of dimensions dx_1, dx_2, and dx_3. Figure 2.2.1 shows the stresses acting on the various faces of the parallelepiped along coordinate lines (x_1, x_2, x_3). The sum of all forces in the x_1-direction is given by

$$\left(\sigma_{11} + \frac{\partial \sigma_{11}}{\partial x_1} dx_1\right) dx_2 dx_3 - \sigma_{11} dx_2 dx_3 + \left(\sigma_{21} + \frac{\partial \sigma_{21}}{\partial x_2} dx_2\right) dx_1 dx_3$$

$$- \sigma_{21} dx_1 dx_3 + \left(\sigma_{31} + \frac{\partial \sigma_{31}}{\partial x_3} dx_3\right) dx_1 dx_2 - \sigma_{31} dx_1 dx_2 + \rho f_1 dx_1 dx_2 dx_3$$

$$= \left(\frac{\partial \sigma_{11}}{\partial x_1} + \frac{\partial \sigma_{21}}{\partial x_2} + \frac{\partial \sigma_{31}}{\partial x_3} + \rho f_1\right) dx_1 dx_2 dx_3.$$

By Newton's second law of motion, the sum of the forces is equal to the time rate of change of linear momentum in the x_1-direction:

$$(\rho \, dx_1 dx_2 dx_3) \frac{\partial^2 u_1}{\partial t^2},$$

where ρ is the density (assumed to be independent of time t). Thus, upon dividing throughout by $dx_1 dx_2 dx_3$, we obtain

$$\frac{\partial \sigma_{11}}{\partial x_1} + \frac{\partial \sigma_{21}}{\partial x_2} + \frac{\partial \sigma_{31}}{\partial x_3} + \rho f_1 = \rho \frac{\partial^2 u_1}{\partial t^2} \quad \text{or} \quad \frac{\partial \sigma_{j1}}{\partial x_j} + \rho f_1 = \rho \frac{\partial^2 u_1}{\partial t^2}.$$

Similarly, the application of Newton's second law in the x_2- and x_3-directions yields

$$\frac{\partial \sigma_{j2}}{\partial x_j} + \rho f_2 = \rho \frac{\partial^2 u_2}{\partial t^2}, \quad \frac{\partial \sigma_{j3}}{\partial x_j} + \rho f_3 = \rho \frac{\partial^2 u_3}{\partial t^2},$$

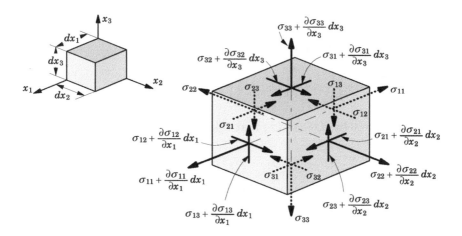

Fig. 2.2.1 Stresses on a parallelepiped element.

or, in index notation,

$$\frac{\partial \sigma_{ji}}{\partial x_j} + \rho f_i = \rho \frac{\partial^2 u_i}{\partial t^2} \quad (i = 1, 2, 3),$$

which is the same as that in Eq. (2.2.8). For static equilibrium, we set the time derivative terms to zero and obtain the equilibrium equations

$$\frac{\partial \sigma_{ji}}{\partial x_j} + \rho f_i = 0 \quad (i = 1, 2, 3).$$

2.2.2 Symmetry of Stress Tensors

The principle of conservation of angular momentum states that *the time rate of change of the total moment of momentum for a continuum is equal to the vector sum of the moments of external forces acting on the continuum*. In the absence of body couples (i.e., volume-dependent couples), the principle leads to the symmetry of the stress tensor. That is, the matrix of the stress components is symmetric:

$$\boldsymbol{\sigma}^{\mathrm{T}} = \boldsymbol{\sigma} \quad (\sigma_{ij} = \sigma_{ji}), \tag{2.2.17}$$

and, more explicitly,

$$\sigma_{23} = \sigma_{32}, \quad \sigma_{31} = \sigma_{13}, \quad \sigma_{12} = \sigma_{21}.$$

Thus, there are only six stress components that are independent. Since the Cauchy stress tensor $\boldsymbol{\sigma}$ is symmetric, it follows from Eq. (2.2.13) that the second Piola–Kirchhoff stress tensor is also symmetric, $\mathbf{S} = \mathbf{S}^{\mathrm{T}}$ (the reader is asked to prove it to herself/himself).

The symmetry of the stress tensor can also be established using Newton's second law for moments. Consider the moment of all forces acting on the parallelepiped about the x_3-axis (see Fig. 2.2.1). Using the right-handed screw rule for positive moment, we obtain

$$\left[\left(\sigma_{12} + \frac{\partial \sigma_{12}}{\partial x_1} dx_1 \right) dx_2 dx_3 \right] dx_1 - \left[\left(\sigma_{21} + \frac{\partial \sigma_{21}}{\partial x_2} dx_2 \right) dx_1 dx_3 \right] dx_2 = 0.$$

Dividing throughout by $dx_1 dx_2 dx_3$ and taking the limit $dx_1 \to 0$ and $dx_2 \to 0$, we obtain

$$\sigma_{12} - \sigma_{21} = 0.$$

Similar considerations of moments about the x_1-axis and x_2-axis give, respectively, the relations

$$\sigma_{23} - \sigma_{32} = 0, \quad \sigma_{13} - \sigma_{31} = 0.$$

Inclosing this section we note that the components of $\boldsymbol{\sigma}$, being a second-order tensor, transform according to Eq. (1.3.19) (we note that $[\beta]^{-1} = [\beta]^{\mathrm{T}}$):

$$\bar{\sigma}_{ij} = \beta_{im}\beta_{jn}\sigma_{mn} \quad \text{or} \quad [\bar{\sigma}] = [\beta][\sigma][\beta]^{\mathrm{T}} \quad \text{and} \quad [\sigma] = [\beta]^{\mathrm{T}}[\bar{\sigma}][\beta]. \tag{2.2.18}$$

In particular, the transformation between components σ_{ij} referred to the rectangular Cartesian coordinate system (x_1, x_2, x_3) and components $\bar{\sigma}_{ij}$ referred to another rectangular Cartesian coordinate system $(\bar{x}_1, \bar{x}_2, \bar{x}_3)$; when the $\bar{x}_1$-axis is oriented at an angle of θ counterclockwise to the x_1-axis and $x_3 = \bar{x}_3$, as shown in Fig. 2.2.2 (i.e., the $\bar{x}_i$ coordinates are obtained by the rotation of the x_1x_2-plane about the x_3-axis by θ in counterclockwise direction), the transformation equations can be obtained using Eq. (2.2.18). The transformation matrix $[\beta]$ in this case is given by (note that $\det[\beta] = 1$ and $[\beta]^{\mathrm{T}} = [\beta]^{-1}$)

$$[\beta] = \begin{bmatrix} \cos\theta & \sin\theta & 0 \\ -\sin\theta & \cos\theta & 0 \\ 0 & 0 & 1 \end{bmatrix}, \quad [\beta]^{\mathrm{T}} = \begin{bmatrix} \cos\theta & -\sin\theta & 0 \\ \sin\theta & \cos\theta & 0 \\ 0 & 0 & 1 \end{bmatrix}. \tag{2.2.19}$$

The relations between σ_{ij} and $\bar{\sigma}_{ij}$ are then given by carrying out the matrix multiplications indicated in Eq. (2.2.18):

$$\begin{Bmatrix} \sigma_{11} \\ \sigma_{22} \\ \sigma_{33} \\ \sigma_{23} \\ \sigma_{13} \\ \sigma_{12} \end{Bmatrix} = \begin{bmatrix} \cos^2\theta & \sin^2\theta & 0 & 0 & 0 & -\sin2\theta \\ \sin^2\theta & \cos^2\theta & 0 & 0 & 0 & \sin2\theta \\ 0 & 0 & 1 & 0 & 0 & 0 \\ 0 & 0 & 0 & \cos\theta & \sin\theta & 0 \\ 0 & 0 & 0 & -\sin\theta & \cos\theta & 0 \\ \sin\theta\cos\theta & -\sin\theta\cos\theta & 0 & 0 & 0 & \cos^2\theta - \sin^2\theta \end{bmatrix} \begin{Bmatrix} \bar{\sigma}_{11} \\ \bar{\sigma}_{22} \\ \bar{\sigma}_{33} \\ \bar{\sigma}_{23} \\ \bar{\sigma}_{13} \\ \bar{\sigma}_{12} \end{Bmatrix}, \tag{2.2.20}$$

$$\begin{Bmatrix} \bar{\sigma}_{11} \\ \bar{\sigma}_{22} \\ \bar{\sigma}_{33} \\ \bar{\sigma}_{23} \\ \bar{\sigma}_{13} \\ \bar{\sigma}_{12} \end{Bmatrix} = \begin{bmatrix} \cos^2\theta & \sin^2\theta & 0 & 0 & 0 & \sin2\theta \\ \sin^2\theta & \cos^2\theta & 0 & 0 & 0 & -\sin2\theta \\ 0 & 0 & 1 & 0 & 0 & 0 \\ 0 & 0 & 0 & \cos\theta & -\sin\theta & 0 \\ 0 & 0 & 0 & \sin\theta & \cos\theta & 0 \\ -\sin\theta\cos\theta & \sin\theta\cos\theta & 0 & 0 & 0 & \cos^2\theta - \sin^2\theta \end{bmatrix} \begin{Bmatrix} \sigma_{11} \\ \sigma_{22} \\ \sigma_{33} \\ \sigma_{23} \\ \sigma_{13} \\ \sigma_{12} \end{Bmatrix}. \tag{2.2.21}$$

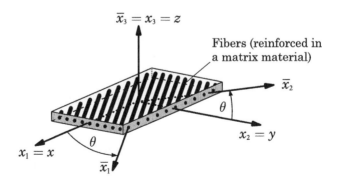

Fig. 2.2.2 Rotation transformation of coordinate system (x_1, x_2, x_3) to $(\bar{x}_1, \bar{x}_2, \bar{x}_3)$.

2.3   Kinematics of Deformation

2.3.1   Green–Lagrange Strain Tensor

Kinematics is a study of the geometry of motion and deformation without consideration of the forces causing them. Under the action of applied loads, a material body gets displaced and deformed or strained. The word *deformation* refers to changes in the geometry of the body. A body is said to undergo *rigid-body motion* if the distance between any two arbitrary points and angle between any two infinitesimal line segments in the body remains unchanged. The rigid-body motion does not alter the shape of the body. Thus, a measure of the deformation is provided by the change in the distance between points and angle between line segments in the body. Here we develop a measure of straining of the body and introduce strain tensor using the material description.

Consider a fixed mass occupying a region with volume V and closed surface S. For simplicity, we also denote the region with V. A typical material particle in V is labeled as X, and its location with respect to a rectangular Cartesian coordinate system is denoted as $\mathbf{X}$. Under the action of applied forces (and subjected to geometric constraints), the body undergoes deformation and the particle X moves to a new location $\mathbf{x}$. Since all material particles of the body move, possibly by different amounts, the body occupies a new region, as shown in Fig. 2.3.1.

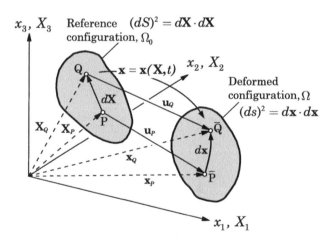

Fig. 2.3.1 Deformation of a material body.

To determine if the motion involved is a deformation (or straining), we must compute the distance between two neighboring points after deformation. Let $P : (X_1, X_2, X_3)$ and $Q : (X_1 + dX_1, X_2 + dX_2, X_3 + dX_3)$ denote two arbitrary material points that are $d\mathbf{X}$ apart in the undeformed configuration of the body at time $t = 0$. Under the action of externally applied forces, the body deforms (i.e., the geometry of the body changes), and the points P and Q move to new places: $\bar{P}$: (x_1, x_2, x_3) and $\bar{Q}$: $(x_1 + dx_1, x_2 + dx_2, x_3 + dx_3)$, respectively, in the deformed body. During this deformation, all material particles that were neighbors to each other in the undeformed body remain neighbors in the deformed body. The motion of a particle X occupying position $\mathbf{X}$ in the undeformed body to point $\mathbf{x}$ in the deformed body can be expressed by the displacement vector

$$\mathbf{u} = \mathbf{x} - \mathbf{X} \quad \text{or} \quad u_i = x_i - X_i. \tag{2.3.1}$$

In the material description used here, the current position $\mathbf{x}$ of a material point is expressed in terms of its position $\mathbf{X}$ in the undeformed body:

$$\mathbf{x} = \mathbf{x}(\mathbf{X}, t), \quad x_i = x_i(X_1, X_2, X_3, t) \quad (i = 1, 2, 3). \tag{2.3.2}$$

Equation (2.3.2) is called the *deformation mapping* because it describes the geometry of the deformed body. By assumption, a continuous medium cannot have gaps or overlaps. Therefore, a one-to-one correspondence exists between points in the undeformed body and points in the deformed body. Consequently, a unique inverse to Eq. (2.3.2) exists, although not easy, in general, to construct it.

Next, we wish to determine, in the interest of measuring the deformation, the change in the distance between points $\bar{P}$ and $\bar{Q}$ in the deformed body and compare with the distance between points P and Q in the undeformed body. The distances between points P and Q and points $\bar{P}$ and $\bar{Q}$ are given, respectively, by

$$(dS)^2 = d\mathbf{X} \cdot d\mathbf{X} = dX_i dX_i, \quad (ds)^2 = d\mathbf{x} \cdot d\mathbf{x} = dx_i dx_i, \tag{2.3.3}$$

where summation on repeated subscripts is implied. If $dS \neq ds$, then we know that the body has strained. Since the length of a vector is obtained by computing the square of its length and then taking the square root of it, it is convenient to consider the square of the distance between points. We have

$$(ds)^2 - (dS)^2 = d\mathbf{x} \cdot d\mathbf{x} - d\mathbf{X} \cdot d\mathbf{X} = dx_m dx_m - dX_i dX_i$$
$$= \left(\frac{\partial x_m}{\partial X_i} dX_i \right) \left(\frac{\partial x_m}{\partial X_j} dX_j \right) - dX_i dX_i$$
$$= \frac{\partial x_m}{\partial X_i} \frac{\partial x_m}{\partial X_j} dX_i dX_j - \delta_{ij} dX_i dX_j = d\mathbf{X} \cdot \left(\mathbf{F}^{\mathrm{T}} \cdot \mathbf{F} - \mathbf{I} \right) \cdot d\mathbf{X}$$
$$\equiv 2 E_{ij} dX_i dX_j = 2 d\mathbf{X} \cdot \mathbf{E} \cdot d\mathbf{X}, \tag{2.3.4}$$

where E_{ij} are the rectangular Cartesian components of the *Green strain tensor* or *Green–Lagrange strain tensor* $\mathbf{E}$ at point $\mathbf{X}$:

$$\mathbf{E} = E_{ij} \hat{\mathbf{E}}_i \hat{\mathbf{E}}_j \tag{2.3.5}$$

with

$$\mathbf{E} = \tfrac{1}{2}\left(\mathbf{F}^{\mathrm{T}} \cdot \mathbf{F} - \mathbf{I}\right), \quad E_{ij} = \tfrac{1}{2}\left(\frac{\partial x_m}{\partial X_i}\frac{\partial x_m}{\partial X_j} - \delta_{ij}\right). \tag{2.3.6}$$

Here $\hat{\mathbf{E}}_i$ denote the unit basis vectors in the rectangular Cartesian system (X_1, X_2, X_3). The fact that E_{ij} are components of a second-order tensor can be established by means of coordinate transformations. We also note that the Green strain tensor components are symmetric by definition. Also, the change in the length of a line element is zero if and only if $E_{ij} = 0$.

The strain components can be expressed in terms of the displacement components at point $\mathbf{X}$ using Eq. (2.3.1). We note

$$\mathbf{x} = \mathbf{u} + \mathbf{X}, \quad \mathbf{F}^{\mathrm{T}} = \boldsymbol{\nabla}_0 \mathbf{x} = \boldsymbol{\nabla}_0 \mathbf{u} + \mathbf{I}, \tag{2.3.7}$$

or, in component form,

$$x_m = u_m + X_m, \quad \frac{\partial x_m}{\partial X_i} = \frac{\partial u_m}{\partial X_i} + \delta_{mi}. \tag{2.3.8}$$

Here $\mathbf{F}$ denotes the deformation gradient defined in Eq. (2.2.11). Consequently, we can express the Green strain tensor in terms of the displacement gradient as

$$\mathbf{E} = \tfrac{1}{2}\left[(\boldsymbol{\nabla}_0 \mathbf{u}) + (\boldsymbol{\nabla}_0 \mathbf{u})^{\mathrm{T}} + (\boldsymbol{\nabla}_0 \mathbf{u}) \cdot (\boldsymbol{\nabla}_0 \mathbf{u})^{\mathrm{T}}\right], \tag{2.3.9}$$

$$E_{ij} = \tfrac{1}{2}\left(\frac{\partial u_i}{\partial X_j} + \frac{\partial u_j}{\partial X_i} + \frac{\partial u_m}{\partial X_i}\frac{\partial u_m}{\partial X_j}\right), \tag{2.3.10}$$

where summation on repeated index (m) is implied. Clearly, the last term in Eqs. (2.3.9) and (2.3.10) is nonlinear in the displacement gradients.

In expanded notation, the Green strain tensor components in rectangular Cartesian coordinate system (X_1, X_2, X_3) are given by

$$E_{11} = \frac{\partial u_1}{\partial X_1} + \frac{1}{2}\left[\left(\frac{\partial u_1}{\partial X_1}\right)^2 + \left(\frac{\partial u_2}{\partial X_1}\right)^2 + \left(\frac{\partial u_3}{\partial X_1}\right)^2\right],$$

$$E_{22} = \frac{\partial u_2}{\partial X_2} + \frac{1}{2}\left[\left(\frac{\partial u_1}{\partial X_2}\right)^2 + \left(\frac{\partial u_2}{\partial X_2}\right)^2 + \left(\frac{\partial u_3}{\partial X_2}\right)^2\right],$$

$$E_{33} = \frac{\partial u_3}{\partial X_3} + \frac{1}{2}\left[\left(\frac{\partial u_1}{\partial X_3}\right)^2 + \left(\frac{\partial u_2}{\partial X_3}\right)^2 + \left(\frac{\partial u_3}{\partial X_3}\right)^2\right], \tag{2.3.11}$$

$$E_{12} = \frac{1}{2}\left(\frac{\partial u_1}{\partial X_2} + \frac{\partial u_2}{\partial X_1} + \frac{\partial u_1}{\partial X_1}\frac{\partial u_1}{\partial X_2} + \frac{\partial u_2}{\partial X_1}\frac{\partial u_2}{\partial X_2} + \frac{\partial u_3}{\partial X_1}\frac{\partial u_3}{\partial X_2}\right),$$

$$E_{13} = \frac{1}{2}\left(\frac{\partial u_1}{\partial X_3} + \frac{\partial u_3}{\partial X_1} + \frac{\partial u_1}{\partial X_1}\frac{\partial u_1}{\partial X_3} + \frac{\partial u_2}{\partial X_1}\frac{\partial u_2}{\partial X_3} + \frac{\partial u_3}{\partial X_1}\frac{\partial u_3}{\partial X_3}\right),$$

$$E_{23} = \frac{1}{2}\left(\frac{\partial u_2}{\partial X_3} + \frac{\partial u_3}{\partial X_2} + \frac{\partial u_1}{\partial X_2}\frac{\partial u_1}{\partial X_3} + \frac{\partial u_2}{\partial X_2}\frac{\partial u_2}{\partial X_3} + \frac{\partial u_3}{\partial X_2}\frac{\partial u_3}{\partial X_3}\right).$$

The components E_{11}, E_{22}, and E_{33} are the normal (i.e., extensional) strains, and E_{12}, E_{23}, and E_{13} are the shear strains. The Green strain tensor **E** can be shown to be dual (or energetically conjugate) to the second Piola–Kirchhoff stress tensor **S** (see Reddy [84]). That is, to compute the strain energy stored in the deformed body, one must evaluate the volume integral of the product of **E** and **S**.

When the displacement gradients are small, that is, $|\nabla_0 \mathbf{u}| \ll 1$,

$$\frac{\partial u_i}{\partial X_j} \ll 1, \quad \left(\frac{\partial u_i}{\partial X_j}\right)^2 \approx 0,$$

we may neglect the nonlinear terms in the definition of the Green strain tensor **E** and obtain the linearized strain tensor $\boldsymbol{\varepsilon}$, called the *infinitesimal strain tensor*. When infinitesimal strains are used, it is common to use x_i in place of X_i, with the understanding that x_i now refer to the material coordinates. The infinitesimal strain tensor $\boldsymbol{\varepsilon}$ is given by

$$\boldsymbol{\varepsilon} = \tfrac{1}{2}\left[(\nabla_0 \mathbf{u}) + (\nabla_0 \mathbf{u})^{\mathrm{T}}\right]; \quad \varepsilon_{ij} = \tfrac{1}{2}\left(\frac{\partial u_i}{\partial X_j} + \frac{\partial u_j}{\partial X_i}\right) = \tfrac{1}{2}\left(\frac{\partial u_i}{\partial x_j} + \frac{\partial u_j}{\partial x_i}\right).$$

$$(2.3.12)$$

Example 2.3.1 _____

Consider a rectangular block of dimensions $a \times b \times h$, where the thickness h is very small compared to a and b. Suppose that the block is deformed (with negligible change in its thickness) into the diamond shape shown in Fig. 2.3.2. Assuming that the deformation mapping is a linear function of X_1 and X_2, determine (a) the deformation mapping $\mathbf{x} = \mathbf{x}(\mathbf{X})$ from the given geometry, (b) the displacement vector, (c) infinitesimal strain components ε_{ij}, (d) the Green strain components, and (e) the strains in the diagonals of the rectangular block.

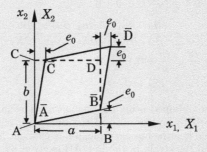

Fig. 2.3.2 Undeformed and deformed rectangular block.

Solution: (a) Considering the deformation mapping in the form (as suggested in the problem statement)

$$x_1 = a_0 + a_1 X_1 + a_2 X_2 + a_3 X_1 X_2, \quad x_2 = b_0 + b_1 X_1 + b_2 X_2 + b_3 X_1 X_2, \quad x_3 = X_3,$$

where the constants a_i and b_i are to be determined from the geometry of the deformed body. Since the point $(X_1, X_2) = (0, 0)$ did not move, we have $a_0 = b_0 = 0$. The point $(X_1, X_2) = (0, b)$ now occupies $(x_1, x_2) = (e_0, b)$, giving

$$e_0 = a_2 b, \quad b = b_2 b \;\rightarrow\; a_2 = \frac{e_0}{b} \text{ and } b_2 = 1.$$

The point $(X_1, X_2) = (a, 0)$ moved to $(x_1, x_2) = (a, e_0)$; so, we have $a_1 = 1$ and $b_1 = e_0/a$. The point $(X_1, X_2) = (a, b)$ now occupies $(x_1, x_2) = (a + e_0, b + e_0)$; then we have

$$a + e_0 = a + e_0 + a_3 ab, \quad b + e_0 = e_0 + b + b_3 ab \;\rightarrow\; a_3 = b_3 = 0.$$

Thus, we have the deformation mapping

$$x_1 = X_1 + \frac{e_0}{b} X_2, \quad x_2 = \frac{e_0}{a} X_1 + X_2, \quad x_3 = X_3.$$

(b) The displacements are given by

$$u_1 = x_1 - X_1 = \frac{e_0}{b} X_2, \quad u_2 = x_2 - X_2 = \frac{e_0}{a} X_1, \quad u_3 = x_3 - X_3 = 0.$$

(c) The infinitesimal strains are

$$\varepsilon_{11} = \frac{\partial u_1}{\partial X_1} = 0, \quad \varepsilon_{22} = \frac{\partial u_2}{\partial X_2} = 0, \quad \varepsilon_{33} = \frac{\partial u_3}{\partial X_3} = 0,$$

$$2\varepsilon_{12} = \frac{\partial u_1}{\partial X_2} + \frac{\partial u_2}{\partial X_1} = \frac{e_0}{b} + \frac{e_0}{a}, \quad 2\varepsilon_{13} = \frac{\partial u_1}{\partial X_3} + \frac{\partial u_3}{\partial X_1} = 0, \quad 2\varepsilon_{23} = \frac{\partial u_2}{\partial X_3} + \frac{\partial u_3}{\partial X_2} = 0.$$

(d) The Green strains are given by

$$E_{11} = \varepsilon_{11} + \frac{1}{2}\left(\frac{\partial u_1}{\partial X_1}\right)^2 + \frac{1}{2}\left(\frac{\partial u_2}{\partial X_1}\right)^2 + \frac{1}{2}\left(\frac{\partial u_3}{\partial X_1}\right)^2$$

$$= 0 + 0 + \frac{1}{2}\left(\frac{e_0}{a}\right)^2 + 0,$$

$$E_{22} = \varepsilon_{22} + \frac{1}{2}\left(\frac{\partial u_1}{\partial X_2}\right)^2 + \frac{1}{2}\left(\frac{\partial u_2}{\partial X_2}\right)^2 + \frac{1}{2}\left(\frac{\partial u_3}{\partial X_2}\right)^2$$

$$= 0 + \frac{1}{2}\left(\frac{e_0}{b}\right)^2 + 0 + 0,$$

$$2E_{12} = 2\varepsilon_{12} + \frac{\partial u_1}{\partial X_1}\frac{\partial u_1}{\partial X_2} + \frac{\partial u_2}{\partial X_1}\frac{\partial u_2}{\partial X_2} + \frac{\partial u_3}{\partial X_1}\frac{\partial u_1}{\partial X_2}$$

$$= \frac{e_0}{b} + \frac{e_0}{a} + 0 + 0 + 0$$

$$E_{33} = E_{13} = E_{23} = 0.$$

The same results can be obtained using the elementary mechanics of materials approach, where the strains are defined to be the ratio of the difference between the final length and original length to the original length. For example, a line element AB in the undeformed body moves to position $\bar{\text{AB}}$. Then the strain in the line AB is given by

$$E_{11} = E_{AB} = \frac{\bar{\text{AB}} - \text{AB}}{\text{AB}}$$

$$= \frac{1}{a}\sqrt{a^2 + e_0^2} - 1 = \sqrt{1 + \left(\frac{e_0}{a}\right)^2} - 1$$

$$= \left[1 + \frac{1}{2}\left(\frac{e_0}{a}\right)^2 + \cdots\right] - 1 \approx \frac{1}{2}\left(\frac{e_0}{a}\right)^2.$$

(e) Now consider a line element oriented at an angle from the X_1-axis. For example, the diagonal line AD deforms and becomes $\bar{\text{AD}}$ in the deformed body. The normal strain in AD is given by

$$\varepsilon'_{xx} = \frac{\sqrt{(a+e_0)^2 + (b+e_0)^2} - \sqrt{a^2+b^2}}{\sqrt{a^2+b^2}}$$

$$= \sqrt{1 + 2\left[\frac{e_0^2 + e_0(a+b)}{a^2+b^2}\right]} - 1 \approx \frac{e_0^2 + e_0(a+b)}{a^2+b^2}.$$

Since ε is a second-order tensor, we can use the transformation relation in Eq. (1.3.19) and obtain the same result. We have

$$\varepsilon'_{11} = \beta_{1i}\beta_{1j}\varepsilon_{ij}$$
$$= \beta_{11}\beta_{11}\varepsilon_{11} + \beta_{12}\beta_{12}\varepsilon_{22} + \beta_{11}\beta_{12}\varepsilon_{12} + \beta_{12}\beta_{11}\varepsilon_{21},$$

where

$$\beta_{ij} = \hat{\mathbf{e}}_i \cdot \hat{\mathbf{e}}_j.$$

In particular, we have $\beta_{11} = \cos\theta$, $\beta_{12} = \sin\theta$, $\beta_{22} = \cos\theta$, $\beta_{21} = -\sin\theta$, and $\theta = \tan^{-1}(b/a)$. Hence,

$$\varepsilon'_{11} = \cos^2\theta\varepsilon_{11} + \sin^2\theta\varepsilon_{22} + 2\sin\theta\cos\theta\varepsilon_{12} \equiv \varepsilon_\theta.$$

Substituting for $\cos\theta$, $\sin\theta$, and ε_{ij} into ε'_{ij}, we obtain

$$\varepsilon'_{11} = \frac{a^2}{a^2+b^2}\frac{1}{2}\left(\frac{e_0}{a}\right)^2 + \frac{b^2}{a^2+b^2}\frac{1}{2}\left(\frac{e_0}{b}\right)^2 + 2\frac{ab}{a^2+b^2}\frac{e_0(a+b)}{2ab}$$
$$= \frac{e_0^2 + e_0(a+b)}{a^2+b^2}.$$

Relations between the components of strain tensors **E** and ε referred to two different rectangular Cartesian coordinate systems can be obtained using equations of the form in Eq. (1.3.19) [see, also, Eq. (2.2.18)]. The next example illustrates this for the specific coordinate transformation shown in Fig. 2.2.2.

Example 2.3.2

Derive the transformation equations between the Green strain components E_{IJ} referred to (X_1, X_2, X_3) coordinate system and $\bar{E}_{IJ}$ in the new coordinate system $(\bar{X}_1, \bar{X}_2, \bar{X}_3)$, which is obtained by rotating the former about the X_3-axis counterclockwise by the angle θ, as shown in Fig. 2.2.2.

Solution: The two coordinate systems are related by

$$\begin{Bmatrix} \bar{X}_1 \\ \bar{X}_2 \\ \bar{X}_3 \end{Bmatrix} = \begin{bmatrix} \cos\theta & \sin\theta & 0 \\ -\sin\theta & \cos\theta & 0 \\ 0 & 0 & 1 \end{bmatrix} \begin{Bmatrix} X_1 \\ X_2 \\ X_3 \end{Bmatrix} \equiv [\beta] \begin{Bmatrix} X_1 \\ X_2 \\ X_3 \end{Bmatrix}. \tag{1}$$

Transformation of strain tensor components follows those of a second-order tensor [see Eq. (2.2.18)]:

$$[\bar{E}] = [\beta][E][\beta]^{\mathrm{T}}; \quad [E] = [\beta]^{\mathrm{T}}[\bar{E}][\beta]. \tag{2}$$

Carrying out the indicated matrix multiplications and expressing the result in single-column format, we have

$$
\left\{
\begin{array}{c}
E_{11} \\
E_{22} \\
E_{33} \\
2E_{23} \\
2E_{13} \\
2E_{12}
\end{array}
\right\}
=
\left[
\begin{array}{cccccc}
\cos^2\theta & \sin^2\theta & 0 & 0 & 0 & -\frac{1}{2}\sin 2\theta \\
\sin^2\theta & \cos^2\theta & 0 & 0 & 0 & \frac{1}{2}\sin 2\theta \\
0 & 0 & 1 & 0 & 0 & 0 \\
0 & 0 & 0 & \cos\theta & \sin\theta & 0 \\
0 & 0 & 0 & -\sin\theta & \cos\theta & 0 \\
\sin 2\theta & -\sin 2\theta & 0 & 0 & 0 & \cos 2\theta
\end{array}
\right]
\left\{
\begin{array}{c}
\bar{E}_{11} \\
\bar{E}_{22} \\
\bar{E}_{33} \\
2\bar{E}_{23} \\
2\bar{E}_{13} \\
2\bar{E}_{12}
\end{array}
\right\}.
\tag{3}
$$

The inverse relations are

$$
\left\{
\begin{array}{c}
\bar{E}_{11} \\
\bar{E}_{22} \\
\bar{E}_{33} \\
2\bar{E}_{23} \\
2\bar{E}_{13} \\
2\bar{E}_{12}
\end{array}
\right\}
=
\left[
\begin{array}{cccccc}
\cos^2\theta & \sin^2\theta & 0 & 0 & 0 & \frac{1}{2}\sin 2\theta \\
\sin^2\theta & \cos^2\theta & 0 & 0 & 0 & -\frac{1}{2}\sin 2\theta \\
0 & 0 & 1 & 0 & 0 & 0 \\
0 & 0 & 0 & \cos\theta & -\sin\theta & 0 \\
0 & 0 & 0 & \sin\theta & \cos\theta & 0 \\
-\sin 2\theta & \sin 2\theta & 0 & 0 & 0 & \cos 2\theta
\end{array}
\right]
\left\{
\begin{array}{c}
E_{11} \\
E_{22} \\
E_{33} \\
2E_{23} \\
2E_{13} \\
2E_{12}
\end{array}
\right\}.
\tag{4}
$$

2.3.2 Strain Compatibility Equations

When a sufficiently differentiable displacement field is given, the computation of the strains is straightforward: one can make use of Eqs (2.3.11) and (2.3.12) to compute the strains. However, when the strain components are given, the determination of the displacements is not always possible because there are six strain components related to three displacement components. Stated in other words, there are six differential equations involving three unknowns. Thus, the six equations should be compatible with one another in the sense that any three equations should give the same displacement field.

To further understand the situation, let us consider the two-dimensional case of linearized strains. The linearized strain–displacement relations are given by

$$
\frac{\partial u_1}{\partial X_1} = \varepsilon_{11}, \quad \frac{\partial u_2}{\partial X_2} = \varepsilon_{22}, \quad \frac{\partial u_1}{\partial X_2} + \frac{\partial u_2}{\partial X_1} = 2\varepsilon_{12}.
\tag{2.3.13}
$$

If ε_{ij} are given as functions of X_1 and X_2, they cannot be arbitrary: $\varepsilon_{11}, \varepsilon_{22}$, and ε_{12} should have a relationship that the three equations are compatible with each other. This relation can be derived as follows: differentiate the third equation with respect to X_1 and X_2 to obtain

$$
\frac{\partial^3 u_1}{\partial X_1 \partial X_2^2} + \frac{\partial^3 u_2}{\partial X_1^2 \partial X_2} = 2\frac{\partial^2 \varepsilon_{12}}{\partial X_1 \partial X_2}.
\tag{2.3.14}
$$

The third derivatives of u_1 and u_2 needed in Eq. (2.3.14) can be computed from the first two equations in Eq. (2.3.13):

$$
\frac{\partial^3 u_1}{\partial X_1 \partial X_2^2} = \frac{\partial^2 \varepsilon_{11}}{\partial X_2^2}, \quad \frac{\partial^3 u_2}{\partial X_1^2 \partial X_2} = \frac{\partial^2 \varepsilon_{22}}{\partial X_1^2}.
\tag{2.3.15}
$$

Substituting these relations into Eq. (2.3.14), we obtain a relationship between the derivatives of the strain components:

$$\frac{\partial^2 \varepsilon_{11}}{\partial X_2^2} + \frac{\partial^2 \varepsilon_{22}}{\partial X_1^2} = 2\frac{\partial^2 \varepsilon_{12}}{\partial X_1 \partial X_2}. \tag{2.3.16}$$

Equation (2.3.16) is called the compatibility equation for a two-dimensional strain field.

The discussion given above can be generalized to infinitesimal strains in three dimensions. There are six equations of strain compatibility in three dimensions, and they are given by

$$
\begin{aligned}
\frac{\partial^2 \varepsilon_{11}}{\partial X_2^2} + \frac{\partial^2 \varepsilon_{22}}{\partial X_1^2} &= 2\frac{\partial^2 \varepsilon_{12}}{\partial X_1 \partial X_2}, \\
\frac{\partial^2 \varepsilon_{22}}{\partial X_3^2} + \frac{\partial^2 \varepsilon_{33}}{\partial X_2^2} &= 2\frac{\partial^2 \varepsilon_{23}}{\partial X_2 \partial X_3}, \\
\frac{\partial^2 \varepsilon_{11}}{\partial X_3^2} + \frac{\partial^2 \varepsilon_{33}}{\partial X_1^2} &= 2\frac{\partial^2 \varepsilon_{13}}{\partial X_1 \partial X_3}, \\
\frac{\partial}{\partial X_1}\left(-\frac{\partial \varepsilon_{23}}{\partial X_1} + \frac{\partial \varepsilon_{13}}{\partial X_2} + \frac{\partial \varepsilon_{12}}{\partial X_3}\right) &= \frac{\partial^2 \varepsilon_{11}}{\partial X_2 \partial X_3}, \\
\frac{\partial}{\partial X_2}\left(-\frac{\partial \varepsilon_{13}}{\partial X_2} + \frac{\partial \varepsilon_{12}}{\partial X_3} + \frac{\partial \varepsilon_{23}}{\partial X_1}\right) &= \frac{\partial^2 \varepsilon_{22}}{\partial X_1 \partial X_3}, \\
\frac{\partial}{\partial X_3}\left(-\frac{\partial \varepsilon_{12}}{\partial X_3} + \frac{\partial \varepsilon_{23}}{\partial X_1} + \frac{\partial \varepsilon_{13}}{\partial X_2}\right) &= \frac{\partial^2 \varepsilon_{33}}{\partial X_1 \partial X_2}.
\end{aligned}
\tag{2.3.17}
$$

In index notation, the compatibility conditions for the infinitesimal strains can be expressed as (not derived here)

$$\varepsilon_{ij,k\ell} + \varepsilon_{k\ell,ij} = \varepsilon_{\ell j,ki} + \varepsilon_{ki,\ell j}, \tag{2.3.18}$$

or in vector form

$$\boldsymbol{\nabla} \times (\boldsymbol{\nabla} \times \boldsymbol{\varepsilon})^{\mathrm{T}} = \mathbf{0} \quad \text{(zero tensor)}. \tag{2.3.19}$$

Equation (2.3.19) constitutes the necessary and sufficient conditions for the existence of a single-valued displacement field (when the strains are given). Although Eq. (2.3.19) yields 81 equations for the three-dimensional case, only six of them, as given in Eq. (2.3.17), are different from each other, and the remaining are either trivial or linear combinations of the six equations. For the two-dimensional case, the six equations reduce to the single equation in Eq. (2.3.16).

In the case of finite strains, the compatibility conditions in terms of the deformation gradient[2] are derived from the mathematical requirement that the

---

[2] The derivation of compatibility conditions in terms of the Green strain tensor $\mathbf{E}$ is quite involved and not attempted here.

curl of a gradient be zero. Since $\mathbf{F}$ is the gradient of $\mathbf{x}$ with respect to $\mathbf{X}$, we require that

$$\boldsymbol{\nabla}_0 \times \mathbf{F}^{\mathrm{T}} = \mathbf{0}. \tag{2.3.20}$$

It should be noted that the strain compatibility equations are satisfied automatically when the strains are computed from a displacement field. Thus, one needs to verify the compatibility conditions only when the strains are computed from stresses that are in equilibrium.

Example 2.3.3

Check to see if the following two-dimensional strain field is compatible:

$$\varepsilon_{11} = c_1 X_1 \left(X_1^2 + X_2^2 \right), \quad \varepsilon_{22} = \frac{1}{3} c_2 X_1^3, \quad \varepsilon_{12} = c_3 X_1^2 X_2,$$

where c_1, c_2, and c_3 are nonzero constants.

Solution: Using Eq. (2.3.16) we obtain

$$\frac{\partial^2 \varepsilon_{11}}{\partial X_2^2} + \frac{\partial^2 \varepsilon_{22}}{\partial X_1^2} - 2 \frac{\partial^2 \varepsilon_{12}}{\partial X_1 \partial X_2} = 2c_1 X_1 + 2c_2 X_1 - 4c_3 X_1.$$

Thus the strain field is *not* compatible, unless $c_1 + c_2 - 2c_3 = 0$.

Example 2.3.4

Given the strain tensor $\mathbf{E} = E_{rr} \hat{\mathbf{e}}_r \hat{\mathbf{e}}_r + E_{\theta\theta} \hat{\mathbf{e}}_\theta \hat{\mathbf{e}}_\theta$ in an axisymmetric body (i.e., E_{rr} and $E_{\theta\theta}$ are functions of r and z only), determine the compatibility conditions on E_{rr} and $E_{\theta\theta}$.

Solution: Toward using the vector form of the compatibility conditions, Eq. (2.3.19), we let

$$\mathbf{G} \equiv \boldsymbol{\nabla} \times \mathbf{E} = \left(\frac{\partial E_{\theta\theta}}{\partial r} + \frac{E_{\theta\theta} - E_{rr}}{r} \right) \hat{\mathbf{e}}_z \hat{\mathbf{e}}_\theta + \frac{\partial E_{rr}}{\partial z} \hat{\mathbf{e}}_\theta \hat{\mathbf{e}}_r - \frac{\partial E_{\theta\theta}}{\partial z} \hat{\mathbf{e}}_r \hat{\mathbf{e}}_\theta.$$

Then

$$
\begin{aligned}
\boldsymbol{\nabla} \times (\mathbf{G})^{\mathrm{T}} = {}& \frac{\partial}{\partial r} \left(\frac{\partial E_{\theta\theta}}{\partial r} + \frac{E_{\theta\theta} - E_{rr}}{r} \right) (\hat{\mathbf{e}}_r \times \hat{\mathbf{e}}_\theta) \hat{\mathbf{e}}_z \\
& - \frac{\partial^2 E_{\theta\theta}}{\partial r \partial z} (\hat{\mathbf{e}}_r \times \hat{\mathbf{e}}_\theta) \hat{\mathbf{e}}_r + \frac{1}{r} \left(\frac{\partial E_{\theta\theta}}{\partial r} + \frac{E_{\theta\theta} - E_{rr}}{r} \right) \left(\hat{\mathbf{e}}_\theta \times \frac{\partial \hat{\mathbf{e}}_\theta}{\partial \theta} \right) \hat{\mathbf{e}}_z \\
& + \frac{1}{r} \frac{\partial E_{rr}}{\partial z} \left(\hat{\mathbf{e}}_\theta \times \frac{\partial \hat{\mathbf{e}}_r}{\partial \theta} \right) \hat{\mathbf{e}}_\theta + \frac{1}{r} \frac{\partial E_{rr}}{\partial z} (\hat{\mathbf{e}}_\theta \times \hat{\mathbf{e}}_r) \frac{\partial \hat{\mathbf{e}}_\theta}{\partial \theta} \\
& - \frac{1}{r} \frac{\partial E_{\theta\theta}}{\partial z} \left(\hat{\mathbf{e}}_\theta \times \frac{\partial \hat{\mathbf{e}}_\theta}{\partial \theta} \right) \hat{\mathbf{e}}_r + \frac{\partial}{\partial z} \left(\frac{\partial E_{\theta\theta}}{\partial r} + \frac{E_{\theta\theta} - E_{rr}}{r} \right) (\hat{\mathbf{e}}_z \times \hat{\mathbf{e}}_\theta) \hat{\mathbf{e}}_z \\
& + \frac{\partial^2 E_{rr}}{\partial z^2} (\hat{\mathbf{e}}_z \times \hat{\mathbf{e}}_r) \hat{\mathbf{e}}_\theta - \frac{\partial^2 E_{\theta\theta}}{\partial z^2} (\hat{\mathbf{e}}_z \times \hat{\mathbf{e}}_\theta) \hat{\mathbf{e}}_r = \mathbf{0}.
\end{aligned}
$$

Noting that

$$\frac{\partial \hat{\mathbf{e}}_r}{\partial \theta} = \hat{\mathbf{e}}_\theta, \quad \frac{\partial \hat{\mathbf{e}}_\theta}{\partial \theta} = -\hat{\mathbf{e}}_r, \quad \hat{\mathbf{e}}_r \times \hat{\mathbf{e}}_\theta = \hat{\mathbf{e}}_z, \quad \hat{\mathbf{e}}_\theta \times \hat{\mathbf{e}}_z = \hat{\mathbf{e}}_r, \quad \hat{\mathbf{e}}_z \times \hat{\mathbf{e}}_r = \hat{\mathbf{e}}_\theta,$$

and that a tensor is zero only when all its components are zero, we obtain

$$\hat{e}_z\hat{e}_z : \quad \frac{\partial}{\partial r}\left(\frac{\partial E_{\theta\theta}}{\partial r} + \frac{E_{\theta\theta} - E_{rr}}{r}\right) + \frac{1}{r}\left(\frac{\partial E_{\theta\theta}}{\partial r} + \frac{E_{\theta\theta} - E_{rr}}{r}\right) = 0,$$

$$\hat{e}_z\hat{e}_r : \quad -\frac{\partial^2 E_{\theta\theta}}{\partial r\partial z} + \frac{1}{r}\frac{\partial}{\partial z}(E_{rr} - E_{\theta\theta}) = 0,$$

$$\hat{e}_r\hat{e}_z : \quad -\frac{\partial}{\partial z}\left(\frac{\partial E_{\theta\theta}}{\partial r} + \frac{E_{\theta\theta} - E_{rr}}{r}\right) = 0,$$

$$\hat{e}_\theta\hat{e}_\theta : \quad \frac{\partial^2 E_{rr}}{\partial z^2} = 0. \qquad \hat{e}_r\hat{e}_r : \quad \frac{\partial^2 E_{\theta\theta}}{\partial z^2} = 0.$$

2.4 Constitutive Equations

2.4.1 Introduction

The kinematic relations and the mechanical and thermodynamic principles are applicable to any continuum irrespective of its physical constitution. Here we consider equations characterizing the individual material and its reaction to applied loads. These equations are called the *constitutive equations*. The formulation of the constitutive equations for a given material is guided by certain rules (i.e., constitutive axioms). We will not discuss them here but will review the linear constitutive relations for solids undergoing small deformations.

A material body is said to be *homogeneous* if the material properties are the same throughout the body (i.e., independent of position). In a *heterogeneous* body, the material properties are a function of position. An *anisotropic* body is one that has different values of a material property in different directions at a point, that is, material properties are direction dependent. An *isotropic* body is one for which every material property in all directions at a point is the same. An isotropic or anisotropic material can be nonhomogeneous or homogeneous.

A material body is said to be *ideally elastic* when, under isothermal conditions, the body recovers its original form completely upon removal of the forces causing deformation, and there is a one-to-one relationship between the state of stress and the state of strain. The constitutive equations described here do not include creep at constant stress and stress relaxation at constant strain. Thus, the material coefficients that specify the constitutive relationship between the stress and strain components are assumed to be constant during the deformation. This does not automatically imply that we neglect temperature effects on deformation. We account for the thermal expansion of the material, which can produce strains or stresses as large as those produced by the applied mechanical forces. Here, we review the basic constitutive equations of linear elasticity (i.e., generalized Hooke's law) for small displacements.

2.4.2 Generalized Hooke's Law

The generalized Hooke's law relates the six components of stress to the six components of strain with respect to the coordinate axes (X_1, X_2, X_3) aligned with the material planes as

$$S_{ij} = C_{ijk\ell} E_{k\ell}, \qquad (2.4.1)$$

where S_{ij} are the components of the second Piola–Kirchhoff stress tensor $\mathbf{S}$, E_{ij} are the components of the Green strain tensor $\mathbf{E}$, and $C_{ijk\ell}$ are the components of the fourth-order elasticity tensor $\mathbf{C}$, all referred to the material coordinates (X_1, X_2, X_3) in the undeformed body.

For the linearized elasticity (i.e., for infinitesimal deformations), we shall replace $\mathbf{S}$ with $\boldsymbol{\sigma}$ and $\mathbf{E}$ with $\boldsymbol{\varepsilon}$. Then Eq. (2.4.1) can be expressed in matrix form as (symmetry of ε_{ij} is used)

$$\begin{Bmatrix} \sigma_{11} \\ \sigma_{22} \\ \sigma_{33} \\ \sigma_{23} \\ \sigma_{13} \\ \sigma_{12} \end{Bmatrix} = \begin{bmatrix} C_{1111} & C_{1122} & C_{1133} & C_{1123} & C_{1113} & C_{1112} \\ C_{2211} & C_{2222} & C_{2233} & C_{2223} & C_{2213} & C_{2212} \\ C_{3311} & C_{3322} & C_{3333} & C_{3323} & C_{3313} & C_{3312} \\ C_{2311} & C_{2322} & C_{2333} & C_{2323} & C_{2313} & C_{2312} \\ C_{1311} & C_{1322} & C_{1333} & C_{1323} & C_{1313} & C_{1312} \\ C_{1211} & C_{1222} & C_{1233} & C_{1223} & C_{1213} & C_{1212} \end{bmatrix} \begin{Bmatrix} \varepsilon_{11} \\ \varepsilon_{22} \\ \varepsilon_{33} \\ 2\varepsilon_{23} \\ 2\varepsilon_{13} \\ 2\varepsilon_{12} \end{Bmatrix}.$$

By virtue of the symmetry of $\boldsymbol{\sigma}$ and $\boldsymbol{\varepsilon}$, we have $C_{ijkl} = C_{jikl} = C_{ijlk} = C_{jilk}$. We can use single-subscript notation for stresses and strains and double-subscript notation for the elasticity tensor components, called the *contracted notation* or the *Voigt–Kelvin notation*, using the following correspondence:

$$11 \rightarrow 1 \quad 22 \rightarrow 2 \quad 33 \rightarrow 3 \quad 23 \rightarrow 4 \quad 13 \rightarrow 5 \quad 12 \rightarrow 6$$

$$\sigma_1 = \sigma_{11}, \quad \sigma_2 = \sigma_{22}, \quad \sigma_3 = \sigma_{33}, \quad \sigma_4 = \sigma_{23}, \quad \sigma_5 = \sigma_{13}, \quad \sigma_6 = \sigma_{12}, \quad (2.4.2)$$

$$\varepsilon_1 = \varepsilon_{11}, \quad \varepsilon_2 = \varepsilon_{22}, \quad \varepsilon_3 = \varepsilon_{33}, \quad \varepsilon_4 = 2\varepsilon_{23}, \quad \varepsilon_5 = 2\varepsilon_{13}, \quad \varepsilon_6 = 2\varepsilon_{12}.$$

The four-subscripted C can be converted to the two-subscripted C by the change of subscripts shown in Eq. (2.4.2). The matrix equation in Eq. (2.4.2) can be expressed as

$$\begin{Bmatrix} \sigma_1 \\ \sigma_2 \\ \sigma_3 \\ \sigma_4 \\ \sigma_5 \\ \sigma_6 \end{Bmatrix} = \begin{bmatrix} C_{11} & C_{12} & C_{13} & C_{14} & C_{15} & C_{16} \\ C_{21} & C_{22} & C_{23} & C_{24} & C_{25} & C_{26} \\ C_{31} & C_{32} & C_{33} & C_{34} & C_{35} & C_{36} \\ C_{41} & C_{42} & C_{43} & C_{44} & C_{45} & C_{46} \\ C_{51} & C_{52} & C_{53} & C_{54} & C_{55} & C_{56} \\ C_{61} & C_{62} & C_{63} & C_{64} & C_{65} & C_{66} \end{bmatrix} \begin{Bmatrix} \varepsilon_1 \\ \varepsilon_2 \\ \varepsilon_3 \\ \varepsilon_4 \\ \varepsilon_5 \\ \varepsilon_6 \end{Bmatrix}. \qquad (2.4.3)$$

The coefficients C_{ijkl} are symmetric (i.e., $C_{ijkl} = C_{klij} = C_{jikl} = C_{ijlk}$) by virtue of the assumption that there exists a potential function $U_0 = U_0(\varepsilon_{ij})$, called the *strain energy density function*, whose derivative with respect to a strain component determines the corresponding stress component (see Section 3.2 for details)

$$\sigma_{ij} = \frac{\partial U_0}{\partial \varepsilon_{ij}}. \tag{2.4.4}$$

Such materials are termed *hyperelastic* materials. In view of the aforementioned symmetries of C_{ijkl}, C_{ij} is also symmetric $C_{ij} = C_{ji}$. For hyperelastic materials, there are only 21 independent coefficients of the matrix $[C]$.

When three mutually orthogonal planes of material symmetry exist, the number of independent elastic coefficients is reduced to nine, and such materials are called *orthotropic*. The stress–strain relations for an orthotropic material take the form

$$\begin{Bmatrix} \sigma_1 \\ \sigma_2 \\ \sigma_3 \\ \sigma_4 \\ \sigma_5 \\ \sigma_6 \end{Bmatrix} = \begin{bmatrix} C_{11} & C_{12} & C_{13} & 0 & 0 & 0 \\ C_{12} & C_{22} & C_{23} & 0 & 0 & 0 \\ C_{13} & C_{23} & C_{33} & 0 & 0 & 0 \\ 0 & 0 & 0 & C_{44} & 0 & 0 \\ 0 & 0 & 0 & 0 & C_{55} & 0 \\ 0 & 0 & 0 & 0 & 0 & C_{66} \end{bmatrix} \begin{Bmatrix} \varepsilon_1 \\ \varepsilon_2 \\ \varepsilon_3 \\ \varepsilon_4 \\ \varepsilon_5 \\ \varepsilon_6 \end{Bmatrix}. \tag{2.4.5}$$

The inverse relations, strain–stress relations, are given by

$$\begin{Bmatrix} \varepsilon_1 \\ \varepsilon_2 \\ \varepsilon_3 \\ \varepsilon_4 \\ \varepsilon_5 \\ \varepsilon_6 \end{Bmatrix} = \begin{bmatrix} C_{11}^* & C_{12}^* & C_{13}^* & 0 & 0 & 0 \\ C_{12}^* & C_{22}^* & C_{23}^* & 0 & 0 & 0 \\ C_{13}^* & C_{23}^* & C_{33}^* & 0 & 0 & 0 \\ 0 & 0 & 0 & C_{44}^* & 0 & 0 \\ 0 & 0 & 0 & 0 & C_{55}^* & 0 \\ 0 & 0 & 0 & 0 & 0 & C_{66}^* \end{bmatrix} \begin{Bmatrix} \sigma_1 \\ \sigma_2 \\ \sigma_3 \\ \sigma_4 \\ \sigma_5 \\ \sigma_6 \end{Bmatrix}$$

$$= \begin{bmatrix} \frac{1}{E_1} & -\frac{\nu_{21}}{E_2} & -\frac{\nu_{31}}{E_3} & 0 & 0 & 0 \\ -\frac{\nu_{12}}{E_1} & \frac{1}{E_2} & -\frac{\nu_{32}}{E_3} & 0 & 0 & 0 \\ -\frac{\nu_{13}}{E_1} & -\frac{\nu_{23}}{E_2} & \frac{1}{E_3} & 0 & 0 & 0 \\ 0 & 0 & 0 & \frac{1}{G_{23}} & 0 & 0 \\ 0 & 0 & 0 & 0 & \frac{1}{G_{13}} & 0 \\ 0 & 0 & 0 & 0 & 0 & \frac{1}{G_{12}} \end{bmatrix} \begin{Bmatrix} \sigma_1 \\ \sigma_2 \\ \sigma_3 \\ \sigma_4 \\ \sigma_5 \\ \sigma_6 \end{Bmatrix}, \tag{2.4.6}$$

where C_{ij}^* denote the compliance coefficients, $[C^*] = [C]^{-1}$; E_1, E_2, E_3 are Young's moduli in 1, 2, and 3 material directions, respectively; ν_{ij} is Poisson's ratio, defined as the ratio of transverse strain in the jth direction to the axial strain in the ith direction when stressed in the i-direction; and G_{23}, G_{13}, G_{12} are shear moduli in the 2-3, 1-3, and 1-2 planes, respectively. Since the compliance matrix $[C^*]$ is the inverse of the stiffness matrix $[C]$ and the inverse of a symmetric matrix is symmetric, it follows that $[C^*]$ is also symmetric. Therefore, the following reciprocal relations hold:

$$\frac{\nu_{21}}{E_2} = \frac{\nu_{12}}{E_1}; \quad \frac{\nu_{31}}{E_3} = \frac{\nu_{13}}{E_1}; \quad \frac{\nu_{32}}{E_3} = \frac{\nu_{23}}{E_2} \tag{2.4.7}$$

or, in short,

$$\frac{\nu_{ij}}{E_i} = \frac{\nu_{ji}}{E_j} \quad (\text{no sum on } i, j) \tag{2.4.8}$$

for $i, j = 1, 2, 3$. Thus, there are only nine independent material coefficients for an orthotropic material:

$$E_1, \ E_2, \ E_3, \ G_{23}, \ G_{13}, \ G_{12}, \ \nu_{12}, \ \nu_{13}, \ \nu_{23}. \tag{2.4.9}$$

When there exist no preferred directions in the material (i.e., the material has infinite number of planes of material symmetry), the number of independent elastic coefficients reduces to two. Such materials are called *isotropic*. For isotropic materials we have $E_1 = E_2 = E_3 = E$, $G_{12} = G_{13} = G_{23} \equiv G$, and $\nu_{12} = \nu_{23} = \nu_{13} \equiv \nu$. Of the three constants (E, ν, G), only two are independent and the third one is related to the other two by the relation

$$G = \frac{E}{2(1 + \nu)}. \tag{2.4.10}$$

For isotropic materials, the stress–strain relations can be expressed in terms of the tensor components as

$$\boldsymbol{\sigma} = 2\mu \boldsymbol{\varepsilon} + \lambda \operatorname{tr}(\boldsymbol{\varepsilon})\mathbf{I}, \quad \sigma_{ij} = 2\mu \varepsilon_{ij} + \lambda \varepsilon_{kk} \delta_{ij}, \tag{2.4.11}$$

where $\operatorname{tr}(\cdot)$ denotes the *trace* (sum of the diagonal elements) of the enclosed tensor and μ and λ are the Lamé constants, which are related to E and ν by

$$\mu = \frac{E}{2(1+\nu)}, \quad \lambda = \frac{\nu E}{(1+\nu)(1-2\nu)}, \quad 2\mu + \lambda = \frac{(1-\nu)E}{(1+\nu)(1-2\nu)}. \tag{2.4.12}$$

2.4.3   Plane Stress-Reduced Constitutive Relations

Typically, when analyzing thin bodies that experience significant stresses in the plane of the body, all stresses (especially σ_{33}) on the plane perpendicular to the thickness coordinate (say, x_3) are neglected, as illustrated in Fig. 2.4.1(a). Such bodies are said to be in a state of *plane stress*. The strain–stress relations of an orthotropic body in plane stress state can be expressed as

$$\begin{Bmatrix} \varepsilon_1 \\ \varepsilon_2 \\ \varepsilon_6 \end{Bmatrix} = \begin{bmatrix} C_{11}^* & C_{12}^* & 0 \\ C_{12}^* & C_{22}^* & 0 \\ 0 & 0 & C_{66}^* \end{bmatrix} \begin{Bmatrix} \sigma_1 \\ \sigma_2 \\ \sigma_6 \end{Bmatrix} = \begin{bmatrix} \frac{1}{E_1} & -\frac{\nu_{21}}{E_2} & 0 \\ -\frac{\nu_{12}}{E_1} & \frac{1}{E_2} & 0 \\ 0 & 0 & \frac{1}{G_{12}} \end{bmatrix} \begin{Bmatrix} \sigma_1 \\ \sigma_2 \\ \sigma_6 \end{Bmatrix}. \tag{2.4.13}$$

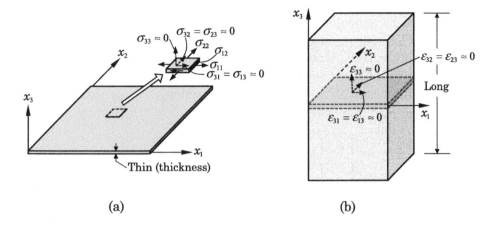

Fig. 2.4.1 (a) Plane stress state. (b) Plane strain state.

The strain–stress relations in Eq. (2.4.13) can be inverted to obtain the stress–strain relations for the plane stress case as

$$\left\{\begin{matrix} \sigma_1 \\ \sigma_2 \\ \sigma_6 \end{matrix}\right\} = \begin{bmatrix} Q_{11} & Q_{12} & 0 \\ Q_{12} & Q_{22} & 0 \\ 0 & 0 & Q_{66} \end{bmatrix} \left\{\begin{matrix} \varepsilon_1 \\ \varepsilon_2 \\ \varepsilon_6 \end{matrix}\right\} \tag{2.4.14}$$

where the Q_{ij}, called the *plane stress-reduced elastic stiffness coefficients*, are given by

$$Q_{11} = \frac{C_{22}^*}{C_{11}^* C_{22}^* - C_{12}^* C_{12}^*} = \frac{E_1}{1 - \nu_{12}\nu_{21}},$$

$$Q_{12} = -\frac{C_{12}^*}{C_{11}^* C_{22}^* - C_{12}^* C_{12}^*} = \frac{\nu_{12} E_2}{1 - \nu_{12}\nu_{21}},$$

$$Q_{22} = \frac{C_{11}^*}{C_{11}^* C_{22}^* - C_{12}^* C_{12}^*} = \frac{E_2}{1 - \nu_{12}\nu_{21}}, \tag{2.4.15}$$

$$Q_{66} = \frac{1}{C_{66}^*} = G_{12}.$$

Note that the reduced stiffness coefficients involve four independent material constants, E_1, E_2, ν_{12}, and G_{12}.

In theories where only the transverse normal stress σ_{33} is neglected but the transverse shear stresses are accounted for, the transverse shear stresses in an orthotropic material are related to the transverse shear stresses by the relations

$$\left\{\begin{matrix} \sigma_4 \\ \sigma_5 \end{matrix}\right\} = \begin{bmatrix} Q_{44} & 0 \\ 0 & Q_{55} \end{bmatrix} \left\{\begin{matrix} \varepsilon_4 \\ \varepsilon_5 \end{matrix}\right\}, \tag{2.4.16}$$

where $Q_{44} = C_{44} = G_{23}$ and $Q_{55} = C_{55} = G_{13}$.

A *plane strain state* is one in which the all three strains in certain plane are assumed to be negligible [see Fig 2.4.1(b)]. The plane strain assumption is used in dealing with problems in which all planes in certain (long) direction essentially have the same strain field. The stress–strain relation for the plane strain case (with x_3 being the direction perpendicular to the plane of the body being analyzed), when $E_3 = E_2$, $\nu_{12} = \nu_{13}$, and $\nu_{23} = \nu_{32}$ (known as the *transversely isotropic material*), are given by

$$\left\{ \begin{array}{c} \sigma_1 \\ \sigma_2 \\ \sigma_6 \end{array} \right\} = \left[\begin{array}{ccc} \hat{Q}_{11} & \hat{Q}_{12} & 0 \\ \hat{Q}_{12} & \hat{Q}_{22} & 0 \\ 0 & 0 & \hat{Q}_{66} \end{array} \right] \left\{ \begin{array}{c} \varepsilon_1 \\ \varepsilon_2 \\ \varepsilon_6 \end{array} \right\} \tag{2.4.17}$$

where

$$\hat{Q}_{11} = \frac{(1 - \nu_{23})E_1}{1 - \nu_{23} - 2\nu_{12}\nu_{21}},$$

$$\hat{Q}_{12} = \frac{\nu_{12}E_2}{1 - \nu_{23} - 2\nu_{12}\nu_{21}},$$

$$\hat{Q}_{22} = \frac{(1 - \nu_{12}\nu_{21})E_2}{(1 + \nu_{23})(1 - \nu_{23} - 2\nu_{12}\nu_{21})},$$

$$\hat{Q}_{66} = G_{12}. \tag{2.4.18}$$

2.4.4   Thermoelastic Constitutive Relations

When temperature changes occur in the elastic body, we account for the thermal expansion of the material, even though the variation of elastic constants with temperature is neglected. When the strains, geometric changes, and temperature variations are sufficiently small, all governing equations are linear, and superposition of mechanical and thermal effects is possible.

The linear thermoelastic constitutive equations have the form

$$\sigma_j = C_{ji}[-\alpha_i(T - T_0) + \varepsilon_i], \quad \varepsilon_j = C_{ji}^*\sigma_i + \alpha_i(T - T_0), \tag{2.4.19}$$

where α_i ($i = 1, 2, 3$) are the linear coefficients of thermal expansion in the x_i-coordinate direction ($\alpha_4 = \alpha_5 = \alpha_6 = 0$), T denotes temperature, and T_0 is the reference temperature of the undeformed body. In writing Eq. (2.4.19), it is assumed that α_i and C_{ij} are independent of strains and temperature. For an isotropic material, we have $\alpha_1 = \alpha_2 = \alpha_3 \equiv \alpha$. The orthotropic plane stress constitutive relations for a thermoelastic case are given by

$$\left\{ \begin{array}{c} \sigma_1 \\ \sigma_2 \\ \sigma_6 \end{array} \right\} = \left[\begin{array}{ccc} Q_{11} & Q_{12} & 0 \\ Q_{12} & Q_{22} & 0 \\ 0 & 0 & Q_{66} \end{array} \right] \left\{ \begin{array}{c} \varepsilon_1 - \alpha_1 \Delta T \\ \varepsilon_2 - \alpha_2 \Delta T \\ \varepsilon_6 \end{array} \right\}, \tag{2.4.20}$$

where $\Delta T = T - T_0$ is the temperature change from a reference temperature T_0. Equation (2.4.16) remains unchanged for the thermoelastic case.

2.5 Theories of Straight Beams

2.5.1 Introduction

Most practical engineering structures, microscale or macroscale, consist of members that can be classified as beams, plates, and shells, called structural members. *Beams* are structural members that have a ratio of length-to-cross-sectional dimensions very large, say, 10 to 100 or more, and subjected to forces both along and transverse to the length and moments that tend to rotate them about an axis perpendicular to their length [see Fig. 2.5.1(a)]. When all applied loads are along the length only, they are often called *bars* (i.e., bars experience only tensile or compressive strains and no bending deformation). *Cables* may be viewed as very flexible form of bars, and they can only take tension and not compression. *Plates* are two-dimensional versions of beams in the sense that one of the cross-sectional dimensions of a plate can be as large as the length. The smallest dimension of a plate is called its thickness. Thus, plates are thin bodies subjected to forces in the plane as well as in the direction normal to the plane and bending moments about either axis in the plane [see Fig. 2.5.1(b)]. A *shell* is a thin structure with curved geometry [see Fig. 2.5.1(c)] and can be subjected to distributed as well as point forces and moments. Because of their geometries and loads applied, beams, plates, and shells are stretched and bent from their original shapes. The difference between structural elements

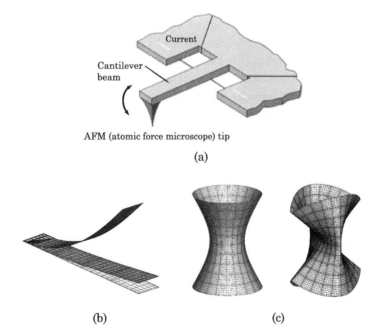

Fig. 2.5.1 (a) An AFM cantilever beam. (b) A narrow plate strip subjected to a vertical force. (c) A hyperboloidal shell. The deformations shown for (b) and (c) are exaggerated.

and three-dimensional solid bodies, such as solid blocks and spheres that have no restrictions on their geometric make up, is that the latter may change their geometries, but they may not show significant "bending" deformation.

Although all solids can be analyzed for stress and deformation using the elasticity equations reviewed in the preceding sections of this chapter, their geometries allow us to develop theories that are simple and yet yield results that are accurate enough for engineering analysis and design. *Structural mechanics* is a branch of solid mechanics that deals with the study of beams (frames), plates, and shells using theories that are derived from three-dimensional elasticity theory by making certain simplifying assumptions concerning the deformation and stress states in these members.

The present section is devoted to the development of structural theories of straight beams in bending for the static case. The equations governing beams will be referenced in the coming chapters, and their development in this chapter will help the reader in understanding the material of the subsequent chapters. Plates will be studied in a chapter devoted entirely for them (see Chapter 7). The geometric description of shells is more involved, and they are not covered in this book to keep the book size within reasonable limits (see, for additional information, Timoshenko and Woinowsky-Krieger [124] and Reddy [50, 51]).

In this section we consider two most commonly used theories of straight beams. They differ from each other in the representation of displacement and strain fields. The first one is *the Bernoulli–Euler beam theory*,[3] a theory that is covered in all undergraduate mechanics of materials books. In the Bernoulli–Euler beam theory, the transverse shear strain is neglected, making the beam infinitely rigid in the transverse direction. The second one is a refinement to the Bernoulli–Euler beam theory, known as *the Timoshenko beam theory*, which accounts for the transverse shear strain. These two beam theories will be developed assuming infinitesimal deformation. Therefore, we use σ_{ij} to mean S_{ij}, the second Piola–Kirchhoff stress tensor, and x_i in place of X_i.

We consider straight beams of length L and symmetric (about the x_2-axis) cross section of area A. The x_1-coordinate is taken along the centroidal axis of the beam with the x_3-coordinate along the thickness (the height) and the x_2-coordinate along the width of the beam (into the plane of the page), as shown in Fig. 2.5.2(a). In general, the cross-sectional area A can be a function of x_1. Suppose that the beam is subjected to distributed axial force $f(x_1)$ and transverse load $q(x_1)$, and let (u_1, u_2, u_3) denote the total displacements along the coordinates $(x_1 = x, x_2 = y, x_3 = z)$. For simplicity of the developments, only stretching along the length of the beam and bending about the x_2-axis are considered here.

---

[3] Jacob Bernoulli (1655–1705) was one of the many prominent Swiss mathematicians in the Bernoulli family. He, and along with his brother Johann Bernoulli, was one of the founders of the calculus of variations. Leonhard Euler (1707–1783) was a pioneering Swiss mathematician and physicist.

2.5.2 The Bernoulli–Euler Beam Theory

The Bernoulli–Euler beam theory is based on certain simplifying assumptions, known as *the Bernoulli–Euler hypothesis*, concerning the kinematics of bending deformation. The hypothesis states that straight lines perpendicular to the beam axis before deformation remain (a) straight, (b) inextensible, and (c) perpendicular to the tangent line to the beam axis after deformation.

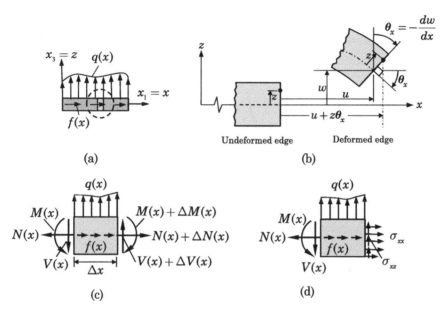

Fig. 2.5.2 (a) A typical beam with loads. (b) Kinematics of deformation of an Bernoulli–Euler beam theory. (c) Equilibrium of a beam element. (d) Definitions (or internal equilibrium) of stress resultants.

The Bernoulli–Euler hypothesis leads to the following displacement field [see Fig. 2.5.2(b)]:

$$u_1(x, y, z) = u(x) - z\frac{dw}{dx}, \quad u_2 = 0, \quad u_3(x, y, z) = w(x), \tag{2.5.1}$$

where (u, v, w) are the displacements of a point on the x-axis in the $x_1 = x$, $x_2 = y$, and $x_3 = z$ coordinate directions, respectively. The infinitesimal strains are

$$\varepsilon_{xx} = \frac{\partial u_1}{\partial x_1} = \frac{du}{dx} - z\frac{d^2w}{dx^2}, \quad 2\varepsilon_{xz} = \frac{\partial u_1}{\partial x_3} + \frac{\partial u_3}{\partial x_1} = -\frac{dw}{dx} + \frac{dw}{dx} = 0, \tag{2.5.2}$$

and all other strains are identically zero.

A free-body diagram of an element of length Δx is shown, with all its forces, in Fig. 2.5.2(c). Summing the forces in the x- and z-directions and summing the moments at the right end of the beam element, we obtain the following equilibrium equations:

$$\sum F_x = 0: \qquad -\frac{dN}{dx} = f(x), \qquad (2.5.3)$$

$$\sum F_z = 0: \qquad -\frac{dV}{dx} = q(x), \qquad (2.5.4)$$

$$\sum M_y = 0: \qquad V - \frac{dM}{dx} = 0, \qquad (2.5.5)$$

where $N(x)$ is the net axial force, $M(x)$ is the net bending moment about the y-axis, and $V(x)$ is the net transverse shear force [see Fig. 2.5.2(d)] on the beam cross section:

$$N(x) = \int_A \sigma_{xx}\, dA, \quad M(x) = \int_A \sigma_{xx} z\, dA, \quad V(x) = \int_A \sigma_{xz}\, dA, \qquad (2.5.6)$$

where dA denotes an area element of the cross section ($dA = dydz$). The set (N, M, V) are known as the stress resultants (because they result from the stresses in the beam).

The stresses in the beam can be computed using the linear elastic constitutive relation for an isotropic but possibly inhomogeneous material (i.e., the material properties may be functions of x and z); we have

$$\sigma_{xx}(x, z) = E(x, z)\varepsilon_{xx}(x, z) = E\left(\frac{du}{dx} - z\frac{d^2 w}{dx^2}\right),$$

$$\sigma_{xz}(x, z) = 2G(x, z)\varepsilon_{xz}(x, z) = 0. \qquad (2.5.7)$$

Although the transverse shear stress computed using constitutive equation is zero (hence, $V = 0$), it cannot be zero in a beam because the force equilibrium, Eq. (2.5.4), is violated. This is the inconsistency resulting from the Bernoulli–Euler beam hypothesis. Therefore, to respect the physical requirement of $V \neq 0$, we do not compute the shear stress from $\sigma_{xz} = 2G\varepsilon_{xz}$; it is computed using the shear force V obtained from the equilibrium condition in Eq. (2.5.5), namely, $V = dM/dx$.

The stress resultants (N, M) now can be related to the displacements u and w and back to the stress σ_{xx}, as discussed next. Using Eq. (2.5.6), we obtain

$$N(x) = \int_A \sigma_{xx}\, dA = A_{xx}\frac{du}{dx} + B_{xx}\left(-\frac{d^2 w}{dx^2}\right),$$

$$M(x) = \int_A z\sigma_{xx}\, dA = B_{xx}\frac{du}{dx} + D_{xx}\left(-\frac{d^2 w}{dx^2}\right), \qquad (2.5.8)$$

where A_{xx} is the axial stiffness, B_{xx} is the bending–stretching stiffness, and D_{xx} is the bending stiffness

$$A_{xx} = \int_A E(x, z)\, dA, \quad B_{xx} = \int_A zE(x, z)\, dA, \quad D_{xx} = \int_A z^2 E(x, z)\, dA. \quad (2.5.9)$$

When E is a function of x only, Eq. (2.5.9) simplifies to

$$A_{xx} = E(x) \int_A dA = E(x)A(x), \quad B_{xx} = E(x) \int_A z\, dA = 0,$$
$$D_{xx} = E(x) \int_A z^2\, dA = E(x)I(x), \quad I = \int_A z^2\, dA, \tag{2.5.10}$$

where we have used the fact that the x-axis coincides with the centroidal axis:

$$\int_A z\, dA = 0. \tag{2.5.11}$$

Then (N, M) of Eq. (2.5.8) reduce to

$$N(x) = EA\frac{du}{dx}, \quad M(x) = -EI\frac{d^2w}{dx^2}. \tag{2.5.12}$$

Equations (2.5.8) and (2.5.9) can be inverted to express du/dx and d^2w/dx^2 in terms of N and M. We obtain

$$\frac{du}{dx} = \frac{D_{xx}N - B_{xx}M}{A_{xx}D_{xx} - B_{xx}^2}, \quad -\frac{d^2w}{dx^2} = \frac{A_{xx}M - B_{xx}N}{A_{xx}D_{xx} - B_{xx}^2}. \tag{2.5.13}$$

Substituting these relations for du/dx and $-d^2w/dx^2$ into the expression for σ_{xx} in Eq. (2.5.7), we obtain

$$\sigma_{xx}(x, z) = \frac{E}{A_{xx}D_{xx} - B_{xx}^2} \left[D_{xx}N - B_{xx}M + z\left(A_{xx}M - B_{xx}N \right) \right]. \tag{2.5.14}$$

When E is not a function of z, we obtain ($A_{xx} = EA$ and $D_{xx} = EI$)

$$\frac{du}{dx} = \frac{N}{EA}, \quad -\frac{d^2w}{dx^2} = \frac{M}{EI}, \tag{2.5.15}$$

and

$$\sigma_{xx} = \frac{N}{A} + \frac{Mz}{I}. \tag{2.5.16}$$

We note that when E is not a function of z, N is independent of w and M is independent of u. Therefore, Eq. (2.5.3) is independent of Eqs (2.5.4) and (2.5.5). In that case the axial deformation of members can be determined by solving Eq. (2.5.3) independent of the bending deformation and vice versa. Also, Eqs (2.5.4) and (2.5.5) can be combined to read

$$-\frac{d^2M}{dx^2} = q(x), \tag{2.5.17}$$

which can be expressed in terms of the transverse deflection w using the second equation in Eq. (2.5.15) as

$$\frac{d^2}{dx^2}\left(EI\frac{d^2w}{dx^2} \right) = q(x). \tag{2.5.18}$$

We now return to the computation of shear stress σ_{xz} in the Bernoulli–Euler beam theory. The shear stress is computed from equilibrium considerations. For beams with $E = E(x)$, the expression for σ_{xz} is given by (see pages 256–261 of Fenner and Reddy [98] for a derivation)

$$\sigma_{xz}(x, z) = \frac{V(x)Q(z)}{I(x)b}, \quad Q(z) = \int_z^{c_1} b(z)z\,dz, \tag{2.5.19}$$

where $Q(z)$ denotes the first moment of the hatched area [see Fig. 2.5.3] about the axis y of the entire cross section and b is the width of the cross section at z where the longitudinal shear stress (σ_{xz}) is computed. The hatched area is only a portion of the total area of cross section that lies below the surface on which the shear force acts; Q is the maximum at the centroid (i.e., $z = 0$), and σ_{xz} is the maximum wherever Q/b is the maximum. For example, for a beam of height $2h$ and width b, $Q(z)$ takes the form

$$Q(z) = b \int_z^h z\,dz = \frac{b}{2} \left(h^2 - z^2 \right), \tag{2.5.20}$$

and the shear stress becomes

$$\sigma_{xz}(x, z) = \frac{V(x)Q(z)}{I(x)b} = \frac{1}{2I} V(x) \left(h^2 - z^2 \right). \tag{2.5.21}$$

Thus, the shear stress varies quadratically through the beam height, vanishing at the top $(z = -h)$ and bottom $(z = h)$. The maximum shear stress is $\sigma_{xz}^{\max} = \sigma_{xz}(x, 0) = \frac{3}{4bh} V(x)$. Thus, the actual maximum occurs wherever V is the maximum.

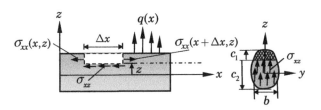

Fig. 2.5.3 Transverse shear stress σ_{xz} due to bending.

2.5.3   The Timoshenko Beam Theory

The Timoshenko beam theory is based on a relaxation of the normality condition, namely, part (c) of the Bernoulli–Euler hypothesis. In other words, the transverse normal has a rotation ϕ_x that is not the same as the slope $-dw/dx$. The difference between these two quantities is the transverse shear strain.

The relaxed Bernoulli–Euler hypothesis leads to the following displacement field for the Timoshenko beam theory (see Fig. 2.5.4):

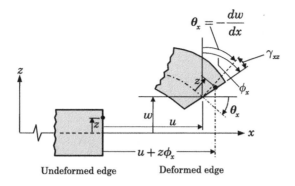

Fig. 2.5.4 Kinematics of deformation in the Timoshenko beam theory.

$$u_1(x, y, z) = u(x) + z\phi_x, \quad u_2 = 0, \quad u_3(x, y, z) = w(x), \qquad (2.5.22)$$

where ϕ_x is the rotation of a transverse normal line. The infinitesimal strains are

$$\varepsilon_{xx} = \frac{\partial u_1}{\partial x_1} = \frac{du}{dx} + z\frac{d\phi_x}{dx}, \quad 2\varepsilon_{xz} = \frac{\partial u_1}{\partial x_3} + \frac{\partial u_3}{\partial x_1} = \phi_x + \frac{dw}{dx}, \qquad (2.5.23)$$

and all other strains are identically zero.

The free-body diagram of a typical element of the beam remains the same as depicted in Fig. 2.5.2(c). Therefore, the equilibrium equations given in Eqs (2.5.3) to (2.5.5) are also valid for the Timoshenko beam theory, except that the stress resultants (N, M, V) are related to (u, w, ϕ_x) in a different way, as discussed in the following paragraphs.

The stresses in the beam are

$$\sigma_{xx}(x, z) = E(x, z)\varepsilon_{xx}(x, z) = E\left(\frac{du}{dx} + z\frac{d\phi_x}{dx}\right),$$
$$\sigma_{xz}(x, z) = 2G(x, z)\varepsilon_{xz}(x, z) = G(x, z)\left(\phi_x + \frac{dw}{dx}\right). \qquad (2.5.24)$$

Note that the shear strain (hence, shear force) is not zero, removing the inconsistency of the Bernoulli–Euler beam theory.

The stress resultants (N, M) in the Timoshenko beam theory have the form

$$N(x) = \int_A \sigma_{xx}\, dA = A_{xx}\frac{du}{dx} + B_{xx}\left(\frac{d\phi_x}{dx}\right), \qquad (2.5.25)$$

$$M(x) = \int_A z\sigma_{xx}\, dA = B_{xx}\frac{du}{dx} + D_{xx}\left(\frac{d\phi_x}{dx}\right), \qquad (2.5.26)$$

$$V(x) = \int_A K_s\sigma_{xz}\, dA = K_s S_{xz}\left(\phi_x + \frac{dw}{dx}\right), \qquad (2.5.27)$$

where K_s is the *shear correction coefficient* introduced to correct the energy loss due to the constant state[4] of transverse shear stress σ_{xz} and S_{xz} is the shear stiffness

$$S_{xz} = \int_A G(x, z)\, dA. \tag{2.5.28}$$

Inverting Eqs (2.5.25) to (2.5.27), we obtain

$$\frac{du}{dx} = \frac{D_{xx}N - B_{xx}M}{A_{xx}D_{xx} - B_{xx}^2}, \quad \frac{d\phi_x}{dx} = \frac{A_{xx}M - B_{xx}N}{A_{xx}D_{xx} - B_{xx}^2}, \quad \phi_x + \frac{dw}{dx} = \frac{V}{K_s S_{xz}}. \tag{2.5.29}$$

Substituting for du/dx, $d\phi_x/dx$, and $\phi_x + dw/dx$ from Eq. (2.5.29) into the expressions for σ_{xx} and σ_{xz} in Eq. (2.5.24), we obtain

$$\sigma_{xx}(x, z) = \frac{E}{A_{xx}D_{xx} - B_{xx}^2} \left[D_{xx}N - B_{xx}M + z\left(A_{xx}M - B_{xx}N\right) \right], \tag{2.5.30}$$

$$\sigma_{xz}(x, z) = \frac{GV}{K_s S_{xz}}.$$

When E is not a function of z, we obtain ($A_{xx} = EA$, $B_{xx} = 0$, and $D_{xx} = EI$)

$$\sigma_{xx}(x, z) = \frac{N}{A} + \frac{Mz}{I}, \quad \sigma_{xz}(x) = \frac{V}{K_s A}, \tag{2.5.31}$$

and the equations of equilibrium in Eqs (2.5.3) to (2.5.5) in terms of the generalized displacements (u, w, ϕ_x) take the form

$$-\frac{d}{dx}\left(EA\frac{du}{dx}\right) = f, \tag{2.5.32}$$

$$-\frac{d}{dx}\left[K_s GA\left(\frac{dw}{dx} + \phi_x\right)\right] = q, \tag{2.5.33}$$

$$K_s GA\left(\frac{dw}{dx} + \phi_x\right) - \frac{d}{dx}\left(EI\frac{d\phi_x}{dx}\right) = 0. \tag{2.5.34}$$

As in the case of the Bernoulli–Euler beam theory, the axial deformation of beams can be determined by solving Eq. (2.5.32) independent of the bending equations, Eqs (2.5.33) and (2.5.34).

Example 2.5.1 illustrates several ideas from this chapter.

Example 2.5.1 ————————————————————————————

Consider the problem of an isotropic cantilever beam of rectangular cross section $2h \times b$ (height $2h$ and width b), bent by an upward transverse load F_0 applied at the free end, as shown in

---

[4]We note that the shear stress in the Timoshenko beam theory is not a function of z, implying that it has a constant value at any section of the beam, violating the condition that it be zero at the top and bottom of the beam.

Fig. 2.5.5. Assume that the beam is isotropic, linearly elastic, and homogeneous. (a) Compute the stresses using Eqs (2.5.16) and (2.5.19), (b) compute the strains using the uniaxial strain–stress relations, (c) show that the strains computed satisfy the compatibility condition in the $x_1 x_3$-plane, and (d) determine the two-dimensional displacement field (u_1, u_2) that satisfies the geometric boundary conditions.

Solution: (a) The stresses in the beam are

$$\sigma_{11} = \frac{M x_3}{I} = -\frac{F_0 x_1 x_3}{I}, \quad \sigma_{13} = \frac{VQ}{Ib} = -\frac{F_0}{2I}(h^2 - x_3^2), \tag{1}$$

where I is the moment of inertia about the x_2 axis, $2h$ is the height of the beam, and b is the width of the beam.

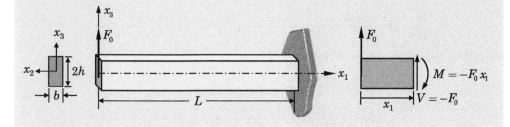

Fig. 2.5.5 Cantilever beam bent by a point load, F_0.

(b) The strains ε_{11} and ε_{13} are known from this section (except for a change of coordinates from X_i to x_i); ε_{11} is the same from the Bernoulli–Euler or the Timoshenko beam theory (TBT), and we use the shear strain obtained from the equilibrium conditions:

$$\begin{aligned}
\varepsilon_{11} &= \frac{\sigma_{11}}{E} = \frac{M(x_1)x_3}{EI} = -\frac{F_0 x_1 x_3}{EI}, \\
\varepsilon_{13} &= \frac{\sigma_{13}}{2G} = \frac{VQ}{2IbG} = -\frac{(1+\nu)F_0}{2EI}(h^2 - x_3^2), \\
\varepsilon_{22} &= \varepsilon_{33} = -\nu\varepsilon_{11} = \frac{\nu F_0 x_1 x_3}{EI},
\end{aligned} \tag{2}$$

where ν is the Poisson ratio, E is Young's modulus, and G is the shear modulus.

(c) Next, we determine if the strains computed are compatible. Substituting ε_{ij} into the third equation in Eq. (2.3.17), we obtain $0 + 0 = 0$. Thus, the strains satisfy the compatibility equations in two dimensions (i.e., in the $x_1 x_3$-plane). Although the two-dimensional strains are compatible, the three-dimensional strains are not compatible. For example, using the additional strains, $\varepsilon_{33} = \varepsilon_{22} = -\nu\varepsilon_{11}$ and $\varepsilon_{12} = \varepsilon_{23} = 0$, one can show that all of the equations except the fifth equation in Eq. (2.3.17) are satisfied. We shall seek the two-dimensional displacement field (u_1, u_3) associated with the $x_1 x_3$-plane.

(d) Integrating the strain–displacement equations, we obtain

$$\frac{\partial u_1}{\partial x_1} = \varepsilon_{11} = -\frac{F_0 x_1 x_3}{EI} \quad \text{or} \quad u_1 = -\frac{F_0 x_1^2 x_3}{2EI} + f(x_3), \tag{3}$$

$$\frac{\partial u_3}{\partial x_3} = \varepsilon_{33} = \frac{\nu F_0 x_1 x_3}{EI} \quad \text{or} \quad u_3 = \frac{\nu F_0 x_1 x_3^2}{2EI} + g(x_1), \tag{4}$$

where $f(x_3)$ and $g(x_1)$ are functions of integration. Substituting u_1 and u_3 into the definition of $2\varepsilon_{13}$, we obtain

$$2\varepsilon_{13} = \frac{\partial u_1}{\partial x_3} + \frac{\partial u_3}{\partial x_1} = -\frac{F_0 x_1^2}{2EI} + \frac{df}{dx_3} + \frac{\nu F_0 x_3^2}{2EI} + \frac{dg}{dx_1}. \tag{5}$$

But this must be equal to the shear strain known from Eq. (2):

$$-\frac{F_0}{2EI}x_1^2 + \frac{df}{dx_3} + \frac{\nu F_0}{2EI}x_3^2 + \frac{dg}{dx_1} = -\frac{(1+\nu)}{EI}F_0(h^2 - x_3^2). \tag{6}$$

Separating the terms that depend only on x_1 and those depend only on x_3 (the constant term can go with either one), we obtain

$$-\frac{dg}{dx_1} + \frac{F_0}{2EI}x_1^2 - \frac{(1+\nu)F_0h^2}{EI} = \frac{df}{dx_3} - \frac{(2+\nu)F_0}{2EI}x_3^2. \tag{7}$$

Since the left side depends only on x_1 and the right side depends only on x_3, and yet the equality must hold, it follows that both sides should be equal to a constant, say, c_0:

$$\frac{df}{dx_3} - \frac{(2+\nu)F_0}{2EI}x_3^2 = c_0, \quad -\frac{dg}{dx_1} + \frac{F_0}{2EI}x_1^2 - \frac{(1+\nu)F_0h^2}{EI} = c_0.$$

Integrating the expressions for f and g, we obtain

$$f(x_3) = \frac{(2+\nu)F_0}{6EI}x_3^3 + c_0 x_3 + c_1,$$

$$g(x_1) = \frac{F_0}{6EI}x_1^3 - \frac{(1+\nu)F_0h^2}{EI}x_1 - c_0 x_1 + c_2, \tag{8}$$

where c_1 and c_2 are constants of integration. Thus, the most general form of displacement field (u_1, u_3) that corresponds to the strains in Eq. (2) is given by

$$u_1(x_1, x_3) = -\frac{F_0}{2EI}x_1^2 x_3 + \frac{(2+\nu)F_0}{6EI}x_3^3 + c_0 x_3 + c_1,$$

$$u_3(x_1, x_3) = -\frac{(1+\nu)F_0h^2}{EI}x_1 + \frac{\nu F_0}{2EI}x_1 x_3^2 + \frac{F_0}{6EI}x_1^3 - c_0 x_1 + c_2. \tag{9}$$

The constants c_0, c_1, and c_2 are determined using suitable boundary conditions. We impose the following boundary conditions that eliminate rigid-body displacements (that is, rigid-body translation and rigid-body rotation):

$$u_1(L,0) = 0, \quad u_3(L,0) = 0, \quad \Omega_{13}\Big|_{x_1=L, x_3=0} = \frac{1}{2}\left(\frac{\partial u_3}{\partial x_1} - \frac{\partial u_1}{\partial x_3}\right)_{x_1=L, x_3=0} = 0. \tag{10}$$

Imposing the boundary conditions from Eq. (10) on the displacement field in Eq. (9), we obtain

$$u_1(L,0) = 0 \quad \Rightarrow \quad c_1 = 0,$$

$$u_3(L,0) = 0 \quad \Rightarrow \quad c_0 L - c_2 = -\frac{(1+\nu)F_0h^2 L}{EI} + \frac{F_0 L^3}{6EI},$$

$$\left(\frac{\partial u_3}{\partial x_1} - \frac{\partial u_1}{\partial x_3}\right)_{x_1=L, x_3=0} = 0, \quad \Rightarrow \quad c_0 = \frac{F_0 L^2}{2EI} - \frac{(1+\nu)F_0h^2}{2EI}. \tag{11}$$

Thus, we have

$$c_0 = \frac{F_0 L^2}{2EI} - \frac{(1+\nu)F_0h^2}{2EI}, \quad c_1 = 0, \quad c_2 = \frac{F_0 L^3}{3EI} + \frac{(1+\nu)F_0h^2 L}{2EI}. \tag{12}$$

Then the final displacement field in Eq. (9) becomes

$$u_1(x_1, x_3) = \frac{F_0 L^2 x_3}{6EI}\left[3\left(1 - \frac{x_1^2}{L^2}\right) + (2+\nu)\frac{x_3^2}{L^2} - 3(1+\nu)\frac{h^2}{L^2}\right],$$

$$u_3(x_1, x_3) = \frac{F_0 L^3}{6EI}\left[2 - 3\frac{x_1}{L}\left(1 - \nu\frac{x_3^2}{L^2}\right) + \frac{x_1^3}{L^3} + 3(1+\nu)\frac{h^2}{L^2}\left(1 - \frac{x_1}{L}\right)\right]. \tag{13}$$

In the Euler–Bernoulli beam theory (EBT), where one assumes that $L >> 2h$ and $\nu = 0$, we have $u_1 = 0$, and u_3 is given by

$$u_3^{\text{EBT}}(x_1, x_3) = \frac{F_0 L^3}{6EI} \left(2 - 3\frac{x_1}{L} + \frac{x_1^3}{L^3} \right), \tag{14}$$

while in the Timoshenko beam theory (TBT) we have $u_1 = 0$ [$E = 2(1+\nu)G$, $I = Ah^2/3$, and $A = 2bh$], and u_3 is given by

$$u_3^{\text{TBT}}(x_1, x_3) = \frac{F_0 L^3}{6EI} \left(2 - 3\frac{x_1}{L} + \frac{x_1^3}{L^3} \right) + \frac{F_0 L}{K_s GA} \left(1 - \frac{x_1}{L} \right). \tag{15}$$

Here K_s denotes the shear correction factor. Thus, the TBT with shear correction factor of $K_s = 4/3$ predicts the same maximum deflection, $u_3(0,0)$, as the two-dimensional elasticity theory. Both beam theory solutions, in general, are in error compared with the plane elasticity solution (primarily because of the Poisson effect).

2.5.4 The von Kármán Theory of Beams

2.5.4.1 Preliminary Discussion

In the preceding sections on beam theories, we made an assumption of infinitesimal strains to write the linear strain–displacement relations [see Eqs (2.5.2) and (2.5.23)]. When one assumes the strain are small but the rotation of the transverse normal line is moderate, the strain fields of the Bernoulli–Euler and Timoshenko beam theories will have additional term, which is nonlinear in the transverse deflection w, in the extensional part of the strain. The nonlinear strain is often referred to as the *von Kármán strain*. To see how this nonlinear term enters the calculation, we first make certain assumptions concerning the magnitude of the various terms in the Green strain tensor components and compute the nonzero strains for beams.

We begin with the assumption that the axial strain (du_1/dx) and the curvature (d^2w/dx^2) are of order ϵ, and (dw/dx) is of order $\sqrt{\epsilon}$, where $\epsilon << 1$ is a small parameter. Then the nonzero Green strain tensor components from Eq. (2.3.11) are (with $X_i \approx x_i$)

$$\varepsilon_{xx} = \frac{\partial u_1}{\partial x_1} + \frac{1}{2} \left(\frac{\partial u_3}{\partial x_1} \right)^2, \tag{2.5.35}$$

$$\varepsilon_{xz} = \frac{1}{2} \left(\frac{\partial u_1}{\partial x_3} + \frac{\partial u_3}{\partial x_1} \right). \tag{2.5.36}$$

Thus, the underlined term in Eq. (2.5.35) is new and it is nonlinear. The beam theory resulting from the inclusion of this nonlinear term is termed the *von Kármán beam theory*. Here we present the complete development for each of the beam theories considered before.

Before we embark on the development of the beam theories, we must consider suitable equations of equilibrium for this nonlinear case. We adopt the equations of equilibrium in terms of the second Piola–Kirchhoff stress tensor $\mathbf{S}$. The vector equation of equilibrium in terms of $\mathbf{S}$ is [see Eqs (2.2.14), (2.2.9), and (2.3.7)]

$$\nabla_0 \cdot \left[\mathbf{S}^{\mathrm{T}} \cdot (\mathbf{I} + \nabla_0 \mathbf{u})\right] + \rho_0 \mathbf{f} = \mathbf{0},$$

which in component form is

$$\frac{\partial}{\partial X_J}\left[\left(\delta_{KI} + \frac{\partial u_I}{\partial X_K}\right) S_{KJ}\right] + \rho_0 f_I = 0, \tag{2.5.37}$$

where f_I are body forces measured per unit mass. In the present case, the equilibrium equations associated with balance of forces in the x- and z-directions are

$$\frac{\partial}{\partial x}\left(S_{xx} + \frac{\partial u_1}{\partial x}S_{xx} + \frac{\partial u_1}{\partial z}S_{xz}\right) + \frac{\partial}{\partial z}\left(S_{xz} + \frac{\partial u_1}{\partial x}S_{xz}\right) + \bar{f}_x = 0, \tag{2.5.38}$$

$$\frac{\partial}{\partial x}\left(S_{xz} + \frac{\partial u_3}{\partial x}S_{xx}\right) + \frac{\partial}{\partial z}\left(\frac{\partial u_3}{\partial x}S_{xz}\right) + \bar{f}_z = 0, \tag{2.5.39}$$

where $\bar{f}_x = \rho_0 f_1$ and $\bar{f}_z = \rho_0 f_3$ are the body forces measured per unit volume. They are assumed, in the present case, to be only functions of x.

2.5.4.2 The Bernoulli–Euler Beam Theory

The displacement field of the Bernoulli–Euler beam theory is given in Eq. (2.5.1). Integration of Eqs (2.5.38) and (2.5.39) over the beam area of cross section yields

$$\frac{d}{dx}\left(N + \frac{du}{dx}N - \frac{d^2 w}{dx^2}M - \frac{dw}{dx}V\right) - \frac{d^2 w}{dx^2}V + f = 0, \tag{2.5.40}$$

$$\frac{d}{dx}\left(V + \frac{dw}{dx}N\right) + q = 0, \tag{2.5.41}$$

where f and q are the axially and transversely distributed forces per unit length.

We note that

$$\int_A z \frac{\partial}{\partial z}\left(S_{xz} + \frac{\partial u_1}{\partial x}S_{xz}\right) dA = -\int_A \left(S_{xz} + \frac{\partial u_1}{\partial x}S_{xz}\right) dA, \tag{2.5.42}$$

where we have used the assumption that $S_{xz} = 0$ on the surface of the beam. Next, multiply Eq. (2.5.38) with z, integrate over the beam area of cross section, and use the identity in Eq. (2.5.42) to obtain

$$\frac{d}{dx}\left(M + \frac{du}{dx}M - \frac{d^2 w}{dx^2}P_x - \frac{dw}{dx}R_x\right) - V - \frac{du}{dx}V + \frac{d^2 w}{dx^2}R_x = 0, \tag{2.5.43}$$

where (P_x, R_x) are higher-order stress resultants

$$P_x = \int_A \sigma_{xx}z^2 \, dA, \quad R_x = \int_A \sigma_{xz}z \, dA. \tag{2.5.44}$$

Note that area integrals of zf_x and zf_z vanish because f_x and f_z are only functions of x and the x-axis coincides with the geometric centroidal axis.

Using the order-of-magnitude assumption of various quantities, Eqs (2.5.40), (2.5.41), and (2.5.43) can be simplified as

$$-\frac{dN}{dx} = f, \qquad (2.5.45)$$

$$-\frac{dV}{dx} - \underline{\frac{d}{dx}\left(\frac{dw}{dx}N\right)} = q, \qquad (2.5.46)$$

$$-\frac{dM}{dx} + V = 0, \qquad (2.5.47)$$

where the underlined expression is the only new term due to the von Kármán nonlinearity when compared with Eqs (2.5.3) to (2.5.5). Of course, the stress resultant N will also contain nonlinear term when it is expressed in terms of the displacements. Since in the Bernoulli–Euler beam theory V cannot be determined independently, we use $V = dM/dx$ from Eq. (2.5.47) and substitute for V into Eq. (2.5.46) the following equations of equilibrium for the Bernoulli–Euler beam theory:

$$-\frac{dN}{dx} = f \qquad (2.5.48)$$

$$-\frac{d^2 M}{dx^2} - \underline{\frac{d}{dx}\left(\frac{dw}{dx}N\right)} = q. \qquad (2.5.49)$$

In view of the displacement field in Eq. (2.5.1), we obtain the following simplified Green strain tensor component:

$$E_{xx} \approx \varepsilon_{xx} = \frac{du}{dx} - z\frac{d^2w}{dx^2} + \frac{1}{2}\left(\frac{dw}{dx}\right)^2 \equiv \varepsilon_{xx}^{(0)} + z\varepsilon_{xx}^{(1)}, \quad \gamma_{xz} = 0, \qquad (2.5.50)$$

where

$$\varepsilon_{xx}^{(0)} = \frac{du}{dx} + \frac{1}{2}\left(\frac{dw}{dx}\right)^2, \quad \varepsilon_{xx}^{(1)} = -\frac{d^2w}{dx^2}, \qquad (2.5.51)$$

and the stress resultants (N, M) can be expressed in terms of the displacements (u, w) (when linear stress–strain relations are used) as [see Eqs (2.5.8) and (2.5.9)]

$$N(x) = \int_A \sigma_{xx}\, dA = A_{xx}\left[\frac{du}{dx} + \frac{1}{2}\left(\frac{dw}{dx}\right)^2\right] + B_{xx}\left(-\frac{d^2w}{dx^2}\right), \qquad (2.5.52)$$

$$M(x) = \int_A z\sigma_{xx}\, dA = B_{xx}\left[\frac{du}{dx} + \frac{1}{2}\left(\frac{dw}{dx}\right)^2\right] + D_{xx}\left(-\frac{d^2w}{dx^2}\right). \qquad (2.5.53)$$

Relations in Eq. (2.5.13) take the form

$$\frac{du}{dx} + \frac{1}{2}\left(\frac{dw}{dx}\right)^2 = \frac{D_{xx}N - B_{xx}M}{A_{xx}D_{xx} - B_{xx}^2}, \qquad -\frac{d^2w}{dx^2} = \frac{A_{xx}M - B_{xx}N}{A_{xx}D_{xx} - B_{xx}^2}. \qquad (2.5.54)$$

Consequently, Eq. (2.5.14) and (2.5.16) remain unchanged.

2.5.4.3   The Timoshenko Beam Theory

Using the displacement field in Eq. (2.5.22) and following the procedure as in Eqs (2.5.40) to (2.5.43), we obtain

$$\frac{d}{dx}\left(N + \frac{du}{dx}N + \frac{d\phi_x}{dx}M + \phi_x V\right) + \frac{d\phi_x}{dx}V + f = 0, \qquad (2.5.55)$$

$$\frac{d}{dx}\left(V + \frac{dw}{dx}N\right) + q = 0, \qquad (2.5.56)$$

$$\frac{d}{dx}\left(M + \frac{du}{dx}M + \frac{d\phi_x}{dx}P_x + \phi_x R_x\right) - V - \frac{du}{dx}V - \frac{d\phi_x}{dx}R_x = 0. \qquad (2.5.57)$$

Using the order-of-magnitude assumption of various quantities, Eqs (2.5.55) to (2.5.57) are simplified to those listed in Eqs (2.5.45) to (2.5.47).

The simplified Green strain tensor components of the Timoshenko beam theory are

$$E_{xx} \approx \varepsilon_{xx} = \frac{du}{dx} + z\frac{d\phi_x}{dx} + \frac{1}{2}\left(\frac{dw}{dx}\right)^2 \equiv \varepsilon_{xx}^{(0)} + z\varepsilon_{xx}^{(1)}, \quad \gamma_{xz} = \phi_x + \frac{dw}{dx} \qquad (2.5.58)$$

where

$$\varepsilon_{xx}^{(0)} = \frac{du}{dx} + \frac{1}{2}\left(\frac{dw}{dx}\right)^2, \quad \varepsilon_{xx}^{(1)} = \frac{d\phi_x}{dx}. \qquad (2.5.59)$$

The stress resultants (N, M, V) are known in terms of the generalized displacements (u, w, ϕ_x) as

$$N(x) = \int_A \sigma_{xx}\, dA = A_{xx}\left[\frac{du}{dx} + \frac{1}{2}\left(\frac{dw}{dx}\right)^2\right] + B_{xx}\left(\frac{d\phi_x}{dx}\right), \qquad (2.5.60)$$

$$M(x) = \int_A z\sigma_{xx}\, dA = B_{xx}\left[\frac{du}{dx} + \frac{1}{2}\left(\frac{dw}{dx}\right)^2\right] + D_{xx}\left(\frac{d\phi_x}{dx}\right), \qquad (2.5.61)$$

$$V(x) = K_s \int_A \sigma_{xz}\, dA = K_s S_{xz}\left(\phi_x + \frac{dw}{dx}\right). \qquad (2.5.62)$$

Relations in Eq. (2.5.29) take the form

$$\frac{du}{dx} + \frac{1}{2}\left(\frac{dw}{dx}\right)^2 = \frac{D_{xx}N - B_{xx}M}{A_{xx}D_{xx} - B_{xx}^2},$$

$$\frac{d\phi_x}{dx} = \frac{A_{xx}M - B_{xx}N}{A_{xx}D_{xx} - B_{xx}^2}, \quad \phi_x + \frac{dw}{dx} = \frac{V}{K_s S_{xz}}. \qquad (2.5.63)$$

Therefore, Eqs (2.5.30) and (2.5.31) remain unchanged.

This completes the development of the Bernoulli–Euler and Timoshenko beam theories with the von Kármán nonlinear strain. The equations governing bars can be obtained as a special case from the equations of beams by setting bending related quantities to zero (i.e., $w = 0$, $\phi_x = 0$, $M = 0$, $V = 0$, and $q = 0$). Thus, the kinematic, constitutive, and equilibrium equations of bars (written in terms of forces as well as displacements) are

$$\varepsilon_{xx} = \frac{du}{dx}, \quad \sigma_{xx} = E\varepsilon_{xx}, \quad N = EA\frac{du}{dx}$$

$$-\frac{dN}{dx} + f = 0 \Rightarrow \left[-\frac{d}{dx}\left(EA\frac{du}{dx} \right) + f = 0 \right], \quad 0 < x < L. \tag{2.5.64}$$

Here u denotes the axial displacement and f is the distributed axial force.

The governing equations for cables (also termed rods, ropes, strings, and so on) in a plane are the same as those for bars, with the exception that axial stiffness EA is replaced with the tension $T(x)$ in the cable. Thus, we have

$$\left[-\frac{d}{dx}\left(T\frac{du}{dx} \right) + f = 0 \right], \quad 0 < x < L, \tag{2.5.65}$$

where u is the transverse displacement and $f(x)$ is the distributed force transverse to the cable.

2.6   Summary

In this chapter a summary of the equations of elasticity are presented. These include the equations of motion, strain–displacement equations, and constitutive (or stress–strain) relations. The stress and strain transformations and compatibility conditions for strains are also discussed. Also, two beam theories, namely, the Bernoulli–Euler and Timoshenko beam theories, are also fully developed in this chapter. Finally, the von Kármán nonlinear beam theories using the Bernoulli–Euler and Timoshenko kinematics are developed. The equations governing the two beam theories will be used extensively in the coming chapters. The main equations of this chapter for *linearized elasticity* and bending of beams are summarized here.

Equations of motion [Eqs (2.2.7) and (2.2.8)]:

$$\nabla \cdot \boldsymbol{\sigma} + \rho\mathbf{f} = \rho\frac{\partial^2 \mathbf{u}}{\partial t^2} \quad \left(\frac{\partial \sigma_{ji}}{\partial x_j} + \rho f_i = \rho\frac{\partial^2 u_i}{\partial t^2} \right). \tag{2.6.1}$$

Strain–displacement relations [Eq. (2.3.12)]:

$$\boldsymbol{\varepsilon} = \tfrac{1}{2}\left[(\nabla\mathbf{u}) + (\nabla\mathbf{u})^{\mathrm{T}} \right] \quad \left[\varepsilon_{ij} = \tfrac{1}{2}\left(\frac{\partial u_i}{\partial x_j} + \frac{\partial u_j}{\partial x_i} \right) \right]. \tag{2.6.2}$$

Strain compatibility equations [Eqs (2.3.18) and (2.3.19)]:

$$\nabla \times (\nabla \times \boldsymbol{\varepsilon})^{\mathrm{T}} = \mathbf{0} \quad \left(\frac{\partial^2 \varepsilon_{ij}}{\partial x_k \partial x_\ell} + \frac{\partial^2 \varepsilon_{k\ell}}{\partial x_i \partial x_j} = \frac{\partial^2 \varepsilon_{\ell j}}{\partial x_k \partial x_i} + \frac{\partial^2 \varepsilon_{ki}}{\partial x_\ell \partial x_j} \right). \tag{2.6.3}$$

Stress–strain relations [modified Eqs (2.4.11) and (2.4.12) to account for thermal strains; see **Problem 2.39**]:

$$\boldsymbol{\sigma} = \frac{E}{1+\nu}\boldsymbol{\varepsilon} + \frac{\nu E}{(1+\nu)(1-2\nu)} \, \mathrm{tr}(\boldsymbol{\varepsilon}) \, \mathbf{I} - \frac{E\alpha}{1-2\nu}(T - T_0) \, \mathbf{I}$$
$$\left(\sigma_{ij} = \frac{E}{1+\nu}\varepsilon_{ij} + \frac{\nu E}{(1+\nu)(1-2\nu)} \, \varepsilon_{kk} \, \delta_{ij} - \frac{E\alpha}{1-2\nu}(T - T_0)\delta_{ij} \right). \tag{2.6.4}$$

Equilibrium equations of the Bernoulli–Euler beam theory [Eqs (2.5.48) and (2.5.49); set the nonlinear terms (underlined) to obtain the linear equations]:

$$-\frac{dN}{dx} = f, \quad -\frac{d^2 M}{dx^2} - \frac{d}{dx}\left(\frac{dw}{dx}N \right) = q. \tag{2.6.5}$$

Kinematic relations of the Bernoulli–Euler beam theory [Eq. (2.5.50)]:

$$\varepsilon_{xx} = \frac{du}{dx} + \frac{1}{2}\left(\frac{dw}{dx} \right)^2 - z\frac{d^2 w}{dx^2}, \quad \gamma_{xz} = 0. \tag{2.6.6}$$

Constitutive relations of the homogeneous Bernoulli–Euler beam theory [Eqs (2.5.52) and (2.5.53)]:

$$N = EA\left[\frac{du}{dx} + \frac{1}{2}\left(\frac{dw}{dx} \right)^2 \right], \quad M = -EI\frac{d^2 w}{dx^2}. \tag{2.6.7}$$

Equilibrium equations of the Timoshenko beam theory [Eqs (2.5.45)– (2.5.47); set the nonlinear terms (underlined) to obtain the linear equations]:

$$-\frac{dN}{dx} = f, \quad -\frac{dV}{dx} - \frac{d}{dx}\left(\frac{dw}{dx}N \right) = q, \quad -\frac{dM}{dx} + V = 0. \tag{2.6.8}$$

Kinematic relations of the Timoshenko beam theory [Eqs (2.5.58) and (2.5.59)]:

$$\varepsilon_{xx} = \frac{du}{dx} + \frac{1}{2}\left(\frac{dw}{dx} \right)^2 + z\frac{d\phi_x}{dx}, \quad \gamma_{xz} = \phi_x + \frac{dw}{dx}. \tag{2.6.9}$$

Constitutive relations of the homogeneous Timoshenko beam theory [Eqs (2.5.60)– (2.5.62)]:

$$N = EA\left[\frac{du}{dx} + \frac{1}{2}\left(\frac{dw}{dx} \right)^2 \right], \quad M = EI\frac{d\phi_x}{dx}, \quad V = GAK_s\left(\phi_x + \frac{dw}{dx} \right). \tag{2.6.10}$$

Table 2.6.1 contains a summary of governing equations of homogeneous, isotropic, and linear elastic bodies in terms of displacements (i.e., equilibrium equations, kinematic relations, and constitutive equations are combined to express the governing equations in terms of displacements). The nonlinear equilibrium equations of beams with the von Kármán nonlinearity can be expressed in terms of displacements with the help of equations presented in Section 2.5. In the later chapters of this book, we shall make reference to the equations developed in this chapter and summarized in this section. Some of the equations developed here may be re-derived using the energy approach in Chapters 3 through 5.

Table 2.6.1 Summary of equations of elasticity, membranes, cables, bars, and beams.

---

1. Linearized elasticity (μ and λ are Lamé constants, **u** is the displacement vector, and **f** is the body force vector measured per unit volume)

$$\mu \nabla^2 \mathbf{u} + (\lambda + \mu) \nabla (\nabla \cdot \mathbf{u}) + \mathbf{f} = \mathbf{0} \text{ in } \Omega. \tag{2.6.11}$$

2. Membranes (a_{11} and a_{22} are tensions in the x- and y-coordinate directions, respectively, u is the transverse deflection, and f is the distributed transverse force measured per unit area)

$$-\frac{\partial}{\partial x} \left(a_{11} \frac{\partial u}{\partial x} \right) - \frac{\partial}{\partial y} \left(a_{22} \frac{\partial u}{\partial y} \right) = f \text{ in } \Omega. \tag{2.6.12}$$

3. Cables (T is the tension in the cable, u is the transverse deflection, and f is the distributed transverse load)

$$-\frac{d}{dx} \left(T \frac{du}{dx} \right) = f, \quad 0 < x < L. \tag{2.6.13}$$

4. Bars (E is Young's modulus, A is the area of cross section, u is the axial displacement, and f is the distributed axial force)

$$-\frac{d}{dx} \left(EA \frac{du}{dx} \right) = f, \quad 0 < x < L. \tag{2.6.14}$$

5. Bernoulli–Euler beam theory (E is Young's modulus; I is the moment of inertia; w is the transverse displacement, and q is the distributed load)

$$\frac{d^2}{dx^2} \left(EI \frac{d^2 w}{dx^2} \right) = q, \quad 0 < x < L. \tag{2.6.15}$$

6. Timoshenko beam theory (E is Young's modulus, G is the shear modulus, A is the area of cross section, K_s shear correction coefficients, I is the moment of inertia; w is the transverse displacement, ϕ_x rotation of a transverse normal line, and q is the distributed load)

$$-\frac{d}{dx} \left(EI \frac{d\phi_x}{dx} \right) + GAK_s \left(\phi_x + \frac{dw}{dx} \right) = 0$$

$$-\frac{d}{dx} \left[GAK_s \left(\phi_x + \frac{dw}{dx} \right) \right] = q. \tag{2.6.16}$$

Problems

CONTINUUM MECHANICS (SECTION 2.1)

2.1 Let an arbitrary region in a continuous medium be denoted by Ω and the bounding, closed surface of this region be continuous and denoted by Γ. Let each point on the bounding surface move with the velocity $\mathbf{v}_s$. It can be shown that the time derivative of the volume integral over some continuous function $Q(\mathbf{r}, t)$ is given by

$$\frac{d}{dt} \int_\Omega Q(\mathbf{r}, t)\, dv \equiv \int_\Omega \frac{\partial Q}{\partial t}\, dv + \oint_\Gamma Q\mathbf{v}_s \cdot \hat{\mathbf{n}}\, ds. \tag{1}$$

This expression for the differentiation of a volume integral with variable limits is sometimes known as the three-dimensional *Leibniz rule*. Let each element of mass in the medium move with the velocity $\mathbf{v}(\mathbf{r}, t)$ and consider a special region Ω such that the bounding surface S is attached to a fixed set of material elements. Then each point of this surface moves itself with the material velocity, that is, $\mathbf{v}_s = \mathbf{v}$, and the region Ω thus contains a fixed total amount of mass since no mass crosses the boundary surface Γ. To distinguish the time rate of change of an integral over this material region, we replace d/dt by D/Dt and write

$$\frac{D}{Dt} \int_\Omega Q(\mathbf{r}, t)\, d\Omega \equiv \int_\Omega \frac{\partial Q}{\partial t}\, d\Omega + \oint_\Gamma Q\mathbf{v} \cdot \hat{\mathbf{n}}\, d\Gamma, \tag{2}$$

which holds for a material region, that is, a region of fixed total mass. Show that the relation between the time derivative following an arbitrary region and the time derivative following a material region (fixed total mass) is

$$\frac{d}{dt} \int_\Omega Q(\mathbf{r}, t) d\Omega \equiv \frac{D}{Dt} \int_\Omega Q(\mathbf{r}, t)\, d\Omega + \oint_\Gamma Q(\mathbf{v}_s - \mathbf{v}) \cdot \hat{\mathbf{n}}\, d\Gamma. \tag{3}$$

The velocity difference $\mathbf{v} - \mathbf{v}_s$ is the velocity of the material measured relative to the velocity of the surface. The surface integral

$$\oint_\Gamma Q(\mathbf{v} - \mathbf{v}_s) \cdot \hat{\mathbf{n}}\, d\Gamma \tag{4}$$

thus measures the total *outflow* of the property Q from the region Ω.

2.2 Let $Q = \rho(\mathbf{r}, t)$ denote the mass density of a continuous region. Then conservation of mass for a *material* region requires that

$$\frac{D}{Dt} \int_\Omega \rho\, d\Omega = 0. \tag{1}$$

Show that for a *fixed region* ($\mathbf{v}_s = 0$), conservation of mass can also be stated as

$$\frac{d}{dt} \int_\Omega \rho\, d\Omega = -\oint_\Gamma \rho\mathbf{v} \cdot \hat{\mathbf{n}}\, d\Gamma \quad \text{or} \quad \int_\Omega \frac{\partial \rho}{\partial t}\, d\Omega = -\oint_\Gamma \rho\mathbf{v} \cdot \hat{\mathbf{n}}\, d\Gamma. \tag{2}$$

Interpret these equations physically.

2.3 In the field description of a continuous variable $\phi = \phi(\mathbf{r}, t)$, let the field position be a function of time such that $\mathbf{r} = \mathbf{r}(t)$. Deduce that the total time derivative of ϕ can be written

$$\frac{d\phi}{dt} = \frac{\partial \phi}{\partial t} + \frac{d\mathbf{r}}{dt} \cdot \operatorname{grad} \phi. \tag{1}$$

This arbitrary total time derivative corresponds to a change in ϕ following a change in $\mathbf{r}$ with time. If we let $\mathbf{r}$ correspond to the position of a fixed material element, then

$d\mathbf{r}/dt = \mathbf{v}$ corresponds to the velocity of the material element. To distinguish the time rates of change following a material element from other arbitrary changes, we write

$$\frac{D\phi}{Dt} \equiv \frac{\partial\phi}{\partial t} + \mathbf{v} \cdot \mathrm{grad}\ \phi. \tag{2}$$

This is the differential time rate of change of a field variable $\phi(\mathbf{r}, t)$ following a material element. It is referred to as the *material derivative*, the *substantial derivative*, or the *Eulerian derivative*. The corresponding material derivative for integrals was defined in **Problem 2.2**. By means of vector identities show that the continuity equation for mass conservation can be written as

$$\frac{D\rho}{Dt} + \rho\,\mathrm{div}\ \mathbf{v} = 0. \tag{3}$$

2.4 The material derivative operator D/Dt corresponds to changes with respect to a fixed mass, that is, $\rho\,dv$ is constant with respect to this operator. Show formally by means of Leibniz's rule, the divergence theorem, and conservation of mass principle that

$$\frac{D}{Dt} \int_\Omega \rho\phi\,d\Omega \equiv \int_\Omega \rho\frac{D\phi}{Dt}\,d\Omega. \tag{1}$$

2.5 Letting a finite volume Ω shrink to an infinitesimal volume dv, show by setting $\phi = 1$ in Leibniz's rule and by use of the divergence theorem that div $\mathbf{v}$ can be interpreted as

$$\mathrm{div}\ \mathbf{v} \equiv \lim_{d\Omega \to 0} \frac{1}{d\Omega}\frac{D}{Dt}(d\Omega). \tag{1}$$

The right-hand side can be interpreted as the rate of volumetric strain following a material particle, called the dilation rate.

2.6 The acceleration of a material element in a continuum is described by

$$\frac{D\mathbf{v}}{Dt} \equiv \frac{\partial\mathbf{v}}{\partial t} + \mathbf{v} \cdot \mathrm{grad}\ \mathbf{v}. \tag{1}$$

Show by means of vector identities that the acceleration can also be written as

$$\frac{D\mathbf{v}}{Dt} \equiv \frac{\partial\mathbf{v}}{\partial t} + \mathrm{grad}\left(\frac{v^2}{2}\right) - \mathbf{v} \times \mathrm{curl}\ \mathbf{v}. \tag{2}$$

This form displays the role of the *vorticity vector*, curl $\mathbf{v}$.

2.7 Deduce that

$$\mathbf{v} \cdot \frac{D\mathbf{v}}{Dt} = \frac{D}{Dt}\left(\frac{\mathbf{v} \cdot \mathbf{v}}{2}\right). \tag{1}$$

Note that $(\mathbf{v} \cdot \mathbf{v})/2 = v^2/2$ is the kinetic energy per unit mass of a material particle.

2.8 Deduce that

$$\mathrm{curl}\left(\frac{D\mathbf{v}}{Dt}\right) \equiv \frac{D\boldsymbol{\omega}}{Dt} + \boldsymbol{\omega}\,\mathrm{div}\mathbf{v} - \boldsymbol{\omega} \cdot \boldsymbol{\nabla}\mathbf{v}, \tag{1}$$

where $\boldsymbol{\omega} \equiv \boldsymbol{\nabla} \times \mathbf{v}$ is the vorticity vector.

2.9 Newton's second law of motion in its elementary form $m\mathbf{a} = \mathbf{F}$ holds strictly for a point particle of fixed mass m. For a material region of continuously distributed mass, Newton's second law reads

$$\frac{D}{Dt} \int_\Omega \rho\mathbf{v}\,d\Omega = \mathbf{F}, \tag{1}$$

where $\mathbf{F}$ is the sum of all the forces acting on the material in R. Deduce by means of the results of **Problem 2.8** that the corresponding law of motion for a region of *variable* mass is

$$\frac{d}{dt} \int_\Omega \rho\mathbf{v}\,d\Omega = \mathbf{F} + \oint_\Gamma \rho\mathbf{v}(\mathbf{v}_s - \mathbf{v}) \cdot \hat{\mathbf{n}}\,d\Gamma, \tag{2}$$

where $\mathbf{v}_s$ is the velocity of points on the surface Γ bounding the variable mass region Ω.

2.10 Newton's second law of motion applied to a continuum states that the rate of change of momentum following a material region of fixed mass is equal to the sum of all the forces on the region. When the forces are divided into surface forces and body forces, Newton's second law reads

$$\frac{D}{Dt} \int_\Omega \rho \mathbf{v} \, d\Omega = \oint_\Gamma \hat{\mathbf{n}} \cdot \boldsymbol{\sigma} \, d\Gamma + \int_\Omega \rho \mathbf{f} d\Omega, \tag{1}$$

where $\boldsymbol{\sigma}$ is the surface stress tensor, $\mathbf{f}$ is the body force per unit mass, ρ is the mass density, and $\mathbf{v}$ is the material velocity. Since the material particle mass ρdv is constant with respect to the material time derivative D/Dt, make use of the divergence theorem and obtain the differential form of Newton's second law of motion for a continuum:

$$\rho \frac{D\mathbf{v}}{Dt} = \text{div } \boldsymbol{\sigma} + \rho \mathbf{f}. \tag{2}$$

2.11 Multiply the continuity equation

$$\frac{\partial \rho}{\partial t} + \text{div } (\rho \mathbf{v}) = 0 \tag{1}$$

by the velocity $\mathbf{v}$, and add the result to the left-hand side of the momentum equation (Newton's second law) in **Problem 2.10**. After use of vector identities, obtain the result

$$\frac{\partial}{\partial t} (\rho \mathbf{v}) + \text{div } (\rho \mathbf{vv} - \boldsymbol{\sigma}) = \rho \mathbf{f}, \tag{2}$$

which is called the conservation form of the momentum equation. The combination ρ $\mathbf{vv}$ is called the momentum-flux tensor.

2.12 Take the scalar product with $\mathbf{v}$ of the equation for Newton's second law in **Problem 2.10**, and obtain the equation of change for the kinetic energy of a material particle in a continuum:

$$\rho \frac{D}{Dt} \left(\frac{v^2}{2} \right) = \mathbf{v} \cdot \text{div} \boldsymbol{\sigma} + \rho \mathbf{v} \cdot \mathbf{f}. \tag{1}$$

How do you interpret the rate-of-work terms on the right-hand side?

2.13 Let e denote the thermodynamic internal energy per unit mass of a material. Then the equation of change for total energy of a material region can be written as

$$\frac{D}{Dt} \int_\Omega \rho \left(e + \frac{v^2}{2} \right) d\Omega = \oint_\Gamma \hat{\mathbf{n}} \cdot \boldsymbol{\sigma} \cdot \mathbf{v} \, d\Gamma + \int_R \rho \mathbf{f} \cdot \mathbf{v} \, d\Omega - \oint_\Gamma \mathbf{q} \cdot \hat{\mathbf{n}} \, d\Gamma. \tag{1}$$

The first two terms on the right-hand side describe the rate of work done on the material region by the surface stresses and the body forces. The third integral describes the net *outflow* of heat from the region, causing a decrease of energy inside the region. The heat-flux vector $\mathbf{q}$ describes the magnitude and direction of the flow of heat energy per unit time and per unit area.

By suitable operations obtain the differential form of the energy equation:

$$\rho \frac{D}{Dt} \left(e + \frac{v^2}{2} \right) = \text{div } (\boldsymbol{\sigma} \cdot \mathbf{v}) + \rho \mathbf{f} \cdot \mathbf{v} - \text{div } \mathbf{q}. \tag{2}$$

Subtract the contribution from kinetic energy and obtain

$$\rho \frac{De}{Dt} = \text{div } (\boldsymbol{\sigma} \cdot \mathbf{v}) - \mathbf{v} \cdot \text{div } \boldsymbol{\sigma} - \text{div } \mathbf{q}. \tag{3}$$

This is called the *thermodynamic form* of the energy equation for a continuum.

2.14 The total rate of work done by the surface stresses per unit volume is given by div $(\boldsymbol{\sigma} \cdot \mathbf{v})$. The rate of work done by the resultant of the surface stresses per unit volume is given by $\mathbf{v} \cdot \text{div } \boldsymbol{\sigma}$. The difference between these two terms yields the rate of work done by the surface stresses in deformation of the material particle, per unit volume. Show that this can be written as

$$\text{div}\,(\boldsymbol{\sigma}\cdot\mathbf{v}) - \mathbf{v}\cdot\text{div}\boldsymbol{\sigma} = \boldsymbol{\sigma} : (\nabla\mathbf{v})^{\mathrm{T}}$$

$$= \boldsymbol{\sigma} : \nabla\mathbf{v} \quad (\boldsymbol{\sigma}\ \text{symmetric})$$

$$= \frac{1}{2}\boldsymbol{\sigma} : \left[\nabla\mathbf{v} + (\nabla\mathbf{v})^{\mathrm{T}}\right] \quad (\boldsymbol{\sigma}\ \text{symmetric}). \tag{1}$$

SOLID MECHANICS (SECTIONS 2.2–2.5)

2.15 Derive the transformation relations relating the normal and shear stresses σ_n and σ_s on a plane whose normal is $\hat{\mathbf{n}} = \cos\theta\hat{\mathbf{e}}_1 + \sin\theta\hat{\mathbf{e}}_2$ to the stress components σ_{11}, σ_{22}, and $\sigma_{12} = \sigma_{21}$ on the $\hat{\mathbf{e}}_1$ and $\hat{\mathbf{e}}_2$ planes (see Fig. P2.15):

$$\sigma_n = \sigma_{11}\cos^2\theta + \sigma_{22}\sin^2\theta + 2\sigma_{12}\sin\theta\cos\theta,$$

$$\sigma_s = (\sigma_{22} - \sigma_{11})\sin\theta\cos\theta + \sigma_{12}(\cos^2\theta - \sin^2\theta). \tag{1}$$

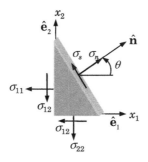

Fig. P2.15

Note that θ is the angle measured from the positive x_1-axis to the normal to the inclined plane. Then show that the principal stresses at a point in a two-dimensional body are given by

$$\sigma_{\max} = \frac{\sigma_{11} + \sigma_{22}}{2} + \sqrt{\left(\frac{\sigma_{11} - \sigma_{22}}{2}\right)^2 + \sigma_{12}^2}\,,$$

$$\sigma_{\min} = \frac{\sigma_{11} + \sigma_{22}}{2} - \sqrt{\left(\frac{\sigma_{11} - \sigma_{22}}{2}\right)^2 + \sigma_{12}^2}, \tag{2}$$

and that the orientation of the principal planes is given by

$$\theta_p = \frac{1}{2}\tan^{-1}\left[\frac{2\sigma_{12}}{\sigma_{11} - \sigma_{22}}\right].$$

2.16 Determine whether the following stress fields are possible in a structural member free of body forces:

(a) $\sigma_{11} = c_1 x_1 + c_2 x_2 + c_3 x_1 x_2$, $\sigma_{12} = -c_3 \frac{x_2^2}{2} - c_1 x_2$, $\sigma_{22} = c_4 x_1 + c_1 x_2$.

(b) $\sigma_{11} = x_1^2 - 2x_1 x_2 + cx_3$, $\sigma_{12} = -x_1 x_2 + x_2^2$, $\sigma_{13} = -x_1 x_3$, $\sigma_{22} = x_2^2$,
  $\sigma_{23} = -x_2 x_3$, $\sigma_{33} = (x_1 + x_2)x_3$.

(c) $\sigma_{11} = 3x_1 + 5x_2$, $\sigma_{12} = 4x_1 - 3x_2$, $\sigma_{22} = 2x_1 - 4x_2$.

2.17 Given the following state of stress ($\sigma_{ij} = \sigma_{ji}$),

$$\sigma_{11} = -2x_1^2, \quad \sigma_{12} = -7 + 4x_1x_2 + x_3, \quad \sigma_{13} = 1 + x_1 - 3x_2,$$
$$\sigma_{22} = 3x_1^2 - 2x_2^2 - 5x_3, \quad \sigma_{23} = 0, \quad \sigma_{33} = -5 + x_1 + 3x_2 + 3x_3,$$

determine the body force vector for which the stress field is in a state of equilibrium.

2.18 For the cantilever beam bent by a point load at the free end (see Fig. 2.5.5), the bending moment M_2 about the x_2-axis is given by $M_2 = -F_0x_1$. The bending stress σ_{11} is given by $\sigma_{11} = M_2x_3/I_2$, where I_2 is the moment of inertia of the cross section about the x_2-axis. Starting with this equation, use the two-dimensional equilibrium equations to determine the stresses σ_{33} and σ_{13} as functions of x_1 and x_3.

2.19 Repeat **Problem 2.18** for the case in which the cantilever beam is bent by uniformly distributed load q_0/unit length applied at the top (i.e., at $x_3 = -h$).

2.20 For the state of stress given in **Problem 2.17**, determine the stress vector at point (x_1, x_2, x_3) on the plane $x_1 + x_2 + x_3 = $ constant. What are the normal and shearing components of the stress vector at point $(1, 1, 3)$?

2.21 Find the principal stresses and their orientation at point $(1,2,1)$ for the state of stress given in **Problem 2.17**.

2.22 Find the maximum principal stress and its orientation for the state of stress

$$[\sigma] = \begin{bmatrix} 3 & 5 & 8 \\ 5 & 1 & 0 \\ 8 & 0 & 2 \end{bmatrix} \text{ MPa.}$$

2.23 Find the linear strains associated with the displacements

$$u_1 = \left[X_1 X_2 (2 - X_1) - c_1 X_2 + c_2 X_2^3 \right],$$
$$u_2 = -\left[c_3 X_2^2 (1 - X_1) + (3 - X_1)\frac{X_1^2}{3} + c_1 X_1 \right].$$

2.24 The two-dimensional displacement field in a body is given by

$$u_1 = X_1 \left[X_1^2 X_2 + c_1 \left(2c_2^3 + 3c_2^2 X_2 - X_2^3 \right) \right],$$
$$u_2 = -X_2 \left(2c_2^3 + \frac{3}{2}c_2^2 X_2 - \frac{1}{4}X_2^3 + \frac{3}{2}c_1 X_1^2 X_2 \right),$$

where c_1 and c_2 are constants. Find the linear and nonlinear strains.

2.25 Determine the displacements and strains in the (x_1, x_2) system for the deformed body shown in Fig. P2.25.

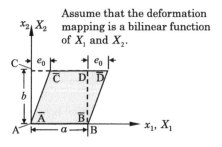

Fig. P2.25

2.26 Determine the displacements and strains in the (x_1, x_2) system for the deformed body shown in Fig. P2.26.

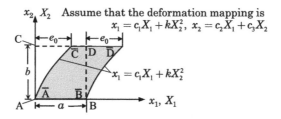

x_2, X_2 Assume that the deformation mapping is
$$x_1 = c_1 X_1 + k X_2^2, \quad x_2 = c_2 X_1 + c_3 X_2$$

$$x_1 = c_1 X_1 + k X_2^2$$

Fig. P2.26

2.27 Find the linear strains associated with the displacement field
$$u_1 = u_1^0(X_1, X_2) + X_3 \phi_1(X_1, X_2),$$
$$u_2 = u_2^0(X_1, X_2) + X_3 \phi_2(X_1, X_2),$$
$$u_3 = u_3^0(X_1, X_2).$$

2.28 Determine whether the following strain fields are possible in a continuous body:

(a)
$$[e] = \begin{bmatrix} (X_1^2 + X_2^2) & X_1 X_2 \\ X_1 X_2 & X_2^2 \end{bmatrix},$$

(b)
$$[e] = \begin{bmatrix} X_3(X_1^2 + X_2^2) & 2X_1 X_2 X_3 & X_3 \\ 2X_1 X_2 X_3 & X_2^2 & x_1 \\ X_3 & X_1 & X_3^2 \end{bmatrix}.$$

2.29 Find the normal and shear strains in the diagonal element of the rectangular block in **Problem 2.25**.

2.30 The biaxial state of strain at a point is given by $\varepsilon_{11} = 800 \times 10^{-6}$, $\varepsilon_{22} = 200 \times 10^{-6}$, $\varepsilon_{12} = 400 \times 10^{-6}$. Find the principal strains and their directions.

2.31 Find the axial strain in the diagonal element of **Problem 2.25** (see Fig. P2.31) using (a) the basic definition of normal strain and (b) the strain transformation equations.

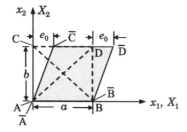

Fig. P2.31

2.32 Using the definition of ∇ and the nonion form of σ, show that the equations of motion in the cylindrical coordinate system are given by
$$\frac{\partial \sigma_{rr}}{\partial r} + \frac{1}{r}\frac{\partial \sigma_{r\theta}}{\partial \theta} + \frac{\partial \sigma_{rz}}{\partial z} + \frac{1}{r}(\sigma_{rr} - \sigma_{\theta\theta}) + \rho_0 f_r = \rho_0 \frac{\partial^2 u_r}{\partial t^2},$$
$$\frac{\partial \sigma_{r\theta}}{\partial r} + \frac{1}{r}\frac{\partial \sigma_{\theta\theta}}{\partial \theta} + \frac{\partial \sigma_{\theta z}}{\partial z} + \frac{2\sigma_{r\theta}}{r} + \rho_0 f_\theta = \rho_0 \frac{\partial^2 u_\theta}{\partial t^2},$$
$$\frac{\partial \sigma_{rz}}{\partial r} + \frac{1}{r}\frac{\partial \sigma_{\theta z}}{\partial \theta} + \frac{\partial \sigma_{zz}}{\partial z} + \frac{\sigma_{rz}}{r} + \rho_0 f_z = \rho_0 \frac{\partial^2 u_z}{\partial t^2}.$$

2.33 Consider the displacement vector in a polar axisymmetric problem,

$$\mathbf{u} = u_r \hat{\mathbf{e}}_r + u_z \hat{\mathbf{e}}_z.$$

The strain tensor in the cylindrical coordinate system can be written in the dyadic form as

$$\varepsilon = \varepsilon_{rr} \hat{\mathbf{e}}_r \hat{\mathbf{e}}_r + \varepsilon_{\theta\theta} \hat{\mathbf{e}}_\theta \hat{\mathbf{e}}_\theta + \varepsilon_{zz} \hat{\mathbf{e}}_z \hat{\mathbf{e}}_z + \varepsilon_{r\theta} \left(\hat{\mathbf{e}}_r \hat{\mathbf{e}}_\theta + \hat{\mathbf{e}}_\theta \hat{\mathbf{e}}_r \right)$$
$$+ \varepsilon_{rz} \left(\hat{\mathbf{e}}_r \hat{\mathbf{e}}_z + \hat{\mathbf{e}}_z \hat{\mathbf{e}}_r \right) + \varepsilon_{\theta z} \left(\hat{\mathbf{e}}_\theta \hat{\mathbf{e}}_z + \hat{\mathbf{e}}_z \hat{\mathbf{e}}_\theta \right).$$

Show that the only nonzero linear strains are given by

$$\varepsilon_{rr} = \frac{\partial u_r}{\partial r}, \quad \varepsilon_{\theta\theta} = \frac{u_r}{r}, \quad \varepsilon_{rz} = \frac{1}{2}\left(\frac{\partial u_r}{\partial z} + \frac{\partial u_z}{\partial r} \right), \quad \varepsilon_{zz} = \frac{\partial u_z}{\partial z}.$$

2.34 Using the definitions of ∇ and $\mathbf{u}$ in the cylindrical coordinate system, show that the linear strains are given by

$$\varepsilon_{rr} = \frac{\partial u_r}{\partial r}, \quad \varepsilon_{r\theta} = \frac{1}{2}\left(\frac{1}{r}\frac{\partial u_r}{\partial \theta} + \frac{\partial u_\theta}{\partial r} - \frac{u_\theta}{r} \right), \quad \varepsilon_{rz} = \frac{1}{2}\left(\frac{\partial u_r}{\partial z} + \frac{\partial u_z}{\partial r} \right),$$

$$\varepsilon_{\theta\theta} = \frac{u_r}{r} + \frac{1}{r}\frac{\partial u_\theta}{\partial \theta}, \quad \varepsilon_{z\theta} = \frac{1}{2}\left(\frac{\partial u_\theta}{\partial z} + \frac{1}{r}\frac{\partial u_z}{\partial \theta} \right), \quad \varepsilon_{zz} = \frac{\partial u_z}{\partial z}.$$

2.35 Establish the vector form of the strain compatibility conditions

$$\nabla \times (\nabla \times \varepsilon)^{\mathrm{T}} = \mathbf{0},$$

where ε is the infinitesimal strain tensor.

2.36 Given the following second Piola–Kirchhoff stress components ($S_{IJ} = S_{JI}$),

$$S_{11} = -2X_1^2, \quad S_{12} = -7 + 4X_1 X_2 + X_3, \quad S_{13} = 1 + X_1 - 3X_2,$$
$$S_{22} = 3X_1^2 - 2X_2^2 + 5X_3, \quad S_{23} = 0, \quad S_{33} = -5 + X_1 + 3X_2 + 3X_3,$$

and components of the displacement vector in a deformed body,

$$u_1 = AX_2, \quad u_2 = BX_1, \quad u_3 = 0,$$

where A and B are arbitrary constants, determine the body force components for which the second Piola–Kirchhoff stress field describes a state of equilibrium:

$$\nabla_0 \cdot \left[\mathbf{S}^{\mathrm{T}} \cdot (\mathbf{I} + \nabla_0 \mathbf{u}) \right] + \rho_0 \mathbf{f} = \mathbf{0}.$$

2.37 Using the free-body-diagram of a typical element of the cable, as shown in Fig. P2.37, derive the equation of equilibrium of a cable subjected to distributed transverse load $f(x)$ by summing the forces in the x- and y-directions. In Fig. P2.37, T denotes the tension in the cable and u is the transverse deflection. Assume small displacements.

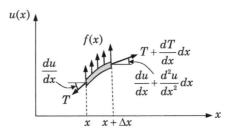

Fig. P2.37

2.38 Use Eq. (2.4.4) to show that

$$\frac{\partial \sigma_{ij}}{\partial \varepsilon_{kl}} = \frac{\partial \sigma_{kl}}{\partial \varepsilon_{ij}}$$

and hence that

$$C_{ijkl} = C_{klij} = C_{jikl} = C_{ijlk}.$$

2.39 For isotropic material C_{ijkl} has the form

$$C_{ijkl} = \lambda \delta_{ij}\delta_{kl} + \mu(\delta_{ik}\delta_{jl} + \delta_{il}\delta_{jk}), \qquad (1)$$

where μ and λ are the Lamé constants, and summation on repeated indices is implied. Using the generalized Hooke's law for the thermoelastic case (see Section 2.4.4)

$$\sigma_{ij} = C_{ijkl}\left[\varepsilon_{kl} - \alpha(T - T_0)\delta_{kl}\right], \qquad (2)$$

deduce that

$$\sigma_{ij} = 2\mu\,\varepsilon_{ij} + \lambda\,\varepsilon_{kk}\,\delta_{ij} - \alpha(2\mu + 3\lambda)(T - T_0)\delta_{ij}, \qquad (3)$$

and then

$$\sigma_{ij} = \frac{E}{1+\nu}\varepsilon_{ij} + \frac{\nu E}{(1+\nu)(1-2\nu)}\,\varepsilon_{kk}\,\delta_{ij} - \frac{E\alpha}{(1-2\nu)}\,(T - T_0)\delta_{ij}. \qquad (4)$$

2.40 Show for an isotropic material the strain–stress relations are given by

$$\varepsilon_{ij} = \left(\frac{1+\nu}{E}\right)\sigma_{ij} - \frac{\nu}{E}\,\sigma_{kk}\,\delta_{ij} + \alpha(T - T_0)\delta_{ij}. \qquad (1)$$

2.41 Establish the following stress–strain relations for an isotropic material with temperature change $\Delta T = T - T_0$:

$$\sigma_{11} = \frac{E}{1-\nu^2}\,(\varepsilon_{11} + \nu\varepsilon_{22}) - \frac{E\alpha}{1-\nu}\,\Delta T$$

$$\sigma_{22} = \frac{E}{1-\nu^2}\,(\nu\varepsilon_{11} + \varepsilon_{22}) - \frac{E\alpha}{1-\nu}\,\Delta T \qquad (1)$$

$$\sigma_{12} = \frac{E}{2(1+\nu)}\,2\varepsilon_{12}, \quad \sigma_{13} = \frac{E}{2(1+\nu)}\,2\varepsilon_{13}, \quad \sigma_{23} = \frac{E}{2(1+\nu)}\,2\varepsilon_{23}$$

in the plane stress state and

$$\sigma_{11} = \frac{E}{(1+\nu)(1-2\nu)}\,\left[(1-\nu)\varepsilon_{11} + \nu\varepsilon_{22} - (1+\nu)\alpha\Delta T\right]$$

$$\sigma_{22} = \frac{E}{(1+\nu)(1-2\nu)}\,\left[\nu\varepsilon_{11} + (1-\nu)\varepsilon_{22} - (1+\nu)\alpha\Delta T\right] \qquad (2)$$

$$\sigma_{12} = \frac{E}{2(1+\nu)}\,2\varepsilon_{12}.$$

for plane strain state.

"You can get into a habit of thought in which you enjoy making fun of all those other people who don't see things as clearly as you do. We have to guard carefully against it."

    – *Carl Sagan* (American astronomer, cosmologist, and science communicator)

"It is really quite amazing by what margins competent but conservative scientists and engineers can miss the mark, when they start with the preconceived idea that what they are investigating is impossible. When this happens, the most well-informed men become blinded by their prejudices and are unable to see what lies directly ahead of them."

    – *Arthur C. Clarke* (British science fiction writer famous for the 1968 film *2001: A Space Odyssey*)

Work, Energy, and Variational Calculus

3.1 Concepts of Work and Energy

3.1.1 Preliminary Comments

The concept of *work done* was introduced in Chapter 1 in connection with the scalar product of vectors. Work done was defined to be the product of displacement, linear or angular, with the force or moment *in the direction* of respective displacement. The total work done is a scalar (i.e., a real number) with units of N-m (Newton meters). Here we generalize the concept to forces and displacements that are possibly functions of position and time.

Consider a material particle moving from point A to point B along some path in space under the influence of a force $\mathbf{F}$, which can be time dependent. The position of the particle is measured from a fixed origin by position vector $\mathbf{r} = \mathbf{x}$. Then the work dW performed by the force $\mathbf{F}$ in moving the particle by an *infinitesimal distance* $d\mathbf{r} = d\mathbf{u}$ along the path over an interval of time dt is defined as [see Fig. 3.1.1(a)]

$$dW = \mathbf{F} \cdot d\mathbf{u} = F_1 du_1 + F_2 du_2 + F_3 du_3 = F_i \, du_i. \qquad (3.1.1)$$

The total work done, W, by force $\mathbf{F}$ in moving the particle from point A to point B is given by

$$W = \int_A^B \mathbf{F} \cdot d\mathbf{u}. \qquad (3.1.2)$$

By definition, work done is a scalar quantity, and it is positive whenever both displacement and force have the same direction and negative if they are in the opposite directions. Since $d\mathbf{u}$ depends on the chosen reference frame, W also depends on the choice of the reference frame. Thus, work is a relative quantity. However, work done does not depend on the path but only the end points, $W = W_B - W_A$, unless the path has constraints. If the reference frame is chosen such that $W_A = 0$, then $W = W_B$.

Analogous to the work done by a force, we can also define the work done by a couple (i.e., a pair of equal, opposite forces) $\mathbf{M}$ in moving through an

Energy Principles and Variational Methods in Applied Mechanics, Third Edition. J.N. Reddy.
©2017 John Wiley & Sons Ltd. Published 2017 by John Wiley & Sons Ltd.
Companion Website: www.wiley.com/go/reddy/applied_mechanics_EPVM

infinitesimal rotation $d\boldsymbol{\theta}$ as [see Fig. 3.1.1(b)]

$$dW = \mathbf{M} \cdot d\boldsymbol{\theta} = M_1 d\theta_1 + M_2 d\theta_2 + M_3 d\theta_3 = M_i \, d\theta_i. \qquad (3.1.3)$$

The total work done in a finite rotation from point A to point B is

$$W = \int_A^B \mathbf{M} \cdot d\theta. \qquad (3.1.4)$$

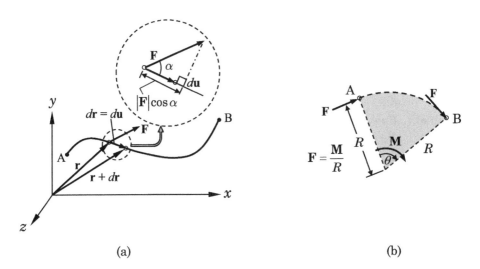

(a)   (b)

Fig. 3.1.1 (a) Motion of a particle under the action of a force. (b) Motion of a particle under the action of a moment.

The rate of work done by force $\mathbf{F}$ in moving the particle through an infinitesimal distance $d\mathbf{r}$ is $\mathbf{F} \cdot d\mathbf{r}/dt$. But $d\mathbf{r}/dt = \mathbf{v}$ is the velocity of the particle. Hence, the rate of work done is equal to $\mathbf{F} \cdot \mathbf{v}$. Thus, the work done by a force $\mathbf{F}$ in moving through the infinitesimal distance $d\mathbf{r}$ is

$$dW = \mathbf{F} \cdot d\mathbf{r} = \mathbf{F} \cdot \mathbf{v} \, dt.$$

Hence, the total work done during the time interval (t_1, t_2) is

$$W = \int_{t_1}^{t_2} \mathbf{F} \cdot \mathbf{v} \, dt. \qquad (3.1.5)$$

3.1.2   External and Internal Work Done

The definition of work done can be extended to material bodies, which can be viewed as a collection of material particles. If work is done on all particles of the body, the total work on the body is the sum of all such works. A deformable body subjected to external forces develops internal forces that move through displacements produced by geometric changes in the body. Thus in a deformable body, work is done by external forces and internal forces. The total work done is the sum of the work done by internal and external forces:

$$W = W_I + W_E, \tag{3.1.6}$$

where W_I denotes the work done by internal forces (stresses) and W_E is the work done by external forces. If a body is subjected to external point forces $\mathbf{F}^1$, $\mathbf{F}^2$, ..., $\mathbf{F}^n$ that displace the points of action by displacements $\Delta\mathbf{r}_1$, $\Delta\mathbf{r}_2$, ... $\Delta\mathbf{r}_n$, respectively, then the work done by the forces on the body during the time interval Δt is the sum of the work done by individual forces in moving through their respective displacements:

$$W_E = -\sum_{i=1}^{n} \mathbf{F}^i \cdot \Delta\mathbf{r}_i = -\sum_{i=1}^{n} \mathbf{F}^i \cdot \mathbf{v}_i \, \Delta t,$$

where $\mathbf{v}_i$ denotes the velocity $\Delta\mathbf{r}_i/\Delta t$. The minus sign indicates that the work is expended on the body as opposed to the work stored in the body. If the time increment Δt approaches zero, the sum approaches a limiting value that is represented by an integral, as in Eq. (3.1.5).

If $\mathbf{f}(\mathbf{x})$ is the body force (measured per unit volume) acting on a material particle occupying position $\mathbf{x}$ in a body with volume Ω, $\mathbf{t}$ is the surface force (measured per unit area) on the boundary Γ, and $\mathbf{u}$ is the displacement of the particle, then the work done on an interior particle is $\mathbf{f} \cdot \mathbf{u}$ and work done on a particle on the boundary is $\mathbf{t} \cdot \mathbf{u}$. The work done on all particles occupying the elemental volume $d\Omega$ is $\mathbf{f} \cdot \mathbf{u} \, d\Omega$, and the work done on all particles on the surface of the body is $\mathbf{t} \cdot \mathbf{u} \, ds$. Hence, the total work done on the body is the sum of the work done on all particles interior and on the surface of the body:

$$W_E = -\left[\int_\Omega \mathbf{f}(\mathbf{x}) \cdot \mathbf{u}(\mathbf{x}) \, d\Omega + \oint_\Gamma \mathbf{t}(s) \cdot \mathbf{u}(s) \, d\Gamma \right]. \tag{3.1.7}$$

In calculating the external work done, the applied (external) forces (or moments) are assumed to be independent of the displacements (or rotations) they cause in a body. Sometimes, W_E is called the potential energy due to the applied loads and is denoted by V. On the other hand, the internal forces generated inside the body due to the application of the external loads are proportional to the displacements they move through, and they can be expressed in terms of the displacements via the constitutive and the strain–displacement relations. In general, the work done by internal forces in moving through their respective finite displacements is *not* equal to the work done by (or potential energy due to) external forces in moving through their respective displacements (i.e., $-W_E \neq W_I$).

To further understand the difference between work done by external forces and internal forces, consider a spring–mass system in static equilibrium [see

Fig.3.1.2(a)]. Suppose that the mass m is placed slowly (to eliminate dynamic effects) at the end of a linear elastic spring. The spring will elongate by an amount e_0, measured from its undeformed state. In this case, the force is due to gravity and it is equal to $F = mg$. Clearly, F is independent of the extension e_0 in the spring, and F does not change during the course of the extension e going from 0 to its final value e_0 [see Fig.3.1.2(b)]. The work done by F in moving through de is $F \, de$. Then the total external work done by F is

$$W_E = - \int_0^{e_0} F \, de = -F \, e_0.$$

Next consider the force in the spring F_s. The force in the spring is proportional to the displacement in the spring. Therefore, the value of F_s goes from zero when $e = 0$ to its final value F_s^f when $e = e_0$ [see Fig.3.1.2(c)]. The work done by F_s in moving through de is $F_s \, de$. The total work done by F_s is

$$W_I = \int_0^{e_0} F_s(e) \, de.$$

If the spring is assumed to be linearly elastic with spring constant k, we have $F_s = ke$, and the work done by the spring force (internal) is

$$W_I = \frac{1}{2} k e_0^2.$$

At equilibrium we have $F_s^f = F$, where F_s^f denotes the final force in the spring.

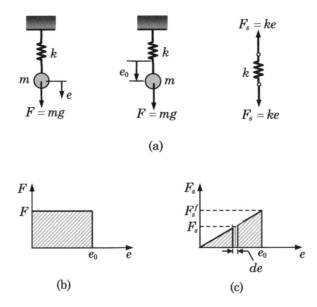

(a)

(b)   (c)

Fig. 3.1.2 A linear spring–mass system in equilibrium. (a) Elongation due to weight mg. (b) External work done. (c) Internal work done.

Now suppose that the spring is imagined to be elongated from e_0 to $e_0 + \Delta e$, Δe being infinitesimally small. Then the additional (or incremental) work done by $F = mg$ and $F_s = F_s(e_0)$ are simply

$$\Delta W_E = -F\Delta e = -mg\Delta e, \quad \Delta W_I = F_s\Delta e = ke_0\Delta e,$$

which shows that $\Delta W_I = -\Delta W_E$ since $F = F_s$.

Example 3.1.1

Figure 3.1.3 shows a rigid link in equilibrium. The initial, no load position corresponds to $\theta = 0$, and the displacement to the equilibrium position is small (i.e., θ is small). Determine the work done by external forces and internal forces when the link is given an infinitesimal rotation $\Delta\theta$ from its equilibrium position. Assume that the springs are linearly elastic.

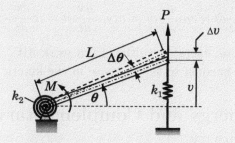

Fig. 3.1.3 A rigid link in equilibrium.

Solution: The work done by external forces is given by

$$\Delta W_E = -(P\Delta v + M\Delta\theta)$$
$$= -(PL\left[\sin(\theta + \Delta\theta) - \sin\theta\right] + M\Delta\theta)$$
$$\approx -PL(\theta + \Delta\theta - \theta) - M\Delta\theta = -(PL + M)\Delta\theta.$$

Once again, the negative sign indicates that the work is done on the body.
    The work done by internal force F_s in the extensional spring k_1 and internal moment M_s in the torsional spring k_2 is given by

$$\Delta W_I = F_s\Delta v + M_s\Delta\theta$$
$$= k_1 v\Delta v + k_2\theta\Delta\theta$$
$$\approx (k_1 L^2 + k_2)\theta\Delta\theta.$$

Note that $\Delta W_I = -\Delta W_E$ (because $P = k_1 L\theta$ and $M = k_2\theta$) or $\Delta W_I + \Delta W_E = 0$. As we shall see later, this is known as the principle of virtual displacements.

Energy is defined as the capacity to do work. The energy stored in a body endows it with a capacity to do work. Therefore, the internal work done is stored as energy. For example, in the case of a linear elastic spring, the internal work done is nothing but the strain energy stored in the spring, which is available to restore the spring back to its original state when the external force causing the deformation is removed. Thus, energy is a measure of the capacity of all forces that can be associated with matter to perform work. Work is also performed on a body through a change in energy. For example, consider Newton's second law of motion for a material particle, $\mathbf{F} = m(d\mathbf{v}/dt)$, where m is the mass and $\mathbf{v}$ is the velocity of the particle. If $\mathbf{r}$ is the position vector of the particle with respect to a fixed frame, the work done by the force $\mathbf{F}$ in moving the particle through an infinitesimal distance $d\mathbf{r}$ is

$$dW = \mathbf{F} \cdot d\mathbf{r} = m\frac{d\mathbf{v}}{dt} \cdot d\mathbf{r} = m\frac{d\mathbf{v}}{dt} \cdot \mathbf{v}\, dt,$$

where $v = |\mathbf{v}|$, the magnitude of the velocity vector $\mathbf{v} = d\mathbf{r}/dt$. Since the kinetic energy of the particle is $K = (1/2)mv^2$, it follows that

$$dK = \frac{dK}{dt}dt = \frac{\partial K}{\partial v}\frac{dv}{dt}dt = mv\frac{dv}{dt}dt = m\mathbf{v} \cdot \frac{d\mathbf{v}}{dt}dt = m\frac{d\mathbf{v}}{dt} \cdot \mathbf{v}\, dt \ \text{ or } \ dW = dK,$$

where dK is the kinetic energy resulting from work dW. In other words, the work done by a force is equal to the increase in the kinetic energy.

3.2 Strain Energy and Complementary Strain Energy

3.2.1 General Development

The first law of thermodynamics (i.e., balance of energy) gives rise to the following energy equation in the material description [84]:

$$\rho_0 \frac{\partial e}{\partial t} = (\mathbf{S} \cdot \mathbf{F}^{\mathrm{T}}) : \boldsymbol{\nabla}_0 \mathbf{v} - \boldsymbol{\nabla}_0 \cdot \mathbf{q}_0 + \rho_0 r_0, \tag{3.2.1}$$

where e is the internal energy per unit mass, $\mathbf{q}_0$ is the heat flux vector, $\mathbf{S}$ is the second Piola–Kirchhoff stress tensor, $\mathbf{F}$ is the deformation gradient, $\mathbf{v}$ is the velocity vector, and r_0 is the internal heat generation per unit mass. For elastic bodies under isothermal conditions ($\mathbf{q}_0 = \mathbf{0}$ and $r_0 = 0$) and infinitesimal deformations ($\mathbf{S} \approx \boldsymbol{\sigma}$, $\mathbf{E} \approx \boldsymbol{\varepsilon}$, and $\boldsymbol{\nabla}_0 \approx \boldsymbol{\nabla}$), the internal energy consists of only stored elastic strain energy, denoted $\rho_0 e = U_0$ and measured per unit volume. In this case (i.e., elastic bodies under isothermal conditions), the energy equation takes the form

$$\frac{\partial U_0}{\partial t} = \boldsymbol{\sigma} : \boldsymbol{\nabla}\mathbf{v} \tag{3.2.2}$$

or (since $\mathbf{v} = \partial\mathbf{u}/\partial t$)

$$dU_0 = \boldsymbol{\sigma} : \boldsymbol{\nabla}(d\mathbf{u}). \tag{3.2.3}$$

Note that

$$\boldsymbol{\nabla}(d\mathbf{u}) = d(\boldsymbol{\nabla}\mathbf{u}) = d\left\{\tfrac{1}{2}\left[\boldsymbol{\nabla}\mathbf{u} + (\boldsymbol{\nabla}\mathbf{u})^{\mathrm{T}}\right] + \tfrac{1}{2}\left[\boldsymbol{\nabla}\mathbf{u} - (\boldsymbol{\nabla}\mathbf{u})^{\mathrm{T}}\right]\right\}$$

$$= d\varepsilon + d\Omega,$$

where ε is the infinitesimal strain tensor and Ω is the rotation tensor. Since the stress tensor is symmetric and Ω is skew–symmetric (i.e., $\Omega^{\mathrm{T}} = -\Omega$ or $\Omega_{ij} = -\Omega_{ji}$), the double-dot product $\sigma : \Omega = \sigma_{ij}\Omega_{ij}$ is zero, giving

$$dU_0 = \sigma : d\varepsilon = \sigma_{ij}\, d\varepsilon_{ij}. \tag{3.2.4}$$

Integrating Eq. 3.2.4, we obtain the expression for the strain energy density of an elastic body:

$$U_0 = \int_0^{\varepsilon_{ij}} \sigma_{ij}\, d\varepsilon_{ij}. \tag{3.2.5}$$

The existence of a scalar function U_0 of strains such that the stresses are derivable from U_0 is of special importance. Such stresses satisfy the energy equation, and consequently they are said to be *conservative*. We have from Eq. (3.2.4) the result

$$\frac{\partial U_0}{\partial \varepsilon_{ij}} = \sigma_{ij} \quad \text{or} \quad \frac{\partial U_0}{\partial \varepsilon} = \sigma. \tag{3.2.6}$$

For large deformation problems, a relation similar to that in Eq. (3.2.6) holds between the second Piola–Kirchhoff stress tensor $\mathbf{S}$ and the deformation gradient $\mathbf{F}$:

$$\mathbf{S} = \mathbf{F}^{-1} \cdot \frac{\partial U_0(\mathbf{F})}{\partial \mathbf{F}}.$$

Equation (3.2.5) can also be arrived by considering the work done by internal forces (i.e., stresses) in moving through displacements. First, we consider axial deformation of a bar of area of cross section A. The free-body diagram of an element of length dx_1 of the bar is shown in Fig. 3.2.1(a). Note that the element is in static equilibrium, and we wish to determine the work done by the internal force associated with stress σ_{11}^f, where the superscript f indicates that it is the final value of the quantity. Suppose that the element is deformed slowly so that axial strain varies from 0 to its final value ε_{11}^f. At any instant during the strain variation from ε_{11} to $\varepsilon_{11} + d\varepsilon_{11}$, we assume that σ_{11} (due to ε_{11}) is kept constant so that equilibrium is maintained. Then the work done by $A\sigma_{11}$ in moving through the displacement $d\varepsilon_{11}dx_1$ is

$$A\sigma_{11}\, d\varepsilon_{11}dx_1 = \sigma_{11}\, d\varepsilon_{11}(Adx_1) = dU_0(Adx_1),$$

where dU_0 denotes the work done per unit volume.

Referring to the stress–strain diagram in Fig. 3.2.1(b), dU_0 represents the elemental area *under* the stress–strain curve. The elemental area in the complement (in the rectangle formed by ε_{11} and σ_{11}) is given by

$$dU_0^* = \varepsilon_{11}\, d\sigma_{11}.$$

The total area under the curve is obtained by integrating from zero to the final value of the strain (during which the stress changes according to its relation to the strain):

$$U_0 = \int_0^{\varepsilon_{11}} \sigma_{11}\, d\varepsilon_{11},$$

where the superscript f is omitted as the expression holds for any value of ε. The quantity U_0 is known as the *strain energy density*.

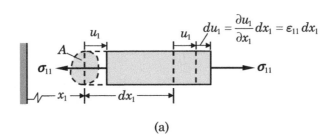

(a)

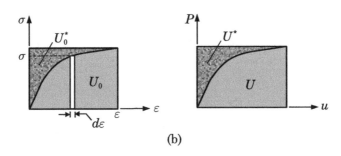

(b)

Fig. 3.2.1 Computation of strain energy for the uniaxially loaded member. (a) Work done by force $\sigma_{11} A$. (b) Definitions of strain energy density and complementary strain energy density and associated energies.

Similarly, the total area above the stress–strain curve in Fig. 3.2.1(b) is (during which the strain changes according to its relation to the stress)

$$U_0^* = \int_0^{\sigma_{11}} \varepsilon_{11} d\sigma_{11},$$

which is termed the *complementary strain energy density*. Note that the strain energy density is expressed in terms of strains, while the complementary strain energy is expressed in terms of stresses. In general, $U_0^* \neq U_0$. They are equal in value when the stress–strain relations are linear (but they are still expressed in terms of different quantities).

The internal work done (or strain energy stored) by $A\sigma_{11}$ over the whole element of length dx during the entire deformation is

$$dU = \int_0^{\varepsilon_{11}} \sigma_{11} d\varepsilon_{11}(A dx_1) = U_0(A dx_1),$$

and the total work done or strain energy stored in the entire body is obtained by integrating over the length of the bar:

$$U = \int_0^L A U_0 \, dx_1.$$

This is the internal strain energy stored in the body and is called the *strain energy*. Similarly, the *complementary strain energy* is given by

$$U^* = \int_0^L A U_0^* \, dx_1.$$

Next, we extend the discussion to the three-dimensional case. Consider the rectangular parallelepiped element of sides dx_1, dx_2, and dx_3 taken from inside an elastic body Ω, as shown in Fig. 3.2.2(a). Suppose that the element is subjected to a system of stresses that vary slowly until they reach their final values, so that the equilibrium is maintained at all times. The forces due to the normal components of stresses are

$$\sigma_{11} dx_2 dx_3, \quad \sigma_{22} dx_3 dx_1, \quad \sigma_{33} dx_1 dx_2,$$

and the forces due to the shear stresses on the six faces are

$$\sigma_{12} dx_2 dx_3, \quad \sigma_{13} dx_3 dx_2, \quad \sigma_{21} dx_1 dx_3, \quad \sigma_{23} dx_3 dx_1, \quad \sigma_{31} dx_1 dx_2, \quad \sigma_{32} dx_2 dx_1.$$

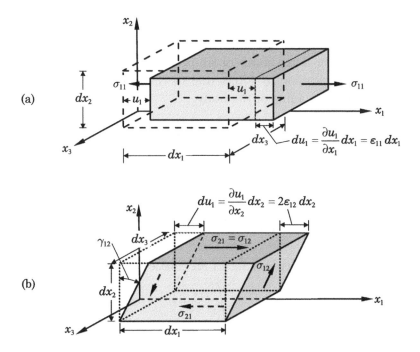

Fig. 3.2.2 Computation of strain energy for the three-dimensional case.

At any stage during the action of these forces, the faces of the parallelepiped will undergo displacements in the normal directions by the amounts $d\varepsilon_{11}dx_1$, $d\varepsilon_{22}dx_2$, and $d\varepsilon_{33}dx_3$, and distort by the amounts $2d\varepsilon_{12}dx_1$, $2d\varepsilon_{23}dx_2$, and $2d\varepsilon_{31}dx_3$. As examples, the deformations caused by normal force $\sigma_{11}dx_2dx_3$ and shear force $\sigma_{21}dx_1dx_3$, each acting alone, are shown in Fig. 3.2.2(a) and (b), respectively. The work done by individual forces can be summed to obtain the total work done by the simultaneous application of all of the forces, because, for example, an x_1-directed force does no work in the x_2 or x_3 directions. The work done, for instance, during the application of the force $\sigma_{11}dx_2dx_3$ is given by (force times displacement)

$$(\sigma_{11}dx_2dx_3)(d\varepsilon_{11}dx_1) = \sigma_{11}\,d\varepsilon_{11}\,d\Omega,$$

where $d\Omega = dx_1dx_2dx_3$. Similarly, the work done by the shear force $\sigma_{21}dx_1dx_3$ is given by

$$(\sigma_{21}dx_1dx_3)(2d\varepsilon_{21}dx_2) = 2\sigma_{21}d\varepsilon_{21}\,d\Omega.$$

The internal work done by all forces in varying slowly from zero to their final values is given by

$$dU = \left(\int_0^{\varepsilon_{11}} \sigma_{11}\,d\varepsilon_{11} + \int_0^{\varepsilon_{12}} 2\sigma_{12}\,d\varepsilon_{12} + \cdots \right)d\Omega = \int_0^{\varepsilon_{ij}} \sigma_{ij}\,d\varepsilon_{ij}\,d\Omega,$$

where sum on the repeated indices is implied but the integral limit corresponds to a fixed i and j. The expression

$$U_0 = \int_0^{\varepsilon_{ij}} \sigma_{ij}\,d\varepsilon_{ij} \tag{3.2.7}$$

is the strain energy per unit volume or simply the strain energy density. Thus, the total internal work done by forces due to the stresses σ_{ij} is given by the integral of U_0 over the volume of the body:

$$U = \int_\Omega U_0\,d\Omega. \tag{3.2.8}$$

Here U represents the mechanical energy stored in the body, and it is known as the strain energy of the body.

The complementary strain energy density, U_0^*, can be computed from

$$U_0^* = \int_0^{\sigma_{ij}} \varepsilon_{ij}\,d\sigma_{ij}. \tag{3.2.9}$$

The complementary strain energy U^* of an elastic body is defined by

$$U^* = \int_\Omega U_0^*\,d\Omega. \tag{3.2.10}$$

Analogous to Eq. (3.2.6), the strains may be derived from the complementary strain energy density function

$$\varepsilon_{ij} = \frac{\partial U_0^*}{\partial \sigma_{ij}} \quad \text{or} \quad \varepsilon = \frac{\partial U_0^*}{\partial \sigma}.$$

For linear elastic solids (i.e., when the material obeys Hooke's law), both the strain energy and complementary strain energy can be expressed as

$$U = \frac{1}{2} \int_\Omega \boldsymbol{\varepsilon} : \boldsymbol{\sigma}(\boldsymbol{\varepsilon}) \, d\Omega, \quad U^* = \frac{1}{2} \int_\Omega \boldsymbol{\varepsilon}(\boldsymbol{\sigma}) : \boldsymbol{\sigma} \, d\Omega, \quad (3.2.11)$$

where ":" denotes the "double-dot product." We note that, although the value of U and U^* is the same for linear elastic solids, U is expressed in terms of strains (or displacements), while U^* is expressed in terms of stresses (or forces).

In general, the internal work done W_I and internal complementary work done W_I^* in a deformable solid consist of energies due to stress and deformation caused by mechanical, thermal, magnetic, and other stimuli. Thus, the strain energy U and complementary strain energy U^* are only a part of W_I and W_I^*, respectively. **In the present study,** *unless stated otherwise,* **we only consider energies due to mechanical forces only.** Therefore, we have

$$W_I = U, \quad W_I^* = U^*. \quad (3.2.12)$$

3.2.2 Expressions for Strain Energy and Complementary Strain Energy Densities of Isotropic Linear Elastic Solids

3.2.2.1 Strain Energy Density

The expression for strain energy density of an elastic solid in three dimensions can be expressed in terms of strains once a stress–strain relation is adopted. For an isotropic linear elastic solid subjected to temperature change of ΔT from a room temperature T_0 (i.e., $\Delta T = T - T_0$, where T is the temperature of the body), the stress–strain relations are

$$\sigma_{ij} = 2\mu \, \varepsilon_{ij} + \lambda \, \varepsilon_{kk} \, \delta_{ij} - \alpha(2\mu + 3\lambda)\Delta T \delta_{ij} \quad (3.2.13)$$

where μ and λ are the Lamé constants and α is the coefficient of thermal expansion. The Lamé constants are related to the engineering constants, namely, Young's modulus E and Poisson's ratio ν by

$$\mu = \frac{E}{2(1+\nu)}, \quad \lambda = \frac{\nu E}{(1+\nu)(1-2\nu)}, \quad 2\mu + \lambda = \frac{(1-\nu)E}{(1+\nu)(1-2\nu)}. \quad (3.2.14)$$

The stress–strain relations in terms of the engineering constants are

$$\sigma_{ij} = \frac{E}{1+\nu}\varepsilon_{ij} + \frac{\nu E}{(1+\nu)(1-2\nu)} \, \varepsilon_{kk} \, \delta_{ij} - \frac{E\alpha}{(1-2\nu)}\Delta T \delta_{ij}. \quad (3.2.15)$$

Then the strain energy density of a linear isotropic elastic solid becomes

$$U_0 = \frac{1}{2} \left[2\mu\varepsilon_{ij}\varepsilon_{ij} + \lambda\varepsilon_{ii}\varepsilon_{jj} - \alpha(2\mu + 3\lambda)\Delta T\varepsilon_{jj} \right] \quad (3.2.16)$$

$$= \frac{1}{2} \left[\frac{E}{1+\nu}\varepsilon_{ij} \, \varepsilon_{ij} + \frac{\nu E}{(1+\nu)(1-2\nu)} \, \varepsilon_{ii} \, \varepsilon_{jj} - \frac{E\alpha}{(1-2\nu)}\Delta T\varepsilon_{jj} \right]$$

$$= \frac{1}{2} \left[\frac{E}{1+\nu} \left(\varepsilon_{11}^2 + \varepsilon_{22}^2 + \varepsilon_{33}^2 + 2\varepsilon_{12}^2 + 2\varepsilon_{13}^2 + 2\varepsilon_{23}^2 \right) \right.$$

$$\left. + \frac{\nu E}{(1+\nu)(1-2\nu)} \left(\varepsilon_{11} + \varepsilon_{22} + \varepsilon_{33} \right)^2 - \frac{E\alpha}{(1-2\nu)}\Delta T \left(\varepsilon_{11} + \varepsilon_{22} + \varepsilon_{22} \right) \right]. \quad (3.2.17)$$

For two-dimensional problems involving isotropic linear elastic materials, Eq. (3.2.17) can be specialized to plane-stress or plane-strain problems. For plane-stress problems ($\sigma_{33} = 0$), the stress–strain relations are (see Section 2.4.3)

$$\sigma_{11} = \frac{E}{1-\nu^2}(\varepsilon_{11} + \nu\varepsilon_{22}) - \frac{E\alpha}{1-\nu}\Delta T$$

$$\sigma_{22} = \frac{E}{1-\nu^2}(\nu\varepsilon_{11} + \varepsilon_{22}) - \frac{E\alpha}{1-\nu}\Delta T \tag{3.2.18}$$

$$\sigma_{12} = \frac{E}{2(1+\nu)}2\varepsilon_{12}, \quad \sigma_{13} = \frac{E}{2(1+\nu)}2\varepsilon_{13}, \quad \sigma_{23} = \frac{E}{2(1+\nu)}2\varepsilon_{23}.$$

The strain energy density for the plane-stress state becomes

$$U_0 = \frac{1}{2}\left[\frac{E}{1-\nu^2}\left(\varepsilon_{11}^2 + 2\nu\varepsilon_{22}\varepsilon_{11} + \varepsilon_{22}^2\right) + \frac{2E}{1+\nu}\left(\varepsilon_{12}^2 + \varepsilon_{13}^2 + \varepsilon_{23}^2\right)\right.$$

$$\left. - \frac{E\alpha}{(1-\nu)}\Delta T\left(\varepsilon_{11} + \varepsilon_{22}\right)\right] \tag{3.2.19}$$

For plane-strain problems ($\varepsilon_{33} = \varepsilon_{13} = \varepsilon_{23} = 0$), the stress–strain relations are

$$\sigma_{11} = \frac{E}{(1+\nu)(1-2\nu)}\left[(1-\nu)\varepsilon_{11} + \nu\varepsilon_{22} - (1+\nu)\alpha\Delta T\right]$$

$$\sigma_{22} = \frac{E}{(1+\nu)(1-2\nu)}\left[\nu\varepsilon_{11} + (1-\nu)\varepsilon_{22} - (1+\nu)\alpha\Delta T\right] \tag{3.2.20}$$

$$\sigma_{12} = \frac{E}{2(1+\nu)}2\varepsilon_{12}.$$

The strain energy density for the plane–strain case becomes

$$U_0 = \frac{E}{2(1+\nu)(1-2\nu)}\left[(1-\nu)\varepsilon_{11}\varepsilon_{11} + 2\nu\varepsilon_{22}\varepsilon_{11} + (1-\nu)\varepsilon_{22}\varepsilon_{22}\right]$$

$$+ \left[\frac{2E}{1+\nu}\varepsilon_{12}^2 - (1+\nu)\alpha\Delta T\left(\varepsilon_{11} + \varepsilon_{22}\right)\right]. \tag{3.2.21}$$

3.2.2.2 Complementary Strain Energy Density

The complementary strain energy density is expressed in terms of stresses. For an isotropic linear elastic solid subjected to temperature change of ΔT from a room temperature, the strain–stress relations in terms of the engineering constants E and ν are

$$\varepsilon_{ij} = \left(\frac{1+\nu}{E}\right)\sigma_{ij} - \frac{\nu}{E}\sigma_{kk}\delta_{ij} + \alpha\Delta T\delta_{ij}. \tag{3.2.22}$$

The complementary strain energy of a three-dimensional linear elastic, isotropic solid is

$$U_0^* = \frac{1}{2}\left[\left(\frac{1+\nu}{E}\right)\sigma_{ij}\sigma_{ij} - \frac{\nu}{E}\sigma_{ii}\sigma_{jj} + \alpha\Delta T\sigma_{jj}\right]$$

$$= \frac{1}{2}\left[\left(\frac{1+\nu}{E}\right)\left(\sigma_{11}^2 + \sigma_{22}^2 + \sigma_{33}^2 + 2\sigma_{12}^2 + 2\sigma_{13}^2 + 2\sigma_{23}^2\right)\right.$$

$$\left. - \frac{\nu}{E}\left(\sigma_{11} + \sigma_{22} + \sigma_{33}\right)^2 + \alpha\Delta T\left(\sigma_{11} + \sigma_{22} + \sigma_{33}\right)\right]. \tag{3.2.23}$$

For plane-stress state, the strain–stress relations are given by

$$\varepsilon_{11} = \tfrac{1}{E}\left(\sigma_{11} - \nu\sigma_{22}\right) + \alpha\Delta T$$
$$\varepsilon_{22} = \tfrac{1}{E}\left(\sigma_{22} - \nu\sigma_{11}\right) + \alpha\Delta T \tag{3.2.24}$$
$$\varepsilon_{12} = \tfrac{(1+\nu)}{E}\sigma_{12}, \quad \varepsilon_{13} = \tfrac{(1+\nu)}{E}\sigma_{13}, \quad \varepsilon_{23} = \tfrac{(1+\nu)}{E}\sigma_{23}.$$

The complementary strain energy density is given by

$$U_0^* = \tfrac{1}{2}\left[\tfrac{1}{E}\left(\sigma_{11}^2 + \sigma_{22}^2 - 2\nu\sigma_{11}\sigma_{22}\right) + \tfrac{2(1+\nu)}{E}\left(\sigma_{12}^2 + \sigma_{13}^2 + \sigma_{23}^2\right)\right.$$
$$\left. + \alpha\Delta T \left(\sigma_{11} + \sigma_{22}\right)\right], \tag{3.2.25}$$

which can also be obtained directly from Eq. (3.2.23) by setting $\sigma_{33} = 0$.

For plane-strain problems, the strain–stress relations are

$$\varepsilon_{11} = \tfrac{1-\nu^2}{E}\left(\sigma_{11} - \tfrac{\nu}{1-\nu}\sigma_{22}\right) + (1+\nu)\alpha\Delta T$$
$$\varepsilon_{22} = \tfrac{1-\nu^2}{E}\left(\sigma_{22} - \tfrac{\nu}{1-\nu}\sigma_{11}\right) + (1+\nu)\alpha\Delta T \tag{3.2.26}$$
$$\varepsilon_{12} = \tfrac{(1+\nu)}{E}\sigma_{12}.$$

The strain energy density for the plane-strain case is

$$U_0^* = \tfrac{1}{2}\left[\tfrac{1-\nu^2}{E}\left(\sigma_{11}^2 + \sigma_{22}^2 - \tfrac{2\nu}{1-\nu}\sigma_{11}\sigma_{22}\right) + \tfrac{2(1+\nu)}{E}\sigma_{12}^2\right.$$
$$\left. + (1+\nu)\alpha\Delta T \left(\sigma_{11} + \sigma_{22}\right)\right]. \tag{3.2.27}$$

3.2.3 Strain Energy and Complementary Strain Energy for Trusses

A *truss* consists of axially load-carrying members connected to each other through pins. Pin connections do not offer resistance to rotations (i.e., members connected through pins are free to rotate about the axis of the pin). The members of a truss are modeled as discrete elements in the sense that we only calculate net axial force (tensile or compressive) and elongation in each member of the truss. This amounts to approximating the axial displacement in the member as linear (and hence the axial strain is a constant). Thus, forces and displacements in truss members are not functions of position. Consequently, the strain energy or complementary strain energy calculation is vastly simplified. Therefore, the strain energy and complementary strain energy of a truss are computed using the following equations:

$$U = \sum_{i=1}^{N} \int_{\Omega_i} U_0^{(i)}\, d\Omega = \sum_{i=1}^{N} A_i L_i U_0^{(i)}, \quad U^* = \sum_{i=1}^{N} \int_{\Omega_i} U_0^{*(i)}\, d\Omega = \sum_{i=1}^{N} A_i L_i U_0^{*(i)}$$
$$\tag{3.2.28}$$

where A_i is the area of cross section, L_i is the length of ith element of the truss, and N is the total number of elements in the truss. The strain energy density and complementary strain energy density are calculated for each member of a structure. They are not to be summed. Only the strain energy and complementary strain energy of members of a structure are summed to obtain the total strain energy or complementary strain energy.

Example 3.2.1 ───

Consider the pin-connected structure shown in Fig. 3.2.3(a). The members of the truss are made of an elastic material whose uniaxial stress–strain behavior in tension and compression are given by

$$\sigma = \begin{cases} K\sqrt{\varepsilon}, & \varepsilon \geq 0, \\ -K\sqrt{-\varepsilon}, & \varepsilon \leq 0, \end{cases} \tag{3.2.29}$$

where K is a material constant. Compute the strain energy density U_0, the complementary strain energy density U_0^*, the strain energy U, and the complementary strain energy U^* of the structure. Assume that the displacements of the point O are small compared with a (i.e., neglect squares of the ratios of the displacements to the lengths of the members).

Solution: To compute U_0, we must find the strains in each member of the structure. On the other hand, U_0^* is computed using the stresses in each member. In both cases, the ultimate result can be expressed in terms of the applied (known) load P. Of course, because of the nonlinear stress–strain relation in Eq. (3.2.29), $U_0 \neq U_0^*$. However, $U_0^{(i)} + U_0^{*(i)}$ for the ith member should equal the product (no sum on i) $\sigma_i \, \varepsilon_i$.

In order to compute U_0 in each of the two members, we first compute the strains in each member by neglecting $(u/a)^2$ and $(v/a)^2$. From Fig. 3.2.3(b), the strains in each member are given by

$$\varepsilon^{(1)} = \frac{\mathrm{B\bar{O}} - \mathrm{BO}}{\mathrm{BO}} = \left[\frac{(a-u)^2 + v^2}{a^2} \right]^{\frac{1}{2}} - 1$$

$$= \left[\frac{a^2 + u^2 + v^2 - 2au}{a^2} \right]^{\frac{1}{2}} - 1 \approx \left(1 - 2\frac{u}{a} \right)^{\frac{1}{2}} - 1$$

$$\approx -\frac{u}{a}, \tag{1}$$

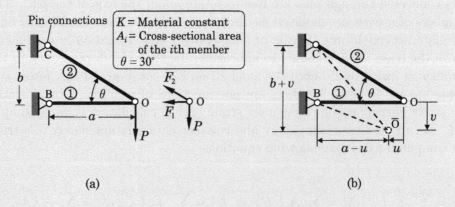

(a)   (b)

Fig. 3.2.3 A pin-connected two-member truss.

where squares of u/a and v/a are neglected, and the binomial series

$$(1 \pm x)^n = 1 \pm nx + \frac{n(n-1)x^2}{2!} \pm \frac{n(n-1)(n-2)x^3}{3!} + \cdots, \quad \text{for } x^2 < 1$$

is used to simplify the expression (with $x = u/a$ and $n = 1/2$). Similarly, we have $(b = a/\sqrt{3})$

$$
\begin{aligned}
\varepsilon^{(2)} &= \frac{\overline{CO} - CO}{CO} = \left[\frac{(a-u)^2 + (b+v)^2}{a^2 + b^2} \right]^{\frac{1}{2}} - 1 \\
&= \left[\frac{a^2 + b^2 + u^2 + v^2 + 2(bv - au)}{a^2 + b^2} \right]^{\frac{1}{2}} - 1 \approx \left(1 + 2\frac{bv - au}{a^2 + b^2} \right)^{\frac{1}{2}} - 1 \\
&\approx \frac{bv - au}{a^2 + b^2} = \frac{\sqrt{3}v - 3u}{4a}.
\end{aligned}
\tag{2}
$$

We note that it is not obvious from the values of strains computed if each one is compressive or tensile. When we compute the forces in the members, it is possible to determine if they are tensile or compressive, which in turn will tell us the nature of the strains. From the free-body diagram of joint O, the axial forces in each member can be calculated as ($F_2 \sin\theta = P$ and $F_1 + F_2 \cos\theta = 0$ with $\theta = 30°$)

$$F_1 = -\sqrt{3}P, \quad F_2 = 2P. \tag{3}$$

From these forces, it is clear that the stress in member 1 is compressive and it is tensile in member 2 (provided the load P is downward). Therefore, the strain $\varepsilon^{(1)}$ is compressive and $\varepsilon^{(2)}$ is tensile.

The strain energy density of each member is computed from

$$
\begin{aligned}
U_0^{(1)} &= \int_0^{\varepsilon^{(1)}} \sigma^{(1)} \, d\varepsilon^{(1)} = \int_0^{\varepsilon^{(1)}} \left(-K\sqrt{-\varepsilon^{(1)}} \right) d\varepsilon^{(1)} = \frac{2K}{3} \left(-\varepsilon^{(1)} \right)^{\frac{3}{2}} = \frac{2K}{3} \left(\frac{u^3}{a^3} \right)^{\frac{1}{2}}, \\
U_0^{(2)} &= \int_0^{\varepsilon^{(2)}} \sigma^{(2)} \, d\varepsilon^{(2)} = \int_0^{\varepsilon^{(2)}} \left(K\sqrt{\varepsilon^{(2)}} \right) d\varepsilon^{(2)} = \frac{2K}{3} \left(\varepsilon^{(2)} \right)^{\frac{3}{2}} = \frac{2K}{3} \left(\frac{\sqrt{3}v - 3u}{4a} \right)^{\frac{3}{2}}.
\end{aligned}
\tag{4}
$$

The strain energy of the structure is ($L_1 = a$ and $L_2 = 2a/\sqrt{3}$)

$$
\begin{aligned}
U &= \int_{V_1} U_0^{(1)} \, d\Omega + \int_{V_2} U_0^{(2)} \, d\Omega = A_1 L_1 U_0^{(1)} + A_2 L_2 U_0^{(2)} \\
&= \frac{2K A_1 a}{3} \left(\frac{u^3}{a^3} \right)^{\frac{1}{2}} + \frac{4K A_2 a}{3\sqrt{3}} \left(\frac{\sqrt{3}v - 3u}{4a} \right)^{\frac{3}{2}},
\end{aligned}
\tag{5}
$$

where A_1 and A_2 denote the cross-sectional areas and V_1 and V_2 are the volumes of members 1 and 2, respectively.

In order to compute the complementary strain energy density in each of the two members, we first compute the stresses (with $\sigma = F/A$). The stress in each member is calculated as

$$\sigma^{(1)} = -\frac{\sqrt{3}P}{A_1}, \quad \sigma^{(2)} = \frac{2P}{A_2}. \tag{6}$$

Then the strain in each member is

$$\varepsilon^{(1)} = -\left(\frac{\sigma^{(1)}}{K} \right)^2, \quad \varepsilon^{(2)} = \left(\frac{\sigma^{(2)}}{K} \right)^2. \tag{7}$$

Then the complementary strain energy density of each member of the structure is

$$
U_0^{*(1)} = \int_0^{\sigma^{(1)}} \varepsilon^{(1)} \, d\sigma^{(1)} = \int_0^{\sigma^{(1)}} \left[-\left(\frac{\sigma^{(1)}}{K} \right)^2 \right] d\sigma^{(1)} = -\frac{1}{3K^2} \left(\sigma^{(1)} \right)^3 = \frac{\sqrt{3}P^3}{A_1^3 K^2},
$$

$$
U_0^{*(2)} = \int_0^{\sigma^{(2)}} \varepsilon^{(2)} \, d\sigma^{(2)} = \int_0^{\sigma^{(2)}} \left[\frac{\left(\sigma^{(2)} \right)^2}{K^2} \right] d\sigma^{(2)} = \frac{1}{3K^2} \left(\sigma^{(2)} \right)^3 = \frac{1}{3} \left(\frac{8P^3}{A_2^3 K^2} \right). \tag{8}
$$

The complementary strain energy of the structure is

$$
U^* = U_0^{*(1)} A_1 L_1 + U_0^{*(2)} A_2 L_2 = \frac{1}{3} \left[\left(\frac{3\sqrt{3}P^3 a}{A_1^2 K^2} \right) + \left(\frac{16P^3 a}{\sqrt{3}A_2^2 K^2} \right) \right]. \tag{9}
$$

It is possible to express the strain energy density of each member in terms of the load P to see if they are equal in value to the corresponding complementary strain energy density. We have

$$
\varepsilon^{(1)} = -\frac{3P^2}{A_1^2 K^2}, \quad \varepsilon^{(2)} = \frac{4P^2}{A_2^2 K^2}. \tag{10}
$$

Then the strain energy density of each member of the structure is

$$
U_0^{(1)} = \int_0^{\varepsilon^{(1)}} \sigma^{(1)} \, d\varepsilon^{(1)} = \int_0^{\varepsilon^{(1)}} \left(-K\sqrt{-\varepsilon^{(1)}} \right) d\varepsilon^{(2)} = \frac{2K}{3} \left(-\varepsilon^{(1)} \right)^{\frac{3}{2}} = \frac{2\sqrt{3}P^3}{A_1^3 K^2},
$$

$$
U_0^{(2)} = \int_0^{\varepsilon^{(2)}} \sigma^{(2)} \, d\varepsilon^{(2)} = \int_0^{\varepsilon^{(2)}} \left(K\sqrt{\varepsilon^{(2)}} \right) d\varepsilon^{(2)} = \frac{2K}{3} \left(\varepsilon^{(2)} \right)^{\frac{3}{2}} = \frac{2}{3} \left(\frac{8P^3}{A_2^3 K^2} \right). \tag{11}
$$

The strain energy of the structure is

$$
U = U_0^{(1)} A_1 L_1 + U_0^{(2)} A_2 L_2 = \frac{2}{3} \left[\left(\frac{3\sqrt{3}P^3 a}{A_1^2 K^2} \right) + \left(\frac{16P^3 a}{\sqrt{3}A_2^2 K^2} \right) \right]. \tag{12}
$$

Clearly, $U_0^{(i)} \neq U_0^{*(i)}$ and $U \neq U^*$.

Example 3.2.2

Consider a uniaxial bar of length L, constant area of cross section A, and constant Young's modulus E. Suppose that the axial displacement $\bar{u}(\bar{x})$ is expressed in terms of the end displacements $\bar{u}_1$ and $\bar{u}_2$ along the length of the member as [see Fig. 3.2.4(a)],

$$
\bar{u}(\bar{x}) = \left(1 - \frac{\bar{x}}{L} \right) \bar{u}_1 + \frac{\bar{x}}{L} \bar{u}_2. \tag{1}
$$

Determine the strain energy density and strain energy in terms of (i) displacements $\bar{u}_1$ and $\bar{u}_2$ and (ii) displacements u_1, u_2, v_1, and v_2 [see Fig. 3.2.4(b)], and (iii) apply the results to determine the strain energy of the truss shown in Fig. 3.2.3 when the material of the truss obey's linear stress–strain relation, $\sigma = E\varepsilon$.

Solution: The only strain the bar experiences is the axial strain $\bar{\varepsilon}$, which is calculated from

$$
\bar{\varepsilon} = \frac{d\bar{u}}{d\bar{x}} = \frac{\bar{u}_2 - \bar{u}_1}{L}. \tag{2}
$$

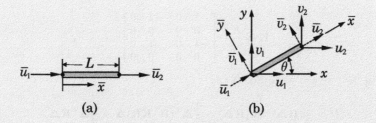

Fig. 3.2.4 Axial deformation of a bar.

(i) The strain energy density is

$$U_0 = \frac{E}{2}(\bar{\varepsilon})^2 = \frac{E}{2}\left(\frac{\bar{u}_2 - \bar{u}_1}{L}\right)^2 = \frac{E}{2L^2}\left\{\begin{matrix}\bar{u}_1\\\bar{u}_2\end{matrix}\right\}^{\mathrm{T}}\begin{bmatrix}1 & -1\\-1 & 1\end{bmatrix}\left\{\begin{matrix}\bar{u}_1\\\bar{u}_2\end{matrix}\right\}. \tag{3}$$

The strain energy stored in the bar is

$$U = \frac{EAL}{2}(\bar{\varepsilon})^2 = \frac{EA}{2L}(\bar{u}_2 - \bar{u}_1)^2 = \frac{EA}{2L}\left\{\begin{matrix}\bar{u}_1\\\bar{u}_2\end{matrix}\right\}^{\mathrm{T}}\begin{bmatrix}1 & -1\\-1 & 1\end{bmatrix}\left\{\begin{matrix}\bar{u}_1\\\bar{u}_2\end{matrix}\right\}. \tag{4}$$

(ii) To express the strain energy in terms of the displacements referred to the x and y axes, first we must set up $\bar{y}$ axis transverse to the bar and add associated displacements $\bar{v}_1$ and $\bar{v}_2$, which are zero (as there are no displacements transverse to the member), and relate the two sets of coordinates to each other:

$$\left\{\begin{matrix}\bar{x}\\\bar{y}\end{matrix}\right\} = \begin{bmatrix}\cos\theta & \sin\theta\\-\sin\theta & \cos\theta\end{bmatrix}\left\{\begin{matrix}x\\y\end{matrix}\right\}, \tag{5}$$

where θ is measured from the positive x-axis to the positive $\bar{x}$-axis. Since the same transformation holds between $(\bar{u}_1, \bar{v}_1)$ and (u_1, v_1) and between $(\bar{u}_2, \bar{v}_2)$ and (u_2, v_2), we can write

$$\left\{\begin{matrix}\bar{u}_1\\\bar{v}_1\\\bar{u}_2\\\bar{v}_2\end{matrix}\right\} = \begin{bmatrix}\cos\theta & \sin\theta & 0 & 0\\-\sin\theta & \cos\theta & 0 & 0\\0 & 0 & \cos\theta & \sin\theta\\0 & 0 & -\sin\theta & \cos\theta\end{bmatrix}\left\{\begin{matrix}u_1\\v_1\\u_2\\v_2\end{matrix}\right\} \quad \text{or} \quad \bar{\boldsymbol{\Delta}} = \mathbf{R}\boldsymbol{\Delta}, \tag{6}$$

where

$$\bar{\boldsymbol{\Delta}} = \left\{\begin{matrix}\bar{u}_1\\\bar{v}_1\\\bar{u}_2\\\bar{v}_2\end{matrix}\right\}, \quad \boldsymbol{\Delta} = \left\{\begin{matrix}u_1\\v_1\\u_2\\v_2\end{matrix}\right\}, \quad \mathbf{R} = \begin{bmatrix}\cos\theta & \sin\theta & 0 & 0\\-\sin\theta & \cos\theta & 0 & 0\\0 & 0 & \cos\theta & \sin\theta\\0 & 0 & -\sin\theta & \cos\theta\end{bmatrix}. \tag{7}$$

Returning to Eqs. (3) and (4), we can rewrite them as

$$U_0 = \frac{E}{2L^2}\left\{\begin{matrix}\bar{u}_1\\\bar{v}_1\\\bar{u}_2\\\bar{v}_2\end{matrix}\right\}^{\mathrm{T}}\begin{bmatrix}1 & 0 & -1 & 0\\0 & 0 & 0 & 0\\-1 & 0 & 1 & 0\\0 & 0 & 0 & 0\end{bmatrix}\left\{\begin{matrix}\bar{u}_1\\\bar{v}_1\\\bar{u}_2\\\bar{v}_2\end{matrix}\right\}, \tag{8}$$

$$U = \frac{EA}{2L}\left\{\begin{matrix}\bar{u}_1\\\bar{v}_1\\\bar{u}_2\\\bar{v}_2\end{matrix}\right\}^{\mathrm{T}}\begin{bmatrix}1 & 0 & -1 & 0\\0 & 0 & 0 & 0\\-1 & 0 & 1 & 0\\0 & 0 & 0 & 0\end{bmatrix}\left\{\begin{matrix}\bar{u}_1\\\bar{v}_1\\\bar{u}_2\\\bar{v}_2\end{matrix}\right\} \quad \text{or} \quad U = \frac{1}{2}\bar{\boldsymbol{\Delta}}^{\mathrm{T}}\bar{\mathbf{K}}\bar{\boldsymbol{\Delta}} \tag{9}$$

with

$$\bar{\mathbf{K}} = \frac{EA}{L} \begin{bmatrix} 1 & 0 & -1 & 0 \\ 0 & 0 & 0 & 0 \\ -1 & 0 & 1 & 0 \\ 0 & 0 & 0 & 0 \end{bmatrix}. \tag{10}$$

Using Eq. (6), we can write Eq. (9) as

$$U = \frac{1}{2} (\mathbf{R\Delta})^{\mathrm{T}} \bar{\mathbf{K}} (\mathbf{R\Delta}) = \frac{1}{2} \mathbf{\Delta}^{\mathrm{T}} (\mathbf{R}^{\mathrm{T}} \bar{\mathbf{K}} \mathbf{R}) \mathbf{\Delta} \equiv \frac{1}{2} \mathbf{\Delta}^{\mathrm{T}} \mathbf{K\Delta}, \tag{11}$$

where

$$\mathbf{K} = \mathbf{R}^{\mathrm{T}} \bar{\mathbf{K}} \mathbf{R} = \frac{EA}{L} \begin{bmatrix} \cos^2\theta & \cos\theta\sin\theta & -\cos^2\theta & -\cos\theta\sin\theta \\ \cos\theta\sin\theta & \sin^2\theta & -\sin\theta\cos\theta & -\sin^2\theta \\ -\cos^2\theta & -\cos\theta\sin\theta & \cos^2\theta & \cos\theta\sin\theta \\ -\cos\theta\sin\theta & -\sin^2\theta & \cos\theta\sin\theta & \sin^2\theta \end{bmatrix}. \tag{12}$$

When $\theta = 0$, we obtain $\bar{\mathbf{K}}$ from $\mathbf{K}$.

(iii) For member 1, we have $\theta_1 = 0$, $L_1 = a$, $u_1 = v_1 = 0$, $u_2 = -u$, and $v_2 = -v$ (the x-axis is horizontal to the right, and the y-axis is vertically up). Hence, from Eq. (11) we have

$$U^{(1)} = \frac{EA_1}{2a} \begin{Bmatrix} -u \\ -v \end{Bmatrix}^{\mathrm{T}} \begin{bmatrix} 1 & 0 \\ 0 & 0 \end{bmatrix} \begin{Bmatrix} -u \\ -v \end{Bmatrix} = \frac{EA_1}{2a} u^2. \tag{13}$$

For member 2 (the $\bar{x}$-axis goes from point C to point O), we have $\theta_1 = -30°$, $L_2 = 2a/\sqrt{3}$, $u_1 = v_1 = 0$, $u_2 = -u$, and $v_2 = -v$. Then from Eq. (11) we obtain

$$U^{(2)} = \frac{\sqrt{3}EA_2}{4a} \begin{Bmatrix} -u \\ -v \end{Bmatrix}^{\mathrm{T}} \frac{1}{4\sqrt{3}} \begin{bmatrix} 3\sqrt{3} & -3 \\ -3 & \sqrt{3} \end{bmatrix} \begin{Bmatrix} -u \\ -v \end{Bmatrix} = \frac{EA_2}{16a} \left(3\sqrt{3}u^2 + \sqrt{3}v^2 - 6uv \right). \tag{14}$$

The total strain energy of the truss is equal to $U = U^{(1)} + U^{(2)}$.

3.2.4   Strain Energy and Complementary Strain Energy for Torsional Members

A circular cylindrical member subjected to torque T about its longitudinal axis develops shear stress $\sigma_{x\theta}$ in its cross section, as shown in Fig. 3.2.5. It can be shown that the stress is related to the applied torque by the relation [98]:

$$\sigma_{x\theta}(r) = \frac{Tr}{J}, \tag{3.2.30}$$

where r is the radial distance from the axis of the member to the location where $\sigma_{x\theta}$ acts and J is the polar moment of inertia of the entire cross section. For a solid circular member of diameter d, the polar moment of inertia is $J = \pi d^4/32$. The stress is the maximum at the outer surface (i.e., at $r = d/2$) of the member.

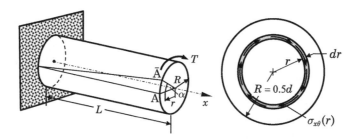

Fig. 3.2.5 Torsion of a solid circular cylindrical shaft.

The strain energy density of a solid circular shaft subjected to torque T is given by

$$U_0 = 2 \int_0^{\varepsilon_{x\theta}} \sigma_{x\theta} \, d\varepsilon_{x\theta} = 4 \int_0^{\varepsilon_{x\theta}} G \varepsilon_{x\theta} \, d\varepsilon_{x\theta} = 2G\varepsilon_{x\theta}^2 \equiv \frac{G}{2} \gamma_{x\theta}^2, \qquad (3.2.31)$$

where G is the shear modulus and $\gamma_{x\theta}$ is the shear strain:

$$\gamma_{x\theta} = 2\varepsilon_{x\theta} = \frac{r\alpha}{L}. \qquad (3.2.32)$$

Here α is the angle of twist, which is related to the applied torque by

$$\alpha = \frac{TL}{GJ}. \qquad (3.2.33)$$

The strain energy of a solid circular shaft subjected to torque T is given by

$$U = \int_v \frac{G}{2} \gamma_{x\theta}^2 \, d\Omega = \frac{1}{2} \int_0^L \int_0^{2\pi} \int_0^{d/2} G \left(\frac{r\alpha}{L}\right)^2 dr \, r d\theta \, dx = \frac{1}{2} \int_0^L GJ \frac{\alpha^2}{L^2} \, dx. \qquad (3.2.34)$$

The complementary strain energy density of a solid circular shaft subjected to torque T is given by

$$U_0^* = 2 \int_0^{\sigma_{x\theta}} \varepsilon_{x\theta} \, d\sigma_{x\theta} = \frac{1}{2G} \sigma_{x\theta}^2. \qquad (3.2.35)$$

The complementary strain energy of a solid circular shaft subjected to torque T is given by

$$U^* = \int_v \frac{1}{2G} \sigma_{x\theta}^2 \, d\Omega = \frac{1}{2} \int_0^L \int_0^{2\pi} \int_0^{d/2} \frac{1}{G} \left(\frac{Tr}{J}\right)^2 dr \, r d\theta \, dx$$
$$= \frac{1}{2} \int_0^L \frac{T^2}{GJ} \, dx. \qquad (3.2.36)$$

Example 3.2.3

Consider a solid shaft made of a linear, isotropic elastic material with shear modulus G held between two rigid walls and subjected to torque, T, as shown in Fig. 3.2.6(a). Determine the complementary strain energy and strain energy in the shaft.

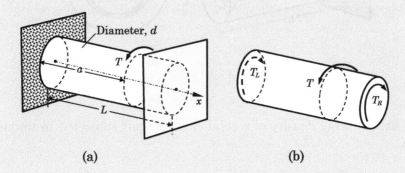

(a)   (b)

Fig. 3.2.6 A solid circular cylindrical shaft held between two rigid walls and subjected to a torque.

Solution: This is an indeterminate structure because we have only one equation of equilibrium, while there are two unknown torques, T_L and T_R [see Fig. 3.2.6(b)], one on the left end and the other on the right end:

$$T - T_R - T_L = 0. \tag{1}$$

Therefore, we need a kinematic relation to determine the unknown torques in terms of the applied torque. The kinematic relation is provided by the requirement that the angle of twist be the same at the point where T is applied. This gives

$$\alpha_L = \alpha_R \ \rightarrow \ \frac{T_L a}{GJ} = \frac{T_R(L-a)}{GJ} \ \text{ or } \ T_L = T_R \frac{L-a}{a}. \tag{2}$$

From Eqs (1) and (2), we obtain

$$T_L = \left(1 - \frac{a}{L}\right) T, \quad T_R = \frac{a}{L} T. \tag{3}$$

Therefore, the complementary strain energy stored as a result of the torsional deformation in the shaft is

$$
\begin{aligned}
U^* &= \frac{1}{2}\left[\int_0^a \frac{T_L^2}{GJ}dx + \int_0^{L-a} \frac{T_R^2}{GJ}dx\right] \\
&= \frac{T^2}{2GJ}\left[a\left(1-\frac{a}{L}\right)^2 + (L-a)\left(\frac{a}{L}\right)^2\right] \\
&= \frac{a}{2GJ}\left(1-\frac{a}{L}\right)T^2.
\end{aligned} \tag{4}
$$

The strain energy stored in the shaft is obtained by applying Eq. (3.2.34) to the portions of the shaft ($\alpha_L = \alpha_R = \alpha$):

$$U = \frac{GJ}{2}\left(\frac{\alpha_L^2}{a} + \frac{\alpha_R^2}{L-a}\right) = \frac{GJL}{2a(L-a)}\alpha^2 = \frac{a}{2GJ}\left(1-\frac{a}{L}\right)T^2. \tag{5}$$

3.2.5 Strain Energy and Complementary Strain Energy for Beams

In this section we derive the strain energy, and complementary strain energy due to extension, bending, and transverse shear in beams. Both beam theories that were developed in Section 2.5 are considered. They include bars (slender members subjected to only axial forces) as a special case. For the strain energy computation, the strain will be expressed in terms of the axial and transverse displacements, whereas for the complementary strain energy computation, the stress will be expressed in terms of the axial and transverse forces and bending moment. The three modes of deformation are separable for straight members with small strains. For beams with a torque, the strain energy or complementary strain energy due to torsion must be added to obtain the total strain energy or complementary strain energy.

3.2.5.1 The Bernuolli–Euler Beam Theory

The strain energy density U_0 of a Bernoulli–Euler beam subjected to axial and bending loads is given by

$$U_0 = \int_0^{\varepsilon_{xx}} \sigma_{xx} \, d\varepsilon_{xx} = \int_0^{\varepsilon_{xx}} E(x,z)\varepsilon_{xx} \, d\varepsilon_{xx}$$

$$= \frac{E}{2}\varepsilon_{xx}^2 = \frac{E}{2}\left(\frac{du}{dx} - z\frac{d^2w}{dx^2}\right)^2. \tag{3.2.37}$$

The strain energy density due to transverse shear strain is zero in the Bernoulli–Euler beam theory. The strain energy U due to axial strain in a beam of length L, area of cross section $A(x)$, and modulus $E = E(x,z)$ is

$$U = \int_v U_0 \, d\Omega = \int_0^L \int_A \frac{E}{2}\left(\frac{du}{dx} - z\frac{d^2w}{dx^2}\right)^2 dA dx$$

$$= \frac{1}{2}\int_0^L \left[A_{xx}\left(\frac{du}{dx}\right)^2 + D_{xx}\left(\frac{d^2w}{dx^2}\right)^2 - 2B_{xx}\frac{du}{dx}\frac{d^2w}{dx^2}\right] dx, \tag{3.2.38}$$

where A_{xx}, B_{xx}, and D_{xx} are as defined in Eq. (2.5.10):

$$A_{xx} = \int_A E(x,z)\,dA, \quad B_{xx} = \int_A zE(x,z)\,dA, \quad D_{xx} = \int_A z^2E(x,z)\,dA.$$

For the case in which E is not a function of z, we have

$$U = \frac{1}{2}\int_0^L \left[EA\left(\frac{du}{dx}\right)^2 + EI\left(\frac{d^2w}{dx^2}\right)^2\right] dx. \tag{3.2.39}$$

The complementary strain energy density of a linear elastic beam with axial and transverse stresses is

$$
U_0^* = \int_0^{\sigma_{xx}} \varepsilon_{xx}\, d\sigma_{xx} + 2 \int_0^{\sigma_{xz}} \varepsilon_{xz}\, d\sigma_{xz}
$$

$$
= \frac{\sigma_{xx}^2}{2E} + \frac{\sigma_{xz}^2}{2G}. \tag{3.2.40}
$$

The complementary strain energy for the case $E = E(x, z)$ [see Eqs (2.5.14) and (2.5.19)] is

$$
U^* = \int_0^L \int_A \frac{1}{2E} \left(\frac{E}{D_e} \right)^2 [D_{xx}N - B_{xx}M + z\,(A_{xx}M - B_{xx}N)]^2 \, dA dx
$$

$$
+ \int_0^L \int_A \frac{1}{2G} \left(\frac{VQ}{Ib} \right)^2 dA dx
$$

or

$$
U^* = \int_0^L \frac{1}{2D_e^2} \Big[A_{xx}\,(D_{xx}N - B_{xx}M)^2 + D_{xx}\,(A_{xx}M - B_{xx}N)^2
$$

$$
+ 2B_{xx}\,(D_{xx}N - B_{xx}M)\,(A_{xx}M - B_{xx}N) \Big] dx
$$

$$
+ \int_0^L \frac{f_s}{2S_{xz}} V^2 \, dx, \tag{3.2.41}
$$

where [see Eqs (2.5.19) and (2.5.28)]

$$
Q(z) = \int_z^{c_1} b(z)z\,dz, \quad S_{xz} = \int_A G(x, z)\, dA \tag{3.2.42}
$$

and

$$
D_e = A_{xx}D_{xx} - B_{xx}^2, \quad f_s = \frac{S_{xz}}{I^2 b^2} \int_A \frac{Q^2(z)}{G(x, z)} dA. \tag{3.2.43}
$$

For a homogeneous beam of rectangular cross section (height h and width b), we have

$$
Q(z) = b \int_z^{h/2} z\, dz = \frac{b}{2} \left(\frac{h^2}{4} - z^2 \right), \quad f_s = \frac{6}{5}.
$$

For the case in which E is not a function of z, we have

$$
U^* = \frac{1}{2} \int_0^L \left(\frac{N^2}{EA} + \frac{M^2}{EI} + \frac{f_s V^2}{GA} \right) dx, \tag{3.2.44}
$$

where N, M, and V are expressed in terms of the externally applied loads.

3.2.5.2 The Timoshenko Beam Theory

The strain energy density U_0 of a Timoshenko beam subjected to axial and bending loads is given by

$$U_0 = \int_0^{\varepsilon_{xx}} \sigma_{xx}\, d\varepsilon_{xx} + \int_0^{\gamma_{xz}} \sigma_{xz}\, d\gamma_{xz} = \frac{E}{2}\varepsilon_{xx}^2 + \frac{G}{2}\gamma_{xz}^2$$

$$= \frac{E}{2}\left(\frac{du}{dx} + z\frac{d\phi_x}{dx}\right)^2 + \frac{G}{2}\left(\phi_x + \frac{dw}{dx}\right)^2. \tag{3.2.45}$$

We note that the strain energy density due to transverse shear strain is *not* zero in the Timoshenko beam theory. The strain energy U due to axial strain in a beam of length L, area of cross section $A(x)$, and modulus $E = E(x, z)$ is[1]

$$U = \int_v U_0\, d\Omega = \int_0^L \int_A \left[\frac{E}{2}\left(\frac{du}{dx} + z\frac{d\phi_x}{dx}\right)^2 + \frac{K_s G}{2}\left(\phi_x + \frac{dw}{dx}\right)^2\right] dA dx$$

$$= \frac{1}{2}\int_0^L \left[A_{xx}\left(\frac{du}{dx}\right)^2 + D_{xx}\left(\frac{d\phi_x}{dx}\right)^2 + 2B_{xx}\frac{du}{dx}\frac{d\phi_x}{dx}\right.$$

$$\left. + K_s S_{xz}\left(\phi_x + \frac{dw}{dx}\right)^2\right] dx, \tag{3.2.46}$$

where K_s is the shear correction coefficient introduced [see Eq. (2.5.26)] to correct the energy loss due to the constant state of transverse shear stress σ_{xz}. When E is not a function of z, we have ($A_{xx} = EA$, $D_{xx} = EI$, and $S_{xz} = GA$)

$$U = \frac{1}{2}\int_0^L \left[EA\left(\frac{du}{dx}\right)^2 + EI\left(\frac{d\phi_x}{dx}\right)^2 + GAK_s\left(\phi_x + \frac{dw}{dx}\right)^2\right] dx. \tag{3.2.47}$$

The complementary strain energy density of a linear elastic beam with axial and transverse stresses is

$$U_0^* = \int_0^{\sigma_{xx}} \varepsilon_{xx}\, d\sigma_{xx} + 2\int_0^{\sigma_{xz}} \varepsilon_{xz}\, d\sigma_{xz} = \frac{\sigma_{xx}^2}{2E} + \frac{\sigma_{xz}^2}{2G}. \tag{3.2.48}$$

The complementary strain energy, for the case $E = E(x, z)$ [see Eq. (2.5.30)], is expressed as

$$U^* = \int_0^L \int_A \frac{1}{2E}\left(\frac{E}{D_e}\right)^2 \left[D_{xx}N - B_{xx}M + z\left(A_{xx}M - B_{xx}N\right)\right]^2 dA dx$$

$$+ \int_0^L \int_A \frac{f_s}{2G}\left(\frac{GV}{S_{xz}}\right)^2 dA dx$$

---

[1] The variables (u, w, ϕ_x) are termed *generalized displacements* to indicate that displacements as well as rotations are displacements in a generalized sense.

or

$$U^* = \int_0^L \frac{1}{2D_e^2} \Big[A_{xx} \left(D_{xx}N - B_{xx}M \right)^2 + D_{xx} \left(A_{xx}M - B_{xx}N \right)^2$$

$$+ 2B_{xx} \left(D_{xx}N - B_{xx}M \right) \left(A_{xx}M - B_{xx}N \right) \Big] dx$$

$$+ \int_0^L \frac{f_s}{2S_{xz}} V^2 \, dx, \tag{3.2.49}$$

where f_s [see Eq. (3.2.43) for its definition] is introduced, in analogy to U^* in Eq. (3.2.41), to correct for the constant state of shear stress predicted by the Timoshenko beam theory [the inverse of f_s is the shear correction coefficient, K_s, introduced in Eq. (3.2.46)]. For the case in which E is not a function of z, we have

$$U^* = \frac{1}{2} \int_0^L \left(\frac{N^2}{EA} + \frac{M^2}{EI} + \frac{V^2}{GAK_s} \right) dx. \tag{3.2.50}$$

Example 3.2.4 ————————————————————————————————

Consider a homogeneous, isotropic linear elastic beam of length L, area of cross-section A, moment of inertia about the axis of bending I, and Young's modulus E. Assuming that the axial displacement $u(x)$ and transverse displacement $w(x)$ of the Bernoulli–Euler beam are of the form

$$u(x) = \left(1 - \frac{x}{L} \right) u_1 + \frac{x}{L} u_2 \equiv \psi_1(x)u_1 + \psi_2(x)u_2, \tag{1}$$

$$w(x) = \left[1 - 3\left(\frac{x}{L}\right)^2 + 2\left(\frac{x}{L}\right)^3 \right] w_1 - x\left(1 - \frac{x}{L} \right)^2 \theta_1$$

$$+ \left[3\left(\frac{x}{L}\right)^2 - 2\left(\frac{x}{L}\right)^3 \right] w_2 - x\left[\left(\frac{x}{L}\right)^2 - \frac{x}{L} \right] \theta_2$$

$$\equiv \varphi_1(x)w_1 + \varphi_2(x)\theta_1 + \varphi_3(x)w_2 + \varphi_4(x)\theta_2, \tag{2}$$

where (u_1, u_2, w_1, w_2) and (θ_1, θ_2) are the displacements and rotations of the two ends of the beam, as shown in Fig. 3.2.7, compute the strain energy of the Bernoulli–Euler beam theory in terms of the parameters (u_1, w_1, θ_1) and (u_2, w_2, θ_2).

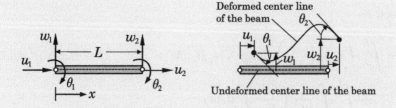

Fig. 3.2.7 A Bernoulli–Euler beam element with generalized displacement degrees of freedom.

Solution: The strain energy for the problem at hand is given by the expression in Eq. (3.2.39). First we compute du/dx and d^2w/dx^2 in terms of the generalized displacements:

$$\frac{du}{dx} = \frac{u_2 - u_1}{L},$$

$$\frac{d^2w}{dx^2} = -\frac{6}{L^2}\left(1 - 2\frac{x}{L}\right)w_1 - \frac{2}{L}\left(3\frac{x}{L} - 2\right)\theta_1 + \frac{6}{L^2}\left(1 - 2\frac{x}{L}\right)w_2 - \frac{2}{L}\left(3\frac{x}{L} - 1\right)\theta_2. \tag{3}$$

Then

$$\frac{EA}{2}\int_0^L \left(\frac{du}{dx}\right)^2 dx = \frac{EA}{2L}(u_2 - u_1)^2 = \frac{EA}{2L}\left\{\begin{matrix} u_1 \\ u_2 \end{matrix}\right\}^T \begin{bmatrix} 1 & -1 \\ -1 & 1 \end{bmatrix}\left\{\begin{matrix} u_1 \\ u_2 \end{matrix}\right\}. \tag{4}$$

Next, we evaluate the integrals ($\xi = x/L$ and $dx = L\,d\xi$),

$$\int_0^L (1 - 2\xi)^2 L\,d\xi = \frac{L}{3}, \quad \int_0^L (1 - 2\xi)(3\xi - 2)L\,d\xi = -\frac{L}{2}$$

$$\int_0^L (3\xi - 2)^2 L\,d\xi = L, \quad \int_0^L (1 - 2\xi)(3\xi - 1)L\,d\xi = -\frac{L}{2}, \tag{5}$$

$$\int_0^L (3\xi - 1)^2 L\,d\xi = L, \quad \int_0^L (3\xi - 2)(3\xi - 1)L\,d\xi = \frac{L}{2}.$$

Then we have

$$\frac{EI}{2}\int_0^L \left(\frac{d^2w}{dx^2}\right)^2 dx = \frac{EI}{L^3}\big(6w_1^2 - 6Lw_1\theta_1 - 12w_1w_2 - 6Lw_1\theta_2 + 2L^2\theta_1^2 + 6L\theta_1w_2$$

$$+ 2L^2\theta_1\theta_2 + 6w_2^2 + 6Lw_2\theta_2 + 2L^2\theta_2^2\big)$$

$$= \frac{EI}{L^3}\left\{\begin{matrix} w_1 \\ \theta_1 \\ w_2 \\ \theta_2 \end{matrix}\right\}^T \begin{bmatrix} 6 & -3L & -6 & -3L \\ -3L & 2L^2 & 3L & L^2 \\ -6 & 3L & 6 & 3L \\ -3L & L^2 & 3L & 2L^2 \end{bmatrix}\left\{\begin{matrix} w_1 \\ \theta_1 \\ w_2 \\ \theta_2 \end{matrix}\right\}. \tag{6}$$

Thus, the strain energy in the beam is

$$U = \left\{\begin{matrix} u_1 \\ u_2 \end{matrix}\right\}^T \frac{EA}{2L}\begin{bmatrix} 1 & -1 \\ -1 & 1 \end{bmatrix}\left\{\begin{matrix} u_1 \\ u_2 \end{matrix}\right\} + \left\{\begin{matrix} w_1 \\ \theta_1 \\ w_2 \\ \theta_2 \end{matrix}\right\}^T \frac{EI}{L^3}\begin{bmatrix} 6 & -3L & -6 & -3L \\ -3L & 2L^2 & 3L & L^2 \\ -6 & 3L & 6 & 3L \\ -3L & L^2 & 3L & 2L^2 \end{bmatrix}\left\{\begin{matrix} w_1 \\ \theta_1 \\ w_2 \\ \theta_2 \end{matrix}\right\}. \tag{7}$$

The first part of the expression is the strain energy due to extensional deformation and the second part corresponds to the strain energy due to bending deformation.

Example 3.2.5

Assuming that the axial displacement $u(x)$, transverse displacement $w(x)$, and rotation ϕ_x of the Timoshenko beam theory are of the form

$$u(x) = \left(1 - \frac{x}{L}\right)u_1 + \frac{x}{L}u_2, \tag{1}$$

$$w(x) = \left[\left(1 - \frac{x}{L}\right)\left(1 - \frac{2x}{L}\right)\right]w_1 + \frac{4x}{L}\left(1 - \frac{x}{L}\right)w_c + \left[-\frac{x}{L}\left(1 - \frac{2x}{L}\right)\right]w_2 \tag{2}$$

$$\phi_x(x) = \left(1 - \frac{x}{L}\right)\phi_1 + \frac{x}{L}\phi_2, \tag{3}$$

where $(u_1, u_2, w_1, w_c, w_2)$ and (ϕ_1, ϕ_2) are the displacements and rotations of the beam, as shown in Fig. 3.2.8, compute the strain energy of the Timoshenko beam theory in terms of the parameters $(u_1, u_2, w_1, w_c, w_2, \phi_1, \phi_2)$.

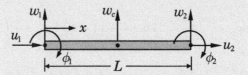

Fig. 3.2.8 A straight beam element with end displacement degrees of freedom.

Solution: The strain energy due to axial deformation is the same as in Eq. (4) of **Example 3.2.4**. The remaining terms of the strain energy expression U from Eq. (3.2.47) are presented next. We have

$$\frac{EI}{2} \int_0^L \left(\frac{d\phi_x}{dx} \right)^2 dx = \frac{EI}{2L} (\phi_2 - \phi_1)^2 = \frac{EI}{2L} \left\{ \begin{matrix} \phi_1 \\ \phi_2 \end{matrix} \right\}^{\mathrm{T}} \left[\begin{matrix} 1 & -1 \\ -1 & 1 \end{matrix} \right] \left\{ \begin{matrix} \phi_1 \\ \phi_2 \end{matrix} \right\}. \tag{4}$$

$$\frac{GAK_s}{2} \int_0^L \phi_x^2 \, dx = \frac{GAK_s L}{12} \left\{ \begin{matrix} \phi_1 \\ \phi_2 \end{matrix} \right\}^{\mathrm{T}} \left[\begin{matrix} 2 & 1 \\ 1 & 2 \end{matrix} \right] \left\{ \begin{matrix} \phi_1 \\ \phi_2 \end{matrix} \right\}. \tag{5}$$

$$\frac{GAK_s}{2} \int_0^L \left(\frac{dw}{dx} \right)^2 dx = \frac{GAK_s}{6L} \left\{ \begin{matrix} w_1 \\ w_c \\ w_2 \end{matrix} \right\}^{\mathrm{T}} \left[\begin{matrix} 7 & -8 & 1 \\ -8 & 16 & -8 \\ 1 & -8 & 7 \end{matrix} \right] \left\{ \begin{matrix} w_1 \\ w_c \\ w_2 \end{matrix} \right\}. \tag{6}$$

$$GAK_s \int_0^L \phi_x \frac{dw}{dx} \, dx = \frac{GAK_s}{6} \left\{ \begin{matrix} \phi_1 \\ \phi_2 \end{matrix} \right\}^{\mathrm{T}} \left[\begin{matrix} -5 & 4 & 1 \\ -1 & -4 & 5 \end{matrix} \right] \left\{ \begin{matrix} w_1 \\ w_c \\ w_2 \end{matrix} \right\}. \tag{7}$$

Sum of these expressions along with the expression in Eq. (4) of **Example 3.2.4** gives the desired result.

Example 3.2.6 deals with the complementary energy of a frame[2] structure.

Example 3.2.6

Consider the frame structure shown in Fig. 3.2.9. Each member of the structure has the same cross-sectional area A, moment of inertia I, and constant Young's modulus E. The material of the frame is assumed to be linearly elastic. Compute the complementary strain energy of the frame structure using the Bernoulli–Euler beam theory and the Timoshenko beam theory.

Solution: This is a determinate structure. Hence, all of the external reactions can be easily determined from the free-body diagram of the whole structure. The expressions for the axial and shear forces and bending moments as functions of position for the two members of the frame can be easily determined, as illustrated in Fig. 3.2.9:

---

[2] A *frame* structure consists of several uniaxial members (beams) that are connected rigidly so that they cannot freely rotate with respect to each other. Consequently, they develop forces and moments at the connecting points when subjected to external loads. When the connections are pin connections that allow free rotation at the joints, the structure is called a *truss*.

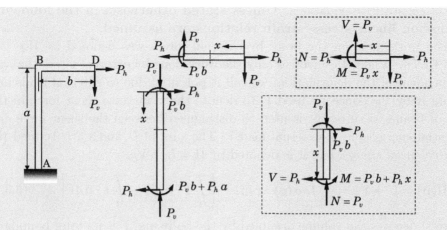

Fig. 3.2.9 A two-member frame structure.

$$N_{\text{BA}} = P_v, \quad V_{\text{BA}} = P_h, \quad M_{\text{BA}} = P_v \cdot b + P_h \cdot x,$$
$$N_{\text{DB}} = P_h, \quad V_{\text{DB}} = P_v, \quad M_{\text{DB}} = P_v \cdot x. \tag{1}$$

The complementary strain energy is $U^* = U_{DB}^* + U_{BA}^*$ with

$$U_{\text{DB}}^* = \int_0^b \left[\frac{P_h^2}{2EA} + \frac{1}{2EI}(P_v x)^2 + \frac{f_s P_v^2}{2GA} \right] dx = \frac{P_h^2 b}{2EA} + \frac{P_v^2 b^3}{6EI} + \frac{f_s P_v^2 b}{2GA}, \tag{2}$$

$$U_{\text{BA}}^* = \int_0^a \left[\frac{P_v^2}{2EA} + \frac{1}{2EI}(P_v b + P_h x)^2 + \frac{f_s P_h^2}{2GA} \right] dx$$
$$= \frac{P_v^2 a}{2EA} + \frac{1}{2EI}\left(P_v^2 ab^2 + P_v P_h a^2 b + \frac{1}{3}P_h^2 a^3 \right) + \frac{f_s P_h^2 a}{2GA}. \tag{3}$$

The complementary strain energy for the Bernoulli–Euler beam theory and the Timoshenko beam theory is the same with $f_s = 1/K_s$, where K_s is the shear correction coefficient.

3.3 Total Potential Energy and Total Complementary Energy

3.3.1 Introduction

As discussed in the previous sections, strain energy U is the internal energy stored in the body due to the deformation of the body, and it is the same as the work done by internal forces in moving through their respective displacements. The strain energy is expressed in terms of strains or displacements. This requires the use of constitutive relations to express the internal stresses in terms of strains. The complementary strain energy U^* is also the energy stored in the body due to the deformation of the body, and it is the same as the work done by internal forces in moving through their respective displacements. However, the complementary strain energy is expressed in terms of stresses or forces with the help of constitutive equations. **Unless stated otherwise, in the following discussion linear stress–strain relations are assumed.**

The work done on the body by external forces was denoted by W_E [see Eq. (3.1.7)]. When W_E involves only the work done by external forces that are independent of the deformation, we call it potential due to external loads and denote it by V_E (since we used V to denote the transverse shear force in the case of beams, a subscript is used to distinguish between the shear force and potential energy due to external loads). The sum of U and V_E is termed the *total potential energy*, and it is denoted by $\Pi = U + V_E$:

$$\Pi(\mathbf{u}) = U + V_E = \frac{1}{2} \int_\Omega \boldsymbol{\sigma}(\varepsilon) : \varepsilon \, d\Omega - \left[\int_\Omega \mathbf{f} \cdot \mathbf{u} \, d\Omega + \oint_\Gamma \mathbf{t} \cdot \mathbf{u} \, d\Gamma \right], \quad (3.3.1)$$

where Ω denotes the volume occupied by the body and Γ is its total boundary. Similarly, the sum of U^* and V_E is termed the *total complementary energy*:

$$\Pi^*(\boldsymbol{\sigma}) = U^* + V_E = \frac{1}{2} \int_\Omega \boldsymbol{\sigma} : \varepsilon(\boldsymbol{\sigma}) \, d\Omega - \left[\int_\Omega \mathbf{f} \cdot \mathbf{u} \, d\Omega + \oint_\Gamma \mathbf{t} \cdot \mathbf{u} \, d\Gamma \right]. \quad (3.3.2)$$

3.3.2 Total Potential Energy of Beams

Here we present the total potential energy expressions for beams, which include bars (i.e., members subjected to uniaxial forces) as a special case. First, we compute the potential energy, V_E, due to the applied loads. Suppose that the beam is subjected to distributed axial force $f(x)$ along the line $z = 0$ and transverse load $q(x)$ at $z = h/2$, both measured per unit length. Then the work done by the applied forces is

$$V_E = - \left\{ \int_0^L [f(x)u_1(x, 0) + q(x)u_3(x, h/2)] \, dx \right\}$$

$$= - \left\{ \int_0^L [f(x)u(x) + q(x)w(x)] \, dx \right\}. \quad (3.3.3)$$

This expression should be added to U that was already derived for beams to obtain Π.

In the case of the Bernoulli–Euler beam theory, the total potential energy is obtained by adding V_E from Eq. (3.3.3) to U from Eq. (3.2.38):

$$\Pi(u, w) = \frac{1}{2} \int_0^L \left[A_{xx} \left(\frac{du}{dx} \right)^2 + D_{xx} \left(\frac{d^2w}{dx^2} \right)^2 - 2B_{xx} \frac{du}{dx} \frac{d^2w}{dx^2} \right] dx$$

$$- \int_0^L (f\,u + q\,w) \, dx, \quad (3.3.4)$$

where A_{xx}, B_{xx}, and D_{xx} are defined in Eq. (2.5.10). The expression in Eq. (3.3.4) must be appended with any terms corresponding to the work done by external point forces and moments.

The total potential energy for the Timoshenko beam theory is obtained by adding V_E from Eq. (3.3.3) to U from (3.2.46):

$$\Pi(u, w, \phi_x) = \frac{1}{2} \int_0^L \left[A_{xx} \left(\frac{du}{dx} \right)^2 + D_{xx} \left(\frac{d\phi_x}{dx} \right)^2 + 2B_{xx} \frac{du}{dx} \frac{d\phi_x}{dx} \right.$$

$$\left. + K_s S_{xz} \left(\phi_x + \frac{dw}{dx} \right)^2 \right] dx - \int_0^L (f\, u + q\, w)\, dx, \qquad (3.3.5)$$

where K_s is the shear correction coefficient.

3.3.3 Total Complementary Energy of Beams

The total complementary energy is denoted with Π^*. The expression for the total complementary energy is the same for both the Bernoulli–Euler beam theory and the Timoshenko beam theory, and it is constructed from Eqs (3.3.3) and (3.2.49):

$$\Pi^*(N, M, V) = \int_0^L \frac{1}{2D_e^2} \left[A_{xx} \left(D_{xx} N - B_{xx} M \right)^2 + D_{xx} \left(A_{xx} M - B_{xx} N \right)^2 \right.$$

$$\left. + 2B_{xx} \left(D_{xx} N - B_{xx} M \right) \left(A_{xx} M - B_{xx} N \right) \right] dx$$

$$+ \int_0^L \frac{f_s}{2S_{xz}} V^2\, dx - \int_0^L (f\, u + q\, w)\, dx. \qquad (3.3.6)$$

Example 3.3.1 ————————————————————————————

Derive the total potential energy and complementary energy expressions for a three-dimensional linear elastic solid that occupies volume Ω with boundary Γ and obeys Hooke's law in Eq. (2.4.1). Assume linearized strain–displacement relations.

Solution: For total potential energy, we use Eqs (2.4.1), (3.1.7), and (3.2.11) in index notation:

$$\Pi(\mathbf{u}) = \frac{1}{2} \int_\Omega \boldsymbol{\sigma} : \boldsymbol{\varepsilon}\, d\Omega - \int_\Omega \mathbf{f} \cdot \mathbf{u}\, d\Omega - \oint_\Gamma \mathbf{t} \cdot \mathbf{u}\, d\Gamma \qquad (1)$$

$$= \frac{1}{2} \int_\Omega \sigma_{ij}\, \varepsilon_{ij}\, d\Omega - \int_\Omega f_i u_i\, d\Omega - \oint_\Gamma t_i u_i\, d\Gamma$$

$$= \frac{1}{2} \int_\Omega C_{ijkl}\, \varepsilon_{kl}\, \varepsilon_{ij}\, d\Omega - \int_\Omega f_i u_i\, d\Omega - \oint_\Gamma t_i u_i\, d\Gamma$$

$$= \frac{1}{8} \int_\Omega C_{ijkl}\, (u_{k,l} + u_{l,k})\, (u_{i,j} + u_{j,i})\, d\Omega - \int_\Omega f_i u_i\, d\Omega - \oint_\Gamma t_i u_i\, d\Gamma$$

$$= \frac{1}{8} \int_\Omega C_{ijkl}\, (u_{k,l} u_{i,j} + u_{l,k} u_{i,j} + u_{k,l} u_{j,i} + u_{l,k} u_{j,i})\, d\Omega - \int_\Omega f_i u_i\, d\Omega - \oint_\Gamma t_i u_i\, d\Gamma$$

$$= \frac{1}{2} \int_\Omega C_{ijkl}\, u_{k,l} u_{i,j}\, d\Omega - \int_\Omega f_i u_i\, d\Omega - \oint_\Gamma t_i u_i\, d\Gamma, \qquad (2)$$

where $u_{i,j} = \partial u_i / \partial x_j$, C_{ijkl} are the components of the fourth-order elasticity tensor, and summation on repeated indices is implied over the range of 1, 2, and 3. The last step was arrived by using the symmetry of C_{ijkl} with respect to its subscripts.

For total complementary energy, we have

$$\Pi^*(\sigma_{ij}) = \frac{1}{2} \int_\Omega \sigma_{ij}\,\varepsilon_{ij}\,d\Omega - \int_\Omega f_i u_i\,d\Omega - \oint_\Gamma t_i u_i\,d\Gamma$$

$$= \frac{1}{2} \int_\Omega C^*_{ijkl}\,\sigma_{kl}\,\sigma_{ij}\,d\Omega - \int_\Omega f_i u_i\,d\Omega - \oint_\Gamma t_i u_i\,d\Gamma, \tag{3}$$

where C^*_{ijkl} are the components of the fourth-order compliance tensor, which is the inverse of the elasticity tensor.

3.4   Virtual Work

3.4.1   Virtual Displacements

From purely geometrical considerations, a given mechanical system can take many possible configurations. The set of configurations that satisfy the geometric constraints (e.g., geometric boundary conditions) of the system is called the *set of admissible configurations*. Of all admissible configurations only one of them corresponds to the equilibrium configuration under the applied loads, and it is this configuration that also satisfies Newton's second law (i.e., equilibrium of forces and moments). The admissible configurations are restricted to a neighborhood of the true configuration so that they are obtained from infinitesimal *variations* of the true configuration (i.e., infinitesimal movement of the material points). During such variations, the geometric constraints of the system are not violated, and all applied forces are fixed at their actual equilibrium values. When a mechanical system experiences such variations in its equilibrium configuration, it is said to undergo *virtual displacements*. These displacements need not have any relationship with the actual displacements, which might occur due to a change in the loads and/or boundary conditions. The displacements are called virtual because they are *imagined* to take place (i.e., hypothetical) with the actual loads acting at their fixed values.

For example, consider a beam fixed at $x = 0$ and subjected to any arbitrary loading (e.g., distributed as well as point loads), as shown in Fig. 3.4.1. The possible geometric configurations the beam can take under the loads may be expressed in terms of the transverse deflection $w(x)$ and axial displacement $u(x)$. The support conditions require that

$$w(0) = \hat{w}, \quad \left(-\frac{dw}{dx}\right)_{x=0} = \hat{\theta}, \quad u(0) = \hat{u}, \tag{3.4.1}$$

where $\hat{w}$, $\hat{\theta}$, and $\hat{u}$ are constants. These are called the *geometric* or *displacement boundary conditions*. Boundary conditions that involve specifying the forces applied on the beam are called *force boundary conditions*.

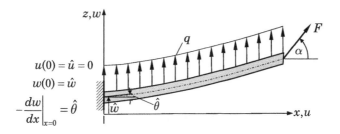

Fig. 3.4.1 A cantilever beam with a set of arbitrary loads.

The set of all functions $w(x)$ and $u(x)$ that satisfy the geometric boundary conditions in Eq. (3.4.1) is the space of admissible configurations for this case. This space consists of pairs of elements $\{(u_i, w_i)\}$ of the form

$$u_1(x) = \hat{u} + a_1 x, \quad w_1(x) = \hat{w} + \hat{\theta} x + b_1 x^2,$$
$$u_2(x) = \hat{u} + a_1 x + a_2 x^2, \quad w_2(x) = \hat{w} + \hat{\theta} x + b_1 x^2 + b_2 x^3,$$

where a_i and b_i are constants. The pair (u, w) that also satisfies, in addition to the geometric boundary conditions, the equilibrium equations and force boundary conditions (which require the precise nature of the applied loads) of the problem is the equilibrium solution. The virtual displacements, $\delta u(x)$ and $\delta w(x)$, must be necessarily of the form

$$\delta u_1 = a_1 x, \quad \delta w_1 = b_1 x^2; \quad \delta u_2 = a_1 x + a_2 x^2, \quad \delta w_2 = b_1 x^2 + b_2 x^3, \quad \text{and so on,}$$

which satisfy the homogeneous form of the specified geometric boundary conditions:

$$\delta w(0) = 0, \quad \left(\frac{d\delta w}{dx}\right)_{x=0} = 0, \quad \delta u(0) = 0. \tag{3.4.2}$$

Thus, the virtual displacements at the boundary points at which the geometric conditions are specified (independent of the specified values) are necessarily zero.

The work done by the actual forces through a virtual displacement of the actual configuration is called *virtual work* done by actual forces. If we denote the virtual displacement by $\delta \mathbf{u}$, the virtual work done by a constant force $\mathbf{F}$ is given by

$$\delta W = \mathbf{F} \cdot \delta \mathbf{u}.$$

The virtual work done by actual forces in moving through virtual displacements in a deformable body consists of two parts: virtual work done by internal forces, δW_I, and virtual work done by external forces, δW_E. These may be computed as discussed next.

Consider a deformable body of volume Ω and closed surface area Γ. Suppose that the body is subjected to a body force $\mathbf{f}(\mathbf{x})$ per unit volume of the body, a specified surface force $\mathbf{t}(s)$ per unit area on a portion Γ_σ of the boundary, and specified displacement $\mathbf{u}$ on the remaining portion Γ_u of the boundary such that $\Gamma = \Gamma_u \cup \Gamma_\sigma$ and $\Gamma_u \cap \Gamma_\sigma$ is empty. For this case, the virtual displacement $\delta\mathbf{u}(\mathbf{x})$ is any function, small in magnitude (to keep the system in equilibrium), that satisfies the requirement

$$\delta\mathbf{u} = \mathbf{0} \quad \text{on} \quad \Gamma_u.$$

The virtual work done by actual forces $\mathbf{f}$ and $\mathbf{t}$ in moving through the virtual displacement $\delta\mathbf{u}$ is

$$\delta W_E = -\left(\int_\Omega \mathbf{f} \cdot \delta\mathbf{u} \, d\Omega + \int_{\Gamma_\sigma} \mathbf{t} \cdot \delta\mathbf{u} \, d\Gamma \right). \tag{3.4.3}$$

The negative sign in front of the expression indicates that work is performed *on* the body.

As a result of the application of the loads, the body develops internal forces in the form of stresses. These stresses also perform work when the body is given a virtual displacement. In this study, we shall be concerned mainly with only *ideal systems*. An ideal system is one in which no work is dissipated by friction. We assume that the virtual displacement $\delta\mathbf{u}$ is applied slowly from zero to its final value. Associated with the virtual displacement is the virtual strain, which can be computed according to the strain–displacement relation (for the linear case):

$$\delta\varepsilon = \frac{1}{2} \left[\boldsymbol{\nabla}(\delta\mathbf{u}) + (\boldsymbol{\nabla}(\delta\mathbf{u}))^\mathrm{T} \right]. \tag{3.4.4}$$

The internal virtual work stored in the body per unit volume, analogous to the calculation of the strain energy density, is the virtual strain energy density:

$$\delta U_0 = \int_0^{\delta\varepsilon} \boldsymbol{\sigma} : d(\delta\varepsilon) = \int_0^{\delta\varepsilon_{ij}} \sigma_{ij} d(\delta\varepsilon_{ij}) = \sigma_{ij}\delta\varepsilon_{ij} \tag{3.4.5}$$

We note that the result in Eq. (3.4.5) is arrived without the use of a constitutive equation, because $\boldsymbol{\sigma}$ is not a function of $\delta\varepsilon$. The total internal virtual work stored in the body is denoted by δW_I, and it is equal to

$$\delta W_I = \int_\Omega \delta U_0 \, d\Omega = \int_\Omega \sigma_{ij} \, \delta\varepsilon_{ij} \, d\Omega. \tag{3.4.6}$$

Example 3.4.1 ―――――――――――――――――――――――――――――

Consider the rigid link of **Example 3.1.1** (see Fig. 3.4.2). The initial, no load position, corresponds to $\theta = 0$, and the displacement to the equilibrium position is small (i.e., θ is small). Determine the virtual work done by external forces and internal forces when the link is given an infinitesimal rotation $\delta\theta$ from its equilibrium position.

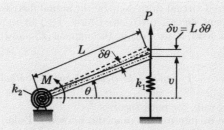

Fig. 3.4.2 A rigid link in equilibrium.

Solution: The virtual work done by actual external forces M and P in moving through the virtual displacements $\delta\theta$ and δv, respectively, is given by

$$\delta W_E = -(M\delta\theta + P\delta v) = -(PL + M)\delta\theta. \tag{1}$$

The work done by internal force F_s in the extensional spring and internal moment M_s in the torsional spring is given by

$$\begin{aligned}
\delta W_I &= F_s\,\delta v + M_s\,\delta\theta \\
&= k_1 v\,\delta v + k_2\theta\,\delta\theta \\
&\approx \left(k_1 L^2 + k_2\right)\theta\,\delta\theta.
\end{aligned} \tag{2}$$

Example 3.4.2

Consider the cantilever beam shown in Fig. 3.4.3. Assume virtual displacements $\delta u(x)$ and $\delta w(x)$ in the Bernoulli–Euler beam theory to be such that

$$\delta w(0) = 0, \quad \left.\frac{d\delta w}{dx}\right|_{x=0} = 0, \quad \delta u(0) = 0, \tag{1}$$

and virtual displacements $\delta u(x)$, $\delta w(x)$, and $\delta\phi_x$ in the Timoshenko beam theory as

$$\delta w(0) = 0, \quad \delta\phi_x(0) = 0, \quad \delta u(0) = 0. \tag{2}$$

Determine the external and internal virtual works done by actual forces in moving through virtual displacements for both the theories. Express the results in terms of the usual stress resultants, applied external loads, and virtual displacements.

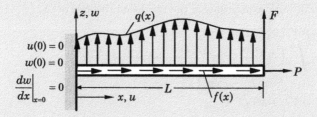

Fig. 3.4.3 A cantilever beam subjected to distributed and point loads.

Solution: The total external virtual work done by the actual distributed force $q(x)$ in moving through virtual displacement $\delta w(x)$, distributed force $f(x)$ in moving through the virtual displacement $\delta u(x)$, point load F in moving through $\delta w(L)$, and point load P in moving through $\delta u(L)$ is given by

$$\delta W_E = - \left[\int_0^L (q\delta w + f\delta u)\, dx + F\delta w(L) + P\delta u(L) \right]. \tag{3}$$

The expression for δW_E is valid for both theories.

The virtual work done by internal forces in the Bernoulli–Euler beam theory is

$$\delta W_I^B = \int_0^L \int_A \sigma_{xx} \delta\varepsilon_{xx}\, dA\,dx, \tag{4}$$

where the virtual strain is

$$\delta\varepsilon_{xx}^B = \frac{d\delta u}{dx} - z\frac{d^2\delta w}{dx^2}. \tag{5}$$

The internal virtual work done in the Bernoulli–Euler beam theory is

$$\delta W_I^B = \int_0^L \int_A \sigma_{xx} \left(\frac{d\delta u}{dx} - z\frac{d^2\delta w}{dx^2} \right) dA\,dx$$

$$= \int_0^L \left(N^B\frac{d\delta u}{dx} - M^B\frac{d^2\delta w}{dx^2} \right) dx, \tag{6}$$

where the stress resultants (N^B, M^B) are

$$N^B = \int_A \sigma_{xx}\, dA, \quad M^B = \int_A \sigma_{xx} z\, dA. \tag{7}$$

In the Timoshenko beam theory, the internal virtual work is given by

$$\delta W_I^T = \int_0^L \int_A (\sigma_{xx}\,\delta\varepsilon_{xx} + 2K_s\sigma_{xz}\,\delta\varepsilon_{xz})\, dA\,dx, \tag{8}$$

where the virtual strains are

$$\delta\varepsilon_{xx}^T = \frac{d\delta u}{dx} + z\frac{d\delta\phi_x}{dx}, \quad 2\delta\varepsilon_{xz}^T = \delta\phi_x + \frac{d\delta w}{dx}. \tag{9}$$

Hence, the expression for the virtual work takes the form

$$\delta W_I^T = \int_0^L \int_A \left[\sigma_{xx} \left(\frac{d\delta u}{dx} + z\frac{d\delta\phi_x}{dx} \right) + K_s\sigma_{xz} \left(\delta\phi_x + \frac{d\delta w}{dx} \right) \right] dA\,dx$$

$$= \int_0^L \left[N^T\frac{d\delta u}{dx} + M^T\frac{d\delta\phi_x}{dx} + V^T \left(\delta\phi_x + \frac{d\delta w}{dx} \right) \right] dx, \tag{10}$$

where

$$N^T = \int_A \sigma_{xx}\, dA, \quad M^T = \int_A \sigma_{xx} z\, dA, \quad V^T = K_s \int_A \sigma_{xz}\, dA. \tag{11}$$

The total virtual work expressions in the two beam theories are

$$\delta W^B = \delta W_I^B + \delta W_E^B$$

$$= \int_0^L \left(N^B\frac{d\delta u}{dx} - M^B\frac{d^2\delta w}{dx^2} - q\delta w - f\delta u \right) dx$$

$$\quad - F\delta w(L) - P\delta u(L), \tag{12}$$

$$\delta W^T = \delta W_I^T + \delta W_E^T$$

$$= \int_0^L \left[N^T\frac{d\delta u}{dx} + M^T\frac{d\delta\phi_x}{dx} + V^T \left(\delta\phi_x + \frac{d\delta w}{dx} \right) - q\delta w - f\delta u \right] dx$$

$$\quad - F\delta w(L) - P\delta u(L). \tag{13}$$

Note that no constitutive law is used in arriving at the results in Eqs (12) and (13), but it is understood that N and M are known in terms of the kinematic variables u, w, and ϕ_x. If we assume linear elastic behavior $\sigma_{xx} = E\varepsilon_{xx}$ and $\sigma_{xz} = G\gamma_{xz}$ with E and (hence G) being independent of the thickness coordinate z, then

$$N^{\mathrm{B}} = \int_A \sigma_{xx}\, dA = \int_A E\left(\frac{du}{dx} - z\frac{d^2w}{dx^2}\right) dx = EA\frac{du}{dx}, \tag{14}$$

$$M^{\mathrm{B}} = \int_A \sigma_{xx} z\, dA = \int_A E\left(\frac{du}{dx} - z\frac{d^2w}{dx^2}\right) z\, dx = -EI\frac{d^2w}{dx^2} \tag{15}$$

for the Bernoulli–Euler beam theory and

$$N^{\mathrm{T}} = \int_A \sigma_{xx}\, dA = \int_A E\left(\frac{du}{dx} + z\frac{d\phi_x}{dx}\right) dx = EA\frac{du}{dx}, \tag{16}$$

$$M^{\mathrm{T}} = \int_A \sigma_{xx} z\, dA = \int_A E\left(\frac{du}{dx} + z\frac{d\phi_x}{dx}\right) z\, dx = EI\frac{d\phi_x}{dx}, \tag{17}$$

$$V^{\mathrm{T}} = K_s \int_A \sigma_{xz}\, dA = K_s GA\left(\phi_x + \frac{dw}{dx}\right) \tag{18}$$

for the Timoshenko beam theory. Note that u is decoupled from the bending variables (w, ϕ_x) because of the fact that the x-axis is taken along the geometric centroid of the beam (so that $\int_A z\, dA = 0$). When E is a function of z (like in a through-thickness functionally graded beam), Eqs (14)–(17) will exhibit coupling between u and w and u and ϕ_x even for linearized beam theories.

In this case the expressions for the total virtual work done, which are denoted here by $\delta\Pi$ to distinguish it from δW (which is independent of the constitutive relation), for the Bernoulli–Euler beam theory (B) and Timoshenko beam theory (T) are

$$\delta\Pi^{\mathrm{B}} = \int_0^L \left(EA\frac{du}{dx}\frac{d\delta u}{dx} + EI\frac{d^2w}{dx^2}\frac{d^2\delta w}{dx^2} - q\delta w - f\delta u\right) dx$$
$$\qquad - F\delta w(L) - P\delta u(L), \tag{19}$$

$$\delta\Pi^{\mathrm{T}} = \int_0^L \left[EA\frac{du}{dx}\frac{d\delta u}{dx} + EI\frac{d\phi_x}{dx}\frac{d\delta\phi_x}{dx} + GAK_s\left(\delta\phi_x + \frac{d\delta w}{dx}\right)\left(\phi_x + \frac{dw}{dx}\right) - q\delta w - f\delta u\right] dx$$
$$\qquad - F\delta w(L) - P\delta u(L), \tag{20}$$

where K_s is the shear correction coefficient introduced in writing the energy expression for the Timoshenko beam theory.

3.4.2   Virtual Forces

Analogous to virtual displacements, we can also think of virtual forces. If a body is imagined to be subjected to a set of *self-equilibrating force system* $\delta\mathbf{F}$, the virtual work done by the virtual forces in moving through actual displacements $\mathbf{u}$ is given by

$$\delta W^* = \delta\mathbf{F} \cdot \mathbf{u},$$

and it is called *complementary virtual work*. The complementary internal virtual work and complementary external virtual work for a deformable body can be expressed, following the ideas already presented, as

$$\delta W_E^* = -\left(\int_\Omega \delta \mathbf{f} \cdot \mathbf{u} \, d\Omega + \int_\Gamma \delta \mathbf{t} \cdot \mathbf{u} \, d\Gamma \right), \tag{3.4.7}$$

$$\delta W_I^* = \int_\Omega \delta U_0^* \, d\Omega = \int_\Omega \boldsymbol{\varepsilon} : \delta \boldsymbol{\sigma} \, d\Omega, \tag{3.4.8}$$

where U_0^* is the complementary strain energy density

$$\delta U_0^* = \int_0^\sigma \boldsymbol{\varepsilon} : d(\delta \boldsymbol{\sigma}) = \boldsymbol{\varepsilon} : \delta \boldsymbol{\sigma} = \varepsilon_{ij} \, \delta \sigma_{ij}. \tag{3.4.9}$$

Note that in selecting a virtual force system, one must make sure that the the virtual forces are in equilibrium among themselves.

Example 3.4.3

Consider the cantilever beam of **Example 3.4.2** [see Fig. 3.4.4(a)]. Define a self-equilibrating virtual force system, and write the complementary external and internal virtual works done by the virtual force system in moving through the actual displacements. Assume that the strains in the beam are of the form

$$\varepsilon_{xx} = \varepsilon_{xx}^{(0)}(x) + z\varepsilon_{xx}^{(1)}(x), \quad \varepsilon_{xz} = \varepsilon_{xz}^{(0)}(x). \tag{1}$$

Express the result in terms of the stress resultants only by assuming the following linear strain–stress relations:

$$\varepsilon_{xx}^{(0)} = \frac{N}{EA}, \quad \varepsilon_{xx}^{(1)} = \frac{M}{EI}, \quad 2\varepsilon_{xz}^{(0)} = \frac{V}{K_s GA}, \tag{2}$$

where K_s is the shear correction coefficient.

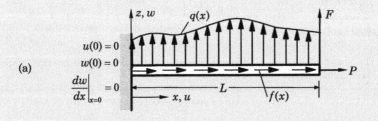

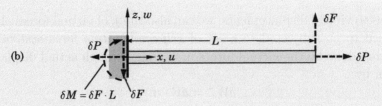

Fig. 3.4.4 (a) A cantilever beam subjected to distributed and point loads. (b) A virtual force system that is in self-equilibrium.

Solution: A self-equilibrating virtual force system is shown in Fig. 3.4.4(b). The complementary virtual work done by the external virtual forces is

$$\delta W_E^* = -\left[\delta F\, w(L) + \delta P\, u(L) - \delta F\, w(0) + (\delta F \cdot L)\left(\frac{dw}{dx}\right)_{x=0} - \delta P\, u(0)\right]$$

$$= -\left[\delta F\, w(L) + \delta P\, u(L)\right]. \tag{3}$$

where we used the fact that the work done by the reactions due to the virtual forces is zero because the corresponding actual displacements are zero.

Then the complementary virtual internal (strain) energy is

$$\delta W_I^* = \int_\Omega \left(\varepsilon_{xx}\delta\sigma_{xx} + 2\varepsilon_{xz}\delta\sigma_{xz}\right) d\Omega$$

$$= \int_0^L \int_A \left[\left(\varepsilon_{xx}^{(0)} + z\varepsilon_{xx}^{(1)}\right)\delta\sigma_{xx} + 2\varepsilon_{xz}^{(0)}\delta\sigma_{xz}\right] dA\,dx$$

$$= \int_0^L \left(\varepsilon_{xx}^{(0)}\,\delta N + \varepsilon_{xx}^{(1)}\,\delta M + 2\varepsilon_{xz}^{(0)}\,\delta V\right) dx. \tag{4}$$

The total complementary virtual work done is

$$\delta W^* = \int_0^L \left(\varepsilon_{xx}^{(0)}\,\delta N + \varepsilon_{xx}^{(1)}\delta M + 2\varepsilon_{xz}^{(1)}\delta V\right) dx - \left[\delta F w(L) + \delta P u(L)\right]. \tag{5}$$

If we use linear constitutive relations in Eq. (2), we obtain

$$\delta\Pi^* = \int_0^L \left(\frac{N}{EA}\delta N + \frac{M}{EI}\delta M + \frac{V}{K_sGA}\delta V\right) dx - \left[\delta F w(L) + \delta P u(L)\right]. \tag{6}$$

Equation (6) is valid for both Bernoulli–Euler beam theory and the Timoshenko beam theory. All three generalized forces (N, M, V) are computed from the equilibrium of the structure under consideration.

Example 3.4.4

Consider the frame structure of Fig. 3.2.9. Compute the total complementary virtual work done, $\delta W^* = \delta W_I^* + \delta W_E^*$. Assume linear elastic constitutive relation to write the total virtual complementary strain energy in terms of the stress resultants.

Solution: From **Example 3.4.3**, we have

$$\delta W_I^* = \delta U^* = \int_0^L \left[\frac{N}{EA}\delta N + \frac{M}{EI}\delta M + \frac{V}{K_sGA}\delta V\right] dx, \tag{1}$$

where (N, M, V) are the axial force, bending moment, and shear force due to actual internal forces, and $(\delta N, \delta M, \delta V)$ are the virtual axial force, virtual moment, and virtual transverse force in the structure. The stress resultants (N, M, V) in the two segments of the structure can be expressed in terms of the applied external forces, P_v and P_h, as

$$N_{DB} = P_h, \quad M_{DB} = P_v \cdot x, \quad V_{DB} = P_v,$$
$$N_{BA} = P_v, \quad M_{BA} = P_v \cdot b + P_h \cdot x \quad V_{BA} = P_h. \tag{2}$$

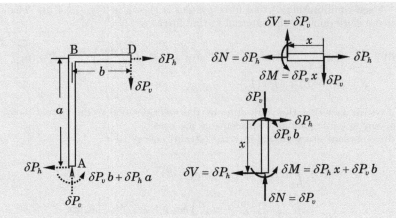

Fig. 3.4.5 A frame structure with virtual forces.

If we apply self-equilibrating virtual forces δP_v and δP_h at point D as shown in Fig. 3.4.5, the axial force, shear force, and bending moment expressions due to the *virtual* forces δP_v and δP_h for the two parts of the beam are

$$\delta N_{\text{DB}} = \delta P_h, \quad \delta N_{\text{BA}} = \delta P_v, \quad \delta V_{\text{DB}} = \delta P_v, \quad \delta V_{\text{BA}} = \delta P_h, \tag{3}$$
$$\delta M_{\text{DB}} = \delta P_v \cdot x, \quad \delta M_{\text{BA}} = \delta P_v \cdot b + \delta P_h \cdot x. \tag{4}$$

In general, the virtual forces need not be the same as the actual forces.

The virtual complementary strain energies due to virtual forces δP_v and δP_h are

$$
\begin{aligned}
\delta U^*_{\text{DB}} &= \int_0^b \left[\frac{N_{\text{DB}}}{EA} \delta N_{\text{DB}} + \frac{M_{\text{DB}}}{EI} \delta M_{\text{DB}} + \frac{V_{DB}}{K_s GA} \delta V_{\text{DB}} \right] dx \\
&= \frac{P_h b}{EA} \delta P_h + \left(\frac{P_v b^3}{3EI} + \frac{P_v b}{K_s GA} \right) \delta P_v,
\end{aligned}
\tag{5}
$$

$$
\begin{aligned}
\delta U^*_{\text{BA}} &= \int_0^b \left[\frac{N_{\text{BA}}}{EA} \delta N_{\text{BA}} + \frac{M_{\text{BA}}}{EI} \delta M_{\text{BA}} + \frac{V_{BA}}{K_s GA} \delta V_{\text{BA}} \right] dx \\
&= \int_0^a \left[\frac{P_v}{EA} \delta P_v + \frac{1}{EI} (P_v b + P_h x)(\delta P_v b + \delta P_h x) + \frac{P_h}{K_s GA} \delta P_h \right] dx \\
&= \left[\frac{P_v a}{EA} + \frac{1}{EI} \left(P_v b^2 a + P_h b \frac{a^2}{2} \right) \right] \delta P_v \\
&\quad + \left[\frac{1}{EI} \left(P_v b \frac{a^2}{2} + P_h \frac{a^3}{3} \right) + \frac{P_h a}{K_s GA} \right] \delta P_h.
\end{aligned}
\tag{6}
$$

The total virtual complementary strain energy of the beam is the sum of δU^*_{DB} and δU^*_{BA}:

$$\delta W^*_I = \delta U^*_{\text{DB}} + \delta U^*_{\text{BA}}. \tag{7}$$

The complementary virtual work done by external virtual forces is

$$\delta W^*_E = -\left(v\, \delta P_v + u\, \delta P_h \right), \tag{8}$$

where u and v are the horizontal and vertical displacements in the direction of the virtual forces at point D.

3.5 Calculus of Variations

3.5.1 The Variational Operator

The delta operator δ used in conjunction with virtual quantities has special importance in variational methods. The operator is called the *variational operator* because it is used to denote a variation (or change) in a given quantity. In this section, we discuss certain operational properties of δ and elements of variational calculus. With these tools in hand, we can study energy and variational principles of solid and structural mechanics.

Let $\mathbf{u}(\mathbf{x})$ be the true displacement (i.e., the one corresponding to equilibrium) at a point $\mathbf{x}$ in the deformed configuration of a body Ω, and suppose that $\mathbf{u} = \hat{\mathbf{u}}$ on the boundary Γ_u of the total boundary Γ. Then an admissible configuration is of the form

$$\bar{\mathbf{u}} = \mathbf{u} + \alpha\mathbf{v} \tag{3.5.1}$$

everywhere in the body, where $\mathbf{v}$ is an arbitrary function that satisfies the homogeneous geometric boundary condition of the system:

$$\mathbf{v} = \mathbf{0} \text{ on } \Gamma_u. \tag{3.5.2}$$

Here $\alpha\mathbf{v}$ represents a variation of the actual displacement vector $\mathbf{u}$. It should be understood that *the variations are small enough* (i.e., α is small) *not to disturb the equilibrium of the system, and the variation is consistent with the geometric constraint of the system.* Equation (3.5.1) defines a set of varied configurations; an infinite number of configurations $\bar{\mathbf{u}}$ can be generated for a fixed $\mathbf{v}$ by assigning values to α. All of these configurations satisfy the specified geometric boundary conditions on boundary Γ_u, and therefore they constitute the set of admissible configurations for the system. For any $\mathbf{v}$, all configurations reduce to the actual one when α is set to zero. Therefore, for any *fixed* point $\mathbf{x}$, $\alpha\mathbf{v}$ can be viewed as a change or *variation* in the actual displacement $\mathbf{u}$. This variation is often denoted by $\delta\mathbf{u}$:

$$\delta\mathbf{u} = \alpha\mathbf{v}. \tag{3.5.3}$$

and $\delta\mathbf{u}$ is called the *first variation* of $\mathbf{u}$. We can show that a differential operator ∇ and the variational operator δ can be interchanged without changing the value:

$$\delta(\nabla\mathbf{u}) = \alpha(\nabla\mathbf{v}) = \nabla(\alpha\mathbf{v}) = \nabla(\delta\mathbf{u}). \tag{3.5.4}$$

For simplicity of discussion, consider a function of the dependent variable $u(x)$ and its derivative $u' \equiv du/dx$ in a one-dimensional problem:

$$F = F(x, u, u'). \tag{3.5.5}$$

For a fixed x, the change in F associated with a variation in u (and hence u') is

$$\Delta F = F(x, u + \alpha v, u' + \alpha v') - F(x, u, u')$$

$$= F(x, u, u') + \frac{\partial F}{\partial u}\alpha v + \frac{\partial F}{\partial u'}\alpha v' + \frac{(\alpha v)^2}{2!}\frac{\partial^2 F}{\partial u^2}$$

$$+ \frac{2(\alpha v)(\alpha v')}{2!}\frac{\partial^2 F}{\partial u \partial u'} + \cdots - F(x, u, u')$$

$$= \frac{\partial F}{\partial u}\alpha v + \frac{\partial F}{\partial u'}\alpha v' + O(\alpha^2), \tag{3.5.6}$$

where $O(\alpha^2)$ denotes terms of order α^2 and higher. The first total variation of $F(x, u, u')$ is defined by

$$\delta F \equiv \alpha \left[\lim_{\alpha \to 0} \frac{\Delta F}{\alpha} \right]$$

$$= \alpha \left(\frac{\partial F}{\partial u}v + \frac{\partial F}{\partial u'}v' \right)$$

$$= \frac{\partial F}{\partial u}\alpha v + \frac{\partial F}{\partial u'}\alpha v'$$

$$= \frac{\partial F}{\partial u}\delta u + \frac{\partial F}{\partial u'}\delta u'. \tag{3.5.7}$$

Alternatively, the first variation may be defined as

$$\delta F = \alpha \left[\frac{dF(u + \alpha v, u' + \alpha v')}{d\alpha} \right]_{\alpha=0}$$

$$= \alpha \left[\frac{\partial F}{\partial(u + \alpha v)}\frac{\partial(u + \alpha v)}{\partial \alpha} + \frac{\partial F}{\partial(u' + \alpha v')}\frac{\partial(u' + \alpha v')}{\partial \alpha} \right]_{\alpha=0}$$

$$= \frac{\partial F}{\partial u}\alpha v + \frac{\partial F}{\partial u'}\alpha v' = \frac{\partial F}{\partial u}\delta u + \frac{\partial F}{\partial u'}\delta u'. \tag{3.5.8}$$

There is an analogy between the first variation δF of F and the total differential dF of F. The total differential of F is

$$dF = \frac{\partial F}{\partial x}dx + \frac{\partial F}{\partial u}du + \frac{\partial F}{\partial u'}du'. \tag{3.5.9}$$

Since x is fixed during the variation of u to $u + \delta u$, we have $dx = 0$, and the analogy between δF in Eq. (3.5.8) and dF in Eq. (3.5.9) becomes apparent. That is, the variational operator δ acts like a differential operator with respect to the dependent variables. Indeed, the laws of variation of sums, products, ratios, powers, and so forth are completely analogous to the corresponding laws of differentiation; that is, the variational calculus resembles the differential calculus. For example, if $F_1 = F_1(u)$ and $F_2 = F_2(u)$, we have

$$
\begin{array}{ll}
(1) & \delta(F_1 \pm F_2) = \delta F_1 \pm \delta F_2. \\[4pt]
(2) & \delta(F_1\, F_2) = \delta F_1\, F_2 + F_1 \delta F_2. \\[4pt]
(3) & \delta\left(\dfrac{F_1}{F_2}\right) = \dfrac{\delta F_1\, F_2 - F_1\, \delta F_2}{F_2^2}. \\[4pt]
(4) & \delta\left[(F_1)^n\right] = n(F_1)^{n-1}\, \delta F_1.
\end{array}
\qquad (3.5.10)
$$

If $G = G(u, v, w)$ is function of several dependent variables u, v, and w (and possibly their derivatives), the total variation of G is the sum of partial variations with respect to u, v, and w:

$$
\delta G = \delta_u G + \delta_v G + \delta_w G, \qquad (3.5.11)
$$

where, for example, δ_u denotes the partial variation with respect to u. The variational operator can be interchanged with differential (as already shown) and integral operators:

$$
\begin{array}{ll}
(1) & \delta\left(\dfrac{du}{dx}\right) = \alpha \dfrac{dv}{dx} = \dfrac{d}{dx}(\alpha v) = \dfrac{d}{dx}(\delta u). \\[10pt]
(2) & \delta\left(\displaystyle\int_0^a u\, dx\right) = \alpha \displaystyle\int_0^a v\, dx = \displaystyle\int_0^a \alpha v\, dx = \displaystyle\int_0^a \delta u\, dx.
\end{array}
\qquad (3.5.12)
$$

All of the discussion can be extended to functions F that depend on more than one dependent variable in two or three dimensions. First, consider a function of two dependent variables u and v and their derivatives in two dimensions:

$$
F = F(x, y, u, v, u_x, v_x, u_y, v_y),
$$

where $u_x = \partial u / \partial x$, $u_y = \partial u / \partial y$, and so on. The first variation of F is given by

$$
\delta F = \frac{\partial F}{\partial u} \delta u + \frac{\partial F}{\partial v} \delta v + \frac{\partial F}{\partial u_x} \delta u_x + \frac{\partial F}{\partial v_x} \delta v_x + \frac{\partial F}{\partial u_y} \delta u_y + \frac{\partial F}{\partial v_y} \delta v_y \qquad (3.5.13)
$$

and

$$
\begin{aligned}
\delta\left[F_1(u, v) F_2(u, v)\right] &= \delta F_1\, F_2 + F_1\, \delta F_2 \\
&= \left(\frac{\partial F_1}{\partial u} \delta u + \frac{\partial F_1}{\partial v} \delta v\right) F_2 + F_1 \left(\frac{\partial F_2}{\partial u} \delta u + \frac{\partial F_2}{\partial v} \delta v\right).
\end{aligned}
$$

Example 3.5.1 ————————————————————————————

Consider the following functions of dependent variables:

$$(1) \quad F(u, u', u'') = c_1 u^2 + c_2 \left(\frac{du}{dx}\right)^2 + c_3 \left(\frac{d^2 u}{dx^2}\right)^2 + c_4 \, u \frac{du}{dx},$$

$$(2) \quad G(u, v, u', v') = c_1 u^2 + c_2 v^2 + c_3 uv + c_4 \left(\frac{du}{dx}\right)^2 + c_5 \left(\frac{dv}{dx}\right)^2 + c_6 \frac{du}{dx}\frac{dv}{dx},$$

where c_i denote functions of x only. Compute their first variations.

Solution: The first variations are

$$(1) \quad \delta F = \frac{\partial F}{\partial u}\delta u + \frac{\partial F}{\partial u'}\delta u' + \frac{\partial F}{\partial u''}\delta u''$$

$$= \left(2c_1 u + c_4 \frac{du}{dx}\right)\delta u + \left(2c_2 \frac{du}{dx} + c_4 u\right)\left(\frac{d\delta u}{dx}\right) + 2c_3 \frac{d^2 u}{dx^2}\left(\frac{d^2 \delta u}{dx^2}\right),$$

$$(2) \quad \delta G = \delta_u G + \delta_v G,$$

$$\delta_u G = \frac{\partial G}{\partial u}\delta u + \frac{\partial G}{\partial u'}\delta u'$$

$$= (2c_1 u + c_3 v)\delta u + \left(2c_4 \frac{du}{dx} + c_6 \frac{dv}{dx}\right)\left(\frac{d\delta u}{dx}\right),$$

$$\delta_v G = \frac{\partial G}{\partial v}\delta v + \frac{\partial G}{\partial v'}\delta v'$$

$$= (2c_2 v + c_3 u)\delta v + \left(2c_5 \frac{dv}{dx} + c_6 \frac{du}{dx}\right)\left(\frac{d\delta v}{dx}\right).$$

3.5.2 Functionals

In the study of variational formulations of continuum problems, we encounter integrals of functions of dependent variables that are themselves functions of other parameters, such as position, time, and so on. Examples of such integral statements are provided by the strain energy and complementary strain energy expressions of preceding sections (see, in particular, **Example 3.3.1**). Such integral expressions are termed functionals. A formal mathematical definition of a functional requires concepts from functional analysis. Here we attempt to present the formal definition without a detailed review of functional analysis [49], as it is outside the scope of the present study.

A *functional* $\mathcal{F}$ is a mapping (or operator) from a vector space $\mathcal{U}$ into the real number field $\Re$. Thus, if $u \in \mathcal{U}$ (i.e., u is an element of $\mathcal{U}$), then $\mathcal{F}(u)$ is a real number:

$$\mathcal{F} : \mathcal{U} \to \Re. \tag{3.5.14}$$

Note that $\mathcal{F}(\cdot)$ is an operator and $\mathcal{F}(u)$ is a functional. An example of a functional is provided by the integral expression

$$\mathcal{F}(u) = \int_0^L \left[a(x)u(x) + b(x)u'(x) + c(x)u''(x) \right] dx, \quad u' \equiv \frac{du}{dx}, \quad u'' \equiv \frac{d^2u}{dx^2},$$

for all integrable and square-integrable functions $u(x)$ in the interval $(0, L)$ with their first and second derivatives:

$$\int_0^L [u(x)]^2 \, dx < \infty, \quad \int_0^L [u'(x)]^2 \, dx < \infty, \quad \int_0^L [u''(x)]^2 \, dx < \infty.$$

We note that the value of the functional $\mathcal{F}(u)$ depends, for a given set of functions $a(x)$, $b(x)$, and $c(x)$, on the dependent variable u.

A functional is said to be *linear* if

$$\mathcal{F}(\alpha u + \beta v) = \alpha \mathcal{F}(u) + \beta \mathcal{F}(v) \tag{3.5.15}$$

for all real numbers $\alpha, \beta \in \Re$ and dependent variables $u, v \in \mathcal{U}$. A *quadratic functional* $Q(u)$ is a functional that satisfies the relation

$$Q(\alpha u) = \alpha^2 \, Q(u) \tag{3.5.16}$$

for all real numbers $\alpha \in \Re$ and the dependent variable $u \in \mathcal{U}$.

3.5.3   The First Variation of a Functional

The first variation of a functional of u (and possibly its derivatives) can be calculated as follows. Let $I(u)$ denote the integral defined in the interval (a, b):

$$I(u) = \int_a^b F(x, u, u') \, dx, \quad u' = \frac{du}{dx}, \tag{3.5.17}$$

where F is a function, in general, of x, u, and du/dx. When the limits a and b are independent of u, the first variation of the functional $I(u)$ is

$$\delta I(u; \delta u) = \delta \int_a^b F(x, u, u') \, dx = \int_a^b \delta F \, dx = \int_a^b \left(\frac{\partial F}{\partial u} \delta u + \frac{\partial F}{\partial u'} \delta u' \right) dx. \tag{3.5.18}$$

When the limits of integration, a and b, depend on the dependent variable, that is, $a = a(u)$ and $b = b(u)$, the Leibniz rule is used to calculate the first variation:

$$\delta I(u; v) = \alpha \left[\frac{d}{d\alpha} \int_{a(u+\alpha v)}^{b(u+\alpha v)} F(x, u + \alpha v, u' + \alpha v') dx \right]_{\alpha=0}. \tag{3.5.19}$$

Using the Leibniz rule, we obtain

$$\delta I(u; \delta u) = \int_a^b \delta F(x, u, u') dx + F(b, u(b), u'(b)) \delta b - F(a, u(a), u'(a)) \delta a. \tag{3.5.20}$$

Thus the variational operator can be used to compute the variation of any functional. Next we consider several examples of finding variations of functionals.

Example 3.5.2 ――

Determine the first variation of the following functionals:

(1) $\Pi_1(u) = \displaystyle\int_0^L \left[\frac{a}{2} \left(\frac{du}{dx} \right)^2 + \frac{b}{2} u^2 - fu \right] dx,$

(2) $\Pi_2(u) = \displaystyle\int_\Omega \left(\frac{1}{2} \boldsymbol{\nabla} u \cdot \boldsymbol{\nabla} u - fu \right) d\Omega - \int_{\Gamma_2} qu \, d\Gamma,$

(3) $\Pi_3(u, w) = \displaystyle\int_0^L \left[\frac{EA}{2} \left(\frac{du}{dx} \right)^2 + \frac{EI}{2} \left(\frac{d^2 w}{dx^2} \right)^2 + \frac{k}{2} w^2 - fu - qw \right] dx,$

where a, b, k, EA, EI, f, and q are functions of x.
Solution: (1) We have

$$\delta\Pi_1(u; \delta u) = \int_0^L \left(a \frac{du}{dx} \frac{d\delta u}{dx} + bu\delta u - f\delta u \right) dx.$$

(2) In this case, we have

$$\delta\Pi_2(u; \delta u) = \int_\Omega \left(\boldsymbol{\nabla} u \cdot \boldsymbol{\nabla}\delta u - f\,\delta u \right) d\Omega - \int_{\Gamma_2} q\,\delta u \, d\Gamma.$$

(3) We have

$$\delta\Pi_3(u, w; \delta u, \delta w) = \int_0^L \left(EA \frac{du}{dx} \frac{d\delta u}{dx} + EI \frac{d^2 w}{dx^2} \frac{d^2 \delta w}{dx^2} + kw\,\delta w \right.$$
$$\left. - f\,\delta u - q\,\delta w \right) dx.$$

3.5.4   Fundamental Lemma of Variational Calculus

The *fundamental lemma of calculus of variations* is useful in obtaining differential equations from variational principles involving integral statements. The lemma can be stated as follows.
Lemma 3.1: *For any integrable function G, if the statement*

$$\int_a^b G(x) \cdot \eta(x) \, dx = 0 \tag{3.5.21}$$

holds for any arbitrary continuous function $\eta(x)$ for all x in (a, b), then it follows that

$$G(x) = 0 \quad x \in (a, b).$$

Proof: A mathematical proof of the lemma can be found in most books on variational calculus. A simple proof of the lemma is presented here. Since $\eta(x)$ is arbitrary, it can be replaced by G. We have

$$\int_a^b G^2(x)\,dx = 0.$$

Since an integral of a positive function, G^2, is positive, the aforementioned statement holds only if $G = 0$.

A more general statement of the fundamental lemma is as follows. If η is arbitrary in $a < x < b$ and $\eta(a)$ is arbitrary, then the statement

$$\int_a^b G(x)\,\eta(x)\,dx + B(a)\eta(a) = 0 \qquad (3.5.22)$$

implies that

$$G(x) = 0 \ \ a < x < b \ \ \text{and} \ \ B(a) = 0, \qquad (3.5.23)$$

because $\eta(x)$ is independent of $\eta(a)$.

3.5.5 Extremum of a Functional

From elementary calculus it is known that a differentiable function $f(x)$, $a < x < b$, has an extremum (i.e., minimum or maximum) at a point x_0 in the interval (a, b) only if (necessary condition)

$$\frac{df}{dx}\Big|_{x=x_0} = 0. \qquad (3.5.24)$$

The sufficient condition for maximum (or minimum) is that

$$\frac{d^2 f}{dx^2} < 0 \ \ \left(\text{or} \ \frac{d^2 f}{dx^2} > 0\right). \qquad (3.5.25)$$

Similarly, for a differential function $f(x, y)$ in two dimensions, the necessary condition for an extremum at the point (x_0, y_0) is that its total differential be zero at (x_0, y_0) :

$$df \equiv \frac{\partial f}{\partial x}dx + \frac{\partial f}{\partial y}dy = 0 \ \ \text{at} \ x = x_0 \ \text{and} \ y = y_0. \qquad (3.5.26)$$

If x and y are linearly independent (i.e., x and y are not constrained), dx and dy are independently arbitrary, and Eq. (3.5.26) implies that

$$\frac{\partial f}{\partial x} = 0 \ \ \text{and} \ \ \frac{\partial f}{\partial y} = 0 \qquad (3.5.27)$$

at $x = x_0$ and $y = y_0$.

In the study of problems by variational principles and methods, we seek the extremum of integrals of functions, $F = F(x, u(x), u'(x))$, or functionals. Consider a simple, but typical, variational problem. Find a function $u = u(x)$ such that $u(a) = u_a$ and $u(b) = u_b$, and

$$I(u) = \int_a^b F(x, u(x), u'(x)) \, dx \qquad (3.5.28)$$

is a minimum.

In analyzing the problem we are not interested in all functions u, but only in those functions that satisfy the stated boundary (or end) conditions. The set of all such functions is called, for obvious reasons, the *set of competing functions* (or set of admissible functions). We shall denote the set by $\mathcal{C}$. The problem is to seek an element u from $\mathcal{C}$, which renders $I(u)$ a minimum. If $u \in \mathcal{C}$, then $(u + \alpha v) \in \mathcal{C}$ for every v satisfying the conditions $v(a) = v(b) = 0$. The space of all such elements is called the space of admissible variations, as already mentioned. Figure 3.5.1 shows a typical competing function $\bar{u}(x) = u(x) + \alpha v(x)$ and a typical admissible variation $v(x)$.

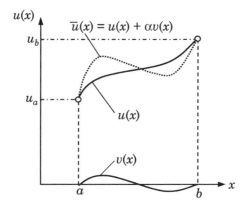

Fig. 3.5.1 The admissible variations of $u(x)$.

Let $I(u)$ be a differentiable functional in the sense that

$$\frac{dI(u + \alpha v, u' + \alpha v')}{d\alpha}$$

exists, and let $\mathcal{C}$ denote the space of competing functions. Then, an element $u \in \mathcal{C}$ is said to yield a *relative minimum (maximum)* for $I(\bar{u})$ in $\mathcal{C}$ if

$$I(\bar{u}) - I(u) \geq 0 \quad (\leq 0). \qquad (3.5.29)$$

If $I(\bar{u})$ assumes a relative minimum (maximum) at $u \in \mathcal{C}$ relative to elements $\bar{u} \in \mathcal{C}$, then it follows from the definition of the space of admissible variations

$\mathcal{H}$ and Eq. (3.5.29) that

$$I(u + \alpha v) - I(u) \geq 0 \quad (\leq 0) \tag{3.5.30}$$

for all $v \in \mathcal{H}, ||v|| < \epsilon$ (a small number), and $\alpha \in \Re$. Since u is the minimizer, any other function $u \in \mathcal{C}$ is of the form $\bar{u} = u + \alpha v$, and the actual minimizer is determined by setting $\alpha = 0$. Once $u(x)$ and $v(x)$ are assigned, $I(\bar{u})$ is a function of α alone, say $\bar{I}(\alpha)$. Now a necessary condition for $I(\bar{u}) = \bar{I}(\alpha)$ to attain a minimum is that

$$\frac{d\bar{I}(\alpha)}{d\alpha} = \frac{d}{d\alpha}[I(u + \alpha v)] = 0. \tag{3.5.31}$$

On the other hand, $I(\bar{u})$ attains its minimum at u, that is, when $\alpha = 0$. These two conditions together imply $(d\bar{I}(\alpha)/d\alpha)|_{\alpha=0} = 0$, which is nothing but

$$\delta I(u) = 0. \tag{3.5.32}$$

Thus, a necessary condition for $I(u)$ to be a minimum at u is that its first variation be zero.

Analogous to the sufficient condition for ordinary functions, the sufficient condition for a functional to assume a relative minimum (maximum) is that the second variation $\delta^2 I(u)$ is greater (less) than zero. The second variation $\delta^2 I(u)$ of a functional $I(u)$ is given by

$$\delta^2 I(u) \equiv \frac{\alpha^2}{2}\left[\frac{d^2}{d\alpha^2} I(u + \alpha v)\right]_{\alpha=0} \tag{3.5.33}$$

for all $v \in \mathcal{H}$ and $\alpha \in \Re$.

3.5.6 The Euler Equations

We now return to the problem of determining the minimum of a functional

$$I(u) = \int_a^b F(x, u, u') \, dx \tag{3.5.34}$$

subject to the end conditions:

$$u(a) = u_a, \quad u(b) = u_b. \tag{3.5.35}$$

It is clear that any candidate for the minimizing functional should satisfy the end conditions in Eq. (3.5.35) and be sufficiently differentiable (twice in the present case, as we shall see shortly). The set of all such functions is the set of admissible functions or competing functions for the present case. Functions from the admissible set can be viewed as smooth (i.e., differentiable twice) functions passing through points (a, u_a) and (b, u_b), as shown in Fig. 3.5.1. Clearly, any element $\bar{u} \in \mathcal{C}$ (the set of competing functions) has the form

$$\bar{u} = u + \alpha v, \tag{3.5.36}$$

where α is a small number and v is a sufficiently differentiable function that satisfies the *homogeneous form* of the end conditions (because $\bar{u}$ must satisfy the specified end conditions) in Eq. (3.5.35):

$$v(a) = v(b) = 0, \tag{3.5.37}$$

and u is the function that minimizes the functional in Eq. (3.5.34). The set of all functions v is the set of admissible variations, $\mathcal{H}$. Now assuming that, for each admissible function $\bar{u}$, $F(x, \bar{u}, \bar{u}')$ exists and is continuously differentiable with respect to its arguments and $I(\bar{u})$ takes one and only one real value, we seek the particular function $u(x)$ that makes the integral a minimum.

The necessary condition for $I(u)$ to attain a minimum is [see Eq. (3.5.32)]

$$
\begin{aligned}
0 &= \frac{dI(u + \alpha v)}{d\alpha}\bigg|_{\alpha=0} = \left[\frac{d}{d\alpha} \int_a^b F(x, \bar{u}, \bar{u}') dx\right]_{\alpha=0} \\
&= \int_a^b \left(\frac{\partial F}{\partial \bar{u}} \frac{\partial \bar{u}}{\partial \alpha} + \frac{\partial F}{\partial \bar{u}'} \frac{\partial \bar{u}'}{\partial \alpha}\right)\bigg|_{\alpha=0} dx = \int_a^b \left(\frac{\partial F}{\partial u} v + \frac{\partial F}{\partial u'} v'\right) dx, \tag{3.5.38}
\end{aligned}
$$

where $\bar{u} = u + \alpha v$. The statement in Eq. (3.5.38) is of no use unless we can deduce some result from the statement. The integrand has two terms, one with v (or δu) and the other with v' (or $\delta u'$), which are not independent of each other. If we can transfer the differentiation from v' to its coefficient, then the two terms in the integrand can be combined into one expression with v multiplying it, and we can use the fundamental lemma of variational calculus to deduce that this expression, because v is arbitrary, is necessarily zero. Integrating the second term in the last equation by parts to transfer differentiation from v' to $\partial F/\partial u'$, we obtain[3]

$$0 = \int_a^b v\left[\frac{\partial F}{\partial u} - \frac{d}{dx}\left(\frac{\partial F}{\partial u'}\right)\right] dx + \left(\frac{\partial F}{\partial u'} v\right)\bigg|_a^b. \tag{3.5.39}$$

The boundary term vanishes because v is zero at $x = a$ and $x = b$ [see Eq. (3.5.37)]. The fact that v is arbitrary inside the internal (a, b) and yet Eq. (3.5.39) hold implies, by the fundamental lemma of the calculus of variations, the expression in the square brackets is zero identically:

$$\frac{\partial F}{\partial u} - \frac{d}{dx}\left(\frac{\partial F}{\partial u'}\right) = 0 \text{ in } a < x < b. \tag{3.5.40}$$

---

[3]The integration by parts involves recognizing that

$$\frac{d}{dx}\left[G(x, u, u')v\right] = G'v + Gv' \text{ or } Gv' = (Gv)' - G'v \text{ with } G = \frac{\partial F}{\partial u'}.$$

Equation (3.5.40) is called the *Euler equation* of the functional in Eq. (3.5.34). Of all the admissible functions, the one that satisfies (i.e., the solution of) Eq. (3.5.40) is the true minimizer of the functional $I(u)$.

Next consider the problem of finding (u, v), defined on a two-dimensional region Ω, such that the following functional is to be minimized:

$$I(u, v) = \int_\Omega F(x, y, u, v, u_x, v_x, u_y, v_y) \, dxdy, \qquad (3.5.41)$$

where $u_x = \partial u/\partial x$, $u_y = \partial u/\partial y$, and so on. For the moment we assume that u and v are specified on the boundary Γ of Ω. The vanishing of the first variation of $I(u, v)$ is written as

$$\delta I(u, v) = \delta_u I(u, v) + \delta_v I(u, v) = 0.$$

Here δ_u and δ_v denote partial variations with respect to u and v, respectively. We have

$$\delta I(u, v; \delta u, \delta v) = \int_\Omega \left(\frac{\partial F}{\partial u} \delta u + \frac{\partial F}{\partial u_x} \delta u_x + \frac{\partial F}{\partial u_y} \delta u_y + \frac{\partial F}{\partial v} \delta v + \frac{\partial F}{\partial v_x} \delta v_x + \frac{\partial F}{\partial v_y} \delta v_y \right) dxdy. \qquad (3.5.42)$$

The next step in the development involves the use of integration by parts or the gradient theorem on the second, third, fifth, and sixth terms in Eq. (3.5.42). Consider the second term. We have

$$\int_\Omega \frac{\partial F}{\partial u_x} \frac{\partial \delta u}{\partial x} \, dxdy = \int_\Omega \left[\frac{\partial}{\partial x} \left(\frac{\partial F}{\partial u_x} \delta u \right) - \frac{\partial}{\partial x} \left(\frac{\partial F}{\partial u_x} \right) \delta u \right] dxdy$$

$$= \oint_\Gamma \frac{\partial F}{\partial u_x} \delta u \, n_x \, d\Gamma - \int_\Omega \frac{\partial}{\partial x} \left(\frac{\partial F}{\partial u_x} \right) \delta u \, dxdy. \qquad (3.5.43)$$

Using a similar procedure on the other terms and collecting the coefficients of δu and δv separately, we obtain

$$0 = \int_\Omega \left\{ \left[\frac{\partial F}{\partial u} - \frac{\partial}{\partial x} \left(\frac{\partial F}{\partial u_x} \right) - \frac{\partial}{\partial y} \left(\frac{\partial F}{\partial u_y} \right) \right] \delta u \right.$$

$$+ \left. \left[\frac{\partial F}{\partial v} - \frac{\partial}{\partial x} \left(\frac{\partial F}{\partial v_x} \right) - \frac{\partial}{\partial y} \left(\frac{\partial F}{\partial v_y} \right) \right] \delta v \right\} dxdy$$

$$+ \oint_\Gamma \left[\left(\frac{\partial F}{\partial u_x} n_x + \frac{\partial F}{\partial u_y} n_y \right) \delta u + \left(\frac{\partial F}{\partial v_x} n_x + \frac{\partial F}{\partial v_y} n_y \right) \delta v \right] d\Gamma. \qquad (3.5.44)$$

Since (u, v) are specified on Γ, $\delta u = \delta v = 0$ and the boundary expressions in Eq. (3.5.44) vanish. Then, since δu and δv are arbitrary and independent of each other in Ω, the fundamental lemma yields the Euler equations

$$\delta u: \qquad \frac{\partial F}{\partial u} - \frac{\partial}{\partial x} \left(\frac{\partial F}{\partial u_x} \right) - \frac{\partial}{\partial y} \left(\frac{\partial F}{\partial u_y} \right) = 0, \qquad (3.5.45)$$

$$\delta v: \qquad \frac{\partial F}{\partial v} - \frac{\partial}{\partial x} \left(\frac{\partial F}{\partial v_x} \right) - \frac{\partial}{\partial y} \left(\frac{\partial F}{\partial v_y} \right) = 0. \qquad (3.5.46)$$

All of the foregoing discussion can be applied to the general case of a functional involving p dependent variables with mth-order partial derivatives with respect to n independent variables. In this case there will be p number of Euler equations involving $2m$th-order derivatives in the n independent variables.

3.5.7 Natural and Essential Boundary Conditions

First consider the problem of minimizing the functional in Eq. (3.5.34) subject to no end conditions (hence $\delta v \neq 0$ at $x = a$ and $x = b$). The necessary condition for I to attain minimum yields, as before, the result in Eq. (3.5.39). Now suppose that $\frac{\partial F}{\partial u'}$ and v are selected such that

$$\frac{\partial F}{\partial u'} v = 0 \text{ for } x = a \text{ and } x = b. \tag{3.5.47}$$

Then using the fundamental lemma of the calculus of variations, we obtain the Euler equation in Eq. (3.5.40).

Equation (3.5.47) is satisfied identically for any of the following combinations:

$$
\begin{aligned}
&(1) &&v(a) = 0, \quad v(b) = 0. \\
&(2) &&v(a) = 0, \quad \frac{\partial F}{\partial u'}(b) = 0. \\
&(3) &&\frac{\partial F}{\partial u'}(a) = 0, \quad v(b) = 0. \\
&(4) &&\frac{\partial F}{\partial u'}(a) = 0, \quad \frac{\partial F}{\partial u'}(b) = 0.
\end{aligned}
\tag{3.5.48}
$$

The requirement that $v = 0$ (or $\delta u = 0$) at an end point is equivalent to the requirement that u be specified (to be some value) at that point. The end conditions in Eq. (3.5.48) are classified into two types: *essential boundary conditions*, which require u and possibly its derivatives to vanish at the boundary, and *natural boundary conditions*, which require the specification of the coefficients of v and its derivatives. Thus we have the following types of boundary conditions:

Essential boundary conditions:

$$\text{Specify } v = 0 \text{ or } u = \hat{u} \text{ on the boundary.} \tag{3.5.49}$$

Natural boundary conditions:

$$\frac{\partial F}{\partial u'} = 0 \text{ on the boundary.} \tag{3.5.50}$$

In a given problem only one of the four combinations given in Eq. (3.5.48) can be specified. Problems in which all of the boundary conditions are of essential type are called *Dirichlet boundary-value problems*, and those in which all of the boundary conditions are of natural type are called *Neumann boundary-value problems. Mixed boundary-value problems* are those in which both essential and natural boundary conditions are specified. Essential boundary conditions are also known as *Dirichlet* or *geometric* boundary conditions, and natural boundary conditions are known as *Neumann* or *force* boundary conditions.

As a general rule, the vanishing of the variation v (or δu) – equivalently, the specification of u – on the boundary constitutes the essential boundary condition, and vanishing of the coefficient of the variation constitutes the natural boundary condition. This rule applies to any functional in one, two, and three dimensions and integrands that are functions of one or more dependent variables and their derivatives of any order.

Next consider the functional in Eq. (3.5.41) involving two dependent variables (u, v), and suppose that (u, v) are arbitrary on the boundary Γ for the moment. It is easy to identify the natural and essential boundary conditions of the problem from Eq. (3.5.44): in each of the pairings on boundary Γ, specifying the first element (which contains no variations of the dependent variables) constitutes the natural boundary condition, and vanishing of the second element (or, equivalently, specifying the quantity in front of the variational operator) constitutes the essential boundary condition. Thus we have either

$$u = \hat{u} \ \text{(specified)} \ \text{ so that } \ \delta u = 0 \quad \text{on} \quad \Gamma \tag{3.5.51}$$

or

$$\frac{\partial F}{\partial u_x} n_x + \frac{\partial F}{\partial u_y} n_y = 0 \quad \text{on} \quad \Gamma \tag{3.5.52}$$

and

$$v = \hat{v} \ \text{(specified)} \ \text{ so that } \ \delta v = 0 \quad \text{on} \quad \Gamma \tag{3.5.53}$$

or

$$\frac{\partial F}{\partial v_x} n_x + \frac{\partial F}{\partial v_y} n_y = 0 \quad \text{on} \quad \Gamma. \tag{3.5.54}$$

Equations (3.5.51) and (3.5.53) represent the essential boundary conditions, and Eqs (3.5.52) and (3.5.54) the natural boundary conditions. The pair of elements (u, v) are called the *primary variables*, and

$$Q_x \equiv \frac{\partial F}{\partial u_x} n_x + \frac{\partial F}{\partial u_y} n_y \ \text{ and } \ Q_y \equiv \frac{\partial F}{\partial v_x} n_x + \frac{\partial F}{\partial v_y} n_y$$

are called the *secondary variables*. Thus, specification of the primary variables constitutes essential boundary conditions, and specification of the secondary variables constitutes natural boundary conditions. In general, one element of each pair (u, Q_x) and (v, Q_y) (but not both elements of the same pair) may be specified at any point of the boundary. Thus there are four possible combinations of natural and essential boundary conditions for the problem under discussion.

Example 3.5.3

Consider a linear elastic bar of length L, area of cross section A, and Young's modulus E. Assume that it is fixed at the left end, supported axially by a linear elastic spring with spring constant k and subjected to distributed axial load $f(x)$ and a point load P at $x = L$, as shown in Fig. 3.5.2. Supposing that the total potential energy is a minimum at equilibrium, derive the equation of equilibrium (i.e., the Euler equation) and natural boundary conditions for the problem.

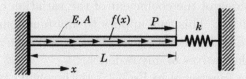

Fig. 3.5.2 A bar fixed at the left end and spring-supported at the right end.

Solution: The total potential energy functional for the stretching of a linear elastic bar is

$$\Pi(u) = \int_0^L \left[\frac{EA}{2} \left(\frac{du}{dx} \right)^2 \right] dx + \frac{k}{2} [u(L)]^2 - \int_0^L fu\, dx - Pu(L), \tag{1}$$

where u denotes the axial displacement of the bar. The first term in $\Pi(u)$ represents the strain energy U stored in the bar, the second term corresponds to the strain energy stored in the spring at $x = L$, and the third and fourth terms denote the external work done V_E on the bar by the distributed load f and point load P, respectively.

In order that Π be a minimum, its first variation must be zero: $\delta\Pi = 0$. The first variation is

$$\delta\Pi(u) = \int_0^L \left(EA \frac{du}{dx} \frac{d\delta u}{dx} - f\delta u \right) dx + ku(L)\, \delta u(L) - P\delta u(L), \tag{2}$$

where δu is arbitrary in $0 < x < L$ and at $x = L$ but satisfies the condition $\delta u(0) = 0$. To use the fundamental lemma of variational calculus, we must relieve δu of any differentiation. Integrating the first term by parts, we obtain

$$\delta\Pi(u; \delta u) = \int_0^L \left[-\frac{d}{dx} \left(EA \frac{du}{dx} \right) - f \right] \delta u\, dx + \left[EA \frac{du}{dx} \delta u \right]_0^L + [ku(L) - P]\, \delta u(L)$$

$$= \int_0^L \delta u \left[-\frac{d}{dx} \left(EA \frac{du}{dx} \right) - f \right] dx + \delta u(L) \left[\left(EA \frac{du}{dx} \right)_{x=L} + ku(L) - P \right]$$

$$- \delta u(0) \left(EA \frac{du}{dx} \right)_{x=0}. \tag{3}$$

The last term is zero because $\delta u(0) = 0$. Also, from the boundary term

$$\delta u(0) \cdot \left(EA \frac{du}{dx} \right)_{x=0},$$

we immediately identify that u is the primary variable and $EA(du/dx)$, which is the axial force, is the secondary variable.

Since $\delta\Pi = 0$, setting the coefficients of δu in $(0, L)$ and δu at $x = L$ from Eq. (3) to zero separately, we obtain the Euler equation and the natural (or force) boundary condition of the problem:

Euler equation:

$$-\frac{d}{dx}\left(EA\frac{du}{dx}\right) - f = 0, \quad 0 < x < L. \tag{4}$$

Natural boundary condition:

$$EA\frac{du}{dx} + ku - P = 0 \text{ at } x = L. \tag{5}$$

Thus, the solution u of Eqs (4) and (5) that satisfies $u(0) = 0$ is the minimizer of the energy functional $\Pi(u)$ in Eq. (1). Equation (4) can be obtained directly from Eq. (3.5.40) by substituting $(u' = du/dx)$

$$F(x, u, u') = \frac{EA}{2}\left(\frac{du}{dx}\right)^2 - fu, \quad \frac{\partial F}{\partial u} = -f, \quad \frac{\partial F}{\partial u'} = EA\frac{du}{dx}. \tag{6}$$

Example 3.5.4

Consider a linear elastic beam of length L, area of cross section A, moment of inertia I, and Young's modulus E. Assume that it is fixed at the left end, supported vertically by a linear elastic spring with spring constant k, and subjected to distributed axial load $f(x)$, transverse load $q(x)$, and generalized forces P, F, and M_0 at $x = L$, as shown in Fig. 3.5.3. Supposing that the total potential energy according to the Timoshenko beam theory, Eq. (3.3.5) with $B_{xx} = 0$, is a minimum at equilibrium, derive the equations of equilibrium and natural boundary conditions associated with the Timoshenko beam theory for the problem.

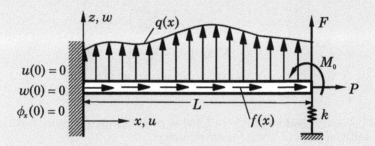

Fig. 3.5.3 A beam fixed at the left end and spring-supported vertically at the right end.

Solution: The total potential energy functional from Eq. (3.3.5) with $B_{xx} = 0$ and accounting for the point loads is

$$\Pi(u, w, \phi_x) = \frac{1}{2}\int_0^L\left[EA\left(\frac{du}{dx}\right)^2 + EI\left(\frac{d\phi_x}{dx}\right)^2 + K_sS_{xz}\left(\phi_x + \frac{dw}{dx}\right)^2\right]dx$$

$$+ \frac{k}{2}[w(L)]^2 - \int_0^L(f\,u + q\,w)\,dx - Pu(L) - Fw(L) - M_0\phi_x(L), \tag{1}$$

where u denotes the axial displacement, w is the transverse displacement, and ϕ_x is the rotation of a transverse normal in the beam. The first variation $\delta\Pi(u, w, \phi_x; \delta u, \delta w, \delta\phi_x)$ is

$$\delta\Pi = \int_0^L\left[EA\frac{du}{dx}\frac{d\delta u}{dx} + EI\frac{d\phi_x}{dx}\frac{d\delta\phi_x}{dx} + K_sS_{xz}\left(\phi_x + \frac{dw}{dx}\right)\left(\delta\phi_x + \frac{d\delta w}{dx}\right)\right]dx$$

$$+ kw(L)\,\delta w(L) - \int_0^L(f\,\delta u + q\,\delta w)\,dx - P\delta u(L) - F\delta w(L) - M_0\delta\phi_x(L). \tag{2}$$

This statement is equivalent to the following statements, obtained by collecting terms involving δu, δw, and $\delta\phi_x$, separately:

$$0 = \int_0^L \left(EA\frac{du}{dx}\frac{d\delta u}{dx} - f\delta u \right) dx - P\delta u(L), \tag{3}$$

$$0 = \int_0^L \left[K_sS_{xz}\left(\phi_x + \frac{dw}{dx}\right)\frac{d\delta w}{dx} - q\delta w \right] dx + kw(L)\,\delta w(L) - F\,\delta w(L), \tag{4}$$

$$0 = \int_0^L \left[EI\frac{d\phi_x}{dx}\frac{d\delta\phi_x}{dx} + K_sS_{xz}\left(\phi_x + \frac{dw}{dx}\right)\delta\phi_x \right] dx - M_0\,\delta\phi_x(L). \tag{5}$$

To obtain the Euler equations (there will be three equations because there are three dependent unknowns), we relieve the varied quantities $(\delta u, \delta w, \delta\phi_x)$ of any derivatives by carrying out integration by parts. We obtain

$$0 = \int_0^L \left[-\frac{d}{dx}\left(EA\frac{du}{dx}\right) - f \right]\delta u\,dx + \left[EA\frac{du}{dx}\delta u \right]_0^L - P\delta u(L), \tag{6}$$

$$0 = \int_0^L \left\{ -\frac{d}{dx}\left[K_sS_{xz}\left(\phi_x + \frac{dw}{dx}\right) \right] - q \right\}\delta w\,dx + \left[K_sS_{xz}\left(\phi_x + \frac{dw}{dx}\right)\delta w \right]_0^L$$
$$+ \left[kw(L) - F\right]\delta w(L), \tag{7}$$

$$0 = \int_0^L \left[-\frac{d}{dx}\left(EI\frac{d\phi_x}{dx}\right) + K_sS_{xz}\left(\phi_x + \frac{dw}{dx}\right) \right]\delta\phi_x\,dx + \left[EI\frac{d\phi_x}{dx}\delta\phi_x \right]_0^L - M_0\,\delta\phi_x(L). \tag{8}$$

Thus, the Euler equations in $(0, L)$ are

$$-\frac{d}{dx}\left(EA\frac{du}{dx}\right) - f = 0, \tag{9}$$

$$-\frac{d}{dx}\left[K_sS_{xz}\left(\phi_x + \frac{dw}{dx}\right) \right] - q = 0, \tag{10}$$

$$-\frac{d}{dx}\left(EI\frac{d\phi_x}{dx}\right) + K_sS_{xz}\left(\phi_x + \frac{dw}{dx}\right) = 0. \tag{11}$$

The natural boundary conditions at $x = L$ follow from Eqs (6)–(8), since $\delta u(0) = \delta w(0) = \delta\phi_x(0) = 0$, and $(\delta u(L), \delta w(L), \delta\phi_x(L))$ are arbitrary:

$$EA\frac{du}{dx} - P = 0, \tag{12}$$

$$K_sS_{xz}\left(\phi_x + \frac{dw}{dx}\right) + kw - F = 0, \tag{13}$$

$$EI\frac{d\phi_x}{dx} - M_0 = 0. \tag{14}$$

Thus, the primary and secondary variables of the theory are (note the pairings):

u axial displacement; $\qquad\qquad EA\dfrac{du}{dx}$, axial force, N,

w vertical displacement; $\qquad\qquad K_sS_{xz}\left(\phi_x + \dfrac{dw}{dx}\right)$, vertical force, V, $\qquad$ (15)

ϕ_x rotation about the y-axis; $\qquad EI\dfrac{d\phi_x}{dx}$, bending moment, M.

3.5.8 Minimization of Functionals with Equality Constraints

In the preceding sections, we devoted our attention to the determination of a function that minimizes a given functional. The necessary condition leads to a differential equation (the Euler equation) and (natural) boundary conditions governing the function. It turns out that it is also a sufficient condition for linear problems, not requiring the evaluation of the second variation of the functional. In the present section, we consider ways to determine the minimum of quadratic functionals (i.e., functionals that involve quadratic terms of the dependent variables) with linear equality constraints.

Consider the problem of finding the minimum of a functional (assumed to be quadratic in u and v),

$$I(u, v) = \int_a^b F(x, u, u', v, v')dx, \qquad (3.5.55)$$

subject to the constraint

$$G(u, u', v, v') = 0. \qquad (3.5.56)$$

In addition u and v should satisfy, as before, certain end conditions (or essential boundary conditions) of the problem.

In the constrained minimization problems, the admissible functions should not only satisfy the specified end conditions and be sufficiently continuous, but they should satisfy the constraint conditions. Furthermore, the admissible variations should be such that the constraint conditions are not violated. Here we consider two separate methods of including the constraints in the (modified) functional.

3.5.8.1 The Lagrange Multiplier Method

The necessary condition for the minimum of $I(u, v)$ in Eq. (3.5.55) is $\delta I = 0$. We have

$$0 = \delta I = \int_a^b \left(\frac{\partial F}{\partial u} \delta u + \frac{\partial F}{\partial u'} \delta u' + \frac{\partial F}{\partial v} \delta v + \frac{\partial F}{\partial v'} \delta v' \right) dx. \qquad (3.5.57)$$

Since u and v must satisfy the constraint condition in Eq. (3.5.56), the variations δu and δv are related by

$$0 = \delta G = \frac{\partial G}{\partial u} \delta u + \frac{\partial G}{\partial u'} \delta u' + \frac{\partial G}{\partial v} \delta v + \frac{\partial G}{\partial v'} \delta v'. \qquad (3.5.58)$$

The Lagrange multiplier method consists of multiplying the right-hand side of Eq. (3.5.58) with an arbitrary parameter λ, integrating over the interval (a, b) and adding the result to Eq. (3.5.57). The multiplier λ is called the *Lagrange multiplier.* We obtain

$$
0 = \int_a^b \left[\frac{\partial F}{\partial u} \delta u + \frac{\partial F}{\partial u'} \delta u' + \frac{\partial F}{\partial v} \delta v + \frac{\partial F}{\partial v'} \delta v' \right.
$$
$$
\left. + \lambda \left(\frac{\partial G}{\partial u} \delta u + \frac{\partial G}{\partial u'} \delta u' + \frac{\partial G}{\partial v} \delta v + \frac{\partial G}{\partial v'} \delta v' \right) \right] dx
$$
$$
= \int_a^b \left\{ \left[\frac{\partial F}{\partial u} + \lambda \frac{\partial G}{\partial u} - \frac{d}{dx} \left(\frac{\partial F}{\partial u'} + \lambda \frac{\partial G}{\partial u'} \right) \right] \delta u \right.
$$
$$
\left. + \left[\frac{\partial F}{\partial v} + \lambda \frac{\partial G}{\partial v} - \frac{d}{dx} \left(\frac{\partial F}{\partial v'} + \lambda \frac{\partial G}{\partial v'} \right) \right] \delta v \right\} dx. \qquad (3.5.59)
$$

The boundary terms vanish because $\delta v(a) = \delta v(b) = \delta u(a) = \delta u(b) = 0$. Since the variations δu and δv are not both independent, we cannot conclude that their coefficients are zero. Suppose the δu is independent and δv is related to δu by Eq. (3.5.58). We choose λ such that the coefficient of δv is zero. Then by the fundamental lemma of variational calculus, it follows that (because δu is arbitrary) the coefficient of δu is also zero. Thus we have

$$
\frac{\partial}{\partial u}(F + \lambda G) - \frac{d}{dx}\left[\frac{\partial}{\partial u'}(F + \lambda G) \right] = 0, \qquad (3.5.60)
$$

$$
\frac{\partial}{\partial v}(F + \lambda G) - \frac{d}{dx}\left[\frac{\partial}{\partial v'}(F + \lambda G) \right] = 0. \qquad (3.5.61)
$$

Equations (3.5.60), (3.5.61), and (3.5.56) furnish three equations for the determination of u, v, and λ. If G also contains u' and v', then G also appears in the second term of Eqs (3.5.60) and (3.5.61).

It is clear from Eq. (3.5.59) that Lagrange's method can be viewed as one of determining u, v, and λ by setting the first variation of the *modified* functional

$$
L(u, v, \lambda) \equiv I(u, v) + \int_a^b \lambda G(u, u', v, v') dx
$$
$$
= \int_a^b (F + \lambda G) dx \qquad (3.5.62)
$$

to zero. The Euler equations of the functional are precisely Eqs (3.5.60), (3.5.61), and (3.5.56). Indeed, we have

$$
0 = \delta L = \int_a^b \delta(F + \lambda G) dx
$$
$$
= \int_a^b (\delta F + \delta \lambda G + \lambda \delta G) dx
$$

$$
= \int_a^b \left[\left(\frac{\partial F}{\partial u} + \lambda \frac{\partial G}{\partial u} \right) \delta u + \left(\frac{\partial F}{\partial u'} + \lambda \frac{\partial G}{\partial u'} \right) \delta u' \right.
$$
$$
\left. + \left(\frac{\partial F}{\partial v} + \lambda \frac{\partial G}{\partial v} \right) \delta v + \left(\frac{\partial F}{\partial v'} + \lambda \frac{\partial G}{\partial v'} \right) \delta v' + \delta \lambda G \right] dx
$$
$$
= \int_a^b \left\{ \left[\frac{\partial F}{\partial u} + \lambda \frac{\partial G}{\partial u} - \frac{d}{dx} \left(\frac{\partial F}{\partial u'} + \lambda \frac{\partial G}{\partial u'} \right) \right] \delta u \right.
$$
$$
\left. + \left[\frac{\partial F}{\partial v} + \lambda \frac{\partial G}{\partial v} - \frac{d}{dx} \left(\frac{\partial F}{\partial v'} + \lambda \frac{\partial G}{\partial v'} \right) \right] \delta v + G \delta \lambda \right\} dx, \qquad (3.5.63)
$$

from which we obtain Eqs (3.5.60), (3.5.61), and (3.5.56) by setting, respectively, the coefficients of δu, δv, and $\delta \lambda$ to zero.

3.5.8.2 The Penalty Function Method

The penalty function method involves the reduction of conditional extremum problems to extremum problems without constraints by the introduction of a penalty function associated with the constraints. As applied to the problem in Eqs (3.5.55) and (3.5.56), the technique involves seeking the minimum of a modified functional obtained by adding a quadratic term associated with the constraint in Eq. (3.5.56) (i.e., the constraint is approximately satisfied in the least-squares sense):

$$
P(u, v) = I(u, v) + \frac{\gamma}{2} \int_a^b \left[G(u, u', v, v') \right]^2 dx, \qquad (3.5.64)
$$

where γ is the *penalty parameter* (a preassigned positive parameter). Setting the first variation of P to zero, we obtain the Euler equations for the functional:

$$
\frac{\partial F}{\partial u} - \frac{d}{dx} \left(\frac{\partial F}{\partial u'} \right) + \gamma \left[G \frac{\partial G}{\partial u} - \frac{d}{dx} \left(G \frac{\partial G}{\partial u'} \right) \right] = 0, \qquad (3.5.65)
$$
$$
\frac{\partial F}{\partial v} - \frac{d}{dx} \left(\frac{\partial F}{\partial v'} \right) + \gamma \left[G \frac{\partial G}{\partial v} - \frac{d}{dx} \left(G \frac{\partial G}{\partial v'} \right) \right] = 0. \qquad (3.5.66)
$$

For successively large values of γ, the solution of Eqs (3.5.65) and (3.5.66) gets closer to the true solution. We note that in the penalty function method, the constraint is satisfied only approximately and that no additional unknowns are introduced into the variational formulation. Further, in the penalty function method an approximation to the Lagrange multiplier can be computed from the solution $(u(\gamma), v(\gamma))$ of Eqs (3.5.65) and (3.5.66) by the formula [compare Eqs (3.5.65) and (3.5.66) with Eqs (3.5.60) and (3.5.61)]:

$$
\lambda_\gamma = \gamma G(u_\gamma, u'_\gamma, v_\gamma, v'_\gamma). \qquad (3.5.67)
$$

Both the Lagrange multiplier method and the penalty function methods can be generalized for functionals with many variables and constraints. Note that when the constraint equation is a relationship involving constants, the Lagrange multiplier is also a constant. For example, the relations

$$c_1 u - c_2 v = c_3, \qquad \int_a^b F(x, u(x), u'(x)) \, dx = c_1,$$

where c_i ($i = 1, 2, 3$) are constants, are constraint equations among numbers (note that a definite integral of a function is a number).

Example 3.5.5

Given the problem of minimizing the functional

$$I(x, y) = \frac{1}{2} \int_{t_1}^{t_2} (x\dot{y} - y\dot{x}) \, dt, \tag{1}$$

subject to the constraint

$$L(x, y) = \int_{t_1}^{t_2} (\dot{x}^2 + \dot{y}^2)^{\frac{1}{2}} \, dt, \tag{2}$$

where t is a parameter, $\dot{x} = dx/dt$, $\dot{y} = dy/dt$, L is a constant, and x and y are dependent variables, determine the Euler equations using the Lagrange multiplier method.

Solution: The constraint condition in Eq. (2) is an algebraic equation in the sense that the integral quantity is a number. Therefore, the Lagrange multiplier, λ, is also a number. Thus, the functional in the Lagrange multiplier method is

$$I_L = \frac{1}{2} \int_{t_1}^{t_2} (x\dot{y} - y\dot{x}) \, dt + \lambda \left(\int_{t_1}^{t_2} \sqrt{\dot{x}^2 + \dot{y}^2} \, dt - L \right)$$

$$\equiv \int_{t_1}^{t_2} F(x, y, \lambda, \dot{x}, \dot{y}) dt. \tag{3}$$

The three Euler equations are given by

$$\delta x : \quad \frac{\partial F}{\partial x} - \frac{d}{dt}\left(\frac{\partial F}{\partial \dot{x}}\right) = 0,$$

$$\delta y : \quad \frac{\partial F}{\partial y} - \frac{d}{dt}\left(\frac{\partial F}{\partial \dot{y}}\right) = 0,$$

$$\delta \lambda : \quad \frac{\partial F}{\partial \lambda} - \frac{d}{dt}\left(\frac{\partial F}{\partial \dot{\lambda}}\right) = 0,$$

or

$$\delta x : \quad \frac{1}{2}\dot{y} - \frac{d}{dt}\left[-\frac{1}{2}y + \frac{\lambda \dot{x}}{\sqrt{\dot{x}^2 + \dot{y}^2}}\right] = 0, \tag{4}$$

$$\delta y : \quad -\frac{1}{2}\dot{x} - \frac{d}{dt}\left[\frac{1}{2}x + \frac{\lambda \dot{y}}{\sqrt{\dot{x}^2 + \dot{y}^2}}\right] = 0, \tag{5}$$

and the third equation is the same as that listed in Eq. (2).

Example 3.5.6

Consider the problem of minimizing the functional

$$\Pi(w, \phi_x) = \int_0^L \left[\frac{EI}{2} \left(\frac{d\phi_x}{dx} \right)^2 + qw \right] dx - Fw(L) \tag{1}$$

subject to the constraint

$$G(w_0', \phi_x) \equiv \frac{dw}{dx} + \phi_x = 0. \tag{2}$$

The functional Π represents the total potential energy of a cantilever beam (according to the Euler–Bernoulli beam theory) subjected to uniformly distributed load q and point load F at its free end. The constraint in Eq. (2) represents the relation between the transverse deflection w and rotation ϕ_x of a transverse normal line. The essential boundary conditions for the problem are

$$w(0) = \phi_x(0) = 0. \tag{3}$$

Determine the Euler equations using the Lagrange multiplier method and the penalty method.

Solution: In the Lagrange multiplier method, the modified functional for the problem is given by

$$\Pi_L(w, \phi_x, \lambda) = \Pi(w, \phi_x) + \int_0^L \lambda \left(\frac{dw}{dx} + \phi_x \right) dx, \tag{4}$$

where $\lambda(x)$ denotes the Lagrange multiplier. The Euler equations of the problem are obtained by setting the partial variations of Π_L with respect to w, ϕ_x, and λ to zero:

$$\delta_w \Pi_L = 0: \qquad -\frac{d\lambda}{dx} + q = 0, \; 0 < x < L. \tag{5}$$

$$\lambda - F = 0, \quad \text{at } x = L. \tag{6}$$

$$\delta_{\phi_x} \Pi_L = 0: \qquad -\frac{d}{dx} \left(EI \frac{d\phi_x}{dx} \right) + \lambda = 0, \; 0 < x < L. \tag{7}$$

$$EI \frac{d\phi_x}{dx} = 0, \quad \text{at } x = L. \tag{8}$$

$$\delta_\lambda \Pi_L = 0: \qquad \phi_x + \frac{dw}{dx} = 0, \; 0 < x < L. \tag{9}$$

The Lagrange multiplier $\lambda(x)$ in the present case turns out to be the shear force, $V(x)$.

In the penalty function method, the modified functional is given by

$$\Pi_p(w, \phi_x) = \Pi(w, \phi_x) + \frac{\gamma}{2} \int_0^L \left(\phi_x + \frac{dw}{dx} \right)^2 dx. \tag{10}$$

The Euler equations are given by

$$\delta_w \Pi_p = 0: \qquad -\gamma \frac{d}{dx} \left(\phi_x + \frac{dw}{dx} \right) + q = 0, \; 0 < x < L. \tag{11}$$

$$\gamma \left(\phi_x + \frac{dw}{dx} \right) - F = 0, \quad \text{at } x = L. \tag{12}$$

$$\delta_{\phi_x} \Pi_p = 0: \qquad -\frac{d}{dx} \left(EI \frac{d\phi_x}{dx} \right) + \gamma \left(\phi_x + \frac{dw}{dx} \right) = 0, \; 0 < x < L. \tag{13}$$

$$EI \frac{d\phi_x}{dx} = 0, \quad \text{at } x = L. \tag{14}$$

A comparison of Eq. (6) with Eq. (12) shows that the Lagrange multiplier is related to ϕ_x and w by

$$\lambda(x) = \gamma \left(\phi_x + \frac{dw}{dx} \right). \tag{15}$$

For the choice of $\gamma = GAK_s$, Eq. (15) represents the shear force and shear strain relation in the Timoshenko beam theory

$$\lambda(x) \sim V(x) = GAK_s\gamma_{xz} = GAK_s \left(\phi_x + \frac{dw}{dx} \right), \tag{16}$$

where G the shear modulus, A the area of cross section, and K_s is the shear correction factor of the beam.

3.6   Summary

In this chapter the concepts of work and energy are introduced, and expressions for strain energy and complementary strain energy densities and strain energy and complementary strain energies are derived. The expressions are then specialized for bars and beams. Total potential energy and total complementary energy are discussed, and their expressions for beams are developed. Then concepts of virtual displacements and virtual forces are introduced, and expressions for virtual work done by actual forces in moving through virtual displacements and virtual complementary work done by virtual forces in moving through actual displacements are derived. Finally, elements from calculus variations are presented. In particular, properties of the variational operator, computation of the first variation of functionals, and the derivation of Euler equations and natural boundary conditions associated with a given functional, without and with constraints, are presented. Thus, the elements of this chapter provide the basis for various energy principles (i.e., principles based on the total energy of the system) to be discussed in Chapter 4. The main equations of this chapter are summarized next.

Strain energy and complementary strain energy densities [see Eqs (3.2.7) and (3.2.9)]:

$$U_0 = \int_0^{\varepsilon_{ij}} \sigma_{ij}(\boldsymbol{\varepsilon}) \, d\varepsilon_{ij}, \quad U_0^* = \int_0^{\sigma_{ij}} \varepsilon_{ij}(\boldsymbol{\sigma}) \, d\sigma_{ij}. \tag{3.6.1}$$

External and internal works done for elastic solids [see Eqs (3.1.7), (3.2.8), (3.2.10), and (3.2.12)]:

$$W_E = W_E^* = -\left[\int_\Omega \mathbf{f}(\mathbf{x}) \cdot \mathbf{u}(\mathbf{x}) \, d\Omega + \oint_\Gamma \mathbf{t}(s) \cdot \mathbf{u}(s) \, d\Gamma \right]$$

$$W_I = U = \int_\Omega U_0 \, d\Omega, \quad W_I^* = U^* = \int_\Omega U_0^* \, d\Omega. \tag{3.6.2}$$

Strain energy and complementary strain energy expressions for linear elastic solids [see Eq. (3.2.11)]:

$$U = \frac{1}{2} \int_\Omega \sigma_{ij}(\boldsymbol{\varepsilon})\, \varepsilon_{ij}\, d\Omega; \quad U^* = \frac{1}{2} \int_\Omega \sigma_{ij}\, \varepsilon_{ij}(\boldsymbol{\sigma})\, d\Omega. \tag{3.6.3}$$

Strain energy expressions (for straight, homogeneous beams made of linear isotropic elastic material axial displacement u, transverse displacement w, and rotation ϕ_x):

The Bernoulli–Euler beam theory [Eq. (3.2.39)]:

$$U = \frac{1}{2} \int_0^L \left[EA \left(\frac{du}{dx} \right)^2 + EI \left(\frac{d^2 w}{dx^2} \right)^2 \right] dx. \tag{3.6.4}$$

The Timoshenko beam theory [Eqs (3.2.47)]:

$$U = \frac{1}{2} \int_0^L \left[EA \left(\frac{du}{dx} \right)^2 + EI \left(\frac{d\phi_x}{dx} \right)^2 + GAK_s \left(\phi_x + \frac{dw}{dx} \right)^2 \right] dx. \tag{3.6.5}$$

Complementary strain energy expressions (for straight, homogeneous beams made of linear isotropic elastic material and subjected to axial force N, bending moment M, transverse shear force V, and torque T)

The Bernoulli–Euler beam theory [Eq. (3.2.44)]:

$$U^* = \frac{1}{2} \int_0^L \left(\frac{N^2}{EA} + \frac{M^2}{EI} + \frac{f_s V^2}{GA} + \frac{T^2}{GJ} \right) dx. \tag{3.6.6}$$

The Timoshenko beam theory [Eq. (3.2.50)]:

$$U^* = \frac{1}{2} \int_0^L \left(\frac{N^2}{EA} + \frac{M^2}{EI} + \frac{V^2}{GAK_s} + \frac{T^2}{GJ} \right) dx. \tag{3.6.7}$$

Total potential energy expressions (for straight, homogeneous beams made of linear isotropic elastic material: axial displacement u, transverse displacement w, and rotation ϕ_x; for bars, omit the expressions involving w and ϕ_x)

The Bernoulli–Euler beam theory [simplified Eq. (3.3.4)]:

$$\Pi(u, w) = \frac{1}{2} \int_0^L \left[EA \left(\frac{du}{dx} \right)^2 + EI \left(\frac{d^2 w}{dx^2} \right)^2 \right] dx - \int_0^L (fu + qw)\, dx. \tag{3.6.8}$$

The Timoshenko beam theory [simplified Eq. (3.3.5)]:

$$\Pi(u, w, \phi_x) = \frac{1}{2} \int_0^L \left[EA \left(\frac{du}{dx} \right)^2 + EI \left(\frac{d\phi_x}{dx} \right)^2 + GAK_s \left(\phi_x + \frac{dw}{dx} \right)^2 \right] dx$$

$$- \int_0^L (fu + qw)\, dx. \tag{3.6.9}$$

Total potential energy expression for plane elasticity and membranes

Plane elasticity (for linear isotropic elastic materials). u_x and u_y are the x and y components, respectively, of the displacement vector $\mathbf{u}$; μ and λ are the Lamé constants; (f_x, f_y) are the components of the body force vector $\mathbf{f}$ (measured per unit area); and $(\hat{t}_x, \hat{t}_y)$ are the specified components of the surface traction vector $\mathbf{t}$ (measured per unit length).

$$
\Pi(u_x, u_y) = \frac{1}{2} \int_\Omega \left\{ (2\mu + \lambda) \left[\left(\frac{\partial u_x}{\partial x} \right)^2 + \left(\frac{\partial u_y}{\partial y} \right)^2 \right] + \mu \left(\frac{\partial u_x}{\partial y} + \frac{\partial u_y}{\partial x} \right)^2 \right.
$$
$$
\left. + 2\lambda \frac{\partial u_x}{\partial x} \frac{\partial u_y}{\partial y} \right\} dx\, dy - \int_\Omega (f_x u_x + f_y u_y)\, dx\, dy
$$
$$
- \int_{\Gamma_\sigma} (\hat{t}_x u_x + \hat{t}_y u_y)\, d\Gamma. \tag{3.6.10}
$$

Membranes (for an orthotropic membrane with the principal material axes coinciding with the x and y axes). u is the transverse deflection, a_{xx} and a_{yy} are the tensions in the x and y coordinate directions, f is the distributed transverse load, and $\hat{t}$ is the specified boundary flux normal to the boundary:

$$
\Pi(u) = \frac{1}{2} \int_\Omega \left[a_{xx} \left(\frac{\partial u}{\partial x} \right)^2 + a_{yy} \left(\frac{\partial u}{\partial y} \right)^2 \right] dx\, dy
$$
$$
- \int_\Omega f u\, dx\, dy - \int_{\Gamma_\sigma} \hat{t} u\, d\Gamma. \tag{3.6.11}
$$

Total complementary potential energy expressions for straight, homogeneous beams made of linear isotropic elastic material and subjected to axial force N, bending moment M, transverse shear force V, and torque T (for both beam theories):

$$
\Pi^*(N, M, V, T) = \frac{1}{2} \int_0^L \left(\frac{N^2}{EA} + \frac{M^2}{EI} + \frac{f_s V^2}{GA} + \frac{T^2}{GJ} \right) dx
$$
$$
- \int_0^L (fu + qw)\, dx. \tag{3.6.12}
$$

External and internal virtual work expressions for elastic solids [Eqs (3.4.3) and (3.4.6)]:

$$
\delta W_E = - \left(\int_\Omega \mathbf{f} \cdot \delta \mathbf{u}\, d\Omega + \int_{\Gamma_\sigma} \mathbf{t} \cdot \delta \mathbf{u}\, d\Gamma \right). \tag{3.6.13}
$$

$$
\delta W_I = \int_\Omega \boldsymbol{\sigma}(\varepsilon) : \delta \varepsilon\, d\Omega. \tag{3.6.14}
$$

External and internal complementary virtual work expressions for elastic solids [Eqs (3.4.7) and (3.4.8)]

$$\delta W_E^* = -\left(\int_\Omega \delta \mathbf{f} \cdot \mathbf{u} \, d\Omega + \int_\Gamma \delta \mathbf{t} \cdot \mathbf{u} \, d\Gamma \right), \tag{3.6.15}$$

$$\delta W_I^* = \int_\Omega \varepsilon(\boldsymbol{\sigma}) : \delta \boldsymbol{\sigma} \, d\Omega. \tag{3.6.16}$$

Problems

WORK AND ENERGY (SECTIONS 3.2 AND 3.3)

3.1 Consider the double pendulum shown in Fig. P3.1 at some instant of time. Write the potential energy (only due to gravity) of the system assuming that the energy is zero when $\theta_1 = \theta_2 = 0$. Neglect the weight of the pendulum bars and assume that they are rigid.

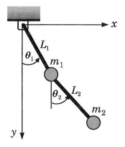

Fig. P3.1

3.2 Consider the rigid-body assemblage shown in Fig. P3.2 at some instant of time. The bars are interconnected by hinges, and their relative rotations are resisted by torsional springs located at each hinge. Write the potential energy of the system assuming that the energy is zero when $\theta_1 = \theta_2 = \theta_3 = 0$. Include the weight W of the rigid bars, and assume that the displacements are small and the torsional springs are linear.

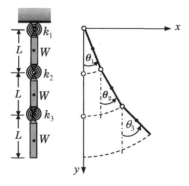

Fig. P3.2

3.3 The rigid link shown in Fig. P3.3 is in equilibrium when pinned at one end and subjected to forces F_x and F_y and moment M as shown. Find the incremental work done ΔW^* when the link is subjected to increments $\Delta F_x, \Delta F_y$, and ΔM. Express the answer only in terms of the increments ΔF_x and ΔF_y.

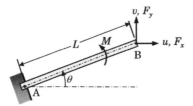

Fig. P3.3

3.4 Determine the strain energy and external work done for the bar with an end linear elastic spring shown in Fig. P3.4.

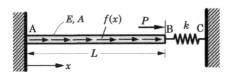

Fig. P3.4

3.5 Determine the strain energy and external work done for the beam with an end linear elastic spring shown in Fig. P3.5. Neglect the energy due to shear.

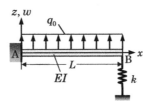

Fig. P3.5

3.6 Determine the total complementary strain energy due to bending and shear for the beam shown in Fig. P3.6. Take $f_s = 6/5$.

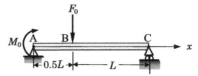

$M_0 = 6$ kN-m, $F_0 = 12$ kN, $L = 1$ m, $A = 1.25 \times 10^{-6}$ m^2

$I = 0.2604 \times 10^{-6}$ m^4, $E = 205$ GPa, $\nu = 0.3$

Fig. P3.6

3.7 Repeat **Problem 3.5** when the spring behaves according to the relation $F_s = kw^n$, w being the elongation of the spring.

3.8 Determine the complementary strain energy of the structure shown in Fig. P3.8. Include energies due to bending and torsion, but neglect that due to transverse shear forces.

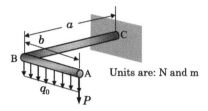

Units are: N and m

Fig. P3.8

3.9 Consider the pin-connected structure shown in Fig. P3.9. Suppose that the material of all members obeys the stress–strain relation given in Eq. (3.2.29). All members have the same cross-sectional area A and material constant K. Write the complementary strain energy densities of the members and the total complementary strain energy of the structure.

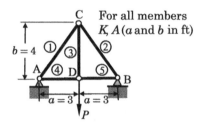

For all members
K, A (a and b in ft)

Fig. P3.9

3.10 Determine the (a) strain energy and (b) complementary strain energy of the truss shown in Fig. P3.10. Assume elastic behavior of the form $\sigma = K\sqrt{\varepsilon}$. Express your answer in terms of displacements in the former case and applied loads in the latter case. *Hint:* You may have to use a displacement constraint to determine the reactions.

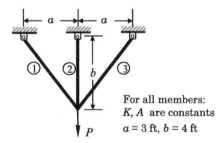

For all members:
K, A are constants
$a = 3$ ft, $b = 4$ ft

Fig. P3.10

3.11 Identify admissible virtual displacements for the problem in **Problem 3.5**.

3.12 Identify admissible virtual displacements for the problem in **Problem 3.8**.

3.13 Identify admissible virtual forces for the problem in **Problem 3.4**.

3.14 Identify admissible virtual forces for the problem in **Problem 3.6**.

3.15 Identify admissible virtual forces for the problem in **Problem 3.9**.

3.16 Write the total virtual work expression for the problem in **Problem 3.4**.

3.17 Write the total virtual work expression for the problem in **Problem 3.5**.

3.18 Write the total complementary virtual work expression for the problem in **Problem 3.3**

3.19 Write the total complementary virtual work expression for the problem in **Problem 3.4**

3.20 Write the total complementary virtual work expression for the problem in **Problem 3.5**.

3.21 Write the total complementary virtual work expression for the problem in **Problem 3.8**.

3.22 Write the total complementary virtual work expression for the problem in **Problem 3.9**.

3.23 Two rigid links are pin-connected at point A and to a rigid support at point O as shown in Fig. P3.23. Compute the virtual work done by the forces due to the virtual changes $\delta\theta_1$ and $\delta\theta_2$ from the equilibrium configuration shown in the figure.

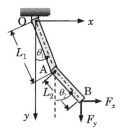

Fig. P3.23

3.24 A rigid bar supported by three linear elastic springs and subjected to load P and moment M occupies the equilibrium configuration shown in Fig. P3.24. Write the total complementary virtual work done, δW^*, and write the equilibrium equations that the virtual forces δP, δM, and δF_3 (virtual force in spring k_3) must satisfy.

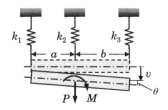

Fig. P3.24

3.25 Write the virtual strain (internal) energy expression for the Bernoulli–Euler beam theory with the following von Kármán nonlinear strain–displacement relation:

$$\varepsilon_{xx} = \frac{du}{dx} + \frac{1}{2}\left(\frac{dw}{dx}\right)^2 - z\frac{d^2w}{dx^2}.$$

Express your solution in terms of the stress resultants in Eq. (2.5.6). In particular, do not employ any stress–strain relations.

3.26 The *Timoshenko beam theory* is based on the displacement field:

$$u_1 = u(x) + z\phi_x(x), \qquad u_2 = 0, \quad u_3 = w(x), \tag{1}$$

where u is the inplane displacement and w is the transverse deflection of a point on the midplane (i.e., $z = 0$) of the plate and ϕ_x is the rotation of a transverse normal line about the y-axis.

(1) Compute von Kármán strains using the displacement field.

(2) Write the expression for the total virtual work done by actual forces in moving through the virtual displacements $(\delta u, \delta w, \delta\phi)$, assuming that the beam is loaded with a distributed axial load $f(x)$ and transverse force $q(x)$. Express the internal virtual work in terms of the stress resultants:

$$N = \int_A \sigma_{xx}\, dA, \quad M = \int_A z\sigma_{xx}\, dA, \quad V = K_s \int_A \sigma_{xz}\, dA. \tag{2}$$

3.27 The classical *Kirchhoff plate theory* for axisymmetric bending of circular plates is based on the displacement field:

$$u_r(r, z) = u(r) - z\frac{dw}{dr}, \quad u_\theta = 0, \quad u_z(r, z) = w(r), \tag{1}$$

where u is the radial displacement and w is the transverse deflection of a point on the midplane (i.e., $z = 0$) of the plate, the r-coordinate is taken radially outward from the center of the plate, z-coordinate along the thickness (or height) of the plate, and the θ-coordinate is taken along the circumference of the plate (see Fig. P3.27). In a general case where applied loads and geometric boundary conditions are not axisymmetric, the displacements (u_r, u_θ, u_z) along the coordinates (r, θ, z) are functions of r, θ, and z coordinates. Assume that the applied loads and boundary conditions are independent of θ coordinate (i.e., axisymmetric) so that the displacement u_θ is identically zero and (u_r, u_z) are only functions of r and z.

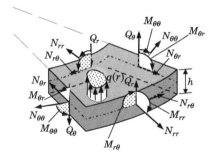

Fig. P3.27

(1) Compute linear strains using the displacement field in Eq. (1).

(2) Write the expression for the total virtual work done $\delta W = \delta W_I + \delta W_E$ assuming that the plate is subjected to transversely distributed load $q = q(r)$. Express the internal virtual work in terms of the moments (see Fig. P3.27):

$$M_{rr} = \int_{-\frac{h}{2}}^{\frac{h}{2}} \sigma_{rr} z \, dz, \quad M_{\theta\theta} = \int_{-\frac{h}{2}}^{\frac{h}{2}} \sigma_{\theta\theta} z \, dz, \tag{2}$$

where σ_{rr} and $\sigma_{\theta\theta}$ are the radial and circumferential stresses, respectively.

VARIATIONAL CALCULUS (SECTION 3.5)

Calculate the first variation of the functionals in **Problems 3.28–3.36**. The following notation is used:

$$u' \equiv \frac{du}{dx}, \quad u'' \equiv \frac{d^2 u}{dx^2}, \quad u_{,x} \equiv \frac{\partial u}{\partial x}, \quad u_{i,j} \equiv \frac{\partial u_i}{\partial x_j}.$$

All variables other than those listed in the argument of the functional are assumed to be functions of position.

3.28 $I(u) = \int_a^b \sqrt{1 + (u')^2} dx, \quad a < x < b.$

3.29 $I(u) = \int_a^b u\sqrt{1 + (u')^2} dx, \quad a < x < b.$

3.30 $I(u) = \int_0^1 \sqrt{\frac{1+(u')^2}{u}} dx, \quad 0 < x < 1..$

3.31 $I(u) = \frac{1}{2} \int_a^b \left[(u'')^2 + 2u'u'' + (u')^2 - 2u \right] dx, \quad a < x < b.$

3.32 The Bernoulli–Euler beam theory with the von Kármán nonlinearity:

$$\Pi(u, w) = \int_0^L \left\{ \frac{EA}{2} \left[\frac{du}{dx} + \frac{1}{2} \left(\frac{dw}{dx} \right)^2 \right]^2 + \frac{EI}{2} \left(\frac{d^2 w}{dx^2} \right)^2 - qw \right\} dx$$
$$- Pu(L) + M_0 \left(\frac{dw}{dx} \right)_{x=L}.$$

The boundary conditions are: $u(0) = 0, w(0) = 0,$ and $\left(\frac{dw}{dx}\right)_0 = 0.$

3.33 Functional with two dependent variables (u, v):

$$I(u, v) = \frac{1}{2} \int_\Omega \left(u_{,x}^2 + 2u_{,x} v_{,x} + 2u_{,y} v_{,y} + v_{,y}^2 + 2fu + 2gv \right) dxdy$$

where $u_{,x} = \partial u / \partial x$, $u_{,y} = \partial u / \partial y$, and so on.

3.34 Functional with two dependent variables, one scalar u and the other a vector **v**:

$$I(u, \mathbf{v}) = \int_\Omega \left\{ \frac{1}{2k} \mathbf{v} \cdot \mathbf{v} + \mathbf{v} \cdot \nabla u + Qu \right\} d\Omega - \int_{\Gamma_q} \hat{q}u \, d\Gamma,$$

where Γ_q is the portion of the boundary on which $\mathbf{v} \cdot \hat{\mathbf{n}}$ is specified; on the remaining portion of the boundary, $\Gamma_u = \Gamma - \Gamma_q$, u is specified.

3.35 Functional with three dependent unknowns (u_1, u_2, P):

$$I(u_i, P) = \int_\Omega \left\{ \frac{\mu}{2}(u_{i,j} + u_{j,i})u_{i,j} - Pu_{i,i} \right\} d\Omega - \int_\Omega f_i u_i \, d\Omega - \int_{\Gamma_\sigma} \hat{t}_i u_i \, d\Gamma,$$

with $\mathbf{u} = \mathbf{0}$ on the boundary Γ_u, where $\Gamma_u \cup \Gamma_\sigma = \Gamma$, the total boundary.

3.36 Functional in a single dependent variable but with second-order derivatives in two dimensions:

$$\Pi(w) = \frac{D}{2} \int_\Omega \left[\left(\frac{\partial^2 w}{\partial x^2}\right)^2 + \left(\frac{\partial^2 w}{\partial y^2}\right)^2 + 2\nu \frac{\partial^2 w}{\partial x^2}\frac{\partial^2 w}{\partial y^2} + 2(1-\nu)\left(\frac{\partial^2 w}{\partial x \partial y}\right)^2 \right] dx dy$$

$$+ \frac{1}{2}\int_\Omega kw^2 \, dx dy - \int_\Omega qw \, dx dy,$$

with $w = 0$ and $\frac{\partial w}{\partial n} = 0$ on the boundary Γ of the domain Ω.

Express the next three problems in mathematical terms as one of minimizing appropriate quantities, subjected to certain end conditions, and possibly some constraints.

3.37 *The Brachistochrone problem.* Determine the curve $y = y(x)$ connecting two points, A: $(0,0)$ and B: (x_b, y_b), in a vertical plane such that a material particle, sliding without friction under its own weight, travels from point A to point B along the curve in the *shortest time* (see Fig. P3.37).

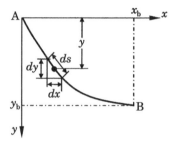 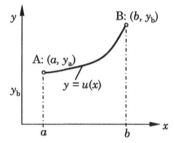

Fig. P3.37   Fig. P3.38

3.38 *Geodesic problem.* Find the curve $y = y(x)$ of *minimum length* joining two given points A: (a, y_a) and B: (b, y_b) in the xy-plane (see Fig. P3.38).

3.39 *Isoperimetric problem.* Among all curves with a continuous derivative that join two given points, A: (a, y_a) and B: (b, y_b), and have the given length L, find the one that encompasses the *largest possible area*.

Determine (a) the Euler equations and (b) the natural boundary conditions associated with the functionals identified in **Problems 3.40–3.44**. Specified essential boundary conditions are shown if there are any.

3.40 Functional in **Problem 3.31**.

3.41 Functional in **Problem 3.32**.

3.42 Functional in **Problem 3.33**.

3.43 Functional in **Problem 3.35**.

3.44 Functional in **Problem 3.36**; assume that Ω is a rectangular region occupying $0 < x < a$ and $0 < y < b$.

3.45 Obtain the Euler equations and the natural boundary conditions associated with the functional

$$\Pi(w) = \frac{\pi}{2}\int_{r_i}^{r_0}\left\{D_{11}\left(\frac{d^2 w}{dr^2}\right)^2 + \frac{2D_{12}}{r}\frac{dw}{dr}\frac{d^2 w}{dr^2} + D_{22}\left(\frac{1}{r}\frac{dw}{dr}\right)^2\right\}r \, dr,$$

which arises in connection with axisymmetric bending of polar orthotropic annular plates with inner radius r_i and outer radius r_0. Assume that the deflection $w = 0$ at $r = r_0$, the outer radius of the plate.

3.46 Derive the Euler equations and the natural and essential boundary conditions associated with the Lagrange multiplier functional of the following problem: minimize the functional

$$
\Pi(w, \phi_x, \phi_y) = \frac{D}{2} \int_{\Omega} \left[\left(\frac{\partial \phi_x}{\partial x} \right)^2 + \left(\frac{\partial \phi_y}{\partial y} \right)^2 + 2\nu \frac{\partial \phi_x}{\partial x} \frac{\partial \phi_y}{\partial y} \right.
$$
$$
\left. + (1 - \nu) \left(\frac{\partial \phi_x}{\partial y} + \frac{\partial \phi_y}{\partial x} \right)^2 \right] dx dy - \int_{\Omega} qw \, dx dy
$$

subject to the constraints

$$
\frac{\partial w}{\partial x} + \phi_x = 0, \quad \frac{\partial w}{\partial y} + \phi_y = 0.
$$

Use two different Lagrange multipliers, say λ_x and λ_y, with the two constraints.

3.47 Repeat **Problem 3.46** with the penalty functional of the problem (i.e., construct the penalty functional and derive the Euler equations). Use two different penalty parameters, say γ_x and γ_y, with the two constraints. Note that the penalty functional represents the total potential energy of the first-order shear deformation plate theory (FSDT) discussed in Chapter 7 (for $\gamma_x = \gamma_y = K_s G h$, where K_s is the shear correction coefficient, G the shear modulus, and h the thickness of the plate).

3.48 Consider the problem of maximizing the functional

$$
I(u) = \int_a^b u(x) dx, \quad u(a) = u_a, \quad u(b) = u_b,
$$

subject to the constraint

$$
G(u') \equiv \int_a^b \sqrt{1 + \left(\frac{du}{dx} \right)^2} \, dx - L = 0.
$$

Use (a) the Lagrange multiplier method and (b) penalty method to introduce the constraint into $I(u)$, and derive associated Euler equations. Note that the constraint is an integral constraint.

3.49 Consider the problem of minimizing the functional

$$
I(u) = \int_a^b u \sqrt{1 + \left(\frac{du}{dx} \right)^2} \, dx, \quad u(a) = u_a, \quad u(b) = u_b,
$$

subject to the constraint

$$
\int_a^b \sqrt{1 + \left(\frac{du}{dx} \right)^2} \, dx = L.
$$

Use (a) the Lagrange multiplier method and (b) penalty function method to introduce the constraint into the functional, and derive the Euler equations.

3.50 Consider the problem of minimizing the functional

$$
\Pi(w, \varepsilon_{xx}^{(1)}) = \frac{EI}{2} \int_0^L \left(\varepsilon_{xx}^{(1)} \right)^2 dx - \int_0^L qw \, dx,
$$

subject to the constraint

$$
\varepsilon_{xx}^{(1)} - \frac{d^2 w}{dx^2} = 0,
$$

where w is the transverse deflection, q the distributed load, EI is the flexural stiffness, and L is the length of the beam. Construct a functional using the Lagrange multiplier method, and determine the Euler equations.

4

Virtual Work and Energy Principles of Mechanics

4.1 Introduction

In this chapter we study the virtual work and energy principles of solid mechanics. These include the principles of virtual displacements and virtual forces, the principle of minimum total potential energy, and the principle of minimum total complementary energy. These principles can also be used to derive the unit dummy-displacement and unit dummy-load methods and Castigliano's theorems I and II of structural mechanics [99, 100]. All of these principles and theorems are used to determine either continuous solutions or point-wise solutions for displacements and forces.

The virtual work principles can be used, as already illustrated in Chapter 3, to derive the equations of equilibrium of deformable solids, including structural elements like bars, beams, plates, and shells. The use of energy methods (i.e., unit dummy-displacement and unit dummy-load methods and Castigliano's theorems) in the determination of point displacements and forces will be illustrated with bars, beams, trusses, and frames. The Betti and Maxwell reciprocity theorems [101, 102] will be presented and their use in finding solutions point-wise displacements in beams. Both direct variational methods (i.e., Ritz, Galerkin, and weighted-residual methods) and the finite element method that make use of virtual work principles will be discussed in the subsequent chapters.

4.2 The Principle of Virtual Displacements

4.2.1 Rigid Bodies

Recall from Section 3.4 that virtual work is the work done on a deformable body by actual forces in moving through hypothetical or virtual displacements that are consistent with the geometric constraints. The applied forces are kept constant during the virtual displacements.

Suppose that a rigid body is in equilibrium under the action of n concurrent forces $\mathbf{F}_1, \mathbf{F}_2, \ldots, \mathbf{F}_n$. Now suppose that the rigid body is given an arbitrary

Energy Principles and Variational Methods in Applied Mechanics, Third Edition. J.N. Reddy.
©2017 John Wiley & Sons Ltd. Published 2017 by John Wiley & Sons Ltd.
Companion Website: www.wiley.com/go/reddy/applied_mechanics_EPVM

virtual displacement $\delta\mathbf{u}$ during which all forces along with their directions are fixed. The total virtual work done by all forces in moving through the virtual displacement $\delta\mathbf{u}$ is given by

$$\delta W = \mathbf{F}_1 \cdot \delta\mathbf{u} + \mathbf{F}_2 \cdot \delta\mathbf{u} + \cdots + \mathbf{F}_n \cdot \delta\mathbf{u}$$

$$= \left(\sum_{i=1}^{n} \mathbf{F}_i\right) \cdot \delta\mathbf{u}. \tag{4.2.1}$$

We note that each force moves by the same displacement because all points in a rigid body move by the same amount, but the work done by each force is different because the component of each force in the direction of $\delta\mathbf{u}$ is different. Since the expression in the parenthesis is the vector sum of all forces acting on the particle and from vector mechanics, we know that the sum is zero if the particle is in equilibrium; thus we have $\delta W = 0$. Conversely, if $\delta W = 0$ and $\delta\mathbf{u}$ is arbitrary, it follows that $\sum_{i=1}^{n} \mathbf{F}_i = 0$, that is, the particle is in equilibrium. In other words, the particle is in equilibrium if and only if $\delta W = 0$ for *any* choice of $\delta\mathbf{u}$. The statement $\delta W = 0$ is the mathematical statement of *the principle of virtual displacements* for a rigid body (equally valid for a single material particle).

4.2.2 Deformable Solids

Now we consider a generalization of the principle of virtual displacements to deformable bodies. In deformable bodies material points can move relative to one another and do internal work in addition to the work done by external forces. Thus we should consider the virtual work done by internal forces (i.e., stresses developed within the body) as well as the work done by external forces in moving through virtual displacements that are consistent with the geometric constraints.

Consider a continuum occupying the volume Ω (which is an open-bounded set of points in $\Re^3$, each point occupied by a material particle) and is in equilibrium under the action of body forces $\mathbf{f}$ and surface tractions $\mathbf{t}$. Suppose that displacements are specified to be $\hat{\mathbf{u}}$, over portion Γ_u of the total boundary Γ of Ω; on the remaining boundary $\Gamma - \Gamma_u \equiv \Gamma_\sigma$, tractions are specified to be $\hat{\mathbf{t}}$. The boundary portions Γ_u and Γ_σ are disjoint (i.e., do not overlap), and their sum is the total boundary, Γ. Let $\mathbf{u} = (u_1, u_2, u_3)$ be the displacement vector corresponding to the equilibrium configuration of the body, and let σ_{ij} and ε_{ij} be the associated stress and strain components, respectively, referred to rectangular Cartesian system (x_1, x_2, x_3). Throughout this discussion, we assume that the strains are infinitesimal and rotations are possibly moderate so that no distinction between the Cauchy stress tensor and second Piola–Kirchhoff stress tensor and between the infinitesimal strain tensor and the Green–Lagrange strain tensor is made. We make no assumption concerning the constitutive behavior of the material body at the moment.

The set of admissible configurations is defined by sufficiently differentiable displacement fields that satisfy the geometric boundary conditions: $\mathbf{u} = \hat{\mathbf{u}}$ on Γ_u. Of all such admissible configurations, the actual one corresponds to the equilibrium configuration with the prescribed loads. In order to determine the displacement field $\mathbf{u}$ corresponding to the equilibrium configuration, we let the body experience a virtual displacement $\delta\mathbf{u}$ from the equilibrium configuration. The virtual displacements are arbitrary, continuous with continuous derivatives as dictated by the strain energy, and satisfy the homogeneous form of the specified geometric boundary conditions, $\delta\mathbf{u} = \mathbf{0}$ on Γ_u. The principle of virtual displacements states that *a continuous body is in equilibrium if and only if the virtual work done by all forces, internal and external, acting on the body is zero in a virtual displacement:*

$$\delta W = \delta W_I + \delta W_E = 0, \tag{4.2.2}$$

where δW_I is the virtual work due to the internal forces and δW_E is the virtual work due to the external forces. For problems involving mechanical stimuli, the internal virtual work is the same as the virtual strain energy, $\delta U = \delta W_I$.

The principle of virtual work is independent of any constitutive law. The principle may be used to derive the equilibrium equations of deformable solids in terms of stresses or stress resultants, as already illustrated by **Examples 3.5.3** and **3.5.4**. Here we derive the stress equilibrium equations of three-dimensional elasticity using the principle of virtual displacements.

We begin with the principle of virtual displacements:

$$\begin{aligned}
0 = \delta W &= \int_\Omega \sigma_{ij}\, \delta\varepsilon_{ij}\, d\Omega - \left[\int_\Omega \mathbf{f} \cdot \delta\mathbf{u}\, d\Omega + \int_{\Gamma_\sigma} \hat{\mathbf{t}} \cdot \delta\mathbf{u}\, d\Gamma \right] \\
&= \tfrac{1}{2} \int_\Omega \left[\sigma_{ij} \left(\delta u_{i,j} + \delta u_{j,i} \right) \right] d\Omega - \left[\int_\Omega f_i \delta u_i\, d\Omega + \int_{\Gamma_\sigma} \hat{t}_i \delta u_i\, d\Gamma \right] \\
&= - \int_\Omega \left(\sigma_{ji,j} + f_i \right) \delta u_i\, d\Omega - \int_{\Gamma_\sigma} \hat{t}_i u_i\, ds + \int_\Gamma \sigma_{ji} n_j \delta u_i\, d\Gamma, \quad (4.2.3)
\end{aligned}$$

where integration by parts (or the Green–Gauss theorem) is used to arrive at the last step. Since $\delta u_i = 0$ on Γ_u, the last boundary integral reduces to one on Γ_σ. Thus, we have

$$0 = - \int_\Omega \left(\sigma_{ji,j} + f_i \right) \delta u_i\, d\Omega + \int_{\Gamma_\sigma} \left(\sigma_{ij} n_j - \hat{t}_i \right) \delta u_i\, d\Gamma. \tag{4.2.4}$$

Since δu_i is arbitrary in Ω and independently on Γ_σ, we obtain the Euler equations

$$\sigma_{ji,j} + f_i = 0 \quad \text{or} \quad \nabla \cdot \boldsymbol{\sigma} + \mathbf{f} = \mathbf{0} \quad \text{in } \Omega \tag{4.2.5}$$

and

$$\sigma_{ji} n_j - \hat{t}_i = 0 \quad \text{or} \quad \hat{\mathbf{n}} \cdot \boldsymbol{\sigma} - \hat{\mathbf{t}} = \mathbf{0} \quad \text{on } \Gamma_\sigma. \tag{4.2.6}$$

Thus, the principle of virtual displacements yields the stress equilibrium equation and stress boundary condition as the Euler equations. Next we consider a specific example from theory of beams.

Example 4.2.1

In Section 2.5.2, we have discussed the kinematic hypothesis of the Bernoulli–Euler beam theory, and used a displacement field consistent with the hypothesis [also see Fig. 2.5.2(b)]. The von Kámán nonlinear theory of beams was discussed in Section 2.5.4. Use the principle of virtual displacements to derive the governing equations of the Bernoulli–Euler beam theory using the von Kámán strain in Eqs (2.5.43) and (2.5.44). Assume that the beam rests on a linear elastic foundation with foundation modulus k and subjected to a distributed longitudinal load $f(x)$ and distributed transverse load $q(x)$ at the top (see Fig. 4.2.1).

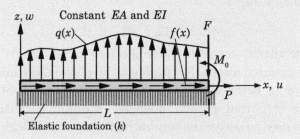

Fig. 4.2.1 A beam on elastic foundation.

Solution: The displacement field of a beam under the Bernoulli–Euler kinematic hypothesis is given by [see Eq. (2.5.1)]

$$u_1(x, y, z) = u(x) + z\theta_x, \quad u_2 = 0, \quad u_3(x, y, z) = w(x); \quad \theta_x \equiv -\frac{dw}{dx}. \tag{1}$$

If we assume that the strains are small in the sense that the following terms are negligible compared to du/dx (as discussed in Section 2.5.4),

$$\left(\frac{du}{dx}\right)^2, \quad \varepsilon_{zz}, \quad \frac{du}{dx}\frac{du}{dz}, \tag{2}$$

the only nonzero nonlinear strain is given by

$$\varepsilon_{xx} = \frac{du}{dx} + \frac{1}{2}\left(\frac{dw}{dx}\right)^2 + z\frac{d\theta_x}{dx}. \tag{3}$$

Let the virtual displacements be δu and δw, which are completely arbitrary because there are no specified geometric boundary conditions for the problem at hand. Then the virtual strain $\delta\varepsilon_{xx}$ is given by

$$\delta\varepsilon_{xx} = \delta\left[\frac{du}{dx} + \frac{1}{2}\left(\frac{dw}{dx}\right)^2 + z\frac{d\theta_x}{dx}\right] = \frac{d\delta u}{dx} + \frac{dw}{dx}\frac{d\delta w}{dx} + z\frac{d\delta\theta_x}{dx}. \tag{4}$$

Then the internal and external virtual works due to the virtual displacements δu and δw are given by

$$\delta W_E = -\left\{\int_0^L [f(x)\delta u + q(x)\delta w(x, h_t)]\, dx + \int_0^L (-F_s)\delta w(x, h_b)\, dx \right.$$
$$\left. + P\delta u(L) + (-F)\delta w(L) + (-M_0)\left(-\frac{d\delta w}{dx}\right)_{x=L}\right\}$$

$$= - \left\{ \int_0^L [f(x)\delta u + q(x)\delta w(x)] \, dx - \int_0^L kw(x)\delta w(x) \, dx \right.$$

$$\left. + P\delta u(L) - F\delta w(L) + M_0 \left(\frac{d\delta w}{dx} \right)_{x=L} \right\}, \tag{5}$$

$$\delta W_I = \int_0^L \int_A \sigma_{xx} \delta \varepsilon_{xx} \, dx dA$$

$$= \int_0^L \int_A \sigma_{xx} \left(\frac{d\delta u}{dx} + \frac{dw}{dx} \frac{d\delta w}{dx} - z \frac{d^2 \delta w}{dx^2} \right) dx dA, \tag{6}$$

where L is the length, h_t is the distance from the x-axis to the top of the beam, h_b is the distance from the x-axis to the bottom of the beam, and A is the cross-sectional area of the beam. The foundation reaction force F_s (acting downward) is replaced with $F_s = kw(x)$ using the linear elastic constitutive equation for the foundation.

The principle of virtual displacements requires that $\delta W = \delta W_I + \delta W_E = 0$, which gives

$$0 = \int_0^L \int_A \sigma_{xx} \left(\frac{d\delta u}{dx} + \frac{dw}{dx} \frac{d\delta w}{dx} - z \frac{d^2 \delta w}{dx^2} \right) dA dx$$

$$- \int_0^L [f\delta u + (q - kw)\delta w] \, dx - P\delta u(L) + F\delta w(L) - M_0 \left(\frac{d\delta w}{dx} \right)_{x=L}$$

$$= \int_0^L \left[N \left(\frac{d\delta u}{dx} + \frac{dw}{dx} \frac{d\delta w}{dx} \right) - M \frac{d^2 \delta w}{dx^2} \right] dx$$

$$- \int_0^L [f\delta u + (q - kw)\delta w] \, dx - P\delta u(L) + F\delta w(L) - M_0 \left(\frac{d\delta w}{dx} \right)_{x=L}$$

$$= \int_0^L \left[-\frac{dN}{dx} \delta u - \frac{d}{dx} \left(\frac{dw}{dx} N \right) \delta w - \frac{d^2 M}{dx^2} \delta w \right] dx$$

$$- \int_0^L [f\delta u + (q - kw)\delta w] \, dx$$

$$- P\delta u(L) + F\delta w(L) - M_0 \left(\frac{d\delta w}{dx} \right)_{x=L}$$

$$+ \left[N\delta u + \left(\frac{dw}{dx} N + \frac{dM}{dx} \right) \delta w - M \frac{d\delta w}{dx} \right]_0^L, \tag{7}$$

where N and M are the stress resultants defined by

$$N = \int_A \sigma_{xx} dA, \quad M = \int_A \sigma_{xx} z \, dA. \tag{8}$$

The Euler equations are obtained by setting the coefficients of δu and δw under the integral separately to zero:

$$\delta u : \quad -\frac{dN}{dx} = f(x), \tag{9}$$

$$\delta w : \quad -\frac{d^2 M}{dx^2} - \frac{d}{dx} \left(\frac{dw}{dx} N \right) + kw - q = 0, \tag{10}$$

in $0 < x < L$.

We note that δu, δw, and $d\delta w/dx = \delta(dw/dx)$ appear in the boundary terms. Hence, u, w, and (dw/dx) are the primary variables of the theory, and their specification constitutes the essential boundary conditions. Since nothing is said about the primary variables being specified, δu, δw, and $d\delta w/dx$ are arbitrary at $x = 0$ and $x = L$. Therefore, the natural boundary conditions are:

$$N = 0, \quad V \equiv \frac{dM}{dx} + \frac{dw}{dx}N = 0, \quad M = 0 \text{ at } x = 0, \tag{11}$$

and

$$N - P = 0, \quad \frac{dM}{dx} + \frac{dw}{dx}N + F = 0, \quad M + M_0 = 0 \text{ at } x = L. \tag{12}$$

A close examination of the steps involved in **Example 4.2.1** shows that the principle of virtual displacements can be used to derive the governing equations and associated boundary conditions of higher-order theories of beams, plates, and shells by (a) assuming a displacement expansion in powers of thickness coordinate with unknown generalized displacements, (b) computing the actual and virtual strains using a suitable measure, (c) using the principle of virtual displacements in an appropriate description of motion, (d) introducing stress resultants, and using the fundamental lemma of the calculus of variations (which requires integration by parts to relieve variations of the generalized displacements of any derivatives). These steps will be used in the coming chapters to derive the equations of equilibrium or motion of plates.

4.2.3   Unit Dummy-Displacement Method

The principle of virtual displacements can also be used, in addition to deriving equations of equilibrium, to directly determine reaction forces and displacements in structural problems. If $\mathbf{F}_0$ is the force at point O in a structure occupying the volume Ω, we can prescribe a virtual displacement $\delta\mathbf{u}_0$ at the point. The virtual displacements at all other points are zero. The virtual strains $\delta\boldsymbol{\varepsilon}^0$ due to the virtual displacement $\delta\mathbf{u}_0$ are determined from the strain–displacement relations. Then the principle of virtual displacements reads as

$$\mathbf{F}_0 \cdot \delta\mathbf{u}_0 = \int_{\Omega} \boldsymbol{\sigma} : \delta\boldsymbol{\varepsilon}^0 \, d\Omega = \int_{\Omega} \sigma_{ij} \, \delta\varepsilon^0_{ij} \, d\Omega \tag{4.2.7}$$

where σ_{ij} are the actual stresses (assumed to be known in terms of the actual displacements) and $\delta\varepsilon^0_{ij}$ are the virtual strains in the structure derived from the virtual displacement $\delta\mathbf{u}_0$, consistent with the geometric constraints. Since $\delta\mathbf{u}_0$ is arbitrary, one can take $\delta\mathbf{u}_0 = \hat{\mathbf{e}}_F$, the unit vector along the force $\mathbf{F}_0$. If $\delta\varepsilon^0_{ij}$ are the strains due to the unit virtual displacement at point O, then

$$F_0 = \int_{\Omega} \sigma_{ij} \, \delta\varepsilon^0_{ij} \, d\Omega, \tag{4.2.8}$$

where F_0 is the magnitude of the force $\mathbf{F}_0$. This procedure is called the *unit dummy-displacement method*. The word "dummy" is used historically to mean "virtual." Equation (4.2.8) is also valid for the case in which F_0 is replaced with a bending moment M_0:

$$M_0 = \int_\Omega \sigma_{ij}\, \delta\varepsilon_{ij}^0\, d\Omega. \qquad (4.2.9)$$

Equation (4.2.7) is more general than Eq. (4.2.8), in the sense that it is valid for arbitrary virtual displacement vector (i.e., the virtual displacement need not be a unit vector).

Equations (4.2.8) and (4.2.9) can be used to determine point loads and moments or deflections and rotations of structures. However, in order to use the unit dummy-displacement method for the determination of point displacements, rotations, forces, and/or moments in bar, truss, beam, or frame structures, it is necessary that the stresses and strains be expressed in terms of point displacements. Two examples of application of the unit dummy-displacement method are presented next.

Example 4.2.2

Consider the two-member truss shown in Fig. 4.2.2(a). Use the unit dummy-displacement method to determine the horizontal displacement u and vertical displacement v of point O. Use small strain assumption and obtain the displacements (u, v) when the material is (a) linearly elastic, $\sigma = E\varepsilon$, and (b) material is nonlinearly elastic with the stress–strain relation:

$$\sigma = \begin{cases} K\sqrt{\varepsilon}, & \varepsilon \geq 0, \\ -K\sqrt{-\varepsilon}, & \varepsilon \leq 0, \end{cases} \qquad (1)$$

where K is a material constant.

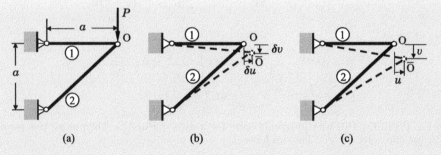

| (a) | (b) | (c) |

Fig. 4.2.2 (a) A two-member truss. (b) Virtual displacements of point O. (c) Actual displacements of point O.

Solution: Assume virtual (or dummy) displacements of δu to the right and δv downward at point O, as shown in Fig. 4.2.2(b). Thus the virtual displacement vector at point O is $\delta\mathbf{u}_0 = \delta u\,\hat{\mathbf{e}}_x + \delta v\,\hat{\mathbf{e}}_y$, where $\hat{\mathbf{e}}_y$ and $\hat{\mathbf{e}}_x$ are unit basis vectors. The load in this case is $\mathbf{F}_0 = 0\hat{\mathbf{e}}_x + P\hat{\mathbf{e}}_y$. Then we have

$$\mathbf{F}_0 \cdot \delta\mathbf{u}_0 = 0 \cdot \delta u + P \cdot \delta v = \int_\Omega \sigma_{ij}\, \delta\varepsilon_{ij}\, dv$$

$$= \int_0^{L_1} A^{(1)}\sigma^{(1)}\, \delta\varepsilon^{(1)}\, dx + \int_0^{L_2} A^{(2)}\sigma^{(2)}\, \delta\varepsilon^{(2)}\, dx$$

$$= L_1 A^{(1)}\sigma^{(1)}\, \delta\varepsilon^{(1)} + L_2 A^{(2)}\sigma^{(2)}\, \delta\varepsilon^{(2)}, \qquad (2)$$

where $A^{(1)} = A^{(2)} = A$, $L_1 = a$, and $L_2 = \sqrt{2}a$. The actual strains are computed as [see Fig. 4.2.2(c)]

$$\varepsilon^{(1)} = \frac{1}{a}\left(\sqrt{(a+u)^2 + v^2} - a\right) \approx \frac{u}{a}$$

$$\varepsilon^{(2)} = \frac{1}{\sqrt{2}a}\left(\sqrt{(a+u)^2 + (a-v)^2} - \sqrt{2}a\right) \approx \frac{u-v}{2a}, \tag{3}$$

where the binomial theorem was used and the squares of u/a and v/a were neglected in computing the strains (small strain assumption). Similarly, the virtual strains $\delta\varepsilon^{(1)}$ and $\delta\varepsilon^{(2)}$ can be computed as (alternatively, the virtual strains can be computed by taking the variation of the actual strains)

$$\delta\varepsilon^{(1)} = \frac{\delta u}{a} , \quad \delta\varepsilon^{(2)} = \frac{\delta u - \delta v}{2a}. \tag{4}$$

Substituting Eq. (4) into Eq. (2), we obtain

$$0 \cdot \delta u + P \cdot \delta v = aA\sigma^{(1)}\frac{\delta u}{a} + \sqrt{2}aA\sigma^{(2)}\frac{\delta u - \delta v}{2a}$$

$$= A\left(\sigma^{(1)} + \frac{1}{\sqrt{2}}\sigma^{(2)}\right)\delta u + A\left(-\frac{1}{\sqrt{2}}\sigma^{(2)}\right)\delta v. \tag{5}$$

Since the variations δu and δv are arbitrary, we can set them to (1) $\delta u = 1$ and $\delta v = 0$, and then to (2) $\delta u = 0$ and $\delta v = 1$ to obtain the two relations for the two displacements. Alternatively, equating the coefficients of δu and δv on two sides separately, we arrive at

$$0 = A\sigma^{(1)} + \frac{1}{\sqrt{2}}A\sigma^{(2)}, \quad P = -\frac{1}{\sqrt{2}}A\sigma^{(2)}. \tag{6}$$

The actual stresses are computed in terms of the actual displacements u and v of point O using the given stress–strain relations.

(a) For the linear stress–strain relation ($E^{(1)} = E^{(2)} = E$), we obtain

$$\sigma^{(1)} = E^{(1)}\varepsilon^{(1)} = \frac{uE}{a} , \quad \sigma^{(2)} = E^{(2)}\varepsilon^{(2)} = \frac{(u-v)E}{2a}. \tag{7}$$

Hence, we have

$$0 = A\sigma^{(1)} + \frac{1}{\sqrt{2}}A\sigma^{(2)} = AE\frac{u}{a} + \frac{1}{\sqrt{2}}AE\left(\frac{u-v}{2a}\right), \tag{8}$$

$$P = -\frac{1}{\sqrt{2}}A\sigma^{(2)} = -\frac{1}{\sqrt{2}}AE\left(\frac{u-v}{2a}\right). \tag{9}$$

Using Eq. (8) in Eq. (9), we can readily solve for u as $u = Pa/AE$. Then using this result in Eq. (9), we can solve for v. Thus, we have

$$u = \frac{Pa}{AE} , \quad v = \frac{2\sqrt{2}Pa}{AE} + u = \frac{Pa}{AE}\left(1 + 2\sqrt{2}\right). \tag{10}$$

(b) Using the stress–strain relation in Eq. (1), we have (noting that member 2 is in compression)

$$\sigma^{(1)} = K\sqrt{\frac{u}{a}} , \quad \sigma^{(2)} = -K\sqrt{\frac{(v-u)}{2a}}. \tag{11}$$

Hence, we have

$$0 = A\sigma^{(1)} + \frac{1}{\sqrt{2}}A\sigma^{(2)} = AK\sqrt{\frac{u}{a}} - \frac{1}{\sqrt{2}}AK\sqrt{\frac{v-u}{2a}}, \tag{12}$$

$$P = -\frac{1}{\sqrt{2}}A\sigma^{(2)} = \frac{1}{\sqrt{2}}AK\sqrt{\frac{v-u}{2a}}. \tag{13}$$

Using Eq. (12) in Eq. (13), we can readily solve for u as $u = P^2 a / A^2 K^2$. Then using this result in Eq. (13), we can solve for v. Thus, we have

$$u = \frac{P^2 a}{A^2 K^2}, \quad v = \frac{4P^2 a}{A^2 K^2} + u = \frac{5P^2 a}{A^2 K^2}. \tag{14}$$

Example 4.2.3

Consider a cantilever beam of length L and constant axial and bending stiffnesses EA and EI, respectively. Determine the transverse deflection and rotation of the free end of the cantilever beam when it is subjected to transverse load F_0, bending moment M_0, and axial load P_0 at its free end, as shown in Fig. 4.2.3. Use (a) the Bernoulli–Euler beam theory and (b) the Timoshenko beam theory.

Solution: First we must write the axial displacement $u(x)$ and transverse deflection $w(x)$ in terms of suitable quantities, called the *generalized coordinates*. Then we apply the unit dummy-displacement method to determine the required generalized displacements in terms of the applied loads. For this linear problem, the axial and transverse displacements are decoupled and, hence, they can be determined independent of each other.

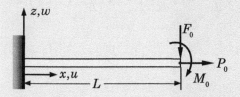

Fig. 4.2.3 A cantilever beam with applied loads at the free end.

The axial displacement $u(x)$ is governed by the equation [see Eq. (4) of **Example 3.5.3** with $f = 0$]

$$\frac{d}{dx}\left(EA\frac{du}{dx}\right) = 0 \;\rightarrow\; u(x) = a_1 + a_2 x. \tag{1}$$

Using the only displacement boundary condition $u(0) = 0$, we determine $a_1 = 0$. Then a_2 can be used as the generalized coordinate; it is useful to express a_2 in terms of the axial displacement at the free end, u_L: $u(L) = a_2 L$ or $a_2 = u(L)/L = u_L/L$. Thus, we have ($c_1 = u_L$)

$$u(x) = \left(\frac{x}{L}\right)c_1, \quad \varepsilon = \frac{du}{dx} = \frac{1}{L}c_1, \quad \delta\varepsilon^0 = \frac{1}{L}\delta c_1. \tag{2}$$

Then the unit dummy-displacement method gives ($\sigma = E\varepsilon$; $\delta c_1 = 1$)

$$P_0 \delta c_1 = \int_0^L \int_A \sigma\,\delta\varepsilon\,dA dx = \int_0^L EA\frac{1}{L^2}c_1\delta c_1\,dx = \frac{EA}{L}c_1\delta c_1, \tag{3}$$

or

$$c_1 = u_L = u(L) = \frac{P_0 L}{EA}. \tag{4}$$

(a) For the Bernoulli–Euler beam theory, we proceed to express $w(x)$ in terms of generalized coordinates, which must be identified in terms of suitable parameters. The equation governing transverse deflection $w(x)$ when $q = 0$ is

$$\frac{d^2}{dx^2}\left(EI\frac{d^2 w}{dx^2}\right) = 0 \;\Rightarrow\; w(x) = a_1 + a_2 x + a_3 x^2 + a_4 x^3. \tag{5}$$

Using the two known boundary conditions $w = 0$ and $(dw/dx) = 0$ at $x = 0$, we obtain $a_1 = 0$ and $a_2 = 0$. Next, we express the remaining constants a_3 and a_4 in terms of the displacement and rotation at the free end, $x = L$:

$$w(L) = w_L \quad \rightarrow \quad w_L = a_3 L^2 + a_4 L^3; \quad \left(-\frac{dw}{dx} \right)_{x=L} = \theta_L \quad \rightarrow \quad \theta_L = -2a_3 L - 3a_4 L^2. \quad (6)$$

Let $w_L \equiv c_2$ and $\theta_L \equiv c_3$ be the generalized coordinates. Solving Eq. (6) for a_3 and a_4 in terms of $c_2 = w_L$ and $c_3 = \theta_L$, we obtain

$$a_3 = \frac{1}{L^2} \left(3c_2 + Lc_3 \right), \quad a_4 = -\frac{1}{L^3} \left(2c_2 + Lc_3 \right). \quad (7)$$

Substituting for a_3 and a_4 into the expression for $w(x)$ in Eq. (5), we obtain

$$w(x) = \left(3\frac{x^2}{L^2} - 2\frac{x^3}{L^3} \right) c_2 + L \left(\frac{x^2}{L^2} - \frac{x^3}{L^3} \right) c_3 \equiv c_2 \phi_2(x) + c_3 \phi_3(x). \quad (8)$$

The bending strain according to the Bernoulli–Euler beam theory is

$$\varepsilon_{xx} = -z \frac{d^2 w}{dx^2} = -z \left(c_2 \frac{d^2 \phi_1}{dx^2} + c_3 \frac{d^2 \phi_2}{dx^2} \right) = -z \left[\frac{6}{L^2} \left(1 - 2\frac{x}{L} \right) c_2 + \frac{2}{L} \left(1 - 3\frac{x}{L} \right) c_3 \right] \quad (9)$$

and

$$\sigma_{xx} = E\varepsilon_{xx} = -zE \left[\frac{6}{L^2} \left(1 - 2\frac{x}{L} \right) c_2 + \frac{2}{L} \left(1 - 3\frac{x}{L} \right) c_3 \right]. \quad (10)$$

The virtual strain is

$$\delta \varepsilon_{xx}^0 = -z \left[\frac{6}{L^2} \left(1 - 2\frac{x}{L} \right) \delta c_2 + \frac{2}{L} \left(1 - 3\frac{x}{L} \right) \delta c_3 \right]. \quad (11)$$

Now using the unit dummy-load method with $\delta c_2 = 1$ and $\delta c_3 = 0$, we obtain

$$F_0 \left(-\delta c_2 \right) = \int_0^L \int_A \sigma_{xx} \left(\delta \varepsilon_{xx}^0 \right)_{\delta c_3 = 0} dAdx = \int_0^L EI \frac{d^2 w}{dx^2} \left(\frac{d^2 \delta w}{dx^2} \right)_{\delta c_3 = 0} dx$$

$$= \int_0^L EI \left[\frac{6}{L^2} \left(1 - 2\frac{x}{L} \right) c_2 + \frac{2}{L} \left(1 - 3\frac{x}{L} \right) c_3 \right] \frac{6}{L^2} \left(1 - 2\frac{x}{L} \right) \delta c_2 dx. \quad (12)$$

Next, using the unit dummy-load method with $\delta c_2 = 0$ and $\delta c_3 = 1$, we obtain

$$M_0 \delta c_3 = \int_0^L \int_A \sigma_{xx} \left(\delta \varepsilon_{xx}^0 \right)_{\delta c_2 = 0} dAdx = \int_0^L EI \frac{d^2 w}{dx^2} \left(\frac{d^2 \delta w}{dx^2} \right)_{\delta c_2 = 0} dx$$

$$= \int_0^L EI \left[\frac{6}{L^2} \left(1 - 2\frac{x}{L} \right) c_2 + \frac{2}{L} \left(1 - 3\frac{x}{L} \right) c_3 \right] \frac{2}{L} \left(1 - 3\frac{x}{L} \right) \delta c_3 \, dx. \quad (13)$$

Upon carrying out the indicated integrations in Eqs (12) and (13), we obtain

$$-F_0 = \frac{12EI}{L^3} c_2 + \frac{6EI}{L^2} c_3, \quad M_0 = \frac{6EI}{L^2} c_2 + \frac{4EI}{L} c_3, \quad (14)$$

whose solution gives the deflection and rotation of the free end

$$c_2 = w_L = w(L) = -\left(\frac{F_0 L^3}{3EI} + \frac{M_0 L^2}{2EI} \right), \quad c_3 = \theta_L = \left(-\frac{dw}{dx} \right)_{x=L} = \frac{F_0 L^2}{2EI} + \frac{M_0 L}{EI}. \quad (15)$$

These are the exact solutions (one can verify the results with those available from a book on mechanics of materials), as we have not introduced any approximations.

(b) For the Timoshenko beam theory, we must express $w(x)$ and ϕ_x in terms of the generalized coordinates, namely, the values of w and ϕ_x at $x = L$. The equations governing transverse deflection $w(x)$ and rotation ϕ_x are [see Eqs (10) and (11) of **Example 3.5.4** with $q = 0$]

$$\frac{d}{dx}\left[K_sGA\left(\phi_x + \frac{dw}{dx}\right)\right] = 0, \quad -\frac{d}{dx}\left(EI\frac{d\phi_x}{dx}\right) + K_sGA\left(\phi_x + \frac{dw}{dx}\right) = 0, \quad (16)$$

where K_s is the shear correction factor. The solution to these two equations is

$$w(x) = -\frac{1}{EI}\left(b_1\frac{x^3}{6} + b_2\frac{x^2}{2} + b_3x + b_4\right) + \frac{b_1}{GAK_s}x, \quad (17)$$

$$\phi_x(x) = \frac{1}{EI}\left(b_1\frac{x^2}{2} + b_2x + b_3\right). \quad (18)$$

Using the two known boundary conditions $w = 0$ and $\phi_x = 0$ at $x = 0$, we obtain $b_4 = 0$ and $b_3 = 0$. Next, we express the remaining constants b_1 and b_2 in terms of the displacement and rotation at the free end, $x = L$:

$$w(L) = w_L \quad \to \quad w_L = -\frac{1}{EI}\left(\frac{L^3}{6}b_1 + \frac{L^2}{2}b_2\right) + \frac{b_1L}{GAK_s}$$

$$\phi_x(L) = \phi_L \quad \to \quad \phi_L = \frac{1}{EI}\left(\frac{L^2}{2}b_1 + b_2L\right). \quad (19)$$

Let $w_L \equiv \hat{c}_2$ and $\phi_L \equiv \hat{c}_3$ be the generalized coordinates. Solving for b_1 and b_2 and substituting into Eqs (17) and (18), we obtain

$$w(x) = \hat{c}_2\hat{\phi}_3(x) + \hat{c}_3\hat{\phi}_4(x), \quad \phi_x(x) = \hat{c}_2\varphi_3(x) + \hat{c}_3\varphi_4(x), \quad (20)$$

where

$$\hat{\phi}_3 = \frac{1}{\mu}\left[\left(3 - 2\frac{x}{L}\right)\left(\frac{x}{L}\right)^2 + 12\Lambda\frac{x}{L}\right],$$

$$\hat{\phi}_4 = \frac{L}{\mu}\left[\left(1 - \frac{x}{L}\right)\left(\frac{x}{L}\right)^2 + 6\Lambda\left(1 - \frac{x}{L}\right)\frac{x}{L}\right],$$

$$\varphi_3 = -\frac{6}{L\mu}\left(1 - \frac{x}{L}\right)\frac{x}{L}, \quad (21)$$

$$\varphi_4 = \frac{1}{\mu}\left[3\left(\frac{x}{L}\right)^2 - 2\frac{x}{L} + 12\Lambda\frac{x}{L}\right]$$

and

$$\mu = 1 + 12\Lambda, \quad \Lambda = \frac{EI}{GAK_sL^2}. \quad (22)$$

The bending strains according to the Timoshenko beam theory are

$$\varepsilon_{xx} = z\frac{d\phi_x}{dx} = z\left(\hat{c}_2\frac{d\varphi_3}{dx} + \hat{c}_3\frac{d\varphi_4}{dx}\right) = z\left[-\frac{6}{\mu L^2}\left(1 - 2\frac{x}{L}\right)\hat{c}_2 - \frac{2}{\mu L}\left(1 - 3\frac{x}{L} - 6\Lambda\right)\hat{c}_3\right],$$

$$\gamma_{xz} = \phi_x + \frac{dw}{dx} = \hat{c}_2\left(\varphi_3 + \frac{d\hat{\phi}_3}{dx}\right) + \hat{c}_3\left(\varphi_4 + \frac{d\hat{\phi}_4}{dx}\right) = \frac{12\Lambda}{\mu L}\hat{c}_2 + \frac{6\Lambda}{\mu}\hat{c}_3, \quad (23)$$

and

$$\sigma_{xx} = E\varepsilon_{xx} = Ez\left[-\frac{6}{\mu L^2}\left(1 - 2\frac{x}{L}\right)\hat{c}_2 - \frac{2}{\mu L}\left(1 - 3\frac{x}{L} - 6\Lambda\right)\hat{c}_3\right],$$

$$\sigma_{xz} = G\gamma_{xz} = G\left(\frac{12\Lambda}{\mu L}\hat{c}_2 + \frac{6\Lambda}{\mu}\hat{c}_3\right) = G\frac{6\Lambda}{\mu L}\left(2\hat{c}_2 + L\hat{c}_3\right). \quad (24)$$

The virtual strains are

$$
\delta\varepsilon_{xx}^0 = z\left[-\frac{6}{\mu L^2}\left(1 - 2\frac{x}{L}\right)\delta\hat{c}_2 - \frac{2}{\mu L}\left(1 - 3\frac{x}{L} - 6\Lambda\right)\delta\hat{c}_3\right],
$$

$$
\delta\gamma_{xz}^0 = \frac{6\Lambda}{\mu L}\left(2\,\delta\hat{c}_2 + L\,\delta\hat{c}_3\right).
$$

$$(25)$$

Now using the unit dummy-load method with $\delta\hat{c}_2 = 1$ and $\delta\hat{c}_3 = 0$, we obtain

$$
F_0\left(-\delta\hat{c}_2\right) = \int_0^L \int_A \left[\sigma_{xx}\left(\delta\varepsilon_{xx}^0\right)_{\delta\hat{c}_3=0} + K_s\sigma_{xz}\left(\delta\gamma_{xz}^0\right)_{\delta\hat{c}_3=0}\right] dA\,dx
$$

$$
= \int_0^L \left[EI\frac{d\phi_x}{dx}\left(\frac{d\delta\phi_x}{dx}\right)_{\delta\hat{c}_3=0} + GAK_s\left(\phi_x + \frac{dw}{dx}\right)\left(\delta\phi_x + \frac{d\delta w}{dx}\right)_{\delta\hat{c}_3=0}\right] dx
$$

or

$$
-F_0 = \int_0^L EI\left[\frac{6}{\mu L^2}\left(1 - 2\frac{x}{L}\right)\hat{c}_2 + \frac{2}{\mu L}\left(1 - 3\frac{x}{L} - 6\Lambda\right)\hat{c}_3\right]\frac{6}{\mu L^2}\left(1 - 2\frac{x}{L}\right) dx
$$

$$
+ \int_0^L GAK_s\left(\frac{12\Lambda}{\mu L}\hat{c}_2 + \frac{6\Lambda}{\mu}\hat{c}_3\right)\frac{12\Lambda}{\mu L} dx
$$

$$
= \frac{12EI}{\mu^2 L^3}\hat{c}_2 + \frac{6EI}{\mu^2 L^2}\hat{c}_3 + GAK_s\left(\frac{12\Lambda}{\mu L}\hat{c}_2 + \frac{6\Lambda}{\mu}\hat{c}_3\right)\frac{12\Lambda}{\mu}
$$

$$
= \frac{12EI}{\mu L^3}\hat{c}_2 + \frac{6EI}{\mu L^2}\hat{c}_3.
$$

$$(26)$$

Next, using the unit dummy-load method with $\delta\hat{c}_2 = 0$ and $\delta\hat{c}_3 = 1$, we obtain

$$
M_0\,\delta\hat{c}_3 = \int_0^L \int_A \left[\sigma_{xx}\left(\delta\varepsilon_{xx}^0\right)_{\delta\hat{c}_2=0} + \sigma_{xz}\left(\delta\gamma_{xz}^0\right)_{\delta\hat{c}_2=0}\right] dA\,dx
$$

$$
= \int_0^L \left[EI\frac{d\phi_x}{dx}\left(\frac{d\delta\phi_x}{dx}\right)_{\delta\hat{c}_2=0} + GAK_s\left(\phi_x + \frac{dw}{dx}\right)\left(\delta\phi_x + \frac{d\delta w}{dx}\right)_{\delta\hat{c}_2=0}\right] dx
$$

or

$$
M_0 = \int_0^L EI\left[\frac{6}{\mu L^2}\left(1 - 2\frac{x}{L}\right)\hat{c}_2 + \frac{2}{\mu L}\left(1 - 3\frac{x}{L} - 6\Lambda\right)\hat{c}_3\right]\frac{2}{\mu L}\left(1 - 3\frac{x}{L} - 6\Lambda\right) dx
$$

$$
+ \int_0^L GAK_s\left(\frac{12\Lambda}{\mu L}\hat{c}_2 + \frac{6\Lambda}{\mu}\hat{c}_3\right)\frac{6\Lambda}{\mu} dx
$$

$$
= \frac{6EI}{\mu^2 L^2}\hat{c}_2 + \frac{4EI}{\mu^2 L}\left(1 + 6\Lambda + 36\Lambda^2\right)\hat{c}_3 + GAK_s\left(\frac{12\Lambda}{\mu L}\hat{c}_2 + \frac{6\Lambda}{\mu}\hat{c}_3\right)\frac{6\Lambda L}{\mu}
$$

$$
= \frac{6EI}{\mu L^2}\hat{c}_2 + \frac{4EI}{\mu L}\left(1 + 3\Lambda\right)\hat{c}_3.
$$

$$(27)$$

The solution of Eqs (26) and (27) yields the deflection and rotation of the free end:

$$
\hat{c}_2 = w(L) = -\left[\frac{F_0 L^3}{3EI}\left(1 + 3\Lambda\right) + \frac{M_0 L^2}{2EI}\right], \quad \hat{c}_3 = \phi_x(L) = \frac{F_0 L^2}{2EI} + \frac{M_0 L}{EI}.
$$

$$(28)$$

Thus, the deflection $w(L)$ predicted by the Timoshenko beam theory is larger than that predicted by the Bernoulli–Euler beam theory (and it is the same if only M_0 is applied), while the rotation predicted by the two theories is the same, $\phi_x(L) = -(dw/dx)(L)$, for the present problem. Equations (8) and (20) can be used to determine the solution at any x along the length of the beam.

The procedure discussed in **Example 4.2.3** to express the displacements $u(x)$ and $w(x)$ in terms of unknown generalized displacement degrees of freedom holds for bar and beam structures. When distributed load $q(x) \neq 0$, one may convert it to a set of statically equivalent point loads F_i acting at the same points and directions as the generalized displacements c_i of the beam. For bars and beams with constant EA and EI but arbitrary load $q(x)$, this procedure results in exact values of the generalized displacements c_i. The procedure can be generalized to bars and beams with arbitrary geometric and material properties. However, in such cases, the solutions obtained are approximate because the polynomial expansions used in Eqs (1), (5), (17), and (18) of **Example 4.2.3** are based on constant EA and EI, and they become approximate when EA and EI are functions of x.

It is also informative to note that the procedure can be generalized to a bar or beam by expressing the unknown parameters a_i in Eq. (1) or (5) of **Example 4.2.3** in terms of the generalized displacements at the end points, as discussed in **Example 3.2.4** [see, in particular, Eqs (1) and (2) of **Example 3.2.4**]. This will be discussed further in **Examples 4.3.3** and **4.3.5**. As will be shown in Chapter 8, this is the same procedure that is used to develop the finite element solutions. When the homogeneous solution to the governing equations is algebraic, this procedure yields the exact solutions for the generalized displacements and forces at the ends of the bars and beams.

4.3 The Principle of Minimum Total Potential Energy and Castigliano's Theorem I

4.3.1 The Principle of Minimum Total Potential Energy

The principle of virtual work discussed in the previous section is applicable to any continuous body with arbitrary constitutive behavior (e.g., linear or nonlinear elastic materials). The principle of minimum total potential energy is obtained as a special case from the principle of virtual displacements when the constitutive relations can be obtained from a potential function. Here we restrict our discussion to materials that admit existence of a strain energy potential such that the stress is derivable from it. Such materials are termed *hyperelastic*.

For elastic bodies (in the absence of temperature variations), there exists a strain energy potential U_0 such that [see Eq. (3.2.6)]

$$\sigma_{ij} = \frac{\partial U_0}{\partial \varepsilon_{ij}}. \tag{4.3.1}$$

The strain energy density U_0 is a function of strains at a point and is assumed to be positive definite. The statement of the principle of virtual displacements,

$\delta W = 0$, can be expressed in terms of the strain energy density U_0 as

$$0 = \delta W = \int_\Omega \sigma_{ij}\,\delta\varepsilon_{ij}\,d\Omega - \left[\int_\Omega \mathbf{f}\cdot\delta\mathbf{u}\,d\Omega + \int_{\Gamma_\sigma} \hat{\mathbf{t}}\cdot\delta\mathbf{u}\,ds\right]$$

$$= \int_\Omega \frac{\partial U_0}{\partial\varepsilon_{ij}}\delta\varepsilon_{ij}\,d\Omega + \delta V_E$$

$$= \int_\Omega \delta U_0\,d\Omega + \delta V_E = \delta(U + V_E) \equiv \delta\Pi, \qquad (4.3.2)$$

where

$$V_E = -\left[\int_\Omega \mathbf{f}\cdot\mathbf{u}\,d\Omega + \int_{\Gamma_\sigma} \hat{\mathbf{t}}\cdot\mathbf{u}\,ds\right] \qquad (4.3.3)$$

is the potential energy due to external loads and U is the strain energy potential

$$U = \int_\Omega U_0\,d\Omega. \qquad (4.3.4)$$

Note that we have used a constitutive equation, Eq. (4.3.1), in arriving at Eq. (4.3.2). As already defined in Section 3.3, the sum $V_E + U \equiv \Pi$ is called the *total potential energy*, and the statement

$$\delta\Pi \equiv \delta(U + V_E) = 0 \qquad (4.3.5)$$

is known as the *principle of minimum total potential energy*. It is a statement of the fact that the energy of the system is the minimum only at its equilibrium configuration $\mathbf{u}$:

$$\Pi(\mathbf{u} + \alpha\mathbf{v}) \geq \Pi(\mathbf{u}) \quad \text{for all scalars } \alpha \text{ and admissible variations } \mathbf{v}. \qquad (4.3.6)$$

The equality holds only for $\alpha = 0$.

Next, we show that the principle of minimum total potential energy, $\delta\Pi = 0$, as applied to an isotropic linear elastic solid in three dimensions yields the equations of equilibrium and force boundary conditions in terms of displacements. Suppose that the body occupying the volume Ω is subjected to body force $\mathbf{f}$ (measured per unit volume), prescribed surface traction $\hat{\mathbf{t}}$ (measured per unit area) on portion Γ_σ, and specified displacements $\hat{\mathbf{u}}$ on portion Γ_u of the total surface Γ of Ω.

The total potential energy for the problem at hand is given by [see Eq. (3.3.1)]

$$\Pi(\mathbf{u}) = \int_\Omega \left[\tfrac{1}{2}\boldsymbol{\sigma}(\boldsymbol{\varepsilon}) : \boldsymbol{\varepsilon} - \mathbf{f}\cdot\mathbf{u}\right]d\Omega - \int_{\Gamma_\sigma} \hat{\mathbf{t}}\cdot\mathbf{u}\,d\Gamma$$

$$= \int_\Omega \left(\tfrac{1}{2}\sigma_{ij}\varepsilon_{ij} - f_i u_i\right)d\Omega - \int_{\Gamma_\sigma} \hat{t}_i u_i\,d\Gamma, \qquad (4.3.7)$$

where u_i denote the components of the displacement vector $\mathbf{u}$, σ_{ij} are the components of the stress tensor $\boldsymbol{\sigma}$, ε_{ij} are the components of the strain tensor

ε, f_i are the components of the body force vector $\mathbf{f}$, and t_i are the components of the stress vector (often called traction vector) $\mathbf{t}$ – all referred to a rectangular Cartesian coordinate system. The stress vector $\mathbf{t}$ is related to the stress tensor $\boldsymbol{\sigma}$ by Cauchy's formula [see Eq. (2.2.5)]:

$$\mathbf{t} = \hat{\mathbf{n}} \cdot \boldsymbol{\sigma}. \tag{4.3.8}$$

Here $\hat{\mathbf{n}}$ denotes the unit normal vector to the surface Γ of the domain Ω. The first term under the volume integral represents the strain energy density of the elastic body, the second term represents the work done by the body force $\mathbf{f}$, and the surface integral denotes the work done by the specified traction $\hat{\mathbf{t}}$.

In order to express the strain energy in terms of the displacements, we must employ the stress–strain relations and strain–displacement equations. For an isotropic body, the stress–strain relations are given by

$$\boldsymbol{\sigma} = 2\mu\boldsymbol{\varepsilon} + \lambda\,\mathrm{tr}\,(\boldsymbol{\varepsilon})\,\mathbf{I} \quad (\sigma_{ij} = 2\mu\varepsilon_{ij} + \lambda\delta_{ij}\varepsilon_{kk}), \tag{4.3.9}$$

where μ and λ are the Lamé constants and $\mathrm{tr}\,(\boldsymbol{\varepsilon})$ denotes the trace of the strain tensor (i.e., sum of the diagonal elements, $\varepsilon_{kk} = \varepsilon_{11} + \varepsilon_{22} + \varepsilon_{33}$). Hence,

$$\boldsymbol{\sigma} : \boldsymbol{\varepsilon} = \sigma_{ij}\varepsilon_{ij} = 2\mu\varepsilon_{ij}\varepsilon_{ij} + \lambda\varepsilon_{ii}\varepsilon_{kk}, \tag{4.3.10}$$

wherein sum on repeated indices is assumed.

The strain-displacement relations of the linearized elasticity are

$$\varepsilon_{ij} = \tfrac{1}{2}\left(u_{i,j} + u_{j,i}\right), \tag{4.3.11}$$

where $u_{i,j} = (\partial u_i/\partial x_j)$. Substituting Eqs (4.3.9) and (4.3.11) into Eq. (4.3.7), we obtain the final expression for the total potential energy in terms of the displacements:

$$\Pi(\mathbf{u}) = \int_{\Omega}\left[\frac{\mu}{4}\left(u_{i,j} + u_{j,i}\right)\left(u_{i,j} + u_{j,i}\right) + \frac{\lambda}{2}u_{i,i}u_{k,k} - f_i u_i\right]d\Omega$$
$$- \int_{\Gamma_\sigma}\hat{t}_i u_i\,d\Gamma. \tag{4.3.12}$$

Now we use the principle of minimum total potential energy, $\delta\Pi = 0$, and derive the Euler equations associated with the total potential energy functional Π in Eq. (4.3.12). Since the body is subjected to specified displacements on the portion Γ_u of the surface, the first variation of the displacement vector is zero on Γ_u:

$$\mathbf{u} = \hat{\mathbf{u}} \quad \text{and} \quad \delta\mathbf{u} = \mathbf{0} \quad \text{on } \Gamma_u. \tag{4.3.13}$$

Setting the first variation of Π to zero, we obtain

$$0 = \int_{\Omega}\left[\frac{\mu}{2}\left(\delta u_{i,j} + \delta u_{j,i}\right)\left(u_{i,j} + u_{j,i}\right) + \lambda\delta u_{i,i}u_{k,k} - f_i\delta u_i\right]d\Omega - \int_{\Gamma_\sigma}\hat{t}_i\delta u_i\,d\Gamma,$$
$$\tag{4.3.14}$$

wherein the product rule of variation is used and similar terms are combined. Using integration by parts (or the Green–Gauss theorem), we obtain

$$\int_\Omega \delta u_{i,j} \left(u_{i,j} + u_{j,i}\right) d\Omega = -\int_\Omega \delta u_i \left(u_{i,j} + u_{j,i}\right)_{,j} d\Omega + \oint_\Gamma \delta u_i \left(u_{i,j} + u_{j,i}\right) n_j \, d\Gamma,$$

where n_j denotes the jth direction cosine of the unit normal to the surface. Hence, Eq. (4.3.14) becomes

$$0 = \int_\Omega \left[-\frac{\mu}{2} \left(u_{i,j} + u_{j,i}\right)_{,j} \delta u_i - \frac{\mu}{2} \left(u_{i,j} + u_{j,i}\right)_{,i} \delta u_j - \lambda u_{k,ki} \delta u_i - f_i \delta u_i \right] d\Omega$$

$$+ \oint_\Gamma \left[\frac{\mu}{2} \left(u_{i,j} + u_{j,i}\right) \left(n_j \delta u_i + n_i \delta u_j\right) + \lambda u_{k,k} n_i \delta u_i \right] d\Gamma - \int_{\Gamma_\sigma} \delta u_i \hat{t}_i \, d\Gamma$$

$$= \int_\Omega \left[-\mu \left(u_{i,j} + u_{j,i}\right)_{,j} - \lambda u_{k,ki} - f_i \right] \delta u_i \, d\Omega$$

$$+ \oint_\Gamma \left[\mu \left(u_{i,j} + u_{j,i}\right) + \lambda u_{k,k} \delta_{ij} \right] n_j \delta u_i \, d\Gamma - \int_{\Gamma_\sigma} \delta u_i \hat{t}_i \, d\Gamma. \tag{4.3.15}$$

In arriving at the last step, change of dummy indices is made to combine terms. Recognizing that the expression inside the square brackets of the closed surface integral is nothing but σ_{ij} and $[\mu \left(u_{i,j} + u_{j,i}\right) + \lambda u_{k,k} \delta_{ij}] n_j = \sigma_{ij} n_j = t_i$ by Cauchy's formula, we can write

$$\oint_\Gamma t_i \delta u_i \, d\Gamma = \int_{\Gamma_u} t_i \delta u_i \, d\Gamma + \int_{\Gamma_\sigma} t_i \delta u_i \, d\Gamma = \int_{\Gamma_\sigma} t_i \delta u_i \, d\Gamma.$$

The integral over Γ_u is zero by virtue of Eq. (4.3.13). Hence, Eq. (4.3.15) becomes

$$0 = \int_\Omega \left[-\mu \left(u_{i,j} + u_{j,i}\right)_{,j} - \lambda u_{k,ki} - f_i \right] \delta u_i \, d\Omega + \int_{\Gamma_\sigma} \left(t_i - \hat{t}_i\right) \delta u_i \, d\Gamma. \tag{4.3.16}$$

Using the fundamental lemma of calculus of variations, we set the coefficients of δu_i in Ω and δu_i on Γ_σ to zero separately and obtain the Euler equations

$$\mu \left(u_{i,jj} + u_{j,ij}\right) + \lambda u_{k,ki} + f_i = 0 \text{ in } \Omega \quad \text{and} \quad t_i - \hat{t}_i = 0 \text{ on } \Gamma_\sigma, \tag{4.3.17}$$

for $i = 1, 2, 3$. Equation (4.3.17) is the well-known Navier equation of elasticity (contains three equations), which can be expressed in vector form as

$$\mu \nabla^2 \mathbf{u} + (\lambda + \mu) \nabla (\nabla \cdot \mathbf{u}) + \mathbf{f} = \mathbf{0} \text{ in } \Omega, \quad \text{and} \quad \mathbf{t} - \hat{\mathbf{t}} = \mathbf{0} \text{ on } \Gamma_\sigma. \tag{4.3.18}$$

The principle of virtual displacements and the principle of minimum total potential energy give, when applied to an elastic body, the equilibrium equations as the Euler equations. The main difference between them is that the principle of virtual displacements gives the equilibrium equations in terms of stresses or stress resultants, whereas the principle of minimum total potential energy gives them in terms of the displacements because, in the latter, constitutive and kinematic relations are used to replace the stresses (or stress resultants) in terms of the displacements. **Examples 3.5.3** and **3.5.4** illustrate these ideas.

Next, we establish the minimum character of the total potential energy functional $\Pi(u)$ from Eq. (4.3.12). We begin with $\Pi(\bar{\mathbf{u}})$, where $\bar{\mathbf{u}}$ is an admissible configuration, which is of the form $\bar{\mathbf{u}} = \mathbf{u} + \alpha\mathbf{v}$, where $\mathbf{u}$ is the actual solution of Navier's equations [see Eq. (4.3.18)], $\mathbf{v}$ is the admissible variation that vanishes on Γ_u, and α is an arbitrary real number. We have

$$
\Pi(\mathbf{u} + \alpha\mathbf{v}) = \int_\Omega \left[\frac{\mu}{4} \left(u_{i,j} + \alpha v_{i,j} + u_{j,i} + \alpha v_{j,i}\right) \left(u_{i,j} + \alpha v_{i,j} + u_{j,i} + \alpha v_{j,i}\right) \right.
$$

$$
\left. + \frac{\lambda}{2} \left(u_{i,i} + \alpha v_{i,i}\right) \left(u_{k,k} + \alpha v_{k,k}\right) - f_i \left(u_i + \alpha v_i\right) \right] d\Omega
$$

$$
- \int_{\Gamma_\sigma} \hat{t}_i (u_i + \alpha v_i) \, d\Gamma
$$

$$
= \Pi(\mathbf{u}) + \alpha^2 \int_\Omega \left[\frac{\mu}{4} \left(v_{i,j} + v_{j,i}\right)\left(v_{i,j} + v_{j,i}\right) + \frac{\lambda}{2} v_{i,i} v_{k,k} \right] d\Omega
$$

$$
+ \alpha \int_\Omega \left[\frac{\mu}{2} \left(u_{i,j} + u_{j,i}\right)\left(v_{i,j} + v_{j,i}\right) + \lambda\, u_{k,k} v_{i,i} - f_i v_i \right] d\Omega
$$

$$
- \alpha \int_{\Gamma_\sigma} \hat{t}_i v_i \, d\Gamma \tag{4.3.19}
$$

$$
= \Pi(\mathbf{u}) + \alpha^2 \int_\Omega \left[\frac{\mu}{4} \left(v_{i,j} + v_{j,i}\right)\left(v_{i,j} + v_{j,i}\right) + \frac{\lambda}{2} v_{i,i} v_{k,k} \right] d\Omega
$$

$$
- \alpha \int_\Omega \left[\mu \left(u_{i,jj} + u_{j,ij}\right) + \lambda\, u_{k,ki} + f_i \right] v_i \, d\Omega - \alpha \int_{\Gamma_\sigma} \hat{t}_i v_i \, d\Gamma
$$

$$
+ \alpha \oint_\Gamma \left[\mu \left(u_{i,j} + u_{j,i}\right) + \lambda\, u_{k,k} \delta_{ij} \right] n_j v_i \, d\Gamma, \tag{4.3.20}
$$

where integration by parts is used in arriving at the last step. Since u_i is the exact solution, the volume integral involving α is zero. Now consider the surface integrals

$$
-\alpha \int_{\Gamma_\sigma} \hat{t}_i v_i \, d\Gamma + \alpha \oint_\Gamma \left[\mu \left(u_{i,j} + u_{j,i}\right) + \lambda\, u_{k,k} \delta_{ij} \right] n_j v_i \, d\Gamma
$$

$$
= \alpha \int_{\Gamma_\sigma} \left\{ \left[\mu \left(u_{i,j} + u_{j,i}\right) + \lambda\, u_{k,k} \delta_{ij} \right] n_j - \hat{t}_i \right\} v_i \, d\Gamma = 0,
$$

where the surface integral over Γ_u is set to zero because $v_i = 0$ there and the remaining expression is zero because u_i satisfies the force boundary condition $\left[\mu \left(u_{i,j} + u_{j,i}\right) + \lambda u_{k,k} \delta_{ij} \right] n_j - \hat{t}_i = 0$. Thus, we have the result from Eq. (4.3.19):

$$
\Pi(\mathbf{u} + \alpha\mathbf{v}) = \Pi(\mathbf{u}) + \alpha^2 \int_\Omega \left[\frac{\mu}{4} \left(v_{i,j} + v_{j,i}\right)\left(v_{i,j} + v_{j,i}\right) + \frac{\lambda}{2} v_{i,i} v_{k,k} \right] d\Omega
$$

$$
\geq \Pi(\mathbf{u}).
$$

We close this section with a specific example of the Bernoulli–Euler beam theory with the von Kármán nonlinearity.

Example 4.3.1

Consider the bending of a beam according to the Bernoulli–Euler beam theory with the von Kármán nonlinear strain (see Fig. 4.3.1 and **Example 4.2.1**). Construct the total potential energy functional and then determine the governing equation and boundary conditions. Also, examine if the functional $\Pi(u, w)$ attains its minimum at equilibrium.

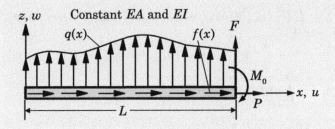

Fig. 4.3.1 A beam with applied loads.

Solution: From Eq. (6) of **Example 4.2.1** that the virtual strain energy of the Bernoulli–Euler beam theory for the linear elastic case (i.e., obeys Hooke's law, $\sigma_{xx} = E\varepsilon_{xx}$) is given by

$$
\delta W_I = \delta U = \int_0^L \int_A \sigma_{xx}\delta\varepsilon_{xx}\,dAdx = \int_0^L \int_A E\varepsilon_{xx}\,\delta\varepsilon_{xx}\,dAdx
$$

$$
= \frac{1}{2}\delta\left[\int_0^L \int_A E\left(\varepsilon_{xx}\right)^2 dxdA\right]
$$

$$
= \frac{1}{2}\delta\left\{\int_0^L \int_A E\left[\frac{du}{dx} + \frac{1}{2}\left(\frac{dw}{dx}\right)^2 - z\frac{d^2w}{dx^2}\right]^2 dxdA\right\}, \tag{1}
$$

where L is the length, A is the cross-sectional area, I is the second moment of area about the axis (y) of bending, and E is Young's modulus of the beam material. Thus, we have

$$
U = \frac{1}{2}\left\{\int_0^L \int_A E\left[\frac{du}{dx} + \frac{1}{2}\left(\frac{dw}{dx}\right)^2 - z\frac{d^2w}{dx^2}\right]^2 dxdA\right\}
$$

$$
= \frac{1}{2}\int_0^L \left\{EA\left[\frac{du}{dx} + \frac{1}{2}\left(\frac{dw}{dx}\right)^2\right]^2 + EI\left(\frac{d^2w}{dx^2}\right)^2\right\} dx \tag{2}
$$

where the fact that the x-axis coincides with the centroidal axis is used; that is,

$$
\int_A z\,dA = 0. \tag{3}
$$

To compute the potential energy due to applied load, suppose that the beam is subjected to distributed axial force $f(x)$, distributed transverse load $q(x)$, horizontal point load P at

$x = L$, transverse point load F at $x = L$, and bending moment M_0 at $x = L$, as shown in Fig. 4.3.1. Then the potential energy of applied forces is given by

$$V_E = -\left[\int_0^L (fu + qw)\,dx + Pu(L) + Fw(L) + M_0 \left(-\frac{dw}{dx}\right)_{x=L}\right]. \tag{4}$$

Note that $\theta_x = -dw/dx$ is the rotation in the counterclockwise sense.

The total potential energy of the beam is $\Pi = U + V_E$:

$$\Pi(u,w) = \frac{1}{2}\int_0^L \left\{ EA \left[\frac{du}{dx} + \frac{1}{2}\left(\frac{dw}{dx}\right)^2\right]^2 + EI\left(\frac{d^2w}{dx^2}\right)^2 \right\} dx$$

$$- \int_0^L (fu + qw)\,dx - \left[Pu(L) + Fw(L) - M_0 \frac{dw}{dx}\Big|_{x=L}\right]. \tag{5}$$

Applying the principle of minimum total potential energy, $\delta\Pi = 0$, and using the tools of variational calculus, we obtain

$$0 = \delta\Pi = \int_0^L \left\{ EA \left[\frac{du}{dx} + \frac{1}{2}\left(\frac{dw}{dx}\right)^2\right]\left(\frac{d\delta u}{dx} + \frac{dw}{dx}\frac{d\delta w}{dx}\right) + EI\frac{d^2w}{dx^2}\frac{d^2\delta w}{dx^2} \right\} dx$$

$$- \int_0^L (f\delta u + q\delta w)\,dx - \left[P\,\delta u(L) + F\,\delta w(L) - M_0 \frac{d\delta w}{dx}\Big|_{x=L}\right]$$

$$= \int_0^L \left[-\frac{dN}{dx}\delta u - \frac{d}{dx}\left(\frac{dw}{dx}N\right)\delta w + \frac{d^2}{dx^2}\left(EI\frac{d^2w}{dx^2}\right)\delta w \right] dx$$

$$+ [N\,\delta u]_0^L + \left[EI\frac{d^2w}{dx^2}\frac{d\delta w}{dx}\right]_0^L + \left[\frac{dw}{dx}N\,\delta w - \frac{d}{dx}\left(EI\frac{d^2w}{dx^2}\right)\delta w\right]_0^L$$

$$- \left[\int_0^L (f\delta u + q\delta w)\,dx + P\delta u(L) + F\delta w(L) - M_0\frac{d\delta w}{dx}\Big|_{x=L}\right], \tag{6}$$

where

$$N \equiv EA \left[\frac{du}{dx} + \frac{1}{2}\left(\frac{dw}{dx}\right)^2\right]. \tag{7}$$

An examination of the boundary terms resulting from integration by parts shows that the primary and secondary variables of the theory are

$$\text{Primary variables:} \quad u, \quad\quad w, \quad\quad -\frac{dw}{dx}$$

$$\text{Secondary variables:} \quad N, \quad V \equiv \frac{dM}{dx} + N\frac{dw}{dx}, \quad M \equiv -EI\frac{d^2w}{dx^2}. \tag{8}$$

The Euler equations resulting from the principle of the minimum total potential energy are

$$\delta u: \quad\quad -\frac{d}{dx}\left\{EA\left[\frac{du}{dx} + \frac{1}{2}\left(\frac{dw}{dx}\right)^2\right]\right\} - f = 0, \tag{9}$$

$$\delta w: \quad -\frac{d}{dx}\left\{EA\frac{dw}{dx}\left[\frac{du}{dx} + \frac{1}{2}\left(\frac{dw}{dx}\right)^2\right]\right\} + \frac{d^2}{dx^2}\left(EI\frac{d^2w}{dx^2}\right) - q = 0. \tag{10}$$

Equations (9) and (10) are the same as Eqs (2.5.45) and (2.5.46), except they are expressed here in terms of the displacements [see Eqs (7) and (8) for the definitions of N, M, and V in terms of the displacements u and w].

The natural boundary conditions at the ends of the beam are dictated by the duality listed in Eq. (8). For example, when u is *not* specified at a boundary point, N must be known (or specified) at the point; when w is not specified, then V must be specified; and when $-(dw/dx)$ is not specified, M should be known. For the beam shown in Fig. 4.3.1, there are no specified geometric (or essential) boundary conditions. The known force (or natural) boundary conditions are [see Eqs (11) and (12) of **Example 4.2.1**]

$$N(0) = 0, \quad M(0) = 0, \quad V(0) = 0; \quad N(L) = P, \quad M(L) = M_0, \quad V(L) = F, \qquad (11)$$

where N, M, and V are defined in Eqs (7) and (8) in terms of the displacements u and w. Thus, *the principle of minimum total potential energy yields the equations of equilibrium as well as the natural boundary conditions in terms of the displacements.*

Now we examine if the total potential energy functional Π attains its minimum value at (u, w), where u and w satisfy the equations of equilibrium in Eqs (9) and (10). We begin with Π at $\bar{u} = u + \alpha u_0$ and $\bar{w} = w + \beta w_0$, where α and β are real numbers and u_0 and w_0 are admissible variations of u and w (i.e., u_0 and w_0 are arbitrary except that they vanish at the points where u and w are specified) and check if $\Pi(\bar{u}, \bar{w}) \geq \Pi(u, w)$ holds. We have

$$\Pi(\bar{u}, \bar{w}) = \frac{1}{2} \int_0^L \left\{ EA \left[\frac{du}{dx} + \alpha \frac{du_0}{dx} + \frac{1}{2} \left(\frac{dw}{dx} + \beta \frac{dw_0}{dx} \right)^2 \right]^2 + EI \left(\frac{d^2w}{dx^2} + \beta \frac{d^2w_0}{dx^2} \right)^2 \right\} dx$$

$$- \int_0^L [f(u + \alpha u_0) + q(w + \beta w_0)] \, dx - \left[Pu(L) + \alpha Pu_0(L) + Fw(L) \right.$$

$$\left. + \beta Fw_0(L) - M_0 \frac{dw}{dx} \bigg|_{x=L} - \beta M_0 \frac{dw_0}{dx} \bigg|_{x=L} \right]$$

$$= \frac{1}{2} \int_0^L \left\{ EA \left[\frac{du}{dx} + \frac{1}{2} \left(\frac{dw}{dx} \right)^2 \right]^2 + EI \left(\frac{d^2w}{dx^2} \right)^2 \right\} dx$$

$$- \int_0^L (fu + qw) \, dx - \left[Pu(L) + Fw(L) - M_0 \frac{dw}{dx} \bigg|_{x=L} \right] + \int_0^L \beta EI \frac{d^2w}{dx^2} \frac{d^2w_0}{dx^2} \, dx$$

$$+ \frac{1}{2} \int_0^L \left\{ EA \left[\alpha \frac{du_0}{dx} + \beta \frac{dw}{dx} \frac{dw_0}{dx} + \frac{1}{2} \beta^2 \left(\frac{dw_0}{dx} \right)^2 \right]^2 + \beta^2 EI \left(\frac{d^2w_0}{dx^2} \right)^2 \right\} dx$$

$$+ \int_0^L EA \left[\frac{du}{dx} + \frac{1}{2} \left(\frac{dw}{dx} \right)^2 \right] \left[\alpha \frac{du_0}{dx} + \beta \frac{dw}{dx} \frac{dw_0}{dx} + \frac{1}{2} \beta^2 \left(\frac{dw_0}{dx} \right)^2 \right] dx$$

$$- \int_0^L (\alpha f u_0 + \beta q w_0) \, dx - \left[\alpha Pu_0(L) + \beta Fw_0(L) - \beta M_0 \frac{dw_0}{dx} \bigg|_{x=L} \right]$$

or

$$\Pi(\bar{u}, \bar{w}) = \Pi(u, w) + \mathcal{P} + \int_0^L EA \left[\frac{du}{dx} + \frac{1}{2} \left(\frac{dw}{dx} \right)^2 \right] \left[\alpha \frac{du_0}{dx} + \beta \frac{dw}{dx} \frac{dw_0}{dx} + \frac{1}{2} \beta^2 \left(\frac{dw_0}{dx} \right)^2 \right] dx$$

$$+ \int_0^L \beta EI \frac{d^2w}{dx^2} \frac{d^2w_0}{dx^2} \, dx - \int_0^L (\alpha f u_0 + \beta q w_0) \, dx$$

$$- \left[\alpha Pu_0(L) + \beta Fw_0(L) - \beta M_0 \frac{dw_0}{dx} \bigg|_{x=L} \right] \qquad (12)$$

where $\mathcal{P}$ is the positive quantity:

$$\mathcal{P} = \frac{1}{2} \int_0^L \left\{ EA \left[\alpha \frac{du_0}{dx} + \beta \frac{dw}{dx} \frac{dw_0}{dx} + \frac{1}{2} \beta^2 \left(\frac{dw_0}{dx} \right)^2 \right]^2 + \beta^2 EI \left(\frac{d^2w_0}{dx^2} \right)^2 \right\} dx. \qquad (13)$$

First, we consider the following terms from Eq. (12):

$$\int_0^L \left\{ EA \left[\frac{du}{dx} + \frac{1}{2} \left(\frac{dw}{dx} \right)^2 \right] \alpha \frac{du_0}{dx} - \alpha f u_0 \right\} dx - \alpha P u_0(L)$$

$$= \alpha \int_0^L \left(-\frac{dN}{dx} - f \right) u_0 \, dx - \alpha P u_0(L) + \alpha \left[N u_0 \right]_0^L$$

$$= \alpha \left[N(L) - P \right] + N(0) u_0(0) = 0, \tag{14}$$

where the equation of equilibrium from Eq. (9), boundary conditions from Eq. (11), and

$$N = EA \left[\frac{du}{dx} + \frac{1}{2} \left(\frac{dw}{dx} \right)^2 \right]$$

are used in arriving at the last result.

Next, we consider the following terms from Eq. (12):

$$\beta \int_0^L \left\{ EA \left[\frac{du}{dx} + \frac{1}{2} \left(\frac{dw}{dx} \right)^2 \right] \frac{dw}{dx} \frac{dw_0}{dx} + EI \frac{d^2 w}{dx^2} \frac{d^2 w_0}{dx^2} - q w_0 \right\} dx$$

$$- \beta \left[F w_0(L) - M_0 \frac{dw_0}{dx} \bigg|_{x=L} \right]$$

$$= \beta \int_0^L \left[-\frac{d}{dx} \left(\frac{dw}{dx} N \right) - \frac{d^2 M}{dx^2} - q \right] w_0 \, dx$$

$$+ \beta \left[\left(\frac{dw}{dx} N + \frac{dM}{dx} \right) w_0 \right]_0^L + \beta \left[M \left(-\frac{dw_0}{dx} \right) \right]_0^L$$

$$- \beta \left[F w_0(L) - M_0 \frac{dw_0}{dx} \bigg|_{x=L} \right] = 0, \tag{15}$$

where the equilibrium equation in Eq. (10), the boundary conditions from Eq. (11), and

$$M = -EI \frac{d^2 w}{dx^2}$$

are used in arriving at the result.

In view of the results in Eqs (14) and (15), Eq. (12) takes the form

$$\Pi(\bar{u}, \bar{w}) = \Pi(u, w) + \mathcal{P} + \frac{1}{2} \beta^2 \int_0^L N \left(\frac{dw_0}{dx} \right)^2 dx$$

or

$$\Pi(\bar{u}, \bar{w}) \geq \Pi(u, w) + \frac{1}{2} \beta^2 \int_0^L N \left(\frac{dw_0}{dx} \right)^2 dx, \tag{16}$$

where $\mathcal{P}$ is a positive number defined in Eq. (13). Thus, the minimum character of Π cannot be established because the integral expression on the right-hand side is not always positive. The minimum character of Π is obvious for the case in which either the von Kármán nonlinear term is neglected or N is positive:

$$\Pi(\bar{u}, \bar{w}) \geq \Pi(u, w) \tag{17}$$

for any arbitrary admissible displacements $(\bar{u}, \bar{w})$.

4.3.2   Castigliano's Theorem I

Like the unit dummy-displacement method, which can be used to determine
the unknown point loads and displacements of structural systems composed of
discrete structural members, Castigliano's theorem allows one to compute displacements or loads. Carlo Alberto Castigliano (1847–1884), an Italian mathematician and railroad engineer, was mainly concerned with linear elastic materials. The generalization of Castigliano's original theorem I to the case in
which displacements are nonlinear functions of external forces is attributed to
Friedrich Engesser (1848–1931), a German engineer. In the present study we
consider Castigliano's theorems in a generalized form that are applicable to
both linear and nonlinear elastic materials.

Suppose that the displacement field of a structure can be expressed in terms
of the displacements (and possibly rotations) of a finite number of points $\mathbf{x}_i$
$(i = 1, 2, \cdots N)$ in the body

$$\mathbf{u}(\mathbf{x}) = \sum_{i=1}^{N} \mathbf{u}_i \phi_i(\mathbf{x}), \qquad (4.3.21)$$

where $\mathbf{u}_i$ are unknown displacement parameters, called *generalized displacements*, and ϕ_i are known functions of position, called *interpolation functions*
with the property that ϕ_i is unity at the ith point (i.e., $\mathbf{x} = \mathbf{x}_i$) and zero at all
other points $(\mathbf{x}_j, \ j \neq i)$. Then it is possible to represent the strain energy U
and potential V_E due to applied loads in terms of the generalized displacements
$\mathbf{u}_i$. The principle of minimum total potential energy can be expressed as

$$\delta \Pi = \delta U + \delta V_E = 0 \ \Rightarrow \ \delta U = -\delta V_E, \quad \text{or} \quad \frac{\partial U}{\partial \mathbf{u}_i} \cdot \delta \mathbf{u}_i = -\frac{\partial V_E}{\partial \mathbf{u}_i} \cdot \delta \mathbf{u}_i$$

where sum on repeated indices is implied. Since (only for conservative systems)

$$\frac{\partial V_E}{\partial \mathbf{u}_i} = -\mathbf{F}_i,$$

it follows that

$$\left(\frac{\partial U}{\partial \mathbf{u}_i} - \mathbf{F}_i \right) \cdot \delta \mathbf{u}_i = 0$$

or, because $\delta \mathbf{u}_i$ are arbitrary,

$$\frac{\partial U}{\partial \mathbf{u}_i} = \mathbf{F}_i \quad \left(\frac{\partial U}{\partial u_{ij}} = F_{ij} \right), \qquad (4.3.22)$$

where (u_{ij}, F_{ij}) are the displacement and force, respectively, at the ith point in
the jth direction. Equation (4.3.22) is known as the *Castigliano theorem I*. For
an elastic body, it states that the rate of change of strain energy with respect to
a generalized displacement is equal to the associated generalized force. When

applied to a structure with point loads F_i (or moment M_i) moving through displacements u_i (or rotation θ_i), both having the same sense, the Castigliano theorem I takes the form

$$\frac{\partial U}{\partial u_i} = F_i, \quad \text{or} \quad \frac{\partial U}{\partial \theta_i} = M_i. \tag{4.3.23}$$

It is clear from the derivation that Castigliano's theorem I is a special case of the principle of minimum total potential energy and hence that of the principle of virtual displacements. Castigliano's theorem I is also equivalent to the unit dummy-displacement method, which is derived using the principle of virtual displacements. Next, we consider examples of application of Castigliano's theorem I to trusses and beams.

Example 4.3.2

Use Castigliano's theorem I to determine the displacements u and v of the point O of the truss problem of **Example 4.2.2** [see Fig. 4.3.2(a)].

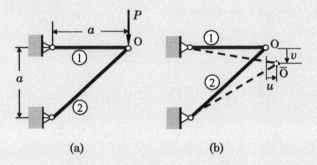

(a)   (b)

Fig. 4.3.2 (a) A two-member truss. (b) Displacements of point O.

Solution: The strain energy of the structure can be expressed in terms of the displacements u and v of point O, as shown in Fig. 4.3.2(b). From Eq. (3) of **Example 4.2.2**, we have

$$\varepsilon^{(1)} = \frac{u}{a}, \quad \varepsilon^{(2)} = \frac{u-v}{2a}. \tag{1}$$

(a) The strain energy is given by ($\sigma^{(i)} = E_i \varepsilon^{(i)}$):

$$U(u,v) = \frac{1}{2} \sum_{i=1}^{2} A_i L_i E_i \left(\varepsilon^{(i)} \right)^2$$

$$= \frac{EA}{2} \left[a \left(\frac{u}{a} \right)^2 + \sqrt{2} a \left(\frac{u-v}{2a} \right)^2 \right]. \tag{2}$$

Then by Castigliano's theorem I, we have

$$0 = \frac{\partial U}{\partial u} = EA \left(\frac{u}{a} + \frac{\sqrt{2}}{2} \frac{u-v}{2a} \right),$$

$$P = \frac{\partial U}{\partial v} = EA \left(0 - \frac{\sqrt{2}}{2} \frac{u-v}{2a} \right), \tag{3}$$

which gives the same result as in Eq. (10) of **Example 4.2.2**:

$$u = \frac{Pa}{EA}, \qquad v = (1 + 2\sqrt{2})\frac{Pa}{EA}. \tag{4}$$

(b) The stresses for this case are

$$\sigma^{(1)} = K\sqrt{\varepsilon^{(1)}}, \quad \sigma^{(2)} = -K\sqrt{-\varepsilon^{(2)}}. \tag{5}$$

The total strain energy becomes

$$
\begin{aligned}
U(u,v) &= A_1 L_1 K \int_0^{\varepsilon^{(1)}} \sqrt{\varepsilon}\, d\varepsilon - A_2 L_2 K \int_0^{\varepsilon^{(2)}} \sqrt{-\varepsilon}\, d\varepsilon \\
&= \frac{2}{3} A_1 L_1 K \left(\varepsilon^{(1)}\right)^{\frac{3}{2}} + \frac{2}{3} A_2 L_2 K \left(-\varepsilon^{(2)}\right)^{\frac{3}{2}} \\
&= \frac{2AK}{3} \left[a\left(\frac{u}{a}\right)^{\frac{3}{2}} + \sqrt{2}a\left(\frac{v-u}{2a}\right)^{\frac{3}{2}} \right].
\end{aligned}
\tag{6}
$$

Using Castigliano's theorem I, we obtain

$$
\begin{aligned}
0 &= \frac{\partial U}{\partial u} = AK\left(\sqrt{\frac{u}{a}} - \frac{\sqrt{2}}{2}\sqrt{\frac{v-u}{2a}} \right), \\
P &= \frac{\partial U}{\partial v} = AK\left(0 + \frac{\sqrt{2}}{2}\sqrt{\frac{v-u}{2a}} \right).
\end{aligned}
\tag{7}
$$

Thus, we obtain the same result as in Eq. (14) of **Example 4.2.2**:

$$u = \frac{P^2 a}{A^2 K^2}, \qquad v = \frac{4P^2 a}{A^2 K^2} + u = \frac{5P^2 a}{A^2 K^2}. \tag{8}$$

Example 4.3.3

Use of Castigliano's theorem I to determine displacements or forces in trusses is made easy by the fact that the axial strain in a truss element (or member) can be expressed solely in terms of the displacements at the two ends of the element. If $\bar{u}_1$ and $\bar{u}_2$ denote the axial displacements at the left and right ends (with respect to the coordinate $\bar{x}$ that is aligned with the length), respectively, then the axial strain in the truss element is $\bar{\varepsilon} = (\bar{u}_2 - \bar{u}_1)/L$, where L is the length of the element (see **Example 3.2.2** and **3.2.4**). If $\bar{P}_1$ and $\bar{P}_2$ are the axial forces at the left and right ends of the truss element in the direction of the displacements $\bar{u}_1$ and $\bar{u}_2$, respectively (see Fig. 4.3.3), use Castigliano's theorem I to establish the relationship between $(\bar{P}_1, \bar{P}_2)$ and $(\bar{u}_1, \bar{u}_2)$.

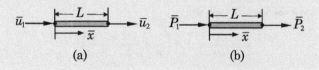

Fig. 4.3.3 A truss element with (a) end displacements and (b) end forces.

Solution: When there is no body force, the exact solution to the second-order equation governing the axial deformation of a truss element with constant cross-sectional area A and modulus E is $\bar{u}(\bar{x}) = a_1 + a_2\bar{x}$ (see **Example 4.2.3**). The two constants of integration, a_1 and a_2, can be expressed in terms of the end displacements $\bar{u}_1$ and $\bar{u}_2$ to give

$$\bar{u}(x) = \left(1 - \frac{\bar{x}}{L}\right)\bar{u}_1 + \frac{\bar{x}}{L}\bar{u}_2 \text{ giving } \bar{\varepsilon} = \frac{d\bar{u}}{d\bar{x}} = \frac{\bar{u}_2 - \bar{u}_1}{L}. \tag{1}$$

The strain energy can be computed, as explained in **Example 3.2.2**, as

$$U = \frac{EAL}{2}\bar{\varepsilon}^2 = \frac{EA}{2L}\left(\bar{u}_2 - \bar{u}_1\right)^2. \tag{2}$$

Then Castigliano's theorem I gives

$$\begin{aligned}
\bar{P}_1 &= \frac{\partial U}{\partial \bar{u}_1} = -\frac{EA}{L}\left(\bar{u}_2 - \bar{u}_1\right), \\
\bar{P}_2 &= \frac{\partial U}{\partial \bar{u}_2} = \frac{EA}{L}\left(\bar{u}_2 - \bar{u}_1\right).
\end{aligned} \tag{3}$$

Expressing the linear relations of Eq. (3) in matrix form, we obtain

$$\left\{\begin{array}{c}\bar{P}_1 \\ \bar{P}_2\end{array}\right\} = \frac{EA}{L}\left[\begin{array}{cc}1 & -1 \\ -1 & 1\end{array}\right]\left\{\begin{array}{c}\bar{u}_1 \\ \bar{u}_2\end{array}\right\}. \tag{4}$$

As we shall see in Chapter 8, Eq. (4) represents the finite element equations of a uniaxial bar finite element when linear approximation of the displacement $u(x)$ is used. The coefficient matrix is called the stiffness matrix; note that EA/L represents the equivalent spring constant (force per unit displacement).

The application of Castigliano's theorem I to beams is relatively more complicated than to trusses. The reason is that in the case of straight beams subjected to bending loads only, the transverse displacement $w(x)$ must be expressed in terms of chosen generalized displacement variables (called coordinates) so that Castigliano's theorem I can be used. The next example is based on the expansion of the displacement field in terms of a Fourier trigonometric series where the role of the generalized displacement coordinates is played by the Fourier coefficients.

Example 4.3.4 ────────────────────────────────────

Determine the center deflection of a simply supported beam with a point load F_0 (acting downward) at the center of the beam using Castigliano's theorem I.

Solution: Let w be expanded in the Fourier sine series as

$$w(x) = \sum_{n=1}^{\infty} c_n \sin\frac{n\pi x}{L}, \tag{1}$$

which satisfies the following boundary conditions of simply supported beam for any arbitrary

$c_n, n = 1, 2, \ldots,$:

$$w = 0, \quad M = -EI\frac{d^2 w}{dx^2} = 0 \quad \text{at} \quad x = 0, L. \tag{2}$$

We wish to determine c_n so that we can determine $w(L/2)$ using the expression in Eq. (1). The strain energy U of the beam can be expressed in terms of the parameters c_n as

$$U = \frac{EI}{2} \int_0^L \left(\frac{d^2 w}{dx^2}\right)^2 dx = \sum_{n=1}^{\infty} \sum_{m=1}^{\infty} \frac{EI}{2} \left(\frac{m\pi}{L}\right)^2 \left(\frac{n\pi}{L}\right)^2 c_m c_n \int_0^L \sin\frac{m\pi x}{L} \sin\frac{n\pi x}{L} dx$$

$$= \sum_{n=1}^{\infty} \frac{EIL}{4} \left(\frac{n\pi}{L}\right)^4 c_n^2, \tag{3}$$

where we have used the orthogonality of the sine functions:

$$\int_0^L \sin\frac{m\pi}{L} \sin\frac{n\pi}{L} dx = \begin{cases} \frac{L}{2}, & \text{for } m = n, \\ 0, & \text{for } m \neq n. \end{cases} \tag{4}$$

Using Castigliano's theorem I (or the principle of virtual displacements), we can write

$$F_0 \delta w(L/2) = \delta U \;\Rightarrow\; F_0 \sum_{n=1}^{\infty} \sin\frac{n\pi}{2}\delta c_n = \frac{EIL}{2} \sum_{n=1}^{\infty} \left(\frac{n\pi}{L}\right)^4 c_n\, \delta c_n, \tag{5}$$

or

$$c_n = -(-1)^{\frac{n+1}{2}} \frac{2F_0 L^3}{n^4\pi^4 EI}, \quad n = 1, 3, 5, \ldots \tag{6}$$

The deflection at any point is given by

$$w(x) = -\frac{2F_0 L^3}{\pi^4 EI} \sum_{n=1}^{\infty} (-1)^{\frac{n+1}{2}} \frac{1}{n^4} \sin\frac{n\pi x}{L}. \tag{7}$$

Since a complete cubic polynomial contains four parameters, we have to identify four generalized displacement variables and corresponding generalized force variables. Such duality between displacement and force variables is provided by Eq. (8) of **Example 4.3.1** for the Bernoulli–Euler beam theory; for pure bending, transverse deflection (w) and slope ($-(dw/dx)$) are the generalized displacement variables, and shear force (V) and bending moment (M) are the generalized force variables. Thus, the four parameters of the cubic polynomial for $w(x)$ should be expressed in terms of the deflection and slope at the two ends of the beam element. The next example is based on expansion of the displacement field in terms of algebraic polynomials with generalized displacements (i.e., displacement and rotation) as the generalized coordinates.

Example 4.3.5

Consider a straight beam of length L and constant bending stiffness EI with generalized displacements ($\Delta_1, \Delta_2, \Delta_3, \Delta_4$), as shown in Fig. 4.3.4(a); the corresponding generalized forces (F_1, F_2, F_3, F_4) are shown in Fig. 4.3.4(b). (a) Establish the relations between the generalized displacements and forces. (b) Use the result to determine the compression in the linear elastic spring of the beam shown in Fig. 4.3.5. (c) Determine the reaction force at $x = L$ when the spring in Fig. 4.3.5 is replaced by a rigid support.

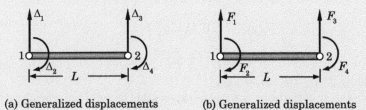

(a) Generalized displacements   (b) Generalized displacements

Fig. 4.3.4 A beam element with (a) generalized displacements and (b) generalized forces.

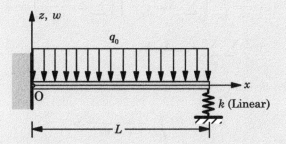

Fig. 4.3.5 A beam fixed at the left end and spring supported at the right end.

Solution: (a) From the discussion presented in **Example 4.2.3**, the solution to the fourth-order equation governing the transverse deflection of a homogeneous beam is given by

$$w(x) = a_1 + a_2 x + a_3 x^2 + a_4 x^3, \tag{1}$$

where a_i ($i = 1, 2, 3, 4$) are constants of integration. First, we express these constants in terms of the generalized displacements at the two ends of the beam element:

$$\begin{aligned}
\Delta_1 &\equiv w(0) = a_1, \\
\Delta_2 &\equiv \left(-\frac{dw}{dx}\right)_{x=0} = -a_2, \\
\Delta_3 &\equiv w(L) = a_1 + a_2 L + a_3 L^2 + a_4 L^3, \\
\Delta_4 &\equiv \left(-\frac{dw}{dx}\right)_{x=L} = -a_2 - 2a_3 L - 3a_4 L^2.
\end{aligned} \tag{2}$$

We note that Δ_1 and Δ_3 are the values of the transverse deflection w at $x = 0$ and $x = L$, respectively, and Δ_2 and Δ_4 are the slopes $-(dw/dx)$, measured clockwise, at $x = 0$ and $x = L$, respectively, as shown in Fig. 4.3.4(a). The four relations in Eq. (2) can be inverted to solve for a_i in terms of Δ_i. Then substituting the result into Eq. (1) yields

$$w(x) = \varphi_1(x)\Delta_1 + \varphi_2(x)\Delta_2 + \varphi_3(x)\Delta_3 + \varphi_4(x)\Delta_4 = \sum_{i=1}^{4} \varphi_i(x)\Delta_i, \tag{3}$$

where

$$\varphi_1(x) = 1 - 3\left(\frac{x}{L}\right)^2 + 2\left(\frac{x}{L}\right)^3,$$

$$\varphi_2(x) = -x\left[1 - 2\left(\frac{x}{L}\right) + \left(\frac{x}{L}\right)^2\right],$$

$$\varphi_3(x) = \left(\frac{x}{L}\right)^2\left(3 - 2\frac{x}{L}\right),$$

(4)

$$\varphi_4(x) = x\frac{x}{L}\left(1 - \frac{x}{L}\right)$$

are called the cubic *Hermite polynomials*.

The strain energy of the beam now can be expressed in terms of the generalized coordinates Δ_i as

$$
\begin{aligned}
U &= \frac{EI}{2}\int_0^L \left(\frac{d^2w}{dx^2}\right)^2 dx \\
&= \frac{EI}{2}\int_0^L \left(\sum_{m=1}^4 \Delta_m \frac{d^2\varphi_m}{dx^2}\right)\left(\sum_{n=1}^4 \Delta_n \frac{d^2\varphi_n}{dx^2}\right) dx \\
&= \frac{1}{2}\sum_{m=1}^4\sum_{n=1}^4 K_{mn}\Delta_m\Delta_n,
\end{aligned}
$$

(5)

where K_{mn} are the coefficients of the stiffness matrix $\mathbf{K}$:

$$K_{mn} = EI\int_0^L \frac{d^2\varphi_m}{dx^2}\frac{d^2\varphi_n}{dx^2}\, dx.$$

(6)

Evaluating the integral in Eq. (6) with φ_i from Eq. (4), we obtain

$$\mathbf{K} = \frac{2EI}{L^3}\begin{bmatrix} 6 & -3L & -6 & -3L \\ -3L & 2L^2 & 3L & L^2 \\ -6 & 3L & 6 & 3L \\ -3L & L^2 & 3L & 2L^2 \end{bmatrix}.$$

(7)

Note that K_{mn} is symmetric ($K_{mn} = K_{nm}$). The work done by applied forces is given by

$$
\begin{aligned}
V_E &= -\left[\int_0^L q(x)w(x)dx + \sum_{m=1}^4 Q_m\Delta_m\right] \\
&= -\sum_{m=1}^4 (q_m\,\Delta_m + Q_m\,\Delta_m) \equiv -\sum_{m=1}^4 F_m\,\Delta_m,
\end{aligned}
$$

(8)

where

$$q_m = \int_0^L q(x)\varphi_m(x)\, dx$$

(9)

and Q_m are the generalized point forces associated with the generalized displacements Δ_m:

$$Q_1 = -V(0), \quad Q_3 = V(L); \quad Q_2 = -M(0), \quad Q_4 = M(L).$$

(10)

Thus, $F_1 = Q_1+q_1$ and $F_3 = Q_3+q_3$ are the transverse forces at $x = 0$ and $x = L$, respectively, and $F_2 = Q_2+q_2$ and $F_4 = Q_4+q_4$ are the bending moments at $x = 0$ and $x = L$, respectively, as shown in Fig. 4.3.4(b). The transverse forces q_1 and q_3 and bending moments q_2 and q_4 together are statically equivalent to the distributed load $q(x)$ on the beam.

Using Castigliano's theorem I $(-\partial V_E/\partial \Delta_i = \partial U/\partial \Delta_i)$, we obtain [removing the summation symbols and adopting the summation convention in Eqs (5) and (8)]

$$\frac{\partial}{\partial \Delta_i}(F_m \Delta_m) = \frac{\partial}{\partial \Delta_i}\left(\frac{1}{2}K_{mn}\Delta_m \Delta_n\right)$$

$$F_m \delta_{im} = \frac{1}{2}K_{mn}(\delta_{im}\Delta_n + \Delta_m \delta_{in})$$

$$F_i = \frac{1}{2}(K_{in}\Delta_n + K_{mi}\Delta_m)$$

$$= K_{ij}\Delta_j \quad \text{because } K_{mi} = K_{im} \tag{11}$$

In matrix form we have $\mathbf{F} = \mathbf{K}\Delta$:

$$\begin{Bmatrix} F_1 \\ F_2 \\ F_3 \\ F_4 \end{Bmatrix} \equiv \begin{Bmatrix} Q_1 \\ Q_2 \\ Q_3 \\ Q_4 \end{Bmatrix} + \begin{Bmatrix} q_1 \\ q_2 \\ q_3 \\ q_4 \end{Bmatrix} = \frac{2EI}{L^3}\begin{bmatrix} 6 & -3L & -6 & -3L \\ -3L & 2L^2 & 3L & L^2 \\ -6 & 3L & 6 & 3L \\ -3L & L^2 & 3L & 2L^2 \end{bmatrix}\begin{Bmatrix} \Delta_1 \\ \Delta_2 \\ \Delta_3 \\ \Delta_4 \end{Bmatrix}. \tag{12}$$

(b) We now consider a beam fixed at $x = 0$ and spring supported at $x = L$ and subjected to uniformly distributed load of intensity q_0 (see Fig. 4.3.5).

First, we compute the load vector corresponding to the distributed load $q = -q_0$:

$$\begin{Bmatrix} q_1 \\ q_2 \\ q_3 \\ q_4 \end{Bmatrix} = -\frac{q_0 L}{12}\begin{Bmatrix} 6 \\ -L \\ 6 \\ L \end{Bmatrix}. \tag{13}$$

Next, we identify the boundary conditions of the problem. We have

Geometric BC: $\quad w(0) = \Delta_1 = 0, \quad \left(-\dfrac{dw}{dx}\right)_{x=0} = \Delta_2 = 0;$

Force BC: $\quad V(L) = Q_3 = -kw(L) = -k\Delta_3, \quad M(L) = Q_4 = 0.$ $\tag{14}$

Thus we have

$$\frac{2EI}{L^3}\begin{bmatrix} 6 & -3L & -6 & -3L \\ -3L & 2L^2 & 3L & L^2 \\ -6 & 3L & 6 & 3L \\ -3L & L^2 & 3L & 2L^2 \end{bmatrix}\begin{Bmatrix} 0 \\ 0 \\ \Delta_3 \\ \Delta_4 \end{Bmatrix} = -\frac{q_0 L}{12}\begin{Bmatrix} 6 \\ -L \\ 6 \\ L \end{Bmatrix} + \begin{Bmatrix} Q_1 \\ Q_2 \\ -k\Delta_3 \\ 0 \end{Bmatrix}, \tag{15}$$

Condensing out the equations for the unknown generalized displacements Δ_3 and Δ_4, we obtain

$$\frac{2EI}{L^3}\begin{bmatrix} 6 & 3L \\ 3L & 2L^2 \end{bmatrix}\begin{Bmatrix} \Delta_3 \\ \Delta_4 \end{Bmatrix} = -\frac{q_0 L}{12}\begin{Bmatrix} 6 \\ L \end{Bmatrix} + \begin{Bmatrix} -k\Delta_3 \\ 0 \end{Bmatrix} \tag{16}$$

or

$$\begin{bmatrix} \frac{12EI}{L^3} + k & \frac{6EI}{L^2} \\ \frac{6EI}{L^2} & \frac{4EI}{L} \end{bmatrix}\begin{Bmatrix} \Delta_3 \\ \Delta_4 \end{Bmatrix} = -\frac{q_0 L}{12}\begin{Bmatrix} 6 \\ L \end{Bmatrix}. \tag{17}$$

Solving for $\Delta_3 = w(L)$ and $\Delta_4 = -(dw/dx)(L)$ by Cramer's rule, we obtain

$$w(L) = -\frac{q_0 L^4}{8EI}\left(1 + \frac{kL^3}{3EI}\right)^{-1}, \quad -\frac{dw}{dx}\bigg|_{x=L} = \frac{q_0 L^3}{6EI}\left(1 - \frac{kL^3}{24EI}\right)\left(1 + \frac{kL^3}{3EI}\right)^{-1}. \tag{18}$$

(c) From Eq. (18), we can obtain deflections and slopes at $x = L$ for the following special cases:

$$\text{When } k = 0, \quad \Delta_3 = w(L) = -\frac{q_0 L^4}{8EI}, \quad \Delta_4 = -\frac{dw}{dx}\bigg|_{x=L} = \frac{q_0 L^3}{6EI}. \tag{19}$$

$$\text{When } k \to \infty, \quad \Delta_3 = w(L) = 0, \quad \Delta_4 = -\frac{dw}{dx}\bigg|_{x=L} = -\frac{q_0 L^3}{48EI}. \tag{20}$$

For the case in which the beam is fixed at the left end and simply supported at the right end (i.e., $k \to \infty$), the reaction force Q_3 at $x = L$ can be determined from Eq. (15) by replacing $-k\Delta_3$ on the right-hand side with Q_3 and setting Δ_3 to zero on the left side of the equality [alternatively, one can go to Eq. (12) and apply the boundary conditions, $\Delta_1 = \Delta_2 = \Delta_3 = 0$ and $Q_4 = 0$]. Then solving for Δ_3, we obtain

$$\Delta_3 = -\frac{q_0 L^2}{12}\frac{L}{4EI} = -\frac{q_0 L^3}{48EI}. \tag{21}$$

Then from the third row of Eq. (15), we determine Q_3, the reaction of the rigid support, as

$$Q_3 = \frac{q_0 L}{2} + \frac{6EI}{L^2}\Delta_3 = \frac{q_0 L}{2} - \frac{6EI}{L^2}\frac{q_0 L^3}{48EI} = \frac{3q_0 L}{8}. \tag{22}$$

4.4   The Principle of Virtual Forces

4.4.1   Deformable Solids

In Section 4.2 we discussed the principle of virtual displacements. Naturally, the virtual work done by virtual forces that are in self-equilibrium in moving through actual displacements should have similar use.

Consider single-valued, differentiable variations of a stress field $\delta\boldsymbol{\sigma}$ and variations of body forces $\delta\mathbf{f}$ and tractions $\delta\mathbf{t}$ that satisfy the linear equilibrium equations both within the body and on its boundaries:

$$\boldsymbol{\nabla} \cdot \delta\boldsymbol{\sigma} + \delta\mathbf{f} = \mathbf{0} \quad \text{in } \Omega, \tag{4.4.1}$$

$$\delta\mathbf{t} = \hat{\mathbf{n}} \cdot \delta\boldsymbol{\sigma} = 0 \quad \text{on } \Gamma_\sigma, \tag{4.4.2}$$

where Γ_σ is the portion of the boundary Γ on which the traction vector $\mathbf{t}$ is specified (to be, say, $\mathbf{t} = \hat{\mathbf{t}}$); the virtual tractions $\delta\mathbf{t}$ that are consistent with the virtual stresses by Cauchy's formula are arbitrary on Γ_u. We shall call such a stress field a *statically admissible field of variation*. These virtual stresses, body forces, and tractions (i.e., surface stresses or forces), except for *self-equilibrating* and $\delta\mathbf{t} = \mathbf{0}$ on Γ_σ, are completely arbitrary and independent of the true stresses and forces.

The external *complementary virtual work* is defined by

$$\delta W_E^* = \delta V_E^* = -\int_\Omega \mathbf{u} \cdot \delta\mathbf{f} \, d\Omega - \int_{\Gamma_u} \hat{\mathbf{u}} \cdot \delta\mathbf{t} \, ds, \tag{4.4.3}$$

where $\delta \mathbf{f}$ is the virtual body force that satisfies the equilibrium equation [see Eq. (4.4.1)] and $\delta \mathbf{t}$ is the virtual surface force that satisfies Eq. (4.4.2).

The internal complementary virtual work is given by

$$\delta W_I^* = \int_\Omega \boldsymbol{\varepsilon}(\boldsymbol{\sigma}) : \delta \boldsymbol{\sigma} \, d\Omega = \int_\Omega \varepsilon_{ij} \, \delta \sigma_{ij} \, d\Omega, \qquad (4.4.4)$$

where $\boldsymbol{\varepsilon}$ is the actual infinitesimal strain tensor and $\delta \boldsymbol{\sigma}$ is the virtual stress tensor field that satisfies Eqs (4.4.1) and (4.4.2).

The principle of complementary virtual work (or virtual forces) states that *the strains and displacements in a deformable body are compatible and consistent with the constraints if and only if the total complementary virtual work is zero:*

$$\delta W_I^* + \delta W_E^* = 0. \qquad (4.4.5)$$

We now show that the principle of virtual forces gives the kinematic relations and geometric boundary conditions as the Euler equations.

Consider a three-dimensional deformable solid occupying the volume Ω with specified displacements $\hat{u}_i$ on the boundary Γ_u of Ω. Let $\delta \mathbf{f}$ and $\delta \mathbf{t}$ be the virtual forces on the solid and let $\delta \boldsymbol{\sigma}$ be the virtual stresses developed in the body. The internal and external virtual works done are (in index notation)

$$\delta W_I^* = \int_\Omega \varepsilon_{ij} \, \delta \sigma_{ij} \, d\Omega, \quad \delta W_E^* = - \left[\int_\Omega u_i \, \delta f_i \, d\Omega + \int_{\Gamma_u} \hat{u}_i \, \delta t_i \, d\Gamma \right]. \qquad (4.4.6)$$

where sum on repeated indices is assumed. Using the principle of virtual forces, namely, Eq. (4.4.5), we can write

$$0 = \delta W_I^* + \delta W_E^* = \int_\Omega \varepsilon_{ij} \, \delta \sigma_{ij} \, d\Omega - \int_\Omega u_i \, \delta f_i \, d\Omega - \int_{\Gamma_u} \hat{u}_i \, \delta t_i \, d\Gamma$$

$$= \int_\Omega \varepsilon_{ij} \, \delta \sigma_{ij} \, d\Omega - \int_\Omega u_i (-\delta \sigma_{ij,j}) d\Omega - \int_{\Gamma_u} \hat{u}_i n_j \, \delta \sigma_{ij} \, d\Gamma, \qquad (4.4.7)$$

where equations of equilibrium and natural boundary conditions from Eqs (4.4.1) and (4.4.2) are used in arriving at the second line of Eq. (4.4.7). Using the Green–Gauss theorem to trade differentiation from σ_{ij} to u_i, we can rewrite the integral

$$\int_\Omega u_i \, (-\delta \sigma_{ij,j}) \, d\Omega = \int_\Omega u_{i,j} \, \delta \sigma_{ij} \, d\Omega - \oint_\Gamma u_i \, \delta \sigma_{ij} n_j \, d\Gamma$$

$$= \int_\Omega \tfrac{1}{2} \left(u_{i,j} + u_{j,i} \right) \delta \sigma_{ij} \, d\Omega - \int_{\Gamma_u} u_i \, \delta \sigma_{ij} n_j \, d\Gamma, \qquad (4.4.8)$$

where $u_{i,j}$ is expressed as a sum of symmetric and antisymmetric parts:

$$u_{i,j} = \tfrac{1}{2} \left(u_{i,j} + u_{j,i} \right) + \tfrac{1}{2} \left(u_{i,j} - u_{j,i} \right),$$

recognizing that the antisymmetric part times symmetric $\delta\sigma_{ij}$ is zero in Ω and $\delta\sigma_{ij}n_j = 0$ on Γ_σ in arriving at the result in Eq. (4.4.8). Using the result from Eq. (4.4.8) in Eq. (4.4.7), we obtain

$$0 = \int_\Omega \left[\varepsilon_{ij} - \tfrac{1}{2}\left(u_{i,j} + u_{j,i}\right)\right]\delta\sigma_{ij}\,d\Omega + \int_{\Gamma_u}\left(u_i - \hat{u}_i\right)n_j\delta\sigma_{ij}\,d\Gamma. \qquad (4.4.9)$$

Because $\delta\sigma_{ij}$ is arbitrary in Ω and on Γ_u independently, we obtain the strain–displacement equations and the displacement boundary conditions as the Euler equations from Eq. (4.4.9):

$$\begin{aligned}\varepsilon_{ij} - \tfrac{1}{2}(u_{i,j} + u_{j,i}) &= 0 \quad \text{in } \Omega \\ u_i - \hat{u}_i &= 0 \quad \text{on } \Gamma_u.\end{aligned} \qquad (4.4.10)$$

4.4.2 Unit Dummy-Load Method

The unit dummy-load method is a special case of the principle of complementary virtual work, and it can be used to determine point displacements and forces in structures. The basic idea can be described in analogy with unit dummy-displacement method. If u_0 is the true displacement at point O in an elastic structure, we can prescribe a virtual force δF_0 at the point in the direction of u_0. The application of virtual force induces a system of virtual stresses $\delta\sigma_{ij}$ that satisfy the equilibrium equations. Then from the principle of virtual forces, namely, Eq. (4.4.5), we have $(\delta W_E^* = -u_0\,\delta F_0)$:

$$u_0\,\delta F_0 = \int_\Omega \varepsilon_{ij}\,\delta\sigma_{ij}^0\,d\Omega. \qquad (4.4.11)$$

Once again, one can take $\delta F_0 = 1$ and calculate the corresponding virtual internal stresses $\delta\sigma_{ij}^0$. Then Eq. (4.4.11) reduces to the statement of the unit dummy-load method:

$$u_0 = \int_\Omega \varepsilon_{ij}\,\delta\sigma_{ij}^0\Big|_{\delta F_0=1}\,d\Omega. \qquad (4.4.12)$$

We now consider several examples of the application of the principle of virtual forces or the unit dummy-load method. It proves to be more convenient not to set the virtual forces to unity, especially when there is more than one (i.e., use the principle of virtual forces). Collecting terms as coefficients of an independent set of virtual forces and setting the coefficients to zero is equivalent to setting a virtual force to unity and the other virtual forces to zero at a time.

Example 4.4.1 ————————————————————————————————————

Consider the truss structure of **Examples 4.2.2** and **4.3.2** [see Fig. 4.4.1(a)]. Determine the vertical displacement v and horizontal displacement u of point O using the principle of virtual forces or unit dummy-load method when the stress–strain relation is (a) linear and (b) nonlinear as given in Eq. (1) of **Example 4.2.2**.

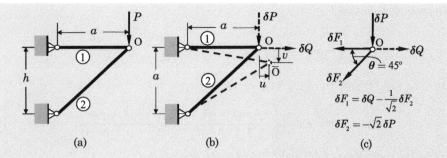

Fig. 4.4.1 (a) Original truss. (b) Applied virtual forces. (c) Member virtual forces.

Solution: Assume virtual (or dummy) forces of δP downward and δQ horizontal, as shown in Fig. 4.4.1(b). Then using the principle of virtual forces, we obtain

$$u \cdot \delta Q + v \cdot \delta P = \int_\Omega \varepsilon_{ij} \, \delta\sigma_{ij} \, d\Omega,$$

$$= \int_0^{L_1} A^{(1)} \varepsilon^{(1)} \, \delta\sigma^{(1)} \, dx + \int_0^{L_2} A^{(2)} \varepsilon^{(2)} \, \delta\sigma^{(2)} \, dx, \qquad (1)$$

where $A^{(1)} = A^{(2)} = A$, $L_1 = a$, and $L_2 = \sqrt{2}a$. The actual stresses are computed in terms of the actual forces in each member due to the applied load P.

From the static equilibrium of the structure, the actual member forces are found to be

$$F^{(1)} = P, \quad F^{(2)} = -\sqrt{2}P. \qquad (2)$$

The virtual forces in the members are computed from the virtual loads δP and δQ on the structure [see Fig. 4.4.1(c)]. We obtain

$$\delta F^{(1)} = \delta Q + \delta P, \quad \delta F^{(2)} = -\sqrt{2} \, \delta P. \qquad (3)$$

The actual stresses and virtual stresses are

$$\sigma^{(1)} = \frac{F^{(1)}}{A} = \frac{P}{A}, \quad \sigma^{(2)} = \frac{F^{(2)}}{A} = -\frac{\sqrt{2}\,P}{A}, \qquad (4)$$

$$\delta\sigma^{(1)} = \frac{\delta Q + \delta P}{A}, \quad \delta\sigma^{(2)} = -\frac{\sqrt{2}\,\delta P}{A}. \qquad (5)$$

(a) The actual strains can be expressed in terms of member forces as

$$\varepsilon^{(1)} = \frac{\sigma^{(1)}}{E} = \frac{P}{EA}, \quad \varepsilon^{(2)} = \frac{\sigma^{(2)}}{E} = -\frac{\sqrt{2}P}{EA}. \qquad (6)$$

Substituting Eqs (5) and (6) in Eq. (1), we obtain

$$\delta Q \cdot u + \delta P \cdot v = Aa \frac{\delta Q + \delta P}{A} \frac{P}{EA} + \sqrt{2}Aa \frac{\sqrt{2}\,\delta P}{A} \frac{\sqrt{2}\,P}{EA}, \qquad (7)$$

or, collecting the coefficients of δQ and δP separately, we obtain

$$u = \frac{Pa}{EA}, \quad v = \frac{Pa}{EA} + 2\sqrt{2}\frac{Pa}{EA} = \frac{Pa}{AE}\left(1 + 2\sqrt{2}\right). \qquad (8)$$

(b) The actual strains are

$$\varepsilon^{(1)} = \left(\frac{\sigma^{(1)}}{K}\right)^2 = \frac{P^2}{K^2 A^2}, \quad \varepsilon^{(2)} = -\left(\frac{\sigma^{(2)}}{K}\right)^2 = -\frac{2P^2}{K^2 A^2}. \tag{9}$$

Substituting Eqs (5) and (9) in Eq. (1), we obtain

$$\delta Q \cdot u + \delta P \cdot v = Aa\frac{\delta Q + \delta P}{A}\frac{P^2}{K^2 A^2} + \sqrt{2}Aa\frac{\sqrt{2}\delta P}{A}\frac{2P^2}{K^2 A^2}. \tag{10}$$

Collecting the coefficients of δQ and δP separately, we obtain

$$u = \frac{P^2 a}{K^2 A^2}, \quad v = \frac{P^2 a}{K^2 A^2} + \frac{4P^2 a}{K^2 A^2} = \frac{5P^2 a}{K^2 A^2}. \tag{11}$$

Example 4.4.2

Consider the overhang simply supported beam shown in Fig. 4.4.2(a). Determine the rotation (or slope) at the left end of the beam using the unit dummy-load method. Neglect the energy due to transverse shear force.

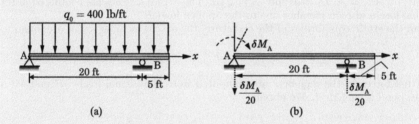

Fig. 4.4.2 (a) Given beam. (b) Virtual force system.

Solution: Since the slope at the left end is desired, we use a virtual moment of δM_A (clockwise) at the left end, as shown in Fig. 4.4.2(b). The virtual moment should be in self-equilibrium. A convenient set of self-equilibrating forces consist of a force of $\delta M_A/20$ downward at the left end and another force of $\delta M_A/20$ upward at the other support. Using the principle of virtual forces, we obtain

$$\theta_A \cdot \delta M_A = \frac{1}{EI}\int_0^L M(x)\,\delta M(x)\,dx, \tag{1}$$

where $M(x)$ is the actual bending moment, $\delta M(x)$ is the virtual bending moment at any location x along the beam due to the virtual forces, and θ_A is the rotation (clockwise) at the left support. To compute $M(x)$, first we must compute the reaction forces at the supports using static equilibrium of forces. We obtain

$$R_A = \frac{(400 \times 25) \times 7.5}{20} = 3,750\,\text{lb}; \quad R_B = \frac{(400 \times 25) \times 12.5}{20} = 6,250\,\text{lb}. \tag{2}$$

Then the actual and virtual bending moments (taken counterclockwise positive) are

$$M(x) = \begin{cases} -3750x + 200x^2, & 0 \le x \le 20', \\ -3750x + 200x^2 - 6250(x - 20), & 20' \le x \le 25', \end{cases} \tag{3}$$

$$\delta M(x) = \begin{cases} -\delta M_{\mathrm{A}} + (\frac{1}{20}\delta M_{\mathrm{A}})x, & 0 \le x \le 20, \\ 0, & 20' \le x \le 25'. \end{cases} \tag{4}$$

Using Eqs (3) and (4) in Eq. (1), canceling δM_{A} on both sides, and evaluating the integral (we note that the integral over the interval [20–25] is zero), we obtain

$$\begin{aligned} \theta_{\mathrm{A}} &= \frac{1}{EI} \int_0^{20} \left(-3750x + 200x^2\right)\left(-1 + \frac{x}{20}\right) dx \\ &= \frac{1}{EI} \int_0^{20} \left(3750x - 200x^2 - 187.5x^2 + 10x^3\right) dx = -\frac{7 \times 10^5}{6EI}, \end{aligned} \tag{5}$$

where EI assumed to be in lb-ft^2.

Example 4.4.3

Consider the beam shown in Fig. 4.3.5, which is shown again in Fig. 4.4.3(a). Determine the displacement and rotation of the right end of the beam, which is supported by a linear elastic spring, using the principle of virtual forces. Specialize the result for the case of rigid spring (i.e., the right end does not move vertically).

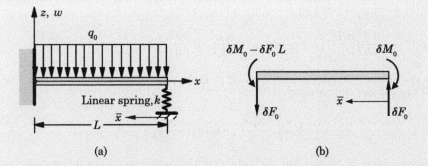

(a)   (b)

Fig. 4.4.3 (a) Given beam. (b) Virtual force system of the beam.

Solution: The virtual force system is shown in Fig. 4.4.3. The principle of virtual forces for this case can be expressed as (we note that the force and moment at the left end do no work because the corresponding actual displacement and rotation are zero)

$$w(0) \cdot \delta F_0 + \theta(0) \cdot \delta M_0 = \frac{1}{EI} \int_0^L M(\bar{x}) \, \delta M(\bar{x}) \, d\bar{x}, \tag{1}$$

where the $\bar{x}$-coordinate is taken from the right to left and $M(\bar{x})$ is the actual bending moment. Let $F_s = kw(0)$ denote the force in the spring acting downward on the beam at $x = L$. Then

$M(\bar{x})$ and $\delta M(\bar{x})$ are given by (taken positive counterclockwise)

$$M(\bar{x}) = F_s\bar{x} + \frac{q_0}{2}\bar{x}^2, \quad 0 \le \bar{x} \le L, \tag{2}$$

$$\delta M(\bar{x}) = \delta M_0 - \delta F_0\bar{x}, \quad 0 \le \bar{x} \le L. \tag{3}$$

Substituting the expressions for M and δM from Eqs (2) and (3) into Eq. (1) and carrying out the indicated integration, we obtain

$$
\begin{aligned}
w(0) \cdot \delta F_0 + \theta(0) \cdot \delta M_0 &= \frac{1}{EI} \int_0^L M(\bar{x})\,\delta M(\bar{x})\,d\bar{x} \\
&= \frac{1}{EI} \int_0^L \left(F_s\bar{x} + \frac{q_0}{2}\bar{x}^2\right)(\delta M_0 - \delta F_0\bar{x})\,dx \\
&= -\frac{1}{EI}\left(\frac{F_sL^3}{3} + \frac{q_0L^4}{8}\right)\delta F_0 + \frac{1}{EI}\left(\frac{F_sL^2}{2} + \frac{q_0L^3}{6}\right)\delta M_0
\end{aligned}
\tag{4}
$$

from which we obtain

$$w(0) = -\frac{1}{EI}\left(\frac{F_sL^3}{3} + \frac{q_0L^4}{8}\right) = -\frac{1}{EI}\left(\frac{kL^3}{3}w(0) + \frac{q_0L^4}{8}\right) \tag{5}$$

$$\theta(0) = \frac{1}{EI}\left(\frac{F_sL^2}{2} + \frac{q_0L^3}{6}\right) = \frac{1}{EI}\left(\frac{kL^2}{2}w(0) + \frac{q_0L^3}{6}\right). \tag{6}$$

Solving Eq. (5) for $w(0)$ first and substituting the result into Eq. (6), the deflection and rotation at the right end (i.e., at $\bar{x} = 0$) are

$$w(0) = -\frac{q_0L^4}{8EI}\left(1 + \frac{kL^3}{3EI}\right)^{-1}, \tag{7}$$

$$\theta(0) = -\frac{kL^2}{2EI}\frac{q_0L^4}{8EI}\left(1 + \frac{kL^3}{3EI}\right)^{-1} + \frac{q_0L^3}{6EI} = \frac{q_0L^3}{6EI}\left(1 - \frac{kL^3}{24EI}\right)\left(1 + \frac{kL^3}{3EI}\right)^{-1}. \tag{8}$$

For the case of rigid spring, $w(0) = 0$ and F_s is the reaction of the support. From the first parts of Eqs (5) and (6), we obtain

$$F_s = -\frac{3q_0L}{8}, \quad \theta(0) = -\frac{q_0L^3}{48EI}. \tag{9}$$

The negative signs of F_s and $\theta(0)$ indicate that the actual quantities are opposite to the assumed directions of F_s and $\theta(0)$. Thus, the actual reaction is upward and the rotation is counterclockwise at the right end of the beam. The results obtained here agree, as expected, with those obtained in **Example 4.3.5**.

Example 4.4.4

Use the unit dummy-load method to determine the horizontal and vertical displacements of point A of the frame structure shown in Fig. 4.4.4(a).

Solution: We use the unit dummy-load method for one direction at a time. First consider the vertical direction. We use a dummy vertical load of $\delta Q = 1$ kN at point A to determine the vertical deflection. The self-equilibrating virtual force system is shown in Fig. 4.4.4(b). For this determinate frame structure, the actual bending moment and the virtual bending moment due to unit load are (moments are taken clockwise as positive)

$$M(x) = \begin{cases} 0, & \text{in AB} \\ F \cdot (x - a), & \text{in BC.} \\ F \cdot a, & \text{in CD} \end{cases} \tag{1}$$

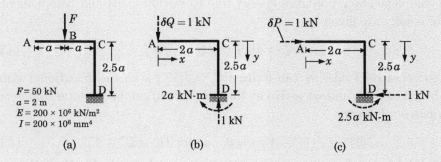

Fig. 4.4.4 (a) Given frame structure. (b) Virtual force system for v. (c) Virtual force system for u.

$$\delta M(x) = \begin{cases} 1 \cdot x, & \text{in AC} \\ 1 \cdot 2a, & \text{in CD} \end{cases} \tag{2}$$

Then by the unit dummy-load method, we have

$$1 \cdot v = \frac{1}{EI} \int M \, \delta M \, dx$$

$$= \frac{1}{EI} \int_0^a (0)(x) \, dx + \frac{1}{EI} \int_a^{2a} [F(x - a)](x) \, dx + \frac{1}{EI} \int_0^{2.5a} (Fa)(2a) \, dy$$

$$v = 0 + \frac{5Fa^3}{6EI} + \frac{5Fa^3}{EI} = \frac{35Fa^3}{6EI} = 0.0583 \text{ (m).} \tag{3}$$

Similarly, we use a dummy horizontal load of $\delta P = 1$ kN at point A to determine the horizontal deflection. The self-equilibrating virtual force system is shown in Fig. 4.4.4(c). The virtual bending moment in this case is (clockwise is positive)

$$\delta M(x) = \begin{cases} 0, & \text{in AC} \\ 1 \cdot (-y), & \text{in CD} \end{cases}. \tag{4}$$

The unit dummy-load method gives

$$1 \cdot u = \frac{1}{EI} \int M \, \delta M \, dx$$

$$= \frac{1}{EI} \int_0^a (0)(0) \, dx + \frac{1}{EI} \int_a^{2a} [F(x - a)](0) \, dx + \frac{1}{EI} \int_0^{2.5a} (Fa)(-y) \, dy$$

$$u = 0 + 0 - \frac{1}{2EI}(2.5)^2 Fa^3 = -0.0313 \text{ (m).} \tag{5}$$

The negative sign for u indicates that the displacement is in the opposite direction to the assumed dummy load. Thus, point A moves downward and to the left.

4.5   Principle of Minimum Total Complementary Potential Energy and Castigliano's Theorem II

4.5.1   The Principle of the Minimum Total Complementary Potential Energy

The complementary potential energy due to virtual loads and complementary strain energy are given by

$$\delta V_E^* = \delta W_E^*, \quad \delta U^* = \delta W_I^*. \tag{4.5.1}$$

For conservative loads, we can write $\delta V_E^* = \delta(V_E^*)$ (i.e., the variational symbol can be taken out), and we arrive at the *principle of minimum total complementary potential energy*:

$$\delta \Pi^* \equiv \delta \left(U^* + V_E^* \right) = 0 \quad \text{with} \quad \Pi^* \equiv U^* + V_E^*, \tag{4.5.2}$$

where Π^* denotes the total complementary potential energy. For three-dimensional linear elasticity, the total complementary potential energy functional has the form

$$\Pi^*(\boldsymbol{\sigma}) = \frac{1}{2} \int_\Omega C_{ijkl}^* \, \sigma_{ij} \, \sigma_{kl} \, d\Omega - \left[\int_\Omega u_i \, f_i \, d\Omega + \int_{\Gamma_u} \hat{u}_i \, t_i \, d\Gamma \right], \tag{4.5.3}$$

where C_{ijkl}^* are the compliance coefficients (i.e., $\mathbf{C}^*$ is the inverse of $\mathbf{C}$).

We now show that the principle of minimum total complementary energy gives combined kinematic and constitutive relations (i.e., stress–displacement equations) and the geometric boundary conditions as the Euler equations. Taking the first variation of Π^* and setting it to zero, we obtain

$$
\begin{aligned}
0 = \delta \Pi^* &= \int_\Omega C_{ijkl}^* \, \sigma_{kl} \, \delta\sigma_{ij} \, d\Omega - \left[\int_\Omega u_i \, \delta f_i \, d\Omega + \int_{\Gamma_u} \hat{u}_i \, \delta t_i \, d\Gamma \right] \\
&= \int_\Omega C_{ijkl}^* \, \sigma_{kl} \, \delta\sigma_{ij} \, d\Omega + \int_\Omega u_i \, \delta\sigma_{ij,j} \, d\Omega - \int_{\Gamma_u} \hat{u}_i \, \delta t_i \, d\Gamma \\
&= \int_\Omega \left[C_{ijkl}^* \, \sigma_{kl} - \tfrac{1}{2} \left(u_{i,j} + u_{j,i} \right) \right] \delta\sigma_{ij} \, d\Omega + \oint_\Gamma u_i \, \delta\sigma_{ij} n_j \, d\Gamma - \int_{\Gamma_u} \hat{u}_i \, \delta t_i \, d\Gamma \\
&= \int_\Omega \left[C_{ijkl}^* \, \sigma_{kl} - \tfrac{1}{2} \left(u_{i,j} + u_{j,i} \right) \right] \delta\sigma_{ij} \, d\Omega + \int_{\Gamma_u} u_i \, \delta t_i \, d\Gamma - \int_{\Gamma_u} \hat{u}_i \, \delta t_i \, d\Gamma \\
&= \int_\Omega \left[C_{ijkl}^* \, \sigma_{kl} - \tfrac{1}{2} \left(u_{i,j} + u_{j,i} \right) \right] \delta\sigma_{ij} \, d\Omega + \int_{\Gamma_u} \left(u_i - \hat{u}_i \right) \delta t_i \, d\Gamma, \tag{4.5.4}
\end{aligned}
$$

where we have used Eqs (4.4.1) and (4.4.2) as well as the Green–Gauss theorem to arrive at the final result in Eq. (4.5.4). Thus the Euler equations resulting from the principle of minimum total complementary potential energy are

$$C_{ijkl}^* \, \sigma_{kl} - \tfrac{1}{2} \left(u_{i,j} + u_{j,i} \right) = 0 \quad \text{in} \quad \Omega, \tag{4.5.5}$$

$$u_i - \hat{u}_i = 0 \quad \text{on} \quad \Gamma_u. \tag{4.5.6}$$

Next we establish the minimum character of Π^*. Let $\boldsymbol{\sigma} + \alpha \boldsymbol{\lambda}$ be an arbitrary admissible stress field, where α is a scalar and $\boldsymbol{\lambda}$ is an arbitrary variation of $\boldsymbol{\sigma}$, which satisfies the homogeneous form of the equilibrium equations and traction boundary conditions:

$$\nabla \cdot \boldsymbol{\lambda} = \mathbf{0} \quad \text{in } \Omega, \tag{4.5.7}$$

$$\hat{\mathbf{n}} \cdot \boldsymbol{\lambda} = \mathbf{0} \quad \text{on } \Gamma_\sigma. \tag{4.5.8}$$

We begin with

$$\Pi^*(\boldsymbol{\sigma} + \alpha \boldsymbol{\lambda}) = \frac{1}{2} \int_\Omega C^*_{ijkl} \left(\sigma_{ij} + \alpha \lambda_{ij}\right) \left(\sigma_{kl} + \alpha \lambda_{kl}\right) \, d\Omega$$

$$+ \int_\Omega u_i \left(\sigma_{ij,j} + \alpha \lambda_{ij,j}\right) \, d\Omega - \int_{\Gamma_u} \hat{u}_i \left(\sigma_{ij} + \alpha \lambda_{ij}\right) n_j \, d\Gamma$$

$$= \Pi^*(\boldsymbol{\sigma}) + \frac{\alpha^2}{2} \int_\Omega C^*_{ijkl} \lambda_{ij} \lambda_{kl} \, d\Omega + \alpha \int_\Omega C^*_{ijkl} \sigma_{ij} \lambda_{kl} \, d\Omega$$

$$+ \alpha \int_\Omega u_i \lambda_{ij,j} \, d\Omega - \alpha \int_{\Gamma_u} \hat{u}_i \lambda_{ij} n_j \, d\Gamma \tag{4.5.9}$$

where $(\cdot)_{,j}$ indicates differentiation of the enclosed quantity with respect to x_j and symmetries of σ_{ij} and C^*_{ijkl} are utilized in arriving at the last step:

$$\sigma_{ij} = \sigma_{ji}, \quad \lambda_{ij} = \lambda_{ji}, \quad C^*_{ijkl} = C^*_{klij} = C^*_{lkij} = C^*_{klji}.$$

Now consider the term

$$\alpha \int_\Omega u_i \lambda_{ij,j} \, d\Omega = -\alpha \int_\Omega u_{i,j} \lambda_{ij} \, d\Omega + \alpha \oint_\Gamma u_i \lambda_{ij} n_j \, d\Gamma$$

$$= -\frac{\alpha}{2} \int_\Omega \left(u_{i,j} + u_{j,i}\right) \lambda_{ij} \, d\Omega + \alpha \oint_\Gamma u_i \lambda_{ij} n_j \, d\Gamma$$

$$= -\frac{\alpha}{2} \int_\Omega \left(u_{i,j} + u_{j,i}\right) \lambda_{ij} \, d\Omega + \alpha \int_{\Gamma_u} u_i \lambda_{ij} n_j \, d\Gamma. \tag{4.5.10}$$

Substituting the result from Eq. (4.5.10) into Eq. (4.5.9), we obtain

$$\Pi^*(\boldsymbol{\sigma} + \alpha \boldsymbol{\lambda}) = \Pi^*(\boldsymbol{\sigma}) + \frac{\alpha^2}{2} \int_\Omega C^*_{ijkl} \lambda_{ij} \lambda_{kl} \, d\Omega + \alpha \int_{\Gamma_u} \left(u_i - \hat{u}_i\right) \lambda_{ij} n_j \, d\Omega$$

$$+ \alpha \int_\Omega \left[C^*_{ijkl} \sigma_{kl} - \tfrac{1}{2} \left(u_{i,j} + u_{j,i}\right)\right] \lambda_{ij} \, d\Omega$$

$$= \Pi^*(\boldsymbol{\sigma}) + \frac{\alpha^2}{2} \int_\Omega C^*_{ijkl} \lambda_{ij} \lambda_{kl} \, d\Omega \tag{4.5.11}$$

from which we conclude that

$$\Pi^*(\boldsymbol{\sigma} + \alpha \boldsymbol{\lambda}) \geq \Pi^*(\boldsymbol{\sigma}) \tag{4.5.12}$$

for all real numbers α and $\boldsymbol{\lambda}$ that satisfies the conditions in Eqs (4.5.7) and (4.5.8). Thus, Π^* attains its minimum at σ_{ij} that satisfies Eq. (4.5.5).

4.5.2   Castigliano's Theorem II

Analogous to the unit dummy-load method, we can derive Castigliano's theorem II from the principle of minimum total complementary potential energy. We begin with

$$\delta\Pi^* \equiv \delta U^* + \delta V^* = 0 \quad \rightarrow \quad \delta U^* = -\delta V^*.$$

If U^* and V^* can be expressed in terms of point loads F_i, then we have

$$\delta U^* = \frac{\partial U^*}{\partial F_i}\,\delta F_i, \quad \delta V^* = \frac{\partial V^*}{\partial F_i}\,\delta F_i = -u_i\,\delta F_i,$$

where u_i are the displacements at the same points and in the same direction as the forces F_i. Thus, we have

$$\left(\frac{\partial U^*}{\partial F_i} - u_i\right)\delta F_i = 0 \quad \text{or} \quad \frac{\partial U^*}{\partial F_i} = u_i. \tag{4.5.13}$$

Equation (4.5.13) is known as *Castigliano's theorem II*. Equation (4.5.13) is valid for structures that are linearly elastic as well as nonlinearly elastic. When the material of the structure is linearly elastic, we have $U_0 = U_0^*$ and $U = U^*$ in value. However, U is always expressed in terms of displacements, while U^* in terms of forces, and Castigliano's theorem I is based on U, while theorem II is based on U^*. Some authors use U in place of U^*, express U in terms of the loads, and apply Castigliano's theorem II. In some books Eq. (4.5.13) is referred to as the *Crotti–Engesser theorem* and state that it is also valid for nonlinear problems (implying that Castigliano's theorem is valid only for linear problems).

Castigliano's theorem II, which is essentially the same as the unit dummy-load method or the principle of virtual forces, is used to determine displacements and rotations under applied point forces and moments. However, when a structure does not have a load (or moment) at the point at which displacement (or rotation) is required, the determination of the displacement (or rotation) at that point requires the use of a fictitious (or dummy) load. For example, consider a beam that is subjected to distributed load $q(x)$ and n point loads $F_1, F_2, \ldots, F_n$. Suppose that we wish to determine the vertical deflection w_0 at a point O at which there is no point load. We introduce a fictitious vertical load F_0 at point O and then write the complementary strain energy in terms of q, F_0, and $F_1, F_2, \ldots, F_n$. Then we use Castigliano's theorem II to determine the desired displacement at the point:

$$w_0 = \left.\frac{\partial U^*}{\partial F_0}\right|_{F_0=0}. \tag{4.5.14}$$

For a frame structure (which includes bars, trusses, and beams as special cases) subjected to distributed and point loads and moments (applied possibly at different points), Castigliano's theorem II can be used to determine the displacement v_A in the direction of the force F_A applied at point A and rotation θ_B in the direction of the moment M_B applied at point B of the frame structure using the following formulas (when the load F_A and moment M_B are fictitious, set them to zero after differentiation with respect to F_A and M_B):

$$v_A = \int \frac{N}{EA} \frac{\partial N}{\partial F_A} dx + \int \frac{M}{EI} \frac{\partial M}{\partial F_A} dx + \int \frac{V}{K_s GA} \frac{\partial V}{\partial F_A} dx + \int \frac{T}{GJ} \frac{\partial T}{\partial F_A} dx,$$

(4.5.15)

$$\theta_B = \int \frac{N}{EA} \frac{\partial N}{\partial M_B} dx + \int \frac{M}{EI} \frac{\partial M}{\partial M_B} dx + \int \frac{V}{K_s GA} \frac{\partial V}{\partial M_B} dx + \int \frac{T}{GJ} \frac{\partial T}{\partial M_B} dx.$$

(4.5.16)

Equations (4.5.15) and (4.5.16) can also be used to determine force F_A and moment M_B when the corresponding displacement v_A and rotation θ_B are known.

We shall use Castigliano's theorem II to solve some of the example problems solved by the other methods of this chapter and some new problems. Castigliano's theorem II is a simpler method to solve truss, beam, and frame problems.

Example 4.5.1

Consider the two-member truss of **Example 4.4.1** [see Fig. 4.4.1(a)]. Determine the vertical displacement v and horizontal displacement u of point O using Castigliano's theorem II when the strain–stress relation is (a) linear $\varepsilon = \sigma/E$ and (b) nonlinear:

$$\varepsilon = \begin{cases} \frac{\sigma^2}{K^2}, & \sigma \geq 0, \\ -\frac{\sigma^2}{K^2}, & \sigma \leq 0. \end{cases}$$

(1)

where E is Young's modulus and K is a material constant.

Solution: First, assume that there is a fictitious (or virtual) load of Q applied in the horizontal direction at point O. This is necessary because unless U^* is a function of Q, we cannot use Castigliano's theorem II to compute u. The member forces can be computed as [see Fig. 4.4.1(c)] $F_1 = Q + P$ and $F_2 = -\sqrt{2}P$.

(a) The complementary strain energy of the structure for the case of linear strain–stress relation is

$$U^* = \frac{1}{2} \sum_{i=1}^{2} A_i L_i \frac{1}{E^{(i)}} \left(\sigma^{(i)}\right)^2$$

$$= \frac{A}{2E} \left[a \left(\frac{Q+P}{A}\right)^2 + \sqrt{2}a \left(-\frac{\sqrt{2}P}{A}\right)^2 \right].$$

(2)

Then, by Castigliano's theorem II, we have

$$u = \left(\frac{\partial U^*}{\partial Q}\right)\bigg|_{Q=0} = \frac{Pa}{EA},$$

$$v = \frac{\partial U^*}{\partial P} = \frac{Pa}{EA} \left(1 + 2\sqrt{2}\right).$$

(3)

(b) Next we consider the nonlinear strain–stress relation in Eq. (1). The complementary strain energy is

$$U^* = \sum_{i=1}^{2} \frac{A_i L_i}{K^2} \int_0^{\sigma^{(i)}} (\sigma)^2 \, d\sigma = \sum_{i=1}^{2} (-1)^{i+1} \frac{A_i L_i}{3K^2} \left(\sigma^{(i)} \right)^3$$

$$= \frac{A}{3K^2} \left[a \left(\frac{Q+P}{A} \right)^3 - \sqrt{2} a \left(-\frac{\sqrt{2}P}{A} \right)^3 \right]. \tag{4}$$

By Castigliano's theorem II, we have

$$u = \left(\frac{\partial U^*}{\partial Q} \right) \Big|_{Q=0} = \frac{P^2 a}{A^2 K^2},$$

$$v = \frac{\partial U^*}{\partial P} = \frac{P^2 a}{A^2 K^2} + 4 \frac{P^2 a}{A^2 K^2} = \frac{5 P^2 a}{A^2 K^2}. \tag{5}$$

Example 4.5.2

Consider the beam shown in Fig. 4.5.1 (same as Fig. 4.3.5). Determine (a) the compression in the linear elastic spring and (b) the reaction force and the rotation at $x = L$ when the spring is replaced by a rigid support. Include the energy due to transverse shear force.

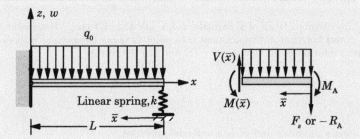

Fig. 4.5.1 A beam fixed at the left end and spring supported at the right end.

Solution: The complementary strain energy for the Bernoulli–Euler beam theory and the Timoshenko beam theory are the same, and it is given by [see Eq. (3.2.50)]

$$U^* = \frac{1}{2} \int_0^L \left(\frac{M^2}{EI} + \frac{V^2}{GAK_s} \right) d\bar{x}, \tag{1}$$

where the axial force (N) term is omitted because it is zero in the present problem. The bending moment M and shear force V for the beam at hand are

$$M(\bar{x}) = \frac{q_0 \bar{x}^2}{2} + F_s \bar{x}, \quad V(\bar{x}) = F_s + q_0 \bar{x}. \tag{2}$$

Substituting the expressions for M and V from Eq. (2) into Eq. (1),

$$U^* = \frac{1}{2EI} \int_0^L \left(\frac{q_0 \bar{x}^2}{2} + F_s \bar{x} \right)^2 d\bar{x} + \frac{1}{2GAK_s} \int_0^L (F_s + q_0 \bar{x})^2 \, d\bar{x}. \tag{3}$$

(a) Using Castigliano's theorem II, we obtain the compression $-w(0)$ in the spring:

$$-w(0) = \frac{\partial U^*}{\partial F_s} = \frac{1}{EI} \int_0^L \left(\frac{q_0 \bar{x}^2}{2} + F_s \bar{x} \right) (\bar{x}) d\bar{x} + \frac{1}{GAK_s} \int_0^L (F_s + q_0 \bar{x}) \, d\bar{x}$$

$$= \frac{q_0 L^4}{8EI} + \frac{F_s L^3}{3EI} + \frac{F_s L}{GAK_s} + \frac{q_0 L^2}{2GAK_s}. \tag{4}$$

Using the relation $F_s = kw(0)$, we obtain

$$-\left(1 + \frac{kL^3}{3EI} + \frac{kL}{GAK_s} \right) w(0) = \frac{q_0 L^4}{8EI} + \frac{q_0 L^2}{2GAK_s}$$

or

$$w(0) = -\left(\frac{q_0 L^4}{8EI} + \frac{q_0 L^2}{2GAK_s} \right) \left(1 + \frac{kL^3}{3EI} + \frac{kL}{GAK_s} \right)^{-1}. \tag{5}$$

(b) To determine the reaction and slope when the spring is replaced by a rigid support at $x = L$ (or $\bar{x} = 0$), we use a fictitious moment M_A (clockwise) at $x = L$ and take the support reaction to be R_A. The bending moment and shear force for this case are

$$M(\bar{x}) = \frac{q_0 \bar{x}^2}{2} - R_A \bar{x} + M_A, \quad V(\bar{x}) = q_0 \bar{x} - R_A. \tag{6}$$

Substituting the expressions for M and V from Eq. (6) into Eq. (1), we obtain

$$U^* = \frac{1}{2EI} \int_0^L \left(\frac{q_0 \bar{x}^2}{2} - R_A \bar{x} + M_A \right)^2 d\bar{x} + \frac{1}{2GAK_s} \int_0^L (q_0 \bar{x} - R_A)^2 \, d\bar{x}. \tag{7}$$

Using Castigliano's theorem II twice, first with respect to R_A and then with respect to M_A,

$$0 = \left(\frac{\partial U^*}{\partial R_A} \right) \bigg|_{M_A=0} = \frac{1}{EI} \int_0^L \left(\frac{q_0 \bar{x}^2}{2} - R_A \bar{x} \right) (-\bar{x}) d\bar{x} + \frac{1}{GAK_s} \int_0^L (R_A - q_0 \bar{x}) \, d\bar{x}$$

$$= -\frac{q_0 L^4}{8EI} + \frac{R_A L^3}{3EI} + \frac{R_A L}{GAK_s} - \frac{q_0 L^2}{2GAK_s}. \tag{8}$$

$$\theta_A = \left(\frac{\partial U^*}{\partial M_A} \right) \bigg|_{M_A=0} = \frac{1}{EI} \int_0^L \left(\frac{q_0 \bar{x}^2}{2} - R_A \bar{x} \right) (1) d\bar{x}$$

$$= \frac{q_0 L^3}{6EI} - \frac{R_A L^2}{2EI}. \tag{9}$$

Thus, we have

$$R_A = \frac{3q_0 L}{8} (1 + 4\Lambda)(1 + 3\Lambda)^{-1}, \quad \theta_A = -\frac{q_0 L^3}{48EI} \left[-8 + 9(1 + 4\Lambda)(1 + 3\Lambda)^{-1} \right], \tag{10}$$

where $\Lambda = EI/GAK_s L^2$. When Λ is set to zero (i.e., the effect of transverse shear is neglected), we obtain the results of **Example 4.3.4**.

Example 4.5.3

Consider a homogeneous beam with an elastic support at its center and subjected to uniformly distributed load, as shown in Fig. 4.5.2. Use Castigliano's theorem II to determine the reaction force and amount of compression in the elastic support at the center of the beam. Assume linear elastic behavior of the beam as well as the support.

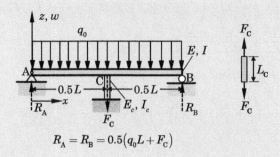

$$R_A = R_B = 0.5(q_0L + F_c)$$

Fig. 4.5.2 A simply–supported beam with a center support.

Solution: Suppose F_c is the reaction in the center support (which can be viewed as an elastic spring with spring constant $k = E_cA_c/L_c$), acting downward . Since the center support is replaced with its reaction force acting on the beam, we must consider the energy of the beam only. Due to symmetry of the problem about the center of the beam, we can consider only one-half of the beam to compute U^* and then double it.

The complementary strain energy of the whole beam is given by

$$U^* = 2\int_0^{L/2} \left(\frac{M^2}{2EI} + \frac{V^2}{2GAK_s} \right) dx, \tag{1}$$

where the bending moment M and shear force V are given by $[R_A = 0.5\,(q_0L - F_c)]$:

$$M(x) = \frac{1}{2}\left[-(q_0L + F_c)\,x + q_0x^2 \right], \quad V(x) = -\frac{1}{2}\,(q_0L + F_c) + q_0x, \quad 0 \le x \le \frac{L}{2}. \tag{2}$$

Using Castigliano's theorem II, we obtain

$$
\begin{aligned}
-w(L/2) &= \frac{\partial U^*}{\partial F_c} = 2\int_0^{L/2} \left(\frac{M}{EI}\frac{\partial M}{\partial F_c} + \frac{V}{GAK_s}\frac{\partial V}{\partial F_c} \right) dx \\
&= \frac{2}{4EI}\int_0^{L/2} \left[-(q_0L + F_c)\,x + q_0x^2 \right](-x)\,dx \\
&\quad + \frac{2}{4GAK_s}\int_0^{L/2} \left[-(q_0L + F_c) + 2q_0x \right](-1)\,dx \\
&= \frac{1}{2EI}\left(\frac{q_0L^4}{24} + \frac{F_cL^3}{24} - \frac{q_0L^4}{64} \right) + \frac{1}{2GAK_s}\left(\frac{q_0L^2}{2} + \frac{F_cL}{2} - \frac{q_0L^2}{4} \right) \\
&= \frac{1}{2EI}\left(\frac{5q_0L^4}{192} + \frac{F_cL^3}{24} \right) + \frac{1}{2GAK_s}\left(\frac{q_0L^2}{4} + \frac{F_cL}{2} \right). \tag{3}
\end{aligned}
$$

The minus sign indicates that $w(L/2)$ is taken (by the sign convention adopted) opposite in sense to that of the reaction F_c. From the uniaxial deformation of the central support, it is clear that

$$F_c = w(L/2)\,k_c \quad \text{or} \quad w(L/2) = \frac{F_c}{k_c}, \quad k_c = \frac{E_cA_c}{L_c}. \tag{4}$$

Hence, we have

$$-F_c\frac{1}{k_c} = \frac{1}{EI}\left(\frac{F_cL^3}{48} + \frac{5q_0L^4}{384} \right) + \frac{1}{GAK_s}\left(\frac{q_0L^2}{8} + \frac{F_cL}{4} \right), \tag{5}$$

from which we obtain

$$F_c = \frac{5q_0 L^4}{384EI}(1+9.6\Lambda)\left[\frac{1}{k_c}+\frac{L^3}{48EI}(1+12\Lambda)\right]^{-1}. \tag{6}$$

The center deflection is then given by

$$w(L/2) = -\frac{5q_0 L^4}{384EI}(1+9.6\Lambda)\left[1+\frac{k_c L^3}{48EI}(1+12\Lambda)\right]^{-1}. \tag{7}$$

When *transverse shear is neglected*, that is, $\Lambda = 0$, we obtain from Eqs (6) and (7)

$$F_c = -\frac{5q_0 L^4}{384EI}\left(\frac{1}{k_c}+\frac{L^3}{48EI}\right)^{-1}, \quad w(L/2) = -\frac{5q_0 L^4}{384EI}\left(1+\frac{k_c L^3}{48EI}\right)^{-1}. \tag{8}$$

If the support is rigid, that is, $k_c = \infty$, Eq. (8) gives

$$R_C = -F_c = \frac{5q_0 L}{8}, \quad w(L/2) = 0. \tag{9}$$

If there is no support (i.e., $k_c = 0$), then

$$F_c = 0, \quad w(L/2) = -\frac{5q_0 L^4}{384EI}. \tag{10}$$

When *transverse shear in **not** neglected* but the support is rigid, we have the solution

$$R_C = -F_c = \frac{5q_0 L}{8}\left(\frac{1+9.6\Lambda}{1+12\Lambda}\right), \quad w(L/2) = 0. \tag{11}$$

If the central support is not present (i.e., $k_c = 0$ and the beam is simply supported), then

$$R_C = 0, \quad w(L/2) = -\frac{5q_0 L^4}{384EI}(1+9.6\Lambda). \tag{12}$$

Clearly, shear deformation has the effect of increasing the deflection.

As can be seen from the examples of beams considered so far that Castigliano's theorem II simplifies the determination of reaction forces and moments of indeterminate beams when compared to the traditional method of using the statics and kinematics. The next example illustrates this point directly.

Example 4.5.4

Consider a clamped beam with linearly varying transverse load, as shown in Fig. 4.5.3. Determine the reactions at the right-end support of the beam.

Solution: First we write the expression for the shear force and bending moment,

$$V(x) = -R_A + \left(q_0\frac{x}{L}\right)\frac{x}{2}, \quad M(x) = M_A - R_A x + \left(q_0\frac{x}{L}\right)\frac{x}{2}\frac{x}{3}, \quad 0 < x < L \tag{1}$$

and compute their derivatives with respect to R_A and M_A as required in the Castigliano's

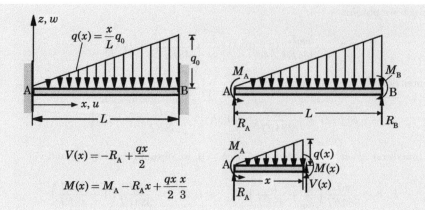

Fig. 4.5.3 A clamped beam with linearly varying load.

theorem II:

$$\frac{\partial M}{\partial R_A} = -x, \quad \frac{\partial M}{\partial M_A} = 1, \quad \frac{\partial V}{\partial R_A} = -1, \quad \frac{\partial V}{\partial M_A} = 0. \tag{2}$$

Using Castigliano's theorem II, we obtain $[\theta_x \equiv -(dw/dx)]$

$$
\begin{aligned}
w(0) &= \frac{\partial U^*}{\partial R_A} = \frac{1}{EI}\int_0^L M\frac{\partial M}{\partial R_A}\,dx + \frac{1}{GAK_s}\int_0^L V\frac{\partial V}{\partial R_A}\,dx \\
&= \frac{1}{EI}\int_0^L \left(M_A - R_A x + \frac{q_0}{6L}x^3\right)(-x)\,dx + \frac{1}{GAK_s}\int_0^L\left(-R_A + \frac{q_0}{2L}x^2\right)(-1)\,dx \\
&= \frac{1}{EI}\left(-\frac{M_A L^2}{2} + \frac{R_A L^3}{3} - \frac{q_0 L^4}{30}\right) + \frac{1}{GAK_s}\left(R_A L - \frac{q_0 L^2}{6}\right)
\end{aligned}
\tag{3}
$$

$$
\begin{aligned}
\theta_x(0) &= \frac{\partial U^*}{\partial M_A} = \frac{1}{EI}\int_0^L M\frac{\partial M}{\partial M_A}\,dx + \frac{1}{GAK_s}\int_0^L V\frac{\partial V}{\partial M_A}\,dx \\
&= \frac{1}{EI}\int_0^L \left(M_A - R_A x + \frac{q_0}{6L}x^3\right)(1)\,dx + \frac{1}{GAK_s}\int_0^L\left(-R_A + \frac{q_0}{2L}x^2\right)(0)\,dx \\
&= \frac{1}{EI}\left(M_A L - \frac{R_A L^2}{2} + \frac{q_0 L^3}{24}\right).
\end{aligned}
\tag{4}
$$

Since the deflection and rotation are zero at $x = 0$ [i.e., $w(0) = 0$ and $\theta_x(0) = 0$], we obtain

$$0 = -\frac{M_A}{2} + \frac{R_A L}{3} - \frac{q_0 L^2}{30} + \Lambda\left(R_A L - \frac{q_0 L^2}{6}\right), \tag{5}$$

$$0 = M_A - \frac{R_A L}{2} + \frac{q_0 L^2}{24}, \tag{6}$$

where $\Lambda = EI/(GAK_s L^2)$. Writing in matrix form for the unknowns R_A and M_A,

$$
\begin{bmatrix} \frac{1}{3}+\Lambda & -\frac{1}{2} \\ -\frac{1}{2} & 1 \end{bmatrix}
\begin{Bmatrix} R_A L \\ M_A \end{Bmatrix}
= \frac{q_0 L^2}{120}
\begin{Bmatrix} 4(1+5\Lambda) \\ -5 \end{Bmatrix},
\tag{7}
$$

and solving the equations using Cramer's rule, we arrive at

$$R_A = \frac{q_0 L}{20}\left(\frac{3+40\Lambda}{1+12\Lambda}\right), \qquad M_A = \frac{q_0 L^2}{30}\left(\frac{1+15\Lambda}{1+12\lambda}\right). \tag{8}$$

To obtain the result in Eq. (8) by the conventional mechanics of materials approach, one must use the Timoshenko beam theory, and it would be algebraically very complicated. When the effect of transverse shear deformation is neglected (i.e., $\Lambda = 0$), we obtain

$$R_A = \frac{3q_0 L}{20}, \qquad M_A = \frac{q_0 L^2}{30}. \tag{9}$$

Example 4.5.5

Consider the frame structure shown in Fig. 4.5.4(a). Determine the transverse deflections of points A and B, assuming linear elastic behavior with E, G, I, and J as the constant material and geometric parameters.

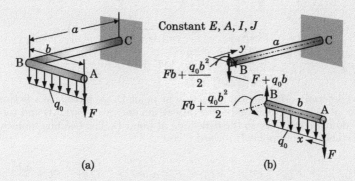

(a)   (b)

Fig. 4.5.4 (a) A frame structure, ABC. (b) Forces on parts AB and BC.

Solution: An inspection of the structure in Fig. 4.5.4(a) shows that portions AB and BC undergo bending deformation while BC also experiences torsional deformation. Therefore, the strain energy of the structure involves computing the bending and transverse shear energies of parts AB and BC and torsional energy of part BC of the frame structure. The total complementary strain energy of the structure is the sum of the energies of its parts: $U^* = U^*_{AB} + U^*_{BC}$, where

$$U^*_{AB} = \int_0^b \left(\frac{M^2}{2EI} + \frac{V^2}{2GAK_s} \right) dx, \quad U^*_{BC} = \int_0^a \left(\frac{M^2}{2EI} + \frac{V^2}{2GAK_s} + \frac{T^2}{2GJ} \right) dx. \tag{1}$$

The expressions for M, V, and T are calculated as follows [see Fig. 4.5.4(b)]:

$$\text{Part AB:} \quad M(x) = Fx + \frac{q_0 x^2}{2}, \quad V(x) = F + q_0 x, \quad T = 0 \quad \text{for } 0 < x < b, \tag{2}$$

$$\text{Part BC:} \quad M(y) = (F + q_0 b)\, y, \quad V(y) = F + q_0 b, \quad T = Fb + \frac{q_0 b^2}{2} \quad \text{for } 0 < y < a. \tag{3}$$

Thus, the total complementary strain energy of the whole structure is

$$U^* = \frac{1}{2EI} \int_0^b M^2\, dx + \frac{1}{2GAK_s} \int_0^b V^2\, dx$$
$$+ \frac{1}{2EI} \int_0^a M^2\, dy + \frac{1}{2GAK_s} \int_0^a V^2\, dy + \frac{1}{2GJ} \int_0^a T^2\, dy$$

$$U^* = \frac{1}{2EI} \int_0^b \left(Fx + \frac{q_0 x^2}{2}\right)^2 dx + \frac{1}{2GAK_s} \int_0^b (F + q_0 x)^2 dx$$

$$+ \frac{1}{2EI} \int_0^a (F + q_0 b)^2 y^2 dy + \frac{1}{2GAK_s} \int_0^a (F + q_0 b)^2 dy + \frac{1}{2GJ} \int_0^a \left(Fb + \frac{q_0 b^2}{2}\right)^2 dy$$

$$= \frac{1}{2EI} \left(\frac{F^2 b^3}{3} + \frac{q_0^2 b^5}{20} + \frac{Fq_0 b^4}{4}\right) + \frac{1}{2GAK_s} \left(F^2 b + \frac{q_0^2 b^3}{3} + Fq_0 b^2\right)$$

$$+ \frac{1}{2EI} (F + q_0 b)^2 \frac{a^3}{3} + \frac{1}{2GAK_s} (F + q_0 b)^2 a + \frac{1}{2GJ} \left(Fb + \frac{q_0 b^2}{2}\right)^2 a. \tag{4}$$

The deflection at point A by Castigliano's theorem II is given by

$$w_A = \frac{\partial U^*}{\partial F} = \frac{1}{EI} \left(\frac{Fb^3}{3} + \frac{q_0 b^4}{8}\right) + \frac{1}{GAK_s} \left(Fb + \frac{q_0 b^2}{2}\right)$$

$$+ \frac{1}{EI} (F + q_0 b) \frac{a^3}{3} + \frac{1}{GAK_s} (F + q_0 b) a + \frac{1}{GJ} \left(Fb + \frac{q_0 b^2}{2}\right) ab$$

$$= \left(\frac{a^3 + b^3}{3EI} + \frac{a + b}{GAK_s} + \frac{ab^2}{GJ}\right) F + \left[\frac{b}{24EI}(8a^3 + 3b^3) + \frac{b}{2GAK_s}(b + 2a) + \frac{b^2}{2GJ}\right] q_0. \tag{5}$$

In order to determine the vertical deflection at point B, we introduce a fictitious load Q, vertically down, at point B. The complementary strain energy of part AB and torque created by the force F do not contribute to the deflection at point B. The bending moment and shear force at any distance y from point B toward C are

$$M(y) = (F + q_0 b + Q)y, \quad V(y) = F + q_0 b + Q, \quad T(y) = Fb + \frac{q_0 b^2}{2}. \tag{6}$$

Using Castigliano's theorem II, we can write

$$w_B = \left.\frac{\partial U^*_{BC}}{\partial Q}\right|_{Q=0} = \left[\frac{1}{EI} \int_0^a M \frac{\partial M}{\partial Q} dy + \frac{1}{GAK_s} \int_0^a V \frac{\partial V}{\partial Q} dy + \int_0^a T \frac{\partial T}{\partial Q} dy\right]_{Q=0}$$

$$= \left[\frac{1}{EI} \int_0^a (F + q_0 b + Q) y(y) dy + \frac{1}{GAK_s} \int_0^a (F + q_0 b + Q)(1) dy\right]_{Q=0}$$

$$= \frac{(F + q_0 b)a^3}{3EI} + \frac{(F + q_0 b)a}{GAK_s} \tag{7}$$

or

$$w_B = \frac{(F + q_0 b)a^3}{3EI} (1 + 3\Lambda_a), \quad \Lambda_a = \frac{EI}{GAK_s a^2}. \tag{8}$$

Example 4.5.6

Consider the cable-supported beam shown in Fig. 4.5.5(a). The beam and the cable are made of homogeneous, linear elastic, isotropic materials, with constant geometric properties. Determine the horizontal and vertical displacements at the left end and the force in the cable using Castigliano's theorem II. Use the Bernoulli–Euler beam theory and neglect the energy due to the transverse shear deformation.

Solution: Figure 4.5.5(b) shows the effect of the cable force on the beam. We note that a cable can only take a tensile force. Because the load on the beam is transversely down, the cable is indeed in tension. Thus, the cable experiences elongation. Using Castigliano's theorem II

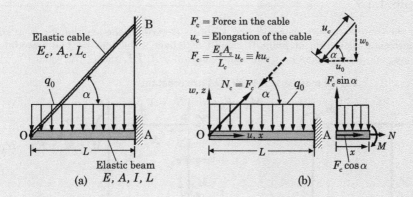

Fig. 4.5.5 A cable-supported beam.

for the *whole* structure (i.e., beam and the cable), we can write (the cable displacement at point B is zero)

$$0 = \frac{\partial U^*}{\partial F_c} = \int_0^L \left(\frac{N}{EA} \frac{\partial N}{\partial F_c} + \frac{M}{EI} \frac{\partial M}{\partial F_c} \right) dx + \int_0^{L_c} \frac{N_c}{E_c A_c} \frac{\partial N_c}{\partial F_c} dx, \tag{1}$$

where

$$N = -F_c \cos \alpha, \quad M = -(F_c \sin \alpha)x + \frac{q_0 x^2}{2}, \quad N_c = F_c, \tag{2}$$

and

$$\frac{\partial N}{\partial F_c} = -\cos \alpha, \quad \frac{\partial M}{\partial F_c} = -x \sin \alpha, \quad \frac{\partial N_c}{\partial F_c} = 1. \tag{3}$$

Substituting the expressions for N, M, and N_c and their derivatives into Eq. (1) and evaluating the integrals, we obtain

$$0 = \frac{F_c L_c}{E_c A_c} + \frac{F_c L}{EA} \cos^2 \alpha + \frac{F_c L^3}{3EI} \sin^2 \alpha - \frac{q_0 L^4}{8EI} \sin \alpha. \tag{4}$$

The cable force is given by

$$F_c = \frac{qL^4}{8EI} \sin \alpha \left(\frac{L_c}{E_c A_c} + \frac{L}{EA} \cos^2 \alpha + \frac{L^3}{3EI} \sin^2 \alpha \right)^{-1}. \tag{5}$$

Then the cable elongation u_c is determined from

$$u_c \equiv \frac{F_c L_c}{E_c A_c} = \frac{qL^4}{8EI} \sin \alpha \left[1 + \frac{E_c A_c}{L_c} \left(\frac{L}{EA} \cos^2 \alpha + \frac{L^3}{3EI} \sin^2 \alpha \right) \right]^{-1}. \tag{6}$$

The horizontal and vertical displacements (u_0, w_0) of the left end of the beam are determined from

$$u_0 = -u_c \cos \alpha, \quad w_0 = -u_c \sin \alpha. \tag{7}$$

When $\alpha = 90°$ (i.e., the cable is vertical), we have

$$w_0 = -\frac{qL^4}{8EI} \left(1 + \frac{kL^3}{3EI} \right)^{-1}, \quad k = \frac{E_c A_c}{L_c}. \tag{8}$$

Example 4.5.7

Consider the frame structure shown in Fig. 4.5.6(a). Determine the reaction forces at support A and horizontal deflection at point B using Castigliano's theorem II. The material obeys Hooke's law, and all members have the same constant EA, EI, and GAK_s values.

For all members: EA, EI, GA are constant

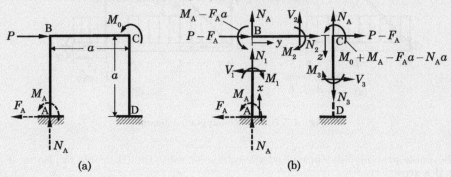

(a)   (b)

Fig. 4.5.6 (a) A frame structure. (b) Coordinate direction and sense of axial and transverse forces and bending moment in each member.

Solution: The expressions for the bending moment and axial force in each member can be expressed in terms of the reactions at point A. The sense of the local coordinate, bending moment, and shear force for each member is indicated in Fig. 4.5.6(b).

Member AB: $N_1(x) = -N_A$, $V_1(x) = -F_A$, $M_1(x) = M_A - F_A x$

$$\frac{\partial N_1}{\partial N_A} = -1, \quad \frac{\partial V_1}{\partial N_A} = 0, \quad \frac{\partial M_1}{\partial N_A} = 0,$$

$$\frac{\partial N_1}{\partial F_A} = 0, \quad \frac{\partial V_1}{\partial F_A} = -1, \quad \frac{\partial M_1}{\partial F_A} = -x, \tag{1}$$

$$\frac{\partial N_1}{\partial M_A} = 0, \quad \frac{\partial V_1}{\partial M_A} = 0, \quad \frac{\partial M_1}{\partial M_A} = 1.$$

Member BC: $N_2(y) = F_A - P$, $V_2(y) = -N_A$, $M_2(y) = M_A - F_A a - N_A y$

$$\frac{\partial N_2}{\partial N_A} = 0, \quad \frac{\partial V_2}{\partial N_A} = -1, \quad \frac{\partial M_2}{\partial N_A} = -y,$$

$$\frac{\partial N_2}{\partial F_A} = 1, \quad \frac{\partial V_2}{\partial F_A} = 0, \quad \frac{\partial M_2}{\partial F_A} = -a, \tag{2}$$

$$\frac{\partial N_2}{\partial M_A} = 0, \quad \frac{\partial V_2}{\partial M_A} = 0, \quad \frac{\partial M_2}{\partial M_A} = 1.$$

Member CD: $N_3(z) = N_A$, $V_3(z) = F_A - P$, $M_3(z) = M_A + M_0 - F_A a - N_A a + (F_A - P)z$

$$\frac{\partial N_3}{\partial N_A} = 1, \quad \frac{\partial V_3}{\partial N_A} = 0, \quad \frac{\partial M_3}{\partial N_A} = -a,$$

$$\frac{\partial N_3}{\partial F_A} = 0, \quad \frac{\partial V_3}{\partial F_A} = 1, \quad \frac{\partial M_3}{\partial F_A} = -a + z, \tag{3}$$

$$\frac{\partial N_3}{\partial M_A} = 0, \quad \frac{\partial V_3}{\partial M_A} = 0, \quad \frac{\partial M_3}{\partial M_A} = 1.$$

The complementary strain energy of the ith member is

$$U_i^* = \frac{1}{2} \int_0^a \left(\frac{N_i^2}{EA} + \frac{M_i^2}{EI} + \frac{V_i^2}{GAK_s} \right) dx, \tag{4}$$

and the total complementary strain energy is the sum of the energies of all members of the structure, $U^* = U_1^* + U_2^* + U_3^*$. Since the support at point A is rigid, all generalized displacements are zero there. Therefore, Castigliano's theorem II gives

$$0 = \frac{\partial U^*}{\partial N_A} = \sum_{i=1}^{3} \int_0^a \left(\frac{N_i}{EA} \frac{\partial N_i}{\partial N_A} + \frac{M_i}{EI} \frac{\partial M_i}{\partial N_A} + \frac{V_i}{GAK_s} \frac{\partial V_i}{\partial N_A} \right) dx, \tag{5}$$

$$0 = \frac{\partial U^*}{\partial F_A} = \sum_{i=1}^{3} \int_0^a \left(\frac{N_i}{EA} \frac{\partial N_i}{\partial F_A} + \frac{M_i}{EI} \frac{\partial M_i}{\partial F_A} + \frac{V_i}{GAK_s} \frac{\partial V_i}{\partial F_A} \right) dx, \tag{6}$$

$$0 = \frac{\partial U^*}{\partial M_A} = \sum_{i=1}^{3} \int_0^a \left(\frac{N_i}{EA} \frac{\partial N_i}{\partial M_A} + \frac{M_i}{EI} \frac{\partial M_i}{\partial M_A} + \frac{V_i}{GAK_s} \frac{\partial V_i}{\partial M_A} \right) dx. \tag{7}$$

Substituting for (N_i, M_i, V_i) and carrying out the integration, we obtain

$$0 = \frac{2N_A a}{EA} + \frac{1}{EI} \left(\frac{4N_A a^3}{3} + F_A a^3 - \frac{3M_A a^2}{2} - M_0 a^2 + \frac{Pa^3}{2} \right) + \frac{N_A a}{GAK_s} \tag{8}$$

$$0 = \frac{1}{EI} \left(\frac{5F_A a^3}{3} - 2M_A a^2 + N_A a^3 + \frac{Pa^3}{6} - \frac{M_0 a^2}{2} \right)$$
$$+ \frac{1}{EA} (F_A a - Pa) + \frac{1}{GAK_s} (2F_A a - Pa) \tag{9}$$

$$0 = \frac{1}{EI} \left(-2F_A a^2 + 3M_A a - \frac{3N_A a^2}{2} - \frac{Pa^2}{2} + M_0 a \right), \tag{10}$$

or, in matrix form, we have

$$\frac{a^3}{EI} \begin{bmatrix} \frac{4}{3} + 2\Lambda_1 + \Lambda_2 & 1 & -\frac{3}{2a} \\ 1 & \frac{5}{3} + \Lambda_1 + 2\Lambda_2 & -\frac{2}{a} \\ -\frac{3}{2a} & -\frac{2}{a} & \frac{3}{a^2} \end{bmatrix} \begin{Bmatrix} N_A \\ F_A \\ M_A \end{Bmatrix} = \frac{a^3}{EI} \begin{Bmatrix} -\frac{P}{2} + \frac{M_0}{a} \\ P(\Lambda_1 + \Lambda_2) - \frac{P}{6} + \frac{M_0}{2a} \\ \frac{P}{2a} - \frac{M_0}{a^2} \end{Bmatrix}, \tag{11}$$

where

$$\Lambda_1 = \frac{EI}{EAa^2}, \qquad \Lambda_2 = \frac{EI}{GAK_s a^2}. \tag{12}$$

We note that the coefficient matrix is symmetric. Solution of these equations will yield N_A, F_A, and M_A.

4.6 Clapeyron's, Betti's, and Maxwell's Theorems

4.6.1 Principle of Superposition for Linear Problems

The principle of superposition is said to hold for a problem if the solution (e.g., displacements) under two sets of boundary conditions and loads are equal to the sum of the solutions obtained by applying each set of boundary conditions and loads separately. The principle can be used to solve, for example, indeterminate beam problems as a superposition of two or more determinate beam problems. It plays a significant role in elasticity to establish some important results, as discussed in this section.

To express the principle in analytical terms, consider the following two sets of boundary conditions and loads in an elasticity problem:

Set 1: $\mathbf{u} = \hat{\mathbf{u}}^{(1)}$ on Γ_u; $\mathbf{t} = \hat{\mathbf{t}}^{(1)}$ on Γ_σ; $\mathbf{f} = \mathbf{f}^{(1)}$ in Ω,   (4.6.1)

Set 2: $\mathbf{u} = \hat{\mathbf{u}}^{(2)}$ on Γ_u; $\mathbf{t} = \hat{\mathbf{t}}^{(2)}$ on Γ_σ; $\mathbf{f} = \mathbf{f}^{(2)}$ in Ω,   (4.6.2)

where the specified data $(\hat{\mathbf{u}}^{(1)}, \hat{\mathbf{t}}^{(1)}, \mathbf{f}^{(1)})$ and $(\hat{\mathbf{u}}^{(2)}, \hat{\mathbf{t}}^{(2)}, \mathbf{f}^{(2)})$ are independent of the displacements $\mathbf{u}$. Suppose that the solution to the two problems be $\mathbf{u}^{(1)}(\mathbf{x})$ and $\mathbf{u}^{(2)}(\mathbf{x})$, respectively. The superposition of the two sets of boundary conditions is

$$\mathbf{u} = \hat{\mathbf{u}}^{(1)} + \hat{\mathbf{u}}^{(2)} \text{ on } \Gamma_u; \quad \mathbf{t} = \hat{\mathbf{t}}^{(1)} + \hat{\mathbf{t}}^{(2)} \text{ on } \Gamma_\sigma; \quad \mathbf{f} = \mathbf{f}^{(1)} + \mathbf{f}^{(2)} \text{ in } \Omega. \quad (4.6.3)$$

If the solution of the problem with the superposed data is

$$\mathbf{u}(\mathbf{x}) = \mathbf{u}^{(1)}(\mathbf{x}) + \mathbf{u}^{(2)}(\mathbf{x}) \text{ for all } \mathbf{x} \text{ in } \Omega, \quad (4.6.4)$$

then the *principle of superposition* is said to hold. For all linear (geometrically and materially) problems, the principle of superposition holds.

For example, the transverse deflection $w(x)$ of a cantilever beam with uniformly distributed load of intensity q_0 and concentrated load F_0 at the free end can be obtained as the superposition (i.e., sum) of the transverse deflection $w^q(x)$ of a cantilever beam with uniformly distributed load of intensity q_0 and the transverse deflection $w^F(x)$ of a cantilever beam with concentrated load F_0 at the free end (provided the geometry and material properties of the two problems being superposed are the same as the original problem and the same coordinate system is used in all cases), as shown in Fig. 4.6.1. From mechanics of material books,[1] the deflections $w^q(x)$ and $w^F(x)$ according to the Bernoulli–Euler beam theory are given by

$$w^q(x) = \frac{q_0 L^4}{24EI}\left[3 - 4\frac{x}{L} + \left(\frac{x}{L}\right)^4\right], \quad (4.6.5)$$

$$w^F(x) = \frac{F_0 L^3}{6EI}\left[2 - 3\frac{x}{L} + \left(\frac{x}{L}\right)^3\right]. \quad (4.6.6)$$

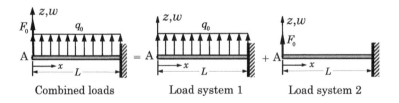

Fig. 4.6.1 Illustration of the principle of superposition for a beam.

---

[1] One may integrate the equations $d^4w/dx^4 = q_0/EI$ and $d^4w/dx^4 = 0$, respectively, for the problems and apply the associated boundary conditions to obtain these solutions.

Then the deflection of the original problem is given by

$$w(x) = \frac{q_0 L^4}{24EI} \left[3 - 4\frac{x}{L} + \left(\frac{x}{L}\right)^4 \right] + \frac{F_0 L^3}{6EI} \left[2 - 3\frac{x}{L} + \left(\frac{x}{L}\right)^3 \right]. \tag{4.6.7}$$

The principle of superposition can be used to solve complicated problems, provided that the given problem can be represented as a superposition (in terms of loads applied) of a set of simpler problems for which solutions are readily available. The next example illustrates the use of the superposition principle in determining the deflection of an indeterminate beam that can be represented as a superposition of two determinate beams.

Example 4.6.1

Consider the indeterminate beam shown in Fig. 4.6.2(a). Determine the transverse deflection of the beam using the principle of superposition.

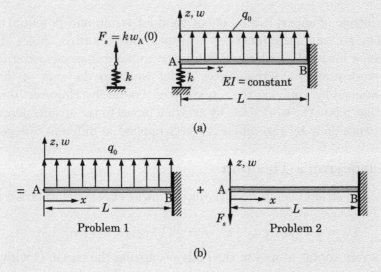

(a)

Problem 1   (b)   Problem 2

Fig. 4.6.2 Representation of an indeterminate beam as a superposition of two determinate beams.

Solution: The indeterminate beam shown in Fig. 4.6.2(a) is equivalent to the two beam problems shown in Fig. 4.6.2(b). At point A the beam experiences transverse deflections w^q and w^F, due, respectively, to the distributed load q_0 and spring force F_s. Within the restrictions of the linear Bernoulli–Euler beam theory, the deflections are linear functions of the loads [see Eqs (4.6.5) and (4.6.6)]. Therefore, the principle of superposition is valid, and the deflection of the original problem is the superposition of the deflections of the two problems:

$$w(x) = \frac{q_0 L^4}{24EI} \left[3 - 4\frac{x}{L} + \left(\frac{x}{L}\right)^4 \right] + \frac{(-F_s)L^3}{6EI} \left[2 - 3\frac{x}{L} + \left(\frac{x}{L}\right)^3 \right], \tag{1}$$

where the spring force, F_s, is yet to be determined. The force F_s is equal to kw_A, where

$$w_A = w(0) = \frac{q_0 L^4}{8EI} - \frac{F_s L^3}{3EI} \tag{2}$$

or

$$w_A = \frac{q_0 L^4}{8EI} - \frac{kL^3}{3EI} w_A \Rightarrow w_A = \frac{q_0 L^4}{8EI} \left(1 + \frac{kL^3}{3EI}\right)^{-1}. \tag{3}$$

Thus, the transverse deflection of the original problem is

$$w(x) = \frac{q_0 L^4}{24EI} \left[3 - 4\frac{x}{L} + \left(\frac{x}{L}\right)^4\right] - \frac{kL^3}{6EI} \left[2 - 3\frac{x}{L} + \left(\frac{x}{L}\right)^3\right] w_A \tag{4}$$

$$= \frac{q_0 L^4}{24EI} \left[3 - 4\frac{x}{L} + \left(\frac{x}{L}\right)^4\right] - \frac{q_0 L^4}{16EI} \left[2 - 3\frac{x}{L} + \left(\frac{x}{L}\right)^3\right] \left(1 + \frac{3EI}{kL^3}\right)^{-1}. \tag{5}$$

When $k = 0$, the original problem and Problem 1 (see Fig. 4.6.2(b)) becomes the same and Eq. (4) gives $w(x) = w^q(x)$. If $k \to \infty$, then Eq. (5) yields the deflection of a beam with uniformly distributed load q_0 and simply supported at the left end and clamped at the right end:

$$w(x) = \frac{q_0 L^4}{48EI} \left[\frac{x}{L} - 3\left(\frac{x}{L}\right)^3 + 2\left(\frac{x}{L}\right)^4\right]. \tag{6}$$

The principle of superposition is not valid for strain and potential energies because they are quadratic functions of displacements and/or forces. In other words, when a linear elastic body is subjected to more than one external force, the total work due to external forces is not equal to the sum of the works that are obtained by applying the single forces separately. However, there exist theorems that relate the work done by external forces to the strain energy stored and relate work done by two different forces applied in different orders.

4.6.2   Clapeyron's Theorem

Recall that the strain energy density due to linear elastic deformation is given by

$$U_0 = \tfrac{1}{2}\boldsymbol{\varepsilon} : \mathbf{C} : \boldsymbol{\varepsilon} = \tfrac{1}{2}\boldsymbol{\varepsilon} : \boldsymbol{\sigma} = \tfrac{1}{2}C_{ijkl}\,\varepsilon_{kl}\,\varepsilon_{ij} = \tfrac{1}{2}\sigma_{ij}\,\varepsilon_{ij}.$$

The total strain energy stored in the body occupying the region Ω with surface Γ is equal to

$$U = \int_\Omega U_0 \, d\Omega = \tfrac{1}{2}\int_\Omega \boldsymbol{\sigma} : \boldsymbol{\varepsilon} \, d\Omega = \tfrac{1}{2}\int_\Omega \sigma_{ij}\,\varepsilon_{ij} \, d\Omega. \tag{4.6.8}$$

The total work done by the body force $\mathbf{f}$ (measured per unit volume) and surface traction $\mathbf{t}$ (measured per unit area) in moving through their respective displacements $\mathbf{u}$ is given by

$$V_E = - \left[\int_\Omega \mathbf{f} \cdot \mathbf{u} \, d\Omega + \oint_\Gamma \mathbf{t} \cdot \mathbf{u} \, d\Gamma\right]. \tag{4.6.9}$$

When $\mathbf{u} = \mathbf{0}$ on a portion Γ_u of the boundary Γ, the surface integral in Eq. (4.6.9) becomes

$$\int_{\Gamma_\sigma} \mathbf{t} \cdot \mathbf{u} \, d\Gamma, \quad \text{where} \quad \Gamma_\sigma = \Gamma - \Gamma_u.$$

Theorem 4.6.1: The strain energy stored in a linear elastic body is equal to one-half of the work done by external forces on the body:

$$U = -\tfrac{1}{2}V_E$$

or

$$\tfrac{1}{2}\int_\Omega \boldsymbol{\sigma} : \boldsymbol{\varepsilon}\, d\Omega = \tfrac{1}{2}\left[\int_\Omega \mathbf{f}\cdot\mathbf{u}\, d\Omega + \oint_\Gamma \mathbf{t}\cdot\mathbf{u}\, d\Gamma\right]. \tag{4.6.10}$$

Proof: We begin with the left-hand side of the Eq.(4.6.10) and arrive at the right-hand side. Owing to the symmetry of the stress tensor $\boldsymbol{\sigma}$, $\boldsymbol{\sigma}^{\mathrm{T}} = \boldsymbol{\sigma}$ ($\sigma_{ij} = \sigma_{ji}$), we can write $\boldsymbol{\sigma} : \boldsymbol{\varepsilon} = \boldsymbol{\sigma} : \nabla\mathbf{u} = \sigma_{ij}u_{i,j}$. Consequently, the strain energy U can be expressed as

$$\begin{aligned}
U &= \tfrac{1}{2}\int_\Omega \boldsymbol{\sigma} : \boldsymbol{\varepsilon}\, d\Omega = \tfrac{1}{4}\int_\Omega \sigma_{ij}\left(u_{i,j} + u_{j,i}\right) d\Omega = \tfrac{1}{2}\int_\Omega \sigma_{ij}\, u_{i,j}\, d\Omega \\
&= -\tfrac{1}{2}\int_\Omega \sigma_{ij,j}\, u_i\, d\Omega + \tfrac{1}{2}\oint_\Gamma n_j\sigma_{ij}u_i\, d\Gamma \\
&= \tfrac{1}{2}\int_\Omega f_i\, u_i\, d\Omega + \tfrac{1}{2}\oint_\Gamma t_i u_i\, d\Gamma \\
&= \tfrac{1}{2}\int_\Omega \mathbf{f}\cdot\mathbf{u}\, d\Omega + \tfrac{1}{2}\oint_\Gamma \mathbf{t}\cdot\mathbf{u}\, d\Gamma,
\end{aligned}$$

where, in arriving at the last line, we have used the stress equilibrium equation $\sigma_{ij,j} + f_i = 0$, Cauchy's formula $t_i = \sigma_{ij}n_j$, and the divergence theorem. Thus, the proof is complete.

Equation (4.6.10) is known as *Clapeyron's theorem*. Although the derivation of the result in Eq. (4.6.10) is based on U, the result holds for U^* because $U = U^*$ (in value) for linear elastic solids.[2] We next consider two examples of the usefulness of the theorem.

Example 4.6.2 ————————————————————————————————

Consider a cantilever beam of length L and flexural rigidity EI and bent by a point load F_0 at a distance a from the free end, as shown in Fig. 4.6.3. Use Clapeyron's theorem to determine the transverse deflection at the point where the load is applied. Use the Timoshenko beam theory.

Solution: Let w_a be the transverse displacement at $x = a$ in the direction of the applied load. By Clapeyron's theorem we have

$$\frac{1}{2}F_0 w_a = \frac{1}{2}\int_A\int_0^L \left(\sigma_{xx}\varepsilon_{xx} + 2K_s\sigma_{xz}\varepsilon_{xz}\right) dx\, dA. \tag{1}$$

---

[2]For Bernoulli–Euler beams, this equality does not hold because U does not contain energy due to shear.

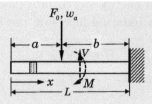

Fig. 4.6.3 A cantilever beam subjected to a point at a distance a from the free end.

But according to the Timoshenko beam theory, the strains and stresses in the beam are given by

$$\varepsilon_{xx} = z\frac{d\phi_x}{dx}, \quad \sigma_{xx} = E\varepsilon_{xx} = Ez\frac{d\phi_x}{dx}, \quad 2\varepsilon_{xz} = \phi_x + \frac{dw}{dx}, \quad \sigma_{xz} = 2G\varepsilon_{xz} = G\left(\phi_x + \frac{dw}{dx}\right).$$

(2)

Then we have

$$\begin{aligned}
\frac{1}{2}F_0 w_a &= \frac{1}{2}\int_A\int_0^L \left(E\varepsilon_{xx}^2 + 4K_s G\varepsilon_{xz}^2\right) dx\, dA \\
&= \frac{1}{2}\int_A\int_0^L \left[Ez^2\left(\frac{d\phi_x}{dx}\right)^2 + GK_s\left(\phi_x + \frac{dw}{dx}\right)^2\right] dA\, dx \\
&= \frac{1}{2}\int_0^L \left[EI\left(\frac{d\phi_x}{dx}\right)^2 + GAK_s\left(\phi_x + \frac{dw}{dx}\right)^2\right] dx \\
&= \frac{1}{2}\int_0^L \left(\frac{M^2}{EI} + \frac{V^2}{GAK_s}\right) dx,
\end{aligned}$$

(3)

where $M(x)$ is the bending moment and $V(x)$ is the shear force at a distance x from the free end (note that we switched from U to U^*; otherwise, it is necessary to express $w(x)$ in terms of w_a and other generalized displacements; see **Examples 3.2.4** and **4.2.3**). The nonzero values of M and V exist only in the span $a \le x \le a+b = L$:

$$M(x) = F_0(x-a), \quad V(x) = F_0 \quad \text{for } a \le x \le a+b = L.$$

(4)

Hence, we have

$$\begin{aligned}
F_0 w_a &= \frac{1}{EI}\int_a^{a+b} F_0^2(x-a)^2\, dx + \frac{1}{GAK_s}\int_a^{a+b} F_0^2\, dx \\
&= \frac{F_0^2 b^3}{3EI} + \frac{F_0^2 b}{GAK_s} \\
&= \frac{F_0^2 b^3}{3EI}\left(1 + 3\Lambda_b\right),
\end{aligned}$$

(5)

where $\Lambda_b = EI/GAK_s b^2$. Thus, the deflection is

$$w_a = \frac{F_0 b^3}{3EI}\left(1 + 3\Lambda_b\right), \quad \Lambda_b = \frac{EI}{GAK_s b^2}.$$

(6)

When Λ_b is set to zero, we obtain the deflection according to the Bernoulli–Euler beam theory.

Example 4.6.3

Consider a beam of length L and flexural rigidity EI, simply supported at the left end and clamped at the right end, and subjected to bending moment M_0 at the simple support and uniformly distributed load of intensity q_0 on the entire span, as shown in Fig. 4.6.4. Use Clapeyron's theorem with the strain energy (not the complementary strain energy) and the Bernoulli–Euler beam theory to determine the rotation at the simple support. *Hint:* Make use of Eqs (2) and (6) of **Example 3.2.4**.

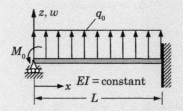

Fig. 4.6.4 A beam simply supported at the left end and fixed at the right end and subjected to bending moment at the simple support and uniformly distributed load on the entire span.

Solution: By Clapeyron's theorem we have

$$\frac{1}{2}\left[\int_0^L q_0\, w(x)\, dx + (-M_0)\left(-\frac{dw}{dx}\right)_{x=0}\right] = \frac{EI}{2}\int_0^L \left(\frac{d^2w}{dx^2}\right)^2 dx. \tag{1}$$

This statement is useful only if $w(x)$ can be expressed in terms of the generalized displacements at the ends of the beam, like we did in **Example 3.2.4**. From Eq. (2) of **Example 3.2.4**, we have

$$w(x) = \sum_{j=1}^{4} \Delta_j \varphi_j(x), \tag{2}$$

where

$$\Delta_1 = w(0), \quad \Delta_2 = -\frac{dw}{dx}\bigg|_{x=0}, \quad \Delta_3 = w(L), \quad \Delta_4 = -\frac{dw}{dx}\bigg|_{x=L} \tag{3}$$

and

$$\varphi_1(x) = \left[1 - 3\left(\frac{x}{L}\right)^2 + 2\left(\frac{x}{L}\right)^3\right], \quad \varphi_2(x) = -x\left(1 - \frac{x}{L}\right)^2$$

$$\varphi_3(x) = \left[3\left(\frac{x}{L}\right)^2 - 2\left(\frac{x}{L}\right)^3\right], \quad \varphi_4(x) = -x\left[\left(\frac{x}{L}\right)^2 - \frac{x}{L}\right]. \tag{4}$$

Substituting for w from Eq. (2) into Eq. (1) and noting that

$$\frac{d\varphi_i}{dx}\bigg|_{x=0} = 0, \; i = 1,3,4 \quad \text{and} \quad -\frac{d\varphi_2}{dx}\bigg|_{x=0} = 1, \tag{5}$$

we obtain

$$\sum_{i=1}^{4} \Delta_i q_i - M_0 \Delta_2 = \sum_{i=1}^{4}\left(\sum_{j=1}^{4} K_{ij}\Delta_j\right)\Delta_i, \tag{6}$$

where

$$q_i = \int_0^L q_0\, \varphi_i(x)\, dx, \quad K_{ij} = EI\int_0^L \frac{d^2\varphi_i}{dx^2}\frac{d^2\varphi_j}{dx^2}\, dx. \tag{7}$$

Carrying out the indicated integrations in Eq. (7), we obtain

$$
\mathbf{q} = \frac{q_0 L}{12} \left\{ \begin{array}{c} 6 \\ -L \\ 6 \\ L \end{array} \right\}, \quad
\mathbf{K} = \frac{2EI}{L^3} \begin{bmatrix} 6 & -3L & -6 & -3L \\ -3L & 2L^2 & 3L & L^2 \\ -6 & 3L & 6 & 3L \\ -3L & L^2 & 3L & 2L^2 \end{bmatrix}. \tag{8}
$$

Since the boundary conditions require

$$
\Delta_1 = \Delta_3 = \Delta_4 = 0, \tag{9}
$$

we obtain from Eq. (5), in view of Eq. (9), the result

$$
q_2 \Delta_2 - M_0 \Delta_2 = K_{22} \Delta_2 \Delta_2
$$

or

$$
\Delta_2 = -\frac{dw}{dx}\bigg|_{x=0} = \frac{q_2 - M_0}{K_{22}} = \left(-\frac{q_0 L^2}{12} - M_0 \right) \frac{L}{4EI} = -\frac{q_0 L^3}{48EI} - \frac{M_0 L}{4EI}. \tag{10}
$$

The negative sign indicates that dw/dx at $x = 0$ is positive for the loads applied.

4.6.3   Types of Elasticity Problems and Uniqueness of Solutions

The static equilibrium problems of elasticity are called boundary value problems (BVPs) because the solution, for a given geometry and material, depends only on the boundary conditions. They can be classified into three types on the basis of the specified boundary conditions. They are defined as follows:

Type I. Boundary value problems in which all specified boundary conditions are of the displacement type

$$
\mathbf{u} = \hat{\mathbf{u}} \quad \text{on} \quad \Gamma \tag{4.6.11}
$$

are called boundary value problems of type I or *displacement or Dirichlet boundary value problems.*

Type II. Boundary value problems in which all specified boundary conditions are of the force type,

$$
\mathbf{t} = \hat{\mathbf{t}} \quad \text{on} \quad \Gamma \tag{4.6.12}
$$

are called boundary value problems of type II or *stress or Neumann boundary value problems.* Such boundary value problems are not common because most practical problems involve specifying displacements at some points that eliminate rigid-body motions.

Type III. Boundary value problems in which all specified boundary conditions are of the mixed type, that is, both displacements and forces are specified:

$$
\mathbf{u} = \hat{\mathbf{u}} \quad \text{on} \quad \Gamma_u \quad \text{and} \quad \mathbf{t} = \hat{\mathbf{t}} \quad \text{on} \quad \Gamma_\sigma \tag{4.6.13}
$$

are called boundary value problems of type III or *mixed boundary value problems*. Most practical problems, including contact problems, fall into this category.

Although the existence of solutions is a difficult task to establish, the uniqueness of solutions is rather easy to prove for linear boundary value problems of elasticity. Consider the problem of finding the solution to Navier's equations of linearized elasticity [see Eq. (4.3.18)] for a given body force $\mathbf{f}$ and boundary conditions:

$$\mathbf{u} = \hat{\mathbf{u}} \ \text{ on } \ \Gamma_u, \tag{4.6.14}$$

$$\mathbf{t} = \hat{\mathbf{t}} \ \text{ on } \ \Gamma_\sigma. \tag{4.6.15}$$

Now suppose that for this set of loads and boundary conditions, there exist two distinct solutions, $\mathbf{u}^{(1)}(\mathbf{x}, t)$ and $\mathbf{u}^{(2)}(\mathbf{x}, t)$. Associated with the two displacement fields, we can compute the strains and stress fields $(\boldsymbol{\varepsilon}^{(1)}, \boldsymbol{\sigma}^{(1)})$ and $(\boldsymbol{\varepsilon}^{(2)}, \boldsymbol{\sigma}^{(2)})$. Then the difference $\bar{\mathbf{u}}(\mathbf{x}, t) \equiv \mathbf{u}^{(1)}(\mathbf{x}, t) - \mathbf{u}^{(2)}(\mathbf{x}, t)$ satisfies the homogeneous form of Navier's equation (with $\bar{\mathbf{f}} = \mathbf{f}^{(1)} - \mathbf{f}^{(2)} = \mathbf{0}$, because the applied forces and boundary values are the same for both solutions),

$$\mu \boldsymbol{\nabla}^2 \bar{\mathbf{u}} + (\lambda + \mu) \boldsymbol{\nabla} \left(\boldsymbol{\nabla} \cdot \bar{\mathbf{u}} \right) = \mathbf{0} \ \text{ in } \ \Omega, \tag{4.6.16}$$

as well as the homogeneous forms of the boundary conditions,

$$\bar{\mathbf{u}} = \mathbf{0} \ \text{ on } \ \Gamma_u, \tag{4.6.17}$$

$$\bar{\mathbf{t}} = \mathbf{0} \ \text{ on } \ \Gamma_\sigma. \tag{4.6.18}$$

Because no work is done on the body by external forces (because $\bar{\mathbf{f}}$ and $\bar{\mathbf{t}}$ are zero), the strain energy density U_0 stored in the body is zero (by Clapeyron's theorem). Noting that the strain energy density U_0 (for the isothermal case) is [see Eqs (3.2.11) and (3.2.13)],

$$U_0(\boldsymbol{\varepsilon}) = \frac{1}{2}\boldsymbol{\sigma} : \boldsymbol{\varepsilon} = \mu\varepsilon_{ij}\varepsilon_{ij} + \tfrac{1}{2}\lambda(\varepsilon_{kk})^2, \tag{4.6.19}$$

which is a positive-definite function of the strains, that is,

$$U_0(\boldsymbol{\varepsilon}) > 0 \text{ whenever } \boldsymbol{\varepsilon} \neq \mathbf{0}, \text{ and } U_0(\boldsymbol{\varepsilon}) = 0 \text{ only when } \boldsymbol{\varepsilon} = \mathbf{0}, \tag{4.6.20}$$

we conclude that the strain field $\bar{\boldsymbol{\varepsilon}}$ due to $\bar{\mathbf{u}}$ is zero and, hence, the stress field $\bar{\boldsymbol{\sigma}}$ is also zero:

$$\bar{\boldsymbol{\varepsilon}} = \boldsymbol{\varepsilon}^{(1)} - \boldsymbol{\varepsilon}^{(2)} = \mathbf{0}, \quad \bar{\boldsymbol{\sigma}} = \boldsymbol{\sigma}^{(1)} - \boldsymbol{\sigma}^{(2)} = \mathbf{0}, \tag{4.6.21}$$

implying that the strain and stress fields associated with the two distinct displacements $\mathbf{u}^{(1)}$ and $\mathbf{u}^{(2)}$ (both being the solutions of the same problem) are the same, that is, they are unique. Since $\bar{\boldsymbol{\varepsilon}} = \mathbf{0}$, it follows that $\boldsymbol{\nabla}\bar{\mathbf{u}} = \mathbf{0}$ or

$\bar{\mathbf{u}} = \mathbf{c}$, where $\mathbf{c}$ is a constant (represents rigid-body motion). Thus, the displacements are unique, in general, only within an arbitrary constant. For type I and type III problems, which involve the specification of displacement boundary conditions, we must have $\bar{\mathbf{u}} = \mathbf{0}$ on the boundary Γ_u, implying that $\mathbf{c} = \mathbf{0}$. Therefore, the displacements are unique for type I and type III problems. For boundary value problems of type II, the displacements are determined within the quantities representing rigid-body motions.

4.6.4 Betti's Reciprocity Theorem

Consider a linear elastic solid that is in equilibrium under the action of two external forces, $\mathbf{F}_1$ and $\mathbf{F}_2$, as shown in Fig. 4.6.5. Since the order of application of the forces is arbitrary, we suppose that force $\mathbf{F}_1$ is applied first. Let W_1 be the work produced by $\mathbf{F}_1$. Then, we apply force $\mathbf{F}_2$, which produces work W_2. This work is the same as that produced by force $\mathbf{F}_2$, if it alone were acting on the body. When force $\mathbf{F}_2$ is applied, force $\mathbf{F}_1$ (which is already acting on the body) does additional work because its point of application is displaced due to the deformation caused by force $\mathbf{F}_2$. Let us denote this work by W_{12}, which is the work done by force $\mathbf{F}_1$ in moving through the displacement produced (at the point of application of load $\mathbf{F}_1$) by force $\mathbf{F}_2$. Thus, the total work done by the application of forces $\mathbf{F}_1$ and $\mathbf{F}_2$, $\mathbf{F}_1$ first and $\mathbf{F}_2$ next, is

$$W = W_1 + W_2 + W_{12}. \tag{4.6.22}$$

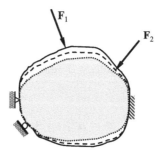

Fig. 4.6.5 Configurations of an elastic body due to the application of load $\mathbf{F}_1$ and load $\mathbf{F}_2$. — Undeformed configuration. - - - Deformed configuration after the application of load $\mathbf{F}_1$. ..... Deformed configuration after the application of load $\mathbf{F}_2$.

Work W_{12}, which can be positive or negative, is zero if and only if the displacement produced by force $\mathbf{F}_2$ at the point of application of force $\mathbf{F}_1$ is zero or perpendicular to the direction of force $\mathbf{F}_1$.

Now suppose that we change the order of application of the forces $\mathbf{F}_1$ and $\mathbf{F}_2$: apply force $\mathbf{F}_2$ first and $\mathbf{F}_1$ next. Then the total work done is equal to

$$\bar{W} = W_1 + W_2 + W_{21}, \tag{4.6.23}$$

where W_{21} is the work done by force $\mathbf{F}_2$ due to the application of force $\mathbf{F}_1$. The work done in both cases should be the same because at the end the elastic body is loaded by the same pair of external forces. Thus we have

$$W = \bar{W} \quad \text{or} \quad W_{12} = W_{21}. \tag{4.6.24}$$

Equation (4.6.24) is a mathematical statement of Betti's (1823–1892) reciprocity theorem. The formal statement of Betti's (1823–1892) reciprocity theorem and its proof as applied to a three-dimensional linear elastic body is presented next.

Theorem 4.6.2: If a linear elastic body is subjected to two different sets of forces, the work done by the first system of forces in moving through the displacements produced by the second system of forces is equal to the work done by the second system of forces in moving through the displacements produced by the first system of forces:

$$\int_\Omega \mathbf{f}^{(1)} \cdot \mathbf{u}^{(2)} \, d\Omega + \int_{\Gamma_\sigma} \mathbf{t}^{(1)} \cdot \mathbf{u}^{(2)} \, d\Gamma = \int_\Omega \mathbf{f}^{(2)} \cdot \mathbf{u}^{(1)} \, d\Omega + \int_{\Gamma_\sigma} \mathbf{t}^{(2)} \cdot \mathbf{u}^{(1)} \, d\Gamma, \tag{4.6.25}$$

where $\mathbf{u}^{(1)}$ is the displacement produced by body forces $\mathbf{f}^{(1)}$ and surface forces $\mathbf{t}^{(1)}$ and $\mathbf{u}^{(2)}$ is the displacement produced by body forces $\mathbf{f}^{(2)}$ and surface forces $\mathbf{t}^{(2)}$. The left-hand side of Eq. (4.6.25), for example, denotes the work done by forces $\mathbf{f}^{(1)}$ and $\mathbf{t}^{(1)}$ in moving through the displacement $\mathbf{u}^{(2)}$ produced by forces $\mathbf{f}^{(2)}$ and $\mathbf{t}^{(2)}$.

Proof: The proof of Betti's reciprocity theorem is straightforward. Let W_{12} denote the work done by forces $(\mathbf{f}^{(1)}, \mathbf{t}^{(1)})$ acting through the displacement $\mathbf{u}^{(2)}$ produced by the forces $(\mathbf{f}^{(2)}, \mathbf{t}^{(2)})$. Then

$$
\begin{aligned}
W_{12} &= \int_\Omega \mathbf{f}^{(1)} \cdot \mathbf{u}^{(2)} \, d\Omega + \oint_\Gamma \mathbf{t}^{(1)} \cdot \mathbf{u}^{(2)} \, d\Gamma \\
&= \int_\Omega f_i^{(1)} u_i^{(2)} \, d\Omega + \oint_\Gamma t_i^{(1)} u_i^{(2)} \, d\Gamma \\
&= \int_\Omega f_i^{(1)} u_i^{(2)} \, d\Omega + \oint_\Gamma n_j \sigma_{ji}^{(1)} u_i^{(2)} \, d\Gamma \\
&= \int_\Omega f_i^{(1)} u_i^{(2)} \, d\Omega + \int_\Omega \left(\sigma_{ji}^{(1)} u_i^{(2)} \right)_{,j} \, d\Omega \\
&= \int_\Omega \left(\sigma_{ij,j}^{(1)} + f_i^{(1)} \right) u_i^{(2)} \, d\Omega + \int_\Omega \sigma_{ij}^{(1)} u_{i,j}^{(2)} \, d\Omega \\
&= \int_\Omega \sigma_{ij}^{(1)} u_{i,j}^{(2)} \, d\Omega = \int_\Omega \sigma_{ij}^{(1)} \varepsilon_{ij}^{(2)} \, d\Omega.
\end{aligned}
$$

Using Hooke's law $\sigma_{ij}^{(1)} = C_{ijk\ell} \, \varepsilon_{k\ell}^{(1)}$, we obtain

$$W_{12} = \int_\Omega C_{ijk\ell} \, \varepsilon_{k\ell}^{(1)} \, \varepsilon_{ij}^{(2)} \, d\Omega. \tag{4.6.26}$$

Since $C_{ijk\ell} = C_{k\ell ij}$, it follows that

$$W_{12} = \int_{\Omega} C_{ijk\ell}\, \varepsilon_{k\ell}^{(1)}\, \varepsilon_{ij}^{(2)}\, d\Omega = \int_{\Omega} C_{k\ell ij}\, \varepsilon_{ij}^{(2)}\, \varepsilon_{k\ell}^{(1)}\, d\Omega = \int_{\Omega} \sigma_{k\ell}^{(2)}\, \varepsilon_{k\ell}^{(1)}\, d\Omega = W_{21}.$$

One can trace back to show that W_{21} is equal to the right-hand side of Eq. (4.6.25). This completes the proof.

During the proof we have also established the equality

$$\int_{\Omega} \sigma_{ij}^{(1)}\, \varepsilon_{ij}^{(2)}\, d\Omega = \int_{\Omega} \sigma_{ij}^{(2)}\, \varepsilon_{ij}^{(1)}\, d\Omega,$$

$$\int_{\Omega} \boldsymbol{\sigma}^{(1)} : \boldsymbol{\varepsilon}^{(2)}\, d\Omega = \int_{\Omega} \boldsymbol{\sigma}^{(2)} : \boldsymbol{\varepsilon}^{(1)}\, d\Omega. \tag{4.6.27}$$

Example 4.6.4

Consider a cantilever beam of length L subjected to two sets of loads: a uniformly distributed load of intensity q_0 throughout the span, as shown in Fig. 4.6.6(a), and a concentrated load F at the free end, as shown in Fig. 4.6.6(b). Verify Betti's reciprocity theorem, that is, the work done by the point load F in moving through the displacement $w^q(0)$ produced by q_0 is equal to the work done by the distributed force q_0 in moving through the displacement $w^F(x)$ produced by the point load F.

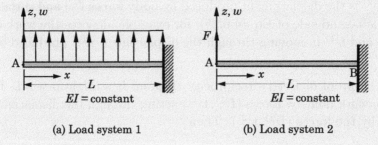

(a) Load system 1   (b) Load system 2

Fig. 4.6.6 (a) A cantilever beam under uniformly distributed load. (b) A cantilever beam with a point load at its free end.

Solution: From Eqs (4.6.5) and (4.6.6), the expression for deflection of the cantilever beam with uniformly distributed load q_0 alone is

$$w_0^q(x) = \frac{q_0 L^4}{24EI}\left[3 - 4\left(\frac{x}{L}\right) + \left(\frac{x}{L}\right)^4\right] \tag{1}$$

and the expression for deflection of the cantilever beam with the point load F at the free end alone is

$$w_0^F(x) = \frac{FL^3}{6EI}\left[2 - 3\left(\frac{x}{L}\right) + \left(\frac{x}{L}\right)^3\right]. \tag{2}$$

The work done by the load F in moving through the displacement due to the application of the uniformly distributed load q_0 is

$$W_{Fq} = Fw_0^q(0) = \frac{Fq_0 L^4}{8EI}.$$

The work done by the uniformly distributed q_0 in moving through the displacement field $w^F(x)$ due to the application of point load F is

$$W_{qF} = \int_0^L \frac{FL^3}{6EI} \left[2 - 3\left(\frac{x}{L}\right) + \left(\frac{x}{L}\right)^3 \right] q_0 \, dx$$

$$= \frac{F q_0 L^4}{8EI},$$

which is in agreement with W_{Fq}.

Example 4.6.5

Use Betti's reciprocity theorem to determine the deflection at the free end of a cantilever beam with distributed load of intensity q_0 in the span between $x = a$ and $x = L$, as shown in Fig. 4.6.7. The deflection $w^F(x)$ due to a point load F at the free end (acting upward) is

$$w^F(x) = \frac{FL^3}{6EI} \left[3\left(\frac{x}{L}\right)^2 - \left(\frac{x}{L}\right)^3 \right]. \tag{1}$$

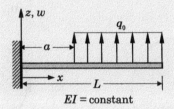

Fig. 4.6.7 A cantilever beam with uniformly distributed load on a portion of the beam.

Solution: The work done by the point load F in moving through the displacement due to the application of the uniformly distributed load q_0 is

$$W_{Fq} = F w_0^q(L). \tag{2}$$

The work done by the uniformly distributed load q_0 in moving through the displacement field $w^F(x)$ due to the application of point load F is

$$W_{qF} = \int_a^L \frac{FL^3}{6EI} \left[3\left(\frac{x}{L}\right)^2 - \left(\frac{x}{L}\right)^3 \right] q_0 \, dx$$

$$= \frac{F q_0}{24EI} \left(3L^4 - 4La^3 + a^4 \right). \tag{3}$$

By Betti's reciprocity theorem, we have $W_{qF} = W_{Fq}$. Hence, the deflection at the free end of the beam due to the distributed load is

$$w_0^q(L) = \frac{q_0}{24EI} \left(3L^4 - 4La^3 + a^4 \right). \tag{4}$$

4.6.5   Maxwell's Reciprocity Theorem

An important special case of Betti's reciprocity theorem is given by Maxwell's (1831–1879) reciprocity theorem. Maxwell's theorem was presented in 1864, whereas Betti's theorem was given in 1872. Therefore, it may be considered that Betti generalized the work of Maxwell. For the sake of completeness, Maxwell's reciprocity theorem is presented here, although it is a special case of Betti's theorem.

Consider a linear elastic solid subjected to force $\mathbf{F}^1$ of unit magnitude[3] acting at point 1 and force $\mathbf{F}^2$ of unit magnitude acting at a different point 2 of the body. Let $\mathbf{u}_{12}$ be the displacement of point 1 produced by unit force $\mathbf{F}^2$ in the direction of force $\mathbf{F}^1$ and $\mathbf{u}_{21}$ be the displacement of point 2 produced by unit force $\mathbf{F}^1$ in the direction of force $\mathbf{F}^2$, as shown in Fig. 4.6.8. Then from Betti's theorem, it follows that

$$\mathbf{F}^1 \cdot \mathbf{u}_{12} = \mathbf{F}^2 \cdot \mathbf{u}_{21} \quad \text{or} \quad u_{12} = u_{21}, \tag{4.6.28}$$

where u_{12} and u_{21} are the magnitudes of the vectors $\mathbf{u}_{12}$ and $\mathbf{u}_{21}$, respectively.

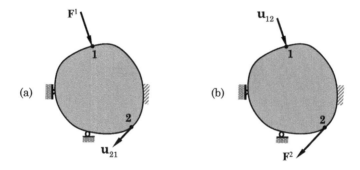

Fig. 4.6.8 Configurations of the body discussed in Maxwell's theorem.

Equation (4.6.28) is a statement of Maxwell's theorem. Maxwell's theorem states that the displacement of point 1 produced in the direction of force $\mathbf{F}_1$ by unit force acting at point 2 is equal to the displacement of point 2 in the direction of force $\mathbf{F}_2$ produced by unit force acting at point 1. We now consider two examples of the use of Maxwell's theorem.

Example 4.6.6 ———————————————————————————

Consider a cantilever beam of length L and constant EI and subjected to a point load F_0 at the free end [see Fig. 4.6.9(a)]. Use Maxwell's theorem to determine the deflection at $x = a$ from the free end. Use the following data: $E = 24 \times 10^6$ psi, $I = 120$ in^4, $F_0 = 1000$ lb, $a = 36$ in, and $b = 108$ in.

---

[3]It is not necessary to make forces to be unity, because they appear on both sides of the equality and thus get canceled.

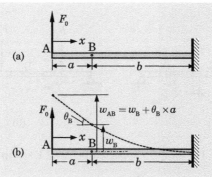

Fig. 4.6.9 A cantilever beam with a point load at the free end.

Solution: By Maxwell's theorem, the displacement w_{BA} at point B ($x = a$) produced by unit load at point A ($x = 0$) is equal to the deflection w_{AB} at point A produced by unit load at point B. We are required to find $w(0) = w_{BA}F_0$. Thus, we must determine w_{AB} (which, presumably, is easier to compute by some way than to compute $w(0)$ directly). Let w_B and θ_B denote the deflection and slope, respectively, at point B owing to a load $F = 1$ applied at point B. Then the deflection at point A due to load $F = 1$ is [see Fig. 4.6.9(b)]

$$w_{AB} = w_B + \theta_B a, \tag{1}$$

and the required solution is

$$w(0) = w_{BA}F_0 = w_{AB}F_0 = F_0 \left(w_B + a\,\theta_B \right). \tag{2}$$

The values of w_B and θ_B can be computed using Eq. (4.6.6) as

$$w_B = \frac{b^3}{6EI}\left[2 - 3\frac{\bar{x}}{b} + \left(\frac{\bar{x}}{b}\right)^3 \right]_{\bar{x}=0} = \frac{b^3}{3EI}, \tag{3}$$

$$\theta_B = -\frac{dw}{dx}\bigg|_{\bar{x}=0} = \frac{b^2}{2EI}\left[1 - \left(\frac{\bar{x}}{b}\right)^2 \right]_{\bar{x}=0} = \frac{b^2}{2EI}. \tag{4}$$

Therefore, we have

$$w(0) = F_0 \left(w_B + a\,\theta_B \right) = F_0 \left(\frac{b^3}{3EI} + \frac{b^2 a}{2EI} \right) = \frac{F_0 b^2}{6EI}(3a + 2b)$$

$$= \frac{1000 \times (108)^2}{6 \times 24 \times 10^6 \times 120}(3 \times 36 + 2 \times 108) = 0.2187 \text{ in.} \tag{5}$$

Example 4.6.7

Consider a simply supported beam subjected to a point load F_B at point B. The load produces deflections w_{AB}, w_{CB}, and w_{DB} at points A, C, and D, respectively [see Fig. 4.6.10(a)]. Determine the deflection w_B at point B produced by loads F_A, F_C, and F_D [see Fig. 4.6.10(b)].

Solution: A unit load acting at point B produces displacements w_{AB}/F_B, w_{CB}/F_B, and w_{DB}/F_B at points A, C, and D, respectively. But, by Maxwell's theorem, w_{AB}/F_B, w_{CB}/F_B, and w_{DB}/F_B are the displacements of point B caused by a unit load acting at points A, C, and

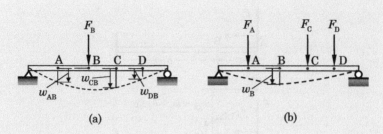

Fig. 4.6.10 The simply supported beam of **Example 4.6.7**.

D, respectively [see Fig. 4.6.10(b)]. Therefore, the displacement at point B owing to forces F_A, F_C, and F_D is given by

$$w_B = F_A \left(\frac{w_{AB}}{F_B} \right) + F_C \left(\frac{w_{CB}}{F_B} \right) + F_D \left(\frac{w_{DB}}{F_B} \right)$$

$$= \frac{F_A \, w_{AB} + F_C \, w_{CB} + F_D \, w_{DB}}{F_B}.$$

4.7 Summary

In this chapter, a host of energy principles and energy theorems were presented and their utility in the determination of displacements and forces in bars, beams, and frames is illustrated through a number of examples. Four different principles are presented: (1) the principle of virtual displacements, (2) the principle of virtual forces, (3) the principle of minimum total potential energy, and (4) the principle of minimum complementary strain energy. The energy theorems developed in this chapter include Clapeyron's theorem, Betti's reciprocity theorem, and Maxwell's reciprocity theorem.

For deformable solids, the principle of virtual displacements yields the equations of equilibrium and the force boundary conditions in terms of stress resultants and thus does not require the knowledge of the constitutive relations. Virtual displacements must be such that they satisfy the homogeneous form of specified geometric boundary conditions. In contrast, the principle of virtual forces yields the kinematic (or strain–displacement) relations and geometric boundary conditions. Virtual forces selected must be such that they satisfy the force equilibrium among themselves and satisfy Cauchy's formula and vanish on the portion on which the force boundary conditions are specified.

The unit dummy-displacement method is a special case of the principle of virtual displacements, and the unit dummy-load method is derived as a special case of the principle of virtual forces. Both can be used to determine point displacements or forces in bar, beam, and frame structures. They are used in structural dynamics to determine the stiffness and flexibility matrices [19].

The principle of minimum total potential energy is derived from the principles of virtual displacements by assuming that the constitutive and strain–displacement relations hold. The total potential energy is expressed in terms of the displacements of the solid continuum, and the principle of minimum total potential energy yields the equations of equilibrium and force boundary conditions as the Euler equations in terms of the displacements. In contrast, the principle of minimum total complementary energy is derived from the principles of virtual forces using the equilibrium and strain–stress relations. The total complementary energy is expressed in terms of the stresses of the solid continuum, and the principle of minimum complementary energy gives the kinematic relations linking the stresses to displacements and geometric boundary conditions as the Euler equations.

Castigliano's theorem I is derived directly from the principle of minimum total potential energy, and Castigliano's theorem II is derived from the principle of minimum complementary energy. They are complementary to each other, although both can be used to determine point displacements and forces in bars, beams, and frames. The energy theorems of Betti and Maxwell are derived independent of the principles of virtual work or the minimum principles. They can also be used to determine point displacements and forces in plates (see **Example 7.2.8**).

Figure 4.7.1 shows the inter-relationship between various principles and methods presented in this chapter. A catalog of important formulas from this chapter is also presented here:

Principle of virtual displacements and the associated Euler equations [see Eqs (4.2.2), (4.2.5), and (4.2.6)]

$$\delta W = \delta W_I + \delta W_E = 0. \tag{4.7.1}$$

$$\boldsymbol{\nabla} \cdot \boldsymbol{\sigma} + \mathbf{f} = \mathbf{0} \text{ in } \Omega; \quad \hat{\mathbf{n}} \cdot \boldsymbol{\sigma} - \hat{\mathbf{t}} = \mathbf{0} \text{ on } \Gamma_\sigma. \tag{4.7.2}$$

Unit dummy-displacement method [see Eq. (4.2.7)]

$$\mathbf{F}_0 \cdot \delta \mathbf{u}_0 = \int_\Omega \boldsymbol{\sigma} : \delta \boldsymbol{\varepsilon}^0 \, d\Omega. \tag{4.7.3}$$

The principle of minimum total potential energy and the associated Euler equations [see Eqs (4.3.5) and (4.3.18)]

$$\delta \Pi \equiv \delta(U + V_E) = 0. \tag{4.7.4}$$

$$\mu \nabla^2 \mathbf{u} + (\lambda + \mu) \boldsymbol{\nabla}(\boldsymbol{\nabla} \cdot \mathbf{u}) + \mathbf{f} = \mathbf{0} \text{ in } \Omega, \quad \text{and} \quad \mathbf{t} - \hat{\mathbf{t}} = \mathbf{0} \text{ on } \Gamma_\sigma. \tag{4.7.5}$$

Castigliano's theorem I [see Eq. (4.3.22)]

$$\frac{\partial U}{\partial \mathbf{u}_i} = \mathbf{F}_i. \tag{4.7.6}$$

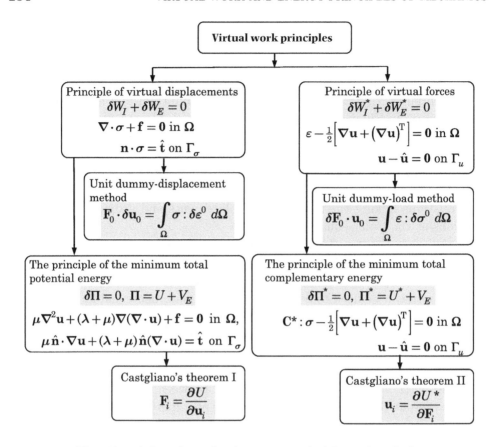

Fig. 4.7.1 A flow chart of various energy principles and methods.

Principle of virtual forces and the associated Euler equations [see Eqs (4.4.5) and (4.4.10)]

$$\delta W^* = \delta W_I^* + \delta W_E^* = 0. \tag{4.7.7}$$

$$\varepsilon - \tfrac{1}{2}\left[\nabla \mathbf{u} + (\nabla \mathbf{u})^{\mathrm{T}}\right] = \mathbf{0} \;\; \text{in } \Omega; \quad \mathbf{u} - \hat{\mathbf{u}} = \mathbf{0} \;\; \text{on } \Gamma_u. \tag{4.7.8}$$

Unit dummy-load method [see Eq. (4.4.11)]

$$\delta \mathbf{F}_0 \cdot \mathbf{u}_0 = \int_\Omega \varepsilon : \delta \boldsymbol{\sigma}^0 \, d\Omega. \tag{4.7.9}$$

The principle of minimum total complementary energy and the associated Euler equations [see Eqs (4.5.2) and (4.5.5)]

$$\delta \Pi^* \equiv \delta(U^* + V_E^*) = 0. \tag{4.7.10}$$

$$\mathbf{C}^* : \boldsymbol{\sigma} - \tfrac{1}{2}\left[\nabla \mathbf{u} + (\nabla \mathbf{u})^{\mathrm{T}}\right] = \mathbf{0} \;\; \text{in } \Omega; \quad \mathbf{u} - \hat{\mathbf{u}} = \mathbf{0} \;\; \text{on } \Gamma_u. \tag{4.7.11}$$

Castigliano's theorem II [see Eq. (4.5.13)]

$$\frac{\partial U^*}{\partial \mathbf{F}_i} = \mathbf{u}_i. \tag{4.7.12}$$

Clapeyron's theorem [see Eq. (4.6.10)]

$$\frac{1}{2}\int_\Omega \boldsymbol{\sigma} : \boldsymbol{\varepsilon}\, d\Omega = \frac{1}{2}\left[\int_\Omega \mathbf{f}\cdot \mathbf{u}\, d\Omega + \oint_\Gamma \mathbf{t}\cdot \mathbf{u}\, d\Gamma\right]. \tag{4.7.13}$$

Betti's reciprocity theorem [see Eqs (4.6.25) and (4.6.27)]

$$\int_\Omega \mathbf{f}^{(1)}\cdot \mathbf{u}^{(2)}\, d\Omega + \int_{\Gamma_\sigma} \mathbf{t}^{(1)}\cdot \mathbf{u}^{(2)}\, d\Gamma = \int_\Omega \mathbf{f}^{(2)}\cdot \mathbf{u}^{(1)}\, d\Omega + \int_{\Gamma_\sigma} \mathbf{t}^{(2)}\cdot \mathbf{u}^{(1)}\, d\Gamma. \tag{4.7.14}$$

$$\int_\Omega \boldsymbol{\sigma}^{(1)} : \boldsymbol{\varepsilon}^{(2)}\, d\Omega = \int_\Omega \boldsymbol{\sigma}^{(2)} : \boldsymbol{\varepsilon}^{(1)}\, d\Omega. \tag{4.7.15}$$

Maxwell's reciprocity theorem [see Eqs (4.6.28)]

$$\mathbf{F}^1\cdot \mathbf{u}_{12} = \mathbf{F}^2\cdot \mathbf{u}_{21} \quad \text{or} \quad u_{12} = u_{21}. \tag{4.7.16}$$

Problems

PRINCIPLE OF VIRTUAL DISPLACEMENTS (SECTIONS 4.2 AND 4.3)

4.1 Use the principle of virtual displacements to determine the governing equations and natural boundary conditions associated with the problem in **Problem 3.4**.

4.2 Use the principle of virtual displacements to determine the governing equations and natural boundary conditions associated with the problem in **Problem 3.5**.

4.3 Use the principle of virtual displacements to derive the equations governing the Timoshenko beam theory with the von Kármán nonlinear strain. Use a distributed axial load $f(x)$ and transverse load of $q(x)$, and assume that the displacement field is given by

$$u_1(x, z) = u(x) + z\phi_x, \quad u_2 = 0, \quad u_3(x, y, z) = w(x), \tag{1}$$

and the only non-zero strains are

$$\varepsilon_{xx}(x, z) = \frac{du}{dx} + \frac{1}{2}\left(\frac{dw}{dx}\right)^2 + z\frac{d\phi_x}{dx}, \quad 2\varepsilon_{xz}(x) = \phi_x + \frac{dw}{dx}. \tag{2}$$

Here u and w are the axial and transverse displacements of a point $(x, y, 0)$ in the beam, and ϕ_x is the rotation of a transverse normal line about the y-axis. Your answer should be expressed in terms of the area-integrated quantities, that is, the stress resultants of Eq. (11) of **Example 3.4.2**:

$$N_{xx} = \int_A \sigma_{xx}\, dA, \quad M_{xx} = \int_A \sigma_{xx}z\, dA, \quad Q_{xz} = K_s\int_A \sigma_{xz}\, dA. \tag{3}$$

4.4 (The third-order beam theory of Reddy) Use the displacement field

$$u_1(x, z) = u(x) + z\phi_x(x) + z^2\psi_x(x) + z^3\theta_x(x), \quad u_2 = 0, \quad u_3 = w(x) \tag{1}$$

where (u_1, u_2, u_3) are the displacements of a point along the (x, y, z) coordinates, (u, w) are the displacements of a point on the midplane of an undeformed beam, and ϕ_x, ψ_x, and θ_x are various order rotations of a transverse normal about the y-axis (into the plane of the paper). (a) Compute the linearized strains and determine ψ_x and θ_x in terms of other dependent variables by requiring that the transverse shear stress vanishes at the top $(z = h/2)$ and bottom $(z = -h/2)$ of the beam. (b) Derive the equations of equilibrium of the resulting third-order beam theory using the principle of virtual displacements for the case of infinitesimal strains. Also derive the natural boundary conditions associated with the theory.

4.5 Use unit dummy-displacement method to determine the displacement in the direction of the applied load P in the structure shown in Fig. P4.5. Assume that the material of all members obeys the following stress–strain relation:

$$\sigma = \begin{cases} K\sqrt{\varepsilon}, & \varepsilon \geq 0, \\ -K\sqrt{-\varepsilon}, & \varepsilon \leq 0, \end{cases} \tag{1}$$

where K is a material constant.

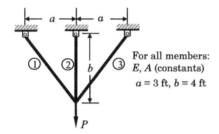

For all members:
E, A (constants)
$a = 3$ ft, $b = 4$ ft

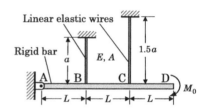

Linear elastic wires

Rigid bar

E, A

M_0

Fig. P4.5   **Fig. P4.6**

4.6 Use Castigliano's theorem I to determine the axial forces in the wires of the structure shown in Fig. P4.6. Assume small deformation. *Hint:* Use the kinematic constraint to relate the forces in the two wires.

4.7 Use Castigliano's theorem I to determine the unknown generalized displacements and forces of the beam shown in Fig. P4.7. Use the Bernoulli–Euler beam theory (see the results of **Example 4.3.5**).

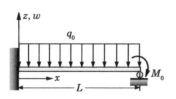

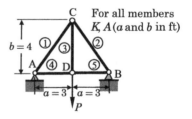

For all members
K, A (a and b in ft)

Fig. P4.7   **Fig. P4.8**

PRINCIPLE OF VIRTUAL FORCES (SECTIONS 4.4 AND 4.5)

4.8 Use unit dummy-load method to determine the horizontal and vertical displacements at point D of the structure shown in Fig. P4.8. Assume that the material of all members obeys the stress–strain relation in Eq. (1) of **Problem 4.5**. Determine the horizontal and vertical displacements at the point of load application using the unit dummy-load method.

4.9 Use the unit dummy-load method to determine the center deflection of a simply supported beam under uniformly distributed transverse load of magnitude q_0. Consider two cases: (a) neglect the energy due to shear and (b) account for the shear energy.

4.10 Use the unit dummy-load method to determine the reaction and the rotation at $x = L$ in **Problem 4.7**.

4.11 Use the unit dummy-load method to determine a relationship between the end deflection w_e and deflection w_c at the center of the beam shown in Fig. P4.11. Neglect the energy due to shear. *Hint:* Use a dummy load at the center that is one-half of the dummy load at the end of the beam.

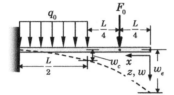

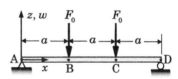

Fig. P4.11   Fig. P4.12

4.12 Use the unit dummy-load method to determine the deflection at point B of the beam shown in Fig. P4.12. Neglect the energy due to shear.

4.13 Use the unit dummy-load method to determine the force in the cable of the cable supported cantilever beam shown in Fig. P4.13. Neglect the energy due to shear. *Hint:* Replace the cable with its tension at point C.

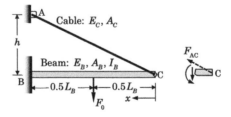

Fig. P4.13

4.14 Use the unit dummy-load method to determine the rotation of joint B of the frame shown in Fig. P4.14. Neglect the energy due to shear.

4.15 Consider the frame structure shown in Fig. P4.15. Each member of the structure has the same geometric and material properties. Assume linear elastic behavior and neglect the energy due to shear. Determine the horizontal and vertical displacements at the point of load application. Use the unit dummy-load method.

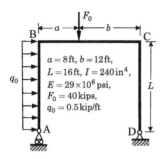

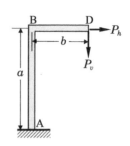

Fig. P4.14   Fig. P4.15

4.16 Use unit dummy-load method to determine the vertical deflection of point A of the structure in Fig. P4.16. Neglect the energy due to shear.

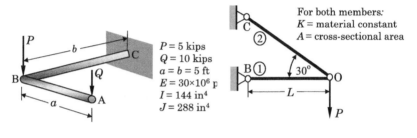

Fig. P4.16   Fig. P4.17

4.17 Consider the pin connected structure shown in Fig. P4.17. Suppose that the material of all members obeys the following stress–strain relation:

$$\sigma = \begin{cases} K\sqrt{\varepsilon}, & \varepsilon \geq 0, \\ -K\sqrt{-\varepsilon}, & \varepsilon \leq 0, \end{cases} \tag{1}$$

where K is a material constant. All members have the same cross-sectional area A and constant K. Neglect the energy due to shear. Determine the horizontal and vertical displacements at the point of load application using the unit dummy-load method.

4.18 Obtain the solution to **Problem 4.6** (see Fig. P4.6) using Castigliano's theorem II.

4.19 Use Castigliano's theorem II to determine the compressive force and displacements in the linear elastic spring (spring constant, k) supporting the free end of a cantilever beam under triangular distributed load (see Fig. P4.19). Neglect the energy due to shear.

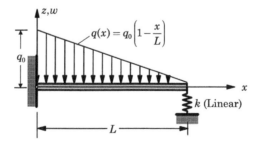

Fig. P4.19

4.20 The thin curved beam shown in Fig. P4.20 has a radius of curvature R, modulus E, and moment of inertia I about the axis of bending. Use Castigliano's theorem II to determine the vertical and horizontal deflections at the free end. Include only energy due to bending.

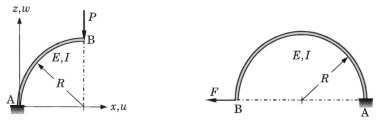

Fig. P4.20   Fig. P4.21

4.21 The thin curved beam shown in Fig. P4.21 has a radius of curvature R, modulus E, and moment of inertia I about the axis of bending. Use Castigliano's theorem II to determine the vertical and horizontal deflections at the free end. Include only energy due to bending.

4.22 The thin curved beam shown in Fig. P4.22 has a radius of curvature R, modulus E, and moment of inertia I about the axis of bending. The radially distributed load (i.e., pressure) is constant, p per unit arc. Use Castigliano's theorem II to determine the horizontal and vertical deflections. Include only energy due to bending.

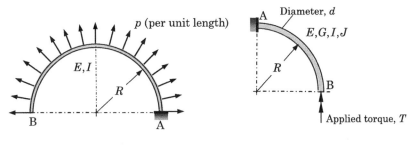

Fig. P4.22   Fig. P4.23

4.23 Consider a cylindrical bar bent into a quarter circle of radius R. The bar is fixed at one end and loaded at the free end by a twisting moment (or torque) T, as shown in Fig. P4.23. Determine the angle of twist using Castigliano's theorem II.

4.24 Repeat **Problem 4.20** by accounting for the energy due to extension as well as shear.

4.25 Use Castigliano's theorem II to determine the force in the spring and the deflection and the rotation at point A of a spring supported cantilever beam under uniformly distributed load (see Fig. P4.25). Neglect the energy due to shear.

4.26 Determine the transverse deflection of the free end of the cantilever beam shown in Fig. P4.26 using Castigliano's theorem II. Solve the problem (a) neglecting the energy due to shear force $V(x)$ and (b) including the shear energy.

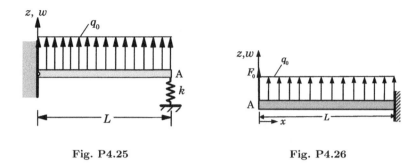

Fig. P4.25   Fig. P4.26

4.27 Determine the spring force of the spring supported cantilever beam shown in Fig. P4.27 using Castigliano's theorem II. Solve the problem (a) neglecting the energy due to shear force $V(x)$ and (b) including the shear energy.

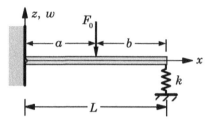

Fig. P4.27

4.28 Use Castigliano's theorem II to determine the vertical and horizontal deflections at point C of the statically determinate structure shown in Fig. P4.28. Assume that all connections are pin connections and that all members have the same area of cross section (A) and modulus (E).

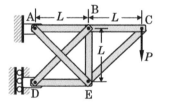

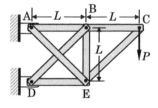

Fig. P4.28   Fig. P4.29

4.29 Repeat **Problem 4.28** when the structure is made indeterminate by replacing the roller at point D with a connection identical to that at point A, as shown in Fig. P4.29.

CLAPEYRON'S, BETTI'S, AND MAXWELL'S THEOREMS (SECTION 4.6)

4.30 Use Clapeyron's theorem to determine the transverse deflection of the free end of a cantilever beam with a transverse load F_0 at its end. Include energy due to transverse shear. Assume linear elastic behavior with constant EI and GA, and take $f_s = 6/5$. *Hint:* The strain energy and complementary strain energy are equal for linear elastic materials (i.e., $U^* = U$).

4.31 Prove Betti's theorem for a three-dimensional elastic solid.

4.32 Consider a simply supported beam of length L subjected to a concentrated load F_0 at the midspan and a bending moment M_0 at the left end, as shown in Fig. P4.32. Verify that Betti's reciprocity theorem holds. Determine the required displacements using energy theorems.

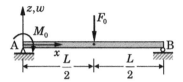

Fig. P4.32

4.33 Determine the slope at point A of the beam shown in Fig. P4.32 using Maxwell's reciprocal relationship and any suitable energy theorems.

4.34 Verify Maxwell's reciprocal relationship for the cantilever problem with two different loadings shown in Fig. P4.34. You must determine the required deflections and slopes to verify Maxwell's reciprocal relationship.

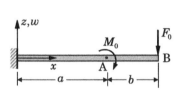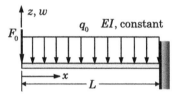

Fig. P4.34   Fig. P4.35

4.35 Determine the deflection at the midspan of a cantilever beam subjected to uniformly distributed load q_0 throughout the span and a point load F_0 at the free end (see Fig. P4.35). Use Maxwell's theorem and superposition.

4.36 A load $F_A = 4000$ lb acting at a point A of a beam produces $w_{BA} = 0.25$ in at point B and $w_{CA} = 0.75$ in at point C of the beam. Find the deflection of point A (w_A) produced by loads $F_B = 4500$ lb and $F_C = 2000$ lb acting at points B and C, respectively.

"Many medical practices are not soundly based. They are sustained, as is true of other human pursuits, by an inertia supported by fashion, custom, and the word of authority. The security provided by a long-held belief system, even when poorly founded, is a strong impediment to progress. General acceptance of a practice becomes the proof of its validity, though it lacks all other merit."   – *Bernard Lown* (developer of heart defibrillator)

"Almost all really new ideas have a certain aspect of foolishness when they are first produced."
                    – *Alfred North Whitehead* (English mathematician and philosopher)

Dynamical Systems: Hamilton's Principle

5.1 Introduction

The principles of virtual work and their special cases discussed in Chapter 4 were limited to static equilibrium of solids. Hamilton's principle is a generalization of the principle of virtual displacements to dynamics of systems of particles, rigid bodies, or deformable solids. The principle assumes that the system under consideration is characterized by two energy functions: a *kinetic energy K* and a *potential energy* Π. For *discrete* systems (i.e., systems with a finite number of degrees of freedom), these energies can be described in terms of a finite number of generalized coordinates and their derivatives with respect to time t. For *continuous* systems (i.e., systems whose exact response cannot be described by a finite number of generalized coordinates), the energies can be expressed in terms of the dependent variables (which are functions of position) of the problem. Hamilton's principle reduces to the principle of virtual displacements for systems that are in static equilibrium.

The present chapter is devoted to the extension of the principle of virtual displacements to dynamical systems and the discussion of Hamilton's principle and its application to dynamics of discrete systems as well as continuous systems. We begin the discussion of Hamilton's principle for particles.

5.2 Hamilton's Principle for Discrete Systems

Consider a single particle of mass m moving under the influence of a force $\mathbf{F} = \mathbf{F}(\mathbf{r})$. The path $\mathbf{r}(t)$ followed by the particle is related to the force $\mathbf{F}$ and mass m by Newton's second law of motion:

$$\mathbf{F}(\mathbf{r}) = \frac{d}{dt}\left(m\frac{d\mathbf{r}}{dt}\right). \tag{5.2.1}$$

A path that differs from the actual path is expressed as $\mathbf{r} + \delta\mathbf{r}$, where $\delta\mathbf{r}$ is the variation of the path for any *fixed* time t. We suppose that the actual path

Energy Principles and Variational Methods in Applied Mechanics, Third Edition. J.N. Reddy.
©2017 John Wiley & Sons Ltd. Published 2017 by John Wiley & Sons Ltd.
Companion Website: www.wiley.com/go/reddy/applied_mechanics_EPVM

$\mathbf{r}$ and the *varied* path differ except at two distinct times t_1 and t_2, that is, $\delta\mathbf{r}(t_1) = \delta\mathbf{r}(t_2) = \mathbf{0}$. Taking the scalar product of Eq. (5.2.1) with the variation $\delta\mathbf{r}$ and integrating with respect to time between times t_1 and t_2 ($t_2 > t_1$), we obtain

$$\int_{t_1}^{t_2} \left[\frac{d}{dt}\left(m\frac{d\mathbf{r}}{dt} \right) - \mathbf{F}(\mathbf{r}) \right] \cdot \delta\mathbf{r}\, dt = 0. \tag{5.2.2}$$

Integration by parts of the first term in Eq. (5.2.2) yields

$$-\int_{t_1}^{t_2} \left(m\frac{d\mathbf{r}}{dt} \cdot \frac{d\delta\mathbf{r}}{dt} + \mathbf{F}(\mathbf{r}) \cdot \delta\mathbf{r} \right) dt + \left(m\frac{d\mathbf{r}}{dt} \cdot \delta\mathbf{r} \right)\Bigg|_{t_1}^{t_2} = 0. \tag{5.2.3}$$

The last term in Eq. (5.2.3) vanishes because $\delta\mathbf{r}(t_1) = \delta\mathbf{r}(t_2) = \mathbf{0}$. Also, note that

$$m\frac{d\mathbf{r}}{dt} \cdot \frac{d\delta\mathbf{r}}{dt} = \delta\left[\frac{m}{2}\frac{d\mathbf{r}}{dt} \cdot \frac{d\mathbf{r}}{dt} \right] \equiv \delta K, \tag{5.2.4}$$

where K is the kinetic energy of the particle,

$$K = \frac{1}{2}m\frac{d\mathbf{r}}{dt} \cdot \frac{d\mathbf{r}}{dt} = \frac{1}{2}m\mathbf{v} \cdot \mathbf{v}, \tag{5.2.5}$$

and $\mathbf{v}$ is the velocity vector. Equation (5.2.3) now takes the form

$$\int_{t_1}^{t_2} \left(\delta K + \mathbf{F} \cdot \delta\mathbf{r} \right) dt = 0, \tag{5.2.6}$$

which is known as the *general form of Hamilton's principle* for a single particle. This can also be viewed as the principle of virtual displacements applied to motion of a particle.

Now suppose that the force $\mathbf{F}$ is conservative, that is, the sum of the potential and kinetic energies is conserved, such that it can be replaced by the gradient of a potential

$$\mathbf{F} = -\boldsymbol{\nabla} V_E, \tag{5.2.7}$$

where $V_E = V_E(\mathbf{r})$ is the potential energy due to the forces on the particle. Then Eq. (5.2.6) can be expressed in the form

$$\delta\int_{t_1}^{t_2} (K - V_E)\, dt = 0, \tag{5.2.8}$$

because

$$\boldsymbol{\nabla} V_E \cdot \delta\mathbf{r} = \frac{\partial V_E}{\partial x_1}\delta x_1 + \frac{\partial V_E}{\partial x_2}\delta x_2 + \frac{\partial V_E}{\partial x_3}\delta x_3 = \delta V_E(x_1, x_2, x_3).$$

The difference between the kinetic and potential energies is called the *Lagrangian* function:

$$L \equiv K - V_E. \tag{5.2.9}$$

Note that for particles and rigid bodies considered here, the internal energy W_I is assumed to be zero and $W_E = V_E$.

Equation (5.2.8) represents Hamilton's principle for the conservative motion of a particle. Hamilton's principle can be stated as *the motion of a particle acted on by conservative forces between two arbitrary instants of time t_1 and t_2 is such that the line integral over the Lagrangian function is an extremum for the path motion.* Stated in other words, of all possible paths that the particle could travel from its position at time t_1 to its position at time t_2, its actual path will be one for which the integral

$$I \equiv \int_{t_1}^{t_2} L \, dt \tag{5.2.10}$$

is an extremum (i.e., a minimum, a maximum, or an inflection).

If the path $\mathbf{r}$ can be expressed in terms of some generalized coordinates g_i ($i = 1, 2, 3$), the Lagrangian function can be written in terms of g_i and their time derivatives

$$L = L(g_1, g_2, g_3, \dot{g}_1, \dot{g}_2, \dot{g}_3). \tag{5.2.11}$$

Then from Sections 3.5.5 and 3.5.6, the condition for the extremum of I results in the equation (note that $\delta g_i = 0$ at t_1 and t_2)

$$\delta I = \delta \int_{t_1}^{t_2} L(g_1, g_2, g_3, \dot{g}_1, \dot{g}_2, \dot{g}_3) \, dt = 0$$

$$= \int_{t_1}^{t_2} \sum_{i=1}^{3} \left[\frac{\partial L}{\partial g_i} - \frac{d}{dt}\left(\frac{\partial L}{\partial \dot{g}_i} \right) \right] \delta g_i \, dt. \tag{5.2.12}$$

When all g_i are linearly independent (i.e., no constraints among g_i), the variations δg_i are independent for all t, except $\delta g_i = 0$ at t_1 and t_2. Therefore, the coefficients of $\delta g_1, \delta g_2$, and δg_3 vanish separately:

$$\frac{\partial L}{\partial g_i} - \frac{d}{dt}\left(\frac{\partial L}{\partial \dot{g}_i} \right) = 0, \quad i = 1, 2, 3. \tag{5.2.13}$$

These equations are called the *Lagrange equations of motion.* In Section 3.5.6 on static problems, these equations were also called the *Euler equations.* In the discussions to follow, we shall refer to these equations as the *Euler–Lagrange equations,* whenever the functional includes kinetic energy of the system.

When the forces are not conservative, we must deal with the general form of Hamilton's principle in Eq. (5.2.6). Recognizing that $\mathbf{F} \cdot \delta \mathbf{r}$ represents the virtual work $-\delta W_E$ due to forces $\mathbf{F}$, we can write Eq. (5.2.6) in the form

$$\int_{t_1}^{t_2} (\delta K - \delta W_E) \, dt = 0. \tag{5.2.14}$$

In this case, there may not exist a functional I that must be an extremum. If the virtual work can be expressed in terms of the generalized coordinates g_i by

$$\delta W_E = -\left(F_1 \delta g_1 + F_2 \delta g_2 + F_3 \delta g_3\right), \tag{5.2.15}$$

where F_i are the *generalized forces*, then we can write Eq. (5.2.14) as

$$\int_{t_1}^{t_2} \sum_{i=1}^{3} \left[\frac{\partial K}{\partial g_i} - \frac{d}{dt}\left(\frac{\partial K}{\partial \dot{g}_i}\right) + F_i\right] \delta g_i \, dt = 0, \tag{5.2.16}$$

and the Euler–Lagrange equations for the nonconservative forces are given by

$$\frac{\partial K}{\partial g_i} - \frac{d}{dt}\left(\frac{\partial K}{\partial \dot{g}_i}\right) + F_i = 0, \quad i = 1, 2, 3. \tag{5.2.17}$$

Example 5.2.1 ————————————————————————————————

Consider the spring mass system shown in Fig. 5.2.1. Determine the equations of motion for (a) the case in which the mass is sliding on a smooth, frictionless surface [see Fig. 5.2.1(a)] and (b) for the case where the surface offers friction.

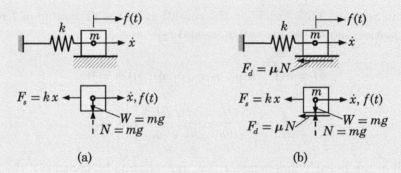

(a)   (b)

Fig. 5.2.1 A spring mass system, with the mass sliding on a (a) smooth, frictionless, surface and (b) dry surface with friction.

Solution: (a) The kinetic and potential energies are (since spring is a discrete element, the strain energy stored in the spring is not labeled here as the internal energy stored, W_I)

$$K = \frac{1}{2}m(\dot{x})^2, \quad V_E = \frac{k}{2}x^2 - f(t)x, \quad L = K - V_E. \tag{1}$$

Hence, the equation of motion is

$$\frac{\partial L}{\partial x} - \frac{d}{dt}\left(\frac{\partial L}{\partial \dot{x}}\right) = 0, \tag{2}$$

which yields

$$-kx + f(t) - m\ddot{x} = 0. \tag{3}$$

(b) In the case of a surface with friction, for motion to begin, there must be a force acting on the body that overcomes the resistance to motion. The friction force is parallel to the surface in the direction opposite to the motion, and its magnitude is proportional to the force normal to the surface. The force normal to the surface in this case is the weight $W = mg$, and the constant of proportionality is known as the static coefficient of friction μ_s. The value of μ_s varies from 0 (when the surface is frictionless) to 1, depending on the surface finish and material. Once motion ensues, the value of the friction coefficient drops to $\mu_k < \mu_s$, known as the kinematic friction coefficient. The friction force remains constant in magnitude during the motion as long as the inertia force and force in the spring are sufficiently large to overcome the static friction.

Thus, in this case, the system involves a force that is nonconservative (i.e., not derivable from a potential function). Denoting the magnitude of the friction force by $F_d = \mu_k N = \mu_k\, mg$, we can write the equation of motion, using Eq. (5.2.17), as

$$-\frac{d}{dt}\left(\frac{\partial K}{\partial \dot{x}}\right) + F = 0, \quad F = -kx - \text{sign}(\dot{x})F_d + f(t), \tag{4}$$

where $\text{sign}(\dot{x})$ denotes the sign of $\dot{x}$ [i.e., $\text{sign}(\dot{x})$ is $+1$ when $\dot{x}$ is positive and -1 if $\dot{x}$ is negative]. The above equation of motion can be separated into two equations as

$$-m\ddot{x} - kx - F_d + f = 0 \ \text{ when } \dot{x} > 0, \quad m\ddot{x} + kx - F_d - f = 0 \ \text{ when } \dot{x} < 0. \tag{5}$$

Example 5.2.2

Consider the motion of a pendulum in a plane. Ideally speaking, the pendulum is imagined to be a mass m (called bob) attached at the end of a rigid massless rod of length l that pivots about a fixed point O, as shown in Fig. 5.2.2. Determine the equations of motion when (a) there is no resistance offered by the medium (i.e., surroundings) to the motion and (b) the mass experiences a resistance force $\mathbf{F}^*$ proportional to its speed (i.e., the pendulum is suspended in a medium that offers resistance to motion). Take the force $\mathbf{F}^*$ according to Stoke's law as

$$\mathbf{F}^* = -6\pi\mu a l\, \dot{\theta}\, \hat{\mathbf{e}}_\theta, \tag{1}$$

where μ is the viscosity of the surrounding medium, a is the radius of the pendulum bob, and $\hat{\mathbf{e}}_\theta$ is the unit vector tangential to the circular path. Neglect the resistance of the massless rod supporting the bob.

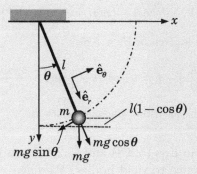

Fig. 5.2.2 Geometry and forces acting on the pendulum.

Solution: (a) The motion can be described using $g_1 = l$ and $g_2 = \theta$ generalized coordinates. However, l is fixed and $\dot{g}_1 = 0$. Hence, $g_2 = \theta$ is the only generalized coordinate, which is measured from the vertical position. The force $\mathbf{F}$ acting on the bob is the component of the gravitational force:

$$\mathbf{F} = mg(-\sin\theta\,\hat{\mathbf{e}}_\theta + \cos\theta\,\hat{\mathbf{e}}_r). \tag{2}$$

The component along $\hat{\mathbf{e}}_r$ does no work (because rod is rigid and hence does not change its length). The first component is derivable from the potential $(\nabla V = -\mathbf{F})$

$$V_E = -[-mg\ell(1-\cos\theta)] = mg\,l(1-\cos\theta). \tag{3}$$

The expression signifies the potential energy of the pendulum bob at any instant of time with respect to the static equilibrium position $\theta = 0$. The gradient operator in the present case is given by $(r = l)$

$$\nabla = \hat{\mathbf{e}}_r\,\frac{\partial}{\partial r} + \hat{\mathbf{e}}_\theta\,\frac{1}{r}\frac{\partial}{\partial\theta}. \tag{4}$$

Thus the kinetic and potential energies are given by

$$K = \frac{m}{2}(l\,\dot{\theta})^2, \quad \delta K = ml^2\dot{\theta}\,\delta\dot{\theta}; \quad V_E = mgl(1-\cos\theta), \quad \delta V_E = mgl\sin\theta\,\delta\theta. \tag{5}$$

Therefore, the Lagrangian function L is a function of θ and $\dot{\theta}$. The Euler–Lagrange equation is given by

$$\frac{\partial L}{\partial\theta} - \frac{d}{dt}\left(\frac{\partial L}{\partial\dot{\theta}}\right) = 0, \tag{6}$$

which yields

$$-mg\,l\sin\theta - \frac{d}{dt}(ml^2\dot{\theta}) = 0 \quad \text{or} \quad \ddot{\theta} + \frac{g}{l}\sin\theta = 0. \tag{7}$$

Equation (7) represents a second-order nonlinear differential equation governing θ. For small angular motions, Eq. (7) can be linearized by replacing $\sin\theta \approx \theta$:

$$\ddot{\theta} + \frac{g}{l}\theta = 0. \tag{8}$$

(b) The resistance force $\mathbf{F}^*$ is not derivable from a potential function. Thus, we have one part of the force (i.e., gravitational force) conservative and the other (i.e., viscous force) nonconservative. Hence, we use Hamilton's principle expressed by Eq. (5.2.14) or Eq. (5.2.16) with

$$\delta W_E = \delta V_E - \mathbf{F}^* \cdot (l\delta\theta\hat{\mathbf{e}}_\theta) = \left(mg\,l\sin\theta + 6\pi\mu a l^2\dot{\theta}\right)\delta\theta \equiv -F\,\delta\theta. \tag{9}$$

Then the equation of motion is given by $[K = K(\dot{\theta})]$

$$-\frac{d}{dt}\left(\frac{\partial K}{\partial\dot{\theta}}\right) + F = 0 \quad \text{or} \quad \ddot{\theta} + \frac{g}{l}\sin\theta + \frac{6\pi a\mu}{m}\dot{\theta} = 0. \tag{10}$$

The coefficient $c = 6\pi a\mu/m$ is called the *damping* coefficient.

Hamilton's principle in Eqs (5.2.8) and (5.2.14) can be easily extended to a system of N particles, hence to rigid bodies. We have

$$\delta\int_{t_1}^{t_2} L(q_i, \dot{q}_i)\,dt = 0 \quad \text{for conservative forces,} \tag{5.2.18}$$

$$\int_{t_1}^{t_2}(\delta K - \delta W_E)\,dt = 0 \quad \text{for nonconservative forces,} \tag{5.2.19}$$

where

$$K = \sum_{i=1}^{N} \frac{m_i}{2} \frac{d\mathbf{r}_i}{dt} \cdot \frac{d\mathbf{r}_i}{dt}, \quad \delta W_E = -\sum_{i=1}^{N} \mathbf{F}_i \cdot \delta\mathbf{r}_i, \tag{5.2.20}$$

and m_i and $\mathbf{r}_i$ denote the mass and path of the ith particle. Of course, Eqs (5.2.18) and (5.2.19) apply to a rigid body, because a rigid body can be viewed as a collection of particles with fixed positions $\mathbf{r}_i$ relative to the mass center.

5.3   Hamilton's Principle for a Continuum

Following essentially the same procedure as that used for discrete systems, we can develop Hamilton's principle for the dynamics of deformable bodies. The main difference between rigid bodies and deformable bodies is the presence of internal energy W_I for deformable bodies. Newton's second law of motion applied to deformable bodies expresses the global statement of the principle of conservation of linear momentum. However, it should be noted that Newton's second law of motion for continuous media is not sufficient to determine its motion $\mathbf{u} = \mathbf{u}(\mathbf{x}, t)$; the kinematic conditions and constitutive equations derived in Chapter 2 are needed to completely determine the motion.

Newton's second law of motion for a continuous body can be written in general terms as

$$\mathbf{F} - m\mathbf{a} = \mathbf{0}, \tag{5.3.1}$$

where m is the mass, $\mathbf{a}$ is the acceleration vector, and $\mathbf{F}$ is the resultant of *all* forces acting on the body. The actual path $\mathbf{u} = \mathbf{u}(\mathbf{x}, t)$ followed by a material particle in position $\mathbf{x}$ in the body is varied, consistent with kinematic (essential) boundary conditions, to $\mathbf{u} + \delta\mathbf{u}$, where $\delta\mathbf{u}$ is the admissible variation (or virtual displacement) of the path. We suppose that the varied path differs from the actual path except at initial and final times, t_1 and t_2, respectively. Thus, an admissible variation $\delta\mathbf{u}$ satisfies the conditions

$$\delta\mathbf{u} = \mathbf{0} \text{ on } \Gamma_u \text{ for all } t,$$
$$\delta\mathbf{u}(\mathbf{x}, t_1) = \delta\mathbf{u}(\mathbf{x}, t_2) = \mathbf{0} \text{ for all } \mathbf{x}, \tag{5.3.2}$$

where Γ_u denotes the portion of the boundary of the body where the displacement vector $\mathbf{u}$ is specified. Note that the scalar product of Eq. (5.3.1) with $\delta\mathbf{u}$ gives work done at point $\mathbf{x}$, because $\mathbf{F}, \mathbf{a}$, and $\mathbf{u}$ are vector functions of position (whereas the work is a scalar). Integration of the product over the volume (and surface) of the body gives the total work done by all points.

The *work done on the body* at time t by the *resultant force* in moving through the virtual displacement $\delta\mathbf{u}$ is given by

$$\int_{\Omega} \mathbf{f} \cdot \delta\mathbf{u} \, d\Omega + \int_{\Gamma_\sigma} \hat{\mathbf{t}} \cdot \delta\mathbf{u} \, d\Gamma - \int_{\Omega} \boldsymbol{\sigma} : \delta\varepsilon \, d\Omega, \tag{5.3.3}$$

where $\mathbf{f}$ is the body force vector, $\hat{\mathbf{t}}$ is the specified surface traction vector, and $\boldsymbol{\sigma}$ and $\boldsymbol{\varepsilon}$ are the stress and strain tensors. The last term in Eq. (5.3.3) represents the *virtual work* of internal forces *stored in the body*. The strains $\delta\boldsymbol{\varepsilon}$ are assumed to be compatible, in the sense that the strain–displacement relations in Eq. (2.3.12) are satisfied. The work done by the inertia force ma in moving through the virtual displacement $\delta\mathbf{u}$ is given by

$$\int_\Omega \rho \frac{\partial^2 \mathbf{u}}{\partial t^2} \cdot \delta\mathbf{u} \, d\Omega \tag{5.3.4}$$

where ρ is the mass density (can be a function of position) of the medium. We have, analogous to Eq. (5.2.2) for discrete systems, the result

$$\int_{t_1}^{t_2} \left\{ \int_\Omega \rho \frac{\partial^2 \mathbf{u}}{\partial t^2} \cdot \delta\mathbf{u} \, d\Omega - \left[\int_\Omega (\mathbf{f} \cdot \delta\mathbf{u} - \boldsymbol{\sigma} : \delta\boldsymbol{\varepsilon}) \, d\Omega + \int_{\Gamma_\sigma} \hat{\mathbf{t}} \cdot \delta\mathbf{u} \, d\Gamma \right] \right\} dt = 0,$$

or

$$-\int_{t_1}^{t_2} \left[\int_\Omega \rho \frac{\partial \mathbf{u}}{\partial t} \cdot \frac{\partial \delta\mathbf{u}}{\partial t} \, d\Omega + \int_\Omega (\mathbf{f} \cdot \delta\mathbf{u} - \boldsymbol{\sigma} : \delta\boldsymbol{\varepsilon}) \, d\Omega + \int_{\Gamma_\sigma} \hat{\mathbf{t}} \cdot \delta\mathbf{u} \, d\Gamma \right] dt = 0. \tag{5.3.5}$$

In arriving at the expression in Eq. (5.3.5), integration by parts is used on the first term; the integrated terms vanish because of the initial and final conditions in Eq. (5.3.2). Equation (5.3.5) is known as *the general form of Hamilton's principle for a continuous medium* (conservative or not, and elastic or not).

We recall from the previous discussions that for an ideal elastic body, the forces $\mathbf{f}$ and $\mathbf{t}$ are conservative (i.e., they are derivable from a potential function called the potential energy due to applied loads),

$$\delta V_E = -\left(\int_\Omega \mathbf{f} \cdot \delta\mathbf{u} \, d\Omega + \int_{\Gamma_\sigma} \hat{\mathbf{t}} \cdot \delta\mathbf{u} \, d\Gamma \right), \tag{5.3.6}$$

and that there exists a strain energy density potential $U_0 = U_0(\varepsilon_{ij})$ such that

$$\boldsymbol{\sigma} = \frac{\partial U_0}{\partial \boldsymbol{\varepsilon}} \quad \left(\sigma_{ij} = \frac{\partial U_0}{\partial \varepsilon_{ij}} \right) \quad \text{or} \quad \boldsymbol{\sigma} : \delta\boldsymbol{\varepsilon} = \delta U_0. \tag{5.3.7}$$

Substituting Eqs (5.3.6) and (5.3.7) into Eq. (5.3.5), we obtain

$$\int_{t_1}^{t_2} [\delta K - (\delta U + \delta V_E)] dt = 0, \tag{5.3.8}$$

where K and U are the kinetic and strain energies, respectively,

$$K = \frac{1}{2} \int_\Omega \rho \frac{\partial \mathbf{u}}{\partial t} \cdot \frac{\partial \mathbf{u}}{\partial t} \, d\Omega, \quad U = \int_\Omega U_0 \, d\Omega. \tag{5.3.9}$$

Equation (5.3.8) represents Hamilton's principle for an elastic (linear or nonlinear) medium. We recall that the sum of the strain energy and potential energy of external forces, $U + V_E$, is called the total potential energy, Π, of the body. For bodies involving no motion (i.e., forces are applied sufficiently slowly such that the motion is independent of time, and the inertia forces are negligible), Hamilton's principle in Eq. (5.3.8) reduces to the principle of virtual displacements for static equilibrium. Indeed, Eq. (5.3.8) may be viewed as the dynamics version of the principle of virtual displacements.

The Euler–Lagrange equations associated with the Lagrangian, $L = K - \Pi$, can be obtained from Eq. (5.3.8) as follows:

$$0 = \delta \int_{t_1}^{t_2} L(\mathbf{u}, \boldsymbol{\nabla}\mathbf{u}, \dot{\mathbf{u}}) \, dt$$

$$= \int_{t_1}^{t_2} \left[\int_{\Omega} \left(\rho \frac{\partial^2 \mathbf{u}}{\partial t^2} - \boldsymbol{\nabla} \cdot \boldsymbol{\sigma} - \mathbf{f} \right) \cdot \delta \mathbf{u} \, d\Omega + \int_{\Gamma_\sigma} (\mathbf{t} - \hat{\mathbf{t}}) \cdot \delta \mathbf{u} \, d\Gamma \right] dt, \quad (5.3.10)$$

where integration by parts (or the divergence theorem) and Eq. (5.3.7) were used in arriving at Eq. (5.3.10) from Eq. (5.3.8). Because $\delta \mathbf{u}$ is arbitrary for t, $t_1 < t < t_2$, and for $\mathbf{x}$ in Ω and also on Γ_σ, it follows that

$$\rho \frac{\partial^2 \mathbf{u}}{\partial t^2} - \boldsymbol{\nabla} \cdot \boldsymbol{\sigma} - \mathbf{f} = \mathbf{0} \quad \text{in } \Omega; \quad \mathbf{t} - \hat{\mathbf{t}} = \mathbf{0} \quad \text{on } \Gamma_\sigma. \quad (5.3.11)$$

Equation (5.3.11) is the Euler–Lagrange equation for a linearized elastic body.

Using the results of this section and the previous sections, one can derive statements of Hamilton's principle corresponding to the principle of virtual forces (or the complementary potential energy) and stationary principles. These are left as exercises for the reader.

Example 5.3.1

Consider the axial motion of an elastic bar of length L, area of cross section A, modulus of elasticity E, and mass density ρ, and subjected to distributed force f per unit length and an end load P. Determine the equations of motion of the bar when (a) no nonconservative forces are applied and (b) when the bar also experiences a nonconservative (viscous damping) force proportional to the velocity,

$$F^* = -\mu \frac{\partial u}{\partial t}, \quad (1)$$

where μ is the damping coefficient (a constant).

Solution: (a) The kinetic and total potential energies of the system are

$$K = \int_{\Omega} \frac{\rho}{2} \left(\frac{\partial u}{\partial t} \right)^2 d\Omega = \int_0^L \frac{\rho A}{2} \left(\frac{\partial u}{\partial t} \right)^2 dx, \quad (2)$$

$$\Pi(u) = \int_{\Omega} \frac{1}{2} \sigma_{ij} \varepsilon_{ij} \, d\Omega - \int_0^L fu \, dx - Pu(L)$$

$$= \int_0^L \frac{A}{2} \sigma_{xx} \varepsilon_{xx} \, dx - \int_0^L fu \, dx - Pu(L), \quad (3)$$

wherein u, σ_{xx}, and ε_{xx} are assumed to be functions of x only, and

$$u(0, t) = 0 \quad \text{(bar is fixed at } x = 0), \quad (4)$$

$$\varepsilon_{xx} = \frac{\partial u}{\partial x} \quad \text{(strain–displacement relation)}. \tag{5}$$

Substituting for K and Π from Eqs (2) and (3) into Eq. (5.3.8), we obtain

$$
\begin{aligned}
0 &= \int_{t_1}^{t_2} \left\{ \int_0^L \left[A\rho \frac{\partial u}{\partial t} \frac{\partial \delta u}{\partial t} - A\sigma_{xx}\delta\left(\frac{\partial u}{\partial x}\right) + f\delta u \right] dx + P\delta u(L) \right\} dt \\
&= \int_0^L \left[\int_{t_1}^{t_2} -\frac{\partial}{\partial t}\left(\rho A \frac{\partial u}{\partial t}\right)\delta u\, dt + \rho A \frac{\partial u}{\partial t}\delta u\Big|_{t_1}^{t_2} \right] dx \\
&\quad + \int_{t_1}^{t_2} \left\{ \int_0^L \left[\frac{\partial}{\partial x}(A\sigma_{xx}) + f\right]\delta u\, dx - (A\sigma_{xx}\delta u)\Big|_0^L + P\delta u(L) \right\} dt \\
&= -\int_{t_1}^{t_2} \left\{ \int_0^L \left[\frac{\partial}{\partial t}\left(\rho A \frac{\partial u}{\partial t}\right) - \frac{\partial}{\partial x}(A\sigma_{xx}) - f \right]\delta u\, dx \right. \\
&\qquad\qquad \left. - (A\sigma_{xx} - P)\big|_{x=L}\delta u(L) \right\} dt, \tag{6}
\end{aligned}
$$

where $\delta u(0,t) = 0$ and $\delta u(x,t_1) = \delta u(x,t_2) = 0$ are used to simplify the expression. The Euler–Lagrange equations are obtained by setting the coefficients of δu in $(0,L)$ and at $x = L$ to zero separately:

$$\frac{\partial}{\partial t}\left(\rho A \frac{\partial u}{\partial t}\right) - \frac{\partial}{\partial x}\left(A\sigma_{xx}\right) - f = 0, \quad 0 < x < L, \tag{7}$$

$$(A\sigma_{xx})\big|_{x=L} - P = 0, \tag{8}$$

for all $t, t_1 < t < t_2$. For linear elastic materials, we have $\sigma_{xx} = E\varepsilon_{xx} = E(\partial u/\partial x)$, and Eqs (7) and (8) become

$$\frac{\partial}{\partial t}\left(\rho A \frac{\partial u}{\partial t}\right) - \frac{\partial}{\partial x}\left(EA \frac{\partial u}{\partial x}\right) - f = 0, \quad 0 < x < L, \tag{9}$$

$$\left(AE \frac{\partial u}{\partial x}\right)\Big|_{x=L} - P = 0. \tag{10}$$

(b) Now suppose that the bar also experiences a nonconservative (viscous damping) force given in Eq. (1). Then the Euler–Lagrange equations from Eq. (5.3.8) are given by

$$\frac{\partial}{\partial t}\left(\rho A \frac{\partial u}{\partial t}\right) - \frac{\partial}{\partial x}\left(EA \frac{\partial u}{\partial x}\right) - f + \mu \frac{\partial u}{\partial t} = 0, \quad 0 < x < L, \tag{11}$$

$$AE \frac{\partial u}{\partial x}\Big|_{x=L} - P = 0. \tag{12}$$

The next example illustrates the use of Hamilton's principle in deriving the equations of motion and boundary conditions associated with the assumed displacement field of an higher-order theory of straight beams. For all higher-order theories, the vector approach (i.e., setting the vector sum of forces and moments to zero) cannot be used to derive the governing equations of motion or boundary conditions. The displacement field of such higher-order theories is arrived at by placing certain kinematic hypotheses (see Reddy [50, 51]).

Example 5.3.2

Consider the displacement field of the Reddy beam theory:

$$
u_1(x, z, t) = u(x, t) + z\phi_x(x, t) - c_1 z^3 \left(\phi_x + \frac{\partial w}{\partial x} \right),
$$

$$
u_3(x, z, t) = w(x, t),
$$

(1)

where $c_1 = 4/(3h^2)$, h is the height of the beam (of rectangular cross section), u is the axial displacement, w is the transverse displacement, and ϕ_x is the rotation of a point on the centroidal axis x of the beam. The displacement field is arrived by (a) relaxing the Bernoulli–Euler hypotheses to let the straight lines normal to the beam axis before deformation to become (cubic) curves with arbitrary slope at $z = 0$ and (b) requiring the transverse shear stress to vanish at the top and bottom of the beam. Thus, the only restriction from the Bernoulli–Euler beam theory that is kept is the extensibility, $u_3(x, z, t) = w(x, t)$ (i.e., transverse deflection is independent of the thickness coordinate z). Assuming that the beam is subjected to a distributed transverse load of $q(x, t)$ along the length of the beam and the strains are linear, derive the equations of motion of the Reddy third-order beam theory.

Solution: Since we are primarily interested in deriving the equations of motion and the nature of the boundary conditions of the beam that experiences a displacement field of the form in Eq. (1), there is no need to consider specific geometric or force boundary conditions. The procedure to obtain the equations of motion and boundary conditions involves the following three steps: (1) compute the strains, (2) compute the virtual energies required in Hamilton's principle, and (3) use Hamilton's principle and derive the Euler–Lagrange equations of motion and identify the primary and secondary variables of the theory (which in turn help identify the nature of the boundary conditions).

Although one can use the general nonlinear strain–displacement relations, here we restrict the development to small strains and displacements. The linear strains associated with the displacement field are

$$
\varepsilon_{xx} = \varepsilon_{xx}^{(0)} + z\varepsilon_{xx}^{(1)} + z^3 \varepsilon_{xx}^{(3)}, \quad \gamma_{xz} = \gamma_{xz}^{(0)} + z^2 \gamma_{xz}^{(2)},
$$

(2)

where

$$
\varepsilon_{xx}^{(0)} = \frac{\partial u}{\partial x}, \quad \varepsilon_{xx}^{(1)} = \frac{\partial \phi_x}{\partial x}, \quad \gamma_{xz}^{(0)} = \phi_x + \frac{\partial w}{\partial x},
$$

$$
\varepsilon_{xx}^{(3)} = -c_1 \left(\frac{\partial \phi_x}{\partial x} + \frac{\partial^2 w}{\partial x^2} \right),
$$

(3)

$$
\gamma_{xz}^{(2)} = -c_2 \left(\phi_x + \frac{\partial w}{\partial x} \right),
$$

and $c_2 = 4/h^2$. Note that $\gamma_{xz} = 2\varepsilon_{xz}$ is a quadratic function of z. Hence, $\sigma_{xz} = G\gamma_{xz}$ varies quadratically through the beam's height.

From the dynamic version of the principle of virtual displacements (i.e., Hamilton's principle for deformable bodies), we have

$$
0 = \int_0^T \int_0^L \int_A \left[\sigma_{xx} \left(\delta\varepsilon_{xx}^{(0)} + z\delta\varepsilon_{xx}^{(1)} + z^3 \delta\varepsilon_{xx}^{(3)} \right) + \sigma_{xz} \left(\delta\gamma_{xz}^{(0)} + z^2 \delta\gamma_{xz}^{(2)} \right) \right] dA\,dx\,dt
$$

$$
- \int_0^T \int_0^L \int_A \rho \left\{ \left[\dot{u} + z\dot{\phi}_x - c_1 z^3 \left(\dot{\phi}_x + \frac{\partial \dot{w}}{\partial x} \right) \right] \times \right.
$$

$$
\left[\delta\dot{u} + z\delta\dot{\phi}_x - c_1 z^3 \left(\delta\dot{\phi}_x + \frac{\partial \delta\dot{w}}{\partial x} \right) \right] + \dot{w}\delta\dot{w} \bigg\} dA\,dx\,dt
$$

$$
- \int_0^T \int_0^L q\delta w \, dx\,dt
$$

$$= \int_0^T \int_0^L \left(N_{xx} \delta \varepsilon_{xx}^{(0)} + M_{xx} \delta \varepsilon_{xx}^{(1)} + P_{xx} \delta \varepsilon_{xx}^{(3)} + V_x \delta \gamma_{xz}^{(0)} + R_x \delta \gamma_{xz}^{(2)} \right) dx \, dt$$

$$- \int_0^T \int_0^L \left\{ I_0 \dot{u} \delta \dot{u} + \left[I_2 \dot{\phi}_x - c_1 I_4 \left(\dot{\phi}_x + \frac{\partial \dot{w}}{\partial x} \right) \right] \delta \dot{\phi}_x + q \delta w \right\} dx \, dt$$

$$- \int_0^T \int_0^L \left\{ -c_1 \left[I_4 \dot{\phi}_x - c_1 I_6 \left(\dot{\phi}_x + \frac{\partial \dot{w}}{\partial x} \right) \right] \left(\delta \dot{\phi}_x + \frac{\partial \delta \dot{w}}{\partial x} \right) + I_0 \dot{w} \delta \dot{w} \right\} dx \, dt$$

$$= \int_0^T \int_0^L \int_A \left\{ \left(-\frac{\partial N_{xx}}{\partial x} + I_0 \frac{\partial^2 u}{\partial t^2} \right) \delta u + \left(-\frac{\partial \bar{M}_{xx}}{\partial x} + \bar{Q}_x + K_2 \frac{\partial^2 \phi_x}{\partial t^2} - c_1 J_4 \frac{\partial^3 w}{\partial x \partial t^2} \right) \delta \phi_x \right.$$

$$\left. + \left[-c_1 \frac{\partial^2 P_{xx}}{\partial x^2} - \frac{\partial \bar{Q}_x}{\partial x} - q + c_1 \left(J_4 \frac{\partial^3 \phi_x}{\partial x \partial t^2} - c_1 I_6 \frac{\partial^4 w}{\partial x^2 \partial t^2} \right) + I_0 \frac{\partial^2 w}{\partial t^2} \right] \delta w \right\} dx \, dt$$

$$+ \int_0^T \left\{ N_{xx} \delta u + \bar{M}_{xx} \delta \phi_x - c_1 P_{xx} \frac{\partial \delta w}{\partial x} + \left[\bar{Q}_x + c_1 \left(\frac{\partial P_{xx}}{\partial x} - J_4 \frac{\partial^2 \phi_x}{\partial t^2} + c_1 I_6 \frac{\partial^3 w}{\partial x \partial t^2} \right) \right] \delta w \right\}_0^L dt,$$

$$(4)$$

where all the terms involving $[\cdot]_0^T$ vanish on account of the assumption that all variations and their derivatives are zero at time $t = 0$ and $t = T$, and the following stress resultants are introduced in arriving at the last expression:

$$\left\{ \begin{matrix} N_{xx} \\ M_{xx} \\ P_{xx} \end{matrix} \right\} = b \int_{-h/2}^{h/2} \left\{ \begin{matrix} 1 \\ z \\ z^3 \end{matrix} \right\} \sigma_{xx} \, dz \,, \quad \left\{ \begin{matrix} V_x \\ R_x \end{matrix} \right\} = b \int_{-h/2}^{h/2} \left\{ \begin{matrix} 1 \\ z^2 \end{matrix} \right\} \sigma_{xz} \, dz, \qquad (5)$$

$$\bar{M}_{xx} = M_{xx} - c_1 P_{xx}, \quad \bar{Q}_x = V_x - c_2 R_x, \quad c_1 = \frac{4}{3h^2}, \quad c_2 = \frac{4}{h^2}, \qquad (6)$$

$$J_4 = I_4 - c_1 I_6, \quad K_2 = I_2 - 2 c_1 I_4 + c_1^2 I_6, \quad I_i = b \int_{-h/2}^{h/2} \rho(z)^i \, dz, \qquad (7)$$

where b is the width of the beam. Note that the mass inertias I_i are zero for odd values of i (i.e., $I_1 = I_3 = I_5 = 0$).

From Eq. (4), we obtain the following Euler–Lagrange equations:

$$\delta u: \quad \frac{\partial N_{xx}}{\partial x} = I_0 \frac{\partial^2 u}{\partial t^2}, \qquad (8)$$

$$\delta w: \quad \frac{\partial \bar{Q}_x}{\partial x} + c_1 \frac{\partial^2 P_{xx}}{\partial x^2} + q = I_0 \frac{\partial^2 w}{\partial t^2} + c_1 \left(J_4 \frac{\partial^3 \phi_x}{\partial x \partial t^2} - c_1 I_6 \frac{\partial^4 w}{\partial x^2 \partial t^2} \right), \qquad (9)$$

$$\delta \phi_x: \quad \frac{\partial \bar{M}_{xx}}{\partial x} - \bar{Q}_x = K_2 \frac{\partial^2 \phi_x}{\partial t^2} - c_1 J_4 \frac{\partial^3 w}{\partial x \partial t^2}. \qquad (10)$$

The last line of Eq. (4) includes boundary terms, which indicates that the primary variables of the Reddy beam theory are (those with the variational symbol) u, w, ϕ_x, and $\partial w / \partial x$. The corresponding secondary variables are the coefficients of δu, δw, $\delta \phi_x$, and $\partial \delta w / \partial x$ in the boundary expression of Eq. (4):

$$N_{xx}, \quad \bar{Q}_x + c_1 \left(\frac{\partial P_{xx}}{\partial x} - J_4 \frac{\partial^2 \phi_x}{\partial t^2} + c_1 I_6 \frac{\partial^3 w}{\partial x \partial t^2} \right), \quad \bar{M}_{xx}, \quad -c_1 P_{xx}. \qquad (11)$$

This completes the development of the Reddy third-order beam theory. The stress resultants $(N_{xx}, M_{xx}, P_{xx}, V_x, R_x)$ can be related to the generalized displacements $(u, w, \phi_x, \partial w / \partial x)$ once the stress–strain relation are identified.

The Timoshenko beam theory can be obtained as a special case from the Reddy third-order beam theory. For $c_1 = 0$, Eq. (1) corresponds to the displacement field of the Timoshenko beam theory. Equation (6) gives $\bar{M}_{xx} = M_{xx}$ and $\bar{Q}_x = V_x$. Thus, the equations of motion of the Timoshenko beam theory can be obtained directly from Eqs (8)–(10) by setting $c_1 = c_2 = 0$:

$$\frac{\partial N_{xx}}{\partial x} = I_0 \frac{\partial^2 u}{\partial t^2}, \tag{12}$$

$$\frac{\partial V_x}{\partial x} + q = I_0 \frac{\partial^2 w}{\partial t^2}, \tag{13}$$

$$\frac{\partial M_{xx}}{\partial x} - V_x = I_2 \frac{\partial^2 \phi_x}{\partial t^2}. \tag{14}$$

The primary and secondary variables of the Timoshenko beam theory are: (u, w, ϕ_x) and (N_{xx}, V_x, M_{xx}).

5.4 Hamilton's Principle for Constrained Systems

The kinematic relations that restrict a motion are called *constraints*. When the relations are between generalized coordinates, and possible position and time, in the form

$$H(t, \mathbf{g}) = 0, \quad \text{for a discrete system,} \tag{5.4.1}$$

$$H(\mathbf{x}, t, \mathbf{u}, \boldsymbol{\nabla}\mathbf{u}) = 0, \quad \text{for a continuous system,} \tag{5.4.2}$$

they are called *holonomic constraints*. When the constraints are of inequality type or relate g_i to $\dot{g}_i$, they are called *nonholonomic constraints*. We only consider holonomic constraints here.

As discussed in Section 3.5.8, constraints can be included into the Lagrangian function either by means of the Lagrange multiplier method or by the penalty function method. To illustrate the ideas for a dynamical system, we consider the motion of a single particle whose motion is constrained by holonomic constraints:

$$H_1(t, g_1, g_2, g_3) = 0, \quad H_2(t, g_1, g_2, g_3) = 0. \tag{5.4.3}$$

The variation of these equations yields

$$\frac{\partial H_1}{\partial g_1} \delta g_1 + \frac{\partial H_1}{\partial g_2} \delta g_2 + \frac{\partial H_1}{\partial g_3} \delta g_3 = 0, \tag{5.4.4}$$

$$\frac{\partial H_2}{\partial g_1} \delta g_1 + \frac{\partial H_2}{\partial g_2} \delta g_2 + \frac{\partial H_2}{\partial g_3} \delta g_3 = 0. \tag{5.4.5}$$

Since there are two constraints and three degrees of freedom, only one of the variations $\delta g_1, \delta g_2$, and δg_3 is independent, and the other two are related to the independent ones by Eq. (5.4.3). We multiply Eq. (5.4.4) by the Lagrange multiplier λ_1 and Eq. (5.4.5) by λ_2, integrate each equation over t_1 to t_2, and add the results to Eq. (5.2.16):

$$
\int_{t_1}^{t_2} \sum_{i=1}^{3} \left[\frac{\partial K}{\partial g_i} - \frac{d}{dt}\left(\frac{\partial K}{\partial \dot{g}_i}\right) + F_i + \lambda_1 \frac{\partial H_1}{\partial g_i} + \lambda_2 \frac{\partial H_2}{\partial g_i} \right] \delta g_i \, dt = 0. \qquad (5.4.6)
$$

Since the variations δg_1, δg_2, and δg_3 are not all independent, we cannot use the usual argument to set the coefficients of $\delta g_i (i = 1, 2, 3)$ to zero. However, the Lagrange multipliers λ_1 and λ_2 are arbitrary. Therefore, we choose λ_1 and λ_2 such that the coefficients of two of the variations out of δg_1, δg_2, and δg_3 vanish. Since the remaining variation is linearly independent, its coefficient should be zero. Thus we obtain

$$
\frac{\partial K}{\partial g_i} - \frac{d}{dt}\left(\frac{\partial K}{\partial \dot{g}_i}\right) + F_i + \lambda_1 \frac{\partial H_1}{\partial g_i} + \lambda_2 \frac{\partial H_2}{\partial g_i} = 0 \qquad (5.4.7)
$$

for $i = 1, 2, 3$. Equations (5.4.7) and (5.4.3) together provide five equations for the five unknowns $(g_1, g_2, g_3, \lambda_1, \lambda_2)$.

In the penalty function method, Hamilton's principle in Eq. (5.2.14) is modified to read

$$
\int_{t_1}^{t_2} \delta K_p \, dt - \int_{t_1}^{t_2} \delta W_E \, dt + \delta \int_{t_1}^{t_2} \left(\frac{\gamma_1}{2} H_1^2 + \frac{\gamma_2}{2} H_2^2 \right) dt = 0, \qquad (5.4.8)
$$

where γ_1 and γ_2 are the penalty parameters and K_p is the kinetic energy of the constrained system. Since the constraint conditions (5.4.3) are now included, in the least squares sense, we suppose that the variations δg_1, δg_2, and δg_3 are independent of each other. Performing the variation indicated in Eq. (5.4.8), we obtain (after the use of the fundamental lemma) the Euler–Lagrange equations

$$
\frac{\partial K_p}{\partial g_i} - \frac{d}{dt}\left(\frac{\partial K_p}{\partial \dot{g}_i}\right) + F_i + \gamma_1 H_1 \frac{\partial H_1}{\partial g_i} + \gamma_2 H_2 \frac{\partial H_2}{\partial g_i} = 0. \qquad (5.4.9)
$$

A comparison of Eq. (5.4.9) with Eq. (5.4.7) ($K_p = K$) shows that the Lagrange multipliers λ_1 and λ_2 can be computed from

$$
\lambda_1 = \gamma_1 H_1, \quad \lambda_2 = \gamma_2 H_2, \qquad (5.4.10)
$$

where H_1 and H_2 are given by Eq. (5.4.3) for the generalized coordinates g_i computed from Eq. (5.4.9); H_1 and H_2 thus computed are not identically zero because g_i computed from Eq. (5.4.9) are different from the true values, and the error is inversely proportional to the penalty parameters γ_1 and γ_2.

Example 5.4.1

To illustrate the ideas presented in the preceding paragraphs, we reconsider the damped motion of a pendulum of **Example 5.2.2**. The generalized coordinates are

$$
g_1 = r, \quad g_2 = \theta, \qquad (1)
$$

and the constraint is

$$H_1(g_1) \equiv g_1 - l = 0. \tag{2}$$

The work done by external forces in moving through virtual displacement $\delta \mathbf{g}$ is given by

$$\delta W_E = -\mathbf{F} \cdot \delta \mathbf{g} = -\left[mg(-\sin\theta\,\hat{\mathbf{e}}_\theta + \cos\theta\,\hat{\mathbf{e}}_r) - 6\pi\mu a l\dot{\theta}\hat{\mathbf{e}}_\theta\right] \cdot \left(\delta g_1 \hat{\mathbf{e}}_r + r\delta g_2 \hat{\mathbf{e}}_\theta\right)$$

$$= -\left[mg\cos\theta\delta g_1 - r(mg\sin\theta + 6\pi\mu a l\dot{\theta})\delta g_2\right] \equiv -(F_1\delta g_1 + F_2\delta g_2). \tag{3}$$

The kinetic energy is given by

$$K = \frac{m}{2}\left[(\dot{r})^2 + (l\,\dot{\theta})^2\right] = \frac{m}{2}\left(\dot{g}_1^2 + l^2\,\dot{g}_2^2\right). \tag{4}$$

For the Lagrange multiplier method, the Euler–Lagrange equations are obtained by substituting Eq. (4) and F_i from (3) into Eq. (5.4.7):

$$-m\ddot{r} + mg\cos\theta + \lambda_1 = 0, \tag{5}$$

$$-ml^2\ddot{\theta} - r(mg\sin\theta + 6\pi\mu a l\dot{\theta}) = 0. \tag{6}$$

In view of the constraint condition (2), we have $\ddot{r} = 0$, and Eqs (5) and (6) become

$$\lambda_1 = -mg\cos\theta, \quad \ddot{\theta} + \frac{6\pi\mu a}{m}\dot{\theta} + \frac{g}{\ell}\sin\theta = 0. \tag{7}$$

The Lagrange multiplier λ_1 can be interpreted as the force exerted on the pendulum bob by the massless rod. It is the force necessary to oppose gravity and maintain the motion of the bob in a circular arc.

In the penalty function method, the Euler–Lagrange equations are obtained by substituting Eqs (3) and (4) into Eq. (5.4.9) (note that $\ddot{r} = 0$):

$$-m\ddot{r} + mg\cos\theta + \gamma_1(r - \ell) = 0, \tag{8}$$

$$-ml^2\ddot{\theta} - r(mg\sin\theta + 6\pi\mu a l\dot{\theta}) = 0. \tag{9}$$

A comparison of Eqs (8) and (9) with Eqs (5) and (6) shows that the approximate Lagrange multiplier is given by

$$\lambda_1(\gamma_1) = \gamma_1(r - l). \tag{10}$$

The error in the constraint is given by [from Eq. (8) with $\ddot{r} = 0$]

$$r - l = -\frac{mg\cos\theta}{\gamma_1}, \tag{11}$$

which goes to zero as γ_1 goes to infinity.

The ideas discussed in this section for particles and rigid bodies can be readily extended to deformable bodies by accounting for the internal energy W_I due to deformation. In particular, Eqs (5.4.7) and (5.4.9) take the following forms:

Lagrange multiplier method:

$$\frac{\partial L_c}{\partial g_i} - \frac{\partial}{\partial t}\left(\frac{\partial L_c}{\partial \dot{g}_i}\right) - \frac{\partial}{\partial x}\left(\frac{\partial L_c}{\partial g_{i,x}}\right) + F_i + \lambda_1\frac{\partial H_1}{\partial g_i} + \lambda_2\frac{\partial H_2}{\partial g_i} = 0. \tag{5.4.11}$$

Penalty method:

$$\frac{\partial L_c}{\partial g_i} - \frac{\partial}{\partial t}\left(\frac{\partial L_c}{\partial \dot{g}_i}\right) - \frac{\partial}{\partial x}\left(\frac{\partial L_c}{\partial g_{i,x}}\right) + F_i + \gamma_1 H_1 \frac{\partial H_1}{\partial g_i} + \gamma_2 H_2 \frac{\partial H_2}{\partial g_i} = 0. \quad (5.4.12)$$

where $g_{i,x} = (\partial g_i/\partial x)$ and $L_c = K_c - (U_c + V_{Ec})$. The quantities K_c, U_c, and V_{Ec} are the kinetic energy, strain energy, and the potential energy due to conservative loads, respectively, of the constrained system. The next example illustrates the use of Eqs (5.4.11) and (5.4.12).

Example 5.4.2

The Lagrange function associated with the dynamics of an Bernoulli–Euler beam is given by $L = K - (U + V_E)$, where

$$K = \int_0^L \int_A \left[\frac{\rho}{2}\left(\frac{\partial u}{\partial t} - z\frac{\partial^2 w}{\partial x \partial t}\right)^2 + \frac{\rho}{2}\left(\frac{\partial w}{\partial t}\right)^2\right] dA\, dx$$

$$= \int_0^L \left[\frac{\rho A}{2}\left(\frac{\partial u}{\partial t}\right)^2 + \frac{\rho I}{2}\left(\frac{\partial^2 w}{\partial x \partial t}\right)^2 + \frac{\rho A}{2}\left(\frac{\partial w}{\partial t}\right)^2\right] dx, \quad (1)$$

$$U = \int_0^L \int_A \frac{E}{2}\left(\frac{\partial u}{\partial x} - z\frac{\partial^2 w}{\partial x^2}\right)^2 dA\, dx$$

$$= \int_0^L \left[\frac{EA}{2}\left(\frac{\partial u}{\partial x}\right)^2 + \frac{EI}{2}\left(\frac{\partial^2 w}{\partial x^2}\right)^2\right] dx, \quad (2)$$

$$V_E = -\int_0^L [f(x,t)u + q(x,t)w]\, dx. \quad (3)$$

Here, u denotes the axial displacement and w the transverse displacement, which are functions of x and t, and f and q are the axial and transverse distributed loads. In arriving at the expressions for K and U, we have used the fact that the x-axis coincides with the geometric centroidal axis, $\int_A z\, dA = 0$.

Note that there is no constraint among the two dependent variables $g_1 = u$ and $g_2 = w$, and $L = L(u, w, u', w'', \dot{u}, \dot{w}, \dot{w}')$. Hence, the Euler–Lagrange equations of the Bernoulli–Euler beam theory are given by

$$\delta u: \quad \frac{\partial L}{\partial u} - \frac{\partial}{\partial t}\left(\frac{\partial L}{\partial \dot{u}}\right) - \frac{\partial}{\partial x}\left(\frac{\partial L}{\partial u'}\right) = 0, \quad (4)$$

$$\delta w: \quad \frac{\partial L}{\partial w} - \frac{\partial}{\partial t}\left(\frac{\partial L}{\partial \dot{w}}\right) + \frac{\partial^2}{\partial x^2}\left(\frac{\partial L}{\partial w''}\right) + \frac{\partial^2}{\partial x \partial t}\left(\frac{\partial L}{\partial \dot{w}'}\right) = 0. \quad (5)$$

Note that

$$\frac{\partial L}{\partial u} = f, \quad \frac{\partial L}{\partial \dot{u}} = \rho A \dot{u}, \quad \frac{\partial L}{\partial u'} = -EAu',$$

$$\frac{\partial L}{\partial w} = q, \quad \frac{\partial L}{\partial \dot{w}} = \rho A \dot{w}, \quad \frac{\partial L}{\partial \dot{w}'} = \rho I \dot{w}', \quad \frac{\partial L}{\partial w''} = -EIw''. \quad (6)$$

Hence, we have

$$\delta u: \quad \frac{\partial}{\partial t}\left(\rho A \frac{\partial u}{\partial t}\right) - \frac{\partial}{\partial x}\left(EA\frac{\partial u}{\partial x}\right) = f, \quad (7)$$

$$\delta w: \quad \frac{\partial}{\partial t}\left(\rho A \frac{\partial w}{\partial t}\right) - \frac{\partial^2}{\partial x^2}\left(EI\frac{\partial^2 w}{\partial x^2}\right) + \frac{\partial^2}{\partial x \partial t}\left(\rho I \frac{\partial^2 w}{\partial x \partial t}\right) = q. \quad (8)$$

Now, suppose that we introduce the function $\theta_x(x,t)$ such that

$$\theta_x + \frac{\partial w}{\partial x} = 0. \tag{9}$$

Now, the problem can be looked at as one of minimizing L subject to the constraint in Eq. (9). We wish to determine the equations of motion of the beam using (a) the Lagrange multiplier method and (b) the penalty method.

First we write out the expressions K, U, and V for the constrained problem. We have $(V = V_{Ec})$

$$K_c = \int_0^L \left[\frac{\rho A}{2} \left(\frac{\partial u}{\partial t} \right)^2 + \frac{\rho I}{2} \left(\frac{\partial \theta_x}{\partial t} \right)^2 + \frac{\rho A}{2} \left(\frac{\partial w}{\partial t} \right)^2 \right] dx, \tag{10}$$

$$U_c = \int_0^L \left[\frac{EA}{2} \left(\frac{\partial u}{\partial x} \right)^2 + \frac{EI}{2} \left(\frac{\partial \theta_x}{\partial x} \right)^2 \right] dx, \tag{11}$$

and $L_c = K_c - (U_c + V_{Ec})$.

(a) *Lagrange multiplier method:* We have

$$L_\ell = L_c + \int_0^L \lambda \left(\theta_x + \frac{\partial w}{\partial x} \right) dx. \tag{12}$$

Hence, the Euler-Lagrange equations are

$$
\begin{aligned}
\delta u : && \frac{\partial L_\ell}{\partial u} - \frac{\partial}{\partial x} \left(\frac{\partial L_\ell}{\partial u'} \right) - \frac{\partial}{\partial t} \left(\frac{\partial L_\ell}{\partial \dot{u}} \right) &= 0, \\
\delta w : && \frac{\partial L_\ell}{\partial w} - \frac{\partial}{\partial x} \left(\frac{\partial L_\ell}{\partial w'} \right) - \frac{\partial}{\partial t} \left(\frac{\partial L_\ell}{\partial \dot{w}} \right) &= 0, \\
\delta \theta_x : && \frac{\partial L_\ell}{\partial \theta_x} - \frac{\partial}{\partial x} \left(\frac{\partial L_\ell}{\partial \theta_x'} \right) - \frac{\partial}{\partial t} \left(\frac{\partial L_\ell}{\partial \dot{\theta}_x} \right) &= 0, \\
\delta \lambda : && \frac{\partial L_\ell}{\partial \lambda} &= 0.
\end{aligned}
\tag{13}
$$

We have

$$\delta u : \qquad f + \frac{\partial}{\partial x} \left(EA \frac{\partial u}{\partial x} \right) - \frac{\partial}{\partial t} \left(\rho A \frac{\partial u}{\partial t} \right) = 0, \tag{14}$$

$$\delta w : \qquad q - \frac{\partial \lambda}{\partial x} - \frac{\partial}{\partial t} \left(\rho A \frac{\partial w}{\partial t} \right) = 0, \tag{15}$$

$$\delta \theta_x : \qquad \lambda + \frac{\partial}{\partial x} \left(EI \frac{\partial \theta_x}{\partial x} \right) - \frac{\partial}{\partial t} \left(\rho I \frac{\partial \theta_x}{\partial t} \right) = 0, \tag{16}$$

$$\delta \lambda : \qquad \theta_x + \frac{\partial w}{\partial x} = 0. \tag{17}$$

(b) *Penalty method:* In this case we have

$$L_p = L_c + \frac{1}{2} \int_0^L \gamma \left(\theta_x + \frac{\partial w}{\partial x} \right)^2 dx. \tag{18}$$

Note that

$$L_p = L_p(u, w, \theta_x, u', w', \theta_x', \dot{u}, \dot{w}, \dot{\theta}_x).$$

Hence, the Euler–Lagrange equations are

$$\delta u : \qquad f + \frac{\partial}{\partial x}\left(EA\frac{\partial u}{\partial x}\right) - \frac{\partial}{\partial t}\left(\rho A\frac{\partial u}{\partial t}\right) = 0, \tag{19}$$

$$\delta w : \qquad q - \frac{\partial}{\partial x}\left[\gamma\left(\theta_x + \frac{\partial w}{\partial x}\right)\right] - \frac{\partial}{\partial t}\left(\rho A\frac{\partial w}{\partial t}\right) = 0, \tag{20}$$

$$\delta\theta_x : \quad \gamma\left(\theta_x + \frac{\partial w}{\partial x}\right) + \frac{\partial}{\partial x}\left(EI\frac{\partial\theta_x}{\partial x}\right) - \frac{\partial}{\partial t}\left(\rho I\frac{\partial\theta_x}{\partial t}\right) = 0. \tag{21}$$

5.5 Rayleigh's Method

For natural vibration of conservative systems, the Lagrangian function consists of only the strain energy and kinetic energy, $L = K - U$. For an ideal mechanical system (i.e., without thermal and other dissipative effects), the kinetic energy is the maximum when the strain energy is zero, and vice versa. Thus, in the absence of other energies, the principle of conservation of energy requires that

$$K_{\max} = U_{\max}. \tag{5.5.1}$$

Equation (5.5.1) merely implies that the maximum kinetic energy has the same value as the maximum strain energy. However, at any given instant, the two energies are *not* equal to each other, but their sum is constant. For example, for the spring mass system shown in Fig. 5.5.1, we have

$$K_{\max} = \frac{1}{2}m(\dot{x}_{\max})^2, \quad U_{\max} = \frac{1}{2}k(x_{\max})^2. \tag{5.5.2}$$

If the system is in a state of natural vibration, then $x = A\sin\omega t$ and $x_{\max} = A$, where A is the amplitude and ω is the frequency of natural vibration. Therefore, $\dot{x} = A\omega\cos\omega t$ and $\dot{x}_{\max} = A\omega$. Then

$$\frac{1}{2}mA^2\omega^2 = \frac{1}{2}kA^2 \quad \Rightarrow \quad \omega^2 = \frac{k}{m}.$$

Thus, given the stiffness of the spring and mass attached to it, we can calculate the natural frequency.

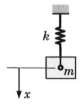

Fig. 5.5.1 A spring mass system.

Extending the discussion to natural vibration of a continuous system, say to an elastic bar, we have

$$K = \frac{1}{2} \int_0^L \rho A \left(\frac{\partial u}{\partial t}\right)^2 dx, \quad U = \frac{1}{2} \int_0^L EA \left(\frac{\partial u}{\partial x}\right)^2 dx, \tag{5.5.3}$$

where ρ is the mass density (measured per unit volume), E is the Young's modulus, and A is the area of cross section of the bar. For natural vibration, we have

$$u(x,t) = u_n(x)\sin(\omega_n t + \theta_n), \quad \dot{u}(x,t) = u_n(x)\omega\cos(\omega_n t + \theta_n).$$

Hence,

$$K = \frac{1}{2}\int_0^L \rho A u_n^2 \omega_n^2 \cos^2(\omega_n t + \theta_n)dx, \quad U = \frac{1}{2}\int_0^L EA\left(\frac{du_n}{dx}\right)^2 \sin^2(\omega_n t + \theta_n)dx.$$

Equating the maximum values of K and U ($K_{\max} = U_{\max}$), we obtain

$$\omega_n^2 \int_0^L \rho A u_n^2 dx = \int_0^L EA\left(\frac{du_n}{dx}\right)^2 dx \quad \text{or} \quad \omega_n^2 = \frac{\int_0^L EA\left(\frac{du_n}{dx}\right)^2 dx}{\int_0^L \rho A u_n^2 dx}. \tag{5.5.4}$$

The right-hand side of Eq. (5.5.4) is known as the *Rayleigh quotient* applied to a bar. With an appropriate choice of the eigenfunctions $u_n(x)$, the Rayleigh quotient allows us to compute the eigenvalue ω_n^2. A suitable candidate for $u_n(x)$ is the function that is sufficiently differentiable as required in Eq. (5.5.4), which satisfies the geometric boundary conditions of the problem. As we shall see in Chapter 6, Rayleigh quotient is a special case of the Ritz approximation. The Rayleigh quotient is good for the estimation of fundamental frequencies.

Example 5.5.1

We wish to use the Rayleigh quotient to estimate the fundamental frequency of vibration of a cantilever beam. For this case, the Rayleigh quotient takes the form

$$\omega_n^2 = \frac{\int_0^L EI\left(\frac{d^2 w_n}{dx^2}\right)^2 dx}{\int_0^L \rho A w_n^2 dx}, \tag{1}$$

where $w_n(x)$ is the eigenmode associated with the transverse motion of the beam.

A suitable candidate for the eigenfunction w_1 is provided by the deflection of the beam under its own weight. For a cantilever beam, the deflection form is given by

$$f(x) = \left(x^4 - 4x^3 L + 6x^2 L^2\right), \tag{2}$$

Therefore, we choose $w_1(x) = f(x) = x^4 - 4x^3 L + 6x^2 L^2$, and compute the Rayleigh quotient

$$\omega_1^2 = \frac{\int_0^L EI\left[12(x-L)^2\right]^2 dx}{\int_0^L \rho A\left(x^4 - 4x^3 L + 6x^2 L^2\right)^2 dx}$$

$$= \frac{EI}{\rho A L^4}\frac{144 \times 63}{728}, \tag{3}$$

or

$$\omega_1 = \frac{3.53}{L^2}\sqrt{\frac{EI}{\rho A}}. \tag{4}$$

The exact fundamental frequencies can be obtained by solving the transcendental equation

$$\cos\lambda\cosh\lambda + 1 = 0, \quad \lambda^2 = \omega L^2\sqrt{\frac{\rho A}{EI}}. \tag{5}$$

The first root of this equation is $\lambda_1 = 1.875$, giving $\omega_1 = (3.516/L^2)\sqrt{EI/\rho A}$. Thus, the solution estimated by the Rayleigh quotient is very accurate.

A higher-order frequency can be obtained only if we can find an appropriate candidate for the corresponding eigenfunction. For example, in the case of a simply supported beam, the eigenfunction associated with the nth mode is $w_n(x) = \sin(n\pi x/L)$, which can be used in the Rayleigh quotient to determine $\omega_n = (n\pi/L)^2$. This turns out to be the exact value of the frequency because of the selection of the exact eigenfunction for the simply supported beam.

5.6   Summary

In this chapter, Hamilton's principle for rigid bodies and continuous systems is presented. Hamilton's principle can be viewed as an extension of the principle of virtual displacements to dynamical systems. The extension involves the addition of the kinetic energy (algebraically, the total potential energy is subtracted from the kinetic energy) to the system's total energy. Analogous to the principle of virtual displacements, Hamilton's principle gives the equations of motion and the force boundary conditions in terms of the stress resultants, which are assumed to be known in terms of the displacements of the system. Hamilton's principle is a pseudo-variational principle because it requires the variations to vanish at the initial as well as final times. The main equations of this chapter are summarized here.

Hamilton's principle for rigid bodies [see Eq. (5.2.14)]:

$$\int_{t_1}^{t_2} (\delta K - \delta W_E)\, dt = 0. \tag{5.6.1}$$

The Euler–Lagrange equations for the nonconservative forces for rigid bodies [see Eq. (5.2.17)]:

$$\frac{\partial K}{\partial g_i} - \frac{d}{dt}\left(\frac{\partial K}{\partial \dot{g}_i}\right) + F_i = 0, \quad i = 1, 2, 3. \tag{5.6.2}$$

Hamilton's principle for rigid bodies with nonconservative forces [see Eq. (5.2.19)]:

$$\int_{t_1}^{t_2} (\delta K - \delta W_E)\, dt = 0 \quad \text{for nonconservative forces.} \tag{5.6.3}$$

Hamilton's principle for a continuum with nonconservative forces [see Eq. (5.3.8)]:

$$\int_{t_1}^{t_2} [\delta K - (\delta U + \delta V_E)]dt = 0. \tag{5.6.4}$$

The Euler–Lagrange equations for 3-D elasticity [see Eq. (5.3.11)]:

$$\rho \frac{\partial^2 \mathbf{u}}{\partial t^2} - \nabla \cdot \boldsymbol{\sigma} - \mathbf{f} = \mathbf{0} \quad \text{in } \Omega; \quad \mathbf{t} - \hat{\mathbf{t}} = \mathbf{0} \quad \text{on } \Gamma_\sigma. \tag{5.6.5}$$

Hamilton's principle for constrained systems [see Eqs (5.4.3), (5.4.7), and (5.4.9)]:
Holonomic constraints [see Eq. (5.4.3)]:

$$H_1(t, g_1, g_2, g_3) = 0, \quad H_2(t, g_1, g_2, g_3) = 0. \tag{5.6.6}$$

The Euler–Lagrange equations using the Lagrange multiplier method [see Eq. (5.4.7)]:

$$\frac{\partial K}{\partial g_i} - \frac{d}{dt}\left(\frac{\partial K}{\partial \dot{g}_i}\right) + F_i + \lambda_1 \frac{\partial H_1}{\partial g_i} + \lambda_2 \frac{\partial H_2}{\partial g_i} = 0. \tag{5.6.7}$$

The Euler–Lagrange equations using the penalty function method [see Eq. (5.4.9)]:

$$\frac{\partial K_p}{\partial g_i} - \frac{d}{dt}\left(\frac{\partial K_p}{\partial \dot{g}_i}\right) + F_i + \gamma_1 H_1 \frac{\partial H_1}{\partial g_i} + \gamma_2 H_2 \frac{\partial H_2}{\partial g_i} = 0. \tag{5.6.8}$$

Problems

HAMILTON'S PRINCIPLE FOR DISCRETE SYSTEMS (SECTION 5.2)

Derive the Lagrangian function for the linear spring, mass, and linear dashpot systems shown in Figs P5.1–P5.3. Assume that motion starts from rest and the contact surfaces are free of friction. The constitutive equations for the spring and dashpot are given by equations of the form Force $= k \times$ displacement, for springs ($k =$ spring constant) and Force $= \eta \times$ velocity, for dashpots ($\eta =$ dashpot constant).

5.1 A system of springs.

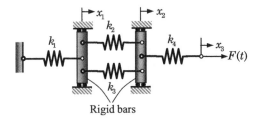

Fig. P5.1

5.2 A spring dashpot system (Kelvin model).

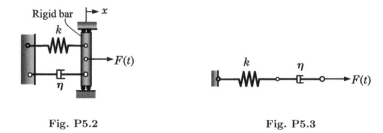

Fig. P5.2   Fig. P5.3

5.3 A spring dashpot system (Maxwell model) shown in Fig. P5.3.

5.4 Derive the equations of motion for the double pendulum of **Problem 3.1**.

5.5 Derive the equations of motion for the rigid body assemblage of **Problem 3.2**.

5.6 Consider a pendulum of mass m_1 with a flexible suspension, as shown in Fig. P5.6. The hinge of the pendulum is in a block of mass m_2, which can move up and down between the frictionless guides. The block is connected by a linear spring (k) to an immovable support. The coordinate x is measured from the position of the block in which the system remains stationary. Derive the Euler–Lagrange equations for the system.

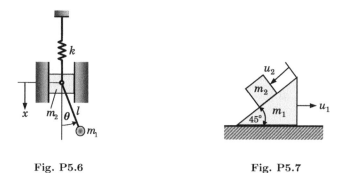

Fig. P5.6   Fig. P5.7

5.7 Consider a block of mass m_2 sliding on another block of mass m_1, which in turn slides on a horizontal surface, as shown in Fig. P5.7. Using u_1 and u_2 as coordinates, obtain the equations of motion. Assume that all surfaces are frictionless.

5.8 Figure P5.8 represents a double pendulum that is suspended from a block that moves horizontally with a prescribed motion, $x(t)$. Derive the Euler–Lagrange equations of motion of the pendulum when it oscillates in the (x, y)-plane under the action of gravity and prescribed motion.

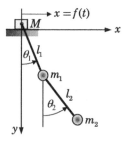

Fig. P5.8

5.9 Two masses m_1 and m_2 are attached to the ends of an inextensible cord that is suspended over a frictionless stationary pulley, as shown in Fig. P5.9. Find the equations of motion of the system.

Fig. P5.9   Fig. P5.10

5.10 Repeat **Problem 5.9** for the case in which a monkey of mass m_3 is climbing up the cord above mass m_1 with a speed v_0 relative to mass m_1 (see Fig. P5.10).

5.11 Determine the motion of all masses in the suspended double-pulley system represented in Fig. P5.11. Assume that the mass of the second pulley is m.

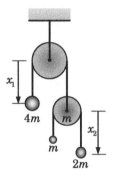

Fig. P5.11

5.12 Derive the equations of motion of the system shown in Fig. P5.12. Assume that the mass moment of inertia of the link about its mass center is $J = m\Omega^2$, where Ω is the radius of gyration.

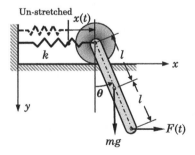

Fig. P5.12

5.13 Consider a cantilever beam supporting a lumped mass m at its end (J is the mass moment of inertia), as shown in Fig. P5.13. Derive the equations of motion and associated natural boundary conditions for the problem. Use the Bernoulli–Euler beam theory.

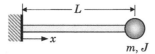

Fig. P5.13

HAMILTON'S PRINCIPLE FOR CONTINUOUS SYSTEMS (SECTIONS 5.3 AND 5.5)

5.14 Derive the equations of motion of the Timoshenko beam theory, starting with the displacement field

$$u_1(x, z, t) = u(x, t) + z\phi_x(x, t), \quad u_2 = 0, \quad u_3(x, z, t) = w(x, t).$$

Assume that the beam is subjected to distributed axial load $f(x, t)$ and transverse load $q(x, t)$, and that the x-axis coincides with the geometric centroidal axis. Account for the von Kármán nonlinear strains.

5.15 Express the equations of motion of the Timoshenko beam theory with the von Kármán nonlinear strains (from **Problem 5.14**) in terms of the generalized displacements (u, w, ϕ_x). Assume linear elastic behavior.

5.16 Rewrite the linearized bending equations of motion of the Timoshenko beam theory from **Problem 5.15** solely in terms of the transverse deflection w. Assume that the stiffness coefficients (EI, GAK_s) and mass inertias (m_0, m_2) are independent of position x and time t.

5.17 Consider the Bernoulli–Euler beam theory, whose displacement field is given by

$$u_1(x, z, t) = u(x, t) - z\frac{\partial w}{\partial x}, \quad u_2 = 0, \quad u_3(x, z, t) = w(x, t).$$

Assume that the beam has two types of viscous (velocity-dependent) damping: (1) viscous resistance to transverse displacement of the beam and (2) a viscous resistance to straining of the beam material. If the resistance to transverse velocity is denoted by $c(x)$, the corresponding damping force is given by $q_D(x, t) = c(x)\dot{w}$. If the resistance to strain velocity is c_s, the damping stress is $\sigma_{xx}^D = c_s\dot{\varepsilon}_{xx}$. Derive the linearized equations of motion of the beam with both types of damping.

5.18 Derive the equations of motion of the Bernoulli–Euler beam theory when the following nonlinear strain ε_{xx} is used:

$$\varepsilon_{xx} = \frac{\partial u}{\partial x} + \frac{1}{2}\left(\frac{\partial w}{\partial x}\right)^2 - z\frac{\partial^2 w}{\partial x^2}.$$

5.19 Suppose that the displacements (u_1, u_2, u_3) along the three coordinate axes $x_1 = x$, $x_2 = y$, and $x_3 = z$ of a point (x, y, z) in the beam can be expressed as

$$u_1(x, z, t) = u(x, t) + z\left[c_1\frac{\partial w}{\partial x} + c_2\phi_x(x, t)\right],$$

$$u_2(x, z, t) = 0, \tag{1}$$

$$u_3(x, z, t) = w(x, t),$$

where (c_1, c_2) are constants, (u_1, u_2, u_3) denote the total displacements along the $x = x_1$, $y = x_2$, and $z = x_3$ coordinates, respectively, of a point (x, y, z) in the beam, $u(x, t)$ is the displacement of a point on the midplane, $w(x, t)$ is the transverse deflection of a point on the midplane, and $\phi_x(x, t)$ is the rotation of a transverse normal about the y-axis. Assume that the beam is subjected to distributed transverse load $q(x, t)$ and axial load $f(x, t)$, both measured per unit length.

(1) Compute the *linear* strains, using the strain–displacement relations.

(2) Use the dynamics version of the principle of virtual displacements and the following definitions of stress resultants to (a) derive the governing equations of motion (i.e., the Euler–Lagrange equations) and (b) identify the primary and secondary variables of the theory:

$$N_{xx} = \int_A \sigma_{xx} \, dA \,, \quad V_x = K_s \int_A \sigma_{xz} \, dA \,, \quad M_{xx} = \int_A z\,\sigma_{xx} \, dA. \qquad (2)$$

Here N_{xx} denotes the axial force, Q_x is the transverse (shear) force, and M_{xx} is the bending moment. Your answer should be in terms of the stress resultants defined in Eq. (2).

5.20 Suppose that the displacements (u_1, u_2, u_3) along the three coordinate axes $x_1 = x$, $x_2 = y$, and $x_3 = z$ of a point (x, y, z) in the beam can be expressed as

$$u_1(x, z, t) = u(x, t) + z\phi_x(x, t) + z^2\psi_x(x, t),$$
$$u_2(x, z, t) = 0,$$
$$u_3(x, z, t) = w(x, t),$$

where (u, w) denote the displacements of a point $(x, y, 0)$ along the x and z directions, respectively, and ϕ_x and ψ_x are unknown functions (i.e., dependent unknowns). Compute the linear strains and write the virtual Lagrangian (omit V) $\delta L = \delta K - \delta U$ in terms of u, w, ϕ_x, ψ_x, and area-integrated quantities.

5.21 Given the following displacement field for the axisymmetric bending of a circular plate

$$u_r(r, z, t) = u(r, t) - z\frac{\partial w}{\partial r} \,, \quad u_\theta = 0, \quad u_z(r, z, t) = w(r, t), \qquad (1)$$

derive the equations of motion. Include the von Kármán nonlinearity. *Hint:* The strains are given by

$$\varepsilon_{rr} = \varepsilon_{rr}^{(0)} + z\varepsilon_{rr}^{(1)}, \quad \varepsilon_{\theta\theta} = \varepsilon_{\theta\theta}^{(0)} + z\varepsilon_{\theta\theta}^{(1)}, \qquad (2)$$

where

$$\varepsilon_{rr}^{(0)} = \frac{\partial u}{\partial r} + \frac{1}{2}\left(\frac{\partial w}{\partial r}\right)^2, \quad \varepsilon_{rr}^{(1)} = -\frac{\partial^2 w}{\partial r^2}, \quad \varepsilon_{\theta\theta}^{(0)} = \frac{u}{r}, \quad \varepsilon_{\theta\theta}^{(1)} = -\frac{1}{r}\frac{\partial w}{\partial r}. \qquad (3)$$

5.22 Repeat **Problem 5.21** when transverse shear deformation is accounted. The displacement field and the strains are given by

$$u_r(r, z, t) = u(r, t) + z\phi_r(r, t), \quad u_\theta = 0, \quad u_z(r, z, t) = w(r, t) \qquad (1)$$

and

$$\varepsilon_{rr} = \varepsilon_{rr}^{(0)} + z\varepsilon_{rr}^{(1)}, \quad \varepsilon_{\theta\theta} = \varepsilon_{\theta\theta}^{(0)} + z\varepsilon_{\theta\theta}^{(1)}, \quad \varepsilon_{rz} = \varepsilon_{rz}^{(0)}, \qquad (2)$$

where

$$\varepsilon_{rr}^{(0)} = \frac{\partial u}{\partial r} + \frac{1}{2}\left(\frac{\partial w}{\partial r}\right)^2, \quad \varepsilon_{rr}^{(1)} = \frac{\partial\phi_r}{\partial r},$$
$$\varepsilon_{\theta\theta}^{(0)} = \frac{u}{r}, \quad \varepsilon_{\theta\theta}^{(1)} = \frac{\phi_r}{r}, \quad 2\varepsilon_{rz}^{(0)} = \phi_r + \frac{\partial w}{\partial r}. \qquad (3)$$

5.23 Derive the equations of motion for the axisymmetric bending of a circular plate according to the Reddy third-order plate theory. The displacement field is given by

$$u_r(r, z, t) = u(r, t) + z\phi(r, t) - c_1 z^3 \left(\phi_r + \frac{\partial w}{\partial r} \right), \quad u_z(r, z, t) = w(r, t), \qquad (1)$$

where $c_1 = 4/(3h^2)$ and h denotes the total thickness of the plate. Include the von Kármán nonlinearity. *Hint:* The strains are given by

$$\varepsilon_{rr} = \varepsilon_{rr}^{(0)} + z\varepsilon_{rr}^{(1)} + z^3\varepsilon_{rr}^{(3)}, \quad \varepsilon_{\theta\theta} = \varepsilon_{\theta\theta}^{(0)} + z\varepsilon_{\theta\theta}^{(1)} + z^3\varepsilon_{\theta\theta}^{(3)}, \quad \varepsilon_{rz} = \varepsilon_{rz}^{(0)} + z^2\varepsilon_{rz}^{(2)}, \qquad (2)$$

where

$$\varepsilon_{rr}^{(0)} = \frac{\partial u}{\partial r} + \frac{1}{2}\left(\frac{\partial w}{\partial r} \right)^2, \quad \varepsilon_{rr}^{(1)} = \frac{\partial \phi_r}{\partial r}, \quad \varepsilon_{rr}^{(3)} = -c_1 \left(\frac{\partial \phi_r}{\partial r} + \frac{\partial^2 w}{\partial r^2} \right),$$

$$\varepsilon_{\theta\theta}^{(0)} = \frac{u}{r}, \quad \varepsilon_{\theta\theta}^{(1)} = \frac{\phi_r}{r}, \quad \varepsilon_{\theta\theta}^{(3)} = -\frac{c_1}{r}\left(\phi_r + \frac{\partial w}{\partial r} \right), \qquad (3)$$

$$2\varepsilon_{rz}^{(0)} = \phi_r + \frac{\partial w}{\partial r}, \quad 2\varepsilon_{rz}^{(2)} = -c_2 \left(\phi_r + \frac{\partial w}{\partial r} \right), \quad c_2 = \frac{4}{h^2}.$$

5.24 Consider a rectangular membrane of dimensions a by b and mass density ρ, and subjected to distributed transverse load $f(x, t)$. The membrane is fixed at all points of its boundary. If the strain energy stored in the membrane is given by

$$U = \int_0^a \int_0^b \frac{T_0}{2} \left[\left(\frac{\partial u}{\partial x} \right)^2 + \left(\frac{\partial u}{\partial y} \right)^2 \right] dx\, dy, \qquad (1)$$

where T_0 is the tension in the membrane and u is the transverse displacement, determine the equation of motion governing u.

5.25 Use the Rayleigh quotient to estimate the fundamental natural frequencies of the following beam problems:

    (a) A simply supported beam. Use the static deflection of the beam under uniform load as the eigenfunction.

    (b) A fixed-simply supported beam. Use $W_1(x) = x^2(2x^2 - 5xL + 3L^2)$.

    (c) A fixed-fixed beam. Use $W_1(x) = x^2(L - x)^2$.

"What we need is not the will to believe but the will to find out."

"The fact that an opinion has been widely held is no evidence whatever that it is not utterly absurd; indeed in view of the silliness of the majority of mankind, a widespread belief is more likely to be foolish than sensible."

                                        – Bertrand Russell (British philosopher and Nobel laureate)

6

Direct Variational Methods

6.1 Introduction

In Chapters 4 and 5, we have seen how energy principles can be used to obtain governing equations, associated boundary conditions, and, in certain simple cases, solutions for displacements and forces at selective points of a structure. However the energy methods considered in Chapters 4 and 5 cannot be used, in general, to determine continuous solutions to complex problems.

The present chapter deals with approximate methods that employ the variational statements (i.e., either variational principles or weak formulations) to determine continuous solutions of problems of mechanics. Recall that the energy principles contain, in a single statement, the governing equation(s) and the natural boundary condition(s) of the problem. The energy principles involved setting the first variation of an appropriate functional with respect to the dependent variables to zero. The procedures of the calculus of variations were then used to obtain the governing (Euler–Lagrange) equations of the problem. In contrast, the methods described in this chapter seek a solution in terms of adjustable parameters that are determined by substituting the assumed solution into the functional and finding its extremum or stationary value with respect to the parameters. Such solution methods are called *direct methods*, because the approximate solutions are obtained directly by using the same variational principle that was used to derive the governing equations.

The assumed solutions in the variational methods are in the form of a finite linear combination of *undetermined parameters* with appropriately chosen functions. This amounts to representing a continuous function by a finite linear combination of functions. Since the solution of a continuum problem in general cannot be represented by a finite set of functions, error is introduced into the solution. Therefore, the solution obtained is an *approximation* of the true solution for the equations describing a physical problem. As the number of linearly independent terms in the assumed solution is increased, the error in the approximation will be reduced, and the assumed solution converges to the desired solution.

Energy Principles and Variational Methods in Applied Mechanics, Third Edition. J.N. Reddy.
©2017 John Wiley & Sons Ltd. Published 2017 by John Wiley & Sons Ltd.
Companion Website: www.wiley.com/go/reddy/applied_mechanics_EPVM

The equations governing a physical problem themselves are approximate. The approximations are introduced via several sources, including the geometry, the representation of specified loads and displacements, and the material behavior. In the present study, our primary interest is to determine accurate approximate solutions to appropriate analytical descriptions of physical problems.

The variational methods of approximation described here include the classical methods of Ritz, Galerkin, and Petrov–Galerkin (weighted-residual methods). Examples of applications of these methods are drawn from the problems of bars, beams, torsion, and membranes, while those to circular and rectangular plates are considered in Chapter 7. We begin with some mathematical preliminaries.

6.2   Concepts from Functional Analysis

6.2.1   General Introduction

Before we discuss the variational methods of approximation, it is useful to equip ourselves with certain mathematical concepts. These include the vector spaces, norm, inner product, linear independence, orthogonality, and linear and bilinear forms of functions. Since the objective of the present study is to learn about variational methods, we limit our discussion here only to concepts that are pertinent in the context. The reader already familiar with these concepts may browse through the section to gain familiarity with the notation. Others who do not wish to burden themselves with the formalism of functional analysis may skip this section; it would not prevent them from gaining an understanding of the main ideas of variational methods.

A set X is any well-defined collection of things, which are called *members* or *elements* of X. In the present study, we are concerned with the collection of numbers, sequences, functions, and functions of functions. Examples of sets are provided below.

1. The set $\Re$ of all real numbers

2. The set $C[0, L]$ of all real-valued continuous functions $f(x)$ defined on the closed interval $0 \leq x \leq L$

3. The collection of all closed intervals, $I_i = [x_i, x_{i+1}]$, on the real line

The following notation, very standard in mathematics, is adopted here:

$$\subset \text{ means "a subset of"}$$

$$\not\subset \text{ means "not a subset of"}$$

$$\in \text{ means "an element of"}$$

$\notin$ means "not an element of"

$\forall$ means "for all"

$\exists$ means "there exists"

$\ni$ means "such that." $\hspace{2cm}$ (6.2.1)

One way of defining a set S is to specify two pieces of information: (1) assume that each element of S is an element of a universal set (i.e., a well-known set), say X, and (2) list the properties that elements of the universal set must satisfy in order to be in S. For example, let X be the set of all sequences of complex numbers $\mathbf{x} = \{x_1, x_2, x_3, \ldots\}$ and S be all elements of X possessing the property

$$\sum_{n=1}^{\infty} |x_n| < \infty.$$

We shall use the following notation:

$$S = \left\{ \mathbf{x} = (x_1, x_2, \ldots, x_n) \in X : \sum_{n=1}^{\infty} |x_n| < \infty \right\}, \hspace{1cm} (6.2.2)$$

which is read "S is the set of all elements $\mathbf{x} = (x_1, x_2, \ldots, x_n)$ of the set X such that (the colon stands for 'such that') $\sum_{n=1}^{\infty} |x_n| < \infty$."

A set $A \subset \Re$ is said to be *bounded from above* if there exists a real number μ such that $a \leq \mu$ for all $a \in A$. The real number μ is said to be an *upper bound* of the set A. Similarly, a set A is said to be *bounded from below* if there exists a real number γ such that $a \geq \gamma$ for all $a \in A$. The real number γ is said to be *lower bound* of the set A. If a set A is bounded from above and from below, we say that A is bounded. An upper (lower) bound $M(m)$ for A is said to be the maximum (minimum) of $A \subset \Re$ if $M \in A$ ($m \in A$). It should be noted that even a bounded set need not have a maximum or a minimum. Every nonempty set of real numbers bounded from above has a "least upper bound," and every nonempty set of real numbers bounded from below has a "greatest lower bound." The least upper bound of a set A is denoted by sup A ("supremum of A"), and the greatest lower bound of A is denoted by inf A ("infimum of A").

6.2.2 Linear Vector Spaces

As we have seen in Chapter 1, the term *vector* is used often to imply a *physical* vector that has "magnitude and direction" and obeys certain rules of vector addition and scalar multiplication. These ideas can be extended to functions, which are also called vectors, provided that the rules of vector addition and scalar multiplication are defined. While the definition of a vector "from a linear vector space" does not require the vector to have a magnitude, in nearly all cases of practical interest, the vector is endowed with a magnitude, called *norm*. In such cases the vector is said to belong to a normed vector space. We begin with a formal definition of an abstract vector space.

A collection of vectors, $u, v, w, \ldots$, is called a *real linear vector space* $\mathcal{V}$ over the real number field $\Re$ if the following rules of vector addition and scalar multiplication of a vector are satisfied by the elements of the vector space.

Vector addition. To every pair of vectors u and v, there corresponds a unique vector $u + v \in \mathcal{V}$, called the *sum* of u and v, with the following properties:

(1a) $\quad u + v = v + u \quad$ (commutative).

(1b) $\quad (u + v) + w = u + (v + w) \quad$ (associative).

(1c) $\quad$ There exists a unique vector, Θ, independent of u such that

$\qquad u + \Theta = u \quad$ for every $\ u \in \mathcal{V}$ (existence of an identity element).

(1d) $\quad$ To every u there exists a unique vector, $-u$,

$\qquad$ (that depends on u) such that $u + (-u) = \Theta \quad$ for every $\ u \in \mathcal{V}$

$\qquad$ (existence of the additive inverse element). $\hfill$ (6.2.3)

Scalar multiplication. To every vector u and every real number $\alpha \in \Re$, there corresponds a unique vector $\alpha u \in \mathcal{V}$, called the *product* of u and α, such that the following properties hold:

(2a) $\quad \alpha(\beta u) = (\alpha\beta)u \quad$ (associative).

(2b) $\quad (\alpha + \beta)u = \alpha u + \beta u \quad$ (distributive w.r.t. the scalar addition).

(2c) $\quad \alpha(u + v) = \alpha u + \alpha v \quad$ (distributive w.r.t. the vector addition).

(2d) $\quad 1 \cdot u = u \cdot 1$. $\hfill$ (6.2.4)

Note that in order to prove that a set of vectors qualifies as a vector space, one must define the identity and inverse elements and prove the "closure property" $u + v \in \mathcal{V}$ and $\alpha u \in \mathcal{V}$ for all $u, v \in \mathcal{V}$, and $\alpha \in \Re$.

A subset S of a vector space $\mathcal{V}$ is called a *subspace* of $\mathcal{V}$, denoted $S \subset \mathcal{V}$, if S itself is a vector space with respect to vector addition and scalar multiplication defined over $\mathcal{V}$. Examples of some common linear vector spaces are listed here.

1. The set of ordered n-tuples $(x_1, x_2, x_3, \ldots, x_n)$ of real numbers $x_1, x_2, \ldots, x_n$ is called the *Cartesian space*, denoted by $\Re^n$. A typical element of $\Re^n$ is denoted by $\mathbf{x} = (x_1, x_2, x_3, \ldots, x_n)$. The Cartesian space is a linear vector space with respect to the usual rules of addition and scalar multiplication:

- *Vector addition:* $\mathbf{x} + \mathbf{y} = (x_1 + y_1, x_2 + y_2, \ldots, x_n + y_n) \ \forall \mathbf{x}, \mathbf{y} \in \Re^n$.

- *Scalar multiplication:* $\alpha \mathbf{x} = (\alpha x_1, \alpha x_2, \ldots, \alpha x_n) \ \forall \mathbf{x} \in \Re^n$ and $\alpha \in \Re$.

The identity element is $\mathbf{0} = (0, 0, 0, \ldots)$ (n zeros), and the inverse element is the negative of the vector.

2. Let $\mathcal{P}$ be the set of *all* polynomials in x with real coefficients. A typical element of $\mathcal{P}$ is of the form

$$p(x) = a_0 + a_1 x + a_2 x^2 + \dots,$$

where $a_0, a_1, \dots$ are real numbers. Then $\mathcal{P}$ is a linear vector space with respect to the usual rules of addition and scalar multiplication. Also, the set $\mathcal{P}_n$ of polynomials of degree *less than or equal to* degree n is also a linear vector space, as can be verified (by the closure property). Moreover, $\mathcal{P}_n \subset \mathcal{P}$. However, the set of polynomials of degree *equal to* n is not a vector space as the closure property is violated. For example, consider the set of all cubic polynomials. The sum of $p_1(x) = 1 - 2x + 3x^2 + 6x^3$ and $p_2 = -3 + 5x + 2x^2 - 6x^3$ is not a cubic polynomial.

3. Let $C^n[a, b]$, where $n \geq 0$ is an integer, denote the set of all real-valued functions $u(x)$ defined on the interval $a \leq x \leq b$ such that u is continuous, and the derivatives $d^k u / dx^k$ of order $k \leq n$ exist and are continuous on $[a, b]$. It can be shown that $C^n[a, b]$ is a linear vector space with respect to the usual rules of vector addition and scalar multiplication.

4. The set

$$S_0 = \left\{ u : u(x) \in C^2[0, L], -\frac{d}{dx}\left(a(x)\frac{du}{dx}\right) + c(x)u = 0, \quad 0 < x < L \right\}$$

is a vector space with respect to the usual addition and scalar multiplication. However, the set

$$S = \left\{ u : u(x) \in C^2[0, L], -\frac{d}{dx}\left(a(x)\frac{du}{dx}\right) + c(x)u = f(x), \quad 0 < x < L \right\}$$

is not a linear vector space.

5. Consider the transverse motion of a cable with length L, fixed at its ends (see Fig. 6.2.1). Let $C[0, L]$ denote the set of all real-valued, continuous functions $u(x, t)$ defined on the closed interval $0 \leq x \leq L$ for any time t. The transverse deflection $u(\cdot, t)$ (i.e., configuration) of the cable at any time t can be viewed as an element of $C[0, L]$. However, not every element of $C[0, L]$ is a possible configuration of the cable because all possible configurations must pass through the points $x = 0$ and $x = L$; that is, the boundary conditions $u(0, t) = 0$ and $u(L, t) = 0$ must be satisfied. Let S be the subset of $C[0, L]$ made up of all real-valued, continuous functions $u(x, t)$ such that $u(0, t) = 0$ and $u(L, t) = 0$,

$$S = \left\{ u : u(x) \in C[0, L], \quad u(0, t) = 0, u(L, t) = 0 \right\}.$$

Then S is a subspace of $C[0, L]$, and all possible configurations (i.e., deflections) are contained in this space.

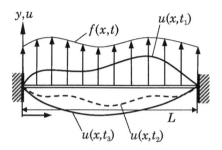

Fig. 6.2.1 Transverse motion of a cable fixed at both ends.

Let $\mathcal{U}$ and $\mathcal{V}$ be each a linear vector space. An *ordered pair* is a pair of elements $u \in \mathcal{U}$ and $v \in \mathcal{V}$ where one of the elements is designated as the first member of the pair and the other is designated as the second. We denote ordered pairs by (u, v) with the obvious order. Then $\mathcal{U} \times \mathcal{V}$ is called a *product space* $\mathcal{W}$ with elements $w = (u, v)$, $u \in \mathcal{U}$, and $v \in \mathcal{V}$:

$$\mathcal{W} = \{w : w = (u, v), u \in \mathcal{U}, v \in \mathcal{V}\},$$

which is also a linear vector space with respect to the following definitions of vector addition and scalar multiplication of a vector in the product space $\mathcal{U} \times \mathcal{V}$:

$$(u_1, v_1) + (u_2, v_2) = (u_1 + u_2, v_1 + v_2),$$
$$\alpha(u, v) = (\alpha u, \alpha v), \quad \alpha \in \Re \tag{6.2.5}$$

for $(u_1, v_1), (u_2, v_2) \in \mathcal{W} = \mathcal{U} \times \mathcal{V}$ with $u_1, u_2 \in \mathcal{U}$ and $v_1, v_2 \in \mathcal{V}$.

Consider an open bounded domain $\Omega \subset \Re^3$. Note that Ω is a set of points $\mathbf{x} = (x_1, x_2, x_3)$. A real-valued function $u(\mathbf{x})$ is said to be *square integrable* in the domain Ω if the integrals

$$\int_\Omega u(\mathbf{x})d\Omega \quad \text{and} \quad \int_\Omega |u(\mathbf{x})|^2 d\Omega \tag{6.2.6}$$

exist and are finite. The space of square-integrable functions u defined over a domain Ω is called the L_2 vector space:

$$L_2(\Omega) = \left\{ u(\mathbf{x}) : \int_\Omega u(\mathbf{x})d\Omega < \infty, \int_\Omega |u(\mathbf{x})|^2 d\Omega < \infty \right\}. \tag{6.2.7}$$

Example 6.2.1 ──

Determine if the following functions belong to $L_2(0, 1)$:

$$\text{(a) } u(x) = \frac{1}{x^{\frac{1}{3}}}, \quad \text{(b) } u(x) = \frac{1}{\sqrt{x}}. \tag{1}$$

Solution: To show that a function u belongs to $L_2(0,1)$, we must verify if the conditions given in Eq. (6.2.7) are met.

(a) We have

$$\int_0^1 u(x)\,dx = \int_0^1 x^{-\frac{1}{3}}\,dx = \left[\frac{3}{2}x^{\frac{2}{3}}\right]_0^1 = \frac{3}{2},$$

$$\int_0^1 [u(x)]^2\,dx = \int_0^1 x^{-\frac{2}{3}}\,dx = \left[3x^{\frac{1}{3}}\right]_0^1 = 3. \tag{2}$$

Thus, $u(x) = x^{-\frac{1}{3}} \in L_2(0,1)$.

(b) We have

$$\int_0^1 u(x)\,dx = \int_0^1 x^{-\frac{1}{2}}\,dx = \left[2x^{\frac{1}{2}}\right]_0^1 = 2,$$

$$\int_0^1 [u(x)]^2\,dx = \int_0^1 x^{-1}\,dx = [\log x]_0^1 = +\infty. \tag{3}$$

Thus, $u(x) = \frac{1}{\sqrt{x}} \notin L_2(0,1)$.

There is a corresponding space $L_\infty(\Omega)$, which consists of all real-valued functions $u(\mathbf{x})$ defined in the domain Ω such that there exists a number Λ with the property

$$|u(\mathbf{x})| \leq \Lambda.$$

Linear independence. Recall the concepts of coplanar and collinear vectors in Euclidean space from Chapter 1. These concepts can be generalized to function spaces. An expression of the form

$$\alpha_1 u_1 + \alpha_2 u_2 + \cdots + \alpha_N u_N = \sum_{i=1}^N \alpha_i u_i \tag{6.2.8}$$

for all functions $u_i(x)$ and scalars $\alpha_i \in \Re$ (real number field) is called a *linear combination* of u_i. The equation $\sum_i^N \alpha_i u_i = 0$ is called a *linear relation* among the functions u_i. A set of N vectors, $u_1, u_2, \ldots, u_N$, is said to be *linearly dependent* if a set of N numbers, $\alpha_1, \alpha_2, \ldots, \alpha_N$, not all of which are zero, can be found such that the following linear relation holds:

$$\sum_{i=1}^N \alpha_i u_i = 0. \tag{6.2.9}$$

If there does not exist at least one nonzero number among α_i such that the above relation is satisfied, the vectors are said to be *linearly independent*.

Example 6.2.2 ———————————————————————————

1. Consider the following set of polynomials, $\{p_i\} = \{p_1, p_2, p_3\}$, with

$$p_1(x) = 1 + x, \quad p_2(x) = 1 + x^2, \quad p_3(x) = 1 + x + x^3.$$

Determine if the set is linearly independent.

2. Determine if the set $\{p_i\} = \{p_1, p_2, p_3\}$, when p_3 is replaced by $p_4 = 2 + x + x^2$, is linearly independent.

Solution:

1. Consider the linear relation

$$\alpha_1 p_1 + \alpha_2 p_2 + \alpha_3 p_3 = 0$$

for $\alpha_i \in \Re$. Since the above relation must hold for all x, it follows that the coefficients of powers of x must be zero separately. Collecting the coefficients of various powers of x and setting them to zero, we obtain

$$\alpha_1 + \alpha_2 + \alpha_3 = 0, \quad \alpha_1 + \alpha_3 = 0, \quad \alpha_2 = 0, \quad \alpha_3 = 0.$$

The solution to these equations is trivial (i.e., all $\alpha_i = 0$); hence, the set $\{p_1, p_2, p_3\}$ is linearly independent.

2. If p_3 is replaced by $p_4 = 2 + x + x^2$, we see that the linear relation

$$\alpha_1 p_1 + \alpha_2 p_2 + \alpha_4 p_4 = 0$$

requires that

$$\alpha_1 + \alpha_2 + 2\alpha_4 = 0, \quad \alpha_1 + \alpha_4 = 0, \quad \alpha_2 + \alpha_4 = 0.$$

An infinite number of solutions to the above set of equations exists. For example,

$$\alpha_4 = 1, \quad \alpha_1 = \alpha_2 = -1$$

is a solution. Hence, the set $\{p_1, p_2, p_4\}$ is linearly dependent. Indeed, p_4 can be expressed as a linear combination of p_1 and p_2 :

$$p_4 = \alpha_1 p_1 + \alpha_2 p_2, \quad \alpha_1 = \alpha_2 = 1.$$

6.2.3   Normed and Inner Product Spaces

6.2.3.1   Norm

The concepts of distance between two points and length of a physical vector can be generalized to abstract vectors, that is, vectors that are functions. Let $\mathcal{V}$ be a linear vector space over the real number field $\Re$. We shall use the notation $\| \cdot \|$ to denote the norm of real-valued functions $u(\mathbf{x})$, $\mathbf{x} \in \Omega \subset \Re^3$. Then, associated with every vector $u \in \mathcal{V}$, there exists a real number $\| \cdot \| \in \Re$, called *norm*, that satisfies certain rules, as discussed below. Thus norm is the operation $\| \cdot \| : \mathcal{V} \to \Re$.

(1) Nonnegative: (a) $\|u\| \geq 0$ for all u. (b) $\|u\| = 0$ only if $u = 0$.

(2) Homogeneous: $\|\alpha u\| = |\alpha| \, \|u\|.$ (6.2.10)

(3) Triangle inequality: $\|u + v\| \leq \|u\| + \|v\|.$

If $\|u\|$ satisfies (1a), (2), and (3), it is called a *seminorm*, and is denoted by $|u|$.

A linear vector space endowed with this norm is called a *normed vector space*. A linear subspace S of a normed vector space $\mathcal{V}$ is a linear subspace equipped with the norm of $\mathcal{V}$.

A norm $\|\cdot\|$ can be used to define a notion of distance between vectors, called *natural metric*,

$$d(u, v) \equiv \|u - v\| \quad \text{for} \quad u, v \in \mathcal{V}. \tag{6.2.11}$$

Examples of norms will be given shortly.

For $1 \leq p \leq \infty$, we define the *Lebesgue spaces*

$$L_p(\Omega) = \{u(\mathbf{x}) : \|u\|_p < \infty, \ \mathbf{x} \in \Omega\}, \tag{6.2.12}$$

where

$$\|u\|_{L_p(\Omega)} \equiv \|u\|_p = \left[\int_\Omega |u(\mathbf{x})|^p \, d\mathbf{x} \right]^{1/p} < \infty. \tag{6.2.13}$$

For $p = \infty$ we set

$$\|u\|_{L_\infty(\Omega)} \equiv \|u\|_\infty = \sup \{|u(\mathbf{x})| : \mathbf{x} \in \Omega\}. \tag{6.2.14}$$

The norm defined by Eq. (6.2.14) is called the "sup-norm."

Two norms $\|\cdot\|_1$ and $\|\cdot\|_2$ on a normed vector space $\mathcal{V}$ are said to be *equivalent* if there exist positive numbers c_1 and c_2, independent of $u \in \mathcal{V}$, such that the following double inequality holds:

$$c_1\|u\|_1 \leq \|u\|_2 \leq c_2\|u\|_1. \tag{6.2.15}$$

A normed space $\mathcal{V}$ is called *complete* if every Cauchy sequence $\{u_j\}$ of elements of $\mathcal{V}$ has a limit $u \in \mathcal{V}$. For a normed vector space, a *Cauchy sequence* is one such that

$$\|u_j - u_k\| \to 0 \quad \text{as} \quad j, k \to \infty,$$

and completeness means that

$$\|u - u_j\| \to 0 \quad \text{as} \quad j \to \infty.$$

A normed vector space, which is complete in its natural metric, is called a *Banach space*. A linear subspace of a Banach space is itself a Banach space if and only if the subspace is complete. Every finite-dimensional subspace of a normed vector space is complete. Examples of Banach spaces are presented next:

1. The n-dimensional *Euclidean space* $\Re^n$ is a Banach space with respect to the *Euclidean norm*:

$$\|\mathbf{x}\| \equiv \sqrt{\sum_{i=1}^n x_i^2}. \tag{6.2.16}$$

2. The space $\mathcal{C}[0,1]$ of real-valued continuous functions $f(x)$ defined on the closed interval $[0,1]$ with the sup-norm defined in Eq. (6.2.14) is a Banach space. It is a linear vector space with respect to the vector addition and scalar multiplication defined as

$$(f+g)(x) = f(x) + g(x), \quad (\alpha f)(x) = \alpha f(x), \quad \alpha \in \mathfrak{R}.$$

Further, it is complete with respect to the sup-norm in Eq. (6.2.14):

$$\|f\|_\infty \equiv \max |f(x)|.$$

3. *Sobolev space, $W^{m,p}(\Omega)$.* Let $\mathcal{C}^m(\Omega)$ denote the set of all real-valued functions with m continuous derivatives defined in $\Omega \in \mathfrak{R}^3$, and let $\mathcal{C}^\infty(\Omega)$ denote the set of infinitely differentiable continuous functions. We define on $\mathcal{C}^m(\Omega)$ the norm, called the *Sobolev norm*,

$$\|u\|_{m,p} = \left[\int_\Omega \sum_{|\alpha| \le m} |D^\alpha u(\mathbf{x})|^p d\mathbf{x} \right]^{1/p}, \tag{6.2.17}$$

for $1 \le p \le \infty$ and for all $u \in \mathcal{C}^m(\Omega)$. In Eq. (6.2.17), α denotes an n-tuple of integers:

$$\alpha = (\alpha_1, \alpha_2, \ldots, \alpha_n), \quad |\alpha| = \sum_{i=1}^n \alpha_i, \quad \alpha_i \ge 0,$$

$$D^\alpha = \frac{\partial^{|\alpha|}}{\partial x_1^{\alpha_1} \partial x_2^{\alpha_2} \ldots \partial x_n^{\alpha_n}}. \tag{6.2.18}$$

For $m = 1, n = 2$, and $1 \le p < \infty$, we have $[\alpha = (\alpha_1, \alpha_2), \alpha_1, \alpha_2 = 0, 1]$, and

$$\|u\|_{1,p} = \left\{ \int_{\Omega \subset \mathfrak{R}^2} \left[|u|^p + \left| \frac{\partial u}{\partial x} \right|^p + \left| \frac{\partial u}{\partial y} \right|^p \right] dx dy \right\}^{1/p}. \tag{6.2.19}$$

The space $\mathcal{C}^m(\Omega)$ is not complete with respect to the Sobolev norm $\| \cdot \|_{m,p}$. The completion of $\mathcal{C}^m(\Omega)$ with respect to the norm $\| \cdot \|_{m,p}$ is called the *Sobolev space of order* (m,p), denoted by $W^{m,p}(\Omega)$. The completion of $\mathcal{C}(\Omega)$ is the $L_2(\Omega)$ space. Hence the Sobolev space is a Banach space. Of course, the Lebesque space $L_p(\Omega)$ is a special case of the Sobolev space $W^{m,p}$ for $m = 0$, and $L_2(\Omega)$ is a special case of $L_p(\Omega)$ for $p = 2$, with the norms defined in Eq. (6.2.13).

If $\mathcal{U}$ and $\mathcal{V}$ are each normed vector spaces, we can define a norm on the product space $\mathcal{U} \times \mathcal{V}$ in one of the following ways:

(1) $\|(u,v)\| = \|u\|_U + \|v\|_V.$

(2) $\|(u,v)\| = \left(\|u\|_U^p + \|v\|_V^p \right)^{1/p}, \quad p \ge 1.$ \hfill (6.2.20)

(3) $\|(u,v)\| = \max \left(\|u\|_U, \|v\|_V \right).$

Then $\mathcal{U} \times \mathcal{V}$ is a normed vector space with respect to any one of the above norms.

6.2.3.2 Inner Product

Analogous to the scalar product of physical vectors, the *inner product* of a pair of vectors u and v from an abstract vector space V is defined to be a real number, denoted by $(u, v)_V$ (i.e., $(\cdot, \cdot)_V : V \times V \to \Re$), which satisfies the following rules for every $u_1, u_2, u, v \in V$ and $\alpha \in \Re$:

(1) Symmetry: $(u, v)_V = (v, u)_V$.

(2a) Homogeneous: $(\alpha u, v)_V = \alpha (u, v)_V$.

(2b) Additive: $(u_1 + u_2, v)_V = (u_1, v)_V + (u_2, v)_V$.

(3) Positive–definite: $(u, u)_V > 0$ for all $u \neq 0$. $\qquad\qquad$ (6.2.21)

One can define a number of inner products and associated *natural norms* for pairs of functions that, along with their derivatives, are square-integrable. In particular, the Sobolev space $W^{m,2}(\Omega) \equiv H^m(\Omega)$, which is also known as the *Hilbert space* of order m, is endowed with the inner product

$$(u, v)_m = \int_\Omega \sum_{|\alpha| \leq m} D^\alpha u(\mathbf{x}) D^\alpha v(\mathbf{x}) d\Omega, \qquad (6.2.22)$$

for all $u, v \in H^m(\Omega)$. We note that for $m = 0$, we have $H^0(\Omega) = L_2(\Omega)$. Some special cases of Eq. (6.2.22) are provided by

$$(u, v)_0 = \int_\Omega uv \, dxdy, \quad \|u\|_0 = \sqrt{(u, u)_0}, \qquad (6.2.23)$$

$$(u, v)_1 = \int_\Omega \left(uv + \frac{\partial u}{\partial x}\frac{\partial v}{\partial x} + \frac{\partial u}{\partial y}\frac{\partial v}{\partial y} \right) dxdy, \quad \|u\|_1 = \sqrt{(u, u)_1}, \qquad (6.2.24)$$

$$(u, v)_2 = \int_\Omega \left(uv + \frac{\partial u}{\partial x}\frac{\partial v}{\partial x} + \frac{\partial u}{\partial y}\frac{\partial v}{\partial y} + \frac{\partial^2 u}{\partial x \partial y}\frac{\partial^2 v}{\partial x \partial y} + \frac{\partial^2 u}{\partial x^2}\frac{\partial^2 v}{\partial x^2} \right.$$
$$\left. + \frac{\partial^2 u}{\partial y^2}\frac{\partial^2 v}{\partial y^2} \right) dxdy, \quad \|u\|_2 = \sqrt{(u, u)_2}. \qquad (6.2.25)$$

A linear vector space on which an inner product can be defined is called an *inner product space*. A linear subspace S of an inner product space V is a subspace with the inner product of V. Note that the square root of the inner product of a vector with itself satisfies the axioms of a norm. Consequently, one can associate with every inner product in vector space V a norm

$$\|u\|_V = \sqrt{(u, u)_V}. \qquad (6.2.26)$$

The norm thus obtained is called the *norm induced by the inner product*. Since we can associate with each inner product a norm, every inner product space is also a normed vector space. It should be obvious to the reader that the converse does not hold in general.

Example 6.2.3

Consider the functions
$$u(x) = \cos x, \quad v(x) = x \quad \text{in } 0 \le x \le \pi \tag{1}$$
and
$$w(x, y) = \sin \pi x \sin \pi y \quad \text{in } 0 \le x \le 1, \ 0 \le y \le 1. \tag{2}$$

Compute $\|u\|_0$, $\|v\|_0$, $\|w\|_0$, $(u, v)_0$, $\|u + v\|_0$, and $d(u, v) = \|u - v\|_0$.

Solution: We have

$$\|u\|_0^2 = \int_0^\pi \cos^2 x \, dx = \frac{1}{2} \int_0^\pi (1 + \cos 2x) \, dx = \frac{\pi}{2}, \tag{3}$$

$$\|v\|_0^2 = \int_0^\pi x^2 \, dx = \frac{\pi^3}{3}, \tag{4}$$

$$\|w\|_0^2 = \int_0^1 \int_0^1 \sin^2 \pi x \sin^2 \pi y \, dx dy = \int_0^1 \sin^2 \pi x \, dx \int_0^1 \sin^2 \pi y \, dy,$$
$$= \frac{1}{2} \int_0^1 (1 - \cos 2\pi x) \, dx \, \frac{1}{2} \int_0^1 (1 - \cos 2\pi y) \, dy = \frac{1}{2} \frac{1}{2} = \frac{1}{4}, \tag{5}$$

$$(u, v)_0 = \int_0^\pi x \cos x \, dx = [x \sin x]_0^\pi - \int_0^\pi \sin x \, dx = [x \sin x + \cos x]_0^\pi = -2, \tag{6}$$

$$\|u + v\|_0^2 = \int_0^\pi (\cos x + x)^2 \, dx = \int_0^\pi (\cos^2 x + x^2 + 2x \cos x) \, dx = \frac{\pi}{2} + \frac{\pi^3}{3} - 4, \tag{7}$$

$$\|u - v\|_0^2 = \int_0^\pi (\cos x - x)^2 \, dx = \frac{\pi}{2} + \frac{\pi^3}{3} + 4. \tag{8}$$

6.2.3.3   Orthogonality

Two vectors $u, v \in \mathcal{V}$ are said to be *orthogonal* if

$$(u, v)_V = 0, \tag{6.2.27}$$

where $(\cdot, \cdot)_V$ denotes an inner product in $\mathcal{V}$. Note that the concept of orthogonality is a generalization of the familiar notion of perpendicularity of one vector to another in Euclidean space. A set of mutually orthogonal vectors is called an orthogonal set.

A sequence of functions $\{\phi_i\}$ in $L_2(\Omega)$ is called *orthonormal* if

$$(\phi_i, \phi_j)_V = \delta_{ij} = \begin{cases} 1, & \text{if } i = j, \\ 0, & \text{if } i \ne j. \end{cases} \tag{6.2.28}$$

Here δ_{ij} denotes the Kronecker delta. It can be shown that every orthonormal system is linearly independent. If two vectors u and v of an inner product space $\mathcal{V}$ are orthogonal, then the Pythagorean theorem holds even in function spaces:

$$\|u + v\|_V^2 = (u + v, u + v)_V = (u, u)_V + 2(u, v)_V + (v, v)_V = \|u\|_V^2 + \|v\|_V^2.$$

A set of functions $\{\phi_i\}$ is said to be *complete* in $L_2(\Omega)$ if every piecewise continuous function f can be approximated in Ω by the sum $\sum_{j=1}^{N} c_j \phi_j$ in such a way that

$$\mathcal{E}_n \equiv \int_\Omega \left(f - \sum_{j=1}^{N} c_j \phi_j \right)^2 d\Omega \qquad (6.2.29)$$

can be made as small as we wish (by increasing N). This property of the coordinate functions is the key to the proof of the convergence of the Ritz and Galerkin approximations to be discussed in the later sections of this chapter.

A complete (in its natural metric) inner product space is called a *Hilbert space*. We mention without proof the fact that every inner product space (hence a normed space) has a completion.

The following lemma, referred to as the *fundamental lemma of variational calculus*, which was already introduced in Chapter 3, plays an important role in variational theory.

Lemma 6.2.1: Let $\mathcal{V}$ be an inner product space. If $(u, v)_V = 0$ for all $v \in \mathcal{V}$, then $u = 0$.

Proof: Since $(u, v)_V = 0$ for all v, it must also hold for $v = u$. Then $(u, u)_V = 0$ implies that $u = 0$.

6.2.4 Transformations, and Linear and Bilinear Forms

A transformation T from a linear vector space $\mathcal{U}$ into another linear vector space $\mathcal{V}$ (both vector spaces are defined on the same field of scalars) is a correspondence that assigns to each element $u \in \mathcal{U}$ a unique element $v = Tu \in \mathcal{V}$. We use the terms *transformation, mapping*, and *operator* interchangeably, and the transformation is expressed as $T : \mathcal{U} \to V$.

A transformation $T : \mathcal{U} \to V$, where $\mathcal{U}$ and $\mathcal{V}$ are vector spaces that have the same scalar field, is said to be *linear* if

1. $T(\alpha u) = \alpha T(u)$, for all $u \in \mathcal{U}$, $\alpha \in \Re$ (homogeneous).
2. $T(u_1 + u_2) = T(u_1) + T(u_2)$, for all $u_1, u_2 \in \mathcal{U}$, (additive).

The two statements can be combined into a single statement as follows:

$$T(\alpha u_1 + \beta u_2) = \alpha T(u_1) + \beta T(u_2), \quad \text{for all } u_1, u_2 \in \mathcal{U}, \ \alpha, \beta \in \Re \quad (6.2.30)$$

Otherwise, it is said to be a *nonlinear transformation*.

Transformations that map vectors (functions) from a linear vector space into real numbers are of special interest in the present study. Such transformations (linear or not) are called *functionals*. A linear transformation $\ell : V \to \Re$ that maps a linear vector space $\mathcal{V}$ into the real number field $\Re$ is called a *linear functional*. Similarly, a linear transformation that maps pairs of vectors

$(u, v) \in V \times V$ into real number field $\Re$, or $B(\cdot, \cdot) : V \times V \to \Re$, is called a *bilinear form.* Examples of linear and bilinear forms are provided by

$$\ell(u) = \int_a^b f u \, dx, \quad B(u, v) = \int_a^b \left(\frac{du}{dx} \frac{dv}{dx} + uv \right) dx.$$

A bilinear form is said to be *symmetric* if it is symmetric in its arguments:

$$B(u, v) = B(v, u). \tag{6.2.31}$$

6.2.5   Minimum of a Quadratic Functional

Consider an operator equation of the form

$$Au = f, \tag{6.2.32}$$

where A is a certain operator (often a differential operator), $A : \mathcal{D}_A \to H$, and $f \in H$ is a given function. Here $\mathcal{D}_A$ denotes a set of elements from a Hilbert space H. The denseness of $\mathcal{D}_A$ in H is often assumed but we will not discuss this topic in the present study.

The differential equations (familiar equations from previous chapters)

$$-\frac{d}{dx}\left(EA_0 \frac{du}{dx} \right) = f(x), \quad EA_0 > 0, \quad 0 < x < L,$$

$$\frac{d^2}{dx^2}\left(EI \frac{d^2 u}{dx^2} \right) = f(x), \quad EI > 0, \quad 0 < x < L$$

are special cases of the operator equation in Eq. (6.2.32) with

$$A = -\frac{d}{dx}\left(EA_0 \frac{d(\cdot)}{dx} \right), \quad A = \frac{d^2}{dx^2}\left(EI \frac{d^2(\cdot)}{dx^2} \right),$$

respectively. In these cases $H = L_2(0, L)$. The set $\mathcal{D}_A$ for the first equation consists of functions from $\mathcal{C}^2(0, L)$ and for the second equation, the functions are from $\mathcal{C}^4(0, L)$.

An operator $A : \mathcal{D}_A \to H$ is called *symmetric* (or *self-adjoint*) if

$$(Au, v)_H = (u, Av)_H \tag{6.2.33}$$

holds for all $u, v \in \mathcal{D}_A$, where $(\cdot, \cdot)_H$ is the inner product in H. An operator A is called *strictly positive* in $\mathcal{D}_A$ if it is symmetric in $\mathcal{D}_A$ and if

$$(Au, u)_H > 0 \text{ holds for all } u \in \mathcal{D}_A \text{ and } u \neq 0,$$
$$(Au, u)_H = 0 \text{ if and only if } u \in \mathcal{D}_A \text{ and } u = 0. \tag{6.2.34}$$

A *quadratic functional* $Q : H \to \Re$ is one that is quadratic in its arguments, $Q(\alpha u) = \alpha^2 Q(u)$ for $\alpha \in \Re$. Every bilinear form $B(\cdot, \cdot)$ can be used to generate a quadratic form Q by setting

$$Q(u) = B(u, u), u \in H. \tag{6.2.35}$$

The following results are of fundamental importance for the present study.

Theorem 6.1: (*uniqueness*) If A is a strictly positive operator in $\mathcal{D}_A$, then

$$Au = f \text{ in } H$$

has at most one solution $u \in \mathcal{D}_A$ in H.

Proof: Suppose that there exist two solutions $u_1, u_2 \in \mathcal{D}_A$. Then

$$Au_1 = f \text{ and } Au_2 = f \;\to\; A(u_1 - u_2) = 0 \text{ in } H,$$

and

$$(A(u_1 - u_2), u_1 - u_2)_H = 0 \;\to\; u_1 - u_2 = 0 \text{ or } u_1 = u_2 \in \mathcal{D}_A,$$

which was to be proved.

Theorem 6.2: (*Minimum of a quadratic functional*) Let A be a positive operator in $\mathcal{D}_A$, $f \in H$. Let Eq. (6.2.32) have a solution $u_0 \in \mathcal{D}_A$. Then the quadratic functional

$$I(u) = \frac{1}{2}(Au, u)_H - (f, u)_H \tag{6.2.36}$$

assumes its minimal value in $\mathcal{D}_A$ for the element u_0, that is,

$$I(u) \geq I(u_0), \quad \text{and} \quad I(u) = I(u_0) \text{ only for } u = u_0.$$

Conversely, if $I(u)$ assumes its minimal value, among all $u \in \mathcal{D}_A$, for the element u_0, then u_0 is the solution of Eq. (6.2.32) (i.e., $Au_0 = f$).

Proof: First note that $I(u)$ is defined for all $u \in \mathcal{D}_A$. Let u_0 be the solution of Eq. (6.2.32). Then $f = Au_0$. Substituting for f into Eq. (6.2.36), we obtain for $u \in \mathcal{D}_A$

$$
\begin{aligned}
I(u) &= \tfrac{1}{2}(Au, u)_H - (Au_0, u)_H \\
&= \tfrac{1}{2}\left[(Au, u)_H - (Au_0, u)_H - (u, Au_0)_H\right] \\
&= \tfrac{1}{2}\left[(Au, u)_H - (Au_0, u)_H - (Au, u_0)_H\right] \\
&= \tfrac{1}{2}\left[(Au, u)_H - (Au_0, u)_H - (Au, u_0)_H + (Au_0, u_0)_H - (Au_0, u_0)_H\right] \\
&= \tfrac{1}{2}\left[(A(u - u_0), u - u_0)_H - (Au_0, u_0)_H\right], \tag{6.2.37}
\end{aligned}
$$

where the linearity and symmetry of A, as well as the symmetry of the bilinear form, are used in arriving at the last step. From Eq. (6.2.37) it follows that

$$I(u_0) = -\tfrac{1}{2}(Au_0, u_0)_H. \qquad (6.2.38)$$

Next, we use the strictly positive property of A to conclude that

$$I(u) \geq I(u_0) \quad \text{for} \quad u \in \mathcal{D}_A, \quad \text{and} \quad I(u) = I(u_0)$$

if and only if $u = u_0$ in $\mathcal{D}_A$. Consequently, if the equation $Au_0 = f$ is satisfied, then the functional $I(u)$ assumes its minimal value in $\mathcal{D}_A$ precisely for the element $u = u_0$.

Now suppose that $I(u)$ assumes its minimal value in $\mathcal{D}_A$ for the element u_0. This implies that

$$I(u_0 + \alpha v) \geq I(u_0) \text{ for } \alpha \in \mathfrak{R}, v \in \mathcal{D}_A. \qquad (6.2.39)$$

Using again the symmetry of A and the symmetry of the inner product, one obtains

$$\begin{aligned}
I(u_0 + \alpha v) &= \tfrac{1}{2}(A(u_0 + \alpha v), u_0 + \alpha v)_H - (f, u_0 + \alpha v)_H \\
&= \tfrac{1}{2}\left[(Au_0 + \alpha Av, u_0 + \alpha v)_H - 2(f, u_0)_H - 2\alpha(f, v)_H\right] \\
&= \tfrac{1}{2}\left[(Au_0, u_0)_H + \alpha(Av, u_0)_H + \alpha(Au_0, v)_H \right. \\
&\quad \left. + \alpha^2(Av, v)_H - 2(f, u_0)_H - 2\alpha(f, v)_H\right] \\
&= \tfrac{1}{2}\left[(Au_0, u_0)_H + 2\alpha(Au_0, v)_H + \alpha^2(Av, v)_H \right. \\
&\quad \left. - 2(f, u_0)_H - 2\alpha(f, v)_H\right]. \qquad (6.2.40)
\end{aligned}$$

Since $u_0 \in \mathcal{D}_A$ and $f \in H$ are fixed elements, it is obvious that for arbitrarily fixed $v \in \mathcal{D}_A$ the function $I(u_0 + \alpha v)$ is a quadratic function in the variable α. From Eq. (6.2.39) it follows that the function has a local minimum at $\alpha = 0$, which implies that its first derivative is equal to zero at $\alpha = 0$ [or, equivalently, the first variation of I is zero; see Eq. (3.5.31)]

$$\left[\frac{d}{d\alpha} I(u_0 + \alpha v)\right]_{\alpha=0} = 0,$$

or by Eq. (6.2.40) that

$$(Au_0, v)_H - (f, v)_H = 0 \quad \text{or} \quad (Au_0 - f, v)_H = 0.$$

Since $v \in H$ is arbitrary, by Lemma 6.2.1 it follows that $Au_0 - f = 0$ in H. This completes the proof.

Example 6.2.4

Consider the differential equation

$$-\frac{d}{dx}\left(a(x)\frac{du}{dx}\right) = f(x), \quad a(x) > 0, \quad 0 < x < L, \tag{1}$$

subjected to the boundary conditions

$$u(0) = 0, \quad u(L) = 0, \tag{2}$$

which arises in connection with the transverse deflection of cables. Here $u(x)$ denotes the deflection of a cable of original length L, tension $a = a(x)$, and subjected to distributed transverse load $f(x)$ (see Fig. 6.2.2). The boundary conditions in Eq. (2) indicate that the cable is fixed at $x = 0$ and $x = L$.

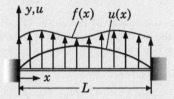

Fig. 6.2.2 Transverse deflection of a cable fixed at its ends (steady-state configuration).

Let us choose $H = L_2(0, L)$, define $\mathcal{D}_A$ as the linear set of functions that are continuous with their derivatives up to the second-order inclusive in the interval $[0, L]$, and satisfy the end conditions in Eq. (2). Define the operator A on $\mathcal{D}_A$ by

$$Au = -\frac{d}{dx}\left(a(x)\frac{du}{dx}\right). \tag{3}$$

We now set out to prove that A is strictly positive on $\mathcal{D}_A$. First, we note that A is symmetric in $\mathcal{D}_A$. For every $u \in \mathcal{D}_A$ and $v \in \mathcal{D}_A$, we have

$$
\begin{aligned}
(Au, v)_H &= \int_0^L \left[-\frac{d}{dx}\left(a\frac{du}{dx}\right)\right] v\, dx \\
&= -\left[a\frac{du}{dx}v\right]_0^L + \int_0^L \frac{dv}{dx}\left(a\frac{du}{dx}\right) dx \\
&= \int_0^L a(x)\frac{dv}{dx}\frac{du}{dx}\, dx \tag{4} \\
&= \left[a\frac{dv}{dx}u\right]_0^L + \int_0^L \left[-\frac{d}{dx}\left(a\frac{dv}{dx}\right)\right] u\, dx \\
&= \int_0^L \left[-\frac{d}{dx}\left(a\frac{dv}{dx}\right)\right] u\, dx = (u, Av)_H, \tag{5}
\end{aligned}
$$

where we have used the fact that $u(0) = u(L) = v(0) = v(L) = 0$. Thus, A is symmetric on $\mathcal{D}_A$. From Eq. (4), it follows that

$$(Au, u)_H = \int_0^L a(x)\left(\frac{du}{dx}\right)^2 dx \quad \text{for all } u \in \mathcal{D}_A. \tag{6}$$

Due to the fact that $a(x) > 0$ in $[0, L]$, it follows from (6) that

$$(Au, u)_H \geq 0 \quad \text{for every } u \in \mathcal{D}_A. \tag{7}$$

Moreover, if $(Au, u)_H = 0$, it follows that

$$\frac{du}{dx} = 0 \quad \text{in } [0, L].$$

This in turn implies that

$$u(x) = c, \text{ constant in } [0, L].$$

Since $u(0) = 0$, it follows that $c = 0$ or $u(x) = 0$ in $[0, L]$. This proves that A is strictly positive in $\mathcal{D}_A$. Further, $(u, u)_A \equiv (Au, u)_H$ is the energy norm.

The quadratic functional associated with Eqs (1) and (2) is, according to Eq. (6.2.36),

$$\Pi(u) = \tfrac{1}{2}(Au, u)_H - (f, u)_H = \frac{1}{2} \int_0^L a(x) \left(\frac{du}{dx} \right)^2 dx - \int_0^L f u \, dx, \tag{8}$$

which represents the total potential energy of the cable. The first term is the elastic strain energy and the second term is the potential energy of the external load $f(x)$.

Let $u_0(x)$ be the solution of Eqs (1) and (2). Then, $f = Au_0$ and we have

$$\begin{aligned}
\Pi(u) &= \tfrac{1}{2}(Au, u)_H - (Au_0, u)_H \\
&= \frac{1}{2} \int_0^L a(x) \left(\frac{du}{dx} \right)^2 dx - \int_0^L \left[-\frac{d}{dx} \left(a(x) \frac{du_0}{dx} \right) \right] u \, dx \\
&= \frac{1}{2} \int_0^L a(x) \left(\frac{du}{dx} \right)^2 dx - \int_0^L a(x) \frac{du_0}{dx} \frac{du}{dx} \, dx \\
&= \frac{1}{2} \int_0^L a(x) \left(\frac{du}{dx} - \frac{du_0}{dx} \right)^2 dx - \frac{1}{2} \int_0^L a(x) \left(\frac{du_0}{dx} \right)^2 dx. \tag{9}
\end{aligned}$$

Since $a(x)(u' - u_0')^2 \geq 0$, it is clear from Eq. (9) that $\Pi(u)$ is minimal in $\mathcal{D}_A$ if and only if $u' = u_0'$ in $\mathcal{D}_A$. Thus, if $u_0 \in \mathcal{D}_A$ is the solution of Eqs (1) and (2), then the functional $\Pi(u)$ assumes its minimum for $u_0 \in \mathcal{D}_A$.

Conversely, let $u_0 \in \mathcal{D}_A$ be the element minimizing the functional $\Pi(u)$ in Eq. (8). Let $v \in \mathcal{D}_A$ be an arbitrary element from $\mathcal{D}_A$ and let α be an arbitrary real number. Then the minimum of $\Pi(u)$ implies that $(u = u_0 + \alpha v)$

$$\begin{aligned}
0 &= \frac{d}{d\alpha} \Pi(u_0 + \alpha v) \Big|_{\alpha=0} \\
&= \frac{d}{d\alpha} \left[\frac{1}{2} \int_0^L a(x) \left(\frac{du}{dx} \right)^2 dx - \int_0^L f u \, dx \right]_{\alpha=0} \\
&= \int_0^L a(x) \frac{du_0}{dx} \frac{dv}{dx} \, dx - \int_0^L f v \, dx \\
&= \int_0^L \left[-\frac{d}{dx} \left(a(x) \frac{du_0}{dx} \right) - f \right] v \, dx.
\end{aligned}$$

Since this result must hold for every v, it follows that

$$-\frac{d}{dx} \left[a(x) \frac{du_0}{dx} \right] - f(x) = 0 \quad \text{in } \Omega = (0, L), \tag{10}$$

which is the Euler equation associated with the functional $\Pi(u)$ in Eq. (8). These arguments are equivalent to the principle of minimum total potential energy discussed in Chapter 4.

Recall that the operator $A : \mathcal{D}_A \to H$ is symmetric,

$$(Au, v)_H = (u, Av)_H \quad \text{for} \quad u, v \in \mathcal{D}_A,$$

and positive definite on $\mathcal{D}_A$, that is, there exists a constant $\mu > 0$ so that

$$(Au, u)_H \geq \mu \|u\|^2 \text{ holds for every } u \in \mathcal{D}_A. \tag{6.2.41}$$

Hence, we can define a new inner product $(u, v)_A$ on $\mathcal{D}_A$ as follows:

$$(u, v)_A = (Au, v)_H \quad \text{for all } u, v \in \mathcal{D}_A. \tag{6.2.42}$$

It can be easily verified that $(u, v)_A$ satisfies the axioms (1)–(3) in Eq. (6.2.21) of an inner product. The linear set $\mathcal{D}_A$ with the inner product in Eq. (6.2.42) constitutes a linear vector space, called the *energy space*, and denoted by H_A. The norm and natural metric follow from the definition in Eq. (6.2.42):

$$\|u\|_A^2 = (u, u)_A, \quad d(u, v) = \|u - v\|_A. \tag{6.2.43}$$

The norm $\|u\|_A^2$ is called the *energy norm*.

The energy space H_A can be shown to be complete with respect to the metric defined in Eq. (6.2.43), and hence it is a Hilbert space. Moreover, it can be shown that the functional $\Pi(u)$ can be extended to all elements of H_A, that the functional assumes its minimum at $u_0 \in H_A$, and that the element u_0 is uniquely determined by the element $f \in H$. These proofs are beyond the scope of the present study, and interested readers may consult References [39–41, 49, 52].

6.3 The Ritz Method

6.3.1 Introduction

As discussed in Chapters 4 and 5, the principles of virtual displacements and forces as applied to continuous systems can be used to determine the governing equations and natural boundary conditions of the problem. The energy methods (e.g., unit dummy-load and unit dummy-displacement methods and Castigliano's theorems I and II) derived from these principles were used to determine deflections and forces at selected points. Here we consider a powerful method of determining approximate solutions to the governing equations of a problem by directly using the variational statements (i.e., the virtual work principles, the principle of total potential energy, or the principle of complementary energy). The method bypasses the derivation of the Euler equations and goes directly from a variational statement of the problem to the solution of the Euler equations. One such direct method was proposed by German engineer W. Ritz (1878–1909).

6.3.2   Description of the Method

Consider the linear operator equation

$$Au = f \quad \text{in} \quad \Omega, \tag{6.3.1}$$

where A is a strictly positive operator from H_A into H and $f \in H$. The solution u_0 of Eq. (6.3.1) is the element $u_0 \in H_A$ that minimizes the quadratic functional

$$I(u) = \tfrac{1}{2}(u, u)_A - \ell(u), \tag{6.3.2}$$

where $\ell(\cdot)$ denotes a linear functional.

In structural mechanics problems, the functional $I(u)$ represents the total potential energy, and the principle of minimum total potential energy, $\delta I(u) = 0$, yields Eq. (6.3.1) as the Euler equation. Of course, one should have either $I(u)$ or the governing equation $Au = f$ for the problem at hand. The Ritz method [4, 103] is based on the idea that $I(u)$ exists. It is already established that minimization of $I(u)$ is equivalent to seeking solution to the equation $Au = f$.

We seek an approximation $U_N(x)$ of the solution $u(x)$ of Eq. (6.3.1) with associated boundary conditions, in the form

$$u(x) \approx U_N(x) = \sum_{i=1}^{N} c_i \phi_i(x) + \phi_0(x), \tag{6.3.3}$$

where $\phi_i(x)$ are the elements of a basis in H_A and c_i are yet unknown real parameters. The number of parameters, N, is preselected, and one expects to obtain better and better solution as N is increased (a property of the approximation known as the *convergence*). The parameters c_i $(i = 1, 2, \ldots, N)$ are determined by the condition that $I(U_N)$ is the minimum (i.e., parameters c_i are adjusted or *varied* to make $I(U_N)$ a minimum). Since $\{\phi_i\}$ is a basis in H_A, the solution $u_0 \in H_A$ can be approximated to arbitrary accuracy by a suitable linear combination of its elements. Therefore, it can be expected that the approximate solution U_N, with constants determined by minimizing the functional $I(U_N)$, will be sufficiently close in H_A to the actual solution u_0 if N is selected sufficiently large. This process of determining U_N by minimizing $I(U_N)$ (or requiring $\delta I(U_N) = 0$) is known as the Ritz method.

Equation (6.3.3) can be viewed as a representation of u in a component form. The parameters c_i are the *components* (or coordinates) and $\{\phi_i\}$ are the *coordinate functions*. Another interpretation of Eq. (6.3.3) is provided by the finite Fourier series, in which c_i are known as the *Fourier coefficients*.

The basic idea of the Ritz method is illustrated here with a specific operator equation $Au = f$. We consider the axial deformation of a bar of nonuniform cross-sectional area $A_0 = A_0(x)$, length a, and modulus of elasticity E; the bar is fixed at the left end and axially supported at the right end by a linearly elastic spring of stiffness coefficient k. The governing equation is

$$-\frac{d}{dx}\left(EA_0\frac{du}{dx}\right) = f(x) \quad \left[A \equiv -\frac{d}{dx}\left(EA_0\frac{d}{dx}\right)\right], \quad 0 < x < a, \quad (6.3.4)$$

subjected to the boundary conditions

$$u(0) = 0, \quad \left[EA_0\frac{du}{dx} + ku\right]_{x=a} = P \quad (6.3.5)$$

where $f(x)$ is the distributed axial load and P is the point axial load at $x = a$ (see Fig. 6.3.1). The space H_A in this case is the completion of the set $\mathcal{D}_A$ of functions that are continuous with their derivatives up to the second order in $[0, a]$.

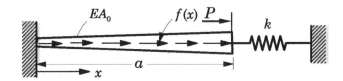

Fig. 6.3.1 Axial deformation of a non-uniform bar fixed at the left end and supported axially by an end spring.

As discussed in **Example 3.5.3**, the problem of solving Eqs (6.3.4) and (6.3.5) is equivalent to minimizing the total potential energy functional Π [i.e., Eqs (6.3.4) and (6.3.5) are the Euler equations of the functional $\Pi(u)$]:

$$\Pi(u) = \int_0^a \left[\frac{EA_0}{2}\left(\frac{du}{dx}\right)^2 - fu\right] dx + \frac{k}{2}[u(a)]^2 - Pu(a). \quad (6.3.6)$$

The exercise of deriving Π from U and V_E (i.e., $\Pi = U + V_E$) for solid mechanics problems was already discussed in Chapters 4 and 5. The construction of functional $\Pi(u)$ from the governing equations, Eqs (6.3.4) and (6.3.5), will be discussed in Section 6.3.6.

The necessary condition for the minimum of Π is

$$0 = \delta\Pi = \int_0^a \left(EA_0\frac{du}{dx}\frac{d\delta u}{dx} - f\,\delta u\right) dx + ku(a)\delta u(a) - P\,\delta u(a)$$

$$\equiv B(\delta u, u) - \ell(\delta u) \quad (6.3.7)$$

or

$$B(\delta u, u) = \ell(\delta u), \quad (6.3.8)$$

where $B(\cdot, \cdot)$ is the bilinear functional and $\ell(\cdot)$ is the linear functional:

$$B(\delta u, u) = \int_0^a E A_0 \frac{du}{dx} \frac{d\delta u}{dx} \, dx + ku(a)\delta u(a),$$

$$\ell(\delta u) = \int_0^a f \delta u \, dx + P \delta u(a). \qquad (6.3.9)$$

The fact that both $B(\cdot, \cdot)$ and $\ell(\cdot)$ are linear in δu is obvious. The linearity of $B(\cdot, \cdot)$ in u depends on the governing equation. Whenever the governing equation is linear in u, $B(\cdot, \cdot)$ is linear u.

Equation (6.3.8) is known as the *variational problem* associated with Eqs (6.3.4) and (6.3.5). The inner product and the energy norm in H_A are defined by

$$(u, v)_A = B(u, v), \quad \|u\|_A = (u, u)_A. \qquad (6.3.10)$$

Returning to the Ritz method, we seek an approximate solution U_N to the problem as a finite linear combination of the form in Eq. (6.3.3). The particular form of the approximate solution resembles that of the exact solution to any differential equation: the homogeneous solution and the particular solution. That is, the first part $\sum_{j=1}^N c_j \phi_j(x)$ can be viewed as the homogeneous part and the second term $\phi_0(x)$ as the particular part. This analogy helps understand the requirements on ϕ_0 and ϕ_j ($j = 1, 2, \ldots, N$). Since the total potential energy functional in Eq. (6.3.6) or the variational problem in Eq. (6.3.8) already account for the specified natural boundary condition in Eq. (6.3.5), the approximate solution U_N must only satisfy the specified essential boundary condition, $u(0) = 0$. This requirement on U_N will in turn place certain conditions on ϕ_0 and ϕ_j, as will be discussed shortly.

Substitution of U_N from Eq. (6.3.3) into the total potential energy functional Π in Eq. (6.3.6) and its minimization with respect to c_i ($i = 1, 2, \ldots, N$) or substitution into the variational problem in Eq. (6.3.8) will result in a set of linear algebraic equations among the parameters c_i ($i = 1, 2, \ldots, N$). First, consider substitution of U_N into the total potential energy functional Π. Then Π becomes a function of the parameters $c_1, c_2, \ldots, c_N$:

$$\Pi(c_1, c_2, \ldots, c_N) = \int_0^a \left[\frac{E A_0}{2} \left(\sum_{j=1}^N c_j \frac{d\phi_j}{dx} + \frac{d\phi_0}{dx} \right)^2 - f \left(\sum_{j=1}^N c_j \phi_j + \phi_0 \right) \right] dx$$

$$+ \frac{k}{2} \left(\sum_{j=1}^N c_j \phi_j(a) + \phi_0(a) \right)^2 - P \left(\sum_{j=1}^N c_j \phi_j(a) + \phi_0(a) \right). \qquad (6.3.11)$$

Then the necessary condition for the minimum of a function of several variables is that its first derivative with respect to each of the variables be zero:

$$\frac{\partial \Pi}{\partial c_i} = 0 \quad \text{for} \quad i = 1, 2, \ldots, N. \qquad (6.3.12)$$

Carrying out the indicated differentiation of Π with respect to c_1, c_2, ..., c_N, we obtain N algebraic equations:

$$0 = \frac{\partial \Pi}{\partial c_1} = \int_0^a \left[EA_0 \frac{d\phi_1}{dx} \left(\sum_{j=1}^N c_j \frac{d\phi_j}{dx} + \frac{d\phi_0}{dx} \right) - f\phi_1 \right] dx$$

$$+ k\phi_1(a) \left(\sum_{j=1}^N c_j \phi_j(a) + \phi_0(a) \right) - P\phi_1(a), \qquad \text{(Eq.1)}$$

$$0 = \frac{\partial \Pi}{\partial c_2} = \int_0^a \left[EA_0 \frac{d\phi_2}{dx} \left(\sum_{j=1}^N c_j \frac{d\phi_j}{dx} + \frac{d\phi_0}{dx} \right) - f\phi_2 \right] dx$$

$$+ k\phi_2(a) \left(\sum_{j=1}^N c_j \phi_j(a) + \phi_0(a) \right) - P\phi_2(a), \qquad \text{(Eq.2)}$$

$$\cdots\cdots\cdots\cdots\cdots\cdots\cdots\cdots\cdots\cdots$$

$$0 = \frac{\partial \Pi}{\partial c_i} = \int_0^a \left[EA_0 \frac{d\phi_i}{dx} \left(\sum_{j=1}^N c_j \frac{d\phi_j}{dx} + \frac{d\phi_0}{dx} \right) - f\phi_i \right] dx$$

$$+ k\phi_i(a) \left(\sum_{j=1}^N c_j \phi_j(a) + \phi_0(a) \right) - P\phi_i(a), \qquad \text{(Eq.}i)$$

$$\cdots\cdots\cdots\cdots\cdots\cdots\cdots\cdots\cdots\cdots$$

$$0 = \frac{\partial \Pi}{\partial c_N} = \int_0^a \left[EA_0 \frac{d\phi_N}{dx} \left(\sum_{j=1}^N c_j \frac{d\phi_j}{dx} + \frac{d\phi_0}{dx} \right) - f\phi_N \right] dx$$

$$+ k\phi_N(a) \left(\sum_{j=1}^N c_j \phi_j(a) + \phi_0(a) \right) - P\phi_N(a). \qquad \text{(Eq.}N)$$

In particular, the ith equation (numbered as Eq. i) of the set of N equations is

$$0 = \int_0^a \left[EA_0 \frac{d\phi_i}{dx} \left(\sum_{j=1}^N c_j \frac{d\phi_j}{dx} + \frac{d\phi_0}{dx} \right) - f\phi_i \right] dx + k\phi_i(a) \left(\sum_{j=1}^N c_j \phi_j(a) + \phi_0(a) \right)$$

$$- P\phi_i(a)$$

$$0 = \sum_{j=1}^{N} \left\{ \int_0^a EA_0 \frac{d\phi_i}{dx} \frac{d\phi_j}{dx} \, dx + k\phi_i(a)\phi_j(a) \right\} c_j$$

$$- \int_0^a \left(-EA_0 \frac{d\phi_0}{dx} \frac{d\phi_i}{dx} + f\phi_i \right) dx - P\phi_i(a) + k\phi_0(a)\phi_i(a)$$

$$\equiv \sum_{j=1}^{N} A_{ij} c_j - b_i \tag{6.3.13}$$

where

$$A_{ij} = B(\phi_i, \phi_j) = \int_0^a EA_0 \frac{d\phi_i}{dx} \frac{d\phi_j}{dx} \, dx + k\phi_i(a)\phi_j(a),$$

$$b_i = \ell(\phi_i) = - \int_0^a EA_0 \frac{d\phi_0}{dx} \frac{d\phi_i}{dx} \, dx - k\phi_0(a)\phi_i(a) + \int_0^a f\phi_i \, dx + P\phi_i(a). \tag{6.3.14}$$

In matrix form, the N linear equations can be expressed as

$$\mathbf{Ac} = \mathbf{b}. \tag{6.3.15}$$

Equation (6.3.15) can also be arrived by substituting the Ritz approximation (6.3.3) and its variation

$$\delta u \approx \delta U_N = \sum_{i=1}^{N} \delta c_i \, \phi_i(x) \tag{6.3.16}$$

into the variational statement $\delta\Pi = 0$, instead of substituting Eq. (6.3.3) in Π and then taking its variation with respect to c_i. This results in

$$0 = \sum_{i=1}^{N} \delta c_i \left\{ \int_0^a \left[EA_0 \frac{d\phi_i}{dx} \left(\sum_{j=1}^{N} c_j \frac{d\phi_j}{dx} + \frac{d\phi_0}{dx} \right) - f\phi_i \right] dx \right.$$

$$\left. + k\phi_i(a) \left(\sum_{j=1}^{N} c_j \phi_j(a) + \phi_0(a) \right) - P\phi_i(a) \right\}. \tag{6.3.17}$$

Since δc_i are arbitrary and linearly independent of each other, the expression in the curly brackets should vanish and we obtain the ith equation, Eq. (6.3.13), of the set. We note that, in view of the fact that $\sum_{i=1}^{N} \delta c_i$ in Eq. (6.3.17) does not enter the final result, the ith equation of the system can be obtained directly from the statement $\delta\Pi = 0$ by substituting Eq. (6.3.3) for u and ϕ_i for δu.

A specific example of the application of the equations derived here will be presented shortly.

6.3.3 Properties of Approximation Functions

6.3.3.1 Preliminary Comments

The most challenging part of using the Ritz method (and the weighted residual methods to be discussed in the next section) is the selection of the *approximation functions* ϕ_0 and ϕ_i. While we do not present any formulas (because such formulas do not exist) for their derivation, certain guidelines are provided here that help the user to derive them. In fact, this (i.e., not having a systematic procedure to derive ϕ_0 and ϕ_i) is the main reason for giving birth to the finite element method and its generalizations.

Two most important considerations in selecting ϕ_0 and ϕ_i are the following: (1) U_N satisfies the specified essential boundary conditions of the problem and (2) the substitution of U_N into $\delta\Pi = 0$ (or its equivalent) results in N linearly independent set of equations for the parameters c_j ($j = 1, 2, \ldots, N$) so that the system has a solution.

6.3.3.2 Boundary Conditions

We recall from the earlier discussions that every engineering problem has a duality between certain variables (e.g., force and displacement) that enter the mathematical model (i.e., the governing equations) of the problem. This duality in a problem can be identified when the Euler or Euler–Lagrange equations are derived in Chapters 3–5. To recap the discussion of Section 3.5.7 (see **Examples 3.5.3** and **3.5.4**), during the integration by parts to relieve the varied quantities of any differentiations (so that the fundamental lemma of the variational calculus can be used), we obtained boundary terms in the form (the boundary integral signs are valid only for two- or three-dimensional problems; for one-dimensional problems the integral sign is omitted but limits are added)

$$\oint_\Gamma [\text{Expression 2}] \cdot \delta[\text{Expression 1}] \, d\Gamma, \tag{6.3.18}$$

where δ is the variational symbol. Then we defined that **Expression 1** is the *primary variable* and **Expression 2** is the *secondary variable* (and their product often defines "work done"). The specification of a primary variable on the boundary is termed the *essential boundary condition*, and the specification of the secondary variable is called the *natural boundary condition*.

Since the natural boundary conditions of the problem are included in the variational statement, whether it is a functional like Π or an integral statement of the form $\delta\Pi = 0$, we require the approximate solution U_N to satisfy only the essential boundary conditions. In order that U_N satisfies the essential boundary conditions *for any* c_i, it is convenient to choose the approximate solution in the form of Eq. (6.3.3) and require $\phi_0(x)$ to satisfy the specified essential boundary conditions (i.e., the particular solution). For instance, if $u(x)$ is specified to be

$u(0) = \hat{u}$, then $U_N(0) = \hat{u}$. If we require $\phi_0(x)$ to be the lowest-order function that meets the condition $\phi_0(0) = \hat{u}$, then from Eq. (6.3.3) we have

$$U_N(0) = \sum_{i=1}^{N} c_i \phi_i(0) + \phi_0(0) = \sum_{i=1}^{N} c_i \phi_i(0) + \hat{u}. \tag{6.3.19}$$

Since $U_N(0) = \hat{u}$, it follows for arbitrary and nonzero c_i ($i = 1, 2, \ldots, N$) that

$$\sum_{i=1}^{N} c_i \phi_i(0) = 0 \;\Rightarrow\; \phi_i(0) = 0 \text{ for all } i = 1, 2, \ldots, N. \tag{6.3.20}$$

Thus, $\phi_i(x)$ must satisfy the *homogeneous form* of the specified essential boundary condition, $\phi_i(0) = 0$.

To ensure that the Ritz approximation $U_N(x)$ converges to the true solution $u(x)$ of the problem as the value of N is increased, ϕ_i ($i = 1, 2, \ldots, N$) and ϕ_0 must satisfy certain additional requirements. Before we list these requirements, it is informative to discuss the concepts of completeness of a set of functions and convergence of a sequence of approximations. To make the ideas presented simple to understand, mathematical rigor of the statements made is sacrificed.

6.3.3.3   Convergence

A sequence $\{U_N\}$ of functions is said to *converge* to u if for each $\epsilon > 0$ there is a number $M > 0$, depending on ϵ (a small number), such that

$$\|U_N(\mathbf{x}) - u(\mathbf{x})\| < \epsilon \quad \text{for all } N > M, \tag{6.3.21}$$

where $\| \cdot \|$ denotes a norm of the enclosed quantity and u is called the limit (function) of the sequence. In the above statement U_N is the approximate solution and u is the true solution. The statement implies that the N-parameter solution U_N can be made as close to u as we wish, say within ϵ, by choosing N to be greater than $M = M(\epsilon)$, provided that the approximate solution is convergent. While there is no formula to determine M (but we may expect M to increase with a decrease in ϵ), a series of trials will help determine the value of N for which the error between the approximate solution U_N and the true solution u is within the specified tolerance ϵ.

6.3.3.4   Completeness

The concept of convergence of a sequence involves a limit of the sequence. If the limit is not a part of the sequence $\{U_N\}_{N=1}^{\infty}$, then there is no hope of attaining convergence. For example, if the true solution to a certain problem is of the form $u(x) = ax^2 + bx^3 + cx^5$, where a, b, and c are constants, then the sequence of approximations

$$U_1 = c_1 x^3, \; U_2 = c_1 x^3 + c_2 x^4, \; \ldots, \; U_N = c_1 x^3 + c_2 x^4 + \cdots + c_N x^{N+2}$$

will not converge to the true solution because the sequence does not contain the x^2 term. The sequence is said to be incomplete. As a rule, in selecting an approximate solution, one should include all terms up to the highest-order term in the expression used for U_N (even if individually ϕ_i are incomplete). If a certain term is not a part of the true solution, like the x^4 term, its coefficient will turn out to be zero by the time all terms of the true solution are included in the approximation.

6.3.3.5 Requirements on ϕ_0 and ϕ_i

We now list the requirements on ϕ_0 and ϕ_i of the Ritz approximation given in Eq. (6.3.3):

1. ϕ_0 must satisfy the actual *specified* essential boundary conditions. It is identically zero if all of the specified essential boundary conditions are homogeneous. One must only select the lowest-order function that meets the specified nonzero essential boundary condition. For example, if the specified essential boundary condition is $u(0) = \hat{u} \neq 0$, then $\phi_0(x) = (1 + ax + bx^2 + \cdots +)\hat{u}$ satisfies the condition $\phi_0(0) = \hat{u}$ for any a, b, and so on. However, the lowest-order function is the constant term. Other terms should not be included in $\phi_0(x)$.

2. $\phi_i \in H_A$ $(i = 1, 2, \ldots, N)$ must satisfy the following three conditions:

   (a) $\phi_i(x)$ be continuous, as required by the variational statement being used;

   (b) $\phi_i(x)$ satisfy the *homogeneous form* of the specified essential boundary conditions; and

   (c) the set $\{\phi_i\}$ be linearly independent and complete. (6.3.22)

The properties listed in Eq. (6.3.22) provide guidelines for selecting the coordinate functions $\phi_0(x)$ and $\phi_i(x)$; they do not, however, give any formulas for generating the functions. Thus, apart from the guidelines, the selection of the coordinate functions is largely arbitrary (i.e., they can be algebraic, trigonometric, or other functions). For reasons of the simplicity of evaluating integrals of algebraic functions, often we prefer algebraic functions for ϕ_0 and ϕ_i. As a general rule, the coordinate functions ϕ_i should be selected from an admissible set [i.e., those meeting the conditions in Eq. (6.3.22)], from the lowest order to a desirable order, without missing any intermediate terms (i.e., satisfying the completeness property). Also, ϕ_0 should be any lowest order (including zero) that satisfied the specified essential boundary conditions of the problem; it has no continuity (differentiability) requirement. We will illustrate the selection of ϕ_0 and ϕ_i for specific problems considered in the examples.

Example 6.3.1

For the problem governed by Eqs (6.3.4) and (6.3.5) with $EA_0 = a_0[2 - (x/L)]$ and $f = f_0$, where L is the length of the bar, formulate the N-parameter Ritz solution using algebraic functions for $\phi_i(x)$. Then obtain the numerical values of the Ritz parameters

$$\bar{c}_i = \frac{a_0 L^{i-1}}{P} c_i \times 10^2 \ (N = 1, 2, \ldots, 8), \quad \text{for} \quad f_0 = 0, \quad k = 0. \tag{1}$$

Solution: For the problem described by Eqs (6.3.4) and (6.3.5), the specified boundary conditions contain one essential, $u(0) = 0$, and the other one is natural (actually mixed), $EA_0(du/dx) + ku = P$ at $x = L$, which is already accounted for in the total potential energy functional of Eq. (6.3.6). The specified essential boundary condition is homogeneous. Therefore, we select

$$\phi_0 = 0. \tag{2}$$

Next, we seek $\phi_i(x)$ such that $\phi_i(0) = 0$ (i.e., satisfy the homogeneous form of the specified essential boundary condition) and differentiable at least once with respect to x because Π involves its first derivatives of $\phi_i(x)$. We begin with $\phi_1(x)$. If an algebraic polynomial is to be selected, the lowest order polynomial that has a non-zero first derivative is

$$\phi_1(x) = \alpha_0 + \alpha_1 x, \tag{3}$$

where α_0 and α_1 are constants to be determined using the requirement $\phi_1(0) = 0$. The condition $\phi_1(0) = 0$ gives $\alpha_0 = 0$. Since α_1 is arbitrary, we can set it to unity (any nonzero constant will be absorbed into c_1). When $N > 1$, property 2(c) in Eq. (6.3.22) requires that ϕ_i, $i > 1$, should be selected such that the set $\{\phi_i\}_{i=1}^N$ is linearly independent and makes the set complete. In the present case, this is done by choosing $\phi_2(x)$ to be x^2. Alternatively, one can begin with $\phi_2(x) = \alpha_0 + \alpha_1 x + \alpha_2 x^2$ and find that $\alpha_0 = 0$; then α_1 and α_2 are arbitrary, except $\alpha_2 \neq 0$. Since x is already accounted for in ϕ_1, one may take $\alpha_1 = 0$. Clearly, $\phi_2(x) = x^2$ meets the conditions $\phi_2(0) = 0$, linearly independent of $\phi_1(x) = x$ (i.e., ϕ_2 is not a constant multiple of ϕ_1), and the set $\{x, x^2\}$ is complete (i.e., no other admissible term up to and including x^2 is omitted). Of course, one can also choose $\phi_2(x) = \alpha_1 x + \alpha_2 x^2$, with any values of α_1 and α_2. Then we have

$$U_2(x) = c_1 x + c_2(\alpha_1 x + \alpha_2 x^2) = \hat{c}_1 x + \hat{c}_2 x^2, \quad \hat{c}_1 = c_1 + \alpha_1 c_2, \quad \hat{c}_2 = c_2 \alpha_2 \tag{4}$$

This amounts to renaming the Ritz parameters c_1 and c_2 as $\hat{c}_1$ and $\hat{c}_2$. Generalizing the process, we arrive at $\phi_i = x^i$, $i = 1, 2, \ldots, N$.

For the choice of algebraic polynomials, the N-parameter Ritz approximation for the bar problem is

$$U_N(x) = \sum_{i=1}^N c_i \phi_i(x), \quad \phi_i(x) = x^i. \tag{5}$$

Then the coefficients A_{ij} and b_i of Eq. (6.3.14) for $EA_0 = a_0[2 - (x/L)]$, $\phi_i(x) = x^i$, $f = f_0$ (a constant), and $a = L$ are

$$A_{ij} = a_0 \int_0^L \left(2 - \frac{x}{L}\right) \frac{d\phi_i}{dx} \frac{d\phi_j}{dx} \, dx + k\phi_i(L)\phi_j(L)$$

$$= a_0 \, ij \int_0^L \left(2 - \frac{x}{L}\right) x^{i+j-2} \, dx + k(L)^{i+j}$$

$$= a_0 \frac{ij(1 + i + j)}{(i + j - 1)(i + j)} (L)^{i+j-1} + k(L)^{i+j}, \tag{6}$$

$$b_i = -\int_0^L EA \frac{d\phi_0}{dx} \frac{d\phi_i}{dx} \, dx - k\phi_0(L)\phi_i(L) + \int_0^L f\phi_i \, dx + P\phi_i(L)$$

$$= \int_0^L f_0 \phi_i \, dx + P\phi_i(L) = \frac{f_0}{i + 1}(L)^{i+1} + P(L)^i. \tag{7}$$

For $k = 0$ and $f_0 = 0$, we can express the Ritz equations as

$$\sum_{j=1}^{N} A_{ij}c_j - b_i = 0 \quad \text{or} \quad \sum_{j=1}^{N} \bar{A}_{ij}\bar{c}_j - \bar{b}_i = 0, \quad \bar{c}_j = \frac{a_0 L^{j-1}}{P}c_j, \tag{8}$$

where

$$\bar{A}_{ij} = \frac{ij(1+i+j)}{(i+j-1)(i+j)}, \quad \bar{b}_i = 1. \tag{9}$$

For a one-term approximation ($N = 1$) with $f_0 = k = 0$, we obtain

$$\bar{A}_{11} = \frac{3}{2}, \quad \bar{b}_1 = 1, \Rightarrow \bar{c}_1 = \frac{\bar{b}_1}{\bar{A}_{11}} = \frac{2}{3}. \tag{10}$$

The one-term Ritz approximation (for $f_0 = 0$) is ($c_1 = \bar{c}_1 P/a_0$)

$$U_1(x) = \frac{2P}{3a_0}x = \frac{2PL}{3a_0}\frac{x}{L} \tag{11}$$

For $N = 2$ and $f_0 = k = 0$, we have

$$\bar{A}_{11} = \frac{3}{2}, \quad \bar{A}_{12} = \bar{A}_{21} = \frac{4}{3}, \quad \bar{A}_{22} = \frac{5}{3}, \quad \bar{b}_1 = 1, \quad \bar{b}_2 = 1.$$

The Ritz equations can be written in matrix form $\bar{\mathbf{A}}\bar{\mathbf{c}} = \bar{\mathbf{b}}$ as

$$\frac{1}{6}\begin{bmatrix} 9 & 8 \\ 8 & 10 \end{bmatrix}\begin{Bmatrix} \bar{c}_1 \\ \bar{c}_2 \end{Bmatrix} = \begin{Bmatrix} 1 \\ 1 \end{Bmatrix},$$

whose solution by Cramer's rule is

$$\bar{c}_1 = \frac{6}{13}, \quad \bar{c}_2 = \frac{3}{13}. \tag{12}$$

Hence, the two-parameter Ritz solution becomes

$$U_2(x) = c_1 x + c_2 x^2 = \frac{3PL}{13a_0}\left(2\frac{x}{L} + \frac{x^2}{L^2}\right). \tag{13}$$

For $N = 3$ and $f_0 = k = 0$, we have (other coefficients are the same as those for $N = 2$)

$$\bar{A}_{13} = \bar{A}_{31} = \frac{5}{4}, \quad \bar{A}_{23} = \bar{A}_{32} = \frac{9}{5}, \quad \bar{A}_{33} = \frac{21}{10}, \quad \bar{b}_3 = 1,$$

and equations for the Ritz coefficients $\bar{c}_1$, $\bar{c}_2$, and $\bar{c}_3$ are

$$\frac{1}{60}\begin{bmatrix} 90 & 80 & 75 \\ 80 & 100 & 108 \\ 75 & 108 & 126 \end{bmatrix}\begin{Bmatrix} \bar{c}_1 \\ \bar{c}_2 \\ \bar{c}_3 \end{Bmatrix} = \begin{Bmatrix} 1 \\ 1 \\ 1 \end{Bmatrix}, \tag{14}$$

and the solution is

$$\bar{c}_1 = 50.7937 \times 10^{-2}, \quad \bar{c}_2 = 7.9365 \times 10^{-2}, \quad \bar{c}_3 = 10.5820 \times 10^{-2}. \tag{15}$$

Hence, the three-parameter Ritz solution becomes

$$U_3(x) = c_1 x + c_2 x^2 + c_3 x^3 = \frac{PL}{a_0}\left(50.7937\frac{x}{L} + 7.9365\frac{x^2}{L^2} + 10.5820\frac{x^3}{L^3}\right)10^{-2}. \tag{16}$$

The exact solution of Eqs (6.3.4) and (6.3.5) with $u(0) = 0$, $k = 0$, $EA = a_0[2 - (x/L)]$, and $f = f_0$ is

$$u(x) = \frac{f_0 L}{a_0} x + \frac{(f_0 L - P)L}{a_0} \log\left(1 - \frac{x}{2L}\right) \tag{17}$$

$$\approx \frac{f_0 L}{a_0} x + \frac{PL - f_0 L^2}{a_0} \left[\left(\frac{x}{2L}\right) + \frac{1}{2}\left(\frac{x}{2L}\right)^2 + \frac{1}{3}\left(\frac{x}{2L}\right)^3 + \frac{1}{4}\left(\frac{x}{2L}\right)^4 + \cdots \right]. \tag{18}$$

Table 6.3.1 contains a comparison of the Ritz coefficients $\bar{c}_i$ for $N = 1, 2, \ldots, 8$ with the exact coefficients in Eq. (18). The first two Ritz coefficients $\bar{c}_i$ have converged to the exact ones for $N = 8$. However, the Ritz solution for $N < 8$ is reasonably close to the exact solution because coefficients of the higher-order terms contribute less and less to the solution, as can be seen from Table 6.3.2 (the numerical results are obtained for $L = 10$ m, $a_0 = 180 \times 10^6$ N, $P = 10^4$ N, and $f_0 = 0$). In fact, the eight-parameter solution matches with the exact solution up to the fourth decimal for $x = 0, 0.1, 0.2, \ldots, 1$.

Table 6.3.1 The Ritz coefficients $(\bar{c}_i = (a_0 L^{i-1}/P)c_i \times 10^2)$ for the axial deformation of an isotropic, linearly elastic, nonuniform bar fixed at one end and subjected to a point force P at the other end.

N	$\bar{c}_1$	$\bar{c}_2$	$\bar{c}_3$	$\bar{c}_4$	$\bar{c}_5$	$\bar{c}_6$	$\bar{c}_7$	$\bar{c}_8$
1	66.667							
2	46.154	23.077						
3	50.794	7.936	10.582					
4	49.844	14.019	0.000	5.452				
5	50.029	12.062	6.100	−1.872	2.994			
6	49.994	12.615	3.426	3.621	−2.506	1.713		
7	50.001	12.471	4.416	0.576	2.605	−1.764	1.008	
8	50.000	12.506	4.090	1.968	−0.517	2.045	−1.384	0.605
Exact	50.000	12.500	4.167	1.562	0.625	0.260	0.112	0.049

Table 6.3.2 Comparison of the Ritz and Petrov–Galerkin solutions (multiplied by 10^4) with the analytical solution of the bar problem described by Eqs (6.3.4) and (6.3.5).

x/L	$N = 1$	$N = 2$	$N = 3$	$N = 8$	Exact
0.00	0.0000	0.0000	0.0000	0.0000	0.0000
0.10	0.3703	0.2692	0.2872	0.2849	0.2849
0.20	0.7407	0.5641	0.5867	0.5853	0.5853
0.30	1.1111	0.8846	0.9021	0.9029	0.9029
0.40	1.4815	1.2308	1.2369	1.2397	1.2397
0.50	1.8519	1.6026	1.5947	1.5982	1.5982
0.60	2.2222	2.0000	1.9788	1.9815	1.9815
0.70	2.5926	2.4231	2.3930	2.3932	2.3932
0.80	2.9630	2.8718	2.8407	2.8379	2.8379
0.90	3.3333	3.3462	3.3254	3.3213	3.3213
1.00	3.7037	3.8462	3.8507	3.8508	3.8508

If trigonometric functions are to be selected, one may be tempted to select $\phi_1 = \sin(\pi x/L)$, which satisfies the condition $\phi_1(0) = 0$. However, this choice also gives $U_1(L) = 0$ unless $N > 1$ and $\phi_2(L) \neq 0$. A better choice would be to select $\phi_1(x) = \sin(\pi x/2L)$ so that $U_1(L) \neq 0$. Even this choice, when used alone, makes $U_1(L/2) = 0$. However, for $N > 1$, the choice

$$\phi_i(x) = \sin\frac{(2i-1)\pi x}{2L}, \quad i = 1, 2, \ldots, N \tag{19}$$

will produce a good Ritz solution.

6.3.4 General Features of the Ritz Method

Some general features of the Ritz method are discussed here.

1. The Ritz method applies to all problems, linear or nonlinear, as long as one can cast the governing equations in the form of a variational statement [see Eq. (6.3.8)]: find $u \in H$ such that

$$B(v, u) = \ell(v)$$

holds for $v \in H$. In general, $B(v, u)$ and $\ell(v)$ are linear in v, but $B(v, u)$ may be nonlinear in u [and $B(\cdot, \cdot)$ may not, in general, be symmetric]. The procedure to develop variational forms of the form in Eq. (6.3.8), also called *weak forms*, from governing differential equations will be discussed in Section 6.3.6.

2. If the variational problem $B(v, u) = \ell(v)$ used in the Ritz approximation is such that $B(v, u)$ is symmetric in v and u, the coefficient matrix $\mathbf{A}$ of the resulting discrete problem $(\mathbf{Ac} = \mathbf{b})$ is also symmetric and, therefore, only elements above or below the diagonal of the coefficient matrix $\mathbf{A}$ need to be computed.

3. If $B(v, u)$ is nonlinear in u, the resulting algebraic equations $\mathbf{A(c)c} = \mathbf{b}$ will also be nonlinear in the parameters c_i. To solve such nonlinear equations, a variety of numerical methods are available (e.g., Newton's method). Generally, there is more than one solution to the set of nonlinear equations, and one of them is selected as the solution based on some physical or mathematical criterion.

4. If the approximation functions ϕ_0 and $\phi_i(x)$ are selected to satisfy the requirements in Eq. (6.3.22), the assumed approximation $U_N(x)$, when substituted into the minimum total potential energy principle, normally converges to the actual solution $u(x)$ with an increase in the number of parameters (i.e., as $N \to \infty$). A mathematical proof of such an assertion is not given here, but interested readers can consult the references at the end of the book.

5. For increasing values of N, the previously computed coefficients of the algebraic equations in Eq. (6.3.14) remain unchanged, provided the previously selected approximation functions are not changed; thus, one must only compute the new coefficients A_{ij} and b_i.

6. If the set of approximation functions $\{\phi_i\}$ chosen is an orthogonal set in the energy sense $B(\phi_i, \phi_j) = A_{ij}\delta_{ij}$ (no sum on i and j) and $k = 0$, then one need not invert the system of equations, and the solution is obtained as $c_i = b_i / A_{ii}$ (no sum on i).

7. When strains (in structural problems) or gradients are computed from the approximate solution by differentiation, the strains, stresses, and gradients are generally less accurate than the solution computed.

8. In general, the governing equation(s) and natural boundary conditions of the problem are satisfied only in the variational (or integral) sense, and not pointwise, by the computed Ritz solution. Therefore, the solution(s) obtained from the Ritz approximation generally does not satisfy the governing equations in a pointwise sense.

9. Since a continuous system is approximated by a finite number of coordinates (or degrees of freedom), the approximate system is less flexible than the actual system (recall that $\Pi(u) < \Pi(U)$ for any U that is not equal to the exact solution u). Consequently, the displacements obtained by the Ritz method using the principle of minimum total potential energy converge to the exact displacement from the following:

$$U_1 < U_2 < \ldots < U_n < U_m \cdots < u, \quad \text{for } m > n,$$

where U_N denotes the N-parameter Ritz approximation of u obtained using the principle of minimum total potential energy. The displacements obtained from the Ritz approximations based on the total complementary energy principle provide an upper bound for the exact solution.

10. Although the discussion of the Ritz method in this section thus far is confined to a linear solid mechanics problem, the method can be used for any equation that admits a variational formulation (in the sense discussed in feature 1 above), as will be illustrated through several examples shortly. However, the bounds mentioned above do not hold unless the variational problem is based on a minimum variational principle.

6.3.5 Examples

In this section we illustrate the application of the Ritz method to a variety of problems. These include static, eigenvalue, and transient problems. As will be shown in Section 6.3.6, the Ritz method can be applied to all problems whose governing equations are derivable (as the Euler equations) from a variational principle or problems for which their governing equations can be cast into an equivalent integral form. In the latter case, a way to develop the integral form, called the *weak form*, is discussed in Section 6.3.6.

To construct algebraic polynomials for ϕ_0 and ϕ_i ($i = 1, 2, \ldots, N$), we follow certain general procedure. For ϕ_0, we begin with a complete polynomial that has the same number of parameters as the number of specified essential boundary conditions and determine parameters by applying the specified conditions. Obviously, if all of the specified boundary conditions are homogeneous (i.e., the boundary conditions contain no nonzero specified values), $\phi_0 = 0$. For ϕ_1 we begin with a complete polynomial that has one more term than the number of specified essential boundary conditions and determine all but one parameter by applying the homogeneous form of the specified essential boundary conditions. For example, if there are n specified essential boundary conditions, we begin with a complete polynomial with $n + 1$ parameters and determine the first n parameters in terms of the $n + 1$st parameter (which is then set to an arbitrary value). The functions $\phi_2, \phi_3, \ldots, \phi_N$ are constructed (sometimes it is obvious to write the ith function by inspection) using the same procedure but by taking complete polynomials with $n + 2$, $n + 3$, and so on parameters. The parameter associated with the highest-order term in a polynomial should never be set to zero, while the other free parameters can be arbitrarily selected (they can even be set to zero). These ideas will be illustrated in the examples discussed here.

Example 6.3.2 ———————————————————————————————

Consider a uniform cross-section bar of length a, with the left end fixed and the right end connected to a rigid support via a linear elastic spring (with a spring constant k), as shown in Fig. 6.3.2. Determine the first two natural axial frequencies of the bar with constant EA and $ka/EA = 1$ using the Ritz method, and compare your results with the analytical solution for the frequencies.

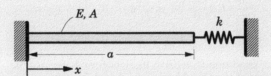

Fig. 6.3.2 Natural vibrations of a bar with an end spring.

Solution: This is a problem of dynamics of bars. Hence, we first write the $L = K - U$ for the problem and then reduce it to a problem of natural vibration (V_E is zero for this case). The kinetic energy K and the strain energy U associated with the axial motion of the member are given by

$$K = \frac{1}{2} \int_0^a \rho A \left(\frac{\partial u}{\partial t} \right)^2 dx, \quad U = \frac{1}{2} \int_0^a EA \left(\frac{\partial u}{\partial x} \right)^2 dx + \frac{1}{2} k[u(a,t)]^2. \tag{1}$$

Substituting for K and U from Eq. (1) and $V_E = 0$ into Hamilton's principle, Eq. (5.3.8), and integrating by parts with respect to time t (necessary for a free vibration problem to relieve δu of any derivatives with respect to time), we obtain $[\delta u(x,t_1) = \delta u(x,t_2) = 0$ and $\delta u(0,t) = 0]$

$$
\begin{aligned}
0 &= \int_{t_1}^{t_2} \delta(K - U)\, dt \\
&= \int_{t_1}^{t_2} \left[\int_0^a \left(\rho A \frac{\partial u}{\partial t} \frac{\partial \delta u}{\partial t} - EA \frac{\partial u}{\partial x} \frac{\partial \delta u}{\partial x} \right) dx - k u(a,t) \delta u(a,t) \right] dt \\
&= \int_{t_1}^{t_2} \left[\int_0^a \left(-\rho A \frac{\partial^2 u}{\partial t^2} \delta u - EA \frac{\partial u}{\partial x} \frac{\partial \delta u}{\partial x} \right) dx - k u(a,t) \delta u(a,t) \right] dt.
\end{aligned} \tag{2}
$$

Next, for natural vibration, we seek the periodic motion of the form

$$u(x, t) = u_0(x) \, e^{i\omega t}, \quad i = \sqrt{-1}, \tag{3}$$

where ω is the frequency of natural vibration and $u_0(x)$ is the mode of vibration. Substituting Eq. (3) into Eq. (2), we obtain (ρa is assumed to be independent of t)

$$0 = \int_{t_1}^{t_2} e^{2i\omega t} \, dt \left[\int_0^a \left(\rho A \omega^2 u_0 \delta u_0 - EA \frac{du_0}{dx} \frac{d\delta u_0}{dx} \right) dx - k u_0(a) \delta u_0(a) \right]. \tag{4}$$

Since

$$\int_{t_1}^{t_2} e^{2i\omega t} \, dt \neq 0,$$

we obtain

$$0 = \int_0^a \left(\rho A \omega^2 u_0 \delta u_0 - EA \frac{du_0}{dx} \frac{d\delta u_0}{dx} \right) dx - k u_0(a) \delta u_0(a). \tag{5}$$

We use Eq. (5) for the Ritz method to determine the values of ω. Note that the Rayleigh quotient for the problem at hand is given by

$$\omega^2 = \frac{\int_0^a EA \frac{du_0}{dx} \frac{d\delta u_0}{dx} \, dx + k u_0(a) \delta u_0(a)}{\int_0^a \rho A u_0 \delta u_0 dx}. \tag{6}$$

The Euler equation and natural boundary condition associated with the variational statement in Eq. (5) are

$$\frac{d}{dx} \left(EA \frac{du_0}{dx} \right) + \rho A \omega^2 u_0 = 0, \quad 0 < x < a,$$

$$EA \frac{du_0}{dx} + k u_0 = 0 \quad \text{at } x = a. \tag{7}$$

The essential boundary condition is $u_0(0) = 0$. Hence, we have $\phi_0 = 0$.

A dimensionless form of the variables is introduced for convenience and simplicity:

$$\xi = \frac{x}{a}, \quad v = \frac{u_0}{a}, \quad \alpha = \frac{ka}{EA}, \quad \lambda = \frac{\rho a^2}{E} \omega^2. \tag{8}$$

Then the dimensionless form of the governing equations in Eq. (7) is

$$\frac{d^2 v}{d\xi^2} + \lambda v = 0, \quad 0 < \xi < 1, \quad v(0) = 0, \quad \frac{dv}{d\xi} + \alpha v = 0 \quad \text{at } \xi = 1, \tag{9}$$

and Eq. (5) becomes

$$0 = \int_0^1 \left(\lambda v \delta v - \frac{dv}{d\xi} \frac{d\delta v}{d\xi} \right) d\xi - \alpha v(1) \delta v(1). \tag{10}$$

Substituting an N-parameter Ritz approximation (with $\phi_0 = 0$)

$$v(\xi) \approx V_N(\xi) = \sum_{i=1}^N c_i \phi_i(\xi) \tag{11}$$

into Eq. (10), we obtain

$$0 = \sum_{i=1}^N \left\{ \sum_{j=1}^N \left[\lambda \int_0^1 \phi_i \phi_j \, d\xi - \left(\int_0^1 \frac{d\phi_i}{d\xi} \frac{d\phi_j}{d\xi} \, d\xi + \alpha \phi_i(1) \phi_j(1) \right) \right] c_j \right\} \delta c_i. \tag{12}$$

Because of the independent nature of δc_i, we obtain the following ith equation of the system of N equations

$$0 = \sum_{j=1}^{N} \left[\lambda \int_0^1 \phi_i \phi_j \, d\xi - \left(\int_0^1 \frac{d\phi_i}{d\xi} \frac{d\phi_j}{d\xi} \, d\xi + \alpha \phi_i(1)\phi_j(1) \right) \right] c_j, \tag{13}$$

and in matrix form we have

$$(\mathbf{A} - \lambda \mathbf{M}) \, \mathbf{c} = \mathbf{0}, \tag{14}$$

where

$$A_{ij} = \int_0^1 \frac{d\phi_i}{d\xi} \frac{d\phi_j}{d\xi} \, d\xi + \alpha \phi_i(1)\phi_j(1), \quad M_{ij} = \int_0^1 \phi_i \phi_j \, d\xi. \tag{15}$$

Equation (14) represents a matrix eigenvalue problem, which involves finding the eigenvalues of λ and associated eigenvectors $\mathbf{v}$. Under the necessary condition for non-trivial solution of Eq. (14) (i.e., not all $c_i = 0$), we set the determinant of the coefficient matrix to zero and we obtain an Nth degree polynomial in λ, giving N eigenvalues λ_i, $i = 1, 2, \ldots, N$. Analytical method for finding eigenvalues and eigenvectors was discussed in Chapter 1 (see Section 1.3.4).

Following the arguments presented in **Example 6.3.1**, we can select $\phi_i(x)$ as

$$\phi_i(\xi) = \xi^i. \tag{16}$$

Substituting $\phi_i = \xi^i$ into Eq. (15), we obtain

$$M_{ij} = \int_0^1 \phi_i \phi_j \, d\xi = \frac{1}{i+j+1},$$

$$A_{ij} = \int_0^1 \frac{d\phi_i}{d\xi} \frac{d\phi_j}{d\xi} d\xi + \alpha \phi_i(1)\phi_j(1) = \frac{ij}{i+j-1} + \alpha. \tag{17}$$

Since we are required to determine two eigenvalues for $\alpha = 1$, we take $N = 2$ and obtain

$$M_{11} = \frac{1}{3}, \quad M_{12} = M_{21} = \frac{1}{4}, \quad M_{22} = \frac{1}{5}, \quad A_{11} = 2, \quad A_{12} = A_{21} = 2, \quad A_{22} = \frac{7}{3}, \tag{18}$$

and the matrix eigenvalue problem (14) becomes

$$\left(\begin{bmatrix} 2 & 2 \\ 2 & \frac{7}{3} \end{bmatrix} - \lambda \begin{bmatrix} \frac{1}{3} & \frac{1}{4} \\ \frac{1}{4} & \frac{1}{5} \end{bmatrix} \right) \begin{Bmatrix} c_1 \\ c_2 \end{Bmatrix} = \begin{Bmatrix} 0 \\ 0 \end{Bmatrix}. \tag{19}$$

For a nontrivial solution (i.e., $c_1 \neq 0$ and $c_2 \neq 0$), the determinant of the coefficient matrix in Eq. (19) is set to zero:

$$\begin{vmatrix} 2 - \frac{\lambda}{3} & 2 - \frac{\lambda}{4} \\ 2 - \frac{\lambda}{4} & \frac{7}{3} - \frac{\lambda}{5} \end{vmatrix} = 15\lambda^2 - 640\lambda + 2400 = 0. \tag{20}$$

This quadratic equation has two roots:

$$\lambda_1 = 4.1545, \quad \lambda_2 = 38.5121 \Rightarrow \omega_1 = \frac{2.0383}{L} \sqrt{\frac{E}{\rho}}, \quad \omega_2 = \frac{6.2058}{L} \sqrt{\frac{E}{\rho}}. \tag{21}$$

The eigenvectors are given by

$$U_2^{(i)} = c_1^{(i)} \xi + c_2^{(i)} \xi^2, \tag{22}$$

where $c_1^{(i)}$ and $c_2^{(i)}$ can be calculated from Eq. (19) for $\lambda = \lambda_i$, $i = 1, 2$ (see **Example 1.3.2**).

The general solution of the second-order equation in Eq. (9) is

$$v(\xi) = A \cos \sqrt{\lambda}\xi + B \sin \sqrt{\lambda}\xi, \quad 0 < \xi < 1, \tag{23}$$

where the constants of integration A and B are determined using the boundary conditions. The condition $v(0) = 0$ gives $A = 0$. The second condition, $(dv/d\xi) + \alpha v = 0$ at $\xi = 1$, gives (for $B \neq 0$)

$$\sqrt{\lambda} \cos \sqrt{\lambda} + \alpha \sin \sqrt{\lambda} = 0, \quad v(\xi) = B \sin \sqrt{\lambda}\xi, \tag{24}$$

which provides the required values of λ. For $\alpha = 0$ (i.e., right end is free), we have

$$\cos \sqrt{\lambda} = 0 \implies \lambda_n^2 = \frac{(2n-1)\pi}{2}. \tag{25}$$

The first two roots of the transcendental equation in Eq. (24) for $\alpha = 1$ are

$$\lambda_1 = \frac{\rho a^2}{E}\omega_1^2 = 4.1158, \quad \lambda_2 = \frac{\rho a^2}{E}\omega_2^2 = 24.1393. \tag{26}$$

or

$$\omega_1 = \frac{2.02875}{a}\sqrt{(E/\rho)}, \quad \omega_2 = \frac{4.91318}{a}\sqrt{(E/\rho)}. \tag{27}$$

Note that the first approximate frequency is closer to the exact than the second.

If one selects ϕ_i (we still have $\phi_0 = 0$) to satisfy the natural boundary condition also, the degree of polynomials inevitably goes up. For example, the lowest-order function that satisfies the homogeneous form of the natural boundary condition $v'(1) + \alpha v(1) = 0$, with $\alpha = 1$, is

$$\hat{\phi}_1 = 3\xi - 2\xi^2. \tag{28}$$

The one-parameter solution with the choice of $\hat{\phi}_1$ in Eq. (28) gives $\lambda_1 = 50/12 = 4.1667$ (or $\omega_1 = (2.04124/a)\sqrt{E/\rho}$), which is no better than the two-parameter solution computed using $\phi_1 = \xi$ and $\phi_2 = \xi^2$. Of course, the one-parameter solution $c_1\hat{\phi}_1$ would yield a more accurate value for λ_1 than the solution $c_1\phi_1$. Although $c_1\hat{\phi}_1$ and $c_1\phi_1 + c_2\phi_2$ are of the same degree (polynomials), the latter gives better accuracy for λ_1 because the number of parameters is greater, which provides more freedom to adjust the parameters in minimizing the total potential energy.

The dimensionless natural frequencies, $\bar{\omega} = \omega(a\sqrt{\rho/E})$, computed using the Ritz method for various values of N, are compared with the exact frequencies for $\alpha = ka/EA = 0$ (i.e., bar is fixed at the left end and the right end is free) and $\alpha = ka/EA = 1$ (i.e., bar is fixed at the left end and the right end is supported by a linear elastic spring with spring stiffness $k = EA/a$) in Tables 6.3.3 and 6.3.4, respectively. We note that the number of eigenvalues

Table 6.3.3 Comparison of dimensionless natural frequencies $(\bar{\omega}_i = \omega_i a\sqrt{\rho/E})^a$ computed using the Ritz method with the exact frequencies of a uniform isotropic bar fixed at one end and free at the other end ($\alpha = 0$).

N	$\bar{\omega}_1$	$\bar{\omega}_2$	$\bar{\omega}_3$	$\bar{\omega}_4$	$\bar{\omega}_5$	$\bar{\omega}_6$	$\bar{\omega}_7$	$\bar{\omega}_8$
1	1.7321							
2	1.5767	5.6728						
3	1.5709	4.8365	10.477					
4	1.5708	4.7426	8.331	16.304				
5	1.5708	4.7132	7.939	12.174	23.361			
6	1.5708	4.7124	7.865	11.279	16.480	31.665		
7	1.5708	4.7124	7.855	11.053	14.813	21.335	41.229	
8	1.5708	4.7124	7.854	11.004	14.315	18.621	26.762	52.089
Exact	1.5708	4.7124	7.854	10.995	14.137	17.279	20.420	23.562

[a]The eigenvalue problem is solved using the Jacobi iteration method with a tolerance of 10^{-12}.

Table 6.3.4 Comparison of dimensionless natural frequencies ($\bar{\omega}_i = \omega_i a \sqrt{\rho/E}$)[a] computed using the Ritz method with the exact frequencies of a uniform isotropic bar fixed at one end and spring-supported at the other end ($\alpha = 1$).

N	$\bar{\omega}_1$	$\bar{\omega}_2$	$\bar{\omega}_3$	$\bar{\omega}_4$	$\bar{\omega}_5$	$\bar{\omega}_6$	$\bar{\omega}_7$	$\bar{\omega}_8$
1	2.4495							
2	2.0383	6.2058						
3	2.0295	5.0387	10.977					
4	2.0288	4.9322	8.452	16.856				
5	2.0288	4.9141	8.083	12.256	23.934			
6	2.0288	4.9132	7.990	11.405	16.541	32.252		
7	2.0288	4.9132	7.980	11.143	14.939	21.382	41.827	
8	2.0288	4.9132	7.979	11.096	14.385	18.754	26.799	52.695
Exact	2.0288	4.9132	7.979	11.086	14.207	17.336	20.469	23.604

[a]The eigenvalue problem is solved using the Jacobi iteration method with a tolerance of 10^{-12}.

one can compute is equal to the value of N, and with the increasing value of N, the previously computed eigenvalues get refined and converge to the exact. Thus, even if one is interested in the smallest (called fundamental) frequency, one has to use $N > 1$ (in the present case, $N = 4$) to obtain an accurate value of the fundamental frequency.

Example 6.3.3

Consider a bar of uniform cross-sectional area A and length a, with the left end fixed and the right end connected to a rigid support via a linear elastic spring with spring constant k (see Fig. 6.3.2). Suppose that the bar (with constant axial stiffness EA) is subjected to a body force $f(x,t)$. Use the N-parameter Ritz approximation from **Example 6.3.2** to formulate the Ritz equations of motion of the bar under the assumption that the motion starts from rest, that is, the initial conditions of the problem are

$$u(x,0) = 0, \quad \dot{u}(x,0) = 0. \tag{1}$$

In particular, solve the resulting equations of motion for one-parameter Ritz approximation.

Solution: The kinetic and strain energies associated with the axial motion of the bar are given in Eq. (1) of **Example 6.3.2**. The potential energy due to $f(x,t)$ is

$$V_E = -\int_{t_1}^{t_2} \int_0^a fu \, dx. \tag{2}$$

Then the statement of Hamilton's principle for the problem at hand is

$$0 = \int_{t_1}^{t_2} \left\{ \int_0^a \left[-\frac{\partial}{\partial t}\left(\rho A \frac{\partial u}{\partial t}\right)\delta u - EA\frac{\partial u}{\partial x}\frac{\partial \delta u}{\partial x} + f\delta u \right] dx - ku(a,t)\delta u(a,t) \right\} dt. \tag{3}$$

The Euler–Lagrange equations resulting from Eq. (3) are

$$\frac{\partial}{\partial x}\left(EA\frac{\partial u}{\partial x}\right) - \frac{\partial}{\partial t}\left(\rho A\frac{\partial u}{\partial t}\right) + f = 0, \quad 0 < x < a; \ t > 0, \tag{4}$$

$$EA\frac{\partial u}{\partial x} + ku = 0, \quad \text{at } x = a; \ t \geq 0. \tag{5}$$

By introducing the dimensionless variables (f_0 is the magnitude of f)

$$\xi = \frac{x}{a}, \quad v = \frac{u}{(f_0 a^2/EA)}, \quad g = \frac{f}{f_0}, \quad \tau = \frac{t}{a\sqrt{\rho/E}}, \quad \alpha = \frac{ka}{EA}, \tag{6}$$

Eqs (4), (5), and (1) can be expressed as

$$\frac{\partial^2 v}{\partial \xi^2} - \frac{\partial^2 v}{\partial \tau^2} + g = 0, \quad 0 < \xi < 1; \ \tau > 0, \tag{7}$$

$$v(0,\tau) = 0, \quad \left[\frac{\partial v}{\partial \xi} + \alpha v\right]_{(1,\tau)} = 0, \quad \tau \geq 0. \tag{8}$$

$$v(\xi,0) = 0, \quad \left[\frac{\partial v}{\partial \tau}\right]_{(\xi,0)} = 0, \quad \text{for any } \xi. \tag{9}$$

When one is interested in determining the time-dependent solution $v(\xi, \tau)$ under applied load $g(\xi, \tau)$, the Ritz solution is sought in the form (with $\phi_0 = 0$)

$$v(\xi,\tau) \approx V_N(\xi,\tau) = \sum_{j=1}^{N} c_j(\tau)\phi_j(\xi), \quad \phi_i = \xi^i, \tag{10}$$

where c_j are now time-dependent parameters to be determined for all times $\tau > 0$. Substituting Eq. (10) into the dimensionless form of Eq. (3), we obtain

$$0 = -\sum_{j=1}^{N} \left[\left(\int_0^1 \phi_i\phi_j \, d\xi\right)\frac{d^2 c_j}{d\tau^2} + \left(\int_0^1 \frac{d\phi_i}{d\xi}\frac{d\phi_j}{d\xi} d\xi + \alpha\phi_i(1)\phi_j(1)\right)c_j\right] + \int_0^1 \phi_i g \, d\xi$$

$$= -\sum_{j=1}^{N} \left(M_{ij}\frac{d^2 c_j}{d\tau^2} + A_{ij}c_j\right) + b_i \quad \text{or} \quad \mathbf{M\ddot{c}} + \mathbf{Ac} = \mathbf{b}, \tag{11}$$

where

$$M_{ij} = \int_0^1 \phi_i\phi_j \, d\xi, \quad A_{ij} = \int_0^1 \frac{d\phi_i}{d\xi}\frac{d\phi_j}{d\xi} d\xi + \alpha\phi_i(1)\phi_j(1), \quad b_i = \int_0^1 \phi_i g(\xi,\tau) \, d\xi. \tag{12}$$

For the choice of $\phi_i = \xi^i$, we obtain (for $g = g_0$, a constant)

$$M_{ij} = \frac{1}{i+j+1}, \quad A_{ij} = \frac{ij}{i+j-1} + \alpha, \quad b_i = \frac{g_0}{i+1}. \tag{13}$$

For $N = 1$, we have

$$M_{11}\frac{d^2 c_1}{d\tau^2} + A_{11}c_1 = b_1 \quad \text{or} \quad \frac{d^2 c_1}{d\tau^2} + 3(1+\alpha)c_1 = \frac{3g_0}{2}. \tag{14}$$

The solution to this second-order differential equation is

$$c_1(\tau) = K_1 \sin\sqrt{\lambda}\tau + K_2 \cos\sqrt{\lambda}\tau + \frac{g_0}{2(1+\alpha)}, \quad \lambda = 3(1+\alpha), \tag{15}$$

and the one-parameter Ritz solution is given by

$$V_1(\xi,\tau) = \left(K_1 \sin\sqrt{\lambda}\tau + K_2 \cos\sqrt{\lambda}\tau + \frac{g_0}{2(1+\alpha)}\right)\xi, \tag{16}$$

where K_1 and K_2 are constants to be determined using the initial conditions in Eq. (9). Using the initial condition $V_1(\xi,0) = 0$, we obtain $K_2 = -g_0/2(1+\alpha)$; and $\dot{V}(\xi,0) = 0$ yields $K_1 = 0$. Then the solution in Eq. (16) becomes

$$V_1(\xi,\tau) = \frac{g_0}{2(1+\alpha)}\left(1 - \cos\sqrt{\lambda}\tau\right)\xi. \tag{17}$$

Next, consider the two-parameter Ritz approximation:

$$v(\xi,\tau) \approx V_2(\xi,\tau) = c_1(\tau)\phi_1(\xi) + c_2(\tau)\phi_2(\xi) = c_1(\tau)\,\xi + c_2(\tau)\,\xi^2. \tag{18}$$

From Eqs (10) and (12), we have (for $\alpha = 0$)

$$\frac{1}{60}\begin{bmatrix} 20 & 15 \\ 15 & 12 \end{bmatrix}\begin{Bmatrix} \ddot{c}_1 \\ \ddot{c}_2 \end{Bmatrix} + \frac{1}{3}\begin{bmatrix} 3 & 3 \\ 3 & 4 \end{bmatrix}\begin{Bmatrix} c_1 \\ c_2 \end{Bmatrix} = \frac{g_0}{6}\begin{Bmatrix} 3 \\ 2 \end{Bmatrix}, \tag{19}$$

which must be solved subjected to the initial conditions in Eq. (9), that is,

$$v(\xi,0) \approx c_1(0)\,\xi + c_2(0)\,\xi^2 = 0, \quad \dot{v}(\xi,0) \approx \dot{c}_1(0)\xi + \dot{c}_2(0)\xi^2 = 0. \tag{20}$$

These conditions can be satisfied, for arbitrary ξ, only if

$$c_1(0) = 0, \quad c_2(0) = 0, \quad \dot{c}_1(0) = 0, \quad \dot{c}_2(0) = 0. \tag{21}$$

We note that nonhomogeneous initial conditions with constant specified values may not be satisfied exactly.

For $N \geq 2$, the resulting system of ordinary differential equations in time, Eq. (11), can be solved for $c_i(\tau)$ using either the Laplace transform method or a numerical method (such as the finite difference or finite element method), the latter being more practical. In the solution by a numerical method, the complete solution is obtained only for discrete values of time. For example, at time $\tau_n = n\Delta\tau$, that is, the total time interval is divided into a finite number (n) of time steps of size $\Delta\tau$, we will have the solution

$$V_N(\xi,\tau_n) = \sum_{j=1}^{N} c_j(\tau_n)\phi_j(\xi). \tag{22}$$

In the present example, we shall use the Laplace transform method to solve the ordinary differential equations in Eq. (19). The procedure is detailed next.

Let $\bar{c}_i(s)$ denote the Laplace transform of $c_i(\tau)$:

$$\bar{c}_i = \mathcal{L}[c_i(\tau)] \equiv \int_{-\infty}^{\infty} c_i(\tau)e^{-s\tau}\,d\tau, \tag{23}$$

where s is the Laplace transform variable (replaces τ). The Laplace transform of the first and second derivatives of $c_i(\tau)$ are given by (see Table 6.3.5)

$$\mathcal{L}\left[\frac{dc_i}{d\tau}\right] = s\bar{c}_i(s) - c_i(0), \quad \mathcal{L}\left[\frac{d^2 c_i}{d\tau^2}\right] = s^2\bar{c}_i(s) - sc_i(0) - \dot{c}_i(0). \tag{24}$$

Using Eqs (23) and (24), Eq. (19) can be transformed to

$$\left(\frac{1}{3}s^2 + 1\right)\bar{c}_1 + \left(\frac{1}{4}s^2 + 1\right)\bar{c}_2 = \frac{g_0}{2s},$$
$$\left(\frac{1}{4}s^2 + 1\right)\bar{c}_1 + \left(\frac{1}{5}s^2 + \frac{4}{3}\right)\bar{c}_2 = \frac{g_0}{3s}. \tag{25}$$

The equations can be solved for $\bar{c}_1(s)$ and $\bar{c}_2(s)$ as

$$\bar{c}_1(s) = \frac{60g_0}{s\Delta}\left(s^2 + 20\right), \quad \bar{c}_2(s) = -\frac{50g_0}{s\Delta}\left(s^2 + 12\right), \tag{26}$$

Table 6.3.5 The Laplace transforms of some standard functions.

$f(t)$	$\bar{f}(s) \equiv \mathcal{L}[f(t)]$
$f(t)$	$\int_0^\infty e^{-st} f(t)\, dt$
$\dot{f} \equiv \frac{df}{dt}$	$s\bar{f}(s) - f(0)$
$\ddot{f} \equiv \frac{d^2 f}{dt^2}$	$s^2 \bar{f}(s) - sf(0) - \dot{f}(0)$
$f^{(n)}(t) \equiv \frac{d^n f}{dt^n}$	$s^n \bar{f}(s) - s^{n-1} f(0) - s^{n-2}\dot{f}(0) - \cdots - f^{(n-1)}(0)$
$\int_0^t f(\xi)\, d\xi$	$\frac{1}{s}\bar{f}(s)$
$\int_0^t \int_0^\tau f(\xi)\, d\xi\, d\tau$	$\frac{1}{s^2}\bar{f}(s)$
$\int_0^t f_1(\xi) f_2(t - \xi)\, d\xi$	$\bar{f}_1(s)\bar{f}_2(s)$
$H(t)$	$\frac{1}{s}$
$\delta(t) = \dot{H}(t)$	1
$\dot{\delta}(t) = \ddot{H}(t)$	s
$\delta^{(n)}(t)$	s^n
t	$\frac{1}{s^2}$
t^n	$\frac{n!}{s^{n+1}}$
$t f(t)$	$-\bar{f}'(s)$
$t^n f(t)$	$(-1)^n \bar{f}^{(n)}(s)$
$\frac{1}{t} f(t)$	$\int_s^\infty f(\xi)\, d\xi$
$e^{at} f(t)$	$\bar{f}(s - a)$
e^{at}	$\frac{1}{s-a}$
$t e^{at}$	$\frac{1}{(s-a)^2}$
$t^n e^{at}$	$\frac{n!}{(s-a)^{n+1}}, \quad n = 0, 1, 2, \ldots$
$\sin at$	$\frac{a}{s^2 + a^2}$
$\cos at$	$\frac{s}{s^2 + a^2}$
$\sinh at$	$\frac{a}{s^2 - a^2}$
$\cosh at$	$\frac{s}{s^2 - a^2}$
$t \sin at$	$\frac{2as}{(s^2 + a^2)^2}$
$t \cos at$	$\frac{s^2 - a^2}{(s^2 + a^2)^2}$
$e^{bt} \sin at$	$\frac{a}{(s-b)^2 + a^2}$
$e^{bt} \cos at$	$\frac{s - b}{(s-b)^2 + a^2}$
$\sqrt{t}$	$\frac{\sqrt{\pi}}{2} s^{-3/2}$
$\frac{1}{\sqrt{\pi t}}$	$\frac{1}{\sqrt{s}}$
$\frac{e^{bt} - e^{at}}{t}$	$\log \frac{s-a}{s-b}$
$\frac{1}{t}(1 - \cos at)$	$\frac{1}{2} \log \frac{s^2 + a^2}{s^2}$
$\frac{1}{t}(1 - \cosh at)$	$\frac{1}{2} \log \frac{s^2 - a^2}{s^2}$
$\frac{1}{t} \sin kt$	$\arctan \frac{k}{s}$

where Δ is the determinant of the coefficient matrix of $\bar{c}$'s in Eq. (25):

$$\Delta = 15s^4 + 520s^2 + 1200 = 15\left(s^4 + \frac{104}{3}s^2 + 80\right) = 15\left(s^2 + \alpha^2\right)\left(s^2 + \beta^2\right),$$

$$\alpha^2 = \frac{52}{3} + \frac{8\sqrt{31}}{3} = 32.1807, \quad \beta^2 = \frac{52}{3} - \frac{8\sqrt{31}}{3} = 2.4860,$$

(27)

where we have used the identity $x^4 + bx^2 + c = (x^2 + 0.5b + 0.5\sqrt{b^2 - 4c})(x^2 + 0.5b - 0.5\sqrt{b^2 - 4c})$. Hence, Eq. (26) becomes

$$\bar{c}_1(s) = 4g_0\frac{(s^2 + 20)}{s\left(s^2 + \alpha^2\right)\left(s^2 + \beta^2\right)}, \quad \bar{c}_2(s) = -\frac{10g_0}{3}\frac{(s^2 + 12)}{s\left(s^2 + \alpha^2\right)\left(s^2 + \beta^2\right)}.$$

(28)

To obtain the inverse transform of Eq. (28) to determine $c_1(\tau)$ and $c_2(\tau)$, first we must express $\bar{c}_1(s)$ and $\bar{c}_2(s)$ in a form that allows ready Laplace inversions. Consider the identity, which is dictated by the solution in Eq. (28):

$$\frac{s^2 + A}{s\left(s^2 + \alpha^2\right)\left(s^2 + \beta^2\right)} = \frac{1}{\alpha^2 - \beta^2}\left[\frac{(A - \alpha^2)}{\alpha^2}\frac{s}{(s^2 + \alpha^2)} - \frac{(A - \beta^2)}{\beta^2}\frac{s}{(s^2 + \beta^2)}\right] + \frac{A}{\alpha^2\beta^2}\frac{1}{s}. \quad (29)$$

Using the results in Eqs (28) and (29) and the fact

$$\alpha^2 - \beta^2 = \frac{16\sqrt{31}}{3} = 29.6947, \quad \alpha^2\beta^2 = \frac{240}{3} = 80, \quad \alpha = 5.6728, \quad \beta = 1.5767, \quad (30)$$

we obtain ($A = 20$ for $\bar{c}_1$ and $A = 12$ for $\bar{c}_2$)

$$\bar{c}_1(s) = 4g_0\frac{s^2 + 20}{s\left(s^2 + \alpha^2\right)\left(s^2 + \beta^2\right)}$$

$$= g_0\left[\frac{1}{s} - 0.0509868\frac{s}{s^2 + \alpha^2} - 0.9489979\frac{s}{s^2 + \beta^2}\right]$$

(31)

$$\bar{c}_2(s) = -\frac{10g_0}{3}\frac{s^2 + 12}{s\left(s^2 + \alpha^2\right)\left(s^2 + \beta^2\right)}$$

$$= -g_0\left[\frac{1}{2s} - 0.0703948\frac{s}{s^2 + \alpha^2} - 0.4295976\frac{s}{s^2 + \beta^2}\right].$$

(32)

Using the Laplace inverse transformation (see Table 6.3.5), we obtain

$$c_1(\tau) = g_0\left[1 - 0.051\cos(5.6728\tau) - 0.949\cos(1.5767\tau)\right],$$
$$c_2(\tau) = -g_0\left[0.5 - 0.07\cos(5.6728\tau) - 0.4296\cos(1.5767\tau)\right].$$

(33)

Equations (18) and (33) together give the two-parameter Ritz solution:

$$V_2(\xi, \tau) = g_0\left[1 - 0.051\cos(5.6728\tau) - 0.949\cos(1.5767\tau)\right]\xi$$
$$- g_0\left[0.5 - 0.07\cos(5.6728\tau) - 0.4296\cos(1.5767\tau)\right]\xi^2$$

(34)

Figure 6.3.3 contains plots of $v(1, \tau)$ versus τ for the one-parameter, Eq. (16), and two-parameter, Eq. (32), Ritz solutions for $\alpha = 0$ and $g_0 = 1$. Clearly, the one-parameter and two-parameter solutions deviate more and more from each other with time. Even the two-parameter solution may not be close to the actual solution. We note that Eqs (7)–(9) do not admit exact solution in order for us to assess the accuracy of the Ritz solutions.

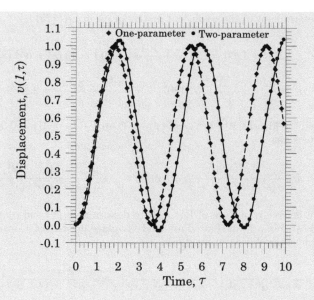

Fig. 6.3.3 Deflection $v(1,\tau)$ versus τ for the transient response of a bar fixed at the left end and free at the other end and subjected to a constant body force.

Example 6.3.4

Consider a simply supported beam of length L and constant bending stiffness EI, loaded with uniformly distributed transverse load of intensity q_0 (see Fig. 6.3.4). Use the N-parameter Ritz approximation with algebraic polynomials to set up the algebraic equations for the problem. Use Bernoulli–Euler beam theory. Compute the one-, two-, and three-parameter solutions, and compare with the exact solution of the problem.

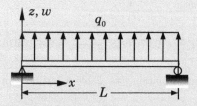

Fig. 6.3.4 A simply supported beam under uniform load.

Solution: The principle of minimum total potential energy for the problem can be expressed as

$$0 = \int_0^L \left(EI \frac{d^2 \delta w}{dx^2} \frac{d^2 w}{dx^2} - \delta w q_0 \right) dx. \tag{1}$$

The boundary conditions of the beam at hand are

$$w(0) = w(L) = 0, \quad M = -EI \frac{d^2 w}{dx^2} = 0 \text{ at } x = 0, L. \tag{2}$$

Of the four boundary conditions, only the first two are the specified essential boundary conditions and they are homogeneous. Hence, we have $\phi_0 = 0$. To construct ϕ_1, we note that $\phi_i(0) = 0$ and $\phi_i(L) = 0$. It is obvious that ϕ_0 and ϕ_i $(i = 1, 2, \ldots, N)$ should be as follows:

$$\phi_0 = 0, \quad \phi_1 = x(L - x), \quad \phi_2 = x^2(L - x), \quad \phi_i = x^i(L - x), \ldots, \phi_N = x^N(L - x). \quad (3)$$

To verify this choice, we begin with a three-parameter (since there are two conditions) complete polynomial for ϕ_1

$$\phi_1(x) = \alpha_0 + \alpha_1 x + \alpha_2 x^2$$

and find that $\alpha_0 = 0$ and $L\alpha_1 + L^2\alpha_2 = 0$, which gives $\alpha_1 = -L\alpha_2$. Then $\phi_1(x)$ becomes $\phi_1(x) = (-Lx + x^2)\alpha_2 = -x(L - x)\alpha_2$. By taking $\alpha_2 = -1$, we obtain $\phi_1(x) = x(L - x)$.

Next, we select a four-parameter polynomial for ϕ_2

$$\phi_2(x) = \alpha_0 + \alpha_1 x + \alpha_2 x^2 + \alpha_3 x^3$$

and obtain [using the conditions $w(0) = w(L) = 0$] $\alpha_0 = 0$ and $L\alpha_1 + L^2\alpha_2 + L^3\alpha_3 = 0$. We now have a choice to set either α_1 or α_2 to zero (of course, there are infinite number of possibilities to select them), and solve the other one in terms of α_3 (which cannot be set to zero because x^3 is the term that makes ϕ_2 independent of ϕ_1). Let us take $\alpha_1 = 0$ and solve for α_2 in terms of α_3. We obtain $\alpha_2 = -L\alpha_3$. Thus, we have $\phi_2(x) = (-Lx^2 + x^3)\alpha_3$. By setting $\alpha_3 = -1$, we obtain $\phi_2(x) = x^2(L - x)$. This procedure of setting $\alpha_1, \alpha_2, \ldots, \alpha_{i-1}$ to zero and solving α_i in terms of α_{i+1} results in $\phi_i(x) = x^i(L - x)$. Another possible set is $\{\phi_i\} = \{x(L^i - x^i)\}$. These sets are equivalent in the sense that the final N-parameter Ritz solution produced by these sets is the same.

Substituting the N-parameter Ritz approximation

$$w(x) \approx W_N(x) = \sum_{j=1}^{N} c_j\phi_j(x) + \phi_0(x) = \sum_{j=1}^{N} c_j x^j(L - x) \quad (4)$$

for w and ϕ_i for δw into Eq. (1), we obtain

$$0 = \int_0^L \left[EI\phi_i'' \left(\sum_{j=1}^{N} c_j\phi_j'' \right) - \phi_i q_0 \right] dx$$

$$= \sum_{j=1}^{N} A_{ij}c_j - b_i \quad \text{or} \quad \mathbf{Ac} = \mathbf{b}, \quad (5)$$

where

$$A_{ij} = \int_0^L EI \frac{d^2\phi_i}{dx^2} \frac{d^2\phi_j}{dx^2} dx$$

$$= EI \int_0^L ij \left[L(i-1)x^{i-2} - (i+1)x^{i-1} \right] \left[L(j-1)x^{j-2} - (j+1)x^{j-1} \right] dx$$

$$= ij \left[\frac{(i-1)(j-1)}{i+j-3} - \frac{2(ij-1)}{i+j-2} + \frac{(i+1)(j+1)}{i+j-1} \right] EIL^{i+j-1} \quad (6)$$

$$b_i = \int_0^L q_0\phi_i \, dx = \int_0^L q_0 \left[Lx^i - x^{i+1} \right] dx = \frac{q_0 L^{i+2}}{(i+1)(i+2)}. \quad (7)$$

For $N = 1$, we have

$$A_{11} = 4EIL, \quad b_1 = \frac{q_0 L^3}{6} \Rightarrow c_1 = \frac{b_1}{A_{11}} = \frac{q_0 L^2}{24EI}. \quad (8)$$

The Ritz solution is

$$W_1(x) = \frac{q_0 L^4}{24EI}\left(\frac{x}{L} - \frac{x^2}{L^2}\right), \quad W_1(0.5L) = \frac{q_0 L^4}{96EI} = 0.01042\frac{q_0 L^4}{EI}. \tag{9}$$

For $N = 2$, we obtain

$$EIL\begin{bmatrix} 4 & 2L \\ 2L & 4L^2 \end{bmatrix}\begin{Bmatrix} c_1 \\ c_2 \end{Bmatrix} = \frac{q_0 L^3}{12}\begin{Bmatrix} 2 \\ L \end{Bmatrix}. \tag{10}$$

The solution of these equations is

$$c_1 = \frac{q_0 L^2}{24EI}, \quad c_2 = 0. \tag{11}$$

Thus, the two-parameter Ritz solution is the same as the one-parameter Ritz solution. Since $c_2 = 0$, one should not conclude that the Ritz approximation has already converged, as it is obvious that it has not.

The three-parameter Ritz approximation yields the following system of algebraic equations:

$$EIL\begin{bmatrix} 4 & 2L & 2L^2 \\ 2L & 4L^2 & 4L^3 \\ 2L^2 & 4L^3 & 4.8L^4 \end{bmatrix}\begin{Bmatrix} c_1 \\ c_2 \\ c_3 \end{Bmatrix} = \frac{q_0 L^3}{12}\begin{Bmatrix} 2 \\ L \\ 0.6L^2 \end{Bmatrix}, \tag{12}$$

whose solution is

$$c_1 = c_2 L = -c_3 L^2 = \frac{q_0 L^2}{24EI}, \tag{13}$$

and the three-parameter Ritz solution becomes

$$W_3(x) = \frac{q_0 L^4}{24EI}\left[\left(\frac{x}{L} - \frac{x^2}{L^2}\right) + \left(\frac{x^2}{L^2} - \frac{x^3}{L^3}\right) - \left(\frac{x^3}{L^3} - \frac{x^4}{L^4}\right)\right] = \frac{q_0 L^4}{24EI}\left(\frac{x}{L} - 2\frac{x^3}{L^3} + \frac{x^4}{L^4}\right). \tag{14}$$

The three-parameter Ritz solution coincides with the exact solution

$$w(x) = \frac{q_0 L^4}{24EI}\left(\frac{x}{L} - 2\frac{x^3}{L^3} + \frac{x^4}{L^4}\right). \tag{15}$$

The maximum deflection occurs at the center of the beam and it is

$$w(0.5L) = \frac{5}{384}\frac{q_0 L^4}{EI} = 0.01302\frac{q_0 L^4}{EI}. \tag{16}$$

The maximum deflection predicted by the one-parameter Ritz approximation is 20% in error.

Several comments concerning the beam problem in **Example 6.3.4** are in order.

1. If the beam is subjected to a point load F_0 at its center, $x = L/2$, or a distributed load on a portion of the beam, instead of a uniform load throughout the span of the beam (see Fig. 6.3.5), the exact solution to the problem is not a single expression, but it would be in two parts because of the load discontinuity. For the case of the beam subjected to a point

load F_0 (downward) at its center, the exact solution is given by

$$w(x) = \begin{cases} -\frac{F_0 L^3}{48EI}\left(3\frac{x}{L} - 4\frac{x^3}{L^3}\right), & 0 \le x \le 0.5L, \\ -\frac{F_0 L^3}{48EI}\left(-1 + 9\frac{x}{L} - 12\frac{x^2}{L^2} + 4\frac{x^3}{L^3}\right), & 0.5L \le x \le L. \end{cases} \quad (6.3.23)$$

Then it is clear that we must also seek the Ritz approximation separately in the two segments of the beam.

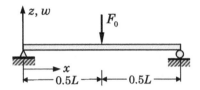

Fig. 6.3.5 A simply supported beam under center point load.

Suppose that we use the principle of minimum total potential energy (i.e., $\delta\Pi = 0$)

$$0 = \int_0^L EI\frac{d^2\delta w}{dx^2}\frac{d^2 w}{dx^2}dx + F_0\delta w(0.5L) \quad (6.3.24)$$

with the three-parameter approximation for the entire span of the beam

$$W_3(x) = c_1 x(L - x) + c_2 x^2(L - x) + c_3 x^3(L - x), \quad (6.3.25)$$

we obtain the same coefficient matrix as in Eq. (12) of **Example 6.3.4** because the bilinear form did not change but the linear form has changed. Therefore, we have

$$EIL\begin{bmatrix} 4 & 2L & 2L^2 \\ 2L & 4L^2 & 4L^3 \\ 2L^2 & 4L^3 & 4.8L^4 \end{bmatrix}\begin{Bmatrix} c_1 \\ c_2 \\ c_3 \end{Bmatrix} = -\frac{F_0 L^2}{4}\begin{Bmatrix} 1 \\ \frac{L}{2} \\ \frac{L^2}{4} \end{Bmatrix}.$$

The solution for the Ritz coefficients yields

$$c_1 = -\frac{4F_0 L}{64EI}, \quad c_2 = -\frac{5F_0}{64EIL}, \quad c_3 = \frac{5F_0}{64EIL}, \quad (6.3.26)$$

and the Ritz solution becomes

$$W_3(x) = \frac{F_0 L^3}{64EI}\left[-4\left(\frac{x}{L} - \frac{x^2}{L^2}\right) - 5\left(\frac{x^2}{L^2} - \frac{x^3}{L^3}\right) + 5\left(\frac{x^3}{L^3} - \frac{x^4}{L^4}\right)\right].$$

The Ritz solution *does not* coincide with the exact solution. In particular, the maximum deflection predicted by the three-parameter Ritz approximation is

$$W_3(0.5L) = -\frac{21}{64 \times 16}\frac{F_0 L^3}{EI} = -\frac{F_0 L^3}{48.762EI},$$

whereas the exact value from Eq. (6.3.23) is $w(0.5L) = -F_0 L^3/48EI$. The reason for the Ritz solution based on the variational statement in Eq. (6.3.24) not being exact even for a point load is that the variational statement does not account for the discontinuity in the load. Note that the exact shear force, $V = -EI(d^3 w/dx^3)$, is discontinuous at $x = L/2$, whereas the Ritz solution obtained above is continuous. Thus, we must apply the variational problem in each segment to obtain a good solution, as illustrated in **Example 6.3.5**.

2. One can also use trigonometric polynomials in place of algebraic polynomials for the approximation functions ϕ_i. For instance, the deflection of a simply supported beam subjected to continuously distributed load $q(x)$ can be represented by

$$w \approx W_N(x) = c_1 \sin\frac{\pi x}{L} + c_2 \sin\frac{2\pi x}{L} + \cdots + c_N \sin\frac{N\pi x}{L}. \qquad (6.3.27)$$

The functions $\phi_i = \sin(i\pi x/L)$ are linearly independent, and are complete if all lower functions up to $\sin(N\pi x/L)$ are included. If the applied loads on the simply supported beam are such that the solution has symmetry about $x = L/2$, then the even functions $\sin(2i\pi x/L)$ may be skipped, as they do not contribute to the solution.

When the load is sinusoidal,

$$q(x) = q_0 \sin\frac{m\pi x}{L} \quad \text{(for fixed } m\text{)}, \qquad (6.3.28)$$

we obtain the exact solution ($c_1 = c_2 = \cdots = c_{m-1} = c_{m+1} = \cdots = c_N = 0$)

$$w(x) = c_m \sin\frac{m\pi x}{L}, \quad c_m = \frac{q_0 L^4}{EI m^4 \pi^4}. \qquad (6.3.29)$$

This solution cannot be represented using a finite set of algebraic functions [e.g., $\{\phi_i\} = x^i(L-x)$], although the Ritz solution with a finite number of terms may be very close to the exact solution when evaluated at a point.

On the other hand, if the load is representable by an algebraic polynomial (e.g., $q(x)$ is a constant, linear, or higher-order polynomial in x), then the Ritz solution in Eq. (6.3.27) will not coincide with the exact solution for any finite value of N, because the sine-series representation of such load is an infinite series. For example, a point load F_0, applied at the center of the beam, can be represented by the sine series

$$F_0\delta(x - 0.5L) = \sum_{n=1}^{\infty} F_n \sin \frac{n\pi x}{L}, \quad F_n = \frac{2F_0}{L} \sin \frac{n\pi}{2} dx, \quad (6.3.30)$$

where $\delta(x)$ is the Dirac delta function. However, the Ritz solution in Eq. (6.3.27) converges rapidly, giving an accurate solution, especially away from the ends, for a finite value of N. Thus, in general, a judicious choice of approximation functions ϕ_i based on the source term $q(x)$ will not only make the computational effort minimal but may give accurate or even exact solution.

Example 6.3.5

Consider a beam with constant EI and discontinuous loading. Divide the beam into as many parts as there are regions with continuous geometry and loading. Use the results of **Example 4.3.5** (based on the Bernoulli–Euler beam theory) for a typical part of the beam and the continuity of the primary variables and balance of the secondary variables to determine the center deflection of the beam problem shown in Fig. 6.3.5.

Solution: Consider a typical part e of a beam located between $x = x_a^e$ and $x = x_b^e$ and let $h_e = x_b^e - x_a^e$, as shown in Fig. 6.3.6(a). The governing differential equation as per the Bernoulli–Euler beam theory is

$$\frac{d^2}{dx^2}\left(EI\frac{d^2w}{dx^2}\right) = q(x), \quad x_a^e < x < x_b^e, \quad (1)$$

where $q(x)$ is any distributed load in the part. A free-body diagram depicting the shear forces $(V_1^{(e)}, V_2^{(e)})$ and moments $(M_1^{(e)}, M_2^{(e)})$ at the ends is shown in Fig. 6.3.6(b), where the left end is labeled as 1 and the right end as 2. The total potential energy for the eth part is

$$\Pi^e(w) = \int_{x_a^e}^{x_b^e}\left[\frac{EI}{2}\left(\frac{d^2w}{dx^2}\right)^2 - qw\right]dx - V_1^{(e)}w(x_a^e) - V_2^{(e)}w(x_b^e)$$
$$- M_1^{(e)}\left(-\frac{dw}{dx}\right)_{x_a^e} - M_2^{(e)}\left(-\frac{dw}{dx}\right)_{x_b^e}, \quad (2)$$

where the x-coordinate is taken positive from the left end of the beam toward the right. The left end of the eth part is at a distance $x = x_a^e$ from the origin of the coordinate system. One may interpret the work done by the generalized forces in moving through their respective generalized displacements at points 1 and 2 as either due to "known" forces or "known" displacements.

If we presume that the displacements are specified at the two ends of the eth part and forces are unknown there, then the Ritz approximation W_N must satisfy the "specified" essential boundary conditions at the ends of the eth part. Suppose that the four essential boundary conditions, two at each end, are

$$w(x_a^e) = \Delta_1^e, \quad -\frac{dw}{dx}\bigg|_{x=x_a^e} = \Delta_2^e, \quad w(x_b^e) = \Delta_3^e, \quad -\frac{dw}{dx}\bigg|_{x=x_b^e} = \Delta_4^e. \quad (3)$$

Since there are four conditions, we select the following four-parameter Ritz approximation in algebraic form:

$$w^e(\bar{x}) \approx W_4^e(\bar{x}) = c_1^e + c_2^e\bar{x} + c_3^e\bar{x}^2 + c_4^e\bar{x}^3, \quad (4)$$

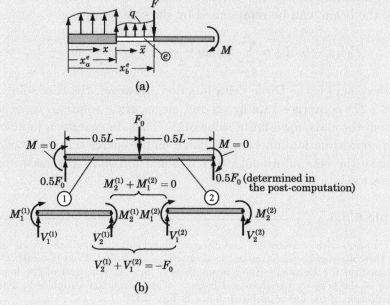

Fig. 6.3.6 Boundary, displacement continuity, and force equilibrium conditions for the beam.

where $\bar{x}$ is the local coordinate with its origin at the left end of the eth part (i.e., $x = \bar{x} + x_a^e$). Then c_i^e ($i = 1, 2, 3, 4$) are determined such that $W_4^e(\bar{x})$ satisfies the four end conditions in Eq. (3) (see **Example 4.3.5**):

$$
\begin{aligned}
\Delta_1 &\equiv W_4^e(0) = c_1^e, \\
\Delta_2 &\equiv \left(-\frac{dW_4^e}{dx}\right)_{x=0} = -c_2^e, \\
\Delta_3 &\equiv W_4^e(h_e) = c_1^e + c_2^e h_e + c_3^e h_e^2 + c_4^e h_e^3, \\
\Delta_4 &\equiv \left(-\frac{dW_4^e}{dx}\right)_{x=h_e} = -c_2^e - 2c_3^e h_e - 3c_4^e h_e^2.
\end{aligned}
\tag{5}
$$

Solving for c_i^e in terms of Δ_i^e and substituting them into Eq. (4) results in the expression

$$
W_4^e(\bar{x}) = \varphi_1^e(\bar{x})\,\Delta_1^e + \varphi_2^e(\bar{x})\,\Delta_2^e + \varphi_3^e(\bar{x})\,\Delta_3^e + \varphi_4^e(\bar{x})\,\Delta_4^e = \sum_{i=1}^{4} \varphi_i^e(\bar{x})\,\Delta_i^e,
\tag{6}
$$

where

$$
\begin{aligned}
\varphi_1^e(\bar{x}) &= 1 - 3\left(\frac{\bar{x}}{h_e}\right)^2 + 2\left(\frac{\bar{x}}{h_e}\right)^3, \\
\varphi_2^e(\bar{x}) &= -\bar{x}\left[1 - 2\left(\frac{\bar{x}}{h_e}\right) + \left(\frac{\bar{x}}{h_e}\right)^2\right], \\
\varphi_3^e(\bar{x}) &= \left(\frac{\bar{x}}{h_e}\right)^2\left(3 - 2\frac{\bar{x}}{h_e}\right), \\
\varphi_4(x) &= \bar{x}\frac{\bar{x}}{h_e}\left(1 - \frac{\bar{x}}{h_e}\right).
\end{aligned}
\tag{7}
$$

Substitution of Eq. (6) into Π^e (we note $x = \bar{x} + x_a^e$ and $dx = d\bar{x}$)

$$\Pi^e(\boldsymbol{\Delta}^e) = \int_0^{h_e} \left[\frac{EI}{2} \left(\sum_{j=1}^4 \Delta_j^e \frac{d^2\varphi_j^e}{d\bar{x}^2} \right)^2 - q \left(\sum_{j=1}^4 \Delta_j^e \varphi_j^e \right) \right] d\bar{x} - \sum_{j=1}^4 Q_j^e \Delta_j^e, \tag{8}$$

where

$$Q_1^e = V_1^{(e)}, \quad Q_2^e = M_1^{(e)}, \quad Q_3^e = V_2^{(e)}, \quad Q_4^e = M_2^{(e)}, \tag{9}$$

and setting $(\partial \Pi / \partial \Delta_i^e) = 0$, we obtain

$$0 = \frac{\partial \Pi^e}{\partial \Delta_i^e} = \int_0^{h_e} \frac{d^2\varphi_i^e}{d\bar{x}^2} \left[EI \left(\sum_{j=1}^4 \Delta_j^e \frac{d^2\varphi_j^e}{d\bar{x}^2} \right) - q\varphi_i^e \right] d\bar{x} - Q_i^e. \tag{10}$$

Carrying out the integration yields the following Ritz equations for the eth part:

$$\frac{2E_e I_e}{h_e^3} \begin{bmatrix} 6 & -3h_e & -6 & -3h_e \\ -3h_e & 2h_e^2 & 3h_e & h_e^2 \\ -6 & 3h_e & 6 & 3h_e \\ -3h_e & h_e^2 & 3h_e & 2h_e^2 \end{bmatrix} \begin{Bmatrix} \Delta_1^e \\ \Delta_2^e \\ \Delta_3^e \\ \Delta_4^e \end{Bmatrix} = \begin{Bmatrix} q_1^e \\ q_2^e \\ q_3^e \\ q_4^e \end{Bmatrix} + \begin{Bmatrix} Q_1^e \\ Q_2^e \\ Q_3^e \\ Q_4^e \end{Bmatrix}, \tag{11}$$

where

$$q_j^{(e)} = \int_0^{h_e} q(\bar{x}) \varphi_j^e(\bar{x}) \, d\bar{x}. \tag{12}$$

For uniformly distributed load $q = q_0$ on part e, we have

$$\begin{Bmatrix} q_1^e \\ q_2^e \\ q_3^e \\ q_4^e \end{Bmatrix} = \frac{q_0 h_e}{12} \begin{Bmatrix} 6 \\ -h_e \\ 6 \\ h_e \end{Bmatrix}. \tag{13}$$

Equation (11) relates the four generalized displacements (Δ_1^e, Δ_2^e, Δ_3^e, Δ_4^e) to the four generalized forces (Q_1^e, Q_2^e, Q_3^e, Q_4^e). Obviously, four of the eight variables should be known in order to solve the four equations. If there are n parts, there will be a total of $4n$ equations in $8n$ variables. Hence, the remaining $4n$ variables should be eliminated through known conditions (e.g., boundary conditions, continuity of the generalized displacements, and equilibrium generalized forces).

To illustrate the ideas, we consider the simply supported beam with a point load F_0 at the center (see Fig. 6.3.5). The beam must be divided into two parts, $h_1 = h_2 = L/2$. The Ritz equations for the two parts are given below ($E_1 = E_2 = E$, $I_1 = I_2 = I$, and $q(x) = 0$).

Part 1 ($e = 1$):

$$\frac{16EI}{L^3} \begin{bmatrix} 6 & -1.5L & -6 & -1.5L \\ -1.5L & 0.5L^2 & 1.5L & 0.25L^2 \\ -6 & 1.5L & 6 & 1.5L \\ -1.5L & 0.25L^2 & 1.5L & 0.5L^2 \end{bmatrix} \begin{Bmatrix} \Delta_1^{(1)} \\ \Delta_2^{(1)} \\ \Delta_3^{(1)} \\ \Delta_4^{(1)} \end{Bmatrix} = \begin{Bmatrix} Q_1^{(1)} \\ Q_2^{(1)} \\ Q_3^{(1)} \\ Q_4^{(1)} \end{Bmatrix}. \tag{14}$$

Part 2 ($e = 2$):

$$\frac{16EI}{L^3} \begin{bmatrix} 6 & -1.5L & -6 & -1.5L \\ -1.5L & 0.5L^2 & 1.5L & 0.25L^2 \\ -6 & 1.5L & 6 & 1.5L \\ -1.5L & 0.25L^2 & 1.5L & 0.5L^2 \end{bmatrix} \begin{Bmatrix} \Delta_1^{(2)} \\ \Delta_2^{(2)} \\ \Delta_3^{(2)} \\ \Delta_4^{(2)} \end{Bmatrix} = \begin{Bmatrix} Q_1^{(2)} \\ Q_2^{(2)} \\ Q_3^{(2)} \\ Q_4^{(2)} \end{Bmatrix}. \tag{15}$$

There are a total of eight equations in 16 variables. However, some of the variables are duplicative and others are related. In particular, we have the following eight conditions [see Fig. 6.3.6(b)]:

Boundary conditions:

$$\Delta_1^{(1)} = 0, \quad \Delta_3^{(2)} = 0, \quad Q_2^{(1)} = 0, \quad Q_4^{(2)} = 0. \tag{16}$$

Displacement continuity conditions:

$$\Delta_3^{(1)} = \Delta_1^{(2)}, \quad \Delta_4^{(1)} = \Delta_2^{(2)}. \tag{17}$$

Force equilibrium conditions:

$$Q_3^{(1)} + Q_1^{(2)} = -F_0, \quad Q_4^{(1)} + Q_2^{(2)} = 0. \tag{18}$$

Imposition of conditions in Eq. (17) is straightforward. We introduce the notation

$$U_1 = \Delta_1^{(1)}, \quad U_2 = \Delta_2^{(1)}, \quad U_3 = \Delta_3^{(1)} = \Delta_1^{(2)}, \quad U_4 = \Delta_4^{(1)} = \Delta_2^{(2)}, \quad U_5 = \Delta_3^{(2)}, \quad U_6 = \Delta_4^{(2)} \tag{19}$$

so that the conditions in Eq. (17) are satisfied. The first two equations of Eq. (14), expressed in terms of U_I, remain unchanged:

$$\frac{16EI}{L^3} \left(6U_1 - 1.5LU_2 - 6U_3 - 1.5LU_4 \right) = Q_1^{(1)}, \tag{20}$$

$$\frac{16EI}{L^3} \left(-1.5LU_1 + 0.5L^2U_2 + 1.5LU_3 + 0.25L^2U_4 \right) = Q_2^{(1)}, \tag{21}$$

To impose the first equilibrium condition of Eq. (18), we must add the third equation of element 1 [see Eq. (14)] to the first equation of element 2 [see Eq. (14)], so that we have

$$\frac{16EI}{L^3} \left[-6U_1 + 1.5LU_2 + (6+6)U_3 + (1.5L - 1.5L)U_4 - 6U_5 - 1.5LU_6 \right] = -F_0, \tag{22}$$

Similarly, the second equilibrium condition in Eq. (18) requires the addition of the fourth equation of element 1 [see Eq. (14)] to the second equation of element 2 [see Eq. (15)]:

$$\frac{16EI}{L^3} \left[-1.5LU_1 + 0.25LU_2 + (1.5L - 1.5L)U_3 + (0.5L^2 + 0.5L^2)U_4 + 1.5LU_5 + 0.25L^2U_6 \right) = 0. \tag{23}$$

The remaining two relations in Eq. (15) remain unchanged:

$$\frac{16EI}{L^3} \left(-6U_3 + 1.5LU_4 + 6U_5 + 1.5LU_6 \right) = Q_3^{(2)}, \tag{24}$$

$$\frac{16EI}{L^3} \left(-1.5LU_3 + 0.25L^2U_4 + 1.5LU_5 + 0.5L^2U_6 \right) = Q_4^{(2)}. \tag{25}$$

Thus, we have a total of six equations: Eqs (20)–(25). Writing these equations in matrix form, we have

$$\frac{16EI}{L^3} \begin{bmatrix} 6 & -1.5L & -6 & -1.5L & 0 & 0 \\ -1.5L & 0.5L^2 & 1.5L & 0.25L^2 & 0 & 0 \\ -6 & 1.5L & 6+6 & 1.5L - 1.5L & -6 & -1.5L \\ -1.5L & 0.25L^2 & 1.5L - 1.5L & 0.5L^2 + 0.5L^2 & 1.5L & 0.25L^2 \\ 0 & 0 & -6 & 1.5L & 6 & 1.5L \\ 0 & 0 & -1.5L & 0.25L^2 & 1.5L & 0.5L^2 \end{bmatrix} \begin{Bmatrix} U_1 \\ U_2 \\ U_3 \\ U_4 \\ U_5 \\ U_6 \end{Bmatrix} = \begin{Bmatrix} Q_1^{(1)} \\ Q_2^{(1)} \\ -F_0 \\ 0 \\ Q_3^{(2)} \\ Q_4^{(2)} \end{Bmatrix}. \tag{26}$$

Now we are ready to impose the boundary conditions from Eq. (16) (we note that $U_1 = U_5 = 0$ and $Q_2^{(1)} = Q_4^{(2)} = 0$). The first and fifth of Eq. (26) can be used to compute the reaction force at the left and right supports, respectively, provided we have the displacements (U_2, U_3, U_4, U_6). Thus, equations (2)–(4), and 6 from Eq. (26) can be written for the unknown generalized displacements (called the *condensed equations* by omitting the first and fifth rows and columns:

$$\frac{16EI}{L^3} \begin{bmatrix} 0.5L^2 & 1.5L & 0.25L^2 & 0 \\ 1.5L & 12 & 0 & -1.5L \\ 0.25L^2 & 0 & L^2 & 0.25L^2 \\ 0 & -1.5L & 0.25L^2 & 0.5L^2 \end{bmatrix} \begin{Bmatrix} U_2 \\ U_3 \\ U_4 \\ U_6 \end{Bmatrix} = \begin{Bmatrix} 0 \\ -F_0 \\ 0 \\ 0 \end{Bmatrix}. \tag{27}$$

Solving the four equations for (U_2, U_3, U_4, U_6), we obtain

$$U_2 = \frac{F_0 L^2}{16EI}, \quad U_3 = -\frac{F_0 L^3}{48EI}, \quad U_4 = 0.0, \quad U_6 = -\frac{F_0 L^2}{16EI}, \tag{28}$$

The rotation $U_4 = -(dw/dx)$ at the center of the beam is correctly predicted to be zero. The four-parameter Ritz solution becomes

$$W_4(x) = \begin{cases} U_2\varphi_2(x) + U_3\varphi_3(x), & 0 \le x \le 0.5L, \\ U_3\varphi_1(\bar{x}) + U_6\varphi_4(\bar{x}) & 0 \le \bar{x} \le 0.5L, \end{cases}$$

$$W_4(x) = \begin{cases} -\frac{F_0 L^3}{16EI}\left(\frac{x}{L} - 4\frac{x^2}{L^2} + 4\frac{x^3}{L^3}\right) - \frac{F_0 L^3}{48EI}\left(12\frac{x^2}{L^2} - 16\frac{x^3}{L^3}\right), & 0 \le x \le 0.5L, \\ -\frac{F_0 L^3}{48EI}\left[1 - 12\left(\frac{\bar{x}}{L}\right)^2 + 16\left(\frac{\bar{x}}{L}\right)^3\right] - \frac{F_0 L^3}{16EI}\left[2\left(\frac{\bar{x}}{L}\right)^2 - 4\left(\frac{\bar{x}}{L}\right)^3\right], & 0 \le \bar{x} \le 0.5L, \end{cases} \tag{29}$$

where $\bar{x} = x - 0.5L$. Simplification of Eq. (29) gives the expression in Eq. (6.3.23). Thus, the Ritz solution based on the cubic polynomial in Eq. (4) matches with the exact solution.

We can use an alternative procedure that does not require dividing the beam at hand into two parts. Suppose that we can represent the point load F_0 as a distributed load by means of Dirac delta:

$$q(x) = -F_0\delta(x - x_0), \quad x_0 = \frac{L}{2}. \tag{30}$$

Then we have from Eq. (12) the result

$$q_i^e = -F_0\varphi_i(x_0), \quad i = 1, 2, 3, 4, \tag{31}$$

or

$$\begin{Bmatrix} q_1^e \\ q_2^e \\ q_3^e \\ q_4^e \end{Bmatrix} = -\frac{F_0}{8} \begin{Bmatrix} 4 \\ -L \\ 4 \\ L \end{Bmatrix}. \tag{32}$$

Thus, a point load applied inside a typical part can be replaced with a set of statically equivalent loads and moments at the ends of the part. Then we can use a single element to formulate the problem:

$$\frac{2EI}{L^3} \begin{bmatrix} 6 & -3L & -6 & -3L \\ -3L & 2L^2 & 3L & L^2 \\ -6 & 3L & 6 & 3L \\ -3L & L^2 & 3L & 2L^2 \end{bmatrix} \begin{Bmatrix} \Delta_1 \\ \Delta_2 \\ \Delta_3 \\ \Delta_4 \end{Bmatrix} = -\frac{F_0}{8} \begin{Bmatrix} 4 \\ -L \\ 4 \\ L \end{Bmatrix} + \begin{Bmatrix} Q_1 \\ Q_2 \\ Q_3 \\ Q_4 \end{Bmatrix}. \tag{33}$$

Applying the boundary conditions [see Eq. (16)], we obtain

$$\frac{2EI}{L^3} \begin{bmatrix} 2L^2 & L^2 \\ L^2 & 2L^2 \end{bmatrix} \begin{Bmatrix} \Delta_2 \\ \Delta_4 \end{Bmatrix} = -\frac{F_0}{8} \begin{Bmatrix} -L \\ L \end{Bmatrix}, \tag{34}$$

whose solution is

$$\Delta_2 = -\Delta_4 = \frac{F_0 L^2}{16EI}.$$

(35)

Although these slopes at the ends of the beam are exact, the center deflection, post-computed using Eq. (6), is

$$W_4(0.5L) = \Delta_2 \varphi_2(0.5L) + \Delta_4 \varphi_4(0.5L) = -\frac{F_0 L^3}{64EI},$$

(36)

which is in 25% error.

The procedure described in **Example 6.3.5** is essentially that of the finite element method, which will be discussed in detail in Chapter 8. The procedure is valid for *all* beam problems irrespective of the nature of the loading and boundary conditions. In addition, the procedure gives exact values of the deflection $w(x)$ and rotation $-(dw/dx)$ at the end points of each part for all loads and boundary conditions, *provided that the flexural rigidity EI is constant within each part.* We note that the solution is not exact, in general, at an interior point.

Example 6.3.6

Consider a square, isotropic membrane with uniform tension T, occupying the domain $\Omega = -a < x < a, \ -a < y < a$, as shown in Fig. 6.3.7. Suppose that the membrane is fixed on all four sides, that is, $u = 0$ on the boundary Γ and subjected to uniformly distributed load of intensity f_0. Formulate an N-parameter Ritz solution, and obtain a one- and three-parameter Ritz solutions using (a) algebraic polynomials and (b) trigonometric functions. Use one quadrant of the domain to exploit the biaxial symmetry of the problem.

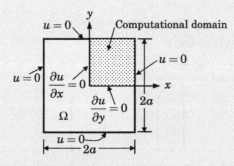

Fig. 6.3.7 A square membrane with boundary conditions of a quadrant.

Solution: In view of the biaxial symmetry present in the problem, the total potential energy functional associated with a square membrane is [see Eq. (3.6.11) with $a_{xx} = a_{yy} = T$ and $f = f_0$]

$$\Pi(u) = \int_0^a \int_0^a \left\{ \frac{T}{2} \left[\left(\frac{\partial u}{\partial x} \right)^2 + \left(\frac{\partial u}{\partial y} \right)^2 \right] - f_0 u \right\} dx \, dy.$$

(1)

The boundary conditions of the computational domain (i.e., quadrant) are

$$u = 0 \text{ at } x = a; \quad u = 0 \text{ at } y = a; \quad \frac{\partial u}{\partial x} = 0 \text{ at } x = 0; \quad \frac{\partial u}{\partial y} = 0 \text{ at } y = 0.$$

(2)

For this problem, specifying u is the essential boundary condition, and specifying the first derivative of u with respect to x or y is a natural boundary condition. Thus, we have $\phi_0 = 0$, and ϕ_i must be such that it vanishes at $x = a$ for any y and at $y = a$ for any x: $\phi_i(a, y) = \phi_i(x, a) = 0$ for $i = 1, 2, \ldots, N$.

Using the N-parameter approximation

$$U_N(x, y) = \sum_{j=1}^{N} c_j \phi_j(x, y) \tag{3}$$

and substituting into the total potential energy functional $\Pi(u)$ in Eq. (1) and setting its first derivative with respect to c_i to zero, we obtain

$$0 = \sum_{j=1}^{N} c_j \int_0^a \int_0^a T \left(\frac{\partial \phi_i}{\partial x} \frac{\partial \phi_j}{\partial x} + \frac{\partial \phi_i}{\partial y} \frac{\partial \phi_j}{\partial y} \right) dx dy - \int_0^a \int_0^a f_0 \phi_i \, dx dy$$

$$= \sum_{j=1}^{N} A_{ij} c_j - b_i, \tag{4}$$

where

$$A_{ij} = \int_0^a \int_0^a T \left(\frac{\partial \phi_i}{\partial x} \frac{\partial \phi_j}{\partial x} + \frac{\partial \phi_i}{\partial y} \frac{\partial \phi_j}{\partial y} \right) dx dy, \quad b_i = \int_0^a \int_0^a f_0 \phi_i \, dx dy \tag{5}$$

(a) For a choice of algebraic polynomials, we can use

$$\phi_1(x, y) = (a - x)(a - y), \quad \phi_2(x, y) = (a^2 - x^2)(a - y), \quad \phi_3(x, y) = (a - x)(a^2 - y^2), \quad \ldots \tag{6}$$

For two-dimensional problems, we must also follow the *equi-presence* rule of having the same degree of polynomials in both x and y in representing U_N. Thus, we can use only $N = 1$ and $N = 3$ but not $N = 2$ as it violates the equi-presence rule.

For $N = 1$, we obtain

$$A_{11} = \int_0^a \int_0^a T \left(\frac{\partial \phi_1}{\partial x} \frac{\partial \phi_1}{\partial x} + \frac{\partial \phi_1}{\partial y} \frac{\partial \phi_1}{\partial y} \right) dx dy$$

$$= T \int_0^a \int_0^a \left[(a - y)^2 + (a - x)^2 \right] dx dy = \frac{2Ta^4}{3} \tag{7}$$

$$b_1 = f_0 \int_0^a \int_0^a (a - x)(a - y) dx dy = \frac{f_0 a^4}{4},$$

and the solution becomes

$$U_1(x, y) = \frac{3 f_0 a^2}{8T} \left(1 - \frac{x}{a} \right) \left(1 - \frac{y}{a} \right). \tag{8}$$

For $N = 3$, we obtain

$$T \begin{bmatrix} \frac{2}{3}a^4 & \frac{3}{4}a^5 & \frac{3}{4}a^5 \\ \frac{3}{4}a^5 & \frac{44}{45}a^6 & \frac{5}{6}a^6 \\ \frac{3}{4}a^5 & \frac{5}{6}a^6 & \frac{44}{45}a^6 \end{bmatrix} \begin{Bmatrix} c_1 \\ c_2 \\ c_3 \end{Bmatrix} = \frac{f_0 a^4}{12} \begin{Bmatrix} 3 \\ 4a \\ 4a \end{Bmatrix}, \tag{9}$$

whose solution is

$$c_1 = -\frac{51 f_0}{89T}, \quad c_2 = \frac{75 f_0}{178aT}, \quad c_3 = \frac{75 f_0}{178aT}. \tag{10}$$

The three-parameter Ritz solution becomes

$$U_3(x, y) = -\frac{51 f_0 a^2}{89T} \left(1 - \frac{x}{a}\right) \left(1 - \frac{y}{a}\right) + \frac{75 f_0 a^2}{178T} \left[\left(1 - \frac{y}{a}\right) \left(1 - \frac{x^2}{a^2}\right) + \left(1 - \frac{x}{a}\right) \left(1 - \frac{y^2}{a^2}\right)\right].$$
(11)

For $N = 4$ with $\phi_4(x, y) = (a^2 - x^2)(a^2 - y^2)$, we obtain

$$T \begin{bmatrix} \frac{2}{3}a^4 & \frac{3}{4}a^5 & \frac{3}{4}a^5 & \frac{5}{6}a^6 \\ \frac{3}{4}a^5 & \frac{44}{45}a^6 & \frac{5}{6}a^6 & \frac{49}{45}a^7 \\ \frac{3}{4}a^5 & \frac{5}{6}a^6 & \frac{44}{45}a^6 & \frac{49}{45}a^7 \\ \frac{5}{6}a^6 & \frac{49}{45}a^7 & \frac{49}{45}a^7 & \frac{64}{45}a^8 \end{bmatrix} \begin{Bmatrix} c_1 \\ c_2 \\ c_3 \\ c_4 \end{Bmatrix} = \frac{f_0 a^4}{36} \begin{Bmatrix} 9 \\ 12a \\ 12a \\ 16a^2 \end{Bmatrix},$$
(12)

whose solution is

$$c_1 = \frac{4 f_0}{39T}, \quad c_2 = -\frac{5 f_0}{39aT}, \quad c_3 = -\frac{5 f_0}{39aT}, \quad c_4 = \frac{35 f_0}{78a^2 T}.$$
(13)

The four-parameter Ritz solution becomes

$$U_4(x, y) = \frac{4 f_0 a^2}{39T} \left(1 - \frac{x}{a}\right) \left(1 - \frac{y}{a}\right) - \frac{5 f_0 a^2}{39T} \left[\left(1 - \frac{y}{a}\right) \left(1 - \frac{x^2}{a^2}\right) + \left(1 - \frac{x}{a}\right) \left(1 - \frac{y^2}{a^2}\right)\right]$$
$$+ \frac{35 f_0}{78a^2 T} \left(1 - \frac{x^2}{a^2}\right) \left(1 - \frac{y^2}{a^2}\right).$$
(14)

The Euler equation associated with the total potential energy functional in Eq. (1) is

$$-T \left(\frac{\partial^2 u}{\partial x^2} + \frac{\partial^2 u}{\partial y^2}\right) = f_0 \quad \text{in } \Omega.$$
(15)

The exact solution to Eq. (15) subjected to the boundary condition $u = 0$ on Γ can be obtained using the separation of variables method, and it is given by

$$u(x, y) = \frac{16 f_0 a^2}{\pi^3 T} \sum_{n=1,3,5\ldots}^{\infty} \frac{1}{n^3} (-1)^{(n-1)/2} \left[1 - \frac{\cosh(n\pi y/2a)}{\cosh(n\pi/2)}\right] \cos \frac{n\pi x}{2a}.$$
(16)

The exact solution at the center of the region is

$$u(0, 0) = 0.2947 \frac{f_0 a^2}{T},$$
(17)

whereas the one-, three-, and four-parameter Ritz solutions from Eqs (8), (11), and (14) are

$$U_1(0, 0) = 0.375 \frac{f_0 a^2}{T}, \quad U_3(0, 0) = 0.2697 \frac{f_0 a^2}{T}, \quad U_4(0, 0) = 0.2948 \frac{f_0 a^2}{T}.$$
(18)

If the full domain is used, the first admissible function is $\phi_4 = (a^2 - x^2)(a^2 - y^2)$. The one-parameter Ritz solution in this case is

$$U_1(x, y) = \frac{5 f_0 a^2}{16T} \left(1 - \frac{x^2}{a^2}\right) \left(1 - \frac{y^2}{a^2}\right),$$
(19)

which is more accurate than the one-parameter solution in Eq. (8).

(b) For a choice of trigonometric functions, we can use

$$\phi_1(x,y) = \cos\frac{\pi x}{2a}\cos\frac{\pi y}{2a}, \quad \phi_2(x,y) = \cos\frac{3\pi x}{2a}\cos\frac{\pi y}{2a}, \quad \phi_3(x,y) = \cos\frac{\pi x}{2a}\cos\frac{3\pi y}{2a}, \quad \dots \quad (20)$$

In closing this example, we note that Eq. (15) arises, with a different interpretation of the variable u and the data (T, f_0), in a number of other fields. For example, it arises in the study of the torsion of a square cross-section prismatic bar where $T = 1$ and $f_0 = 2G\theta$, where G denotes the shear modulus and θ is the angle of twist per unit length. Variable u is the Prandtl stress function. It also arises in the study of conduction heat transfer in a square isotropic medium with conductivity k (i.e., $T = k$) and internal heat generation of f_0 per unit area, and u denotes the temperature.

6.3.6 The Ritz Method for General Boundary-Value Problems

6.3.6.1 Preliminary Comments

In the previous sections, the Ritz method was introduced as an approximate method that utilizes a variational principle, such as the principle of minimum total potential energy. Since the variational principle is equivalent to a set of governing equations, the Ritz method provides approximate solution of the underlying equations. That is why the Ritz method is termed a *direct* variational method. For problems outside the field of solid and structural mechanics, the construction of an analog of the minimum total potential energy principle or its equivalent is needed to use the Ritz method. In this section a procedure for constructing an integral form that is equivalent to a differential equation, called *weak form*, is presented, so that the Ritz method can be used for problems outside of solid and structural mechanics. Indeed, for structural mechanics problems, these integral statements are the same as the statements of principles of virtual displacements. Although the weak forms of equations outside solid mechanics provide a means to compute the approximate solutions, there is no underlying physical principle to guarantee that the approximate solution is the "best approximation" for problems outside solid mechanics.

6.3.6.2 Weak Forms

Typically, the weak forms are integral statements that involve weaker (i.e., lesser) differentiability of the dependent variables and thus allow the use of lower-order approximations. The steps involved in the weak formulation of differential equations are described with the aid of three *model* equations that cover 1-D and 2-D problems: (1) a second-order equation in one dimension, (2) a fourth-order equation in one dimension, and (3) a second-order equation in two dimensions. These equations are quite general, and they arise in a number of fields of engineering and applied sciences. The weak forms for structural mechanics problems are equivalent, in most cases, to the principle of virtual displacements.

6.3.6.3　Model Equation 1

Consider the problem of finding the function $u(x)$ that satisfies the differential equation

$$-\frac{d}{dx}\left(a(x)\frac{du}{dx}\right) + c(x)u - f(x) = 0 \quad \text{for} \quad 0 < x < L, \tag{6.3.31}$$

and the boundary conditions

$$u(0) = u_0, \quad \left(a\frac{du}{dx}\right)\Big|_{x=L} = Q, \tag{6.3.32}$$

where $a = a(x)$, $c = c(x)$, $f = f(x)$, u_0, and Q are the problem data (i.e., known quantities). Equation (6.3.31) arises in connection with the analytical description of many physical processes. For example, conduction and convection heat transfer in a plane wall or fin (1-D heat transfer), flow through channels and pipes, transverse deflection of cables, axial deformation of bars, and many others. Table 6.3.6 contains a list of several field problems described by Eq. (6.3.31) when $c(x) = 0$. The mathematical structure common to different fields is brought out in this table. Thus, if we can develop a numerical procedure by which Eq. (6.3.31) can be solved for all physically possible boundary conditions, the procedure can be used to solve all field problems listed in Table 6.3.6, as well as many others. This fact provides us the motivation to use Eq. (6.3.31) as a model second-order equation in one dimension.

As discussed before, we seek an N-parameter approximation of $u(x)$ in the form

$$u(x) \approx U_N = \sum_{j=1}^{N} c_j\phi_j(x) + \phi_0(x), \tag{6.3.33}$$

where c_j are the parameters to be determined and ϕ_j and ϕ_0 are selected such that Eqs (6.3.31) and (6.3.32) are satisfied in some sense, as already discussed in Section 6.3.2.

The necessary and sufficient number of algebraic relations among the c_js can be obtained by recasting the differential equation (6.3.31) in a weighted-integral form (see Section 6.3.2):

$$0 = \int_0^L w(x)\left[-\frac{d}{dx}\left(a\frac{du}{dx}\right) + cu - f\right]dx, \tag{6.3.34}$$

where $w(x)$ denotes a weight function that can take a set of independent values, which for the moment are arbitrary. For each choice of $w(x)$, we obtain an independent algebraic equation relating all c_j, and a total of N independent equations are obtained to solve for the N parameters c_j. When a weighted-integral statement like the one in Eq. (6.3.34) is used to obtain the N equations among c_j, the method is known as the *weighted-residual method*, which will be

Table 6.3.6 List of fields in which the model equation in Eq. (6.3.31) arises, with meaning of various parameters and variables; see the bottom of the table for the meanings of the parameters[a].

Field of study	Primary variable u	Coefficient a	Coefficient c	Source term f	Secondary variable Q
1. Heat transfer	Temperature $T - T_\infty$	Thermal conductance kA	Surface convection $p\beta$	Heat generation f	Heat Q
2. Flow through porous medium	Fluid head ϕ	Permeability μ	– 0	Infiltration f	Point source Q
3. Flow through pipes	Pressure P	Pipe resistance $1/R$	– 0	– 0	Point source Q
4. Flow of viscous fluids	Velocity v_z	Viscosity μ	– 0	Pressure gradient $-dP/dx$	Shear stress σ_{xz}
5. Elastic cables	Displacement u	Tension T	– 0	Transverse force f	Point force P
6. Elastic bars	Displacement u	Axial stiffness EA	– 0	Axial force f	Point load P
7. Torsion of bars	Angle of twist θ	Shear stiffness GJ	– 0	– 0	Torque T
8. Electro-statics	Electrical potential ϕ	Dielectric constant ϵ	– 0	Charge density ρ	Electric flux E

[a] k = thermal conductance; β = convective film conductance; p = perimeter; P = pressure or force; T_∞ = ambient temperature of the surrounding fluid medium; $R = 128\mu h/(\pi d^4)$ with μ being the viscosity; h, the length and d the diameter of the pipe; E = Young's modulus; A = area of cross-section; J = polar moment of inertia.

discussed in Section 6.4. Note that the use of Eq. (6.3.34) precludes that $u(x)$ or its approximation satisfies *all* specified boundary conditions and differentiable as many times as required in the differential equation. That is, ϕ_0 meets all specified boundary conditions of the problem, and ϕ_j satisfies the homogeneous form of the specified boundary conditions.

To weaken the continuity (i.e., differentiability) required of $u \approx U_N(x)$ and therefore of $\phi_j(x)$, we trade the differentiation in Eq. (6.3.34) from u to w such that both u and w are differentiated equally – once each in the present case. The resulting integral form is termed the *weak form* of Eq. (6.3.31). This form not only is equivalent to Eq. (6.3.31) but also contains the natural boundary conditions of the problem, and therefore, $u(x)$ need not satisfy the natural boundary conditions. The three-step procedure of constructing the weak form of Eq. (6.3.31) is discussed next.

The first step is to multiply the governing differential equation with a weight function w and integrate over domain $(0, L)$, giving Eq. (6.3.34). The second step is to trade differentiation from u to w, using integration by parts. This is achieved as follows. Consider the identity

$$-w\left[\frac{d}{dx}\left(a\frac{du}{dx}\right)\right] = -\frac{d}{dx}\left(w\,a\frac{du}{dx}\right) + a\frac{dw}{dx}\frac{du}{dx}, \qquad (6.3.35)$$

which is simply the product rule of differentiation applied to the product of two functions, $a(du/dx)$ and w. Integrating this identity over the domain, we obtain

$$-\int_0^L w\left[\frac{d}{dx}\left(a\frac{du}{dx}\right)\right] dx = -\int_0^L \frac{d}{dx}\left(w\,a\frac{du}{dx}\right) dx + \int_0^L a\frac{dw}{dx}\frac{du}{dx}\,dx$$

$$= -\left[w\,a\frac{du}{dx}\right]_0^L + \int_0^L a\frac{dw}{dx}\frac{du}{dx}\,dx. \qquad (6.3.36)$$

Using Eq. (6.3.36) in Eq. (6.3.34), we arrive at the result

$$0 = \int_0^L \left(a\frac{dw}{dx}\frac{du}{dx} + cwu - wf\right) dx - \left[w\,a\frac{du}{dx}\right]_0^L. \qquad (6.3.37)$$

The third and last step is to identify the so-called primary and secondary variables of the variational (or weak) form. This allows us to classify the boundary conditions of each differential equation into *essential* (or geometric) and *natural* (or force) boundary conditions. The classification is made uniquely by examining the boundary term appearing in the second step of the weak form development, namely, Eq. (6.3.37):

$$\left[w\,a\frac{du}{dx}\right]_0^L.$$

We define the coefficient of the weight function appearing in the boundary expression, namely, $a(du/dx)$, as the *secondary variable*. The specification of a secondary variable on the boundary constitutes the *natural* or *Neumann* boundary condition. The dependent unknown *in the same form as the weight function*

w in the boundary expression is termed the *primary variable*. The specification of a primary variable on the boundary is termed the *essential* or *Dirichlet* boundary condition. Thus, for the model equation at hand, the primary and secondary variables are

$$u \quad \text{and} \quad a\frac{du}{dx} \equiv Q.$$

Next, we denote the secondary variables at the end points by some symbols,

$$-Q_0 = \left(a\frac{du}{dx}\right)\bigg|_0, \quad Q_L = \left(a\frac{du}{dx}\right)\bigg|_L = P. \tag{6.3.38}$$

With the notation in Eq. (6.3.38), Eq. (6.3.37) becomes the final weak form for the problem:

$$0 = \int_0^L \left(a\frac{dw}{dx}\frac{du}{dx} + cwu - wf\right) dx - w(0)Q_0 - w(L)Q_L. \tag{6.3.39}$$

Now is the time to discuss the conditions on the weight function $w(x)$. Clearly, it should be differentiable at least once (like $u(x)$ is). The Ritz method uses the weak form, with $w = \phi_i$ to obtain the ith equation of the set of N relations among c_js. Thus, we require $w(x)$ to satisfy the homogeneous form of specified essential boundary conditions (i.e., belonging to the set of admissible variations). In the present case, $w(x)$ should be once differentiable and vanish at $x = 0$. Hence, the final weak form is

$$0 = \int_0^L \left(a\frac{dw}{dx}\frac{du}{dx} + cwu - wf\right) dx - w(L)P. \tag{6.3.40}$$

This completes the three-step procedure of constructing the weak form. The weak form in Eq. (6.3.40) contains two types of expressions: those containing both w and u and those containing only w. We group the former type into a single expression,

$$B(w, u) \equiv \int_0^L \left(a\frac{dw}{dx}\frac{du}{dx} + cwu\right) dx. \tag{6.3.41}$$

We denote all terms containing only w (but not u) by $\ell(w)$, called the *linear form*:

$$\ell(w) = \int_0^L wf \, dx + w(L)P. \tag{6.3.42}$$

The statement in Eq. (6.3.40) can now be expressed as one of finding u from the set of admissible functions (i.e., differentiable at least once and satisfies the essential boundary conditions) such that the variational problem

$$B(w, u) = \ell(w) \tag{6.3.43}$$

is satisfied for all w from the set of admissible variations. As seen before, $B(w, u)$ results directly in the coefficient matrix and while the linear form $\ell(w)$ gives rise to the right-hand-side column vector $\mathbf{B}$ of the Ritz equations. Since $B(w, u)$ is symmetric (in the present case), the functional associated with the variational problem (6.3.43) is given by Eq. (6.2.36) (which represents the total potential energy in the case of bars or cables):

$$I(u) = \tfrac{1}{2} B(u, u) - \ell(u) \tag{6.3.44}$$

$$= \int_0^L \left[\frac{a}{2} \left(\frac{du}{dx} \right)^2 + \frac{c}{2} u^2 \right] dx - \int_0^L uf \, dx - u(L)P. \tag{6.3.45}$$

The weak form development up to Eq. (6.3.43) is valid even for the case in which the coefficients a and c are functions of the dependent variable u (and hence U_N), making the problem nonlinear. In that case, $B(w, u)$ will be no longer a bilinear form and the functional $I(u)$ may not exist. However, to use the Ritz method, one only needs the variational statement in Eq. (6.3.43), which can be constructed using the three-step procedure discussed in this section.

6.3.6.4   Model Equation 2

Here we consider a fourth-order differential equation of the form

$$\frac{d^2}{dx^2} \left[b(x) \frac{d^2 u}{dx^2} \right] + c(x)u = f, \quad 0 < x < L, \tag{6.3.46}$$

We will not be concerned with any specific boundary conditions, as the weak form development naturally leads to the classification of the variables into primary and secondary types, and their specification constitutes the essential and natural type, respectively. This equation arises, for example, in connection with the bending of straight beams, where $u(x)$ denotes the transverse deflection; $b(x) = E(x)I(x)$, the bending rigidity; $c(x) = k$, the foundation modulus (if any); and $f(x)$ distributed transverse load.

The first two steps in the development of the weak form of the equation is summarized below. Note that, in this case we must transfer two derivatives to the weight function so that both w and u are required to have the same order of continuity (or differentiability).

Step 1. *Write the weighted-residual statement of the equation.*

$$0 = \int_0^L w(x) \left[\frac{d^2}{dx^2} \left(b \frac{d^2 u}{dx^2} \right) + cu - f \right] dx. \tag{6.3.47}$$

The equation requires nonzero fourth-order derivative of the approximation used for u.

Step 2. *Carry out integrations by parts to distribute the differentiations equally between u and w.*

$$
\begin{aligned}
0 &= \int_0^L \left[-\frac{dw}{dx}\frac{d}{dx}\left(b\frac{d^2u}{dx^2}\right) + cwu - wf \right] dx + \left[w\frac{d}{dx}\left(b\frac{d^2u}{dx^2}\right) \right]_0^L \\
&= \int_0^L \left(b\frac{d^2w}{dx^2}\frac{d^2u}{dx^2} + cwu - wf \right) dx + \left[w\frac{d}{dx}\left(b\frac{d^2u}{dx^2}\right) \right]_0^L \\
&\quad - \left[\frac{dw}{dx}b\frac{d^2u}{dx^2} \right]_0^L .
\end{aligned}
\tag{6.3.48}
$$

The integral statement now requires only continuous second derivatives of both u and w.

Step 3. *Identify the secondary variables at the boundary points of the domain and give them specific symbols.*

It is clear from the boundary expressions that the secondary variables are

$$
\frac{d}{dx}\left(b\frac{d^2u}{dx^2}\right), \quad b\frac{d^2u}{dx^2}.
\tag{6.3.49}
$$

In the case of beams, they represent the shear force and bending moment, respectively, in the beam. The primary variables are [$w \to u$ and $(dw/dx) \to (du/dx)$]

$$
u, \quad \frac{du}{dx}.
\tag{6.3.50}
$$

Denoting the shear forces and bending moments at the two ends of the beam as (proper signs are inserted to make all of the Qs and Ms to have the negative sign in the weak form; this also happens to be the correct definition of the bending moments and shear forces on the left and right ends of the beam)

$$
\begin{aligned}
\left[\frac{d}{dx}\left(b\frac{d^2u}{dx^2}\right) \right]_{x=0} = Q_0, \quad & \left[b\frac{d^2u}{dx^2} \right]_{x=0} = M_0, \\
\left[-\frac{d}{dx}\left(b\frac{d^2u}{dx^2}\right) \right]_{x=L} = Q_L, \quad & \left[-b\frac{d^2u}{dx^2} \right]_{x=L} = M_L.
\end{aligned}
\tag{6.3.51}
$$

Then Eq. (6.3.48) takes the final form

$$
\begin{aligned}
0 &= \int_0^L \left(b\frac{d^2w}{dx^2}\frac{d^2u}{dx^2} + cwu - wf \right) dx - w(0)Q_0 - w(L)Q_L \\
&\quad - \left(-\frac{dw}{dx} \right)\bigg|_0 M_0 - \left(-\frac{dw}{dx} \right)\bigg|_L M_L.
\end{aligned}
\tag{6.3.52}
$$

The bilinear form, linear form, and functional for the problem are

$$B(w,u) = \int_0^L \left(b\frac{d^2w}{dx^2}\frac{d^2u}{dx^2} + cwu \right) dx,$$

$$\ell(w) = \int_0^L wf\ dx + w(L)Q_L + \left(-\frac{dw}{dx} \right)\bigg|_0 M_0 + \left(-\frac{dw}{dx} \right)\bigg|_L M_L,$$

$$I(u) = \int_0^L \left[\frac{b}{2}\left(\frac{d^2u}{dx^2} \right)^2 + \frac{c}{2}u^2 - fu \right] dx$$

$$- u(0)Q_0 - u(L)Q_L - \left(-\frac{du}{dx} \right)\bigg|_0 M_0 - \left(-\frac{du}{dx} \right)\bigg|_L M_L. \quad (6.3.53)$$

In the case of beam bending, $I(u)$ represents the total potential energy $\Pi(u)$, where u is the transverse deflection, b is the flexural rigidity EI, and c is the foundation modulus. The weight function $w(x)$ in this case is required to be twice differentiable and vanish at the points where u and du/dx are specified. Equation (6.3.53) can be specialized to any beam with specific boundary conditions and loads.

6.3.6.5   Model Equation 3

Lastly, we consider the problem of determining the solution $u(x,y)$ to the partial differential equation,

$$-\frac{\partial}{\partial x}\left(a_1\frac{\partial u}{\partial x} \right) - \frac{\partial}{\partial y}\left(a_2\frac{\partial u}{\partial y} \right) + a_0 u = f \quad \text{in } \Omega \qquad (6.3.54)$$

in a two-dimensional (2-D) domain Ω. Here a_0, a_1, a_2, and f are known functions of position (x,y) in Ω. The function u is required to satisfy, in addition to the differential equation (6.3.54), certain boundary conditions on the boundary Γ of Ω. The variational formulation to be presented tells us the precise form of the boundary conditions that can be specified for problems described by the equation. Equation (6.3.54) arises in many fields of engineering, including in 2-D heat transfer, stream function or velocity potential formulation of inviscid flows, transverse deflections of a membrane, and torsion of a cylindrical member.

Next, the three-step procedure is used to develop the weak form associated with Eq. (6.3.54). The weight function is denoted by $w(x,y)$.

Step 1. *The weighted-integral statement.*

$$0 = \int_\Omega w\left[-\frac{\partial}{\partial x}\left(a_1\frac{\partial u}{\partial x} \right) - \frac{\partial}{\partial y}\left(a_2\frac{\partial u}{\partial y} \right) + a_0 u - f \right] dxdy. \qquad (6.3.55)$$

Step 2. *Weakening the differentiability of u.*

$$0 = \int_\Omega \left(a_1 \frac{\partial w}{\partial x}\frac{\partial u}{\partial x} + a_2 \frac{\partial w}{\partial y}\frac{\partial u}{\partial y} + a_0 wu - wf \right) dxdy$$

$$- \oint_\Gamma w \left(a_1 \frac{\partial u}{\partial x} n_x + a_2 \frac{\partial u}{\partial y} n_y \right) ds, \qquad (6.3.56)$$

where we used integration by parts (or the Green–Gauss theorem) to transfer differentiation to w so that both u and w have the same order derivatives. The boundary term shows that u is the primary variable, while

$$q_n \equiv a_1 \frac{\partial u}{\partial x} n_x + a_2 \frac{\partial u}{\partial y} n_y$$

is the secondary variable.

Step 3. *Identification of the secondary variables and writing of the final weak form.*

The last step in the procedure is to impose the specified boundary conditions. Suppose that u is specified on portion Γ_1 and q_n is specified on the remaining portion Γ_2 of the boundary Γ:

$$u = \hat{u} \ \text{ on } \Gamma_1, \qquad q_n = \hat{g} \ \text{ on } \Gamma_2. \qquad (6.3.57)$$

Then w is arbitrary on Γ_2 and equal to zero on Γ_1. Consequently, Eq. (6.3.56) simplifies to

$$0 = \int_\Omega \left(a_1 \frac{\partial w}{\partial x}\frac{\partial u}{\partial x} + a_2 \frac{\partial w}{\partial y}\frac{\partial u}{\partial y} + a_0 wu - wf \right) dxdy - \int_{\Gamma_2} w\hat{g}\, ds. \qquad (6.3.58)$$

The bilinear form, linear form, and functionals are

$$B(w, u) = \int_\Omega \left(a_1 \frac{\partial w}{\partial x}\frac{\partial u}{\partial x} + a_2 \frac{\partial w}{\partial y}\frac{\partial u}{\partial y} + a_0 wu \right) dxdy,$$

$$\ell(w) = \int_\Omega wf\, dxdy + \int_{\Gamma_2} w\hat{g}\, ds, \qquad (6.3.59)$$

$$I(u) = \frac{1}{2} \int_\Omega \left[a_1 \left(\frac{\partial u}{\partial x}\right)^2 + a_2 \left(\frac{\partial u}{\partial y}\right)^2 + a_0 u^2 \right] dxdy$$

$$- \int_\Omega uf\, dxdy - \int_{\Gamma_2} \hat{g}u\, ds.$$

In the case of a membrane, $I(u)$ represents the total potential energy and u is the transverse deflection.

6.3.6.6   Ritz Approximations

The Ritz method can be applied directly to the weak forms developed for the model problems of this section. In all cases, the variational problem is of the form

$$B(w, u) = \ell(w). \tag{6.3.60}$$

Here we consider the case in which $B(\cdot, \cdot)$ is a bilinear form (i.e., $B(w, u)$ is linear in both w and u). Although for this case a quadratic functional $I(u)$ can be constructed using the formula in Eq. (6.3.44), it is not necessary to have $I(u)$ to use the Ritz method; we only need $\delta I = 0$, which is equivalent to the variational statement in Eq. (6.3.60). The discrete form of Eq. (6.3.60) is given by

$$B(W_i, U_N) = \ell(W_i). \tag{6.3.61}$$

Substituting

$$U_N = \sum_{j=1}^{N} c_j \phi_j + \phi_0 \tag{6.3.62}$$

and $W_i = \phi_i$ into Eq. (6.3.61), we obtain the ith algebraic equation out of a set of N algebraic relations among $c_1, c_2, \ldots, c_N$:

$$\sum_{j=1}^{N} A_{ij} c_j = b_i, \quad j = 1, 2, \cdots, N, \quad \Rightarrow \quad \mathbf{Ac} = \mathbf{b}, \tag{6.3.63}$$

where

$$A_{ij} = B(\phi_i, \phi_j), \quad b_i = \ell(\phi_i) - B(\phi_i, \phi_0). \tag{6.3.64}$$

The specific expressions of A_{ij} and b_i of each model equation can be written using the respective bilinear and linear forms. For example, for the third model equation, we have

$$A_{ij} = \int_{\Omega} \left(a_1 \frac{\partial \phi_i}{\partial x} \frac{\partial \phi_j}{\partial x} + a_2 \frac{\partial \phi_i}{\partial y} \frac{\partial \phi_j}{\partial y} + a_0 \phi_i \phi_j \right) dx dy,$$

$$b_i = \int_{\Omega} \left(\phi_i f - a_1 \frac{\partial \phi_i}{\partial x} \frac{\partial \phi_0}{\partial x} - a_2 \frac{\partial \phi_i}{\partial y} \frac{\partial \phi_0}{\partial y} - a_0 \phi_i \phi_0 \right) dx dy + \int_{\Gamma_2} \phi_i \hat{g} \, ds. \tag{6.3.65}$$

Once ϕ_i and ϕ_0 are selected, subjected to the conditions stated in Section 6.2, the coefficients of matrix $\mathbf{A}$ and column vector $\mathbf{b}$ can be computed, and the linear algebraic equations in Eq. (6.3.63) can be solved for the Ritz coefficients, c_j. Then the N-parameter Ritz solution is given by Eq. (6.3.62).

Next, we consider couple of specific examples of application of the Ritz method to general boundary-value problems.

Example 6.3.7

Solve the partial differential equation (Laplace's equation)

$$-\left(\frac{\partial^2 u}{\partial x^2} + \frac{\partial^2 u}{\partial y^2}\right) = 0 \quad \text{in} \quad 0 < (x,y) < 1, \tag{1}$$

subject to the boundary conditions (see Fig. 6.3.8)

$$u = 0 \quad \text{on} \quad x = 0,1 \text{ and } y = 0,$$

$$u = \sin \pi x \quad \text{on} \quad y = 1 \tag{2}$$

using the Ritz method. Use trigonometric approximation functions.

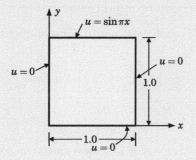

Fig. 6.3.8 Domain and boundary conditions of the problem in **Example 6.3.7**.

Solution: The weak form of the equation can be obtained as a special case from Eq. (6.3.58) by setting $f = 0$, $a_1 = a_2 = 1$, $a_0 = 0$, and $\Gamma_1 = \Gamma$ ($\Gamma_2 = 0$):

$$0 = \int_0^1 \int_0^1 \left(\frac{\partial \delta u}{\partial x}\frac{\partial u}{\partial x} + \frac{\partial \delta u}{\partial y}\frac{\partial u}{\partial y}\right) dx dy. \tag{3}$$

The Ritz equations are given by Eq. (6.3.63) with A_{ij} and b_i given by

$$A_{ij} = \int_0^1 \int_0^1 \left(\frac{\partial \phi_i}{\partial x}\frac{\partial \phi_j}{\partial x} + \frac{\partial \phi_i}{\partial y}\frac{\partial \phi_j}{\partial y}\right) dx \, dy,$$

$$b_i = -\int_0^1 \int_0^1 \left(\frac{\partial \phi_i}{\partial x}\frac{\partial \phi_0}{\partial x} + \frac{\partial \phi_i}{\partial y}\frac{\partial \phi_0}{\partial y}\right) dx \, dy. \tag{4}$$

Next we select ϕ_0 and ϕ_i for the problem. The function ϕ_0 is required to satisfy the boundary conditions in Eq. (2) because all of them are of the essential type. The following choice for ϕ_0 meets the conditions:

$$\phi_0 = y \sin \pi x. \tag{5}$$

The functions $\phi_i (i = 1, 2, \ldots, N)$ are required to satisfy the homogenous form of Eq. (2) and be linearly independent. We choose

$$\phi_1 = \sin \pi x \sin \pi y, \quad \phi_2 = \sin \pi x \sin 2\pi y, \quad \phi_3 = \sin 2\pi x \sin \pi y, \quad \text{etc.} \tag{6}$$

From Eq. (4) we obtain

$$A_{ij} = \begin{cases} \frac{\pi^2}{2} & \text{if } i = j, \\ 0 & \text{if } i \neq j, \end{cases}, \quad b_i = \begin{cases} -\frac{\pi}{2} & \text{if } i = 1, \\ 0 & \text{if } i \neq 1. \end{cases} \tag{7}$$

That is, $\mathbf{A}$ is a diagonal matrix. Hence, the Ritz solution becomes ($c_1 = -1/\pi$ and all other $c_i = 0$)

$$U_1(x,y) = c_1\phi_1(x,y) + \phi_0(x,y)$$

$$= y\sin\pi x - \frac{1}{\pi}\sin\pi x\sin\pi y = \sin\pi x\left(y - \frac{1}{\pi}\sin\pi y\right). \tag{8}$$

The exact solution of Eq. (1) is given by

$$u(x,y) = \frac{\sin\pi x\sinh\pi y}{\sinh\pi}. \tag{9}$$

Example 6.3.8

Consider the following pair of coupled differential equations, which arise in connection with the Timoshenko beam theory:

$$\frac{\partial}{\partial x}\left[S\left(\frac{\partial w}{\partial x} + \phi_x\right)\right] + q = m_0\frac{\partial^2 w}{\partial t^2}, \tag{1}$$

$$\frac{\partial}{\partial x}\left(D\frac{\partial\phi_x}{\partial x}\right) - S\left(\frac{\partial w}{\partial x} + \phi_x\right) = m_2\frac{\partial^2\phi_x}{\partial t^2}, \tag{2}$$

where S is the shear stiffness, $S = K_sGA$ (K_s shear correction coefficient, G shear modulus, and A area of cross section), D is the bending stiffness ($D = EI$, where E is Young's modulus, and I is the moment of inertia), w is the transverse deflection, ϕ_x is the rotation, q is the distributed transverse load, and $m_0 = \rho A$ and $m_2 = \rho I$ are the mass inertias. Assume that D, S, m_0, and m_2 are constants. The "specified" boundary conditions are of the form (as will be clear from the third step of the weak form)

$$-\left(D\frac{\partial\phi_x}{\partial x}\right)_{x=0} = M_1, \qquad \left(D\frac{\partial\phi_x}{\partial x}\right)_{x=L} = M_2, \tag{3}$$

$$-S\left(\frac{\partial w}{\partial x} + \phi_x\right)_{x=0} = Q_1, \quad S\left(\frac{\partial w}{\partial x} + \phi_x\right)_{x=L} = Q_2. \tag{4}$$

Assuming the Ritz approximations of the form

$$w(x,t) \approx \sum_{j=1}^{M} d_j(t)\psi_j(x) + \psi_0(x), \quad \phi_x(x,t) \approx \sum_{j=1}^{N} c_j(t)\theta_j(x) + \theta_0(x), \tag{5}$$

derive the Ritz equations (ordinary differential equations in time) of the problem.

Solution: First we develop the weak form of the equations using the three-step procedure. Multiply the first equation with weight function v_1 and the second one with weight function v_2, integrate over the length of the beam, and use the three-step procedure to obtain the weak forms. We obtain

$$0 = \int_0^L v_1\left\{\frac{\partial}{\partial x}\left[S\left(\frac{\partial w}{\partial x} + \phi_x\right)\right] + q - m_0\frac{\partial^2 w}{\partial t^2}\right\}dx$$

$$= \int_0^L\left\{-S\frac{\partial v_1}{\partial x}\left(\frac{\partial w}{\partial x} + \phi_x\right) + v_1q - m_0v_1\frac{\partial^2 w_0}{\partial t^2}\right\}dx$$

$$+ \left[S\left(\frac{\partial w}{\partial x} + \phi_x\right)v_1\right]_0^L$$

$$= \int_0^L \left\{ -S\frac{\partial v_1}{\partial x}\left(\frac{\partial w}{\partial x} + \phi_x\right) + v_1 q - m_0 v_1 \frac{\partial^2 w}{\partial t^2} \right\} dx$$
$$+ Q_1 v_1(0) + Q_2 v_1(L), \tag{6}$$

$$0 = \int_0^L v_2 \left\{ \frac{\partial}{\partial x}\left(D\frac{\partial \phi_x}{\partial x}\right) - S\left(\frac{\partial w}{\partial x} + \phi_x\right) - m_2\frac{\partial^2 \phi_x}{\partial t^2} \right\} dx$$

$$= \int_0^L \left\{ -D\frac{\partial v_2}{\partial x}\frac{\partial \phi_x}{\partial x} - Sv_2\left(\frac{\partial w}{\partial x} + \phi_x\right) - m_2 v_2\frac{\partial^2 \phi_x}{\partial t^2} \right\} dx$$

$$+ \left[D\frac{\partial \phi_x}{\partial x} v_2 \right]_0^L$$

$$= \int_0^L \left\{ -D\frac{\partial v_2}{\partial x}\frac{\partial \phi_x}{\partial x} - Sv_2\left(\frac{\partial w}{\partial x} + \phi_x\right) - m_2 v_2\frac{\partial^2 \phi_x}{\partial t^2} \right\} dx$$

$$+ M_1 v_2(0) + M_2 v_2(L), \tag{7}$$

where the notation given in Eqs (3) and (4) is utilized in writing the final weak forms. Note that integration by parts was used such that the expression $w_{,x} + \phi_x$ is preserved, as it enters the boundary term representing the shear force. Such considerations can only be used by knowing the physics behind the problem at hand. Also, note that the pair of weight functions (v_1, v_2) (from a product space of admissible variations) satisfy the homogeneous form of specified essential boundary conditions on the pair (w, ϕ_x) (with the correspondence $v_1 \sim w$ and $v_2 \sim \phi_x$). Writing the bilinear form for the problem is a bit involved; one may treat $u = (w, \phi_x)$ and $v = (v_1, v_2)$ as vectors (from a vector space) and then write the bilinear form $B(v, u)$.

From weak forms in Eqs (6) and (7), it is clear that all of the specified boundary conditions are of the natural type. Hence $\psi_0(x) = 0$ and $\theta_0(x) = 0$ in Eq. (5). Next we substitute the approximations from Eq. (5) into the weak forms for $w(x, t)$ and $\phi_x(x, t)$, and $v_1(x) = \psi_i(x)$ and $v_2(x) = \theta_i(x)$ and obtain

$$0 = \int_0^L \left[-S\frac{d\psi_i}{dx}\left(\sum_{j=1}^M \frac{d\psi_j}{dx}d_j + \sum_{j=1}^N c_j\theta_j\right) + \psi_i q - m_0\psi_i\left(\sum_{j=1}^M \frac{d^2 d_j}{dt^2}\psi_j\right) \right] dx$$

$$+ Q_1\psi_i(0) + Q_2\psi_i(L)$$

$$= -\sum_{j=1}^M A_{ij}d_j - \sum_{j=1}^N B_{ij}c_j - \sum_{j=1}^M M_{ij}^1\frac{d^2 d_j}{dt^2} + F_i^1, \tag{8}$$

$$0 = \int_0^L \left[-D\frac{d\theta_i}{dx}\left(\sum_{j=1}^N \frac{d\theta_j}{dx}c_j\right) - S\theta_i\left(\sum_{j=1}^M \frac{d\psi_j}{dx}d_j + \sum_{j=1}^N \theta_j c_j\right) \right.$$

$$\left. - m_2\theta_i\left(\sum_{j=1}^N \frac{d^2 c_j}{dt^2}\theta_j\right) \right] dx + M_1\theta_i(0) + M_2\theta_i(L)$$

$$= -\sum_{j=1}^M C_{ij}d_j - \sum_{j=1}^N D_{ij}c_j - \sum_{j=1}^N M_{ij}^2\frac{d^2 c_j}{dt^2} + F_i^2. \tag{9}$$

In matrix form, Eqs (8) and (9) can be expressed as

$$\begin{bmatrix} \mathbf{A} & \mathbf{B} \\ \mathbf{C} & \mathbf{D} \end{bmatrix}\begin{Bmatrix} \mathbf{d} \\ \mathbf{c} \end{Bmatrix} + \begin{bmatrix} \mathbf{M}^1 & \mathbf{0} \\ \mathbf{0} & \mathbf{M}^2 \end{bmatrix}\begin{Bmatrix} \ddot{\mathbf{d}} \\ \ddot{\mathbf{c}} \end{Bmatrix} = \begin{Bmatrix} \mathbf{F}^1 \\ \mathbf{F}^2 \end{Bmatrix}, \tag{10}$$

or

$$\mathbf{K\Delta} + \mathbf{M\ddot{\Delta}} = \mathbf{F}, \tag{11}$$

where the coefficients A_{ij}, B_{ij}, and so on are given by

$$
A_{ij} = \int_0^L S \frac{d\psi_i}{dx} \frac{d\psi_j}{dx} dx, \quad B_{ij} = \int_0^L S \frac{d\psi_i}{dx} \theta_j dx,
$$

$$
M_{ij}^1 = \int_0^L m_0 \psi_i \psi_j \, dx, \quad F_i^1 = \int_0^L \psi_i q \, dx + Q_1 \psi_i(0) + Q_2 \psi_i(L),
$$

$$
C_{ij} = \int_0^L S \theta_i \frac{d\psi_j}{dx} dx, \quad D_{ij} = \int_0^L \left(D \frac{d\theta_i}{dx} \frac{d\theta_j}{dx} + S \theta_i \theta_j \right) dx,
$$

$$
M_{ij}^2 = \int_0^L m_2 \theta_i \theta_j \, dx, \quad F_i^2 = M_1 \theta_i(0) + M_2 \theta_i(L).
$$

(10)

In closing this section, we make a couple of additional comments on the Ritz method. In developing the weak form, one should bear in mind that the boundary terms obtained from the integration by parts should be physically meaningful. The variational form used for the Ritz method does not have to be a quadratic functional, but it should be a form that includes the natural boundary conditions of the problem. Therefore, the Ritz method can be applied even to nonlinear problems. Of course, the resulting simultaneous algebraic equations are nonlinear, and there can be more than one solution to the equations (see **Example 6.4.18**). The selection of approximation functions becomes increasingly difficult with the dimension and shape of the domain.

The weak form of the operator equation in Eq. (6.3.1) can be constructed whenever the operator A permits the use of integration by parts [or gradient/divergence theorems in Eqs (1.2.55)–(1.2.57)] to transfer a part of the differentiation from the dependent variable u to the weight function w and incorporate the natural boundary conditions of the problem. In general, the weak form can be constructed if the operator A is expressible as a product (or composition) of two operators

$$
A = T^*(aT), \tag{6.3.66}
$$

where operator T^* is called the *adjoint* of T and is related to T by

$$
\int_\Omega T(u) v \, d\Omega = \int_\Omega u T^*(v) \, d\Omega + \oint_\Gamma C_1(u) C_2(v) \, d\Gamma \tag{6.3.67}
$$

for all u and v. Here Ω is a domain with boundary Γ and a is a function of position in Ω. The boundary operators C_1 and C_2 depend on the operator A. For example, when $A = -(d/dx)[a(du/dx)]$, operators T, T^*, C_1, and C_2 are given by

$$
T = \frac{d}{dx}, \quad T^* = -\frac{d}{dx}, \quad C_1 = 1, \quad C_2 = 1.
$$

If $A = -\nabla^2 = -\nabla \cdot \nabla$, then we have ($u$ must be a scalar and $\mathbf{v}$ a vector function of position)

$$
T = \nabla \ (\text{grad}), \quad T^* = -\nabla \cdot \ (\text{div.}), \quad C_1 = 1, \quad C_2 = \hat{\mathbf{n}} \cdot \ . \tag{6.3.68}
$$

The weak form of Eq. (6.3.1), with A given by Eq. (6.3.66), can be derived as follows:

$$
\begin{aligned}
0 &= \int_\Omega w\left[A(u) - f\right] d\Omega \\
&= \int_\Omega w\left[T^*(aT(u)) - f\right] d\Omega \\
&= \int_\Omega [T(w)(aT(u)) - wf]\, d\Omega - \oint_\Gamma B_1(w)B_2(u)\, d\Gamma,
\end{aligned} \qquad (6.3.69)
$$

where $C_1 = B_1$ and $B_2 = C_2(aT(u))$. Since $B_1(w) = 0$ on portion Γ_1 (where $B_1(u)$ is specified), and $B_2(u) = \hat{g}$ on Γ_2, with $\Gamma_1 + \Gamma_2 = \Gamma$, we have

$$
0 = \int_\Omega [aT(w)T(u) - wf]\, d\Omega - \int_{\Gamma_2} B_1(w)\hat{g}\, d\Gamma, \qquad (6.3.70)
$$

which is the weak form we set to derive.

6.4 Weighted-Residual Methods

6.4.1 Introduction

Consider a boundary-value problem described by the operator equation [see Eq. (6.3.1)]

$$
A(u) + C(u) = f \quad \text{in} \ \ \Omega, \qquad (6.4.1)
$$

where u is the dependent unknown, f is a source term, and A and C are differential operators (in spatial coordinates) defined over Ω, an open bounded region with total boundary Γ. In some cases, the operator C is an identity operator. Often A is a linear or nonlinear differential operator (or transformation) from a linear vector space V to $L_2(\Omega)$, $u \in V$ is the dependent variable, and $f \in L_2(\Omega)$ is a given source term in the domain Ω.

Equation (6.4.1) is subjected to boundary conditions

$$
B_1(u) = \hat{u} \ \ \text{on} \ \ \Gamma_u, \quad B_2(u) = \hat{g} \ \ \text{on} \ \ \Gamma_g, \qquad (6.4.2)
$$

where B_1 and B_2 are boundary operators associated with the essential and natural boundary conditions of the operator A, and $\hat{u}$ and $\hat{g}$ are specified values on the portions Γ_u and Γ_g of the boundary Γ of the domain. An example of Eqs (6.4.1) and (6.4.2) is provided by

$$
A(u) = -\frac{d}{dx}\left(a\frac{du}{dx}\right), \quad C = 1, \quad B_1(u) = u, \quad B_2(u) = a\frac{du}{dx},
$$

Here Γ_u is the point $x = 0$ and Γ_g is the point $x = L$. The operator A is linear whenever the parameter a is independent of u. If a is a function of the dependent variable u, $a = a(x, u)$, then A is nonlinear [see Section 6.2.4 and Eq. (6.2.30) for the definition of linear transformations].

We seek N-parameter solution to Eq. (6.4.1) in the form

$$U_N(x) = \sum_{j=1}^{N} c_j \phi_j(x) + \phi_0(x),   \tag{6.4.3}$$

and determine the unknown parameters c_j such that Eqs (6.4.1) and (6.4.2) are satisfied by the N-parameter solution U_N. Often, we select ϕ_0 and ϕ_i such that U_N satisfies the boundary conditions in Eq. (6.4.2) and then determine c_j such that Eq. (6.4.1) is satisfied in some sense.

For example, consider the two-point boundary-value problem

$$A(u) + C(u) \equiv -\frac{d}{dx}\left(x\frac{du}{dx}\right) + u = f_0 \text{ in } \Omega = (0, 1)$$

subjected to the boundary conditions

$$u(0) = u_0, \quad \left[x\frac{du}{dx}\right]_{x=1} = 0.$$

Suppose that the two-parameter solution is selected to be of the form

$$u \approx U_2 = c_1\phi_1(x) + c_2\phi_2(x) + \phi_0(x) = c_1(x^2 - 2x) + c_2(x^3 - 3x) + u_0,$$

which satisfies the boundary conditions for any values of c_1 and c_2. Then the constants c_1 and c_2 may be determined such that the approximate solution U_2 satisfies $A(U_2) + U_2 - f_0 = 0$ exactly at every point of the domain. That is,

$$\begin{aligned}
0 &= -\frac{dU_2}{dx} - x\frac{d^2U_2}{dx^2} + U_2 - f_0 \\
&= -2c_1(x - 1) - 3c_2(x^2 - 1) - x(2c_1 + 6c_2x) \\
&\quad + \left[c_1(x^2 - 2x) + c_2(x^3 - 3x) + u_0\right] - f_0.
\end{aligned}$$

Since this expression must be zero for any x, the coefficients of the various powers of x must be zero. We have

$$\begin{aligned}
x^0 : &\quad 2c_1 + 3c_2 = f_0 - u_0, \\
x^1 : &\quad -6c_1 - 3c_2 = 0, \\
x^2 : &\quad c_1 - 9c_2 = 0, \\
x^3 : &\quad c_2 = 0.
\end{aligned}$$

The above set of linear relations among c_1 and c_2 are inconsistent for given f_0 and u_0 (both c_1 and c_2 cannot be zero because the solution is trivial; also, the first equation cannot hold for $c_1 = c_2 = 0$ for arbitrary f_0 and u_0); hence, there is *no solution* to the set of equations. If we were able to find a unique solution to these equations for nonzero data, f_0 and u_0, then $U_2 = c_1(x^2 - 2x) + c_2(x^3 - 3x) + u_0$ is the exact solution of the problem. In general, such a procedure does not yield the solution for arbitrary data of the problem.

An alternative procedure must be found to make $A(U_2) + U_2 - f_0 = 0$ in some acceptable sense. For example, we may require the error in the differential equation to be orthogonal to a set of *weight functions* that are linearly independent, $\{w_1, w_2\}$, with respect to the $L_2(0,1)$ inner product:

$$(w_1, \mathbb{R})_0 = \int_0^1 w_1(x)\mathbb{R}\,dx = 0 \quad \text{and} \quad (w_2, \mathbb{R})_0 = \int_0^1 w_2(x)\mathbb{R}\,dx = 0, \quad (6.4.4)$$

where $\mathbb{R}$ is the *residual* (i.e., error) in the differential equation,

$$\mathbb{R} \equiv A(U_2) + U_2 - f_0 = -\frac{dU_2}{dx} - x\frac{d^2U_2}{dx^2} + U_2 - f_0 \neq 0.$$

For example, if we take the choices $w_1(x) = 1$ and $w_2(x) = x$, which are linearly independent, we obtain (for $f_0 = 0$ and $u_0 = 1$)

$$0 = \int_0^1 1 \cdot \mathbb{R}\ dx = (1 + 2c_1 + 3c_2) + \frac{1}{2}(-6c_1 - 3c_2) + \frac{1}{3}(c_1 - 9c_2) + \frac{1}{4}c_2,$$

$$0 = \int_0^1 x \cdot \mathbb{R}\ dx = \frac{1}{2}(1 + 2c_1 + 3c_2) + \frac{1}{3}(-6c_1 - 3c_2) + \frac{1}{4}(c_1 - 9c_2) + \frac{1}{5}c_2,$$

or

$$\frac{2}{3}c_1 + \frac{5}{4}c_2 = 1, \quad \frac{3}{4}c_1 + \frac{31}{20}c_2 = \frac{1}{2}.$$

These equations provide two linearly independent relations for c_1 and c_2 that can be solved to obtain the unique solution, $c_1 = \frac{222}{23}$ and $c_2 = -\frac{100}{23}$, and the approximate solution $U_2(x)$ becomes

$$U_2(x) = c_1(x^2 - 2x) + c_2(x^3 - 3x) + 1 = \frac{222}{23}(x^2 - 2x) - \frac{100}{23}(x^3 - 3x) + 1.$$

6.4.2 The General Method of Weighted Residuals

Thus, *weighted-integral statements* of the type in Eq. (6.4.4) provide means for obtaining as many algebraic equations as there are unknown coefficients c_j in the approximate solution, as in Eq. (6.4.3). There are several variational methods, in addition to the Ritz method that we have already discussed, in which approximate solutions of the type in Eq. (6.4.3) are sought and the coefficients c_j are determined. These methods differ from each other in the choice of the weight functions $w_i(x)$ and the integral statement used, which in turn dictates the choice of the approximation functions ϕ_j and ϕ_0.

Returning to the operator equation in Eq. (6.4.1), the N-parameter approximation of u yields the following residual in the operator equation $A(u)+C(u) = f$ in Ω:

$$\mathbb{R}_N \equiv A\left(\sum_{j=1}^N c_j\phi_j + \phi_0\right) + C\left(\sum_{j=1}^N c_j\phi_j + \phi_0\right) - f \neq 0 \text{ in } \Omega. \qquad (6.4.5)$$

We require that $\mathbb{R}$ be orthogonal to a set of N linearly independent set of functions, called the *weight functions* $\psi_i(x)$, which, in general, are different from the approximation functions ϕ_i:

$$\int_\Omega \psi_i \mathbb{R}_N(\mathbf{x}, \{c\}, \{\phi\}, f)d\Omega = 0, \ (i = 1, 2, \ldots, N). \qquad (6.4.6)$$

Equation (6.4.6) provides N linearly independent equations for the determination of the parameters c_i. If A is a nonlinear operator, the resulting algebraic equations are nonlinear. When A and C are linear, Eq. (6.4.5) is simplified to

$$\mathbb{R}_N = \sum_{j=1}^N c_j \left[A(\phi_j) + C(\phi_j)\right] + A(\phi_0) + C(\phi_0) - f \neq 0. \qquad (6.4.7)$$

Unless stated otherwise, we shall only consider cases where A and C are linear differential operators in the rest of this section.

Substituting $\mathbb{R}_N$ from Eq. (6.4.7) into Eq. (6.4.6), we obtain

$$0 = \int_\Omega \psi_i \left\{\sum_{j=1}^N [A(\phi_j) + C(\phi_j)]\, c_j + A(\phi_0) + C(\phi_0) - f \right\} d\Omega$$

$$= \sum_{j=1}^N A_{ij}c_j - b_i = 0 \Rightarrow \mathbf{Ac} = \mathbf{b} \qquad (6.4.8)$$

for $i = 1, 2, \ldots, N$, where

$$A_{ij} = \int_\Omega \psi_i \left[A(\phi_j) + C(\phi_j)\right] d\Omega$$

$$b_i = \int_\Omega \psi_i \left[-A(\phi_0) - C(\phi_0) + f\right] d\Omega. \qquad (6.4.9)$$

We note that the coefficient matrix $\mathbf{A}$ is unsymmetric, unless $\psi_i = A(\phi_i) + C(\phi_i)$ (i.e., the matrix $\mathbf{A}$ is symmetric for the least-squares method, as will be discussed shortly).

Since the weighted-residual statement in Eq. (6.4.6) does not account for the boundary conditions in Eq. (6.4.2) of the problem, we must make sure that U_N satisfies all of the specified boundary conditions. This in turn places certain requirements on the approximation functions ϕ_0 and ϕ_j. The approximation functions (ϕ_0, ϕ_i) and weight functions ψ_i in a weighted-residual method must satisfy the following conditions:

1. ϕ_0 has the main purpose of satisfying *all* specified boundary conditions associated with Eq. (6.4.1). It is necessarily zero when the specified boundary conditions are all homogeneous.

2. $\phi_j (j = 1, 2, \ldots, N)$ should satisfy three conditions:

    (a) Each ϕ_j is *continuous* as required in the weighted-residual statement; that is, ϕ_j should be such that U_N yields a nonzero value of $A(U_N) + C(U_N)$ (i.e., matrix $\mathbf{A}$ is invertible).

    (b) Each ϕ_j satisfies the *homogeneous form* of *all* specified (i.e., essential as well as natural) boundary conditions.

    (c) The set $\{\phi_j\}$ is *linearly independent* and *complete*.

3. ψ_i should be linearly independent. (6.4.10)

There are two main differences between the approximation functions used in the Ritz method and those used in a weighted-residual method:

(1) *Continuity.* The approximation functions used in a weighted-residual method are required to have the same differentiability as in the differential equation, whereas those used in the Ritz method must be differentiable as required by the total potential energy functional. Typically, for a $2m$th-order differential equation, the total potential energy functional contains only the mth-order derivatives of the dependent variable. Thus, the approximation functions ϕ_j used in a weighted-residual method are required to be twice as differentiable as in the Ritz method.

(2) *Boundary Conditions.* The approximation functions used in the weighted-residual method must satisfy the homogeneous form of the specified geometric as well as force boundary conditions, whereas those used in the Ritz method must satisfy the homogeneous form of only the specified essential boundary conditions (because the natural boundary conditions are already included in the total potential energy).

Both of these differences require ϕ_i used in a weighted-residual method to be of higher order than those used in the Ritz method. In an effort to satisfy the homogeneous form of all of the specified boundary conditions, the order of the functions will automatically go up.

Various special cases of the weighted-residual method differ from each other due to the choice of the weight function, ψ_i. The most commonly used weight functions are:

- *The Petrov–Galerkin method:* $\psi_i \neq \phi_i.$

- *Galerkin's method* (see [5, 104, 105]): $\psi_i = \phi_i.$ (6.4.11)

- *Least-squares method:* $\psi_i = A(\phi_i) + C(\phi_i).$

- *Collocation method:* $\psi_i = \delta(\mathbf{x} - \mathbf{x}_i).$

Here $\delta(\cdot)$ denotes the Dirac delta function. Although the least-squares method is listed as a special case of the weighted-residual method here, it is based on the concept of minimizing an integral statement. In general, the least-squares method is *not* a special case of the weighted-residual method. In addition to the methods listed above, there are other variational methods (methods in which the unknown parameters c_i are adjusted such the governing equations are satisfied in an integral sense). These include the subdomain method and *Trefftz method* [106, 107]. These methods will also be discussed briefly later in this chapter.

Example 6.4.1

Consider the bar problem of **Example 6.3.1**, which is concerned with solving Eq. (6.3.4) subjected to boundary conditions in Eq. (6.3.5) (with $f = k = 0$). Solve the problem using the two-parameter Petrov–Galerkin method with algebraic approximation functions and weight functions $\psi_1 = 1$ and $\psi_2 = x$.

Solution: The operator equation in this case is [see Eq. (6.3.4)]

$$A(u) = f, \quad A = -\frac{d}{dx}\left(EA\frac{du}{dx}\right), \quad EA = a_0\left(2 - \frac{x}{L}\right). \tag{1}$$

First, we determine ϕ_0, ϕ_1, and ϕ_2. The function ϕ_0 is required to satisfy the conditions

$$\phi_0(0) = 0, \quad \left[a_0\left(2 - \frac{x}{L}\right)\frac{d\phi_0}{dx}\right]_{x=L} = P. \tag{2}$$

We begin with the linear polynomial [the two conditions in Eq. (2) require the use of a polynomial with two parameters]

$$\phi_0(x) = \alpha_0 + \alpha_1 x$$

and determine α_0 and α_1 to satisfy the conditions in Eq. (2). The first condition gives $\alpha_0 = 0$. The second condition leads to ($a_0 \neq 0$)

$$a_0\alpha_1 = P \implies \alpha_1 = \frac{P}{a_0}.$$

Thus, $\phi_0 = (P/a_0)x$.

Next, we begin with the three-parameter complete polynomial for ϕ_1, $\phi_1(x) = \alpha_0 + \alpha_1 x + \alpha_2 x^2$ and require

$$\phi_1(0) = 0, \quad \left[a_0\left(2 - \frac{x}{L}\right)\frac{d\phi_1}{dx}\right]_{x=L} = 0. \tag{3}$$

We obtain $\alpha_0 = 0$ and

$$a_0(\alpha_1 + 2L\alpha_2) = 0 \implies \alpha_1 = -2L\alpha_2.$$

Thus, we can select $\phi_1(x)$ as (by setting $\alpha_2 = -1$)

$$\phi_1(x) = 2Lx - x^2 = x(2L - x). \tag{4}$$

Following the same procedure for $\phi_2(x) = \alpha_0 + \alpha_1 x + \alpha_2 x^2 + \alpha_3 x^3$, we obtain

$$\alpha_0 = 0, \quad \alpha_1 + 2L\alpha_2 + 3L^2\alpha_3 = 0. \tag{5}$$

Now we have an infinite number of choices, with the condition that $\alpha_3 \neq 0$ (otherwise, we will have a linearly dependent set), in selecting the values of α_1 and α_2 so that Eq. (5) holds. Two simple choices are to (a) set $\alpha_1 = 0$ and solve for α_2 in terms of α_3 or (b) set $\alpha_2 = 0$ and solve for α_1 in terms of α_3. These choices only change the values of c_1 and c_2 we compute at the end, but the total expression for $U_2(x)$ will not change (one may check this). We select here $\alpha_2 = 0$ and solve for α_1 in terms of α_3 (and then set $\alpha_3 = -1$). Thus, we have

$$\phi_2(x) = 3L^2 x - x^3 = x(3L^2 - x^2). \tag{6}$$

Thus, the two-parameter weighted-residual approximation is

$$U_2(x) = c_1\phi_1(x) + c_2\phi_2(x) + \phi_0(x) = c_1\left(2Lx - x^2\right) + c_2\left(3L^2 x - x^3\right) + \frac{P}{a_0}x. \tag{7}$$

Now we can apply the Petrov–Galerkin method to obtain two equations in two unknowns, c_1 and c_2. We have

$$A(\phi_i) = -\frac{d}{dx}\left[a_0\left(2 - \frac{x}{L}\right)\frac{d\phi_i}{dx}\right] = -a_0\left(2 - \frac{x}{L}\right)\frac{d^2\phi_i}{dx^2} + \frac{a_0}{L}\frac{d\phi_i}{dx},$$

$$A(\phi_0) = -a_0\left(2 - \frac{x}{L}\right)\frac{d^2\phi_0}{dx^2} + \frac{a_0}{L}\frac{d\phi_0}{dx} = \frac{P}{L}. \tag{8}$$

and (with $\psi_1 = 1$ and $\psi_2 = x$)

$$A_{11} = \int_0^L \psi_1 A(\phi_1)\, dx = \int_0^L \left[a_0\left(2 - \frac{x}{L}\right)2 + \frac{a_0}{L}(2L - 2x)\right] dx = 4a_0 L,$$

$$A_{21} = \int_0^L \psi_2 A(\phi_1)\, dx = \int_0^L x\left[a_0\left(2 - \frac{x}{L}\right)2 + \frac{a_0}{L}(2L - 2x)\right] dx = \frac{5}{3}a_0 L^2,$$

$$A_{12} = \int_0^L \psi_1 A(\phi_2)\, dx = \int_0^L \left[a_0\left(2 - \frac{x}{L}\right)6x + \frac{a_0}{L}(3L^2 - 3x^2)\right] dx = 6a_0 L^2,$$

$$A_{22} = \int_0^L \psi_2 A(\phi_2)\, dx = \int_0^L x\left[a_0\left(2 - \frac{x}{L}\right)6x + \frac{a_0}{L}(3L^2 - 3x^2)\right] dx = \frac{13}{4}a_0 L^3,$$

$$B_1 = -\int_0^L \frac{P}{L}\, dx = -P, \quad B_2 = -\int_0^L x\frac{P}{L}\, dx = -\frac{PL}{2}. \tag{9}$$

In matrix form we have

$$a_0 \begin{bmatrix} 4L & 6L^2 \\ \frac{5}{3}L^2 & \frac{13}{4}L^3 \end{bmatrix} \begin{Bmatrix} c_1 \\ c_2 \end{Bmatrix} = -P \begin{Bmatrix} 1 \\ \frac{L}{2} \end{Bmatrix}, \tag{10}$$

whose solution is

$$c_1 = -\frac{P}{12a_0 L}, \quad c_2 = -\frac{P}{9a_0 L^2}. \tag{11}$$

The two-parameter Petrov–Galerkin solution becomes

$$U_2(x) = \frac{PL}{36a_0}\left(18\frac{x}{L} + 3\frac{x^2}{L^2} + 4\frac{x^3}{L^3}\right). \tag{12}$$

We recall from Eq. (18) of **Example 6.3.1** that the analytical solution is

$$u(x) = -\frac{PL}{a_0} \log\left(1 - \frac{x}{2L}\right).$$ (13)

Table 6.4.1 contains a comparison of the three-parameter Ritz solution [from Eq. (16) of **Example 6.3.1**] and the two-parameter Petrov–Galerkin solution [from Eq. (12) of this example], both contain terms up to and including order x^3, with the analytical solution in Eq. (13). The numerical results are obtained for $L = 10$ m, $a_0 = 180 \times 10^6$ N, and $P = 10^4$ N. Table 6.4.1 also contains percentage of errors computed from

$$\text{Error} = \left|\frac{U_N - u}{u}\right| \times 100.$$

Table 6.4.1 Comparison of the Ritz and Petrov–Galerkin (P–Galerkin) solutions (multiplied by 10^4) with the analytical solution of the bar problem described by Eqs (6.3.4) and (6.3.5) with $k = 0$ and $f = 0$.

x/L	Analytical	Ritz $(N = 3)$	Error	P–Galerkin $(N = 2)$	Error
0.00	0.0000	0.0000	0.00	0.0000	0.00
0.10	0.2849	0.2872	0.78	0.2830	0.68
0.20	0.5853	0.5867	0.24	0.5790	1.08
0.30	0.9029	0.9021	0.08	0.8916	1.24
0.40	1.2397	1.2369	0.22	1.2247	1.21
0.50	1.5982	1.5947	0.22	1.5818	1.03
0.60	1.9815	1.9788	0.14	1.9667	0.75
0.70	2.3932	2.3930	0.01	2.3830	0.43
0.80	2.8379	2.8407	0.10	2.8346	0.12
0.90	3.3213	3.3254	0.12	3.3250	0.11
1.00	3.8508	3.8507	0.00	3.8580	0.19

We note that the solutions in the Ritz and Petrov–Gakerkin methods are obtained by minimizing the errors in integral sense. Both solutions are comparable in terms of accuracy with the three-parameter Ritz solution being slightly more accurate compared with the two-parameter Petrov–Galerkin solution.

6.4.3 The Galerkin Method

As discussed in the previous section, the Galerkin method is one in which the residual $\mathbb{R}_N$ is made orthogonal (with respect to the $L_2(\Omega)$ inner product) to the set of approximation functions $\{\phi_i\}_{i=1}^N$ used to approximate the solution u: Substituting from Eq. (6.4.7) into Eq. (6.4.6), we obtain [see Eqs (6.4.8) and (6.4.9)]

$$0 = \int_\Omega \phi_i \left\{ \sum_{j=1}^N c_j \left[A(\phi_j) + C(\phi_j) \right] + A(\phi_0) + C(\phi_0) - f \right\} d\Omega$$

$$= \sum_{j=1}^N A_{ij} c_j - b_i = 0 \quad \text{or} \quad \mathbf{Ac} = \mathbf{B}$$ (6.4.12)

for $i = 1, 2, \ldots, N$, where

$$A_{ij} = \int_\Omega \phi_i \left[A(\phi_j) + C(\phi_j) \right] d\Omega$$

$$\text{(6.4.13)}$$

$$b_i = \int_\Omega \phi_i \left[-A(\phi_0) - C(\phi_0) + f \right] d\Omega.$$

Thus, in the Galerkin method the error is made orthogonal to every element of the finite-dimensional vector space $\mathcal{V}_N$ in which the solution is sought. The set $\{\phi_j\}$ constitutes a basis of the finite-dimensional subspace $\mathcal{V}_N$. As in the Ritz method, the previously computed coefficients A_{ij} and b_i remain unaltered as we increase the value of N, provided the same approximation functions are retained.

The phrase "Galerkin's method" is often misused in the literature (especially, in the finite element literature). Some authors have used the phrase even when the integral statement in Eq. (6.4.12) is modified by carrying out integration by parts (when allowed) to reduce the order of differentiation on U_N and including the specified natural boundary conditions as a part of the variational statement. Such modification essentially amounts to changing the original idea of Galerkin and making it the same as the Ritz method.

The Ritz and Galerkin methods yield the same set of algebraic equations for the following three cases:

1. The specified boundary conditions of the problem are all essential type, and therefore the requirements on ϕ_i and ϕ_0 in both methods are the same.

2. The problem has both essential and natural boundary conditions, but the approximate functions used in the Galerkin method are also used in the Ritz method.

3. The governing equations of the problems are even order, allowing integration by parts to reduce the Galerkin statement to that used in the Ritz method and using the same approximation functions in both cases.

Example 6.4.2

Solve the bar problem of **Example 6.4.1** using the two-parameter Galerkin method with algebraic approximation functions.

Solution: The functions $\phi_0(x)$, $\phi_1(x)$, and $\phi_2(x)$ derived in **Example 6.4.1** are valid here. Hence, the two-parameter Galerkin approximation is

$$U_2(x) = c_1\phi_1(x) + c_2\phi_2(x) + \phi_0(x) = c_1\left(2Lx - x^2\right) + c_2\left(3L^2x - x^3\right) + \frac{P}{a_0}x. \quad (1)$$

The coefficients A_{ij} and b_i for the Galerkin model are (note that $C = 0$)

$$A_{11} = \int_0^L \phi_1 A(\phi_1)\,dx = \int_0^L (2Lx - x^2)\left[2a_0\left(2 - \frac{x}{L}\right) + \frac{a_0}{L}(2L - 2x)\right]dx = \frac{7}{3}a_0L^3,$$

$$A_{21} = \int_0^L \phi_2 A(\phi_1)\,dx = \int_0^L (3L^2x - x^3)\left[2a_0\left(2 - \frac{x}{L}\right) + \frac{a_0}{L}(2L - 2x)\right]dx = \frac{43}{10}a_0L^4,$$

$$A_{12} = \int_0^L \phi_1 A(\phi_2)\,dx = \int_0^L (2Lx - x^2)\left[a_0\left(2 - \frac{x}{L}\right)6x + \frac{a_0}{L}(3L^2 - 3x^2)\right]dx = \frac{43}{10}a_0L^4,$$

$$A_{22} = \int_0^L \psi_2 A(\phi_2)\,dx = \int_0^L (3L^2x - x^3)\left[a_0\left(2 - \frac{x}{L}\right)6x + \frac{a_0}{L}(3L^2 - 3x^2)\right]dx = \frac{81}{10}a_0L^5,$$

$$B_1 = -\int_0^L (2Lx - x^2)\frac{P}{L}\,dx = -\frac{2}{3}PL^2, \quad B_2 = -\int_0^L (3L^2x - x^3)\frac{P}{L}\,dx = -\frac{5}{4}PL^3. \quad (2)$$

In matrix form we have

$$\frac{a_0L^3}{30}\begin{bmatrix} 70 & 129L \\ 129L & 243L^2 \end{bmatrix}\begin{Bmatrix} c_1 \\ c_2 \end{Bmatrix} = -\frac{PL^2}{12}\begin{Bmatrix} 8 \\ 15L \end{Bmatrix}, \quad (3)$$

whose solution is

$$c_1 = -\frac{5}{82}\frac{P}{a_0L}, \quad c_2 = -\frac{10}{82}\frac{P}{a_0L^2}. \quad (4)$$

The two-parameter Galerkin solution becomes

$$U_2(x) = \frac{PL}{82a_0}\left(40\frac{x}{L} + 5\frac{x^2}{L^2} + 10\frac{x^3}{L^3}\right). \quad (5)$$

Table 6.4.2 contains a comparison of the three-parameter Ritz solution [from Eq. (16) of **Example 6.3.1**] and the two-parameter Galerkin solution [from Eq. (5) of this example], both containing terms up to and including order x^3, with the analytical solution in Eq. (13) of **Example 6.4.1**. Both solutions are comparable in terms of accuracy with the three-parameter Ritz solution being more accurate compared with the two-parameter Galerkin solution, the latter being less accurate than the Petrov–Galerkin solution presented in **Example 6.4.1**.

Table 6.4.2 Comparison of the Ritz and Galerkin solutions (multiplied by 10^4) with the analytical solution of the bar problem described by Eqs (6.3.4) and (6.3.5) with $k = 0$ and $f = 0$.

x/L	Analytical	Ritz ($N = 3$)	Error	Galerkin ($N = 2$)	Error
0.00	0.0000	0.0000	0.00	0.0000	0.00
0.10	0.2849	0.2872	0.78	0.2751	3.47
0.20	0.5853	0.5867	0.24	0.5610	4.16
0.30	0.9029	0.9021	0.08	0.8618	4.55
0.40	1.2397	1.2369	0.22	1.1816	4.69
0.50	1.5982	1.5947	0.22	1.5244	4.62
0.60	1.9815	1.9788	0.14	1.8943	4.40
0.70	2.3932	2.3930	0.01	2.2954	4.09
0.80	2.8379	2.8407	0.10	2.7317	3.74
0.90	3.3213	3.3254	0.12	3.2073	3.43
1.00	3.8508	3.8507	0.00	3.7263	3.23

Example 6.4.3 ————————————————————————————

Determine the one-parameter Galerkin solution for the transverse deflection of the simply supported beam problem of **Example 6.3.4**.

Solution: The approximation functions for the simply supported beam shown in Fig. 6.3.4 are determined by satisfying all of the conditions in Eq. (2) of **Example 6.3.4** ($\phi_0 = 0$):

$$\phi_i(0) = \phi_i(L) = 0, \quad \left.\frac{d^2\phi_i}{dx^2}\right|_{x=0,L} = 0. \tag{1}$$

We begin with a fourth-degree polynomial for $\phi_1(x)$:

$$\phi_1(x) = \alpha_0 + \alpha_1 x + \alpha_2 x^2 + \alpha_3 x^3 + \alpha_4 x^4$$

and find $\alpha_0 = 0, \alpha_2 = 0$ and

$$\alpha_1 + \alpha_3 L^2 + \alpha_4 L^3 = 0, \quad 6\alpha_3 + 12\alpha_4 L = 0,$$

from which we have

$$\alpha_3 = -2\alpha_4 L, \quad \alpha_1 = \alpha_4 L^3.$$

Thus, we have

$$\phi_1(x) = L^4 \left(\frac{x}{L} - 2\frac{x^3}{L^3} + \frac{x^4}{L^4} \right). \tag{2}$$

Substituting ϕ_1 into the Galerkin integral for the problem, we obtain

$$0 = \int_0^L \phi_1(x) \left(EI\frac{d^4 w}{dx^4} - q_0 \right) dx = \int_0^L \phi_1(x) \left(24EI c_1 - q_0 \right) dx \;\Rightarrow\; c_1 = \frac{q_0}{24EI}. \tag{3}$$

Thus, the one-parameter solution is

$$W_1(x) = \frac{q_0 L^4}{24EI} \left(\frac{x}{L} - 2\frac{x^3}{L^3} + \frac{x^4}{L^4} \right), \tag{4}$$

which coincides with the exact solution of the problem. The maximum deflection occurs at the center of the beam

$$w_{\max} = \frac{5q_0 L^4}{384EI}. \tag{5}$$

We note that the solution obtained here is independent of the particular weighted-residual method because $A(W_1) - q_0$ is a constant. It can be shown that a one-parameter Ritz solution with ϕ_1 given by Eq. (2) also yields the same exact solution in Eq. (4).

If we were to exploit the solution symmetry about $x = L/2$, we can use one-half of the beam to solve the problem. In using the first half of the beam, we must address the boundary conditions at $x = L/2$. The boundary conditions at this point are that the slope is zero, $(dw/dx) = 0$, and the shear force is zero, $EI(d^3 w/dx^3) = 0$. Hence, for this case, the approximation functions ϕ_i of the weighted-residual method must satisfy the conditions

$$\text{at } x = 0: \;\; \phi_i = 0, \;\; \frac{d^2\phi_i}{dx^2} = 0, \;\; \text{and at } x = \frac{L}{2}: \;\; \frac{d\phi_i}{dx} = 0, \;\; \frac{d^3\phi_i}{dx^3} = 0.$$

Obviously, the function $\phi_1(x)$ of Eq. (2) satisfies the conditions above. Hence, we obtain the same exact solution using any of the weighted-residual methods. If a trigonometric function is used for ϕ_i, we can use

$$\phi_i(x) = \sin \frac{(2i-1)\pi x}{L}.$$

However, it will not yield an exact solution with a finite N. This is because the representation of a uniform load q_0 by trigonometric functions is an infinite series.

Example 6.4.4

Consider the membrane of **Example 6.3.6** (see Fig. 6.3.7). Find the one-parameter Galerkin solution using algebraic polynomials. As in **Example 6.3.6** use one quadrant of the domain to exploit the biaxial symmetry of the problem.

Solution: The boundary conditions of the computational domain are

$$u = 0 \text{ at } x = a; \quad u = 0 \text{ at } y = a; \quad \frac{\partial u}{\partial x} = 0 \text{ at } x = 0; \quad \frac{\partial u}{\partial y} = 0 \text{ at } y = 0. \tag{1}$$

Thus, $\phi_0 = 0$. Using the boundary conditions, we can take a educated guess and write

$$\phi_1(x, y) = (a^2 - x^2)(a^2 - y^2), \tag{2}$$

which satisfies all of the boundary conditions in Eq. (1). We note that the same function is valid for the full domain.

From Eq. (15) of **Example 6.3.6**, we can write

$$A(U_1) - f_0 = -T \left(\frac{\partial^2 \phi_1}{\partial x^2} + \frac{\partial^2 \phi_1}{\partial y^2} \right) c_1 - f_0 = 2T(2a^2 - y^2 - x^2)c_1 - f_0. \tag{3}$$

Thus, we have

$$0 = \int_0^a \int_0^a (a^2 - x^2)(a^2 - y^2) \left[2T(2a^2 - y^2 - x^2)c_1 - f_0 \right] dx dy, \tag{4}$$

which gives

$$A_{11} = 2T \int_0^a \int_0^a (a^2 - x^2)(a^2 - y^2)(2a^2 - y^2 - x^2) \, dx dy = \frac{64}{45} a^8 T$$

$$B_1 = f_0 \int_0^a \int_0^a (a^2 - x^2)(a^2 - y^2) \, dx dy = \frac{4}{9} f_0 a^6. \tag{5}$$

The one-parameter Galerkin solution becomes

$$U_1(x, y) = \frac{5 f_0 a^2}{16T} \left(1 - \frac{x^2}{a^2} \right) \left(1 - \frac{y^2}{a^2} \right), \tag{6}$$

which coincides with the one-parameter Ritz solution with $\phi_1 = (a^2 - x^2)(a^2 - y^2)$ [see Eq. (19) of **Example 6.3.6**].

For the two-parameter Galerkin solution, we assume ($\phi_0 = 0$)

$$U_2(x, y) = c_1 \phi_1(x) + c_2 \phi_2(x) \tag{7}$$

with

$$\phi_1(x) = (a^2 - x^2)(a^2 - y^2), \quad \phi_2(x) = (x^2 + y^2)\phi_1. \tag{8}$$

The matrix equations obtained are

$$a^8 T \begin{bmatrix} \frac{256}{45} & \frac{1024}{525} a^2 \\ \frac{1024}{525} a^2 & \frac{11264}{4725} a^4 \end{bmatrix} \begin{Bmatrix} c_1 \\ c_2 \end{Bmatrix} = f_0 a^6 \begin{Bmatrix} \frac{16}{9} \\ \frac{32}{45} a^2 \end{Bmatrix}, \tag{9}$$

whose solution is

$$c_1 = \frac{1295 f_0}{4432 a^2 T}, \quad c_2 = \frac{525 f_0}{8864 a^4 T}. \tag{10}$$

The two-parameter Galerkin solution is given by

$$U_2(x,y) = \frac{f_0 a^2}{8864T} \left[2590 + 525 \left(\frac{x^2}{a^2} + \frac{y^2}{a^2} \right) \right] \left(1 - \frac{x^2}{a^2} \right) \left(1 - \frac{y^2}{a^2} \right). \tag{11}$$

We would have obtained the same solution by the Ritz method when Eqs (7) and (8) are used. The analytical solution of the problem is given in Eq. (16) of **Example 6.3.6**. The two-parameter Galerkin/Ritz solution at the center of the domain is $0.2922(f_0a^2/T)$, which is only 0.68% in error compared to the analytical solution.

6.4.4 The Least-Squares Method

The least-squares method is based on the idea of minimizing the integral of the sums of the squares of the residual in the approximations of the operator equations, Eqs (6.4.1) and (6.4.2) (when A, C, and B_2 are linear operators):

$$\mathbb{R}_N = \sum_{j=1}^{N} c_j \left[A(\phi_j) + C(\phi_j) \right] + A(\phi_0) + C(\phi_0) - f \neq 0 \text{ in } \Omega, \tag{6.4.14}$$

$$\mathbb{B}_N = \sum_{j=1}^{N} c_j B_2(\phi_j) + B_2(\phi_0) - \hat{g} \neq 0 \text{ on } \Gamma_g. \tag{6.4.15}$$

The least-squares functional is

$$I(c_1, c_2, \ldots, c_N) = \int_\Omega \mathbb{R}_N^2 \, d\Omega + \gamma \int_{\Gamma_g} \mathbb{B}_N^2 \, d\Gamma, \tag{6.4.16}$$

where γ is a weight parameter to make the squares of the two residuals, $\mathbb{R}_N$ and $\mathbb{B}_N$, to have the same physical dimensions (or units). By requiring that I to be the minimum, we obtain

$$0 = \int_\Omega \mathbb{R}_N \frac{\partial \mathbb{R}_N}{\partial c_i} \, d\Omega + \gamma \int_{\Gamma_g} \mathbb{B}_N \frac{\partial \mathbb{B}_N}{\partial c_i} \, d\Gamma, \tag{6.4.17}$$

or

$$0 = \sum_{j=1}^{N} c_j \int_\Omega \left[A(\phi_i) + C(\phi_i) \right] \left[A(\phi_j) + C(\phi_j) \right] d\Omega$$

$$+ \int_\Omega \left[A(\phi_i) + C(\phi_i) \right] \left[A(\phi_0) + C(\phi_0) - f \right] d\Omega$$

$$+ \gamma \sum_{j=1}^{N} c_j \int_{\Gamma_g} B_2(\phi_i) B_2(\phi_j) \, d\Gamma + \gamma \int_{\Gamma_g} B_2(\phi_i) \left[B_2(\phi_0) - \hat{g} \right] d\Gamma$$

$$= \sum_{j=1}^{N} c_j L_{ij} - F_i, \tag{6.4.18}$$

where

$$L_{ij} = \int_{\Omega} [A(\phi_i) + C(\phi_i)] [A(\phi_j) + C(\phi_j)] \, d\Omega + \gamma \int_{\Gamma_g} B_2(\phi_i) B_2(\phi_j) \, d\Gamma,$$

$$F_i = \int_{\Omega} [A(\phi_i) + C(\phi_i)] [-A(\phi_0) - C(\phi_0) + f] \, d\Omega$$

$$+ \gamma \int_{\Gamma_g} B_2(\phi_i) [-B_2(\phi_0) + \hat{g}] \, d\Gamma. \tag{6.4.19}$$

When $\gamma = 0$ (i.e., $\mathbb{B}_N$ is not included in the least-squares functional), ϕ_0 and ϕ_i are required to satisfy the conditions in Eq. (6.4.10). When $\mathbb{B}_N$ is included in the least-squares functional, ϕ_0 and ϕ_i are not required to meet the natural boundary conditions, and the requirements on ϕ_0 and ϕ_i become the same as those in the Ritz method. The least-squares method is the only other method, in addition to the Ritz method, that is based on the minimization of a functional and results in a symmetric coefficient matrix for linear operator problems. It also results in a positive-definite coefficient matrix.

Because $\mathbb{R}_N$ contains operators A and C, the least-squares method requires the same order of differentiability of ϕ_i as operators A and C. Therefore, it is desirable to express higher-order equations as a set of first-order equations and then apply the least-squares method. Suppose that $C = c$, a constant, and operator A is a differential operator that can be expressed as the product of two operators

$$A = S(aT) \tag{6.4.20}$$

where S and T are lower-order differential operators and a and c are functions of position. When A is a self-adjoint operator, $S = T^*$ is the adjoint of operator T. For example, we have

$$A(u) = -\frac{d}{dx}\left(a\frac{du}{dx}\right), \quad S = -\frac{d}{dx}, \quad T = \frac{d}{dx}.$$

We have

$$(v, Tu) = \int_0^L v \frac{du}{dx} \, dx = \int_0^L \left(-\frac{dv}{dx}\right) u \, dx + [u\,v]_0^L = (T^*v, u) + [u\,v]_0^L.$$

Then $S = T^*$ is said to be a formal adjoint of T.

When A is of the form in Eq. (6.4.1) with $C = c$, we can rewrite the original equation (6.4.1) as a pair of two lower-order equations

$$v = aT(u), \quad S(v) + cu = f \quad \text{or} \quad T(u) - \frac{v}{a} = 0, \quad S(v) + cu - f = 0. \tag{6.4.21}$$

Let u and v be approximated independently by M- and N-parameter expansions

$$u \approx U_M = \sum_{j=1}^{M} a_j \phi_j(x) + \phi_0(x), \quad v \approx V_N = \sum_{j=1}^{N} b_j \varphi_j + \varphi_0(x), \tag{6.4.22}$$

where $\phi_0(x)$ and $\varphi_0(x)$ satisfy the actual boundary conditions on u and v, respectively, and $\phi_i(x)$ and $\varphi_i(x)$ satisfy their homogeneous form, respectively. Then the residuals in the two equations are

$$\mathbb{R}_M \equiv T(U_M) - \frac{1}{a}V_N, \quad \mathbb{R}_N \equiv S(V_N) + cU_N - f. \tag{6.4.23}$$

The least-squares functional is

$$J(U_M, V_N) = \int_\Omega \left(\mathbb{R}_M^2 + \gamma \mathbb{R}_N^2\right) d\Omega, \tag{6.4.24}$$

where γ is a positive weight function that can be selected to equalize the magnitudes of the two residuals (to make both residual to approach zero at the same rate). Setting the first variation of J to zero and noting that the variations δa_i and δb_i are independent of each other, we obtain

$$0 = \frac{\partial J}{\partial a_i} = 2 \int_\Omega \left(\mathbb{R}_M \frac{\partial \mathbb{R}_M}{\partial a_i} + \gamma \mathbb{R}_N \frac{\partial \mathbb{R}_N}{\partial a_i}\right) d\Omega,$$

$$0 = \frac{\partial J}{\partial b_i} = 2 \int_\Omega \left(\mathbb{R}_M \frac{\partial \mathbb{R}_M}{\partial b_i} + \gamma \mathbb{R}_N \frac{\partial \mathbb{R}_N}{\partial b_i}\right) d\Omega,$$

or

$$0 = \int_0^L \left\{ T(\phi_i) \left[\sum_{j=1}^M T(\phi_j)a_j + T(\phi_0) - \frac{1}{a}\left(\sum_{j=1}^N \varphi_j b_j - \phi_0 \right) \right] \right.$$
$$\left. + \gamma c\phi_i \left[\sum_{j=1}^N S(\varphi_j)b_j + S(\phi_0) + c\left(\sum_{j=1}^M \phi_j a_j + \phi_0 \right) - f \right] \right\} dx, \tag{6.4.25}$$

$$0 = \int_0^L \left\{ -\frac{1}{a}\varphi_i \left[\sum_{j=1}^M T(\phi_j)a_j + T(\phi_0) - \frac{1}{a}\left(\sum_{j=1}^N \varphi_j b_j + \phi_0 \right) \right] \right.$$
$$\left. + \gamma S(\varphi_i) \left[\sum_{j=1}^N S(\varphi_j)b_j + S(\phi_0) + c\left(\sum_{j=1}^M \phi_j a_j + \phi_0 \right) - f \right] \right\} dx.$$
$$\tag{6.4.26}$$

Equations (6.4.25) and (6.4.26) can be expressed in matrix form as

$$\begin{bmatrix} \mathbf{A} & \mathbf{B} \\ \mathbf{B}^T & \mathbf{D} \end{bmatrix} \begin{Bmatrix} \mathbf{a} \\ \mathbf{b} \end{Bmatrix} = \begin{Bmatrix} \mathbf{F}^1 \\ \mathbf{F}^2 \end{Bmatrix}, \tag{6.4.27}$$

where the coefficients of matrices $\mathbf{A}$, $\mathbf{B}$, and $\mathbf{D}$ and column vectors $\mathbf{F}^1$ and $\mathbf{F}^2$

are defined by

$$A_{ij} = \int_0^L \left[T(\phi_i)T(\phi_j) + \gamma c^2 \phi_i \phi_j \right] dx,$$

$$B_{ij} = \int_0^L \left[-\frac{1}{a} T(\phi_i)\varphi_j + \gamma c \phi_i S(\varphi_j) \right] dx,$$

$$D_{ij} = \int_0^L \left[\frac{1}{a^2} \varphi_i \varphi_j + \gamma S(\varphi_i) S(\varphi_j) \right] dx, \tag{6.4.28}$$

$$F_i^1 = \int_0^L \left[\gamma c \phi_i f - T(\phi_i)T(\phi_0) + \frac{1}{a} T(\phi_i)\phi_0 - \gamma c \phi_i S(\phi_0) - \gamma c^2 \phi_i \phi_0 \right] dx,$$

$$F_i^2 = \int_0^L \left[\gamma S(\varphi_i) f + \frac{1}{a} \varphi_i T(\phi_0) - \frac{1}{a^2} \varphi_i \phi_0 - \gamma S(\varphi_i) S(\phi_0) - \gamma c S(\varphi_i)\phi_0 \right] dx.$$

Clearly, the coefficient matrix in Eq. (6.4.27) is symmetric. The least-squares model in Eq. (6.4.27) is termed a mixed least-squares model.

Example 6.4.5

Consider the problem of axial deformation of a bar from **Example 6.3.1**, **Example 6.4.1**, and **Example 6.4.2**. The governing equations are (with operator $C = 0$)

$$-\frac{d}{dx}\left[a_0 \left(2 - \frac{x}{L}\right) \frac{du}{dx} \right] = 0, \quad A \equiv -\frac{d}{dx}\left[a_0 \left(2 - \frac{x}{L}\right) \frac{d}{dx} \right], \quad 0 < x < L, \tag{1}$$

subjected to the boundary conditions (see Fig. 6.3.1)

$$u(0) = 0, \quad \left[a_0 \left(2 - \frac{x}{L}\right) \frac{du}{dx} \right]_{x=L} = P, \quad B_2 \equiv \left[a_0 \left(2 - \frac{x}{L}\right) \frac{d}{dx} \right]_{x=L}. \tag{2}$$

Determine (a) a two-parameter least-squares (LS-2) solution by minimizing the square of the residual in the differential equation and (b) a three-parameter least-squares (LS-3) solution by minimizing the sum of the squares of the residuals in the differential equation as well as in the natural boundary condition.

Solution: (a) For the problem at hand, we have ($f = 0$, $\hat{g} = 0$, and $\gamma = 0$)

$$L_{ij} = \int_0^L A(\phi_i) A(\phi_j) dx, \quad F_i = -\int_0^L A(\phi_i) A(\phi_0) dx, \tag{3}$$

where ϕ_0 must satisfy all of the actual specified boundary conditions and ϕ_i must satisfy the corresponding homogeneous boundary conditions. These functions are readily available from **Example 6.4.1**:

$$\phi_0(x) = \frac{Px}{a_0}, \quad \phi_1(x) = 2Lx - x^2, \quad \phi_2(x) = 3L^2 x - x^3. \tag{4}$$

Evaluating L_{ij} and F_i, we obtain

$$a_0^2 L \begin{bmatrix} \frac{52}{3} & 23L \\ 23L & \frac{186}{5}L^2 \end{bmatrix} \begin{Bmatrix} c_1 \\ c_2 \end{Bmatrix} = -a_0 P \begin{Bmatrix} 4 \\ 6L \end{Bmatrix}, \tag{5}$$

whose solution is

$$c_1 = -\frac{18}{193}\frac{P}{a_0 L}, \quad c_2 = -\frac{20}{193}\frac{P}{a_0 L^2}.$$

(6)

Hence the LS-2 solution based on Eq. (3) is

$$U_2(x) = \frac{PL}{193 a_0}\left(97\frac{x}{L} + 18\frac{x^2}{L^2} + 20\frac{x^3}{L^3}\right).$$

(7)

(b) Next, we consider the problem in which the sum of the squares of the residuals from Eq. (1) and the force (natural) boundary condition in Eq. (2) is minimized. In this case, the ϕ_0 and ϕ_i must satisfy the same conditions as in the Ritz method. Thus, we can take

$$\phi_0 = 0, \quad \phi_i = x^i.$$

(8)

Then we have

$$L_{ij} = \int_0^L A\left(\phi_i\right) A\left(\phi_j\right) dx + \gamma B_2(\phi_i) B_2(\phi_j), \quad F_i = B_2\left(\phi_i\right)_{x=L} P,$$

(9)

where

$$A(\phi_i) = -\frac{d}{dx}\left[a_0\left(2 - \frac{x}{L}\right)\frac{d\phi_i}{dx}\right] = -a_0\left(2 - \frac{x}{L}\right)\frac{d^2\phi_i}{dx^2} + \frac{a_0}{L}\frac{d\phi_i}{dx}$$

$$= -a_0\left(2 - \frac{x}{L}\right)i(i - 1)x^{i-2} + \frac{a_0}{L}ix^{i-1}$$

(10)

$$B_2(\phi_i) = \left[a_0\left(2 - \frac{x}{L}\right)\frac{d\phi_i}{dx}\right]_{x=L} = a_0 i L^{i-1}.$$

Then

$$L_{ij} = \int_0^L \frac{a_0^2}{L^2}\left[-(2L - x)i(i - 1)x^{i-2} + ix^{i-1}\right]\left[-(2L - x)j(j - 1)x^{j-2} + jx^{j-1}\right] dx$$

$$+ \gamma a_0^2 ij L^{i+j-2}$$

$$= a_0^2 L^{i+j-3}\left[\left(\frac{4}{i+j-3} + \frac{1}{i+j-1} - \frac{4}{i+j-2}\right)ij(i - 1)(j - 1) + \frac{ij}{i+j-1}\right.$$

$$\left. + \left(\frac{1}{i+j-1} - \frac{2}{i+j-2}\right)ij(i + j - 2)\right] + \gamma a_0^2 ij L^{i+j-2}$$

(11)

$$F_i = \gamma a_0 i L^{i-1} P.$$

It is clear from the two expressions of L_{ij} that the value of the parameter γ should be equal to $\gamma = 1/L$. Care should be taken in using various terms in the expression for L_{ij}. In particular, a term comes into effect only when the denominator is greater than zero [e.g., $1/(i+j-3)$ is valid only when $i + j - 3 > 0$]. Evaluating L_{ij} and F_i for $N = 3$, we obtain

$$\frac{a_0^2}{15L}\begin{bmatrix} 30 & 0 & 0 \\ 0 & 140L^2 & 165L^3 \\ 0 & 165L^3 & 288L^4 \end{bmatrix}\begin{Bmatrix} c_1 \\ c_2 \\ c_3 \end{Bmatrix} = \frac{a_0 P}{L}\begin{Bmatrix} 1 \\ 2L \\ 3L^2 \end{Bmatrix},$$

(12)

whose solution is

$$c_1 = \frac{P}{2a_0}, \quad c_2 = \frac{9}{97}\frac{P}{a_0 L}, \quad c_3 = \frac{10}{97}\frac{P}{a_0 L^2}.$$

(13)

Hence the three-parameter least-squares solution based on Eq. (9) is

$$U_3(x) = \frac{PL}{194 a_0}\left(97\frac{x}{L} + 18\frac{x^2}{L^2} + 20\frac{x^3}{L^3}\right).$$

(14)

Table 6.4.3 contains a comparison of the LS-2 and the LS-3 solutions from Eqs (7) and (14), respectively, with the analytical solution and the three-parameter Ritz solution from Eq. (16) of **Example 6.3.1**. Although the errors presented in Table 6.4.3 are pointwise errors, the variational methods try to minimize the errors in the integral sense over the entire domain.

Table 6.4.3 Comparison of the LS-2 and LS-3 solutions (multiplied by 10^4) with the analytical and three-parameter Ritz (Ritz-3) solutions of the bar problem described by Eqs (1) and (2).

x/L	Analytical	Ritz-3	Error	LS-2	Error	LS-3	Error
0.00	0.0000	0.0000	0.00	0.0000	0.00	0.0000	0.00
0.10	0.2849	0.2872	0.78	0.2849	0.00	0.2835	0.51
0.20	0.5853	0.5867	0.24	0.5838	0.27	0.5807	0.78
0.30	0.9029	0.9021	0.08	0.8998	0.34	0.8952	0.85
0.40	1.2397	1.2369	0.22	1.2366	0.25	1.2302	0.76
0.50	1.5982	1.5947	0.22	1.5976	0.04	1.5893	0.56
0.60	1.9815	1.9788	0.14	1.9862	0.23	1.9759	0.28
0.70	2.3932	2.3930	0.01	2.4059	0.53	2.3935	0.01
0.80	2.8379	2.8047	0.10	2.8601	0.78	2.8454	0.26
0.90	3.3213	3.3254	0.12	3.3523	0.93	3.3351	0.41
1.00	3.8508	3.8507	0.00	3.8860	0.91	3.8660	0.39

Example 6.4.6

Use the least-squares formulations of (a) the original equation and (b) two second-order equations to determine the solution for the transverse deflection of the simply supported beam problem of **Example 6.3.4**.

Solution: (a) The approximation function needed in a weighted-residual method for the simply supported beam shown in Fig. 6.3.4 was determined as ($\phi_0 = 0$):

$$\phi_1(x) = L^4 \left(\frac{x}{L} - 2\frac{x^3}{L^3} + \frac{x^4}{L^4} \right). \tag{1}$$

The least-squares functional for the original equation is ($A(w) = (d^2/dx^2)[EI(d^2w/dx^2)]$)

$$I_1(c_1) = \frac{1}{2} \int_0^L \left(c_1 EI \frac{d^4\phi_1}{dx^4} - q_0 \right)^2 dx, \tag{2}$$

and the necessary condition for its minimum with respect to c_1 is

$$0 = \int_0^L \left(c_1 EI \frac{d^4\phi_1}{dx^4} - q_0 \right) EI \frac{d^4\phi_1}{dx^4} dx$$
$$= 24EIL \left(24c_1 EI - q_0 \right)$$

Thus, c_1 is given by

$$c_1 = \frac{q_0}{24EI}. \tag{3}$$

Thus, the one-parameter least-squares solution based on the least-squares functional in Eq. (2) is the same as the one-parameter Galerkin solution that coincides with the exact solution of the problem:

$$W_1(x) = \frac{q_0 L^4}{24EI} \left(\frac{x}{L} - 2\frac{x^3}{L^3} + \frac{x^4}{L^4} \right). \tag{4}$$

(b) We consider the following two lower-order equations equivalent to the original fourth-order equation $[A = S \cdot EI \cdot T \text{ and } S = T^* = T = d^2/dx^2]$:

$$\frac{d^2 M}{dx^2} + q_0 = 0, \quad M = -EI \frac{d^2 w}{dx^2}, \tag{5}$$

or in dimensionless form

$$\frac{d^2 v}{d\xi^2} + \hat{q} = 0, \quad v + \frac{d^2 u}{d\xi^2} = 0, \tag{6}$$

where

$$\xi = \frac{x}{L}, \quad u = \frac{w}{L}, \quad v = \frac{ML}{EI}, \quad \hat{q} = \frac{q_0 L^3}{EI}. \tag{7}$$

The least-squares functional associated with the two equations in Eq. (6) is

$$J(u, v) = \int_0^1 \left[\left(\frac{d^2 v}{d\xi^2} + \hat{q} \right)^2 + \left(v + \frac{d^2 u}{d\xi^2} \right)^2 \right] d\xi. \tag{8}$$

The necessary condition for a minimum is

$$\delta J = 2 \int_0^1 \left[\frac{d^2 \delta v}{d\xi^2} \left(\frac{d^2 v}{d\xi^2} + \hat{q} \right) + \left(\delta v + \frac{d^2 \delta u}{d\xi^2} \right) \left(v + \frac{d^2 u}{d\xi^2} \right) \right] d\xi = 0, \tag{9}$$

which is equivalent to the following two integral statements:

$$0 = \int_0^1 \frac{d^2 \delta u}{d\xi^2} \left(v + \frac{d^2 u}{d\xi^2} \right) d\xi. \tag{10}$$

$$0 = \int_0^1 \left[\frac{d^2 \delta v}{d\xi^2} \left(\frac{d^2 v}{d\xi^2} + \hat{q} \right) + \delta v \left(v + \frac{d^2 u}{d\xi^2} \right) \right] d\xi \tag{11}$$

Let us assume approximations of the form

$$u(\xi) = \sum_{j=1}^{M} a_j \phi_j(\xi), \quad v(\xi) = \sum_{j=1}^{N} b_j \varphi_j(\xi). \tag{12}$$

Substituting the approximations from Eq. (12) into the integral statements in Eqs (10) and (11), we obtain

$$0 = \int_0^1 \frac{d^2 \phi_i}{d\xi^2} \left(\sum_{j=1}^{M} a_j \frac{d^2 \phi_j}{d\xi^2} + \sum_{j=1}^{N} b_j \varphi_j \right) d\xi$$

$$= \left(\sum_{j=1}^{M} A_{ij} a_j + \sum_{j=1}^{N} B_{ij} b_j \right)$$

$$\mathbf{0} = \mathbf{A}\mathbf{a} + \mathbf{B}\mathbf{b}, \tag{13}$$

and

$$0 = \sum_{j=1}^{M} a_j \int_0^1 \varphi_i \frac{d^2 \phi_j}{d\xi^2} d\xi + \sum_{j=1}^{N} b_j \int_0^1 \left(\frac{d^2 \varphi_i}{d\xi^2} \frac{d^2 \varphi_j}{d\xi^2} + \varphi_i \varphi_j \right) d\xi + \int_0^1 \frac{d^2 \varphi_i}{d\xi^2} \hat{q} \, d\xi$$

$$= \sum_{j=1}^{M} B_{ji} a_j + \sum_{j=1}^{N} D_{ij} b_j + F_i$$

$$\mathbf{0} = \mathbf{B}^{\mathsf{T}} \mathbf{a} + \mathbf{D}\mathbf{b} + \mathbf{F}, \tag{14}$$

where

$$A_{ij} = \int_0^1 \frac{d^2\phi_i}{d\xi^2} \frac{d^2\phi_j}{d\xi^2} \, d\xi,$$

$$B_{ij} = \int_0^1 \frac{d^2\phi_i}{d\xi^2} \varphi_j \, d\xi,$$

$$D_{ij} = \int_0^1 \left(\frac{d^2\varphi_i}{d\xi^2} \frac{d^2\varphi_j}{d\xi^2} + \varphi_i\varphi_j \right) d\xi \qquad (15)$$

$$F_i = \int_0^1 \frac{d^2\varphi_i}{d\xi^2} \hat{q} \, d\xi.$$

Equation (14) can be used to solve for **b** (because matrix **D** is invertible) and substituted into Eq. (13) to obtain an equation for **a**:

$$\mathbf{b} = -\mathbf{D}^{-1}\left(\mathbf{F} + \mathbf{B}^{\mathrm{T}}\mathbf{a}\right) \;\Rightarrow\; \left(\mathbf{A} - \mathbf{B}\mathbf{D}^{-1}\mathbf{B}^{\mathrm{T}}\right)\mathbf{a} = \mathbf{B}\mathbf{D}^{-1}\mathbf{F}. \qquad (16)$$

Since both $u(x)$ and $v(x)$ must satisfy similar conditions, $u(0) = u(1) = 0$ and $v(0) = v(1) = 0$, one may take $\phi_j(\xi) = \varphi_j(\xi) = \xi^i(1 - \xi)$.

6.4.5   The Collocation Method

In the collocation method we require the residual to vanish at a selected number of points $\mathbf{x}^i$ in the domain:

$$\mathbb{R}_N(\mathbf{x}^i, \{c\}, \{\phi\}, f) = 0 \;\Rightarrow\; \sum_{i=1}^N A_{ij}c_j = b_i, \quad (i = 1, 2, \ldots, N),$$

which can be written, with the help of the Dirac delta function, as

$$\int_\Omega \delta(\mathbf{x} - \mathbf{x}^i)\, \mathbb{R}_N(\mathbf{x}, \{c\}, \{\phi\}, f)\, d\mathbf{x} = 0, \quad (i = 1, 2, \ldots, N). \qquad (6.4.29)$$

Thus, the collocation method is a special case of the weighted-residual method given by Eq (6.4.6) with $\psi_i(\mathbf{x}) = \delta(\mathbf{x} - \mathbf{x}^i)$. The coefficients A_{ij} and b_i are defined as

$$A_{ij} = A(\phi_j(\mathbf{x}^i)) + C(\phi_j(\mathbf{x}^i)), \quad b_i = f(\mathbf{x}^i) - A(\phi_0(\mathbf{x}^i)) - C(\phi_0(\mathbf{x}^i)). \qquad (6.4.30)$$

In the collocation method, one must choose as many collocation points as there are undetermined parameters. In general, these points should be distributed uniformly in the domain, excluding the boundary points. Otherwise, ill-conditioned equations among c_j may result. Next, we consider two examples to illustrate the idea. Unlike in the Ritz and Galerkin methods, the previously computed coefficients A_{ij} and b_i in the collocation method *do not* remain unaltered as we increase the value of N, unless the collocation points $\mathbf{x}^i$ as well as the approximation functions ϕ_i from the previous approximation are retained.

Example 6.4.7

Consider the bar problem of **Example 6.3.1**, which is concerned with solving

$$-\frac{d}{dx}\left[a_0\left(2-\frac{x}{L}\right)\frac{du}{dx}\right] = 0, \quad 0 < x < L \tag{1}$$

subjected to boundary conditions

$$u(0) = 0, \quad \left[a_0\left(2-\frac{x}{L}\right)\frac{du}{dx}\right]_{x=L} = 0. \tag{2}$$

Solve the problem using the two-parameter collocation method with algebraic approximation functions.

Solution: From **Example 6.4.1**, we have the following approximation functions that are valid for any weighted-residual method:

$$\phi_0(x) = \frac{Px}{a_0}, \quad \phi_1(x) = x(2L - x), \quad \phi_2(x) = x(3L^2 - x^2). \tag{3}$$

Thus, the two-parameter approximation is

$$U_2(x) = c_1\left(2Lx - x^2\right) + c_2\left(3L^2x - x^3\right) + \frac{P}{a_0}x. \tag{4}$$

The residual of the approximation is

$$\mathbb{R}_N(x, c_1, c_2) = -\frac{d}{dx}\left[a_0\left(2-\frac{x}{L}\right)\frac{dU_2}{dx}\right] = -a_0\left(2-\frac{x}{L}\right)\frac{d^2U_2}{dx^2} + \frac{a_0}{L}\frac{dU_2}{dx}$$

$$= a_0c_1\left[\left(2-\frac{x}{L}\right)2 + \frac{1}{L}(2L - 2x)\right] + a_0c_2\left[\left(2-\frac{x}{L}\right)6x + \frac{1}{L}(3L^2 - 3x^2)\right] + \frac{P}{L}.$$

We select the collocation points to be $x = L/3$ and $x = 2L/3$. This choice is dictated by the fact that we have two points to choose and the obvious choice to place them is at one-third and two-third points of the domain. We obtain

$$\mathbb{R}_N(L/3, c_1, c_2) = a_0c_1\left[\left(2-\frac{1}{3}\right)2 + \left(2-\frac{2}{3}\right)\right] + a_0c_2L\left[2\left(2-\frac{1}{3}\right) + \left(3-\frac{1}{3}\right)\right] + \frac{P}{L},$$

$$\mathbb{R}_N(2L/3, c_1, c_2) = a_0c_1\left[\left(2-\frac{2}{3}\right)2 + \left(2-\frac{4}{3}\right)\right] + a_0c_2L\left[4\left(2-\frac{2}{3}\right) + \left(3-\frac{4}{3}\right)\right] + \frac{P}{L}.$$

In matrix form we have

$$\frac{a_0}{3}\begin{bmatrix} 14 & 18L \\ 10 & 21L \end{bmatrix}\begin{Bmatrix} c_1 \\ c_2 \end{Bmatrix} = -\frac{P}{L}\begin{Bmatrix} 1 \\ 1 \end{Bmatrix}, \tag{5}$$

whose solution is

$$c_1 = -\frac{3P}{38a_0L}, \quad c_2 = -\frac{4P}{38a_0L^2}. \tag{6}$$

Thus, the two-parameter collocation solution is

$$U_2(x) = \frac{PL}{38a_0}\left(20\frac{x}{L} + 3\frac{x^2}{L^2} + 4\frac{x^3}{L^3}\right). \tag{7}$$

The analytical solution from Eq. (18) of **Example 6.3.1** is

$$u(x) = -\frac{PL}{a_0}\log\left(1 - \frac{x}{2L}\right) \approx \frac{PL}{a_0}\left(\frac{1}{2}\frac{x}{L} + \frac{1}{8}\frac{x^2}{L^2} + \frac{1}{24}\frac{x^3}{L^3} + \cdots\right). \tag{8}$$

Example 6.4.8

Consider a simply supported beam of length L and constant flexural rigidity EI, subjected to uniformly distributed transverse load of intensity q_0 (see **Example 6.4.3**). Determine the two-parameter collocation solution of the form (the origin of the coordinate x is taken at the left end of the beam and the positive to the right)

$$W_2(x) = c_1 \sin \frac{\pi x}{L} + c_2 \sin \frac{3\pi x}{L},$$

which satisfies the boundary conditions $w = 0$ and $M = 0$ at $x = 0, L$ of the Bernoulli–Euler beam theory.

Solution: Due to the symmetry of the solution about $x = L/2$, it is sufficient to take the two points in the first half of the beam. The residual of approximation in the differential equation is

$$\mathbb{R}_N(x, c_1, c_2) = EI\frac{d^4 W_2}{dx^4} - q_0$$

$$= EI\left[c_1\left(\frac{\pi}{L}\right)^4 \sin\frac{\pi x}{L} + c_2\left(\frac{3\pi}{L}\right)^4 \sin\frac{3\pi x}{L}\right] - q_0. \tag{1}$$

Using collocation points at $x = L/4$ and $x = L/2$, we obtain

$$\mathbb{R}_N(L/4, c_1, c_2) = EI\left[c_1\left(\frac{\pi}{L}\right)^4 \sin\frac{\pi}{4} + c_2\left(\frac{3\pi}{L}\right)^4 \sin\frac{3\pi}{4}\right] - q_0 = 0,$$

$$\mathbb{R}_N(L/2, c_1, c_2) = EI\left[c_1\left(\frac{\pi}{L}\right)^4 \sin\frac{\pi}{2} + c_2\left(\frac{3\pi}{L}\right)^4 \sin\frac{3\pi}{2}\right] - q_0 = 0, \tag{2}$$

which yield

$$EI\left(\frac{\pi}{L}\right)^4 \begin{bmatrix} 1 & 81 \\ 1 & -81 \end{bmatrix} \begin{Bmatrix} c_1 \\ c_2 \end{Bmatrix} = q_0 \begin{Bmatrix} \sqrt{2} \\ 1 \end{Bmatrix}. \tag{3}$$

Solving for c_1 and c_2, we obtain

$$c_1 = \frac{1 + \sqrt{2}}{2}\frac{q_0 L^4}{EI\pi^4}, \quad c_2 = \frac{-1 + \sqrt{2}}{162}\frac{q_0 L^4}{EI\pi^4}. \tag{4}$$

The two-parameter collocation solution becomes

$$W_2(x) = \frac{q_0 L^4}{162 EI\pi^4}\left[81(1 + \sqrt{2}) \sin\frac{\pi x}{L} + (-1 + \sqrt{2}) \sin\frac{3\pi x}{L}\right]. \tag{5}$$

The maximum deflection is

$$W_2(L/2) = \frac{82 + 80\sqrt{2}}{162}\frac{q_0 L^4}{EI\pi^4} = \frac{q_0 L^4}{80.8676 EI} = \frac{4.7485}{384}\frac{q_0 L^4}{EI}, \tag{6}$$

which is about 5% lower compared with the exact value, $(5q_0 L^4/384EI) = (q_0 L^4/76.8EI)$.

If we have used $x = L/2$ and $x = 3L/4$, we would have obtained the same solution. We also note that we cannot use $x = L/4$ and $x = 3L/4$ because they both yield the same relations among c_1 and c_2. Thus, a judicious choice of the collocation points is necessary to obtain accurate solutions.

6.4.6 The Subdomain Method

To describe the subdomain method, we consider the problem of approximately solving the operator equation in Eq. (6.4.1):

$$A(u) + C(u) = f \quad \text{in} \quad \Omega. \tag{6.4.31}$$

As in Eq. (6.4.3), we seek approximate solution in the form

$$U_N(x) = \sum_{j=1}^{N} c_j \phi_j(x) + \phi_0(x) \quad \text{for} \quad x \in \Omega. \tag{6.4.32}$$

The residual in the approximation of Eq. (6.4.1) is given by Eq. (6.4.5)

$$\mathbb{R}_N \equiv A \left(\sum_{j=1}^{N} c_j \phi_j + \phi_0 \right) + C \left(\sum_{j=1}^{N} c_j \phi_j + \phi_0 \right) - f \neq 0 \quad \text{in} \quad \Omega. \tag{6.4.33}$$

Then the total domain Ω is subdivided into a finite number N subdomains,

$$\Omega = \cup_{i=1}^{N} \Omega^{(i)}, \tag{6.4.34}$$

and the integral of the residual $\mathbb{R}_N$ over each subdomain is set to zero:

$$0 = \int_{\Omega^{(i)}} \mathbb{R}_N \, d\Omega^{(i)} = \sum_{j=1}^{N} A_{ij} c_j - b_i, \quad i = 1, 2, \dots, N, \tag{6.4.35}$$

where

$$A_{ij} = \int_{\Omega^{(i)}} [A(\phi_j) + C(\phi_j)] \, d\Omega^{(i)}, \quad b_i = \int_{\Omega^{(i)}} [f - A(\phi_0) - C(\phi_0)] \, d\Omega^{(i)}. \tag{6.4.36}$$

Thus, in the subdomain method, the domain Ω is divided into as many subdomains $\Omega^{(i)}$ $(i = 1, 2, \dots, N)$ as there are the number of unknown parameters c_i in the approximation of the variable u and then the integral of the residual $\mathbb{R}_N$ over each subdomain is set to zero (i.e., the weight function on each subdomain is equal to unity). Equation (6.4.35) provides the required N equations for the determination of c_i, $i = 1, 2, \dots, N$. Obviously, in this method, negative errors can cancel positive errors within each subdomain to give zero net error (unless $\mathbb{R}_N$ is a positive function) in $\Omega^{(i)}$, although the sum of the absolute values of the errors may be very large. We note that the previously computed coefficients A_{ij} and b_i do *not* remain unaltered as we increase the value of N, unless the subdomains as well as the approximation functions ϕ_i from the previous approximation are retained. We also remark that the subdomain method is the same as the *finite volume method* used in computational fluid mechanics literature.

Example 6.4.9

Consider the bar problem of **Example 6.4.7**. Determine the two-parameter solution with algebraic approximation functions, using the subdomain method.

Solution: From **Example 6.4.7**, we have the following residual of the approximation:

$$\mathbb{R}_N(x, c_1, c_2) = a_0 c_1 \left(6 - 4\frac{x}{L}\right) + a_0 c_2 \left(3L + 12x - 9\frac{x^2}{L}\right) + \frac{P}{L}. \tag{1}$$

We divide the domain into two subdomains, $\Omega^1 = (0, L/2)$ and $\Omega^2 = (L/2, L)$. We obtain

$$\int_0^{L/2} \mathbb{R}_N \, dx = a_0 c_1 \left(3 - \frac{1}{2}\right) L + a_0 c_2 \left(3 - \frac{3}{8}\right) L^2 + \frac{P}{2},$$

$$\int_{L/2}^{L} \mathbb{R}_N \, dx = a_0 c_1 \left(4 - \frac{5}{2}\right) L + a_0 c_2 \left(6 - \frac{21}{8}\right) L^2 + \frac{P}{2}. \tag{2}$$

In matrix form we have

$$\frac{a_0 L}{8} \begin{bmatrix} 20 & 21L \\ 12 & 27L \end{bmatrix} \begin{Bmatrix} c_1 \\ c_2 \end{Bmatrix} = -\frac{P}{2} \begin{Bmatrix} 1 \\ 1 \end{Bmatrix}, \tag{3}$$

whose solution is

$$c_1 = -\frac{P}{12 a_0 L}, \quad c_2 = -\frac{P}{9 a_0 L^2}. \tag{4}$$

Thus, the two-parameter solution by the subdomain method is

$$U_2(x) = \frac{PL}{108 a_0} \left(54\frac{x}{L} + 9\frac{x^2}{L^2} + 12\frac{x^3}{L^3}\right) = \frac{PL}{a_0} \left(\frac{1}{2}\frac{x}{L} + \frac{1}{12}\frac{x^2}{L^2} + \frac{1}{9}\frac{x^3}{L^3}\right). \tag{5}$$

Example 6.4.10

Consider a simply supported beam of **Example 6.4.8**. Determine the two-parameter solution of the form

$$W_2(x) = c_1 \sin \frac{\pi x}{L} + c_2 \sin \frac{3\pi x}{L},$$

using the subdomain method.

Solution: Due to the symmetry of the solution about $x = L/2$, it is sufficient to subdivide the first half of the beam into two subdomains (i.e., intervals), say, $(0, L/4)$ and $(L/4, L/2)$. The residual is

$$\mathbb{R}_N(x, c_1, c_2) = EI\left[c_1 \left(\frac{\pi}{L}\right)^4 \sin \frac{\pi x}{L} + c_2 \left(\frac{3\pi}{L}\right)^4 \sin \frac{3\pi x}{L}\right] - q_0. \tag{1}$$

Using the subdomain method, we obtain

$$\int_0^{L/4} \mathbb{R}_N(x, c_1, c_2) \, dx = -EI\left[c_1 \left(\frac{\pi}{L}\right)^3 \cos \frac{\pi x}{L} + c_2 \left(\frac{3\pi}{L}\right)^3 \cos \frac{3\pi x}{L}\right]_0^{L/4} - \frac{q_0 L}{4} = 0,$$

$$\int_{L/4}^{L/2} \mathbb{R}_N(x, c_1, c_2) \, dx = -EI\left[c_1 \left(\frac{\pi}{L}\right)^3 \cos \frac{\pi x}{L} + c_2 \left(\frac{3\pi}{L}\right)^3 \cos \frac{3\pi x}{L}\right]_{L/4}^{L/2} - \frac{q_0 L}{4} = 0, \tag{2}$$

or

$$EI\left(\frac{\pi}{L}\right)^3 \begin{bmatrix} 1 - \frac{1}{\sqrt{2}} & 27(1 + \frac{1}{\sqrt{2}}) \\ \frac{1}{\sqrt{2}} & -27\frac{1}{\sqrt{2}} \end{bmatrix} \begin{Bmatrix} c_1 \\ c_2 \end{Bmatrix} = \frac{q_0 L}{4} \begin{Bmatrix} 1 \\ 1 \end{Bmatrix}. \tag{3}$$

Solving for c_1 and c_2, we obtain

$$c_1 = \frac{1 + \sqrt{2}}{4\sqrt{2}} \frac{q_0 L^4}{EI\pi^3}, \quad c_2 = \frac{-1 + \sqrt{2}}{108\sqrt{2}} \frac{q_0 L^4}{EI\pi^3}. \tag{4}$$

The two-parameter collocation solution becomes

$$W_2(x) = \frac{q_0 L^4}{108\sqrt{2}EI\pi^3} \left[27(1 + \sqrt{2}) \sin \frac{\pi x}{L} + (-1 + \sqrt{2}) \sin \frac{3\pi x}{L} \right]. \tag{5}$$

The maximum deflection is

$$W_2(L/2) = \frac{28 + 26\sqrt{2}}{108\sqrt{2}} \frac{q_0 L^4}{EI\pi^3} = \frac{q_0 L^4}{73.1169EI} = \frac{5.252}{384} \frac{q_0 L^4}{384EI}, \tag{6}$$

which is about 5% higher compared with the exact value, $(5q_0 L^4/384EI) = (q_0 L^4/76.8EI)$.

6.4.7 Eigenvalue and Time-Dependent Problems

In the previous subsections, we consider operator equations involving boundary value problems. Here we consider operator equations for eigenvalue problems and time-dependent problems and discuss their variational formulations.

6.4.7.1 Eigenvalue Problems

Consider operator equations of the form [see Eqs (6.4.1) and (6.4.2)]

$$A(u) + \lambda C(u) = 0 \text{ in } \Omega \quad B_1(u) = 0 \text{ on } \Gamma_u, \quad B_2(u) = 0 \text{ on } \Gamma_g \quad (6.4.37)$$

where A is a linear spatial differential operator, C is either a linear spatial differential operator or an algebraic operator, and λ is an unknown parameter called the eigenvalue, which is to be determined along with the eigenvector $u(\mathbf{x})$. The boundary operators B_1 and B_2 have the same meaning as before. In form Eq. (6.4.36) is similar but not the same as the equilibrium problem in Eqs (6.4.1) and (6.4.2) in view of the additional unknown λ in the problem and the homogeneous data $f = 0$, $\hat{u} = 0$, and $\hat{g} = 0$.

All linear structural dynamics problems can be reformulated as eigenvalue problems to determine natural frequencies (see **Example 6.3.2**), where λ denotes the square of the natural frequency. In addition, problems of buckling of structures can be formulated as an eigenvalue problem. An example of Eq. (6.4.37) is provided by the buckling of a beam column subjected to a compressive load P:

$$\frac{d^2}{dx^2}\left(EI\frac{d^2u}{dx^2}\right) + P\frac{d^2u}{dx^2} = 0, \quad 0 < x < L \tag{6.4.38}$$

where u denotes the lateral deflection and P is the axial compressive load. The problem involves determining the value of P and mode shape $u(x)$ such that the governing equation (6.4.38) and certain end conditions of the beam are satisfied. The minimum value of P is called the *critical buckling load*. Comparing the above equation with Eq. (6.4.37), we note that

$$A(u) = \frac{d^2}{dx^2}\left(EI\frac{d^2u}{dx^2}\right), \quad \lambda = P, \quad C(u) = \frac{d^2u}{dx^2}.$$

The N-parameter approximation for eigenvalue problems remains the same as in Eq. (6.4.3) with $\phi_0 = 0$ because the boundary conditions are homogeneous. However, the discrete problem described by Eqs (6.4.5) to (6.4.9) is different because of the presence of λ. The residual in the approximation of Eq. (6.4.37) is

$$\mathbb{R}_N \equiv A\left(\sum_{j=1}^{N} c_j\phi_j\right) + \lambda C\left(\sum_{j=1}^{N} c_j\phi_j\right) \neq 0 \text{ in } \Omega. \tag{6.4.39}$$

Then, as in the equilibrium problems, we set the integral of the weighted-residual to zero,

$$\int_\Omega \psi_i \, \mathbb{R}_N(\mathbf{x}, \{c\}, \{\phi\}, \lambda)d\Omega = 0, \ (i = 1, 2, \ldots, N), \tag{6.4.40}$$

and obtain (assuming that A and C are linear operators)

$$\int_\Omega \psi_i \left\{\sum_{j=1}^{N} [A(\phi_j) + \lambda C(\phi_j)] \, c_j\right\} d\Omega = \sum_{j=1}^{N} (A_{ij} + \lambda C_{ij}) \, c_j = 0$$

for $i = 1, 2, \ldots, N$. In matrix form we have

$$(\mathbf{A} + \lambda\mathbf{C})\mathbf{c} = \mathbf{0}, \tag{6.4.41}$$

where

$$A_{ij} = \int_\Omega \psi_i \, A(\phi_j) \, d\Omega, \quad C_{ij} = \int_\Omega \psi_i \, C(\phi_j) \, d\Omega. \tag{6.4.42}$$

Equation (6.4.41) is a standard eigenvalue problem (either $\mathbf{A}$ or $\mathbf{C}$ will contain a negative sign), which can be solved for the values of λ and corresponding $u(\mathbf{x})$.

6.4.7.2   Time-Dependent Problems

Consider operator equations of the form

$$A_t(u) + A(u) = f(\mathbf{x}, t) \text{ in } \Omega, \tag{6.4.43}$$

where A is a spatial differential operator and A_t is a temporal differential operator. Equation (6.4.43) is supplemented by boundary conditions of the form in Eq. (6.4.2) and initial conditions on u. Examples of Eq. (6.4.43) are provided by equations of heat transfer in a plane wall and axial motion of a bar, respectively:

$$\rho c_v \frac{\partial u}{\partial t} - \frac{\partial}{\partial x}\left(k\frac{\partial u}{\partial x}\right) = f(x, t), \tag{6.4.44}$$

$$-\rho A_0 \frac{\partial^2 u}{\partial t^2} - \frac{\partial}{\partial x}\left(EA_0 \frac{\partial u}{\partial x}\right) = f(x, t). \tag{6.4.45}$$

In Eq. (6.4.44), u is the temperature, ρ is the density, c_v is the specific heat at constant volume, k is the conductivity, and f is the internal heat generation. Clearly, we have

$$A_t(u) = \rho c_v \frac{\partial u}{\partial t}, \quad A(u) = -\frac{\partial}{\partial x}\left(k\frac{\partial u}{\partial x}\right).$$

In Eq. (6.4.45), u is the axial displacement, ρ is the density, E is Young's modulus, A_0 is the area of cross section, and f is the body force per unit length. In this case, we have

$$A_t(u) = -\rho A_0 \frac{\partial^2 u}{\partial t^2}, \quad A(u) = -\frac{\partial}{\partial x}\left(EA_0 \frac{\partial u}{\partial x}\right).$$

The N-parameter weighted residual approximation is assumed in the form

$$U_N(\mathbf{x}, t) = \sum_{j=1}^{N} c_j(t)\phi_j(\mathbf{x}) + \phi_0(\mathbf{x}), \tag{6.4.46}$$

where ϕ_0 and ϕ_i satisfy the same conditions as in the static problems and they are functions of only the spatial coordinate $\mathbf{x}$. However, we now made c_j to be a function of time t. The residual of the approximation is (assuming that A and A_t are linear operators)

$$\mathbb{R}_N \equiv A_t\left(\sum_{j=1}^{N} c_j\phi_j\right) + A\left(\sum_{j=1}^{N} c_j\phi_j\right) + A(\phi_0) - f(\mathbf{x}, t)$$

$$= \sum_{j=1}^{N} A_t(c_j)\phi_j + \sum_{j=1}^{N} c_j A(\phi_j) + A(\phi_0) - f(\mathbf{x}, t) \neq 0 \quad \text{in } \Omega. \tag{6.4.47}$$

Then, we set the integral of the weighted-residual to zero over the domain:

$$\int_{\Omega} \psi_i \, \mathbb{R}_N(\mathbf{x}, t, \{c\}, \{\phi\}) d\Omega = 0, \quad (i = 1, 2, \dots, N). \tag{6.4.48}$$

Substitution of the approximation (6.4.47) in Eq. (6.4.48), we have

$$\sum_{j=1}^{N} \left\{ \int_{\Omega} \psi_i \left[A(\phi_j)c_j + \phi_j A_t(c_j) \right] d\mathbf{x} \right\} - \int_{\Omega} \psi_i \left[f - A(\phi_0) \right] d\mathbf{x} = 0, \qquad (6.4.49)$$

or

$$\sum_{j=1}^{N} (A_{ij}c_j + M_{ij}A_t(c_j)) = F_i \quad (i = 1, 2, \ldots, N), \quad \mathbf{Ac} + \mathbf{MA_t(c)} = \mathbf{F}, \quad (6.4.50)$$

where A_{ij}, M_{ij}, and F_i are defined by

$$A_{ij} = \int_{\Omega} \psi_i A(\phi_j) \, d\mathbf{x}, \quad M_{ij} = \int_{\Omega} \psi_i \phi_j \, d\mathbf{x}, \quad F_i = \int_{\Omega} \psi_i \left[f - A(\phi_0) \right] d\mathbf{x}. \quad (6.4.51)$$

Equation (6.4.50) represents a set of N ordinary differential equations in time. They must be either further numerically approximated using a suitable time-integration scheme or solved by some analytical means. Three comments are in order.

1. In the case of time-dependent problems, the integral in Eq. (6.4.6) can still be taken over the spatial domain and the weight function ψ to be a function of spatial coordinate only. Then, Eq. (6.4.6) leads to a set of ordinary differential equations in time among the undetermined parameters $c_j(t)$. These ordinary differential equations in time need to be further approximated using suitable time approximation schemes.

2. In the case of the least-squares method, question arises as to what should be the weight function ψ_i for eigenvalue as well as time-dependent problems. Let us examine the least-squares method first for the linear eigenvalue problem (i.e., A and C are linear operators). We have

$$0 = \delta \int_{\Omega} \mathbb{R}_N^2 d\mathbf{x}, \quad \text{where} \quad \mathbb{R}_N = A(U_N) - \lambda C(U_N)$$

$$= 2 \int_{\Omega} \left[A(\phi_i) + \lambda C(\phi_i) \right] \mathbb{R}_N \, d\mathbf{x},$$

$$= 2 \sum_{j=1}^{N} \left\{ \int_{\Omega} \left[A(\phi_i) + \lambda C(\phi_i) \right] \left[A(\phi_j) + \lambda C(\phi_j) \right] d\mathbf{x} \right\} c_j.$$

Clearly, the eigenvalue problem becomes quadratic in λ, which is not desirable from a computational view point.

In the case of time-dependent problems, we have

$$0 = \delta \int_{\Omega} \mathbb{R}_N^2 d\mathbf{x}, \quad \mathbb{R}_N = A(U_N) + A_t(U_N)$$

$$= \int_{\Omega} \left[A(\phi_i) + \sum_{k=1}^{N} \phi_k \frac{\partial}{\partial c_i} A_t(c_k) \right] \mathbb{R}_N \, d\mathbf{x}, \quad (i = 1, 2, \ldots, N),$$

which is complicated due to the time derivative term. Thus, the least-squares method leads to complicated systems of equations for eigenvalue or time-dependent problems. An alternative is to use $\psi_i = A(\phi_i)$ in all cases. This avoids the problems seen above, although it deviates from the true least-squares idea of minimizing a quadratic functional.

3. It is possible to develop the so-called space–time approximations, that is, using the variational methods, treating time as an additional coordinate. It is found that such approaches are complicated, and they are not commonly used in practice. Therefore, we will not consider the space–time approximations in the present study.

Next, we consider a number of examples to illustrate the use of various weighted-residual methods, with the exception of the least-squares method due to its drawback in modeling eigenvalue and initial-value problems, in the solution of eigenvalue and time-dependent problems.

Example 6.4.11

Consider the eigenvalue problem described by Eq. (9) of **Example 6.3.2** (in a nondimensional form) for $\alpha = 1$:

$$\frac{d^2v}{d\xi^2} + \lambda v = 0, \quad v(0) = 0, \quad \frac{dv}{d\xi} + v = 0 \text{ at } \xi = 1.$$

Determine the first eigenvalue using various weighted-residual methods discussed in this section. Use (a) one-parameter and (b) two-parameter approximations.

Solution: In a weighted-residual method, ϕ_i must satisfy both of the boundary conditions:

$$\phi_i(0) = 0, \quad \phi_i'(1) + \phi_i(1) = 0. \tag{1}$$

(a) The lowest-order function that satisfies the two boundary conditions is

$$\phi_1(\xi) = 3\xi - 2\xi^2. \tag{2}$$

Galerkin's method. The one-parameter Galerkin's solution for the natural frequency can be computed using

$$0 = c_1 \int_0^1 \phi_1 \left(\frac{d^2\phi_1}{d\xi^2} + \lambda\phi_1 \right) d\xi \quad \text{or} \quad \left(-\frac{10}{3} + \frac{4}{5}\lambda \right) c_1 = 0, \tag{3}$$

which gives (for nonzero c_1) $\lambda = 50/12 = 4.1667$. If the same function is used for ϕ_1 in the one-parameter Ritz solution, we obtain, as discussed in **Example 6.3.2**, the same result as in the one-parameter Galerkin solution.

Least-squares method. The one-parameter least-squares approximation with $\psi_1 = A(\phi_1)$ gives

$$0 = c_1 \int_0^1 \frac{d^2\phi_1}{d\xi^2} \left(\frac{d^2\phi_1}{d\xi^2} + \lambda\phi_1 \right) d\xi \quad \text{or} \quad \left(-4 + \frac{5}{6}\lambda \right) c_1 = 0 \tag{4}$$

gives $\lambda = 4.8$. If we use $\psi_1 = A(\phi_1) + \lambda\phi_1$, we obtain

$$0 = c_1 \int_0^1 \left(\frac{d^2\phi_1}{d\xi^2} + \lambda\phi_1 \right) \left(\frac{d^2\phi_1}{dx^2} + \lambda\phi_1 \right) d\xi$$

$$= \left(\frac{4}{5}\lambda^2 - \frac{20}{3}\lambda + 16 \right) c_1, \tag{5}$$

whose roots are

$$\lambda_{1,2} = \frac{25}{6} \pm \frac{1}{6}\sqrt{445} \ \rightarrow \ \lambda_1 = 0.6508, \quad \lambda_2 = 7.6825. \tag{6}$$

Neither root is closer to the exact value of 4.1158. This indicates that the least-squares method with $\psi_i = A(\phi_i)$ is perhaps more suitable than $\psi_i = A(\phi_i) + \lambda C(\phi_i)$.

Collocation method. If we use one-parameter collocation method with the collocation point at $x = 0.5$, we obtain $[\phi_1(0.5) = 1.0$ and $(d^2\phi_1/d\xi^2) = -4.0]$

$$0 = c_1\phi_1(0.5) \left[\left. \left(\frac{d^2\phi_1}{dx^2} \right) \right|_{x=0.5} + \lambda\phi_1(0.5) \right] \quad \text{or} \quad (-4 + \lambda)c_1 = 0, \tag{7}$$

which gives $\lambda = 4$.

Subdomain method. The one-parameter subdomain method gives

$$0 = c_1 \int_0^1 \left(\frac{d^2\phi_1}{d\xi^2} + \lambda\phi_1 \right) d\xi \quad \text{or} \quad \left[-4 + \left(\frac{3}{2} - \frac{2}{3} \right)\lambda \right] c_1 = 0, \tag{8}$$

which gives $\lambda = 24/5 = 4.8$.

(b) Let us consider a two-parameter weighted-residual approximation. To determine $\phi_2(\xi)$, we begin with a polynomial that is one degree higher than that used for ϕ_1,

$$\phi_2(\xi) = a + b\xi + c\xi^2 + d\xi^3,$$

and obtain

$$\phi_2(0) = 0 \ \rightarrow \ a = 0; \quad \phi_2'(1) + \phi_2(1) = 0 \ \rightarrow \ 2b + 3c + 4d = 0 \text{ or } d = -\frac{2}{4}b - \frac{3}{4}c.$$

We can arbitrarily pick the values of b and c, except that not both equal to zero (for obvious reasons). Thus, we have infinite number of possibilities. Let us consider two such possibilities. First let us choose $b = 0$ and $c = 4$ and obtain $d = -3$. Then ϕ_2 becomes

$$\phi_2(\xi) = 4\xi^2 - 3\xi^3. \tag{9}$$

Alternatively, choose $c = 0$ and $b = 4$ and obtain $d = -2$. We denote this new ϕ_2 as φ_2,

$$\varphi_2(\xi) = 4\xi - 2\xi^3. \tag{10}$$

We can show that the set $\{\phi_1, \phi_2\}$ is equivalent to the set $\{\phi_1, \varphi_2\}$ in the sense that both yield the same solution with different values of their coefficients. Consider

$$V_2(\xi) = c_1\phi_1(\xi) + c_2\phi_2(\xi)$$
$$= c_1(3\xi - 2\xi^2) + c_2(4\xi^2 - 3\xi^3)$$
$$= 3c_1\xi + (-2c_1 + 4c_2)\xi^2 - 3c_2\xi^3,$$

and

$$V_2(\xi) = c_1\phi_1(\xi) + c_2\varphi_2(\xi)$$
$$= \bar{c}_1(3\xi - 2\xi^2) + \bar{c}_2(4\xi - 2\xi^3)$$
$$= (3\bar{c}_1 + 4\bar{c}_2)\,\xi - 2\bar{c}_1\xi^2 - 2\bar{c}_2\xi^3.$$

Comparing the two relations we find that

$$\bar{c}_1 = c_1 - 2c_2, \quad \bar{c}_2 = 1.5c_2.$$

Hence, either set will yield the same final solution for $V_2(\xi)$ or λ.

The residual of the approximation using (ϕ_1, ϕ_2) is

$$\mathbb{R}_N = \frac{d^2 V_2}{d\xi^2} + \lambda V_2 = c_1\frac{d^2\phi_1}{d\xi^2} + c_2\frac{d^2\phi_1}{d\xi^2} + \lambda\,(c_1\phi_1 + c_2\phi_2)$$

$$= c_1\left(\frac{d^2\phi_1}{d\xi^2} + \lambda\phi_1\right) + c_2\left(\frac{d^2\phi_2}{d\xi^2} + \lambda\phi_2\right)$$

$$= c_1\left[-4 + \lambda(3\xi - 2\xi^2)\right] + c_2\left[(8 - 18\xi) + \lambda(4\xi^2 - 3\xi^3)\right]. \tag{11}$$

Galerkin's method. For the Galerkin method, we set the integral of the weighted residual to zero and obtain

$$0 = \int_0^1 \phi_1(\xi)\mathbb{R}_N\,d\xi = \int_0^1 \phi_1(\xi)\left[c_1\frac{d^2\phi_1}{d\xi^2} + c_2\frac{d^2\phi_2}{d\xi^2} + \lambda\,(c_1\phi_1 + c_2\phi_2)\right]d\xi$$

$$= A_{11}c_1 + A_{12}c_2 + \lambda\,(C_{11}c_1 + C_{12}c_2),$$

$$0 = \int_0^1 \phi_2(\xi)\mathbb{R}_N\,d\xi = \int_0^1 \phi_2(\xi)\left[c_1\frac{d^2\phi_1}{d\xi^2} + c_2\frac{d^2\phi_1}{d\xi^2} + \lambda\,(c_1\phi_1 + c_2\phi_2)\right]d\xi$$

$$= A_{21}c_1 + A_{22}c_2 + \lambda\,(C_{21}c_1 + C_{22}c_2).$$

In matrix form, we have

$$\mathbf{Ac} + \lambda\mathbf{Cc} = \mathbf{0},$$

where

$$A_{ij} = \int_0^1 \phi_i\frac{d^2\phi_j}{d\xi^2}\,dx, \quad C_{ij} = \int_0^1 \phi_i\phi_j\,d\xi.$$

First, for the choice of functions in Eqs (2) and (9), we have

$$\frac{d^2\phi_1}{d\xi^2} = -4, \quad \frac{d^2\phi_2}{d\xi^2} = 8 - 18\xi.$$

Evaluating the integrals, we obtain

$$A_{11} = \int_0^1 \phi_1\frac{d^2\phi_1}{d\xi^2}\,d\xi = \int_0^1 (3\xi - 2\xi^2)(-4)d\xi = -\frac{10}{3},$$

$$A_{12} = \int_0^1 \phi_1\frac{d^2\phi_2}{d\xi^2}\,dx = \int_0^1 (3\xi - 2\xi^2)(8 - 18\xi)d\xi = -\frac{7}{3},$$

$$A_{21} = \int_0^1 \phi_2\frac{d^2\phi_1}{d\xi^2}\,dx = \int_0^1 (4\xi^2 - 3\xi^3)(-4)d\xi = -\frac{7}{3},$$

$$A_{22} = \int_0^1 \phi_2\frac{d^2\phi_2}{d\xi^2}\,dx = \int_0^1 (4\xi^2 - 3\xi^3)(8 - 18\xi)d\xi = -\frac{38}{15},$$

$$C_{11} = \int_0^1 \phi_1\phi_1\,d\xi = \int_0^1 (3\xi - 2\xi^2)(3\xi - 2\xi^2)d\xi = \frac{4}{5},$$

$$C_{12} = \int_0^1 \phi_1\phi_2\,d\xi = \int_0^1 (3\xi - 2\xi^2)(4\xi^2 - 3\xi^3)d\xi = \frac{3}{5} = C_{21},$$

$$C_{22} = \int_0^1 \phi_2\phi_2\,d\xi = \int_0^1 (4\xi^2 - 3\xi^3)(4\xi^2 - 3\xi^3)d\xi = \frac{17}{35},$$

and in matrix form, we have

$$\left(-\frac{1}{15}\begin{bmatrix} 50 & 35 \\ 35 & 38 \end{bmatrix} + \frac{\lambda}{35}\begin{bmatrix} 28 & 21 \\ 21 & 17 \end{bmatrix}\right)\begin{Bmatrix} c_1 \\ c_2 \end{Bmatrix} = \begin{Bmatrix} 0 \\ 0 \end{Bmatrix}. \tag{12}$$

For nontrivial solution, $c_1 \neq 0$ and $c_2 \neq 0$, we set the determinant of the coefficient matrix to zero to obtain the characteristic polynomial

$$675 - \frac{1332}{7}\lambda + \frac{315}{49}\lambda^2 = 0 \quad \text{or} \quad 525 - 148\lambda + 5\lambda^2 = 0, \tag{13}$$

which gives

$$\lambda_1 = 4.121, \quad \lambda_2 = 25.479. \tag{14}$$

Clearly, the value of λ_1 has improved over that computed using the one-parameter approximation. The exact value of the second eigenvalue is 24.139.

Collocation method. We select $\xi = 1/3$ and $\xi = 2/3$ as the collocation points and obtain

$$\mathbb{R}_N(\xi, c_1, c_2) = c_1\left[-4 + \lambda(3\xi - 2\xi^2)\right] + c_2\left[(8 - 18\xi) + \lambda(4\xi^2 - 3\xi^3)\right],$$

$$\mathbb{R}_N(1/3, c_1, c_2) = c_1\left(-4 + \frac{7}{9}\lambda\right) + c_2\left(2 + \frac{3}{9}\lambda\right) = 0, \tag{15}$$

$$\mathbb{R}_N(2/3, c_1, c_2) = c_1\left(-4 + \frac{10}{9}\lambda\right) + c_2\left(-4 + \frac{8}{9}\lambda\right) = 0,$$

and in matrix form, we have

$$\left(-\begin{bmatrix} 4 & -2 \\ 4 & 4 \end{bmatrix} + \frac{\lambda}{9}\begin{bmatrix} 7 & 3 \\ 10 & 8 \end{bmatrix}\right)\begin{Bmatrix} c_1 \\ c_2 \end{Bmatrix} = \begin{Bmatrix} 0 \\ 0 \end{Bmatrix}. \tag{16}$$

Setting the determinant of the coefficient matrix to zero, we obtain the following characteristic polynomial:

$$26\bar{\lambda}^2 - 68\bar{\lambda} + 24 = 0, \tag{17}$$

where $\bar{\lambda} = \lambda/9$. The roots of the characteristic polynomial are

$$\bar{\lambda}_1 = 0.4206, \quad \bar{\lambda}_2 = 2.1948 \quad \text{or} \quad \lambda_1 = 3.7851, \quad \lambda_2 = 19.7533. \tag{18}$$

The first eigenvalue is about -4.81% in error while the second eigenvalue is about 17.21% in error.

Subdomain method. Using the subdomain method with two equal intervals $(0, 0.5)$ and $(0.5, 1.0)$, we obtain

$$\int_0^{0.5} \mathbb{R}_N(\xi, c_1, c_2)\, d\xi = c_1\left[-4\xi + \lambda\left(\frac{3}{2}\xi^2 - \frac{2}{3}\xi^3\right)\right]_0^{0.5} + c_2\left[(8\xi - 9\xi^2) + \lambda\left(\frac{4}{3}\xi^3 - \frac{3}{4}\xi^4\right)\right]_0^{0.5}$$

$$= c_1\left[-2 + \lambda\left(\frac{3}{8} - \frac{2}{24}\right)\right] + c_2\left[\left(4 - \frac{9}{4}\right) + \lambda\left(\frac{1}{6} - \frac{3}{64}\right)\right]$$

$$= c_1\left(-2 + \frac{7}{24}\lambda\right) + c_2\left(\frac{7}{4} + \frac{23}{192}\lambda\right),$$

$$\int_{0.5}^1 \mathbb{R}_N(\xi, c_1, c_2)\, d\xi = c_1\left[-4\xi + \lambda\left(\frac{3}{2}\xi^2 - \frac{2}{3}\xi^3\right)\right]_{0.5}^1 + c_2\left[(8\xi - 9\xi^2) + \lambda\left(\frac{4}{3}\xi^3 - \frac{3}{4}\xi^4\right)\right]_{0.5}^1$$

$$= c_1\left(-4 + \frac{5}{6}\lambda\right) + c_2\left(-1 + \frac{7}{12}\lambda\right) - c_1\left(-2 + \frac{7}{24}\lambda\right) - c_2\left(\frac{7}{4} + \frac{23}{192}\lambda\right)$$

$$= c_1\left(-2 + \frac{13}{24}\lambda\right) + c_2\left(-\frac{11}{4} + \frac{89}{192}\lambda\right),$$

and we have the eigenvalue problem

$$\left(-\frac{1}{24}\begin{bmatrix} 48 & -42 \\ 48 & 66 \end{bmatrix} + \frac{\lambda}{192}\begin{bmatrix} 56 & 23 \\ 104 & 89 \end{bmatrix}\right)\begin{Bmatrix} c_1 \\ c_2 \end{Bmatrix} = \begin{Bmatrix} 0 \\ 0 \end{Bmatrix}. \tag{19}$$

The characteristic polynomial is

$$0 = (48 - 56\bar{\lambda})(66 - 89\bar{\lambda}) + (48 - 104\bar{\lambda})(42 + 23\bar{\lambda})$$
$$= 5184 - 11232\,\bar{\lambda} + 2592\,\bar{\lambda}^2$$

or

$$648 - 1404\,\bar{\lambda} + 324\,\bar{\lambda}^2 = 0$$

where $\bar{\lambda} = \lambda/8$. The roots of this equation are

$$\bar{\lambda}_1 = 0.5252 \quad \bar{\lambda}_2 = 3.8081 \quad \text{or} \quad \lambda_1 = 4.2015, \quad \lambda_2 = 30.4651. \tag{20}$$

Clearly, the first root has improved; the second root is in more error than that computed using the other weighted-residual methods.

Example 6.4.12

Formulate the problem of natural vibration considered in **Examples 6.3.2** (with $\alpha = 1$) and **6.4.11** using the N-parameter Galerkin approximation, and compare the frequencies obtained with the Ritz method and the analytical solutions for $N = 1, 2, \ldots, 8$. Note that $\omega = \sqrt{\lambda}\sqrt{E/\rho}/L$ [see Eq. (8) of **Example 6.3.2**].

Solution: We begin with the dimensionless form of the governing equations

$$\frac{d^2v}{d\xi^2} + \lambda v = 0, \quad 0 < \xi < 1, \quad v(0) = 0, \quad \frac{dv}{d\xi} + \alpha v = 0 \quad \text{at } \xi = 1. \tag{1}$$

Here the operator equation is of the form

$$A(v) + \lambda C(v) = 0, \quad A = \frac{d^2}{d\xi^2}, \quad C = 1. \tag{2}$$

The Galerkin integral is given by ($\phi_0 = 0$)

$$0 = \int_0^1 \phi_i \left[A(V_N) + \lambda C(V_N) \right] d\xi$$
$$= \sum_{j=1}^{N} \int_0^1 \phi_i \left[A(\phi_j) + \lambda C(\phi_j) \right] d\xi$$
$$= \sum_{j=1}^{N} (A_{ij} + \lambda C_{ij})\, c_j, \tag{3}$$

where

$$A_{ij} = \int_0^1 \phi_i \frac{d^2\phi_j}{d\xi^2}\, d\xi, \quad C_{ij} = \int_0^1 \phi_i \phi_j\, d\xi. \tag{4}$$

Next we derive $\phi_i(x)$ using the conditions

$$\phi_i(0) = 0, \quad \left[\frac{d\phi_i}{d\xi} + \alpha\phi_i \right]_{\xi=1} = 0. \tag{5}$$

Beginning with $\phi_1 = a_0 + a_1\xi + a_2\xi^2$ and using the conditions in Eq. (5), we obtain

$$a_0 = 0, \quad a_1 + 2a_2 + \alpha(a_1 + a_2) = 0 \quad \text{or} \quad a_1 = -\frac{2+\alpha}{1+\alpha}. \tag{6}$$

For $\alpha = 1$, we have

$$\phi_1(\xi) = (2+\alpha)\xi - (1+\alpha)\xi^2 = 3\xi - 2\xi^2.$$

Similarly, for ϕ_2 we begin with $\phi_2 = a_0 + a_1\xi + a_3\xi^3$, where we chose $a_2 = 0$. We obtain

$$\phi_2(\xi) = (3+\alpha)\xi - (1+\alpha)\xi^3 = 4\xi - 2\xi^3.$$

Thus, the ith approximation function for any weighted-residual method can be expressed as

$$\phi_i(\xi) = (2+i)\xi - 2\xi^{i+1}, \quad \frac{d^2\phi_i}{d\xi^2} = -2i(i+1)\xi^{i-1}. \tag{7}$$

Then we have

$$
\begin{aligned}
A_{ij} &= \int_0^1 \phi_i \frac{d^2\phi_j}{d\xi^2}\, d\xi = -\left[2(2+i)j - 4j\frac{j+1}{i+j+1}\right], \\
C_{ij} &= \int_0^1 \phi_i \phi_j\, d\xi = \frac{(2+i)(2+j)}{3} - 2\frac{2+i}{3+j} - 2\frac{2+j}{3+i} + \frac{4}{i+j+3}.
\end{aligned}
\tag{8}
$$

For $N = 1$, we obtain

$$-\frac{10}{3} + \frac{4}{5}\lambda = 0 \;\Rightarrow\; \lambda = \frac{50}{12} = 4.1667 \quad \text{or} \quad \omega = \frac{2.0412}{L}\sqrt{\frac{E}{\rho}}.$$

For $N = 2$, we obtain the following matrix eigenvalue problem:

$$\left(-\begin{bmatrix} 3.3333 & 6.0000 \\ 6.0000 & 11.2000 \end{bmatrix} + \lambda \begin{bmatrix} 0.8000 & 1.4667 \\ 1.4667 & 2.7047 \end{bmatrix}\right)\begin{Bmatrix} c_1 \\ c_2 \end{Bmatrix} = \begin{Bmatrix} 0 \\ 0 \end{Bmatrix}.$$

The eigenvalues and eigenfrequencies are

$$\lambda_1 = 4.1211, \quad \omega_1 = \frac{2.0300}{L}\sqrt{\frac{E}{\rho}}; \quad \lambda_2 = 25.4790, \quad \omega_2 = \frac{5.0477}{L}\sqrt{\frac{E}{\rho}}.$$

The dimensionless natural frequencies, $\bar{\omega} = \omega(L\sqrt{\rho/E})$, computed using the Galerkin method for various values of N are compared with the exact frequencies for $\alpha = kL/EA = 1$ (uniform bar fixed at the left end and supported by a linear elastic spring with spring stiffness $k = EA/L$ at the other end) in Table 6.4.4. The frequencies computed using the Galerkin method are very close to those computed using the Ritz method (see Table 6.3.3), knowing that $N+1$-parameter Ritz solution is equivalent to N-parameter Galerkin solution.

Table 6.4.4 Comparison of dimensionless natural frequencies ($\bar{\omega}_i = \omega_i L\sqrt{\rho/E}$)[a] computed with the Galerkin method for various values of N with the exact frequencies of a uniform isotropic bar fixed at one end and spring supported at the other end ($\alpha = 1$).

N	$\bar{\omega}_1$	$\bar{\omega}_2$	$\bar{\omega}_3$	$\bar{\omega}_4$	$\bar{\omega}_5$	$\bar{\omega}_6$	$\bar{\omega}_7$	$\bar{\omega}_8$
1	2.0412							
2	2.0300	5.0477						
3	2.0288	4.9439	8.458					
4	2.0288	4.9142	8.134	12.259				
5	2.0288	4.9133	7.991	11.532	16.542			
6	2.0288	4.9132	7.981	11.145	15.181	21.382		
7	2.0288	4.9132	7.979	11.099	14.388	19.139	26.826	
8	2.0288	4.9132	7.979	11.087	14.261	17.760	23.464	32.901
Exact	2.0288	4.9132	7.979	11.086	14.207	17.336	20.469	23.604

[a]The eigenvalue problem is solved using the Jacobi iteration method with a tolerance of 10^{-12}.

Example 6.4.13

Consider the equation of motion governing a membrane fixed on all its edges

$$T\left(\frac{\partial^2 u}{\partial x^2} + \frac{\partial^2 u}{\partial y^2}\right) - \rho\frac{\partial^2 u}{\partial t^2} = f(x,y,t), \quad \text{in } \Omega, \quad u = 0 \text{ on } \Gamma, \quad \text{for any } t > 0 \qquad (1)$$

where Ω is the triangular domain shown in Fig. 6.4.1, Γ is its boundary, ρ is the mass density, and T is the tension. Determine the fundamental frequency of natural vibration using a one-parameter Galerkin approximation of the problem.

Solution: For natural vibration, we set $f = 0$ and

$$u(x,y,t) = u_0(x,y)e^{i\omega t} \qquad (2)$$

and obtain

$$T\left(\frac{\partial^2 u_0}{\partial x^2} + \frac{\partial^2 u_0}{\partial y^2}\right) + \rho\omega^2 u_0 = 0, \quad \text{in } \Omega, \quad u_0 = 0 \text{ on } \Gamma. \qquad (3)$$

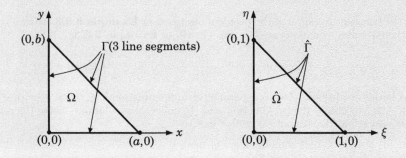

Fig. 6.4.1 Geometry and boundary conditions of a triangular membrane.

Introducing the following dimensionless variables,

$$\xi = \frac{x}{a}, \quad \eta = \frac{y}{b}, \quad v = \frac{u_0}{a}, \quad \lambda = \frac{\rho a^2 \omega^2}{T}, \qquad (4)$$

we obtain the dimensionless form of the governing equation

$$\left(\frac{\partial^2 v}{\partial \xi^2} + \frac{\partial^2 v}{\partial \eta^2}\right) + \lambda v = 0, \quad \text{in } \hat{\Omega}, \quad v = 0 \text{ on } \hat{\Gamma}. \qquad (5)$$

The one-parameter Galerkin approximation is based on the weighted-integral statement

$$0 = \int_{\hat{\Omega}} \phi_1 c_1 \left(\frac{\partial^2 \phi_1}{\partial \xi^2} + \frac{\partial^2 \phi_1}{\partial \eta^2} + \lambda\phi_1\right) d\xi d\eta. \qquad (6)$$

The function $\phi_1(\xi,\eta)$ must vanish on the boundary $\hat{\Gamma}$. Thus, we have $(a = b)$

$$\phi_1(\xi,0) = 0, \quad \phi_1(0,\eta) = 0, \quad \phi_1(\xi,\eta) = 0 \text{ on the line } \xi + \eta - 1 = 0. \qquad (7)$$

Hence, the choice for $\phi_1(\xi,\eta)$ is

$$\phi_1(\xi,\eta) = (\xi - 0)(\eta - 0)(\xi + \eta - 1), \quad \frac{\partial^2 \phi_1}{\partial \xi^2} + \frac{\partial^2 \phi_1}{\partial \eta^2} = 2(\xi + \eta). \qquad (8)$$

Hence, we have

$$0 = c_1 \int_0^1 \int_0^{1-\eta} [2(\xi + \eta) + \lambda \xi \eta(\xi + \eta - 1)] \, \xi \eta(\xi + \eta - 1) \, d\xi d\eta,$$

or, for $c_1 \neq 0$, we have the result

$$\lambda = -\frac{\int_0^1 \int_0^{1-\eta} 2(\xi + \eta) \xi \eta(\xi + \eta - 1) \, d\xi d\eta}{\int_0^1 \int_0^{1-\eta} \xi^2 \eta^2 (\xi + \eta - 1)^2 \, d\xi d\eta} = 56. \tag{9}$$

The natural frequency of vibration is

$$\omega = \frac{7.4833}{a} \sqrt{\frac{T}{\rho}}. \tag{10}$$

Example 6.4.14

Consider the transient response of the problem discussed in **Example 6.3.3** (with $g = 1$ and $\alpha = 1$). The governing equations are [see Eqs (7)–(9) of **Example 6.3.3**]

$$-\frac{\partial^2 v}{\partial \xi^2} + \frac{\partial^2 v}{\partial \tau^2} = 1, \quad v(0, \tau) = 0, \quad \frac{\partial v}{\partial \xi} + v = 0 \quad \text{at } \xi = 1 \text{ for all } \tau \geq 0, \tag{1}$$

with zero initial conditions. Use one-parameter approximation $v(\xi, \tau) \approx c_1(\tau)\phi_1(\xi)$ with $\phi_1(\xi) = 3\xi - 2\xi^2$ to obtain the (a) Galerkin and (b) collocation solutions. Solve the resulting ordinary equations in (dimensionless) time τ exactly.

Solution: (a) For the Galerkin method, we obtain

$$0 = \int_0^1 \phi_1 \left(-c_1 \frac{d^2 \phi_1}{d\xi^2} + \frac{d^2 c_1}{d\tau^2} \phi_1 - 1 \right) dx \quad \text{or} \quad \frac{4}{5} \frac{d^2 c_1}{d\tau^2} + \frac{10}{3} c_1 = \frac{5}{6}, \tag{2}$$

whose (exact) solution is ($\sqrt{50/12} \approx 2.0412$)

$$c_1(\tau) = A \sin 2.04\tau + B \cos 2.04\tau + \frac{1}{4}.$$

For zero initial conditions, $v(\xi, 0) = \dot{v}(\xi, 0) = 0$ [or $c_1(0) = 0$ and $\dot{c}_1(0) = 0$], the total solution becomes ($A = 0$ and $B = -1/4$)

$$V_1(\xi, \tau) = \frac{1}{4}(1 - \cos 2.04\tau)\left(3\xi - 2\xi^2\right). \tag{3}$$

(b) For one-parameter collocation method with collocation point at $\xi = 0.5$, we obtain

$$0 = \phi_1(0.5)\left[-c_1\left(\frac{d^2 \phi_1}{d\xi^2} \right)\bigg|_{0.5} + \frac{d^2 c_1}{d\tau^2}\phi_1(0.5) - 1 \right] \quad \text{or} \quad \frac{d^2 c_1}{d\tau^2} + 4c_1 = 1, \tag{4}$$

so that

$$c_1(\tau) = A \sin 2\tau + B \cos 2\tau + \frac{1}{4}.$$

The zero initial conditions give $A = 0$ and $B = -1/4$, and the one-parameter collocation solution becomes

$$V_1(\xi, \tau) = \frac{1}{4}(1 - \cos 2\tau)\left(3\xi - 2\xi^2\right). \tag{5}$$

Example 6.4.15

Consider the following partial differential equation (such as the one arising in transient heat transfer) in dimensionless form:

$$\frac{\partial u}{\partial t} - \frac{\partial^2 u}{\partial x^2} = 0, \quad 0 < x < 2, \tag{1}$$

$$u(0,t) = u(2,t) = 0 \quad \text{for } t > 0, \quad u(x,0) = 1.0 \quad \text{for } 0 < x < 2. \tag{2}$$

Use the symmetry about $x = 1$ and obtain (a) a one-parameter Galerkin solution and (b) a two-parameter Ritz solution to the time-dependent problem.

Solution: Owing to the symmetry about $x = 1$, we can solve the following equivalent problem:

$$\frac{\partial u}{\partial t} - \frac{\partial^2 u}{\partial x^2} = 0, \quad 0 < x < 1, \tag{3}$$

$$u(0,t) = 0, \quad \frac{\partial u}{\partial x}(1,t) = 0 \quad \text{for } t > 0, \quad u(x,0) = 1.0 \quad \text{for } 0 < x < 1. \tag{4}$$

(a) We first consider the following form of a one-parameter Galerkin approximation:

$$U_1(x,t) = c_1(t)\phi_1(x) = c_1(t)(2x - x^2). \tag{5}$$

The function $\phi_1(x) = 2x - x^2$ satisfies the boundary conditions in Eq. (4). We should determine $c_1(t)$ such that the initial condition in Eq. (4) is also satisfied. This requires that $c_1(t)$ must satisfy the condition

$$c_1(0) = 1. \tag{6}$$

Substituting Eq. (5) into Eq. (3) and setting up the Galerkin integral, we obtain

$$\begin{aligned}
0 &= \int_0^1 \left(\frac{\partial U_1}{\partial t} - \frac{\partial^2 U_1}{\partial x^2} \right) \phi_1 \, dx \\
&= \int_0^1 \left[\frac{dc_1}{dt}(2x - x^2) - c_1(-2) \right] (2x - x^2) \, dx \\
&= \frac{8}{15} \frac{dc_1}{dt} + \frac{4}{3} c_1.
\end{aligned} \tag{7}$$

The exact solution of Eq. (7) is

$$c_1(t) = Ae^{-(5/2)t}, \tag{8}$$

where the constant A is determined using the condition in Eq. (6)

$$c_1(0) = 1 \rightarrow A = 1.$$

Thus the one-parameter Galerkin solution becomes

$$U_{1G}(x,t) = e^{-2.5t}(2x - x^2). \tag{9}$$

The exact solution of Eqs (3) and (4) is given by

$$u(x,t) = 2 \sum_{n=0}^{\infty} \frac{e^{-\lambda_n^2 t} \sin \lambda_n x}{\lambda_n}, \quad \lambda_n = \frac{(2n+1)\pi}{2}. \tag{10}$$

(b) For a two-parameter Ritz approximation, we seek the solution in the form

$$u(x,t) \approx U_2(x,t) = c_1(t)\phi_1(x) + c_2(t)\phi_2(x), \tag{11}$$

with

$$\phi_1(x) = x, \quad \phi_2(x) = x^2, \tag{12}$$

which satisfy (the homogeneous form of) the essential boundary condition, $u(0) = 0$.
  Substituting the two-parameter approximation (11) into the weak form

$$0 = \int_0^1 \phi_i \left(\frac{\partial u}{\partial t} - \frac{\partial^2 u}{\partial x^2} \right) dx = \int_0^1 \left(\phi_i \frac{\partial u}{\partial t} + \frac{\partial \phi_i}{\partial x} \frac{\partial u}{\partial x} \right) dx, \tag{13}$$

we obtain the following two ordinary differential equations:

$$0 = \frac{1}{3} \frac{dc_1}{dt} + \frac{1}{4} \frac{dc_2}{dt} + c_1 + c_2, \tag{14}$$

$$0 = \frac{1}{4} \frac{dc_1}{dt} + \frac{1}{5} \frac{dc_2}{dt} + c_1 + \frac{4}{3} c_2, \tag{15}$$

which must be solved subject to the initial condition in Eq. (4). It is clear that there is no way the initial condition can be satisfied exactly by the selected approximation in Eqs (11) and (12). Therefore, we try to satisfy the initial condition in the Galerkin-integral sense:

$$\int_0^1 [u(x,0) - 1]\phi_1 \, dx = 0, \quad \int_0^1 [u(x,0) - 1]\phi_2 \, dx = 0, \tag{16}$$

which give

$$\frac{1}{3} c_1(0) + \frac{1}{4} c_2(0) = \frac{1}{2}, \quad \frac{1}{4} c_1(0) + \frac{1}{5} c_2(0) = \frac{1}{3}. \tag{17}$$

Then we use the Laplace transform method to solve Eqs (14), (15), and (17).
  Let $\bar{c}_i(s)$ denote the Laplace transform of $c_i(t)$:

$$\bar{c}_i = L[c_i(t)] \equiv \int_{-\infty}^{\infty} c_i(t) e^{-st} dt. \tag{18}$$

The Laplace transform of the derivative of $c_i(t)$ is given by

$$L\left[\frac{dc_i}{dt} \right] = s\bar{c}_i(s) - c_i(0). \tag{19}$$

Using Eq. (19), Eqs (14) and (15) can be transformed to

$$\left(\frac{1}{3} s + 1 \right) \bar{c}_1 + \left(\frac{1}{4} s + 1 \right) \bar{c}_2 = \frac{1}{3} c_1(0) + \frac{1}{4} c_2(0), \tag{20}$$

$$\left(\frac{1}{4} s + 1 \right) \bar{c}_1 + \left(\frac{1}{5} s + \frac{4}{3} \right) \bar{c}_2 = \frac{1}{4} c_1(0) + \frac{1}{5} c_2(0). \tag{21}$$

The right-hand sides of Eqs (20) and (21) can be replaced using Eq. (17), and the resulting equations can be solved for $\bar{c}_1(s)$ and $\bar{c}_2(s)$:

$$\bar{c}_1(s) = \frac{\frac{1}{2}\left(\frac{1}{5}s + \frac{4}{3}\right) - \frac{1}{3}\left(\frac{1}{4}s + 1\right)}{\left(\frac{1}{5}s + \frac{4}{3}\right)\left(\frac{1}{3}s + 1\right) - \left(\frac{1}{4}s + 1\right)^2} = \frac{12s + 240}{3s^2 + 104s + 240}, \tag{22}$$

$$\bar{c}_2(s) = \frac{\frac{1}{3}\left(\frac{1}{3}s + 1\right) - \frac{1}{2}\left(\frac{1}{4}s + 1\right)}{\left(\frac{1}{5}s + \frac{4}{3}\right)\left(\frac{1}{3}s + 1\right) - \left(\frac{1}{4}s + 1\right)^2} = -\frac{10s + 120}{3s^2 + 104s + 240}. \tag{23}$$

The inverse Laplace transform of Eqs (22) and (23) can be obtained by using the identity

$$L^{-1}\left[\frac{s+a_1}{(s+a)(s+b)}\right] = \frac{a_1-a}{b-a}e^{-at} + \frac{a_1-b}{a-b}e^{-bt}. \qquad (24)$$

Hence, we obtain

$$c_1(t) = 1.6408e^{-32.1807t} + 2.3592e^{-2.486t},$$

$$c_2(t) = -\left(2.265e^{-32.1807t} + 1.068e^{-2.486t}\right). \qquad (25)$$

The complete two-parameter Ritz solution is

$$U_{2R}(t) = x\left(1.6408e^{-32.1807t} + 2.3592e^{-2.486t}\right)$$

$$- x^2\left(2.265e^{-32.1807t} + 1.068e^{-2.486t}\right). \qquad (26)$$

The two variational solutions in Eqs (9) and (26) are compared, for various values of (x, t), with the series solution in Eq. (10) in Table 6.4.5. The two-parameter Ritz solution is more accurate than the one-parameter Galerkin solution and agrees more closely with the series solution for larger values of time.

Table 6.4.5 Comparison of the variational solutions with the series solution of a parabolic equation with initial condition, $u(x, 0) = 1$ [see Eqs (3) and (4)].

t	x	Series solution Eq. (10)	Variational solutions Eq. (9)	Eq. (26)
	0.2	0.4727	0.3177	0.4265
	0.4	0.7938	0.5648	0.7413
0.05	0.6	0.9418	0.7413	0.9443
	0.8	0.9880	0.8472	1.0357
	1.0	0.9965	0.8825	1.0154
	0.2	0.2135	0.1927	0.2306
	0.4	0.4052	0.3426	0.4152
0.25	0.6	0.5560	0.4496	0.5538
	0.8	0.6520	0.5139	0.6466
	1.0	0.6848	0.5353	0.6934
	0.2	0.1145	0.1031	0.1238
	0.4	0.2177	0.1834	0.2230
0.50	0.6	0.2996	0.2407	0.2975
	0.8	0.3522	0.2751	0.3473
	1.0	0.3703	0.2865	0.3725
	0.2	0.0617	0.0552	0.0665
	0.4	0.1174	0.0981	0.1198
0.75	0.6	0.1616	0.1288	0.1600
	0.8	0.1900	0.1472	0.1866
	1.0	0.1997	0.1534	0.2001
	0.2	0.0333	0.0296	0.0357
	0.4	0.0633	0.0525	0.0643
1.00	0.6	0.0872	0.0690	0.0858
	0.8	0.1025	0.0788	0.1002
	1.0	0.1077	0.0821	0.1075
	0.2	0.0097	0.0085	0.0103
	0.4	0.0184	0.0151	0.0186
1.50	0.6	0.0254	0.0198	0.0248
	0.8	0.0298	0.0226	0.0289
	1.0	0.0313	0.0235	0.0310

Example 6.4.16

Kantorovich method [34]. Consider the membrane/torsion problem considered in **Examples 6.3.6** and **6.4.4** (see Fig. 6.3.7). Find a one-parameter approximation of the form

$$U_1(x,y) = c_1(x)\phi_1(y) = c_1(x)(y^2 - a^2).$$ (1)

and obtain the equation governing $c_1(x)$ and solve it analytically. This procedure in which the unknown parameter is taken as a function of one of the coordinates is known as the *Kantorovich method*.

Solution: Since $u = 0$ on $x = \pm a$ and on $y = \pm a$, it follows that the parameter $c_1(x)$ must satisfy the conditions

$$c_1(-a) = c_1(a) = 0.$$ (2)

Substituting Eq. (1) into the Galerkin integral [with $T = 1$ in Eq. (15) of **Example 6.3.6**]

$$
\begin{aligned}
0 &= \int_{-a}^{a}\int_{-a}^{a}\left[-\left(\frac{\partial^2 U_1}{\partial x^2} + \frac{\partial^2 U_1}{\partial y^2}\right) - f_0\right]\phi_1\,dxdy \\
&= \int_{-a}^{a}\int_{-a}^{a}\left(-\frac{d^2 c_1}{dx^2}\phi_1 - c_1\frac{d^2\phi_1}{dy^2} - f_0\right)\phi_1\,dxdy \\
&= -\int_{-a}^{a}\left\{\int_{-a}^{a}\left[(y^2 - a^2)\frac{d^2 c_1}{dx^2} + 2c_1 + f_0\right](y^2 - a^2)dy\right\}dx
\end{aligned}
$$ (3)

and carrying out the integration with respect to y, we obtain

$$0 = \int_{-a}^{a}\left(\frac{d^2 c_1}{dx^2} - \frac{5}{2a^2}c_1 - \frac{5f_0}{4a^2}\right)dx.$$ (4)

An examination of the integrand in Eq. (4) shows that $c_1(x)$ is *not* a periodic function. Hence, the integral vanishes only if the integrand is identically zero:

$$\frac{d^2 c_1}{dx^2} - \frac{5}{2a^2}c_1 - \frac{5f_0}{4a^2} = 0.$$ (5)

This completes the application of the Galerkin method. We can solve the ordinary differential equation (5) either exactly or by an approximate method, such as the Ritz or Galerkin method. We consider the exact solution of Eq. (5) subjected to the boundary conditions in Eq. (2).

The general solution of Eq. (5) is given by

$$c_1(x) = A\cosh kx + B\sinh kx - \frac{f_0}{2}, \quad k = \sqrt{\frac{5}{2a^2}},$$ (6)

where the constants of integration, A and B, are determined using the conditions in Eq. (2). We obtain

$$A = \frac{f_0}{2\cosh ka}, \quad B = 0.$$

Then the solution in Eq. (1) becomes

$$U_1(x,y) = \frac{f_0 a^2}{2}\left(1 - \frac{y^2}{a^2}\right)\left(1 - \frac{\cosh kx}{\cosh ka}\right).$$ (7)

Using a two-parameter Galerkin approximation of the form

$$U_2(x,y) = (y^2 - a^2)[c_1(x) + c_2(x)y^2],$$ (8)

we obtain the differential equations

$$\frac{8}{15}a^2\frac{d^2c_1}{dx^2} + \frac{8}{105}a^4\frac{d^2c_2}{dx^2} - \left(\frac{4}{3}c_1 + \frac{4}{15}a^2c_2\right) = \frac{2}{3}f_0$$

$$\frac{8}{105}a^4\frac{d^2c_1}{dx^2} + \frac{8}{315}a^6\frac{d^2c_2}{dx^2} - \left(\frac{4a^2}{15}c_1 + \frac{44}{105}a^4c_2\right) = \frac{2}{15}f_0a^2,$$

(9)

with the boundary conditions on c_1 and c_2:

$$c_1(-a) = c_1(a) = c_2(-a) = c_2(a) = 0. \tag{10}$$

The simultaneous differential equations in Eq. (9) can be solved as follows: let $D = d/dx$, $D^2 = d^2/dx^2$, and so on. Then we can write Eq. (9) in the operator form

$$\begin{bmatrix} \frac{8a^2}{15}D^2 - \frac{4}{3} & \frac{8a^4}{105}D^2 - \frac{4a^2}{15} \\ \frac{8a^4}{105}D^2 - \frac{4a^2}{15} & \frac{8a^6}{315}D^2 - \frac{44a^4}{105} \end{bmatrix} \begin{Bmatrix} c_1 \\ c_2 \end{Bmatrix} = \frac{2}{15}f_0 \begin{Bmatrix} 5 \\ a^2 \end{Bmatrix}. \tag{11}$$

Using Cramer's rule, but keeping in mind that D's are differential operators that operate on the quantities in front of them, we obtain

$$\mathcal{L}(c_1) = \begin{vmatrix} \frac{2}{3}f_0 & \frac{8a^4}{105}D^2 - \frac{4a^2}{15} \\ \frac{2}{15}f_0a^2 & \frac{8a^6}{315}D^2 - \frac{44a^4}{105} \end{vmatrix}$$

$$= \left(\frac{8}{315}a^6D^2 - \frac{44}{105}a^4\right)\frac{2}{3}f_0 - \left(\frac{8}{105}a^4D^2 - \frac{4a^2}{15}\right)\frac{2}{15}f_0a^2,$$

$$= -\frac{384}{1575}f_0a^4, \tag{12}$$

$$\mathcal{L}(c_2) = \begin{vmatrix} \frac{8a^2}{15}D^2 - \frac{4}{3} & \frac{2}{3}f_0 \\ \frac{8}{105}a^4D^2 - \frac{4a^2}{15} & \frac{2}{15}f_0a^2 \end{vmatrix} = 0, \tag{13}$$

where $\mathcal{L}$ is the determinant of the operator matrix in Eq. (11),

$$\mathcal{L} = \frac{256a^8}{33075}\left(D^4 - \frac{28}{a^2}D^2 + \frac{63}{a^4}\right). \tag{14}$$

The general solutions of Eqs (12) and (13) are

$$c_1(x) = -\frac{f_0}{2} + A_1\cosh k_1x + A_2\sinh k_1x + A_3\cosh k_2x + A_4\sinh k_2x,$$

$$c_2(x) = B_1\cosh k_1x + B_2\sinh k_1x + B_3\cosh k_2x + B_4\sinh k_2x,$$

(15)

where A_i and $b_i(i = 1, 2, 3, 4)$ are constants to be determined, and k_1^2 and k_2^2 are the roots of the quadratic equation

$$k^4 - \frac{28}{a^2}k^2 + \frac{63}{a^4} = 0, \quad k_1^2 = 14 - \sqrt{133}, \quad k_2^2 = 14 + \sqrt{133}. \tag{16}$$

Substituting Eq. (15) into the first equation in (9), we obtain the relationships

$$\left(\frac{8a^2}{15}k_1^2 - \frac{4}{3}\right)A_i = \left(\frac{4a^2}{15} - \frac{8}{105}a^4k_1^2\right)b_i, \quad i = 1, 2,$$

$$\left(\frac{8a^2}{15}k_2^2 - \frac{4}{3}\right)A_i = \left(\frac{4a^2}{15} - \frac{8}{105}a^4k_2^2\right)b_i, \quad i = 3, 4.$$

(17)

Use of the boundary conditions in Eq. (10) gives $A_2 = A_4 = B_2 = B_4 = 0$, and

$$A_1 \cosh k_1 a + A_3 \cosh k_2 a = 0.5 f_0, \quad B_1 \cosh k_1 a + B_3 \cosh k_2 a = 0, \tag{18}$$

which can be solved along with Eq. (17) for A_1, B_1, A_3, and B_3, and we have

$$c_1(x) = 0.5 f_0 + 0.516 f_0 \frac{\cosh k_1 x}{\cosh k_1 a} - 0.0156 f_0 \frac{\cosh k_2 x}{\cos h k_2 a},$$
$$c_2(x) = -0.1138 f_0 \frac{\cosh k_1 x}{\cosh k_1 a} + 0.1138 f_0 \frac{\cosh k_2 x}{\cosh k_2 a}. \tag{19}$$

Equations (8) and (19) together define the two-parameter solution of the torsion problem.

A comparison of the solutions obtained using the Galerkin method [see Eqs (6) and (11) of **Example 6.4.4**] and the present solution (with one and two parameters) with the series solution [from Eq. (16) of **Example 6.3.6**] of the torsion problem is presented in Table 6.4.6. The dimensionless shearing stress, $\bar{\sigma}_{yz} = \sigma_{yz}/f_0 a$ [$\sigma_{yz} = -(\partial u/\partial x)$] at $x = a$ and $y = 0$ is found to be: 0.675 (exact), 0.625 (one-parameter Galerkin), 0.70284 (two-parameter Galerkin), 0.72636 (one-parameter Kantorovich), and 0.66416 (two-parameter Kantorovich). It is clear from the results that the two-parameter solutions for $\bar{u}(x, 0)$ are more accurate than the one-parameter solutions.

Table 6.4.6 Comparison of the solution $\bar{u} = u(x, 0)/q_0 a^2$ of Eq. (15) of **Example 6.3.6** obtained by the Ritz–Galerkin[a] and the Kantorovich methods with the series solution.

| x/a | Series | Ritz–Galerkin solution[a] | | Kantorovich solution | |
(y = 0)	solution	One-parameter	Two-parameter	One-parameter	Two-parameter
0.0	0.29445	0.31250	0.29219	0.30261	0.29473
0.1	0.29194	0.30937	0.28986	0.30014	0.29222
0.2	0.28437	0.30000	0.28278	0.29266	0.28462
0.3	0.27158	0.28437	0.27075	0.27999	0.27177
0.4	0.25328	0.26250	0.25340	0.26180	0.25339
0.5	0.22909	0.23437	0.23025	0.23765	0.22909
0.6	0.19854	0.20000	0.20065	0.20693	0.19839
0.7	0.16104	0.15937	0.16382	0.16886	0.16072
0.8	0.11591	0.11250	0.11884	0.12250	0.11546
0.9	0.06239	0.05937	0.06463	0.06667	0.06203
1.0	−0.00038	0.00000	0.00000	−0.00447	0.00000

[a] The Ritz and Galerkin solutions are the same for the same choice of approximation functions.

The Trefftz method. In all the variational methods discussed in this chapter, the coordinate functions were selected such that they satisfied the boundary conditions of the problem, and the unknown parameters were determined using a variational procedure, such as the minimization of a quadratic functional or setting the weighted residual to zero. In the Tefftz method [106, 107], the approximation functions are selected to satisfy the governing differential equation, and the unknown parameters are determined such that the boundary conditions are satisfied in a variational/integral sense. The application of the method is illustrated in **Example 6.4.17** via the torsion problem of **Example 6.4.16**.

Example 6.4.17 ————————————————————————

The torsion of cylindrical members can be formulated alternatively in terms of the conjugate function Ψ, which is related to the Prandtl stress function Φ,

$$\Phi = \Psi - \frac{f_0}{4}(x^2 + y^2) \quad \text{in } \Omega, \tag{1}$$

where $f_0 = 2G\theta$. Consequently, Ψ is governed by the Laplace equation

$$\nabla^2 \Psi = 0 \quad \text{in} \quad \Omega = \{(x, y) : -a < x, y < a\}. \tag{2}$$

subjected to the boundary condition

$$\Psi = \frac{f_0}{4}(x^2 + y^2) \quad \text{on} \quad \Gamma. \tag{3}$$

Note that a nonhomogeneous equation (i.e., Poisson's equation) with homogeneous boundary conditions is transformed to a homogeneous equation (i.e., Laplace's equation) with nonhomogeneous boundary conditions). For the square domain of **Example 6.4.16**, formulate an N-parameter approximation of Eqs (2) and (3) using the Trefftz method and then give the numerical solution for $N = 1$.

Solution: We select an N-parameter approximation of the form

$$\Psi \approx \Psi_N = \sum_{j=1}^{N} c_j \phi_j(x, y), \tag{4}$$

where ϕ_j are selected such that Ψ_N satisfies the governing equation $\nabla^2 \Psi_N = 0$ (i.e., ϕ_j should be harmonic functions). The parameters c_j are determined by the requirement that the boundary condition (3) is satisfied in an integral sense

$$
\begin{aligned}
0 &= \oint_{\Gamma} \left[\Psi_N - \frac{f_0}{4}(x^2 + y^2) \right] \frac{\partial \phi_i}{\partial n}\, ds \\
&= \oint_{\Gamma} \left[\sum_{j=1}^{N} c_j \phi_j - \frac{f_0}{4}(x^2 + y^2) \right] \frac{\partial \phi_i}{\partial n}\, ds \\
&\equiv \sum_{j=1}^{N} T_{ij} c_j - b_i,
\end{aligned}
\tag{5}
$$

where

$$T_{ij} = \oint_{\Gamma} \frac{\partial \phi_i}{\partial n} \phi_j\, ds, \quad b_i = \oint_{\Gamma} \frac{f_0}{4}(x^2 + y^2) \frac{\partial \phi_i}{\partial n}\, ds. \tag{6}$$

For $N = 1$, we choose

$$\phi_1 = x^4 - 6x^2 y^2 + y^4, \tag{7}$$

which satisfies the equation $\nabla^2 \phi_1 = 0$. We then have

$$
\begin{aligned}
T_{11} &= 4 \int_{-a}^{a} (4a^3 - 12ax^2)(a^4 - 6a^2 x^2 + x^4)\, dx = \frac{1536}{35} a^8, \\
B_1 &= -\int_{-a}^{a} f_0(-12x^2 a + 4a^3)(x^2 + a^2)\, dx = -\frac{64}{30} a^6 f_0,
\end{aligned}
\tag{8}
$$

and the solution is given by

$$\Psi_1 = -\frac{7f_0}{144a^2}(x^4 - 6x^2 y^2 + y^4). \tag{9}$$

The Prandtl stress function Φ is given by

$$\Phi = \Psi - \frac{f_0}{4}(x^2+y^2) = -\frac{7f_0}{144a^2}(x^4 - 6x^2y^2 + y^4) - \frac{f_0}{4}(x^2+y^2). \tag{10}$$

The maximum shear stress σ_{yz} is given by

$$\sigma_{yz}(a,0) = -\left(\frac{\partial\Phi_1}{\partial x}\right)_{(a,0)} = \frac{25}{36}f_0 a = 0.6944 f_0 a.$$

The maximum shear stress is about 2.88% greater than the exact solution.

The Trefftz method has limited use because it can only be utilized for Dirichlet type boundary-value problems (i.e., boundary-value problems in which the function is specified on the boundary). Furthermore, it is not easy to find approximation functions that satisfy the governing differential equations. Some of the hybrid finite element models are based on ideas similar to the Trefftz method.

Example 6.4.18

Consider the nonlinear differential equation

$$-\frac{d}{dx}\left(u\frac{du}{dx}\right) + 1 = 0, \quad 0 < x < 1, \tag{1}$$

subject to the boundary conditions

$$u(1) = \sqrt{2}, \quad \left(\frac{du}{dx}\right)\bigg|_{x=0} = 0. \tag{2}$$

Determine a one-parameter Ritz solution to the problem.

Solution: The weak form of Eq. (1) is given by

$$0 = \int_0^1 \left[u\frac{dw}{dx}\frac{du}{dx} + w\right]dx.$$

The boundary term is zero because $w(1) = 0$ and $\frac{du}{dx}(0) = 0$. Let

$$u \approx U_1 = c_1\phi_1 + \phi_0, \quad \text{with } \phi_0 = \sqrt{2}, \quad \phi_1 = 1 - x.$$

Substituting into the weak form, we obtain

$$0 = \int_0^1 \left\{\left[c_1(1-x) + \sqrt{2}\right](-1)(-c_1) + (1-x)\right\}dx$$

$$= c_1\left[c_1\cdot\frac{1}{2} + \sqrt{2}\right] + \frac{1}{2},$$

or

$$c_1^2 + 2\sqrt{2}c_1 + 1 = 0, \quad (c_1)_{1,2} = -\sqrt{2} \pm \sqrt{2-1} = -\sqrt{2} \pm 1.$$

Thus, there are two approximate solutions to the nonlinear problem. We must choose one value of c_i using some meaningful criterion. We shall take the value of c_1 that yields the smallest integral of the residual in the differential equation:

$$\int_0^1 \left[-\frac{d}{dx}\left(u\frac{du}{dx}\right) + 1\right]dx = (c_1^2 + 1).$$

Clearly, the smaller (in absolute value) root gives the smaller value of the residual. Hence, we choose $(c_1)_1 = -\sqrt{2} + 1$. The solution becomes

$$U_1 = (1 - \sqrt{2})(1 - x) + \sqrt{2} = 1 + (\sqrt{2} - 1)x.$$

The exact solution is $u(x) = \sqrt{1 + x^2}$. At $x = 0$ the approximate solution matches with the exact.

6.5 Summary

In this chapter, the Ritz and weighted-residual (e.g., Galerkin, least-squares, subdomain, and collocation) methods were presented, and their application to simple problems was illustrated. The Ritz method makes use of the weak form provided by the principle of virtual displacements, the principles of minimum total potential energy, or the one developed from the governing equations of the problem as discussed in Section 6.3.6. The key feature of the weak form is that it includes the governing equation(s) as well as the natural boundary condition(s) of the problem. Hence, the Ritz approximation is not required to satisfy the natural boundary conditions.

All weighted-residual methods, except the least-squares method, are based on a weighted-integral statement of the governing equation(s), whereas the least-squares method is based on the minimization of the square of the residuals in the governing equation(s). Thus, the integral statements used in all weighted-residual methods do not include any boundary conditions as a part of the statements. Hence, the approximations chosen for the weighted-residual methods, including the least-squares method, are required to satisfy *all*, both natural and essential, specified boundary conditions of the problem. Consequently, the approximation functions are of higher order than those used in the Ritz method.

The Ritz as well as the weighted-residual methods may be used for linear and nonlinear differential equations. However, in the case of least-squares method, there are some limitations, as discussed in Sections 6.4.4 and 6.4.7. Although the least-squares method can be interpreted as a special case of the weighted-residual methods for linear static problems, it is based on an integral statement whose Euler equations, if derived, are *not* the same as the governing equations. Thus, the least-squares method is quite different from the other weighted-residual methods. Also, the subdomain method is a weighted-residual method with weight function equal to unity when the domain is subdivided into the same number of subdomains as there are unknown coefficients. Figure 6.5.1 summarizes various variational methods. Overall, the Ritz method is the most efficient method, especially for solid and structural mechanics problems, and it is the most commonly used technique in the development of displacement-based finite element models.

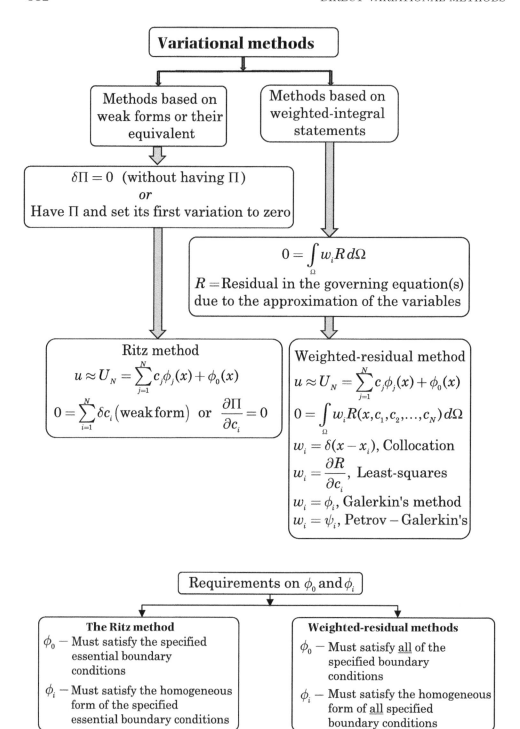

Fig. 6.5.1 A flow chart summarizing the various variational methods.

The single most difficult step in using the classical variational methods presented in this chapter is the selection of the approximation functions. The requirements in Eqs (6.3.22) and (6.4.10) on the approximation functions merely provide some guidelines for their selection. The selection of the approximation functions becomes even more difficult for problems with irregular domains (i.e., noncircular and nonrectangular) or discontinuous data (loading as well as geometry). Further, the generation of coefficient matrices for the resulting algebraic equations cannot be automated for a *class* of problems that differ from each other only in the geometry of the domain, boundary conditions, or loading. These limitations of the classical variational methods can be overcome by representing a given domain as a collection of geometrically simple subdomains for which we can systematically generate the approximation functions (see **Example 6.3.5**). One such technique, namely, the finite element method, is discussed in Chapter 8. The finite element method is based on ideas similar to the classical variational methods, especially in developing the system of algebraic equations for the unknown coefficients, but the method views a given domain as a collection of conveniently chosen subdomains that allows a systematic generation of the approximation functions.

Problems

ELEMENTS OF FUNCTIONAL ANALYSIS (SECTIONS 6.1 AND 6.2)

In **Problems 6.1–6.5**, determine whether the sets of functions are linearly independent.

6.1 $\{\sin \frac{n\pi x}{L}\}_{n=1}^{3}$, $0 \le x \le L$.

6.2 $\{\cos \frac{n\pi x}{L}\}_{n=1}^{3}$, $0 \le x \le L$.

6.3 $\{x^n(1-x)\}_{n=0}^{3}$, $0 \le x \le 1$.

6.4 $\{x, 5x, x^2\}$, $0 \le x \le 1$.

6.5 $\{1 + x + x^2, 1 + 2x + 3x^2, 2 - x + x^2\}$.

6.6 Let $\mathbf{u} = \mathbf{u}(x)$ and $\mathbf{v} = \mathbf{v}(x)$ be two vector functions of x. Define the products

    (a) $(\mathbf{u}, \mathbf{v}) \equiv \int_0^L \mathbf{u} \cdot \mathbf{v} \, dx$.

    (b) $(\mathbf{u}, \mathbf{v}) \equiv \int_0^L \mathbf{u}(L - x) \cdot \mathbf{v}(x) \, dx$.

    Which one of these products qualifies as an inner product?

Compute the norm of the functions in **Problems 6.7–6.9** using the inner products indicated.

6.7 $u = x^3 - 3L^2x + 2L^3$, $\Omega = (0, L)$, using the one-dimensional version of the norm in Eq. (6.2.23).

6.8 $u = x^3 - 3L^2x + 2L^3$, $\Omega = (0, L)$, using the one-dimensional version of the norm in Eq. (6.2.24).

6.9 $u = (x^2 - a^2)(y^2 - b^2)$, $0 \le x \le a$ and $0 \le y \le b$, using Eq. (6.2.23).

6.10 Find the distance between the following pair of functions using the inner product in Eq. (6.2.23): $u = x^3 - 3x + 2$ and $v = (x - 1)^2$, $0 \le x \le 1$.

6.11 Check whether the following pair of functions are orthogonal in the $L_2(0,1)$-space: $u(x) = 2 + 3x^2 - x$ and $v(x) = \frac{1}{3} + 3x - 5x^2$.

6.12 Check whether the following pair of functions in **Problem 6.11** are orthogonal in the $H^1(0,1)$-space.

6.13 Determine the constants a and b such that $w(x) = a + bx + 3x^2$ is orthogonal to both $u(x)$ and $v(x)$ of **Problem 6.11**.

In **Problems 6.14–6.18**, which operators qualify as linear operators? Identify the linear and bilinear functionals.

6.14 $A(u) = -\frac{d}{dx}\left(a\frac{du}{dx}\right)$.

6.15 $T(u) = \nabla^2 u + 1$.

6.16 $I(u) = \int_0^a K(x)\frac{du}{dx}\,dx - u(0)$.

6.17 $I(u,v) = \int_\Omega \nabla u \cdot \nabla v \, dxdy$.

6.18 $I(u,v) = \int_0^a \left(b\frac{d^2 u}{dx^2}\frac{d^2 v}{dx^2} + fv\right)dx$.

6.19 Show that if $I(u)$ is a quadratic functional, its first variation is a bilinear functional of u and δu.

6.20 Show that if $I(u)$ is any functional, its first variation is a linear functional of δu.

THE RITZ METHOD (SECTION 6.3)   Give admissible approximation functions, either algebraic or trigonometric, for a *two-parameter* Ritz approximation of **Problems 6.21–6.27**. Assume that the total potential energy principle is used to construct the Ritz approximation.

6.21 A cable suspended between points $A : (0,0)$ and $B : (L,h)$ and subjected to uniformly distributed transverse load of intensity f_0.

6.22 A cantilever beam subjected to uniformly distributed load of intensity q_0.

6.23 The symmetric half of the simply supported beam problem considered in **Example 6.3.4** $(0 \le x \le L/2)$.

6.24 A beam clamped at the left end and simply supported at the right end and subjected to point load F_0 at $x = L/2$.

6.25 A simply supported beam with a spring support at $x = L/2$.

6.26 A square elastic membrane, $-a < x < a$ and $-a < y < a$, fixed on all its sides and subjected to a uniformly distributed load of intensity f_0. Use the full domain.

6.27 Consider the differential equation

$$\frac{d^2}{dx^2}\left[(x+2)\frac{d^2 w}{dx^2}\right] + w = 3x, \quad 0 < x < 1$$

subject to the boundary conditions

$$w(1) = 0, \quad \frac{dw}{dx}\bigg|_{x=1} = 0, \quad \left[(x+2)\frac{d^2w}{dx^2}\right]_{x=0} = 0, \quad \left\{\frac{d}{dx}\left[(x+2)\frac{d^2w}{dx^2}\right]\right\}_{x=0} = 0.$$

Give (a) an algebraic approximation function ϕ_1 for a one-parameter Ritz approximation and (b) an algebraic approximation function ϕ_1 that satisfies all of the boundary
conditions.

6.28–6.33 Find the *two-parameter* Ritz approximation of the problems in **Problems 6.21–
6.26**, and compare the solutions with the exact solutions when possible.

6.34 Determine two-parameter Ritz solution of the problem

$$-\frac{d}{dx}\left[(1+x)\frac{du}{dx}\right] = 0 \quad \text{for} \quad 0 < x < 1,$$

$$u(0) = 0, \quad u(1) = 1.$$

Use algebraic polynomials for the approximation functions.

6.35 Set up the two-parameter Ritz equations to determine the first two natural frequencies
of a cantilever beam (length L and bending stiffness EI) using the Bernoulli–Euler
beam theory. Take $EI = (a + bx)^{-1}$, where a and b are constants.

6.36 Find a two-parameter Ritz approximation of the transverse deflection of a simply supported beam (of length L and bending stiffness EI) on an elastic foundation (foundation modulus k) that is subjected to uniformly distributed load (of intensity q_0). Use
the Bernoulli–Euler beam theory and (a) algebraic polynomials and (b) trigonometric
polynomials.

6.37 Derive the matrix equations corresponding to the N-parameter Ritz approximation

$$W_N = c_1 x^2 + c_2 x^3 + \cdots + c_N x^{N+1}$$

of a cantilever beam with a uniformly distributed load, q_0. Use the Bernoulli–Euler
beam theory, and compute a_{ij} and b_j in explicit form in terms of i, j, L, EI, and q_0.

6.38 Use a two-parameter Ritz approximation with trigonometric functions to determine
the critical buckling load of a simply supported beam. Use the Bernoulli–Euler beam
theory.

6.39 Consider the buckling of a uniform beam according to the Timoshenko beam theory.
The total potential energy functional for the problem is

$$\Pi(w, \phi_x) = \frac{1}{2}\int_0^L \left[D\left(\frac{d\phi_x}{dx}\right)^2 + S\left(\frac{dw}{dx} + \phi_x\right)^2 - \hat{N}\left(\frac{dw}{dx}\right)^2\right]dx,$$

where $w(x)$ is the transverse deflection, ϕ_x is the rotation, $D = EI$ is the flexural stiffness, $S = GAK_s$ is the shear stiffness, and $\hat{N}$ is the axial compressive load. Determine
the critical (i.e., the minimum) buckling load of a simply supported beam using the
Ritz method. Assume a one-parameter approximations of both w and ϕ_x.

6.40 Determine the N-parameter Ritz solution for the transient response of a simply supported beam under step loading $q(x,t) = q_0 H(t-t_0)$, where $H(t)$ denotes the Heaviside
step function. Use trigonometric functions for $\phi_i(x)$ and the Bernoulli–Euler beam theory, and neglect the rotatory inertia.

6.41 Show that the two-parameter Ritz solution for the transient response of the bar considered in **Example 6.3.3** with $\alpha = 1$ yields the equations

$$\begin{bmatrix} \frac{1}{3} & \frac{1}{4} \\ \frac{1}{4} & \frac{1}{5} \end{bmatrix} \begin{Bmatrix} \ddot{c}_1 \\ \ddot{c}_2 \end{Bmatrix} + \begin{bmatrix} 2 & 2 \\ 2 & \frac{7}{3} \end{bmatrix} \begin{Bmatrix} c_1 \\ c_2 \end{Bmatrix} = \frac{g_0}{6} \begin{Bmatrix} 3 \\ 2 \end{Bmatrix}. \tag{1}$$

Set $g_0 = 1$ and use the Laplace transform method to determine the solution of these equations.

6.42 Derive the weak forms of the following nonlinear equations [see Eq. (2.6.8)]:

$$-\frac{dN_{xx}}{dx} = f(x), \tag{1}$$

$$-\frac{d}{dx}\left(\frac{dw}{dx}N_{xx}\right) - \frac{d^2 M_{xx}}{dx^2} = q(x), \tag{2}$$

where

$$N_{xx} = EA\left[\frac{du}{dx} + \frac{1}{2}\left(\frac{dw}{dx}\right)^2\right], \quad M_{xx} = -EI\frac{d^2 w}{dx^2}. \tag{3}$$

6.43 Construct the weak form of the equation governing buckling of a simply supported beam,

$$\frac{d^2}{dx^2}\left(EI\frac{d^2 w}{dx^2}\right) + P\frac{d^2 w}{dx^2} = 0, \quad 0 < x < L,$$

with boundary conditions

$$w = 0, \quad EI\frac{d^2 w}{dx^2} = 0 \text{ at } x = 0, L,$$

where P is the axial compressive load on the beam column. Develop a N-parameter Ritz formulation for the determination of the buckling load P. Assume that $EI =$ constant and use algebraic polynomials for $\phi_i(x)$. In particular, determine the buckling load using $N = 1$, $N = 2$, and $N = 3$.

6.44 Consider the problem of finding the fundamental frequency of a circular membrane of radius a, fixed at its edge. The governing equation for axisymmetric vibration is

$$-\frac{1}{r}\frac{d}{dr}\left(r\frac{du}{dr}\right) - \lambda u = 0, \quad 0 < r < a,$$

where λ is the frequency parameter and u is the deflection of the membrane. (a) Construct the weak form, (b) use one-parameter Ritz approximation to determine λ, and (c) use two-parameter Ritz approximation to determine λ. Select trigonometric functions for $\phi_i(r)$.

6.45 Consider the problem of finding the solution of the equation

$$\frac{d^2 u}{dx^2} + u + x = 0, \quad 0 < x < 1; \quad u(0) = u(1) = 0.$$

(a) Develop the weak form, (b) assume $N-$parameter Ritz approximation of the form

$$U_N(x) = x(1 - x)\left(c_1 + c_2 x + \cdots + c_N x^{N-1}\right)$$

and obtain the Ritz equations for the unknown coefficients, and (c) determine the two-parameter solution and compare with the exact solution

$$u(x) = \frac{\sin x}{\sin 1} - x.$$

6.46 Consider the Bessel equation

$$x^2 \frac{d^2 u}{dx^2} + x \frac{du}{dx} + (x^2 - 1)u = 0, \quad 1 < x < 2$$

subject to the boundary conditions $u(1) = 1$ and $u(2) = 2$. Assume $u = w + x$ and reduce the equation to

$$-\frac{d}{dx}\left(x \frac{dw}{dx}\right) - \frac{x^2 - 1}{x} w = x^2, \quad 1 < x < 2$$

subject to the boundary conditions $w(1) = w(2) = 0$. Determine a one-parameter Ritz approximation of the problem, and compare it with the analytical solution

$$u(x) = 3.6072 J_1(x) + 0.75195 Y_1(x)$$

where J_1 and Y_1 are Bessel functions of the first and second kind, respectively.

6.47 Derive the weak form and obtain a one- and two-parameter Ritz solutions of the problem

$$-\left(\frac{\partial^2 u}{\partial x^2} + \frac{\partial^2 u}{\partial y^2}\right) = 1 \quad \text{in a unit square,}$$

$$u(1, y) = u(x, 1) = 0,$$

$$\frac{\partial u}{\partial x}(0, y) = \frac{\partial u}{\partial y}(x, 0) = 0.$$

The origin of the coordinate system is taken at the lower left corner of the unit square.

WEIGHTED-RESIDUAL METHODS (SECTION 6.4)

6.48 Determine the two-parameter Petrov–Galerkin solution (with $\psi_1 = 1$ and $\psi_2 = x$) of the following differential equation, and compare, if possible, with the exact solution.

$$-\frac{d^2 u}{dx^2} = \frac{1}{1 + x}, \quad 0 < x < 1; \quad u(0) = u'(1) = 0.$$

6.49 Determine the two-parameter Petrov–Galerkin solution (with $\psi_1 = 1$ and $\psi_2 = x$) of the following differential equation:

$$-\frac{d}{dx}\left[(1 + x^2) \frac{du}{dx}\right] + u = \sin \pi x + 3x - 1, \quad 0 < x < 1; \quad u(0) = u'(1) = 0.$$

6.50 Determine the one-parameter Galerkin solution of the equation

$$\frac{d^2}{dx^2}\left[\left(2 + \frac{x}{L}\right)\frac{d^2 w}{dx^2}\right] = q_0 \left(1 - \frac{x}{L}\right)$$

that governs a cantilever beam subjected to linearly varying load (from q_0 at the fixed end, $x = 0$, to zero at the free end, $x = L$). Use an algebraic polynomial for $\phi_1(x)$.

6.51 Determine the two-parameter Petrov–Galerkin solution of the equation

$$-\nabla^2 u = 0 \quad \text{in } 0 < (x, y) < 1,$$

subjected to the boundary conditions

$$u = \sin \pi x \quad \text{on } y = 0,$$

$$u = 0 \text{ on all other sides.}$$

Use $\psi_1 = 1$ and $\psi_2 = xy$ as the weight functions to determine the solution. Use trigonometric functions.

6.52 Solve the nonlinear differential equation

$$-\frac{d}{dx}\left(u\frac{du}{dx}\right) + 1 = 0, \quad 0 < x < 1,$$

$$u(1) = \sqrt{2}, \quad u'(0) = 0.$$

Use a one-parameter Petrov–Galerkin approximation with (a) $\psi_1 = 1$ and (b) $\psi_1 = x$. Compare the results with the exact solution, $u(x) = \sqrt{1+x^2}$ (see **Example 6.4.18**). Select the value of c_1 (when multiple values are found) that makes the integral of the residual the smallest.

6.53 Find the first two eigenvalues associated with the differential equation

$$-\frac{d^2u}{dx^2} = \lambda u, \quad 0 < x < 1,$$

$$u(0) = 0, \quad u(1) + u'(1) = 0.$$

Use the Galerkin method and compare the solution with the results from **Example 6.3.2**.

6.54 Solve the Poisson equation

$$-\nabla^2 u = f_0 \quad \text{in a unit square,}$$

$$u = 0 \quad \text{on the boundary.}$$

using the following N-parameter Galerkin approximation:

$$U_N = \sum_{i,j=1}^{N} c_{ij} \sin i\pi x \sin j\pi y.$$

Compare the solution obtained with the Ritz method when the same approximation is used.

6.55 Solve the nonlinear equation in **Problem 6.52** by the Galerkin method.

6.56 Find a one-parameter Galerkin solution of a beam clamped at both ends and under uniformly distributed transverse load of intensity q_0. Use the Bernoulli–Euler beam theory.

6.57 Solve the equation in **Problem 6.49** using the least-squares method.

6.58 Solve the problem of **Problem 6.52** using the least-squares method.

6.59 Solve the equation in **Problem 6.53** using the least-squares method. Use the operator definition to be $A = -(d^2/dx^2)$ to avoid increasing the degree of the characteristic polynomial for λ.

6.60 Solve the equation in **Problem 6.54** using the least-squares method.

6.61 Consider a cantilever beam (fixed at $x = 0$ and free at $x = L$) of variable flexural rigidity, $EI = a_0[2 - (x/L)^2]$ and carrying a distributed load, $q = q_0[1 - (x/L)]$. Find a three-parameter solution using the collocation method.

6.62 Solve **Problem 6.52** using the collocation method with the collocation point at $x = 0.5$.

6.63 Solve **Problem 6.53** by the collocation method. Use the collocation points at $x = \frac{1}{3}$ and $x = \frac{2}{3}$.

6.64 Solve **Problem 6.54** using one-parameter collocation method. Use the collocation point $(x, y) = (0.5, 0.5)$.

6.65 Consider the Poisson equation

$$-\nabla^2 u = 0, \quad 0 < x < 1, \quad 0 < y < \infty,$$

$$u(0, y) = u(1, y) = 0 \quad \text{for } y > 0,$$

$$u(x, 0) = x(1 - x), \quad u(x, \infty) = 0, \quad 0 \le x \le 1.$$

Assuming one-parameter Galerkin approximation of the form

$$U_1(x, y) = c_1(y)x(1 - x),$$

find the differential equation for $c_1(y)$ and solve it exactly.

6.66 Use the semidiscretization method and a two-parameter Galerkin approximation of the form

$$u(x, y) = c_1(y) \cos \frac{\pi x}{2a} + c_2(y) \cos \frac{3\pi x}{2a},$$

to determine an approximation of the membrane problem in **Example 6.3.6**. Note that u satisfies the boundary conditions on $x = \pm a$ for any y.

6.67 Use the semidiscretization method and a two-parameter Galerkin approximation of the form

$$W_1(x, t) = c_1(t)(1 - \cos 2\pi x),$$

to determine an approximate solution of the equation

$$\frac{\partial^4 w}{\partial x^4} + \frac{\partial^2 w}{\partial t^2} = 0, \quad 0 < x < 1, \quad t > 0,$$

$$w = \frac{\partial w}{\partial x} = 0 \quad \text{at } x = 0, 1 \quad \text{and } t > 0,$$

$$w(x, 0) = \sin \pi x - \pi x(1 - x), \quad \dot{w}(x, 0) = 0, \quad 0 < x < 1,$$

where $\dot{w} = \frac{\partial w}{\partial t}$. Use the Galerkin method to satisfy the initial conditions.

6.68 Consider the problem of finding the eigenvalues associated with the equation

$$\nabla^2 u + \lambda u = 0 \quad \text{in } \Omega; \quad u = 0 \quad \text{on } \Gamma,$$

where Ω is the rectangle, $-a < x < a$ and $-b < y < b$, and Γ is its boundary (see **Example 6.4.16**). Assuming one-parameter Galerkin approximation of the form

$$U_1(x, y) = c_1(x)\phi_1(y) = c_1(x)(y^2 - b^2),$$

determine the first eigenvalue λ_1.

6.69 Repeat **Problem 6.68** with two-parameter Galerkin approximation of the form

$$U_2(x, y) = c_1(x)\phi_1(y) + c_2(x)\phi_2(y) = [c_1(x) + c_2(x)y^2](y^2 - b^2).$$

6.70 Consider the Eigenvalue problem for the Laplace operator, $-\nabla^2$,

$$-\nabla^2 u - \lambda u = 0 \quad \text{in } \Omega, \quad u = 0 \quad \text{on } \Gamma,$$

where Ω is the domain represented by the first quadrant of the ellipse (see Fig. P6.70)

$$\frac{x^2}{a^2} + \frac{y^2}{b^2} = 1$$

and Γ is the boundary of Ω. Determine the one-parameter Galerkin solution (i.e., find the first value of λ). What is the one-parameter Ritz solution?

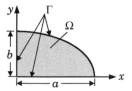

Fig. P6.70

"Imagination is more important than knowledge."

"The important thing is not to stop questioning. Curiosity has its own reason for existing."

"Anyone who has never made a mistake has never tried anything new."
     – *Albert Einstein* (German-born American Nobel laureate of physics who developed the theory of relativity)

"If I have seen further than others, it is by standing upon the shoulders of giants."

"Tact is the knack of making a point without making an enemy."

"I can calculate the motion of heavenly bodies, but not the madness of people."
     – *Isaac Newton* (English mathematician, astronomer, and physicist)

Theory and Analysis of Plates

7.1 Introduction

7.1.1 General Comments

In this chapter, we study bending, buckling, and natural vibration of plate structures undergoing bending deformation. A *plate* is initially a flat structural element with planform dimensions much larger than its thickness, h, and is subjected to loads transverse to the plane that cause bending deformation in addition to stretching (see Fig. 7.1.1). Street manhole covers, tabletops, side panels and roofs of buildings and transportation vehicles, glass window panels, turbine disks, bulkheads, and tank bottoms provide familiar examples of plate structures. A *shell* is a curved structural element with thickness much smaller than the other dimensions. Like a plate, shell is subjected to loads that cause stretching and bending deformations. Examples of shell structures are provided by pressure vessels, pipes, curved panels of a variety of structures including automobiles and aerospace vehicles, tires, and roof domes and sheds.

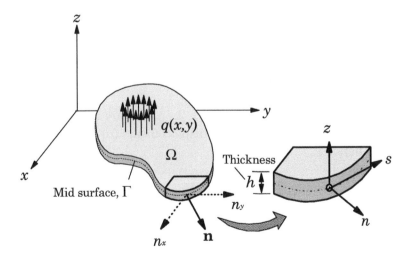

Fig. 7.1.1 Geometry of a plate with curved boundary.

Energy Principles and Variational Methods in Applied Mechanics, Third Edition. J.N. Reddy.
©2017 John Wiley & Sons Ltd. Published 2017 by John Wiley & Sons Ltd.
Companion Website: www.wiley.com/go/reddy/applied_mechanics_EPVM

In most cases, the thickness of plate and shell structures is about one-tenth or less of the smallest in-plane (i.e., within the plane) dimension. When the ratio of the thickness to the smallest in-plane dimension of a plate is 20 or greater, it is termed a *thin* plate; otherwise, the plate is said to be *thick*. Because of the smallness of the thickness dimension, it is often not necessary to model plate and shell structures using three-dimensional (3-D) elasticity equations. Simple two-dimensional theories (i.e., theories in which the governing equations are only functions of the in-plane coordinates, x and y) derived using some kinematic assumptions can be used to study deformation, stresses, natural frequencies, and global buckling loads of plate and shell structures. In the present study, the governing equations are derived using the principle of virtual displacements (or the principle of minimum total potential energy) for circular plates in cylindrical coordinates and for noncircular plates in rectangular Cartesian coordinates. Bending, buckling, and vibration solutions are obtained for particular cases of geometry, boundary conditions, and loads using the energy and variational methods developed in the earlier chapters. When possible, exact analytical solutions are also presented.

The two-dimensional (2-D) theories of plates are developed from an assumed displacement or stress field. They are often taken as a finite linear combination of unknown functions and the thickness coordinate z. For example, the ith displacement component u_i is expanded as

$$u_i(x,y,z,t) = \varphi_i^{(0)}(x,y,t) + z\varphi_i^{(1)} + \cdots + (z)^N \varphi_i^{(N)} = \sum_{j=0}^{N} (z)^j \varphi_i^{(j)}(x,y,t), \quad (7.1.1)$$

where (x,y) are the coordinates in the plane of the plate, z is the thickness coordinate, t is the time, and $\varphi_i^{(j)}$ are functions that describe the deformation and are to be determined. The meaning of functions $\varphi_i^{(j)}$ for $j > 1$ is difficult to physically interpret.

The equations governing $\varphi_i^{(j)}$ $(j = 0, 1, 2, \ldots, N)$ are derived using the dynamic version of the principle of virtual displacements (or Hamilton's principle; see Chapter 5 details):

$$0 = \int_0^T (\delta U + \delta V - \delta K)\, dt, \quad (7.1.2)$$

where $\delta U, \delta V$, and δK denote the virtual strain energy, virtual potential energy due to external applied forces, and virtual kinetic energy, respectively. These quantities are determined in terms of actual stresses and virtual strains, which depend on the assumed displacement expansion in Eq. (7.1.1). The integration over the domain of the plate structure is represented as the tensor product of the integration over the midplane (or mid surface) of the plate and integration over the thickness. This is possible due to the explicit nature of the assumed displacement field in the thickness coordinate z. Thus, we can write

$$\int_{Vol.} (\cdot)\, dV = \int_{-\frac{h}{2}}^{\frac{h}{2}} \int_{\Omega} (\cdot)\, d\Omega\, dz, \qquad (7.1.3)$$

where h denotes the total thickness of the plate and Ω denotes the undeformed midplane of the plate, which is assumed to coincide with the xy-plane. Since all undetermined variables $\phi_i^{(j)}$ are independent of the thickness coordinate z and u_i are explicit functions of z, the integration over plate thickness is carried out explicitly, reducing the problem to a two-dimensional one. This is completely analogous to the procedure used in deriving the beam theories in Chapters 3 and 4. Consequently, the Euler–Lagrange equations resulting from Eq. (7.1.2) consist of partial differential equations involving the dependent variables $\varphi_i^{(j)}(x, y, t)$ and thickness-averaged stress resultants, $M_{ij}^{(m)}$:

$$M_{ij}^{(m)} = \int_{-\frac{h}{2}}^{\frac{h}{2}} (z)^m \sigma_{ij}\, dz. \qquad (7.1.4)$$

The stress resultants $M_{ij}^{(m)}$ can be written in terms of $\varphi_i^{(j)}$ with the help of the assumed constitutive equations (stress–strain relations) and strain–displacement relations. More complete development of this procedure for plates is forthcoming in this chapter.

The same approach is used when φ_i denote stress components, except that the basis of the derivation of the governing equations is the principle of virtual forces. In the present book, the stress-based theories will receive very little attention.

7.1.2 An Overview of Plate Theories

There exist a number of plate, and they differ from each other in two principal ways:

(1) Choice of the assumed field (i.e., displacement or stress)

(2) The choice of terms in the expansion [i.e., number of terms and powers of the thickness coordinate z retained in Eq. (7.1.1)]

The choice of the field is often restricted to either displacements or stresses, although a mixed approach is possible. The number of terms retained in the displacement or stress expansions is often limited, at most, to cubic in thickness coordinate in the interest of keeping the theory simple but adequate to represent transverse shear strains. Thus, a plate or shell theory can be developed for any combination of the field variables and number of terms in the expansion of the variables. The number of theories further multiply if different order expansions are used for different components of a field variable and types of nonlinear terms included in the strain–displacement relations.

A general but brief review of various theories of plates is presented before closing this section. The ideas are also applicable to shells. No review of plate and shell theories is complete, as there are thousands of papers dealing with one aspect or the other of the many combinations mentioned previously.

7.1.2.1   The Classical Plate Theory

The simplest plate theory is known as the *Kirchhoff plate theory* or simply the *classical plate theory* (CPT), which is an extension of the Bernoulli–Euler beam theory (see Section 3.2.5.1 and **Example 4.2.1**) to two dimensions. The CPT is based on the following assumptions, known as the *Kirchhoff hypotheses* [108]:

(1) Straight lines perpendicular to the midplane (i.e., transverse normals) before deformation remain straight after deformation.

(2) The transverse normals do not experience elongation (i.e., they are inextensible).

(3) The transverse normals rotate such that they remain perpendicular to the mid surface after deformation.

These assumptions allow us to describe the plate deformation in terms of certain displacement quantities. Assumption (1) requires that the displacement field $(u_1 = u_x, u_2 = u_y, u_3 = u_z)$ be a linear function of the thickness coordinate z:

$$
\begin{aligned}
u_x(x, y, z, t) &= u(x, y, t) + zF_1(x, y, t), \\
u_y(x, y, z, t) &= v(x, y, t) + zF_2(x, y, t), \\
u_z(x, y, z, t) &= w(x, y, t) + zF_3(x, y, t),
\end{aligned}
\tag{7.1.5}
$$

where (u, v, w, F_1, F_2, F_3) are functions to be determined such that the remaining two assumptions of the Kirchhoff hypothesis are satisfied. The inextensibility assumption (2) requires that

$$
\frac{\partial u_z}{\partial z} = 0 \ \rightarrow \ F_3 = 0 \text{ for all } x, y, \text{ and } t.
$$

Thus, w is independent of z, that is, $u_z = w(x, y, t)$. Assumption (3) requires that

$$
\begin{aligned}
\frac{\partial u_x}{\partial z} &= -\frac{\partial u_z}{\partial x} \ \rightarrow \ F_1 = -\frac{\partial w}{\partial x}, \\
\frac{\partial u_y}{\partial z} &= -\frac{\partial u_z}{\partial y} \ \rightarrow \ F_2 = -\frac{\partial w}{\partial y}.
\end{aligned}
\tag{7.1.6}
$$

Hence, the displacement field in Eq. (7.1.5) takes the form

$$
\begin{aligned}
u_x(x, y, z, t) &= u(x, y, t) + z\theta_x, \quad \theta_x \equiv -\frac{\partial w}{\partial x}, \\
u_y(x, y, z, t) &= v(x, y, t) + z\theta_y, \quad \theta_y \equiv -\frac{\partial w}{\partial y} \\
u_z(x, y, z, t) &= w(x, y, t).
\end{aligned}
\tag{7.1.7}
$$

Thus, the displacement field of CPT is completely determined by the functions (u, v, w), which denote the displacements of a point on the midplane along the three coordinate directions (x, y, z). Note that the displacement field in Eq. (7.1.7) will result, as will be seen shortly, in the neglect of all transverse strains, that is, $\varepsilon_{zz} = \varepsilon_{xz} = \varepsilon_{yz} = 0$.

7.1.2.2 The First-Order Plate Theory

The simplest displacement-based plate theory that accounts for nonzero transverse shear strains $\gamma_{xz} = 2\varepsilon_{xz}$ and $\gamma_{yz} = 2\varepsilon_{yz}$ is one based on the displacement field

$$
\begin{aligned}
u_x(x, y, z, t) &= u(x, y, t) + z\phi_x(x, y, t), \\
u_y(x, y, z, t) &= v(x, y, t) + z\phi_y(x, y, t), \\
u_z(x, y, z, t) &= w(x, y, t),
\end{aligned}
\tag{7.1.8}
$$

where ϕ_x and $-\phi_y$ denote rotations about the y and x axes, respectively:

$$
\phi_x(x, y, t) = \left(\frac{\partial u_x}{\partial z}\right)_{z=0}, \quad \phi_y(x, y, t) = \left(\frac{\partial u_y}{\partial z}\right)_{z=0}.
\tag{7.1.9}
$$

In this theory, the normality restriction is removed by allowing for independent and arbitrary rotations (ϕ_x, ϕ_y) of a transverse normal line. The theory is known in the literature as the *Mindlin plate theory* although the use of the displacement field in Eq. (7.1.8) and associated plate theory were developed much earlier by Basset [109], Hildebrand *et al.* [110], and Hencky [111]. Mindlin [112] extended the theory developed by Hencky [111] to the vibration of crystal plates. The theory is now being referred to as the *first-order shear deformation plate theory* (FSDT); see Reddy [50, 51].

Note that the CPT is also a first-order theory in the sense that the first-order terms in the thickness coordinate (z) are included; however, the coefficients of the thickness coordinate in the displacement field of the CPT are not independent of the other variables (θ_x and θ_y are the derivatives of the transverse deflection w) and it is not a shear deformation theory. As will be seen later in this chapter, the transverse shear strains in FSDT are represented as constants through the plate thickness (but they are functions of x and y). Therefore, the transverse shear stresses are also constant through the plate thickness, whereas the stress equilibrium equations predict them to be quadratic. To make the shear forces ($M_{xz}^{(0)} = Q_x$ and $M_{yz}^{(0)} = Q_y$) computed in the FSDT to be equal to those obtained using the transverse shear stresses from the stress-equilibrium equations, shear correction factors were introduced. In both CPT and FSDT, the plane-stress state assumption is used and plane-stress reduced form of the constitutive law is used (see Section 2.4.3). In both theories, the inextensibility (i.e., $\varepsilon_{zz} = 0$) and/or straightness of transverse normals can be removed. Such extensions lead to second- and higher-order theories of plates.

7.1.2.3 The Third-Order Plate Theory

Second- and higher-order plate theories[1] use higher-order polynomials [i.e., $N > 1$ in Eq. (7.1.1)] in the expansion of the displacement components through the thickness of the plate. The higher-order theories introduce additional unknowns that are often difficult to interpret in physical terms.

A third-order plate theory with transverse inextensibility is based on the displacement field

$$
\begin{aligned}
u_x(x, y, z, t) &= u(x, y, t) + f(z)\varphi_x(x, y, t), \\
u_y(x, y, z, t) &= v(x, y, t) + f(z)\varphi_y(x, y, t), \\
u_z(x, y, z, t) &= w(x, y, t),
\end{aligned}
\tag{7.1.10}
$$

where $f(z)$ is a cubic function of z. There are number of people who have used a displacement field of the form in Eq. (7.1.10), but they differ in actual form because of the choice of variables; consequently, the resulting governing equations have different looks. However, it can be shown that all third-order theories are special cases of that derived by Reddy [113–115] for laminated plates and shells using the displacement field in Eq. (7.1.10) and the principle of virtual displacements. The displacement field in Eq. (7.1.10) accommodates quadratic variation of transverse shear strains (and hence stresses) and vanishing of transverse shear stresses on the top and bottom of a plate. This particular third-order shear deformation plate theory is known in the literature as the *Reddy third-order plate theory*. Thus, there is no need to use shear correction factors in a third-order plate theory.

The third-order plate theory of Levinson [116] is based on the same displacement field as in Eq. (7.1.10), but he used the equilibrium equations of the FSDT, that is, the principle of virtual displacements was not used to derive the equilibrium equations. This amounts to not including certain terms (e.g., higher-order stress resultants) in the strain energy of the plate. The Levinson plate theory results in much simpler equations, but the theory leads to an unsymmetric stiffness matrix even for linear problems.

---

[1]The order referred here is to the degree n of the thickness coordinate in the displacement expansion and not the order of the governing differential equations.

7.1.2.4 Stress-Based Theories

The plate theories based on the expansion of the stress field are due to Reissner [117–119] and Kromm [120, 121], and the book by Panc [122] contains chapters devoted to these theories and their extensions. In the *Reissner plate theory*, the distribution of the stress components through the plate thickness, for the static case, is assumed to be of the form

$$\sigma_{xx} = z\frac{12M_{xx}}{h^3}, \quad \sigma_{yy} = z\frac{12M_{yy}}{h^3}, \quad \sigma_{xy} = z\frac{12M_{xy}}{h^3},$$

$$\sigma_{xz} = \frac{3Q_x}{2h}\left[1 - \left(\frac{2z}{h}\right)^2\right], \quad \sigma_{yz} = \frac{3Q_y}{2h}\left[1 - \left(\frac{2z}{h}\right)^2\right], \tag{7.1.11}$$

$$\sigma_{zz} = -\frac{q}{4}\left[2 - 3\left(\frac{2z}{h}\right) + \left(\frac{2z}{h}\right)^3\right],$$

where ($M_{xx}^{(1)} = M_{xx}$, $M_{yy}^{(1)} = M_{yy}$, $M_{xy}^{(1)} = M_{xy}$) are the bending moments per unit length, ($M_{xz}^{(0)} = Q_x$, $M_{yz}^{(0)} = Q_y$) shear forces per unit length, q is the distributed transverse load per unit area, and h is the plate thickness. The stress field in Eq. (7.1.11) is the same as that of the CPT for the pure bending case (to be shown shortly); the transverse shear stress field in Eq. (7.1.11) is the same as that obtained from 3-D stress equilibrium equations after using the in-plane stress field ($\sigma_{xx}, \sigma_{yy}, \sigma_{xy}$). Thus, the inplane stresses are linear functions of z, the transverse shear stresses are quadratic functions of z, and transverse normal stress is cubic in z. Obviously, these stress components satisfy the stress equilibrium equations of 3-D elasticity [see Eq. (2.2.9)]. The transverse displacement w of Reissner's theory is a function of (x, y, z), and this complicates its determination. To make the theory tractable, Reissner introduced the thickness–integrated transverse displacement (a "mean deflection with respect to the plate thickness")

$$\bar{w}(x, y) = \frac{3}{2h}\int_{-\frac{h}{2}}^{\frac{h}{2}} w\left[1 - \left(\frac{2z}{h}\right)^2\right] dz. \tag{7.1.12}$$

The refined theory of Kromm [120, 121] is based on a more general stress distribution, especially for the transverse shear stress components across the thickness of the plate. For a complete description of the theory and its governing equations, the reader may consult the book by Panc [122].

The stress-based theories have not received as much attention as those based on the displacement expansions. This might be due to the fact that the stress-based theories are relatively more complicated and inconsistencies between the actual (i.e., consistent with the assumed stress distributions) and adopted displacements may exist.

In this chapter, we consider only small strain and small displacement bending deformation of plates; that is, the stretching deformation is omitted in the derivation of the governing equations, as the equations governing stretching deformation are not coupled to bending deformation. Therefore, in the absence of loads that cause stretching, the extensional displacement field (u, v) is identically zero. We begin with an assumed displacement field, compute strains, and then use Hamilton's principle (or the dynamic version of the principle of virtual displacements) to derive the governing equations of motion.

7.2   The Classical Plate Theory

7.2.1   Governing Equations of Circular Plates

Consider a circular plate on a linear elastic foundation, with foundation modulus k. Let r denote the radial coordinate outward from the center of the plate, z the thickness coordinate along the height of the plate, and θ the coordinate along a circumference of the plate, as shown in Fig. 7.2.1. In a general case where applied loads and boundary conditions are not axisymmetric, the displacements (u_r, u_θ, u_z) along the coordinates (r, θ, z) are functions of r, θ, and z coordinates. We begin with the following displacement field implied by the Kirchhoff assumptions:

$$u_r(r, \theta, z, t) = -z\frac{\partial w}{\partial r},$$

$$u_\theta(r, \theta, z, t) = -z\left(\frac{1}{r}\frac{\partial w}{\partial \theta}\right), \qquad (7.2.1)$$

$$u_z(r, \theta, z, t) = w(r, \theta, t),$$

where w is the transverse displacement at point (x, y) in the plate. The linear strain components referred to the cylindrical coordinate system are given by (see **Problem 2.34**)

$$\varepsilon_{rr} = \frac{\partial u_r}{\partial r}, \qquad \varepsilon_{\theta\theta} = \frac{u_r}{r} + \frac{1}{r}\frac{\partial u_\theta}{\partial \theta},$$

$$\varepsilon_{zz} = \frac{\partial u_z}{\partial z}, \qquad \varepsilon_{r\theta} = \frac{1}{2}\left(\frac{1}{r}\frac{\partial u_r}{\partial \theta} + \frac{\partial u_\theta}{\partial r} - \frac{u_\theta}{r}\right), \qquad (7.2.2)$$

$$\varepsilon_{z\theta} = \frac{1}{2}\left(\frac{\partial u_\theta}{\partial z} + \frac{1}{r}\frac{\partial u_z}{\partial \theta}\right), \quad \varepsilon_{rz} = \frac{1}{2}\left(\frac{\partial u_r}{\partial z} + \frac{\partial u_z}{\partial r}\right).$$

For the choice of the displacement field in Eq. (7.2.1), the only nonzero strains are

$$\varepsilon_{rr} = z\varepsilon_{rr}^{(1)}, \quad \varepsilon_{\theta\theta} = z\varepsilon_{\theta\theta}^{(1)}, \quad 2\varepsilon_{r\theta} = z\gamma_{r\theta}^{(1)}, \qquad (7.2.3)$$

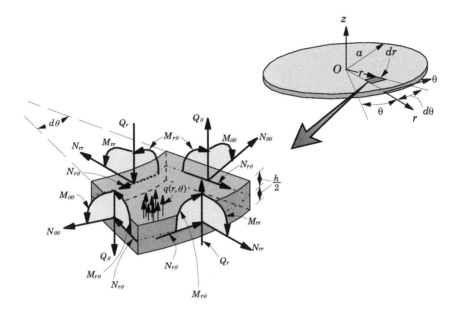

Fig. 7.2.1 A circular plate with stress resultants.

where

$$\varepsilon_{rr}^{(1)} = -\frac{\partial^2 w}{\partial r^2},$$

$$\varepsilon_{\theta\theta}^{(1)} = -\frac{1}{r}\left(\frac{\partial w}{\partial r} + \frac{1}{r}\frac{\partial^2 w}{\partial \theta^2}\right), \tag{7.2.4}$$

$$\gamma_{r\theta}^{(1)} = -\frac{2}{r}\left(\frac{\partial^2 w}{\partial r\partial\theta} - \frac{1}{r}\frac{\partial w}{\partial \theta}\right).$$

Next we use Hamilton's principle,

$$\int_{t_1}^{t_2} (\delta U + \delta V - \delta K)\,dt = 0, \tag{7.2.5}$$

to derive the Euler–Lagrange equations. The virtual strain energy δU is given by

$$\delta U = \int_{\Omega}\int_{-\frac{h}{2}}^{\frac{h}{2}} (\sigma_{rr}\delta\varepsilon_{rr} + \sigma_{\theta\theta}\delta\varepsilon_{\theta\theta} + \sigma_{r\theta}\delta\gamma_{r\theta})\,dz\,rdrd\theta$$

$$= \int_{\Omega}\left[-M_{rr}\frac{\partial^2\delta w}{\partial r^2} - \frac{1}{r}M_{\theta\theta}\left(\frac{\partial\delta w}{\partial r} + \frac{1}{r}\frac{\partial^2\delta w}{\partial\theta^2}\right)\right.$$

$$\left. -\frac{2}{r}M_{r\theta}\left(\frac{\partial^2\delta w}{\partial r\partial\theta} - \frac{1}{r}\frac{\partial\delta w}{\partial\theta}\right)\right]rdrd\theta, \tag{7.2.6}$$

where h is the plate thickness, and $(M_{rr}, M_{r\theta}, M_{\theta\theta})$ are the moments per unit length:

$$M_{rr} = \int_{-\frac{h}{2}}^{\frac{h}{2}} \sigma_{rr} z \, dz,$$

$$M_{\theta\theta} = \int_{-\frac{h}{2}}^{\frac{h}{2}} \sigma_{\theta\theta} z \, dz, \qquad (7.2.7)$$

$$M_{r\theta} = \int_{-\frac{h}{2}}^{\frac{h}{2}} \sigma_{r\theta} z \, dz.$$

The virtual total potential energy due to applied loads is calculated as follows. Suppose that $q = q(r, \theta)$ is the distributed transverse load, F_s is the reaction force of a linear elastic foundation, and $(\hat{N}_{rr}, \hat{N}_{\theta\theta}, \hat{N}_{r\theta})$ are the compressive and shear forces, measured per unit length, applied in the midplane of the plate. We have

$$\delta V = -\int_{\Omega} (q + F_s)\delta w \; r dr d\theta$$

$$-\int_{\Omega} \left[\hat{N}_{rr} \frac{\partial w}{\partial r} \frac{\partial \delta w}{\partial r} + \hat{N}_{\theta\theta} \frac{1}{r} \frac{\partial w}{\partial \theta} \frac{1}{r} \frac{\partial \delta w}{\partial \theta} \right.$$

$$\left. + \hat{N}_{r\theta} \frac{1}{r} \left(\frac{\partial w}{\partial \theta} \frac{\partial \delta w}{\partial r} + \frac{\partial w}{\partial r} \frac{\partial \delta w}{\partial \theta} \right) \right] r dr d\theta, \qquad (7.2.8)$$

where the linearly elastic foundation force is given by $F_s = -kw$ and k is the modulus of the elastic foundation.

The virtual kinetic energy δK is

$$\delta K = \int_{\Omega} \int_{-\frac{h}{2}}^{\frac{h}{2}} \rho \left[z^2 \frac{\partial \dot{w}}{\partial r} \frac{\partial \delta \dot{w}}{\partial r} + \frac{z^2}{r} \frac{\partial \dot{w}}{\partial \theta} \frac{1}{r} \frac{\partial \delta \dot{w}}{\partial \theta} + \dot{w} \delta \dot{w} \right] dz r dr d\theta$$

$$= \int_{\Omega} \left[m_2 \left(\frac{\partial \dot{w}}{\partial r} \frac{\partial \delta \dot{w}}{\partial r} + \frac{1}{r^2} \frac{\partial \dot{w}}{\partial \theta} \frac{\partial \delta \dot{w}}{\partial \theta} \right) + m_0 \dot{w} \delta \dot{w} \right] r dr d\theta, \qquad (7.2.9)$$

where m_0 is the principal mass inertia and m_2 is the rotatory inertia

$$m_0 = \int_{-\frac{h}{2}}^{\frac{h}{2}} \rho dz = \rho h, \quad m_2 = \int_{-\frac{h}{2}}^{\frac{h}{2}} z^2 \rho dz = \frac{\rho h^3}{12}, \qquad (7.2.10)$$

and superposed dot on w denotes differentiation with respect to t.

Next, we substitute the expressions for δU, δV, and δK into Eq. (7.2.5), and carry out integration by parts with respect to r, θ, and t to relieve δw of any derivatives. In Hamilton's principle, we require $\delta w(r, \theta, t_1) = 0$ and $\delta w(r, \theta, t_2) = 0$. The mixed derivative term $\partial^2 \delta w/\partial r \partial \theta$ in Eq. (7.2.6) requires special attention. Instead of integrating by parts with respect to r and then with respect to θ, or vice versa, it should be split into two terms so that the integration by parts is carried in a symmetric way:

$$\int_\Omega \frac{2}{r} M_{r\theta} \frac{\partial^2 \delta w}{\partial r \partial \theta} \, r dr d\theta = \int_\Omega M_{r\theta} \left(\frac{\partial^2 \delta w}{\partial r \partial \theta} + \frac{\partial^2 \delta w}{\partial \theta \partial r} \right) dr d\theta. \tag{7.2.11}$$

We obtain

$$
\begin{aligned}
0 = \int_{t_1}^{t_2} \int_\Omega \Bigg\{ & -\frac{1}{r} \left[\frac{\partial^2}{\partial r^2}(r M_{rr}) - \frac{\partial M_{\theta\theta}}{\partial r} + \frac{1}{r}\frac{\partial^2 M_{\theta\theta}}{\partial \theta^2} + 2\frac{\partial^2 M_{r\theta}}{\partial r \partial \theta} + \frac{2}{r}\frac{\partial M_{r\theta}}{\partial \theta} \right] \\
& + \frac{1}{r} \left[\frac{\partial}{\partial r}\left(r\hat{N}_{rr}\frac{\partial w}{\partial r} \right) + \frac{\partial}{\partial \theta}\left(\hat{N}_{\theta\theta}\frac{1}{r}\frac{\partial w}{\partial \theta} \right) \right. \\
& \left. + \frac{\partial}{\partial r}\left(\hat{N}_{r\theta}\frac{\partial w}{\partial \theta} \right) + \frac{\partial}{\partial \theta}\left(\hat{N}_{r\theta}\frac{\partial w}{\partial r} \right) \right] - q + kw \\
& + \frac{1}{r}\left[-m_2\frac{\partial}{\partial r}\left(r\frac{\partial \ddot{w}}{\partial r} \right) - m_2\frac{1}{r}\frac{\partial^2 \ddot{w}}{\partial \theta^2} + r m_0\ddot{w} \right] \Bigg\} \delta w \, r dr d\theta dt \\
- \int_{t_1}^{t_2} \oint_\Gamma \Bigg[& (r M_{rr}n_r + M_{r\theta}n_\theta)\frac{\partial \delta w}{\partial r} + \frac{1}{r}(r M_{r\theta}n_r + M_{\theta\theta}n_\theta)\frac{\partial \delta w}{\partial \theta} \\
& + \left\{ \left[\frac{\partial}{\partial r}(r M_{rr}) + \frac{\partial M_{r\theta}}{\partial \theta} - M_{\theta\theta} \right]n_r + \left(\frac{1}{r}\frac{\partial M_{\theta\theta}}{\partial \theta} + \frac{\partial M_{r\theta}}{\partial r} + \frac{2}{r}M_{r\theta} \right)n_\theta \right. \\
& - \frac{1}{r}\left(r\hat{N}_{rr}\frac{\partial w}{\partial r} + \hat{N}_{r\theta}\frac{\partial w}{\partial \theta} \right)n_r - \frac{1}{r}\left(\hat{N}_{\theta\theta}\frac{1}{r}\frac{\partial w}{\partial \theta} + \hat{N}_{r\theta}\frac{\partial w}{\partial r} \right)n_\theta \\
& + m_2\left(\frac{\partial \ddot{w}}{\partial r}n_r + \frac{1}{r^2}\frac{\partial \ddot{w}}{\partial \theta}n_\theta \right) \Bigg\} \delta w \Bigg] dS dt, \tag{7.2.12}
\end{aligned}
$$

where (n_r, n_θ) are the direction cosines of unit normal $\hat{\mathbf{n}} = n_r\hat{\mathbf{e}}_r + n_\theta\hat{\mathbf{e}}_\theta$ to the boundary Γ of the domain (midplane) Ω of the plate. Setting the coefficient of δw to zero inside Ω, we obtain the Euler–Lagrange equation

$$
\begin{aligned}
-\frac{1}{r}\Bigg[& \frac{\partial^2}{\partial r^2}(r M_{rr}) - \frac{\partial M_{\theta\theta}}{\partial r} + \frac{1}{r}\frac{\partial^2 M_{\theta\theta}}{\partial \theta^2} + 2\frac{\partial^2 M_{r\theta}}{\partial r \partial \theta} + \frac{2}{r}\frac{\partial M_{r\theta}}{\partial \theta} \Bigg] \\
& + \frac{1}{r}\left[\frac{\partial}{\partial r}\left(r\hat{N}_{rr}\frac{\partial w}{\partial r} \right) + \frac{\partial}{\partial \theta}\left(\hat{N}_{\theta\theta}\frac{1}{r}\frac{\partial w}{\partial \theta} \right) \right. \\
& \left. + \frac{\partial}{\partial r}\left(\hat{N}_{r\theta}\frac{\partial w}{\partial \theta} \right) + \frac{\partial}{\partial \theta}\left(\hat{N}_{r\theta}\frac{\partial w}{\partial r} \right) \right] - q + kw \\
& + \frac{1}{r}\left[-m_2\frac{\partial}{\partial r}\left(r\frac{\partial \ddot{w}}{\partial r} \right) - m_2\frac{1}{r}\frac{\partial^2 \ddot{w}}{\partial \theta^2} + r m_0\ddot{w} \right] = 0. \tag{7.2.13}
\end{aligned}
$$

The natural boundary conditions follow from the boundary expressions in Eq. (7.2.12). Assuming that δw, $\partial \delta w / \partial r$, and $\partial \delta w / \partial \theta$ are arbitrary, which means that w, $\partial w / \partial r$, and $\partial w / \partial \theta$ are not specified on the boundary, we obtain

$$\delta w: \qquad r V_r n_r + V_\theta n_\theta = 0, \tag{7.2.14}$$

$$\frac{\partial \delta w}{\partial r}: \qquad r M_{rr} n_r + M_{r\theta} n_\theta = 0, \tag{7.2.15}$$

$$\frac{\partial \delta w}{\partial \theta}: \qquad r M_{r\theta} n_r + M_{\theta\theta} n_\theta = 0, \tag{7.2.16}$$

where V_r and V_θ are the effective shear forces per unit length:

$$V_r = Q_r - \frac{1}{r} \left(r \hat{N}_{rr} \frac{\partial w}{\partial r} + \hat{N}_{r\theta} \frac{\partial w}{\partial \theta} \right) + m_2 \frac{\partial \ddot{w}}{\partial r}, \tag{7.2.17}$$

$$V_\theta = Q_\theta - \frac{1}{r} \left(\hat{N}_{\theta\theta} \frac{1}{r} \frac{\partial w}{\partial \theta} + \hat{N}_{r\theta} \frac{\partial w}{\partial r} \right) + m_2 \frac{1}{r} \frac{\partial \ddot{w}}{\partial \theta}, \tag{7.2.18}$$

$$Q_r \equiv \frac{1}{r} \left[\frac{\partial}{\partial r} (r M_{rr}) + \frac{\partial M_{r\theta}}{\partial \theta} - M_{\theta\theta} \right], \tag{7.2.19}$$

$$Q_\theta \equiv \frac{1}{r} \left[\frac{\partial}{\partial r} (r M_{r\theta}) + \frac{\partial M_{\theta\theta}}{\partial \theta} + M_{r\theta} \right]. \tag{7.2.20}$$

The equation of motion, Eq. (7.2.13), can be expressed, in view of the definitions in Eqs (7.2.17) to (7.2.20), as

$$-\frac{1}{r} \left[\frac{\partial}{\partial r} (r Q_r) + \frac{\partial Q_\theta}{\partial \theta} \right] + kw$$

$$+ \frac{1}{r} \left[\frac{\partial}{\partial r} \left(r \hat{N}_{rr} \frac{\partial w}{\partial r} \right) + \frac{\partial}{\partial \theta} \left(\hat{N}_{\theta\theta} \frac{1}{r} \frac{\partial w}{\partial \theta} \right) + \frac{\partial}{\partial r} \left(\hat{N}_{r\theta} \frac{\partial w}{\partial \theta} \right) + \frac{\partial}{\partial \theta} \left(\hat{N}_{r\theta} \frac{\partial w}{\partial r} \right) \right]$$

$$+ \frac{1}{r} \left[-m_2 \frac{\partial}{\partial r} \left(r \frac{\partial \ddot{w}}{\partial r} \right) - m_2 \frac{1}{r} \frac{\partial^2 \ddot{w}}{\partial \theta^2} + r m_0 \ddot{w} \right] = q. \tag{7.2.21}$$

For the axisymmetric case (i.e., when the material properties, loads, and boundary conditions are independent of θ), Eq. (7.2.21) becomes

$$-\frac{1}{r} \frac{\partial}{\partial r} (r Q_r) + kw + \frac{1}{r} \frac{\partial}{\partial r} \left(r \hat{N}_{rr} \frac{\partial w}{\partial r} \right) + \frac{1}{r} \left[-m_2 \frac{\partial}{\partial r} \left(r \frac{\partial \ddot{w}}{\partial r} \right) + r m_0 \ddot{w} \right] = q, \tag{7.2.22}$$

for $b < r < a$, where a and b are the outer and inner radii, respectively, of an annular plate. For the solid circular plate, we set $b = 0$.

This completes the derivation of the governing equations of a circular plate. The equations are valid for any suitable constitutive behavior of the plate material.

In order to express the governing equations in terms of the displacements u and w, the bending moments in Eq. (7.2.7) must be expressed in terms of the displacements through stress-strain relations and strain–displacement relations. For an isotropic linear elastic material, assuming that the elastic stiffnesses are independent of the temperature, the stress–strain relations are [see Eqs (2.4.14) and (2.4.15), which also hold for components in the cylindrical coordinates; see Eq. (2.4.20) for the thermo elastic case]

$$
\left\{ \begin{matrix} \sigma_{rr} \\ \sigma_{\theta\theta} \\ \sigma_{r\theta} \end{matrix} \right\} = \frac{E}{1-\nu^2} \begin{bmatrix} 1 & \nu & 0 \\ \nu & 1 & 0 \\ 0 & 0 & \frac{1-\nu}{2} \end{bmatrix} \left\{ \begin{matrix} \varepsilon_{rr} - \alpha\,\Delta T \\ \varepsilon_{\theta\theta} - \alpha\,\Delta T \\ 2\varepsilon_{r\theta} \end{matrix} \right\}, \tag{7.2.23}
$$

where $\Delta T(r,\theta,z)$ is the temperature rise from a stress-free (or undeformed) state, α is the coefficient of thermal expansion, E is Young's modulus, and ν Poisson's ratio. Then we find that

$$
\begin{aligned}
M_{rr} &= \int_{-\frac{h}{2}}^{\frac{h}{2}} \sigma_{rr}\, z\, dz = \frac{Eh^3}{12(1-\nu^2)} \left(\varepsilon_{rr}^{(1)} + \nu\varepsilon_{\theta\theta}^{(1)} \right) - M_{rr}^T, \\
M_{\theta\theta} &= \int_{-\frac{h}{2}}^{\frac{h}{2}} \sigma_{\theta\theta}\, z\, dz = \frac{Eh^3}{12(1-\nu^2)} \left(\nu\varepsilon_{rr}^{(1)} + \varepsilon_{\theta\theta}^{(1)} \right) - M_{\theta\theta}^T, \\
M_{r\theta} &= \int_{-\frac{h}{2}}^{\frac{h}{2}} \sigma_{r\theta}\, z\, dz = \frac{Gh^3}{12}\gamma_{r\theta}^{(1)},
\end{aligned}
\tag{7.2.24}
$$

where M_{rr}^T and $M_{\theta\theta}^T$ are the thermal moments:

$$
M^T \equiv M_{rr}^T = M_{\theta\theta}^T = \frac{E\alpha}{1-\nu} \int_{-\frac{h}{2}}^{\frac{h}{2}} \Delta T z\, dz. \tag{7.2.25}
$$

Now, the stress resultants $(M_{rr}, M_{\theta\theta}, M_{r\theta})$ can be expressed in terms of the deflection w using the strain–displacement relations in Eq. (7.2.3) as

$$
\begin{aligned}
M_{rr} &= -D \left[\frac{\partial^2 w}{\partial r^2} + \frac{\nu}{r} \left(\frac{\partial w}{\partial r} + \frac{1}{r}\frac{\partial^2 w}{\partial \theta^2} \right) \right] - M^T, \\
M_{\theta\theta} &= -D \left[\nu\frac{\partial^2 w}{\partial r^2} + \frac{1}{r} \left(\frac{\partial w}{\partial r} + \frac{1}{r}\frac{\partial^2 w}{\partial \theta^2} \right) \right] - M^T, \\
M_{r\theta} &= -(1-\nu)D\frac{1}{r} \left(\frac{\partial^2 w}{\partial r\partial\theta} - \frac{1}{r}\frac{\partial w}{\partial\theta} \right).
\end{aligned}
\tag{7.2.26}
$$

The bending stiffness D is given by

$$
D = \frac{Eh^3}{12(1-\nu^2)}. \tag{7.2.27}
$$

The equation of equilibrium in Eq. (7.2.21) can now be written in terms of the deflection w as

$$
D \left[\frac{1}{r} \frac{\partial}{\partial r} \left(r \frac{\partial}{\partial r} \right) + \frac{1}{r^2} \frac{\partial^2}{\partial \theta^2} \right] \left[\frac{1}{r} \frac{\partial}{\partial r} \left(r \frac{\partial w}{\partial r} \right) + \frac{1}{r^2} \frac{\partial^2 w}{\partial \theta^2} \right] + kw
$$

$$
+ \frac{1}{r} \left[\frac{\partial}{\partial r} \left(r \hat{N}_{rr} \frac{\partial w}{\partial r} \right) + \frac{\partial}{\partial \theta} \left(\hat{N}_{\theta\theta} \frac{1}{r} \frac{\partial w}{\partial \theta} \right) + \frac{\partial}{\partial r} \left(\hat{N}_{r\theta} \frac{\partial w}{\partial \theta} \right) + \frac{\partial}{\partial \theta} \left(\hat{N}_{r\theta} \frac{\partial w}{\partial r} \right) \right]
$$

$$
+ \frac{1}{r} \left[-m_2 \frac{\partial}{\partial r} \left(r \frac{\partial \ddot{w}}{\partial r} \right) - m_2 \frac{1}{r^2} \frac{\partial^2 \ddot{w}}{\partial \theta^2} + m_0 \ddot{w} \right]
$$

$$
= q - \left[\frac{1}{r} \frac{\partial}{\partial r} \left(r \frac{\partial M^T}{\partial r} \right) + \frac{1}{r^2} \frac{\partial^2 M^T}{\partial \theta^2} \right]. \tag{7.2.28}
$$

For the axisymmetric case, Eq. (7.2.28) simplifies to

$$
\frac{D}{r} \frac{\partial}{\partial r} \left\{ r \frac{\partial}{\partial r} \left[\frac{1}{r} \frac{\partial}{\partial r} \left(r \frac{\partial w}{\partial r} \right) \right] \right\} + kw + \frac{1}{r} \frac{\partial}{\partial r} \left(r \hat{N}_{rr} \frac{\partial w}{\partial r} \right)
$$

$$
+ \frac{1}{r} \left[-m_2 \frac{\partial}{\partial r} \left(r \frac{\partial \ddot{w}}{\partial r} \right) + m_0 \ddot{w} \right] = q - \frac{1}{r} \frac{\partial}{\partial r} \left(r \frac{\partial M^T}{\partial r} \right). \tag{7.2.29}
$$

For the static case (with applied compressive force $\hat{N}_{rr}$), Eq. (7.2.29) becomes

$$
\frac{D}{r} \frac{d}{dr} \left\{ r \frac{d}{dr} \left[\frac{1}{r} \frac{d}{dr} \left(r \frac{dw}{dr} \right) \right] \right\} + kw + \frac{1}{r} \frac{d}{dr} \left(r \hat{N}_{rr} \frac{dw}{dr} \right) = q - \frac{1}{r} \frac{d}{dr} \left(r \frac{dM^T}{dr} \right). \tag{7.2.30}
$$

The standard boundary conditions for axisymmetric bending of solid circular plates and annular plates (see Fig. 7.2.2) are listed in Table tab7.2.1.

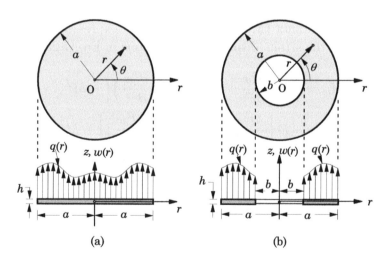

Fig. 7.2.2 Axisymmetric (a) circular and (b) annular plates.

Table 7.2.1 Typical boundary conditions for solid circular and annular plates[a].

Plate type/edge	Free	Hinged	Clamped
Solid circular plate			
$r = 0$ (for all cases)		$\frac{dw}{dr} = 0, \quad 2\pi r Q_r = -Q_0$	
$r = a$	$Q_r = Q_a$	$w = 0$	$w = 0$
	$M_{rr} = M_a$	$M_{rr} = M_a$	$\frac{dw}{dr} = 0$
Annular plate			
$r = b$	$Q_r = Q_b$	$w = 0$	$w = 0$
	$M_{rr} = M_b$	$M_{rr} = M_b$	$\frac{dw}{dr} = 0$
$r = a$	$Q_r = Q_a$	$w = 0$	$w = 0$
	$M_{rr} = M_a$	$M_{rr} = M_a$	$\frac{dw}{dr} = 0$

[a]Q_a, Q_b, M_a, and M_b are distributed edge forces and moments and Q_0 is a concentrated force.

7.2.2 Analysis of Circular Plates

7.2.2.1 Analytical Solutions for Bending

Here, we develop analytical solutions of isotropic circular plates for axisymmetric boundary conditions and loads (see Reddy [51] for additional details). For axisymmetric circular plates with $k = 0$, $M^T = 0$, and $\hat{N}_{rr} = 0$, Eq. (7.2.29) simplifies to

$$\frac{D}{r}\frac{\partial}{\partial r}\left\{r\frac{\partial}{\partial r}\left[\frac{1}{r}\frac{\partial}{\partial r}\left(r\frac{\partial w}{\partial r}\right)\right]\right\} + \frac{1}{r}\left[-m_2\frac{\partial}{\partial r}\left(r\frac{\partial \ddot{w}}{\partial r}\right) + m_0\ddot{w}\right] = q. \quad (7.2.31)$$

For static case, this further simplifies to

$$\frac{D}{r}\frac{d}{dr}\left\{r\frac{d}{dr}\left[\frac{1}{r}\frac{d}{dr}\left(r\frac{dw}{dr}\right)\right]\right\} = q. \quad (7.2.32)$$

The stresses, bending moments, and shear force are related to the deflection w by

$$\sigma_{rr}(r, z) = -\frac{Ez}{(1 - \nu^2)}\left(\frac{d^2w}{dr^2} + \frac{\nu}{r}\frac{dw}{dr}\right), \quad (7.2.33)$$

$$\sigma_{\theta\theta}(r, z) = -\frac{Ez}{(1 - \nu^2)}\left(\nu\frac{d^2w}{dr^2} + \frac{1}{r}\frac{dw}{dr}\right), \quad (7.2.34)$$

$$M_{rr}(r) = -D\left(\frac{d^2w}{dr^2} + \frac{\nu}{r}\frac{dw}{dr}\right), \tag{7.2.35}$$

$$M_{\theta\theta}(r) = -D\left(\nu\frac{d^2w}{dr^2} + \frac{1}{r}\frac{dw}{dr}\right), \tag{7.2.36}$$

$$Q_r(r) = -D\frac{d}{dr}\left[\frac{1}{r}\frac{d}{dr}\left(r\frac{dw}{dr}\right)\right]. \tag{7.2.37}$$

In the case of plates subjected to point loads, the maximum values of stresses can occur near to the point of application of the point load. According to Nadai[2] and Westergaard[3], the maximum bending moments and stresses within a small radius $r_c \ll a$ can be computed using the equivalent radius r_e given by

$$r_e = \sqrt{1.6r_c^2 + h^2} - 0.675h \text{ for } r_c < 0.5h; \quad r_e = r_c \text{ for } r_c \geq 0.5h, \tag{7.2.38}$$

where h is the plate thickness.

The general solution of Eq. (7.2.32) can be obtained by successive integrations:

$$D\frac{d}{dr}\left[\frac{1}{r}\frac{d}{dr}\left(r\frac{dw}{dr}\right)\right] = \frac{1}{r}\int rq\,dr + \frac{c_1}{r} \quad (=-Q_r), \tag{7.2.39}$$

$$D\frac{d}{dr}\left(r\frac{dw}{dr}\right) = r\int\frac{1}{r}\left(\int rq\,dr\right)dr + c_1 r\log r + c_2 r, \tag{7.2.40}$$

$$D\frac{dw}{dr} = F(r) + c_1\frac{r}{4}(2\log r - 1) + c_2\frac{r}{2} + c_3\frac{1}{r}, \tag{7.2.41}$$

$$Dw = G(r) + c_1\frac{r^2}{4}(\log r - 1) + c_2\frac{r^2}{4} + c_3\log r + c_4, \tag{7.2.42}$$

where

$$\int r\log r\,dr = \frac{r^2}{4}(2\log r - 1)$$

$$F(r) = \frac{1}{r}\int r\left[\int\frac{1}{r}\left(\int rq\,dr\right)dr\right]dr, \tag{7.2.43}$$

$$G(r) = \int F(r)\,dr,$$

and c_i $(i = 1, 2, 3, 4)$ are constants of integration that will be evaluated using the boundary conditions. For solid circular plates the requirement that the slope be zero due to symmetry at $r = 0$ requires $c_3 = 0$.

---

[2]A. Nadai, *Die Elastichen Platten*, Springer, Berlin, 1925.
[3]H.M. Westergaard, "Stresses in concrete pavements computed by theoretical analysis," *Public Roads*, U.S. Department of Agriculture, Bureau of Public Roads, Vol. 7, No. 2, 1926.

The bending moments and stresses are given by

$$M_{rr}(r) = -\left[\frac{dF}{dr} + \frac{c_1}{4}(2\log r + 1) + \frac{c_2}{2} - \frac{c_3}{r^2}\right]$$
$$\quad - \frac{\nu}{r}\left[F + \frac{c_1 r}{4}(2\log r - 1) + \frac{c_2 r}{2} + \frac{c_3}{r}\right], \tag{7.2.44}$$

$$M_{\theta\theta}(r) = -\left[\nu\frac{dF}{dr} + \nu\frac{c_1}{4}(2\log r + 1) + \nu\frac{c_2}{2} - \nu\frac{c_3}{r^2}\right]$$
$$\quad - \frac{1}{r}\left[F + \frac{c_1 r}{4}(2\log r - 1) + \frac{c_2 r}{2} + \frac{c_3}{r}\right], \tag{7.2.45}$$

$$\sigma_{rr}(r,z) = \frac{12z}{h^3}M_{rr}(r), \qquad \sigma_{\theta\theta}(r,z) = \frac{12z}{h^3}M_{\theta\theta}(r). \tag{7.2.46}$$

Example 7.2.1

Determine the analytical solutions for deflection, moments, and stresses in a solid circular plate with clamped edge (see Fig. 7.2.3) when the plate is loaded by (a) uniformly distributed load of intensity q_0 and (b) point load Q_0 at the center.

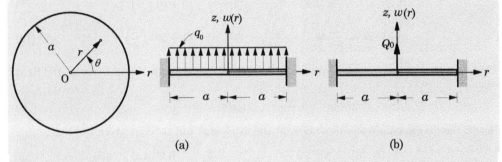

(a)   (b)

Fig. 7.2.3 A clamped circular plate under (a) uniformly distributed load, and (b) point load.

Solution: (a) *Uniform load.* The boundary conditions at $r = 0$ for solid circular plates without a point load at $r = 0$ are

$$(rQ_r) = 0, \quad \frac{dw}{dr} = 0. \tag{1}$$

The first condition, in view of Eq. (7.2.39), yields $c_1 = 0$. The second condition, in light of Eq. (7.2.41), requires $c_3 = 0$.

The boundary conditions associated with the clamped outer edge at $r = a$ are

$$w = 0, \quad \frac{dw}{dr} = 0. \tag{2}$$

These conditions yield

$$c_2 = -\frac{2F(a)}{a}, \quad c_4 = -G(a) - c_2\frac{a^2}{4} = -G(a) + \frac{a}{2}F(a).$$

For uniformly distributed load of intensity q_0, we have

$$F(r) = \frac{q_0 r^3}{16}, \quad G(r) = \frac{q_0 r^4}{64}. \tag{3}$$

Therefore, we have

$$c_1 = 0, \quad c_2 = -\frac{q_0 a^2}{8}, \quad c_3 = 0, \quad c_4 = \frac{q_0 a^4}{64}. \tag{4}$$

Hence, the deflection becomes

$$w(r) = \frac{q_0 a^4}{64D} \left[1 - \left(\frac{r}{a} \right)^2 \right]^2, \tag{5}$$

and the maximum deflection occurs at the center of the plate ($r = 0$):

$$w_{\max} = \frac{q_0 a^4}{64D}. \tag{6}$$

Expressions for the bending moments and stresses from Eqs (7.2.44) to (7.2.46) become

$$M_{rr}(r) = \frac{q_0 a^2}{16} \left[(1 + \nu) - (3 + \nu) \left(\frac{r}{a} \right)^2 \right], \tag{7}$$

$$M_{\theta\theta}(r) = \frac{q_0 a^2}{16} \left[(1 + \nu) - (1 + 3\nu) \left(\frac{r}{a} \right)^2 \right], \tag{8}$$

$$\sigma_{rr}(r, z) = \frac{12z}{h^3} M_{rr} = \frac{3q_0 a^2 z}{4h^3} \left[(1 + \nu) - (3 + \nu) \left(\frac{r}{a} \right)^2 \right], \tag{9}$$

$$\sigma_{\theta\theta}(r, z) = \frac{12z}{h^3} M_{\theta\theta} = \frac{3q_0 a^2 z}{4h^3} \left[(1 + \nu) - (1 + 3\nu) \left(\frac{r}{a} \right)^2 \right]. \tag{10}$$

The maximum (in magnitude) values of the bending moments are found to be at the fixed edge:

$$M_{rr}(a) = -\frac{q_0 a^2}{8}, \quad M_{\theta\theta}(a) = -\frac{\nu q_0 a^2}{8}. \tag{11}$$

Hence, the maximum stresses also occur at the fixed edge and they are given by

$$\sigma_{rr}^{\max}\left(a, -\frac{h}{2}\right) = -\frac{6M_{rr}(a)}{h^2} = \frac{3q_0}{4} \left(\frac{a}{h} \right)^2, \quad \sigma_{\theta\theta}^{\max}\left(a, -\frac{h}{2}\right) = -\frac{6M_{\theta\theta}(a)}{h^2} = \frac{3\nu q_0}{4} \left(\frac{a}{h} \right)^2. \tag{12}$$

(b) *Point load.* For point load Q_0 at the center, the boundary conditions are (Q_r acts downward at the center, whereas Q_0 is applied upward; see Figs. 7.2.1 and 7.2.3)

$$(2\pi r Q_r)_{r=0} = -Q_0, \quad \frac{dw}{dr}\bigg|_{r=0} = 0, \quad w(a) = 0, \quad \frac{dw}{dr}\bigg|_{r=a} = 0. \tag{13}$$

Since $q = 0$, from Eq. (7.2.39) we have $2\pi c_1 = Q_0$. For a finite value of dw/dr at $r = 0$, from Eq. (7.2.41), we infer that $c_3 = 0$. Thus, the first two boundary conditions in Eq. (13) give

$$c_1 = \frac{Q_0}{2\pi}, \quad c_3 = 0. \tag{14}$$

The remaining two boundary conditions in Eq. (13) correspond to the clamped edge that, with the help of Eqs (7.2.41) and (7.2.42), yield

$$\frac{Q_0 a}{8\pi} (2\log a - 1) + \frac{a}{2} c_2 = 0, \quad \frac{Q_0 a^2}{8\pi} (\log a - 1) + \frac{a^2}{4} c_2 + c_4 = 0.$$

Solving for c_2 and c_4, we obtain

$$c_2 = -\frac{Q_0}{4\pi} (2\log a - 1), \quad c_4 = \frac{Q_0 a^2}{16\pi}. \tag{15}$$

Hence, the solution becomes

$$w(r) = \frac{Q_0 a^2}{16\pi D}\left[1 - \left(\frac{r}{a}\right)^2 + 2\left(\frac{r}{a}\right)^2 \log\left(\frac{r}{a}\right)\right], \tag{16}$$

$$M_{rr}(r) = -\frac{Q_0}{4\pi}\left[1 + (1+\nu)\log\left(\frac{r}{a}\right)\right], \qquad M_{\theta\theta}(r) = -\frac{Q_0}{4\pi}\left[\nu + (1+\nu)\log\left(\frac{r}{a}\right)\right], \tag{17}$$

$$\sigma_{rr}(r,z) = -\frac{3Q_0 z}{\pi h^3}\left[1 + (1+\nu)\log\left(\frac{r}{a}\right)\right], \qquad \sigma_{\theta\theta}(r,z) = -\frac{3Q_0 z}{\pi h^3}\left[\nu + (1+\nu)\log\left(\frac{r}{a}\right)\right]. \tag{18}$$

The maximum deflection occurs at $r = 0$ (the last term goes to zero as $r \to 0$) and is given by

$$w_{\max} = \frac{Q_0 a^2}{16\pi D}. \tag{19}$$

We note that M_{rr} and $M_{\theta\theta}$ (and the corresponding stresses) are not defined at $r = 0$; the maximum values occur at $r = r_e$, where r_e is given by Eq. (7.2.38).

Example 7.2.2

Consider a simply supported annular plate with inner radius b and outer radius a, and subjected to uniformly distributed load of intensity q_0. Determine the expressions for the transverse deflection, bending moments, and stresses in the plate. Specialize the results for the case of simply supported solid circular plates under uniformly distributed load and point load at the center.

Solution: The boundary conditions for this case (i.e., simply supported at $r = a$ and free at $r = b$) are

$$\text{At } r = b: \qquad M_{rr} = 0, \quad (rQ_r) = 0, \tag{1}$$
$$\text{At } r = a: \qquad w = 0, \quad M_{rr} = 0. \tag{2}$$

The functions $F(r)$ and $G(r)$ are as given in Eq. (3) of **Example 7.2.1**: $F(r) = q_0 r^3/16$ and $G(r) = q_0 r^4/64$. The second condition in Eq. (1), upon substitution into Eq. (7.2.39), gives $c_1 = -q_0 b^2/2$; the substitution of the remaining boundary conditions into Eqs (7.2.42) and (7.2.44) yield, after some algebraic manipulations, the following relations among c_2, c_3, and c_4:

$$-\left(\frac{3+\nu}{16}\right)q_0 b^2 + \left[\frac{1+\nu}{4}\log b + \frac{1-\nu}{8}\right]q_0 b^2 - \frac{1+\nu}{2}c_2 + \frac{1-\nu}{b^2}c_3 = 0,$$

$$-\left(\frac{3+\nu}{16}\right)q_0 a^2 + \left[\frac{1+\nu}{4}\log a + \frac{1-\nu}{8}\right]q_0 b^2 - \frac{1+\nu}{2}c_2 + \frac{1-\nu}{a^2}c_3 = 0,$$

$$\frac{q_0 a^4}{64} - \frac{a^2}{8}(\log a - 1)q_0 b^2 + \frac{a^2}{4}c_2 + c_3 \log a + c_4 = 0.$$

Solving for the constants c_2, c_3, and c_4, we obtain

$$c_2 = -\left(\frac{1+3\nu}{1+\nu}\right)\frac{q_0 b^2}{8} - \left(\frac{3+\nu}{1+\nu}\right)\frac{q_0 a^2}{8} + \frac{q_0 b^2}{2}\log a - \frac{q_0 b^4}{2(a^2-b^2)}\log\beta,$$

$$c_3 = -\left(\frac{3+\nu}{1-\nu}\right)\frac{q_0 b^2 a^2}{16} - \left(\frac{1+\nu}{1-\nu}\right)\frac{q_0 a^2 b^4}{4(a^2-b^2)}\log\beta,$$

$$c_4 = -\frac{q_0 a^4}{64} + \left(\frac{3+\nu}{1+\nu}\right)\frac{q_0 a^2(a^2-b^2)}{32} + \left(\frac{3+\nu}{1-\nu}\right)\frac{q_0 b^2 a^2}{16}\log a$$

$$+ \frac{q_0 b^4 a^2}{8(a^2-b^2)}\log\beta + \left(\frac{1+\nu}{1-\nu}\right)\frac{q_0 b^4 a^2}{4(a^2-b^2)}\log a \log\beta.$$

The deflection and bending moments in the annular plate are

$$w(r) = \frac{q_0 a^4}{64D} \left\{ -1 + \left(\frac{r}{a}\right)^4 + \frac{2\alpha_1}{1+\nu}\left[1 - \left(\frac{r}{a}\right)^2\right] - \frac{4\alpha_2\beta^2}{1-\nu}\log\left(\frac{r}{a}\right) \right\}, \tag{3}$$

$$M_{rr}(r) = \frac{q_0 a^2}{16}\left\{ (3+\nu)\left[1 - \left(\frac{r}{a}\right)^2\right] + \beta^2\left[3 + \nu + 4(1+\nu)\kappa \right.\right.$$
$$\left.\left. -\alpha_2\left(\frac{a}{r}\right)^2 + 4(1+\nu)\log\left(\frac{r}{a}\right)\right]\right\}, \tag{4}$$

$$M_{\theta\theta}(r) = \frac{q_0 a^2}{16}\left\{ (3+\nu) - (1+3\nu)\left(\frac{r}{a}\right)^2 + \beta^2\left[(5\nu-1) + 4(1+\nu)\kappa \right.\right.$$
$$\left.\left. +\alpha_2\left(\frac{a}{r}\right)^2 + 4(1+\nu)\log\left(\frac{r}{a}\right)\right]\right\}, \tag{5}$$

where

$$\alpha_1 = (3+\nu)(1-\beta^2) - 4(1+\nu)\beta^2\kappa, \quad \alpha_2 = (3+\nu) + 4(1+\nu)\kappa,$$

$$\kappa = \frac{\beta^2}{1-\beta^2}\log\beta, \quad \beta = \frac{b}{a}.$$

The stresses are given by

$$\sigma_{rr}(r,z) = \frac{12z}{h^3}M_{rr}(r), \quad \sigma_{\theta\theta}(r,z) = \frac{12z}{h^3}M_{\theta\theta}(r). \tag{6}$$

For a simply supported solid circular plate under uniform load, we set $b = 0$ (i.e., $\beta = 0$, $\kappa = 0$, and $\alpha_1 = \alpha_2 = 3 + \nu$) in the previous equations and obtain

$$w(r) = \frac{q_0 a^4}{64D}\left[\left(\frac{r}{a}\right)^4 - 2\left(\frac{3+\nu}{1+\nu}\right)\left(\frac{r}{a}\right)^2 + \frac{5+\nu}{1+\nu}\right], \tag{7}$$

$$M_{rr}(r) = \frac{q_0 a^2}{16}(3+\nu)\left[1 - \left(\frac{r}{a}\right)^2\right], \tag{8}$$

$$M_{\theta\theta}(r) = \frac{q_0 a^2}{16}\left[(3+\nu) - (1+3\nu)\left(\frac{r}{a}\right)^2\right]. \tag{9}$$

The maximum deflection, bending moments, and stresses occur at $r = 0$, and they are

$$w_{max} = \left(\frac{5+\nu}{1+\nu}\right)\frac{q_0 a^4}{64D}, \quad M_{max} = (3+\nu)\frac{q_0 a^2}{16}, \quad \sigma_{max} = 3(3+\nu)\frac{q_0 a^2}{8h^2}. \tag{10}$$

Finally, for a simply supported solid circular plate with a central point load Q_0, Eqs (7.2.39) and (7.2.41) yield the constants $c_1 = Q_0/2\pi$ and $c_3 = 0$. Using the boundary conditions $w(a) = 0$ and $M_{rr}(a) = 0$ in Eqs (7.2.42) and (7.2.44), we obtain

$$\frac{Q_0 a^2}{8\pi}(\log a - 1) + \frac{a^2}{4}c_2 + c_4 = 0$$

$$-\frac{Q_0}{2\pi}\left[\frac{1+\nu}{2}\log a + \frac{1-\nu}{4}\right] - \frac{1+\nu}{2}c_2 = 0.$$

Solving for the constants c_2 and c_4, we obtain

$$c_2 = -\frac{Q_0}{4\pi}\left[2\log a + \left(\frac{1-\nu}{1+\nu}\right)\right], \quad c_4 = \left(\frac{3+\nu}{1+\nu}\right)\frac{Q_0 a^2}{16\pi}.$$

The solution for any $r \neq 0$ is given by

$$w(r) = \frac{Q_0 a^2}{16\pi D} \left[\left(\frac{3+\nu}{1+\nu} \right) \left(1 - \frac{r^2}{a^2} \right) + 2 \left(\frac{r}{a} \right)^2 \log \left(\frac{r}{a} \right) \right], \tag{11}$$

$$M_{rr}(r) = -\frac{Q_0(1+\nu)}{4\pi} \log \left(\frac{r}{a} \right), \tag{12}$$

$$M_{\theta\theta}(r) = -\frac{Q_0}{4\pi} \left[(1+\nu) \log \left(\frac{r}{a} \right) - (1-\nu) \right], \tag{13}$$

$$\sigma_{rr}(r) = -\frac{3zQ_0(1+\nu)}{h^3 \pi} \log \left(\frac{r}{a} \right), \tag{14}$$

$$\sigma_{\theta\theta}(r) = -\frac{3zQ_0}{h^3 \pi} \left[(1+\nu) \log \left(\frac{r}{a} \right) - (1-\nu) \right]. \tag{15}$$

The maximum deflection is given by

$$w_{\max} = w(0) = \frac{Q_0 a^2}{16\pi D} \left(\frac{3+\nu}{1+\nu} \right). \tag{16}$$

The stresses and bending moments cannot be calculated at $r = 0$ due to the logarithmic singularity. The maximum finite stresses produced by load Q_0 on a very small circular area of radius r_c can be calculated using the equivalent radius r_e in Eqs (12)–(15) (see Roark and Young [123]):

$$r_e = \sqrt{1.6r_c^2 + h^2} - 0.675h \quad \text{when } r_c < 1.7h, \tag{17}$$

$$r_e = r_c \quad \text{when } r_c \geq 1.7h. \tag{18}$$

7.2.2.2   Analytical Solutions for Buckling

Here we consider exact solutions for buckling of circular plates under in-plane compressive load $\hat{N}_{rr} = N_0$. Consider the shear force–deflection relation in Eq. (7.2.37), which can be expressed as

$$Q_r = -D \left[\frac{1}{r} \frac{d}{dr} \left(r \frac{d^2 w}{dr^2} \right) - \frac{1}{r^2} \frac{dw}{dr} \right]. \tag{7.2.47}$$

We write the equation in terms of $\phi = (dw/dr)$, which represents the angle between the central axis of the plate and the normal to the deflected surface at any point (see Fig. 7.2.4) as

$$Q_r = -D \left[\frac{1}{r} \frac{d}{dr} \left(r \frac{d\phi}{dr} \right) - \frac{\phi}{r^2} \right]. \tag{7.2.48}$$

For a circular plate under the action of uniform radial compressive load $\hat{N}_{rr} = N_0$ per unit length, the shear force at any point is obtained by integrating Eq. (7.2.30) for the case with $q = 0$, $k = 0$, and $M^T = 0$, and setting the constant of integration to zero (because $Q_r = 0$ at $r = 0$ for an axisymmetric mode):

$$Q_r = N_0 \frac{dw}{dr} = N_0 \phi. \tag{7.2.49}$$

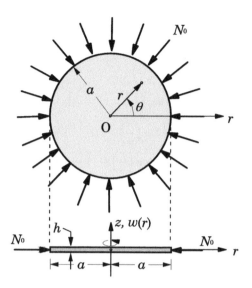

Fig. 7.2.4 Buckling of a circular plate under uniform radial compressive load.

From Eqs (7.2.48) and (7.2.49), we obtain

$$-D\left[\frac{1}{r}\frac{d}{dr}\left(r\frac{d\phi}{dr}\right) - \frac{1}{r^2}\phi\right] = N_0\phi,$$

or

$$r\frac{d}{dr}\left(r\frac{d\phi}{dr}\right) + \left(\frac{N_0}{D}r^2 - 1\right)\phi = 0. \qquad (7.2.50)$$

Equation (7.2.50) can be recast in an alternative form by invoking the transformation

$$\bar{r} = r\alpha, \qquad \alpha^2 = \frac{N_0}{D},$$

and the alternative form is

$$\bar{r}\frac{d}{d\bar{r}}\left(\bar{r}\frac{d\phi}{d\bar{r}}\right) + \left(\bar{r}^2 - n^2\right)\phi = 0, \qquad (7.2.51)$$

with $n = 1$ (corresponds to the critical buckling load), which is recognized as the *Bessel differential equation*.

The general solution of Eq. (7.2.51) is

$$\phi(\bar{r}) = AJ_n(\bar{r}) + BY_n(\bar{r}), \qquad (7.2.52)$$

where J_n is the Bessel function of the first kind of order n, Y_n is the Bessel function of the second kind of order n, and A and B are constants to be determined using the boundary conditions. In the case of buckling, we do not actually find these constants but determine the stability criterion. In the following we consider specific examples of buckling of circular plates.

Example 7.2.3 ────────────────────────────────

Determine the critical buckling loads for solid circular plates under in-plane compressive load when the outer edge is (a) clamped, (b) simply supported, and (c) simply supported with rotational restraint at the outer edge.

Solution:
(a) *Clamped plate.* For a clamped plate, the boundary conditions are (dw/dr is zero at $r = 0, a$)

$$\phi(0) = 0, \quad \phi(a) = 0. \tag{1}$$

Using the general solution in Eq. (7.2.52) for an isotropic plate, we obtain

$$AJ_1(0) + BY_1(0) = 0, \quad AJ_1(\alpha a) + BY_1(\alpha a) = 0, \quad \alpha^2 = \frac{\hat{N}_{rr}}{D}. \tag{2}$$

Since $J_1(0) = 0$ and $Y_1(0)$ is unbounded, we must have $B = 0$. The fact that $B = 0$ reduces the second equation (since $A \neq 0$ for a non-trivial solution) to

$$J_1(\alpha a) = 0, \tag{3}$$

which is the stability criterion. The smallest root of the condition in Eq. (3) is $\alpha a = 3.8317$. Thus we have the following critical buckling load for a solid clamped circular plate:

$$N_{cr} = 14.682 \frac{D}{a^2}. \tag{4}$$

(b) *Simply supported plate.* For a simply supported plate, the boundary conditions are (dw/dr is zero at $r = 0$ and $M_{rr} = 0$ at $r = a$)

$$\phi(0) = 0, \quad \left[\frac{d\phi}{dr} + \nu\frac{1}{r}\phi\right]_{r=a} = 0. \tag{5}$$

The second boundary condition can be written in terms of $\bar{r}$ as

$$\left[\frac{d\phi}{d\bar{r}} + \nu\frac{1}{\bar{r}}\phi\right]_{\bar{r}=\alpha a} = 0. \tag{6}$$

Using the general solution in Eq. (7.2.52), we obtain

$$AJ_1(0) + BY_1(0) = 0, \quad AJ_1'(\alpha a) + BY_1'(\alpha a) + \frac{\nu}{\alpha a}[AJ_1(\alpha a) + BY_1(\alpha a)] = 0. \tag{7}$$

The first equation in Eq. (6) gives $B = 0$, and the second equation, in view of $B = 0$ and the identity

$$\frac{dJ_n}{d\bar{r}} = J_{n-1}(\bar{r}) - \frac{1}{\bar{r}}J_n(\bar{r}), \tag{8}$$

gives

$$\alpha a J_0'(\alpha a) - (1 - \nu)J_1(\alpha a) = 0, \tag{9}$$

which is the stability condition for the simply supported plate. For $\nu = 0.3$, the smallest root of the transcendental equation in Eq. (9) is $\alpha a = 2.05$. Hence, the buckling load for simply supported solid circular plate becomes

$$N_{cr} = 4.198\frac{D}{a^2}. \tag{10}$$

(c)*Simply supported plate with rotational restraint.* For a simply supported plate with rotational restraint (see Fig. 7.2.5), the boundary conditions are

$$\phi(0) = 0, \quad D\left[r\frac{d\phi}{dr} + \nu\phi\right]_{r=a} + ak_R\,\phi(a) = 0, \tag{11}$$

where k_R denotes the rotational spring constant. We obtain

$$\alpha a J_0'(\alpha a) - (1 - \nu - \beta)J_1(\alpha a) = 0, \quad \beta = \frac{ak_R}{D} \tag{12}$$

as the stability condition. When $\beta = 0$, we obtain Eq. (9) and, when $\beta = \infty$, we obtain Eq. (3) as special cases.

Table 7.2.2 contains numerical values of the buckling load factor $\bar{N} = N_{cr}(a^2/D)$ for various values of the parameter β (for $\nu = 0.3$).

Table 7.2.2 Buckling load factors $\bar{N} = \hat{N}_{cr}(a^2/D)$ for rotationally restrained circular plates.

$\beta \rightarrow$	0	0.1	0.5	1	5	10	100	∞
$\bar{N} \rightarrow$	4.198	4.449	5.369	6.353	10.462	12.173	14.392	14.682

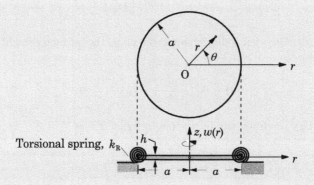

Fig. 7.2.5 Buckling of a rotationally restrained circular plate under uniform radial compressive load.

7.2.2.3 Variational Solutions

In this section we consider the Ritz and Galerkin solutions for axisymmetric bending of circular plates (solid as well as annular), possibly on elastic foundation. Towards this end, we first identify the variational statement (i.e., weak form) needed for the Ritz method. From the virtual work statement in Eq. (7.2.5), with $(\delta U, \delta V)$ from Eqs (7.2.6) and (7.2.8) and $\delta K = 0$ [or directly from Eq. (7.2.12)], and omitting terms involving derivatives with respect to the

θ coordinate, we have ($\hat{N}_{rr} = N_0$)

$$0 = -2\pi \int_b^a \left(M_{rr} \frac{d^2\delta w}{dr^2} + M_{\theta\theta} \frac{1}{r} \frac{d\delta w}{dr} - kw\,\delta w + q\delta w + N_0 \frac{dw}{dr} \frac{d\delta w}{dr} \right) r\,dr$$

$$+ 2\pi \left[a\bar{Q}_a \delta w(a) - b\bar{Q}_b \delta w(b) - aM_a \left(\frac{d\delta w}{dr} \right)_a + bM_b \left(\frac{d\delta w}{dr} \right)_b \right]$$

$$= 2\pi \int_b^a \left\{ D \left[\left(\frac{d^2 w}{dr^2} + \frac{\nu}{r} \frac{dw}{dr} \right) \frac{d^2\delta w}{dr^2} + \left(\nu \frac{d^2 w}{dr^2} + \frac{1}{r} \frac{dw}{dr} \right) \frac{1}{r} \frac{d\delta w}{dr} \right] \right.$$

$$\left. + kw\,\delta w - q\delta w - N_0 \frac{dw}{dr} \frac{d\delta w}{dr} \right\} r\,dr$$

$$+ 2\pi \left[a\bar{Q}_a\,\delta w(a) - b\bar{Q}_b\,\delta w(b) \right]$$

$$+ 2\pi \left[-aM_a \left(\frac{d\delta w}{dr} \right)_a + bM_b \left(\frac{d\delta w}{dr} \right)_b \right], \tag{7.2.53}$$

where a and b denote the outer and inner radii of an annular plate, k is the foundation modulus of the linear elastic foundation, D is the bending stiffness, q is the distributed transverse load, $\bar{Q}_a$ and $\bar{Q}_b$ are the intensities of effective line loads (include in-plane compressive force) at the outer and inner edges, respectively,

$$\bar{Q}_a = \left[Q_r - N_0 \frac{dw}{dr} \right]_{r=a},$$

$$\bar{Q}_b = \left[Q_r - N_0 \frac{dw}{dr} \right]_{r=b}, \tag{7.2.54}$$

and M_a and M_b the distributed edge moments at the outer and inner edges, respectively. When $b = 0$ (for solid circular plate), we have $M_b = 0$ and $2\pi b Q_b = Q_0$, Q_0 being the applied point load at the center of the plate. Equation (7.2.53) is the weak form used in the Ritz method.

The weak form in Eq. (7.2.53) should be modified if the edge $r = a$ of the plate is elastically restrained (see Fig. 7.2.6):

$$\left(r\bar{Q}_r \right)_{r=a} + k_E w(a) = 0,$$

$$(-rM_{rr})_{r=a} + k_R \left(\frac{dw}{dr} \right)_{r=a} = 0, \tag{7.2.55}$$

where k_E and k_R are spring constants associated with extensional and rotational springs, respectively. One can simulate the simply supported boundary condition ($k_E = \infty$ and $k_R = 0$), the clamped boundary condition ($k_E = \infty$ and $k_R = \infty$), and free edge condition ($k_E = 0$ and $k_R = 0$). Of course, similar expressions can be written for the edge at $r = b$. The weak form in Eq. (7.2.53) takes the form

$$0 = 2\pi \int_b^a \left\{ D \left[\left(\frac{d^2w}{dr^2} + \frac{\nu}{r}\frac{dw}{dr} \right) \frac{d^2\delta w}{dr^2} + \left(\nu\frac{d^2w}{dr^2} + \frac{1}{r}\frac{dw}{dr} \right) \frac{1}{r}\frac{d\delta w}{dr} \right] \right.$$

$$\left. + kw\delta w - q\delta w - N_0 \frac{dw}{dr}\frac{d\delta w}{dr} \right\} r\,dr$$

$$+ 2\pi \left[-k_E \delta w(a)w(a) - b\bar{Q}_b\delta w(b) \right]$$

$$+ 2\pi \left[-k_R \left(\frac{d\delta w}{dr}\frac{dw}{dr} \right)_a + bM_b \left(\frac{d\delta w}{dr} \right)_b \right]. \tag{7.2.56}$$

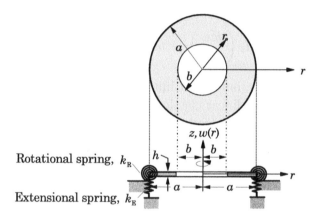

Fig. 7.2.6 An elastically restrained annular plate.

Assume an N-parameter Ritz solution of the form, assuming that all specified geometric boundary conditions are homogeneous so that we can take $\phi_0(r) = 0$:

$$w(r) \approx W_N(r) = \sum_{j=1}^N c_j\phi_j(r). \tag{7.2.57}$$

The approximation functions ϕ_i for the Ritz method must satisfy three conditions (as discussed before): (1) ϕ_i must be continuous as required by the weak form in Eq. (7.2.56), (2) the set $\{\phi_i\}_i^N$ should be linearly independent, and (3) ϕ_i must satisfy the homogeneous form of any specified essential boundary conditions.

Substituting Eq. (7.2.57) into Eq. (7.2.53), we obtain a set of algebraic equations in terms of the undetermined parameters $(c_1, c_2, \ldots, c_N)$

$$\sum_{j=1}^N A_{ij}c_j - b_i = 0 \ (i = 1, 2, \ldots, N); \quad \text{or} \quad \mathbf{Ac} = \mathbf{b}, \tag{7.2.58}$$

where

$$A_{ij} = 2\pi D \int_b^a \left[\frac{d^2\phi_i}{dr^2} \frac{d^2\phi_j}{dr^2} + \frac{\nu}{r} \left(\frac{d\phi_i}{dr} \frac{d^2\phi_j}{dr^2} + \frac{d^2\phi_i}{dr^2} \frac{d\phi_j}{dr} \right) + \frac{1}{r^2} \frac{d\phi_i}{dr} \frac{d\phi_j}{dr} \right] r\, dr$$

$$+ 2k\pi \int_b^a \phi_i \phi_j \, r\, dr - 2\pi \int_b^a N_0 \frac{d\phi_i}{dr} \frac{d\phi_j}{dr} \, r\, dr, \qquad (7.2.59)$$

$$b_i = 2\pi \int_b^a q\phi_i \, r\, dr - 2\pi \left[a\bar{Q}_a \phi_i(a) - b\bar{Q}_b \phi_i(b) - aM_a \frac{d\phi_i}{dr}|_a + bM_b \frac{d\phi_i}{dr}|_b \right].$$

$$(7.2.60)$$

Similar expression can be obtained using the variational statement in Eq. (7.2.56). Once the approximation functions ϕ_i are selected, the coefficients A_{ij} and b_i can be computed by evaluating the integrals in Eqs (7.2.59) and (7.2.60), and Eq. (7.2.58) can be solved for the parameters c_i ($i = 1, 2, \ldots, N$). Then, the N-parameter solution $W_N(r)$ is given by Eq. (7.2.57).

For the weighted-residual family of methods, we use the weighted-integral statement

$$0 = 2\pi \int_b^a \psi_i(r) \left[\frac{D}{r} \frac{d}{dr} \left\{ r \frac{d}{dr} \left[\frac{1}{r} \frac{d}{dr} \left(r \frac{dw}{dr} \right) \right] \right\} + kw \right.$$

$$\left. + \frac{1}{r} \frac{d}{dr} \left(rN_0 \frac{dw}{dr} \right) - q \right] r\, dr, \qquad (7.2.61)$$

where $\psi_i(r)$ is the weight function that takes different forms depending on the method used. Substituting the N-parameter approximation of the form

$$w(r) \approx W_N(r) = \sum_{j=1}^N c_j \phi_j(r) + \phi_0(r) \qquad (7.2.62)$$

into Eq. (7.2.61), we obtain

$$\sum_{j=1}^N A_{ij} c_j - b_i = 0 \ (i = 1, 2, \ldots, N); \quad \text{or} \quad \mathbf{Ac} = \mathbf{b}, \qquad (7.2.63)$$

where

$$A_{ij} = 2\pi \int_b^a \psi_i \left[\frac{D}{r} \frac{d}{dr} \left\{ r \frac{d}{dr} \left[\frac{1}{r} \frac{d}{dr} \left(r \frac{d\phi_j}{dr} \right) \right] \right\} + k\phi_j \right.$$

$$\left. + \frac{1}{r} \frac{d}{dr} \left(rN_0 \frac{d\phi_j}{dr} \right) \right] r\, dr, \qquad (7.2.64)$$

$$b_i = 2\pi \int_b^a q\psi_i \, r\, dr - 2\pi \int_b^a \psi_i \left[\frac{D}{r} \frac{d}{dr} \left\{ r \frac{d}{dr} \left[\frac{1}{r} \frac{d}{dr} \left(r \frac{d\phi_0}{dr} \right) \right] \right\} \right.$$

$$\left. + \frac{1}{r} \frac{d}{dr} \left(rN_0 \frac{d\phi_0}{dr} \right) + k\phi_0 \right] r\, dr. \qquad (7.2.65)$$

The approximation functions ϕ_i must be (1) continuous as required by the weighted-integral statement in Eq. (7.2.61) and (2) linearly independent and (3) satisfy the homogeneous form of *all* specified boundary conditions, whereas ϕ_0 must satisfy *all* specified boundary conditions. Of course, $\phi_0 = 0$ when all specified boundary conditions are homogeneous.

Example 7.2.4

Consider a simply supported solid ($b = 0$) circular plate under uniformly distributed load of intensity q_0, as shown in Fig. 7.2.7. Formulate the N-parameter Ritz solution and determine the solution for $N = 1, 2,$ and 3. Note that $k = 0$ and $\hat{N}_{rr} = N_0 = 0$ for the present problem.

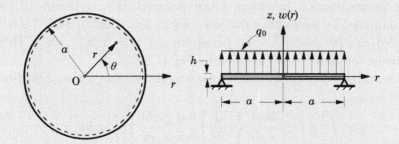

Fig. 7.2.7 A simply supported solid circular plate under uniformly distributed load.

Solution: Recall that for this problem the essential boundary conditions are

$$\frac{dw}{dr}(0) = 0, \quad w(a) = 0, \tag{1}$$

and the natural boundary conditions are given by

$$M_{rr} = 0 \text{ at } r = a, \quad Q_r = \frac{1}{r}\left[\frac{d}{dr}(rM_{rr}) - M_{\theta\theta}\right] = 0 \text{ at } r = 0. \tag{2}$$

The natural boundary conditions will have no bearing on the selection of ϕ_i in the Ritz method. Each ϕ_i must satisfy the homogeneous form of the geometric boundary conditions

$$\frac{d\phi_i}{dr}(0) = 0, \quad \phi_i(a) = 0. \tag{3}$$

For the choice of algebraic polynomials, we assume

$$\phi_1(r) = \alpha_1 + \alpha_2 r + \alpha_3 r^2$$

and determine, using the conditions in Eq. (3), that $\alpha_2 = 0$ and $\alpha_1 + \alpha_3 a^2 = 0$. Thus, we have (for the choice of $\alpha_1 = 1$)

$$\phi_1 = 1 - \frac{r^2}{a^2}. \tag{4}$$

The procedure can be used to obtain a linearly independent and complete set of functions

$$\phi_1 = 1 - \frac{r^2}{a^2}, \quad \phi_2 = 1 - \frac{r^3}{a^3}, \quad \ldots, \quad \phi_j = 1 - \left(\frac{r}{a}\right)^{j+1}. \tag{5}$$

Substituting for ϕ_j from Eq. (5) into Eqs (7.2.56) and (7.2.57) and noting that $k = 0$ and $N_0 = 0$, we obtain

$$A_{ij} = \frac{2\pi D}{a^2}\left(\frac{ij+1}{i+j} + \nu\right)(i+1)(j+1),$$

$$b_i = 2\pi q_0 a^2 \frac{(i+1)}{2(i+3)}. \tag{6}$$

For $N = 3$ we have the algebraic equations (the factor 2π cancels on both sides of the equation)

$$\frac{D}{a^2}\begin{bmatrix} 4(1+\nu) & 6(1+\nu) & 8(1+\nu) \\ 6(1+\nu) & 9(1.25+\nu) & 12(1.4+\nu) \\ 8(1+\nu) & 12(1.4+\nu) & 16(\frac{5}{3}+\nu) \end{bmatrix}\begin{Bmatrix} c_1 \\ c_2 \\ c_3 \end{Bmatrix} = \frac{q_0 a^2}{60}\begin{Bmatrix} 15 \\ 18 \\ 20 \end{Bmatrix}. \tag{7}$$

The one-, two- and three-parameter Ritz solutions are

$$W_1(r) = \frac{q_0 a^4}{16D(1+\nu)}\left(1 - \frac{r^2}{a^2}\right),$$

$$W_2(r) = \frac{q_0 a^4}{80D}\left(\frac{9+4\nu}{1+\nu}\right)\left(1 - \frac{r^2}{a^2}\right) - \frac{q_0 a^4}{30D}\left(1 - \frac{r^3}{a^3}\right),$$

$$W_3(r) = \frac{q_0 a^4}{64D}\left(\frac{6+2\nu}{1+\nu}\right)\left(1 - \frac{r^2}{a^2}\right) - \frac{q_0 a^4}{64D}\left(1 - \frac{r^4}{a^4}\right)$$

$$= \frac{q_0 a^4}{64D}\left[\frac{5+\nu}{1+\nu} - 2\left(\frac{3+\nu}{1+\nu}\right)\left(\frac{r}{a}\right)^2 + \left(\frac{r}{a}\right)^4\right]. \tag{8}$$

Thus, the three-parameter solution in Eq. (8) coincides with the exact solution in Eq. (7) of **Example 7.2.2**. If one tries to use $N \geq 4$, the associated coefficients c_4, c_5, and so on will be computed as zeros because the solution for $N = 3$ is already exact. One can use $W_N(r)$ to compute the stresses, bending moments, and shear force using the relations in Eqs (7.2.33) to (7.2.37).

Example 7.2.5

Consider the bending of a clamped (at $r = a$) solid circular plate under uniformly distributed transverse load of intensity q_0. Determine the two-parameter Ritz and Galerkin solutions.

Solution: The boundary conditions of the problem are

$$\text{Geometric (essential)}: \ w(a) = 0, \quad \frac{dw}{dr} = 0 \quad \text{at } r = 0, a, \tag{1}$$

$$\text{Force (natural)}: Q_r(0) \equiv -D\left\{\frac{d}{dr}\left[\frac{1}{r}\frac{d}{dr}\left(r\frac{dw}{dr}\right)\right]\right\}_{r=0} = 0. \tag{2}$$

The three boundary conditions in Eq. (1) are of the essential type, while that in Eq. (2) is of the natural type.

Approximation functions. First, we discuss the selection of the approximation functions. Obviously, ϕ_0 is zero since all specified boundary conditions are homogeneous. The choice of ϕ_i depends on the method we use. We derive them for both the Ritz and weighted-residual methods.

In the Ritz method the approximation functions are required to satisfy the homogeneous form of only the geometric boundary conditions. Since there are *three* essential boundary conditions, we begin with a four-parameter polynomial,

$$\phi_1(r) = c_0 + c_1 r + c_2 r^2 + c_3 r^3,$$ (3)

and determine the constants (one of them is arbitrary but nonzero) such that the three conditions

$$\phi_i'(0) = 0, \quad \phi_i(a) = 0, \quad \phi_i'(a) = 0$$

are satisfied. We obtain ($c_0 = a^3 c_3/2, c_1 = 0, c_2 = -3ac_3/2$)

$$\phi_1 = 1 - 3\left(\frac{r}{a}\right)^2 + 2\left(\frac{r}{a}\right)^3.$$ (4)

Similarly, we pick the five-term polynomial for ϕ_2 and determine the constants (two of them are arbitrary but the coefficient of the highest-order term, c_4, should never be taken as zero)

$$\phi_2(r) = c_0 + c_1 r + c_2 r^2 + c_3 r^3 + c_4 r^4.$$ (5)

We obtain

$$c_0 = \frac{3}{2}a^3 c_3 + c_4 a^4, \quad c_1 = 0, \quad c_2 = -\frac{3}{2}ac_3 - 2a^2 c_4.$$

For simplicity, we take $c_3 = 0$ and obtain

$$\phi_2 = \left[1 - \left(\frac{r}{a}\right)^2\right]^2.$$ (6)

In the weighted-residual method, the approximation functions are required to satisfy the homogeneous form of all specified boundary conditions. Since there are *four* specified boundary conditions, we begin with a five-term complete polynomial,

$$\phi_1(r) = c_0 + c_1 r + c_2 r^2 + c_3 r^3 + c_4 r^4,$$ (7)

and determine the constants such that the four conditions

$$\phi_i'(0) = 0, \quad \phi_i(a) = 0, \quad \phi_i'(a) = 0, \quad \left\{\frac{d}{dr}\left[\frac{1}{r}\frac{d}{dr}\left(r\frac{d\phi_i}{dr}\right)\right]\right\}_{r=0} = 0$$ (8)

are satisfied. We obtain ($c_0 = a^4 c_4, c_1 = 0, c_2 = -2a^2 c_4, c_3 = 0$)

$$\phi_1(r) = \left[1 - \left(\frac{r}{a}\right)^2\right]^2.$$ (9)

Next, we pick the six-term complete polynomial for ϕ_2 and determine the constants (two of them are arbitrary but the coefficient of the highest-order term, c_5, should not be taken as zero)

$$\phi_2(r) = c_0 + c_1 r + c_2 r^2 + c_3 r^3 + c_4 r^4 + c_5 r^5.$$ (10)

We obtain

$$c_0 = a^4 c_4 + \frac{3}{2}a^5 c_5, \quad c_1 = 0, \quad c_2 = -2a^2 c_4 - \frac{5}{2}a^3 c_5, \quad c_3 = 0,$$

and

$$\phi_2(r) = \left[1 - \left(\frac{r}{a}\right)^2\right]^2 a^4 c_4 + \frac{1}{2}a^5\left[3 - 5\left(\frac{r}{a}\right)^2 + \left(\frac{r}{a}\right)^5\right]c_5.$$ (11)

Since the first part is already represented by ϕ_1, we set $c_4 = 0$ and obtain

$$\phi_2(r) = 3 - 5\left(\frac{r}{a}\right)^2 + \left(\frac{r}{a}\right)^5.$$ (12)

Variational solutions. For the two-parameter Ritz approximation, we use the approximation functions ϕ_1 and ϕ_2 from Eqs (4) and (6) and compute A_{ij} and b_i from Eqs (7.2.59) and (7.2.60) (note that k, $\bar{Q}_a$, $\bar{Q}_b$, M_a, and M_b are zero for the present problem):

$$\frac{1}{r}\frac{d\phi_1}{dr} = -\frac{6}{a^2}\left(1 - \frac{r}{a}\right), \qquad \frac{d^2\phi_1}{dr^2} = -\frac{6}{a^2}\left[1 - 2\left(\frac{r}{a}\right)\right],$$

$$\frac{1}{r}\frac{d\phi_2}{dr} = -\frac{4}{a^2}\left[1 - \left(\frac{r}{a}\right)^2\right], \qquad \frac{d^2\phi_2}{dr^2} = -\frac{4}{a^2}\left[1 - 3\left(\frac{r}{a}\right)^2\right],$$

$$a_{11} = \frac{18\pi D}{3a^2}, \qquad a_{12} = a_{21} = \frac{96\pi D}{5a^2}, \qquad a_{22} = \frac{64\pi D}{3a^2},$$

$$b_1 = \frac{3\pi q_0 a^2}{10}, \qquad b_2 = \frac{\pi q_0 a^2}{3}, \qquad c_1 = 0, \qquad c_2 = \frac{q_0 a^4}{64D}.$$

(13)

Thus, the two-parameter Ritz solution coincides with the exact solution in Eq. (5) of **Example 7.2.1**:

$$w(r) = \frac{q_0 a^4}{64D}\left[1 - \left(\frac{r}{a}\right)^2\right]^2.$$

(14)

Note that the one-parameter solution is given by ($c_1 = b_1/a_{11}$)

$$W_1(r) = c_1\phi_1(r) = \frac{9q_0 a^4}{640D}\left[1 - 3\left(\frac{r}{a}\right)^2 + 2\left(\frac{r}{a}\right)^3\right].$$

(15)

The maximum deflection obtained with the one-parameter Ritz method is in 10% error.

In Galerkin's method, we first compute the residual using the functions in Eqs (9) and (12). We have ($\phi_0 = 0$)

$$\frac{dW_2}{dr} = c_1\frac{d\phi_1}{dr} + c_2\frac{d\phi_2}{dr}$$

$$= \frac{4}{a}\left(\frac{r}{a}\right)\left[1 - \left(\frac{r}{a}\right)^2\right]c_1 + \frac{5}{a}\left[-2\left(\frac{r}{a}\right) + \left(\frac{r}{a}\right)^4\right]c_2,$$

$$\frac{d^2W_2}{dr^2} = \frac{4}{a^2}\left[1 - 3\left(\frac{r}{a}\right)^2\right]c_1 + \frac{10}{a^2}\left[-1 + 2\left(\frac{r}{a}\right)^3\right]c_2,$$

(16)

$$\nabla^2 W_2 \equiv \frac{d^2W_2}{dr^2} + \frac{1}{r}\frac{dW_2}{dr} = \frac{8}{a^2}\left[1 - 2\left(\frac{r}{a}\right)^2\right]c_1 + \frac{5}{a^2}\left[-4 + 5\left(\frac{r}{a}\right)^3\right]c_2,$$

$$\nabla^4 W_2 \equiv \frac{d^2}{dr^2}(\nabla^2 W_2) + \frac{1}{r}\frac{d}{dr}(\nabla^2 W_2) = -\frac{64}{a^4}c_1 + \frac{225}{a^4}\left(\frac{r}{a}\right)c_2.$$

Also, note that

$$Q_r(r) \equiv -D\frac{d}{dr}\left[\frac{1}{r}\frac{d}{dr}\left(r\frac{dW_2}{dr}\right)\right] = D\left[-\frac{32}{a^3}\left(\frac{r}{a}\right)c_1 + \frac{75}{a^3}\left(\frac{r}{a}\right)^2 c_2\right],$$

(17)

$$\frac{d}{dr}(rQ_r) = -D\frac{d}{dr}\left\{r\frac{d}{dr}\left[\frac{1}{r}\frac{d}{dr}\left(r\frac{dW_2}{dr}\right)\right]\right\} = D\left[-\frac{64}{a^3}\left(\frac{r}{a}\right)c_1 + \frac{225}{a^3}\left(\frac{r}{a}\right)^2 c_2\right].$$

(18)

Hence, the coefficients A_{ij} and b_i of Eqs (7.2.61) and (7.2.62) have the values

$$a_{11} = -\frac{64\pi D}{3a^2}, \qquad a_{12} = \frac{240\pi D}{7a^2}, \qquad a_{21} = -\frac{352\pi D}{7a^2},$$

$$a_{22} = \frac{225\pi D}{4a^2}, \qquad b_1 = -\frac{\pi q_0 a^2}{3}, \qquad b_2 = -\frac{11\pi q_0 a^2}{14},$$

(19)

and the solution of the Galerkin equations yields

$$c_1 = \frac{q_0 a^4}{64D}, \qquad c_2 = 0.$$

Therefore, the two-parameter Galerkin solution is the same as the one-parameter Galerkin solution, which coincides with the exact solution [see Eq. (5) of **Example 7.2.1**]:

$$w(r) = \frac{q_0 a^4}{64D} \left[1 - \left(\frac{r}{a} \right)^2 \right]^2. \tag{20}$$

Determination of the natural frequencies of circular plates with various boundary conditions by analytical means leads to complicated equations involving Bessel functions, and the solutions are difficult to obtain, requiring the use of approximate methods of solution. Here, we use the Ritz method to determine the natural frequencies.

The equation of motion of an isotropic plate is given by (see Reddy [51])

$$D\nabla^2\nabla^2 w + kw + m_0 \frac{\partial^2 w}{\partial t^2} - m_2 \frac{\partial^2}{\partial t^2} \left(\nabla^2 w \right) = 0, \tag{7.2.66}$$

where $m_0 = \rho h$ and $m_2 = \rho h^3/12$ are the principal and rotatory inertias, respectively, and the Laplace operator ∇^2 is defined in polar coordinate system by

$$\nabla^2 = \frac{1}{r} \frac{\partial}{\partial r} \left(r \frac{\partial}{\partial r} \right) + \frac{1}{r^2} \frac{\partial^2}{\partial \theta^2}. \tag{7.2.67}$$

For free harmonic motion (i.e., natural vibration), the deflection can be expressed as

$$w(r, \theta, t) = w_0(r, \theta) \cos \omega t, \tag{7.2.68}$$

where ω is the circular frequency of vibration (radians per unit time) and w_0 (mode shape) is a function of only r and θ. Substituting Eq. (7.2.68) into Eq. (7.2.66), we obtain

$$D\nabla^2\nabla^2 w_0 + kw_0 - m_0\omega^2 w_0 + m_2\omega^2\nabla^2 w_0 = 0. \tag{7.2.69}$$

The weak form of Eq. (7.2.69) is given by

$$
\begin{aligned}
0 = \int_\Omega \Bigg\{ & D\frac{\partial^2 w_0}{\partial r^2}\frac{\partial^2 \delta w_0}{\partial r^2} + D\frac{\nu}{r}\left(\frac{\partial w_0}{\partial r}\frac{\partial^2 \delta w_0}{\partial r^2} + \frac{\partial \delta w_0}{\partial r}\frac{\partial^2 w_0}{\partial r^2} \right) \\
& + \frac{1}{r}\frac{\partial^2 w_0}{\partial \theta^2}\frac{\partial^2 \delta w_0}{\partial r^2} + \frac{1}{r}\frac{\partial^2 \delta w_0}{\partial \theta^2}\frac{\partial^2 w_0}{\partial r^2} \Bigg) + kw_0\delta w_0 \\
& + D\left(\frac{1}{r}\frac{\partial w_0}{\partial r} + \frac{1}{r^2}\frac{\partial^2 w_0}{\partial \theta^2} \right)\left(\frac{1}{r}\frac{\partial \delta w_0}{\partial r} + \frac{1}{r^2}\frac{\partial^2 \delta w_0}{\partial \theta^2} \right) \\
& + 2(1-\nu)D\left(\frac{1}{r}\frac{\partial^2 w_0}{\partial r\partial \theta} - \frac{1}{r^2}\frac{\partial w_0}{\partial \theta} \right)\left(\frac{1}{r}\frac{\partial^2 \delta w_0}{\partial r\partial \theta} - \frac{1}{r^2}\frac{\partial \delta w_0}{\partial \theta} \right) \\
& - \omega^2\left[m_0 w_0\delta w_0 + m_2\left(\frac{\partial w_0}{\partial r}\frac{\partial \delta w_0}{\partial r} + \frac{1}{r^2}\frac{\partial w_0}{\partial \theta}\frac{\partial \delta w_0}{\partial \theta} \right) \right] \Bigg\} r\, dr\, d\theta. \tag{7.2.70}
\end{aligned}
$$

Assume an N-parameter Ritz approximation of the form

$$w_0(r, \theta) \approx W_N(r, \theta) = \sum_{j=1}^{N} c_j \phi_j(r, \theta). \tag{7.2.71}$$

Substituting Eq. (7.2.71) into Eq. (7.2.70), we obtain

$$0 = \sum_{j=1}^{N} \left(A_{ij}^{(1)} + A_{ij}^{(2)} - \omega^2 M_{ij} \right) c_j, \tag{7.2.72}$$

where $A_{ij}^{(1)}$, $A_{ij}^{(2)}$, and M_{ij} are defined as

$$
\begin{aligned}
A_{ij}^{(1)} = D \int_\Omega & \left[\frac{\partial^2 \phi_i}{\partial r^2} \frac{\partial^2 \phi_j}{\partial r^2} + \left(\frac{1}{r} \frac{\partial \phi_i}{\partial r} + \frac{1}{r^2} \frac{\partial^2 \phi_i}{\partial \theta^2} \right) \left(\frac{1}{r} \frac{\partial \phi_j}{\partial r} + \frac{1}{r^2} \frac{\partial^2 \phi_j}{\partial \theta^2} \right) \right. \\
& + \frac{\nu}{r} \left(\frac{\partial \phi_i}{\partial r} \frac{\partial^2 \phi_j}{\partial r^2} + \frac{\partial \phi_j}{\partial r} \frac{\partial^2 \phi_i}{\partial r^2} + \frac{1}{r} \frac{\partial^2 \phi_i}{\partial \theta^2} \frac{\partial^2 \phi_j}{\partial r^2} + \frac{1}{r} \frac{\partial^2 \phi_j}{\partial \theta^2} \frac{\partial^2 \phi_i}{\partial r^2} \right) \\
& \left. + 2(1 - \nu) \left(\frac{1}{r} \frac{\partial^2 \phi_i}{\partial r \partial \theta} - \frac{1}{r^2} \frac{\partial \phi_i}{\partial \theta} \right) \left(\frac{1}{r} \frac{\partial^2 \phi_j}{\partial r \partial \theta} - \frac{1}{r^2} \frac{\partial \phi_j}{\partial \theta} \right) \right] r \, dr \, d\theta,
\end{aligned} \tag{7.2.73}
$$

$$A_{ij}^{(2)} = k \int_\Omega \phi_i \phi_j \, r \, dr \, d\theta, \tag{7.2.74}$$

$$M_{ij} = \int_\Omega \left[m_0 \phi_i \phi_j + m_2 \left(\frac{\partial \phi_i}{\partial r} \frac{\partial \phi_j}{\partial r} + \frac{1}{r^2} \frac{\partial \phi_i}{\partial \theta} \frac{\partial \phi_j}{\partial \theta} \right) \right] r \, dr \, d\theta. \tag{7.2.75}$$

In matrix notation, Eq. (7.2.72) has the form of an eigenvalue problem:

$$\left(\mathbf{A}^{(1)} + \mathbf{A}^{(2)} - \omega^2 \mathbf{M} \right) \mathbf{c} = \mathbf{0}. \tag{7.2.76}$$

For a non-trivial solution, the coefficient matrix in Eq. (7.2.76) should be singular, that is,

$$|\mathbf{A}^{(1)} + \mathbf{A}^{(2)} - \omega^2 \mathbf{M}| = 0. \tag{7.2.77}$$

The aforementioned formulation is valid also for annular plates. Next, we now consider couple of examples.

Example 7.2.6

Determine the fundamental (i.e., the lowest) frequency of a clamped solid circular plate using a one-parameter Ritz approximation.

Solution: For a one-parameter ($N = 1$) Ritz solution, let

$$\phi_1(r, \theta) = f_1(r) \cos n\theta \tag{1}$$

and compute $A_{11}^{(1)}$, $A_{11}^{(2)}$, and M_{11} as

$$A_{11}^{(1)} = D \int_0^{2\pi} \int_0^a \left\{ \left[f_1'' f_1'' + \left(\frac{1}{r} f_1' - \frac{n^2}{r^2} f_1 \right)^2 + 2\nu \frac{1}{r} \left(f_1' f_1'' - \frac{n^2}{r} f_1 f_1'' \right) \right] \cos^2 n\theta \right.$$

$$\left. + 2(1-\nu) \left(\frac{n}{r} f_1' - \frac{n}{r^2} f_1 \right)^2 \sin^2 n\theta \right\} r dr d\theta, \tag{2}$$

$$A_{11}^{(2)} = k \int_0^{2\pi} \int_0^a f_1 f_1 \cos^2 n\theta \, r dr d\theta, \tag{3}$$

$$M_{11} = \int_0^{2\pi} \int_0^a \left[(m_0 f_1 f_1 + m_2 f_1' f_1') \cos^2 n\theta + m_2 \frac{n^2}{r^2} f_1 f_1 \sin^2 n\theta \right] r dr d\theta. \tag{4}$$

The conditions on the approximation functions are

$$\phi_i(r,\theta) = \frac{\partial \phi_i}{\partial r} = 0 \text{ at } r = a, \text{ and } \frac{\partial \phi_i}{\partial r} = 0 \text{ at } r = 0, \tag{5}$$

which translate into the conditions $f_1(a) = f_1'(a) = f_1'(0) = 0$. Clearly, the choice

$$f_1(r) = 1 - 3 \left(\frac{r}{a} \right)^2 + 2 \left(\frac{r}{a} \right)^3 \tag{6}$$

satisfies the conditions. Since

$$f_1' = -\frac{6r}{a^2} + \frac{6r^2}{a^3}, \quad f_1'' = -\frac{6}{a^2} + \frac{12r}{a^3},$$

it is clear that the integrals (for $n > 0$)

$$\int_0^a \frac{1}{r} f_1 f_1'' \, dr, \quad \int_0^a \frac{1}{r^3} f_1' f_1 \, dr$$

required in $A_{11}^{(1)}$ of Eq. (2) do not exist because of the logarithmic singularity. Thus $f_1(r)$ defined in Eq. (6) is not admissible for $n > 0$. The next function that satisfies the boundary conditions $f_1(a) = f_1'(a) = f_1'(0) = 0$ is

$$f_1(r) = \left[1 - \left(\frac{r}{a} \right)^2 \right]^2 = 1 - 2 \left(\frac{r}{a} \right)^2 + \left(\frac{r}{a} \right)^4, \tag{7}$$

which is also not admissible for $n > 0$. The next admissible function is

$$f_1(r) = \frac{r}{a} \left[1 - \left(\frac{r}{a} \right)^2 \right]^2 = \frac{r}{a} - 2 \left(\frac{r}{a} \right)^3 + \left(\frac{r}{a} \right)^5, \tag{8}$$

which will not present any problem in evaluating the integrals for $n \geq 0$.

The fundamental frequency corresponding to the axisymmetric mode, $n = 0$. For this case, we use the function in Eq. (6) and obtain

$$A_{11}^{(1)} = 2\pi D \int_0^a \left(f_1'' f_1'' + \frac{1}{r^2} f_1' f_1' + 2\nu \frac{1}{r} f_1' f_1'' \right) r dr = 2\pi \frac{9D}{a^2},$$

$$A_{11}^{(2)} = 2\pi k \int_0^a f_1 f_1 \, r dr = 2\pi \frac{3ka^2}{35},$$

$$M_{11} = 2\pi \int_0^a (m_0 f_1 f_1 + m_2 f_1' f_1') \, r dr = 2\pi \left(\frac{3a^2}{35} m_0 + \frac{3}{5} m_2 \right).$$

Hence,

$$\omega^2 = \left(\frac{9D}{a^2} + \frac{3ka^2}{35}\right)\left[\frac{3a^2}{35}m_0 + \frac{3}{5}m_2\right]^{-1} = \frac{D}{a^4 m_0}\left[\frac{105 + (ka^4/D)}{1 + 7(m_2/m_0 a^2)}\right]. \tag{9}$$

Clearly, rotatory inertia has the effect of reducing the frequency of vibration while the elastic foundation modulus increases it. For a clamped isotropic plate without elastic foundation (i.e., $k = 0$), the frequency parameter becomes $(m_2 = m_0 h^2/12)$

$$\lambda^2 \equiv \omega a^2 \sqrt{\frac{m_0}{D}} = 10.247\left[\frac{1}{1 + 0.583\frac{h^2}{a^2}}\right]^{\frac{1}{2}}. \tag{10}$$

For very thin plates, say, $h/a = 0.01$, the effect of rotatory inertia is negligible. Even for $h/a = 0.1$, the effect is less than 1%. Note that the one-parameter Ritz solution differs from the analytical solution $\omega = 10.216$ listed in Table 5.5.1 (for $m = n = 0$) of Reddy [51] by less than half a percent!

For $n = 1$, the function in Eq. (8) may be used. We obtain

$$a_{11}^{(1)} = \frac{\pi D}{a^2}\left[6 + \frac{2}{3} + 2\left(\frac{1}{2} - \frac{7}{6}\right)\nu + \frac{4}{3}(1 - \nu)\right],$$

$$a_{11}^{(2)} = \frac{\pi a^2}{60}k, \quad m_{11} = \pi\left(\frac{a^2}{60}m_0 + \frac{1}{6}m_2\right).$$

The frequency parameter is given by

$$\lambda^2 = \omega a^2 \sqrt{\frac{m_0}{D}} = \left[\frac{480 + \frac{a^4 k}{60 D}}{1 + \frac{10 m_2}{a^2 m_0}}\right]^{\frac{1}{2}}. \tag{11}$$

For a clamped isotropic plate without elastic foundation, the frequency for the case $m = 0$, $n = 1$ becomes $(m_2 = m_0 h^2/12)$

$$\lambda^2 = \omega a^2 \sqrt{\frac{m_0}{D}} = 21.909\left[\frac{1}{1 + \frac{5}{6}\frac{h^2}{a^2}}\right]^{\frac{1}{2}}. \tag{12}$$

When rotatory inertia is neglected, the frequency predicted by Eq. (12) differs from the analytical solution $\omega = 21.26$ listed in Table 5.5.1 (for $m = 0, n = 1$) of Reddy [51] only by 3%.

For other values of n, one must select functions that are admissible (i.e., allow evaluation of the integrals). Generally, higher values of n require higher-order functions $f_i(r)$.

Example 7.2.7

Determine the critical buckling load of a solid circular plate under uniform compression $\hat{N}_{rr} = N_0$ in the middle plane of the plate (see Fig. 7.2.4). First formulate the N-parameter Ritz approximation and then determine the critical buckling load.

Solution: The governing equation is

$$D\nabla^2\nabla^2 w + N_0\nabla^2 w = 0. \tag{1}$$

The weak form of this equation is a slight modification of that given in Eq. (7.2.70) (omit the foundation modulus term and replace the frequency term with the buckling term). We have

$$
0 = \int_\Omega \left\{ D\frac{\partial^2 w}{\partial r^2}\frac{\partial^2 \delta w}{\partial r^2} + D\frac{\nu}{r}\left(\frac{\partial w}{\partial r}\frac{\partial^2 \delta w}{\partial r^2} + \frac{\partial \delta w}{\partial r}\frac{\partial^2 w}{\partial r^2}\right) \right.
$$
$$
+ \frac{1}{r}\frac{\partial^2 w}{\partial \theta^2}\frac{\partial^2 \delta w}{\partial r^2} + \frac{1}{r}\frac{\partial^2 \delta w}{\partial \theta^2}\frac{\partial^2 w}{\partial r^2}\Bigg)
$$
$$
+ D\left(\frac{1}{r}\frac{\partial w}{\partial r} + \frac{1}{r^2}\frac{\partial^2 w}{\partial \theta^2}\right)\left(\frac{1}{r}\frac{\partial \delta w}{\partial r} + \frac{1}{r^2}\frac{\partial^2 \delta w}{\partial \theta^2}\right)
$$
$$
+ 2(1-\nu)D\left(\frac{1}{r}\frac{\partial^2 w}{\partial r\partial \theta} - \frac{1}{r^2}\frac{\partial w}{\partial \theta}\right)\left(\frac{1}{r}\frac{\partial^2 \delta w}{\partial r\partial \theta} - \frac{1}{r^2}\frac{\partial \delta w}{\partial \theta}\right)
$$
$$
\left. - N_0\left(\frac{\partial \delta w}{\partial r}\frac{\partial w}{\partial r} + \frac{1}{r^2}\frac{\partial \delta w}{\partial \theta}\frac{\partial w}{\partial \theta}\right)\right\} r\,dr\,d\theta. \tag{2}
$$

The N-parameter Ritz approximation in Eq. (7.2.71) results in

$$
([A] - N_0[G])\{c\} = \{0\}, \tag{3}
$$

where A_{ij} are the same as $A_{ij}^{(1)}$ in Eq. (7.2.73) (with $k = 0$) and G_{ij} are given by

$$
G_{ij} = \int_0^{2\pi}\int_0^a \left(\frac{\partial \phi_i}{\partial r}\frac{\partial \phi_j}{\partial r} + \frac{1}{r^2}\frac{\partial \phi_i}{\partial \theta}\frac{\partial \phi_j}{\partial \theta}\right) r\,dr\,d\theta. \tag{4}
$$

Since we are interested in the minimum buckling load, which occurs in the axisymmetric mode $(n = 0)$, we can use the one-parameter approximation $W_1(r) = c_1 f_1(r)$, where $f_1(r)$ is defined in Eq. (6) of **Example 7.2.6**. We obtain

$$
\frac{9D}{a^2} - \frac{3}{5}N_0 = 0
$$

or

$$
N_0 = \frac{15D}{a^2}, \tag{5}
$$

which differs from the exact solution in Eq. (4) of **Example 7.2.3**, namely, $14.682(D/a^2)$, by 2.18%.

The next example illustrates the use of the Betti–Maxwell reciprocity theorem in obtaining the deflections of circular plates under an asymmetric load. The reader may consult Section 4.6 for the details of the theorem.

Example 7.2.8

Consider a circular plate of radius a with an axisymmetric boundary condition, and subjected to an *asymmetric* load of the type shown in Fig. 7.2.8:

$$
q(r,\theta) = q_0 + q_1\frac{r}{a}\cos\theta, \tag{1}
$$

where q_0 represents the uniform part of the load for which the solution is known for various axisymmetric boundary conditions. In particular, the deflection of a clamped circular plate under a point load at the center is given by Eq. (16) of **Example 7.2.1**, and that of a simply supported circular plate under a point load at the center is given by Eq. (11) of **Example 7.2.2**. Use Betti–Maxwell's reciprocity theorem to determine the center deflection of the clamped circular plate under asymmetric distributed load $q(r,\theta)$ given in Eq. (1).

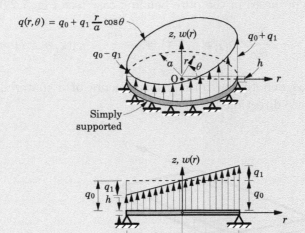

Fig. 7.2.8 A circular plate subjected to an asymmetric loading.

Solution: By Maxwell's theorem, the work done by a point load Q_0 at the center of the plate due to the deflection (at the center) w_c caused by the distributed load $q(r,\theta)$ is equal to the work done by the distributed load $q(r,\theta)$ in moving through the displacement $w(r)$ caused by the point load Q_0 at the center. Hence, the center deflection of a clamped circular plate under asymmetric load in Eq. (1) is

$$w_c = \int_0^{2\pi} \int_0^a q(r,\theta)w(r)\,rdrd\theta = \frac{q_0 a^4}{64D}, \tag{2}$$

where $w(r)$ is given by Eq. (16) of **Example 7.2.1**.

7.2.3   Governing Equations in Rectangular Coordinates

Here we consider the CPT in rectangular Cartesian coordinates. We wish to develop the total Lagrangian functional and use Hamilton's principle to derive the equations of motion and natural boundary conditions for linear bending of a plate according to the Kirchhoff hypotheses. Stretching deformation is not considered as it can be uncoupled from bending deformation for the linear problem.

Let us denote, as before, the undeformed midplane of the plate with the symbol Ω_0, and let it coincide with the xy-plane of the coordinate system with the z coordinate along the thickness of the plate. The total domain of the plate is symbolically expressed as the tensor product $\Omega = \Omega_0 \times (-h/2, h/2)$. The boundary of the total domain Ω consists of the top surface $S_t(z = h/2)$, bottom surface $S_b(z = -h/2)$, and the edge $\bar{\Gamma} \equiv \Gamma \times (-h/2, h/2)$. In general,

$\bar{\Gamma}$ is a curved surface, with outward normal $\hat{\mathbf{n}} = n_x \hat{\mathbf{e}}_x + n_y \hat{\mathbf{e}}_y$, where n_x and n_y are the direction cosines of the unit normal (see Fig. 7.1.1).

We begin with the following displacement field, which is a direct result of the Kirchhoff hypotheses in the pure bending case (see Fig. 7.2.9):

$$u_1(x, y, z, t) = -z\frac{\partial w}{\partial x}, \quad u_2(x, y, z, t) = -z\frac{\partial w}{\partial y}, \quad u_3(x, y, z, t) = w(x, y, t),$$

(7.2.78)

where (u_1, u_2, u_3) denote the total displacements of a material point in the (x, y, z) coordinate directions at time t.

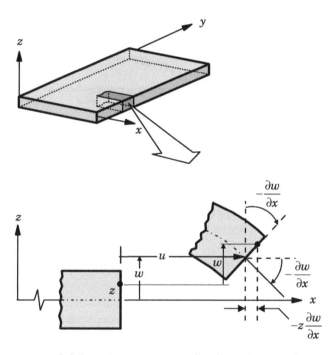

Fig. 7.2.9 Kinematics of deformation in rectangular Cartesian coordinates of the classical plate theory.

Assuming small strains and displacements, the linear strains can be computed using Eq. (2.3.12). For the displacement field in Eq. (7.2.78), the linear strains are

$$\varepsilon_{xx} = -z\frac{\partial^2 w}{\partial x^2}, \quad \varepsilon_{xy} = -z\frac{\partial^2 w}{\partial x \partial y}, \quad \varepsilon_{yy} = -z\frac{\partial^2 w}{\partial y^2},$$

$$\varepsilon_{xz} = 0, \qquad \varepsilon_{yz} = 0, \qquad \varepsilon_{zz} = 0.$$

(7.2.79)

We note that the transverse strains $(\varepsilon_{xz}, \varepsilon_{yz}, \varepsilon_{zz})$ are identically zero in the CPT.

Hamilton's principle can be used in two different forms: One way is to construct the Lagrangian functional $L = (U + V - K)$ and use Hamilton's principle (the dynamic version of the principle of "minimum" potential energy):

$$0 = \delta \int_{t_1}^{t_2} L \, dt = \delta \int_{t_1}^{t_2} (U + V - K) dt, \tag{7.2.80}$$

where (U, V, K) are the total strain energy, potential energy due to applied loads, and kinetic energy, respectively. The other way is to only construct virtual Lagrangian $\delta L = \delta U + \delta V - \delta K$ and use Hamilton's principle in the form (the dynamic version of the principle of virtual displacements)

$$0 = \int_{t_1}^{t_2} \delta L \, dt = \int_{t_1}^{t_2} (\delta U + \delta V - \delta K) dt. \tag{7.2.81}$$

The first way makes use of the constitutive equations to write U and the assumption that the forces are conservative to write V, much the same way as in the principle of minimum total potential energy. The second way is general and does not require the use of constitutive relations or assumption that the forces are derivable from a potential. Here we use the latter but also give the expressions for (U, V, K) for completeness.

The virtual strain energy δU is given by

$$\delta U = \int_{\Omega_0} \int_{-\frac{h}{2}}^{\frac{h}{2}} (\sigma_{xx} \delta \varepsilon_{xx} + \sigma_{yy} \delta \varepsilon_{yy} + 2\sigma_{xy} \delta \varepsilon_{xy}) \, dz dx dy$$

$$= -\int_{\Omega_0} \left(M_{xx} \frac{\partial^2 \delta w}{\partial x^2} + M_{yy} \frac{\partial^2 \delta w}{\partial y^2} + 2M_{xy} \frac{\partial^2 \delta w}{\partial x \partial y} \right) dx dy, \tag{7.2.82}$$

where (M_{xx}, M_{yy}, M_{xy}) are the *moments* per unit length (see Fig. 7.2.10 for a schematic of a plate element with moments and shear forces):

$$\left\{ \begin{array}{c} M_{xx} \\ M_{yy} \\ M_{xy} \end{array} \right\} = \int_{-\frac{h}{2}}^{\frac{h}{2}} \left\{ \begin{array}{c} \sigma_{xx} \\ \sigma_{yy} \\ \sigma_{xy} \end{array} \right\} z \, dz. \tag{7.2.83}$$

To write U, we assume linear elastic behavior of the plate material and write [see Eqs (2.4.14) and (2.4.15)]

$$\left\{ \begin{array}{c} \sigma_{xx} \\ \sigma_{yy} \\ \sigma_{xy} \end{array} \right\} = \frac{E}{1 - \nu^2} \left[\begin{array}{ccc} 1 & \nu & 0 \\ \nu & 1 & 0 \\ 0 & 0 & \frac{1-\nu}{2} \end{array} \right] \left\{ \begin{array}{c} \varepsilon_{xx} \\ \varepsilon_{yy} \\ 2\varepsilon_{xy} \end{array} \right\} \tag{7.2.84}$$

for the isotropic material and

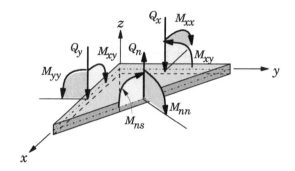

Fig. 7.2.10 Definitions of moments and shear forces.

$$\begin{Bmatrix} \sigma_{xx} \\ \sigma_{yy} \\ \sigma_{xy} \end{Bmatrix} = \begin{bmatrix} Q_{11} & Q_{12} & 0 \\ Q_{12} & Q_{22} & 0 \\ 0 & 0 & Q_{66} \end{bmatrix} \begin{Bmatrix} \varepsilon_{xx} \\ \varepsilon_{yy} \\ 2\varepsilon_{xy} \end{Bmatrix} \tag{7.2.85}$$

for an orthotropic material, where

$$Q_{11} = \frac{E_1}{1 - \nu_{12}\nu_{21}}, \quad Q_{12} = \frac{\nu_{12}E_2}{1 - \nu_{12}\nu_{21}}, \quad Q_{22} = \frac{E_2}{1 - \nu_{12}\nu_{21}},$$
$$Q_{66} = G_{12}, \quad \nu_{21} = \nu_{12}\frac{E_2}{E_1}. \tag{7.2.86}$$

Then substituting Eq. (7.2.79) into Eq. (7.2.85) and the result into Eq. (7.2.83) gives

$$M_{xx} = -\left(D_{11}\frac{\partial^2 w}{\partial x^2} + D_{12}\frac{\partial^2 w}{\partial y^2} \right),$$
$$M_{yy} = -\left(D_{22}\frac{\partial^2 w}{\partial y^2} + D_{12}\frac{\partial^2 w}{\partial x^2} \right), \tag{7.2.87}$$
$$M_{xy} = -2D_{66}\frac{\partial^2 w}{\partial x \partial y},$$

where

$$D_{ij} = \frac{h^3}{12}Q_{ij}, \quad h = \text{plate thickness.} \tag{7.2.88}$$

For an isotropic plate we have $E_1 = E_2 = E$, $\nu_{12} = \nu_{21} = \nu$, and $G_{12} = G = E/[2(1 + \nu)]$ and

$$D_{11} = D_{22} = D = \frac{Eh^3}{12(1 - \nu^2)}, \quad D_{12} = \nu D, \quad 2D_{66} = (1 - \nu)D. \tag{7.2.89}$$

The strain energy of an orthotropic plate is

$$U = \frac{1}{2} \int_V (\sigma_{xx}\varepsilon_{xx} + \sigma_{yy}\varepsilon_{yy} + 2\sigma_{xy}\varepsilon_{xy})dV$$

$$= \frac{1}{2} \int_{\Omega_0} \int_{-\frac{h}{2}}^{\frac{h}{2}} \left[D_{11}\varepsilon_{xx}^2 + D_{22}\varepsilon_{yy}^2 + 2D_{12}\varepsilon_{xx}\varepsilon_{yy} + 4D_{66}\varepsilon_{xy}^2 \right] dzdxdy$$

$$= \frac{1}{2} \int_{\Omega_0} \left[D_{11}\left(\frac{\partial^2 w}{\partial x^2}\right)^2 + D_{22}\left(\frac{\partial^2 w}{\partial y^2}\right)^2 + 2D_{12}\frac{\partial^2 w}{\partial x^2}\frac{\partial^2 w}{\partial y^2} \right.$$

$$\left. + 4D_{66}\left(\frac{\partial^2 w}{\partial x \partial y}\right)^2 \right] dxdy. \tag{7.2.90}$$

The virtual work done by applied distributed load $q(x,y)$ on the top surface, $z = h/2$, applied transverse edge force $\hat{V}_n$, applied normal edge moment $\hat{M}_{nn}$ on boundary Γ_2 (a portion of the total boundary Γ of Ω_0), and applied in-plane compressive and shear forces ($\hat{N}_{xx}$, $\hat{N}_{yy}$, $\hat{N}_{xy}$ per unit length) is

$$\delta V = -\left\{ \int_{\Omega_0} \left[\hat{N}_{xx}\frac{\partial w}{\partial x}\frac{\partial \delta w}{\partial x} + \hat{N}_{yy}\frac{\partial w}{\partial y}\frac{\partial \delta w}{\partial y} \right.\right.$$

$$\left. + \hat{N}_{xy}\left(\frac{\partial w}{\partial x}\frac{\partial \delta w}{\partial y} + \frac{\partial w}{\partial y}\frac{\partial \delta w}{\partial x}\right) \right] dxdy$$

$$\left. + \int_{\Omega_0} q(x,y)\delta w \, dxdy + \int_{\Gamma_2} \left(\hat{V}_n\delta w - \hat{M}_{nn}\frac{\partial \delta w}{\partial n} \right) dS \right\}. \tag{7.2.91}$$

The potential energy due to applied loads is

$$V = -\left\{ \frac{1}{2} \int_{\Omega_0} \left[\hat{N}_{xx}\left(\frac{\partial w}{\partial x}\right)^2 + \hat{N}_{yy}\left(\frac{\partial w}{\partial y}\right)^2 + 2\hat{N}_{xy}\frac{\partial w}{\partial x}\frac{\partial w}{\partial y} \right] dxdy \right.$$

$$\left. + \int_{\Omega_0} q(x,y)w \, dxdy + \int_{\Gamma_2} \left(\hat{V}_n w - \hat{M}_{nn}\frac{\partial w}{\partial n} \right) dS \right\}. \tag{7.2.92}$$

The virtual kinetic energy is given by

$$\delta K = \int_{\Omega_0} \int_{-\frac{h}{2}}^{\frac{h}{2}} \rho\,(\dot{u}\delta\dot{u} + \dot{v}\delta\dot{v} + \dot{w}\delta\dot{w})\,dz\,dxdy$$

$$= \int_{\Omega_0} \int_{-\frac{h}{2}}^{\frac{h}{2}} \rho\left[\left(-z\frac{\partial\dot{w}}{\partial x}\right)\left(-z\frac{\partial\delta\dot{w}}{\partial x}\right) + \left(-z\frac{\partial\dot{w}}{\partial y}\right)\left(-z\frac{\partial\delta\dot{w}}{\partial y}\right) \right.$$

$$\left. + \dot{w}\delta\dot{w} \right] dz\,dxdy,$$

or

$$\delta K = \int_{\Omega_0} \left[m_0 \dot{w} \delta \dot{w} + m_2 \left(\frac{\partial \dot{w}}{\partial x} \frac{\partial \delta \dot{w}}{\partial x} + \frac{\partial \dot{w}}{\partial y} \frac{\partial \delta \dot{w}}{\partial y} \right) \right] dx dy, \qquad (7.2.93)$$

where ρ is the mass density and the superposed dot on a variable indicates time derivative, $\dot{w} = \partial w / \partial t$, and (m_0, m_2) are the mass moments of inertia:

$$\left\{ \begin{matrix} m_0 \\ m_2 \end{matrix} \right\} = \int_{-\frac{h}{2}}^{\frac{h}{2}} \left\{ \begin{matrix} 1 \\ z^2 \end{matrix} \right\} \rho \, dz = \rho \left\{ \begin{matrix} h \\ \frac{h^3}{12} \end{matrix} \right\}. \qquad (7.2.94)$$

The kinetic energy is

$$K = \frac{1}{2} \int_{\Omega_0} \left\{ m_0 (\dot{w})^2 + m_2 \left[\left(\frac{\partial \dot{w}}{\partial x} \right)^2 + \left(\frac{\partial \dot{w}}{\partial y} \right)^2 \right] \right\} dx dy. \qquad (7.2.95)$$

We now have all the elements in place to apply Hamilton's principle in Eq. (7.2.81). We have

$$0 = \int_{t_1}^{t_2} \int_{\Omega_0} \left\{ - \left(M_{xx} \frac{\partial^2 \delta w}{\partial x^2} + M_{yy} \frac{\partial^2 \delta w}{\partial y^2} + 2 M_{xy} \frac{\partial^2 \delta w}{\partial x \partial y} \right) \right.$$

$$- \left[m_0 \dot{w} \delta \dot{w} + m_2 \left(\frac{\partial \dot{w}}{\partial x} \frac{\partial \delta \dot{w}}{\partial x} + \frac{\partial \dot{w}}{\partial y} \frac{\partial \delta \dot{w}}{\partial y} \right) \right]$$

$$- \left[\hat{N}_{xx} \frac{\partial w}{\partial x} \frac{\partial \delta w}{\partial x} + \hat{N}_{yy} \frac{\partial w}{\partial y} \frac{\partial \delta w}{\partial y} \right.$$

$$\left. \left. + \hat{N}_{xy} \left(\frac{\partial w}{\partial x} \frac{\partial \delta w}{\partial y} + \frac{\partial w}{\partial y} \frac{\partial \delta w}{\partial x} \right) \right] \right\} dx dy dt$$

$$- \int_{t_1}^{t_2} \left[\int_{\Omega_0} q \delta w \, dx dy + \int_{\Gamma_2} \left(\hat{V}_n \delta w + \hat{M}_{nn} \frac{\partial \delta w}{\partial n} \right) dS \right] dt. \qquad (7.2.96)$$

Integrating terms by parts to relieve δw of any differentiation, we obtain

$$0 = \int_{t_1}^{t_2} \left\{ \int_{\Omega_0} \left[- \left(\frac{\partial^2 M_{xx}}{\partial x^2} + \frac{\partial^2 M_{yy}}{\partial y^2} + 2 \frac{\partial^2 M_{xy}}{\partial x \partial y} \right) + m_0 \ddot{w} - m_2 \left(\frac{\partial^2 \ddot{w}}{\partial x^2} + \frac{\partial^2 \ddot{w}}{\partial y^2} \right) \right. \right.$$

$$\left. + \frac{\partial}{\partial x} \left(\hat{N}_{xx} \frac{\partial w}{\partial x} + \hat{N}_{xy} \frac{\partial w}{\partial y} \right) + \frac{\partial}{\partial y} \left(\hat{N}_{xy} \frac{\partial w}{\partial x} + \hat{N}_{yy} \frac{\partial w}{\partial y} \right) - q \right] \delta w \, dx dy$$

$$- \oint_{\Gamma} \left[M_{xx} \frac{\partial \delta w}{\partial x} n_x + M_{yy} \frac{\partial \delta w}{\partial y} n_y + M_{xy} \frac{\partial \delta w}{\partial y} n_x + M_{xy} \frac{\partial \delta w}{\partial x} n_y \right.$$

$$\left. - \left(\frac{\partial M_{xx}}{\partial x} n_x + \frac{\partial M_{yy}}{\partial y} n_y + \frac{\partial M_{xy}}{\partial x} n_y + \frac{\partial M_{xy}}{\partial y} n_x \right) \delta w \right.$$

$$- m_2 \left(\frac{\partial \dot{w}}{\partial x} n_x + \frac{\partial \dot{w}}{\partial y} n_y \right) \delta w + \left(\hat{N}_{xx} \frac{\partial w}{\partial x} + \hat{N}_{xy} \frac{\partial w}{\partial y} \right) n_x \delta w$$

$$\left. + \left(\hat{N}_{xy} \frac{\partial w}{\partial x} + \hat{N}_{yy} \frac{\partial w}{\partial y} \right) n_y \delta w \right] dS - \int_{\Gamma_2} \left(\bar{V}_n \delta w - \hat{M}_{nn} \frac{\partial \delta w}{\partial n} \right) dS \Bigg\} dt, \quad (7.2.97)$$

where all terms evaluated at $t = t_1$ and $t = t_2$ are zero (by assumption) and not included in Eq. (7.2.97). The Euler–Lagrange equation (i.e., the equation of motion) is clearly

$$- \left(\frac{\partial^2 M_{xx}}{\partial x^2} + 2 \frac{\partial^2 M_{xy}}{\partial y \partial x} + \frac{\partial^2 M_{yy}}{\partial y^2} \right) - q$$

$$+ \frac{\partial}{\partial x} \left(\hat{N}_{xx} \frac{\partial w}{\partial x} + \hat{N}_{xy} \frac{\partial w}{\partial y} \right) + \frac{\partial}{\partial y} \left(\hat{N}_{xy} \frac{\partial w}{\partial x} + \hat{N}_{yy} \frac{\partial w}{\partial y} \right)$$

$$+ m_0 \ddot{w} - m_2 \left(\frac{\partial^2 \ddot{w}}{\partial x^2} + \frac{\partial^2 \ddot{w}}{\partial y^2} \right) = 0. \quad (7.2.98)$$

Inspection of the boundary terms in Eq. (7.2.94) indicates that w, $\partial w / \partial x$ and $\partial w / \partial y$ are the primary variables and

$$\left(\bar{Q}_x + m_2 \frac{\partial \dot{w}}{\partial x} \right) n_x + \left(\bar{Q}_y + m_2 \frac{\partial \dot{w}}{\partial y} \right) n_y,$$

$$M_{xx} n_x + M_{xy} n_y, \qquad M_{xy} n_x + M_{yy} n_y \quad (7.2.99)$$

are the secondary variables of the theory. Here $\bar{Q}_x$ and $\bar{Q}_y$ are defined as

$$\bar{Q}_x = Q_x - \left(\hat{N}_{xx} \frac{\partial w}{\partial x} + \hat{N}_{xy} \frac{\partial w}{\partial y} \right),$$

$$\bar{Q}_y = Q_y - \left(\hat{N}_{xy} \frac{\partial w}{\partial x} + \hat{N}_{yy} \frac{\partial w}{\partial y} \right),$$

$$Q_x \equiv \frac{\partial M_{xx}}{\partial x} + \frac{\partial M_{xy}}{\partial y}, \quad (7.2.100)$$

$$Q_y \equiv \frac{\partial M_{xy}}{\partial x} + \frac{\partial M_{yy}}{\partial y}.$$

In order to cast the boundary conditions on an edge that is arbitrarily oriented in the xy-plane, we convert the derivatives with respect to x and y to those in terms of the normal (n) and tangential (s) coordinates. If the unit outward normal vector $\hat{\mathbf{n}}$ is oriented at an angle θ counterclockwise from the positive x-axis, then its direction cosines are $n_x = \cos \theta$ and $n_y = \sin \theta$. Hence, the transformation between the coordinate system (n, s) and (x, y) is given by

$$\hat{\mathbf{e}}_x = \cos \theta \, \hat{\mathbf{e}}_n - \sin \theta \, \hat{\mathbf{e}}_s = n_x \, \hat{\mathbf{e}}_n - n_y \, \hat{\mathbf{e}}_s,$$

$$\hat{\mathbf{e}}_y = \sin \theta \, \hat{\mathbf{e}}_n + \cos \theta \, \hat{\mathbf{e}}_s = n_y \, \hat{\mathbf{e}}_n + n_x \, \hat{\mathbf{e}}_s. \quad (7.2.101)$$

Hence, the normal and tangential derivatives $(\partial w/\partial n, \partial w/\partial s)$ are related to the derivatives $(\partial w/\partial x, \partial w/\partial y)$ by

$$\frac{\partial w}{\partial x} = n_x \frac{\partial w}{\partial n} - n_y \frac{\partial w}{\partial s} , \quad \frac{\partial w}{\partial y} = n_y \frac{\partial w}{\partial n} + n_x \frac{\partial w}{\partial s}. \tag{7.2.102}$$

Using Eq. (7.2.102), we can rewrite the boundary expressions in Eq. (7.2.97) as

$$
\begin{aligned}
0 = \oint_{\Gamma} \Bigg\{ & (V_x n_x + V_y n_y)\,\delta w - \left(M_{xx} n_x^2 + 2 M_{xy} n_x n_y + M_{yy} n_y^2 \right) \frac{\partial \delta w}{\partial n} \\
& - \left[(M_{yy} - M_{xx})\, n_x n_y + M_{xy} \left(n_x^2 - n_y^2 \right) \right] \frac{\partial \delta w}{\partial s} \Bigg\} dS \\
& - \int_{\Gamma_2} \left(\hat{V}_n \delta w - \hat{M}_{nn} \frac{\partial \delta w}{\partial n} \right) dS \\
= \int_{\Gamma_2} & \Bigg[\left(V_x n_x + V_y n_y + \frac{\partial M_{ns}}{\partial s} - \hat{V}_n \right) \delta w \\
& + \left(\hat{M}_{nn} - M_{nn} \right) \frac{\partial \delta w}{\partial n} \Bigg] dS,
\end{aligned}
\tag{7.2.103}
$$

where w and $(\partial w/\partial n)$ are assumed to be specified on Γ_1; hence $\delta w = 0$ and $(\partial \delta w/\partial n) = 0$ there, and

$$
\begin{aligned}
V_x &= Q_x - \left(\hat{N}_{xx} \frac{\partial w}{\partial x} + \hat{N}_{xy} \frac{\partial w}{\partial y} \right) + m_2 \frac{\partial \ddot{w}}{\partial x}, \\
V_y &= Q_y - \left(\hat{N}_{xy} \frac{\partial w}{\partial x} + \hat{N}_{yy} \frac{\partial w}{\partial y} \right) + m_2 \frac{\partial \ddot{w}}{\partial x}, \\
M_{nn} &= M_{xx} n_x^2 + 2 M_{xy} n_x n_y + M_{yy} n_y^2, \\
M_{ns} &= (M_{yy} - M_{xx})\, n_x n_y + M_{xy} \left(n_x^2 - n_y^2 \right).
\end{aligned}
\tag{7.2.104}
$$

We note that the expression involving $\partial \delta w/\partial s$ was integrated by parts to give

$$-\oint_{\Gamma} M_{ns} \frac{\partial \delta w}{\partial s}\, dS = \oint_{\Gamma} \frac{\partial M_{ns}}{\partial s} \delta w\, dS - [M_{ns}\delta w]_{\Gamma}, \tag{7.2.105}$$

and the term $[M_{ns}\delta w]_{\Gamma}$ is set to zero since the end points of a closed smooth curve coincide. For plates with corners (i.e., for polygonal plates), concentrated forces of magnitude

$$F_c = -2M_{ns} \tag{7.2.106}$$

will be produced at the corners. The factor of 2 appears because M_{ns} from two sides of the corner are added there. Finally, the natural boundary conditions are

$$V_x n_x + V_y n_y + \frac{\partial M_{ns}}{\partial s} - \hat{V}_n = 0, \quad M_{nn} - \hat{M}_{nn} = 0 \quad \text{on } \Gamma_2. \tag{7.2.107}$$

The first of the two boundary conditions in Eq. (7.2.107) is known as the *Kirch-hoff free edge condition*. Thus, the effective shear force is

$$V_n \equiv V_x n_x + V_y n_y + \frac{\partial M_{ns}}{\partial s}.$$ (7.2.108)

Equation (7.2.98) can be expressed in terms of the transverse deflection with the help of moment–deflection relations in Eq. (7.2.87). We have

$$\left[D_{11}\frac{\partial^4 w}{\partial x^4} + 2(2D_{66} + D_{12})\frac{\partial^4 w}{\partial x^2 \partial y^2} + D_{22}\frac{\partial^4 w}{\partial y^4} \right]$$
$$+ \frac{\partial}{\partial x}\left(\hat{N}_{xx}\frac{\partial w}{\partial x} + \hat{N}_{xy}\frac{\partial w}{\partial y} \right) + \frac{\partial}{\partial y}\left(\hat{N}_{xy}\frac{\partial w}{\partial x} + \hat{N}_{yy}\frac{\partial w}{\partial y} \right)$$
$$= q - m_0\ddot{w} + m_2\left(\frac{\partial^2 \ddot{w}}{\partial x^2} + \frac{\partial^2 \ddot{w}}{\partial y^2} \right)$$ (7.2.109)

for orthotropic plates and

$$D\left(\frac{\partial^4 w}{\partial x^4} + 2\frac{\partial^4 w}{\partial x^2 \partial y^2} + \frac{\partial^4 w}{\partial y^4} \right) + \frac{\partial}{\partial x}\left(\hat{N}_{xx}\frac{\partial w}{\partial x} + \hat{N}_{xy}\frac{\partial w}{\partial y} \right)$$
$$+ \frac{\partial}{\partial y}\left(\hat{N}_{xy}\frac{\partial w}{\partial x} + \hat{N}_{yy}\frac{\partial w}{\partial y} \right) = q - m_0\ddot{w} + m_2\left(\frac{\partial^2 \ddot{w}}{\partial x^2} + \frac{\partial^2 \ddot{w}}{\partial y^2} \right)$$ (7.2.110)

for isotropic plates, where $D = Eh^3/[12(1 - \nu^2)]$.

7.2.4 Navier Solutions of Rectangular Plates

In this section, we briefly discuss Navier solutions of isotropic rectangular plates for bending, natural vibration, buckling, and transient response. Navier solutions can be developed for simply supported rectangular plates in the following cases:

1. Bending solutions under arbitrary transverse load $q(x, y)$.

2. Natural frequencies.

3. Buckling loads under in-plane biaxial compressive loads $\hat{N}_{yy} = \gamma\hat{N}_{xx} = \gamma N_0$.

4. Spatial part of the solution for the transient case.

Keeping the scope of this book in mind, only a brief discussion of analytical solutions is presented. For additional information, the reader may consult the books on plates, especially those by Reddy [51] and Timoshenko and Woinowsky-Krieger [124].

The simply supported boundary conditions for a rectangular plate, for the choice of coordinate system shown in Fig. 7.2.11, are

$$w(0, y, t) = 0, \quad w(a, y, t) = 0, \quad w(x, 0, t) = 0, \quad w(x, b, t) = 0,$$
$$M_{xx}(0, y, t) = 0, \quad M_{xx}(a, y, t) = 0, \quad M_{yy}(x, 0, t) = 0, \quad M_{yy}(x, b, t) = 0. \tag{7.2.111}$$

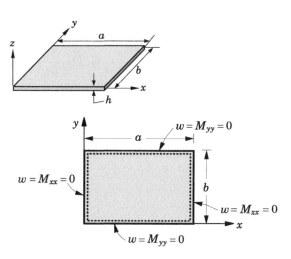

Fig. 7.2.11 Geometry and coordinate system for a rectangular plate.

In Navier's method the displacement w is expanded in double sine series with unknown coefficients so that the boundary conditions are satisfied exactly:

$$w(x, y, t) = \sum_{n=1}^{\infty} \sum_{m=1}^{\infty} W_{mn}(t) \, \sin \alpha_m x \, \sin \beta_n y, \tag{7.2.112}$$

where

$$\alpha_m = \frac{m\pi}{a}, \quad \beta_n = \frac{n\pi}{b}. \tag{7.2.113}$$

Here W_{mn} denote the coefficients (that depend on time t) to be determined such that Eq. (7.2.109) or (7.2.110) is satisfied everywhere in the domain of the plate for all $t > 0$. The choice of expansion in Eq. (7.2.112) is dictated by the fact that the double sine series satisfies the simply supported boundary conditions in Eq. (7.2.111). Substitution of the expansion in Eq. (7.2.112) into the governing equation in Eq. (7.2.109) shows that the mechanical load q also be expanded in double sine series:

$$q(x, y, t) = \sum_{n=1}^{\infty} \sum_{m=1}^{\infty} q_{mn}(t) \sin \alpha_m x \, \sin \beta_n y, \tag{7.2.114}$$

$$q_{mn}(t) = \frac{4}{ab} \int_0^b \int_0^a q(x, y, t) \, \sin \alpha_m x \, \sin \beta_n y \, dx dy, \tag{7.2.115}$$

in order to have a solution. The coefficients q_{mn} for various types of loads can be calculated using Eq. (7.2.115) (see Table tab7.2.3 for typical loads).

Table 7.2.3 Coefficients in the double sine series expansion of loads in Navier's method.

Load type	Coefficients q_{mn}
1. Uniform load, $q(x,y) = q_0$	$q_{mn} = \frac{16q_0}{\pi^2 mn}$ $(m, n = 1, 3, 5, \ldots)$
2. Hydrostatic load, $q(x,y) = q_0 \frac{x}{a}$	$q_{mn} = \frac{8q_0}{\pi^2 mn}(-1)^{m+1}$ $(m = 1, 2, 3, \ldots; n = 1, 3, 5, \ldots)$
3. Point load, Q_0 at (x_0, y_0)	$q_{mn} = \frac{4Q_0}{ab} \sin \frac{m\pi x_0}{a} \sin \frac{n\pi y_0}{b}$ $(m, n = 1, 2, 3, \ldots)$
4. Line load, $q(x,y) = q_0\, \delta(x - x_0)$	$q_{mn} = \frac{8q_0}{\pi a n} \sin \frac{m\pi x_0}{a}$ $(m = 1, 2, 3, \ldots; n = 1, 3, 5, \ldots)$

Substituting Eqs (7.2.112) and (7.2.115) into Eq. (7.2.110), we obtain

$$
0 = \sum_{n=1}^{\infty} \sum_{m=1}^{\infty} \left\{ D \left(\alpha_m^4 + 2\alpha_m^2 \beta_n^2 + \beta_n^4 \right) W_{mn} - (\alpha_m^2 + \gamma\beta_n^2) N_0 W_{mn} \right.
$$

$$
\left. - q_{mn} + \left[m_0 + m_2 \left(\alpha_m^2 + \beta_n^2 \right) \right] \ddot{W}_{mn} \right\} \sin \alpha_m x \, \sin \beta_n y \quad (7.2.116)
$$

for any (x, y) and $t > 0$. Hence, it follows that

$$
\left[m_0 + m_2 \left(\alpha_m^2 + \beta_n^2 \right) \right] \frac{d^2 W_{mn}}{dt^2} + \left[D \left(\alpha_m^2 + \beta_n^2 \right)^2 - (\alpha_m^2 + \gamma\beta_n^2) N_0 \right] W_{mn} = q_{mn}
$$
$$
(7.2.117)
$$

for every pair of m and n and $t > 0$. Here $\hat{N}_{xx} = N_0$ and $\hat{N}_{yy} = \gamma N_0$, and N_0 is applied compressive in-plane force. Equation (7.2.117) can be specialized to static bending, natural vibration, or buckling of simply supported rectangular plates. For the transient case, the ordinary differential equation in time, Eq. (7.2.117), must be solved subject to initial conditions on the deflection w and velocity $\dot{w}$. We discuss these cases next for isotropic plates.

7.2.4.1 Bending

For the static bending response under applied transverse load $q(x, y)$ and in-plane biaxial compressive forces $(\hat{N}_0, \gamma\hat{N}_0)$ per unit length, the coefficients W_{mn} (which are constants) in the displacement expansion, Eq. (7.2.112), can be obtained from Eq. (7.2.117) by setting time derivative terms to zero:

$$
W_{mn} = \frac{q_{mn}}{\left[D \left(\alpha_m^2 + \beta_n^2 \right)^2 - (\alpha_m^2 + \gamma\beta_n^2) N_0 \right]}, \quad (7.2.118)
$$

and the static solution becomes

$$
w(x, y) = \sum_{n=1}^{\infty} \sum_{m=1}^{\infty} W_{mn} \sin \alpha_m x \, \sin \beta_n y. \quad (7.2.119)
$$

Note that the compressive force N_0 has the effect of increasing the deflection (or reduce the deflection when the in-plane forces are tensile).

When the in-plane compressive loads are zero, we have

$$
W_{mn} = \frac{q_{mn}}{D \left(\alpha_m^2 + \beta_n^2 \right)^2} = \frac{b^4}{D\pi^4} \frac{q_{mn}}{(m^2 s^2 + n^2)^2}, \quad (7.2.120)
$$

where s is the plate aspect ratio, $s = b/a$. The solution is given by

$$
w(x, y) = \frac{b^4}{D\pi^4} \sum_{n=1}^{\infty} \sum_{m=1}^{\infty} \frac{q_{mn}}{(m^2 s^2 + n^2)^2} \sin \alpha_m x \, \sin \beta_n y. \quad (7.2.121)
$$

The bending moments can be calculated from

$$M_{xx} = D \sum_{n=1}^{\infty} \sum_{m=1}^{\infty} \frac{\pi^2}{b^2} \left(m^2 s^2 + \nu n^2\right) W_{mn} \sin \alpha_m x \, \sin \beta_n y,$$

$$M_{yy} = D \sum_{n=1}^{\infty} \sum_{m=1}^{\infty} \frac{\pi^2}{b^2} \left(\nu m^2 s^2 + n^2\right) W_{mn} \sin \alpha_m x \, \sin \beta_n y, \qquad (7.2.122)$$

$$M_{xy} = -(1-\nu)\frac{s\pi^2}{b^2} D \sum_{n=1}^{\infty} \sum_{m=1}^{\infty} mn \, W_{mn} \, \cos \alpha_m x \, \cos \beta_n y.$$

The shear forces Q_x and Q_y can be computed using

$$Q_x = \frac{\partial M_{xx}}{\partial x} + \frac{\partial M_{xy}}{\partial y} = \sum_{n=1}^{\infty} \sum_{m=1}^{\infty} S_{xx} W_{mn} \, \cos \alpha_m x \, \sin \beta_n y,$$

$$Q_y = \frac{\partial M_{xy}}{\partial x} + \frac{\partial M_{yy}}{\partial y} = \sum_{n=1}^{\infty} \sum_{m=1}^{\infty} S_{yy} W_{mn} \, \sin \alpha_m x \, \cos \beta_n y, \qquad (7.2.123)$$

where

$$S_{xx} = D\alpha_m \left(\alpha_m^2 + \beta_n^2\right), \quad S_{yy} = D\beta_n \left(\alpha_m^2 + \beta_n^2\right). \qquad (7.2.124)$$

The effective shear forces (i.e., reaction forces) V_x and V_y along the simply supported edges $x = a$ and $y = b$, respectively, can be calculated using

$$V_x(a, y) = Q_x + \frac{\partial M_{xy}}{\partial y} = \frac{\partial M_{xx}}{\partial x} + 2\frac{\partial M_{xy}}{\partial y}$$

$$= \sum_{n=1}^{\infty} \sum_{m=1}^{\infty} (-1)^m \hat{S}_{xx} W_{mn} \, \sin \beta_n y, \qquad (7.2.125)$$

$$V_y(x, b) = Q_y + \frac{\partial M_{xy}}{\partial x} = 2\frac{\partial M_{xy}}{\partial x} + \frac{\partial M_{yy}}{\partial y}$$

$$= \sum_{n=1}^{\infty} \sum_{m=1}^{\infty} (-1)^n \hat{S}_{yy} W_{mn} \, \sin \alpha_m x, \qquad (7.2.126)$$

where

$$\hat{S}_{xx} = D\left[\alpha_m^3 + (2-\nu)\alpha_m \beta_n^2\right], \quad \hat{S}_{yy} = D\left[\beta_n^3 + (2-\nu)\alpha_m^2 \beta_n\right]. \qquad (7.2.127)$$

Thus, the distribution of the reaction forces along the edges follows a sinusoidal form. Similar expressions hold for $V_x(0, y)$ and $V_y(x, 0)$.

In addition to the reactions in Eqs (7.2.125) and (7.2.126) along the edges, the plate experiences concentrated forces at the corners of a rectangular plate due to the twisting moment M_{xy} per unit length (which has the dimensions of a force). The concentrated force at the corners $x = a$ and $y = b$ is given by

$$F_c = -2M_{xy} = 2(1 - \nu)D\frac{\partial^2 w}{\partial x \partial y}$$

$$= 2(1 - \nu)D\sum_{n=1}^{\infty}\sum_{m=1}^{\infty}\left(\frac{m\pi}{a}\right)\left(\frac{n\pi}{b}\right)(-1)^{m+n}W_{mn}. \qquad (7.2.128)$$

For the pure bending case considered here, the stresses in a simply supported rectangular plate are given by

$$\left\{\begin{matrix} \sigma_{xx} \\ \sigma_{yy} \\ \sigma_{xy} \end{matrix}\right\} = -\frac{Ez}{(1 - \nu^2)}\begin{bmatrix} 1 & \nu & 0 \\ \nu & 1 & 0 \\ 0 & 0 & \frac{1-\nu}{2} \end{bmatrix}\left\{\begin{matrix} \frac{\partial^2 w}{\partial x^2} \\ \frac{\partial^2 w}{\partial y^2} \\ 2\frac{\partial^2 w}{\partial x \partial y} \end{matrix}\right\}$$

$$= z\sum_{n=1}^{\infty}\sum_{m=1}^{\infty}W_{mn}\left\{\begin{matrix} R_{xx} \ \sin\alpha_m x \ \sin\beta_n y \\ R_{yy} \ \sin\alpha_m x \ \sin\beta_n y \\ -R_{xy} \ \cos\alpha_m x \ \cos\beta_n y \end{matrix}\right\}, \qquad (7.2.129)$$

where

$$R_{xx} = \frac{\pi^2 E}{b^2(1 - \nu^2)}\left(m^2 s^2 + \nu n^2\right),$$

$$R_{yy} = \frac{\pi^2 E}{b^2(1 - \nu^2)}\left(\nu m^2 s^2 + n^2\right), \qquad (7.2.130)$$

$$R_{xy} = mns\frac{\pi^2 E}{b^2(1 + \nu)}.$$

The maximum stresses occur at $(x, y, z) = (a/2, b/2, \pm h/2)$.

In the CPT, the transverse stresses (σ_{xz}, σ_{yz}, and σ_{zz}) are identically zero when computed from the constitutive equations because the transverse shear strains are zero. However, they can be computed using the 3-D stress equilibrium equations for any $-h/2 \leq z \leq h/2$:

$$\sigma_{xz} = -\int_{-\frac{h}{2}}^{z}\left(\frac{\partial\sigma_{xx}}{\partial x} + \frac{\partial\sigma_{xy}}{\partial y}\right)dz + C_1(x, y),$$

$$\sigma_{yz} = -\int_{-\frac{h}{2}}^{z}\left(\frac{\partial\sigma_{xy}}{\partial x} + \frac{\partial\sigma_{yy}}{\partial y}\right)dz + C_2(x, y), \qquad (7.2.131)$$

$$\sigma_{zz} = -\int_{-\frac{h}{2}}^{z}\left(\frac{\partial\sigma_{xz}}{\partial x} + \frac{\partial\sigma_{yz}}{\partial y}\right)dz + C_3(x, y),$$

where the stresses σ_{xx}, σ_{xy}, and σ_{yy} are known from Eq. (7.2.129) and C_i are functions to be determined using the conditions $\sigma_{xz}(x, y, -h/2) = \sigma_{yz}(x, y, -h/2)$

$= \sigma_{zz}(x, y, -h/2) = 0$. We obtain $C_i = 0$ and

$$
\sigma_{xz} = \frac{h^2}{8}\left[1 - \left(\frac{2z}{h}\right)^2\right] \sum_{n=1}^{\infty}\sum_{m=1}^{\infty} S_{mn}^{xz} \cos\frac{m\pi x}{a} \sin\frac{n\pi y}{b},
$$

$$
\sigma_{yz} = \frac{h^2}{8}\left[1 - \left(\frac{2z}{h}\right)^2\right] \sum_{n=1}^{\infty}\sum_{m=1}^{\infty} S_{mn}^{yz} \sin\frac{m\pi x}{a} \cos\frac{n\pi y}{b}, \tag{7.2.132}
$$

$$
\sigma_{zz} = -\frac{h^3}{48}\left\{\left[1 + \left(\frac{2z}{h}\right)^3\right] - 3\left[1 + \left(\frac{2z}{h}\right)\right]\right\}
$$

$$
\times \sum_{n=1}^{\infty}\sum_{m=1}^{\infty} S_{mn}^{zz} \sin\frac{m\pi x}{a} \sin\frac{n\pi y}{b},
$$

where

$$
S_{mn}^{xz} = S_{13}W_{mn}, \quad S_{mn}^{yz} = S_{23}W_{mn}, \quad S_{mn}^{zz} = S_{33}W_{mn}, \tag{7.2.133}
$$

with S_{ij} defined by

$$
S_{13} = \frac{E}{(1-\nu^2)}\left(\alpha_m^3 + \alpha_m\beta_n^2\right),
$$

$$
S_{23} = \frac{E}{(1-\nu^2)}\left(\beta_n^3 + \alpha_m^2\beta_n\right), \tag{7.2.134}
$$

$$
S_{33} = \frac{E}{(1-\nu^2)}\left(\alpha_m^2 + \beta_n^2\right)^2.
$$

Note that σ_{xz} and σ_{yz} are zero and $\sigma_{zz} = q$ at the top surface of the plate ($z = h/2$). The transverse shear stress σ_{xz} is the maximum at $(x, y, z) = (0, b/2, 0)$, σ_{yz} is the maximum at $(x, y, z) = (a/2, 0, 0)$, and the transverse normal stress σ_{zz} is the maximum at $(x, y, z) = (a/2, b/2, h/2)$.

Example 7.2.9 _____

Determine the deflection $w(x, y)$ and bending moment M_{xx} of an isotropic rectangular plate subjected to (a) uniformly distributed load of intensity q_0 and (b) point load Q_0 at the center.

Solution: The deflection is given by

$$
w(x, y) = \frac{16q_0 b^4}{D\pi^6} \sum_{n=1,3,\ldots}^{\infty}\sum_{m=1,3,\ldots}^{\infty} \frac{1}{mn(m^2 s^2 + n^2)^2} \sin\frac{m\pi x}{a} \sin\frac{n\pi y}{b}. \tag{1}
$$

The maximum values of w and M_{xx} are given by ($s = b/a$)

$$
w_{\max} = w\left(\frac{a}{2}, \frac{b}{2}\right) = \frac{16q_0 b^4}{D\pi^6} \sum_{n=1,3,\ldots}^{\infty}\sum_{m=1,3,\ldots}^{\infty} \frac{(-1)^{\frac{m+n}{2}-1}}{mn(m^2 s^2 + n^2)^2}, \tag{2}
$$

$$
(M_{xx})_{\max} = M_{xx}\left(\frac{a}{2}, \frac{b}{2}\right) = \frac{16q_0 b^2}{D\pi^4} \sum_{n=1,3,\ldots}^{\infty}\sum_{m=1,3,\ldots}^{\infty} (-1)^{\frac{m+n}{2}-1}\frac{(m^2 s^2 + \nu n^2)^2}{mn(m^2 s^2 + n^2)^2}. \tag{3}
$$

For a square plate $(b = a)$ we have

$$w_{\max} = \frac{16q_0 a^4}{D\pi^6} \sum_{n=1,3,\ldots}^{\infty} \sum_{m=1,3,\ldots}^{\infty} \frac{(-1)^{\frac{m+n}{2}-1}}{mn(m^2+n^2)^2}, \tag{4}$$

$$(M_{xx})_{\max} = \frac{16q_0 a^2}{D\pi^4} \sum_{n=1,3,\ldots}^{\infty} \sum_{m=1,3,\ldots}^{\infty} (-1)^{\frac{m+n}{2}-1} \frac{(m^2+\nu n^2)^2}{mn(m^2+n^2)^2}. \tag{5}$$

A one-term solutions for the deflection and bending moment are ($\nu = 0.3$)

$$w_{\max} = \frac{4q_0 a^4}{D\pi^6} = 0.00416 \frac{q_0 a^4}{D}, \tag{6}$$

$$(M_{xx})_{\max} = 0.05338 q_0 a^2, \quad (\sigma_{xx})_{max} = 0.3203 \frac{q_0 a^2}{h^2}. \tag{7}$$

The deflection is about 2.4% in error compared to the solution obtained with $m, n = 1, 3, \ldots, 9$. Thus, the series in Eq. (4) converges rapidly. The expressions for bending moments do not converge as rapidly. For $m, n = 1, 3, \ldots, 29$, we obtain

$$w_{\max} = 0.004062 \frac{q_0 a^4}{D}, \tag{8}$$

$$(M_{xx})_{\max} = 0.04789 q_0 a^2, \quad (\sigma_{xx})_{max} = 0.2816 \frac{q_0 a^2}{h^2}. \tag{9}$$

For an isotropic rectangular plate under a point load Q_0 at (x_0, y_0), the deflection is given by

$$w(x,y) = \frac{4Q_0 b^2 s}{D\pi^4} \sum_{n=1,3,\ldots}^{\infty} \sum_{m=1,3,\ldots}^{\infty} \frac{\sin \frac{m\pi x_0}{a} \sin \frac{n\pi y_0}{b}}{(m^2 s^2 + n^2)^2} \sin \frac{m\pi x}{a} \sin \frac{n\pi y}{b}. \tag{10}$$

The center deflection when the load is applied at the center is given by

$$w\left(\frac{a}{2}, \frac{b}{2}\right) = \frac{4Q_0 b^2 s}{D\pi^4} \sum_{n=1,3,\ldots}^{\infty} \sum_{m=1,3,\ldots}^{\infty} \frac{1}{(m^2 s^2 + n^2)^2}. \tag{11}$$

In the case of a square plate, the center deflection becomes

$$w_{\max} = \frac{4Q_0 a^2}{D\pi^4} \sum_{n=1,3,\ldots}^{\infty} \sum_{m=1,3,\ldots}^{\infty} \frac{1}{(m^2 + n^2)^2}. \tag{12}$$

The first term of the series yields ($\nu = 0.3$)

$$w_{\max} = 0.01027 \frac{Q_0 a^2}{D} = 0.1121 \frac{Q_0 a^2}{Eh^3}, \tag{13}$$

$$(M_{xx})_{\max} = 0.1317 Q_0 a, \quad (\sigma_{xx})_{\max} = 0.7903 \frac{Q_0 a}{h^2}. \tag{14}$$

Taking the first four terms (i.e., $m, n = 1, 3$) of the series, we obtain

$$w_{\max} = 0.01121 \frac{Q_0 a^2}{D} = 0.1225 \frac{Q_0 a^2}{Eh^3}, \tag{15}$$

$$(M_{xx})_{\max} = 0.199 Q_0 a, \quad (\sigma_{xx})_{\max} = 1.194 \frac{Q_0 a}{h^2}. \tag{16}$$

The deflection is about 3.4% in error compared to the solution 0.0116 $(Q_0 a^2/D)$ obtained using $m, n = 1, 3, \ldots, 19$.

Table 7.2.4 contains the dimensionless maximum transverse deflections and stresses of square and rectangular plates under various types of loads. The dimensionless transverse deflection and stresses are defined as follows:

$$\bar{w} = w(0,0) \left(\frac{Eh^3}{a^4 q_0} \right); \quad \bar{\sigma}_{xx} = \sigma_{xx}(a/2, b/2, h/2) \left(\frac{h^2}{a^2 q_0} \right),$$

$$\bar{\sigma}_{yy} = \sigma_{yy}(a/2, b/2, h/2) \left(\frac{h^2}{a^2 q_0} \right); \quad \bar{\sigma}_{xy} = \sigma_{xy}(a, b, -h/2) \left(\frac{h^2}{a^2 q_0} \right), \quad (17)$$

$$\bar{\sigma}_{xz} = \sigma_{xz}(0, b/2, 0) \left(\frac{h}{a q_0} \right); \quad \bar{\sigma}_{yz} = \sigma_{yz}(a/2, 0, 0) \left(\frac{h}{a q_0} \right).$$

The loads considered are sinusoidal (SL), uniform (UL), hydrostatic (HL), or central point load (PL). Since convergence for derivatives of a function is slower than the function itself, the convergence is slower for stresses, which are calculated using the derivatives of the deflection. In the case of the point load, convergence for stresses will not be reached due to the stress singularity at the center of the plate.

Table 7.2.4 Transverse deflections and stresses in isotropic ($\nu = 0.3$) square and rectangular plates subjected to various types of loads.

Load	$\bar{w}$	$\bar{\sigma}_{xx}$	$\bar{\sigma}_{yy}$	$\bar{\sigma}_{xy}$	$\bar{\sigma}_{xz}$	$\bar{\sigma}_{yz}$
Square plates ($b/a = 1$)						
SL	0.0280	0.1976	0.1976	0.1064	0.2387	0.2387
UL (19)[a]	0.0444	0.2873	0.2873	0.1946	0.4909	0.4909
HL (19)	0.0222	0.1436	0.1436	0.0775	0.2455	0.1353
PL (29)	0.1266	2.4350	2.4350	0.3658	1.0010	1.0010
Rectangular plates ($b/a = 3$)						
SL	0.0908	0.5088	0.2024	0.1149	0.4297	0.1432
UL (19)	0.1336	0.7130	0.2433	0.2830	0.7221	0.5110
HL (19)	0.0668	0.3565	0.1217	0.0579	0.3610	0.0636
PL (29)	0.1845	2.3523	1.8828	0.0566	0.9032	0.2257

[a] The number in parenthesis denotes the maximum values of m and n used to evaluate the series.

7.2.4.2 Natural Vibration

For natural vibration, the deflection is assumed to be periodic in time

$$w(x, y, t) = \sum_{m=1}^{\infty} \sum_{n=1}^{\infty} W_{mn} \sin \frac{m\pi x}{a} \sin \frac{n\pi y}{b} e^{i\omega_{mn} t}, \quad (7.2.135)$$

where $i = \sqrt{-1}$ and ω_{mn} is the frequency of natural vibration associated with the (m, n)th mode shape

$$\sin \frac{m\pi x}{a} \sin \frac{n\pi y}{b} \quad (7.2.136)$$

and W_{mn} is the amplitude of vibration of mode (m, n). Substituting Eq. (7.2.135) into Eq. (7.2.110), we obtain, for nonzero W_{mn}, the result

$$D \left(\alpha_m^2 + \beta_n^2\right)^2 - \left(\alpha_m^2 + \gamma\beta_n^2\right) N_0 - \omega_{mn}^2 \left[m_0 + \left(\alpha_m^2 + \beta_n^2\right) m_2\right] = 0, \quad (7.2.137)$$

where $\alpha_m = m\pi/a$ and $\beta_n = n\pi/b$. Solving Eq. (7.2.137) for the natural frequency, we obtain

$$\omega_{mn}^2 = \frac{\pi^2}{\tilde{m}_0 b^2} \left[\frac{\pi^2 D}{b^2} \left(m^2 s^2 + n^2\right)^2 - N_0 \left(m^2 s^2 + \gamma n^2\right)\right], \quad s = \frac{b}{a}, \quad (7.2.138)$$

where (the principal and rotary inertias are $m_0 = \rho h$ and $m_2 = \rho h^3/12$)

$$\tilde{m}_0 = m_0 + m_2 \left[\left(\frac{m\pi}{a}\right)^2 + \left(\frac{n\pi}{b}\right)^2\right]. \quad (7.2.139)$$

For different values of m and n, there corresponds a unique frequency ω_{mn} and mode shape given by Eq. (7.2.136). The in-plane compressive force as well as the rotatory (or rotary) inertia m_2 has the effect of reducing the magnitude of the frequency of vibration, and their relative effect depends on m and n and the plate thickness-to-side ratio h/b. For most plates with $h/b < 0.1$, the rotary inertia may be neglected.

The smallest value of ω_{mn} is called the *fundamental frequency*. When the rotatory inertia m_2 is neglected, the frequency of a rectangular isotropic plate without in-plane forces N_0 reduces to $(m_0 = \rho h)$

$$\omega_{mn} = \frac{\pi^2}{b^2} \sqrt{\frac{D}{\rho h}} \left(m^2 s^2 + n^2\right), \quad s = \frac{b}{a}. \quad (7.2.140)$$

For a square isotropic plate without rotary inertia, the frequency is given by

$$\omega_{mn} = \frac{\pi^2}{a^2} \sqrt{\frac{D}{\rho h}} \left(m^2 + n^2\right). \quad (7.2.141)$$

The fundamental frequency of a square isotropic plate is given by

$$\omega_{11} = \frac{2\pi^2}{a^2} \sqrt{\frac{D}{\rho h}}. \quad (7.2.142)$$

For $\nu = 0.3$ the value is $(D = Eh^3/[12(1 - \nu^2)])$

$$\omega_{11} = 5.973 \frac{h}{a^2} \sqrt{\frac{E}{\rho}}.$$

Table 7.2.5 contains nondimensionalized frequencies $\bar{\omega}_{mn}$ of isotropic ($\nu = 0.25$) plates for modes $(m, n) = (1, 1)$, $(1, 2)$, and $(2, 1)$. Results are presented for various values of the plate aspect ratio (a/b) and side-to-thickness ratio (h/b). The effect of rotary inertia is negligible for plates with $h/b < 0.1$.

Table 7.2.5 Nondimensionalized natural frequencies $\bar{\omega}_{mn}$ of simply supported isotropic plates[a] $[\bar{\omega}_{mn} = \omega_{mn}(b^2/\pi^2)\sqrt{\rho h/D}]$.

	$\bar{\omega}_{11}$			$\bar{\omega}_{12}$			$\bar{\omega}_{21}$		
b/a	w/o	0.01	0.1	w/o	0.01	0.1	w/o	0.01	0.1
0.5	1.250	1.250	1.249	4.250	4.249	4.243	2.000	2.000	1.998
1.0	2.000	2.000	1.998	5.000	4.999	4.990	5.000	4.999	4.990
1.5	3.250	3.250	3.246	6.250	6.248	6.234	10.000	9.996	9.959
2.0	5.000	4.999	4.990	8.000	7.997	7.974	17.000	16.988	16.882
2.5	7.250	7.248	7.228	10.250	10.246	10.207	26.000	25.972	25.726
3.0	10.000	9.996	9.959	13.000	12.993	12.931	37.000	36.944	36.450

[a] w/o = without rotary inertia; the second and third columns contain frequencies when the rotary inertia is included for $h/b = 0.01$ and 0.1, respectively.

7.2.4.3 Buckling Analysis

For buckling of rectangular plates under in-plane biaxial compressive loads (see Fig. 7.2.12), Eq. (7.2.117) reduces, after omitting the time derivative and load terms, to

$$\left[D\left(\alpha_m^2 + \beta_n^2\right)^2 - (\alpha_m^2 + \gamma\beta_n^2)N_0\right]W_{mn} = 0, \tag{7.2.143}$$

which gives, for nonzero W_{mn}, the result

$$N_0(m,n) = \frac{\pi^2 D}{b^2}\frac{\left(s^2 m^2 + n^2\right)^2}{s^2 m^2 + \gamma n^2}, \tag{7.2.144}$$

where s is the plate aspect ratio $s = (b/a)$. For each choice of m and n, there corresponds a unique value of N_0. The *critical buckling load* is the smallest of $N_0(m,n)$. For a given plate this value is dictated by a particular combination of the values of m and n, value of γ, plate geometry, and material properties. Next we present critical buckling loads for various cases.

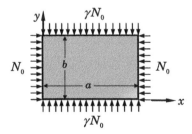

Fig. 7.2.12 Biaxial compression of a rectangular plate ($\hat{N}_{xx} = N_0$ and $\hat{N}_{yy} = \gamma N_0$).

Biaxial compression of a plate

For a rectangular isotropic plate under biaxial compression ($\gamma = 1$), the buckling load is

$$N_0(m,n) = \frac{\pi^2 D}{b^2}\left(m^2 s^2 + n^2\right), \qquad (7.2.145)$$

where s is the plate aspect ratio $s = b/a$. Clearly, the critical buckling load occurs at $m = n = 1$, and it is equal to

$$N_{cr} = (1 + s^2)\frac{\pi^2 D}{b^2}, \qquad s = \frac{b}{a}, \qquad (7.2.146)$$

and for a square plate it reduces to

$$N_{cr} = \frac{2\pi^2 D}{b^2}. \qquad (7.2.147)$$

Biaxial loading of a plate

When the edges $x = 0, a$ are subjected to compressive load $\hat{N}_{xx} = N_0$ and the edges $y = 0, b$ are subjected to tensile load $\hat{N}_{yy} = -\gamma N_0$, Eq. (7.2.144) becomes

$$N_0(m,n) = \frac{\pi^2 D}{b^2}\left[\frac{(s^2 m^2 + n^2)^2}{s^2 m^2 - \gamma n^2}\right], \qquad (7.2.148)$$

when $\gamma n^2 < s^2 m^2$. The minimum buckling load occurs for $n = 1$:

$$N_0(m,1) = \left(\frac{\pi^2 D}{b^2}\right)\frac{(m^2 s^2 + 1)^2}{m^2 s^2 - \gamma}. \qquad (7.2.149)$$

In theory, the minimum of $N_0(m,1)$ occurs when $m^2 s^2 = 1 + 2\gamma$. For a square plate with $\gamma = 0.5$, we find

$$N_0(1,1) = \frac{8\pi^2 D}{a^2}, \qquad N_0(2,1) = 7.1429\frac{\pi^2 D}{a^2} = N_{cr}. \qquad (7.2.150)$$

Uniaxial compression of a plate

When a rectangular plate is subjected to uniform compressive load N_0 on edges $x = 0$ and $x = a$, that is, when $\gamma = 0$, the buckling load is given by

$$N_0(m,n) = \frac{\pi^2 a^2 D}{m^2}\left(\frac{m^2}{a^2} + \frac{n^2}{b^2}\right)^2, \qquad (7.2.151)$$

$$N_0(m,1) = \frac{\pi^2 D}{a^2}\left(m + \frac{1}{m}\frac{a^2}{b^2}\right)^2. \qquad (7.2.152)$$

For a given aspect ratio, two different modes, m_1 and m_2, will have the same buckling load when $\sqrt{m_1 m_2} = a/b$. In particular, the point of intersection of curves m and $m + 1$ occurs for aspect ratios:

$$\frac{a}{b} = \sqrt{2}, \ \sqrt{6}, \ \sqrt{12}, \ \sqrt{20}, \ \ldots, \ \sqrt{m^2 + m}.$$

Thus, there is a mode change at these aspect ratios from m half-waves to $m+1$ half-waves. Putting $m = 1$ in Eq. (7.2.152), we find

$$N_{cr} = \frac{\pi^2 D}{b^2} \left(\frac{a}{b} + \frac{b}{a} \right)^2. \tag{7.2.153}$$

For a square plate we obtain

$$N_{cr} = \frac{4\pi^2 D}{b^2}. \tag{7.2.154}$$

Table 7.2.6 shows the effect of plate aspect ratio and modulus ratio (orthotropy) on the critical buckling loads $\bar{N} = N_{cr} b^2 / (\pi^2 D)$ of rectangular isotropic ($\nu = 0.3$) plates under uniform axial compression ($\gamma = 0$) and biaxial compression ($\gamma = 1$). In all cases, the critical buckling mode is $(m, n) = (1, 1)$, except as indicated.

Table 7.2.6 Effect of plate aspect ratio on the nondimensionalized buckling loads $\bar{N}$ of simply supported (SSSS) rectangular plates under uniform axial compression ($\gamma = 0$) and biaxial compression ($\gamma = 1$). The super script $(\cdot, \cdot)$ denotes the mode numbers.

	$\gamma = 0$				$\gamma = 1$		
$\frac{a}{b}$	$\bar{N}$	$\frac{a}{b}$	$\bar{N}$	$\frac{a}{b}$	$\bar{N}$	$\frac{a}{b}$	$\bar{N}$
0.5	6.250	2.0	$4.000^{(2,1)}$	0.5	5.000	2.0	1.250
1.0	4.000	2.5	$4.134^{(3,1)}$	1.0	2.000	2.5	1.160
1.5	$4.340^{(2,1)}$	3.0	$4.000^{(3,1)}$	1.5	1.444	3.0	1.111

7.2.4.4 Transient Analysis

The determination of the solution $w(x, y, t)$ of Eq. (7.2.109) or (7.2.110) for all times $t > 0$ under an applied load $q(x, y, t)$ and known initial conditions

$$w(x, y, 0) = d_0(x, y), \quad \frac{\partial w}{\partial t}(x, y, 0) = v_0(x, y) \quad \text{for all } x \text{ and } y, \tag{7.2.155}$$

where d_0 and v_0 are the initial displacement and velocity, respectively, is termed the *transient response*. The transient response of simply supported rectangular plates can be determined by assuming solution of the form in Eq. (7.2.112) and determining the solution $W_{mn}(t)$ of the ordinary differential equation in

Eq. (7.2.117). It is necessary to expand the non-zero initial displacement and velocity field in double sine series:

$$d_0(x,y) = \sum_{n=1}^{\infty}\sum_{m=1}^{\infty} D_{mn} \sin\alpha_m x \sin\beta_n y, \qquad (7.2.156)$$

$$v_0(x,y) = \sum_{n=1}^{\infty}\sum_{m=1}^{\infty} V_{mn} \sin\alpha_m x \sin\beta_n y, \qquad (7.2.157)$$

where $\alpha_m = m\pi/a$, $\beta_n = n\pi/b$, and D_{mn} and V_{mn} are given by

$$D_{mn} = \frac{4}{ab}\int_0^b\int_0^a d(x,y)\sin\alpha_m x \sin\beta_n y\, dxdy, \qquad (7.2.158)$$

$$V_{mn} = \frac{4}{ab}\int_0^b\int_0^a v(x,y)\sin\alpha_m x \sin\beta_n y\, dxdy. \qquad (7.2.159)$$

Equation (7.2.117) has the general form

$$K_{mn}W_{mn}(t) + M_{mn}\frac{d^2W_{mn}}{dt^2} = q_{mn}(t), \qquad (7.2.160)$$

where

$$K_{mn} = D\left(\alpha_m^2 + \beta_n^2\right)^2 - N_0\left(\alpha_m^2 + \gamma\beta_n^2\right), \quad M_{mn} = m_0 + m_2\left(\alpha_m^2 + \beta_n^2\right). \qquad (7.2.161)$$

The second-order differential equation in Eq. (7.2.159) can be solved either exactly or numerically. The numerical solutions are often based on the finite difference methods.

To solve Eq. (7.2.160) exactly, we first write it in the form

$$\frac{d^2W_{mn}}{dt^2} + \left(\frac{K_{mn}}{M_{mn}}\right)W_{mn} = \frac{1}{M_{mn}}q_{mn}(t) \equiv \hat{q}_{mn}(t). \qquad (7.2.162)$$

The solution of Eq. (7.2.162) is given by

$$W_{mn}(t) = C_1 e^{\lambda_1 t} + C_2 e^{\lambda_2 t} + W_{mn}^p(t), \qquad (7.2.163)$$

where C_1 and C_2 are constants to be determined using the initial conditions, $W_{mn}^p(t)$ is the particular solution

$$W_{mn}^p(t) = \int_0^t \frac{r_1(\tau)r_2(t) - r_1(t)r_2(\tau)}{r_1(\tau)\dot{r}_2(\tau) - \dot{r}_1(\tau)r_2(\tau)}\,\hat{q}_{mn}(\tau)\,d\tau, \qquad (7.2.164)$$

with $r_1(t) = e^{\lambda_1 t}$ and $r_2(t) = e^{\lambda_2 t}$, and λ_1 and λ_2 are the roots of the equation

$$\lambda^2 + \frac{K_{mn}}{M_{mn}} = 0; \quad \lambda_1 = -i\mu, \ \lambda_2 = i\mu, \ i = \sqrt{-1}, \ \mu = \sqrt{\frac{K_{mn}}{M_{mn}}}. \qquad (7.2.165)$$

The solution becomes

$$W_{mn}(t) = A\cos\mu t + B\sin\mu t + W_{mn}^{p}(t), \tag{7.2.166}$$

$$W_{mn}^{p}(t) = \frac{1}{2i\mu}\left(e^{i\mu t}\int_{0}^{t}e^{-i\mu\tau}\hat{q}_{mn}(\tau)d\tau - e^{-i\mu t}\int_{0}^{t}e^{i\mu\tau}\hat{q}_{mn}(\tau)d\tau\right). \tag{7.2.167}$$

Once the load distribution is known, the solution can be determined from Eq. (7.2.163) or (7.2.166).

For a step loading, $q_{mn}(t) = q_{mn}^{0}H(t)$, where $H(t)$ denotes the Heaviside step function, Eq. (7.2.166) takes the form

$$W_{mn}(t) = A_{mn}\cos\mu t + B_{mn}\sin\mu t + \frac{1}{K_{mn}}q_{mn}^{0}. \tag{7.2.168}$$

Using the initial conditions in Eq. (7.2.155), we obtain

$$A_{mn} = D_{mn} - \frac{1}{K_{mn}}q_{mn}^{0}, \quad B_{mn} = \frac{V_{mn}}{\mu}. \tag{7.2.169}$$

Thus the final solution is given by

$$w(x,y,t) = \sum_{n=1}^{\infty}\sum_{m=1}^{\infty}\left[D_{mn}\cos\mu t + \frac{V_{mn}}{\mu}\sin\mu t + \frac{q_{mn}^{0}}{K_{mn}}(1-\cos\mu t)\right]$$
$$\times \sin\alpha_{m}x \sin\beta_{n}y. \tag{7.2.170}$$

The coefficients q_{mn}^{0} were given in Table 7.2.3 for various types of load distributions. The same holds for D_{mn} and V_{mn}.

The exact solution of the differential equation (7.2.162) can also be obtained using the Laplace transform method.

7.2.5 Lévy Solutions of Rectangular Plates

The deflection of a rectangular plate with two opposite edges $(x = 0, a)$ simply supported and the other two edges $(-b/2 \le y \le b/2)$ having arbitrary boundary conditions (see Figure 7.2.13) can be represented in the form (see the semidiscretization method of **Example 6.4.16** and **Example 7.2.14**)

$$w(x,y) = \sum_{m=1}^{\infty} W_{m}(y)\sin\alpha_{m}x, \tag{7.2.171}$$

which satisfies the simply supported boundary conditions on edges $x = 0$ and $x = a$. Substituting the expansion (7.2.171) into the equilibrium equation of the plate

$$D\left(\frac{\partial^{4}w}{\partial x^{4}} + 2\frac{\partial^{4}w}{\partial x^{2}\partial y^{2}} + \frac{\partial^{4}w}{\partial y^{4}}\right) = q, \tag{7.2.172}$$

we obtain an ordinary differential equation for $W_m(y)$:

$$D\left(\alpha_m^4 W_m - 2\alpha_m^2 \frac{d^2 W_m}{dy^2} + \frac{d^4 W_m}{dy^4}\right) = q_m, \qquad (7.2.173)$$

where q_m is defined by

$$q_m(y) = \frac{2}{a}\int_0^a q(x,y)\,\sin\alpha_m x\,dx, \quad \alpha_m = \frac{m\pi}{a}. \qquad (7.2.174)$$

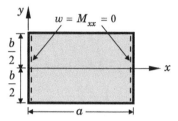

Fig. 7.2.13 Geometry and coordinate system for a rectangular plate with sides $x = 0, a$ simply supported.

Table 7.2.7 contains the coefficients q_m for various types of loads. The solution of Eq. (7.2.173) may be determined exactly or numerically. Here we discuss its exact solution. One may also solve it using the Ritz method (see Reddy [51]).

The solution W_m of Eq. (7.2.173) consists of the homogeneous solution W_m^h and particular solution W_m^p. The homogeneous solution is given by

$$W_m^h(y) = (A_m + B_m y)\cosh\alpha_m y + (C_m + D_m y)\sinh\alpha_m y. \qquad (7.2.175)$$

The particular solution W_m^p is given by the solution of

$$\alpha_m^4 W_m^p - 2\alpha_m^2 \frac{d^2 W_m^p}{dy^2} + \frac{d^4 W_m^p}{dy^4} = \frac{q_m}{D}. \qquad (7.2.176)$$

Assuming solution of Eq. (7.2.176) in the form

$$W_m^p(y) = \sum_{n=1}^{\infty} W_{mn}\sin\beta_n y, \quad \beta_n = \frac{n\pi}{b}, \qquad (7.2.177)$$

and expanding the load q_m also in the same form

$$q_m(y) = \sum_{n=1}^{\infty} q_{mn}\sin\beta_n y, \qquad (7.2.178)$$

Table 7.2.7 Coefficients in the single sine series expansion of loads in Lévy's method.

Load type	Coefficients q_m
1. Uniform load, $q(x,y) = q_0$	$q_m = \frac{4q_0}{m\pi}$ $(m = 1, 3, 5, \ldots)$

2. Hydrostatic load, $q(x,y) = q_0 \frac{x}{a}$	$q_m = \frac{2q_0}{m\pi}(-1)^{m+1}$ $(m = 1, 2, 3, \ldots)$

3. Point load, Q_0 at (x_0, y_0)	$q_m = \frac{2Q_0}{a} \delta(y - y_0) \sin \frac{m\pi x_0}{a}$ $(m = 1, 2, 3, \ldots)$

4. Line load, $q(x,y) = q_0\, \delta(x - x_0)$	$q_m = \frac{2q_0}{a} \sin \frac{m\pi x_0}{a}$ $(m = 1, 2, 3, \ldots)$

and substituting both into Eq. (7.2.176), we obtain

$$W_{mn} = \frac{q_{mn}}{d_{mn}}, \quad d_{mn} = D\left(\alpha_m^2 + \beta_n^2\right)^2, \quad \alpha_m = \frac{m\pi}{a}, \quad \beta_n = \frac{n\pi}{b}. \tag{7.2.179}$$

The particular solution becomes

$$W_m^p(y) = \sum_{n=1}^{\infty} \frac{q_{mn}}{d_{mn}} \sin \beta_n y, \tag{7.2.180}$$

and the complete solution becomes

$$w(x,y) = \sum_{m=1}^{\infty} \left[(A_m + B_m y)\cosh\alpha_m y + (C_m + D_m y)\sinh\alpha_m y + W_m^p \right] \sin\alpha_m x,$$

(7.2.181)

where W_m^p is given by Eq. (7.2.180). The constants A_m, B_m, C_m, and D_m must be determined for the particular set of boundary conditions on edges $y = \pm b/2$.

Example 7.2.10

Consider a rectangular plate simply supported on all four edges and subjected to distributed bending moments along edges $y = \pm b/2$, as shown in Fig. 7.2.14:

$$f(x) = \sum_{m=1}^{\infty} M_m \sin\alpha_m x, \quad \alpha_m = \frac{m\pi}{a},$$

(1)

with the coefficients M_m known from

$$M_m = \frac{2}{a}\int_0^a f(x)\sin\alpha_m x \, dx.$$

(2)

Determine the deflections and bending moments throughout the plate using the Lévy solution procedure.

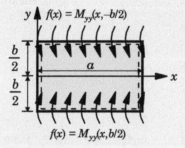

Fig. 7.2.14 A simply supported rectangular plate with distributed moments along edges $y = \pm b/2$.

Solution: The general solution in Eq. (7.2.181) for rectangular plates with simply supported edges $x = 0, a$ is valid here except that the distributed load is zero: $q = 0$. Therefore, the particular solution is zero.

The boundary conditions on edges $y = \pm b/2$ are

$$w = 0, \quad M_{yy} = f(x).$$

(3)

Since $w = 0$ on $y = \pm b/2$, the condition $M_{yy} = f(x)$ on $y = \pm b/2$ reduces to

$$-D\frac{\partial^2 w}{\partial y^2} = f(x).$$

(4)

Due to the symmetry of the problem about the x axis, $w(x,y)$ must be an even function of y. This implies that $B_m = C_m = 0$, and we have

$$w(x,y) = \sum_{m=1}^{\infty}(A_m\cosh\alpha_m y + D_m y\sinh\alpha_m y)\sin\alpha_m x.$$

(5)

Using the boundary conditions in Eq. (3), we obtain

$$A_m = -\frac{b}{2}D_m \tanh \hat{\alpha}_m, \quad D_m = -\frac{aM_m}{2m\pi D \cosh \hat{\alpha}_m}, \quad \hat{\alpha}_m = \alpha_m \frac{b}{2}. \tag{6}$$

The deflection becomes

$$w(x,y) = \frac{a}{2\pi D} \sum_{m=1}^{\infty} M_m \frac{\sin \alpha_m x}{m \cosh \hat{\alpha}_m} \left(\frac{b}{2} \tanh \hat{\alpha}_m \cosh \alpha_m y - y \sinh \alpha_m y \right). \tag{7}$$

The distribution of bending moments for the problem can be computed using the definitions in Eq. (7.2.87).

Dimensionless deflections and bending moments of simply supported rectangular plates with uniformly distributed moments of intensity $f = M_0$ (or $M_m = 4M_0/m\pi$, $m = 1, 3, 5, \ldots$) are presented in Table 7.2.8 for various aspect ratios. The results were computed using $m = 1, 3, \ldots, 9$. The dimensionless quantities used are:

$$\text{For } \frac{b}{a} < 1: \quad \bar{w} = w(a/2, 0)\frac{D}{M_0 b^2}, \quad \bar{M} = M(a/2, 0)\frac{1}{M_0},$$

$$\text{For } \frac{b}{a} \geq 1: \quad \bar{w} = w(a/2, 0)\frac{D}{M_0 a^2}, \quad \bar{M} = M(a/2, 0)\frac{1}{M_0}. \tag{8}$$

Table 7.2.8 Nondimensionalized center deflections and bending moments of simply supported plates with applied edge moment along $y = \pm b/2$.

Variable	a/b					
	0.5	0.75	1.0	1.5	2.0	3.0
$\bar{w}$	0.0965	0.0620	0.0368	0.0280	0.0174	0.0055
$\bar{M}_{xx}$	0.3873	0.4240	0.3938	0.2635	0.1530	0.0446
$\bar{M}_{yy}$	0.7701	0.4764	0.2562	0.0465	-0.0103	-0.0148

Example 7.2.11

Consider a rectangular plate with edges $x = 0, a$ simply supported and edges $y = \pm b/2$ clamped, and subjected to distributed transverse load, as shown in Fig. 7.2.15. Determine the Lévy solution.

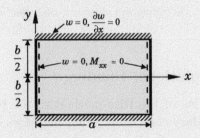

Fig. 7.2.15 Geometry and coordinate system for a rectangular plate with sides $x = 0, a$ simply supported and $y = \pm b/2$ clamped.

Solution: The clamped boundary conditions on edges $y = \pm b/2$ are

$$w = 0, \quad \frac{\partial w}{\partial y} = 0. \tag{1}$$

The general solution is given by

$$w(x,y) = \sum_{m=1}^{\infty} \left(A_m \cosh \alpha_m y + D_m y \sinh \alpha_m y + \sum_{n=1}^{\infty} W_{mn} \sin \beta_n y \right) \sin \alpha_m x, \tag{2}$$

and its derivative is

$$\frac{\partial w}{\partial y} = \sum_{m=1}^{\infty} \Big(\alpha_m A_m \sinh \alpha_m y + D_m \sinh \alpha_m y + D_m \alpha_m y \cosh \alpha_m y$$

$$+ \sum_{n=1}^{\infty} \beta_n W_{mn} \cos \beta_n y \Big) \sin \alpha_m x, \tag{3}$$

where $\alpha_m = m\pi/a$ and $\beta_n = n\pi/b$. Boundary conditions in Eq. (1) give

$$A_m = -\frac{b}{2} D_m \tanh \hat{\alpha}_m - Q_m, \quad \hat{\alpha}_m = \frac{b}{2} \alpha_m,$$

$$D_m = -\frac{a M_m}{2 m \pi D \cosh \hat{\alpha}_m},$$

$$Q_m = \frac{\hat{q}_m}{\cosh \hat{\alpha}_m}, \quad \hat{q}_m = \sum_{n=1}^{\infty} (-1)^n W_{mn}, \tag{4}$$

$$M_m = \hat{\alpha}_m + (1 - \hat{\alpha}_m \tanh \hat{\alpha}_m) \tanh \hat{\alpha}_m,$$

where W_{mn} is defined by

$$W_{mn} = \frac{q_{mn}}{D \left(\alpha_m^2 + \beta_n^2 \right)^2} \tag{5}$$

and q_{mn} are defined in Table 7.2.7 for various types of loads.

Table 7.2.9 contains dimensionless deflections and bending moments:

$$\bar{w} = (wD/q_0 a^4) \times 10^2, \quad \bar{M} = (M/q_0 a^2) \times 10, \tag{6}$$

$$\bar{w} = (wD/q_0 b^4) \times 10^2, \quad \bar{M} = (M/q_0 b^2) \times 10, \tag{7}$$

for various aspect ratios. The dimensionless quantities in Eq. (6) are used for $b \geq a$ (the first three columns of values), and the dimensionless variables in Eq. (7) are used for $a > b$ (the last three columns of values). The values are obtained using the first five terms of the series ($m = 1, 3, \ldots, 9$).

7.2.6   Variational Solutions: Bending

Here we discuss application of the Ritz method to the bending of rectangular plates with various boundary conditions along the edges of the plate. The Ritz approximation functions are selected as products of one-dimensional functions in each coordinate directions that satisfy the required boundary conditions.

Table 7.2.9 Maximum dimensionless deflections and bending moments in isotropic ($\nu = 0.3$) rectangular plates with edges $x = 0, a$ simply supported and edges $y = \pm b/2$ clamped and subjected to distributed loads.

Variable	$\frac{b}{a} = 3$	$\frac{b}{a} = 2$	$\frac{b}{a} = 1$	$\frac{a}{b} = \frac{3}{2}$	$\frac{a}{b} = 2$	$\frac{a}{b} = 3$
Uniform load, $q = q_0$						
$\bar{w}(a/2, 0)$	1.1681	0.8445	0.1917	0.2476	0.2612	0.2619
$\bar{M}_{xx}(a/2, 0)$	1.1442	0.8693	0.2445	0.1794	0.1441	0.1302
$\bar{M}_{yy}(a/2, 0)$	0.4214	0.4738	0.3326	0.4067	0.4214	0.4201
$-\bar{M}_{xx}(a/2, b/2)$	0.3740	0.3574	0.2097	0.2470	0.2535	0.2525
$-\bar{M}_{yy}(a/2, b/2)$	1.2467	1.1915	0.6990	0.8233	0.8451	0.8417
Hydrostatic load, $q = q_0(x/a)$						
$\bar{w}(a/2, 0)$	0.5841	0.4222	0.0959	0.1238	0.1306	0.1309
$\bar{M}_{xx}(a/2, 0)$	0.5721	0.4346	0.1222	0.0897	0.0721	0.0651
$\bar{M}_{yy}(a/2, 0)$	0.2107	0.2369	0.1663	0.2033	0.2107	0.2100
$-\bar{M}_{xx}(a/2, b/2)$	0.1870	0.1787	0.1048	0.1235	0.1267	0.1263
$-\bar{M}_{yy}(a/2, b/2)$	0.6233	0.5957	0.3495	0.4117	0.4225	0.4209
$\bar{w}(3a/4, 0)$	0.4368	0.3218	0.0841	0.1313	0.1614	0.1884
$\bar{M}_{xx}(3a/4, b/2)$	0.5085	0.4097	0.1640	0.1665	0.1509	0.1153
$\bar{M}_{yy}(3a/4, b/2)$	0.1803	0.1997	0.1562	0.2276	0.2705	0.3055
$-\bar{M}_{xx}(a/2, 3b/4)$	0.1635	0.1576	0.1042	0.1443	0.1670	0.1842
$-\bar{M}_{yy}(a/2, 3b/4)$	0.5449	0.5254	0.3475	0.4810	0.5566	0.6140

We begin with the virtual work statement for an orthotropic rectangular plate:

$$
0 = \int_{\Omega} \left[D_{11} \frac{\partial^2 w}{\partial x^2} \frac{\partial^2 \delta w}{\partial x^2} + D_{12} \left(\frac{\partial^2 w}{\partial y^2} \frac{\partial^2 \delta w}{\partial x^2} + \frac{\partial^2 w}{\partial x^2} \frac{\partial^2 \delta w}{\partial y^2} \right) \right.
$$
$$
\left. + 4 D_{66} \frac{\partial^2 w}{\partial x \partial y} \frac{\partial^2 \delta w}{\partial x \partial y} + D_{22} \frac{\partial^2 w}{\partial y^2} \frac{\partial^2 \delta w}{\partial y^2} - q \delta w \right] dx dy
$$
$$
- \oint_{\Gamma} \left(-\hat{M}_{nn} \frac{\partial \delta w}{\partial n} + \hat{V}_n \delta w \right) dS, \tag{7.2.182}
$$

where $\hat{M}_{nn}$ and $\hat{V}_n$ are applied edge moment and effective shear force, respectively, on the boundary Γ of the domain Ω (the midplane of the plate).

We seek an N-parameter Ritz solution in the form

$$
w(x, y) \approx \sum_{j=1}^{N} c_j \phi_j(x, y) \tag{7.2.183}
$$

and substitute it into Eq. (7.2.182) to obtain

$$
\mathbf{Rc} = \mathbf{F}, \tag{7.2.184}
$$

where

$$R_{ij} = \int_{\Omega} \left[D_{11} \frac{\partial^2 \phi_i}{\partial x^2} \frac{\partial^2 \phi_j}{\partial x^2} + D_{12} \left(\frac{\partial^2 \phi_i}{\partial y^2} \frac{\partial^2 \phi_j}{\partial x^2} + \frac{\partial^2 \phi_i}{\partial x^2} \frac{\partial^2 \phi_j}{\partial y^2} \right) \right.$$
$$\left. + 4D_{66} \frac{\partial^2 \phi_i}{\partial x \partial y} \frac{\partial^2 \phi_j}{\partial x \partial y} + D_{22} \frac{\partial^2 \phi_i}{\partial y^2} \frac{\partial^2 \phi_j}{\partial y^2} \right] dx dy, \quad (7.2.185)$$

$$F_i = \int_{\Omega} q\phi_i \, dx dy + \oint_{\Gamma} \left(\hat{V}_n \phi_i - \hat{M}_{nn} \frac{\partial \phi_i}{\partial n} \right) dS. \quad (7.2.186)$$

For rectangular plates, it is convenient to express the Ritz approximation in the form

$$w(x, y) \approx W_N(x, y) = \sum_{i=1}^{N} \sum_{j=1}^{N} c_{ij} \, \phi_{ij}(x, y) = \sum_{i=1}^{N} \sum_{j=1}^{N} c_{ij} X_i(x) Y_j(y), \quad (7.2.187)$$

where $\phi_{ij}(x, y)$ is expressed as a tensor product of the one-dimensional functions X_i and Y_j of x and y, respectively. Substituting Eq. (7.2.187) into Eq. (7.2.182) ($\hat{M}_{nn} = 0$ and $\hat{V}_n = 0$ on Γ), we obtain

$$0 = \sum_{i=1}^{N} \sum_{j=1}^{N} \left\{ \int_{0}^{b} \int_{0}^{a} \left[D_{11} \frac{d^2 X_i}{dx^2} Y_j \frac{d^2 X_p}{dx^2} Y_q + 4D_{66} \frac{dX_i}{dx} \frac{dY_j}{dy} \frac{dX_p}{dx} \frac{dY_q}{dy} \right. \right.$$
$$+ D_{12} \left(X_i \frac{d^2 Y_j}{dy^2} \frac{d^2 X_p}{dx^2} Y_q + \frac{d^2 X_i}{dx^2} Y_j X_p \frac{d^2 Y_q}{dy^2} \right)$$
$$\left. \left. + D_{22} X_i \frac{d^2 Y_j}{dy^2} X_p \frac{d^2 Y_q}{dy^2} \right] dx dy \right\} c_{ij} - \int_{0}^{b} \int_{0}^{a} q X_p Y_q \, dx dy$$

$$\equiv R_{(pq)(ij)} c_{ij} - F_{pq}, \quad (7.2.188)$$

where

$$R_{(pq)(ij)} = \int_{0}^{b} \int_{0}^{a} \left[D_{11} \frac{d^2 X_i}{dx^2} Y_j \frac{d^2 X_p}{dx^2} Y_q + 4D_{66} \frac{dX_i}{dx} \frac{dY_j}{dy} \frac{dX_p}{dx} \frac{dY_q}{dy} \right.$$
$$+ D_{12} \left(X_i \frac{d^2 Y_j}{dy^2} \frac{d^2 X_p}{dx^2} Y_q + \frac{d^2 X_i}{dx^2} Y_j X_p \frac{d^2 Y_q}{dy^2} \right)$$
$$\left. + D_{22} X_i \frac{d^2 Y_j}{dy^2} X_p \frac{d^2 Y_q}{dy^2} \right] dx dy, \quad (7.2.189)$$

$$F_{pq} = \int_{0}^{b} \int_{0}^{a} q X_p Y_q \, dx dy. \quad (7.2.190)$$

One choice for X_i and Y_j is to use algebraic polynomials. A second choice is provided by the characteristic equations of beams. Both sets of functions are given here for typical boundary conditions (see Fig. 7.2.16). The notation CFSF, for example, means that edge $x = 0$ is clamped (C), edge $x = a$ is free

(F), edge $y = 0$ is simply supported (S), and edge $y = b$ is free (F). In the case of characteristic polynomials, the roots λ_i are to be determined by solving a transcendental equation. The first few values of $\lambda_i a$ are given in Table 7.2.10.

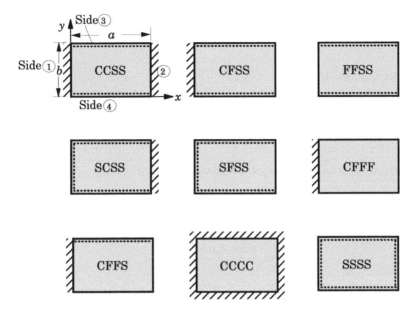

Fig. 7.2.16 Rectangular plates with various boundary conditions.

Table 7.2.10 Values of the constants and eigenvalues for natural vibration of plate strips with various boundary conditions ($\lambda_n^4 \equiv \omega_n^2 m_0/D = (e_n/a)^4$). The classical plate theory without rotary inertia is used.

End conditions at $x = 0$ and $x = a$	Characteristic equation and values of $e_n \equiv \lambda_n a$
• Hinged–hinged (S-S)	$\sin e_n = 0$, $e_n = n\pi$
• Fixed–fixed (C-C)	$\cos e_n \cosh e_n - 1 = 0$ $e_n = 4.730, 7.853, \ldots$
• Fixed–free (C-F)	$\cos e_n \cosh e_n + 1 = 0$ $e_n = 1.875, 4.694, \ldots$
• Free–free (F-F)	$\cos e_n \cosh e_n - 1 = 0$ $e_n = 4.730, 7.853, \ldots$
• Hinged–fixed (S-C)	$\tan e_n = \tanh e_n$ $e_n = 3.927, 7.069, \ldots$
• Hinged–free (S-F)	$\tan e_n = \tanh e_n$ $e_n = 3.927, 7.069, \ldots$

CCSS Plates:

Algebraic polynomials

$$X_i(x) = \left(\frac{x}{a}\right)^{i+1} - 2\left(\frac{x}{a}\right)^{i+2} + \left(\frac{x}{a}\right)^{i+3},$$
$$Y_j(x) = \left(\frac{y}{b}\right)^{j} - \left(\frac{y}{b}\right)^{j+1}. \tag{7.2.191}$$

Characteristic polynomials

$$X_i(x) = \sin \lambda_i x - \sinh \lambda_i x + \alpha_i \left(\cosh \lambda_i x - \cos \lambda_i x\right),$$
$$Y_j(y) = \sin \frac{n\pi y}{b}, \tag{7.2.192}$$

where λ_i are the roots of the characteristic equation

$$\cos \lambda_i a \ \cosh \lambda_i a - 1 = 0, \tag{7.2.193}$$

and α_i are defined by

$$\alpha_i = \frac{\sinh \lambda_i a - \sin \lambda_i a}{\cosh \lambda_i a - \cos \lambda_i a}. \tag{7.2.194}$$

CFSS Plates:

Algebraic polynomials

$$X_i(x) = \left(\frac{x}{a}\right)^{i+1}, \quad Y_j(x) = \left(\frac{y}{b}\right)^{j} - \left(\frac{y}{b}\right)^{j+1}. \tag{7.2.195}$$

Characteristic polynomials

$$X_i(x) = \sin \lambda_i x - \sinh \lambda_i x + \alpha_i \left(\cosh \lambda_i x - \cos \lambda_i x\right),$$
$$Y_j(y) = \sin \frac{n\pi y}{b}, \tag{7.2.196}$$

$$\cos \lambda_i a \ \cosh \lambda_i a + 1 = 0, \quad \alpha_i = \frac{\sinh \lambda_i a + \sin \lambda_i a}{\cosh \lambda_i a + \cos \lambda_i a}. \tag{7.2.197}$$

FFSS Plates:

Algebraic polynomials $(N > 1)$

$$X_i(x) = \left(\frac{x}{a}\right)^{i-1}, \quad Y_j(x) = \left(\frac{y}{b}\right)^{j} - \left(\frac{y}{b}\right)^{j+1}. \tag{7.2.198}$$

Characteristic polynomials

$$X_i(x) = \sin \lambda_i x + \sinh \lambda_i x - \alpha_i \left(\cosh \lambda_i x + \cos \lambda_i x\right),$$
$$Y_j(y) = \sin \frac{n\pi y}{b}, \tag{7.2.199}$$

$$\cos \lambda_i a \, \cosh \lambda_i a - 1 = 0, \quad \alpha_i = \frac{\sinh \lambda_i a - \sin \lambda_i a}{\cosh \lambda_i a - \cos \lambda_i a}. \tag{7.2.200}$$

SCSS Plates:

Algebraic polynomials

$$X_i(x) = \left(\frac{x}{a}\right)\left[1 - \left(\frac{x}{a}\right)\right]^{i+1}, \quad Y_j(x) = \left(\frac{y}{b}\right)^j - \left(\frac{y}{b}\right)^{j+1}. \tag{7.2.201}$$

Characteristic polynomials

$$X_i(x) = \sinh \lambda_i a \sin \lambda_i x + \sin \lambda_i a \sinh \lambda_i x,$$
$$Y_j(y) = \sin \frac{n\pi y}{b}, \tag{7.2.202}$$
$$\tan \lambda_i a - \tanh \lambda_i a = 0.$$

SFSS Plates:

Algebraic polynomials

$$X_i(x) = \left(\frac{x}{a}\right)^i, \quad Y_j(x) = \left(\frac{y}{b}\right)^j - \left(\frac{y}{b}\right)^{j+1}. \tag{7.2.203}$$

Characteristic polynomials

$$X_i(x) = \sinh \lambda_i a \, \sin \lambda_i x - \sin \lambda_i a \, \sinh \lambda_i x,$$
$$Y_j(y) = \sin \frac{n\pi y}{b}, \tag{7.2.204}$$
$$\tan \lambda_i a - \tanh \lambda_i a = 0.$$

CFFF Plates: $(N > 1)$
Algebraic polynomials

$$X_i(x) = \left(\frac{x}{a}\right)^{i+1}, \quad Y_j(y) = \left(\frac{y}{b}\right)^{j-1}. \tag{7.2.205}$$

Characteristic polynomials

$$X_i(x) = \sin \lambda_i x - \sinh \lambda_i x + \alpha_i \left(\cosh \lambda_i x - \cos \lambda_i x\right),$$
$$Y_j(y) = \sin \mu_j y + \sinh \mu_j y - \beta_i \left(\cosh \mu_j y + \cos \mu_j y\right), \tag{7.2.206}$$

$$\cos \lambda_i a \, \cosh \lambda_i a + 1 = 0, \quad \cos \mu_j b \cosh \mu_j b - 1 = 0, \tag{7.2.207}$$

$$\alpha_i = \frac{\sinh \lambda_i a + \sin \lambda_i a}{\cosh \lambda_i a + \cos \lambda_i a}, \quad \beta_j = \frac{\sinh \mu_j b - \sin \mu_i b}{\cosh \mu_j b - \cos \mu_j b}. \tag{7.2.208}$$

CFSF Plates:
Algebraic polynomials

$$X_i(x) = \left(\frac{x}{a}\right)^{i+1}, \quad Y_j(y) = \left(\frac{y}{b}\right)^{j}. \tag{7.2.209}$$

Characteristic polynomials

$$
\begin{aligned}
X_i(x) &= \sin \lambda_i x - \sinh \lambda_i x + \alpha_i \left(\cosh \lambda_i x - \cos \lambda_i x\right), \\
Y_j(y) &= \sinh \mu_j b \, \sin \mu_j y - \sin \mu_j b \, \sinh \mu_j y,
\end{aligned} \tag{7.2.210}
$$

$$\cos \lambda_i a \, \cosh \lambda_i a + 1 = 0, \quad \tan \mu_j b - \tanh \mu_j b = 0, \tag{7.2.211}$$

$$\alpha_i = \frac{\sinh \lambda_i a + \sin \lambda_i a}{\cosh \lambda_i a + \cos \lambda_i a}. \tag{7.2.212}$$

CCCC Plates:
Algebraic polynomials

$$
\begin{aligned}
X_i(x) &= \left(\frac{x}{a}\right)^{i+1} - 2\left(\frac{x}{a}\right)^{i+2} + \left(\frac{x}{a}\right)^{i+3}, \\
Y_j(y) &= \left(\frac{y}{b}\right)^{j+1} - 2\left(\frac{y}{b}\right)^{j+2} + \left(\frac{y}{b}\right)^{j+3}.
\end{aligned} \tag{7.2.213}
$$

Characteristic polynomials

$$
\begin{aligned}
X_i(x) &= \sin \lambda_i x - \sinh \lambda_i x + \alpha_i \left(\cosh \lambda_i x - \cos \lambda_i x\right), \\
Y_j(y) &= \sin \lambda_j y - \sinh \lambda_j y + \alpha_j \left(\cosh \lambda_j y - \cos \lambda_j y\right),
\end{aligned} \tag{7.2.214}
$$

$$\cos \lambda_i a \cosh \lambda_i a - 1 = 0, \tag{7.2.215}$$

$$\alpha_i = \frac{\sinh \lambda_i a - \sin \lambda_i a}{\cosh \lambda_i a - \cos \lambda_i a} = \frac{\cosh \lambda_i a - \cos \lambda_i a}{\sinh \lambda_i a + \sin \lambda_i a}. \tag{7.2.216}$$

The plate types presented in the preceding equations indicate how one can construct the approximation functions for any combination of fixed, hinged, and free boundary conditions on the four edges of a rectangular plate. The difficult task is to evaluate the integrals of these functions as required in Eqs (7.2.189) and (7.2.190). One may use a symbolic manipulator, such as *Mathematica* or *Maple*, to evaluate the integrals. Of course, one may also use trigonometric functions, which have the orthogonality property, for certain boundary conditions. In general, the Ritz method for general rectangular plates with arbitrary boundary conditions is algebraically more complicated than a numerical method, such as the finite element method.

Example 7.2.12

Consider a CCCC plate subjected to uniformly distributed transverse load of intensity q_0. Determine the one-parameter Ritz solution using (a) the algebraic functions in Eq. (7.2.213) and (b) the characteristic polynomials in Eqs (7.2.214) to (7.2.216).

Solution: (a) For the choice of the algebraic functions in Eq. (7.2.213), Eq. (7.2.188) takes the form

$$0 = \left[\left(\frac{4}{5a^3} \right) \left(\frac{b}{630} \right) D_{11} + 4D_{66} \left(\frac{2}{105a} \right) \left(\frac{2}{105b} \right) \right.$$
$$\left. + 2D_{12} \left(-\frac{2}{105a} \right) \left(-\frac{2}{105b} \right) + \left(\frac{a}{630} \right) \left(\frac{4}{5b^3} \right) D_{22} \right] c_{11} - \left(\frac{ab}{900} \right) q_0,$$

or

$$c_{11} = \frac{7q_0 a^4}{8\bar{D}}, \tag{1}$$

where

$$\bar{D} = D_{11} + \frac{4}{7}(D_{12} + 2D_{66})s^2 + D_{22}s^4. \tag{2}$$

and $s = a/b$ denotes the plate aspect ratio. The one-parameter Ritz solution becomes

$$W_{11}(x,y) = \left(\frac{7}{8} \right) \frac{q_0 a^4}{\bar{D}} \left[\frac{x}{a} - \left(\frac{x}{a} \right)^2 \right]^2 \left[\frac{y}{b} - \left(\frac{y}{b} \right)^2 \right]^2. \tag{3}$$

The maximum deflection occurs at $x = a/2$ and $y = b/2$:

$$W_{11} \left(\frac{a}{2}, \frac{b}{2} \right) = 0.00342 \frac{q_0 a^4}{\bar{D}}, \quad \bar{D} = D_{11} + 0.5714(D_{12} + 2D_{66})s^2 + D_{22}s^4. \tag{4}$$

(b) The use of the characteristic functions from Eqs (7.2.214) to (7.2.216)

$$X_1(x) = \sin \frac{4.73x}{a} - \sinh \frac{4.73x}{a} + 1.0178 \left(\cosh \frac{4.73x}{a} - \cos \frac{4.73x}{a} \right),$$
$$Y_1(y) = \sin \frac{4.73y}{b} - \sinh \frac{4.73y}{b} + 1.0178 \left(\cosh \frac{4.73y}{b} - \cos \frac{4.73y}{b} \right). \tag{5}$$

Evaluating the integrals, we obtain

$$\left[537.181D_{11} + 324.829(D_{12} + 2D_{66})s^2 + 537.181D_{22}s^4 \right] c_{11} = 0.715 \, q_0 a^4, \tag{6}$$

or

$$c_{11} = 0.001331 \frac{q_0 a^4}{\hat{D}}, \quad \hat{D} = D_{11}b^4 + 0.604692(D_{12} + 2D_{66})s^2 + D_{22}s^4. \tag{7}$$

The maximum deflection is given by $(X_1(a/2) = Y_1(b/2) = 1.6164)$

$$W_{11} \left(\frac{a}{2}, \frac{b}{2} \right) = c_{11} X_1 \left(\frac{a}{2} \right) Y_1 \left(\frac{b}{2} \right) = 0.003478 \frac{q_0 a^4}{\hat{D}}. \tag{8}$$

For an isotropic square plate, the maximum deflection in Eq. (4) becomes $W_{11}(a/2, b/2) = 0.00134(q_0 a^4/D)$, whereas Eq. (8) gives $W_{11}(a/2, b/2) = 0.00133(q_0 a^4/D)$. The "exact" solution (see Timoshenko and Woinowsky-Krieger [124], p. 202) is $0.00126(q_0 a^4/D)$.

Example 7.2.13

Consider a uniformly loaded, isotropic, rectangular plate $(a \times b)$ that has as its edges $x = 0$ and $x = a$ simply supported, the edge $y = 0$ clamped, and the edge $y = b$ free (i.e., SSCF plate shown in Fig. 7.2.17). The geometric boundary conditions of the problem are

$$w(0,y) = w(a,y) = w(x,0) = \frac{\partial w}{\partial y}\bigg|_{(x,0)} = 0. \tag{1}$$

Determine a one-parameter Ritz solution to the problem.

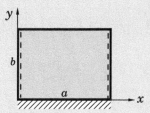

Fig. 7.2.17 A rectangular plate with sides $x = 0, a$ simply supported, side $y = 0$ clamped, and side $y = b$ free.

Solution: We choose an approximation of the form

$$w(x,y) \approx W_1(x,y) = c_1\phi_1(x,y), \quad \phi_1 = \left(\frac{y}{b}\right)^2 \sin\frac{\pi x}{a}, \tag{2}$$

which satisfies the geometric boundary conditions of the problem. Substituting Eq. (2) into the total potential energy expression, $\Pi = U + V$, where U and V are given in Eqs (7.2.90) and (7.2.92)

$$\Pi(w) = \frac{D}{2}\int_0^b\int_0^a\left[\left(\frac{\partial^2 w}{\partial x^2}\right)^2 + \left(\frac{\partial^2 w}{\partial y^2}\right)^2 + 2\nu\frac{\partial^2 w}{\partial x^2}\frac{\partial^2 w}{\partial y^2}\right.$$

$$\left. + 2(1-\nu)\left(\frac{\partial^2 w}{\partial x \partial y}\right)^2\right]dxdy - \int_0^b\int_0^a q_0 w\,dxdy, \tag{3}$$

we obtain

$$\Pi(c_1) = \frac{Dc_1^2}{2}\int_0^a\int_0^b\left\{\left[-\left(\frac{\pi}{a}\right)^2\left(\frac{y}{b}\right)^2\sin\frac{\pi x}{a} + \frac{2}{b^2}\sin\frac{\pi x}{a}\right]^2\right.$$

$$\left. + 2(1-\nu)\left[\frac{2}{b^2}\left(\frac{\pi}{a}\right)^2\left(\frac{y}{b}\right)^2\sin^2\frac{\pi x}{a} + \left(\frac{\pi}{a}\frac{2y}{b^2}\cos\frac{\pi x}{a}\right)^2\right]\right\}dxdy$$

$$- c_1\int_0^a\int_0^b q_0\left(\frac{y}{b}\right)^2\sin\frac{\pi x}{a}\,dxdy$$

$$= \frac{D}{2}c_1^2\left\{\int_0^a\sin^2\frac{\pi x}{a}\,dx\int_0^b\left[\left(\frac{2}{b^2} - \frac{\pi^2}{a^2b^2}y^2\right)^2 + \frac{4(1-\nu)\pi^2}{a^2b^4}y^2\right]dy\right.$$

$$\left. + \frac{8(1-\nu)\pi^2}{a^2b^4}\int_0^a\cos^2\frac{\pi x}{a}\,dx\int_0^b y^2dy\right\} - c_1q_0\int_0^a\sin\frac{\pi x}{a}\,dx\int_0^b\frac{y^2}{b^2}dy$$

$$= \frac{Dc_1^2a}{4}\left[\frac{4}{b^3} + \frac{\pi^4 b}{5a^4} - \frac{4\pi^2}{3a^2b} + \frac{4(1-\nu)\pi^2}{a^2b}\right] - \frac{2abc_1q_0}{3\pi}. \tag{4}$$

Using the principle of minimum total potential energy, $\delta\Pi = (d\Pi/dc_1)\delta c_1 = 0$, we obtain

$$c_1 = \frac{20q_0 a^4}{D\pi \left[3\pi^4 + 20\pi^2 s^2 (2 - 3\nu) + 60s^4\right]}, \quad s = \frac{a}{b}, \tag{5}$$

and the solution is given by

$$W_1(x,y) = \frac{20q_0 a^4}{D\pi \left[3\pi^4 + 20\pi^2 s^2 (2 - 3\nu) + 60s^4\right]} \left(\frac{y}{b}\right)^2 \sin\frac{\pi x}{a}. \tag{6}$$

The maximum deflection is given by

$$W_{\max} = W_1(a/2, b) = c_1, \tag{7}$$

which, for a square plate with $\nu = 0.3$, is equal to $0.01118(q_0 a^4/D)$.

Example 7.2.14

Consider an isotropic rectangular plate of dimensions $2a \times 2b$, and subjected to uniformly distributed load of intensity q_0. Solve the problem using the semidiscrete approximation method of **Example 6.4.16** for the following two boundary conditions: (a) all sides clamped and (b) two opposite sides clamped and the remaining two sides simply supported.

Solution: (a) For the clamped rectangular plate (see Fig. 7.2.18), we seek a one-parameter solution in the form

$$w(x,y) \approx W_1(x,y) = c_1(x)\phi_1(y) = c_1(x)(y^2 - b^2)^2. \tag{1}$$

For convenience, the origin of the coordinate system is taken at the center of the plate. The approximation $W_1(x,y)$ must satisfy the boundary conditions

$$W_1 = 0, \quad \frac{\partial W_1}{\partial x} = 0 \quad \text{at} \quad x = \pm a \text{ and for any } y, \tag{2}$$

$$W_1 = 0, \quad \frac{\partial W_1}{\partial y} = 0 \quad \text{at} \quad y = \pm b \text{ and for any } x. \tag{3}$$

The boundary conditions in Eq. (3) are satisfied for any x because of the choice of ϕ_1. To satisfy the boundary conditions in Eq. (2) for any y, the function $c_1(x)$ is required to satisfy the conditions

$$c_1 = 0, \quad \frac{dc_1}{dx} = 0 \quad \text{at} \quad x = \pm a. \tag{4}$$

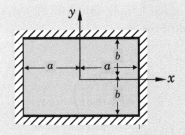

Fig. 7.2.18 A rectangular plate with sides $x = \pm a$ and $y = \pm b$ clamped.

The Galerkin integral of Eq. (7.2.172) (with w replaced by W_1) over the domain $(-b, b)$ is

$$0 = \int_{-b}^{b} \left[D \left(\frac{\partial^4 W_1}{\partial x^4} + 2 \frac{\partial^4 W_1}{\partial x^2 \partial y^2} + \frac{\partial^4 W_1}{\partial y^4} \right) - q_0 \right] \phi_1(y) \, dy$$

$$= D \int_{-b}^{b} \left[(y^2 - b^2)^2 \frac{d^4 c_1}{dx^4} + 2(12y^2 - 4b^2) \frac{d^2 c_1}{dx^2} + 24 c_1 - \frac{q_0}{D} \right] (y^2 - b^2)^2 dy. \tag{5}$$

After performing the integration, we obtain

$$b^5 \left(\frac{256 b^4}{315} \frac{d^4 c_1}{dx^4} - \frac{512 b^2}{105} \frac{d^2 c_1}{dx^2} + \frac{384}{15} c_1 - \frac{16}{15} \frac{q_0}{D} \right) = 0,$$

or

$$\frac{d^4 c_1}{d\xi^4} - 6 \frac{d^2 c_1}{d\xi^2} + \frac{63}{2} c_1 = \frac{21}{16} \frac{q_0}{D}, \tag{6}$$

where $\xi = x/b$. The homogeneous solution of Eq. (6) can be calculated using the roots of the characteristic equation associated with Eq. (6):

$$d^4 - 6 d^2 + \frac{63}{2} = 0 \; \rightarrow \; (d^2)_{1,2} = 3 \pm \sqrt{9 - (63/2)} = 3 \pm i 4.7434,$$

where $i = \sqrt{-1}$. To determine the roots d_1 through d_4, we use the polar form

$$(d^2)_{1,2} = x \pm iy = r(\cos \theta \pm i \sin \theta) = r e^{\pm i\theta},$$

where $x = 3$, $y = 4.7434$, $r = \sqrt{x^2 + y^2} = 5.6125$, and $\theta = \tan^{-1}(y/x) = 57.69°$. Therefore, we have

$$(d)_{1-4} = \pm \sqrt{r} e^{\pm i\theta/2} = \pm \sqrt{r} \left(\cos \frac{\theta}{2} \pm i \sin \frac{\theta}{2} \right) \equiv \pm(\alpha \pm i\beta), \tag{7}$$

where

$$\alpha = \sqrt{r} \cos \frac{\theta}{2} = 2.075, \quad \beta = \sqrt{r} \sin \frac{\theta}{2} = 1.143.$$

Thus, we have

$$\begin{aligned} d_1 &= 2.075 + i1.143, & d_2 &= 2.075 - i1.143 \\ d_3 &= -(2.075 + i1.143), & d_4 &= -(2.075 - i1.143). \end{aligned} \tag{8}$$

Therefore, the homogeneous solution is given by

$$\begin{aligned} c_{1h} &= \bar{A}_1 e^{(\alpha+i\beta)\xi} + \bar{A}_2 e^{(\alpha-i\beta)\xi} + \bar{A}_3 e^{-(\alpha+i\beta)\xi} + \bar{A}_4 e^{-(\alpha-i\beta)\xi} \\ &= A_1 \cosh \alpha\xi \cos \beta\xi + A_2 \cosh \alpha\xi \sin \beta\xi + A_3 \sinh \alpha\xi \sin \beta\xi \\ &\quad + A_4 \sinh \alpha\xi \cos \beta\xi. \end{aligned} \tag{9}$$

The particular solution is given by

$$c_{1p} = \frac{2}{63} \left(\frac{21}{16} \frac{q_0}{D} \right) = \frac{q_0}{24D}.$$

The general solution of Eq. (6) becomes

$$c_1(\xi) = c_{1h} + c_{1p}. \tag{10}$$

Because of the symmetry of the expected solution about the y-axis, it follows that $A_2 = A_4 = 0$. The remaining constants A_1 and A_3 can be determined using boundary conditions in (d) on c_1 at $x = \pm a$. We obtain

$$A_1 \cosh k_1 \cos k_2 + A_3 \sinh k_1 \sin k_2 = -\frac{q_0}{24D},$$

$$A_1 \left(\frac{\alpha}{b} \sinh k_1 \cos k_2 - \frac{\beta}{b} \cosh k_1 \sin k_2 \right)$$

$$+ A_3 \left(\frac{\alpha}{b} \cosh k_1 \sin k_2 + \frac{\beta}{b} \sinh k_1 \cos k_2 \right) = 0,$$

where $k_1 = \alpha a/b$ and $k_2 = \beta a/b$. Solving these equations, we obtain

$$A_1 = \frac{\mu_1}{\mu_0} \left(\frac{q_0}{24D} \right), \quad A_3 = \frac{\mu_2}{\mu_0} \frac{q_0}{24D}, \tag{11}$$

where

$$\mu_0 = \beta \sinh k_1 \cosh k_1 + \alpha \sin k_2 \cos k_2,$$
$$\mu_1 = -(\alpha \cosh k_1 \sin k_2 + \beta \sinh k_1 \cos k_2), \tag{12}$$
$$\mu_2 = \alpha \sinh k_1 \cos k_2 - \beta \cosh k_1 \sin k_2.$$

Thus, the one-parameter solution becomes

$$W_1(x,y) = \frac{q_0}{24D} \left[\left(\frac{\mu_1}{\mu_0} \cosh \frac{\alpha x}{b} \cos \frac{\beta x}{b} + \frac{\mu_2}{\mu_0} \sinh \frac{\alpha x}{b} \sin \frac{\beta x}{b} \right) + 1 \right] (y^2 - b^2)^2. \tag{13}$$

The maximum deflection occurs at the center of the plate; for a square plate of dimensions $2a \times 2a$ the maximum deflection is given by ($k_1 \to \alpha = 2.075$, $k_2 \to \beta = 1.143$)

$$W_{\max} = W_1(0,0) = \frac{q_0 a^4}{24D} \left(1 + \frac{\mu_1}{\mu_0} \right) = 0.4977 \frac{q_0 a^4}{24D} = 0.0207 \frac{q_0 a^4}{D}. \tag{14}$$

For a square plate of dimension $a \times a$, this reduces to (divide the above result with 16)

$$W_{\max} = 0.00129 \frac{q_0 a^4}{D}, \tag{15}$$

which is closer to the "exact" solution $0.00126(q_0 a^4/D)$ than that obtained in **Example 7.2.12**.

(b) Much of the discussion presented in part (a) is also valid for a rectangular plate with edges $y = \pm b$ clamped and $x = \pm a$ simply supported (see Fig. 7.2.19). The general solution in Eq. (10) is valid with the following boundary conditions on c_1:

$$c_1(\pm a) = \left(\frac{d^2 c_1}{dx^2} \right)_{x=\pm a} = 0. \tag{16}$$

These conditions give

$$A_1 \cosh k_1 \cos k_2 + A_3 \sinh k_1 \sin k_2 = -\frac{q_0}{24D},$$

$$A_1 \sinh k_1 \sin k_2 - A_3 \cosh k_1 \cos k_2 = \frac{q_0}{24D} \left(\frac{\beta^2 - \alpha^2}{2\alpha\beta} \right),$$

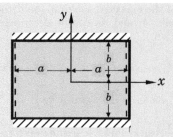

Fig. 7.2.19 A rectangular plate with sides $x = \pm a$ simply supported and sides $y = \pm b$ clamped.

where $\alpha = 2.075$ and $\beta = 1.143$. The solution of these equations is given by

$$
A_1 = -\frac{q_0}{24D} \frac{\cosh k_1 \cos k_2 + a_0 \sinh k_1 \sin k_2}{\cosh^2 k_1 \cos^2 k_2 + \sinh^2 k_1 \sin^2 k_2},
$$

$$
A_3 = \frac{q_0}{24D} \frac{a_0 \cosh k_1 \cos k_2 + \sinh k_1 \sin k_2}{\cosh^2 k_1 \cos^2 k_2 + \sinh^2 k_1 \sin^2 k_2},
\tag{17}
$$

where $a_0 = \left(\beta^2 - \alpha^2\right)/2\alpha\beta$. For a square plate, the maximum deflection is given by

$$
W_{\max} = W_1(0,0) = 0.0248 \frac{q_0 a^4}{D}.
\tag{18}
$$

Example 7.2.15

Show that the Galerkin solution of a simply supported rectangular plate using trigonometric functions coincides with the Navier solution.

Solution: We begin with the approximation [see Eq. (7.2.112)]

$$
w(x,y,t) \approx W_N(x,y,t) = \sum_{i=1}^{N} \sum_{j=1}^{N} c_{ij}(t) \sin \frac{i\pi x}{a} \sin \frac{j\pi y}{b},
\tag{1}
$$

which satisfies all of the boundary conditions in Eq. (7.2.111). Substituting W_N into the (semidiscrete) Galerkin integral

$$
0 = \int_0^b \int_0^a \left[D \left(\frac{\partial^4 W_N}{\partial x^4} + 2 \frac{\partial^4 W_N}{\partial x^2 \partial y^2} + \frac{\partial^4 W_N}{\partial y^4} \right) \right.
$$

$$
\left. - q + m_0 \ddot{W}_N - m_2 \left(\frac{\partial^2 \ddot{W}_N}{\partial x^2} + \frac{\partial^2 \ddot{W}_N}{\partial y^2} \right) \right] \sin \frac{m\pi x}{a} \sin \frac{n\pi y}{b} \, dx dy
$$

for $m, n = 1, 2, \ldots, N$, we obtain ($\alpha_i = i\pi/a$ and $\beta_j = j\pi/b$)

$$
0 = \int_0^b \int_0^a \left\{ \sum_{i=1}^{N} \sum_{j=1}^{N} \left[D \left(\alpha_i^2 + \beta_j^2 \right)^2 c_{ij} + \left[m_0 + m_2 \left(\alpha_i^2 + \beta_j^2 \right) \right] \frac{d^2 c_{ij}}{dt^2} \right] \right.
$$

$$
\left. \times \sin \frac{i\pi x}{a} \sin \frac{j\pi y}{b} - q \right\} \sin \frac{m\pi x}{a} \sin \frac{n\pi y}{b} \, dx dy.
\tag{2}
$$

Since we can write

$$q(x, y, t) = \sum_{i=1}^{N} \sum_{j=1}^{N} q_{ij}(t) \, \sin \frac{i\pi x}{a} \, \sin \frac{j\pi y}{b}, \tag{3}$$

and due to the orthogonality of the trigonometric functions

$$\int_{0}^{a} \sin \frac{i\pi x}{a} \, \sin \frac{m\pi x}{a} \, dx = \begin{cases} \frac{a}{2} & \text{for } i = m, \\ 0 & \text{for } i \neq m, \end{cases} \tag{4}$$

we have

$$\left[m_0 + m_2 \left(\alpha_m^2 + \beta_n^2 \right) \right] \frac{d^2 c_{mn}}{dt^2} + \frac{ab}{4} D \left(\alpha_m^2 + \beta_n^2 \right)^2 c_{mn} = q_{mn}. \tag{5}$$

Equation (5) is identical to Eq. (7.2.117) with $c_{mn} = W_{mn}$ and $N_0 = 0$. Therefore, for $N \to \infty$, the Galerkin solution is the same as the Navier solution. The use of the same approximation, that is, Eq. (1), in the Ritz method will also produce the same result.

Example 7.2.16

Consider an isotropic equilateral triangular plate shown in Fig. 7.2.20 subjected to uniformly distributed transverse load of intensity q_0. Determine the transverse deflection using the classical plate theory and the Ritz method.

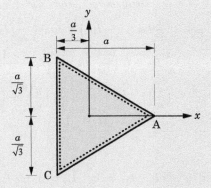

Fig. 7.2.20 An equilateral triangular plate with simply supported edges.

Solution: Since the essential boundary conditions require $w = 0$ on the edges, the approximation functions should vanish on the edges. A natural choice for the first approximation function is the product of the equations of the edges themselves. The equations of the edges, with respect to the coordinate system shown in Fig. 7.2.20, are

$$\frac{x}{a} + \frac{1}{3} = 0 \quad \text{on BC},$$

$$\frac{1}{\sqrt{3}} \frac{x}{a} + \frac{y}{a} - \frac{2}{3\sqrt{3}} = 0 \quad \text{on AC}, \tag{1}$$

$$\frac{1}{\sqrt{3}} \frac{x}{a} - \frac{y}{a} - \frac{2}{3\sqrt{3}} = 0 \quad \text{on AB}.$$

Hence, a one-parameter Ritz approximation is of the form

$$
\begin{aligned}
W_1(x,y) &= c_1 \left(\frac{x}{a} + \frac{1}{3}\right)\left(\frac{1}{\sqrt{3}}\frac{x}{a} + \frac{y}{a} - \frac{2}{3\sqrt{3}}\right)\left(\frac{1}{\sqrt{3}}\frac{x}{a} - \frac{y}{a} - \frac{2}{3\sqrt{3}}\right) \\
&= \frac{1}{3}c_1\left[\frac{4}{27} - \left(\frac{x}{a}\right)^2 - \left(\frac{y}{a}\right)^2 - 3\left(\frac{x}{a}\right)\left(\frac{y}{a}\right)^2 + \left(\frac{x}{a}\right)^3\right] \\
&\equiv c_1\phi_1(x,y),
\end{aligned}
\tag{2}
$$

where

$$
\phi_1(x,y) = \frac{1}{3}\left[\frac{4}{27} - \left(\frac{x}{a}\right)^2 - \left(\frac{y}{a}\right)^2 - 3\left(\frac{x}{a}\right)\left(\frac{y}{a}\right)^2 + \left(\frac{x}{a}\right)^3\right].
$$

The first and second derivatives of ϕ_1 are

$$
\frac{\partial\phi_1}{\partial x} = \frac{1}{3}\left(-\frac{2x}{a^2} - \frac{3y^2}{a^3} + \frac{3x^2}{a^3}\right), \quad \frac{\partial\phi_1}{\partial y} = \frac{1}{3}\left(-\frac{2y}{a^2} - \frac{6xy}{a^3}\right),
$$

$$
\frac{\partial^2\phi_1}{\partial x^2} = -\frac{2}{3a^2} + \frac{2x}{a^3}, \quad \frac{\partial^2\phi_1}{\partial y^2} = -\frac{2}{3a^2} - \frac{2x}{a^3}, \quad \frac{\partial^2\phi_1}{\partial x\partial y} = -\frac{2y}{a^3}.
\tag{3}
$$

Substituting the expressions (3) into Eqs (7.2.185) and (7.2.186), we obtain $[D_{11} = D_{22} = D$, $D_{12} = \nu D$, and $2D_{66} = (1-\nu)D]$

$$
\begin{aligned}
R_{11} &= D \int_T \left[\frac{\partial^2\phi_1}{\partial x^2}\frac{\partial^2\phi_1}{\partial x^2} + \nu\left(\frac{\partial^2\phi_1}{\partial y^2}\frac{\partial^2\phi_1}{\partial x^2} + \frac{\partial^2\phi_1}{\partial x^2}\frac{\partial^2\phi_1}{\partial y^2}\right)\right. \\
&\qquad \left. + 2(1-\nu)\frac{\partial^2\phi_1}{\partial x\partial y}\frac{\partial^2\phi_1}{\partial x\partial y} + \frac{\partial^2\phi_1}{\partial y^2}\frac{\partial^2\phi_1}{\partial y^2}\right]dxdy \\
&= D\left(\frac{4}{9a^4}I_{00} + \frac{4}{a^6}I_{20} - \frac{8}{3a^5}I_{10}\right) + 2\nu D\left(\frac{4}{9a^4}I_{00} - \frac{4}{a^6}I_{20}\right) \\
&\quad + 2(1-\nu)D\frac{4}{a^6}I_{02} + D\left(\frac{4}{9a^4}I_{00} + \frac{4}{a^6}I_{20} + \frac{4}{3a^5}I_{10}\right),
\end{aligned}
\tag{4}
$$

$$
\begin{aligned}
F_1 &= \frac{q_0}{3}\int_T\left[\frac{4}{27} - \left(\frac{x}{a}\right)^2 - \left(\frac{y}{a}\right)^2 - 3\left(\frac{x}{a}\right)\left(\frac{y}{a}\right)^2 + \left(\frac{x}{a}\right)^3\right]dxdy \\
&= \frac{q_0}{3}\left(\frac{4A_0}{27} - \frac{I_{20}}{a^2} - \frac{I_{02}}{a^2} - \frac{3I_{12}}{a^3} + \frac{I_{30}}{a^3}\right),
\end{aligned}
\tag{5}
$$

where

$$
I_{mn} \equiv \int_T x^m y^n \, dxdy, \quad I_{00} = A_0, \quad I_{10} = A_0\bar{x}, \quad I_{01} = A_0\bar{y}.
\tag{6}
$$

We have

$$
\bar{x} = \frac{1}{3}\sum_{i=1}^{3}x_i, \quad \bar{y} = \frac{1}{3}\sum_{i=1}^{3}y_i, \quad I_{11} = \frac{A_0}{12}\left(\sum_{i=1}^{3}x_iy_i + 9\bar{x}\bar{y}\right),
$$

$$
I_{20} = \frac{A_0}{12}\left(\sum_{i=1}^{3}x_i^2 + 9\bar{x}^2\right), \quad I_{02} = \frac{A_0}{12}\left(\sum_{i=1}^{3}y_i^2 + 9\bar{y}^2\right),
\tag{7}
$$

where (x_i, y_i) are the coordinates of the three vertices of a triangle and A_0 is the area of a triangle $(A_0 = a^2/\sqrt{3})$. For the triangle in Fig. 7.2.20, we have

$$
x_1 = \frac{2a}{3}, \quad x_2 = x_3 = -\frac{a}{3}, \quad y_1 = 0, \quad y_2 = -y_3 = \frac{a}{\sqrt{3}},
$$

$$
\bar{x} = \bar{y} = 0, \quad I_{10} = I_{01} = I_{11} = 0, \quad I_{20} = I_{02} = \frac{A_0 a^2}{18},
\tag{8}
$$

$$
I_{30} = \frac{A_0 a^3}{135}, \quad I_{12} = -\frac{A_0 a^3}{135}.
$$

Substituting these values into the expressions in Eqs (4) and (5) for R_{11} and F_1, respectively, we obtain

$$R_{11} = \frac{16DA_0}{9a^4}, \quad F_1 = \frac{q_0 A_0}{45}; \rightarrow c_1 = \frac{F_1}{R_{11}} = \frac{q_0 a^4}{80D}. \tag{9}$$

Thus the one-parameter Ritz solution becomes

$$W_1(x,y) = \frac{q_0 a^4}{240D}\left[\frac{4}{27} - \left(\frac{x}{a}\right)^2 - \left(\frac{y}{a}\right)^2 - 3\left(\frac{x}{a}\right)\left(\frac{y}{a}\right)^2 + \left(\frac{x}{a}\right)^3\right]. \tag{10}$$

For the N-parameter Ritz approximation, one may use

$$\phi_i(x,y) = p_{(i-1)}(x,y)\phi_{(i-1)}(x,y), \quad i = 2, 3, \ldots, N. \tag{11}$$

where $p_i(x,y)$ is a polynomial of degree i:

$$p_1(x,y) = \frac{x}{a} + \frac{y}{a}, \quad p_2(x,y) = \frac{x}{a}\frac{y}{a} + \frac{x^2}{a^2} + \frac{y^2}{a^2}, \ldots. \tag{12}$$

The exact deflection, which can be verified by substitution into the governing equation, Eq. (7.2.172), is given by

$$w(x,y) = \frac{q_0 a^4}{64D}\left[\frac{4}{27} - \left(\frac{x}{a}\right)^2 - \left(\frac{y}{a}\right)^2 - 3\left(\frac{x}{a}\right)\left(\frac{y}{a}\right)^2 + \left(\frac{x}{a}\right)^3\right]$$
$$\times \left[\frac{4}{9} - \left(\frac{x}{a}\right)^2 - \left(\frac{y}{a}\right)^2\right]. \tag{13}$$

Example 7.2.17

Consider a clamped, isotropic elliptic plate with major and minor axes $2a$ and $2b$, respectively (see Fig. 7.2.21). Determine a one-parameter Galerkin solution for the case in which the plate is subjected to uniformly distributed load q_0.

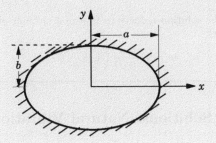

Fig. 7.2.21 A clamped elliptic plate.

Solution: The boundary conditions are

$$w = 0, \quad \frac{\partial w}{\partial n} = 0 \quad \left(\text{or } \frac{\partial w}{\partial x} = 0 \text{ and } \frac{\partial w}{\partial y} = 0\right). \tag{1}$$

As before, we use the equation for the boundary of the ellipse but square it to satisfy the deflection as well as the slope boundary conditions:

$$W_1(x,y) = c_1\phi_1(x,y) = c_1\left(1 - \frac{x^2}{a^2} - \frac{y^2}{b^2}\right)^2. \tag{2}$$

We have

$$\frac{\partial\phi_1}{\partial x} = -\frac{4x}{a^2}\left(1 - \frac{x^2}{a^2} - \frac{y^2}{b^2}\right), \qquad \frac{\partial\phi_1}{\partial y} = -\frac{4y}{b^2}\left(1 - \frac{x^2}{a^2} - \frac{y^2}{b^2}\right),$$

$$\frac{\partial^4\phi_1}{\partial x^4} = \frac{24}{a^4}, \qquad \frac{\partial^4\phi_1}{\partial x^2\partial y^2} = \frac{8}{a^2b^2}, \qquad \frac{\partial^4\phi_1}{\partial y^4} = \frac{24}{b^4}. \tag{3}$$

Clearly, the boundary conditions in Eq. (1) are satisfied exactly by the choice. Substituting into the residual expression,

$$\mathcal{R} = D\left(\frac{\partial^4 W_1}{\partial x^4} + 2\frac{\partial^4 W_1}{\partial x^2\partial y^2} + \frac{\partial^4 W_1}{\partial y^4}\right) - q_0,$$

$$= c_1 D\left(\frac{24}{a^4} + 2\frac{8}{a^2b^2} + \frac{24}{b^4}\right) - q_0, \tag{4}$$

which is independent of x and y. In this case, there is no need to evaluate the weighted-integral of the residual

$$0 = \int_{-b}^{b}\int_{-a}^{a} \mathcal{R}\phi_1(x,y)\,dx\,dy,$$

as it only yields a constant multiple of the expression in Eq. (4). Thus we have

$$c_1 = \frac{q_0 a^4}{8D\left(3 + 2s^2 + 3s^4\right)}, \qquad s = \frac{a}{b},$$

and the Galerkin solution, which coincides with the exact solution, is given by

$$w(x,y) = \frac{q_0 a^4}{8D(3 + 2s^2 + 3s^4)}\left(1 - \frac{x^2}{a^2} - \frac{y^2}{b^2}\right)^2, \qquad s = \frac{a}{b}. \tag{5}$$

When $a = b$ (i.e., $s = 1$), the solution reduces to that of a circular plate (with $x^2 + y^2 = r^2$, r being the radial coordinate):

$$w(r) = \frac{q_0 a^4}{64D}\left(1 - \frac{r^2}{a^2}\right)^2. \tag{6}$$

7.2.7 Variational Solutions: Natural Vibration

This section is dedicated to the study of natural vibrations of rectangular plates by the Ritz method. Equation (7.2.110), without the applied q and $\hat{N}_{xx} = 0$, $\hat{N}_{yy} = 0$, and $\hat{N}_{xy} = 0$, reduces to

$$D\left(\frac{\partial^4 w}{\partial x^4} + 2\frac{\partial^4 w}{\partial x^2\partial y^2} + \frac{\partial^4 w}{\partial y^4}\right)$$

$$+ m_0\frac{\partial^2 w}{\partial t^2} - m_2\left(\frac{\partial^4 w}{\partial t^2\partial x^2} + \frac{\partial^4 w}{\partial t^2\partial y^2}\right) = 0, \tag{7.2.217}$$

where

$$m_0 = \rho_0 h, \quad m_2 = \frac{\rho_0 h^3}{12}. \tag{7.2.218}$$

For natural vibration, the solution is assumed to be periodic:

$$w(x, y, t) = w_0(x, y)e^{i\omega t}, \tag{7.2.219}$$

where $i = \sqrt{-1}$ and ω is the frequency of natural vibration associated with mode shape $w_0(x, y)$. Then Eq. (7.2.217) takes the form

$$D\left(\frac{\partial^4 w_0}{\partial x^4} + 2\frac{\partial^4 w_0}{\partial x^2 \partial y^2} + \frac{\partial^4 w_0}{\partial y^4}\right)$$
$$- \omega^2\left[m_0 w_0 - m_2\left(\frac{\partial^2 w_0}{\partial x^2} + \frac{\partial^2 w_0}{\partial y^2}\right)\right] = 0. \tag{7.2.220}$$

We wish to find values of ω such that Eq. (7.2.220) has a non-trivial solution w_0.

The principle of minimum total potential energy that yields Eq. (7.2.220) as the Euler equation is

$$0 = \int_0^b \int_0^a \left\{ D\left[\frac{\partial^2 w_0}{\partial x^2}\frac{\partial^2 \delta w_0}{\partial x^2} + \nu\left(\frac{\partial^2 w_0}{\partial y^2}\frac{\partial^2 \delta w_0}{\partial x^2} + \frac{\partial^2 w_0}{\partial x^2}\frac{\partial^2 \delta w_0}{\partial y^2}\right)\right.\right.$$
$$\left.+ 2(1-\nu)\frac{\partial^2 w_0}{\partial x \partial y}\frac{\partial^2 \delta w_0}{\partial x \partial y} + \frac{\partial^2 w_0}{\partial y^2}\frac{\partial^2 \delta w_0}{\partial y^2}\right]$$
$$\left.- \omega^2\left[m_0 w_0 \delta w_0 + m_2\left(\frac{\partial w_0}{\partial x}\frac{\partial \delta w_0}{\partial x} + \frac{\partial w_0}{\partial y}\frac{\partial \delta w_0}{\partial y}\right)\right]\right\} dx dy. \tag{7.2.221}$$

We note that the weak form in Eq. (7.2.221) is based on the principle of minimum total potential energy and not obtained by applying the three-step procedure of constructing the weak form from Eq. (7.2.220). The terms containing Poisson's ratio in Eq. (7.2.221) cancel out when deriving the Euler equations and yield the result in Eq. (7.2.220). If we develop the weak form from Eq. (7.2.220) using the three step procedure, the expressions containing Poisson's ratio in Eq. (7.2.221) do not appear !

Using an N-parameter Ritz solution of the form

$$w_0(x, y) \approx \sum_{j=1}^N c_j \phi_j(x, y) \tag{7.2.222}$$

in Eq. (7.2.221), we obtain the eigenvalue problem

$$(\mathbf{R} - \omega^2 \mathbf{B})\, \mathbf{c} = \mathbf{0}, \tag{7.2.223}$$

where

$$R_{ij} = \int_\Omega \left\{ D \left(\frac{\partial^2 \phi_i}{\partial x^2} \frac{\partial^2 \phi_j}{\partial x^2} + \nu \left(\frac{\partial^2 \phi_i}{\partial y^2} \frac{\partial^2 \phi_j}{\partial x^2} + \frac{\partial^2 \phi_i}{\partial x^2} \frac{\partial^2 \phi_j}{\partial y^2} \right) \right. \right.$$
$$\left. \left. + 2(1 - \nu) \frac{\partial^2 \phi_i}{\partial x \partial y} \frac{\partial^2 \phi_j}{\partial x \partial y} + \frac{\partial^2 \phi_i}{\partial y^2} \frac{\partial^2 \phi_j}{\partial y^2} \right] \right\} dx dy, \qquad (7.2.224)$$

$$B_{ij} = \int_\Omega \left[m_0 \phi_i \phi_j + m_2 \left(\frac{\partial \phi_i}{\partial x} \frac{\partial \phi_j}{\partial x} + \frac{\partial \phi_i}{\partial y} \frac{\partial \phi_j}{\partial y} \right) \right] dx dy. \qquad (7.2.225)$$

Equation (7.2.223) represents a standard eigenvalue problem, which can be solved numerically.

As discussed earlier, it is convenient to express the Ritz approximation of rectangular plates in the form

$$w_0(x, y) \approx W_{mn}(x, y) = \sum_{i=1}^{N} \sum_{j=1}^{N} c_{ij} X_i(x) Y_j(y), \qquad (7.2.226)$$

where the functions X_i and Y_j for various boundary conditions were given in Eqs (7.2.191) to (7.2.216). These functions satisfy only the geometric boundary conditions of the problem. Substituting Eq. (7.2.226) into Eqs (7.2.221), we obtain Eq. (7.2.223) in which the coefficients $R_{(ij)(pq)}$ of matrix $\mathbf{R}$ and coefficients $B_{(ij)(pq)}$ of matrix $\mathbf{B}$ are defined by

$$R_{(ij)(pq)} = D \int_0^b \int_0^a \left[\frac{d^2 X_i}{dx^2} Y_j \frac{d^2 X_p}{dx^2} Y_q + 2(1 - \nu) \frac{dX_i}{dx} \frac{dY_j}{dy} \frac{dX_p}{dx} \frac{dY_q}{dy} \right.$$
$$+ \nu \left(X_i \frac{d^2 Y_j}{dy^2} \frac{d^2 X_p}{dx^2} Y_q + \frac{d^2 X_i}{dx^2} Y_j X_p \frac{d^2 Y_q}{dy^2} \right)$$
$$\left. + X_i \frac{d^2 Y_j}{dy^2} X_p \frac{d^2 Y_q}{dy^2} \right] dx dy, \qquad (7.2.227)$$

$$B_{(ij)(pq)} = \int_0^b \int_0^a \left[m_0 X_i X_p Y_j Y_q + m_2 \left(\frac{dX_i}{dx} \frac{dX_p}{dx} Y_j Y_q + X_i X_p \frac{dY_j}{dy} \frac{dY_q}{dy} \right) \right] dx dy. \qquad (7.2.228)$$

The size of the matrices $\mathbf{R}$ and $\mathbf{B}$ is $N^2 \times N^2$, and Eq. (7.2.223) can be used to calculate first N^2 frequencies of the infinite set.

Example 7.2.18

Determine the first four natural frequencies of an isotropic ($\nu = 0.25$) square plate that is simply supported on all four sides. Use the Ritz method with the approximation given in Eq. (7.2.226).

Solution: We select the following approximation functions:

$$\phi_{ij} = X_i(x) Y_j(y), \qquad (1)$$

where

$$X_i(x) = \left(\frac{x}{a}\right)^i - \left(\frac{x}{a}\right)^{i+1}, \quad Y_j(x) = \left(\frac{y}{a}\right)^j - \left(\frac{y}{a}\right)^{j+1}. \tag{2}$$

For square plates with $N = 2$, the matrices $\mathbf{R}$ and $\mathbf{B}$ are given by

$$\mathbf{R} = \begin{bmatrix} 0.4346 & 0.2173 & 0.2173 & 0.1086 \\ 0.2173 & 0.2314 & 0.1086 & 0.1157 \\ 0.2173 & 0.1086 & 0.2314 & 0.1157 \\ 0.1086 & 0.1157 & 0.1157 & 0.0993 \end{bmatrix} \times 10^{-1},$$

$$\mathbf{B} = \begin{bmatrix} 1.1111 & 0.5556 & 0.5556 & 0.2778 \\ 0.5556 & 0.3175 & 0.2778 & 0.1587 \\ 0.5556 & 0.2778 & 0.3175 & 0.1587 \\ 0.2778 & 0.1587 & 0.1587 & 0.0907 \end{bmatrix} \times 10^{-5}, \tag{3}$$

where the following notation is used to store the coefficients

$$\begin{aligned} R_{11} &= R_{(11)(11)}, & R_{12} &= R_{(11)(12)}, & R_{13} &= R_{(11)(21)}, & R_{14} &= R_{(11)(22)}, \\ R_{21} &= R_{(12)(11)}, & R_{22} &= R_{(12)(12)}, & R_{23} &= R_{(12)(21)}, & R_{24} &= R_{(12)(22)}, \\ R_{31} &= R_{(21)(11)}, & R_{32} &= R_{(21)(12)}, & R_{33} &= R_{(21)(21)}, & R_{34} &= R_{(21)(22)}, \\ R_{41} &= R_{(22)(11)}, & R_{42} &= R_{(22)(12)}, & R_{43} &= R_{(22)(21)}, & R_{44} &= R_{(22)(22)}. \end{aligned} \tag{4}$$

The first four dimensionless frequencies, $\bar{\omega}_{mn} = \omega_{mn}a^2\sqrt{\rho h/D}$, are

$$\bar{\omega}_{11} = 20.973, \quad \bar{\omega}_{12} = 58.992, \quad \bar{\omega}_{21} = 59.007, \quad \bar{\omega}_{22} = 92.529. \tag{5}$$

The exact frequencies from Eq. (7.2.217) are

$$\bar{\omega}_{11} = 19.739, \quad \bar{\omega}_{12} = \bar{\omega}_{21} = 49.348, \quad \bar{\omega}_{22} = 88.826. \tag{6}$$

For larger values of N and M, the Ritz method will yield increasingly accurate frequencies.

Example 7.2.19

Determine the first four natural frequencies of an isotropic ($\nu = 0.25$) rectangular plate that is clamped on all four sides. Use the Ritz method with the approximation given in Eq. (7.2.226).

Solution: For a square plate clamped on all sides, we can select the following approximation functions:

$$\phi_{ij} = X_i(x)Y_j(y), \tag{1}$$

$$X_i(x) = \left(\frac{x}{a}\right)^{i+1} - 2\left(\frac{x}{a}\right)^{i+2} + \left(\frac{x}{a}\right)^{i+3}, \tag{2}$$

$$Y_j(y) = \left(\frac{y}{b}\right)^{j+1} - 2\left(\frac{y}{b}\right)^{j+2} + \left(\frac{y}{b}\right)^{j+3}. \tag{3}$$

For isotropic ($\nu = 0.25$) square plates ($a = b$) with $N = 2$, the matrices $\mathbf{R}$ and $\mathbf{B}$ are given by

$$\mathbf{R} = \begin{bmatrix} 0.2902 & 0.1451 & 0.1451 & 0.0725 \\ 0.1451 & 0.1007 & 0.0725 & 0.0503 \\ 0.1451 & 0.0725 & 0.1007 & 0.0503 \\ 0.0725 & 0.0503 & 0.0503 & 0.0335 \end{bmatrix} \times 10^{-3},$$

$$\mathbf{B} = \begin{bmatrix} 0.2520 & 0.1260 & 0.1260 & 0.0630 \\ 0.1260 & 0.0687 & 0.0630 & 0.0344 \\ 0.1260 & 0.0630 & 0.0687 & 0.0344 \\ 0.0630 & 0.0344 & 0.0344 & 0.0187 \end{bmatrix} \times 10^{-7}. \tag{4}$$

The first four dimensionless frequencies, $\bar{\omega}_{mn} = \omega_{mn} a^2 \sqrt{\rho h/D}$, are

$$\bar{\omega}_{11} = 36.000, \quad \bar{\omega}_{12} = 74.296, \quad \bar{\omega}_{21} = 74.297, \quad \bar{\omega}_{22} = 108.592, \tag{5}$$

and the approximate frequencies obtained by Ödman [125] (see Leissa [126]) are

$$\bar{\omega}_{11} = 35.999, \quad \bar{\omega}_{12} = 73.405, \quad \bar{\omega}_{21} = 73.405, \quad \bar{\omega}_{22} = 108.237. \tag{6}$$

Table 7.2.11 contains the first four dimensionless natural frequencies of clamped rectangular plates. The results are obtained with $N = 2$ (compare with Tables 4.28 and 4.29 on pages 63 and 64, respectively, of Leissa [126]; no mention is made of Poisson's ratio used there). The mode shape is of the form

$$w_0(x, y) = X_1 (c_1 Y_1 + c_2 Y_2) + X_2 (c_3 Y_1 + c_4 Y_2). \tag{7}$$

The vectors $\mathbf{c}^{(i)}$ for the four modes in the case of $a/b = 0.25$, for example, are given by

$$\mathbf{c}^{(1)} = \begin{Bmatrix} 1.0 \\ 0.0 \\ 0.0 \\ 0.0 \end{Bmatrix}, \quad \mathbf{c}^{(2)} = \begin{Bmatrix} -0.5 \\ 1.0 \\ 0.0 \\ 0.0 \end{Bmatrix}, \quad \mathbf{c}^{(3)} = \begin{Bmatrix} -0.50 \\ 0.08 \\ 1.00 \\ 0.16 \end{Bmatrix}, \quad \mathbf{c}^{(4)} = \begin{Bmatrix} 0.25 \\ -0.50 \\ -0.50 \\ 1.00 \end{Bmatrix}.$$

The vectors $\mathbf{c}^i$ for the four modes in the case of $a/b = 1$ are given by

$$\mathbf{c}^{(1)} = \begin{Bmatrix} 1.0 \\ 0.0 \\ 0.0 \\ 0.0 \end{Bmatrix}, \quad \mathbf{c}^{(2)} = \begin{Bmatrix} -0.5 \\ 1.0 \\ 0.0 \\ 0.0 \end{Bmatrix}, \quad \mathbf{c}^{(3)} = \begin{Bmatrix} -0.5 \\ 0.0 \\ 1.0 \\ 0.0 \end{Bmatrix}, \quad \mathbf{c}^{(4)} = \begin{Bmatrix} 0.25 \\ -0.50 \\ -0.50 \\ 1.00 \end{Bmatrix}.$$

Table 7.2.11 Effect of the plate aspect ratio a/b on the dimensionless frequencies $\bar{\omega}_{mn} = \omega_{mn} a^2 \sqrt{\rho h/D}$ of clamped isotropic ($\nu = 0.25$) rectangular plates.

Mode				a/b		
(m, n)	0.25	0.5	0.667	1.0	1.5	2.0
$(1, 1)$	22.890	24.647	27.047	36.000	60.856	98.590
$(1, 2)$	24.196	31.867	41.899	74.296	94.273	127.466
$(2, 1)$	63.466	65.234	69.297	74.297	151.419	260.938
$(2, 2)$	64.943	71.941	80.394	108.592	180.887	287.766

Example 7.2.20

Determine the first four natural frequencies of an isotropic ($\nu = 0.25$) rectangular plate with sides $x = 0, a$ and $y = 0$ clamped and side $y = b$ simply supported (CCCS). Use the Ritz method with the approximation given in Eq. (7.2.226).

Solution: For this case, the approximation functions are given by

$$X_i = \left(\frac{x}{a}\right)^{i+1} - 2\left(\frac{x}{a}\right)^{i+2} + \left(\frac{x}{a}\right)^{i+3}, \quad Y_j = \left(\frac{y}{b}\right)^{j+1} - \left(1 - \frac{y}{b}\right)^{j+2}. \tag{1}$$

The first four natural frequencies of rectangular plates are presented in Table 7.2.12.

Table 7.2.12 Effect of the plate aspect ratio a/b on the dimensionless frequencies $\bar{\omega}_{mn} = \omega_{mn}a^2\sqrt{\rho h/D}$ of isotropic ($\nu = 0.25$) rectangular plates (CCCS).

Mode	a/b					
(m, n)	0.25	0.50	0.667	1.0	1.5	2.0
$(1, 1)$	22.840	24.215	25.913	31.849	48.217	73.573
$(1, 2)$	24.654	34.421	46.951	72.025	86.045	108.452
$(2, 1)$	63.411	64.957	66.663	86.497	179.342	310.663
$(2, 2)$	65.444	74.254	84.914	120.087	208.448	337.474

Additional results of the natural vibration of rectangular plates with other boundary conditions can be found in Leissa [126], Liew *et al.* [127], and Reddy [51].

7.2.8 Variational Solutions: Buckling

7.2.8.1 Rectangular Plates Simply Supported along Two Opposite Sides and Compressed in the Direction Perpendicular to Those Sides

In Section 7.2.5, we discussed bending solutions of rectangular plates simply supported along two sides and having arbitrary boundary conditions on the other two sides. There we used the Lévy method of solution. Here we consider variational solutions of the buckling of rectangular plates simply supported along edges $x = 0, a$ and subjected to uniform compression along the same edges (see Fig. 7.2.22). From Eq. (7.2.110), by setting the time derivative and load terms to zero, and taking $\hat{N}_{xx} = N_0$, $\hat{N}_{yy} = 0$, and $\hat{N}_{xy} = 0$, we obtain

$$D\left(\frac{\partial^4 w}{\partial x^4} + 2\frac{\partial^4 w}{\partial x^2 \partial y^2} + \frac{\partial^4 w}{\partial y^4}\right) + \hat{N}_{xx}\frac{\partial^2 w}{\partial x^2} = 0. \tag{7.2.229}$$

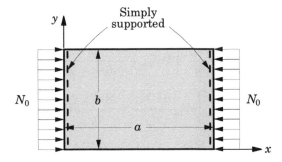

Fig. 7.2.22 Buckling of rectangular plates simply supported along two opposite edges ($x = 0, a$) and subjected to uniform compressive forces on the same edges; other edges may have any combination of boundary conditions.

In the Lévy method, we assumed solution of the form

$$w(x,y) = \sum_{m=1}^{\infty} W_m(y) \sin \alpha_m x, \quad \alpha_m = \frac{m\pi}{a}. \tag{7.2.230}$$

That is, the plate buckles into m sinusoidal half-waves. The assumed solution satisfies the simply supported conditions

$$w = 0, \quad M_{xx} \equiv -D \left(\frac{\partial^2 w}{\partial x^2} + \nu \frac{\partial^2 w}{\partial y^2} \right) = 0 \text{ at } x = 0, a. \tag{7.2.231}$$

Substitution of Eq.(7.2.230) in Eq. (7.2.229) yields

$$\frac{d^4 W_m}{dy^4} - 2\alpha_m^2 \frac{d^2 W_m}{dy^2} + \left(\alpha_m^4 - \frac{\alpha_m^2}{D} N_0 \right) W_m = 0, \tag{7.2.232}$$

which must be solved using the boundary conditions on edges $y = 0, b$. Here we wish to solve Eq. (7.2.232) using the Ritz method.

The weak form of Eq. (7.2.232) is

$$\int_0^b \left[\frac{d^2 W_m}{dy^2} \frac{d^2 \delta W_m}{dy^2} + 2\alpha_m^2 \frac{dW_m}{dy} \frac{d\delta W_m}{dy} + \left(\alpha_m^4 - \bar{N}_0 \alpha_m^2 \right) W_m \delta W_m \right] dy$$

$$+ \left[\left(\frac{d^3 W_m}{dy^3} - 2\alpha_m^2 \frac{dW_m}{dy} \right) \delta W_m - \frac{d^2 W_m}{dy^2} \frac{d\delta W_m}{dy} \right]_0^b = 0, \tag{7.2.233}$$

where $\bar{N}_0 = (N_0/D)$. For a simply supported edge $[W = 0$ and $(d^2 W/dy^2) = 0]$ or a clamped edge $[W = 0$ and $(dW/dy) = 0]$, the boundary terms in Eq. (7.2.233) vanish identically. However, for a free edge, the vanishing of bending moment and effective shear force require

$$\frac{d^2 W}{dy^2} - \nu \alpha_m^2 W_m = 0, \quad \frac{d^3 W_m}{dy^3} - (2 - \nu)\alpha_m^2 \frac{dW_m}{dy} = 0. \tag{7.2.234}$$

Hence, on a free edge, the boundary expression of Eq. (7.2.233) can be simplified to

$$\left[\left(\frac{d^3 W_m}{dy^3} - 2\alpha_m^2 \frac{dW_m}{dy} \right) \delta W_m - \frac{d^2 W_m}{dy^2} \frac{d\delta W_m}{dy} \right]_0^b$$

$$= -\nu \alpha_m^2 \left[W_m \frac{d\delta W_m}{dy} + \frac{dW_m}{dy} \delta W_m \right]_0^b \tag{7.2.235}$$

so that the weak form for the SSSS, SSCC, and SSCS plates becomes

$$0 = \int_0^b \left[\frac{d^2 W_m}{dy^2} \frac{d^2 \delta W_m}{dy^2} + 2\alpha_m^2 \frac{dW}{dy} \frac{d\delta W_m}{dy} + \left(\alpha_m^4 - \bar{N}_0 \alpha_m^2 \right) W_m \delta W_m \right] dy. \tag{7.2.236}$$

For the SSSF and SSCF plates, it is given by

$$
0 = \int_0^b \left[\frac{d^2 W_m}{dy^2} \frac{d^2 \delta W_m}{dy^2} + 2\alpha_m^2 \frac{dW_m}{dy} \frac{d\delta W_m}{dy} + \left(\alpha_m^4 - \bar{N}_0 \alpha_m^2 \right) W_m \delta W_m \right] dy
$$

$$
- \nu \alpha_m^2 \left[W_m \frac{d\delta W_m}{dy} + \frac{dW_m}{dy} \delta W_m \right]_0^b . \tag{7.2.237}
$$

Next we assume an N-parameter Ritz approximation of the form

$$
W_m(y) \approx \sum_{j=1}^N c_j^{(m)} \varphi_j(y), \tag{7.2.238}
$$

where φ_j are the approximation functions [see Eqs (7.2.191) to (7.2.216); replace x with y and a with b]. Substituting the approximation in Eq. (7.2.238) into the weak form in Eq. (7.2.237), we obtain

$$
(\mathbf{R} - N_0 \mathbf{B}) \, \mathbf{c} = \mathbf{0}, \tag{7.2.239}
$$

where

$$
R_{ij} = D \int_0^b \left(\frac{d^2 \varphi_i}{dy^2} \frac{d^2 \varphi_j}{dy^2} + 2\alpha_m^2 \frac{d\varphi_i}{dy} \frac{d\varphi_j}{dy} + \alpha_m^4 \varphi_i \varphi_j \right) dy
$$

$$
- \alpha_m^2 D \nu \left[\varphi_j \frac{d\varphi_i}{dy} + \frac{d\varphi_j}{dy} \varphi_i \right]_0^b , \tag{7.2.240}
$$

$$
B_{ij} = \alpha_m^2 \int_0^b \varphi_i \varphi_j \, dy, \quad \alpha_m = \frac{m\pi}{a}. \tag{7.2.241}
$$

The boundary term in the coefficient R_{ij} is nonzero only for SSSF and SSCF plates. Equation (7.2.240) has a non-trivial solution, $c_i \neq 0$, only if the determinant of the coefficient matrix $\mathbf{R} - N_0 \mathbf{B}$ is zero:

$$
|\mathbf{R} - N_0 \mathbf{B}| = 0. \tag{7.2.242}
$$

Equation (7.2.242) yields an Nth-order polynomial in N_0 that depends on m. Hence, for any given aspect ratio a/b, the critical buckling load is the smallest value of N_0 for all m.

Table 7.2.13 contains numerical values of dimensionless critical buckling loads, $\bar{N}_{cr} = N_{cr} b^2 / (\pi^2 D)$, obtained using the algebraic approximation functions given in Eqs (7.2.191) to (7.2.216) for various boundary conditions. The accuracy of the buckling loads predicted by the three-parameter Ritz approximation is very good and in agreement with the analytical solutions. Note that for certain aspect ratios, there is a change in the buckling mode.

Table 7.2.13 Nondimensionalized buckling loads $\bar{N}$ of rectangular isotropic ($\nu = 0.25$) plates under uniform compression $\hat{N}_{xx} = N_0$ (the Ritz solutions).

$\frac{a}{b}$	N	SSSS	SSCC	SSCS	SSCF	SSSF	SSFF
0.5	1	6.334	7.725	7.915	4.896	4.456	4.000
	3	6.250	7.693	6.860	4.526	4.436	3.958
	E	6.250	7.691	6.853	4.518	4.404	–
1.0	1	4.258	8.606	8.149	2.050	1.456	1.000
	3	4.001	8.605	5.741	1.699	1.445	0.972
	E	4.000	8.604	5.740	1.698	1.434	–
1.5	1	4.497^a	7.120^a	7.040^a	1.751	0.900	0.988
	3	4.341	7.116	5.432	1.339	0.894	0.426
	E	4.340	7.116	5.431	1.339	0.888	–
2.0	1	4.258	8.606	8.149	1.915	0.706	0.250
	3	4.001	8.605	5.741	1.386	0.702	0.238
	E	4.000	8.604	5.740	1.386	0.698	–

^{a} Denotes change to the next higher mode; E denotes the "exact" solution obtained in Section 7.2.4.3.

7.2.8.2 Formulation for Rectangular Plates with Arbitrary Boundary Conditions

In this section we consider buckling of orthotropic rectangular plates using the Ritz method. Buckling problems of plates with combined bending and compression or under pure in-plane shear stress do not permit analytical solutions; therefore, it is useful to consider the Ritz method to solve such problems. The statement of the principle of virtual displacements or the minimum total potential energy for isotropic plates subjected to in-plane edge forces $\hat{N}_{xx}$, $\hat{N}_{yy}$, and $\hat{N}_{xy}$ can be expressed as

$$
0 = \int_0^b \int_0^a \left\{ D \left[\frac{\partial^2 w}{\partial x^2} \frac{\partial^2 \delta w}{\partial x^2} + \nu \left(\frac{\partial^2 w}{\partial y^2} \frac{\partial^2 \delta w}{\partial x^2} + \frac{\partial^2 w}{\partial x^2} \frac{\partial^2 \delta w}{\partial y^2} \right) \right. \right.
$$
$$
\left. + 2(1-\nu) \frac{\partial^2 w}{\partial x \partial y} \frac{\partial^2 \delta w}{\partial x \partial y} + \frac{\partial^2 w}{\partial y^2} \frac{\partial^2 \delta w}{\partial y^2} \right]
$$
$$
+ \hat{N}_{xx} \frac{\partial w}{\partial x} \frac{\partial \delta w}{\partial x} + \hat{N}_{yy} \frac{\partial w}{\partial y} \frac{\partial \delta w}{\partial y}
$$
$$
\left. + \hat{N}_{xy} \left(\frac{\partial w}{\partial y} \frac{\partial \delta w}{\partial x} + \frac{\partial w}{\partial x} \frac{\partial \delta w}{\partial y} \right) \right\} dx dy. \tag{7.2.243}
$$

In general, the applied edge forces are functions of position and they are independent of each other. Let

$$
\hat{N}_{xx} = N_0, \quad \hat{N}_{yy} = \gamma_1 N_0, \quad \hat{N}_{xy} = \gamma_2 N_0, \tag{7.2.244}
$$

where N_0 is a constant and γ_1 and γ_2 are possibly functions of position.

Using an N-parameter Ritz solution of the form

$$w(x,y) \approx \sum_{j=1}^{N} c_j \varphi_j(x,y) \tag{7.2.245}$$

in Eq. (7.2.243), we obtain

$$(\mathbf{R} - N_0\mathbf{B})\,\mathbf{c} = \mathbf{0}, \tag{7.2.246}$$

with

$$R_{ij} = D \int_0^b \int_0^a \left[\frac{\partial^2 \varphi_i}{\partial x^2} \frac{\partial^2 \varphi_j}{\partial x^2} + \nu \left(\frac{\partial^2 \varphi_i}{\partial y^2} \frac{\partial^2 \varphi_j}{\partial x^2} + \frac{\partial^2 \varphi_i}{\partial x^2} \frac{\partial^2 \varphi_j}{\partial y^2} \right) \right.$$
$$\left. + 2(1-\nu) \frac{\partial^2 \varphi_i}{\partial x \partial y} \frac{\partial^2 \varphi_j}{\partial x \partial y} + \frac{\partial^2 \varphi_i}{\partial y^2} \frac{\partial^2 \varphi_j}{\partial y^2} \right] dx\,dy, \tag{7.2.247}$$

$$B_{ij} = \int_0^b \int_0^a \left[\frac{\partial \varphi_i}{\partial x} \frac{\partial \varphi_j}{\partial x} + \gamma_1 \frac{\partial \varphi_i}{\partial y} \frac{\partial \varphi_j}{\partial y} + \gamma_2 \left(\frac{\partial \varphi_i}{\partial x} \frac{\partial \varphi_j}{\partial y} + \frac{\partial \varphi_i}{\partial y} \frac{\partial \varphi_j}{\partial x} \right) \right]. \tag{7.2.248}$$

As discussed earlier, it is convenient to express the Ritz approximation of rectangular plates in the form

$$w(x,y) \approx W_{mn}(x,y) = \sum_{i=1}^{M} \sum_{j=1}^{N} c_{ij}\, X_i(x) Y_j(y). \tag{7.2.249}$$

The functions X_i and Y_j for various boundary conditions are given in Eqs (7.2.191) to (7.2.216). Substituting Eq. (7.2.245) into Eq. (7.2.243), we obtain Eq. (7.2.246) with the following definitions of the coefficients:

$$R_{(ij)(k\ell)} = D \int_0^b \int_0^a \left[\frac{d^2 X_i}{dx^2} \frac{d^2 X_k}{dx^2} Y_j Y_\ell + X_i X_k \frac{d^2 Y_j}{dy^2} \frac{d^2 Y_\ell}{dy^2} \right.$$
$$+ \nu \left(X_i \frac{d^2 X_k}{dx^2} \frac{d^2 Y_j}{dy^2} Y_\ell + \frac{d^2 X_i}{dx^2} X_k Y_j \frac{d^2 Y_\ell}{dy^2} \right)$$
$$\left. + 2(1-\nu) \frac{dX_i}{dx} \frac{dX_k}{dx} \frac{dY_j}{dy} \frac{dY_\ell}{dy} \right] dx\,dy, \tag{7.2.250}$$

$$B_{(ij)(k\ell)} = \int_0^b \int_0^a \left[\frac{dX_i}{dx} \frac{dX_k}{dx} Y_j Y_\ell + \gamma_1 X_i X_k \frac{dY_j}{dy} \frac{dY_\ell}{dy} \right.$$
$$\left. + \gamma_2 \left(\frac{dX_i}{dx} X_k Y_j \frac{dY_\ell}{dy} + X_i \frac{dX_k}{dx} \frac{dY_j}{dy} Y_\ell \right) \right] dx\,dy. \tag{7.2.251}$$

Next we consider an example of the application of Eqs (7.2.246), (7.2.250), and (7.2.251).

Example 7.2.21

Consider a simply supported plate. For the choice of algebraic functions

$$X_i(x) = \left(\frac{x}{a}\right)^i - \left(\frac{x}{a}\right)^{i+1}, \quad Y_j(y) = \left(\frac{y}{b}\right)^i - \left(\frac{y}{b}\right)^{j+1}, \tag{1}$$

and for $M = N = 1$, we obtain

$$R_{(11)(11)} = D\left(\frac{2b}{15a^3} + \frac{2}{9ab} + \frac{2a}{15b^3}\right),$$

$$B_{(11)(11)} = \frac{1}{90} + \frac{\gamma_1}{90} + \gamma_2 \times 0. \tag{2}$$

Note that the buckling load under in-plane shear cannot be determined with one-parameter approximation. The buckling load under uniaxial compression along the x-axis is given by setting $\gamma_1 = 0$:

$$N_{cr}^u = D\left(\frac{12b}{a^3} + \frac{20}{ab} + \frac{12a}{b^3}\right), \tag{3}$$

and the critical buckling load under biaxial compression is given by ($\gamma_1 = 1$)

$$N_{cr}^b = D\left(\frac{6b}{a^3} + \frac{10}{ab} + \frac{6a}{b^3}\right). \tag{4}$$

For square isotropic plates, the expressions in Eqs (3) and (4) become

$$N_{cr}^u = 44\frac{D}{a^2} = 4.458\frac{D\pi^2}{a^2}, \quad N_{cr}^b = 22\frac{D}{a^2} = 2.229\frac{D\pi^2}{a^2}. \tag{5}$$

The exact values for the two cases are $4D\pi^2/a^2$ and $2D\pi^2/a^2$, respectively. The results are in about -11.5% error, and the values do not change for $N = M = 2$.

For a square isotropic plate clamped on all sides and subjected to uniaxial compression (N_{cr}^u) or biaxial compression (N_{cr}^b), the buckling loads obtained using the Ritz method with algebraic polynomials given in Eq. (7.2.213) are (for $N = M = 1$ or 2)

$$N_{cr}^u = 10.943\frac{D\pi^2}{a^2}, \quad N_{cr}^b = 5.471\frac{D\pi^2}{a^2}. \tag{6}$$

Example 7.2.22

Consider a simply supported rectangular plate (see Fig. 7.2.23) with distributed in-plane forces applied in the middle plane of the plate on sides $x = 0, a$. The distribution of the applied forces is assumed to be

$$\hat{N}_{xx} = -N_0\gamma_1 = -N_0\left(1 - c_0\frac{y}{b}\right) \tag{1}$$

where N_0 is the magnitude of the compressive force at $y = 0$ and c_0 is a parameter that defines the relative bending and compression. For example, $c_0 = 0$ corresponds to the case of uniformly distributed compressive force, and for $c_0 = 2$ we obtain the case of pure bending. All other values give a combination of bending and compression or tension. Determine the deflection of the buckled plate, which is simply supported on all sides.

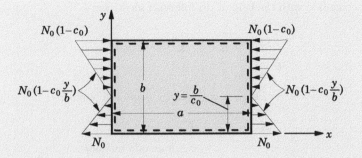

Fig. 7.2.23 Buckling of simply supported plates under combined bending and compression.

Solution: We seek the deflection of the simply supported buckled plate in the form of a double sine series

$$w(x, y) \approx W_{MN} = \sum_{n=1}^{N} \sum_{m=1}^{M} c_{mn} \sin \frac{m\pi x}{a} \sin \frac{n\pi y}{b}, \tag{2}$$

which satisfies the geometric as well as the force boundary conditions of the problem. Substituting Eq. (2) for w and

$$\sin \frac{p\pi x}{a} \sin \frac{q\pi y}{b}$$

for δw into Eq. (7.2.243) with $\hat{N}_{yy} = \hat{N}_{xy} = 0$, we obtain

$$0 = D \sum_{n=1}^{N} \sum_{m=1}^{M} c_{mn} \left\{ \left(\frac{m\pi}{a} \right)^2 \left(\frac{p\pi}{a} \right)^2 + \left(\frac{n\pi}{b} \right)^2 \left(\frac{q\pi}{b} \right)^2 \right.$$

$$\left. + \nu \left[\left(\frac{n\pi}{b} \right)^2 \left(\frac{p\pi}{a} \right)^2 + \left(\frac{m\pi}{a} \right)^2 \left(\frac{q\pi}{b} \right)^2 \right] \right\} I_1$$

$$+ D \sum_{n=1}^{N} \sum_{m=1}^{M} c_{mn} 2(1-\nu) \left(\frac{m\pi}{a} \right) \left(\frac{n\pi}{b} \right) \left(\frac{p\pi}{a} \right) \left(\frac{q\pi}{b} \right) m_2$$

$$- \sum_{n=1}^{N} \sum_{m=1}^{M} \sum_{q=1}^{M} c_{mq} N_0 \left(\frac{m\pi}{a} \right) \left(\frac{p\pi}{a} \right) I_{nq}, \tag{3}$$

where I_{11} and I_{22} are nonzero only when $p = m$ and $q = n$,

$$I_{11} = \int_0^b \int_0^a \sin \frac{m\pi x}{a} \sin \frac{n\pi y}{b} \sin \frac{p\pi x}{a} \sin \frac{q\pi y}{b} \, dxdy = \frac{ab}{4},$$

$$I_{22} = \int_0^b \int_0^a \cos \frac{m\pi x}{a} \cos \frac{n\pi y}{b} \cos \frac{p\pi x}{a} \cos \frac{q\pi y}{b} \, dxdy = \frac{ab}{4}, \tag{4}$$

and I_{nq} is defined as

$$I_{nq} = \int_0^b \int_0^a \left(1 - c_0 \frac{y}{b} \right) \cos \frac{m\pi x}{a} \cos \frac{p\pi x}{a} \sin \frac{n\pi y}{b} \sin \frac{q\pi y}{b} \, dxdy, \tag{5}$$

which can be computed with the help of the following identities:

$$I_{00} \equiv \int_0^b y \sin \frac{n\pi y}{b} \sin \frac{q\pi y}{b} \, dy$$

$$= \frac{b^2}{4} \quad \text{when } n = q$$

$$= 0 \quad \text{when } n \neq q \text{ and } n \pm q \text{ an even number}$$

$$= -\frac{b^2}{\pi^2} \frac{2nq}{(n^2 - q^2)^2} \quad \text{when } n \neq q \text{ and}$$

$$\qquad\qquad\qquad n \pm q \text{ an odd number} . \tag{6}$$

Thus $I_{nq} = 0$ if $p \neq m$ and

$$I_{nq} = \frac{a}{2}\left(\frac{b}{2} - \frac{c_0}{b} I_{00}\right), \tag{7}$$

when $p = m$. For any m and n, Eq. (3) becomes

$$0 = D\left[\left(\frac{m\pi}{a}\right)^4 + \left(\frac{n\pi}{b}\right)^4 + 2\left(\frac{n\pi}{b}\right)^2 \left(\frac{m\pi}{a}\right)^2\right] c_{mn}$$

$$- N_0 \left(\frac{m\pi}{a}\right)^2 \left[c_{mn} - \frac{c_0}{2}\left(c_{mn} - \frac{8}{\pi^2} \sum_{q=1}^{M} \frac{2nq c_{mq}}{(n^2 - q^2)^2}\right)\right], \tag{8}$$

where the summation is taken over all numbers q such that $n \pm q$ is an odd number. Taking $m = 1$ in Eq. (8), we obtain

$$D\left(\bar{s}^2 + n^2\right)^2 c_{1n} = N_0 \frac{\bar{s}^2 b^2}{\pi^2}\left[c_{1n}\left(1 - \frac{c_0}{2}\right) + \frac{8 c_0}{\pi^2} \sum_{q=1}^{M} \frac{nq c_{1q}}{(n^2 - q^2)^2}\right], \tag{9}$$

where $\bar{s}$ denotes the aspect ratio $\bar{s} = b/a$. A nontrivial solution, that is, for nonzero c_{1i}, the determinant of the linear equations in Eq. (9) must be zero if the plate buckles.

For one-parameter approximation ($N = M = 1$), we obtain

$$N_0 = \frac{\pi^2 D}{\bar{s}^2 b^2}(\bar{s}^2 + n^2)^2 \frac{1}{1 - 0.5 c_0}, \tag{10}$$

which gives a satisfactory result only for small values of c_0, that is, in cases where the bending stresses are small compared with the uniform compressive stress. Higher-order approximations yield sufficiently accurate results for the case of pure bending. Table 7.2.14 contains critical buckling loads $\bar{N} = N_0(b^2/\pi^2 D)$ obtained using two-parameter approximation, except for $c_0 = 2$, where the three-parameter approximation was used.

Table 7.2.14 Nondimensionalized critical buckling loads $\bar{N}$ of simply supported rectangular isotropic plates under combined bending and compression $\hat{N}_{xy} = N_0(1 - c_0 y/b)$ (the Ritz solutions).

c_0	$\frac{a}{b} \rightarrow$	0.4	0.5	0.6	$\frac{2}{3}$	0.75	0.8	0.9	1.0	1.5
2		29.1	25.6	24.1	23.9	24.1	24.4	25.6	25.6	24.1
4/3		18.7	–	12.9	–	11.5	11.2	–	11.0	11.5
1		15.1	–	9.7	–	8.4	8.1	–	7.8	8.4
4/5		13.3	–	8.3	–	7.1	6.9	–	6.6	7.1
2/3		10.8	–	7.1	–	6.1	6.0	–	5.8	6.1

Example 7.2.23

When a rectangular plate is simply supported on all its edges and subjected to uniformly distributed in-plane shear force $\hat{N}_{xy} = N_{xy}^0$ (see Fig. 7.2.24), the Navier or Lévy solution procedure cannot be used to determine the critical buckling load. Hence, use the Ritz or Galerkin method with trigonometric functions to determine the buckling loads.

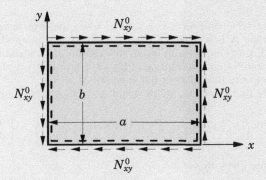

Fig. 7.2.24 Buckling of rectangular plates under the action of shearing stresses.

Solution: Let us seek the solution in the form

$$w(x,y) \approx W_{MN} = \sum_{n=1}^{N} \sum_{m=1}^{M} c_{mn} \, \sin \frac{m\pi x}{a} \, \sin \frac{n\pi y}{b}. \tag{1}$$

The approximate solution satisfies the geometric $(w = 0)$ as well as the natural $(M_{xx} = 0$ on sides $x = 0, a$ and $M_{yy} = 0$ on sides $y = 0, b)$ boundary conditions of the problem. Therefore, both the Ritz and Galerkin methods yield the same solutions.

The Galerkin solution is obtained by substituting Eq. (1) in the weighted-residual statement

$$0 = \int_0^b \int_0^a \left[D \left(\frac{\partial^4 w}{\partial x^4} + 2 \frac{\partial^4 w}{\partial x^2 \partial y^2} + \frac{\partial^4 w}{\partial y^4} \right) - 2N_{xy}^0 \frac{\partial^2 w}{\partial x \partial y} \right] \varphi_{pq} \, dx dy. \tag{2}$$

We obtain

$$0 = \frac{Dab}{4} \left[\left(\frac{p\pi}{a} \right)^4 + \left(\frac{p\pi}{a} \right)^2 \left(\frac{q\pi}{b} \right)^2 + \left(\frac{q\pi}{b} \right)^4 \right] c_{pq}$$

$$- 2N_{xy}^0 \sum_{m=1}^{N} \sum_{n=1}^{M} \left(\frac{m\pi}{a} \right) \left(\frac{n\pi}{b} \right) I_{mp} I_{nq} \, c_{mn}, \tag{3}$$

where

$$I_{mp} = \int_0^a \cos \frac{m\pi x}{a} \sin \frac{p\pi x}{a} \, dx = \left(\frac{2a}{\pi^2} \right) \frac{p}{(p^2 - m^2)} \quad \text{for } p^2 \neq m^2,$$

$$I_{nq} = \int_0^b \cos \frac{n\pi y}{b} \sin \frac{q\pi y}{b} \, dy = \left(\frac{2b}{\pi^2} \right) \frac{q}{(q^2 - n^2)} \quad \text{for } q^2 \neq n^2,$$

and the integral I_{mp} is zero when $p = m$ or $p \pm m$ is an even number, and I_{nq} is zero when $q = n$ or $q \pm n$ is an even number. The set of mn homogeneous equations in Eq. (3) define an eigenvalue problem,

$$\sum_{m=1}^{N} \sum_{n=1}^{M} \left(A_{(mn),(pq)} - N_{xy}^0 G_{(mn)(pq)} \right) c_{mn} = 0$$

$$([A] - N_{xy}^0[G])\{c\} = \{0\},\tag{4}$$

where

$$A_{(mn)(pq)} = D\delta_{mp}\delta_{nq}\frac{ab}{4}\left[\alpha_m^2\alpha_p^2 + 2\alpha_m\alpha_p\beta_n\beta_q + \beta_n^2\beta_q^2\right],$$

$$G_{(mn)(pq)} = 2\alpha_m\beta_n\,I_{mp}I_{nq},\quad \alpha_m = \frac{m\pi}{a},\quad \beta_n = \frac{n\pi}{b}.$$

Equation (4) has a nontrivial solution (i.e., $c_{mn} \neq 0$) when the determinant of the coefficient matrix is zero. Note that $\mathbf{A}$ is a diagonal matrix while $\mathbf{G}$ is a nonpositive-definite matrix; hence, the solution of Eq. (4) requires an eigenvalue routine that is suitable for nonpositive-definite matrices. It is found that the solution of Eq. (4) converges very slowly with increasing values of M and N. We note that $\mathbf{G}$ does not exist for $M = N = 1$.

For $M = N = 2$, we find from Eq. (4) the result ($G_{(11)(12)} = G_{(11)(21)} = 0$)

$$\left(\begin{bmatrix} A_{11} & 0 \\ 0 & A_{22} \end{bmatrix} - N_{xy}^0 \begin{bmatrix} 0 & G_{(11)(22)} \\ G_{(22)(11)} & 0 \end{bmatrix}\right)\begin{Bmatrix} c_{11} \\ c_{22} \end{Bmatrix} = \begin{Bmatrix} 0 \\ 0 \end{Bmatrix},\tag{5}$$

where the coefficients are given by

$$A_{(11)(11)} = \frac{\pi^4 D}{4\bar{s}b^2}(\bar{s}^2 + 1)^2,\quad A_{(22)(22)} = 16A_{(11)(11)},$$

$$G_{(11)(11)} = G_{(22)(22)} = 0,\quad G_{(11)(22)} = G_{(22)(11)} = \frac{32}{9},$$

and $\bar{s} = b/a$ is the plate aspect ratio. Setting the determinant of the coefficient matrix in Eq. (5) to zero, we obtain

$$N_{xy}^0 = \pm\frac{9D\pi^4}{32\bar{s}b^2}(\bar{s}^2 + 1)^2.$$

The two signs indicate that the value of the critical buckling load does not depend on sign.

Timoshenko and Gere [128] obtained the following equation for short isotropic plates ($a/b < 2$) using a five-term ($c_{11}, c_{22}, c_{13}, c_{31}, c_{33}$, and c_{42}) approximation:

$$\lambda^2 = \frac{\bar{s}^4}{81(1+\bar{s}^2)^4}\left[1 + \frac{81}{625} + \frac{81}{25}\left(\frac{1+\bar{s}^2}{9+\bar{s}^2}\right)^2 + \frac{81}{25}\left(\frac{1+\bar{s}^2}{1+9\bar{s}^2}\right)^2\right],$$

where

$$\bar{s} = \frac{b}{a},\quad \lambda = -\frac{\pi^4 D}{32abN_{xy}^0}.$$

For a square plate ($b = a$), the previous equation yields the critical buckling load $(N_{xy}^0)_{cr} = 9.4(\pi^2 D/a^2)$, whereas the value obtained with a larger (than 5) number of equations gives 9.34 in place of 9.4. Table 7.2.15 contains the critical buckling loads $\bar{N} = (N_{xy}^0)_{cr}(a^2/\pi^2 D)$ obtained using a large number of parameters.

Table 7.2.15 Nondimensionalized critical buckling loads $\bar{N}$ of rectangular isotropic ($\nu = 0.25$) plates under uniform shear $\hat{N}_{xy} = N_{xy}^0$.

$s = \frac{a}{b}$	1.0	1.2	1.4	1.5	1.6	1.8	2.0	2.5	3	4
$\bar{N}$	9.34	8.0	7.3	7.1	7.0	6.8	6.6	6.1	5.9	5.7

Example 7.2.24

Determine the critical buckling load of a clamped rectangular plate under in-plane shear using the Ritz method.

Solution: The principle of the minimum total potential energy for this case is

$$
0 = D \int_0^b \int_0^a \left[\frac{\partial^2 w}{\partial x^2} \frac{\partial^2 \delta w}{\partial x^2} + \nu \left(\frac{\partial^2 w}{\partial y^2} \frac{\partial^2 \delta w}{\partial x^2} + \frac{\partial^2 w}{\partial x^2} \frac{\partial^2 \delta w}{\partial y^2} \right) \right.
$$
$$
+ 2(1-\nu) \frac{\partial^2 w}{\partial x \partial y} \frac{\partial^2 \delta w}{\partial x \partial y} + \frac{\partial^2 w}{\partial y^2} \frac{\partial^2 \delta w}{\partial y^2}
$$
$$
\left. - \frac{N_{xy}^0}{D} \left(\frac{\partial \delta w}{\partial x} \frac{\partial w}{\partial y} + \frac{\partial w}{\partial x} \frac{\partial \delta w}{\partial y} \right) \right] dx dy. \tag{1}
$$

We assume a Ritz approximation of the form

$$
w(x,y) \approx W_{MN}(x,y) = \sum_{j=1}^N \sum_{i=1}^M c_{ij} \, X_i(x) \, Y_j(y), \tag{2}
$$

and we obtain

$$
0 = D \sum_{j=1}^N \sum_{i=1}^M \left\{ \int_0^b \int_0^a \left[\frac{d^2 X_i}{dx^2} Y_j \frac{d^2 X_p}{dx^2} Y_q + X_i \frac{d^2 Y_j}{dy^2} X_p \frac{d^2 Y_q}{dy^2} \right. \right.
$$
$$
+ 2 \frac{dX_i}{dx} \frac{dY_j}{dy} \frac{dX_p}{dx} \frac{dY_q}{dy}
$$
$$
\left. \left. - \frac{N_{xy}^0}{D} \left(\frac{dX_i}{dx} Y_j X_p \frac{dY_q}{dy} + X_i \frac{dY_j}{dy} \frac{dX_p}{dx} Y_q \right) \right] dx dy \right\} c_{ij}. \tag{3}
$$

Using the two-parameter approximation

$$
w(x,y) \approx c_{11} X_1(x) Y_1(y) + c_{22} X_2(x) Y_2(y), \tag{4}
$$

with

$$
X_1(x) = \sin \frac{4.73x}{a} - \sinh \frac{4.73x}{a} + 1.0178 \left(\cosh \frac{4.73x}{a} - \cos \frac{4.73x}{a} \right)
$$
$$
X_2(x) = \sin \frac{7.853x}{a} - \sinh \frac{7.853x}{a} + 0.9992 \left(\cosh \frac{7.853x}{a} - \cos \frac{7.853x}{a} \right)
$$
$$
Y_1(y) = \sin \frac{4.73y}{b} - \sinh \frac{4.73y}{b} + 1.0178 \left(\cosh \frac{4.73y}{b} - \cos \frac{4.73y}{b} \right)
$$
$$
Y_2(y) = \sin \frac{7.853y}{b} - \sinh \frac{7.853y}{b} + 0.9992 \left(\cosh \frac{7.853y}{b} - \cos \frac{7.853y}{b} \right), \tag{5}
$$

we obtain

$$
\begin{bmatrix} B_{11} & -N_{xy}^0 B_{12} \\ -N_{xy}^0 B_{12} & B_{22} \end{bmatrix} \begin{Bmatrix} c_{11} \\ c_{22} \end{Bmatrix} = \begin{Bmatrix} 0 \\ 0 \end{Bmatrix}, \tag{6}
$$

where

$$
B_{11} = D \left(\frac{537.181}{a^4} + \frac{324.829}{a^2 b^2} + \frac{537.181}{b^4} \right), \quad B_{12} = \frac{23.107}{ab},
$$
$$
B_{22} = D \left(\frac{3791.532}{a^4} + \frac{4227.255}{a^2 b^2} + \frac{3791.532}{b^4} \right).
$$

For a nontrivial solution, the determinant of the coefficient matrix in Eq. (6) should be zero, $B_{11}B_{22} - B_{12}B_{12}(N_{xy}^0)^2 = 0$. Solving for the buckling load N_{xy}^0, we obtain

$$N_{xy}^0 = \pm \frac{1}{B_{12}}\sqrt{B_{11}B_{22}}.$$

The $\pm$ sign indicates that the shear buckling load may be either positive or negative. For an isotropic square plate, we have $a = b$, and the shear buckling load becomes

$$N_{xy}^0 = \pm 176 \frac{D}{a^2}, \tag{7}$$

whereas the "exact" critical buckling load is

$$N_{xy}^0 = \pm 145 \frac{D}{a^2}. \tag{8}$$

The two-term Ritz solution in Eq. (7) is over 21% in error.

This completes the discussion on the application of the Ritz method to the buckling of rectangular plates.

7.3   The First-Order Shear Deformation Plate Theory

The extension of the Timoshenko beam theory (TBT), which accounts for the transverse shear strain, to two dimensions is termed the first-order shear deformation plate theory (FSDT). In this section, we consider bending, vibration, and buckling of plates according to the FSDT. We consider bending deformation of circular and rectangular plate geometries.

7.3.1   Equations of Circular Plates

We begin with the following displacement field:

$$\begin{aligned}
u_r(r, \theta, z) &= z\phi_r(r, \theta), \\
u_\theta(r, \theta, z) &= z\phi_\theta(r, \theta), \\
u_z(r, \theta, z) &= w(r, \theta),
\end{aligned} \tag{7.3.1}$$

where w is the transverse displacement, and (ϕ_r, ϕ_θ) are rotations of a transverse normal line about the (r, θ) coordinates. The quantities (w, ϕ_r, ϕ_θ) are called the *generalized displacements*. For thin plates, that is, when the plate in-plane characteristic dimension to thickness ratio is on the order of 50 or greater, the rotation functions ϕ_r and ϕ_θ will approach the respective slopes of the transverse deflection:

$$\phi_r = -\frac{\partial w}{\partial r}, \quad \phi_\theta = -\frac{1}{r}\frac{\partial w}{\partial \theta}.$$

The linear strain components referred to the cylindrical coordinate system are given by

$$\varepsilon_{rr} = z\varepsilon_{rr}^{(1)}, \quad \varepsilon_{\theta\theta} = z\varepsilon_{\theta\theta}^{(1)}, \quad 2\varepsilon_{r\theta} = z\gamma_{r\theta}^{(1)},$$

$$2\varepsilon_{z\theta} = \phi_\theta + \frac{1}{r}\frac{\partial w}{\partial \theta}, \quad 2\varepsilon_{zr} = \phi_r + \frac{\partial w}{\partial r}, \tag{7.3.2}$$

where

$$\varepsilon_{rr}^{(1)} = \frac{\partial \phi_r}{\partial r}, \quad \varepsilon_{\theta\theta}^{(1)} = \frac{\phi_r}{r} + \frac{1}{r}\frac{\partial \phi_\theta}{\partial \theta}, \quad \gamma_{r\theta}^{(1)} = \frac{1}{r}\frac{\partial \phi_r}{\partial \theta} + \frac{\partial \phi_\theta}{\partial r} - \frac{\phi_\theta}{r}. \tag{7.3.3}$$

Since we are interested in deriving the equations of motion and the form of the boundary conditions, we assume that the plate is subjected to transverse load q. Using the principle of virtual displacements, $\delta W = \delta W_I + \delta W_E = 0$, we can write

$$0 = \int_{\Omega_0} \int_{-\frac{h}{2}}^{\frac{h}{2}} \left(\sigma_{rr}\delta\varepsilon_{rr} + \sigma_{\theta\theta}\delta\varepsilon_{\theta\theta} + \sigma_{r\theta}\delta\gamma_{r\theta} + K_s\sigma_{rz}\delta\gamma_{rz} + K_s\sigma_{z\theta}\delta\gamma_{z\theta}\right) dz\, rdrd\theta$$

$$- \int_{\Omega_0} q\delta w\, rdrd\theta$$

$$= \int_{\Omega_0} \left[M_{rr}\frac{\partial\delta\phi_r}{\partial r} + M_{r\theta}\left(\frac{1}{r}\frac{\partial\delta\phi_r}{\partial\theta} + \frac{\partial\delta\phi_\theta}{\partial r} - \frac{\delta\phi_\theta}{r}\right) + M_{\theta\theta}\left(\frac{\delta\phi_r}{r} + \frac{1}{r}\frac{\partial\delta\phi_\theta}{\partial\theta}\right) \right.$$

$$\left. + Q_r\left(\delta\phi_r + \frac{\partial\delta w}{\partial r}\right) + Q_\theta\left(\delta\phi_\theta + \frac{1}{r}\frac{\partial\delta w}{\partial\theta}\right) - q\delta w \right] rdrd\theta, \tag{7.3.4}$$

where

$$M_{rr} = \int_{-\frac{h}{2}}^{\frac{h}{2}} \sigma_{rr}z\, dz, \quad M_{\theta\theta} = \int_{-\frac{h}{2}}^{\frac{h}{2}} \sigma_{\theta\theta}z\, dz, \quad M_{r\theta} = \int_{-\frac{h}{2}}^{\frac{h}{2}} \sigma_{r\theta}z\, dz,$$

$$Q_\theta = K_s \int_{-\frac{h}{2}}^{\frac{h}{2}} \sigma_{z\theta}\, dz, \quad Q_r = K_s \int_{-\frac{h}{2}}^{\frac{h}{2}} \sigma_{rz}\, dz, \tag{7.3.5}$$

and K_s is the shear correction factor. The Euler equations are

$$\delta\phi_r : \qquad -\frac{1}{r}\left[\frac{\partial}{\partial r}(rM_{rr}) + \frac{\partial M_{r\theta}}{\partial\theta} - M_{\theta\theta}\right] + Q_r = 0, \tag{7.3.6}$$

$$\delta\phi_\theta : \qquad -\frac{1}{r}\left[\frac{\partial}{\partial r}(rM_{r\theta}) + \frac{\partial M_{\theta\theta}}{\partial\theta} + M_{r\theta}\right] + Q_\theta = 0, \tag{7.3.7}$$

$$\delta w : \qquad -\frac{1}{r}\left[\frac{\partial}{\partial r}(rQ_r) + \frac{\partial Q_\theta}{\partial\theta}\right] - q = 0. \tag{7.3.8}$$

The essential boundary conditions involve specifying (ϕ_r, ϕ_θ, w) and the natural boundary conditions require specifying the expressions

$$
\begin{aligned}
\delta\phi_r : & \quad M_{nn} \equiv r M_{rr} n_r + M_{r\theta} n_\theta = 0, \\
\delta\phi_\theta : & \quad M_{ns} \equiv r M_{r\theta} n_r + M_{\theta\theta} n_\theta = 0, \\
\delta w : & \quad V_r \equiv r Q_r n_r + Q_\theta n_\theta = 0.
\end{aligned}
\tag{7.3.9}
$$

We note that only one element of the following three pairs are specified at a boundary point:

$$
(w, V_r), \quad (\phi_r, M_{nn}), \quad (\phi_\theta, M_{ns}).
\tag{7.3.10}
$$

The bending moments $(M_{rr}, M_{\theta\theta})$ and shear forces (Q_r, Q_θ) are related to the generalized displacements (w, ϕ_r, ϕ_θ) by

$$
M_{rr} = D \left[\frac{\partial \phi_r}{\partial r} + \frac{\nu}{r} \left(\phi_r + \frac{\partial \phi_\theta}{\partial \theta} \right) \right],
$$

$$
M_{\theta\theta} = D \left[\nu \frac{\partial \phi_r}{\partial r} + \frac{1}{r} \left(\phi_r + \frac{\partial \phi_\theta}{\partial \theta} \right) \right],
\tag{7.3.11}
$$

$$
M_{r\theta} = (1 - \nu) D \frac{1}{r} \left(\frac{\partial \phi_r}{\partial \theta} + r \frac{\partial \phi_\theta}{\partial r} - \phi_\theta \right),
$$

$$
Q_r = K_s G h \left(\phi_r + \frac{\partial w}{\partial r} \right),
$$

$$
Q_\theta = K_s G h \left(\phi_\theta + \frac{1}{r} \frac{\partial w}{\partial \theta} \right).
\tag{7.3.12}
$$

where K_s denotes the shear correction factor and $D = Eh^3/[12(1 - \nu^2)]$; E is Young's modulus, G is the shear modulus, ν is Poisson's ratio, and h is the total thickness of the plate.

7.3.2   Exact Solutions of Axisymmetric Circular Plates

For axisymmetric bending, all variables are independent of the angular coorinate θ and $u_\theta = 0$ and $\phi_\theta = 0$. Consequently, $\varepsilon_{r\theta}$ and $\varepsilon_{z\theta}$ are zero; hence, $\sigma_{r\theta}$ and $\sigma_{z\theta}$ and their resultants Q_θ and $M_{r\theta}$ are all zero. Then the equilibrium equations in Eqs (7.3.6) and (7.3.8) [Eq. (7.3.7) is trivially satisfied] become

$$
-\frac{1}{r} \left[\frac{d}{dr} (r M_{rr}) - M_{\theta\theta} \right] + Q_r = 0,
\tag{7.3.13}
$$

$$
-\frac{1}{r} \frac{d}{dr} (r Q_r) - q = 0,
\tag{7.3.14}
$$

where

$$M_{rr} = D\left(\frac{d\phi_r}{dr} + \nu\frac{\phi_r}{r}\right), \tag{7.3.15}$$

$$M_{\theta\theta} = D\left(\nu\frac{d\phi_r}{dr} + \frac{\phi_r}{r}\right), \tag{7.3.16}$$

$$Q_r = K_sGh\left(\phi_r + \frac{dw}{dr}\right), \tag{7.3.17}$$

and $D = Eh^3/[12(1-\nu^2)]$.

Integration of Eq. (7.3.14) gives

$$rQ_r = -\int rq(r)\,dr + c_1. \tag{7.3.18}$$

Use of Eqs (7.3.15), (7.3.16), and (7.3.18) in Eq. (7.3.13) and integration leads to the result

$$D\phi_r(r) = -\frac{dF}{dr} + \frac{r}{4}(2\log r - 1)c_1 + \frac{r}{2}c_2 + \frac{1}{r}c_3, \tag{7.3.19}$$

where

$$F(r) = \int\frac{1}{r}\int r\int\frac{1}{r}\int rq(r)\,dr dr dr dr. \tag{7.3.20}$$

Finally, from Eqs (7.3.17) to (7.3.19), we arrive at

$$K_sGh\frac{dw}{dr} = -\frac{K_sGh}{D}\left[-\frac{dF}{dr} + \frac{r}{4}(2\log r - 1)c_1 + \frac{r}{2}c_2 + \frac{1}{r}c_3\right]$$
$$-\frac{1}{r}\int rq(r)\,dr + \frac{c_1}{r}, \tag{7.3.21}$$

$$K_sGh\,w(r) = -\frac{K_sGh}{D}\left[-F(r) + \frac{r^2}{4}(\log r - 1)c_1 + \frac{r^2}{4}c_2 + c_3\log r\right]$$
$$-\int\frac{1}{r}\int rq(r)\,dr dr + c_1\log r + c_4. \tag{7.3.22}$$

The constants of integration, c_1, c_2, c_3, and c_4 are determined using the boundary conditions. The derivative of ϕ_r is given by

$$D\frac{d\phi_r}{dr} = -\frac{d^2F}{dr^2} + \frac{1}{4}(2\log r + 1)c_1 + \frac{1}{2}c_2 - \frac{1}{r^2}c_3.$$

Then we can compute the bending moments as

$$M_{rr}(r) = -\frac{d^2 F}{dr^2} + \frac{1}{4}(2\log r + 1)c_1 + \frac{1}{2}c_2 - \frac{1}{r^2}c_3$$
$$+ \frac{\nu}{r}\left[-\frac{dF}{dr} + \frac{r}{4}(2\log r - 1)c_1 + \frac{r}{2}c_2 + \frac{1}{r}c_3\right], \qquad (7.3.23)$$

$$M_{\theta\theta}(r) = \nu\left(-\frac{d^2 F}{dr^2} + \frac{1}{4}(2\log r + 1)c_1 + \frac{1}{2}c_2 - \frac{1}{r^2}c_3\right)$$
$$+ \frac{1}{r}\left[-\frac{dF}{dr} + \frac{r}{4}(2\log r - 1)c_1 + \frac{r}{2}c_2 + \frac{1}{r}c_3\right]. \qquad (7.3.24)$$

For solid circular plates, the condition that the rotation ϕ_r be finite at $r = 0$ requires [from Eq. (7.3.19)] that $c_3 = 0$. In addition, if the plate is not subjected to a point load at $r = 0$, the shear force must be zero there. This implies that [from Eq. (7.3.18)] $c_1 = 0$. Thus, for solid circular plates without a point load at the center, we must have

$$c_1 = 0, \quad c_3 = 0. \qquad (7.3.25)$$

If a solid circular plate is subjected to a point load Q_0 at the center, we have

$$2\pi(rQ_r) = -Q_0 \text{ at } r = 0 \rightarrow c_1 = -\frac{Q_0}{2\pi}. \qquad (7.3.26)$$

Obviously, $c_3 \neq 0$ and the condition in Eq. (7.3.26) is not meaningful for annular plates.

Example 7.3.1

Determine the exact solutions of a clamped, solid circular plate of radius a and subjected to (a) uniformly distributed load of intensity q_0 and (b) point load Q_0 at the center of the plate.

Solution: (a) For this case $c_3 = 0$. The boundary conditions associated with the clamped edge $r = a$ are

$$w(a) = 0, \quad \phi_r(a) = 0. \qquad (1)$$

For uniform load of intensity q_0, we have $c_1 = 0$ and

$$c_2 = \frac{q_0 a^2}{8}, \quad c_4 = \frac{q_0 a^2}{4} + \frac{K_s G h}{D}\frac{q_0 a^4}{64}. \qquad (2)$$

Hence, the deflection and rotation becomes

$$w(r) = \frac{q_0 a^4}{64D}\left(1 - \frac{r^2}{a^2}\right)^2 + \frac{q_0 a^2}{4K_s G h}\left(1 - \frac{r^2}{a^2}\right), \qquad (3)$$

$$\phi_r(r) = \frac{q_0 a^3}{16D}\frac{r}{a}\left(1 - \frac{r^2}{a^2}\right). \qquad (4)$$

Note that the deflection in Eq. (3) has two parts, one due to bending $w^b(r)$ and the other due to shear $w^s(r)$:

$$w^b(r) = \frac{q_0 a^4}{64D}\left(1 - \frac{r^2}{a^2}\right)^2, \quad w^s(r) = \frac{q_0 a^2}{4K_s G h}\left(1 - \frac{r^2}{a^2}\right). \qquad (5)$$

The bending deflection $w^b(r)$ is the same as that predicted by the classical plate theory [see Eq. (5) of **Example 7.2.1**]. Thus, the effect of including transverse shear strain in the formulation is to increase the deflection by $w^s(r)$. In other words, the CPT underpredicts deflection compared with the FSDT. We also note that the rotation ϕ_r, in the present case, is equal to

$$\phi_r(r) = -\frac{dw^b}{dr}. \tag{6}$$

That is, the rotation of a transverse normal is not affected by the shear deformation. Also, note that (dw^s/dr) is not zero at $r = a$.

The maximum deflection of the plate is

$$w_{max} = \frac{q_0 a^4}{64D} + \frac{q_0 a^2}{4K_s Gh} = \frac{q_0 a^4}{64D}\left(1 + \frac{8}{3(1-\nu)K_s}\frac{h^2}{a^2}\right). \tag{7}$$

For $\nu = 0.3$, $K_s = 5/6$, and $h/a = 0.01$ (a thin plate), the difference between the maximum deflections predicted by the classical and first-order shear deformation theories is 4.57×10^{-4}, which is negligible. For $h/a = 0.1$ (a moderately thick plate) it is 4.57×10^{-2}, or 4.57%. Thus, the effect of shear deformation on the deflection is more for thick plates.

The expressions for the bending moments and stresses are the same as those in Eqs (7)–(10) of **Example 7.2.1** (why so?). The shear force Q_r and shear stress σ_{rz} are given by

$$Q_r(r) = K_s Gh\left(\phi_r + \frac{dw}{dr}\right)$$

$$= K_s Gh\frac{dw^s}{dr} = -\frac{q_0 r}{2}, \tag{8}$$

$$\sigma_{rz}(r) = G\left(\phi_r + \frac{dw}{dr}\right)$$

$$= -\frac{K_s q_0 r}{2h}. \tag{9}$$

(b) For point load Q_0 at the center, we have

$$c_1 = -\frac{Q_0}{2\pi}, \quad c_3 = 0. \tag{10}$$

The boundary conditions of the clamped edge give

$$c_2 = \frac{Q_0}{4\pi}(2\log a - 1), \quad c_4 = \frac{K_s Gh}{D}\frac{Q_0 a^2}{16\pi} + \frac{Q_0}{2\pi}\log a. \tag{11}$$

Hence, the solution becomes

$$w(r) = \frac{Q_0 a^2}{16\pi D}\left[1 - \frac{r^2}{a^2} + 2\frac{r^2}{a^2}\log\left(\frac{r}{a}\right)\right] - \frac{Q_0}{2\pi K_s Gh}\log\frac{r}{a}, \tag{12}$$

$$\phi_r(r) = -\frac{Q_0 r}{4\pi D}\log\frac{r}{a}. \tag{13}$$

As before, the deflection predicted by the first-order shear deformation plate theory has two parts: one due to bending and the other due to shear. The bending part is the same as that predicted by the classical plate theory, and $\phi_r(r) = -(dw^b/dr)$. However, the shear part $w^s(r)$ is singular at $r = 0$, making it difficult to determine the maximum deflection.

Example 7.3.2

Determine the exact solution of a simply supported solid circular plate under uniformly distributed load of intensity q_0.

Solution: The boundary conditions are

$$\text{At } r = 0: \ (rQ_r) = 0, \quad \text{At } r = a: \ w = 0, \ M_{rr} = 0. \tag{1}$$

The constants of integration are calculated to be $(c_1 = c_3 = 0)$

$$c_2 = \frac{q_0 a^2}{8} \frac{(3 + \nu)}{(1 + \nu)}, \quad c_4 = \frac{q_0 a^2}{4} + \frac{K_s G h}{D} \frac{q_0 a^4}{64} \frac{(5 + \nu)}{(1 + \nu)}. \tag{2}$$

The deflection and rotation become

$$w(r) = \frac{q_0 a^4}{64D} \left(\frac{r^4}{a^4} - 2\frac{3 + \nu}{1 + \nu} \frac{r^2}{a^2} + \frac{5 + \nu}{1 + \nu} \right) + \frac{q_0 a^2}{4K_s G h} \left(1 - \frac{r^2}{a^2} \right), \tag{3}$$

$$\phi_r(r) = \frac{q_0 a^3}{16D} \frac{r}{a} \left[\frac{(3 + \nu)}{(1 + \nu)} - \frac{r^2}{a^2} \right]. \tag{4}$$

Note that the deflection is affected by shear deformation [the second term in Eq. (3)]; however, for this problem, the rotation is not affected by transverse shear deformation.
    The maximum deflection is

$$w_{\text{max}} = w(0) = \left(\frac{5 + \nu}{1 + \nu} \right) \frac{q_0 a^4}{64D} + \frac{q_0 a^2}{4K_s G h}, \tag{5}$$

and the maximum rotation is equal to the negative of the slope at $r = a$:

$$\phi_{\text{max}} = \phi_r(a) = \frac{q_0 a^3}{8D(1 + \nu)}. \tag{6}$$

7.3.3   Equations of Plates in Rectangular Coordinates

We begin with the displacement field for the general case that includes in-plane deformation and time dependency:

$$\begin{aligned}
u_1(x, y, z, t) &= u(x, y, t) + z\phi_x(x, y, t), \\
u_2(x, y, z, t) &= v(x, y, t) + z\phi_y(x, y, t), \\
u_3(x, y, z, t) &= w(x, y, t).
\end{aligned} \tag{7.3.27}$$

As before, (u, v, w) denote the displacements of a point on the plane $z = 0$ and ϕ_x and ϕ_y are the rotations of a transverse normal about the y- and x-axes[4], respectively (see Fig. 7.3.1).

----

[4]To keep the same sign for both ϕ_x and ϕ_y in the displacement expansion (for convenience), we have not followed the vector convention (i.e., the right-hand rule) for defining ϕ_x and ϕ_y.

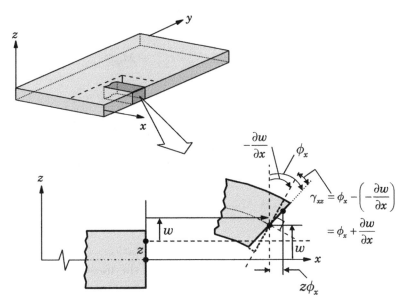

Fig. 7.3.1 Undeformed and deformed geometries of an edge of a plate under the assumptions of the first-order shear deformation plate theory (FSDT).

The linear strains associated with the displacement field in Eq. (7.3.27) are

$$
\begin{Bmatrix} \varepsilon_{xx} \\ \varepsilon_{yy} \\ \gamma_{yz} \\ \gamma_{xz} \\ \gamma_{xy} \end{Bmatrix} = \begin{Bmatrix} \varepsilon_{xx}^{(0)} \\ \varepsilon_{yy}^{(0)} \\ \gamma_{yz}^{(0)} \\ \gamma_{xz}^{(0)} \\ \gamma_{xy}^{(0)} \end{Bmatrix} + z \begin{Bmatrix} \varepsilon_{xx}^{(1)} \\ \varepsilon_{yy}^{(1)} \\ 0 \\ 0 \\ \gamma_{xy}^{(1)} \end{Bmatrix},
$$

$$
\{\varepsilon\} = \{\varepsilon^{(0)}\} + z\{\varepsilon^{(1)}\}, \tag{7.3.28}
$$

where

$$
\{\varepsilon^{(0)}\} = \begin{Bmatrix} \dfrac{\partial u}{\partial x} \\[2mm] \dfrac{\partial v}{\partial y} \\[2mm] \dfrac{\partial w}{\partial y} + \phi_y \\[2mm] \dfrac{\partial w}{\partial x} + \phi_x \\[2mm] \dfrac{\partial u}{\partial y} + \dfrac{\partial v}{\partial x} \end{Bmatrix}, \quad \{\varepsilon^{(1)}\} = \begin{Bmatrix} \dfrac{\partial \phi_x}{\partial x} \\[2mm] \dfrac{\partial \phi_y}{\partial y} \\[2mm] 0 \\[2mm] 0 \\[2mm] \dfrac{\partial \phi_x}{\partial y} + \dfrac{\partial \phi_y}{\partial x} \end{Bmatrix}. \tag{7.3.29}
$$

The governing equations of the first-order plate theory in rectangular Cartesian coordinates can be derived using the dynamic version of the principle of virtual displacements (or Hamilton's principle). The total virtual work done is

$$
\begin{aligned}
0 = \int_0^T \Bigg\{ & \int_{\Omega_0} \Big[N_{xx}\delta\varepsilon_{xx}^{(0)} + M_{xx}\delta\varepsilon_{xx}^{(1)} + N_{yy}\delta\varepsilon_{yy}^{(0)} + M_{yy}\delta\varepsilon_{yy}^{(1)} + N_{xy}\delta\gamma_{xy}^{(0)} \\
& + M_{xy}\delta\gamma_{xy}^{(1)} + Q_x\delta\gamma_{xz}^{(0)} + Q_y\delta\gamma_{yz}^{(0)} + kw\delta w - q\delta w \\
& + \frac{\partial\delta w}{\partial x}\left(\hat{N}_{xx}\frac{\partial w}{\partial x} + \hat{N}_{xy}\frac{\partial w}{\partial y} \right) + \frac{\partial\delta w}{\partial y}\left(\hat{N}_{xy}\frac{\partial w}{\partial x} + \hat{N}_{yy}\frac{\partial w}{\partial y} \right) \\
& - m_0\left(\dot{u}\delta\dot{u} + \dot{v}\delta\dot{v} + \dot{w}\delta\dot{w} \right) - m_2\left(\dot{\phi}_x\delta\dot{\phi}_x + \dot{\phi}_y\delta\dot{\phi}_y \right) \Big] dxdy \\
& - \int_{\Gamma_\sigma}\left(\hat{N}_{nn}\delta u_n + \hat{N}_{ns}\delta u_s + \hat{M}_{nn}\delta\phi_n + \hat{M}_{ns}\delta\phi_s + \hat{Q}_n\delta w \right) dS \Bigg\} dt, \quad (7.3.30)
\end{aligned}
$$

where k denotes the modulus of the elastic foundation; $(\hat{N}_{xx}, \hat{N}_{xy}, \hat{N}_{yy})$ are the applied in-plane edge forces; and (m_0, m_2) are the mass inertias defined in Eq. (7.2.91). The in-plane forces (N_{xx}, N_{xy}, N_{yy}) and moments (M_{xx}, M_{xy}, M_{yy}) are defined by

$$
(N_{xx}, N_{xy}, N_{yy}) = \int_{-\frac{h}{2}}^{\frac{h}{2}} (\sigma_{xx}, \sigma_{xy}, \sigma_{yy})dz,
$$

$$
(M_{xx}, M_{xy}, M_{yy}) = \int_{-\frac{h}{2}}^{\frac{h}{2}} (\sigma_{xx}, \sigma_{xy}, \sigma_{yy})z\, dz, \tag{7.3.31}
$$

and (N_{nn}, N_{ns}) are defined similar to the moments (M_{nn}, M_{ns}) in Eq. (7.2.101), and the transverse shear forces (Q_x, Q_y) are defined by

$$
\begin{Bmatrix} Q_x \\ Q_y \end{Bmatrix} = K_s \int_{-\frac{h}{2}}^{\frac{h}{2}} \begin{Bmatrix} \sigma_{xz} \\ \sigma_{yz} \end{Bmatrix} dz. \tag{7.3.32}
$$

The Euler–Lagrange equations are

$$
\delta u: \quad \frac{\partial N_{xx}}{\partial x} + \frac{\partial N_{xy}}{\partial y} = m_0\frac{\partial^2 u}{\partial t^2}, \tag{7.3.33}
$$

$$
\delta v: \quad \frac{\partial N_{xy}}{\partial x} + \frac{\partial N_{yy}}{\partial y} = m_0\frac{\partial^2 v}{\partial t^2}, \tag{7.3.34}
$$

$$
\delta w: \quad \frac{\partial Q_x}{\partial x} + \frac{\partial Q_y}{\partial y} - kw + q + \mathcal{N}(w, \hat{N}_{xx}, \hat{N}_{xy}, \hat{N}_{yy}) = m_0\frac{\partial^2 w}{\partial t^2}, \tag{7.3.35}
$$

$$
\delta\phi_x: \quad \frac{\partial M_{xx}}{\partial x} + \frac{\partial M_{xy}}{\partial y} - Q_x = m_2\frac{\partial^2 \phi_x}{\partial t^2}, \tag{7.3.36}
$$

$$
\delta\phi_y: \quad \frac{\partial M_{xy}}{\partial x} + \frac{\partial M_{yy}}{\partial y} - Q_y = m_2\frac{\partial^2 \phi_y}{\partial t^2}, \tag{7.3.37}
$$

where

$$\mathcal{N} = \frac{\partial}{\partial x}\left(\hat{N}_{xx}\frac{\partial w}{\partial x} + \hat{N}_{xy}\frac{\partial w}{\partial y}\right) + \frac{\partial}{\partial y}\left(\hat{N}_{xy}\frac{\partial w}{\partial x} + \hat{N}_{yy}\frac{\partial w}{\partial y}\right). \tag{7.3.38}$$

The natural boundary conditions are

$$N_{nn} - \hat{N}_{nn} = 0, \quad N_{ns} - \hat{N}_{ns} = 0,$$
$$M_{nn} - \hat{M}_{nn} = 0, \quad M_{ns} - \hat{M}_{ns} = 0, \tag{7.3.39}$$
$$Q_n - \hat{Q}_n = 0, \quad Q_n \equiv Q_x n_x + Q_y n_y.$$

The stress resultants in an isotropic plate are related to the generalized displacements by

$$\left\{\begin{matrix} N_{xx} \\ N_{yy} \\ N_{xy} \end{matrix}\right\} = A\begin{bmatrix} 1 & \nu & 0 \\ \nu & 1 & 0 \\ 0 & 0 & \frac{1-\nu}{2} \end{bmatrix}\left\{\begin{matrix} \frac{\partial u}{\partial x} \\ \frac{\partial v}{\partial y} \\ \frac{\partial u}{\partial y} + \frac{\partial v}{\partial x} \end{matrix}\right\}, \tag{7.3.40}$$

$$\left\{\begin{matrix} M_{xx} \\ M_{yy} \\ M_{xy} \end{matrix}\right\} = D\begin{bmatrix} 1 & \nu & 0 \\ \nu & 1 & 0 \\ 0 & 0 & \frac{1-\nu}{2} \end{bmatrix}\left\{\begin{matrix} \frac{\partial \phi_x}{\partial x} \\ \frac{\partial \phi_y}{\partial y} \\ \frac{\partial \phi_x}{\partial y} + \frac{\partial \phi_y}{\partial x} \end{matrix}\right\}, \tag{7.3.41}$$

$$\left\{\begin{matrix} Q_y \\ Q_x \end{matrix}\right\} = K_sGh\begin{bmatrix} 1 & 0 \\ 0 & 1 \end{bmatrix}\left\{\begin{matrix} \frac{\partial w}{\partial y} + \phi_y \\ \frac{\partial w}{\partial x} + \phi_x \end{matrix}\right\}, \tag{7.3.42}$$

where the extensional stiffness A and bending stiffness D are defined as

$$A = \frac{Eh}{1-\nu^2}, \qquad D = \frac{Eh^3}{12(1-\nu^2)}. \tag{7.3.43}$$

In the linear theories of orthotropic plates, the equations governing the inplane displacements (u, v), Eqs (7.3.33) and (7.3.34) are decoupled from equations governing the bending variables (w, ϕ_x, ϕ_y), Eqs (7.3.35) to (7.3.37). Hence, for plates with only bending loads, we have $u = v = 0$, and (w, ϕ_x, ϕ_y) can be determined by solving Eqs (7.3.35) to (7.3.37). These equations can be expressed in terms of the generalized displacements (w, ϕ_x, ϕ_y) by means of Eqs (7.3.41) and (7.3.42) (for constant values of D and K_sGh):

$$K_sGh\left(\frac{\partial^2 w}{\partial x^2} + \frac{\partial^2 w}{\partial y^2} + \frac{\partial \phi_x}{\partial x} + \frac{\partial \phi_y}{\partial y}\right) - kw + q(x, y)$$
$$+ \mathcal{N}(w, \hat{N}_{xx}, \hat{N}_{xy}, \hat{N}_{yy}) = m_0\frac{\partial^2 w}{\partial t^2}, \tag{7.3.44}$$

$$D\left[\frac{\partial^2 \phi_x}{\partial x^2} + \nu\frac{\partial^2 \phi_y}{\partial y \partial x} + \frac{(1-\nu)}{2}\left(\frac{\partial^2 \phi_x}{\partial y^2} + \frac{\partial^2 \phi_y}{\partial y \partial x}\right)\right]$$
$$- K_s Gh\left(\frac{\partial w}{\partial x} + \phi_x\right) = m_2\frac{\partial^2 \phi_x}{\partial t^2}, \tag{7.3.45}$$

$$D\left[\frac{(1-\nu)}{2}\left(\frac{\partial^2 \phi_x}{\partial x \partial y} + \frac{\partial^2 \phi_y}{\partial x^2}\right) + \nu\frac{\partial^2 \phi_x}{\partial x \partial y} + \frac{\partial^2 \phi_y}{\partial y^2}\right]$$
$$- K_s Gh\left(\frac{\partial w}{\partial y} + \phi_y\right) = m_2\frac{\partial^2 \phi_y}{\partial t^2}. \tag{7.3.46}$$

7.3.4   Exact Solutions of Rectangular Plates

As in the case of the CPT, analytical solutions of the FSDT can be developed for simply supported rectangular plates using Navier's method and for rectangular plates with two parallel edges simply supported and other two edges having arbitrary boundary conditions by Lévy's (semidiscretization) method. Here we limit our discussion to the Navier method of solution for the pure bending case. The Lévy method of analysis for the FSDT is more involved than the CPT (as there are three equations to be solved even for pure bending case), and details can be found in the books by Reddy [50, 51].

The simply supported boundary conditions for the FSDT can be expressed as (see Fig. 7.3.2)

$$w(x,0,t) = 0, \quad w(x,b,t) = 0, \quad w(0,y,t) = 0, \quad w(a,y,t) = 0, \tag{7.3.47}$$
$$\phi_x(x,0,t) = 0, \quad \phi_x(x,b,t) = 0, \quad \phi_y(0,y,t) = 0, \quad \phi_y(a,y,t) = 0, \tag{7.3.48}$$
$$M_{yy}(x,0,t) = 0, \quad M_{yy}(x,b,t) = 0, \quad M_{xx}(0,y,t) = 0, \quad M_{xx}(a,y,t) = 0. \tag{7.3.49}$$

The boundary conditions in Eqs (7.3.47) to (7.3.49) are satisfied exactly by the following expansions:

$$w(x,y,t) = \sum_{n=1}^{\infty}\sum_{m=1}^{\infty} W_{mn}(t)\sin\frac{m\pi x}{a}\sin\frac{n\pi y}{b},$$

$$\phi_x(x,y,t) = \sum_{n=1}^{\infty}\sum_{m=1}^{\infty} X_{mn}(t)\cos\frac{m\pi x}{a}\sin\frac{n\pi y}{b}, \tag{7.3.50}$$

$$\phi_y(x,y,t) = \sum_{n=1}^{\infty}\sum_{m=1}^{\infty} Y_{mn}(t)\sin\frac{m\pi x}{a}\cos\frac{n\pi y}{b},$$

where a and b denote the dimensions of the rectangular plate. The mechanical load is also expanded in double sine series:

$$q(x,y,t) = \sum_{n=1}^{\infty}\sum_{m=1}^{\infty} Q_{mn}(t)\sin\frac{m\pi x}{a}\sin\frac{n\pi y}{b}, \tag{7.3.51}$$

where

$$Q_{mn}(t) = \frac{4}{ab} \int_0^a \int_0^b q(x,y,t) \sin \frac{m\pi x}{a} \sin \frac{n\pi y}{b} \, dx dy. \tag{7.3.52}$$

Substitution of Eq. (7.3.50) into Eqs (7.3.44) to (7.3.46) yields (assuming that $\hat{N}_{xx}$ and $\hat{N}_{yy}$ are constants and $\hat{N}_{xy} = 0$) the following equations for the coefficients (W_{mn}, X_{mn}, Y_{mn}):

$$\begin{bmatrix} s_{11} + \hat{s}_{11} & s_{12} & s_{13} \\ s_{12} & s_{22} & s_{23} \\ s_{13} & s_{23} & s_{33} \end{bmatrix} \begin{Bmatrix} W_{mn} \\ X_{mn} \\ Y_{mn} \end{Bmatrix} + \begin{bmatrix} \hat{m}_{11} & 0 & 0 \\ 0 & \hat{m}_{22} & 0 \\ 0 & 0 & \hat{m}_{33} \end{bmatrix} \begin{Bmatrix} \ddot{W}_{mn} \\ \ddot{X}_{mn} \\ \ddot{Y}_{mn} \end{Bmatrix} = \begin{Bmatrix} Q_{mn} \\ 0 \\ 0 \end{Bmatrix}, \tag{7.3.53}$$

where s_{ij}, $\hat{s}_{11}$, and $\hat{M}_{ij}$ are defined (for an isotropic plate) by

$$s_{11} = K_s G h(\alpha_m^2 + \beta_n^2) + k, \quad \hat{s}_{11} = \alpha_m^2 \hat{N}_{xx} + \beta_n^2 \hat{N}_{yy},$$

$$s_{12} = K_s G h \, \alpha_m, \quad s_{13} = K_s G h \, \beta_n, \quad s_{22} = D\left(\alpha_m^2 + \frac{1-\nu}{2}\beta_n^2\right) + K_s G h$$

$$s_{23} = D\left(\nu + \frac{1-\nu}{2}\right)\alpha_m \beta_n, \quad s_{33} = D\left(\frac{1-\nu}{2}\alpha_m^2 + \beta_n^2\right) + K_s G h, \tag{7.3.54}$$

$$\hat{m}_{11} = m_0 = \rho h, \quad \hat{m}_{22} = \hat{m}_{33} = m_2 = \frac{\rho h^3}{12},$$

and $\alpha_m = m\pi/a$ and $\beta_n = n\pi/b$.

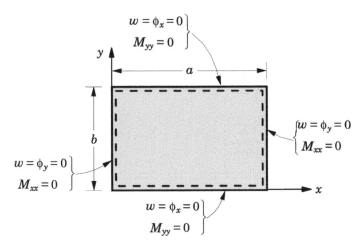

Fig. 7.3.2 The simply supported boundary conditions of the first-order shear deformation theory.

7.3.4.1   Bending Analysis

The static solution can be obtained from Eq. (7.3.53) by setting the time derivative terms and the inplane loads $(\hat{N}_{xx}, \hat{N}_{yy})$ to zero:

$$
\begin{bmatrix}
s_{11} & s_{12} & s_{13} \\
s_{12} & s_{22} & s_{23} \\
s_{13} & s_{23} & s_{33}
\end{bmatrix}
\begin{Bmatrix}
W_{mn} \\
X_{mn} \\
Y_{mn}
\end{Bmatrix}
=
\begin{Bmatrix}
Q_{mn} \\
0 \\
0
\end{Bmatrix}.
\tag{7.3.55}
$$

Solution of Eq. (7.3.55) for each $m, n = 1, 2, \ldots$ gives (W_{mn}, X_{mn}, Y_{mn}), which can then be used to compute the solution (w, ϕ_x, ϕ_y) from Eq. (7.3.50). We obtain

$$
W_{mn} = \frac{b_0}{b_{mn}} Q_{mn}, \ \ X_{mn} = \frac{b_1}{b_{mn}} Q_{mn}, \ Y_{mn} = \frac{b_2}{b_{mn}} Q_{mn},
\tag{7.3.56}
$$

where

$$
b_{mn} = s_{11}b_0 + s_{12}b_1 + s_{13}b_2, \ \ b_0 = s_{22}s_{33} - s_{23}s_{23},
$$
$$
b_1 = s_{23}s_{13} - s_{12}s_{33}, \ \ b_2 = s_{12}s_{23} - s_{22}s_{13}.
\tag{7.3.57}
$$

The bending moments are given by

$$
M_{xx} = -D \sum_{n=1}^{\infty} \sum_{m=1}^{\infty} (\alpha_m X_{mn} + \nu\beta_n Y_{mn}) \sin \alpha_m x \ \sin \beta_n y,
$$
$$
M_{yy} = -D \sum_{n=1}^{\infty} \sum_{m=1}^{\infty} (\nu\alpha_m X_{mn} + \beta_n Y_{mn}) \sin \alpha_m x \sin \beta_n y,
\tag{7.3.58}
$$
$$
M_{xy} = \frac{(1-\nu)D}{2} \sum_{n=1}^{\infty} \sum_{m=1}^{\infty} (\beta_n X_{mn} + \alpha_m Y_{mn}) \ \cos \alpha_m x \ \cos \beta_n y,
$$

and the shear forces can be computed from

$$
Q_x = -K_s Gh \sum_{n=1}^{\infty} \sum_{m=1}^{\infty} (\alpha_m W_{mn} + X_{mn}) \cos \alpha_m x \ \sin \beta_n y,
$$
$$
Q_y = -K_s Gh \sum_{n=1}^{\infty} \sum_{m=1}^{\infty} (\beta_n W_{mn} + Y_{mn}) \sin \alpha_m x \ \cos \beta_n y.
\tag{7.3.59}
$$

The stresses can be computed using the constitutive equations

$$
\begin{Bmatrix}
\sigma_{xx} \\
\sigma_{yy} \\
\sigma_{xy}
\end{Bmatrix}
= -\frac{Ez}{(1-\nu^2)} \sum_{n=1}^{\infty} \sum_{m=1}^{\infty}
\begin{Bmatrix}
(\alpha_m X_{mn} + \nu\beta_n Y_{mn}) \sin \alpha_m x \ \sin \beta_n y \\
(\nu\alpha_m X_{mn} + \beta_n Y_{mn}) \sin \alpha_m x \ \sin \beta_n y \\
-\frac{(1-\nu)}{2} (\beta_n X_{mn} + \alpha_m Y_{mn}) \ \cos \alpha_m x \ \cos \beta_n y
\end{Bmatrix},
\tag{7.3.60}
$$

$$
\begin{Bmatrix}
\sigma_{yz} \\
\sigma_{xz}
\end{Bmatrix}
= Gh \sum_{n=1}^{\infty} \sum_{m=1}^{\infty}
\begin{Bmatrix}
(Y_{mn} + \beta_n W_{mn}) \sin \alpha_m x \ \cos \beta_n y \\
(X_{mn} + \alpha_m W_{mn}) \cos \alpha_m x \ \sin \beta_n y
\end{Bmatrix}.
\tag{7.3.61}
$$

The transverse shear stresses can also be computed using the equilibrium equations of 3-D elasticity expressed in terms of stresses (see Reddy [50, 51] for the details). They are given by

$$
\sigma_{xz} = -\frac{h^2}{8}\left[1 - \left(\frac{2z}{h}\right)^2\right] (T_{11}X_{mn} + T_{12}Y_{mn}) \, \cos\alpha_m x \, \sin\beta_n y,
$$

$$
\sigma_{yz} = -\frac{h^2}{8}\left[1 - \left(\frac{2z}{h}\right)^2\right] (T_{12}X_{mn} + T_{22}Y_{mn}) \, \sin\alpha_m x \, \cos\beta_n y, \qquad (7.3.62)
$$

$$
\sigma_{zz} = \frac{h^3}{48}\left\{\left[1 + \left(\frac{2z}{h}\right)^3\right] - 3\left[1 + \left(\frac{2z}{h}\right)\right]\right\} (T_{31}X_{mn} + T_{32}Y_{mn})
$$
$$
\times \sin\alpha_m x \, \sin\beta_n y,
$$

where

$$
T_{11} = \frac{E}{(1 - \nu^2)}\left(\alpha_m^2 + \frac{1 - \nu}{2}\beta_n^2\right), \quad T_{12} = \frac{E}{2(1 - \nu)}\alpha_m\beta_n,
$$

$$
T_{22} = \frac{E}{(1 - \nu^2)}\left(\frac{1 - \nu}{2}\alpha_m^2 + \beta_n^2\right), \qquad\qquad (7.3.63)
$$

$$
T_{31} = \frac{E}{(1 - \nu^2)}\left(\alpha_m^3 + \alpha_m\beta_n^2\right), \quad T_{32} = \frac{E}{(1 - \nu^2)}\left(\alpha_m^2\beta_n + \beta_n^3\right).
$$

Example 7.3.3

Numerically evaluate the Navier solutions developed for the deflection and stresses in a simply supported, isotropic ($\nu = 0.25$), square plate subjected to uniformly distributed load q_0.

Solution: The following dimensionless quantities are used to report the numerical results:

$$
\bar{w} = w\left(Eh^3/a^4 q_0\right) \text{ for UL}, \quad \bar{w} = w\left(D/a^4 q_0\right) \times 10^2 \text{ for SL},
$$
$$
\bar{\sigma}_{xx} = \sigma_{xx}\left(h^2/a^2 q_0\right), \quad \bar{\sigma}_{yy} = \sigma_{yy}\left(h^2/a^2 q_0\right), \quad \bar{\sigma}_{xy} = \sigma_{xy}\left(h^2/a^2 q_0\right), \qquad (1)
$$
$$
\bar{\sigma}_{xz} = \sigma_{xz}\left(h/a q_0\right), \quad \bar{\sigma}_{yz} = \sigma_{yz}\left(h/a q_0\right),
$$

where a denotes the in-plane dimension and h is the thickness of the square plate. Table 7.3.1 contains the maximum dimensionless deflections ($\bar{w}$) and stresses of simply supported square plates under sinusoidally distributed load (SL) and uniformly distributed load (UL), and for different side-to-thickness ratios, a/h. The stresses were evaluated at the locations indicated below:

$$
\bar{\sigma}_{xx}(a/2, a/2, h/2), \quad \bar{\sigma}_{yy}(a/2, a/2, h/2), \quad \bar{\sigma}_{xy}(a, a, -h/2). \qquad (2)
$$

The transverse shear stresses are calculated using the constitutive equations, Eqs (7.3.60) and (7.3.61), as well as equilibrium equations, Eqs (7.3.62) and (7.3.63). They are the maximum at the locations $\bar{\sigma}_{xz}(0, a/2, h/2)$ and $\bar{\sigma}_{yz}(a/2, 0, h/2)$. Of course, the constitutively derived transverse shear stresses in the FSDT are independent of the z coordinate. The dimensionless quantities in the CPT are independent of the side-to-thickness ratio. The influence of transverse shear deformation is to increase the transverse deflection. The difference between the deflections predicted by the first-order shear deformation theory and classical plate theory decreases with the increase in the ratio a/h (see Fig. 7.3.3).

Table 7.3.1 Effect of the transverse shear deformation on deflections and stresses in isotropic ($\nu = 0.25$) square ($a = b$) plates subjected to distributed loads (see Fig. 7.3.2 for the plate geometry and coordinate system).

Load	$\frac{a}{h}$	$\bar{w}$	$\bar{\sigma}_{xx}$	$\bar{\sigma}_{yy}$	$\bar{\sigma}_{xy}$	$\bar{\sigma}_{xz}$	$\bar{\sigma}_{yz}$
SL	10	0.2702	0.1900	0.1900	0.1140	0.1910	0.1910
						0.2387^a	0.2387^a
	20	0.2600	0.1900	0.1900	0.1140	0.1910	0.1910
	50	0.2572	0.1900	0.1900	0.1140	0.1910	0.1910
	100	0.2568	0.1900	0.1900	0.1140	0.1910	0.1910
	CPT	0.2566	0.1900	0.1900	0.1140	0.2387^a	$0.2387^\dagger$
UL $(19)^b$	10	0.4259	0.2762	0.2762	0.2085	0.3927	0.3927
						0.4909^a	0.4909^a
	20	0.4111	0.2762	0.2762	0.2085	0.3927	0.3927
	50	0.4070	0.2762	0.2762	0.2085	0.3927	0.3927
	100	0.4060	0.2762	0.2762	0.2085	0.3927	0.3927
	CPT	0.4062	0.2762	0.2762	0.2085	$0.4909^\dagger$	$0.4909^\dagger$

[a]Stresses computed using the stress equilibrium equations, Eqs (7.3.62) and (7.3.63); they are the same for all ratios of a/h.

[b]The numbers in parenthesis denotes the maximum values of $m = n$ used to evaluate the series.

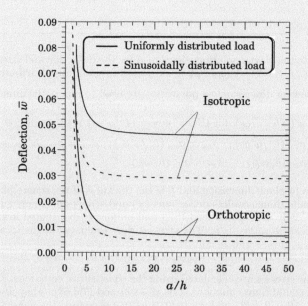

Fig. 7.3.3 Nondimensionalized center transverse deflection ($\bar{w}$) versus side-to-thickness ratio (a/h) for simply supported square plates.

7.3.4.2 Natural Vibration

For natural vibration, we set the mechanical load, q (hence Q_{mn}), and ($\hat{N}_{xx}$, $\hat{N}_{yy}$) to zero and seek periodic solution to Eq. (7.3.53) in the form

$$W_{mn}(t) = W_{mn}^0 e^{i\omega t}, \quad X_{mn}(t) = X_{mn}^0 e^{i\omega t}, \quad Y_{mn}(t) = Y_{mn}^0 e^{i\omega t}, \qquad (7.3.64)$$

and obtain the following matrix eigenvalue problem:

$$\left(\begin{bmatrix} s_{11} & s_{12} & s_{13} \\ s_{12} & s_{22} & s_{23} \\ s_{13} & s_{23} & s_{33} \end{bmatrix} - \omega^2 \begin{bmatrix} \hat{m}_{11} & 0 & 0 \\ 0 & \hat{m}_{22} & 0 \\ 0 & 0 & \hat{m}_{33} \end{bmatrix} \right) \left\{ \begin{array}{c} W_{mn}^0 \\ X_{mn}^0 \\ Y_{mn}^0 \end{array} \right\} = \left\{ \begin{array}{c} 0 \\ 0 \\ 0 \end{array} \right\}, \qquad (7.3.65)$$

where s_{ij} and $\hat{m}_{ij}$ are defined in Eq. (7.3.54). Setting the determinant of the 3×3 coefficient matrix in Eq. (7.3.65) yields a cubic polynomial for ω^2.

If the rotatory inertia m_2 is omitted (i.e., $\hat{m}_{22} = \hat{m}_{33} = 0$), setting the determinant of the coefficient matrix in Eq. (7.3.65) yields the following expression for ω^2:

$$\omega^2 = \frac{1}{\hat{m}_{11}} \left(s_{11} - \frac{s_{13}s_{23} - s_{12}s_{33}}{s_{22}s_{33} - s_{23}s_{23}} s_{12} - \frac{s_{12}s_{23} - s_{13}s_{22}}{s_{22}s_{33} - s_{23}s_{23}} s_{13} \right). \qquad (7.3.66)$$

Example 7.3.4

Determine the natural frequencies of an isotropic, simply supported, square plate using the first-order shear deformation plate theory.

Solution: Using Eq. (7.3.66), we can compute the frequencies. Table 7.3.2 contains the first four natural frequencies of square isotropic plates. The rotary inertia (RI) has the effect of decreasing the frequencies. Table 7.3.3 contains fundamental natural frequencies of square plates for various values of side-to-thickness ratio, a/h.

Table 7.3.2 Effect of the shear deformation, rotatory inertia, and shear correction coefficient on nondimensionalized natural frequencies of simply supported plates ($\bar{\omega} = \omega(a^2/h)\sqrt{\rho/E}$; $\nu = 0.3, a/h = 10$).

m	n	CPT[a] w/o RI	CPT with RI	K_s	FSDT w/o RI	FSDT with RI
1	1	5.973	5.925	1.0	5.838	5.794
				5/6	5.812	5.769
				2/3	5.773	5.732
2	1	14.933	14.635	1.0	14.127	13.899
				5/6	13.980	13.764
				2/3	13.769	13.568
2	2	23.893	23.144	1.0	21.922	21.424
				5/6	21.582	21.121
				2/3	21.103	20.688

[a] w/o RI means without rotary inertia.

Table 7.3.3 Effect of shear deformation, rotatory inertia on dimensionless fundamental frequencies of simply supported square plates $[\bar{\omega} = \omega(a^2/h)\sqrt{\rho/E}, K_s = 5/6, \nu = 0.25]$.

Theory	$a/h \rightarrow$	5	10	20	25	50	100
FSDT	w-RI[a]	5.232	5.694	5.835	5.853	5.877	5.883
	w/o-RI[b]	5.349	5.736	5.847	5.860	5.879	5.883
CPT	(5.885)[c]	5.700	5.837	5.873	5.877	5.883	5.885

[a]w-RI = with rotatory inertia; [b]w/o-RI = without rotatory inertia.
[c]Value in the parenthesis is the frequency without rotatory inertia.

7.3.4.3   Buckling Analysis

For buckling analysis, we assume that the only applied loads are the in-plane compressive forces $(\hat{N}_{xx}, \hat{N}_{yy})$ per unit length:

$$\hat{N}_{xx} = -N_0, \quad \hat{N}_{yy} = -\gamma N_0, \quad \gamma = \frac{\hat{N}_{yy}}{\hat{N}_{xx}}, \tag{7.3.67}$$

and the mechanical load and the foundation term is set to zero (i.e., $q = 0$ and $k = 0$). From Eq. (7.3.53) we have

$$\begin{bmatrix} s_{11} - N_0\left(\alpha_m^2 + \gamma\beta_n^2\right) & s_{12} & s_{13} \\ s_{12} & s_{22} & s_{23} \\ s_{13} & s_{23} & s_{33} \end{bmatrix} \begin{Bmatrix} W_{mn} \\ X_{mn} \\ Y_{mn} \end{Bmatrix} = \begin{Bmatrix} 0 \\ 0 \\ 0 \end{Bmatrix}. \tag{7.3.68}$$

For a nontrivial solution the determinant of the coefficient matrix in Eq. (7.3.68) must be zero. This gives the following expression for the buckling load:

$$N_0 = \left(\frac{1}{\alpha_m^2 + \gamma\beta_n^2}\right)\left[\frac{c_0 + \left(\frac{\alpha_m^2}{K_sGh} + \frac{\beta_n^2}{K_sGh}\right)c_1}{1 + \frac{c_1}{K_s^2G^2h^2} + \frac{c_2}{K_sGh} + \frac{c_3}{K_sGh}}\right] \tag{7.3.69}$$

$$c_0 = D\left(\alpha_m^4 + 2\alpha_m^2\beta_n^2 + \beta_n^4\right),$$

$$c_1 = c_2c_3 - (c_4)^2 > 0, \quad c_2 = D\left(\alpha_m^2 + \frac{1-\nu}{2}\beta_n^2\right), \tag{7.3.70}$$

$$c_3 = D\left(\frac{1-\nu}{2}\alpha_m^2 + \beta_n^2\right), \quad c_4 = \frac{Gh^3}{12}\alpha_m\beta_n.$$

When the effect of transverse shear deformation is neglected (i.e., set K_s to a large value), Eq. (7.3.69) yields the result in Eq. (7.2.141) obtained using the CPT. The expression in Eq. (7.3.69) is of the form

$$\frac{c_0 + k_1}{1 + k_2} \quad \text{with} \quad k_1 < k_2, \quad \text{which implies} \quad c_0 \geq \frac{c_0 + k_1}{1 + k_2},$$

indicating that transverse shear deformation has the effect of *reducing* the buckling load (as long as $c_0 > 1$).

The critical buckling load occurs for $m = n = 1$ and it is given by

$$N_{cr} = 4D \left(\frac{\pi}{a}\right)^2 \frac{\left[1 + \frac{3(1-\nu^2)\pi^2(h/a)^2}{K_s}\right]}{\left[1 + \frac{72(1+\nu)(1-\nu^2)\pi^4(h/a)^4}{K_s^2} + \frac{6(1+\nu)(3-\nu)\pi^2(h/a)^2}{K_s}\right]}. \qquad (7.3.71)$$

Table 7.3.4 contains the critical buckling loads $\bar{N} = N_{cr}b^2/(\pi^2 D)$ as a function of the plate aspect ratio a/b and side-to-thickness ratio b/h for uniaxial ($\gamma = 0$) and biaxial ($\gamma = 1$) compression. The CPT results are also included for comparison. The effect of transverse shear deformation is significant for lower aspect ratios and thick plates. For thin plates, irrespective of the aspect ratio, the buckling loads predicted by the FSDT are very close to those of the CPT.

Table 7.3.4 Nondimensional buckling loads $\bar{N}$ of simply supported plates under in-plane uniform compression ($\gamma = 0$) and biaxial compression ($\gamma = 1$).

γ	$\frac{h}{b}$	$\frac{a}{b} = 0.5$	$\frac{a}{b} = 1.0$	$\frac{a}{b} = 1.5$	$\frac{a}{b} = 3.0$
0	10	5.523	3.800	$4.045^{(2,1)}$	$3.800^{(3,1)a}$
	20	6.051	3.948	$4.262^{(2,1)}$	$3.948^{(3,1)}$
	100	6.242	3.998	$4.337^{(2,1)}$	$3.998^{(3,1)}$
	CPT	6.250	4.000	$4.340^{(2,1)}$	$4.000^{(3,1)}$
1	10	4.418	1.900	1.391	1.079
	20	4.841	1.974	1.431	1.103
	100	4.993	1.999	1.444	1.111
	CPT	5.000	2.000	1.444	1.111

[a]The pair of numbers in parentheses denote mode numbers (m, n) at which the critical buckling load occurred. For all other cases, $(m, n) = (1, 1)$.

7.3.5   Variational Solutions of Circular and Rectangular Plates

In this section, we consider variational formulation of axisymmetric bending of circular plates and plates in rectangular Cartesian coordinates using the FSDT. The variational formulation of the FSDT mirrors that of the CPT. Note that one may use either the Ritz method or the Galerkin method to solve any problem.

7.3.5.1   Axisymmetric Circular Plates

For axisymmetric plates, we set $u_\theta = \phi_\theta = 0$ and omit all terms involving derivatives with respect to θ. In order to use the Ritz method, we must construct the weak form. The weak form based on the virtual work statement in Eq. (7.3.4) is given by

$$0 = 2\pi \int_0^a \left[D \left(\frac{d\phi_r}{dr} + \nu \frac{\phi_r}{r} \right) \frac{d\delta\phi_r}{dr} + D \left(\nu \frac{d\phi_r}{dr} + \frac{\phi_r}{r} \right) \frac{\delta\phi_r}{r} \right.$$

$$\left. + K_s Gh \left(\phi_r + \frac{dw}{dr} \right) \left(\delta\phi_r + \frac{d\delta w}{dr} \right) - q\delta w \right] r \, dr. \qquad (7.3.72)$$

One must add additional terms corresponding to any applied loads and moments to the expression in Eq. (7.3.72).

Next, we seek Ritz approximation of the form

$$w(r) \approx \sum_{j=1}^M a_j \varphi_j(r), \quad \phi_r(r) \approx \sum_{j=1}^N b_j \psi_j(r), \qquad (7.3.73)$$

where we assumed that all specified essential boundary conditions are homogeneous. Substituting Eq. (7.3.73) into the virtual work statement in Eq. (7.3.72) and collecting the coefficients of δa_i and δb_k separately, we obtain

$$0 = \sum_{j=1}^M \left[K_s Gh \int_0^a \frac{d\varphi_i}{dr} \frac{d\varphi_j}{dr} r \, dr \right] a_j + \sum_{j=1}^N \left[K_s Gh \int_0^a \frac{d\varphi_i}{dr} \psi_j \, r \, dr \right] b_j$$

$$- \int_0^a q\varphi_i \, r \, dr, \qquad (7.3.74)$$

$$0 = K_s Gh \sum_{j=1}^M \left[\int_0^a \psi_k \frac{d\varphi_j}{dr} r \, dr \right] a_j + \sum_{j=1}^N \left\{ K_s Gh \int_0^a \psi_k \psi_j \, r \, dr \right.$$

$$\left. + D \int_0^a \left[\frac{d\psi_k}{dr} \frac{d\psi_j}{dr} + \frac{\nu}{r} \left(\psi_k \frac{d\psi_j}{dr} + \frac{d\psi_k}{dr} \psi_j \right) + \frac{1}{r^2} \psi_k \psi_j \right] r \, dr \right\} b_j. \quad (7.3.75)$$

for $k = 1, 2, \ldots, N$. In matrix form, these equations can be written as

$$\begin{bmatrix} \mathbf{A} & \mathbf{B} \\ \mathbf{B}^{\mathrm{T}} & \mathbf{C} \end{bmatrix} \begin{Bmatrix} \mathbf{a} \\ \mathbf{b} \end{Bmatrix} = \begin{Bmatrix} \mathbf{F} \\ \mathbf{0} \end{Bmatrix}, \qquad (7.3.76)$$

where

$$A_{ij} = K_s Gh \int_0^a \frac{d\varphi_i}{dr} \frac{d\varphi_j}{dr} r \, dr, \quad B_{ij} = K_s Gh \int_0^a \frac{d\varphi_i}{dr} \psi_j \, r \, dr,$$

$$C_{kj} = D \int_0^a \left[\frac{d\psi_k}{dr} \frac{d\psi_j}{dr} + \frac{\nu}{r} \left(\psi_k \frac{d\psi_j}{dr} + \frac{d\psi_k}{dr} \psi_j \right) + \frac{1}{r^2} \psi_k \psi_j \right] r \, dr \qquad (7.3.77)$$

$$+ K_s Gh \int_0^a \psi_k \psi_j \, r \, dr,$$

$$F_i = \int_0^a q\phi_i \, r \, dr.$$

Example 7.3.5

Consider a simply supported solid circular plate under uniformly distributed load. Find the Ritz approximation in Eq. (7.3.73) with $M = 2$ and $N = 1$, and

$$\varphi_1(r) = 1 - \frac{r}{a}, \quad \varphi_2(r) = 1 - \frac{r^2}{a^2}, \quad \psi_1(r) = \frac{r}{a}, \quad (1)$$

which meet the essential boundary conditions of the problem: $w(a) = 0$ and $\varphi_r(0) = 0$.

Solution: Evaluating the integrals in Eq. (7.3.77), we obtain ($\Lambda = D/K_sGh$)

$$\frac{K_sGh}{12} \begin{bmatrix} 6 & 8 & -4a \\ 8 & 12 & -6a \\ -4a & -6a & 3a^2 + 12(1+\nu)\Lambda \end{bmatrix} \begin{Bmatrix} a_1 \\ a_2 \\ b_1 \end{Bmatrix} = \frac{q_0a^2}{12} \begin{Bmatrix} 2 \\ 3 \\ 0 \end{Bmatrix}, \quad (2)$$

whose solution is

$$a_1 = 0, \quad a_2 = \frac{q_0a^2}{16} \left[\frac{a^2}{(1+\nu)D} + \frac{4}{K_sGh} \right], \quad b_1 = \frac{q_0a^3}{8(1+\nu)D}. \quad (3)$$

The Ritz solution becomes

$$W_2(r) = \frac{q_0a^4}{16D} \frac{1}{(1+\nu)} \left(1 - \frac{r^2}{a^2} \right) + \frac{q_0a^2}{4K_sGh} \left(1 - \frac{r^2}{a^2} \right), \quad (4)$$

$$\Phi_1(r) = \frac{q_0a^3}{8(1+\nu)D} \left(\frac{r}{a} \right). \quad (5)$$

A close examination of the problem indicates that the deflection is symmetric and the rotation is antisymmetric about $r = 0$. Hence, $w(r)$ is a symmetric (i.e., even) function and ϕ_r is an antisymmetric (or odd) function of r. This also explains why $a_1 = 0$ (only when $M > 1$). This understanding helps in selecting proper functions and reducing the effort. Thus, the solution in Eq. (7.3.73) amounts to using one-parameter ($M = 1$ and $N = 1$) approximations with

$$\varphi_1(r) = 1 - \frac{r^2}{a^2}, \quad \psi_1(r) = \frac{r}{a}. \quad (6)$$

If a two-parameter approximation for w and one-parameter approximation for ϕ_r is used, with the following choice of approximation functions

$$\varphi_1(r) = 1 - \frac{r^2}{a^2}, \quad \varphi_2(r) = 1 - \frac{r^4}{a^4}, \quad \psi_1(r) = \frac{r}{a}, \quad (7)$$

the exact solution would be obtained.

7.3.5.2 Rectangular Plates

Next, we consider the variational formulation of plates in rectangular Cartesian coordinates. While no numerical examples are included, the formulation should prove useful in obtaining Ritz solutions of plates with arbitrary boundary conditions.

We begin with the weak form [which is the same as the virtual work statement in Eq. (7.3.30) specialized to the static, pure bending case]

$$0 = \int_{\Omega_0} \left[M_{xx}\delta\varepsilon_{xx}^{(1)} + M_{yy}\delta\varepsilon_{yy}^{(1)} + M_{xy}\delta\gamma_{xy}^{(1)} + Q_x\delta\gamma_{xz}^{(0)} + Q_y\delta\gamma_{yz}^{(0)} - q\delta w \right] dxdy$$

$$- \oint_{\Gamma} \left(\hat{M}_{nn}\delta\phi_n + \hat{M}_{ns}\delta\phi_s + \hat{Q}_n\delta w \right) dS$$

$$0 = \int_{\Omega_0} \left\{ D\left(\frac{\partial\phi_x}{\partial x} + \nu\frac{\partial\phi_y}{\partial y} \right) \frac{\partial\delta\phi_x}{\partial x} + D\left(\nu\frac{\partial\phi_x}{\partial x} + \frac{\partial\phi_y}{\partial y} \right) \frac{\partial\delta\phi_y}{\partial y} \right.$$

$$+ D\frac{(1-\nu)}{2}\left(\frac{\partial\phi_x}{\partial y} + \frac{\partial\phi_y}{\partial x} \right) \left(\frac{\partial\delta\phi_x}{\partial y} + \frac{\partial\delta\phi_y}{\partial x} \right)$$

$$+ K_sGh\left[\left(\frac{\partial w}{\partial x} + \phi_x \right) \left(\frac{\partial\delta w}{\partial x} + \delta\phi_x \right) + \left(\frac{\partial w}{\partial y} + \phi_y \right) \left(\frac{\partial\delta w}{\partial y} + \delta\phi_y \right) \right]$$

$$\left. - q\delta w \right\} dxdy - \oint_{\Gamma} \left(\hat{M}_{nn}\delta\phi_n + \hat{M}_{ns}\delta\phi_s + \hat{Q}_n\delta w \right) dS.$$

$$(7.3.78)$$

Assuming Ritz appoximation of the form

$$w \approx W_N(x,y) = \sum_{i=1}^{N} a_i\psi_i(x,y),$$

$$\phi_x \approx \Phi_x^M(x,y) = \sum_{i=1}^{M} b_i\varphi_i^{(1)}(x,y), \qquad (7.3.79)$$

$$\phi_y \approx \Phi_y^M(x,y) = \sum_{i=1}^{M} c_i\varphi_i^{(2)}(x,y),$$

and substituting into the weak form in Eq. (7.3.78), we obtain the following Ritz equations:

$$\begin{bmatrix} \mathbf{A} & \mathbf{B} & \mathbf{C} \\ \mathbf{B}^T & \mathbf{D} & \mathbf{E} \\ \mathbf{C}^T & \mathbf{E}^T & \mathbf{G} \end{bmatrix} \begin{Bmatrix} \mathbf{a} \\ \mathbf{b} \\ \mathbf{c} \end{Bmatrix} = \begin{Bmatrix} \mathbf{F}^1 \\ \mathbf{F}^2 \\ \mathbf{F}^3 \end{Bmatrix}, \qquad (7.3.80)$$

where

$$A_{ij} = \int_{\Omega_0} K_sGh\left(\frac{\partial\psi_i}{\partial x}\frac{\partial\psi_j}{\partial x} + \frac{\partial\psi_i}{\partial y}\frac{\partial\psi_j}{\partial y} \right) dxdy,$$

$$B_{ij} = \int_{\Omega_0} K_sGh\frac{\partial\psi_i}{\partial x}\varphi_j^{(1)} dxdy,$$

$$C_{ij} = \int_{\Omega_0} K_sGh\frac{\partial\psi_i}{\partial y}\varphi_j^{(2)} dxdy,$$

$$D_{ij} = \int_{\Omega_0} \left[D \left(\frac{\partial \varphi_i^{(1)}}{\partial x} \frac{\partial \varphi_j^{(1)}}{\partial x} + \frac{1-\nu}{2} \frac{\partial \varphi_i^{(1)}}{\partial y} \frac{\partial \varphi_j^{(1)}}{\partial y} \right) + K_s Gh\, \varphi_i^{(1)} \varphi_j^{(1)} \right] dxdy,$$

$$E_{ij} = \int_{\Omega_0} D \left(\nu \frac{\partial \varphi_i^{(1)}}{\partial x} \frac{\partial \varphi_j^{(2)}}{\partial y} + \frac{1-\nu}{2} \frac{\partial \varphi_i^{(1)}}{\partial y} \frac{\partial \varphi_j^{(2)}}{\partial x} \right) dxdy, \qquad (7.3.81)$$

$$G_{ij} = \int_{\Omega_0} \left[D \left(\frac{\partial \varphi_i^{(2)}}{\partial y} \frac{\partial \varphi_j^{(2)}}{\partial y} + \frac{1-\nu}{2} \frac{\partial \varphi_i^{(2)}}{\partial x} \frac{\partial \varphi_j^{(2)}}{\partial x} \right) + K_s Gh\, \varphi_i^{(2)} \varphi_j^{(2)} \right] dxdy,$$

$$F_i^1 = \int_{\Omega_0} q\psi_i\, dxdy + \oint_\Gamma \hat{Q}_n \psi_i\, dS,$$

$$F_i^2 = \oint_\Gamma \left(M_{xx} n_x + M_{xy} n_y \right) \varphi_i^{(1)}\, dS,$$

$$F_i^3 = \oint_\Gamma \left(M_{xy} n_x + M_{yy} n_y \right) \varphi_i^{(2)}\, dS.$$

This completes the variational formulation of the first-order shear deformation plate theory in rectangular Cartesian coordinates.

7.4 Relationships between Bending Solutions of Classical and Shear Deformation Theories

From the analytical solutions of axisymmetric bending of circular plates, it is clear that there exists a relationship between the solutions of the two theories, namely, the classical (CPT) and first-order shear deformation (FSDT) theories. Indeed, such relationships are developed using the similarity of the equations of the two theories and the load equivalence (see Wang *et al.* [129]). The usefulness of these relationships lies in finding the solution of a beam or plate problem according to the first-order shear deformation theory whenever the corresponding solution according to the classical theory for the same problem exists. The primary objective of this section is to establish relationships between the bending solutions (i.e., deflection, rotation, bending moment, and shear force) of the first-order shear deformation theory in terms of the corresponding quantities of the classical theory. We present a brief discussion of the relationships for the case of bending of beams, circular plates, and rectangular plates.

7.4.1 Beams

First we summarize the governing equations of the classical (i.e., the Bernoulli–Euler) and first-order (i.e., the Timoshenko) beam theories. For details, see Sections 2.5.2 and 2.5.3.

7.4.1.1   Governing Equations

Bernoulli–Euler beam theory (BET) The governing equations are [see Eqs (2.5.4) and (2.5.5)]

$$-\frac{dV^C}{dx} = q, \quad -\frac{dM^C}{dx} + V^C = 0, \tag{7.4.1}$$

where M^C and V^C are the bending moment and shear force, respectively (here the superscript refers to the classical beam theory, that is, the BET). They are related to the generalized displacements $(w^C, \theta_x^C = -dw^C/dx)$ by [see Eq. (2.5.12)]

$$M^C = EI\frac{d\theta_x^C}{dx}, \quad V^C = \frac{d}{dx}\left(EI\frac{d\theta_x^C}{dx}\right), \quad \theta_x^C = -\frac{dw^C}{dx}. \tag{7.4.2}$$

Timoshenko beam theory (TBT) The governing equations for the TBT are the same as those for the BET (here the superscript F is used indicate the TBT):

$$-\frac{dV^F}{dx} = q, \quad -\frac{dM^F}{dx} + V^F = 0, \tag{7.4.3}$$

but M^F and V^F are related to the generalized displacements (w^F, ϕ_x^F) by

$$M^F = EI\frac{d\phi_x^F}{dx}, \quad V^F = GAK_s\left(\phi_x^F + \frac{dw^F}{dx}\right). \tag{7.4.4}$$

7.4.1.2   Relationships between BET and TBT

Comparing the first equation in Eq. (7.4.1) with the first equation in Eq. (7.4.3), we arrive at the relation

$$\frac{dV^F}{dx} = \frac{dV^C}{dx} \;\Rightarrow\; V^F = V^C + C_1.$$

where C_1 is a constant of integration. Similarly, from the second equation in Eq. (7.4.1) and the second equation in Eq. (7.4.3), we obtain

$$-\frac{dM^F}{dx} + V^F = -\frac{dM^C}{dx} + V^C \;\Rightarrow\; M^F = M^C + C_1 x + C_2.$$

Finally, from the first equation in Eq. (7.4.2) and the first equation in Eq. (7.4.4), we obtain

$$EI\frac{d\phi_x^F}{dx} = EI\frac{d\theta_x^C}{dx} + C_1 x + C_2 \;\Rightarrow\; EI\phi_x^F = EI\theta_x^C + C_1\frac{x^2}{2} + C_2 x + C_3.$$

Thus, the shear forces, bending moments, and the slopes of the two beam theories are related by

$$V^F = V^C + C_1 \tag{7.4.5}$$

$$M^F = M^C + C_1 x + C_2 \tag{7.4.6}$$

$$\phi_x^F = \theta_x^C + C_1 \frac{x^2}{2EI} + C_2 \frac{x}{EI} + C_3 \frac{1}{EI} \tag{7.4.7}$$

where C_1, C_2, and C_3 are constants of integration.

To obtain the relationship between w^F and quantities from the classical solution, we begin with Eq. (7.4.5) and use the second equation in Eq. (7.4.1) and the second relation in Eq. (7.4.4) to arrive at the relation

$$V^F = V^C + C_1$$

$$K_S GA \left(\phi_x^F + \frac{dw^F}{dx} \right) = \frac{dM^C}{dx} + C_1 \tag{7.4.8}$$

Using Eq. (7.4.7) for ϕ_x^F in Eq. (7.4.8) and integrating with respect to x, we obtain the following deflection relationship:

$$w^F = w^C + \frac{M^C}{K_s GA} + C_1 \left(\frac{x}{K_s GA} - \frac{x^3}{6EI} \right) - C_2 \frac{x^2}{2EI} - C_3 \frac{x}{EI} - C_4 \frac{1}{EI}, \tag{7.4.9}$$

where C_1, C_2, C_3, C_4 are constants of integration to be determined using the boundary conditions of a particular beam problem of interest. For free (F), simply supported (S), and clamped (C) ends, the boundary conditions in the two beam theories are given by

$$\mathbf{F}: \ M^C = M^F = 0, \ \ V^C = V^F = 0, \tag{7.4.10}$$

$$\mathbf{S}: \ w^C = w^F = 0, \ \ M^C = M^F = 0, \tag{7.4.11}$$

$$\mathbf{C}: \ w^C = w^F = 0, \ \ \frac{dw^C}{dx} = \phi_x^F = 0. \tag{7.4.12}$$

Example 7.4.1 ————————————————————————————————

Determine the deflection, slope, bending moment, and shear force of a Timoshenko beam simply supported at both ends using the relationships developed in Eqs (7.4.5) to (7.4.7) and (7.4.9). Assume that the beam is loaded with uniformly distributed load of intensity q_0.

Solution: The boundary conditions for simply supported beams of length L are given by Eq. (7.4.11) for $x = 0$ and $x = L$. The substitution of these boundary conditions into Eqs (7.4.6) and (7.4.9) gives $C_1 = C_2 = C_3 = C_4 = 0$. Hence, we have

$$w^F = w^C + \frac{M^C}{K_s GA}, \ \ \phi_x^F = \theta_x^C = -\frac{dw^C}{dx}, \ \ M^F = M^C, \ \ V^F = V^C. \tag{1}$$

Thus, for statically determinate beams, the shear force, bending moment, and slope in the two theories remain the same, while the deflection differs.

For a simply supported beam loaded by uniformly distributed load of intensity q_0, the Bernoulli–Euler solution is

$$w^C(x) = \frac{q_0 L^4}{24EI}\left(\frac{x}{L} - 2\frac{x^3}{L^3} + \frac{x^4}{L^4}\right),$$

$$\theta_x^C(x) = -\frac{q_0 L^3}{24EI}\left(1 - 6\frac{x^2}{L^2} + 4\frac{x^3}{L^3}\right),$$

$$M^C(x) = -\frac{q_0 L^2}{2}\left(-\frac{x}{L} + \frac{x^2}{L^2}\right),$$

$$V^C(x) = -\frac{q_0 L}{2}\left(-1 + 2\frac{x}{L}\right).$$

(2)

Therefore, the Timoshenko beam solution is

$$w^F(x) = \frac{q_0 L^4}{24EI}\left(\frac{x}{L} - 2\frac{x^3}{L^3} + \frac{x^4}{L^4}\right) + \frac{1}{K_s GA}\frac{q_0 L^2}{2}\left(\frac{x}{L} - \frac{x^2}{L^2}\right),$$

$$\phi_x^F(x) = -\frac{q_0 L^3}{24EI}\left(1 - 6\frac{x^2}{L^2} + 4\frac{x^3}{L^3}\right),$$

$$M^F(x) = -\frac{q_0 L^2}{2}\left(-\frac{x}{L} + \frac{x^2}{L^2}\right),$$

$$V^F(x) = -\frac{q_0 L}{2}\left(-1 + 2\frac{x}{L}\right).$$

(3)

The maximum deflection occurs at $x = L/2$:

$$w_{\max}^F = \frac{5q_0 L^4}{384EI} + \frac{q_0 L^2}{8K_s GA},$$

(4)

which is larger than the maximum deflection predicted by the Bernoulli–Euler beam theory by $q_0 L^2/(8K_s GA)$. The slope, bending moment, and shear force are the same in the two theories for the simply supported beam loaded by uniformly distributed load.

Example 7.4.2

Determine the deflection, slope, bending moment, and shear force of a Timoshenko beam clamped at both ends using the relationships developed in Eqs (7.4.5) to (7.4.7) and (7.4.9). Assume that the beam is loaded with uniformly distributed load of intensity q_0 as well as a point load F_0 at the center of the beam.

Solution: The boundary conditions for clamped beams are given by Eq. (7.4.12) for $x = 0$ and $x = L$. The substitution of these boundary conditions into Eqs (7.4.7) and (7.4.9) gives

$$C_1 = -\frac{12\Omega}{(1 + 12\Omega)L}M_d, \quad C_2 = \frac{6\Omega}{(1 + 12\Omega)}M_d, \quad C_3 = 0, \quad C_4 = M^C(0)\Omega L^2,$$

(1)

where

$$\Omega = \frac{EI}{K_s GAL^2}, \quad M_d = M^C(L) - M^C(0).$$

(2)

For statically indeterminate beams, in general, the solutions for shear force, bending moment, slope, and deflection predicted by the two theories are different. Of course, the type of load makes a difference in the final form of the solution. We have

$$w^F(x) = w^C + \frac{1}{K_s GA} M^C(x) - \frac{12\Omega}{(1+12\Omega)L} M_d \left(\frac{x}{K_s GA} - \frac{x^3}{6EI} \right)$$

$$- \frac{6\Omega}{(1+12\Omega)} M_d \frac{x^2}{2EI} - M^C(0) \frac{\Omega L^2}{EI},$$

$$\phi_x^F(x) = \theta_x^C - \frac{12\Omega}{(1+12\Omega)L} M_d \frac{x^2}{2EI} + \frac{6\Omega}{(1+12\Omega)} M_d \frac{x}{EI}, \tag{3}$$

$$M^F(x) = M^C(x) - \frac{12\Omega}{(1+12\Omega)L} M_d x + \frac{6\Omega}{(1+12\Omega)} M_d,$$

$$V^F(x) = V^C(x) - \frac{12\Omega}{(1+12\Omega)L} M_d.$$

For a clamped beam loaded by uniformly distributed load of intensity q_0 and point load F_0 at the center, the Bernoulli–Euler solution is

$$w^C(x) = \frac{q_0 L^4}{24EI} \frac{x^2}{L^2} \left(1 - \frac{x}{L}\right)^2 + \frac{F_0 L^3}{48} \left(3\frac{x^2}{L^2} - 4\frac{x^3}{L^3} + 8\langle x - 0.5L \rangle^3\right),$$

$$\theta_x^C(x) = -\frac{q_0 L^3}{24EI} \left(2\frac{x}{L} - 6\frac{x^2}{L^2} + 4\frac{x^3}{L^3}\right) - \frac{F_0 L^2}{8} \left(\frac{x}{L} - 2\frac{x^2}{L^2} + 4\langle x - 0.5L \rangle^2\right),$$

$$M^C(x) = -\frac{q_0 L^2}{24} \left(2 - 12\frac{x}{L} + 12\frac{x^2}{L^2}\right) - \frac{F_0 L}{8} \left(1 - 4\frac{x}{L} + 8\langle x - 0.5L \rangle\right), \tag{4}$$

$$V^C(x) = -\frac{q_0 L^2}{2} \left(-1 + \frac{x}{L}\right) - \frac{F_0 L}{2} \left[-1 + 2\delta(x - 0.5L)\right],$$

where $\delta(x - a)$ is the usual Dirac delta function [$\delta(x - a) = 0$ when $x < a$ and $x > a$ and $\delta(x - a) = 1$ when $x = a$] and $\mathcal{F}(x) = \langle x - a \rangle^n$ is a *step function*, where a is a constant, and the index n is either zero or any positive number. The function $\mathcal{F}(x)$ has the following properties:

1. If $x < a$ then $\mathcal{F}(x) = 0$.

2. If $x \geq a$ then $\mathcal{F}(x) = (x - a)^n$.

3. The function can be integrated with respect to its independent variable as

$$\int \langle x - a \rangle^n dx = \frac{\langle x - a \rangle^{n+1}}{n+1} + \text{(a constant)}. \tag{5}$$

Properties 1 and 2 mean that when the expression within the $\langle \ \rangle$ brackets is negative, the function is zero, but that once the expression becomes positive or zero, the function behaves just as though the brackets are the usual parentheses. We note that this is true for even integer values of n, when the function $\mathcal{F}(x)$ itself is positive for all values of x. Property 3 means that the step function can be integrated like a normal algebraic expression of this type, but that the resulting integral is itself a step function, with the step occurring at the same value of x. We should note that the bracketed expression must be integrated in its entirety rather than be expanded and integrated term by term.

First, we note that

$$M^C(0) = -\frac{q_0 L^2}{24} - \frac{F_0 L}{8}, \quad M^C(L) = -\frac{q_0 L^2}{24} - \frac{F_0 L}{8} \Rightarrow M_d = 0. \tag{6}$$

Thus, the only nonzero constant is C_4. We note that this happened because of the symmetry of the solution. Then the TBT solution for the problem at hand is

$$
w^F(x) = \frac{q_0 L^4}{24EI} \frac{x^2}{L^2} \left(1 - \frac{x}{L}\right)^2 + \frac{F_0 L^3}{48} \left(3\frac{x^2}{L^2} - 4\frac{x^3}{L^3} + 8\langle x - 0.5L \rangle^3\right)
$$
$$
- \frac{1}{K_s GA} \left[\frac{q_0 L^2}{24} \left(1 - 12\frac{x}{L} + 12\frac{x^2}{L^2}\right) + \frac{F_0 L}{8} \left(1 - 4\frac{x}{L} + 8\langle x - 0.5L \rangle\right)\right]
$$
$$
+ \frac{\Omega L^2}{EI} \left(\frac{q_0 L^2}{24} + \frac{F_0 L}{8}\right), \tag{7}
$$

$$
\phi_x^F(x) = -\frac{q_0 L^3}{24EI} \left(2\frac{x}{L} - 6\frac{x^2}{L^2} + 4\frac{x^3}{L^3}\right) - \frac{F_0 L^2}{8} \left(\frac{x}{L} - 2\frac{x^2}{L^2} + 4\langle x - 0.5L \rangle^2\right), \tag{8}
$$

$$
M^F(x) = -\frac{q_0 L^2}{24} \left(2 - 12\frac{x}{L} + 12\frac{x^2}{L^2}\right) - \frac{F_0 L}{8} \left(1 - 4\frac{x}{L} + 8\langle x - 0.5L \rangle\right), \tag{9}
$$

$$
V^F(x) = -\frac{q_0 L^2}{2} \left(-1 + 2\frac{x}{L}\right) - \frac{F_0 L}{2} \left[-1 + 2\delta(x - 0.5L)\right]. \tag{10}
$$

The maximum deflection is

$$
w_{\max} = w(0.5L) = \frac{1}{384EI} \left(q_0 L^4 + 2F_0 L^3\right) + \frac{1}{8K_s GA} \left(q_0 L^2 + 4F_0 L\right). \tag{11}
$$

The relationships between the solutions of the BET and the TBT are summarized in Table 7.4.1. From the results presented in Table 7.4.1, for statically determinate beams, the shear force, bending moment, and slope in the two theories remain the same, while the deflection differs. For statically indeterminate beams the solutions for shear force, bending moment, slope, and deflection predicted by the two theories are, in general, not the same.

7.4.2　Circular Plates

The governing equations of the classical and first-order shear deformation plate theories are summarized in the following for isotropic (E and ν) plates. The superscripts on variables refer to the theory used: C for CPT and F for FSDT.

CPT:

$$
\frac{1}{r} \frac{d}{dr} \left(rQ_r^C\right) = -q, \tag{7.4.13}
$$

where

$$
rQ_r^C \equiv \frac{d}{dr} \left(rM_{rr}^C\right) - M_{\theta\theta}^C, \tag{7.4.14}
$$

$$
M_{rr}^C = -D \left(\frac{d^2 w^C}{dr^2} + \frac{\nu}{r} \frac{dw^C}{dr}\right), \tag{7.4.15}
$$

$$
M_{\theta\theta}^C = -D \left(\nu \frac{d^2 w^C}{dr^2} + \frac{1}{r} \frac{dw^C}{dr}\right). \tag{7.4.16}
$$

Table 7.4.1 Generalized deflection and force relationships between Timoshenko (F) and Bernoulli–Euler (C) beams for various boundary conditions (BC).

BC	Relationships
SS	$w^F(x) = w^C(x) + \frac{\Omega L^2}{EI} M^C(x)$
	$\phi_x^F(x) = -\frac{dw^C}{dx}$
	$M^F(x) = M^C(x)$
	$V^F(x) = V^C(x)$
CF	$w^F(x) = w^C(x) + \frac{\Omega L^2}{EI} \left[M^C(x) - M^C(0) \right]$
	$\phi_x^F(x) = -\frac{dw^C}{dx}$
	$M^F(x) = M^C(x)$
	$V^F(x) = V^C(x)$
FC	$w^F(x) = w^C(x) + \frac{\Omega L^2}{EI} \left[M^C(x) - M^C(L) \right]$
	$\phi_x^F(x) = -\frac{dw^C}{dx}$
	$M^F(x) = M^C(x)$
	$V^F(x) = V^C(x)$
CS	$w^F(x) = w^C(x) + \frac{\Omega L^2}{EI} \left[M^C(x) - M^C(0) \right]$
	$\quad + \frac{3\Omega L^2}{EI(1+3\Omega)} \frac{x}{L} \left(\Omega + \frac{x}{2L} - \frac{x^2}{6L^2} \right) M^C(0)$
	$\phi_x^F(x) = -\frac{dw^C}{dx} + \frac{3\Omega L}{EI(1+3\Omega)} \frac{x}{L} \left(1 - \frac{x}{2L} \right) M^C(0)$
	$M^F(x) = M^C(x) - \frac{3\Omega}{(1+3\Omega)} \left(1 - \frac{x}{L} \right) M^C(0)$
	$V^F(x) = V^C(x) + \frac{3\Omega}{(1+3\Omega)L} M^C(0)$
SC	$w^F(x) = w^C(x) + \frac{\Omega L^2}{EI} M^C(x)$
	$\quad - \frac{3\Omega L^2}{EI(1+3\Omega)} \frac{x}{L} \left(\Omega + \frac{1}{2} - \frac{x^2}{6L^2} \right) M^C(L)$
	$\phi_x^F(x) = -\frac{dw^C}{dx} + \frac{3\Omega L}{2EI(1+3\Omega)} \left(\frac{x^2}{L^2} - 1 \right) M^C(L)$
	$M^F(x) = M^C(x) - \frac{3\Omega}{(1+3\Omega)} \frac{x}{L} M^C(L)$
	$V^F(x) = V^C(x) - \frac{3\Omega}{(1+3\Omega)L} M^C(L)$
CC[a]	$w^F(x) = w^C(x) + \frac{\Omega L^2}{EI} \left[M^C(x) - M^C(0) \right]$
	$\quad + \frac{3\mu L^2}{EI} \frac{x}{L} \left(\frac{2}{3} \frac{x^2}{L^2} - 4\Omega - \frac{x}{L} \right) \left[M^C(L) - M^C(0) \right]$
	$\phi_x^F(x) = -\frac{dw^C}{dx} - \frac{6\mu L}{EI} \frac{x}{L} \left(\frac{x}{L} - 1 \right) \left[M^C(L) - M^C(0) \right]$
	$M^F(x) = M^C(x) - 6\mu \left(2\frac{x}{L} - 1 \right) \left[M^C(L) - M^C(0) \right]$
	$V^F(x) = V^C(x) - \frac{12\mu}{L} \left[M^C(L) - M^C(0) \right]$

[a] $\mu = \frac{\Omega}{(1+12\Omega)}$ and $\Omega = \frac{EI}{K_s G A L^2}$.

FSDT:

$$\frac{1}{r}\frac{d}{dr}\left(rQ_r^F\right) = -q, \tag{7.4.17}$$

$$rQ_r^F = \frac{d}{dr}\left(rM_{rr}^F\right) - M_{\theta\theta}^F \tag{7.4.18}$$

where

$$M_{rr}^F = D\left(\frac{d\phi_r^F}{dr} + \frac{\nu}{r}\phi_r^F\right), \tag{7.4.19}$$

$$M_{\theta\theta}^F = D\left(\nu\frac{d\phi_r^F}{dr} + \frac{1}{r}\phi_r^F\right), \tag{7.4.20}$$

$$Q_r^F = K_s Gh\left(\phi_r^F + \frac{dw^F}{dr}\right). \tag{7.4.21}$$

Next, we introduce the moment sum

$$\mathcal{M} = \frac{M_{rr} + M_{\theta\theta}}{1+\nu}. \tag{7.4.22}$$

Using Eqs (7.4.13) to (7.4.21), we can show that

$$\mathcal{M}^C = -D\left(\frac{d^2 w^C}{dr^2} + \frac{1}{r}\frac{dw^C}{dr}\right) = -D\frac{1}{r}\frac{d}{dr}\left(r\frac{dw^C}{dr}\right), \tag{7.4.23}$$

$$\mathcal{M}^F = D\left(\frac{d\phi_r^F}{dr} + \frac{1}{r}\phi_r^F\right) = D\frac{1}{r}\frac{d}{dr}\left(r\phi_r^F\right), \tag{7.4.24}$$

$$\frac{1}{r}\frac{d}{dr}\left(r\frac{d\mathcal{M}^C}{dr}\right) = -q, \quad \frac{1}{r}\frac{d}{dr}\left(r\frac{d\mathcal{M}^F}{dr}\right) = -q, \tag{7.4.25}$$

$$r\frac{d\mathcal{M}^C}{dr} = \frac{d}{dr}\left(rM_{rr}^C\right) - M_{\theta\theta}^C = rQ_r^C \tag{7.4.26}$$

$$r\frac{d\mathcal{M}^F}{dr} = \frac{d}{dr}\left(rM_{rr}^F\right) - M_{\theta\theta}^F = rQ_r^F. \tag{7.4.27}$$

We are now ready to establish relationships between the solutions of the two theories. From Eqs (7.4.13) and (7.4.17), because the load q being the same, it follows that

$$rQ_r^F = rQ_r^C + C_1, \tag{7.4.28}$$

where C_1 is a constant of integration. From Eqs (7.4.25) to (7.4.28), we have

$$r\frac{d\mathcal{M}^F}{dr} = r\frac{d\mathcal{M}^C}{dr} + C_1 \implies \mathcal{M}^F = \mathcal{M}^C + C_1 \log r + C_2, \tag{7.4.29}$$

where C_1 and C_2 are constants of integration. Next, from Eqs (7.4.23), (7.4.24), and (7.4.29), we have

$$\phi_r^F = -\frac{dw^C}{dr} + \frac{C_1 r}{4D}\left(2\log r - 1\right) + \frac{C_2 r}{2D} + \frac{C_3}{rD}. \tag{7.4.30}$$

Finally, from Eqs (7.4.21), (7.4.26), (7.4.27), (7.4.29), and (7.4.30), we obtain

$$\frac{dw^F}{dr} = -\phi_r^F + \frac{1}{K_s Gh}\left(Q_r^C + \frac{C_1}{r}\right), \tag{7.4.31}$$

and noting that $Q_r^C = d\mathcal{M}^C/dr$, we have

$$w^F = w^C + \frac{\mathcal{M}^C}{K_s Gh} + \frac{C_1 r^2}{4D}(1 - \log r) + \frac{C_1}{K_s Gh}\log r - \frac{C_2 r^2}{4D} - \frac{C_3 \log r}{D} + \frac{C_4}{D}. \tag{7.4.32}$$

The four constants of integration, C_1, C_2, C_3, and C_4, are determined using the boundary conditions. The boundary conditions for various cases are given as follows.

Free edge:

$$rQ_r^F = rQ_r^C = 0, \quad rM_{rr}^F = rM_{rr}^C = 0. \tag{7.4.33}$$

Simply supported edge:

$$w^F = w^C = 0, \quad rM_{rr}^F = rM_{rr}^C = 0. \tag{7.4.34}$$

Clamped edge:

$$w^F = w^C = 0, \quad \phi_r^F = \frac{dw^C}{dr} = 0. \tag{7.4.35}$$

Solid circular plate at $r = 0$ (i.e., at the plate center):

$$\phi_r^F = \frac{dw^C}{dr} = 0, \quad C_1 = 0. \tag{7.4.36}$$

Example 7.4.3

Consider a circular plate of radius a and subjected to a linearly varying axisymmetric load $q = q_0(1 - r/a)$ (set $q_1 = 0$ in Fig. 7.4.1). Determine the relationships between the solutions of the CPT and FSDT.

Solution: For simply supported as well as clamped (at $r = a$) circular plates, the boundary conditions in Eq. (7.4.34) and (7.4.35) give

$$C_1 = C_2 = C_3 = 0 \quad \text{and} \quad C_4 = -\frac{D\mathcal{M}_a^C}{K_s Gh}, \tag{1}$$

where $\mathcal{M}_a^C$ is the moment sum at the simply supported or clamped edge ($r = a$) of the CPT, and it is given as follows for various boundary conditions.

Simply supported edge:

$$\mathcal{M}_a^C = \frac{M_{\theta\theta}^C(a)}{1+\nu} = -\frac{D(1-\nu)}{a}\frac{dw^C}{dr}\bigg|_{r=a} = \frac{D(1-\nu)}{\nu}\frac{d^2 w^C}{dr^2}\bigg|_{r=a}. \tag{2}$$

Clamped edge:

$$\mathcal{M}_a^C = -D\frac{d^2 w^C}{dr^2}\bigg|_{r=a}. \tag{3}$$

Hence, the relationships in Eqs (7.4.30) and (7.4.32) become

$$\phi_r(r) = -\frac{dw^C}{dr}, \quad w^F(r) = w^C + \frac{\mathcal{M}^C(r) - \mathcal{M}_a^C}{K_s Gh}. \tag{4,5}$$

The CPT solution for such plates is given by

$$w^C = \frac{q_0 a^4}{14400D}\left[\frac{3(183+43\nu)}{1+\nu} - \frac{10(71+29\nu)}{1+\nu}\left(\frac{r}{a}\right)^2 + 255\left(\frac{r}{a}\right)^4 - 64\left(\frac{r}{a}\right)^5\right] \tag{6}$$

for simply supported plate and

$$w^C = \frac{q_0 a^4}{14400D}\left[129 - 290\left(\frac{r}{a}\right)^2 + 225\left(\frac{r}{a}\right)^4 - 64\left(\frac{r}{a}\right)^5\right] \tag{7}$$

for clamped plate. For example, substitution of Eq. (6) into Eq. (2), and the result into Eq. (5) yields the deflection of the simply supported plate according to the FSDT:

$$w^F = w^C + \frac{q_0 a^4}{36 K_s Gh}\left[5 - 9\left(\frac{r}{a}\right)^2 + 4\left(\frac{r}{a}\right)^3\right]. \tag{8}$$

The maximum deflection (occurs at $r=0$) is

$$w_{\max}^F = w_{\max}^C + \frac{5q_0 a^4}{36 K_s Gh} = \frac{43q_0 a^4}{4800D} + \frac{5q_0 a^4}{36 K_s Gh}. \tag{9}$$

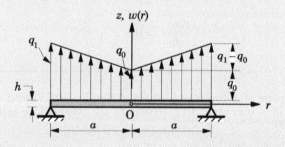

Fig. 7.4.1 Simply supported circular plate under linearly varying axisymmetric load.

7.4.3   Rectangular Plates

Next, we consider the relationships between the bending solutions of the CPT and the FSDT for polygonal plates (i.e., plates with multiple straight edges). The equations of equilibrium of plates according to the CPT and the FSDT can be expressed in terms of the deflection w and the moment sum (or Marcus moment) $\mathcal{M}$ of each theory,

$$\mathcal{M}^C = \frac{M_{xx}^C + M_{yy}^C}{1+\nu}, \quad \mathcal{M}^F = \frac{M_{xx}^F + M_{yy}^F}{1+\nu}, \tag{7.4.37}$$

as

$$\nabla^2 \mathcal{M}^C = -q, \qquad\qquad \nabla^2 w^C = -\frac{\mathcal{M}^C}{D}, \tag{7.4.38}$$

$$\nabla^2 \mathcal{M}^F = -q, \quad \nabla^2\left(w^F - \frac{\mathcal{M}^F}{K_s G h}\right) = -\frac{\mathcal{M}^F}{D}, \tag{7.4.39}$$

where the superscripts C and F refer to quantities of the classical and first-order plate theories, respectively, D is the flexural rigidity, and ν is Poisson's ratio. The moment sum in each theory is related to the generalized displacements of the theory by the relations

$$\mathcal{M}^C = -D\left(\frac{\partial^2 w^C}{\partial x^2} + \frac{\partial^2 w^C}{\partial y^2}\right), \tag{7.4.40}$$

$$\mathcal{M}^F = D\left(\frac{\partial \phi_x}{\partial x} + \frac{\partial \phi_y}{\partial y}\right). \tag{7.4.41}$$

From the first equation in Eqs (7.4.38) and (7.4.39), since q is the same in both equations, it follows that

$$\mathcal{M}^F = \mathcal{M}^C + D\nabla^2\Phi, \tag{7.4.42}$$

where Φ is a function such that it satisfies the biharmonic equation

$$\nabla^4\Phi = 0. \tag{7.4.43}$$

Using this result in the second equation of Eqs (7.4.38) and (7.4.39), we arrive at the relationship

$$w^F = w^C + \frac{\mathcal{M}^C}{K_s G h} + \Psi - \Phi \tag{7.4.44}$$

$$= w^C + \frac{h^2}{6K_s(1-\nu)}\nabla^2 w^C + \Psi - \Phi, \tag{7.4.45}$$

where Ψ is a harmonic function that satisfies the Laplace equation

$$\nabla^2\Psi = 0. \tag{7.4.46}$$

Note that the relationship in Eq. (7.4.45) is valid for all plates, including polygonal plates, with arbitrary boundary conditions and transverse load. One must determine Φ and Ψ from Eqs (7.4.43) and (7.4.46), respectively, subject to the boundary conditions of the plate. It is not simple to determine these functions for arbitrary geometries and boundary conditions.

In cases where $w^F = w^C$ on the boundary of the polygonal plates and $\mathcal{M}^C$ is either zero or equal to a constant $\mathcal{M}^{*C}$ (which can be zero) over the boundary, $\Psi - \Phi$ simply takes on the value of $-\mathcal{M}^{*C}/(K_s G h)$. However, if $\mathcal{M}^C$

varies over the boundary, the functions Ψ and Φ must be determined separately. Restricting our analysis to the former case, Eq. (7.4.44) can be written as

$$w^F(x, y) = w^C(x, y) + \frac{\mathcal{M}^C - \mathcal{M}^{*C}}{K_s Gh}. \tag{7.4.47}$$

Equation (7.4.47) can be used to establish relationships between deflection gradients, bending moments, twisting moment, and shear forces of the CPT and the FSDT, as follows:

$$\frac{\partial w^F}{\partial x} = \frac{\partial w^C}{\partial x} + \frac{Q_x^C}{K_s Gh}, \tag{7.4.48}$$

$$\frac{\partial w^F}{\partial y} = \frac{\partial w^C}{\partial y} + \frac{Q_y^C}{K_s Gh}, \tag{7.4.49}$$

$$
\begin{aligned}
M_{xx}^F &= M_{xx}^C + \frac{D(1-\nu)}{2K_s Gh}\left[\frac{\partial}{\partial x}\left(Q_x^F - Q_x^C\right) - \frac{\partial}{\partial y}\left(Q_y^F - Q_y^C\right)\right] \\
&= M_{xx}^C + \frac{D}{K_s Gh}\left[\frac{\partial}{\partial x}\left(Q_x^F - Q_x^C\right) + \nu\frac{\partial}{\partial y}\left(Q_y^F - Q_y^C\right)\right] \\
&= M_{xx}^C + \frac{D(1-\nu)}{K_s Gh}\frac{\partial}{\partial x}\left(Q_x^F - Q_x^C\right), \tag{7.4.50}
\end{aligned}
$$

$$
\begin{aligned}
M_{yy}^F &= M_{yy}^C + \frac{D(1-\nu)}{2K_s Gh}\left[\frac{\partial}{\partial y}\left(Q_y^F - Q_y^C\right) - \frac{\partial}{\partial x}\left(Q_x^F - Q_x^C\right)\right] \\
&= M_{yy}^C + \frac{D}{K_s Gh}\left[\frac{\partial}{\partial y}\left(Q_y^F - Q_y^C\right) + \nu\frac{\partial}{\partial x}\left(Q_x^F - Q_x^C\right)\right] \\
&= M_{yy}^C + \frac{D(1-\nu)}{K_s Gh}\frac{\partial}{\partial y}\left(Q_y^F - Q_y^C\right), \tag{7.4.51}
\end{aligned}
$$

$$M_{xy}^F = M_{xy}^C + \frac{D(1-\nu)}{2K_s Gh}\left[\frac{\partial}{\partial y}\left(Q_x^F - Q_x^C\right) - \frac{\partial}{\partial x}\left(Q_y^F - Q_y^C\right)\right]. \tag{7.4.52}$$

The relationships given in Eqs (7.4.48) to (7.4.52) are exact if $w^F = w^C$ at the boundaries and the Marcus moments in both theories at the boundary are equal to the same constant.

In the case of simply supported polygonal plates, it can be shown that the Marcus moment (or moment sum) and the deflection in the CPT are zero:

$$w^C = \mathcal{M}^C = 0 \text{ along the straight simply supported edges.} \tag{7.4.53}$$

In the first-order plate theory, the simply supported boundary condition involves specifying, in addition to the deflection and normal bending moment, the tangential rotation on the edge. This boundary condition is known as the simple support of the "hard" type:

$$w^F = 0, \quad M_{nn}^F = 0, \quad \phi_s = 0, \tag{7.4.54}$$

where n is the direction normal to the simply supported edge and s is the direction tangential to the edge. Owing to these conditions, $\partial \phi_s / \partial s = 0$ and the Marcus moment $\mathcal{M}^F$ is thus equal to zero. The boundary conditions of the FSDT for the simply supported plate are therefore

$$w^F = \mathcal{M}^F = 0 \text{ along the straight simply supported edges.} \qquad (7.4.55)$$

Since the Marcus moments at the boundaries of plates with any polygonal shape and simply supported edges are equal to zero, Eq. (7.4.47) applies to such plates with $\mathcal{M}^{*C} = 0$. We have

$$w^F = w^C + \frac{\mathcal{M}^C}{K_s G h}. \qquad (7.4.56)$$

Equation (7.4.56) is an important relationship between the deflections w^F and w^C of a simply supported polygonal plate. That is, if the deflection of a simply supported plate using the CPT is available, we can immediately determine the deflection according to the FSDT for the same problem from Eq. (7.4.56), thus bypassing the task of solving the equations of the FSDT. Using the same reasoning, one may readily deduce that (7.4.44) holds for simply supported polygonal plates under a constant distributed moment $\mathcal{M}^{*K}$ along their edges. For additional results on this topic, the reader may consult the book by Wang *et al.* [129] and references therein. Bending relationships in a canonical form with constants that can be specialized readily for different plate theories can be found in the paper by Lim and Reddy[5].

Example 7.4.4 _____

Consider the simply supported equilateral triangular plate of **Example 7.2.16**. Equation (13) of that example contains the analytical solution for the deflection of the plate, according to the classical plate theory, under uniform load q_0:

$$w^C = \frac{q_0 a^4}{64 D} \left[\bar{x}^3 - 3\bar{y}^2 \bar{x} - (\bar{x}^2 + \bar{y}^2) + \frac{4}{27} \right] \left(\frac{4}{9} - \bar{x}^2 - \bar{y}^2 \right), \qquad (1)$$

where $\bar{x} = x/a$ and $\bar{y} = y/a$ (see Fig. 7.2.20). Determine the deflection of the same plate according to the first-order shear deformation plate theory.

Solution: In view of Eq. (1), the Marcus moment is given by

$$\mathcal{M}^C = -D\nabla^2 w^C = \frac{q_0 L^2}{4} \left[\bar{x}^3 - 3\bar{x}\bar{y}^2 - (\bar{x}^2 + \bar{y}^2) + \frac{4}{27} \right]. \qquad (2)$$

Therefore, the deflection w^F according to the first-order plate theory of the same problem is given by Eq. (7.4.56) as

$$w^F = \frac{q_0 a^4}{4D} \left[\bar{x}^3 - 3\bar{y}^2 - (\bar{x}^2 + \bar{y}^2) + \frac{4}{27} \right] \left[\frac{\frac{4}{9} - \bar{x}^2 - \bar{y}^2}{16} + \frac{D}{K_s G a^2} \right]. \qquad (3)$$

_____

[5]G.T. Lim and J.N. Reddy, "On canonical bending relationships for plates," *International Journal of Solids and Structures*, **40** (2003) 3039–3067.

Example 7.4.5

Consider a simply supported rectangular plate of side lengths $a \times b$, as shown in Fig. 7.2.11 for the classical plate theory and Fig. 7.3.2 for the first-order plate theory. The plate is subjected to a distributed load $q(x,y)$. The deflection of this problem according to the classical plate theory is given by Eqs (7.2.120) and (7.2.121):

$$w^C = \sum_{n=1}^{\infty} \sum_{m=1}^{\infty} \left[\frac{q_{mn}}{\pi^4 D \left(\frac{m^2}{a^2} + \frac{n^2}{b^2} \right)^2} \right] \sin \frac{m\pi x}{a} \sin \frac{n\pi y}{b}, \tag{1}$$

where

$$q_{mn} = \frac{4}{ab} \int_0^b \int_0^a q(x,y) \sin \frac{m\pi x}{a} \sin \frac{n\pi y}{b} \, dx dy. \tag{2}$$

Determine the deflection of the plate according to the first-order plate theory.

Solution: In view of Eq. (7.4.40), the Marcus moment $\mathcal{M}^C$ is given by

$$\mathcal{M}^C = -D\nabla^2 w^C = -\sum_{n=1}^{\infty} \sum_{m=1}^{\infty} \left[\frac{q_{mn}}{\pi^2 \left(\frac{m^2}{a^2} + \frac{n^2}{b^2} \right)} \right] \sin \frac{m\pi x}{a} \sin \frac{n\pi y}{b}. \tag{3}$$

Using Eq. (7.4.56), the deflection w^F according to the first-order plate theory applied to the simply supported rectangular plate under any distributed load q is given by

$$w^F(x,y) = \sum_{n=1}^{\infty} \sum_{m=1}^{\infty} \left[\frac{Q_{mn}}{\pi^4 \left(\frac{m^2}{a^2} + \frac{n^2}{b^2} \right)^2} \right] \sin \frac{m\pi x}{a} \sin \frac{n\pi y}{b} \tag{4}$$

where

$$Q_{mn} = q_{mn} \left[1 + \frac{\pi^2 \left(\frac{m^2}{a^2} + \frac{n^2}{b^2} \right)}{6 K_s G h} \right]. \tag{5}$$

This solution can be verified to be the same as that given for w in Eq. (7.3.50) with W_{mn} defined in Eqs (7.3.56), (7.3.57), and (7.3.54). In particular, for sinusoidally distributed load,

$$q(x,y) = q_0 \sin \frac{\pi x}{a} \sin \frac{\pi y}{b}. \tag{6}$$

The deflection becomes ($q_{mn} = q_0$)

$$w^F(x,y) = \frac{q_0 b^4}{\pi^4 (1 + \hat{s}^2)^2} \left[1 + \frac{\pi^2 (1 + \hat{s}^2)}{6 K_s G b^2} \right] \sin \frac{\pi x}{a} \sin \frac{\pi y}{b}, \tag{7}$$

where $\hat{s}$ is the plate aspect ratio, $\hat{s} = b/a$. The maximum deflection occurs at the center of the rectangular plate, $(x,y) = (0.5a, 0.5b)$, and it is given by

$$w_{\max}^F = \frac{q_0 b^4}{\pi^4 (1 + \hat{s}^2)^2} \left[1 + \frac{\pi^2 (1 + \hat{s}^2)}{6 K_s G b^2} \right]. \tag{8}$$

For a square plate ($b = a$), the maximum deflection becomes

$$w_{\max}^F = \frac{q_0 a^4}{4\pi^4} \left[1 + \frac{2\pi^2}{6 K_s G a^2} \right]. \tag{9}$$

7.5 Summary

This chapter is dedicated to the study of bending of plates. The classical and first-order shear deformation theories of plates (i.e., their governing equations) are developed using the dynamic version of the principle of virtual displacements (or Hamilton's principle). Governing equations for both circular plates and plates in a rectangular Cartesian system are developed. Analytical solutions of bending, natural vibration, and buckling for the axisymmetric circular plates and the Navier as well as the Lévy solutions of rectangular plates are presented. Then variational solutions for circular and rectangular plates are formulated using the Ritz method. In closing the chapter, algebraic relations between the bending solutions of the FSDT (TBT) in terms of the bending solutions of the CPT (BET) are developed for plates (beams). Several numerical examples are included in every section.

Problems

<small>SECTION 7.2.2: ANALYTICAL SOLUTIONS OF CIRCULAR PLATES</small>

7.1 Obtain the Euler equations and the natural boundary conditions associated with the functional

$$\Pi(w) = \frac{\pi}{2} \int_b^a \left\{ D_{11} \left(\frac{d^2 w}{dr^2} \right)^2 + \frac{2D_{12}}{r} \frac{dw}{dr} \frac{d^2 w}{dr^2} + D_{22} \left(\frac{1}{r} \frac{dw}{dr} \right)^2 \right\} r \, dr,$$

which arises in connection with axisymmetric bending of polar orthotropic annular plates with inner radius b and outer radius a. Assume that the deflection w and its slope dw/dr are zero at $r = a$, the outer radius of the plate.

7.2 Use the virtual work statement $\delta W = \delta U + \delta V = 0$, with δU given by Eq. (7.2.6) and δV given by Eq. (7.2.8) for the pure bending case to show that the total potential energy functional for the bending of circular plates is given by

$$\Pi(w) = \frac{D}{2} \int_{\Omega_0} \left[\left(\frac{\partial^2 w}{\partial r^2} + \frac{1}{r} \frac{\partial w}{\partial r} + \frac{1}{r^2} \frac{\partial^2 w}{\partial \theta^2} \right)^2 \right.$$
$$- 2(1 - \nu) \frac{\partial^2 w}{\partial r^2} \left(\frac{1}{r} \frac{\partial w}{\partial r} + \frac{1}{r^2} \frac{\partial^2 w}{\partial \theta^2} \right)$$
$$\left. + 2(1 - \nu) \left(\frac{1}{r} \frac{\partial^2 w}{\partial r \partial \theta} - \frac{1}{r^2} \frac{\partial w}{\partial \theta} \right)^2 \right] r \, dr \, d\theta$$
$$+ \frac{1}{2} \int_{\Omega_0} k w^2 \, r \, dr \, d\theta - \int_{\Omega_0} q w \, r \, dr \, d\theta.$$

7.3 Derive the Euler equation and natural boundary conditions of the functional in **Problem 7.2**.

7.4 Show that the exact deflection of a simply supported circular plate subjected to applied bending moment $M_{rr} = M_a$ at $r = a$ is

$$w(r) = \frac{M_a a^2}{2D(1 + \nu)} \left(1 - \frac{r^2}{a^2} \right).$$

7.5 Determine the deflection and bending moments of a circular plate under uniformly distributed transverse load q_0 when the edge $r = a$ is elastically built-in. The boundary conditions are

$$w(a) = 0, \quad M_{rr} = \beta\frac{dw}{dr} \quad \text{at} \quad r = a$$

where β denotes rotational stiffness constant. *Answer*:

$$w(r) = \frac{q_0 a^4}{64D}\left\{\frac{(5+\nu)D + \beta a}{(1+\nu)D + \beta a} - 2\left[\frac{(3+\nu)D + \beta a}{(1+\nu)D + \beta a}\right]\left(\frac{r}{a}\right)^2 + \left(\frac{r}{a}\right)^4\right\},$$

$$M_{rr} = \frac{q_0 a^2}{16}\left[(1+\nu)\frac{(3+\nu)D + \beta a}{(1+\nu)D + \beta a} - (3+\nu)\left(\frac{r}{a}\right)^2\right],$$

$$M_{\theta\theta} = \frac{q_0 a^2}{16}\left[(1+\nu)\frac{(3+\nu)D + \beta a}{(1+\nu)D + \beta a} - (1+3\nu)\left(\frac{r}{a}\right)^2\right].$$

7.6 Show that the maximum deflection and bending moment of a simply supported circular plate under linearly varying load $q = q_0(1 - r/a)$ are

$$w_{\max} = \frac{q_0 a^4}{4800D}\left(\frac{183 + 43\nu}{1+\nu}\right), \quad M_{\max} = q_0 a^2\left(\frac{71 + 29\nu}{720}\right).$$

7.7 Show that the maximum deflection and bending moment of a simply supported circular plate under linearly varying load $q = q_1(r/a)$ are

$$w_{\max} = \frac{q_1 a^4}{150D}\left(\frac{6+\nu}{1+\nu}\right), \quad M_{\max} = q_1 a^2\left(\frac{4+\nu}{45}\right).$$

7.8 Show that the deflection of a clamped circular plate under the load $q = q_0(r^2/a^2)$ is given by

$$w(r) = \frac{q_0 a^4}{576D}\left[2 - 3\left(\frac{r}{a}\right)^2 + \left(\frac{r}{a}\right)^6\right].$$

7.9 Show that the expression for the deflection of a clamped (at the outer edge) circular plate under linearly varying load, $q = q_0(1 - r/a)$ is

$$w(r) = \frac{q_0 a^4}{14400D}\left(129 - 290\frac{r^2}{a^2} + 225\frac{r^4}{a^4} - 64\frac{r^5}{a^5}\right).$$

7.10 Show that the expressions for the deflection and bending moments of a clamped (at the outer edge) circular plate under linearly varying load $q = q_1(r/a)$ are

$$w(r) = \frac{q_1 a^4}{450D}\left(3 - 5\frac{r^2}{a^2} + 2\frac{r^5}{a^5}\right),$$

$$M_{rr}(r) = \frac{q_1 a^2}{45}\left[(1+\nu) - (4+\nu)\frac{r^3}{a^3}\right],$$

$$M_{\theta\theta}(r) = \frac{q_1 a^2}{45}\left[(1+\nu) - (1+4\nu)\frac{r^3}{a^3}\right].$$

SECTION 7.2.2.3: VARIATIONAL SOLUTIONS OF CIRCULAR PLATES

7.11 Determine the one-parameter Ritz solution for the deflection of a simply supported circular plate under linearly varying load $q = q_1(r/a)$.

7.12 Determine the one-parameter Ritz solution for the deflection of a clamped circular plate under linearly varying load $q = q_1(r/a)$.

7.13 Determine the fundamental frequency of a simply supported circular plate using a one-parameter Ritz approximation.

7.14 Determine the fundamental frequency of a clamped circular plate using a two-parameter Ritz approximation. Use algebraic polynomials.

7.15 The governing equation for axisymmetric vibration of a circular membrane is

$$-\frac{1}{r}\frac{d}{dr}\left(r\frac{du}{dr}\right) - \lambda u = 0, \quad 0 < r < a,$$

where λ is the frequency parameter (square of the frequency) and u is the deflection of the membrane. Determine the fundamental frequency of a clamped circular membrane using a two-parameter Galerkin approximation with trigonometric functions (see **Problem 6.44**).

7.16 Determine the deflection at the center of a simply supported circular plate under asymmetric loading given in Eq. (1) of **Example 7.2.8** using the reciprocity theorem.

7.17 Determine the center deflection of a simply supported circular plate under hydrostatic loading $q(r) = q_0(1 - r/a)$ using the reciprocity theorem.

7.18 Repeat **Problem 7.17** for a clamped circular plate.

7.19 Determine the center deflection of a clamped circular plate subjected to a point load Q_0 at a distance b from the center (and for some θ) using the reciprocity theorem.

7.20 Repeat **Problem 7.19** for a simply supported circular plate.

7.21 The total potential energy functional for the axisymmetric bending of circular plate on an elastic foundation and subjected to concentrated load Q_0 at the center is given by

$$\Pi(w) = \frac{\pi}{2}\int^a \left[D\left(\frac{d^2w}{dr^2}\right)^2 + 2D\nu\frac{1}{r}\frac{dw}{dr}\frac{d^2w}{dr^2} + D\left(\frac{1}{r}\frac{dw}{dr}\right)^2 + kw^2\right]r\,dr - Q_0 w(0),$$

where w is the transverse deflection, a is the radius, D is the flexural rigidity of the plate, and k is the foundation modulus. Determine the two-parameter Ritz solution in the form $w(r) = c_1 + c_2 r^2$.

SECTIONS 7.2.4 AND 7.2.5: THE NAVIER AND LÉVY SOLUTIONS OF RECTANGULAR PLATES

7.22 Consider an isotropic, simply supported rectangular plate subjected to hydrostatic load that varies linearly along the x-axis, as shown in Fig. P7.22 (also see Table 7.2.3). Using the Navier solution method, determine the expressions for deflection $w(x, y)$ and bending moment M_{xx}. What are the maximum values of these quantities for a square plate?

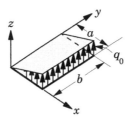

Fig. P7.22

7.23 Consider an isotropic rectangular plate, clamped at $y = 0$, simply supported at $x = 0, a$ and $y = b$ (see Fig. P7.23), and subjected to uniformly distributed load of intensity q_0. Using the Lévy method of solution, determine the expression for deflection $w(x, y)$. What are the maximum values of the deflection and bending moment for a square plate?

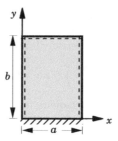

Fig. P7.23

7.24 Consider a square isotropic plate, clamped at $y = \pm a/2$, simply supported at $x = 0, a$ (see Fig. 7.2.15 with $b = a$), and subjected to sinusoidal load:

$$q(x, y) = q_0 \, \sin \frac{\pi x}{a}. \tag{1}$$

Use the Lévy method of solution and determine the expression for deflection $w(x, y)$. What are the maximum values of the deflection and bending moment?

SECTION 7.2.8: VARIATIONAL SOLUTIONS OF RECTANGULAR PLATES

Give the functions X_i and Y_j required in the Ritz approximation in Eq. (7.2.187) for the following rectangular plates:

7.25 Plate with edges $x = 0, a$ and $y = 0$ clamped and edge $y = b$ simply supported.

7.26 Plate with edge $x = 0$ clamped and the remaining edges free.

7.27 Plate with edges $x = 0, a$ and $y = 0$ clamped and edge $y = b$ free.

7.28 Plate with edges $x = 0$ and $y = 0$ clamped and edges $x = a$ and $y = b$ free.

7.29 Determine a one-parameter Ritz solution of an isotropic rectangular plate with edges $x = 0, a$ clamped and edges $y = 0, b$ simply supported, and subjected to uniformly distributed transverse load of intensity q_0. Use the approximation

$$W(x, y) = c_{11} \left(1 - \cos \frac{2\pi x}{a} \right) \sin \frac{\pi y}{b} \tag{1}$$

to determine the maximum deflection of a square plate.

7.30 Determine a one-parameter Ritz solution of an isotropic rectangular plate with edges $x = 0, a$ clamped and edges $y = 0, b$ simply supported and subjected to a center point load Q_0 transverse load. Use the approximation in Eq. (1) of **Problem 7.29**. Determine the maximum deflection of a square plate.

7.31 Determine a one-parameter Ritz solution of an isotropic rectangular plate with all edges clamped and subjected to uniform load of intensity q_0. Use the approximation

$$W(x, y) = c_{11} \left(1 - \cos \frac{2\pi x}{a} \right) \left(1 - \cos \frac{2\pi y}{b} \right), \tag{1}$$

where the origin of the coordinate system is taken at the bottom left corner of the plate with y-axis up (i.e., $0 \le x \le a$ and $0 \le y \le b$). Determine the maximum deflection of a square plate.

7.32 Repeat **Problem 7.31** for a point load Q_0 at the center of the plate.

7.33 Repeat **Problem 7.31** for an orthotropic square plate, and determine the maximum deflection for the case $D_{11} = D_0$, $D_{22} = 0.5 D_0$, and $D_{12} + 2 D_{66} = 1.248 D_0$.

7.34 Determine the deflection surface $w(x, y)$ of a simply supported, isotropic equilateral triangular plate of Fig. 7.2.20 when the plate is subjected to a point load Q_0 at the centroid $(x, y) = (0, 0)$. Use the one-parameter Ritz approximation of **Example 7.2.16**.

7.35 Consider the equilateral triangular plate of Fig. 7.2.20. Suppose that all edges are simply supported and loaded by a uniform moment $\hat{M}_{nn} = M_0$ along its edges. Use a one-parameter Ritz solution of **Example 7.2.16** and determine F_1. The exact solution is

$$w(x, y) = \frac{M_0 a^2}{4D} \left[\frac{4}{27} - \left(\frac{x}{a}\right)^2 - \left(\frac{y}{a}\right)^2 - 3\left(\frac{x}{a}\right)\left(\frac{y}{a}\right)^2 + \left(\frac{x}{a}\right)^3 \right]. \tag{1}$$

Hint: Evaluate the line integral

$$F_1 = \oint_\Gamma M_0 \left(n_x \frac{\partial \phi_1}{\partial x} + n_y \frac{\partial \phi_1}{\partial y} \right) dS, \tag{2}$$

where Γ denotes the boundary of the triangle and (n_x, n_y) are the direction cosines of the line segment.

7.36 Consider a clamped, isotropic elliptic plate with major and minor axes $2a$ and $2b$, respectively. Obtain a one-parameter Galerkin solution for the case in which the plate is subjected to distributed load $q = q_0(x/a)$.

7.37 Determine a one-parameter Ritz solution of a simply supported, isotropic elliptic plate under uniformly distributed load. *Hint:* Use

$$\phi_1(x, y) = \left(1 - \frac{x^2}{a^2} - \frac{y^2}{b^2} \right), \tag{1}$$

which satisfies the geometric boundary condition $w = 0$.

SECTION 7.3: SHEAR DEFORMATION THEORY OF PLATES

7.38 Show that substitution of Eqs (7.3.15) and (7.3.16) for M_{rr} and $M_{\theta\theta}$ into Eq. (7.3.13) yields the result

$$\begin{aligned} rQ_r &= D \left(r\frac{d^2\phi_r}{dr^2} + \frac{d\phi_r}{dr} - \frac{1}{r}\phi_r \right) \\ &= Dr \left[\frac{d^2\phi_r}{dr^2} + \frac{d}{dr}\left(\frac{1}{r}\phi_r\right) \right] \\ &= Dr \frac{d}{dr} \left[\frac{1}{r}\frac{d}{dr}(r\phi_r) \right]. \end{aligned} \tag{1}$$

7.39 Determine the generalized displacements (w, ϕ_r) of a clamped circular plate using a two-parameter Ritz approximation and the first-order shear deformation plate theory. Use **Example 7.3.5** as a guide.

SECTION 7.4: RELATIONSHIPS BETWEEN SOLUTIONS

7.40 Find the constants C_1 through C_4 of Eq. (7.4.9) for clamped–free beams.

7.41 Find the constants C_1 through C_4 of Eq. (7.4.9) for clamped–simply supported beams.

7.42 Use the bending relationships to determine the deflection and bending moment of a cantilever beam under uniformly distributed load according to the Timoshenko beam theory.

7.43 For axisymmetric bending of circular plates (without in-plane load $\hat{N}_{rr}$) according to the classical plate theory, Eq. (7.2.22) reduces to

$$-\frac{1}{r}\frac{d}{dr}\left(rQ_r^C\right) = q; \quad rQ_r^C \equiv \frac{d}{dr}\left(rM_{rr}^C\right) - M_{\theta\theta}^C, \tag{1}$$

where

$$M_{rr}^C = -D\left(\frac{d^2 w^C}{dr^2} + \nu \frac{1}{r}\frac{dw^C}{dr}\right), \quad M_{\theta\theta}^C = -D\left(\nu\frac{d^2 w^C}{dr^2} + \frac{1}{r}\frac{dw^C}{dr}\right), \quad (2)$$

where $D = Eh^3/12(1-\nu^2)$, h being the thickness of the plate. The corresponding equations of the first-order theory are

$$-\frac{1}{r}\frac{d}{dr}\left(rQ_r^F\right) = q, \quad -\frac{d}{dr}\left(rM_{rr}^F\right) + M_{\theta\theta}^F + rQ_r^F = 0, \quad (3)$$

with

$$M_{rr}^F = D\left(\frac{d\phi_r^F}{dr} + \nu\frac{1}{r}\phi_r^F\right), \quad M_{\theta\theta}^F = D\left(\nu\frac{d\phi_r^F}{dr} - \frac{1}{r}\phi_r^F\right),$$

$$Q_r^F = K_s G h\left(\frac{dw^F}{dr} + \phi_r^F\right). \quad (4)$$

Introduce the moment sum

$$\mathcal{M} = \frac{M_{rr} + M_{\theta\theta}}{1+\nu} \quad (5)$$

and show that

$$\mathcal{M}^C = -D\frac{1}{r}\frac{d}{dr}\left(r\frac{dw^C}{dr}\right), \quad (6)$$

$$-\frac{1}{r}\frac{d}{dr}\left(r\frac{d\mathcal{M}^C}{dr}\right) = q, \quad (7)$$

$$r\frac{d\mathcal{M}^C}{dr} = \frac{d}{dr}\left(rM_{rr}^C\right) - M_{\theta\theta}^C = rQ_r^C, \quad (8)$$

$$\mathcal{M}^F = D\frac{1}{r}\frac{d}{dr}\left(r\phi_r^F\right), \quad (9)$$

$$-\frac{1}{r}\frac{d}{dr}\left(r\frac{d\mathcal{M}^F}{dr}\right) = q, \quad (10)$$

$$r\frac{d\mathcal{M}^F}{dr} = \frac{d}{dr}\left(rM_{rr}^F\right) - M_{\theta\theta}^F = rQ_r^F. \quad (11)$$

7.44 Using Eqs (1)–(11) of **Problem 7.43**, establish the following relationships:

$$rQ_r^F = rQ_r^C + C_1, \quad (1)$$

$$\mathcal{M}^F = \mathcal{M}^C + C_1 \log r + C_2, \quad (2)$$

$$\phi_r^F = -\frac{dw^C}{dr} + \frac{C_1}{4D}r\left(2\log r - 1\right) + \frac{C_2}{2D}r + \frac{C_3}{D}\frac{1}{r}, \quad (3)$$

$$w^F = w^C + \frac{C_1}{4D}r^2\left(1 - \log r\right) + \frac{C_1}{K_s G h}\log r - \frac{C_2}{4D}r^2$$

$$- \frac{C_3}{D}\log r + \frac{C_4}{D} + \frac{\mathcal{M}^C}{K_s G h}. \quad (4)$$

"Research is to see what everybody else has seen, and to think what nobody else has thought."
– *Albert Szent-Gyoergi* (Hungarian physiologist who won the Nobel Prize in Medicine)

An Introduction to the Finite Element Method

8.1 Introduction

The traditional variational methods presented in Chapter 6 are very powerful (in fact, they are truly meshless methods), *provided* we can construct the approximation functions $\{\phi_j\}$ needed for any boundary-value problem. The traditional variational methods (i.e., variational methods that are applied to the whole domain) suffer from three major disadvantages: (1) the construction of the approximation functions for geometrically complex domains, problems with discontinuous loads, or for problems with discontinuous material or geometric properties is not possible; (2) even for simple problems, there is no unique way to construct the approximation functions; and (3) the approximation functions are problem dependent. Consequently, even in cases where the approximation functions are available, the computation of associated coefficient matrices cannot be automated for a fixed *class of problems* (e.g., bars, beams, or plates) because the approximation functions are not always algebraic polynomials and they depend on the boundary conditions of the specific problem. Consequently, each time the boundary conditions are changed, for the same differential equation (i.e., the class of problems is the same), the approximation functions are also changed, and the coefficient matrices have to be recomputed. This process is not readily adaptable to modular computer programming.

The finite element method is a procedure that uses the philosophy of the traditional variational methods to derive the equations relating undetermined coefficients. However, the method differs in two ways from the traditional variational methods in generating the algebraic equations of the problem. First, the approximation functions, by choice, are algebraic polynomials that are developed in a unique manner using ideas from the interpolation theory. Second, the approximation functions are developed for subdomains into which a given domain is divided. The subdomains, called *finite elements*, are geometrically simple shapes that permit a systematic and unique construction of the approximation functions over the element. The division of the whole domain into finite

Energy Principles and Variational Methods in Applied Mechanics, Third Edition. J.N. Reddy.
©2017 John Wiley & Sons Ltd. Published 2017 by John Wiley & Sons Ltd.
Companion Website: www.wiley.com/go/reddy/applied_mechanics_EPVM

elements not only simplifies the task of generating the approximation functions but also allows representation of the solution over individual elements. Thus, geometric and/or material discontinuities can be naturally included. Further, since the approximation functions are algebraic polynomials, the computation of the coefficient matrices of resulting from the approximate method can be automated on a computer.

As will be shown shortly, the construction of the approximation functions is systematic, and the process is independent of the boundary conditions and the data of the problem. In short, the finite element method is a piecewise (i.e., element-wise) application of classical variational methods. The undetermined parameters often, but not always, represent the values of the dependent variables and possibly their derivatives at a finite number of preselected points, whose number and location dictates the degree and form of the approximation functions used. This aspect inherently is coupled to the geometric shape of the element, that is, only certain geometries qualify as finite elements. The method is modular and therefore well suited for electronic computation and the development of general-purpose computer programs. A major drawback of the method is the introduction of interfaces between elements where derivatives of the variables being approximated may be discontinuous. The magnitude of these continuities can be reduced by refined meshes.

The present study is intended to expose the reader to some of the basic concepts of the finite element method as applied to types of problems discussed in the previous chapters, namely, bars, beams, and plates. Therefore, the material presented is introductory in nature and confined to the basic finite elements of one- and two-dimensional linear problems of solid mechanics, although the developments presented for these problems apply to all field problems that are governed by the same differential equations (but may have been arrived at using different physical principles). The reader may consult references on the finite element method listed at the end of the book.

The finite element method is described and illustrated here for (1) one-dimensional second-order equations governing bars, (2) one-dimensional fourth-order equation governing Bernoulli–Euler beam theory (BET), (3) pair of coupled one-dimensional equations governing the Timoshenko beam theory (TBT), (4) two-dimensional fourth-order equation governing the classical plate theory (CPT), and (5) set of coupled two-dimensional equations governing the first-order shear deformation plate theory (FSDT). While the general procedure is the same for all five classes of problems, the details differ from each other, and therefore it is best to consider the five classes of problems separately. The equations governing these problems closely resemble the equations governing a variety of field problems, for example, heat transfer, ground water flow, electrostatics, and so on, and thus the developments are readily applicable to those problems as well. For additional details, the reader may consult the textbooks by Reddy [46, 130].

8.2 Finite Element Analysis of Straight Bars

8.2.1 Governing Equation

From earlier discussions, the governing equation of a straight bar is

$$-\frac{d}{dx}\left(EA\frac{du}{dx}\right) + cu = f(x), \quad 0 < x < L, \tag{8.2.1}$$

where $u(x)$ is the axial displacement, E is Young's modulus, A is the area of cross section, $c = c(x)$ the surface resistance (zero in most problems), and $f = f(x)$ is the body force per unit length. We are interested in determining an approximate solution that satisfies Eq. (8.2.1) and appropriate boundary conditions in a variational sense (as discussed in Chapter 6).

In order to account for all possible problem cases, we allow the problem data, namely, E, A, c, and f, to be continuous or discontinuous in the total domain $(0, L)$. For example, a tapered, stepped composite shaft has sudden changes in its area of cross section as well as material properties at a finite number of points along the length (see Fig. 8.2.1)

$$E(x)A(x) = \begin{cases} E_1(x)A_1(x), & 0 < x < x_1, \\ E_2(x)A_2(x), & x_1 < x < x_2, \\ \quad\vdots \\ E_n(x)A_n(x), & x_{n-1} < x < x_n = L. \end{cases} \tag{8.2.2}$$

Thus, the data $(E, A, c, \text{and } f)$ can be element-wise different, and, therefore, we attach the element label e to these parameters when element equations are developed.

In developing the finite element model, we wish to consider all possible boundary conditions that a bar can be subjected to, and their form was already established in earlier chapters.[1] The two possible boundary conditions that we derived for bar problems in Chapters 4–6 are of the form

$$u = \hat{u} \quad \text{or} \quad EA\frac{du}{dx} + ku = \hat{P}, \tag{8.2.3}$$

where k is the spring constant (when the point is connected to a linear elastic spring) and $(\hat{u}, \hat{P})$ are the specified values of the displacement and force, respectively. When $k = 0$, we obtain the force boundary condition $EA(du/dx) = \hat{P}$, and when $k \neq 0$, the mixed boundary condition that relates the unknown force to the unknown displacement at the point.

---

[1]The type of boundary conditions admitted by a given differential equation can be readily determined from the associated variational principle or the weak form (the second step of the weak form development indicates the form of the variables that can be specified on the boundary), as amply illustrated in the preceding chapters.

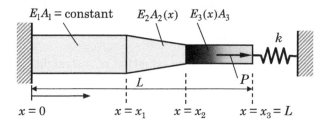

Fig. 8.2.1 A stepped composite bar with an end spring.

Thus, in a general case, we must seek approximate solution to Eq. (8.2.1) in each of the subintervals of $(0, L)$ in which the equation has continuous coefficients (i.e., a, c, and f are constant or continuous functions of x). A step-by-step procedure for the finite element analysis of Eq. (8.2.1) for arbitrary variation of a, c, and f is presented next.

8.2.2 Representation of the Domain by Finite Elements

In the finite element method, the given domain, $\Omega = (0, L)$ is divided into a number of subdomains or intervals, $\Omega_e = (x_a, x_b)$, called finite elements: $\Omega = \cup_{e=1}^{N} \Omega_e$, where N is the total number subdivisions of the domain Ω. This is necessitated by one or both of the following reasons: (1) It is easier to represent the solution $u(x)$ by a polynomial of a desired degree over each element than to use a single polynomial to approximate $u(x)$ over the entire domain. For example, a linear polynomial over each element may be used to adequately represent a cubic or higher degree variation of u with x (see Fig. 8.2.2). Obviously, the more the number of elements used, the less the error between the true solution $u(x)$ and the piecewise finite element solution $u_e(x)$ over the element. (2) The actual solution may be defined in each subdomain because of the geometric and/or material discontinuities.

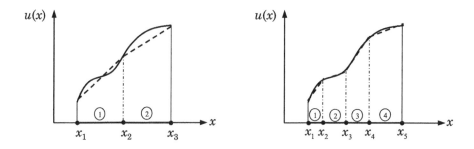

Fig. 8.2.2 Piecewise linear approximation of the solution u that is cubic.

The collection of elements into which the given domain is discretized is called the *finite element mesh* (see Fig. 8.2.3). The number of divisions is analogous (but not equal) to the number of parameters in the traditional variational methods. Therefore, the larger the number of elements, the more accurate the solution will be. The minimum number of subdivisions is equal to the number of subdivisions created by the discontinuities in the data (i.e., loads, material properties, and geometry) of the problem. Selecting the number of elements to be more than the minimum enables improved representation of the solution. An element or interval in one-dimensional problems, like straight bars, is a line of finite length.

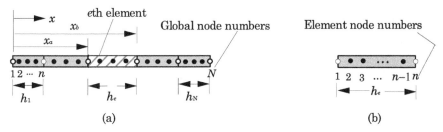

Fig. 8.2.3 (a) Subdivision of the domain into finite elements (mesh), and (b) a typical finite element.

The elements in a finite element mesh are connected to neighboring elements at a finite number of points, called *global nodes* (see Fig. 8.2.3). The word "global" refers to the whole problem as opposed to an element. The end points of individual elements are called *element nodes*, and they match with a pair of global nodes.

8.2.3   Weak Form over an Element

The weak form of the governing equation, Eq. (8.2.1), over the domain of a typical element from the mesh, $\Omega_e = (x_a, x_b) = (x_1^e, x_n^e)$, can be obtained directly from the principle of virtual displacements. A typical element with n nodes and nodal displacements is shown in Fig. 8.2.4(a), while Fig. 8.2.4(b) contains the n-noded element with forces (i.e., a free-body diagram of the element). The forces and displacements at the end nodes are defined by

$$u_e(x_1^e) = u_1^e, \qquad\qquad u(x_n^e) = u_n^e,$$

$$\left(-E_e A_e \frac{du_e}{dx}\right)\Big|_{x=x_1^e} = P_1^e, \qquad \left(E_e A_e \frac{du_e}{dx}\right)\Big|_{x=x_n^e} = P_n^e. \tag{8.2.4}$$

The nodal forces $P_2^e, P_3^e, \ldots, P_{n-1}^e$ are externally applied (i.e., known) forces, if any. Since the forces P_i^e can be included into the variational (or weak) form, we presume that all "specified" boundary conditions at the element level are the essential boundary conditions.

The weak form for a bar element shown in Fig. 8.2.4 is provided by the principle of minimum total potential energy, $\delta\Pi^e = 0$:

$$0 = \delta\left\{\int_{x_a}^{x_b}\left[\frac{E_eA_e}{2}\left(\frac{du_e}{dx}\right)^2 + \frac{c_e}{2}(u_e)^2 - fu_e\right]dx\right.$$

$$\left. - P_1^e u_1^e - P_2^e u_2^e - \cdots - P_n^e u_n^e\right\}$$

$$= \int_{x_a}^{x_b}\left(E_eA_e\frac{d\delta u_e}{dx}\frac{du_e}{dx} + c_e\delta u_e\,u_e - f\delta u_e\right)dx - \sum_{k=1}^n P_k^e\delta u_k^e, \quad (8.2.5)$$

where δ is the variational symbol and the subscript e on the variables indicates that the variables are defined in element Ω_e. Equation (8.2.5) is the same as the *weak form* [because it requires weakened (i.e., reduced) differentiability of u_e than the original equation, Eq. (8.2.1)] that can be derived from the governing differential equation over an element, as described in Chapter 6.

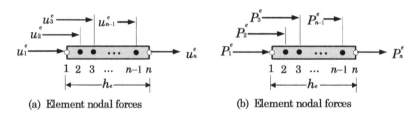

(a) Element nodal forces   (b) Element nodal forces

Fig. 8.2.4 (a) A typical finite element with n nodes and nodal displacements. (b) A typical finite element with n nodes and nodal forces.

8.2.4 Approximation over an Element

Consider an arbitrary element, $\Omega_e = (x_a, x_b)$, located between points $x = x_a \equiv x_1^e$ and $x_b \equiv x_2^e$ in the domain. We isolate the element and study the variational approximation of Eq. (8.2.1) using any one of the variational methods presented in Chapter 6. In the classical variational methods, we seek approximation of the form

$$u(x) \approx \sum_{i=1}^N c_i\phi_i(x) \quad (8.2.6)$$

over the whole domain and determine relations among c_i such that Eq. (8.2.1) is satisfied in the weak form or weighted-integral sense *over the entire domain.* In the finite element method, we seek solution of the form

$$u(x) \approx u_e(x) = \sum_{i=1}^n c_i^e\phi_i^e(x), \quad \text{with} \quad \phi_i^e(x) = x^{i-1} \quad (8.2.7)$$

over an element and determine relations among c_i by satisfying Eq. (8.2.1) in a variational (or weak form) sense *over each element*. Since the elements are connected to each other at the nodes, the solutions from various elements connected at a node must have the same value at that node. In order to enforce this continuity (or uniqueness) of solution, it is convenient to express the solution over each element in terms of its values at n nodes of the element

$$u_e(x) = \sum_{i=1}^{n} c_i^e \phi_i^e(x) = \sum_{i=1}^{n} u_i^e \psi_i^e(x), \tag{8.2.8}$$

where u_i^e is the value of $u_e(x)$ at the ith node (i.e., $u_e(x_i) = u_i^e$). Hence, by definition, the functions ψ_i^e satisfy the interpolation property

$$\psi_i^e(x_j^e) = \delta_{ij}. \tag{8.2.9}$$

That is, ψ_i is unity at its own node (i.e., ith node) and zero at all other nodes. If there are n nodes in the element, a polynomial with n terms is required to fit the solution at the n points. Thus, each ψ_i^e is a $(n-1)$ degree polynomial. For example, when $n = 2$, we have

$$u_e(x) = c_1^e + c_2^e x. \tag{8.2.10}$$

Using the definition $u_e(x_i^e) = u_i^e$, we obtain

$$u_e(x_1^e) = u_1^e = c_1^e + c_2^e x_1^e, \quad u_e(x_2^e) = u_2^e = c_1^e + c_2^e x_2^e.$$

Solving for (c_1^e, c_2^e) in terms of (u_1^e, u_2^e), we obtain

$$\begin{Bmatrix} u_1^e \\ u_2^e \end{Bmatrix} = \begin{bmatrix} 1 & x_1^e \\ 1 & x_2^e \end{bmatrix} \begin{Bmatrix} c_1^e \\ c_2^e \end{Bmatrix} \rightarrow c_1^e = \frac{u_1^e x_2^e - u_2^e x_1^e}{x_2^e - x_1^e}, \quad c_2^e = \frac{u_2^e - u_1^e}{x_2^e - x_1^e}.$$

Substituting for c_1^e and c_2^e into Eq. (8.2.8), we obtain

$$u_e(x) = c_1^e + c_2^e x = \frac{u_1^e x_2^e - u_2^e x_1^e}{x_2^e - x_1^e} + \left(\frac{u_2^e - u_1^e}{x_2^e - x_1^e} \right) x$$

$$= \left(\frac{x_2^e - x}{x_2^e - x_1^e} \right) u_1^e + \left(\frac{x - x_1^e}{x_2^e - x_1^e} \right) u_2^e \equiv \sum_{i=1}^{2} u_i^e \psi_i^e(x), \tag{8.2.11}$$

where

$$\psi_1^e(x) = \left(\frac{x_2^e - x}{x_2^e - x_1^e} \right), \quad \psi_2^e(x) = \left(\frac{x - x_1^e}{x_2^e - x_1^e} \right). \tag{8.2.12}$$

Clearly, $\psi_i^e(x)$ satisfy the interpolation property in Eq. (8.2.9); see Fig. 8.2.5. Thus the nodal values u_i^e take the place of the undetermined parameters, while ψ_i^e take the role of the approximation functions in the traditional variational methods.

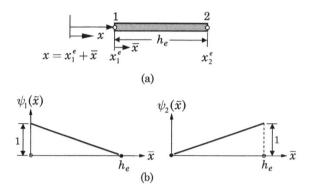

Fig. 8.2.5 Linear interpolation functions.

Alternatively, we can determine $\psi_i^e(x)$ using the interpolation property in Eq. (8.2.9). For example, for linear interpolation, the functions have the properties

$$\psi_1^e(x_1^e) = 1, \quad \psi_1^e(x_2^e) = 0, \quad \psi_2^e(x_1^e) = 0, \quad \psi_2^e(x_2^e) = 1.$$

Since ψ_1^e and ψ_2^e must vanish at x_2^e and x_1^e, respectively, we may write them as

$$\psi_1^e(x) = \alpha_1^e \left(x - x_2^e \right), \quad \psi_2^e(x) = \alpha_2^e \left(x - x_1^e \right),$$

where the constants α_1^e and α_2^e are determined so that ψ_1^e and ψ_2^e are unity at x_1^e and x_2^e, respectively:

$$\psi_1^e(x_1^e) = 1 = \alpha_1^e \left(x_1^e - x_2^e \right), \quad \psi_2^e(x_2^e) = 1 = \alpha_2^e \left(x_2^e - x_1^e \right).$$

These relations give $\alpha_1^e = -1/h_e$ and $\alpha_2^e = 1/h_e$, where h_e denotes the length of the element, $h_e = x_2^e - x_1^e = x_b^e - x_a^e$. Hence, the approximation functions $\psi_i^e(x)$ are given by

$$\psi_1^e(x) = \frac{x_2^e - x}{h_e}, \quad \psi_2^e(x) = \frac{x - x_1^e}{h_e},$$

which are the same as those listed in Eq. (8.2.12).

The interpolation functions ψ_i^e are defined over the element Ω^e, and they are zero outside the element. Hence, they are said to have compact support. The interpolation in which only the function alone is interpolated – and not its derivatives – is known as the Lagrange interpolation, and the corresponding functions are termed *the Lagrange interpolation functions*. It is convenient to express the interpolation functions in terms of a coordinate system fixed in the element. For example, we may take the origin of the element coordinate, $\bar{x}$, at node 1 such that

$$x = \bar{x} + x_1^e. \tag{8.2.13}$$

Then the interpolation functions in Eq. (8.2.12) can be expressed in terms of $\bar{x}$ as

$$\psi_1^e(\bar{x}) = 1 - \frac{\bar{x}}{h_e}, \quad \psi_2^e(x) = \frac{\bar{x}}{h_e}. \tag{8.2.14}$$

On the other hand, if ξ is the element coordinate such that $\xi = -1$ when $\bar{x} = 0$ and $\xi = 1$ when $\bar{x} = h_e$, that is,

$$\bar{x} = \frac{h_e}{2}(1 + \xi), \tag{8.2.15}$$

the $\psi_i(\bar{x})$ of Eq. (8.2.14) take the form

$$\psi_1^e(\xi) = \frac{1}{2}(1 - \xi), \quad \psi_2^e(\xi) = \frac{1}{2}(1 + \xi), \qquad -1 \le \xi \le 1. \tag{8.2.16}$$

The coordinate ξ is known as the *normalized* or *natural coordinate*. Both $\bar{x}$ and ξ are element coordinates, and as such they should have element label e, which is omitted.

For values $n > 2$, we must identify additional points or element nodes in the interior of the element. These points, in principle, can be any distinctly different points of the domain. For example, when $n = 3$, the third node can be placed anywhere other than at the ends. The optimal location is the one that is equidistant from both end nodes, that is, the midpoint of the element. When $n = 4$, the one-third points are selected for the two interior nodes. The $(n - 1)$ degree Lagrange interpolation functions (for an element with n nodes) are defined, for arbitrary location of the nodes, by

$$\psi_j^e = \Pi_{k=1, k \ne j}^{m} \frac{(x - x_k^e)}{(x_j^e - x_k^e)}. \tag{8.2.17}$$

As specific examples, the quadratic and cubic Lagrange interpolation functions are given next.

Quadratic ($n = 3$):

$$\psi_1^e = \frac{x - x_2^e}{x_1^e - x_2^e} \frac{x - x_3^e}{x_1^e - x_3^e}, \quad \psi_2^e = \frac{x - x_1^e}{x_2^e - x_1^e} \frac{x - x_3^e}{x_2^e - x_3^e}, \quad \psi_3^e = \frac{x - x_1^e}{x_3^e - x_1^e} \frac{x - x_2^e}{x_3^e - x_2^e}. \tag{8.2.18}$$

Cubic ($n = 4$):

$$\psi_1^e = \frac{x - x_2^e}{x_1^e - x_2^e} \frac{x - x_3^e}{x_1^e - x_3^e} \frac{x - x_4^e}{x_1^e - x_4^e}, \quad \psi_2^e = \frac{x - x_1^e}{x_2^e - x_1^e} \frac{x - x_3^e}{x_2^e - x_3^e} \frac{x - x_4^e}{x_2^e - x_4^e},$$

$$\psi_3^e = \frac{x - x_1^e}{x_3^e - x_1^e} \frac{x - x_2^e}{x_3^e - x_2^e} \frac{x - x_4^e}{x_3^e - x_4^e}, \quad \psi_4^e = \frac{x - x_1^e}{x_4^e - x_1^e} \frac{x - x_2^e}{x_4^e - x_2^e} \frac{x - x_3^e}{x_4^e - x_3^e}. \tag{8.2.19}$$

For equally spaced nodal points, these functions have the following simpler form, especially when written in terms of the local coordinate $\bar{x}$ (but expressed in terms of the dimensionless quantity $\hat{x} = \bar{x}/h_e$):

Quadratic functions:

$$\psi_1^e(\bar{x}) = (1 - 2\hat{x})(1 - \hat{x}),$$
$$\psi_2^e(\bar{x}) = 4\hat{x}(1 - \hat{x}), \tag{8.2.20}$$
$$\psi_3^e(\bar{x}) = \hat{x}(2\hat{x} - 1).$$

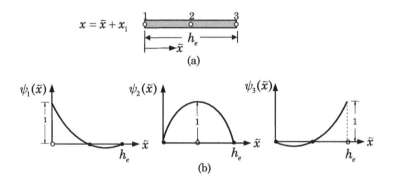

Fig. 8.2.6 Quadratic interpolation functions.

Cubic functions:

$$\psi_1^e(\bar{x}) = (1 - 3\hat{x})(1 - 1.5\hat{x})(1 - \hat{x}),$$
$$\psi_2^e(\bar{x}) = 9\hat{x}(1 - 1.5\hat{x})(1 - \hat{x}),$$
$$\psi_3^e(\bar{x}) = -4.5\hat{x}(1 - 3\hat{x})(1 - \hat{x}), \tag{8.2.21}$$
$$\psi_4^e(\bar{x}) = \hat{x}(1 - 3\hat{x})(1 - 1.5\hat{x}).$$

Note that the interpolation functions $\psi_i^e(x)$ $(i = 1, 2, \ldots, n)$ depend only on the geometry and position of the nodes in element $\Omega^e = (x_a^e, x_b^e) = (x_1^e, x_n^e)$; they do not depend on the solution or the boundary conditions of the problem. The functions ψ_i^e satisfy the interpolation property in Eq. (8.2.9). In addition, we note that

$$\psi_1^e(x) + \psi_2^e(x) + \cdots + \psi_n^e(x) = 1 \quad \text{or} \quad \sum_{i=1}^{n} \psi_i^e(x) = 1, \tag{8.2.22}$$

which is sometimes referred to as *the partition of unity*. This property is the direct result of including the constant term in the approximation, Eq. (8.2.10). For example, if the solution is constant, say, u_0^e, over the entire element, then $u_1^e = u_2^e = \cdots = u_n^e = u_0^e$, and we have

$$u_0^e = \sum_{i=1}^{n} u_i^e \psi_i^e(x) = u_0^e \sum_{i=1}^{n} \psi_i^e(x) \quad \rightarrow \quad 1 = \sum_{i=1}^{n} \psi_i^e(x). \tag{8.2.23}$$

8.2.5 Finite Element Equations

We seek Ritz approximation of $u(x)$ in the form

$$u(x) \approx u_e(x) = \sum_{j=1}^{n} u_j^e \psi_j^e(x), \quad w = \delta u(x) \approx \sum_{i=1}^{n} \delta u_i^e \psi_i^e(x), \qquad (8.2.24)$$

where ψ_j^e are the approximation functions derived earlier; they can be linear $(n = 2)$, quadratic $(n = 3)$, or higher $(n > 3)$.

Substitution of Eq. (8.2.24) for $u_e(x)$ and its variation into Eq. (8.2.5) yields

$$0 = \int_{x_a}^{x_b} \left[E_e A_e \left(\sum_{i=1}^{n} \delta u_i^e \frac{d\psi_i}{dx} \right) \left(\sum_{j=1}^{n} u_j^e \frac{d\psi_j}{dx} \right) + c_e \left(\sum_{i=1}^{n} \delta u_i^e \psi_i \right) \left(\sum_{j=1}^{n} u_j^e \psi_j \right) \right.$$

$$\left. - f_e(x) \left(\sum_{i=1}^{n} \delta u_i^e \psi_i \right) \right] dx - \sum_{i=1}^{n} P_i^e \delta u_i^e$$

$$= \sum_{i=1}^{n} \delta u_i^e \left\{ \sum_{j=1}^{n} \left[\int_{x_a}^{x_b} \left(E_e A_e \frac{d\psi_i}{dx} \frac{d\psi_j}{dx} + c_e \psi_i^e \psi_j^e \right) dx \right] u_j^e - \int_{x_a}^{x_b} f \psi_i \, dx - P_i^e \right\}$$

$$= \sum_{i=1}^{n} \delta u_i^e \left[\sum_{j=1}^{n} K_{ij}^e u_j^e - f_i^e - P_i^e \right]. \qquad (8.2.25)$$

Since δu_1^e, δu_2^e, ..., δu_n^e are arbitrary and independent of each other, it follows that

$$0 = \sum_{j=1}^{n} K_{ij}^e u_j^e - f_i^e - P_i^e \equiv \sum_{j=1}^{n} K_{ij}^e u_j^e - F_i^e, \qquad (8.2.26)$$

where

$$K_{ij}^e = \int_{x_a}^{x_b} \left(E_e A_e \frac{d\psi_i^e}{dx} \frac{d\psi_j^e}{dx} + c_e \psi_i^e \psi_j^e \right) dx$$

$$= \int_0^{h_e} \left(E_e(\bar{x}) A_e(\bar{x}) \frac{d\psi_i^e}{d\bar{x}} \frac{d\psi_j^e}{d\bar{x}} + c^e \psi_i^e \psi_j^e \right) d\bar{x}, \qquad (8.2.27)$$

$$f_i^e = \int_{x_a}^{x_b} f^e(x) \psi_i^e \, dx = \int_0^{h_e} f^e(\bar{x}) \psi_i^e(\bar{x}) \, d\bar{x}.$$

The coefficient matrix $\mathbf{K}^e$ is called the *stiffness matrix*, and $\mathbf{F}^e \equiv \mathbf{f}^e + \mathbf{P}^e$ is the *force vector*. Equation (8.2.26) is often referred to as the *finite element model* of the differential equation, Eq. (8.2.1), and it provides n linear algebraic equations relating n nodal values of the primary variables, u_i^e, to n nodal values of the secondary variables, P_i^e for $i = 1, 2, \ldots, n$.

The stiffness matrix $\mathbf{K}^e$, which is symmetric, and source vector $\mathbf{f}^e$ can be evaluated for a given element type, that is, linear $(n = 2)$, quadratic $(n = 3)$, and so on, and element data $(E_e, A_e, c_e, \text{ and } f_e)$. For element-wise constant values of E_e, A_e, c_e, and f_e, the coefficients K_{ij}^e and f_i^e can easily be evaluated. For linear and quadratic elements, these matrices are presented here.

8.2.5.1   Linear Element

For a typical linear element of length $h_e = x_b^e - x_a^e$, we have

$$\mathbf{K}^e = \frac{E_e A_e}{h_e} \begin{bmatrix} 1 & -1 \\ -1 & 1 \end{bmatrix} + \frac{c_e h_e}{6} \begin{bmatrix} 2 & 1 \\ 1 & 2 \end{bmatrix}, \quad \mathbf{f}^e = \frac{f_e h_e}{2} \begin{Bmatrix} 1 \\ 1 \end{Bmatrix}. \tag{8.2.28}$$

If $A_e(x) = \bar{A}_e \cdot x$, and E_e, $\bar{A}_e$, and $c = c_e$ are constant, then the coefficient matrix $\mathbf{K}^e$ for a linear element can be evaluated as

$$\mathbf{K}^e = \frac{E_e \bar{A}_e}{h_e} \left(\frac{x_1^e + x_2^e}{2} \right) \begin{bmatrix} 1 & -1 \\ -1 & 1 \end{bmatrix} + \frac{c_e h_e}{6} \begin{bmatrix} 2 & 1 \\ 1 & 2 \end{bmatrix}, \tag{8.2.29}$$

where (x_1^e, x_2^e) are global coordinates of node 1 and node 2 of the element $\Omega_e = (x_a^e, x_b^e) = (x_1^e, x_2^e)$. The reader should verify this. Note that $\mathbf{K}^e$ in Eq. (8.2.29) is the same as that in Eq. (8.2.28) with A_e replaced by the average value

$$A_{\text{avg}} = \frac{1}{2}(x_1^e + x_2^e)\bar{A}_e.$$

For example, in the study of bars with linearly varying cross section $A_e(x)$ but constant modulus of elasticity E_e, we have

$$E_e A_e(x) = E_e \left(A_1^e + \frac{A_2^e - A_1^e}{h_e} \bar{x} \right),$$

where $\bar{x}$ is the local or element coordinate with origin at node 1 and A_1^e and A_2^e are areas of cross section at node 1 and 2, respectively. Then the element stiffness matrix is the same as that of a constant cross-sectional bar with the cross-sectional area being the average of the two ends, $A_{\text{avg}} = (A_1^e + A_2^e)/2$.

When $E_e A_e$ and $f(x) = f_e$ are constant, and $c_e = 0$, the finite element equations corresponding to the linear element reduce to

$$\frac{E_e A_e}{h_e} \begin{bmatrix} 1 & -1 \\ -1 & 1 \end{bmatrix} \begin{Bmatrix} u_1^e \\ u_2^e \end{Bmatrix} = \frac{f_e h_e}{2} \begin{Bmatrix} 1 \\ 1 \end{Bmatrix} + \begin{Bmatrix} P_1^e \\ P_2^e \end{Bmatrix}, \tag{8.2.30}$$

or

$$\begin{aligned} \frac{E_e A_e}{h_e} u_1^e - \frac{E_e A_e}{h_e} u_2^e &= \tfrac{1}{2} f_e h_e + P_1^e, \\ -\frac{E_e A_e}{h_e} u_1^e + \frac{E_e A_e}{h_e} u_2^e &= \tfrac{1}{2} f_e h_e + P_2^e. \end{aligned} \tag{8.2.31}$$

8.2.5.2 Quadratic Element

For a quadratic element $\Omega^e = (x_a^e, x_b^e) = (x_1^e, x_3^e)$, $h_e = x_b^e - x_a^e = x_3^e - x_1^e$, with constant values of the data (E_e, A_e, f_e, and c_e), we have

$$\mathbf{K}^e = \frac{E_e A_e}{3h_e} \begin{bmatrix} 7 & -8 & 1 \\ -8 & 16 & -8 \\ 1 & -8 & 7 \end{bmatrix} + \frac{c_e h_e}{30} \begin{bmatrix} 4 & 2 & -1 \\ 2 & 16 & 2 \\ -1 & 2 & 4 \end{bmatrix}, \quad \mathbf{f}^e = \frac{f_e h_e}{6} \begin{Bmatrix} 1 \\ 4 \\ 1 \end{Bmatrix}. \tag{8.2.32}$$

For arbitrary variation of the data E_e, A_e, c_e, and f_e with x, numerical integration may be used to evaluate the coefficients K_{ij}^e and f_i^e.

8.2.6 Assembly (or Connectivity) of Elements

The finite element equations in Eq. (8.2.26) can be specialized to each one of the elements in the mesh by assigning the values of x_a^e and x_b^e and identifying the functions E_e, A_e, c_e, and f_e. Because each of the elements in the mesh is connected to its neighboring elements at the global nodes, and the displacement is continuous from one element to the next, one can relate the nodal values of displacements at the interelement connecting nodes. To this end, let U_I denote the value of the displacement $u(x)$ at the Ith global node. Then, we have the following correspondence between U_I and u_j^e [see Fig. 8.2.7(a)] of a linear element mesh:

$$u_1^1 = U_1, \quad u_2^1 = u_1^2 \equiv U_2, \dots, \quad u_2^e = u_1^{e+1} \equiv U_{e+1}, \dots, u_2^N \equiv U_{N+1}, \tag{8.2.33}$$

where N is the total number of linear elements connected in series. Equation (8.2.33) relates the global displacements to local displacements and enforces interelement continuity of the displacements [see Fig. 8.2.7(a)].

The assembly of element equations is based on the satisfaction of the principle of minimum total potential energy for the whole system:

$$\delta\Pi = \sum_{I=1}^{N+1} \frac{\partial\Pi}{\partial U_I}\delta U_I = 0 \text{ or } \frac{\partial\Pi}{\partial U_I} = 0, \quad I = 1, 2, \dots, N+1, \tag{8.2.34}$$

where Π is the sum of Π^e for $e = 1, 2, \dots, N$ (we note that the existence of Π^e implies that the associated bilinear form is symmetric, i.e., $K_{ij}^e = K_{ji}^e$):

$$\Pi = \sum_{e=1}^{N} \Pi^e = \sum_{e=1}^{N} \sum_{i=1}^{2} u_i^e \left(\frac{1}{2}\sum_{j=1}^{2} K_{ij}^e u_j^e - F_i^e\right)$$

$$= \frac{1}{2}\sum_{e=1}^{N} \left[u_1^e\left(K_{11}^e u_1^e + K_{12}^e u_2^e - 2F_1^e\right) + u_2^e\left(K_{21}^e u_1^e + K_{22}^e u_2^e - 2F_2^e\right)\right]$$

$$= \frac{1}{2}\left[u_1^1\left(K_{11}^1 u_1^1 + K_{12}^1 u_2^1 - 2F_1^1\right) + u_1^2\left(K_{11}^2 u_1^2 + K_{12}^2 u_2^2 - 2F_1^2\right)\right.$$

$$\left. + u_2^1\left(K_{21}^1 u_1^1 + K_{22}^1 u_2^1 - 2F_2^1\right) + u_2^2\left(K_{21}^2 u_1^2 + K_{22}^2 u_2^2 - 2F_2^2\right) + \cdots\right]$$

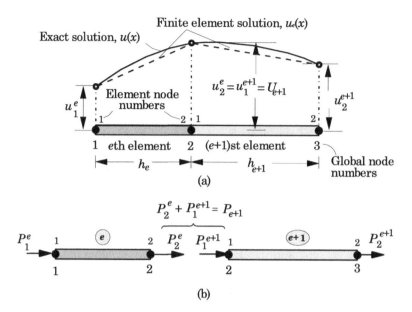

Fig. 8.2.7 Connectivity of elements in one dimension. (a) Element-wise linear approximation of the displacement and interelement continuity of the displacements. (b) Balance of element forces at common nodes.

or

$$\Pi = \tfrac{1}{2}\big[U_1\big(K_{11}^1 U_1 + K_{12}^1 U_2 - 2F_1^1\big) + U_2\big(K_{11}^2 U_2 + K_{12}^2 U_3 - 2F_1^2\big)$$
$$+ U_2\big(K_{21}^1 U_1 + K_{22}^1 U_2 - 2F_2^1\big) + U_3\big(K_{21}^2 U_2 + K_{22}^2 U_3 - 2F_2^2\big) + \cdots\big],$$
$$(8.2.35)$$

wherein arriving at the last line, the correspondence in Eq. (8.2.33) is used to replace the element nodal displacements by the global nodal displacements. Using Eq. (8.2.35) in Eq. (8.2.34), we obtain

$$K_{11}^1 U_1 + \frac{1}{2}\big(K_{12}^1 + K_{21}^1\big)U_2 - F_1^1 = 0,$$
$$\tfrac{1}{2}\big(K_{12}^1 + K_{21}^1\big)U_1 + \big(K_{22}^1 + K_{11}^2\big)U_2 + \tfrac{1}{2}\big(K_{12}^2 + K_{21}^2\big)U_3 - F_2^1 - F_1^2 = 0,$$
$$\tfrac{1}{2}\big(K_{12}^2 + K_{21}^2\big)U_2 + \big(K_{22}^2 + K_{11}^3\big)U_3 + \tfrac{1}{2}\big(K_{12}^3 + K_{21}^3\big)U_2 - F_2^2 - F_1^3 = 0,$$

$$\cdots\cdots\cdots\cdots\cdots\cdots\cdots\cdots\cdots\cdots\cdots\cdots\cdots\cdots$$

$$\tfrac{1}{2}\big(K_{12}^{N-1} + K_{21}^{N-1}\big)U_{N-1} + \big(K_{22}^{N-1} + K_{11}^N\big)U_N$$
$$+ \tfrac{1}{2}\big(K_{12}^N + K_{21}^N\big)U_{N+1} - F_2^{N-1} - F_1^N = 0,$$
$$\tfrac{1}{2}\big(K_{12}^N + K_{21}^N\big)U_N + K_{22}^N U_{N+1} - F_2^N = 0.$$
$$(8.2.36)$$

It is clear from Eq. (8.2.36) that the elements of stiffness matrices of elements Ω_e and Ω_{e+1} add up at the common global node $I = e + 1$ and the global stiffness coefficient K_{IJ} is zero if global nodes I and J do not belong to the same element. Using the symmetry of the element stiffness matrices, $K_{ji}^e = K_{ij}^e$, and the fact that $\mathbf{F}^e = \mathbf{f}^e + \mathbf{P}^e$, we can write the relations in Eq. (8.2.36) in the matrix form as

$$
\begin{bmatrix}
K_{11}^1 & K_{12}^1 & 0 & \cdots & 0 \\
K_{12}^1 & K_{22}^1 + K_{11}^2 & K_{12}^2 & 0 & 0 \\
0 & K_{12}^2 & K_{22}^2 + K_{11}^3 & K_{12}^3 & 0 \\
\cdots & \cdots & \cdots & \cdots & \cdots \\
0 & 0 & \cdots & K_{22}^{N-1} + K_{11}^N & K_{12}^N \\
0 & 0 & \cdots & K_{12}^N & K_{22}^N
\end{bmatrix}
\begin{Bmatrix}
U_1 \\ U_2 \\ U_3 \\ \vdots \\ U_N \\ U_{N+1}
\end{Bmatrix}
$$

$$
= \begin{Bmatrix}
f_1^1 \\ f_2^1 + f_1^2 \\ \vdots \\ f_2^{N-1} + f_1^N \\ f_2^N
\end{Bmatrix}
+ \begin{Bmatrix}
P_1^1 \\ P_2^1 + P_1^2 \\ \vdots \\ P_2^{N-1} + P_1^N \\ P_2^N
\end{Bmatrix}. \tag{8.2.37}
$$

Thus, the resulting global stiffness matrix is not only symmetric but also banded, that is, all entries beyond a diagonal parallel to the main diagonal, below and above, are zero. This is a feature of all finite element equations, irrespective of the differential equation being solved, and is a result of the piecewise definition of the coordinate functions. Keeping in mind the general pattern of the assembled stiffness matrix and force column, one can routinely assemble the element matrices for any number of elements.

The assembly procedure for a general case is based on the following two requirements:

1. Continuity of the primary variable(s) at the inter-element boundary, as expressed by Eq. (8.2.33).

2. Balance of secondary variables; that is, the secondary variables from the elements connected at a global node should add up to the value of the externally applied secondary variable at the node.

The second condition for the mesh of two linear elements shown in Fig. 8.2.7(b) requires

$$
P_2^e + P_1^{e+1} = P_{e+1}, \tag{8.2.38}
$$

where P_{e+1} is the value of externally applied force at node $e+1$ [see Fig. 8.2.7(b)]. These conditions require the addition of the second equation of element Ω_e to the first equation of element Ω_{e+1} so that we can replace $P_2^e + P_1^{e+1}$ with P_{e+1}.

This reduced $2N$ equations to $N+1$ equations, where N is the number of linear elements connected in series, as shown in Fig. 8.2.7(a). In general, if the ith node of element Ω^e is connected to the jth node of element Ω^f, the balance of secondary variables requires

$$P_i^e + P_j^f = F_K, \tag{8.2.39}$$

where K is the global node number of the ith node of element Ω^e, which is the same as the jth node of element Ω^f.

8.2.7 Imposition of Boundary Conditions

Equation (8.2.37) contains more variables than the number of equations, prior to applying the boundary conditions of the problem. In general, for the model problem in Eq. (8.2.1), we must know the value of either the primary variable u or the secondary variable $EA(du/dx)$ at each node, including the boundary nodes. In all bar problems, we have either a displacement u or a force $EA(du/dx)$ specified at a point. If the displacement is specified at a global node, the corresponding nodal displacement should be replaced by the specified value. To impose a specified force, recall that F_i^e consists of contributions due to the distributed force $f(x)$ and internal force $\pm EA(du/dx)$ at the nodes [see Eq. (8.2.4)], $F_i^e = f_i^e + P_i^e$. At any global node where the displacement is unknown, the externally applied point force at the node should be equal to the sum of internal forces from all elements connected at the global node. To illustrate this point further, consider a bar fixed at the left end, $\hat{u} = 0$, and subjected to distributed force f throughout the length of the bar and a point load $\hat{P}$ at the right end. The fact that no point loads are specified at the intermediate nodes implies that [see Fig. 8.2.7(b)]

$$P_2^1 + P_1^2 = 0, \quad P_2^2 + P_1^3 = 0, \quad \ldots, \quad P_2^{N-1} + P_1^N = 0. \tag{8.2.40}$$

Then Eq. (8.2.37) becomes, after omitting the first row and first column corresponding to the specified (first) degree of freedom,

$$\begin{bmatrix} K_{22}^1 + K_{11}^2 & K_{12}^2 & 0 & \cdots & 0 \\ K_{12}^2 & K_{22}^2 + K_{11}^3 & K_{12}^3 & \cdots & 0 \\ \vdots & \vdots & \ddots & \vdots & \vdots \\ 0 & 0 & 0 & K_{22}^{N-1} + K_{11}^N & K_{12}^N \\ 0 & 0 & \cdots & K_{12}^N & K_{22}^N \end{bmatrix} \begin{Bmatrix} U_2 \\ U_3 \\ \vdots \\ U_N \\ U_{N+1} \end{Bmatrix}$$

$$= \begin{Bmatrix} f_2^1 + f_1^2 \\ f_2^2 + f_1^3 \\ \vdots \\ f_2^{N-1} + f_1^N \\ f_2^N + P + 0 \end{Bmatrix} - \begin{Bmatrix} K_{11}^1 u_0 \\ K_{12}^1 u_0 \\ 0 \\ \vdots \\ 0 \end{Bmatrix}, \tag{8.2.41}$$

which consists of N equations for the N unknowns, U_I ($I = 2, \ldots, N+1$). Since the right-hand column is completely known, one can use any standard solution procedure to solve the linear algebraic equations. Now the row that is omitted can be used to determine the unknown quantity on the right side of the equation, namely, Q_1^1.

8.2.8 Postprocessing

The displacement at any point can be computed using the equation

$$
u_e(x) = \begin{cases}
u_1(x), & x_1^{(1)} \le x \le x_n^{(1)}, \\
u_2(x), & x_1^{(2)} \le x \le x_n^{(2)}, \\
\quad \vdots & \\
u_N(x), & x_1^{(N)} \le x \le x_n^{(N)},
\end{cases}
\tag{8.2.42}
$$

where $u_e(x)$ is the finite element solution of Eq. (8.2.1) in element Ω_e and $x_1^{(e)}$ and $x_n^{(e)}$ denotes the global coordinates of the first and the last nodes of element Ω^e. For example, if $u(x_0)$, where $x_1^{(e)} \le x_0 \le x_2^{(e)}$, is desired for a linear element ($n = 2$), we have

$$
u(x_0) = u_1^{(e)} \psi_1^{(e)}(x_0) + u_2^{(e)} \psi_2^{(e)}(x_0) = U_e \psi_1^{(e)}(x_0) + U_{e+1} \psi_2^{(e)}(x_0). \tag{8.2.43}
$$

Similarly, the derivatives of $u_e(x)$ can be computed from

$$
\frac{du_e}{dx} = \begin{cases}
\sum_{j=1}^{n} u_j^{(1)} \frac{d\psi_j^{(1)}}{dx}, & x^{(1)} < x < x_n^{(1)}, \\
\sum_{j=1}^{n} u_j^{(2)} \frac{d\psi_j^{(2)}}{dx}, & x_1^{(2)} < x < x_n^{(2)}, \\
\quad \vdots & \\
\sum_{j=1}^{n} u_j^{(N)} \frac{d\psi_j^{(N)}}{dx}, & x_1^{(N)} < x < x_n^{(N)},
\end{cases}
\tag{8.2.44}
$$

It should be noted that the nodal values of the derivative computed from the two elements sharing that node do not coincide, irrespective of the order of the elements because the continuity of the derivative is not imposed in the procedure. Therefore, one must further process these values to assign a single value at the node (e.g., weighted average of the values of the derivative from all elements connected at the node).

The unknown reaction forces at any global node can be computed after the nodal displacements U_i are computed. They can be computed from either (1) the definition in Eq. (8.2.4) or (2) the element equations Eq. (8.2.30). For the problem at hand, the first equation of Element 1 can be used to determine the reaction force Q_1^1 at global node 1 ($U_1 = u_0$):

$$
\left(P_1^1\right)_{\text{equil}} = -f_1^1 + K_{11}^1 u_0 + K_{12}^1 U_2 = -f_1^1 + \frac{E_1 A_1}{h_1}\left(U_1 - U_2\right). \tag{8.2.45}
$$

Alternatively, P_1^1 can be computed using the definition

$$\left(P_1^1\right)_{\text{def}} \equiv \left(-E_1 A_1 \frac{du}{dx}\right)_{x=0} = -E_1 A_1 \sum_{j=1}^{2} u_j^1 \left(\frac{d\psi_j^1}{dx}\right)_{x=0}$$

$$= -E_1 A_1 \left(\frac{U_2 - U_1}{h_1}\right). \tag{8.2.46}$$

Clearly, the two values, $(P_1^1)_{\text{equil}}$ from Eq. (8.2.45) and $(P_1^1)_{\text{def}}$ from Eq. (8.2.46), are the same only when the distributed force $f(x)$ is zero. The difference between $(P_1^1)_{\text{equil}}$ and $(P_1^1)_{\text{def}}$ reduces as the number of elements is increased (i.e., f_i^e gets smaller in magnitude as the number of elements is increased). The value $(P_1^1)_{\text{equil}}$ is more accurate as it is computed from the equilibrium equation.

This completes the description of the basic steps in the finite element analysis of the model problem. Although the derivation of the approximate functions ψ_i^e is included here to illustrate the procedure, they are readily available in finite element textbooks [46]. One need to select suitable approximate functions from a library of functions of various types for the problem at hand. Although we used the terminology of solid mechanics in the development of the finite element equations, all steps are applicable to any physical problem described by Eq. (8.2.1). In other words, if a computer program is developed to solve Eq. (8.2.1), then the program can be used to analyze all other problems described by the equation (e.g., transverse deflection of cables, heat transfer in fins, flow through pipes, and so on; see Table 6.3.6). In the following example, we illustrate the steps involved in the finite element analysis of a composite bar.

Example 8.2.1 ————————————————————————

Consider the composite bar consisting of a tapered steel bar fastened to an aluminum rod of uniform cross section, and subjected to loads, as shown in Fig. 8.2.8. Determine the deformation in the bar using (a) two linear elements and (b) two quadratic elements. Use the following data:

$$E_s = 30 \times 10^6 \text{ psi}, \ A_s = (c_1 + c_2 x)^2 = \left(1.5 - 0.5\frac{x}{96}\right)^2 \text{ in}^2, \ E_a = 10^7 \text{ psi},$$

$$A_a = 1 \text{ in}^2, \ h_1 = 96 \text{ in}, \ h_2 = 120 \text{ in}, \ L = 216 \text{ in}, \ P_0 = 10000 \text{ lb}. \tag{1}$$

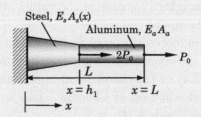

Fig. 8.2.8 Axial deformation of the composite bar of **Example 8.2.1**.

Solution: The governing equations of each part of the structure are given by

$$-\frac{d}{dx}\left(E_s A_s \frac{du}{dx}\right) = 0, \quad 0 < x < h_1,$$

$$-\frac{d}{dx}\left(E_a A_a \frac{du}{dx}\right) = 0, \quad h_1 < x < h_1 + h_2 = L, \tag{2}$$

where the subscripts s and a refer to steel and aluminum, respectively. We will follow the steps in the finite element analysis of the problem. We do not worry about the boundary conditions until all element equations are assembled (Step 4 of the analysis). In actual practice all of the steps can be carried on a digital computer. We begin with the mesh of two linear elements.

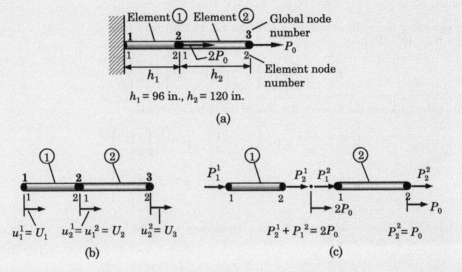

Fig. 8.2.9 Axial deformation of a composite member. (a) Finite element representation. (b) Continuity of displacements. (c) Balance of forces.

1. Finite element mesh. From the discontinuity in the material property, area of cross section, and loading at $x = h_1$, we are required to divide the domain $\Omega = (0, L)$ into at least two elements: $\Omega_1 = (0, h_1)$ and $\Omega_2 = (h_1, L)$. For the mesh of two linear elements, there will be three global nodes; if quadratic elements are used, then there will be five global nodes.

2. Element equations. For each element, whether linear or quadratic, the element stiffness matrix and force vector coefficients are given by Eq. (8.2.27). For the problem at hand, we have

$$c_e = 0, \quad f_e = 0, \quad E_1 = E_s, \quad A_1 = A_s = (c_1 + c_2 x)^2,$$

$$c_1 = \frac{3}{2}, \quad c_2 = -\frac{1}{192}, \quad E_2 = E_a, \quad A_2 = A_a = 1. \tag{3}$$

We now compute $\mathbf{K}^e$ using the linear interpolation functions expressed in terms of the global coordinate x:

$$K_{ij}^{(1)} = \int_{x_a=0}^{x_b=h_1} E_e (c_1 + c_2 x)^2 \frac{d\psi_i^{(1)}}{dx} \frac{d\psi_j^{(2)}}{dx} dx,$$

where

$$\psi_1^e = \frac{x_b - x}{h_e}, \quad \frac{d\psi_1^e}{dx} = -\frac{1}{h_e}, \quad \psi_2^e = \frac{x - x_a}{h_e}, \quad \frac{d\psi_2^e}{dx} = \frac{1}{h_e}.$$

Thus we have

$$K_{11}^{(1)} = \frac{E_s}{h_1^2}\left[(c_1)^2 h_1 + \frac{1}{3}(c_2)^2(h_1^3 - 0) + c_1 c_2(h_1^2 - 0)\right],$$ (4)

$$K_{12}^{(1)} = K_{21}^{(1)} = -K_{11}^{(1)}, \quad K_{22}^{(1)} = K_{11}^{(1)},$$

or

$$\mathbf{K}^{(1)} = \frac{E_s \bar{A}}{h_1}\begin{bmatrix} 1 & -1 \\ -1 & 1 \end{bmatrix}, \quad \mathbf{F}^{(1)} = \begin{Bmatrix} P_1^{(1)} \\ P_2^{(1)} \end{Bmatrix},$$ (5)

where

$$\bar{A} = (c_1)^2 + \frac{1}{3}(c_2 h_1)^2 + c_1 c_2 h_1 = \frac{4.75}{3}.$$

For element 2 the stiffness matrix is given by Eq. (8.2.30) with $c_e = 0$, $A_e = A_a = 1$, and $E_e = E_a = 10^7$. We have

$$\mathbf{K}^{(2)} = \frac{10^7}{120}\begin{bmatrix} 1 & -1 \\ -1 & 1 \end{bmatrix}, \quad \mathbf{F}^{(2)} = \begin{Bmatrix} P_1^{(2)} \\ P_2^{(2)} \end{Bmatrix},$$ (6)

After substitution of the numerical values of the date, we obtain

$$\mathbf{K}^{(1)} = \frac{4.75}{96} \times 10^7 \begin{bmatrix} 1 & -1 \\ -1 & 1 \end{bmatrix}, \quad \mathbf{F}^{(1)} = \begin{Bmatrix} P_1^{(1)} \\ P_2^{(1)} \end{Bmatrix},$$

$$\mathbf{K}^{(2)} = \frac{1}{120} \times 10^7 \begin{bmatrix} 1 & -1 \\ -1 & 1 \end{bmatrix}, \quad \mathbf{F}^{(2)} = \begin{Bmatrix} P_1^{(2)} \\ P_2^{(2)} \end{Bmatrix}.$$ (7)

3. Assembly of element equations. For the two-element mesh, we have from Eq. (8.2.37) the assembled set of equations is

$$\begin{bmatrix} K_{11}^1 & K_{12}^1 & 0 \\ K_{21}^1 & K_{22}^1 + K_{11}^2 & K_{12}^2 \\ 0 & K_{21}^2 & K_{22}^2 \end{bmatrix} \begin{Bmatrix} U_1 \\ U_2 \\ U_3 \end{Bmatrix} = \begin{Bmatrix} P_1^{(1)} \\ P_2^{(1)} + P_1^{(2)} \\ P_2^{(2)} \end{Bmatrix}.$$ (8)

For the problem at hand, these have the explicit form

$$10^7\begin{bmatrix} \frac{4.75}{96} & -\frac{4.75}{96} & 0 \\ -\frac{4.75}{96} & \frac{4.75}{96} + \frac{1}{120} & -\frac{1}{120} \\ 0 & -\frac{1}{120} & \frac{1}{120} \end{bmatrix} \begin{Bmatrix} U_1 \\ U_2 \\ U_3 \end{Bmatrix} = \begin{Bmatrix} P_1^{(1)} \\ P_2^{(1)} + P_1^{(2)} \\ P_2^{(2)} \end{Bmatrix}.$$ (9)

4. Imposition of boundary conditions. From Fig. 8.2.8, we have the boundary and equilibrium conditions

$$u(0) \equiv U_1 = 0, \quad P_2^{(1)} + P_1^{(2)} = 2P_0, \quad P_2^{(2)} = P_0.$$ (10)

Using the boundary conditions in Eq. (10), Eq. (9) can be written as

$$10^4\begin{bmatrix} 49.479 & -49.479 & 0.000 \\ -49.479 & 57.812 & -8.333 \\ 0.000 & -8.333 & 8.333 \end{bmatrix} \begin{Bmatrix} 0 \\ U_2 \\ U_3 \end{Bmatrix} = \begin{Bmatrix} P_1^{(1)} \\ 2P_0 \\ P_0 \end{Bmatrix}.$$ (11)

5. Solution of equations. From Eq. (11), we obtain the following condensed equations for the unknown nodal displacements by deleting the first row and column (because U_1 is specified as zero):

$$10^4\begin{bmatrix} 57.812 & -8.333 \\ -8.333 & 8.333 \end{bmatrix} \begin{Bmatrix} U_2 \\ U_3 \end{Bmatrix} = \begin{Bmatrix} 2P_0 \\ P_0 \end{Bmatrix} = 10^4\begin{Bmatrix} 2 \\ 1 \end{Bmatrix},$$ (12)

or

$$U_2 = 0.06063 \text{ in}, \quad U_3 = 0.18063 \text{ in}. \tag{13}$$

6. Postprocessing. The reaction force at the left end of the bar is given by (since the distributed load is zero in the present problem, the values of P_1^1 by definition and equilibrium are the same)

$$P_1^{(1)} \equiv 10^4(49.479U_1 - 49.479U_2) = -3 \times 10^4 \text{ lb}. \tag{14}$$

The negative sign in Eq. (14) indicates that $P_1^{(1)}$ is acting opposite to the convention used in Fig. 8.2.9(c), and therefore the reaction is acting away from the left end (i.e., a tensile force). The magnitude of $P_1^{(1)}$ is consistent with the static equilibrium of the forces:

$$P_1^{(1)} + 2P_0 + P_0 = 0 \text{ or } P_1^{(1)} = -3P_0 = -3 \times 10^4 \text{ lb}.$$

The axial displacement at any point x along the bar is given by

$$u_e(x) = \begin{cases} u_1^{(1)}\psi_1^{(1)} + u_2^{(1)}\psi_2^{(1)} = 0.06063\frac{x}{96}, & 0 \leq x \leq 96, \\ u_1^{(2)}\psi_1^{(2)} + u_2^{(2)}\psi_2^{(2)} = -0.03537 + 0.001x, & 96 \leq x \leq 216, \end{cases} \tag{15}$$

and its first derivative is given by

$$\frac{du_e}{dx} = \begin{cases} \frac{0.06063}{96}, & 0 \leq x \leq 96, \\ 0.0001, & 96 \leq x \leq 216. \end{cases} \tag{16}$$

The exact solution of Eq. (2) subject to the boundary conditions

$$u(0) = 0, \quad \left[\left(a\frac{du}{dx}\right)_{x=96+} - \left(a\frac{du}{dx}\right)_{x=96-}\right] = 2P_0, \quad \left(a\frac{du}{dx}\right)_{x=216} = P_0 \tag{17}$$

is given by

$$u(x) = \begin{cases} 0.128\frac{x}{288-x}, & 0 \leq x \leq 96, \\ 0.001(x - 32), & 96 \leq x \leq 216, \end{cases} \tag{18a}$$

$$\frac{du}{dx} = \begin{cases} \frac{36.864}{(288-x)^2}, & 0 \leq x \leq 96, \\ 0.001, & 96 \leq x \leq 216. \end{cases} \tag{18b}$$

In particular, the exact solution at nodes 2 and 3 is given by

$$u(96) = 0.064 \text{ in.}, \quad u(216) = 0.1840 \text{ in.} \tag{19}$$

Thus, the two-element solution is about 1.8% off from the maximum displacement.

(b) Next consider a two-element mesh of quadratic elements. The element matrix and force vector for Element 1 are (details of the integration of the coefficients K_{ij}^1 are not included here)

$$\mathbf{K}^1 = 10^4 \begin{bmatrix} 142.19 & -159.37 & 17.18 \\ -159.37 & 266.67 & -107.29 \\ 17.18 & -107.29 & 90.10 \end{bmatrix}, \quad \mathbf{F}^1 = \begin{Bmatrix} P_1^{(1)} \\ P_2^{(1)} \\ P_3^{(1)} \end{Bmatrix}. \tag{20}$$

The element matrix and force vector for Element 2 are [see Eq. (8.2.32)]

$$\mathbf{K}^{(2)} = \frac{E_aA_a}{3h_2}\begin{bmatrix} 7 & -8 & 1 \\ -8 & 16 & -8 \\ 1 & -8 & 7 \end{bmatrix} = 10^4 \begin{bmatrix} 19.444 & -22.222 & 2.778 \\ -22.222 & 44.444 & -22.222 \\ 2.778 & -22.222 & 19.444 \end{bmatrix}, \quad \mathbf{F}^{(2)} = \begin{Bmatrix} P_1^{(2)} \\ P_2^{(2)} \\ P_3^{(2)} \end{Bmatrix}. \tag{21}$$

The assembled stiffness matrix and force vector are of the order 5×5 and 5×1, respectively. The assembled equations for the mesh of two quadratic elements are

$$
\begin{bmatrix}
K_{11}^1 & K_{12}^1 & K_{13}^1 & 0 & 0 \\
K_{21}^1 & K_{22}^1 & K_{23}^1 & 0 & 0 \\
K_{31}^1 & K_{32}^1 & K_{33}^1 + K_{11}^2 & K_{12}^2 & K_{13}^2 \\
0 & 0 & K_{21}^2 & K_{22}^2 & K_{23}^2 \\
0 & 0 & K_{31}^2 & K_{32}^2 & K_{33}^2
\end{bmatrix}
\begin{Bmatrix}
U_1 \\ U_2 \\ U_3 \\ U_4 \\ U_5
\end{Bmatrix}
=
\begin{Bmatrix}
P_1^{(1)} \\
P_2^{(1)} \\
P_3^{(1)} + P_1^{(2)} \\
P_2^{(2)} \\
P_3^{(2)}
\end{Bmatrix}.
\tag{22}
$$

After imposing boundary conditions

$$
U_1 = 0, \; P_2^{(1)} = 0, \; P_3^{(1)} + P_1^{(2)} = 2 \times 10^4, \; P_2^{(2)} = 0, \; P_3^{(2)} = 10^4,
$$

and omitting the first row and first column, we obtain the condensed equations for the unknown generalized displacements

$$
\begin{bmatrix}
K_{22}^1 & K_{23}^1 & 0 & 0 \\
K_{32}^1 & K_{33}^1 + K_{11}^2 & K_{12}^2 & K_{13}^2 \\
0 & K_{21}^2 & K_{22}^2 & K_{23}^2 \\
0 & K_{31}^2 & K_{32}^2 & K_{33}^2
\end{bmatrix}
\begin{Bmatrix}
U_2 \\ U_3 \\ U_4 \\ U_5
\end{Bmatrix}
=
\begin{Bmatrix}
0 \\ 2 \times 10^4 \\ 0 \\ 10^4
\end{Bmatrix},
\tag{23}
$$

whose solution is

$$
U_2 = 0.02572 \text{ in.}, \quad U_3 = 0.06392 \text{ in}, \quad U_4 = 0.12392 \text{ in.}, \quad U_5 = 0.18392 \text{ in.}
\tag{24}
$$

The post-computation of $P_1^{(1)}$ using the definition yields

$$
\left(P_1^{(1)} \right)_{\text{def}} = - \left(E_s A_s \frac{du_1}{dx} \right)\Bigg|_{x=0} = - \left[E_s A_s \left(U_2 \frac{d\psi_2}{dx} + U_3 \frac{d\psi_3}{dx} \right) \right]_{x=0}
$$

$$
= -E_s c_1^2 \left(\frac{4U_2 - U_3}{h_1} \right) = -27386 \text{ lb.}
\tag{25}
$$

This computed value is in 8.71% error compared with the true solution $(-30,000)$ or the solution obtained from the first equation of the system in Eq. (22).

A comparison of the finite element solution obtained by various meshes of linear and quadratic elements with the exact solution is presented in Table 8.2.1. Obviously, the two-element solution obtained using the quadratic element is very accurate, and the four-element solution matches with the exact solution.

Table 8.2.1 Comparison of the finite element solutions with the exact solution at the nodes of the composite bar of **Example 8.2.1**.

x	Exact	Linear[a]				Quadratic[a]	
(in.)	solution	$1+1$	$2+1$	$3+2$	$6+2$	$1+1$	$3+1$
16	0.00753	–	–	–	0.00752	–	0.00753
32	0.01600	–	–	0.01593	0.01598	–	0.01600
48	0.02560	–	0.02532	-	0.02557	0.02572	0.02560
64	0.03657	–	–	0.03638	0.03652	–	0.03657
80	0.04923	–	–	–	0.04916	–	0.04923
96	0.06400	0.06063	0.06309	0.06359	0.06390	0.06392	0.06400
156	0.12400	–	–	0.12359	0.12390	0.12392	0.12400
216	0.18400	0.18063	0.18309	0.18359	0.18390	0.18392	0.18400

[a]$m + n$ means m elements in the interval $(0, 96)$ and n elements in the interval $(96, 216)$; all elements in each interval are of the same size.

8.3 Finite Element Analysis of the Bernoulli–Euler Beam Theory

8.3.1 Governing Equation

Here we consider the finite element formulation of the fourth-order differential equation governing the bending of straight elastic beams according to the Bernoulli–Euler beam theory (see Section 2.5.2) . The equation governing the transverse deflection is given by [see Eq. (10) of **Example 4.3.1**; set the nonlinear term to zero to obtain the fourth-order equation]

$$\frac{d^2}{dx^2}\left(EI\frac{d^2w}{dx^2}\right) - q(x) = 0, \quad 0 < x < L, \tag{8.3.1}$$

where $w(x)$ denotes the transverse deflection, EI is the bending stiffness, and $q(x)$ is the distributed transverse load [see Fig. 8.3.1(a)]. A finite element mesh of the beam is shown in Fig. 8.3.1(b). The basic steps in the formulation are the same as those discussed for bars, but the specific mathematical details differ from those of second-order equations. Since the basic terminology of the finite element method is already introduced in the preceding section, we go straight to the approximation over an element (see **Example 4.3.5**).

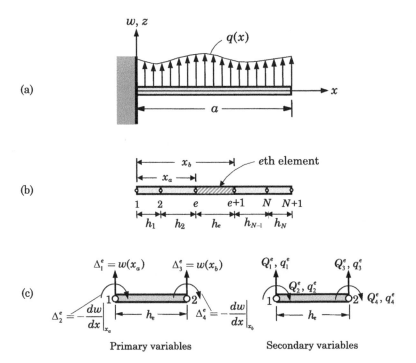

Fig. 8.3.1 Finite element representation of a beam. (a) Geometry and loading on a straight beam. (b) Finite element mesh. (c) Beam element with primary (displacement) and secondary (force) degrees of freedom.

8.3.2 Weak Form over an Element

A finite element for a beam is a free-body diagram of an interval, $\Omega_e = (x_a, x_b)$, of a typical beam, as shown in Fig. 8.3.1(c) with generalized nodal displacements (primary variables) and generalized nodal forces (secondary variables). As discussed earlier, the choice of the nodal variables is dictated by the primary variables resulting from the virtual work statement. From the principle of virtual displacements for beams [see Eq. (7) of **Example 4.2.1**], we know that the primary variables are w and $-\frac{dw}{dx}$. Hence, the generalized displacement degrees of freedom are:

$$
w(x_a) \equiv \Delta_1^e, \quad -\frac{dw}{dx}\bigg|_{x_a} \equiv \Delta_2^e, \quad w(x_b) \equiv \Delta_3^e, \quad -\frac{dw}{dx}\bigg|_{x_b} = \Delta_4^e, \qquad (8.3.2)
$$

and the natural boundary conditions involve the specification of bending moments and shear forces [see Eq. (2.5.12) for moment–deflection relation and Eq. (2.5.5) for the moment–shear force relation; also see Fig. 8.3.1(c) for the sign convention]:

$$
\left[\frac{d}{dx}\left(EI\frac{d^2w}{dx^2}\right)\right]_{x=x_a} \equiv Q_1^e, \qquad \left(EI\frac{d^2w}{dx^2}\right)\bigg|_{x=x_a} \equiv Q_2^e, \qquad (8.3.3)
$$

$$
\left[\frac{d}{dx}\left(EI\frac{d^2w}{dx^2}\right)\right]_{x=x_b} \equiv -Q_3^e, \qquad \left(EI\frac{d^2w}{dx^2}\right)\bigg|_{x=x_b} \equiv -Q_4^e. \qquad (8.3.4)
$$

The weak form of Eq. (8.3.1) is obtained directly from the principle of virtual displacements (or the principle of minimum total potential energy) for the typical element, $\Omega^e = (x_a, x_b)$. The total potential energy functional for the beam element is given by

$$
\Pi^e(w_e) = \int_{x_a}^{x_b} \left[\frac{E_e I_e}{2}\left(\frac{d^2 w_e}{dx^2}\right)^2 - w_e q_e\right] dx - Q_1^e \Delta_1^e - Q_2^e \Delta_2^e - Q_3^e \Delta_3^e - Q_4^e \Delta_4^e,
$$
$$ (8.3.5) $$

where the sum $\sum_{i=1}^4 Q_i^e \Delta_i^e$ represents the work done by the forces (Q_1^e, Q_3^e) and moments (Q_2^e, Q_4^e) in displacing and rotating the ends of the element through displacement (Δ_1^e, Δ_3^e) and rotations (Δ_2^e, Δ_4^e). Setting $\delta\Pi^e = 0$ yields the weak form

$$
0 = \int_{x_a}^{x_b} \left(E_e I_e \frac{d^2 \delta w_e}{dx^2}\frac{d^2 w_e}{dx^2} - \delta w_e q_e\right) dx - Q_1^e \delta w_e(x_a) - Q_2^e \left(-\frac{d\delta w_e}{dx}\right)_{x_a}
$$

$$
- Q_3^e \delta w_e(x_b) - Q_4^e \left(-\frac{d\delta w_e}{dx}\right)_{x_b}. \qquad (8.3.6)
$$

8.3.3 Derivation of the Approximation Functions

To derive the approximation functions for the Ritz approximation of the function w_e, we must first select a polynomial that is continuous and complete up to the degree required. It is clear from the functional Π^e that the function we select should be twice-differentiable to yield nonzero strain energy, and it should be cubic to yield nonzero shear forces Q_1^e and Q_3^e. Another reason for selecting w_e to be a cubic polynomial is that we need four parameters in the polynomial in order to represent the two primary variables at each of the two nodes [see Eq. (8.3.2)]. Thus, a minimum degree polynomial for the Ritz approximation of w_e over the element is cubic:

$$w_e(\bar{x}) \approx \alpha_0 + \alpha_1 \bar{x} + \alpha_2 \bar{x}^2 + \alpha_3 \bar{x}^3, \tag{8.3.7}$$

where $\bar{x}$ denotes the local coordinate (the global coordinate x is related to the local coordinate $\bar{x}$ by $x = \bar{x} + x_a$; hence, $d/dx = d/d\bar{x}$) Next, select the parameters α_0, α_1, α_2, and α_3 such that w satisfies the essential boundary conditions in Eq. (8.3.2):

$$\Delta_1^e \equiv w_e(0) = \alpha_0,$$

$$\Delta_2^e \equiv -\frac{dw_e}{d\bar{x}}(0) = -\alpha_1,$$

$$\Delta_3^e \equiv w_e(h_e) = \alpha_0 + \alpha_1 h_e + \alpha_2 h_e^2 + \alpha_3 h_e^3, \tag{8.3.8}$$

$$\Delta_4^e \equiv -\frac{dw_e}{d\bar{x}}(h_e) = -\alpha_1 - 2\alpha_2 h_e - 3\alpha_3 h_e^2,$$

where $h_e = x_b - x_a$. Solving Eq. (8.3.8) for α_0, α_1, α_2, and α_3 in terms of Δ_i^e ($i = 1, 2, 3, 4$), and substituting the result into Eq. (8.3.7), we obtain

$$w_e(\bar{x}) \approx \sum_{j=1}^{4} \Delta_j^e \phi_j^e(\bar{x}), \tag{8.3.9}$$

where the coordinate functions $\phi_j^e(\bar{x})$, $\bar{x} = x - x_1$, for the beam element are given by

$$\phi_1^e(\bar{x}) = 1 - 3\left(\frac{\bar{x}}{h_e}\right)^2 + 2\left(\frac{\bar{x}}{h_e}\right)^3,$$

$$\phi_2^e(\bar{x}) = -\bar{x}\left[1 - \left(\frac{\bar{x}}{h_e}\right)\right]^2,$$

$$\phi_3^e(\bar{x}) = 3\left(\frac{\bar{x}}{h_e}\right)^2 - 2\left(\frac{\bar{x}}{h_e}\right)^3, \tag{8.3.10}$$

$$\phi_4^e(\bar{x}) = -\bar{x}\left[\left(\frac{\bar{x}}{h_e}\right)^2 - \left(\frac{\bar{x}}{h_e}\right)\right].$$

The approximation functions ϕ_i^e are called the *Hermite cubic interpolation functions*. We note the following properties of ϕ_i^e, $i = 1, 2, 3, 4$ (see Fig. 8.3.2):

$$\phi_{2i-1}^e(\bar{x}_j) = \delta_{ij}, \quad \phi_{2i}^e(\bar{x}_j) = 0, \quad \sum_{i=1}^{2} \phi_{2i-1}^e = 1,$$

$$\frac{d\phi_{2i-1}^e}{dx}(\bar{x}_j) = 0, \quad -\frac{d\phi_{2i}^e}{dx}(\bar{x}_j) = \delta_{ij}, \quad (i, j = 1, 2), \tag{8.3.11}$$

where $\bar{x}_1 = 0$ and $\bar{x}_2 = h_e$ are the local coordinates of nodes 1 and 2 of element $\Omega_e = (x_a, x_b)$ with x_a being the global coordinate of the first node of the element.

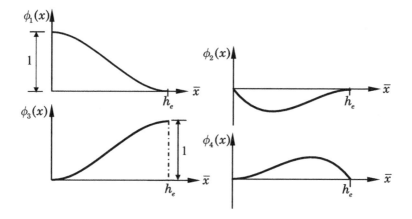

Fig. 8.3.2 Hermite cubic interpolation functions for the beam element.

We note that, by including the rotations Δ_2^e and Δ_4^e as nodal quantities, we will be able to enforce the slope continuity at the interelement boundaries during the assembly process. Of course, the total potential energy formulation of the governing equation yields slopes as part of the essential boundary conditions. All dependent variables and their derivatives that enter the specification of the essential boundary conditions of a problem always end up as the primary nodal variables, and they take the role of the undetermined coefficients of the Ritz method.

8.3.4   Finite Element Model

To derive the finite element equations associated with Eq. (8.3.1), we substitute Eq. (8.3.9) for w_e into Π^e, and set its derivatives with respect to each Δ_j^e, $j = 1, 2, 3, 4$, to zero [alternatively, one can use the weak form in Eq. (8.3.6) to obtain the same result]:

$$
\Pi^e(\Delta_i^e) = \int_{x_a}^{x_b} \left[\frac{E_e I_e}{2} \left(\sum_{j=1}^{4} \Delta_j^e \frac{d^2 \phi_j^e}{dx^2} \right)^2 - \left(\sum_{i=1}^{4} \Delta_i^e \phi_i^e \right) q_e \right] dx - \sum_{i=1}^{4} Q_i^e \Delta_i^e
$$

$$
= \frac{1}{2} \sum_{i=1}^{4} \sum_{j=1}^{4} \Delta_i^e \left(\int_{x_a}^{x_b} E_e I_e \frac{d^2 \phi_i^e}{dx^2} \frac{d^2 \phi_j^e}{dx^2} dx \right) \Delta_j^e
$$

$$
- \sum_{i=1}^{4} \Delta_i^e \left(\int_{x_a}^{x_b} q_e \phi_i^e \, dx + Q_i^e \right) \equiv \sum_{i=1}^{4} \left(\frac{1}{2} \sum_{j=1}^{4} \Delta_i^e K_{ij}^e \Delta_j^e - \Delta_i^e F_i^e \right),
$$

$$
(8.3.12)
$$

and

$$\frac{\partial \Pi^e}{\partial \Delta_i^e} = 0 = \sum_{j=1}^{4} K_{ij}^e \Delta_j^e - F_i^e \quad \text{or} \quad \mathbf{K}^e \mathbf{\Delta}^e = \mathbf{F}^e, \tag{8.3.13}$$

where

$$K_{ij}^e = \int_{x_a}^{x_b} E_e I_e \frac{d^2 \phi_i^e}{dx^2} \frac{d^2 \phi_j^e}{dx^2} \, dx, \quad F_i^e = \int_{x_a}^{x_b} q_e \phi_i^e \, dx + Q_i^e. \tag{8.3.14}$$

In Eq. (8.3.14), it is understood that ϕ_i^e are expressed in terms of the global coordinate x. In the element coordinate $\bar{x}$, Eq. (8.3.14) takes the simple form,

$$K_{ij}^e = \int_{0}^{h_e} \bar{E}_e \bar{I}_e \frac{d^2 \phi_i^e}{d\bar{x}^2} \frac{d^2 \phi_j^e}{d\bar{x}^2} \, d\bar{x}, \quad F_i^e = \int_{0}^{h_e} \bar{q}_e \phi_i^e \, d\bar{x} + Q_i^e, \tag{8.3.15}$$

where $\bar{E}_e$, $\bar{I}_e$, and $\bar{q}_e$ are the transformed functions:

$$\bar{E}_e = E_e(\bar{x}), \quad \bar{I}_e = I_e(\bar{x}), \quad \bar{q}_e = q_e(\bar{x}).$$

For the case in which E_e, I_e, and q_e are constant over the element Ω_e, the stiffness matrix $\mathbf{K}^e$ and force vector $\mathbf{F}^e$ are given by

$$\mathbf{K}^e = \frac{2 E_e I_e}{h_e^3} \begin{bmatrix} 6 & -3h_e & -6 & -3h_e \\ -3h_e & 2h_e^2 & 3h_e & h_e^2 \\ -6 & 3h_e & 6 & 3h_e \\ -3h_e & h_e^2 & 3h_e & 2h_e^2 \end{bmatrix}; \quad \mathbf{F}^e = \frac{q_e h_e}{12} \begin{Bmatrix} 6 \\ -h_e \\ 6 \\ h_e \end{Bmatrix} + \begin{Bmatrix} Q_1^e \\ Q_2^e \\ Q_3^e \\ Q_4^e \end{Bmatrix}. \tag{8.3.16}$$

One can show that the first column of the force vector represents the "statically equivalent" forces and moments at the nodes due to uniformly distributed load $q_e = q_0$ over the element. This element gives, for element-wise constant values of $E_e I_e$, the exact values of w_e and dw_e/dx at the nodes for any load $q_e(x)$, and thus it is called a *superconvergent element* (SCE).

8.3.5 Assembly of Element Equations

The assembly procedure for the beam element is analogous to that described for the bar element. The difference is that in the beam element there are two unknowns, called *degrees of freedom*, per node. Consequently, at every global node that is shared by two elements, the elements in the last two rows and columns of Ith element overlap with the elements in the first two rows and columns of $(I+1)$th element when they are connected in series (see Fig. 8.3.3).

For a three-element case, the assembled system of equations in matrix form is given in Eq. (8.3.17), where the global stiffness coefficients are expressed in terms of the element stiffness coefficients and the global forces are written in terms of element forces:

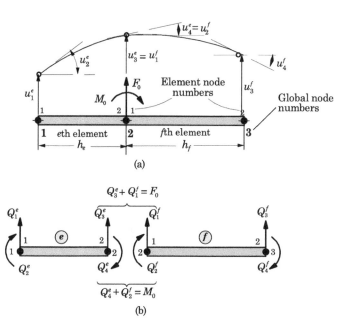

Fig. 8.3.3 Assembly of two beam finite elements. (a) Continuity of generalized displacements. (b) Balance of generalized forces.

$$
\begin{bmatrix}
K_{11}^1 & K_{12}^1 & K_{13}^1 & K_{14}^1 & 0 & 0 & 0 & 0 \\
K_{21}^1 & K_{22}^1 & K_{23}^1 & K_{24}^1 & 0 & 0 & 0 & 0 \\
K_{31}^1 & K_{32}^1 & K_{33}^1 + K_{11}^2 & K_{34}^1 + K_{12}^2 & K_{13}^2 & K_{14}^2 & 0 & 0 \\
K_{41}^1 & K_{42}^1 & K_{43}^1 + K_{21}^2 & K_{44}^1 + K_{22}^2 & K_{23}^2 & K_{24}^2 & 0 & 0 \\
0 & 0 & K_{31}^2 & K_{32}^2 & K_{33}^2 + K_{11}^3 & K_{34}^2 + K_{12}^3 & K_{13}^3 & K_{14}^3 \\
0 & 0 & K_{41}^2 & K_{42}^2 & K_{43}^2 + K_{21}^3 & K_{44}^2 + K_{22}^3 & K_{23}^3 & K_{24}^3 \\
0 & 0 & 0 & 0 & K_{31}^3 & K_{32}^3 & K_{33}^3 & K_{34}^3 \\
0 & 0 & 0 & 0 & K_{41}^3 & K_{42}^3 & K_{43}^3 & K_{44}^3
\end{bmatrix}
\begin{Bmatrix}
U_1 \\ U_2 \\ U_3 \\ U_4 \\ U_5 \\ U_6 \\ U_7 \\ U_8
\end{Bmatrix}
$$

$$
=
\begin{Bmatrix}
q_1^1 \\
q_2^1 \\
q_3^1 + q_1^2 \\
q_4^1 + q_2^2 \\
q_3^2 + q_1^3 \\
q_4^2 + q_2^3 \\
q_3^3 \\
q_4^3
\end{Bmatrix}
+
\begin{Bmatrix}
Q_1^1 \\
Q_2^1 \\
Q_3^1 + Q_1^2 \\
Q_4^1 + Q_2^2 \\
Q_3^2 + Q_1^3 \\
Q_4^2 + Q_2^3 \\
Q_3^3 \\
Q_4^3
\end{Bmatrix},
\qquad (8.3.17)
$$

where U_i are the generalized displacements of the assembled system:

$$
\begin{aligned}
& U_1 = \Delta_1^1, \; U_2 = \Delta_2^1, \; U_3 = \Delta_3^1 = \Delta_1^2, \; U_4 = \Delta_4^1 = \Delta_2^2, \\
& U_5 = \Delta_3^2 = \Delta_1^3, \; U_6 = \Delta_4^2 = \Delta_2^3, \; U_7 = \Delta_3^3, \; U_8 = \Delta_4^3.
\end{aligned}
\qquad (8.3.18)
$$

8.3.6 Imposition of Boundary Conditions

The boundary conditions on the generalized displacements can be imposed in the manner described earlier for bars. The boundary (or equilibrium) conditions on forces are imposed by modifying the second column of the assembled force vector. If, for example, the force at the Ith global node is specified to be F_0, and the moment at Kth global node is specified to be M_0, then the following assembled coefficients get modified:

$$Q_3^{I-1} + Q_1^I = F_0, \quad Q_4^{K-1} + Q_2^K = M_0. \tag{8.3.19}$$

If no force or moment is specified at a nodal point that is unconstrained, it is understood that the force and moment are specified to be zero there.

The solution and post-computation of the displacements and their derivatives at various points of the beam follow the same procedure as described for bar problems. We now consider an example to illustrate the steps involved.

Example 8.3.1 ───────────────────────────────────────

Consider the indeterminate beam shown in Fig. 8.3.4. The beam is made of steel ($E = 30 \times 10^6$ psi) and the cross-sectional dimensions are 2×3 in (i.e., $I = 4.5$ in.^{4}). Find the deflection w and its derivatives using the Bernoulli–Euler beam finite element.

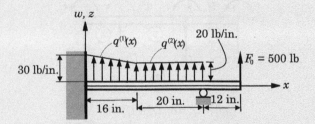

Fig. 8.3.4 The indeterminate beam of **Example 8.3.1**.

Solution: Because of the discontinuity in loading, the beam should be divided into a minimum of three elements: $\Omega_1 = (0, 16)$, $\Omega_2 = (16, 36)$, and $\Omega_3 = (36, 48)$; the element lengths are: $h_1 = 16$ in., $h_2 = 20$ in., and $h_3 = 12$ in.

The element equations for all three elements are readily available from Eq. (8.3.16) only exception is that the load vector for Element 1, which has linearly varying distributed load, must be computed. The load variation is given by

$$q^{(1)}(x) = \left(30 - \frac{10}{16}x\right), \quad q^{(2)}(x) = 20, \quad q^{(3)}(x) = 0, \tag{1}$$

and their contribution to the element nodes can be computed using Eq. (8.3.14). For Element 1, we have

$$q_i^1 = \int_0^{16} \left(30 - \frac{10}{16}x\right)\phi_i^1(x) \, dx, \quad (i = 1, 2, 3, 4). \tag{2}$$

We have

$$
\mathbf{q}^1 = \left\{ \begin{array}{c} 216.00 \\ -554.67 \\ 184.00 \\ 512.00 \end{array} \right\}, \quad \mathbf{q}^2 = \left\{ \begin{array}{c} 200.00 \\ -666.77 \\ 200.00 \\ 666.67 \end{array} \right\}, \quad \mathbf{q}^3 = \mathbf{0}. \tag{3}
$$

The element stiffness matrices can be computed from Eq. (8.3.16) by substituting appropriate values of $h_e, E_e,$ and I_e. For example, for Element 1, we have ($h_1 = 16$ in., $E_1 I_1 = 135 \times 10^6$ lb-in.^{2}), and

$$
\mathbf{K}^1 = 10^6 \begin{bmatrix} 0.3955 & -3.1641 & -0.3955 & -3.1641 \\ -3.1641 & 33.7500 & 3.1641 & 16.8750 \\ -0.3955 & 3.1641 & 0.3955 & 3.1641 \\ -3.1641 & 16.8750 & 3.1641 & 33.750 \end{bmatrix}. \tag{4}
$$

The assembled matrix for this mesh is of the same form as given in Eq. (8.3.17); one only needs to replace the algebraic entries with the numerical ones. The boundary conditions for the problem are (see Fig. 8.3.5):

Specified generalized displacements

$$
w_e(0) = 0 \rightarrow U_1 = 0, \quad \frac{dw_e}{dx}\bigg|_{x=0} = 0 \rightarrow U_2 = 0, \quad w_e(36) = 0 \rightarrow U_5 = 0. \tag{5}
$$

Balance of forces and specified generalized forces

$$
Q_3^{(1)} + Q_1^{(2)} = 0, \quad Q_4^{(1)} + Q_2^{(2)} = 0, \quad Q_4^{(2)} + Q_2^{(3)} = 0, \quad Q_3^{(3)} = 500, \quad Q_4^{(3)} = 0. \tag{6}
$$

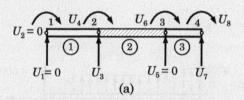

(a)

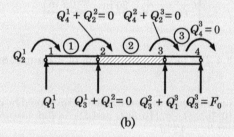

(b)

Fig. 8.3.5 Specified generalized (a) displacements and (b) forces in the mesh.

Note that $Q_1^{(1)}, Q_2^{(1)},$ and $Q_3^{(2)} + Q_1^{(3)}$ are the reactions that are not known, and are to be calculated in the post-computation. Since the specified essential boundary conditions are homogeneous, one can delete the rows and columns corresponding to the specified displacements (i.e., delete rows and columns 1, 2, and 5) and solve the remaining five equations for $U_3, U_4, U_6, U_7,$ and U_8:

$$
\begin{gathered}
U_3 = -0.000322 \text{ in.}, \quad U_4 = 0.0000593 \text{ rad}, \quad U_6 = -0.0002513 \text{ rad}, \\
U_7 = 0.00515 \text{ in.}, \quad U_8 = -0.000518 \text{ rad}.
\end{gathered} \tag{7}
$$

The exact solution to the problem can be obtained by integrating the moment-deflection expression $(d^2w/dx^2) = -M(x)/EI$, in which the bending moment can be readily obtained in terms of the reactions at the supports and applied loads. The deflection and bending moment $(M = -EI\,d^2w/dx^2)$ in the three intervals of the beam are given by (large numbers were rounded to whole numbers)

$$
w(x) = \begin{cases}
\frac{1}{EI}\left(268.543x^2 - \frac{691}{15}x^3 + \frac{5}{4}x^4 - \frac{1}{192}x^5\right), & 0 \le x \le 16, \\[2mm]
\frac{1}{EI}\left(-\frac{16384}{3} + \frac{5120}{3}x + 55.2097x^2 - \frac{491}{15}x^3 + \frac{5}{6}x^4\right), & 16 \le x \le 36, \\[2mm]
\frac{1}{EI}\left(6554368 - 506066x + 12000x^2 - \frac{250}{3}x^3\right), & 36 \le x \le 48.
\end{cases}
\tag{8}
$$

$$
M(x) = \begin{cases}
-\left(537.086 - \frac{1382}{5}x + 15x^2 - \frac{5}{48}x^3\right), & 0 \le x \le 16, \\[2mm]
-\left(110.4194 - \frac{982}{5}x + 10x^2\right), & 16 \le x \le 36, \\[2mm]
-(24000 - 500x), & 36 \le x \le 48.
\end{cases}
\tag{9}
$$

It can be verified that the nodal values in Eq. (7) coincide with the exact values.

A comparison of the finite element results for deflections, slopes, and bending moments (calculated in the post-computation) *at points other than the nodes* are compared with the exact values in Table 8.3.1 for three different meshes. The values of deflections, slopes, and bending moments were computed using the finite element solution and its derivatives.

Table 8.3.1 Comparison of the finite element solution with the exact solution of the beam problem considered in **Example 8.3.1**.

x	N^a	$10^6 \times w$		$10^6(-dw/dx)$		$-10^6(d^2w/dx^2)/EI$	
		FEM	Exact	FEM	Exact	FEM	Exact
	3	−0.8570		1.1553		−1.0250	
2.0	5	4.1405	5.3739	−3.3386	−4.1552	0.4663	0.3219
	6	5.2319		−4.1558		0.4638	
	3	−18.4510		8.8348		−2.8147	
6.0	5	8.3936	9.6048	4.4199	5.2328	−4.3456	−4.4727
	6	9.4751		5.2323		−4.3431	
	3	−138.2300		33.7760		−5.4992	
12.0	5	−122.5900	−120.8100	39.6630	39.6720	−5.4794	−5.9238
	6	−122.5900		39.6630		−5.4794	
	3	−674.3600		71.5620		3.5507	
21.0	5	−643.4900	−639.6300	62.3030	62.3020	3.5507	2.9335
	6	−643.4900		62.3030		3.5507	
	3	−812.9300		−83.7970		27.5210	
31.0	5	−782.0700	−778.1900	−74.5370	−74.5390	27.5210	26.9040
	6	−782.0700		−74.5370		27.5210	
	3	2174.9000		−451.3600		22.2220	
42.0	5	2174.9000	2174.8000	−451.3600	−451.3600	22.2222	22.2222
	6	2174.9000		−451.3600		22.2222	

^{a}3 elements: $h_1 = 16$, $h_2 = 20$, $h_3 = 12$; 5 elements: $h_1 = 8$, $h_2 = 8$, $h_3 = 10$, $h_4 = 10$, $h_5 = 12$; 6 elements: $h_1 = 4$, $h_2 = 4$, $h_3 = 8$, $h_4 = 10$, $h_5 = 10$, $h_6 = 12$.

8.4   Finite Element Analysis of the Timoshenko Beam Theory

8.4.1   Governing Equations

The governing equations of the TBT are [see Eqs (10) and (11) of **Example 3.5.4**; $S_{xz} = GA$]

$$-\frac{d}{dx}\left(EI\frac{d\phi_x}{dx}\right) + GAK_s\left(\phi_x + \frac{dw}{dx}\right) = 0, \qquad (8.4.1)$$

$$-\frac{d}{dx}\left[GAK_s\left(\phi_x + \frac{dw}{dx}\right)\right] = q(x), \qquad (8.4.2)$$

where K_s is the shear correction coefficient and G is the shear modulus. In the following a number of displacement finite element models of these equations are presented for element-wise constant properties, $E = E_e$, $G = G_e$, $A = A_e$, and $I = I_e$.

8.4.2   Weak Forms

The weak forms of Eqs (8.4.1) and (8.4.2) over a typical element $\Omega_e = (x_a^e, x_b^e)$ are provided by the principle of minimum total potential energy [see Eqs (1) and (2) of **Example 3.5.4**]:

$$0 = \int_{x_a^e}^{x_b^e} \left[E_e I_e \frac{d\delta\phi_x^e}{dx}\frac{d\phi_x^e}{dx} + G_e A_e K_s \left(\delta\phi_x^e + \frac{d\delta w_e}{dx}\right)\left(\phi_x^e + \frac{dw_e}{dx}\right) - q(x)\delta w_e \right] dx$$

$$- Q_1^e \delta w_e(x_a^e) - Q_3^e \delta w_e(x_b^e) - Q_2^e \delta\phi_x^e(x_a^e) - Q_4^e \delta\phi_x^e(x_b^e), \qquad (8.4.3)$$

where

$$Q_1^e \equiv -V(x_a^e) = -\left[G_e A_e K_s\left(\frac{dw_e}{dx} + \phi_x^e\right)\right]_{x=x_a^e},$$

$$Q_2^e \equiv -M(x_a^e) = -\left[E_e I_e \frac{d\phi_x^e}{dx}\right]_{x=x_a^e},$$

$$Q_3^e \equiv V(x_b^e) = \left[G_e A_e K_s\left(\frac{dw_e}{dx} + \phi_x^e\right)\right]_{x=x_b^e}, \qquad (8.4.4)$$

$$Q_4^e \equiv M(x_b^e) = \left[E_e I_e \frac{d\phi_x^e}{dx}\right]_{x=x_b^e}.$$

We recall from Eq. (15) of **Example 3.4.5** that (w_e, ϕ_x^e) are the primary variables and (V_e, M_e) are secondary variables associated with the principle of minimum total potential energy.

By collecting terms involving δw_e and $\delta\phi_x^e$ from Eq. (8.4.3) separately, we obtain the following two statements (weak forms):

$$0 = \int_{x_a^e}^{x_b^e} \left[G_e A_e K_s \frac{d\delta w_e}{dx} \left(\phi_x^e + \frac{dw_e}{dx} \right) - q(x)\delta w_e \right] dx$$
$$- Q_1^e \delta w(x_a^e) - Q_3^e \delta w_e(x_b^e), \tag{8.4.5}$$

$$0 = \int_{x_a^e}^{x_b^e} \left[E_e I_e \frac{d\delta\phi_x^e}{dx} \frac{d\phi_x^e}{dx} + G_e A_e K_s \delta\phi_x^e \left(\phi_x^e + \frac{dw_e}{dx} \right) \right] dx$$
$$- Q_2^e \delta\phi_x^e(x_a^e) - Q_4^e \delta\phi_x^e(x_b^e). \tag{8.4.6}$$

8.4.3   Finite Element Models

An examination of the weak forms in Eqs (8.4.5) and (8.4.6) indicate that both w_e and ϕ_x^e should be differentiable once with respect to x, that is, the minimum degree of interpolation of w_e and ϕ_x^e is linear. Because of the fact that the primary variables involve only (w_e, ϕ_x^e) and not their derivatives, Lagrange interpolation of (w_e, ϕ_x^e) is needed. Suppose that w_e and ϕ_x^e are approximated as

$$w_e(x) \approx \sum_{j=1}^{m} \psi_j^e(x) W_j^e, \quad \phi_x^e(x) \approx \sum_{j=1}^{n} \varphi_j^e(x) \Phi_j^e, \tag{8.4.7}$$

where (W_j^e, Φ_j^e) are the nodal values of (w_e, ϕ_x^e) and (ψ_j^e, φ_j^e) are the associated Lagrange interpolation functions. The minimum value of m and n is 2. Substitution of Eq. (8.4.7) for w_e and ϕ_x^e, and $\delta w_e = \psi_i^e$ and $\delta\phi_x^e = \varphi_i^e$ into Eqs (8.4.5) and (8.4.6) yields the finite element model

$$\begin{bmatrix} \mathbf{K}^{11} & \mathbf{K}^{12} \\ (\mathbf{K}^{12})^{\mathrm{T}} & \mathbf{K}^{22} \end{bmatrix} \begin{Bmatrix} \mathbf{W} \\ \mathbf{\Phi} \end{Bmatrix} = \begin{Bmatrix} \mathbf{F}^1 \\ \mathbf{F}^2 \end{Bmatrix}, \tag{8.4.8}$$

where

$$K_{ij}^{11} = \int_{x_a}^{x_b} K_s G_e A_e \frac{d\psi_i^e}{dx} \frac{d\psi_j^e}{dx} \, dx, \quad K_{ij}^{12} = \int_{x_a}^{x_b} K_s G_e A_e \frac{d\psi_i^e}{dx} \varphi_j^e \, dx,$$
$$K_{ij}^{22} = \int_{x_a}^{x_b} \left(E_e I_e \frac{d\varphi_i^e}{dx} \frac{d\varphi_j^e}{dx} + K_s G_e A_e \varphi_i^e \varphi_j^e \right) dx, \tag{8.4.9}$$
$$F_i^1 = \int_{x_a}^{x_b} \psi_i^e q \, dx + Q_1^e \psi_i^e(x_a) + Q_3^e \psi_i^e(x_b), \quad F_i^2 = Q_2^e \varphi_i^e(x_a) + Q_4^e \varphi_i^e(x_b).$$

8.4.4   Reduced Integration Element (RIE)

For the linear interpolation of both w_e and ϕ_x^e (i.e., $m = n = 2$; see Fig. 8.4.1) and exact evaluation of the integrals in Eq. (8.4.9), Eq. (8.4.8) takes the form

$$\frac{G_e A_e K_s}{6h_e} \begin{bmatrix} 6 & -6 & -3h_e & -3h_e \\ -6 & 6 & 3h_e & 3h_e \\ -3h_e & 3h_e & 2h_e^2\lambda_e & h_e^2\xi \\ -3h_e & 3h_e & h_e^2\xi & 2h_e^2\lambda_e \end{bmatrix} \begin{Bmatrix} W_1^e \\ W_2^e \\ \Phi_1^e \\ \Phi_2^e \end{Bmatrix} = \begin{Bmatrix} q_1^e \\ q_2^e \\ 0 \\ 0 \end{Bmatrix} + \begin{Bmatrix} Q_1^e \\ Q_3^e \\ Q_2^e \\ Q_4^e \end{Bmatrix}, \tag{8.4.10}$$

or, rearranging the equations, we obtain

$$\left(\frac{2E_eI_e}{\mu_0^e h_e^3}\right)\begin{bmatrix} 6 & -3h_e & -6 & -3h_e \\ -3h_e & 2h_e^2\lambda_e & 3h_e & h_e^2\xi_e \\ -6 & 3h_e & 6 & 3h_e \\ -3h_e & h_e^2\xi_e & 3h_e & 2h_e^2\lambda_e \end{bmatrix}\begin{Bmatrix} W_1^e \\ \Phi_1^e \\ W_2^e \\ \Phi_2^e \end{Bmatrix} = \begin{Bmatrix} q_1^e \\ 0 \\ q_2^e \\ 0 \end{Bmatrix} + \begin{Bmatrix} Q_1^e \\ Q_2^e \\ Q_3^e \\ Q_4^e \end{Bmatrix}, \quad (8.4.11)$$

where

$$q_i^e = \int_{x_a}^{x_b} \psi_i^e q \, dx, \quad (i = 1, 2) \tag{8.4.12}$$

$$\Lambda_e = \frac{E_eI_e}{G_eA_eK_sh_e^2}, \quad \xi_e = 1 - 6\Lambda_e, \quad \mu_0^e = 12\Lambda_e, \quad \lambda_e = 1 + 3\Lambda_e. \tag{8.4.13}$$

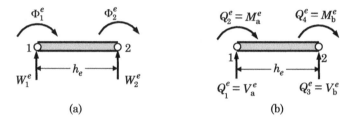

Fig. 8.4.1 Linear finite element model of the TBT. (a) Generalized displacements. (b) Generalized forces.

In the thin beam limit (i.e., h_e/H_e very large, where H_e is the height and b_e is the width of the beam), we have

$$\Lambda_e = \frac{E_eI_e}{G_eA_eK_sh_e^2} = 2(1+\nu)\frac{b_eH_e^3}{12}\frac{1}{b_eH_eK_sh_e^2} = \frac{(1+\nu)}{6K_s}\frac{H_e^2}{h_e^2} \to 0. \tag{8.4.14}$$

Then the first and third equations of Eq. (8.4.11), after multiplying both sides with $\mu_0^e h_e^3/2E_eI_e$ (and $\mu_0^e \to 0$), imply the following relation among $(W_1^e, W_2^e, \Phi_1^e, \Phi_2^e)$:

$$\left(\frac{\Phi_1^e + \Phi_2^e}{2}\right) + \left(\frac{W_2^e - W_1^e}{h_e}\right) = 0, \tag{8.4.15}$$

which is equivalent to the constraint $\phi_x^e + (dw_e/dx) = 0$ (or shear strain $\gamma_{xz} = 0$). The second and fourth equations of Eq. (8.4.11), in view of Eq. (8.4.15), yield the constraint

$$\Phi_1^e - \Phi_2^e = 0. \tag{8.4.16}$$

This is equivalent to $d\phi_x^e/dx = 0$, which forces the curvature and hence the bending energy to zero. Thus, the finite element equations in Eq. (8.4.11), in an attempt to satisfy the constraints in Eqs (8.4.15) and (8.4.16), will yield the trivial solution $W_1^e = W_2^e = \Phi_1^e = \Phi_2^e = 0$. This is known in the finite element literature as the *shear locking*.

The condition in Eq. (8.4.15) suggests that w_e and ϕ_x^e be interpolated such that dw_e/dx is a polynomial of the same order as ϕ_x^e. If w_e is approximated using a linear polynomial (a minimum requirement), then ϕ_x^e should be a constant, which makes the bending energy to vanish. Conversely, if ϕ_x^e is approximated as linear, then it follows that w_e must be approximated as a quadratic polynomial. This is termed a *consistent interpolation*. Alternatively, one may approximate w_e and ϕ_x^e appearing in the bending energy as linear while approximating ϕ_x^e appearing in the shear energy as a constant. This can be achieved by using numerical integration with one-point Gauss rule to evaluate the stiffness terms involving shear coefficient $G_e A_e K_s$. This procedure is known in the literature as the *reduced integration* of the shear stiffness. It amounts to evaluating the second term of K_{ij}^{22} in Eq. (8.4.9) with one-point integration as opposed to two-point integration required to exactly evaluate the integral. This results in the element equations

$$\left(\frac{2EI}{\mu_0 h_e^3}\right) \begin{bmatrix} 6 & -3h_e & -6 & -3h_e \\ -3h_e & h^2(1.5+6\Lambda_e) & 3h_e & h_e^2(1.5-6\Lambda_e) \\ -6 & 3h_e & 6 & 3h_e \\ -3h_e & h_e^2(1.5-6\Lambda_e) & 3h_e & h_e^2(1.5+6\Lambda_e) \end{bmatrix} \begin{Bmatrix} W_1^e \\ \Phi_1^e \\ W_2^e \\ \Phi_2^e \end{Bmatrix} = \begin{Bmatrix} q_1^e \\ 0 \\ q_2^e \\ 0 \end{Bmatrix} + \begin{Bmatrix} Q_1^e \\ Q_2^e \\ Q_3^e \\ Q_4^e \end{Bmatrix}.$$

$$(8.4.17)$$

This element is designated as the *reduced integration element* (RIE). In the thin (or slender) beam limit, these element equations reduce to only one constraint, namely the condition in Eq. (8.4.15). While the element does not lock, it does not yield exact displacements at the nodes. However, with sufficient number of elements in the mesh of a problem, one does get very accurate solution.

8.4.5 Consistent Interpolation Element (CIE)

If we use a quadratic approximation of w_e and linear approximation of ϕ_x^e, Eqs (8.4.5) and (8.4.6) yield the following 5×5 system of equations:

$$\frac{G_e A_e K_s}{6h_e} \begin{bmatrix} 14 & -16 & 2 & -5h_e & -h_e \\ -16 & 32 & -16 & 4h_e & -4h_e \\ 2 & -16 & 14 & h_e & 5h_e \\ -5h_e & 4h_e & h_e & 2h_e^2\lambda_e & h_e^2\xi_e \\ -h_e & -4h_e & 5h_e & h_e^2\xi_e & 2h_e^2\lambda_e \end{bmatrix} \begin{Bmatrix} W_1^e \\ W_c^e \\ W_2^e \\ \Phi_1^e \\ \Phi_2^e \end{Bmatrix} = \begin{Bmatrix} q_1^e \\ q_c^e \\ q_2^e \\ 0 \\ 0 \end{Bmatrix} + \begin{Bmatrix} Q_1^e \\ \hat{Q}_c^e \\ Q_3^e \\ Q_2^e \\ Q_4^e \end{Bmatrix},$$

$$(8.4.18)$$

where $(Q_1^e, Q_3^e, Q_2^e, Q_4^e)$ are the generalized forces defined in Eq. (8.4.4), W_c^e and $\hat{Q}_c^e$ are the deflection and applied external load, respectively, at the center node of the quadratic element, and

$$q_i^e = \int_{x_a^e}^{x_b^e} \psi_i^e q\, dx, \quad (i = 1, 2, c). \qquad (8.4.19)$$

This element is designated as the *consistent interpolation element* (CIE); see Fig. 8.4.2.

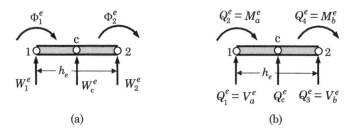

(a)   (b)

Fig. 8.4.2 Consistent finite element model of the TBT. (a) Generalized displacements. (b) Generalized forces.

The center displacement degree of freedom W_c^e may be statically condensed out to reduce Eq. (8.4.18) to a 4×4 system of equations. The second equation of Eq. (8.4.18) can be used to express W_c^e in terms of W_1^e, W_2^e, Φ_1^e, Φ_2^e, q_c^e, and $\hat{Q}_c^e$ as

$$W_c^e = \frac{6h_e}{32 G_e A_e K_s}\left(q_c^e + \hat{Q}_c^e\right) + \left(\frac{W_1^e + W_2^e}{2}\right) + h_e\left(\frac{\Phi_2^e - \Phi_1^e}{8}\right). \qquad (8.4.20)$$

Substituting for W_c^e from Eq. (8.4.20) into the remaining equations of Eq. (8.4.18) [i.e., eliminate W_c^e from Eq. (8.4.18)], we obtain

$$\left(\frac{2E_e I_e}{\mu_0 h_e^3}\right)\begin{bmatrix} 6 & -3h_e & -6 & -3h_e \\ -3h_e & h_e^2(1.5+6\Lambda) & 3h_e & h_e^2(1.5-6\Lambda) \\ -6 & 3h_e & 6 & 3h_e \\ -3h_e & h_e^2(1.5-6\Lambda) & 3h_e & h_e^2(1.5+6\Lambda) \end{bmatrix}\begin{Bmatrix} W_1^e \\ \Phi_1^e \\ W_2^e \\ \Phi_2^e \end{Bmatrix}$$
$$= \begin{Bmatrix} q_1^e + \tfrac{1}{2}\hat{q}_c^e \\ -\tfrac{1}{8}\hat{q}_c^e h_e \\ q_2^e + \tfrac{1}{2}\hat{q}_c^e \\ \tfrac{1}{8}\hat{q}_c^e h_e \end{Bmatrix} + \begin{Bmatrix} Q_1^e \\ Q_2^e \\ Q_3^e \\ Q_4^e \end{Bmatrix}, \qquad (8.4.21)$$

where $\hat{q}_c^e = q_c^e + \hat{Q}_c^e$. For simplicity, but without the loss of generality, we will assume that $\hat{Q}_c^e = 0$ so that $\hat{q}_c^e = q_c^e$. Note that the element has the same stiffness matrix as the RIE but a different load vector. The load vector is equivalent to that of the BET element [see Eq. (8.3.16)]. In fact, for uniform load q, the load vector in Eq. (8.4.21) is identical to that of the BET element. Note that *the elimination of W_c is not possible for the dynamic case* because of the presence of $\ddot{W}_c$.

8.4.6   Superconvergent Element (SCE)

The next choice of consistent interpolation is to use the Lagrange cubic polynomial for w_e and quadratic polynomial for ϕ_x^e. This will lead to a 7×7 system of equations. The displacement degrees of freedom associated with the interior nodes can again be condensed out, for the static case, to obtain a 4×4 system of equations. Here we will not consider this approach. Instead, we consider the Hermite cubic interpolation of w_e and a dependent quadratic approximation of ϕ_x^e (see Reddy [46]). The 4×4 system of equations thus obtained yield exact values of the nodal displacements for the TBT, like in the case of Bernoulli–Euler beam element. Such elements are called *superconvergent elements* (SCEs). The approximation functions for w_e and ϕ_x^e are constructed using the general analytical solutions of the governing differential equations in Eqs (8.4.1) and (8.4.2).

The exact solution of Eqs (8.4.1) and (8.4.2) for the homogeneous case (analogous to what was done in **Examples 4.2.3** and **4.3.5** in connection with the BET) is

$$w_e(x) = -\frac{1}{E_e I_e}\left(a_1 \frac{x^3}{6} + a_2 \frac{x^2}{2} + a_3 x + a_4\right) + \frac{1}{G_e A_e K_s}(a_1 x), \qquad (8.4.22)$$

$$\phi_x^e(x) = \frac{1}{E_e I_e}\left(a_1 \frac{x^2}{2} + a_2 x + a_3\right), \qquad (8.4.23)$$

where a_1 through a_4 are the constants of integration. Note that the constants a_1, a_2, and a_3 appearing in Eq. (8.4.23) are the same as those in Eq. (8.4.22). Thus, Eqs (8.4.22) and (8.4.23) suggest that one may use cubic approximation of w_e and an *interdependent* quadratic approximation of ϕ_x^e. The resulting finite element is termed the *interdependent interpolation element* (IIE).

The constants a_i appearing in Eqs (8.4.22) and (8.4.23) can be expressed in terms of the nodal values of w_e and ϕ_x^e (see **Example 4.3.5**). Then Eqs (8.4.22) and (8.4.23) take the form

$$w_e(x) \approx \sum_{j=1}^{4} \psi_j^e \Delta_j^e, \quad \phi_x^e(x) \approx \sum_{j=1}^{4} \varphi_j^e \Delta_j^e, \qquad (8.4.24)$$

where

$$\Delta_1^e = w_e(x_a^e), \quad \Delta_2^e = \phi_x^e(x_a^e), \quad \Delta_3^e = w_e(x_b^e), \quad \Delta_4^e = \phi_x^e(x_b^e), \qquad (8.4.25)$$

where ψ_i^e and φ_i^e are the approximation functions

$$
\begin{aligned}
&\psi_1^e = \frac{1}{\mu_e}\left[\mu_e - 12\Lambda_e \eta - (3 - 2\eta)\eta^2\right], \quad \psi_2^e = -\frac{h_e}{\mu_e}\left[(1 - \eta)^2 \eta + 6\Lambda_e (1 - \eta)\eta\right], \\
&\psi_3^e = \frac{1}{\mu_e}\left[(3 - 2\eta)\eta^2 + 12\Lambda_e \eta\right], \quad \psi_4^e = \frac{h_e}{\mu_e}\left[(1 - \eta)\eta^2 + 6\Lambda_e (1 - \eta)\eta\right], \\
&\varphi_1^e = \frac{6}{h_e \mu_e}(1 - \eta)\eta, \quad \varphi_2^e = \frac{1}{\mu_e}(\mu_e - 4\eta + 3\eta^2 - 12\Lambda_e \eta), \\
&\varphi_3^e = -\frac{6}{h_e \mu_e}(1 - \eta)\eta, \quad \varphi_4^e = \frac{1}{\mu_e}(3\eta^2 - 2\eta + 12\Lambda_e \eta).
\end{aligned}
\qquad (8.4.26)
$$

Here, η and μ_e are defined by

$$\eta = \frac{x - x_a}{h_e}, \quad \mu_e = 1 + 12\Lambda_e, \quad \Lambda_e = \frac{E_e I_e}{K_s G_e A_e h_e^2}. \tag{8.4.27}$$

Substitution of the approximations in Eqs (8.4.24) into Eqs (8.4.5) and (8.4.6) yields the finite element model

$$\mathbf{K}^e \boldsymbol{\Delta}^e = \mathbf{q}^e + \mathbf{Q}^e, \tag{8.4.28}$$

where

$$K_{ij}^e = \int_{x_a^e}^{x_b^e} \left[E_e I_e \frac{d\varphi_i^e}{dx} \frac{d\varphi_j^e}{dx} + G_e A_e K_s \left(\varphi_i^e + \frac{d\psi_i^e}{dx} \right) \left(\varphi_j^e + \frac{d\psi_j^e}{dx} \right) \right] dx, \tag{8.4.29}$$

$$q_i^e = \int_{x_a^e}^{x_b^e} \psi_i^e q(x) \, dx.$$

Equation (8.4.28) has the explicit form [see Eq. (8.4.13) for the definitions of ξ_e and λ_e]

$$\left(\frac{2E_e I_e}{\mu_e h_e^3} \right) \begin{bmatrix} 6 & -3h_e & -6 & -3h_e \\ -3h_e & 2h_e^2 \lambda_e & 3h_e & h_e^2 \xi_e \\ -6 & 3h_e & 6 & 3h_e \\ -3h_e & h_e^2 \xi_e & 3h_e & 2h_e^2 \lambda_e \end{bmatrix} \begin{Bmatrix} W_1^e \\ \Phi_1^e \\ W_2^e \\ \Phi_2^e \end{Bmatrix} = \begin{Bmatrix} q_1^e \\ q_2^e \\ q_3^e \\ q_4^e \end{Bmatrix} + \begin{Bmatrix} Q_1^e \\ Q_2^e \\ Q_3^e \\ Q_4^e \end{Bmatrix}. \tag{8.4.30}$$

As stated earlier, this element leads to the exact nodal values for any distribution of the transverse load $q(x)$, provided that the bending stiffness $E_e I_e$ and shear stiffness $K_s G_e A_e$ are element-wise constant. In the thin beam limit, the stiffness matrix in Eq. (8.4.30) reduces to that of the Bernoulli–Euler beam equations in Eq. (8.3.16), and the element does not exhibit shear locking.

Example 8.4.1

Consider a simply supported beam. Obtain the solutions using various Timoshenko beam elements developed in this section. Use the following data:

$$E = 10^6, \ \nu = 0.25, \ K = \frac{5}{6}, \ q_0 = 1 \tag{1}$$

uniform and sinusoidally distributed loads. Consider two different beam length-to-height ratios $L/H = 10$ and 100 in presenting the numerical results.

Solution: Table 8.4.1 shows a comparison of the finite element solutions obtained with one, two, and four elements in half beam with the exact beam solutions for two different types of loads, namely, uniformly distributed load (UDL) and sinusoidally distributed load (SDL). Clearly, more than two elements of CIE and RIE are required to obtain acceptable solutions. On the other hand, IIE yields exact nodal values with one element.

Table 8.4.1 Comparison of the finite element solutions with the exact maximum deflection and rotation of a simply supported isotropic beam (N = number of elements used in half beam).

Element	$w \times 10^2$			$-\phi_x \times 10^3$		
	$N = 1$	$N = 2$	$N = 4$	$N = 1$	$N = 2$	$N = 4$
UDL ($L/H = 10$)						
RIE	0.09750	0.14438	0.15609	0.37500	0.46875	0.49219
CIE	0.12875	0.15219	0.15805	0.50000	0.50000	0.50000
IIE[a]	0.16000	0.16000	0.16000	0.50000	0.50000	0.50000
BEE[a]	0.15265	0.15265	0.15265	0.50000	0.50000	0.50000
UDL ($L/H = 100$)						
RIE	0.09379	0.14066	0.15238	0.37500	0.46875	0.49219
CIE	0.12504	0.14847	0.15433	0.50000	0.50000	0.50000
IIE[a]	0.15629	0.15629	0.15629	0.50000	0.50000	0.50000
BEE[a]	0.15265	0.15265	0.15265	0.50000	0.50000	0.50000
SDL ($L/H = 100$)						
RIE	0.07639	0.11079	0.12007	0.30543	0.36702	0.38204
CIE	0.09679	0.11682	0.12163	0.38705	0.38702	0.38702
IIE[a]	0.12322	0.12322	0.12322	0.38702	0.38702	0.38702
BEE[a]	0.12319	0.12319	0.12319	0.38702	0.38702	0.38702

[a]The results obtained using the Bernoulli–Euler element (BEE) and interdependent interpolation element (IIE) are exact compared with the respective beam theories.

8.5 Finite Element Analysis of the Classical Plate Theory

8.5.1 Introduction

Here we develop displacement finite element models of the equations governing the motion of plates according to the CPT. While many of the basic ideas of the finite element method from one-dimensional beam problems carries over to two-dimensional plate problems, finite element models of plates are considerably more complicated due to the fact that two-dimensional problems are described by partial differential equations over geometrically complex regions. The boundary Γ of a two-dimensional domain Ω is, in general, a curve. Therefore, the two-dimensional finite elements must be simple geometric shapes that can be used to approximate the geometry of a given two-dimensional domain as well as the solution over it. Consequently, the finite element solution will have errors due to the approximation of the solution as well as the geometry of the domain.

In two dimensions, there is more than one geometric shape that can be used as a finite element (see Fig. 8.5.1). The interpolation functions depend not only on the number of nodes in the element but also on the shape of the element.

The shape of the element must be such that its geometry is uniquely defined by a set of points, which serves as the element nodes in the development of the interpolation functions. A triangular element is the simplest geometric shape, followed by a rectangle.

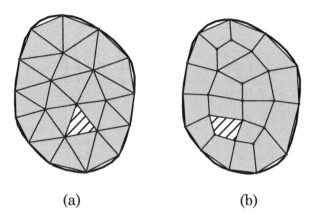

(a)   (b)

Fig. 8.5.1 Finite element discretization of a two-dimensional domain using meshes of (a) triangular and (b) quadrilateral elements.

8.5.2   General Formulation

Here we consider the general case of the dynamic response of plates subjected to in-plane compressive and shear forces. The virtual work statement of the CPT over typical finite element Ω_e is given by [from Eq. (7.2.93), except that the time derivative terms are integrated by parts in time]

$$
\begin{aligned}
0 = \int_{t_1}^{t_2} \int_{\Omega_e} &\left[-\frac{\partial^2 \delta w}{\partial x^2} M_{xx} - 2\frac{\partial^2 \delta w}{\partial x \partial y} M_{xy} - \frac{\partial^2 \delta w}{\partial y^2} M_{yy} \right. \\
&- \frac{\partial \delta w}{\partial x}\left(\hat{N}_{xx}\frac{\partial w}{\partial x} + \hat{N}_{xy}\frac{\partial w}{\partial y} \right) - \frac{\partial \delta w}{\partial y}\left(\hat{N}_{xy}\frac{\partial w}{\partial x} + \hat{N}_{yy}\frac{\partial w}{\partial y} \right) \\
&\left. + m_0 \delta w \ddot{w} + m_2 \left(\frac{\partial \delta w}{\partial x}\frac{\partial \ddot{w}}{\partial x} + \frac{\partial \delta w}{\partial y}\frac{\partial \ddot{w}}{\partial y} \right) - \delta w q \right] dxdydt \\
&+ \int_{t_1}^{t_2} \oint_{\Gamma^e} \left(-\delta w V_n + \frac{\partial \delta w}{\partial x}T_x + \frac{\partial \delta w}{\partial y}T_y \right) dsdt,
\end{aligned}
\tag{8.5.1}
$$

where

$$
\begin{aligned}
T_x &\equiv M_{xx}n_x + M_{xy}n_y, \quad T_y \equiv M_{xy}n_x + M_{yy}n_y, \\
V_n &\equiv \left(\frac{\partial M_{xx}}{\partial x} + \frac{\partial M_{xy}}{\partial y} - \hat{N}_{xx}\frac{\partial w}{\partial x} - \hat{N}_{xy}\frac{\partial w}{\partial y} - m_2\frac{\partial^3 w}{\partial x \partial t^2} \right) n_x \\
&+ \left(\frac{\partial M_{xy}}{\partial x} + \frac{\partial M_{yy}}{\partial y} - \hat{N}_{xy}\frac{\partial w}{\partial x} - \hat{N}_{yy}\frac{\partial w}{\partial y} - m_2\frac{\partial^3 w}{\partial y \partial t^2} \right) n_y.
\end{aligned}
\tag{8.5.2}
$$

Here (n_x, n_y) denote the direction cosines of the unit normal vector on the element boundary Γ_e, (M_{xx}, M_{xy}, M_{yy}) are defined in terms of w by Eq. (7.2.84), and $(\hat{N}_{xx}, \hat{N}_{yy}, \hat{N}_{xy})$ are the in-plane compressive and shear forces. It is clear from the variational statement in Eq. (8.5.1) that the finite elements based on the CPT require continuity of the primary variables, namely, the transverse deflection w and its normal derivative $\partial w/\partial n$ (or $\partial w/\partial x$ and $\partial w/\partial y$) across element boundaries. Also, to satisfy the constant displacement (rigid-body mode) and constant strain requirements, the polynomial expansion for w should be a complete polynomial with a minimum of 9 terms for a triangular element and 12 terms for a quadrilateral element (because there are 3 degrees of freedom per node).

Let us assume finite element approximation of the form

$$w(x, y, t) = \sum_{j=1}^{n} \Delta_j^e(t) \varphi_j^e(x, y), \tag{8.5.3}$$

where Δ_j^e are the values of w and its derivatives at the nodes and φ_j^e are the interpolation functions, the specific form of which will depend on the geometry of the element and the nodal degrees of freedom interpolated. Substituting approximation in Eq. (8.5.3) for w and φ_i^e for the virtual displacement δw into Eq. (8.5.1), we obtain the ith equation of the finite element model

$$\sum_{j=1}^{n} \left[\left(K_{ij}^e - G_{ij}^e \right) \Delta_j^e + M_{ij}^e \ddot{\Delta}_j^e \right] = F_i^e, \tag{8.5.4}$$

where $i, j = 1, 2, \ldots, n$. The coefficients of the stiffness matrix $K_{ij}^e = K_{ji}^e$, mass matrix $M_{ij}^e = M_{ji}^e$, geometric stiffness (or stability) matrix $G_{ij}^e = G_{ji}^e$, and force vectors F_i^e are defined as follows:

$$K_{ij}^e = \int_{\Omega_e} D \left[T_{ij}^{xxxx} + \nu \left(T_{ij}^{xxyy} + T_{ij}^{yyxx} \right) + 2(1 - \nu) T_{ij}^{xyxy} + T_{ij}^{yyyy} \right] dx dy,$$

$$G_{ij}^e = \int_{\Omega_e} \left[\hat{N}_{xx} S_{ij}^{xx} + \hat{N}_{xy} \left(S_{ij}^{xy} + S_{ij}^{yx} \right) + \hat{N}_{yy} S_{ij}^{yy} \right] dx dy,$$

$$\tag{8.5.5}$$

$$M_{ij}^e = \int_{\Omega_e} \left[m_0 S_{ij}^{00} + m_2 \left(S_{ij}^{xx} + S_{ij}^{yy} \right) \right] dx dy,$$

$$F_i^e = \int_{\Omega_e} q \varphi_i^e \, dx dy + \oint_{\Gamma_e} \left(V_n \varphi_i^e - T_x \frac{\partial \varphi_i^e}{\partial x} - T_y \frac{\partial \varphi_i^e}{\partial y} \right) ds,$$

where

$$T_{ij}^{\xi\eta\zeta\mu} = \frac{\partial^2 \varphi_i^e}{\partial \xi \partial \eta} \frac{\partial^2 \varphi_j^e}{\partial \zeta \partial \mu}, \quad S_{ij}^{\xi\eta} = \frac{\partial \varphi_i^e}{\partial \xi} \frac{\partial \varphi_j^e}{\partial \eta}, \quad S_{ij}^{00} = \varphi_i^e \varphi_j^e, \tag{8.5.6}$$

and ξ, η, ζ, and μ take on the symbols x and y. In matrix notation, Eq. (8.5.4) can be expressed as

$$(\mathbf{K}^e - \mathbf{G}^e) \Delta^e + \mathbf{M}^e \ddot{\Delta}^e = \mathbf{F}^e. \tag{8.5.7}$$

This completes the finite element model development of the CPT. The finite element model in Eq. (8.5.7) is called a *displacement finite element model* because it is based on the dynamic version of the principle of virtual displacements and the generalized displacements are the nodal degrees of freedom.

8.5.3   Conforming and Nonconforming Plate Elements

There is a large body of literature on triangular and rectangular plate bending finite elements of isotropic or orthotropic plates based on the CPT (e.g., see [131–139]). The two types of CPT plate bending elements used are as follows:

(1) A *conforming element* is one in which the interelement continuity of w, $\theta_x \equiv \frac{\partial w}{\partial x}$, and $\theta_y \equiv \frac{\partial w}{\partial y}$ (or $\frac{\partial w}{\partial n}$) is satisfied.

(2) A *nonconforming element* is one in which the continuity of the normal slope, $\frac{\partial w}{\partial n}$, is not satisfied.

A conforming rectangular element has w, $\frac{\partial w}{\partial x}$, $\frac{\partial w}{\partial y}$, and $\frac{\partial^2 w}{\partial x \partial y}$ as the nodal variables. The interpolation functions for this element [see Fig. 8.5.2(a)] are

$$\varphi_i^e = g_{i1} \ (i = 1, 5, 9, 13); \quad \varphi_i^e = g_{i2} \ (i = 2, 6, 10, 14),$$
$$\varphi_i^e = g_{i3} \ (i = 3, 7, 11, 15); \quad \varphi_i^e = g_{i4} \ (i = 4, 8, 12, 16), \tag{8.5.8}$$

where

$$g_{i1} = \frac{1}{16}(\xi + \xi_i)^2(\xi_0 - 2)(\eta + \eta_i)^2(\eta_0 - 2),$$

$$g_{i2} = \frac{1}{16}\xi_i(\xi + \xi_i)^2(1 - \xi_0)(\eta + \eta_i)^2(\eta_0 - 2),$$

$$g_{i3} = \frac{1}{16}\eta_i(\xi + \xi_i)^2(\xi_0 - 2)(\eta + \eta_i)^2(1 - \eta_0),$$

$$g_{i4} = \frac{1}{16}\xi_i\eta_i(\xi + \xi_i)^2(1 - \xi_0)(\eta + \eta_i)^2(1 - \eta_0), \tag{8.5.9}$$

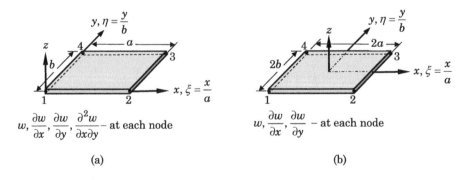

$$w, \frac{\partial w}{\partial x}, \frac{\partial w}{\partial y}, \frac{\partial^2 w}{\partial x \partial y} - \text{at each node}$$

$$w, \frac{\partial w}{\partial x}, \frac{\partial w}{\partial y} - \text{at each node}$$

(a)   (b)

Fig. 8.5.2 (a) Conforming rectangular element. (b) Nonconforming rectangular element.

where $\xi_0 = \xi\xi_i$ and $\eta_0 = \eta\eta_i$. The conforming element has four degrees of freedom per node, and the total number of bending nodal degrees of freedom per element is 16.

A nonconforming rectangular element has w, θ_x, and θ_y as the nodal variables [see Fig. 8.5.2(b)]. The normal slope variation is cubic along an edge, whereas there are only two values of $\frac{\partial w}{\partial n}$ available on the edge. Therefore, the cubic polynomial for the normal derivative of w is not the same on the edge common to two elements. The interpolation functions for this element can be expressed compactly as

$$\varphi_i^e = g_{i1} \ (i = 1, 4, 7, 10); \quad \varphi_i^e = g_{i2} \ (i = 2, 5, 8, 11),$$
$$\varphi_i^e = g_{i3} \ (i = 3, 6, 9, 12), \tag{8.5.10}$$

where $[\xi = (x - x_c)/a, \ \eta = (y - y_c)/b, \ \xi_0 = \xi\xi_i, \ \eta_0 = \eta\eta_i]$

$$g_{i1} = \tfrac{1}{8}(1 + \xi_0)(1 + \eta_0)(2 + \xi_0 + \eta_0 - \xi^2 - \eta^2),$$
$$g_{i2} = \tfrac{1}{8}\xi_i(\xi_0 - 1)(1 + \eta_0)(1 + \xi_0)^2, \tag{8.5.11}$$
$$g_{i3} = \tfrac{1}{8}\eta_i(\eta_0 - 1)(1 + \xi_0)(1 + \eta_0)^2,$$

and $2a$ and $2b$ are the sides of the rectangle, and (x_c, y_c) are the global coordinates of the center of the rectangle. The nonconforming element has three degrees of freedom per node, and the total number of bending degrees of freedom is 12.

8.5.4 Fully Discretized Finite Element Models

8.5.4.1 Static Bending

In the case of static bending under applied mechanical loads, Eq. (8.5.7) reduces to

$$(\mathbf{K}^e - \mathbf{G}^e)\,\mathbf{\Delta}^e = \mathbf{F}^e. \tag{8.5.12}$$

These element equations are assembled and suitable boundary conditions of the problem are implemented. Once the nodal values of generalized displacements $(w, \frac{\partial w}{\partial x}, \frac{\partial w}{\partial y})$ have been obtained by solving the assembled equations of a problem, the strains are evaluated in each element by differentiating the displacements. The strains and stresses are the most accurate if they are computed at the center of the element (i.e., use reduced integration to determine strains and stresses) [140, 141].

8.5.4.2 Buckling

In the case of buckling under applied in-plane compressive $(\hat{N}_{xx}, \hat{N}_{yy})$ and shear $\hat{N}_{xy}$ edge loads, Eq. (8.5.7) reduces to

$$\left(\mathbf{K}^e - \lambda\hat{\mathbf{G}}^e\right)\mathbf{\Delta}^e = \bar{\mathbf{F}}^e. \tag{8.5.13}$$

where

$$\lambda = \hat{N}_{xx}/\hat{N}_{xx}^0 = \hat{N}_{yy}/\hat{N}_{yy}^0 = \hat{N}_{xy}/\hat{N}_{xy}^0,$$

$$\hat{G}_{ij}^e = \int_{\Omega^e} \left[\hat{N}_{xx}^0 S_{ij}^{xx} + \hat{N}_{xy}^0 \left(S_{ij}^{xy} + S_{ij}^{yx} \right) + \hat{N}_{yy}^0 S_{ij}^{yy} \right] dx dy, \qquad (8.5.14)$$

$$\bar{F}_k = \oint_{\Gamma^e} \left(V_n \varphi_k^e - T_x \frac{\partial \varphi_k^e}{\partial x} - T_y \frac{\partial \varphi_k^e}{\partial y} \right) ds.$$

8.5.4.3 Natural Vibration

In the case of natural vibration, the response of the plate is assumed to be periodic, $\boldsymbol{\Delta}^e(t) = \bar{\boldsymbol{\Delta}}^e e^{i\omega t}$, where ω denotes the frequency of natural vibration. Equation (8.5.7) becomes

$$\left(\mathbf{K}^e - \omega^2 \mathbf{M}^e \right) \bar{\boldsymbol{\Delta}}^e = \bar{\mathbf{F}}^e. \qquad (8.5.15)$$

8.5.4.4 Transient Response

In the case of transient response, Eq. (8.5.7) must be integrated with respect to time t to determine the nodal values $\Delta_j^e(t)$ as functions of time. Here we consider the Newmark family of time-integration schemes [46, 130, 142] that are used widely in structural dynamics. In the Newmark method, the first and second time derivatives are approximated as

$$\begin{aligned} \dot{\boldsymbol{\Delta}}_{s+1}^e &= \dot{\boldsymbol{\Delta}}_s^e + a_1 \ddot{\boldsymbol{\Delta}}_s^e + a_2 \ddot{\boldsymbol{\Delta}}_{s+1}^e, \\ \ddot{\boldsymbol{\Delta}}_{s+1}^e &= a_3 \left(\boldsymbol{\Delta}_{s+1}^e - \boldsymbol{\Delta}_s^e \right) - a_4 \dot{\boldsymbol{\Delta}}_s^e - a_5 \ddot{\boldsymbol{\Delta}}_s^e, \end{aligned} \qquad (8.5.16)$$

where the superposed dot denotes differentiation with respect to time and a_1 through a_5 are the parameters defined in terms of the time increment $dt_s = t_{s+1} - t_s$ and the parameters α and γ of the Newmark scheme:

$$a_1 = (1 - \alpha)dt_s, \quad a_2 = \alpha\, dt_s, \quad a_3 = \frac{2}{\gamma (dt_s)^2},$$

$$a_4 = dt_s\, a_3, \quad a_5 = \frac{(1 - \gamma)}{\gamma}, \qquad (8.5.17)$$

where $\{\cdot\}_s$, for example, denotes the value of the enclosed vector at time t_s.

    The parameters α and γ in the Newmark scheme are selected such that the scheme is either stable or conditionally stable; that is, the error introduced through the time approximation in Eq. (8.5.16) does not grow unboundedly with time. All schemes for which $\gamma \geq \alpha \geq 1/2$ are unconditionally stable (i.e., no restriction on dt_s). Schemes for which $\gamma < \alpha$ and $\alpha \geq 0.5$ are conditionally stable, and the stability condition is

$$\Delta t \leq \Delta t_{cr} \equiv \left[\frac{1}{2}\omega_{\max} (\alpha - \gamma) \right]^{-\frac{1}{2}}, \qquad (8.5.18)$$

where $\omega_{\max}$ denotes the maximum frequency associated with the finite element equations in Eq. (8.5.15) for the same mesh.

The Newmark family contains the following widely used schemes:

$$\alpha = \frac{1}{2}, \quad \gamma = \frac{1}{2}, \quad \text{the constant average acceleration method (stable).}$$

$$\alpha = \frac{1}{2}, \quad \gamma = \frac{1}{3}, \quad \text{the linear acceleration method (conditionally stable).}$$
$$\text{(8.5.19)}$$
$$\alpha = \frac{1}{2}, \quad \gamma = 0, \quad \text{the centered difference method (conditionally stable).}$$

$$\alpha = \frac{3}{2}, \quad \gamma = 2, \quad \text{the backward difference method (stable).}$$

Using the time approximations from Eq. (8.5.16) in Eq. (8.5.7) (after some algebraic manipulations), we obtain

$$\hat{\mathbf{K}}^e \mathbf{\Delta}^e_{s+1} = \hat{\mathbf{F}}^e_{s,s+1}, \tag{8.5.20}$$

where

$$\begin{aligned}
\hat{\mathbf{K}}^e &= \left(\mathbf{K}^e_{s+1} - \mathbf{G}^e_{s+1}\right) + a_3 \mathbf{M}^e_{s+1} \\
\hat{\mathbf{F}}^e &= \mathbf{F}^e_{s+1} + \mathbf{M}^e_{s+1}\left(a_3 \mathbf{\Delta}^e_s + a_4 \dot{\mathbf{\Delta}}^e_s + a_5 \ddot{\mathbf{\Delta}}^e_s\right).
\end{aligned} \tag{8.5.21}$$

Note that for the centered difference scheme ($\gamma = 0$), it is necessary to use the following alternative form of Eq. (8.5.20):

$$\mathbf{H}^e_{s+1} \ddot{\mathbf{\Delta}}^e_{s+1} = \mathbf{F}^e_{s+1} - \left(\mathbf{K}^e_s - \mathbf{G}^e_s\right) \mathbf{A}^e_s, \tag{8.5.22}$$

where

$$\begin{aligned}
\mathbf{H}^e_{s+1} &= 0.5\gamma(dt_s)^2\left(\mathbf{K}^e_{s+1} - \mathbf{G}^e_{s+1}\right) + \mathbf{M}^e_{s+1}, \\
\mathbf{A}^e_s &= \mathbf{\Delta}^e_s + dt_s\,\dot{\mathbf{\Delta}}^e_s + 0.5(1-\gamma)(dt_s)^2 \ddot{\mathbf{\Delta}}^e_s.
\end{aligned} \tag{8.5.23}$$

The reader is asked to verify this result.

Equation (8.5.20) represents a system of algebraic equations among the (discrete) values of $\mathbf{\Delta}^e(t)$ at time $t = t_{s+1}$ in terms of known values at time $t = t_s$. At the first time step (i.e., $s = 0$), the values $\mathbf{\Delta}^e_0 = \mathbf{\Delta}^e(0)$ and $\dot{\mathbf{\Delta}}^e_0 = \dot{\mathbf{\Delta}}^e(0)$ are known from the initial conditions of the problem, and Eq. (8.5.7) is used to determine $\ddot{\mathbf{\Delta}}^e_0$ at $t = 0$:

$$\ddot{\mathbf{\Delta}}^e_0 = (\mathbf{M}^e)^{-1}\left[\mathbf{F}^e - (\mathbf{K}^e - \mathbf{G}^e)\,\mathbf{\Delta}^e_0\right]. \tag{8.5.24}$$

The conforming (C) and nonconforming (NC) rectangular finite elements discussed herein are used to analyze plates for bending and natural vibration response, and the results are presented in the next two examples. In the case of conforming element, it is necessary that the cross-derivative $\frac{\partial^2 w}{\partial x \partial y}$ be also set to zero at the center of the plate when a quarter-plate model is used. Otherwise, the results will be less accurate. The stresses in the finite element analysis are computed at the center of the elements. We shall use the notation $m \times n$ mesh to indicate that m elements along the x-axis and n elements along the y-axis are used.

Example 8.5.1 _____

Consider a square isotropic ($\nu = 0.25$) plate subjected to uniformly distributed load of intensity q_0. Consider the two cases of boundary conditions: (a) all edges simply supported and (b) all edges clamped. Determine the bending solutions using 2×2, 4×4, and 8×8 meshes of conforming and nonconforming elements and tabulate the maximum deflections and stresses in the plate. Exploit the biaxial symmetry to model a quadrant of the plate.

Solution: (a) The geometric boundary conditions of the computational domain (see the shaded quadrant in Fig. 8.5.3) are

$$\frac{\partial w}{\partial x} = 0 \text{ at } x = 0; \quad \frac{\partial w}{\partial y} = 0 \text{ at } y = 0, \tag{1}$$

$$w = \frac{\partial w}{\partial y} = 0 \text{ at } x = \frac{a}{2}; \quad w = \frac{\partial w}{\partial x} = 0 \text{ at } y = \frac{b}{2}. \tag{2}$$

In addition, $\frac{\partial^2 w}{\partial x \partial y} = 0$ was used at $x = y = 0$ for the conforming element. The boundary conditions (for CPT) along the symmetry lines are shown in Fig. 8.5.3. Full integration (i.e., 4×4 Gauss rule) was used to evaluate the stiffness coefficients in both elements, and it is found that the nonconforming element gave the same results for 3×3 Gauss rule.

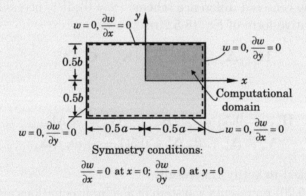

Fig. 8.5.3 Symmetry boundary conditions for rectangular plates.

Table 8.5.1 shows a comparison of finite element solutions with the analytical solutions developed in Chapter 7. The series solutions were evaluated using $m, n = 1, 3, \ldots, 19$. The maximum deflection occurs at $x = y = 0$, maximum stresses σ_{xx} and σ_{yy} occur at $(0, 0, h/2)$, and the maximum shear stress σ_{xy} occurs at $(a/2, b/2, -h/2)$. The maximum normal stresses in the finite element analysis are computed at the (x, y) locations: $(a/8, b/8)$, $(a/16, b/16)$, and $(a/32, b/32)$ for uniform meshes 2×2, 4×4, and 8×8, respectively. For the σ_{xy}, the locations are $(3a/8, 3b/8)$, $(7a/16, 7b/16)$, and $(15a/32, 15b/32)$ for the three meshes.

(b) Similar results are presented in Table 8.5.2 for a clamped square plate. Reduced integration (i.e., 3×3 Gauss rule) was used to evaluate the stiffness coefficients in both elements. The locations of the normal and shear stresses reported for the three meshes are:

Table 8.5.1 A comparison of the maximum transverse deflections and stresses[a] of simply supported (SSSS) square plates under uniformly distributed load.

Variable	Nonconforming			Conforming			Analytical
	2×2	4×4	8×8	2×2	4×4	8×8	
$\bar{w}$	4.8571	4.6425	4.5883	4.7388	4.5952	4.5734	4.5701
$\bar{\sigma}_{xx}$	0.2405	0.2673	0.2740	0.2259	0.2637	0.2732	0.2762
$\bar{\sigma}_{yy}$	0.2405	0.2673	0.2740	0.2259	0.2637	0.2732	0.2762
$\bar{\sigma}_{xy}$	0.1713	0.1964	0.2050	0.1669	0.1935	0.2040	0.2085

[a] $\bar{w} = 10^2[wEh^3/(q_0a^4)]$, $\bar{\sigma} = \sigma h^2/(q_0a^2)$.

$$\sigma \to 2 \times 2: (a/8, b/8); \quad 4 \times 4: (a/16, b/16); \quad 8 \times 8: (a/32, b/32)$$

$$\tau \to 2 \times 2: (3a/8, 3b/8); \quad 4 \times 4: (7a/16, 7b/16); \quad 8 \times 8: (15a/32, 15b/32)$$

These stresses are not necessarily the maximum ones in the plate. For example, for the 8×8 mesh the maximum normal stress in the plate is found to be 0.2316 at $(0.46875a, 0.03125b, -h/2)$, and the maximum shear stress is 0.0225 at $(0.28125a, 0.09375b, -h/2)$ for the nonconforming element.

Table 8.5.2 Maximum transverse deflections and stresses[a] of clamped (CCCC) isotropic ($\nu = 0.25$) square plates under uniformly distributed load.

Variable	Nonconforming			Conforming		
	2×2	4×4	8×8	2×2	4×4	8×8
$\bar{w}^b$	1.5731	1.4653	1.4342	1.7245	1.5327	1.4539
$\bar{\sigma}_{xx}$	0.0987	0.1238	0.1301	0.1115	0.1247	0.1305
$\bar{\sigma}_{yy}$	0.0987	0.1238	0.1301	0.1115	0.1247	0.1305
$\bar{\sigma}_{xy}$	0.0497	0.0222	0.0067	0.0700	0.0297	0.0086

[a] $\bar{w} = 10^2[wEh^3/(q_0a^4)]$, $\bar{\sigma} = \sigma h^2/(q_0a^2)$.
[b] The analytical value of the deflection from **Example 7.2.12** is 1.4231.

Both conforming and nonconforming elements show good convergence. Since the stresses in the finite element analysis are computed at locations different from the analytical solutions, they are expected to be different. Mesh refinement not only improves the accuracy of the solution, but the Gauss point locations also get closer to the true locations of the maximum values.

Example 8.5.2

Determine the fundamental natural frequencies of isotropic ($\nu = 0.25$) rectangular plates for two separate boundary conditions: (a) all edges simply supported and (b) all edges clamped. Use the conforming element with 2×2 and 4×4 meshes. Exploit the biaxial symmetry of the problem to determine the fundamental frequencies.

Solution: Table 8.5.3 contains fundamental frequencies for isotropic ($\nu = 0.25$) simply supported and clamped plates. The results were obtained using the conforming plate bending element. Only a quadrant was modeled and rotary inertia was included ($b/h = 0.1$). Because a quadrant of the plate is used to model the problem (exploiting the biaxial symmetry of symmetric modes), only symmetric vibration modes can be calculated. To obtain all frequencies, one must use the whole domain.

Table 8.5.3 A comparison of the fundamental frequencies[a] of simply supported (SSSS) and clamped (CCCC) rectangular plates.

	SSSS Plates			CCCC Plates		
a/b	2×2	4×4	Exact	2×2	4×4	Ritz[b]
0.5	4.8301	4.9752	4.9003	7.8213	9.1185	9.9891
1.0	1.9736	1.9959	1.9838	3.3103	3.5203	3.6476
1.5	1.4144	1.4395	1.4359	2.3435	2.5861	2.7404
2.0	1.2074	1.2439	1.2436	1.9551	2.2781	2.4973

[a] $\bar{\omega} = \omega \frac{b^2}{\pi^2} \sqrt{\frac{\rho h}{D_{22}}}$; foundation modulus $k = 0$.

[b] Ritz solution discussed in Chapter 7.

8.6 Finite Element Analysis of the First-Order Shear Deformation Plate Theory

8.6.1 Governing Equations and Weak Forms

Here we develop the finite element models of the equations governing the FDST. The weak form resulting from the principle of virtual displacements for the FSDT is given in Eq. (7.3.30). Here we consider the pure bending case only [i.e., terms involving the in-plane displacements u and v are omitted because they are decoupled from the bending variables (w, ϕ_x, ϕ_y)]. In addition, we also omit the in-plane edge forces $(\hat{N}_{xx}, \hat{N}_{yy}, \hat{N}_{xy})$ so that Eq. (7.3.30) simplifies to

$$
\begin{aligned}
0 = \int_0^T \Bigg\{ & \int_{\Omega_e} \Big[M_{xx}\delta\varepsilon_{xx}^{(1)} + M_{yy}\delta\varepsilon_{yy}^{(1)} + M_{xy}\delta\gamma_{xy}^{(1)} + Q_x\delta\gamma_{xz}^{(0)} + Q_y\delta\gamma_{yz}^{(0)} \\
& + kw\,\delta w - q\delta w - m_0\ddot{w}\delta\dot{w} - m_2\left(\dot{\phi}_x\delta\dot{\phi}_x + \dot{\phi}_y\delta\dot{\phi}_y \right) \Big] dxdy \\
& - \oint_{\Gamma_e} \left[T_x\delta\phi_x + T_y\delta\phi_y + (Q_xn_x + Q_yn_y)\,\delta w \right] dS \Bigg\} dt,
\end{aligned}
\tag{8.6.1}
$$

where k is the foundation modulus (if any), $\varepsilon_{xx}^{(1)}$, $\varepsilon_{yy}^{(1)}$, $\varepsilon_{xy}^{(1)}$, $\gamma_{xz}^{(0)}$, and $\gamma_{yz}^{(0)}$ are defined in Eq. (7.3.29):

$$
\{\varepsilon^{(1)}\} = \left\{ \begin{array}{c} \frac{\partial\phi_x}{\partial x} \\ \frac{\partial\phi_y}{\partial y} \\ 0 \\ 0 \\ \frac{\partial\phi_x}{\partial y} + \frac{\partial\phi_y}{\partial x} \end{array} \right\}, \quad \{\gamma^{(0)}\} = \left\{ \begin{array}{c} \gamma_{xz}^{(0)} \\ \gamma_{yz}^{(0)} \end{array} \right\} = \left\{ \begin{array}{c} \phi_x + \frac{\partial w}{\partial x} \\ \phi_y + \frac{\partial w}{\partial y} \end{array} \right\}.
\tag{8.6.2}
$$

and

$$
T_x \equiv M_{xx}n_x + M_{xy}n_y, \quad T_y \equiv M_{xy}n_x + M_{yy}n_y,
\tag{8.6.3}
$$

and (M_{xx}, M_{yy}, M_{xy}) and (Q_x, Q_y) are defined in terms of the generalized displacements (w, ϕ_x, ϕ_y) for an orthotropic plate as

$$
\left\{ \begin{array}{c} M_{xx} \\ M_{yy} \\ M_{xy} \end{array} \right\} = \begin{bmatrix} D_{11} & D_{12} & 0 \\ D_{12} & D_{22} & 0 \\ 0 & 0 & D_{66} \end{bmatrix} \left\{ \begin{array}{c} \frac{\partial \phi_x}{\partial x} \\ \frac{\partial \phi_y}{\partial y} \\ \frac{\partial \phi_x}{\partial y} + \frac{\partial \phi_y}{\partial x} \end{array} \right\}, \tag{8.6.4}
$$

and

$$
\left\{ \begin{array}{c} Q_x \\ Q_y \end{array} \right\} = \begin{bmatrix} A_{55} & 0 \\ 0 & A_{44} \end{bmatrix} \left\{ \begin{array}{c} \phi_x + \frac{\partial w}{\partial x} \\ \phi_y + \frac{\partial w}{\partial y} \end{array} \right\}, \tag{8.6.5}
$$

where the bending stiffness coefficients D_{ij} and shear stiffness coefficients (A_{44}, A_{55}) for an orthotropic material are given by

$$
D_{11} = \frac{E_1 h^3}{12(1 - \nu_{12}\nu_{21})}, \quad D_{22} = \frac{E_2 h^3}{12(1 - \nu_{12}\nu_{21})}, \quad D_{12} = \nu_{12} D_{22} = \nu_{21} D_{11},
$$

$$
D_{66} = \frac{G_{12} h^3}{12}, \quad A_{44} = G_{23} h, \quad A_{55} = G_{13} h, \tag{8.6.6}
$$

where h is the plate thickness and E_1, E_2, G_{12}, G_{13}, G_{23}, and ν_{12} are the engineering material properties.

The statement in Eq. (8.6.1) is equivalent to the following three weak forms:

$$
0 = \int_{\Omega_e} \left[\frac{\partial \delta w}{\partial x} Q_x + \frac{\partial \delta w}{\partial y} Q_y - \delta w q + k w \delta w + m_0 \delta w \frac{\partial^2 w}{\partial t^2} \right]
$$

$$
- \oint_{\Gamma_e} (Q_x n_x + Q_y n_y) \, \delta w \, dS \tag{8.6.7}
$$

$$
0 = \int_{\Omega_e} \left(\frac{\partial \delta \phi_x}{\partial x} M_{xx} + \frac{\partial \delta \phi_x}{\partial y} M_{xy} + \delta \phi_x Q_x + m_2 \delta \phi_x \frac{\partial^2 \phi_x}{\partial t^2} \right) dx dy
$$

$$
- \oint_{\Gamma_e} T_x \delta \phi_x \, dS \tag{8.6.8}
$$

$$
0 = \int_{\Omega_e} \left(\frac{\partial \delta \phi_y}{\partial x} M_{xy} + \frac{\partial \delta \phi_y}{\partial y} M_{yy} + \delta \phi_y Q_y + m_2 \delta \phi_y \frac{\partial^2 \phi_y}{\partial t^2} \right) dx dy
$$

$$
- \oint_{\Gamma_e} T_y \delta \phi_y \, dS \tag{8.6.9}
$$

We note from the boundary terms in Eqs (8.6.7)–(8.6.9) that (w, ϕ_x, ϕ_y) are the primary variables and (Q_n, T_x, T_y) are the secondary variables, where

$$
Q_n = Q_x n_x + Q_y n_y. \tag{8.6.10}
$$

Unlike in the CPT, the rotations (ϕ_x, ϕ_y) are independent of w. Note also that no derivatives of w are in the list of the primary variables, and therefore the finite element to be developed admits Lagrange interpolation functions, the minimum being linear. Such an element is termed a C^0 finite element.

8.6.2   Finite Element Approximations

The primary dependent variables (w, ϕ_x, ϕ_y) can all be approximated, in principle, with differing degrees of Lagrange interpolation functions. Let

$$w(x, y, t) = \sum_{j=1}^{m} w_j(t)\psi_j^e(x, y) \tag{8.6.11}$$

$$\phi_x(x, y, t) = \sum_{j=1}^{n} S_j^1(t)\varphi_j^e(x, y) \tag{8.6.12}$$

$$\phi_y(x, y, t) = \sum_{j=1}^{n} S_j^2(t)\varphi_j^e(x, y) \tag{8.6.13}$$

where (ψ_j^e, φ_j^e) are Lagrange interpolation functions of different degree, and (w_j, S_j^1, S_j^2) are the nodal values of (w, ϕ_x, ϕ_y), respectively. One can use linear, quadratic, or higher-order interpolation functions for $\psi_j^e(x, y)$ and $\varphi_j^e(x, y)$. In the present study we use quadrilateral elements.

The linear and quadratic Lagrange interpolation functions of master elements (see Fig. 8.6.1) that are used to generate the approximation functions for quadrilateral elements are given below in terms of the element coordinates (ξ, η), called the *natural coordinates*. The subscripts of these functions ψ_i correspond to the node numbers shown in Fig. 8.6.1. For a derivation of these functions, see Reddy [46].

Linear interpolation functions

$$\begin{Bmatrix} \psi_1^e \\ \psi_2^e \\ \psi_3^e \\ \psi_4^e \end{Bmatrix} = \frac{1}{4} \begin{Bmatrix} (1 - \xi)(1 - \eta) \\ (1 + \xi)(1 - \eta) \\ (1 + \xi)(1 + \eta) \\ (1 - \xi)(1 + \eta) \end{Bmatrix}. \tag{8.6.14}$$

Eight-node quadratic interpolation functions

$$\begin{Bmatrix} \psi_1^e \\ \psi_2^e \\ \psi_3^e \\ \psi_4^e \\ \psi_5^e \\ \psi_6^e \\ \psi_7^e \\ \psi_8^e \end{Bmatrix} = \frac{1}{4} \begin{Bmatrix} (1 - \xi)(1 - \eta)(-\xi - \eta - 1) \\ (1 + \xi)(1 - \eta)(\xi - \eta - 1) \\ (1 + \xi)(1 + \eta)(\xi + \eta - 1) \\ (1 - \xi)(1 + \eta)(-\xi + \eta - 1) \\ 2(1 - \xi^2)(1 - \eta) \\ 2(1 + \xi)(1 - \eta^2) \\ 2(1 - \xi^2)(1 + \eta) \\ 2(1 - \xi)(1 - \eta^2) \end{Bmatrix}. \tag{8.6.15}$$

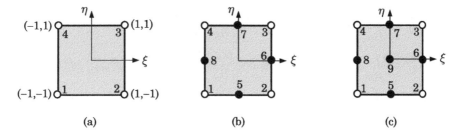

Fig. 8.6.1 Linear and quadratic master rectangular elements. (a) Linear element. (b) Eight-node rectangular element. (c) Nine-node rectangular element.

Nine-node quadratic interpolation functions

$$
\begin{Bmatrix} \psi_1^e \\ \psi_2^e \\ \psi_3^e \\ \psi_4^e \\ \psi_5^e \\ \psi_6^e \\ \psi_7^e \\ \psi_8^e \\ \psi_9^e \end{Bmatrix} = \frac{1}{4} \begin{Bmatrix}
(1-\xi)(1-\eta)(-\xi-\eta-1) + (1-\xi^2)(1-\eta^2) \\
(1+\xi)(1-\eta)(\xi-\eta-1) + (1-\xi^2)(1-\eta^2) \\
(1+\xi)(1+\eta)(\xi+\eta-1) + (1-\xi^2)(1-\eta^2) \\
(1-\xi)(1+\eta)(-\xi+\eta-1) + (1-\xi^2)(1-\eta^2) \\
2(1-\xi^2)(1-\eta) - (1-\xi^2)(1-\eta^2) \\
2(1+\xi)(1-\eta^2) - (1-\xi^2)(1-\eta^2) \\
2(1-\xi^2)(1+\eta) - (1-\xi^2)(1-\eta^2) \\
2(1-\xi)(1-\eta^2) - (1-\xi^2)(1-\eta^2) \\
4(1-\xi^2)(1-\eta^2)
\end{Bmatrix} . \tag{8.6.16}
$$

8.6.3   Finite Element Model

Substituting Eqs (8.6.11)–(8.6.13) for (w, ϕ_x, ϕ_y) into Eqs (8.6.7)–(8.6.9), we obtain the following semidiscrete finite element model of the FSDT:

$$
\begin{bmatrix} \mathbf{K}^{11} & \mathbf{K}^{12} & \mathbf{K}^{13} \\ (\mathbf{K}^{12})^{\mathrm{T}} & \mathbf{K}^{22} & \mathbf{K}^{23} \\ (\mathbf{K}^{13})^{\mathrm{T}} & (\mathbf{K}^{23})^{\mathrm{T}} & \mathbf{K}^{33} \end{bmatrix}^e \begin{Bmatrix} \mathbf{w} \\ \mathbf{S}^1 \\ \mathbf{S}^2 \end{Bmatrix}^e + \begin{bmatrix} \mathbf{M}^{11} & 0 & 0 \\ 0 & \mathbf{M}^{22} & 0 \\ 0 & 0 & \mathbf{M}^{33} \end{bmatrix}^e \begin{Bmatrix} \ddot{\mathbf{w}} \\ \ddot{\mathbf{S}}^1 \\ \ddot{\mathbf{S}}^2 \end{Bmatrix}^e = \begin{Bmatrix} \mathbf{F}^1 \\ \mathbf{F}^2 \\ \mathbf{F}^3 \end{Bmatrix}^e ,
\tag{8.6.17}
$$

or

$$
\mathbf{K}^e \boldsymbol{\Delta}^e + \mathbf{M}^e \ddot{\boldsymbol{\Delta}}^e = \mathbf{F}^e . \tag{8.6.18}
$$

The coefficients of the submatrices $\mathbf{K}^{\alpha\beta}$ and $\mathbf{M}^{\alpha\beta}$ and vectors $\mathbf{F}^{\alpha}$ are defined for $(\alpha, \beta = 1, 2, 3)$ by the expressions

$$
K_{ij}^{11} = \int_{\Omega^e} \left(K_s A_{55} \frac{\partial \psi_i^e}{\partial x} \frac{\partial \psi_j^e}{\partial x} + K_s A_{44} \frac{\partial \psi_i^e}{\partial y} \frac{\partial \psi_j^e}{\partial y} + k \psi_i^e \psi_j^e \right) dx dy,
$$

$$
K_{ij}^{12} = \int_{\Omega^e} K_s A_{55} \frac{\partial \psi_i^e}{\partial x} \varphi_j^e \, dx dy, \quad K_{ij}^{13} = \int_{\Omega^e} K_s A_{44} \frac{\partial \psi_i^e}{\partial y} \varphi_j^e \, dx dy,
$$

$$
K_{ij}^{22} = \int_{\Omega^e} \left(D_{11} \frac{\partial \varphi_i^e}{\partial x} \frac{\partial \varphi_j^e}{\partial x} + D_{66} \frac{\partial \varphi_i^e}{\partial y} \frac{\partial \varphi_j^e}{\partial y} + K_s A_{55} \varphi_i^e \varphi_j^e \right) dx dy,
$$

$$K_{ij}^{23} = \int_{\Omega^e} \left(D_{12} \frac{\partial \varphi_i^e}{\partial x} \frac{\partial \varphi_j^e}{\partial y} + D_{66} \frac{\partial \varphi_i^e}{\partial y} \frac{\partial \varphi_j^e}{\partial x} \right) dxdy,$$

$$K_{ij}^{33} = \int_{\Omega^e} \left(D_{66} \frac{\partial \varphi_i^e}{\partial x} \frac{\partial \varphi_j^e}{\partial x} + D_{22} \frac{\partial \varphi_i^e}{\partial y} \frac{\partial \varphi_j^e}{\partial y} + K_s A_{44} \varphi_i^e \varphi_j^e \right) dxdy,$$

$$M_{ij}^{11} = \int_{\Omega^e} m_0 \psi_i^e \psi_j^e \, dxdy, \quad M_{ij}^{22} = \int_{\Omega^e} m_2 \varphi_i^e \varphi_j^e \, dxdy = M_{ij}^{33},$$

$$F_i^1 = \int_{\Omega^e} q \psi_i^e \, dxdy + \oint_{\Gamma^e} Q_n \psi_i^e \, ds,$$

$$F_i^2 = \oint_{\Gamma^e} T_x \varphi_i^e \, dxdy, \quad F_i^3 = \int_{\Gamma_e} T_y \varphi_i^e \, dxdy. \qquad (8.6.19)$$

When the bilinear interpolation functions are used with $\psi_i^e = \varphi_i^e$ for all generalized displacements (w, ϕ_x, ϕ_y), the element stiffness matrices are of the order 12×12; and for the nine-node quadratic element they are 27×27 (see Fig. 8.6.2).

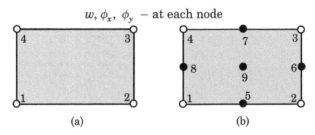

w, ϕ_x, ϕ_y − at each node

(a)   (b)

Fig. 8.6.2 C^0 rectangular plate bending finite elements for the FSDT. (a) Four-node element. (b) Nine-node element.

The C^0-plate bending element of Eq. (8.6.18) is among the simplest available in the literature. It is well known that when lower-order (quadratic or less) equal interpolation of the transverse deflection and rotations is used, the elements experience shear locking (similar to the TBT finite element). A commonly used technique to alleviate the problem of shear locking is to under-integrate the transverse shear parts of stiffness coefficients (i.e., all coefficients in $\mathbf{K}^{\alpha\beta}$ that contain A_{44} and A_{55}). To better understand what is reduced integration in the present context, we visit briefly the concept of numerical integration using the Gauss–Legendre quadrature. For alternative finite element models of the FSDT, the reader may consult References [143–146] at the end of the book.

Equation (8.6.18) can be simplified for static bending, buckling, natural vibration, and transient analysis, as described for the classical plate element in Section 8.5.4. The simplifications are obvious and therefore are not repeated here. Numerical results for bending, buckling, and natural vibration will be discussed next.

8.6.4 Numerical Integration

Numerical integration schemes, such as the Gauss–Legendre numerical integration scheme, require the integral to be evaluated on a specific domain or with respect to a specific coordinate system. Gauss quadrature, for example, requires the integral to be expressed over a square region $\hat{\Omega}$ of dimension 2×2 and the coordinate system (ξ, η) be such that $-1 \leq (\xi, \eta) \leq 1$. The coordinates (ξ, η) are called *normalized* or *natural coordinates*. Thus, the transformation between (x, y) and (ξ, η) of a given integral expression defined over an element Ω_e to one on the domain $\hat{\Omega}$ facilitates the use of Gauss–Legendre quadrature to evaluate integrals. The element $\hat{\Omega}$ is called a *master element* (see Reddy [46], Chapter 9).

The transformation between Ω_e and $\hat{\Omega}$ is accomplished by a coordinate transformation of the form (see Fig. 8.6.3)

$$x = \sum_{j=1}^{m} x_j^e \hat{\psi}_j^e(\xi, \eta), \qquad y = \sum_{j=1}^{m} y_j^e \hat{\psi}_j^e(\xi, \eta), \tag{8.6.20}$$

while a typical dependent variable $u(x, y)$ is approximated by

$$u(x, y) = \sum_{j=1}^{n} u_j^e \psi_j^e(x, y) = \sum_{j=1}^{n} u_j^e \psi_j^e(x(\xi, \eta), y(\xi, \eta)), \tag{8.6.21}$$

where $\hat{\psi}_j^e$ denote the interpolation functions of the master element $\hat{\Omega}$ and ψ_j^e are interpolation functions of a typical element Ω_e over which u is approximated. The transformation in Eq. (8.6.20) maps a point (ξ, η) in the master element $\hat{\Omega}$ to a point (x, y) in a typical element Ω_e of the mesh and vice versa if the Jacobian of the transformation is positive-definite and invertible. The positive definite requirement of the Jacobian dictates admissible geometries of elements in a mesh (see Reddy [46], pp. 421–448).

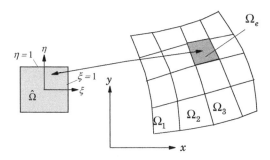

Fig. 8.6.3 Transformation between the domain on which an integral is defined (actual element domain) and the domain on which the integral is evaluated.

The interpolation functions ψ_j^e used for the approximation of the dependent variable are, in general, different from $\hat{\psi}_j^e$ used in the approximation of the geometry. Depending on the relative degree of approximations used for the geometry and the dependent variable(s), the finite element formulations are classified into three categories:

(a) *Superparametric formulation* $(m > n)$: The polynomial degree of approximation used for the geometry is of higher order than that used for the dependent variable.

(b) *Isoparametric formulation* $(m = n)$: Equal degree of approximation is used for both geometry and dependent variables.

(c) *Subparametric formulation* $(m < n)$: Higher-order approximation of the dependent variable is used.

For example, the nonconforming and conforming finite elements are based on higher-order interpolation (e.g., cubic or higher order) of the deflection w, while the geometry is represented by linear or quadratic interpolation functions. Hence, the formulation falls into subparametric category. The first-order shear deformation plate bending elements presented here are based on the isoparametric formulation.

Next, consider the evaluation of the integral

$$I_e = \int_{\Omega_e} F_e(x, y) \, dxdy. \tag{8.6.22}$$

The integrand $F_e(x, y)$ in the present case consists of the elastic coefficients D_{ij}^e and A_{ij}^e and the interpolation functions ψ_i^e and their derivatives. Hence, if D_{ij}^e and A_{ij}^e are constant within the element Ω_e, $F_e(x, y)$ is a polynomial of x and y of a certain degree, depending on the degree of ψ_i^e.

When a Gauss rule is used, the integral in Eq. (8.6.22) is expressed in terms of the natural coordinates (ξ, η) as

$$
\begin{aligned}
I_e &= \int_{\Omega_e} F_e(x, y) \, dxdy \\
&\approx \int_{\hat{\Omega}_e} F_e(x(\xi, \eta), y(\xi, \eta)) \, J^e \, d\xi d\eta \\
&\equiv \int_{-1}^{1} \int_{-1}^{1} \hat{F}_e(\xi, \eta) \, d\xi d\eta,
\end{aligned}
\tag{8.6.23}
$$

where J^e is the determinant of the matrix of coordinate transformation, called the Jacobian:

$$\mathbf{J}^e = \begin{bmatrix} \frac{\partial x}{\partial \xi} & \frac{\partial y}{\partial \xi} \\ \frac{\partial x}{\partial \eta} & \frac{\partial y}{\partial \eta} \end{bmatrix}^e. \tag{8.6.24}$$

For straight-sided elements, J^e is a constant, and the polynomial degree of $\hat{F}_e(\xi, \eta)$ is the same as that of $F_e(x, y)$.

The numerical evaluation of I_e by the Gauss quadrature amounts to evaluating the integrand at a number of points in $\hat{\Omega}$, called Gauss points, multiplying the functions values with so-called Gauss weights, and adding them up:

$$I_e = \int_{-1}^{1} \int_{-1}^{1} \hat{F}_e(\xi, \eta) \, d\xi d\eta = \sum_{I=1}^{M} \sum_{J=1}^{N} \hat{F}_e(\xi_I, \eta_J) W_I W_J, \qquad (8.6.25)$$

where (ξ_I, η_J) are the coordinates of the (I, J) Gauss point in $\hat{\Omega}$, W_I and W_J are the corresponding Gauss weights, and (M, N) are the number of Gauss points in each coordinate direction (see Fig. 8.6.4). If $\hat{F}_e(\xi, \eta)$ is a polynomial of degree p in ξ and degree q in η, then the choice

$$M = \left[\frac{p+1}{2}\right], \quad N = \left[\frac{q+1}{2}\right] \qquad (8.6.26)$$

gives the exact value of the integral I_e. Here $[\cdot]$ denotes the nearest integer equal to or greater than the enclosed value. For example, for the C^0 elements based on isoparametric formulation, the interpolation functions of the same degree are in both ξ and η. Hence, for the evaluation of the stiffness coefficients, we have $M = N$; and $p = q$ are at most equal to twice the degree r of the polynomials (of ξ or η) in $\psi_i^e(\xi, \eta)$. This can be seen by examining the coefficients $K_{ij}^{\alpha\alpha}$ $(\alpha = 1, 2, 3)$, which contain the products $\psi_i^e \psi_j^e$. Thus, we have

$$M = N = \left[\frac{2r+1}{2}\right] = r + 1, \quad r = \text{order of the element.} \qquad (8.6.27)$$

Note that $r = 1$ for the linear element and $r = 2$ for the quadratic element. The Gauss rule $M \times N$ means that M Gauss points in the ξ-coordinate direction and N Gauss points in the η-coordinate direction. If $N \times N$ is the full Gauss rule, then $(N-1) \times (N-1)$ is the reduced Gauss rule.

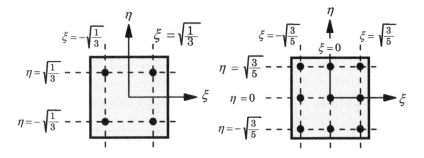

Fig. 8.6.4 Gauss points in a master rectangular element.

8.6.5 Numerical Examples

8.6.5.1 Isotropic Plates

Example 8.6.1 ──

Consider a simply supported isotropic square plate under the (sinusoidal) load

$$q(x,y) = q_0 \cos \frac{\pi x}{a} \cos \frac{\pi y}{a}, \tag{1}$$

where the origin of the coordinate system is taken at the center of the plate, $-a/2 \leq x \leq a/2, -a/2 \leq y \leq a/2$, and $-h/2 \leq z \leq h/2$. Because of the biaxial symmetry of the solution, the quadrant $0 \leq x \leq a/2$ and $0 \leq y \leq a/2$ may be used as the computational domain. Investigate the effect of full and reduced integration Gauss rules with different meshes of linear and quadratic elements on the deflections and stresses.

Solution: We use the notation nL for $n \times n$ uniform mesh of linear rectangular elements, nQ8 for $n \times n$ uniform mesh of eight-node quadratic elements, and nQ9 for $n \times n$ uniform mesh of nine-node quadratic elements in a quarter plate. Three different Gauss rules are used to evaluate the stiffness coefficients:

1. *Full integration* (F). All stiffness coefficients $\mathbf{K}^{\alpha\beta}$ of Eq. (8.6.17) are evaluated using the integration rule $N \times N$ that would yield their exact values.

2. *Selective (mixed) integration* (S). All terms of the stiffness coefficients *except those* containing A_{44} and A_{55} are evaluated using the full integration rule $N \times N$, and the terms involving A_{44} and A_{55} are evaluated using the reduced integration rule $(N-1) \times (N-1)$.

3. *Reduced integration* (R). All stiffness coefficients $\mathbf{K}^{\alpha\beta}$ of Eq. (8.6.17) are evaluated using the reduced integration rule $(N-1) \times (N-1)$.

Various stresses in the finite element analysis are evaluated at the reduced Gauss points as indicated below:

$$\sigma_{xx}(A,A,h/2), \quad \sigma_{yy}(A,A,h/2), \quad \sigma_{xy}(B,B,-h/2)$$

$$\sigma_{xz}(B,A) \text{ at } z = \pm\frac{h}{2}, \quad \sigma_{yz}(A,B) \text{ at } z = 0, \tag{2}$$

where the values of (A,B) are given in Table 8.6.1. The maximum deflection and stresses obtained for the problem are presented in Table 8.6.2. The following dimensionless stresses are used:

$$\bar{w} = w(A,A) \times 10^2 \frac{D}{a^4 q_0}, \quad \bar{\sigma}_{xx} = \sigma_{xx}(A,A,h/2)\frac{h^2}{a^2 q_0},$$

$$\bar{\sigma}_{yy} = \sigma_{yy}(A,A,h/2)\frac{h^2}{a^2 q_0}, \quad \bar{\sigma}_{xy} = \sigma_{xy}(B,B,-h/2)\frac{h^2}{a^2 q_0}, \tag{3}$$

$$\bar{\sigma}_{xz} = -\sigma_{xz}(B,A)\frac{h}{a q_0}, \quad \bar{\sigma}_{yz} = -\sigma_{yz}(A,B)\frac{h}{a q_0}.$$

An examination of the numerical results presented in Table 8.6.2 shows that the FSDT finite element with equal interpolation of all generalized displacements does not experience shear locking for thick plates even when full integration rule is used. Shear locking is evident

Table 8.6.1 The Gauss point locations at which the stresses are computed in the finite element analysis.

Coord.	Exact	2L	4L	8L	2Q8/2Q9	4Q8/4Q9
A	0.0	0.125a	0.0625a	0.03125a	0.05283a	0.02642a
B	0.5a	0.375a	0.4375a	0.46875a	0.44717a	0.47358a

Table 8.6.2 Effect of reduced integration on the nondimensionalized maximum deflections $\bar{w}$ and stresses $\bar{\sigma}$ of simply supported isotropic square plates subjected to sinusoidal loading.

a/h	Source	$\bar{w}$	$\bar{\sigma}_{xx}$	$\bar{\sigma}_{yy}$	$\bar{\sigma}_{xy}$	$\bar{\sigma}_{xz}$	$\bar{\sigma}_{yz}$
	Finite element solutions						
10	2L–F	0.1382	0.0744	0.0744	0.0446	0.1545	0.1545
	2L–R	0.2571	0.1458	0.1458	0.0874	0.1545	0.1545
	2L–S	0.2621	0.1483	0.1483	0.0889	0.1634	0.1634
	1Q9-F	0.2461	0.1402	0.1402	0.0965	0.1626	0.1626
	1Q9-R	0.2720	0.1645	0.1645	0.0985	0.1658	0.1658
	1Q9-S	0.2715	0.1638	0.1638	0.0987	0.405	0.1665
	4L–F	0.2179	0.1436	0.1436	0.0862	0.1537	0.1537
	4L–R	0.2670	0.1781	0.1781	0.1068	0.1814	0.1814
	4L–S	0.2682	0.1788	0.1788	0.1073	0.1837	0.1837
	2Q9-F	0.2682	0.1812	0.1812	0.1100	0.1879	0.1879
	2Q9-R	0.2703	0.1844	0.1844	0.1106	0.1854	0.1854
	2Q9-S	0.2703	0.1844	0.1844	0.1106	0.1855	0.1855
	8L–S	0.2697	0.1871	0.1871	0.1123	0.1892	0.1892
	4Q9-S	0.2702	0.1886	0.1886	0.1132	0.1897	0.1897
	Analytical solutions						
	FSDT	0.2702	0.1900	0.1900	0.1140	0.1910	0.1910
	CPT	0.2566	0.1900	0.1900	0.1140	0.2387	0.2387
	Finite element solutions						
100	2L–F	0.0026	0.0015	0.0015	0.0009	0.1545	0.1545
	2L–R	0.2431	0.1458	0.1458	0.0875	0.1546	0.1546
	2L–S	0.2472	0.1483	0.1483	0.0890	0.1634	0.1634
	1Q9-F	0.2133	0.1223	0.1223	0.0953	0.1567	0.1567
	1Q9-R	0.2583	0.1645	0.1645	0.0985	0.1658	0.1658
	1Q9-S	0.2581	0.1638	0.1638	0.0990	0.1659	0.1659
	4L–F	0.0101	0.0070	0.0070	0.0042	0.1813	0.1813
	4L–R	0.2535	0.1781	0.1781	0.1068	0.1814	0.1814
	4L–S	0.2545	0.1788	0.1788	0.1073	0.1837	0.1837
	2Q9-F	0.2494	0.1731	0.1731	0.1083	0.1846	0.1846
	2Q9-R	0.2569	0.1844	0.1844	0.1106	0.1854	0.1854
	2Q9-S	0.2569	0.1844	0.1844	0.1106	0.1850	0.1850
	8L–S	0.2562	0.1871	0.1871	0.1123	0.1819	0.1819
	4Q9-S	0.2568	0.1886	0.1886	0.1132	0.1896	0.1896
	Analytical solutions						
	FSDT	0.2568	0.1900	0.1900	0.1140	0.1910	0.1910
	CPT	0.2566	0.1900	0.1900	0.1140	0.2387	0.2387

when the element is used to model thin plates ($a/h \geq 100$) with full integration rule (F). Also, higher-order elements show less sensitivity for locking but with slower convergence. The element behaves uniformly well for thin and thick plates when the reduced (R) or selectively reduced integration (S) rule is used. It is clear that the effect of shear deformation is to increase the deflections. It should be noted that the nondimensionalization in Eq. (3) is such that the CPT solution is independent of the side-to-thickness ratio.

Higher-order elements or refined meshes of lower-order elements experience relatively less locking, but sometimes at the expense of rate of convergence. With the suggested Gauss rule, highly distorted elements tend to have slower rates of convergence, but they give reasonably accurate results.

8.6.5.2   Laminated Plates

Plates composed of multiple layers are called *laminated plates* (see Reddy [50]
and Fig. 8.6.5). A laminated plate composed of multiple orthotropic layers,
with the material axes (x_1, x_2) of each layer oriented at either $0°$ or $90°$ to
the plate axes (x, y), has the bending moment–deflection relationships given in
Eq. (8.6.4) with the laminate bending stiffness coefficients D_{ij} defined by

$$D_{ij} = \frac{1}{3} \sum_{k=1}^{N} \bar{Q}_{ij}^{(k)} (z_{k+1}^3 - z_k^3). \tag{8.6.28}$$

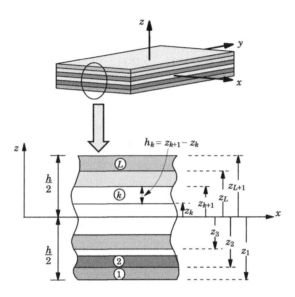

Fig. 8.6.5 The layer numbering and coordinate system used for a typical laminated plate.

Here N denotes the total number of layers in the laminate, z_k is the z-coordinate
of the bottom of the kth layer (counted from the bottom to the top in the
positive z direction), and $\bar{Q}_{ij}^{(k)}$ are the elastic stiffnesses of the kth layer referred
to the plate coordinates (x, y)

$$\bar{Q}_{11}^{(k)} = \frac{E_1^k}{1 - \nu_{12}^k \nu_{21}^k}, \quad \bar{Q}_{12}^{(k)} = \frac{\nu_{21}^k E_1^k}{1 - \nu_{12}^k \nu_{21}^k}, \quad \bar{Q}_{22}^{(k)} = \frac{E_2^k}{1 - \nu_{12}^k \nu_{21}^k}$$

$$\bar{Q}_{16}^{(k)} = 0, \quad \bar{Q}_{26}^{(k)} = 0, \quad \bar{Q}_{66}^{(k)} = G_{12}^k, \quad \bar{Q}_{44}^{(k)} = G_{23}^k, \quad \bar{Q}_{55}^{(k)} = G_{13}^k. \tag{8.6.29}$$

The engineering constants E_1^k, E_2^k, G_{12}^k, G_{23}^k, and ν_{12}^k are referred to the principal material axes of the kth layer. Note that in laminated plates, the principal
material axis x_3 is always taken along the plate axis z that is transverse to the
plane of the laminae.

Example 8.6.2

Consider a square sandwich plate, which is treated as a three-layer laminate with different thickness of the face sheets (layers 1 and 3) and the core (layer 2). The face sheets are assumed to be orthotropic, with the principal material coordinates coinciding with the plate axes (x, y, z), and the material properties are given by

$$E_1 = 25E_2, \ E_2 = 10^6 \text{ psi}, \ G_{12} = G_{13} = 0.5E_2, \ G_{23} = 0.2E_2, \ \nu_{12} = 0.25. \tag{1}$$

The core material is assumed to be characterized by the following material properties:

$$E_1 = E_2 = 10^6 \text{ psi}, \ G_{13} = G_{23} = 0.06 \times 10^6 \text{ psi}, \ \nu_{12} = 0.25,$$

$$G_{12} = \frac{E_1}{2(1 + \nu_{12})} = 0.016 \times 10^6 \text{ psi}. \tag{2}$$

Each face sheet is assumed to be one-tenth of the total thickness of the sandwich plate. Carry out the finite element analysis of the sandwich plate using a uniform 4×4 mesh of eight-node quadrilateral elements with reduced integration (4Q8-R) for the following cases: (a) simply supported on all edges and subjected to sinusoidally distributed transverse load, (b) simply supported on all edges and subjected to uniformly distributed transverse load, and (c) clamped on all edges and subjected to uniformly distributed transverse load.

Solution: (a) For the simply supported sandwich plate under sinusoidally distributed load, the finite element results obtained with a uniform 4×4 mesh of 4Q8-R are compared with the exact solutions based on FSDT and elasticity solution (ELS) of Pagano [147] in Table 8.6.3. The dimensionless deflection is defined as

$$\bar{w} = w(0,0) \times 10^2 \left(\frac{E_2 h^3}{a^4 q_0} \right). \tag{3}$$

The dimensionless stresses are the same as defined in Eq. (3) of **Example 8.6.1**, and their locations with respect to a coordinate system whose origin is at the center of the plate are as follows:

$$\sigma_{xx}(0,0,h/2), \ \sigma_{yy}(0,0,h/2), \ \sigma_{xy}(a/2,b/2,-h/2), \ \sigma_{xz}(0,b/2,0), \ \sigma_{yz}(a/2,0,0). \tag{4}$$

Figure 8.6.6 shows the variation of the transverse shear stresses through the thickness of the sandwich plates for side-to-thickness ratios $a/h = 2, 10$, and 100.

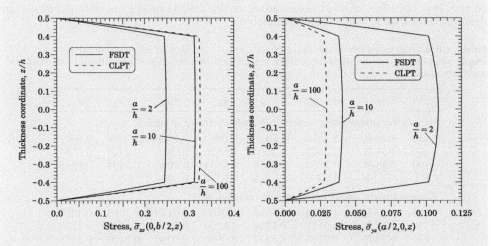

Fig. 8.6.6 Distribution of shear stresses through the thickness of a simply supported sandwich plate under sinusoidal load. (a) σ_{xz}. (b) σ_{yz}.

The results of Table 8.6.3 indicate that the effect of shear deformation on deflections is significant in sandwich plates even at large values of a/h (i.e., in thin plate range). The equilibrium-derived transverse shear stresses are very close to those predicted by the elasticity theory for $a/h \geq 10$, while those computed from constitutive equations are considerably underestimated for small side-to-thickness ratios. The transverse shear stress component σ_{yz} is significantly overestimated by CPT.

Table 8.6.3 Comparison of dimensionless maximum deflections and stresses in simply supported, square, sandwich plates subjected to sinusoidally distributed transverse load ($h_1 = h_3 = 0.1h$, $h_2 = 0.8h$, $K = 5/6$).

a/h	Source	$\bar{w}$	$\bar{\sigma}_{xx}$	$\bar{\sigma}_{yy}$	$\bar{\sigma}_{xy}$	$\bar{\sigma}_{xz}$	$\bar{\sigma}_{yz}$
10	ELS[b]	–	1.1530	0.1104	0.0717	0.3000	0.0527
	Exact[c]	1.5604	1.0457	0.0798	0.0552	0.1374	0.0293
						$(0.3134)^a$	$(0.0408)^a$
	FEM[c]	1.5603	1.0384	0.0792	0.0548	0.1365	0.0278
20	ELS	–	1.1100	0.0700	0.0511	0.3170	0.0361
	Exact	1.0524	1.0831	0.0612	0.0466	0.1409	0.0234
						$(0.3213)^a$	$(0.0325)^a$
	FEM	1.0523	1.0755	0.0608	0.0462	0.1399	0.0233
100	ELS	–	1.0980	0.0550	0.0437	0.3240	0.0297
	Exact	0.8852	1.0964	0.0546	0.0435	0.1422	0.0213
						$(0.3242)^a$	$(0.0296)^a$
	FEM	0.8851	1.0887	0.0542	0.0432	0.1412	0.0161
	CPT	0.8782	1.0970	0.0543	0.0433	$(0.3243)^a$	$(0.0295)^a$

^{a}Values computed from stress equilibrium equations.
^{b}3-D elasticity solution from Ref. [147]. ^{c}Based on the FSDT.

(b) and (c). The same sandwich plate discussed above is analyzed for simply supported and clamped boundary conditions when uniformly distributed transverse load is used. A quarter-plate model is used with 4×4 mesh of nine-node quadrilateral FSDT elements with reduced integration (4Q9-R) and 8×8 mesh of CPT conforming elements with full integration for stiffness coefficient evaluation (8CC-F) and one-point Gauss rule for stresses. From Table 8.6.4, we note that the effect of shear deformation on the deflections is even more significant in clamped plates than in simply supported plates.

Table 8.6.4 Dimensionless maximum deflections and stresses in square sandwich plates with simply supported and clamped boundary conditions and subjected to uniformly distributed transverse load ($h_1 = h_3 = 0.1h$, $h_2 = 0.8h$, $K = 5/6$).

a/h	Source	$\bar{w} \times 10^2$	$\bar{\sigma}_{xx}$	$\bar{\sigma}_{yy}$	$\bar{\sigma}_{xy}$	$\bar{\sigma}_{xz}$	$\bar{\sigma}_{yz}$
Simply supported plate under uniformly distributed load							
10	4Q8-R	2.3370	1.5430	0.0883	0.1136	0.2396	0.0991
50	4Q8-R	1.3671	1.5964	0.0526	0.0916	0.2433	0.0881
100	4Q8-R	1.3359	1.5978	0.0514	0.0906	0.2394	0.0880
CPT	8CC-F	1.3296	1.5830	0.0509	0.0906	–	–
Clamped plate under uniformly distributed load							
10	4Q9-R^a	1.2654	0.5018	0.0550	0.0120	0.2318	0.1445
50	4Q9-R	0.3111	0.5356	0.0108	0.0039	0.2406	0.1160
100	4Q9-R	0.2785	0.5347	0.0094	0.0030	0.2400	0.1148
CPT	8CC-F	0.2951	0.5401	0.0145	0.0605	–	–

^{a}The 4Q9-S element gives the same results as 4Q9-R.

Example 8.6.3

Consider the natural vibration of a simply supported laminated plate. The plate is made of four orthotropic layers. The principal material axes (x_1, x_2) of the first and fourth layers coincide with the plate axes, while those of the middle two layers are oriented at $90°$ to the plate axes (x, y). Such a laminate is designated as $(0/90/90/0)$ or simply $(0/90)_s$, where the latter notation implies that the laminate is symmetric about the midplane. Take the material properties of each layer (or lamina) to be

$$E_1 = 40E_2, \ G_{12} = G_{13} = 0.6E_2, \ G_{23} = 0.5E_2, \ \nu_{12} = 0.25. \tag{1}$$

Use a 2×2 mesh of (a) eight-node and (b) nine-node quadratic elements with reduced and selective integrations in a quarter plate to obtain the fundamental frequency for $a/h = 2, 10$, and 100 (include rotary inertia).

Solution: Effects of side-to-thickness ratio, integration, and type of element on the dimensionless fundamental frequency $\bar{\omega} = \omega(a^2/h)\sqrt{\rho/E_2}$ can be seen from the results presented in Table 8.6.5. From the results obtained, it is clear that both the reduced integration (R) and selective integration (S) rules give good results for a wide range of side-to-thickness ratios.

Table 8.6.5 Effects of side-to-thickness ratio, integration, and type of element on the dimensionless fundamental frequency $\bar{\omega}$ of simply supported square laminates $(0/90/90/0)$.

	Eight-node element			Nine-node element			
a/h	F	R	S	F	R	S	Exact
2	5.502	5.503	5.501	5.502	5.503	5.501	5.500
10	15.174	15.179	15.159	15.182	15.193	15.172	15.143
100	19.171	18.841	18.808	19.225	18.883	18.933	18.836

8.7 Summary

In this chapter an introduction to the finite element method is presented as a piecewise application of the Ritz method (or the weak-form Galerkin method). Displacement finite element models are developed using the principle of virtual displacements. Five different classes of problems are considered for the development of finite element models: (1) axial deformation of straight bars, (2) bending of beams according to the Bernoulli–Euler beam theory (BET), (3) bending of beams according to the Timoshenko beam theory (TBT), (4) bending of plates using the classical plate theory (CPT), and (5) bending of plates using the first-order shear deformation plate theory (FSDT). Detailed discussion of the shear locking in TBT element is presented, and a number of alternative finite element models of the TBT are developed. In the case of the CPT, C^1 conforming and nonconforming rectangular elements are presented. Finally, C^0 finite element model of the FSDT is presented. Illustrative examples throughout the chapter are presented to bring out the effect of integration rule on the degree of locking and shear deformation on bending deflections and natural frequencies (see Reddy [50, 51, 130] for more details and examples).

Problems

8.1 Derive the finite element model of the equation,

$$-\frac{d}{dx}\left[(1+x)\frac{du}{dx}\right] = f, \quad 0 < x < 1$$

for the boundary conditions of the type

$$u(0) = \hat{u}, \quad \left[(1+x)\frac{du}{dx} + u\right]_{x=1} = \hat{q},$$

and solve the problem for the data $f = 0, \hat{u} = 1$, and $\hat{q} = 0$. Use two linear finite elements.

8.2 Use the following procedure to derive the cubic interpolation functions for a line element with equally spaced (four) nodes. Set up the local coordinate $\bar{x}$ at the left-end node (i.e., node 1), and then use the interpolation property of ψ_i,

$$\psi_i(\bar{x}_j) = \delta_{ij}, \quad i, j = 1, 2, 3, 4,$$

to write ψ_i as a polynomial that vanishes at all $\bar{x}_j$, $\bar{x}_j \neq \bar{x}_i$,

$$\psi_1(\bar{x}) = c_1(\bar{x} - \bar{x}_2)(\bar{x} - \bar{x}_3)(\bar{x} - \bar{x}_4),$$
$$\psi_2(\bar{x}) = c_2(\bar{x} - \bar{x}_1)(\bar{x} - \bar{x}_3)(\bar{x} - \bar{x}_4),$$

and determine c_1, c_2, etc. such that $\psi_1(\bar{x}_1) = 1$, $\psi_2(\bar{x}_2) = 1$, etc.

Use the minimum possible number of linear finite elements, unless stated otherwise, to analyze the axially-loaded structures of **Problems 8.3–8.9**.

8.3 Find the stresses and compressions in each section of the composite member shown in Fig. P8.3. Use $E_s = 30 \times 10^6$ psi, $E_a = 10^7$ psi, $E_b = 15 \times 10^6$ psi.

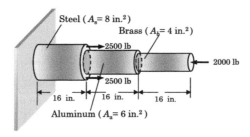

Fig. P8.3

8.4 A solid circular copper cylinder ($E_c = 16 \times 10^6$ psi, $d_c = 0.21$ in.) is encased in a hollow circular steel tube ($E_s = 30 \times 10^6$ psi, $d_s = 0.25$ in.) without any clearance between the two, and the cylinder and tube are compressed between rigid plates by a force $P = 1,250$ lb, as shown in Fig. P8.4. Determine (a) the compression, (b) compressive forces in the steel tube and copper cylinder, and corresponding stresses.

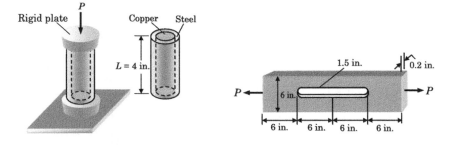

Fig. P8.4   Fig. P8.5

8.5 A rectangular steel bar ($E_s = 30 \times 10^6$ psi) of length 24 in. has a slot in the middle half of its length, as shown in Fig. P8.5. Determine the displacement of the ends due to the axial loads $P = 2000$ lb and stresses in each portion of the bar.

8.6 The two members in Fig. P8.6 are fastened together and to rigid walls. If the members are stress free before they are loaded, what will be the stresses and deformations in each after the two 50,000 lb loads are applied? Use $E_s = 30 \times 10^6$ psi and $E_a = 10^7$ psi; the aluminum rod is 2 in. in diameter and the steel rod is 1.5 in. in diameter.

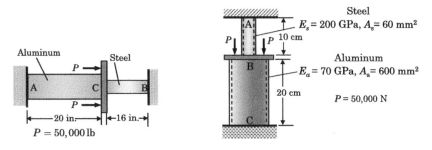

Fig. P8.6   Fig. P8.7

8.7 The aluminum and steel pipes shown in Fig. P8.7 are fastened to rigid supports at ends A and C and to a rigid plate B at their junction. Determine the displacement of point B and stresses in the aluminum and steel pipes. Use the minimum number of linear finite elements.

8.8 A steel bar ABC (assumed to be deformable) is pin-supported at its upper end A to an immovable wall and loaded by a force F_1 at its lower end C, as shown in Fig. P8.8. A *rigid* horizontal beam BDE is pinned to the vertical bar at B, supported at point D, and carries a load F_2 at end E. Determine the axial displacements u_B and u_C of points B and C.

8.9 A steel bar ABC (assumed to de deformable) is pin-supported at its upper end A to an immovable wall and spring-supported (spring is assumed to be linearly elastic) at its lower end C, as shown in Fig. P8.9. A *rigid* horizontal beam BDE is pinned to the vertical bar at B, supported at point D, and carries a load F_2 at end E. Determine the axial displacements u_B and u_C of points B and C.

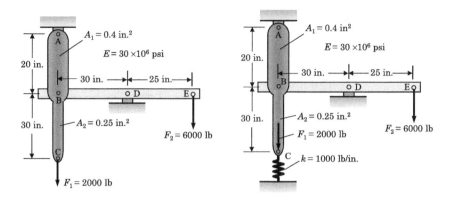

Fig. P8.8   Fig. P8.9

SECTION 8.3: BERNOULLI–EULER BEAMS

8.10–8.16 Use the minimum possible number of Bernoulli–Euler beam elements, unless stated otherwise, to analyze (i.e., determine deflections and slopes at the nodes, and reactions at the supports of) the beam structures shown in Figs. P8.10–P8.16. Use the minimum number of elements and follow the sign convention used in Fig. 8.3.1 for the generalized displacements, generalized forces, and applied transverse load.

8.17 Construct the finite element model, based on the total potential energy functional, for the following differential equation, which arises in connection with the buckling of beam columns:

$$\frac{d^2}{dx^2}\left(EI\frac{d^2w}{dx^2}\right) + P\frac{d^2w}{dx^2} = 0, \quad 0 < x < L$$

where P is the axial compressive load (at which the column buckles) to be determined and EI is the flexural rigidity.

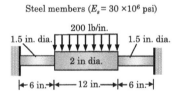

Fig. P8.10

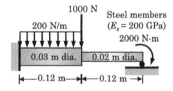

Fig. P8.11

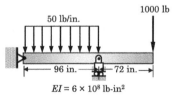

Fig. P8.12

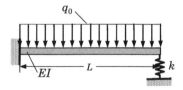

Fig. P8.13

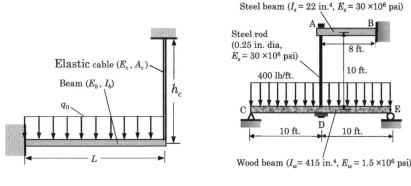

Fig. P8.14   Fig. P8.15

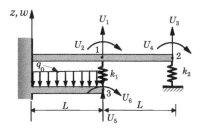

Fig. P8.16

8.18 Compute the stiffness coefficients of the finite element equations in **Problem 8.17**. Take EI as constant. Make use of the results already available in this book.

8.19 Use the finite element model of **Problems 8.17** and **8.18** to determine the critical buckling load P of a simply supported beam. Use one element in a half beam. This problem requires the solution of an eigenvalue problem.

SECTION 8.4: TIMOSHENKO BEAMS

8.20 Analyze the beam shown in Fig. P8.20 using the RIE. Use a value of $\frac{5}{6}$ for the shear correction factor K_s and $\nu = 0.25$.

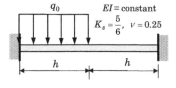

Fig. P8.20

8.21 Repeat **Problem 8.20** using the CIE (quadratic w and linear ϕ_x).

8.22 Repeat **Problem 8.20** using the SCE (cubic w and inter-dependent quadratic ϕ_x).

8.23 Analyze the beam shown in Fig. P8.23 using the RIE. Use a value of $\frac{5}{6}$ for the shear correction factor K_s and $\nu = 0.25$.

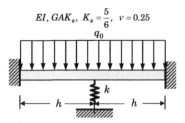

$EI, GAK_s, \ K_s = \dfrac{5}{6}, \ \nu = 0.25$

q_0

k

$\longleftarrow h \longrightarrow$ $\longleftarrow h \longrightarrow$

Fig. P8.23

SECTION 8.5: CLASSICAL THEORY OF PLATES

8.24 The differential equation governing axisymmetric bending of circular plates on elastic foundation is given by

$$-\frac{1}{r}\frac{d}{dr}\left[\frac{d}{dr}(rM_{rr}) - M_{\theta\theta}\right] + kw = q(r), \tag{1}$$

where r is the radial coordinate, k is the modulus of the elastic foundation, q is the transverse distributed load, and $(M_{rr}, M_{\theta\theta})$ are the bending moments defined in terms of the transverse deflection w by

$$M_{rr} = -D\left(\frac{d^2w}{dr^2} + \nu\frac{1}{r}\frac{dw}{dr}\right), \quad M_{\theta\theta} = -D\left(\nu\frac{d^2w}{dr^2} + \frac{1}{r}\frac{dw}{dr}\right), \tag{2}$$

where $D = Eh^3/[12(1-\nu^2)]$ is the bending stiffness. Develop (a) the weak form and identify the primary and secondary variables and (b) the finite element model. Note that the shear force is defined by

$$Q_r = \frac{1}{r}\left[\frac{d}{dr}(rM_{rr}) - M_{\theta\theta}\right]. \tag{3}$$

SECTION 8.6: SHEAR DEFORMATION THEORY OF PLATES

8.25 The governing equations of the first-order shear deformation plate theory for the axisymmetric bending of isotropic circular plates are given by

$$-\frac{d}{dr}(rM_{rr}) + M_{\theta\theta} + rQ_r = 0, \quad -\frac{d}{dr}(rQ_r) = rq, \tag{1}$$

where

$$M_{rr} = D\left(\frac{d\phi_r}{dr} + \frac{\nu}{r}\phi_r\right), \quad M_{\theta\theta} = D\left(\nu\frac{d\phi_r}{dr} + \frac{1}{r}\phi_r\right), \tag{2}$$

$$Q_r = GAK\left(\phi_r + \frac{dw}{dr}\right). \tag{3}$$

Develop the weak form and associated displacement finite element model of equations in Eq. (1).

SECTIONS 8.5 AND 8.6: GENERAL 2-D PROBLEMS

8.26 Give the variational formulation and associated finite element model of the equation

$$-\frac{\partial}{\partial x}\left(k_x\frac{\partial u}{\partial x}\right) - \frac{\partial}{\partial y}\left(k_y\frac{\partial u}{\partial y}\right) = f \text{ in } \Omega, \tag{1}$$

where k_x, k_y, and f are given functions of x and y in a two-dimensional domain Ω, and account for the following boundary condition in the model:

$$\left(k_x \frac{\partial u}{\partial x} n_x + k_y \frac{\partial u}{\partial y} n_y\right) + \beta(u - u_0) = q_n. \tag{2}$$

Here β and u_0 are known constants, and q_n is a given function on boundary of the element.

8.27 Consider the following pair of coupled partial differential equations:

$$-\frac{\partial}{\partial x}\left(a\frac{\partial u}{\partial x}\right) - \frac{\partial}{\partial y}\left[b\left(\frac{\partial u}{\partial y} + \frac{\partial v}{\partial x}\right)\right] - f_x = 0, \tag{1}$$

$$-\frac{\partial}{\partial x}\left[b\left(\frac{\partial u}{\partial y} + \frac{\partial v}{\partial x}\right)\right] - \frac{\partial}{\partial y}\left(c\frac{\partial v}{\partial y}\right) - f_y = 0, \tag{2}$$

where u and v are the dependent variables (unknown functions), a, b f_x, and f_y are known functions of position (x, y). Use the three-step procedure on each equation with a different weight function for each equation (say, w_1 and w_2) to develop the weak forms and finite element model of the equations. Assume approximation of (u, v) in the following form

$$u(x, y) = \sum_{j=1}^{n} \psi_j(x, y)U_j\ ,\quad v(x, y) = \sum_{j=1}^{n} \psi_j(x, y)V_j. \tag{3}$$

The finite element model is of the form

$$0 = \sum_{j=1}^{n} K_{ij}^{11}U_j + \sum_{j=1}^{n} K_{ij}^{12}V_j - F_i^1,$$

$$0 = \sum_{j=1}^{n} K_{ij}^{21}U_j + \sum_{j=1}^{n} K_{ij}^{22}V_j - F_i^2. \tag{4}$$

You must define the algebraic form of the element coefficients K_{ij}^{11}, K_{ij}^{12}, F_i^1, and so on.

"I know that most men, including those at ease with problems of the greatest complexity, can seldom accept even the simplest and most obvious truth if it be such as would oblige them to admit the falsity of conclusions which they have delighted in explaining to colleagues, which they have proudly taught to others, and which they have woven, thread by thread, into the fabric of their lives." – *Leo Tolstoy* (a Russian writer who is regarded as one of the greatest authors of all time)

"What is there that confers the noblest delight? What is that which swells a man's breast with pride above that which any other experience can bring to him? Discovery! To know that you are walking where none others have walked ... ." – *Mark Twain* (American author of *The Adventures of Huckleberry Finn* and humorist)

Mixed Variational and Finite Element Formulations

9.1 Introduction

9.1.1 General Comments

In this chapter we will cover mixed variational principles and their applications. A "mixed" (or sometimes called "hybrid") variational formulation is one where secondary (force-like) variables of the conventional formulation are also treated as dependent variables along with the primary (displacement) variables. Most often, these formulations are developed with the objective of determining the secondary variables directly rather than from post-computations.

We will begin with a brief exposure of mixed variational principles of elasticity, including Reissner's variational principle and the Hellinger–Reissner variational principles [6–10]. Use of variational methods as well as the finite element method to determine solutions based on the mixed variational formulations will also be discussed.

9.1.2 Mixed Variational Principles

To illustrate the basic ideas in the development of mixed variational principles, we shall consider the simple case of the axial deformation of a bar. From equilibrium considerations [see Eq. (2.5.3)], we have

$$-\frac{dN}{dx} = f(x), \quad 0 < x < L, \quad N = \int_A \sigma_{xx}\, dA, \qquad (9.1.1)$$

where L is the length of the bar, f is the body force along the axis x, and σ_{xx} is the axial stress in the bar. The axial force N is related to the axial displacement u through the kinematic and constitutive relations

$$\varepsilon_{xx} = \frac{du}{dx}, \quad \sigma_{xx} = E\varepsilon_{xx} \qquad (9.1.2)$$

by substituting Eq. (9.1.2) into the definition of N in Eq. (9.1.1),

$$N = EA\frac{du}{dx} \quad \text{or} \quad \frac{du}{dx} - \frac{N}{EA} = 0. \qquad (9.1.3)$$

Substituting Eq. (9.1.3) for N in Eq. (9.1.1), we arrive at

$$-\frac{d}{dx}\left(EA\frac{du}{dx}\right) = f(x), \quad 0 < x < L. \tag{9.1.4}$$

The conventional variational formulation is based on Eq. (9.1.4). Consequently, the displacement u is the primary dependent variable of the formulation, and N is post-computed from the known displacement using Eq. (9.1.3). The weak formulation of Eq. (9.1.4) is obtained by either following the steps presented in Section 6.3.6 (model equation 1) or using the principle of minimum total potential energy. The total potential energy functional is given by [see Eq. (3.3.1)]

$$\Pi(u) = \int_0^L \left[\frac{EA}{2}\left(\frac{du}{dx}\right)^2 - fu\right] dx - \text{WPL}, \tag{9.1.5}$$

where WPL stands for work done by applied point loads. Equation (9.1.4) is the Euler equation of this functional.

In a mixed formulation, we treat Eqs (9.1.1) and (9.1.3) as a pair of coupled equations in two independent variables u and N. The mixed variational formulation of this pair can be developed using the same steps as discussed in Section 6.3.6 (or see **Example 6.3.8**). We have

$$0 = \int_0^L w_1\left(-\frac{dN}{dx} - f\right) dx, \tag{9.1.6}$$

$$0 = \int_0^L w_2\left(\frac{du}{dx} - \frac{N}{EA}\right) dx, \tag{9.1.7}$$

where w_1 and w_2 are the weight functions, which may be interpreted as follows. If $w_1 \cdot f\,dx$ were to represent the work done, w_1 must be an axial displacement. Similarly, $w_2 \cdot (du/dx)dx$ represents work done if w_2 is an axial force. Thus, $w_1 \sim \delta u$ and $w_2 \sim \delta N$. Since N appears without differential in Eq. (9.1.7), in the interest of lowering the order of approximation used for N, we carry out the integration by parts of the first term in Eq. (9.1.6). There are other reasons for this that will be apparent in the sequel. One reason is we do not wish to make u a secondary variable by carrying out integration by parts of the first term in Eq. (9.1.7). We obtain

$$0 = \int_0^L w_1\left(-\frac{dN}{dx} - f\right) dx$$

$$= \int_0^L \left(\frac{dw_1}{dx}N - w_1 f\right) dx - [w_1 N]_0^L, \tag{9.1.8}$$

$$0 = \int_0^L w_2\left(\frac{du}{dx} - \frac{N}{EA}\right) dx. \tag{9.1.9}$$

Now it is clear that since $w_1 \sim u$, u is the primary variable and N is the secondary variable. Note that N is also an independent variable of the formulation but it is *not* a primary variable.

Equations (9.1.8) and (9.1.9) can be written, since $w_1 = \delta u$ and $w_2 = \delta N$ are linearly independent, as

$$0 = \int_0^L \left(\frac{d\delta u}{dx} N - \delta u f + \delta N \frac{du}{dx} - \frac{N}{EA} \delta N \right) dx - [\delta u N]_0^L$$

$$= \delta \int_0^L \left(\frac{du}{dx} N - \frac{N^2}{2EA} - fu \right) dx - [\delta u N]_0^L . \qquad (9.1.10)$$

The boundary term can be simplified for a specific problem. Whenever u is specified at $x = 0$ or $x = L$, δu is zero there. If u is not specified at a boundary point, then N is known (either zero or some specified value) at that point. Suppose that u is zero at $x = 0$ (essential boundary condition) and u is not specified at $x = L$, but N is specified to be P at $x = L$. Then we have

$$0 = \delta \left[\int_0^L \left(\frac{du}{dx} N - \frac{N^2}{2EA} - fu \right) dx - u(L)P \right] \equiv \delta \Pi_m . \qquad (9.1.11)$$

This is the mixed formulation associated with Eqs (9.1.1) and (9.1.3). Clearly, the functional

$$\Pi_m(u, N) = \int_0^L \left(\frac{du}{dx} N - \frac{N^2}{2EA} - fu \right) dx - u(L)P \qquad (9.1.12)$$

yields Eqs (9.1.1) and (9.1.3) as its Euler equations along with the natural boundary condition

$$N - P = 0 \quad \text{at} \quad x = L. \qquad (9.1.13)$$

In other words, $\delta \Pi_m = 0$ is equivalent to Eqs (9.1.1), (9.1.3), and (9.1.13). Note that Π_m does not attain a minimum or maximum with respect to u and N. It may be minimum with respect to u but maximum with respect to N. Thus the mixed variational principle can be stated as: find (u, N) such that

$$0 = \delta \Pi_m = \delta_u \Pi_m + \delta_N \Pi_m . \qquad (9.1.14)$$

9.1.3   Extremum and Stationary Behavior of Functionals

It can be shown that $\Pi(u)$ of Eq. (9.1.5) attains a minimum when u is the solution of Eq. (9.1.4) and satisfies the boundary conditions

$$u(0) = 0, \quad \left(EA \frac{du}{dx} \right)_{x=L} = P. \qquad (9.1.15)$$

To establish this, suppose that $\bar{u}(x) = u(x) + \alpha v(x)$ is an arbitrary element from the set of admissible functions (i.e., $\bar{u}$ satisfies the specified geometric boundary conditions and differentiable as required by the functional Π), where u is the solution of Eq. (9.1.4), α is a real number, and v is an element from the space of admissible variations (i.e., v satisfies the homogeneous form of the specified geometric boundary conditions and differentiable as required by Π). Then we have

$$
\begin{aligned}
\Pi(\bar{u}) &= \int_0^L \left[\frac{EA}{2} \left(\frac{d\bar{u}}{dx} \right)^2 - f\bar{u} \right] dx - \bar{u}(L)P \\
&= \int_0^L \left[\frac{EA}{2} \left(\frac{du}{dx} \right)^2 - fu \right] dx - u(L)P + \alpha^2 \int_0^L \frac{EA}{2} \left(\frac{dv}{dx} \right)^2 dx \\
&\quad + \alpha \left[\int_0^L \left(EA \frac{du}{dx} \frac{dv}{dx} - fv \right) dx - v(L)P \right] \\
&= \Pi(u) + \alpha^2 \int_0^L \frac{EA}{2} \left(\frac{dv}{dx} \right)^2 dx \\
&\quad + \alpha \int_0^L \left[-\frac{d}{dx} \left(EA \frac{du}{dx} \right) - f \right] v\,dx + \alpha \left[EA \frac{du}{dx} - P \right] v(L) \\
&= \Pi(u) + \alpha^2 \int_0^L \frac{EA}{2} \left(\frac{dv}{dx} \right)^2 dx \geq \Pi(u), \quad\quad (9.1.16)
\end{aligned}
$$

where in arriving at the last step, we have used the fact that u satisfies Eqs (9.1.4) and (9.1.15). It is clear from Eq. (9.1.16) that $\Pi(\bar{u}) > \Pi(u)$ for any $\alpha \neq 0$ and $\Pi(\bar{u}) = \Pi(u)$ if and only if $\alpha = 0$ (i.e., $\bar{u} = u$). Hence $\Pi(u)$ attains its minimum only at $\bar{u} = u$, and for any other admissible function $\bar{u}$, we have $\Pi(\bar{u}) > \Pi(u)$. This is essentially the statement of the principle of minimum total potential energy [see Eqs (4.3.2)–(4.3.5) and (5.3.10)].

Next, we establish the stationarity of $\Pi_m(u, N)$. Let $\bar{u} = u + \alpha v$ and $\bar{N} = N + \beta Q$, where α and β are real numbers, and (u, N) satisfy Eqs (9.1.1) and (9.1.3) inside the domain, Eqs (9.1.13) at $x = L$, and $u = 0$ at $x = 0$. Then

$$
\begin{aligned}
\Pi_m(\bar{u}, \bar{N}) &= \int_0^L \left(\frac{d\bar{u}}{dx} \bar{N} - \frac{\bar{N}^2}{2EA} - f\bar{u} \right) dx - \bar{u}(L)P \\
&= \int_0^L \left(\frac{du}{dx} N - \frac{N^2}{2EA} - fu \right) dx - u(L)P \\
&\quad + \alpha \left[\int_0^L \left(\frac{dv}{dx} N - fv \right) dx - v(L)P \right] \\
&\quad + \beta \int_0^L \left(\frac{du}{dx} - \frac{N}{EA} \right) Q\,dx - \frac{\beta^2}{2} \int_0^L \frac{Q^2}{EA} dx + \alpha\beta \int_0^L \frac{dv}{dx} Q\,dx \\
&= \Pi_m(u, N) - \frac{\beta^2}{2} \int_0^L \frac{Q^2}{EA} dx + \alpha\beta \int_0^L \frac{dv}{dx} Q\,dx. \quad\quad (9.1.17)
\end{aligned}
$$

The coefficient of α in the third line and the coefficient of β in the fourth line of Eq. (9.1.17) vanish by virtue of Eqs (9.1.1), (9.1.3), and (9.1.13). It is clear that

$\Pi_m(u, \bar{N})$ decreases from its value at equilibrium when u is kept at its actual value u (i.e., $\alpha = 0$) and N is changed to $\bar{N} = N + \beta Q$. That is,

$$\Pi_m(u, \bar{N}) < \Pi_m(u, N). \tag{9.1.18}$$

The value of the functional is unchanged when N is kept at its actual value (i.e., $\beta = 0$), and u is changed to $\bar{u} = u + \alpha v$. In fact, if the end $x = L$ is connected to an elastic spring, one can show that

$$\Pi_m(\bar{u}, N) \leq \Pi_m(u, N). \tag{9.1.19}$$

Thus, Π_m exhibits a stationary behavior and not an extremum (i.e., minimum or maximum) character.

9.2 Stationary Variational Principles

9.2.1 Minimum Total Potential Energy

The principles of minimum total potential energy and maximum total complementary energy are both extremum principles. Recall from Eq. (4.3.6) that a functional $\Pi(\mathbf{u})$ is a minimum at the displacement vector $\mathbf{u}$ if and only if

$$\Pi(\bar{\mathbf{u}}) \geq \Pi(\mathbf{u}) \quad \text{for all} \quad \bar{\mathbf{u}} \tag{9.2.1}$$

and the equality holds only if $\bar{\mathbf{u}} = \mathbf{u}$.

To show that the total potential energy of a linear elasticity body is the minimum at its equilibrium configuration, consider the total potential energy functional of an anisotropic linear elastic body undergoing infinitesimal strains:

$$\Pi(\mathbf{u}) = \int_\Omega \left(\frac{1}{2} C_{ijk\ell} \, \varepsilon_{k\ell} \, \varepsilon_{ij} - f_i u_i \right) dV - \int_{\Gamma_2} \hat{t}_i u_i \, dS, \tag{9.2.2}$$

where $\mathbf{u}$ is the displacement vector, $\mathbf{f}$ is the body force vector, $\hat{\mathbf{t}}$ is the specified traction vector on boundary Γ_2 of the volume region Ω occupied by the body, dV is the differential volume element and dS is the differential surface element, $C_{ijk\ell}$ are the Cartesian components of the fourth-order elasticity tensor, and ε_{ij} are the components of the infinitesimal strain tensor

$$\varepsilon_{ij} = \tfrac{1}{2}(u_{i,j} + u_{j,i}). \tag{9.2.3}$$

In Eqs (9.2.2) and (9.2.3), summation on repeated indices is implied, and comma followed by a subscript j denotes differentiation with respect to x_j: $u_{i,j} = \partial u_i / \partial x_j$. The geometric boundary condition on $\mathbf{u}$ is

$$\mathbf{u} = \hat{\mathbf{u}} \quad \text{on} \quad \Gamma_1, \tag{9.2.4}$$

where $\Gamma_1 + \Gamma_2 = \Gamma$, the total boundary of Ω.

Let $\mathbf{u}$ be the true displacement field and $\bar{\mathbf{u}}$ be an arbitrary but admissible displacement field. Then $\bar{\mathbf{u}}$ is of the form

$$\bar{\mathbf{u}} = \mathbf{u} + \alpha\mathbf{v}, \tag{9.2.5}$$

where $\mathbf{v}$ is a sufficiently differentiable function that satisfies the homogeneous form of the essential boundary condition $\mathbf{v} = \mathbf{0}$ on Γ_1 and α is a real number. From Eq. (9.2.2), we have

$$\Pi(\mathbf{u} + \alpha\mathbf{v}) = \int_\Omega \left[\tfrac{1}{2}C_{ijk\ell}\left(\varepsilon_{k\ell} + \alpha g_{k\ell}\right)\left(\varepsilon_{ij} + \alpha g_{ij}\right) - f_i\left(u_i + \alpha v_i\right)\right]dV$$
$$- \int_{\Gamma_2} \hat{t}_i(u_i + \alpha v_i)dS, \tag{9.2.6}$$

where

$$g_{ij} = \tfrac{1}{2}(v_{i,j} + v_{j,i}). \tag{9.2.7}$$

Collecting the terms, we obtain (because $C_{ijk\ell} = C_{k\ell ij}$)

$$\Pi(\bar{\mathbf{u}}) = \Pi(\mathbf{u}) + \alpha\left[\int_\Omega \left(-f_i v_i + C_{ijk\ell}\,\varepsilon_{k\ell}\,g_{ij} + \tfrac{1}{2}\alpha C_{ijk\ell}\,g_{ij}\,g_{k\ell}\right)dV - \int_{\Gamma_2} \hat{t}_i v_i\,dS\right]. \tag{9.2.8}$$

Using the equilibrium equations, we can write

$$-\int_\Omega f_i v_i\,dV = \int_\Omega \sigma_{ij,j} v_i\,dV = \int_\Omega C_{ijk\ell}\,\varepsilon_{k\ell,j}\,v_i\,dV$$
$$= -\int_\Omega C_{ijk\ell}\,\varepsilon_{k\ell}\,v_{i,j}\,dV + \int_{\Gamma_2} C_{ijk\ell}\,\varepsilon_{k\ell}\,v_i n_j\,dS$$
$$= -\int_\Omega C_{ijk\ell}\,\varepsilon_{k\ell}\,g_{ij}\,dV + \int_{\Gamma_2} \hat{t}_i v_i\,dS, \tag{9.2.9}$$

where the condition $v_i = 0$ on Γ_1 is used in arriving at the last step. Substituting Eq. (9.2.9) into Eq. (9.2.8), we arrive at

$$\Pi(\bar{\mathbf{u}}) = \Pi(\mathbf{u}) + \tfrac{1}{2}\alpha^2 \int_\Omega C_{ijk\ell}\,g_{ij}\,g_{k\ell}\,dV. \tag{9.2.10}$$

In view of the nonnegative nature of the second term on the right-hand side of Eq. (9.2.10), it follows that

$$\Pi(\bar{\mathbf{u}}) \geq \Pi(\mathbf{u}) \tag{9.2.11}$$

and $\Pi(\bar{\mathbf{u}}) = \Pi(\mathbf{u})$ only if the quadratic expression $\tfrac{1}{2}C_{ijk\ell}\,g_{ij}\,g_{k\ell}$ is zero. Due to the positive definiteness of the strain energy density, the quadratic expression is zero only if $v_i = 0$, which in turn implies $\bar{u}_i = u_i$. Thus, Eq. (9.2.11) implies that of all admissible displacement fields the body can assume, the true one is that which makes the total potential energy a minimum. Therefore, the total potential energy principle is a minimum principle. Similar arguments can be made for the total complementary energy. It should be noted that the consideration of whether a functional is minimum or maximum does not enter the calculation of the Euler equations. Such consideration is needed in analyzing the stability of structures.

9.2.2 The Hellinger–Reissner Variational Principle

A stationary principle is one in which the functional attains neither a minimum nor a maximum in its arguments. In fact, a functional can attain a maximum with respect to one set of variables (while others fixed) and a minimum with respect to another set of variables involved in the functional. An example of such functionals is provided by the functional based on the Lagrange multiplier method (see Section 3.5.8). For example, the functional in Eq. (4) of **Example 3.5.6** attains a minimum with respect to (w, ϕ_x) and a maximum with respect to λ. Stationary principles, also called *mixed* variational principles, are of special importance in the analysis of structures. Here we consider the Hellinger–Reissner variational principle (see [6–10, 17, 18, 28] for an elastic body. The stationary functional is constructed from the equations of an elastic continuum by treating all the dependent variables as independent of each other. The stationary conditions of the principle are the strain–displacement equations, stress–strain equations, stress–equilibrium equations, and both natural and essential boundary conditions – in short, all of the governing equations of elasticity (see Oden and Reddy [18]). Here we consider the principle for an elastic body undergoing large displacements (static case).

Recall the total potential energy functional Π from Eq. (4.3.12):

$$\Pi(\mathbf{u}) = \int_\Omega \left[\frac{L_1}{4} \left(u_{i,j} + u_{j,i}\right)\left(u_{i,j} + u_{j,i}\right) + \frac{L_2}{2} u_{i,i} u_{k,k} - f_i u_i \right] dV$$
$$- \int_{\Gamma_2} \hat{t}_i u_i \, dS, \tag{9.2.12}$$

where L_1 and L_2 denote the Lamé constants.[1] In writing the total potential energy functional Π, it was assumed that the displacements and strains are related by a kinematic relationship in Eq. (9.2.3) and that the displacement field satisfies the specified boundary conditions on Γ_1. The principle of minimum total potential energy gives the equilibrium equations and traction boundary conditions as the Euler equations:

$$L_1\left(u_{i,jj} + u_{j,ij}\right) + L_2 u_{k,ki} + f_i = 0 \text{ in } \Omega \tag{9.2.13}$$
$$t_i - \hat{t}_i = 0, \text{ where } t_i \equiv \left[L_1\left(u_{i,j} + u_{j,i}\right) + L_2 u_{k,k}\delta_{ij}\right]n_j \text{ on } \Gamma_2. \tag{9.2.14}$$

Here n_i denote the components of unit normal vector on the boundary Γ of the region Ω occupied by the elastic body.

Now suppose that we wish to include the strain–displacement relations

$$\varepsilon_{ij} = \tfrac{1}{2}\left(u_{i,j} + u_{j,i} + u_{m,i} u_{m,j}\right) \tag{9.2.15}$$

and displacement boundary conditions

$$u_i = \hat{u}_i \quad (i = 1, 2, 3) \quad \text{on } \Gamma_1 \tag{9.2.16}$$

---

[1]The Lamé constants are often denoted as $L_1 = \mu$ and $L_2 = \lambda$.

into the variational statement by treating them as constraints. Toward this end, first we rewrite the total potential energy functional Π of Eq. (9.2.12) in terms of the nonlinear strains

$$\Pi(u_i, \varepsilon_{ij}) = \int_{\Omega} (U_0(\varepsilon_{ij}) - f_i u_i)\, dV - \int_{\Gamma_2} \hat{t}_i u_i\, dS, \qquad (9.2.17)$$

where U_0 is the strain energy density function. Next, let λ_{ij} $(i, j = 1, 2, 3)$ and $\mu_i (i = 1, 2, 3)$ be the Lagrange multipliers associated with the strain–displacement relations in Eq. (9.2.15) and the displacement boundary conditions in Eq. (9.2.16), respectively. Using the Lagrange multiplier method, we introduce the new functional (Hu–Washizu functional)

$$\Pi_{HW}(u_i, \varepsilon_{ij}, \lambda_{ij}, \mu_i)$$

$$= \Pi(u_i, \varepsilon_{ij}) - \int_{\Gamma_1} \mu_i (u_i - \hat{u}_i)\, dS$$

$$+ \int_{\Omega} \lambda_{ij} \left[\tfrac{1}{2}(u_{i,j} + u_{j,i} + u_{m,i} u_{m,j}) - \varepsilon_{ij} \right] dV$$

$$= \int_{\Omega} \left\{ \left[\tfrac{1}{2}(u_{i,j} + u_{j,i} + u_{m,i} u_{m,j}) - \varepsilon_{ij} \right] \lambda_{ij} + U_0(\varepsilon_{ij}) - f_i u_i \right\} dV$$

$$- \int_{\Gamma_1} \mu_i (u_i - \hat{u}_i)\, ds - \int_{\Gamma_2} \hat{t}_i u_i\, dS \qquad (9.2.18)$$

where $\Gamma_1 + \Gamma_2 = \Gamma$ is the total boundary and $\hat{u}_i$ and $\hat{t}_i$ are the specified displacement and traction components, respectively, on Γ_1 and Γ_2.

We now wish to check if the Euler equations of the new functional Π_{HW} are indeed the governing equations of elasticity. Setting the first variation of Π_{HW} to zero

$$0 = \delta\Pi_{HW} \equiv \delta_{\mathbf{u}}\Pi_{HW} + \delta_{\boldsymbol{\varepsilon}}\Pi_{HW} + \delta_{\boldsymbol{\lambda}}\Pi_{HW} + \delta_{\boldsymbol{\mu}}\Pi_{HW} \qquad (9.2.19)$$

and noting that the variations in $\mathbf{u}$, $\boldsymbol{\varepsilon}$, $\boldsymbol{\lambda}$, and $\boldsymbol{\mu}$ are arbitrary, it follows that

$$\delta_{\mathbf{u}}\Pi_{HW} = 0, \;\; \delta_{\boldsymbol{\varepsilon}}\Pi_{HW} = 0, \;\; \delta_{\boldsymbol{\lambda}}\Pi_{HW} = 0, \;\; \delta_{\boldsymbol{\mu}}\Pi_{HW} = 0. \qquad (9.2.20)$$

Carrying out the variations with respect to the dependent variables, we obtain

$$\delta\Pi_{HW} = \int_{\Omega} \left[\tfrac{1}{2}(\delta u_{i,j} + \delta u_{j,i} + \delta u_{m,i} u_{m,j} + u_{m,i}\delta u_{m,j})\lambda_{ij} - f_i \delta u_i \right] dV$$

$$- \int_{\Gamma_1} \mu_i \delta u_i\, dS - \int_{\Gamma_2} \hat{t}_i \delta u_i\, dS + \int_{\Omega} \left(-\delta\varepsilon_{ij}\lambda_{ij} + \frac{\partial U_0}{\partial \varepsilon_{ij}} \delta\varepsilon_{ij} \right) dV$$

$$+ \int_{\Omega} \left[\tfrac{1}{2}(u_{i,j} + u_{j,i} + u_{m,i} u_{m,j}) - \varepsilon_{ij} \right] \delta\lambda_{ij}\, dV - \int_{\Gamma_1} \delta\mu_i (u_i - \hat{u}_i)\, dS$$

$$= \int_{\Omega} \left[\delta u_{i,j}(\delta_{im} + u_{i,m})\lambda_{mj} + \left(\frac{\partial U_0}{\partial \varepsilon_{ij}} - \lambda_{ij} \right) \delta \varepsilon_{ij} - f_i \delta u_i \right] dV$$

$$+ \int_{\Omega} \left[\tfrac{1}{2}(u_{i,j} + u_{j,i} + u_{m,i}u_{m,j}) - \varepsilon_{ij} \right] \delta \lambda_{ij} \, dV$$

$$- \int_{\Gamma_1} \mu_i \delta u_i \, dS - \int_{\Gamma_2} \hat{t}_i \delta u_i \, dS - \int_{\Gamma_1} \delta \mu_i (u_i - \hat{u}_i) dS. \tag{9.2.21}$$

Using the integration by parts, the first term in Eq. (9.2.21) can be expressed as

$$\int_{\Omega} \delta u_{i,j}(\delta_{im} + u_{i,m})\lambda_{mj} \, dV = - \int_{\Omega} \lambda_{mj} \left(\delta_{im} + u_{i,m} \right)_{,j} \delta u_i \, dV$$

$$+ \oint_{\Gamma} n_j \lambda_{mj}(\delta_{im} + u_{i,m})\delta u_i \, dS. \tag{9.2.22}$$

Substituting Eq. (9.2.22) into Eq. (9.2.21) and setting the coefficients of δu_i, $\delta \varepsilon_{ij}$, and $\delta \lambda_{ij}$ in Ω, δu_i on Γ_1 and Γ_2, and $\delta \mu_i$ on Γ_1 to zero, we obtain

$$\delta u_i : \qquad \left[\lambda_{mj}(\delta_{im} + u_{i,m}) \right]_{,j} + f_i = 0 \qquad \text{in } \Omega, \tag{9.2.23}$$

$$\delta \varepsilon_{ij} : \qquad \frac{\partial U_0}{\partial \varepsilon_{ij}} - \lambda_{ij} = 0 \qquad \text{in } \Omega, \tag{9.2.24}$$

$$\delta \lambda_{ij} : \qquad \frac{1}{2}\left(u_{i,j} + u_{j,i} + u_{m,i}u_{m,j}\right) - \varepsilon_{ij} = 0 \qquad \text{in } \Omega, \tag{9.2.25}$$

$$\delta u_i : \qquad n_j \lambda_{mj}\left(\delta_{im} + u_{i,m}\right) - \mu_i = 0 \qquad \text{on } \Gamma_1, \tag{9.2.26}$$

$$\delta u_i : \qquad n_j \lambda_{mj}\left(\delta_{im} + u_{i,m}\right) - \hat{t}_i = 0 \qquad \text{on } \Gamma_2, \tag{9.2.27}$$

$$\delta \mu_i : \qquad u_i - \hat{u}_i = 0 \qquad \text{on } \Gamma_1. \tag{9.2.28}$$

Equations (9.2.24) and (9.2.26) can be interpreted as the definitions of the Lagrange multipliers λ_{ij} and μ_i, respectively. Clearly, they have the meaning of stress and traction components, respectively. Thus, the functional Π_{HW} gives the equilibrium equations in Eq. (9.2.23), constitutive equations in Eq. (9.2.24), strain–displacement relations in Eq. (9.2.25), traction boundary conditions in Eq. (9.2.27), and displacement boundary conditions in Eq. (9.2.28) as the Euler equations. Consequently, finding the stationary values of Π_{HW} is stated as a variational principle, and it is termed *the Hellinger–Reissner variational principle* for an elastic body.

Example 9.2.1

Consider the following equations of the Bernoulli–Euler theory of beams:

Kinematics: $\qquad\qquad \kappa = -\frac{d^2 w}{dx^2}, \qquad w(0) = w_0, \tag{1}$

Constitutive equation: $\qquad M = EI\kappa, \qquad -\frac{dw}{dx}(0) = \theta_0, \tag{2}$

Kinetics (equilibrium): $\quad -\frac{d^2 M}{dx^2} = q, \quad M(L) = M_L, \quad \frac{dM}{dx}(L) = V_L, \tag{3}$

where w_0 and θ_0 denote specified displacement and rotation, respectively, at $x = 0$. Construct the Hellinger–Reissner variational statement that yields Eqs (1) to (3) as the Euler equations.

Solution: The Hellinger–Reissner-type variational principle for the Bernoulli–Euler beams is constructed by beginning with the statement in which each equation is multiplied by an independent variable (λ, μ_1, μ_2):

$$\Pi_{HW}(w, \kappa, \lambda, \mu_1, \mu_2) = \Pi(\kappa) + \int_0^L \left(\frac{d^2 w}{dx^2} + \kappa\right) \lambda \, dx + \mu_1 [w(0) - w_0]$$
$$+ \mu_2 \left[\left(\frac{dw}{dx}\right)\bigg|_0 + \theta_0\right], \tag{4}$$

where $\Pi(\kappa)$ is the total potential energy functional expressed in terms of the curvature κ. The functional Π accounts for the moment and shear force boundary conditions because they are the natural boundary conditions for Π. We have

$$\Pi_{HW}(w, \kappa, \lambda, \mu_1, \mu_2) = \int_0^L \left[\frac{EI}{2}\kappa^2 + \left(\frac{d^2 w}{dx^2} + \kappa\right)\lambda - qw\right] dx - M_L\left(-\frac{dw}{dx}\right)\bigg|_L$$
$$- V_L w(L) + \mu_1 [w(0) - w_0] + \mu_2 \left[\left(\frac{dw}{dx}\right)\bigg|_0 + \theta_0\right]. \tag{5}$$

The Euler equations of functional Π_{HW} are

$$\delta\kappa: \quad EI\kappa + \lambda = 0 \tag{6}$$

$$\delta w: \quad \frac{d^2\lambda}{dx^2} - q = 0, \quad \text{in } 0 < x < L \tag{7}$$

$$\delta\lambda: \quad \frac{d^2 w}{dx^2} + \kappa = 0 \tag{8}$$

$$\frac{d\delta w}{dx}: \quad -\lambda(0) + \mu_2 = 0, \quad \lambda(L) + M_L = 0 \tag{9}$$

$$\delta w: \quad -\left(\frac{d\lambda}{dx}\right)\bigg|_0 + \mu_1 = 0, \quad \left(\frac{d\lambda}{dx}\right)\bigg|_L + V_L = 0, \tag{10}$$

$$\delta\mu_1: \quad w(0) - w_0 = 0, \tag{11}$$

$$\delta\mu_2: \quad \frac{dw}{dx}(0) + \theta_0 = 0. \tag{12}$$

From a comparison of Eq. (2) with Eq. (6) and, subsequently, from Eqs (9) and (10), we note that the Lagrange multipliers λ, μ_1, and μ_2 have the following meaning:

$$\lambda(x) = -M(x), \quad \mu_1 = -\left(\frac{dM}{dx}\right)\bigg|_0, \quad \mu_2 = -M(0). \tag{13}$$

Example 9.2.2

Develop a mixed formulation of the Timoshenko beam theory (TBT) by treating the displacements (w, ϕ_x) and strains $(\kappa_{xx}, \gamma_{xz})$ as the independent dependent variables.

Solution: The variational statement associated with this mixed formulation is given by the stationarity of the following functional (see Oden and Reddy [18], p. 116):

$$\Pi_m = \int_0^L \left[EI \left(\frac{d\phi_x}{dx} - \frac{1}{2}\kappa_{xx} \right) \kappa_{xx} + GAK_s \left(\frac{dw}{dx} + \phi_x - \frac{1}{2}\gamma_{xz} \right) \gamma_{xz} - qw \right] dx$$
$$- V_1 w(0) - V_2 w(L) - M_1 \phi_x(0) - M_2 \phi_x(L), \tag{1}$$

where

$$V_1 = [-GAK_s\gamma_{xz}]_0, \quad V_2 = [GAK_s\gamma_{xz}]_L,$$
$$M_1 = [-EI\kappa_{xx}]_0, \quad M_2 = [EI\kappa_{xx}]_L. \tag{2}$$

The Euler-Lagrange equations associated with the functional in Eq. (1) are

$$\delta w: \qquad -\frac{d}{dx}[GAK_s\gamma_{xz}] - q = 0, \tag{3}$$

$$\delta\phi_x: \qquad -\frac{d}{dx}[EI\kappa_{xx}] + GAK_s\gamma_{xz} = 0, \tag{4}$$

$$\delta\kappa_{xx}: \qquad \frac{d\phi_x}{dx} - \kappa_{xx} = 0, \tag{5}$$

$$\delta\gamma_{xz}: \qquad \left(\frac{dw}{dx} + \phi_x \right) - \gamma_{xz} = 0. \tag{6}$$

Note that Eqs (3) and (4) represent the equilibrium equations in terms of the curvature κ_{xx} and shear strain γ_{xz}, while Eqs (5) and (6) are the kinematic relations.

9.2.3 The Reissner Variational Principle

A special case of the Hellinger–Reissner variational principle is the Reissner principle, which is obtained by eliminating strains ε_{ij} by assuming that the strains are related to the stresses $\sigma_{ij} = \lambda_{ij}$ by the constitutive relations

$$\varepsilon_{ij} = \frac{\partial U_0^*}{\partial \sigma_{ij}}. \tag{9.2.29}$$

Then we obtain from Eq. (9.2.29)

$$\Pi_R(u_i, \sigma_{ij}) = \int_\Omega \left[\frac{1}{2}\left(u_{i,j} + u_{j,i} + u_{m,i}u_{m,j} \right)\sigma_{ij} - U_0^*(\sigma_{ij}) - f_i u_i \right] dV$$

$$- \int_{\Gamma_1} t_i(u_i - \hat{u}_i)ds - \int_{\Gamma_2} \hat{t}_i u_i \, dS. \tag{9.2.30}$$

The Euler equations are given by Eqs (9.2.23), (9.2.26)–(9.2.28), and

$$\delta\sigma_{ij}: \quad \frac{1}{2}\left(u_{i,j} + u_{j,i} + u_{m,i}u_{m,j} \right) - \frac{\partial U_0^*}{\partial \sigma_{ij}} = 0 \quad \text{in } \Omega. \tag{9.2.31}$$

9.3 Variational Solutions Based on Mixed Formulations

The advantage of stationary principles lies in incorporating all of the governing equations into a single functional. Such functionals form the basis of a variety of new finite element models that often yield better accuracies for stresses than the displacement finite element models (which are based on the total potential energy functional).

The variational methods presented in Chapter 6 can be used to determine solutions of problems based on mixed formulations. The variational methods require a variational statement to determine the undetermined parameters of the approximation, and the variational statement can be anything that is equivalent to the governing equations. Thus, the Ritz solution, for example, of an Bernoulli–Euler beam can be determined using the weak form (which is equivalent to the principle of minimum potential energy) or a mixed variational principle/statement. The conditions on the approximation functions are dictated by the variational principle. If all governing equations including the boundary conditions are included in the variational statement, then the approximation functions are required to satisfy the continuity requirements. Here we illustrate these ideas through an example.

Example 9.3.1 ——————————————————————————————————

Consider the bending of a beam using the Bernoulli–Euler beam theory. The governing equations are (see **Example 4.3.1**)

$$-\frac{d^2 M}{dx^2} = q \tag{1}$$

$$-\frac{d^2 w}{dx^2} - \frac{M}{EI} = 0. \tag{2}$$

Develop a mixed variational formulation in which both $w(x)$ and $M(x)$ are treated as independent–dependent variables, each approximated independently. That is, solve Eqs (1) and (2) simultaneously (without eliminating M). As a specific example problem, consider a cantilever beam under uniformly distributed load of intensity q_0. Compare the results of the mixed formulation with the Ritz method based on the principle of minimum total potential energy. Use two-parameter approximations.

Solution: In **Example 4.3.1** we have solved the Bernoulli–Euler beam problem using the principle of minimum total potential energy, whose Euler equation is the following fourth-order differential equation governing the deflection w [set the nonlinear terms to zero in Eq. (10) of **Example 4.3.1**]:

$$\frac{d^2}{dx^2}\left(EI\frac{d^2 w}{dx^2}\right) = q, \tag{3}$$

which is obtained by combining Eqs (1) and (2). In the displacement formulation based on the principle of minimum total potential energy, the primary variables were $(w, -dw/dx)$ and the secondary variables were (V, M), where V is the shear force. As we shall see, in a mixed formulation, the primary variables and secondary variables are not the same as those in a formulation based on the principle of minimum total potential energy.

The boundary conditions for the cantilever beam problem are

$$w(0) = 0, \quad \left(\frac{dw}{dx}\right)\bigg|_{x=0} = 0,$$

$$M(L) = 0, \quad V(L) \equiv \left(\frac{dM}{dx}\right)\bigg|_{x=L} = 0. \tag{4}$$

First, we consider the Ritz approximation based on the minimum total potential energy principle (with uniformly distributed load of intensity q_0):

$$0 = \delta\Pi(w) \equiv \int_0^L \left(EI\frac{d^2\delta w}{dx^2}\frac{d^2w}{dx^2} - q_0\delta w\right)dx. \tag{5}$$

The two-parameter approximation of the form $W_2(x) = c_1 x^2 + c_2 x^3$ satisfies only the geometric boundary conditions, $w(0) = 0$ and $(dw/dx)(0) = 0$; we obtain the solution

$$W_2(x) = c_1 x^2 + c_2 x^3 = \frac{q_0 L^4}{24EI}\left(5\frac{x^2}{L^2} - 2\frac{x^3}{L^3}\right), \quad W_2(L) = \frac{q_0 L^4}{8EI}. \tag{6}$$

The bending moment obtained from this Ritz solution is

$$M_2(x) = -EI\frac{d^2W_2}{dx^2} = -\frac{q_0 L^2}{12}\left(5 - 6\frac{x}{L}\right), \quad M_2(0) = -\frac{5q_0 L^2}{12}. \tag{7}$$

The exact solutions for $w(x)$ and $M(x)$ are given by

$$w(x) = \frac{q_0 L^4}{24EI}\left(6\frac{x^2}{L^2} - 4\frac{x^3}{L^3} + \frac{x^4}{L^4}\right), \quad w(L) = \frac{q_0 L^4}{8EI},$$

$$M(x) = -\frac{q_0 L^2}{2}\left(1 - \frac{x}{L}\right)^2, \quad M(0) = -\frac{q_0 L^2}{2}. \tag{8}$$

Thus the two-parameter Ritz solution based on the total potential energy principle (or displacement model) yields the exact maximum deflection, but the maximum bending moment is in 16.67% error.

Next, we consider the pair in Eqs (1) and (2) and construct their weak forms. Using the three-step procedure, after two steps we obtain the result

$$0 = \int_0^L \left(\frac{dv_1}{dx}\frac{dM}{dx} - v_1 q\right)dx - \left[v_1 \cdot \frac{dM}{dx}\right]_0^L, \tag{9}$$

$$0 = \int_0^L \left(\frac{dv_2}{dx}\frac{dw}{dx} - v_2\frac{M}{EI}\right)dx - \left[v_2 \cdot \frac{dw}{dx}\right]_0^L, \tag{10}$$

where (v_1, v_2) are the weight functions. An examination of the boundary terms shows that the secondary variables are

$$\frac{dM}{dx}, \quad \frac{dw}{dx}, \tag{11}$$

which have the meaning of shear force and rotation (or slope), respectively. Thus, the primary variable, dw/dx, of the conventional formulation became the secondary variable of the mixed formulation. Obviously, w and M are the primary variables of the mixed formulation. Thus, the nature of the boundary conditions (i.e., classification into essential and natural) depends on the equations set used.

To physically interpret the weight functions, we examine the expressions in the integrands. The most obvious one to interpret is the product $v_1 q_0 dx$. In order for it to have the units of "work done," v_1 should have the meaning of the displacement, w; that is, $v_1 \sim \delta w$. Since $(M/EI)dx$ has the units of rotation, v_2 must have the units of a moment, giving $v_2 \sim \delta M$. Thus, w and M are the primary variables. This information is needed in developing the weak forms (and later in applying the Ritz method, where we will replace v_1 and v_2 with the proper approximation functions). It is clear that v_1 must be zero at boundary points where w is specified, and v_2 should be zero where M is specified.

Returning to the weak form development, we use the specified boundary conditions in Eq. (4) to simplify the boundary terms in Eqs (9) and (10). In fact, $(dw/dx)(0) = 0$, $(dM/dx)(L) = 0$, $v_1(0) = 0$, and $v_2(L) = 0$ make all boundary terms vanish, and we obtain the final weak forms:

$$0 = \int_0^L \left(\frac{dv_1}{dx} \frac{dM}{dx} - v_1 q \right) dx, \tag{12}$$

$$0 = \int_0^L \left(\frac{dv_2}{dx} \frac{dw}{dx} - v_2 \frac{M}{EI} \right) dx. \tag{13}$$

Let us consider the two-parameter Ritz approximations of w and M:

$$\begin{aligned} W_2(x) &= a_1 \phi_1(x) + a_2 \phi_2(x) = a_1 x + a_2 x^2, \\ M_2(x) &= b_1 \psi_1(x) + b_2 \psi_2(x) = b_1(L - x) + b_2(L - x)^2, \end{aligned} \tag{14}$$

which satisfy the specified essential boundary conditions, $w(0) = 0$ and $M(L) = 0$. In general, there is a relationship between the number of parameters chosen for $w(x)$ and $M(x)$ to ensure the invertibility of the resulting algebraic equations for the unknown parameters (a_i, b_i).

Substituting approximations from Eq. (14) for w and M, and $v_1 = \phi_i$ and $v_2 = \psi_i$ into Eqs (12) and (13), we obtain

$$0 = \int_0^L \left[\frac{d\phi_i}{dx} \left(\sum_{j=1}^{2} b_j \frac{d\psi_j}{dx} \right) - \phi_i q \right] dx = 0, \quad i = 1, 2, \tag{15}$$

$$0 = \int_0^L \left[\frac{d\psi_i}{dx} \left(\sum_{j=1}^{2} a_j \frac{d\phi_j}{dx} \right) - \frac{1}{EI} \psi_i \left(\sum_{j=1}^{2} b_j \psi_j \right) \right] dx = 0, \quad i = 1, 2, \tag{16}$$

or

$$\sum_{j=1}^{2} K_{ij} b_j = f_i, \quad \sum_{j=1}^{2} K_{ji} a_j - \sum_{j=1}^{2} G_{ij} b_j = 0. \tag{17}$$

In matrix form, we have

$$\begin{bmatrix} \mathbf{0} & \mathbf{K} \\ (\mathbf{K})^{\mathrm{T}} & -\mathbf{G} \end{bmatrix} \begin{Bmatrix} \mathbf{a} \\ \mathbf{b} \end{Bmatrix} = \begin{Bmatrix} \mathbf{f} \\ \mathbf{0} \end{Bmatrix}, \tag{18}$$

where

$$K_{ij} = \int_0^L \frac{d\phi_i}{dx} \frac{d\psi_j}{dx} dx, \quad G_{ij} = \frac{1}{EI} \int_0^L \psi_i \psi_j dx, \quad f_i = \int_0^L q \phi_i \, dx. \tag{19}$$

Evaluating the integrals for the choice of ϕ_i and ψ_i from Eq. (14), we obtain (for $q = q_0$)

$$\mathbf{K} = - \begin{bmatrix} L & L^2 \\ L^2 & \frac{2}{3}L^3 \end{bmatrix}, \quad \mathbf{G} = \frac{1}{EI} \begin{bmatrix} \frac{L^3}{3} & \frac{L^4}{4} \\ \frac{L^4}{4} & \frac{L^5}{5} \end{bmatrix}, \quad \mathbf{f} = \frac{q_0 L^2}{6} \begin{Bmatrix} 3 \\ 2L \end{Bmatrix}.$$

Solving the first equation in Eq. (17) for b's yields $b_1 = 0$ and $b_2 = -q_0/2$. Substituting for b's in the second equation of (17) or (18), and solving for a's, we obtain

$$a_1 = \frac{1}{20} \frac{q_0 L^3}{EI}, \quad a_2 = \frac{3}{40} \frac{q_0 L^2}{EI}.$$

The two-parameter Ritz solutions based on the mixed formulation are

$$W_2(x) = \frac{q_0 L^4}{40 EI}\left(2\frac{x}{L} + 3\frac{x^2}{L^2}\right), \quad W_2(L) = \frac{q_0 L^4}{8EI},$$

$$M_2(x) = -\frac{q_0 L^2}{2}\left(1 - \frac{x}{L}\right)^2, \qquad M_2(0) = -\frac{q_0 L^2}{2}. \tag{20}$$

We note that the maximum values of both deflection and moment coincide with the exact values and that the expression for the bending moment coincides with the exact solution for all values of x. Although the maximum deflection predicted by the mixed Ritz approximation coincides with the exact solution, the deflection obtained violates the slope boundary condition at $x = 0$ (which is a natural boundary condition in the present case). Thus, the mixed variational formulation gives more accurate solutions for the force variable because they are approximated independently; however, the accuracy of the displacements is somewhat degraded. It can be improved by increasing the number of parameters (note that the polynomial degree used for w is only 2 in the mixed formulation, while it is 3 in the displacement formulation).

Table 9.3.1 contains a comparison of point-wise deflections, slope, bending moment, and shear force obtained by the two approaches with the exact solution. The variational solutions are in good agrement with the exact solutions.

Table 9.3.1 Comparison of the scaled[a] generalized displacements and forces computed by the two-parameter Ritz approximations based on the total potential energy formulation (PEF) and mixed variational formulation (MVF).

	Deflection, $\bar{w}(x)$			Bending moment, $\bar{M}(x)$		
x	Exact	PEF	MVF	Exact	PEF	MVF
0.0	0.0000	0.0000	0.0000	−0.500	−0.4167	−0.500
0.1	0.0023	0.0020	0.0057	−0.405	−0.3667	−0.405
0.2	0.0087	0.0077	0.0130	−0.320	−0.3167	−0.320
0.3	0.0183	0.0165	0.0218	−0.245	−0.2667	−0.245
0.4	0.0304	0.0280	0.0320	−0.180	−0.2167	−0.180
0.5	0.0443	0.0417	0.0438	−0.125	−0.1667	−0.125
0.6	0.0594	0.0570	0.0570	−0.080	−0.1167	−0.080
0.7	0.0753	0.0735	0.0717	−0.045	−0.0667	−0.045
0.8	0.0917	0.0907	0.0880	−0.020	−0.0167	−0.020
0.9	0.1083	0.1080	0.1057	−0.005	−0.0333	−0.005
1.0	0.1250	0.1250	0.1250	−0.000	−0.0833	−0.000

	Slope, $\frac{d\bar{w}}{dx}$			Shear force, $\bar{V}(x)$		
x	Exact	PEF	MVF	Exact	PEF	MVF
0.0	0.0000	0.0000	−0.050	1.0	0.5	1.0
0.1	0.0452	0.0392	−0.065	0.9	0.5	0.9
0.2	0.0813	0.0733	−0.080	0.8	0.5	0.8
0.3	0.1095	0.1025	−0.095	0.7	0.5	0.7
0.4	0.1307	0.1267	−0.110	0.6	0.5	0.6
0.5	0.1458	0.1458	−0.125	0.5	0.5	0.5
0.6	0.1560	0.1600	−0.140	0.4	0.5	0.4
0.7	0.1622	0.1692	−0.155	0.3	0.5	0.3
0.8	0.1653	0.1733	−0.170	0.2	0.5	0.2
0.9	0.1665	0.1725	−0.185	0.1	0.5	0.1
1.0	0.1667	0.1667	−0.200	0.0	0.5	0.0

[a] $\bar{w} = w(L)\frac{EI}{q_0 L^4}$; $\frac{d\bar{w}}{dx} = \left(\frac{dw}{dx}\right)_L \frac{EI}{q_0 L^3}$; $\bar{M} = M(0)\frac{1}{q_0 L^2}$; $\bar{V} = \left(\frac{dM}{dx}\right)_0 \frac{1}{q_0 L}$.

9.4   Mixed Finite Element Models of Beams

9.4.1   The Bernoulli–Euler Beam Theory

9.4.1.1   Governing Equations and Weak Forms

Here, we develop a mixed finite element model of the equations governing the Bernoulli–Euler beam theory (BET). Consider the following equations in terms of the transverse deflection w and bending moment M of the BET [same as Eqs (1) and (2) of **Example 9.3.1**]:

$$-\frac{d^2 M}{dx^2} = q, \qquad -\frac{d^2 w}{dx^2} - \frac{M}{EI} = 0. \tag{9.4.1}$$

We can construct a mixed finite element model based on the equations using either (a) the Ritz method (which is the same as the weak-form Galerkin method) or (b) a weighted-residual method applied to the two equations over an element $\Omega_e = (x_a^e, x_b^e)$.

First, we develop the weak forms for the Ritz (or weak-form Galerkin) finite element model over a typical element Ω_e. We have [see Eqs (9) and (10) of **Example 9.3.1**]

$$0 = \int_{x_a^e}^{x_b^e} \left(\frac{dv_1}{dx}\frac{dM}{dx} - v_1 q \right) dx - v_1(x_a^e)\bar{Q}_1^e - v_1(x_b^e)\bar{Q}_2^e, \tag{9.4.2}$$

$$0 = \int_{x_a^e}^{x_b^e} \left(\frac{dv_2}{dx}\frac{dw}{dx} - v_2\frac{M}{E_e I_e} \right) dx - v_2(x_a^e)\Theta_1^e - v_2(x_b^e)\Theta_2^e, \tag{9.4.3}$$

where (v_1, v_2) are the weight functions that have the interpretation of (see **Example 9.3.1**) virtual deflection δw and virtual moment δM, respectively, and

$$\bar{Q}_1^e = -\left(\frac{dM_e}{dx}\right)_{x=x_a^e}, \qquad \bar{Q}_2^e = \left(\frac{dM_e}{dx}\right)_{x=x_b^e},$$

$$\bar{\Theta}_1^e = -\left(\frac{dw_e}{dx}\right)_{x=x_a^e}, \qquad \bar{\Theta}_2^e = \left(\frac{dw_e}{dx}\right)_{x=x_b^e}. \tag{9.4.4}$$

9.4.1.2   Weak-Form Mixed Finite Element Model

The weak forms in Eqs (9.4.2) and (9.4.3) suggest that both w and M may be approximated using Lagrange interpolation functions. Suppose that w_e and M_e over an element Ω_e are approximated as

$$w_e(x) \approx \sum_{j=1}^{m} w_j^e \phi_j^e(x), \qquad M_e(x) \approx \sum_{j=1}^{n} M_j^e \psi_j^e(x), \tag{9.4.5}$$

where (ϕ_i^e, ψ_i^e) are the Lagrange interpolation functions of different degree used for w_e and M_e, respectively. Substituting Eq. (9.4.5) for w_e and M_e, and $v_1 = \phi_i^e$ and $v_2 = \psi_i^e$ into Eqs (9.4.2) and (9.4.3), we obtain

$$0 = \int_{x_a^e}^{x_b^e} \left[\frac{d\phi_i^e}{dx} \left(\sum_{j=1}^{n} M_j^e \frac{d\psi_j^e}{dx} \right) - \phi_i^e q \right] dx - \bar{Q}_i^e, \quad (i = 1, 2, \ldots, m), \tag{9.4.6}$$

$$0 = \int_{x_a^e}^{x_b^e} \left[\frac{d\psi_i^e}{dx} \left(\sum_{j=1}^{m} w_j \frac{d\phi_j^e}{dx} \right) - \frac{1}{E_e I_e} \psi_i^e \left(\sum_{j=1}^{n} M_j^e \psi_j \right) \right] dx - \bar{\Theta}_i^e, \tag{9.4.7}$$

$$(i = 1, 2, \ldots, n).$$

or

$$\sum_{j=1}^{m} K_{ij}^e M_j^e = f_i^e + \bar{Q}_i^e, \quad \sum_{j=1}^{m} K_{ji}^e w_j^e - \sum_{j=1}^{n} G_{ij}^e M_j^e = \bar{\Theta}_i^e. \tag{9.4.8}$$

In matrix form, we have

$$\begin{bmatrix} \mathbf{0} & \mathbf{K}^e \\ (\mathbf{K}^e)^{\mathrm{T}} & -\mathbf{G}^e \end{bmatrix} \left\{ \begin{array}{c} \mathbf{w}^e \\ \mathbf{M}^e \end{array} \right\} = \left\{ \begin{array}{c} \mathbf{f}^e \\ \mathbf{0} \end{array} \right\} + \left\{ \begin{array}{c} \bar{\mathbf{Q}}^e \\ \bar{\Theta}^e \end{array} \right\}, \tag{9.4.9}$$

where

$$K_{ij}^e = \int_{x_a^e}^{x_b^e} \frac{d\phi_i^e}{dx} \frac{d\psi_j^e}{dx} dx, \quad G_{ij}^e = \frac{1}{E_e I_e} \int_{x_a^e}^{x_b^e} \psi_i^e \psi_j^e dx, \quad f_i^e = \int_{x_a^e}^{x_b^e} q \phi_i^e \, dx. \tag{9.4.10}$$

The matrix equations in Eq. (9.4.9) in terms of the nodal values of the displacements $\mathbf{w}^e$ and moments $\mathbf{M}^e$ may be rearranged as matrix equations in terms of $\mathbf{w}^e$ and $\bar{\Theta}^e$ (i.e., move $\bar{\Theta}_e$ to the left side and $\mathbf{M}^e$ to the right side of the equality). Toward this end, solving the second equation in Eq. (9.4.9) for $\mathbf{M}^e$, we obtain

$$-\mathbf{M}^e = -(\mathbf{G}^e)^{-1}(\mathbf{K}^e)^{\mathrm{T}} \mathbf{w}^e + (\mathbf{G}^e)^{-1} \bar{\Theta}^e. \tag{9.4.11}$$

Substituting for $\mathbf{M}^e$ from Eq. (9.4.11) into the first equation in Eq. (9.4.9), we obtain

$$\mathbf{K}^e (\mathbf{G}^e)^{-1} (\mathbf{K}^e)^{\mathrm{T}} \mathbf{w}^e - \mathbf{K}^e (\mathbf{G}^e)^{-1} \bar{\Theta}^e = \mathbf{f}^e + \bar{\mathbf{Q}}^e. \tag{9.4.12}$$

Now we can write Eqs (9.4.12) and (9.4.11) in a single matrix equation as

$$\begin{bmatrix} \mathbf{K}^e (\mathbf{G}^e)^{-1} (\mathbf{K}^e)^{\mathrm{T}} & -\mathbf{K}^e (\mathbf{G}^e)^{-1} \\ -(\mathbf{G}^e)^{-1} (\mathbf{K}^e)^{\mathrm{T}} & (\mathbf{G}^e)^{-1} \end{bmatrix} \left\{ \begin{array}{c} \mathbf{w}^e \\ \bar{\Theta}^e \end{array} \right\} = \left\{ \begin{array}{c} \mathbf{f}^e + \bar{\mathbf{Q}}^e \\ -\mathbf{M}^e \end{array} \right\}, \tag{9.4.13}$$

where $\bar{\Theta}_1^e = \Theta_1^e$ and $\bar{\Theta}_2^e = -\Theta_2^e$. This set of finite element equations is now in the same form as the conventional finite element model in the sense that the nodal variables of the conventional and mixed formulations that multiply the coefficient matrix are the same. However, in general, the coefficient matrices (i.e., the stiffness matrices) may not be the same in the two formulations.

For the choice of linear interpolation of both w_e and M_e, various entries of the stiffness matrix in Eq. (9.4.13) can be computed as

$$\mathbf{K}^e = \frac{1}{h_e}\begin{bmatrix} 1 & -1 \\ -1 & 1 \end{bmatrix}, \quad \mathbf{G}^e = \frac{h_e}{6E_eI_e}\begin{bmatrix} 2 & 1 \\ 1 & 2 \end{bmatrix},$$

$$\mathbf{f}^e = \begin{Bmatrix} f_1^e \\ f_2^e \end{Bmatrix}, \quad (\mathbf{G}^e)^{-1} = \frac{2E_eI_e}{h_e}\begin{bmatrix} 2 & -1 \\ -1 & 2 \end{bmatrix},$$

$$(\mathbf{G}^e)^{-1}(\mathbf{K}^e)^{\mathrm{T}} = \frac{6E_eI_e}{h_e^2}\begin{bmatrix} 1 & -1 \\ -1 & 1 \end{bmatrix}, \tag{9.4.14}$$

$$\mathbf{K}^e(\mathbf{G}^e)^{-1}(\mathbf{K}^e)^{\mathrm{T}} = \frac{12E_eI_e}{h_e^3}\begin{bmatrix} 1 & -1 \\ -1 & 1 \end{bmatrix},$$

$$(\mathbf{G}^e)^{-1}(\mathbf{K}^e)^{\mathrm{T}} = \left(\mathbf{K}^e(\mathbf{G}^e)^{-1}\right)^{\mathrm{T}}.$$

Hence, Eq. (9.4.13) becomes

$$\frac{2E_eI_e}{h_e^3}\begin{bmatrix} 6 & -6 & -3h_e & 3h_e \\ -6 & 6 & 3h_e & -3h_e \\ -3h_e & 3h_e & 2h_e^2 & -h_e^2 \\ 3h_e & -3h_e & -h_e^2 & 2h_e^2 \end{bmatrix}\begin{Bmatrix} w_1^e \\ w_2 \\ \bar{\Theta}_1^e \\ \bar{\Theta}_2^e \end{Bmatrix} = \begin{Bmatrix} f_1^e \\ f_2^e \\ 0 \\ 0 \end{Bmatrix} + \begin{Bmatrix} \bar{Q}_1^e \\ \bar{Q}_2^e \\ -M_1^e \\ -M_2^e \end{Bmatrix}. \tag{9.4.15}$$

After rearranging the coefficients according to the generalized nodal displacements Δ_i^e and generalized nodal forces Q_i^e ($i = 1, 2, 3, 4$)

$$\Delta_1^e = w_1^e, \quad \Delta_3^e = w_2^e, \quad \Delta_2^e = \bar{\Theta}_1^e, \quad \Delta_4^e = -\bar{\Theta}_2^e,$$

$$Q_1^e = \bar{Q}_1^e, \quad Q_3^e = \bar{Q}_2^e, \quad Q_2^e = -M_1^e, \quad Q_4^e = M_2^e, \tag{9.4.16}$$

Eq. (9.4.15) becomes

$$\frac{2E_eI_e}{h_e^3}\begin{bmatrix} 6 & -3h_e & -6 & -3h_e \\ -3h_e & 2h_e^2 & 3h_e & h_e^2 \\ -6 & 3h_e & 6 & 3h_e \\ -3h_e & h_e^2 & 3h_e & 2h_e^2 \end{bmatrix}\begin{Bmatrix} \Delta_1^e \\ \Delta_2^e \\ \Delta_3^e \\ \Delta_4^e \end{Bmatrix} = \begin{Bmatrix} f_1^e \\ 0 \\ f_2^e \\ 0 \end{Bmatrix} + \begin{Bmatrix} Q_1^e \\ Q_2^e \\ Q_3^e \\ Q_4^e \end{Bmatrix}. \tag{9.4.17}$$

We note that the stiffness matrix of the mixed finite element model in Eq. (9.4.17) with the linear interpolation of both w_e and M_e is the same as that of the displacement finite element model derived in Section 8.3.4 using the C^1 (Hermite cubic) interpolation. However, the load vector $\mathbf{q}^e$ differs in the sense that the mixed model does not contain contributions of distributed load $q(x)$ to the nodal moments. Hence the two models will yield the same solution when the distributed load is zero, and the solutions will be different when $q(x) \neq 0$.

9.4.1.3 Weighted-Residual Finite Element Models

It is possible to construct a mixed finite element model of the pair in Eq. (9.4.1) using a weighted-residual formulation. As will be seen shortly, this model requires higher-order approximations of both w and M because they must satisfy both essential and natural boundary conditions. This leads to a complicated set of finite element equations that are not practical. Here we present the main ideas behind the development of the model.

Galerkin finite element model. The weighted-residual statement of the pair of equations in Eq. (9.4.1) is

$$0 = - \int_{x_a^e}^{x_b^e} v_1 \left(\frac{d^2 M_e}{dx^2} + q \right) dx, \tag{9.4.18}$$

$$0 = - \int_{x_a^e}^{x_b^e} v_2 \left(\frac{d^2 w_e}{dx^2} + \frac{M_e}{E_e I_e} \right) dx, \tag{9.4.19}$$

where (v_1, v_2) are the weight functions. A close examination of the statements in Eqs (9.4.18) and (9.4.19) indicates that $v_1 \sim \delta w_e$ and $v_2 \sim \delta M_e$. Since second derivatives of both w_e and M_e appear in the integral statements, we are required to use a minimum of quadratic approximation of the two fields. However, to be able to physically impose continuity of dw_e/dx (slope) and balance of dM_e/dx (shear force), we must use Hermite cubic approximation of w_e and M_e.

Using the Hermite cubic approximations of the form (even though different notation is used, in reality $\varphi_i^{(1)} = \varphi_i^{(2)}$)

$$w_e(x) \approx \sum_{i=1}^{4} \Delta_i^e \varphi_i^{(1)}(x), \quad M_e(x) \approx \sum_{i=1}^{4} \Lambda_i^e \varphi_i^{(2)}(x), \tag{9.4.20}$$

we obtain the following Galerkin (i.e., $v_1 \sim \varphi_i^{(1)}$ and $v_2 \sim \varphi_i^{(2)}$) finite element model:

$$\begin{bmatrix} \mathbf{0} & \mathbf{A}^e \\ \mathbf{B}^e & \mathbf{D}^e \end{bmatrix} \begin{Bmatrix} \mathbf{\Delta}^e \\ \mathbf{\Lambda}^e \end{Bmatrix} = \begin{Bmatrix} \mathbf{f}^e \\ \mathbf{0} \end{Bmatrix}, \tag{9.4.21}$$

where

$$A_{ij}^e = \int_{x_a^e}^{x_b^e} \varphi_i^{(1)} \frac{d^2 \varphi_j^{(2)}}{dx^2} dx, \quad f_i^e = - \int_{x_a}^{x_b} q \varphi_i^{(1)} \, dx,$$

$$B_{ij}^e = \int_{x_a^e}^{x_b^e} \varphi_i^{(2)} \frac{d^2 \varphi_j^{(1)}}{dx^2} \, dx, \quad D_{ij}^e = \int_{x_a^e}^{x_b^e} \frac{1}{E_e I_e} \varphi_i^{(2)} \varphi_j^{(2)} \, dx. \tag{9.4.22}$$

The coefficient matrix in Eq. (9.4.21) is *not* symmetric.

Least-squares finite element model. The least-squares finite element model of the pair in Eq. (9.4.1) is based on the variational statement

$$
0 = \delta \int_{x_a^e}^{x_b^e} \left[\left(\frac{d^2 M_e}{dx^2} + q \right)^2 + \gamma \left(\frac{d^2 w_e}{dx^2} + \frac{M_e}{E_e I_e} \right)^2 \right] dx
$$

$$
= \int_{x_a^e}^{x_b^e} \left[\left(\frac{d^2 M_e}{dx^2} + q \right) \frac{d^2 \delta M_e}{dx^2} + \gamma \left(\frac{d^2 w_e}{dx^2} + \frac{M_e}{E_e I_e} \right) \left(\frac{d^2 \delta w_e}{dx^2} + \frac{\delta M_e}{E_e I_e} \right) \right] dx.
$$

$$(9.4.23)$$

The statement in Eq. (9.4.23) is equivalent to the following two statements (corresponding to the independent variations of δw^e and δM^e):

$$
0 = \int_{x_a^e}^{x_b^e} \gamma \frac{d^2 \delta w_e}{dx^2} \left(\frac{d^2 w_e}{dx^2} + \frac{M_e}{E_e I_e} \right) dx,
$$

$$(9.4.24)$$

$$
0 = \int_{x_a^e}^{x_b^e} \left[\frac{\gamma}{E_e I_e} \delta M_e \left(\frac{d^2 w_e}{dx^2} + \frac{M_e}{E_e I_e} \right) + \frac{d^2 \delta M_e}{dx^2} \left(\frac{d^2 M_e}{dx^2} + q \right) \right] dx,
$$

$$(9.4.25)$$

where γ is the weight used to make the entire statement have the same physical units and possibly the same order of magnitude of the terms. Substituting the approximations from Eq. (9.4.20) into Eqs (9.4.24) and (9.4.25), we obtain

$$
\begin{bmatrix} \mathbf{A}^e & \mathbf{B}^e \\ (\mathbf{B}^e)^{\mathrm{T}} & \mathbf{D}^e \end{bmatrix} \begin{Bmatrix} \boldsymbol{\Delta}^e \\ \boldsymbol{\Lambda}^e \end{Bmatrix} = \begin{Bmatrix} \mathbf{0} \\ \mathbf{q}^e \end{Bmatrix},
$$

$$(9.4.26)$$

where

$$
A_{ij}^e = \int_{x_a^e}^{x_b^e} \gamma \frac{d^2 \varphi_i^{(1)}}{dx^2} \frac{d^2 \varphi_j^{(1)}}{dx^2} \, dx,
$$

$$
B_{ij}^e = \int_{x_a^e}^{x_b^e} \frac{\gamma}{E_e I_e} \frac{d^2 \varphi_i^{(1)}}{dx^2} \varphi_j^{(2)} \, dx,
$$

$$(9.4.27)$$

$$
D_{ij}^e = \int_{x_a^e}^{x_b^e} \left(\frac{d^2 \varphi_i^{(2)}}{dx^2} \frac{d^2 \varphi_j^{(2)}}{dx^2} + \frac{\gamma}{E_e^2 I_e^2} \varphi_i^{(2)} \varphi_j^{(2)} \right) dx,
$$

$$
q_i^e = -\int_{x_a^e}^{x_b^e} \frac{d^2 \varphi_i^{(2)}}{dx^2} q \, dx.
$$

We close this section with the comment that finite element models other than the displacement finite element model are not commonly used in practice. The next section is devoted to the discussion of the mixed finite element models of the TBT.

9.4.2 The Timoshenko Beam Theory

9.4.2.1 Governing Equations

The governing equations of the TBT for pure bending are given by [see Eqs (10) and (11) of **Example 3.5.4**]

$$-\frac{d}{dx}\left[GAK_s\left(\phi_x + \frac{dw}{dx}\right)\right] = q(x). \tag{9.4.28}$$

$$-\frac{d}{dx}\left(EI\frac{d\phi_x}{dx}\right) + GAK_s\left(\phi_x + \frac{dw}{dx}\right) = 0, \tag{9.4.29}$$

Here $q(x)$ denotes the distributed transverse load, E Young's modulus, G the shear modulus, A the area of cross section, I the moment of inertia, and K_s the shear correction factor.

Displacement finite element models of Eqs (9.4.28) and (9.4.29) were presented in Section 8.4. Here we shall discuss a number of mixed finite element models of these equations or their equivalent.

9.4.2.2 General Finite Element Model

The mixed finite element model to be discussed here is based on the stationary functional in Eq. (1) of **Example 9.2.2** in which the displacements (w, ϕ_x) and strains $(\kappa_{xx}, \gamma_{xz})$ are treated as the dependent variables. The Euler equations of this functional are given in Eqs (3) to (6) of **Example 9.2.2**.

The weak forms associated with Eqs (3) to (6) of **Example 9.2.2** over a typical element $\Omega_e = (x_a^e, x_b^e)$ are

$$\int_{x_a^e}^{x_b^e}\left(G_eA_eK_s\frac{d\delta w_e}{dx}\gamma_{xz}^e - \delta w_e q\right)dx - V_a^e\delta w_e(x_a^e) - V_b^e\delta w_e(x_b^e) = 0, \tag{9.4.30}$$

$$\int_{x_a^e}^{x_b^e}\left(E_eI_e\frac{d\delta\phi_x^e}{dx}\kappa_{xx}^e + G_eA_eK_s\delta\phi_x^e\gamma_{xz}^e\right)dx - M_a^e\delta\phi_x^e(x_a^e) - M_b^e\delta\phi_x^e(x_b^e) = 0, \tag{9.4.31}$$

$$\int_{x_a^e}^{x_b^e}E_eI_e\delta\kappa_{xx}^e\left(\frac{d\phi_x^e}{dx} - \kappa_{xx}^e\right)dx = 0, \tag{9.4.32}$$

$$\int_{x_a^e}^{x_b^e}G_eA_eK_s\delta\gamma_{xz}^e\left(\frac{dw_e}{dx} + \phi_x^e - \gamma_{xz}^e\right)dx = 0, \tag{9.4.33}$$

where

$$V_1^e = [-G_eA_aK_s\gamma_{xz}^e]_{x_a^e}, \quad V_2^e = [G_eA_eK_s\gamma_{xz}^e]_{x_b^e},$$

$$M_1^e = [-E_eI_e\kappa_{xx}^e]_{x_a^e}, \qquad M_2^e = [E_eI_e\kappa_{xx}^e]_{x_b^e}. \tag{9.4.34}$$

Let the variables $(w_e, \phi_x^e, \kappa_{xx}^e, \gamma_{xz}^e)$ be approximated as

$$w_e \approx \sum_{j=1}^{m} \psi_j^{(1)} W_j^e, \quad \phi_x^e \approx \sum_{j=1}^{n} \psi_j^{(2)} \Phi_j^e, \quad \kappa_{xx}^e \approx \sum_{j=1}^{p} \psi_j^{(3)} \mathcal{K}_j^e, \quad \gamma_{xz}^e \approx \sum_{j=1}^{q} \psi_j^{(4)} \Gamma_j^e,$$

(9.4.35)

where $(W_j^e, \Phi_j^e, \mathcal{K}_j^e, \Gamma_j^e)$ are the nodal values of $(w_e, \phi_x^e, \kappa_{xx}^e, \gamma_{xz}^e)$ and $\psi_j^{(\alpha)}(x)$ $(\alpha = 1, 2, 3, 4)$ are the associated interpolation functions whose choice is yet to be made. Substituting the approximations from Eq. (9.4.35) into Eqs (9.4.30) to (9.4.33), we obtain the following finite element model:

$$\begin{bmatrix} \mathbf{0} & \mathbf{0} & \mathbf{0} & \mathbf{A}^e \\ \mathbf{0} & \mathbf{0} & \mathbf{B}^e & \mathbf{C}^e \\ \mathbf{0} & (\mathbf{B}^e)^{\mathrm{T}} & -\mathbf{D}^e & \mathbf{0} \\ (\mathbf{A}^e)^{\mathrm{T}} & (\mathbf{C}^e)^{\mathrm{T}} & \mathbf{0} & -\mathbf{G}^e \end{bmatrix} \begin{Bmatrix} \mathbf{W}^e \\ \mathbf{\Phi}^e \\ \mathcal{K}^e \\ \mathbf{\Gamma}^e \end{Bmatrix} = \begin{Bmatrix} \mathbf{F}^e \\ \mathbf{0} \\ \mathbf{0} \\ \mathbf{0} \end{Bmatrix} + \begin{Bmatrix} \mathbf{V}^e \\ \mathbf{M}^e \\ \mathbf{0} \\ \mathbf{0} \end{Bmatrix}, \qquad (9.4.36)$$

where

$$A_{ij}^e = \int_{x_a^e}^{x_b^e} G_e A_e K_s \frac{d\psi_i^{(1)}}{dx} \psi_j^{(4)} \, dx, \quad B_{ij}^e = \int_{x_a^e}^{x_b^e} E_e I_e \frac{d\psi_i^{(2)}}{dx} \psi_j^{(3)} \, dx,$$

$$C_{ij}^e = \int_{x_a^e}^{x_b^e} G_e A_e K_s \psi_i^{(2)} \psi_j^{(4)} \, dx, \quad D_{ij}^e = \int_{x_a^e}^{x_b^e} E_e I_e \psi_i^{(3)} \psi_j^{(3)} \, dx,$$

$$G_{ij}^e = \int_{x_a^e}^{x_b^e} G_e A_e K_s \psi_i^{(4)} \psi_j^{(4)} \, dx, \quad F_i^e = \int_{x_a^e}^{x_b^e} q \psi_i^{(1)} \, dx,$$

(9.4.37)

$$V_1^e = V_a^e, \quad V_m^e = V_b^e, \quad\quad M_1^e = M_a^e, \; M_n^e = M_b^e.$$

Couple of observations are in order concerning the finite element model in Eq. (9.4.36). We note that $\mathbf{A}^e$ is a vector when γ_{xz}^e is approximated as a constant, Γ_0^e. In addition, the first equation of Eq. (9.4.36) has the form

$$G_e A_e K_s \begin{Bmatrix} -1 \\ 0 \\ \vdots \\ 0 \\ 1 \end{Bmatrix} \Gamma_0 = \begin{Bmatrix} F_1^e \\ F_2^e \\ F_3^e \\ \vdots \\ F_m^e \end{Bmatrix} + \begin{Bmatrix} V_1^e \\ 0 \\ \vdots \\ 0 \\ V_m^e \end{Bmatrix} \qquad (9.4.38)$$

when w is interpolated using quadratic or higher-order polynomials. The nonzero entries correspond to the deflection degrees of freedom at node 1 and node m. For linear interpolation of w_e, we have $m = 2$ and Eq. (9.4.38) is alright. However, when $m > 2$, Eq. (9.4.38) implies that $F_i^e = 0$ for $i = 2, \ldots, m-1$, which, in general, is not true. Thus, either the distributed load is zero or it is converted to generalized point forces at the end nodes through Hermite cubic polynomials φ_i^e of Eq. (8.3.10). In the latter case, the force components can be added to V_1^e and V_m^e and the moment components to M_1^e and M_m^e at nodes 1 and m, respectively.

9.4.2.3 ASD-LLCC Element

In the assumed strain-displacement model (ASD), use of linear (L) interpolations of the generalized displacements w_e and ϕ_x^e and constant (C) representation of the generalized strains κ_{xx}^e and γ_{xz}^e ($m = n = 2$ and $p = q = 1$) results in the following finite element equations (when $E_e I_e$ and $G_e A_e K_s$ are assumed to be element-wise constant):

$$G_e A_e K_s \begin{Bmatrix} -1 \\ 1 \end{Bmatrix} \Gamma_0^e = \begin{Bmatrix} q_1^e \\ q_2^e \end{Bmatrix} + \begin{Bmatrix} V_1^e \\ V_2^e \end{Bmatrix}, \qquad (9.4.39)$$

$$E_e I_e \begin{Bmatrix} -1 \\ 1 \end{Bmatrix} \mathcal{K}_0^e + G_e A_e K_s \frac{h_e}{2} \begin{Bmatrix} 1 \\ 1 \end{Bmatrix} \Gamma_0^e = \begin{Bmatrix} M_1^e \\ M_2^e \end{Bmatrix}, \qquad (9.4.40)$$

$$\{-1 \ \ 1\} \begin{Bmatrix} \Phi_1^e \\ \Phi_2^e \end{Bmatrix} - h_e \mathcal{K}_0 = 0, \qquad (9.4.41)$$

$$\{-1 \ \ 1\} \begin{Bmatrix} W_1^e \\ W_2^e \end{Bmatrix} + \frac{h_e}{2} \{1 \ \ 1\} \begin{Bmatrix} \Phi_1^e \\ \Phi_2^e \end{Bmatrix} - h_e \Gamma_0^e = 0. \qquad (9.4.42)$$

Solving Eqs (9.4.41) and (9.4.42) for $\mathcal{K}_0^e$ and Γ_0^e and substituting into Eqs (9.4.39) and (9.4.40) (i.e., condensing out $\mathcal{K}_0^e$ and Γ_0^e), we arrive at the following 4×4 system of equations

$$\frac{G_e A_e K_s}{4 h_e} \begin{bmatrix} 4 & -2h_e & -4 & -2h \\ -2h_e & h_e^2(1 + 4\Lambda_e) & 2h_e & h_e^2(1 - 4\Lambda_e) \\ -4 & 2h_e & 4 & 2h_e \\ -2h_e & h_e^2(1 - 4\Lambda_e) & 2h_e & h_e^2(1 + 4\Lambda_e) \end{bmatrix} \begin{Bmatrix} W_1^e \\ \Phi_1^e \\ W_2^e \\ \Phi_2^e \end{Bmatrix} = \begin{Bmatrix} q_1^e \\ 0 \\ q_2^e \\ 0 \end{Bmatrix} + \begin{Bmatrix} Q_1^e \\ Q_2^e \\ Q_3^e \\ Q_4^e \end{Bmatrix},$$

$$(9.4.43)$$

where $\Lambda_e = E_e I_e / G_e A_e K_s h_e^2$. These are exactly the same equations obtained in the displacement formulation with the linear interpolation of w_e and ϕ_e and using one-point Gauss quadrature to evaluate the shear stiffness coefficients. The element was referred to as the reduced integration element (RIE). Thus, the ASD formulation eliminates the need for reduced integration concepts.

9.4.2.4 ASD-QLCC Element

A quadratic (Q) interpolation of w, linear (L) interpolation of ϕ_x, and constant (C) representation of κ_{xx}^e and γ_{xz}^e will also yield the same stiffness matrix as in Eq. (9.4.43). However, as noted earlier, the distributed load q must be calculated using Hermite (H) cubic polynomials φ_i^e ($i = 1, 2, 3, 4$):

$$q_i^{(H)} = \int_{x_a^e}^{x_b^e} q(x) \varphi_i(x) \, dx, \qquad (9.4.44)$$

and the force components of $q_1^{(H)}$ and $q_3^{(H)}$ must be added to V_1^e and V_2^e, and the moment components $q_2^{(H)}$ and $q_4^{(H)}$ to M_1^e and M_2^e, respectively:

$$
\left\{
\begin{array}{c}
\bar{V}_1^e \\
\bar{M}_1^e \\
\bar{V}_2^e \\
\bar{M}_2^e
\end{array}
\right\}
=
\left\{
\begin{array}{c}
V_1^e + q_1^{(H)} \\
M_1^e + q_2^{(H)} \\
V_2^e + q_3^{(H)} \\
M_2^e + q_4^{(H)}
\end{array}
\right\}.
\tag{9.4.45}
$$

Thus, the load vector is the same as that of the Bernoulli–Euler beam equations. This element was referred to as the consistent interpolation element (CIE).

9.4.2.5   ASD-HQLC Element

Suppose that the distributed load is represented using Eq. (9.4.44). A Lagrange cubic (higher-order, H) interpolation of w_e, quadratic interpolation of ϕ_x^e, linear interpolation of κ_{xx}^e, and constant representation of γ_{xz}^e yield the equations

$$
G_e A_e K_s \left\{ \begin{array}{c} -1 \\ 1 \end{array} \right\} \Gamma_0^e = \left\{ \begin{array}{c} \bar{V}_1^e \\ \bar{V}_2^e \end{array} \right\},
\tag{9.4.46}
$$

$$
\frac{E_e I_e}{6}
\begin{bmatrix}
-5 & -1 \\
4 & -4 \\
1 & 5
\end{bmatrix}
\left\{ \begin{array}{c} \mathcal{K}_1^e \\ \mathcal{K}_2^e \end{array} \right\}
+ \frac{G_e A_e K_s h_e}{6}
\left\{ \begin{array}{c} 1 \\ 4 \\ 1 \end{array} \right\} \Gamma_0^e
= \left\{ \begin{array}{c} \bar{M}_1^e \\ 0 \\ \bar{M}_2^e \end{array} \right\},
\tag{9.4.47}
$$

$$
\frac{E_e I_e}{6}
\begin{bmatrix}
-5 & 4 & 1 \\
-1 & -4 & 5
\end{bmatrix}
\left\{ \begin{array}{c} \Phi_1^e \\ \Phi_c^e \\ \Phi_2^e \end{array} \right\}
- \frac{E_e I_e h_e}{6}
\begin{bmatrix} 2 & 1 \\ 1 & 2 \end{bmatrix}
\left\{ \begin{array}{c} \mathcal{K}_1^e \\ \mathcal{K}_2^e \end{array} \right\}
= \left\{ \begin{array}{c} 0 \\ 0 \end{array} \right\},
\tag{9.4.48}
$$

$$
\{-1 \ \ 1\} \left\{ \begin{array}{c} W_1^e \\ W_2^e \end{array} \right\}
+ \frac{h_e}{6} \{1 \ \ 4 \ \ 1\}
\left\{ \begin{array}{c} \Phi_1^e \\ \Phi_c^e \\ \Phi_2^e \end{array} \right\}
- h_e \Gamma_0^e = 0,
\tag{9.4.49}
$$

where the end nodes of the element are designated as 1 and 2 and the middle node as c, and the interior nodal degrees of freedom associated with w_e are omitted as they do not contribute to the equations.

Solving Eq. (9.4.48) for $\{\mathcal{K}\}^1$ and Eq. (9.4.49) for Γ_0^e and substituting the result into Eqs (9.4.46) and (9.4.47), we obtain

$$
S_e \left(\frac{1}{h_e}
\begin{bmatrix} 1 & -1 \\ -1 & 1 \end{bmatrix}
\left\{ \begin{array}{c} W_1^e \\ W_2^e \end{array} \right\}
+ \frac{1}{6}
\begin{bmatrix} -1 & -1 \\ 1 & 1 \end{bmatrix}
\left\{ \begin{array}{c} \Phi_1^e \\ \Phi_2^e \end{array} \right\}
+ \frac{4}{6}
\left\{ \begin{array}{c} -1 \\ 1 \end{array} \right\} \Phi_c^e \right)
$$

$$
= \left\{ \begin{array}{c} \bar{V}_1^e \\ \bar{V}_2^e \end{array} \right\},
\tag{9.4.50}
$$

$$S_e \left\{ \frac{1}{6} \begin{bmatrix} -1 & 1 \\ -4 & 4 \\ -1 & 1 \end{bmatrix} \begin{Bmatrix} W_1^e \\ W_2^e \end{Bmatrix} + \frac{\bar{\Lambda}}{3h_e} \begin{bmatrix} 7 & 1 \\ -8 & -8 \\ 1 & 7 \end{bmatrix} \begin{Bmatrix} \Phi_1^e \\ \Phi_2^e \end{Bmatrix} + \frac{1}{36} \begin{bmatrix} 1 & 1 \\ 4 & 4 \\ 1 & 1 \end{bmatrix} \begin{Bmatrix} \Phi_1^e \\ \Phi_2^e \end{Bmatrix} \right.$$

$$\left. + \left(\frac{8\bar{\Lambda}}{3h_e} \begin{Bmatrix} -1 \\ 2 \\ -1 \end{Bmatrix} + \frac{h_e}{9} \begin{Bmatrix} 1 \\ 4 \\ 1 \end{Bmatrix} \right) \Phi_c^e \right\} = \begin{Bmatrix} \bar{M}_1^e \\ 0 \\ \bar{M}_2^e \end{Bmatrix}, \tag{9.4.51}$$

where $S_e = G_e A_e K_s$ and $\bar{\Lambda} = (E_e I_e / S_e)$. The second algebraic equation in Eq. (9.4.51) can be used to eliminate Φ_c^e from Eqs (9.4.50) and (9.4.51). Thus, we obtain

$$\frac{2E_e I_e}{\mu h_e^3} \left(\begin{bmatrix} 6 & -6 \\ -6 & 6 \end{bmatrix} \begin{Bmatrix} W_1^e \\ W_2^e \end{Bmatrix} + \begin{bmatrix} -3h_e & -3h_e \\ 3h_e & 3h_e \end{bmatrix} \begin{Bmatrix} \Phi_1^e \\ \Phi_2^e \end{Bmatrix} \right) = \begin{Bmatrix} \bar{V}_1^e \\ \bar{V}_2^e \end{Bmatrix}, \tag{9.4.52}$$

$$\frac{2E_e I_e}{\mu h_e^3} \left(\begin{bmatrix} -3h_e & 3h_e \\ -3h_e & 3h_e \end{bmatrix} \begin{Bmatrix} W_1^e \\ W_2^e \end{Bmatrix} + \begin{bmatrix} 2h^2\lambda & h_e^2\xi \\ h_e^2\xi & 2h_e^2\lambda \end{bmatrix} \begin{Bmatrix} \Phi_1^e \\ \Phi_2^e \end{Bmatrix} \right) = \begin{Bmatrix} \bar{M}_1^e \\ \bar{M}_2^e \end{Bmatrix}. \tag{9.4.53}$$

Adding Eqs (9.4.52) and (9.4.53), we obtain [see Eq. (9.4.45)]

$$\frac{2E_e I_e}{\mu h_e^3} \begin{bmatrix} 6 & -3h_e & -6 & -3h_e \\ -3h_e & 2h_e^2\lambda & 3h_e & h_e^2\xi \\ -6 & 3h_e & 6 & 3h_e \\ -3h_e & h_e^2\xi & 3h_e & 2h_e^2\lambda \end{bmatrix} \begin{Bmatrix} W_1^e \\ \Phi_1^e \\ W_2^e \\ \Phi_2^e \end{Bmatrix} = \begin{Bmatrix} \bar{V}_1^e \\ \bar{M}_1^e \\ \bar{V}_2^e \\ \bar{M}_2^e \end{Bmatrix}.$$

Using Eq. (9.4.45), we obtain

$$\frac{2E_e I_e}{\mu h_e^3} \begin{bmatrix} 6 & -3h_e & -6 & -3h_e \\ -3h_e & 2h_e^2\lambda & 3h_e & h_e^2\xi \\ -6 & 3h_e & 6 & 3h_e \\ -3h_e & h_e^2\xi & 3h_e & 2h_e^2\lambda \end{bmatrix} \begin{Bmatrix} W_1^e \\ \Phi_1^e \\ W_2^e \\ \Phi_2^e \end{Bmatrix} = \begin{Bmatrix} V_1^e + q_1^{(H)} \\ M_1^e + q_2^{(H)} \\ V_2^e + q_3^{(H)} \\ M_2^e + q_4^{(H)} \end{Bmatrix}. \tag{9.4.54}$$

The stiffness matrix is the same as that of the superconvergent element derived in Eq. (8.4.30); however, the load vector is different. It is the same when either the applied load q is element-wise uniform or the load vector is computed using Eq. (8.4.29).

It should be noted that the degree of the polynomial interpolation used for w does not enter the equations presented in all models discussed in this section. However, the load representation implies that w be interpolated with Hermite cubic polynomials φ_i^e of Eq. (8.3.10) or Lagrange cubic polynomials ψ_i^e of Eq. (8.4.26).

9.5 Mixed Finite Element Models of the Classical Plate Theory

9.5.1 Preliminary Comments

As mentioned in Chapter 7, a conforming plate finite element based on the displacement formulation of the classical plate theory is algebraically complex and computationally demanding. A quintic polynomial would satisfy the continuity requirements at element interfaces and results in 21 degrees of freedom per element. The continuity requirements can be relaxed by reformulating the fourth-order equation in Eq. (7.2.109) or Eq. (7.2.110) as a set of second-order equations in terms of the deflection and bending moments [see Eq. (7.2.95)]. Such a formulation is called *mixed formulation*. Here we discuss two such models for the static bending of orthotropic plates (see [143–146]).

9.5.2 Mixed Model I

9.5.2.1 Governing Equations

Consider the following set of equations [set the in-plane forces $\hat{N}_{xx}$, $\hat{N}_{xy}$, and $\hat{N}_{yy}$ as well as the time derivative terms to zero in Eq. (7.2.95), and consider the bending moment–curvature relations in Eq. (7.2.84)]:

$$-\left(\frac{\partial^2 M_x}{\partial x^2} + 2\frac{\partial^2 M_{xy}}{\partial x \partial y} + \frac{\partial^2 M_y}{\partial y^2}\right) = q, \tag{9.5.1}$$

$$M_{xx} = -\left(D_{11}\frac{\partial^2 w}{\partial x^2} + D_{12}\frac{\partial^2 w}{\partial y^2}\right), \tag{9.5.2}$$

$$M_{yy} = -\left(D_{12}\frac{\partial^2 w}{\partial x^2} + D_{22}\frac{\partial^2 w}{\partial y^2}\right), \tag{9.5.3}$$

$$M_{xy} = -2D_{66}\frac{\partial^2 w}{\partial x \partial y}, \tag{9.5.4}$$

where D_{ij} are the bending stiffness coefficients of an orthotropic plate:

$$D_{ij} = Q_{ij}\frac{h^3}{12}, \quad (i = 1, 2, 6), \tag{9.5.5}$$

and the elastic coefficients Q_{ij} are defined in terms of the principal moduli (E_1, E_2), shear modulus G_{12}, and Poisson's ratio ν_{12} (based on plane stress assumption) as

$$Q_{11} = \frac{E_1}{1 - \nu_{12}\nu_{21}}, \quad Q_{12} = \frac{\nu_{12}E_2}{1 - \nu_{12}\nu_{21}}, \quad Q_{22} = \frac{E_2}{1 - \nu_{12}\nu_{21}},$$

$$Q_{66} = G_{12}, \quad \nu_{21} = \nu_{12}\frac{E_2}{E_1}. \tag{9.5.6}$$

Substitution of Eqs (9.5.2) to (9.5.4) into Eq. (9.5.1) results in a fourth-order equation given in Eq. (7.2.109), whose displacement finite element models were discussed in Section 8.3.4. Here we formulate a mixed finite element model of Eqs (9.5.1) to (9.5.4), which involves a set of second-order equations among the dependent variables $(w, M_{xx}, M_{yy}, M_{xy})$. In the mixed formulation we treat these variables as the dependent unknowns.

Toward developing the mixed finite element model, first we express the three equations in Eqs (9.5.2) to (9.5.4) in the following alternative form

$$\frac{\partial^2 w}{\partial x^2} = -\left(\bar{D}_{22}M_{xx} - \bar{D}_{12}M_{yy}\right),$$

$$\frac{\partial^2 w}{\partial y^2} = -\left(\bar{D}_{11}M_{yy} - \bar{D}_{12}M_{xx}\right), \tag{9.5.7}$$

$$2\frac{\partial^2 w}{\partial x \partial y} = -(D_{66})^{-1}M_{xy},$$

where

$$\bar{D}_{ij} = \frac{D_{ij}}{D_0}, \quad D_0 = D_{11}D_{22} - D_{12}^2. \tag{9.5.8}$$

9.5.2.2 Weak Forms

The weak forms of Eqs (9.5.1) and (9.5.7) over a typical element Ω_e can be derived using the procedure described in Section 6.3.6. We have

$$0 = \int_{\Omega_e} \left[\frac{\partial \delta w_e}{\partial x} \left(\frac{\partial M_{xx}^e}{\partial x} + \frac{\partial M_{xy}^e}{\partial y} \right) + \frac{\partial \delta w_e}{\partial y} \left(\frac{\partial M_{yy}^e}{\partial y} + \frac{\partial M_{xy}^e}{\partial x} \right) - q\delta w_e \right] dxdy$$

$$- \oint_{\Gamma_e} \delta w_e \left[\left(\frac{\partial M_{xx}^e}{\partial x} + \frac{\partial M_{xy}^e}{\partial y} \right) n_x + \left(\frac{\partial M_{xy}^e}{\partial x} + \frac{\partial M_{yy}^e}{\partial y} \right) n_y \right] dS, \tag{9.5.9}$$

$$0 = \int_{\Omega_e} \left[\frac{\partial w_e}{\partial x} \frac{\partial \delta M_{xx}^e}{\partial x} - \delta M_{xx}^e \left(\bar{D}_{22}M_{xx}^e - \bar{D}_{12}M_{yy}^e \right) \right] dxdy$$

$$- \oint_{\Gamma_e} \delta M_{xx}^e \frac{\partial w_e}{\partial x} n_x \, dS, \tag{9.5.10}$$

$$0 = \int_{\Omega_e} \left[\frac{\partial w_e}{\partial y} \frac{\partial \delta M_{yy}^e}{\partial y} - \delta M_{yy}^e \left(\bar{D}_{11}M_{yy}^e - \bar{D}_{12}M_{xx}^e \right) \right] dxdy$$

$$- \oint_{\Gamma_e} \delta M_{yy}^e \frac{\partial w_e}{\partial y} n_y \, dS, \tag{9.5.11}$$

$$0 = \int_{\Omega_e} \left(\frac{\partial w_e}{\partial x} \frac{\partial \delta M_{xy}^e}{\partial y} + \frac{\partial w_e}{\partial y} \frac{\partial \delta M_{xy}^e}{\partial x} - (D_{66})^{-1}\delta M_{xy}^e M_{xy}^e \right) dxdy$$

$$- \oint_{\Gamma_e} \delta M_{xy}^e \left(\frac{\partial w_e}{\partial y} n_x + \frac{\partial w_e}{\partial x} n_y \right) dS. \tag{9.5.12}$$

It is clear that the primary variables of the formulation are

$$w_e, \; M_{xx}^e, \; M_{yy}^e, \; M_{xy}^e, \tag{9.5.13}$$

and the secondary variables are

$$Q_n^e \equiv \left(\frac{\partial M_{xx}^e}{\partial x} + \frac{\partial M_{xy}^e}{\partial y} \right) n_x + \left(\frac{\partial M_{xy}^e}{\partial x} + \frac{\partial M_{yy}^e}{\partial y} \right) n_y,$$

$$\frac{\partial w_e}{\partial x} n_x, \quad \frac{\partial w_e}{\partial y} n_y, \quad \frac{\partial w_e}{\partial x} n_y + \frac{\partial w_e}{\partial y} n_x. \tag{9.5.14}$$

The functional associated with the weak forms in Eqs (9.5.9) to (9.5.12) is given by

$$J_1^e(\mathbf{\Lambda}_e) = \int_{\Omega_e} \left[\frac{1}{2} \left(-\bar{D}_{22}(M_{xx}^e)^2 + 2\bar{D}_{12} M_{xx}^e M_{yy}^e - \bar{D}_{11}(M_{yy}^e)^2 - \frac{1}{D_{66}}(M_{xy}^e)^2 \right) \right.$$

$$\left. + \frac{\partial w_e}{\partial x} \left(\frac{\partial M_{xx}^e}{\partial x} + \frac{\partial M_{xy}^e}{\partial y} \right) + \frac{\partial w_e}{\partial y} \left(\frac{\partial M_{xy}^e}{\partial x} + \frac{\partial M_{yy}^e}{\partial y} \right) - q w_e \right] dx dy$$

$$- \oint_{\Gamma_e} \left[M_{xx}^e \theta_x n_x + M_{yy}^e \theta_y n_y + M_{xy}^e (\theta_x^e n_y + \theta_y^e n_x) + Q_n^e w_e \right] ds, \tag{9.5.15}$$

where

$$\theta_x^e = \frac{\partial w_e}{\partial x}, \quad \theta_y^e = \frac{\partial w_e}{\partial y}, \quad \mathbf{\Lambda}_e = (w_e, M_{xx}^e, M_{yy}^e, M_{xy}^e). \tag{9.5.16}$$

The condition $\delta J_1 = 0$ gives Eqs (9.5.1) and (9.5.7) as the Euler equations.

9.5.2.3 Finite Element Model

Let $(w_e, M_{xx}^e, M_{yy}^e, M_{xy}^e)$ be interpolated by expressions of the form

$$w_e = \sum_{i=1}^{r} w_i^e \psi_i^{(1)},$$

$$M_{xx}^e = \sum_{i=1}^{s} M_{xi}^e \psi_i^{(2)}, \quad M_{yy}^e = \sum_{i=1}^{p} M_{yi}^e \psi_i^{(3)}, \quad M_{xy}^e = \sum_{i=1}^{q} M_{xyi}^e \psi_i^{(4)}, \tag{9.5.17}$$

where $\psi_i^{(\alpha)}$ ($\alpha = 1, 2, 3, 4$) are appropriate interpolation functions. Substituting Eq. (9.5.17) into Eqs (9.5.9) to (9.5.12) [or substituting Eq. (9.5.17) into J_1^e and setting the partial variations of J_1^e with respect to $w_j^e, M_{xj}^e, M_{yj}^e$, and M_{xyj}^e to zero separately], we obtain

$$\begin{bmatrix} \mathbf{K}^{11} & \mathbf{K}^{12} & \mathbf{K}^{13} & \mathbf{K}^{14} \\ & \mathbf{K}^{22} & \mathbf{K}^{23} & \mathbf{K}^{24} \\ \text{symm.} & & \mathbf{K}^{33} & \mathbf{K}^{34} \\ & & & \mathbf{K}^{44} \end{bmatrix} \begin{Bmatrix} \mathbf{w} \\ \mathbf{M}_x \\ \mathbf{M}_y \\ \mathbf{M}_{xy} \end{Bmatrix} = \begin{Bmatrix} \mathbf{F}^1 \\ \mathbf{F}^2 \\ \mathbf{F}^3 \\ \mathbf{F}^4 \end{Bmatrix}, \tag{9.5.18}$$

where

$$K_{ij}^{11} = 0 \ (i,j = 1,2,\ldots,r),$$

$$K_{ij}^{12} = \int_{\Omega_e} \frac{\partial \psi_i^{(1)}}{\partial x} \frac{\partial \psi_j^{(2)}}{\partial x} dxdy \ (i = 1,2,\ldots,r; \ j = 1,2,\ldots,s),$$

$$K_{ij}^{13} = \int_{\Omega_e} \frac{\partial \psi_i^{(1)}}{\partial y} \frac{\partial \psi_j^{(3)}}{\partial y} dxdy \ (i = 1,2,\ldots,r; \ j = 1,2,\ldots,p),$$

$$K_{ij}^{14} = \int_{\Omega_e} \left(\frac{\partial \psi_i^{(1)}}{\partial x} \frac{\partial \psi_j^{(4)}}{\partial y} + \frac{\partial \psi_i^{(1)}}{\partial y} \frac{\partial \psi_j^{(4)}}{\partial x} \right) dxdy \ (i = 1,2,\ldots,r; \ j = 1,2,\ldots,q),$$

$$K_{ij}^{22} = \int_{\Omega_e} (-\bar{D}_{22}) \psi_i^{(2)} \psi_j^{(2)} dxdy \ (i,j = 1,2,\ldots,s),$$

$$K_{ij}^{23} = \int_{\Omega_e} (-\bar{D}_{12}) \psi_i^{(2)} \psi_j^{(3)} dxdy \ (i = 1,2,\ldots,s; \ j = 1,2,\ldots,p),$$

$$K_{ij}^{24} = 0 \ (i = 1,2,\ldots,s; \ j = 1,2,\ldots,q), \qquad (9.5.19)$$

$$K_{ij}^{33} = \int_{\Omega_e} (-\bar{D}_{11}) \psi_i^{(3)} \psi_j^{(3)} dxdy \ (i,j = 1,2,\ldots,p),$$

$$K_{ij}^{34} = 0 \ (i = 1,2,\ldots,p; \ j = 1,2,\ldots,q),$$

$$K_{ij}^{44} = -\int_{\Omega_e} (D_{66})^{-1} \psi_i^{(4)} \psi_j^{(4)} dxdy \ (i,j = 1,2,\ldots,q),$$

$$F_i^1 = \int_{\Omega_e} q\psi_i^{(1)} dxdy + \oint_{\Gamma_e} Q_n \psi_i^{(1)} ds \ (i = 1,2,\ldots,r),$$

$$F_i^2 = \oint_{\Gamma_e} \theta_x n_x \ \psi_i^{(2)} \ ds \ (i = 1,2,\ldots,s),$$

$$F_i^3 = \oint_{\Gamma_e} \theta_y n_y \ \psi_i^{(3)} \ ds \ (i = 1,2,\ldots,p),$$

$$F_i^4 = \oint_{\Gamma_e} (\theta_y n_x + \theta_x n_y) \psi_i^{(4)} \ ds \ (i = 1,2,\ldots,q).$$

An examination of the weak forms in Eqs (9.5.9) to (9.5.12) shows that the following minimum continuity conditions are required of the interpolations functions $\psi_i^{(\alpha)}$ ($\alpha = 1,2,3,4$) used in Eq. (9.5.17):

$$\begin{aligned}
\psi_i^{(1)} &= \text{linear in } x \text{ and linear in } y, \\
\psi_i^{(2)} &= \text{linear in } x \text{ and constant in } y, \\
\psi_i^{(3)} &= \text{linear in } y \text{ and constant in } x, \\
\psi_i^{(4)} &= \text{linear in } x \text{ and linear in } y.
\end{aligned} \qquad (9.5.20)$$

The simplest choice of ψ_i^α for a rectangular element is provided by [see Fig. 9.5.1(a)]

$$\psi_i^{(1)} = \psi_i^{(2)} = \psi_i^{(3)} = \psi_i^{(4)} = \text{bilinear functions of a rectangular element.} \tag{9.5.21}$$

Computation of the coefficient matrices is simple and straightforward. The resulting stiffness matrix is of the order 16 by 16.

For a rectangular element, the following choice of interpolation function $\psi_i^{(\alpha)}$ also meets the minimum requirements (a and b are the sides of the rectangular element):

$$\psi_1^{(2)} = 1 - \frac{\overline{x}}{a}, \quad \psi_2^{(2)} = \frac{\overline{x}}{a}, \quad \psi_1^{(3)} = 1 - \frac{\overline{y}}{b}, \quad \psi_2^{(3)} = \frac{\overline{y}}{b}, \tag{9.5.22}$$

and $\psi_i^{(1)}$ and $\psi_i^{(4)}$ are the bilinear functions of a rectangular element. The rectangular element associated with this choice of interpolation functions is shown in Fig. 9.5.1(b). The coefficient matrices in Eq. (9.5.19) can be easily evaluated for this element. For instance, we have (see Sec. 8.2.6 of Reddy [46] for element matrix computations)

$$\mathbf{K}^{12} = \frac{b}{2a} \begin{bmatrix} 1 & -1 \\ -1 & 1 \\ -1 & 1 \\ 1 & -1 \end{bmatrix}, \qquad \mathbf{K}^{13} = \frac{a}{2b} \begin{bmatrix} 1 & -1 \\ 1 & -1 \\ -1 & 1 \\ -1 & 1 \end{bmatrix},$$

$$\mathbf{K}^{22} = -\bar{D}_{22} \frac{a}{6} \begin{bmatrix} 2 & 1 \\ 1 & 2 \end{bmatrix}, \qquad \mathbf{K}^{33} = -\bar{D}_{11} \frac{b}{6} \begin{bmatrix} 2 & 1 \\ 1 & 2 \end{bmatrix}, \tag{9.5.23}$$

$$\mathbf{K}^{14} = \frac{1}{2} \begin{bmatrix} 1 & 0 & -1 & 0 \\ 0 & -1 & 0 & 1 \\ -1 & 0 & 1 & 0 \\ 0 & 1 & 0 & -1 \end{bmatrix}, \qquad \mathbf{K}^{44} = \frac{ab}{36} \begin{bmatrix} 4 & 2 & 1 & 2 \\ 2 & 4 & 2 & 1 \\ 1 & 2 & 4 & 2 \\ 2 & 1 & 2 & 4 \end{bmatrix}.$$

The element stiffness matrix is of the order 12 by 12.

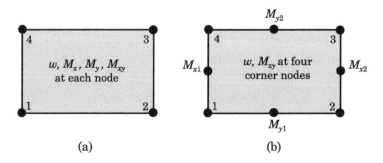

(a)   (b)

Fig. 9.5.1 Mixed plate bending elements based on CPT. (a) MIxed Model IA. (b) Mixed Model IB.

9.5.3 Mixed Model II

9.5.3.1 Governing Equations

A simplified mixed model can be derived by eliminating the twisting moment M_{xy}^e from Eqs (9.5.1) to (9.5.4). We obtain

$$
-\left(\frac{\partial^2 M_{xx}}{\partial x^2} - 4D_{66}\frac{\partial^4 w}{\partial x^2 \partial y^2} + \frac{\partial^2 M_{yy}}{\partial y^2}\right) - \lambda_v w
$$

$$
+ \lambda_b \left(N_1 \frac{\partial^2 w}{\partial x^2} + 2N_6 \frac{\partial^2 w}{\partial x \partial y} + N_2 \frac{\partial^2 w}{\partial y^2}\right) = q, \tag{9.5.24}
$$

$$
\frac{\partial^2 w}{\partial x^2} = -\left(\bar{D}_{22} M_{xx} - \bar{D}_{12} M_{yy}\right), \tag{9.5.25}
$$

$$
\frac{\partial^2 w}{\partial y^2} = -\left(\bar{D}_{11} M_{yy} - \bar{D}_{12} M_{xx}\right), \tag{9.5.26}
$$

where, in the interest of solving the natural vibration frequencies and buckling loads, the contributions of the inertia term (neglecting the rotary inertia) and in-plane compressive and shear forces are added through the parameters λ_v and λ_b, where $(I_0 = \rho h)$

$$
\lambda_v = I_0 \omega^2, \quad \lambda_b = \frac{\hat{N}_{xx}}{N_1} = \frac{\hat{N}_{yy}}{N_2} = \frac{\hat{N}_{xy}}{N_6}. \tag{9.5.27}
$$

Here h denotes the plate thickness and $(\hat{N}_{xx}, \hat{N}_{yy}, \hat{N}_{xy})$ the in-plane compressive and shear forces.

9.5.3.2 Weak Forms

The weak forms of Eqs (9.5.24) to (9.5.26) over a typical element Ω_e are

$$
0 = \int_{\Omega_e} \left(\frac{\partial \delta w_e}{\partial x}\frac{\partial M_{xx}^e}{\partial x} + \frac{\partial \delta w_e}{\partial y}\frac{\partial M_{yy}^e}{\partial y} + 4D_{66}\frac{\partial^2 \delta w_e}{\partial x \partial y}\frac{\partial^2 w_e}{\partial x \partial y} - q\delta w_e\right) dx\,dy
$$

$$
- \int_{\Omega_e} \left\{\lambda_v w_e \delta w_e + \lambda_b \left[N_1 \frac{\partial \delta w_e}{\partial x}\frac{\partial w_e}{\partial x} + N_2 \frac{\partial \delta w_e}{\partial y}\frac{\partial w_e}{\partial y}\right.\right.
$$

$$
\left.\left. + N_6 \left(\frac{\partial \delta w_e}{\partial x}\frac{\partial w_e}{\partial y} + \frac{\partial \delta w}{\partial y}\frac{\partial w}{\partial x}\right)\right]\right\} dx\,dy
$$

$$
- \oint_{\Gamma_e} \left[\delta w_e V_n^e - 2D_{66}\frac{\partial^2 w_e}{\partial x \partial y}\left(\delta \theta_x^e n_y + \delta \theta_y^e n_x\right)\right] dS, \tag{9.5.28}
$$

$$
0 = \int_{\Omega_e} \left[\frac{\partial w_e}{\partial x}\frac{\partial \delta M_{xx}^e}{\partial x} - \delta M_{xx}^e \left(\bar{D}_{22} M_{xx}^e - \bar{D}_{12} M_{yy}^e\right)\right] dx\,dy
$$

$$
- \oint_{\Gamma_e} \delta M_{xx}^e \theta_x^e n_x \, dS, \tag{9.5.29}
$$

$$0 = \int_{\Omega_e} \left[\frac{\partial w_e}{\partial y} \frac{\partial \delta M_{yy}^e}{\partial y} - \delta M_{yy}^e \left(\bar{D}_{11} M_{yy}^e - \bar{D}_{12} M_{xx}^e \right) \right] dxdy$$

$$- \oint_{\Gamma_e} \delta M_{yy}^e \theta_y^e n_y \, dS. \tag{9.5.30}$$

The primary and secondary variables of the formulation are

$$w_e, \qquad M_{xx}^e, \qquad M_{yy}^e, \tag{9.5.31}$$

$$V_n^e, \quad \theta_x^e \equiv \frac{\partial w_e}{\partial x}, \quad \theta_y^e \equiv \frac{\partial w_e}{\partial y}.$$

Here V_n^e denotes the effective shear force (Kirchhoff free-edge condition)

$$V_n^e \equiv \bar{Q}_n^e + \frac{\partial M_{ns}^e}{\partial s}, \qquad \bar{Q}_n^e = \bar{Q}_x^e n_x + \bar{Q}_y^e n_y, \tag{9.5.32}$$

where

$$\bar{Q}_x^e \equiv Q_x^e - \lambda_b \left(N_1 \frac{\partial w_e}{\partial x} + N_6 \frac{\partial w_e}{\partial y} \right), \quad Q_x^e = \frac{\partial M_{xx}}{\partial x} + \frac{\partial M_{xy}}{\partial y}, \tag{9.5.33}$$

$$\bar{Q}_y^e \equiv Q_y^e - \lambda_b \left(N_6 \frac{\partial w_e}{\partial x} + N_2 \frac{\partial w_e}{\partial y} \right), \quad Q_y^e = \frac{\partial M_{xy}}{\partial x} + \frac{\partial M_{yy}}{\partial y}. \tag{9.5.34}$$

The functional associated with Eqs (9.5.28) to (9.5.30) is

$$J_2^e(w_e, M_x^e, M_y^e) = \int_{\Omega_e} \left\{ \frac{1}{2} \left(-\bar{D}_{22}(M_{xx}^e)^2 + 2\bar{D}_{12}M_{xx}^e M_{yy}^e - \bar{D}_{11}(M_{yy}^e)^2 - \lambda_v w_e^2 \right) \right.$$

$$- \frac{1}{2} \lambda_b \left[N_1 \left(\frac{\partial w_e}{\partial x} \right)^2 + N_2 \left(\frac{\partial w_e}{\partial y} \right)^2 + 2 N_6 \frac{\partial w_e}{\partial x} \frac{\partial w_e}{\partial y} \right]$$

$$\left. + \frac{\partial w_e}{\partial x} \frac{\partial M_{xx}^e}{\partial x} + \frac{\partial w_e}{\partial y} \frac{\partial M_{yy}^e}{\partial y} + 2 D_{66} \left(\frac{\partial^2 w_e}{\partial x \partial y} \right)^2 - q w_e \right\} dxdy$$

$$- \oint_{\Gamma_e} \left(M_{xx}^e \theta_x^e n_x + M_{yy}^e \theta_y^e n_y + V_n^e w_e \right) dS. \tag{9.5.35}$$

9.5.3.3   Finite Element Model

An examination of the weak forms in Eqs (9.5.28) to (9.5.30) reveals that the minimum continuity conditions of the interpolation functions $\psi_i^{(\alpha)}$ ($\alpha = 1, 2, 3$) are the same as those listed in Eq. (9.5.20). Therefore, interpolation functions of Eq. (9.5.21) or Eq. (9.5.23) can be used. The corresponding rectangular elements are shown in Figs. 9.5.2 (a)–(c).

The finite element model of Eqs (9.5.24) to (9.5.26) is obtained by substituting the approximations in Eq. (9.5.17) for $(w_e, M_{xx}^e, M_{yy}^e)$. We obtain

$$\begin{bmatrix} \mathbf{K}^{11} & \mathbf{K}^{12} & \mathbf{K}^{13} \\ & \mathbf{K}^{22} & \mathbf{K}^{23} \\ \text{sym.} & & \mathbf{K}^{33} \end{bmatrix} \begin{Bmatrix} \mathbf{w}^e \\ \mathbf{M}_x^e \\ \mathbf{M}_y^e \end{Bmatrix} = \begin{Bmatrix} \mathbf{F}^1 \\ \mathbf{F}^2 \\ \mathbf{F}^3 \end{Bmatrix} + \lambda_v \begin{Bmatrix} \mathbf{S}^e \mathbf{w}^e \\ \mathbf{0} \\ \mathbf{0} \end{Bmatrix} + \lambda_b \begin{Bmatrix} \mathbf{G}^e \mathbf{w}^e \\ \mathbf{0} \\ \mathbf{0} \end{Bmatrix}, \tag{9.5.36}$$

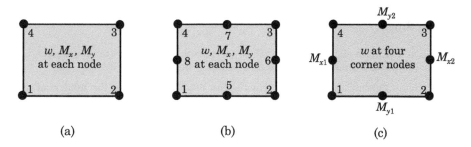

Fig. 9.5.2 Mixed rectangular plate bending elements based on CPT. (a) Mixed model IIA. (b) Mixed model IIB. (c) Mixed model IIC.

where

$$K_{ij}^{11} = 4D_{66} \int_{\Omega_e} \frac{\partial^2 \psi_i^{(1)}}{\partial x \partial y} \frac{\partial^2 \psi_i^{(1)}}{\partial x \partial y} \, dxdy, \quad (i, j = 1, 2, \ldots, r),$$

$$K_{ij}^{12} = \int_{\Omega_e} \frac{\partial \psi_i^{(1)}}{\partial x} \frac{\partial \psi_j^{(2)}}{\partial x} \, dxdy, \quad (i = 1, 2, \ldots, r; \ j = 1, 2, \ldots, s),$$

$$K_{ij}^{13} = \int_{\Omega_e} \frac{\partial \psi_i^{(1)}}{\partial y} \frac{\partial \psi_j^{(3)}}{\partial y} \, dxdy, \quad (i = 1, 2, \ldots, r; \ j = 1, 2, \ldots, p),$$

$$K_{ij}^{22} = \int_{\Omega_e} (-\bar{D}_{22}) \psi_i^{(2)} \psi_j^{(2)} \, dxdy, \quad (i, j = 1, 2, \ldots, s),$$

$$K_{ij}^{23} = \int_{\Omega_e} (\bar{D}_{12}) \psi_i^{(2)} \psi_j^{(3)} \, dxdy, \quad (i = 1, 2, \ldots, s; \ j = 1, 2, \ldots, p), \qquad (9.5.37)$$

$$K_{ij}^{33} = \int_{\Omega_e} (-\bar{D}_{11}) \psi_i^{(3)} \psi_j^{(3)} \, dxdy, \quad (i, j = 1, 2, \ldots, p),$$

$$F_i^1 = \int_{\Omega_e} q \psi_i^{(1)} \, dxdy + \oint_{\Gamma_e} V_n \psi_i^{(1)} \, dS, \quad (i = 1, 2, \ldots, r),$$

$$F_i^2 = \oint_{\Gamma_e} \theta_x n_x \psi_i^{(2)} \, dS, \ (i = 1, 2, \ldots, s), \quad F_i^3 = \oint_{\Gamma_e} \theta_y n_y \psi_i^{(3)} \, dS, \ (i = 1, 2, \ldots, p),$$

$$S_{ij} = \int_{\Omega_e} \psi_i^{(1)} \psi_j^{(1)} \, dxdy, \quad (i, j = 1, 2, \ldots, r),$$

$$G_{ij} = \int_{\Omega_e} \left[N_1 \frac{\partial \psi_i^{(1)}}{\partial x} \frac{\partial \psi_j^{(1)}}{\partial x} + N_6 \left(\frac{\partial \psi_i^{(1)}}{\partial x} \frac{\partial \psi_j^{(1)}}{\partial y} + \frac{\partial \psi_i^{(1)}}{\partial y} \frac{\partial \psi_j^{(1)}}{\partial x} \right) \right. $$
$$\left. + N_2 \frac{\partial \psi_i^{(1)}}{\partial y} \frac{\partial \psi_j^{(1)}}{\partial y} \right] dxdy \quad (i, j = 1, 2, \ldots, r).$$

This completes the development of the mixed finite element models. Next, we consider some numerical examples.

Example 9.5.1

Use various mixed plate elements discussed in Sections 9.5.2 and 9.5.3 to analyze bending of rectangular plates with simply supported and clamped boundary conditions.

Solution: The following mixed finite elements are used in obtaining the numerical results presented in Tables 9.5.1. and 9.5.2:

Mixed model I. Bilinear approximations are used for deflection w, and moments M_x, M_y, and M_{xy} (mixed model IA). Thus, the element has four nodes and four degrees of freedom at each node [see Fig. 9.5.1(a)], resulting in a 16×16 element stiffness matrix.

Mixed model II. This model is subdivided into three models depending on the type of interpolation used [see Fig. 9.5.2(a)–(c)]. These elements have stiffness matrices of order 12×12, 24×24, and 8×8, respectively.

Table 9.5.1 Maximum deflection and bending moment of simply supported, square, isotropic ($\nu = 0.3$) plate under uniformly distributed load.

Mesh	Mixed models				Herrmann
Size	IA	IIA	IIB	IIC	[135]
	Central deflection, $(wD/q_0a^4)10^2$ $(0.4062)^a$				
1×1	0.4613 (16)^c	0.3906 (12)	0.3867 (24)	0.4943 (8)	0.9018 (6)
2×2	0.4237 (36)	0.4082 (27)	0.4053 (63)	0.4289 (21)	0.5127 (15)
4×4	0.4106 (100)	0.4069 (75)	0.4062 (195)	0.4117 (65)	0.4316 (45)
6×6	0.4082 (196)	0.4066 (147)	0.4062 (399)	0.4087 (133)	0.4174 (101)
	Bending moment at the center $(M_{xx}/q_0a^2)10$ $(0.4789)^b$				
1×1	0.7196	0.6094	0.3813	0.3482	0.328
2×2	0.5246	0.5049	0.4818	0.4498	0.446
4×4	0.4891	0.4849	0.4788	0.4721	0.471
6×6	0.4834	0.4851	0.4789	0.4579	0.476

^{a,b}Exact solutions from Eqs (8) and (9), respectively, of **Example 7.2.9**.
^{c}Number of degrees of freedom in the mesh of a quarter plate model.

Table 9.5.2 Maximum deflection and bending moment of clamped, square, isotropic ($\nu = 0.3$) plate under uniformly distributed load.

Mesh	Mixed models				Herrmann
Size	IA	IIA	IIB	IIC	[135]
	Central deflection, $(wD/q_0a^4)10^2(0.1265)^a$				
1×1	0.1664 (16)^b	0.1563 (12)	0.1466 (24)	0.2278 (8)	0.7440 (6)
2×2	0.1529 (36)	0.1480 (27)	0.1260 (63)	0.1627 (21)	0.2854 (15)
4×4	0.1339 (100)	0.1325 (75)	0.1264 (195)	0.1359 (65)	0.1696 (45)
6×6	0.1299 (196)	0.1292 (147)	0.1265 (399)	0.1307 (133)	0.1463 (101)
	Bending moment at the center $(M_{xx}/q_0a^2)10$ $(0.231)^a$				
1×1	0.5193	0.4875	0.2056	0.2487	0.208
2×2	0.3166	0.2899	0.2248	0.2432	0.242
4×4	0.2478	0.2443	0.2287	0.2339	0.235
6×6	0.2374	0.2358	0.2290	0.2313	0.232

^{a}From Timoshenko and Woinowsky-Krieger [124]; also see **Example 7.2.12**.
^{b}Number of degrees of freedom in the mesh of a quarter plate.

The mixed finite element solutions are presented in Tables 9.5.1 and 9.5.2 for isotropic ($\nu = 0.3$) plates and in Table 9.5.3 for orthotropic plates. The orthotropic material properties used are:

Glass-epoxy: $E_1 = 7.8 \times 10^6$ psi, $E_2 = 2.6 \times 10^6$ psi, $\nu_{12} = 0.25$, $G_{12} = 1.3 \times 10^6$ psi.
Graphite-epoxy: $E_1 = 31.8 \times 10^6$ psi, $E_2 = 1.02 \times 10^6$ psi, $\nu_{12} = 0.31$, $G_{12} = 0.96 \times 10^6$ psi.

The present finite element solutions for center deflection and bending moments are compared with various finite element solutions available in the literature. From the numerical results it is clear that mixed models IIA and IIB are the best among the mixed models discussed here.

Table 9.5.3 Maximum deflection and bending moments of simply supported, square, orthotropic plates under uniformly distributed transverse load [$\bar{w} = (wH/q_0 a^4)10^3$, $H = D_{12} + 2D_{66}$; $\bar{M}_{xx} = (M_{xx}/q_0 a^2)10$; $\bar{M}_{yy} = (M_{yy}/q_0 a^2)10^2$].

Mesh	IIA			IIB		
	$\bar{w}$	$\bar{M}_{xx}$	$\bar{M}_{yy}$	$\bar{w}$	$\bar{M}_{xx}$	$\bar{M}_{yy}$
Glass-epoxy plates						
1×1	2.9737	0.9435	3.6290	2.9412	0.5718	2.2297
2×2	3.1041	0.8078	2.8937	3.3081	0.7687	2.7849
4×4	3.0930	0.7737	2.7851	3.0872	0.7627	2.7556
6×6	3.0901	0.7676	2.7685	3.0876	0.7628	2.7556
8×8	3.0890	0.7654	2.7628	3.0876	0.7628	2.7556
Exact[a]	3.0876	0.7628	2.7556	3.0876	0.7628	2.7556
Graphite-epoxy plates						
1×1	0.9348	1.6152	0.6720	0.9207	0.9491	0.3800
2×2	0.9346	1.3570	0.2860	0.9224	1.3094	0.3370
4×4	0.9220	1.2845	0.3143	0.9198	1.2661	0.3274
6×6	0.9208	1.2740	0.3252	0.9199	1.2658	0.3272
8×8	0.9204	1.2704	0.3270	0.9200	1.2658	0.3272
Exact[a]	0.9200	1.2659	0.3271	0.9200	1.2659	0.3271

[a] From Reddy [50].

Example 9.5.2

Determine the fundamental frequencies and buckling loads of plates for the following cases: (1) vibration of simply supported and clamped square plates; (2) vibration of rectangular cantilever plate; and (3) buckling of simply supported and clamped square plates under various edge loads.

Solution: The results are presented in Tables 9.5.4 to 9.5.6.

Table 9.5.4 Natural frequencies ($\lambda_v = \omega^2 \rho h\, a^4/D$) of simply supported (or hinged) and clamped square isotropic ($\nu = 0.3$) plates.

Mesh	Hinged plates (Exact: 389.636)		Clamped plates (Exact: 1295.28)	
	IIA	IIB	IIA	IIB
1×1	576.000	408.439	1440.000	1355.077
2×2	431.529	390.461	1361.283	1303.380
3×3	407.809	389.792	1325.837	1298.051
4×4	399.766	389.685	1312.403	1296.114

Table 9.5.5 Frequency parameters ($\lambda_v^{1/2} = \omega a^2 \sqrt{\rho h/D}$) of cantilever plates.

Mode	Mesh	Present analysis IIA	Present analysis IIB	Anderson [148]	Barton [149][a] Ritz	Barton [149][a] Test	Plunkett[a] [150]
1	2×1	3.27	3.46	3.99	3.47	3.42	3.50
2		14.46	14.60	15.30	14.93	14.52	14.50
3		22.89	22.49	21.16	21.26	20.86	21.70
4		51.28	49.07	49.47	48.71	46.90	48.10
1	4×2	3.41	3.44	3.44	3.47	3.42	3.50
2		14.55	14.76	14.76	14.93	14.52	14.50
3		22.64	21.51	21.60	21.26	20.86	21.70
4		49.93	48.13	48.28	48.71	46.90	48.10

[a]Results are independent of the mesh.

Table 9.5.6 Buckling coefficients (λ_b) for simply supported (S) and clamped (C), isotropic ($\nu = 0.3$) square plates under uniform inplane edge loads.

Load case (Exact)[a]	Mesh	Simply supported IIA	Simply supported IIB	Simply supported Ref. [151]	Clamped IIA	Clamped IIB	Clamped Ref. [151]
Uniaxial	2×2	4.2095	4.0063	4.0158	11.5300	10.2009	–
(S: 4.00)	3×3	4.0922	4.0012	4.0032	10.6880	10.1107	–
(C: 10.07)	4×4	4.0517	4.0004	4.0010	10.4120	10.0866	–
	8×8	4.0129	4.0000	4.0007	10.1560	10.0748	–
Biaxial	2×2	2.1048	2.0031	–	6.0592	5.3746	5.3622
(S: 2.00)	3×3	2.0461	2.0006	–	5.6339	5.3233	5.3660
(C: 5.30)	4×4	2.0258	2.0002	–	5.4868	5.3102	5.3271
	8×8	2.0064	2.0000	–	5.3487	5.3040	5.3054
Shear[b]	2×2	–	14.4124	10.0160	–	29.3967	23.2640
(S: 9.34)	3×3	14.3220	9.6758	9.4180	25.7773	15.4640	15.0430
(C: 14.71)	4×4	11.1499	9.3816	–	18.5902	14.8361	–
	8×8	10.2898	9.3403	–	16.7104	14.7123	–

[a]From Timoshenko and Gere [92] and Reddy [51].
[b]Obtained using the full plate model.

9.6   Summary

In this chapter, mixed variational formulations of problems of solid mechanics are presented. The construction of functionals associated with the Hellinger–Reissner, Reissner, and Hu–Washizu variational principles was discussed for simple problems of beams. The use of the variational methods to determine approximate solutions as well as finite element models based on the mixed variational statements is illustrated via beam theories. Alternative mixed variational formulations of the classical plate theory were also developed, and their finite element models were derived. The mixed finite element models of plates give accurate results for deflections, stresses, buckling loads, and natural frequencies. The mixed models give more accurate results when compared to certain conforming and nonconforming elements. Further, the mixed finite elements are algebraically less complex and involve less "element-forming" efforts.

Mixed formulations and associated finite element models of the first-order shear deformation plate theory are not included here, but they can be found in [145, 146]. Finite element models based on mixed formulations in which the strains are treated as independent variables (e.g., see Section 9.4.2) have enjoyed better success.

Problems

SECTION 9.2: MIXED VARIATIONAL PRINCIPLES
Derive the Euler equations of the stationary functionals in **Problems 9.1–9.4**:

9.1

$$I(u, \mathbf{v}) = \int_\Omega \left(-\frac{1}{2k}\mathbf{v} \cdot \mathbf{v} + \mathbf{v} \cdot \operatorname{grad} u - fu \right) dV - \int_{\Gamma_2} qu\, dS - \int_{\Gamma_1} \hat{\mathbf{n}} \cdot \mathbf{v}(u - \hat{u}) dS$$

9.2

$$I(w, \phi_x, \lambda) = \int_0^L \left[\frac{EI}{2}\left(\frac{d\phi_x}{dx}\right)^2 + fw + \lambda\left(\frac{dw}{dx} + \phi_x\right) \right] dx - F_0 w(L)$$

9.3

$$I(u_1, u_2, P) = \int_\Omega \left\{ \frac{\mu}{2}(u_{i,j} + u_{j,i})u_{i,j} - Pu_{i,i} \right\} dx_1 dx_2 - \int_{\Gamma_2} \hat{t}_i u_i\, dS$$

9.4

$$I(w, \phi_1, \phi_2, \lambda_1, \lambda_2) = \frac{D}{2}\int_\Omega \left[\left(\frac{\partial \phi_1}{\partial x_1}\right)^2 + \left(\frac{\partial \phi_2}{\partial x_2}\right)^2 + 2\nu\frac{\partial \phi_2}{\partial x_1}\frac{\partial \phi_2}{\partial x_2} \right.$$

$$\left. + (1 - \nu)\left(\frac{\partial \phi_1}{\partial x_2} + \frac{\partial \phi_2}{\partial x_1}\right)^2 \right] dx_1 dx_2$$

$$+ \int_\Omega \left[\lambda_1\left(\frac{\partial w}{\partial x_1} + \phi_1\right) + \lambda_2\left(\frac{\partial w}{\partial x_2} + \phi_2\right) - fw \right] dx_1 dx_2.$$

9.5 Derive the Hellinger–Reissner functional for a bar of length L stiffness EA, fixed at the left and right ends subjected to axial load P [see Eqs (9.1.1) to (9.1.4)].

9.6 Derive the stationary functional corresponding to the problem of minimizing the functional

$$I(u_i) = \int_\Omega \left[\frac{\mu}{2}(u_{i,j} + u_{j,i})u_{i,j} - f_i u_i \right] dV - \int_{\Gamma_2} \hat{t}_i u_i\, dS$$

subject to the constraint $u_{i,i} = 0$.

9.7 (*Washizu's principle*). Derive the following generalized Washizu functional:

$$\Pi_W(u_i, \varepsilon_{ij}, \sigma_{ij}) = \int_\Omega \left\{ \left[\frac{1}{2}(u_{i,j} + u_{j,i} + u_{m,i}u_{m,j}) - \varepsilon_{ij} \right]\sigma_{ij} \right.$$

$$\left. + \rho\Psi - f_i u_i \right\} dV - \int_{\Gamma_1} (u_i - \hat{u}_i)t_i\, dS - \int_{\Gamma_2} \hat{t}_i u_i\, dS,$$

where ε_{ij}, σ_{ij}, u_i and t_i are subject to independent variations, and $\Psi = \Psi(\varepsilon_{ij}, T)$ is the Gibbs free-energy potential.

9.8 Derive the Euler equations of the Washizu functional by requiring that the actual state of equilibrium of a body subjected to mechanical and thermal loads is associated with a stationary value of the functional Π_W in **Problem 9.7**.

9.9 Modify the Hellinger–Reissner variational principle for elastic bodies acted upon by mechanical and thermal loads.

9.10 The mixed variational principle for axisymmetric bending of circular plates according to the first-order theory requires $\delta\Pi_m = 0$, where

$$\Pi_m(w, \phi_r, M_r, M_\theta) = \pi \int_0^a \left[K_s G_{13} h \left(\phi_r + \frac{dw}{dr} \right)^2 + 2M_r \frac{d\phi_r}{dr} + \frac{2}{r} M_\theta \phi_r \right.$$
$$\left. + \bar{D}_{22} M_r^2 - 2\bar{D}_{12} M_r M_\theta + \bar{D}_{11} M_\theta^2 - 2qw \right] r\, dr, \quad (1)$$

where a is the radius of the plate, K_s shear correction factor, and

$$\bar{D}_{ij} = \frac{D_{ij}}{D_0}, \quad D_0 = D_{12}^2 - D_{11} D_{22}. \quad (2)$$

Obtain the Euler equations of the functional.

9.11 The principle of minimum total potential energy for axisymmetric bending of polar orthotropic plates according to the first-order shear deformation theory requires $\delta\Pi(w, \phi_r) = 0$, where

$$\delta\Pi_m(w, \phi_r) = 2 \int_b^a \left[\left(D_{11} \frac{d\phi_r}{dr} + D_{12} \frac{\phi_r}{r} \right) \frac{d\delta\phi_r}{dr} + \frac{1}{r} \left(D_{12} \frac{d\phi_r}{dr} + D_{22} \frac{\phi_r}{r} \right) \delta\phi_r \right.$$
$$\left. + A_{55} \left(\phi_r + \frac{dw}{dr} \right) \left(\delta\phi_r + \frac{d\delta w}{dr} \right) - q\delta w \right] r\, dr \quad (1)$$

where b is the inner radius and a the outer radius. Derive the displacement finite element model of the equations. In particular, show that the finite element model is of the form (i.e., define the matrix coefficients of the following equation)

$$\begin{bmatrix} \mathbf{K}^{11} & \mathbf{K}^{12} \\ (\mathbf{K}^{12})^{\mathrm{T}} & \mathbf{K}^{22} \end{bmatrix} \begin{Bmatrix} \mathbf{w} \\ \mathbf{\Phi} \end{Bmatrix} = \begin{Bmatrix} \mathbf{F} \\ \mathbf{0} \end{Bmatrix} \quad (2)$$

SECTIONS 9.4–9.5: MIXED FINITE ELEMENT MODELS

9.12 Derive the mixed finite element model for the annular plates using the following variational statement:

$$0 = 2\pi \int_b^a \left\{ A_{55} \left(\phi_r + \frac{dw}{dr} \right) \frac{d\delta w}{dr} + M_r \frac{d\delta\phi_r}{dr} + \left(\frac{1}{r} M_\theta + Q_r \right) \delta\phi_r \right.$$
$$+ \left(\frac{d\phi_r}{dr} + \bar{D}_{22} M_r - \bar{D}_{12} M_\theta \right) \delta M_r$$
$$\left. + \left[\frac{\phi_r}{r} + \left(\bar{D}_{11} M_\theta - \bar{D}_{12} M_r \right) \right] \delta M_\theta - q\delta w \right\} r\, dr, \quad (1)$$

where $Q_r = A_{55}(\phi_r + \frac{dw}{dr})$ and $\bar{D}_{ij} = D_{ij}/D_0$, $D_0 = D_{12} D_{12} - D_{11} D_{22}$. Show that the finite element model is of the form

$$\begin{bmatrix} \mathbf{K}^{11} & \mathbf{K}^{12} & \mathbf{0} & \mathbf{0} \\ & \mathbf{K}^{22} & \mathbf{K}^{23} & \mathbf{K}^{24} \\ \text{sym.} & & \mathbf{K}^{33} & \mathbf{K}^{34} \\ & & & \mathbf{K}^{44} \end{bmatrix} \begin{Bmatrix} \mathbf{w} \\ \mathbf{\Phi} \\ \mathbf{M}_r \\ \mathbf{M}_\theta \end{Bmatrix} = \begin{Bmatrix} \mathbf{F} \\ \mathbf{0} \\ \mathbf{0} \\ \mathbf{0} \end{Bmatrix}. \quad (2)$$

9.13 Consider the second-order equation

$$-\frac{d}{dx}\left(a\frac{du}{dx}\right) = f \tag{1}$$

and rewrite it as a pair of first-order equations:

$$-\frac{du}{dx} + \frac{P}{a} = 0, \quad -\frac{dP}{dx} - f = 0. \tag{2}$$

Construct the weighted residual finite element model of the equations. Assume approximation of the form

$$u = \sum_{j=1}^{m} u_j \psi_j(x), \quad P = \sum_{j=1}^{n} P_j \phi_j(x) \tag{3}$$

and obtain the finite element model

$$\begin{bmatrix} \mathbf{K}^{11} & \mathbf{K}^{12} \\ \mathbf{K}^{21} & \mathbf{K}^{22} \end{bmatrix} \begin{Bmatrix} \mathbf{u} \\ \mathbf{P} \end{Bmatrix} = \begin{Bmatrix} \mathbf{F}^1 \\ \mathbf{F}^2 \end{Bmatrix}. \tag{4}$$

The model is clearly a mixed model because (u, P) are of different kind of variables.

9.14 Develop the least-squares finite element model of equations in Eq. (2) of **Problem 9.13**, and compute element coefficient matrices and vectors when $\psi_i = \phi_i$ are the linear interpolation functions.

9.15 Consider the pair of equations:

$$\nabla u - \frac{1}{k}\mathbf{q} = 0, \quad \nabla \cdot \mathbf{q} + f = 0 \quad \text{in} \quad \Omega, \tag{1}$$

where u and $\mathbf{q}$ are the dependent variables and k and f are given functions of position (x, y) in a two-dimensional domain Ω. Derive the mixed finite element model of the equations in the form

$$\begin{bmatrix} \mathbf{K}^{11} & \mathbf{K}^{12} & \mathbf{K}^{13} \\ & \mathbf{K}^{22} & \mathbf{K}^{23} \\ \text{symm.} & & \mathbf{K}^{33} \end{bmatrix} \begin{Bmatrix} \mathbf{u} \\ \mathbf{q}^1 \\ \mathbf{q}^2 \end{Bmatrix} = \begin{Bmatrix} \mathbf{F}^1 \\ \mathbf{F}^2 \\ \mathbf{F}^3 \end{Bmatrix}. \tag{2}$$

"Gentleness, self-sacrifice, and generosity are the exclusive possession of no one race or religion."

" You must be the change you wish to see in the world."
– *Mohandas Karamchand Gandhi*

Preeminent leader of the Indian independence movement in British-ruled India, employing nonviolent civil disobedience.

Analysis of Functionally Graded Beams and Plates

10.1 Introduction

On the threshold of the 21st century, new materials that feature *electro-thermo-mechanical coupling, functionality, intelligence,* and *miniaturization* (down to nano-length scales) are being developed. Functionally gradient materials (FGMs) are a class of composite materials that have a gradual variation of material properties in a chosen spatial direction of the structure, resulting in structural components having a smooth variation of mechanical properties; see Hasselman and Youngblood [152], Yamanouchi *et al.* [153], Koizumi [154], Noda [155], and Noda and Tsuji [156]. These novel materials were proposed as thermal barrier coatings for applications in space planes, space structures, nuclear reactors, turbine rotors, flywheels, gears, cams, cutting tools, high temperature chambers, furnace liners, and microelectronics, to name a few. Such enhancements in properties endow FGMs with qualities such as resilience to fracture through reduction in propensity for stress concentration. As conceived and manufactured today, these materials are isotropic but inhomogeneous in the direction of the variation. One reason for increased interest in FGMs is that it is possible to create certain types of FGM structures capable of adapting to operating conditions.

Two-constituent FGMs are most common in practice. For example, in a thermal barrier plate or shell structure made of a mixture of ceramic and metals for use in thermal environments, the mixture varies in its composition from one surface to the other (i.e., through the thickness of the plate or shell); 100% ceramic is on the high temperature side, while 100% metal on the other side, and their composition varies in a predetermined manner through the thickness (see Fig. 10.1.1). The ceramic constituent of the material provides the high temperature resistance due to its low thermal conductivity. The ductile metal constituent, on the other hand, prevents fracture due to high temperature gradient in a very short period of time.

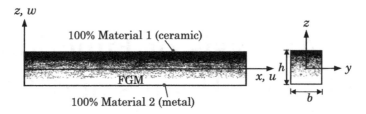

Fig. 10.1.1 A two-constituent beam with ceramic surface at the top and metal at the bottom.

The vast majority of two-constituent FGM studies employed either a power law or exponential distribution of the materials. In the power-law model, which is more commonly used in bending, vibration, and buckling studies, a typical material property $\mathcal{P}$ (e.g., Young's modulus E and mass density ρ) is assumed to vary through the thickness according to the formula (see Praveen and Reddy [157], Praveen *et al.* [158], Reddy and Chin [159], and Reddy [160])

$$\mathcal{P}(z) = (\mathcal{P}_1 - \mathcal{P}_2)\, f(z) + \mathcal{P}_2, \quad f(z) = \left(\frac{1}{2} + \frac{z}{h}\right)^n \qquad (10.1.1)$$

where $\mathcal{P}_1$ and $\mathcal{P}_2$ are the material properties of the top (material 1) and bottom (material 2) faces of the beam or plate, respectively, and n is the *power-law index*. Figure 10.1.2 shows the variation of $f(z)$ through the thickness.

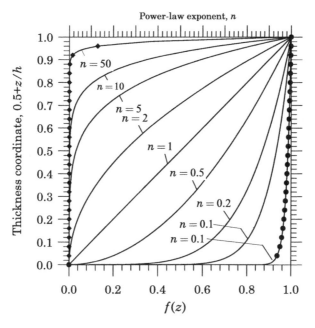

Fig. 10.1.2 Variation of the volume fraction $f(z)$ as a function of the dimensionless thickness coordinate, $\xi = 0.5 + z/h$, for various values of the power-law index, n.

The following integrals are of interest in the sequel:

$$\int_{-\frac{h}{2}}^{\frac{h}{2}} f(z)\,dz = \frac{h}{n+1},$$

$$\int_{-\frac{h}{2}}^{\frac{h}{2}} f(z)z\,dz = \frac{nh^2}{2(n+1)(n+2)}, \qquad (10.1.2)$$

$$\int_{-\frac{h}{2}}^{\frac{h}{2}} f(z)z^2\,dz = \frac{(2+n+n^2)h^3}{4(n+1)(n+2)(n+3)}.$$

Noda [155] presented an extensive review that covers a wide range of topics from thermoelastic to thermoinelastic problems. He also discussed the importance of temperature-dependent properties on thermoelastic problems. He further presented analytical methods to handle transient heat conduction problems and indicated the necessity for the optimization of FGM properties. Tanigawa [161] compiled a comprehensive review on the thermoelastic analyses of FGMs.

A number of other investigations dealing with thermal stresses and deformations in beams, plates, and cylinders had been published in the literature (see, e.g., Noda and Tsuji [156], Praveen and Reddy [157], Praveen *et al.* [158], Reddy and Chin [159], Reddy [160], Vel and Batra [162], Shen [163], and Yang and Shen [164], among others). Among these studies that concern the thermoelastic analysis of plates, beams, or cylinders made of FGMs where the material properties have been considered temperature dependent are Noda and Tsuji [156], Praveen and Reddy [157], Praveen *et al.* [158], Yang and Shen [164], and Kitipornchai *et al.* [165]. The work of Praveen and Reddy [157], Reddy [160], Aliaga and Reddy [166], Reddy [167], Reddy and Kim [168], and Kim and Reddy [169, 170] also considered the von Kármán nonlinearity and transverse shear deformation in plates. In [166–170] the third-order plate theory of Reddy [117–119] was considered and the effect of modified couple stress was also included.

The present chapter is devoted to the development of functionally graded beams and plates. Two-constituent FGMs are considered, and the variation is assumed to be through thickness of the beam or plate. The governing equations of FGM beams and plates are developed using the dynamics version of the principle of virtual displacements (or Hamilton's principle), and their analytical and finite element solutions are presented. The main difference between the developments of the FGM beams and plates and their homogeneous counterparts is in the constitutive equations. Thus, the equations of motion expressed in terms of the stress resultants are the same as for homogeneous beams, but the relations between the stress resultants and the generalized displacements will bring the coupling between the in-plane and out-of-place (i.e., extensional and bending) displacements. The governing equations of beams and plates are summarized without repeating the details covered in the previous chapters. The differences between homogeneous and FGM beams and plates will be detailed (for more details, see [175–181] for FGM beams and [157–160, 167–170, 182–184] for FGM circular and rectangular plates).

10.2 Functionally Graded Beams

10.2.1 The Bernoulli–Euler Beam Theory

10.2.1.1 Displacement and Strain Fields

The Bernoulli–Euler beam theory (BET) is based on the displacement field

$$\mathbf{u}(x, z, t) = [u(x, t) + z\theta_x(x, t)]\,\hat{\mathbf{e}}_x + w(x, t)\,\hat{\mathbf{e}}_z, \quad \theta_x \equiv -\frac{\partial w}{\partial x}, \qquad (10.2.1)$$

where (u, w) are the axial and transverse displacements of the point $(x, 0)$ on the midplane (i.e., $z = 0$) of the beam, and $(\hat{\mathbf{e}}_x, \hat{\mathbf{e}}_z)$ are unit vectors along the (x, z) coordinates.

We assume that the axial strain $(\partial u/\partial x)$ is of order ϵ, $(\partial w/\partial x)$ is of order $\sqrt{\epsilon}$. The simplified Lagrangian strain tensor $\mathbf{E}$ is of the form (omitting terms of order ϵ^2)

$$\mathbf{E} \approx \varepsilon = \left[\frac{\partial u}{\partial x} + z\frac{\partial \theta_x}{\partial x} + \frac{1}{2}\left(\frac{\partial w}{\partial x}\right)^2\right]\hat{\mathbf{e}}_x\hat{\mathbf{e}}_x. \qquad (10.2.2)$$

10.2.1.2 Equations of Motion and Boundary Conditions

The Hamilton principle for the beam has the form

$$\begin{aligned}
0 = \int_0^T \int_0^L &\left[m_0\left(\frac{\partial u}{\partial t}\frac{\partial \delta u}{\partial t} + \frac{\partial w}{\partial t}\frac{\partial \delta w}{\partial t}\right) - m_1\left(\frac{\partial u}{\partial t}\frac{\partial^2 \delta w}{\partial x \partial t} + \frac{\partial^2 w}{\partial x \partial t}\frac{\partial \delta u}{\partial t}\right)\right. \\
&\left. + m_2\frac{\partial^2 w}{\partial x \partial t}\frac{\partial^2 \delta w}{\partial x \partial t} - N_{xx}\,\delta\varepsilon_{xx}^{(0)} - M_{xx}\,\delta\varepsilon_{xx}^{(1)} + f_x\,\delta u + f_z\,\delta w\right] dx\,dt \\
&+ \int_0^T \left[Q_1\delta u(0, t) + Q_2\delta w(0, t) + Q_3\delta\theta_x(0, t) + Q_4\delta u(L, t)\right. \\
&\left. + Q_5\delta w(L, t) + Q_6\delta\theta_x(L, t)\right] dt
\end{aligned} \qquad (10.2.3)$$

where f_x and f_z are the distributed loads along the x and z axes, respectively; $\varepsilon_{xx}^{(0)}$ and $\varepsilon_{xx}^{(1)}$ are the extensional and bending strains, respectively:

$$\varepsilon_{xx}^{(0)} = \frac{\partial u}{\partial x} + \frac{1}{2}\left(\frac{\partial w}{\partial x}\right)^2, \quad \varepsilon_{xx}^{(1)} = \frac{\partial \theta_x}{\partial x} = -\frac{\partial^2 w}{\partial x^2}, \qquad (10.2.4)$$

(N_{xx}, M_{xx}) are the stress resultants (axial force and bending moment, respectively)

$$N_{xx} = \int_A \sigma_{xx}\, dA, \quad M_{xx} = \int_A z\sigma_{xx}\, dA, \tag{10.2.5}$$

and Q_i are the generalized forces at the ends of a beam, as shown in Fig. 10.2.1.

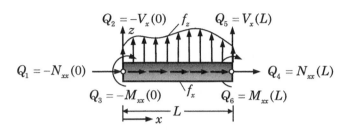

Fig. 10.2.1 External generalized forces at the two ends of the beam.

The Euler–Lagrange equations resulting from the Hamilton principle are

$$\frac{\partial N_{xx}}{\partial x} + f_x = m_0 \frac{\partial^2 u}{\partial t^2} - m_1 \frac{\partial^3 w}{\partial x \partial t^2} \tag{10.2.6}$$

$$\frac{\partial^2 M_{xx}}{\partial x^2} + \frac{\partial}{\partial x}\left(N_{xx} \frac{\partial w}{\partial x}\right) + f_z = m_0 \frac{\partial^2 w}{\partial t^2} + m_1 \frac{\partial^3 u}{\partial x \partial t^2} - m_2 \frac{\partial^4 w}{\partial t^2 \partial x^2}, \tag{10.2.7}$$

where (m_0, m_1, m_2) are the mass inertias

$$(m_0, m_1, m_2) = \int_A \rho(z)(1, z, z^2)\, dA. \tag{10.2.8}$$

The force (or natural) boundary conditions are to specify the following expressions (when the corresponding displacements are not specified)

$$
\begin{aligned}
Q_1 + N_{xx}(0, t) &= 0, & Q_4 - N_{xx}(L, t) &= 0 \\
Q_2 + V_x(0, t) &= 0, & Q_5 - V_x(L, t) &= 0 \\
Q_3 + M_{xx}(0, t) &= 0, & Q_6 - M_{xx}(L, t) &= 0,
\end{aligned}
\tag{10.2.9}
$$

where

$$V_x \equiv N_{xx} \frac{\partial w}{\partial x} + \frac{\partial M_{xx}}{\partial x} - m_1 \frac{\partial^2 u}{\partial t^2} + m_2 \frac{\partial^3 w}{\partial t^2 \partial x} \tag{10.2.10}$$

10.2.2 The Timoshenko Beam Theory

10.2.2.1 Displacement and Strain Fields

The displacement field in the Timoshenko beam theory (TBT) is

$$\mathbf{u}(x, z, t) = [u(x, t) + z\phi_x(x, t)]\,\hat{\mathbf{e}}_x + w(x, t)\,\hat{\mathbf{e}}_z, \tag{10.2.11}$$

where ϕ_x is the independent rotation of the transverse line at x about the y axis.

The Lagrangian strain tensor **E** is of the form (omitting terms of order ϵ^2)

$$\mathbf{E} \approx \varepsilon = \left[\frac{\partial u}{\partial x} + z\frac{\partial \phi_x}{\partial x} + \frac{1}{2}\left(\frac{\partial w}{\partial x}\right)^2 \right] \hat{\mathbf{e}}_x\hat{\mathbf{e}}_x + \frac{1}{2}\left(\phi_x + \frac{\partial w}{\partial x}\right)(\hat{\mathbf{e}}_x\hat{\mathbf{e}}_z + \hat{\mathbf{e}}_z\hat{\mathbf{e}}_x).$$

(10.2.12)

The TBT requires shear correction factors to compensate for the error due to the constant state of shear stress predicted by the kinematics. The shear correction factor depends, in general, not only on the material and geometric parameters but also on the load and boundary conditions.

10.2.2.2 Equations of Motion and Boundary Conditions

The principle of virtual displacements for the TBT is given by

$$0 = \int_0^T \int_0^L \left[m_0\left(\frac{\partial u}{\partial t}\frac{\partial \delta u}{\partial t} + \frac{\partial w}{\partial t}\frac{\partial \delta w}{\partial t}\right) + m_1\left(\frac{\partial u}{\partial t}\frac{\partial \delta \phi_x}{\partial t} + \frac{\partial \phi_x}{\partial t}\frac{\partial \delta u}{\partial t}\right) \right.$$
$$+ m_2\frac{\partial \phi_x}{\partial t}\frac{\partial \delta \phi_x}{\partial t} - N_{xx}\,\delta\varepsilon_{xx}^{(0)} - M_{xx}\,\delta\varepsilon_{xx}^{(1)} - Q_x\,\delta\gamma_{xz}$$
$$\left. + f_x\,\delta u + f_z\,\delta w \right] dx\,dt + \int_0^T \left[Q_1\delta u(0,t) + Q_2\delta w(0,t) \right.$$
$$\left. + Q_3\delta\phi_x(0,t) + Q_4\delta u(L,t) + Q_5\delta w(L,t) + Q_6\delta\phi_x(L,t) \right] dt \quad (10.2.13)$$

where

$$\varepsilon_{xx}^{(0)} = \frac{\partial u}{\partial x} + \frac{1}{2}\left(\frac{\partial w}{\partial x}\right)^2, \quad \varepsilon_{xx}^{(1)} = \frac{\partial \phi_x}{\partial x}, \quad \gamma_{xz} = \phi_x + \frac{\partial w}{\partial x}. \quad (10.2.14)$$

The Euler–Lagrange equations are

$$\frac{\partial N_{xx}}{\partial x} + f_x = m_0\frac{\partial^2 u}{\partial t^2} + m_1\frac{\partial^2 \phi_x}{\partial t^2}, \quad (10.2.15)$$

$$\frac{\partial}{\partial x}\left(Q_x + N_{xx}\frac{\partial w}{\partial x}\right) + f_z = m_0\frac{\partial^2 w}{\partial t^2}, \quad (10.2.16)$$

$$\frac{\partial M_{xx}}{\partial x} - Q_x = m_1\frac{\partial^2 u}{\partial t^2} + m_2\frac{\partial^2 \phi_x}{\partial t^2}. \quad (10.2.17)$$

where (N_{xx}, M_{xx}) are the stress resultants defined in Eq. (10.2.5) and the shear stress resultant Q_x is defined by

$$Q_x = K_s \int_A \sigma_{xz}\,dA, \quad (10.2.18)$$

where K_s is the shear correction factor. The boundary conditions involve specifying the following variables:

$$Q_1 + N_{xx}(0,t) = 0, \qquad\qquad Q_4 - N_{xx}(L,t) = 0$$
$$Q_2 + V_x(0,t) = 0, \qquad\qquad Q_5 - V_x(L,t) = 0 \qquad (10.2.19)$$
$$Q_3 + M_{xx}(0,t) = 0, \qquad\qquad Q_6 - M_{xx}(L,t) = 0$$

where

$$V_x = N_{xx}\frac{\partial w}{\partial x} + Q_x \qquad\qquad (10.2.20)$$

10.2.3 Equations of Motion in Terms of Generalized Displacements

10.2.3.1 Constitutive Equations

For an isotropic linear elastic material, the uniaxial stress–strain relations are

$$\sigma_{xx}(x,z,t) = E(z)\left[\varepsilon_{xx}(x,z,t) - \alpha\Delta T(z,t)\right], \quad \sigma_{xz} = \frac{E(z)}{2(1+\nu)}\gamma_{xz}, \quad (10.2.21)$$

where E is Young's modulus, ν is the Poisson's ratio that is assumed to be constant, α is the coefficient of thermal expansion, and $\Delta T = T - T_0$ is the temperature increment from the room temperature, T_0. The temperature T is assumed to vary only through the thickness.

10.2.3.2 Stress Resultants of BET

The stress resultants in the BET are

$$N_{xx} = \int_A \sigma_{xx}\, dA = A_{xx}\left[\frac{\partial u}{\partial x} + \frac{1}{2}\left(\frac{\partial w}{\partial x}\right)^2\right] + B_{xx}\frac{\partial\theta_x}{\partial x} - N_{xx}^T, \qquad (10.2.22)$$

$$M_{xx} = \int_A z\sigma_{xx}\, dA = B_{xx}\left[\frac{\partial u}{\partial x} + \frac{1}{2}\left(\frac{\partial w}{\partial x}\right)^2\right] + D_{xx}\frac{\partial\theta_x}{\partial x} - M_{xx}^T, \qquad (10.2.23)$$

where A_{xx}, B_{xx}, and D_{xx} are the extensional, extensional–bending, and bending stiffness coefficients

$$(A_{xx}, B_{xx}, D_{xx}) = \int_A (1, z, z^2)E(z)\, dA, \qquad (10.2.24)$$

and N_{xx}^T and M_{xx}^T are the thermal stress resultants defined by

$$N_{xx}^T = \int_A \alpha(z)\,E(z)\,\Delta T\, dA, \quad M_{xx}^T = \int_A z\,\alpha(z)\,E(z)\,\Delta T\, dA. \qquad (10.2.25)$$

10.2.3.3   Stress Resultants of TBT

The stress resultants of the TBT are

$$N_{xx} = \int_A \sigma_{xx}\, dA = A_{xx}\left[\frac{\partial u}{\partial x} + \frac{1}{2}\left(\frac{\partial w}{\partial x}\right)^2\right] + B_{xx}\frac{\partial \phi_x}{\partial x} - N_{xx}^T, \qquad (10.2.26)$$

$$M_{xx} = \int_A z\sigma_{xx}\, dA = B_{xx}\left[\frac{\partial u}{\partial x} + \frac{1}{2}\left(\frac{\partial w}{\partial x}\right)^2\right] + D_{xx}\frac{\partial \phi_x}{\partial x} - M_{xx}^T, \qquad (10.2.27)$$

$$Q_x = K_s \int_A \sigma_{xz}\, dA = K_s S_{xz}\left(\phi_x + \frac{\partial w}{\partial x}\right), \qquad (10.2.28)$$

where K_s is the shear correction factor and S_{xz} is the shear stiffness coefficient

$$S_{xz} = \frac{1}{2(1+\nu)}\int_A E(z)\, dA. \qquad (10.2.29)$$

10.2.3.4   Equations of Motion of the BET

The equations of motion of the BET in terms of the generalized displacements $(u, w, \theta_x = -\partial w/\partial x)$ are

$$\frac{\partial}{\partial x}\left\{A_{xx}\left[\frac{\partial u}{\partial x} + \frac{1}{2}\left(\frac{\partial w}{\partial x}\right)^2\right] + B_{xx}\frac{\partial \theta_x}{\partial x} - N_{xx}^T\right\} + f_x$$
$$= m_0\frac{\partial^2 u}{\partial t^2} - m_1\frac{\partial^3 w}{\partial x \partial t^2}, \qquad (10.2.30)$$

$$\frac{\partial^2}{\partial x^2}\left\{B_{xx}\left[\frac{\partial u}{\partial x} + \frac{1}{2}\left(\frac{\partial w}{\partial x}\right)^2\right] + D_{xx}\frac{\partial \theta_x}{\partial x} - M_{xx}^T\right\}$$
$$+\frac{\partial}{\partial x}\left[\frac{\partial w}{\partial x}\left\{A_{xx}\left[\frac{\partial u}{\partial x} + \frac{1}{2}\left(\frac{\partial w}{\partial x}\right)^2\right] + B_{xx}\frac{\partial \theta_x}{\partial x} - N_{xx}^T\right\}\right] + f_z$$
$$= m_0\frac{\partial^2 w}{\partial t^2} + m_1\frac{\partial^3 u}{\partial x \partial t^2} - m_2\frac{\partial^4 w}{\partial t^2 \partial x^2}. \qquad (10.2.31)$$

10.2.3.5   Equations of Motion of the TBT

The equations of motion of the TBT in terms of the generalized displacements (u, w, ϕ_x) are

$$\frac{\partial}{\partial x}\left\{A_{xx}\left[\frac{\partial u}{\partial x} + \frac{1}{2}\left(\frac{\partial w}{\partial x}\right)^2\right] + B_{xx}\frac{\partial \phi_x}{\partial x} - N_{xx}^T\right\} + f_x$$
$$= m_0\frac{\partial^2 u}{\partial t^2} + m_1\frac{\partial^2 \phi_x}{\partial t^2}, \qquad (10.2.32)$$

$$\frac{\partial}{\partial x}\left[\frac{\partial w}{\partial x}\left\{A_{xx}\left[\frac{\partial u}{\partial x}+\frac{1}{2}\left(\frac{\partial w}{\partial x}\right)^2\right]+B_{xx}\frac{\partial \phi_x}{\partial x}-N_{xx}^T\right\}\right]$$

$$+\frac{\partial}{\partial x}\left[K_s S_{xz}\left(\phi_x+\frac{\partial w}{\partial x}\right)\right]+f_z=m_0\frac{\partial^2 w}{\partial t^2}, \tag{10.2.33}$$

$$\frac{\partial}{\partial x}\left\{B_{xx}\left[\frac{\partial u}{\partial x}+\frac{1}{2}\left(\frac{\partial w}{\partial x}\right)^2\right]+D_{xx}\frac{\partial \phi_x}{\partial x}-M_{xx}^T\right\}$$

$$-\left[K_s S_{xz}\left(\phi_x+\frac{\partial w}{\partial x}\right)\right]=m_1\frac{\partial^2 u}{\partial t^2}+m_2\frac{\partial^2 \phi_x}{\partial t^2}. \tag{10.2.34}$$

10.2.4 Stiffness Coefficients

We assume that the beam is graded through its height using two constituents, whose Young's moduli and mass densities are (E_1,ρ_1) and (E_2,ρ_2), respectively, and the Poisson ratio ν is assumed to be a constant:

$$E(z)=[E_1-E_2]\left(\frac{1}{2}+\frac{z}{h}\right)^n+E_2, \quad \rho(z)=[\rho_1-\rho_2]\left(\frac{1}{2}+\frac{z}{h}\right)^n+\rho_2, \tag{10.2.35}$$

where n is the power-law index that dictates the volume fractions of the two materials.

The stiffness coefficients $(A_{xx},B_{xx},D_{xx},S_{xz})$ and mass inertias (m_0,m_1,m_2), appearing in the beam theories, can be expressed in terms of the modulus ratio $M=E_1/E_2$, ρ_1, ρ_2, and power-law index n as

$$A_{xx}=E_2 A_0\frac{M+n}{1+n}, \quad B_{xx}=E_2 B_0\frac{n(M-1)}{2(1+n)(2+n)}$$

$$D_{xx}=E_2 I_0\left[\frac{(6+3n+3n^2)M+(8n+3n^2+n^3)}{6+11n+6n^2+n^3}\right]$$

$$S_{xz}=\frac{E_2 A_0}{2(1+\nu)}\frac{M+n}{1+n}, \quad A_0=bh, \quad B_0=bh^2, \quad I_0=\frac{bh^3}{12}$$

$$m_0=\frac{A_0}{1+n}(\rho_1+n\rho_2), \quad m_1=\frac{B_0 n}{2(1+n)(2+n)}(\rho_1-\rho_2) \tag{10.2.36}$$

$$m_2=I_0\left[\frac{(6+3n+3n^2)\rho_1+(8n+3n^2+n^3)\rho_2}{6+11n+6n^2+n^3}\right].$$

Figure 10.2.2 contains the variation of the dimensionless axial stiffness ($\bar{A}_{xx}=A_{xx}/E_2 A_0$) and bending stiffness ($\bar{D}_{xx}=D_{xx}/E_2 I_0$) as functions of the power-law index n for various values of the modulus ratio $M=E_1/E_2\geq 1$, and Figure 10.2.3 contains similar plots for the dimensionless axial–bending coupling stiffness ($\bar{B}_{xx}=B_{xx}/E_2 B_0$). It is clear that $\bar{A}_{xx}$ is the maximum at $n=0$ and decreases with increasing value of n; $\bar{B}_{xx}$ is zero at $n=0$, increases to a maximum at $n = \sqrt{2}$ and then decreases with increasing value of n; $\bar{D}_{xx}$ is the maximum at $n = 0$ and decreases with increasing value of n.

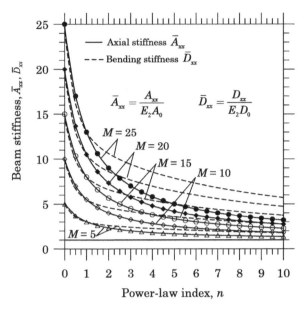

Fig. 10.2.2 Variation of the normalized axial stiffness $\bar{A}_{xx}(n)$ and bending stiffness $\bar{D}_{xx}(n)$ as functions of the power-law index, n, for various values of the modulus ratio, $M = E_1/E_2$.

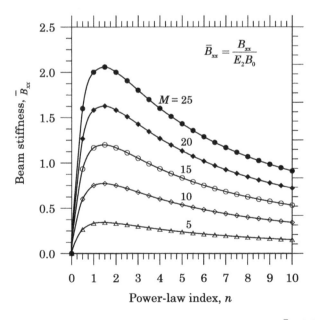

Fig. 10.2.3 Variation of the normalized bending-stretching stiffness $\bar{B}_{xx}(n)$ as a function of the power-law index, n, for various values of the modulus ratio, $M = E_1/E_2$.

10.3 Functionally Graded Circular Plates

10.3.1 Introduction

In this section, we develop governing equations for axisymmetric bending of circular plates, accounting for through-thickness power-law variation of a two-constituent material as per Eq. (10.1.1) and the von Kármán nonlinearity (see Reddy and Berry [182]). The CPT and FSDT are considered. We select the cylindrical coordinate system (r, θ, z) such that r is the radial coordinate outward from the center of the plate $(b \leq r \leq a)$, with a and b being the external and internal radii, respectively; z denotes the transverse coordinate $(-h/2 \leq z \leq h/2)$; and θ is the angular coordinate $(0 \leq \theta \leq 2\pi)$, as shown in Fig. 10.3.1. We assume that the material of the plate is isotropic but varies from one kind of material at the bottom, $z = -h/2$, to another material on the top, $z = h/2$, so that the modulus $E(z)$ and mass density $\rho(z)$ of the material of the plate vary through the plate thickness according to the power law:

$$E(z) = [E_1 - E_2]\, f(z) + E_2,$$
$$\rho(z) = [\rho_1 - \rho_2]\, f(z) + \rho_2, \tag{10.3.1}$$
$$f(z) = \left(\frac{1}{2} + \frac{z}{h}\right)^n,$$

where the subscripts 1 and 2 on E and ρ refer to the material number and n denotes the power-law index. Poisson's ratio ν is assumed to be constant throughout. When $n = 0$, we obtain the single-material plate (with the property of material 1).

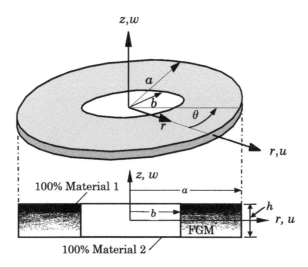

Fig. 10.3.1 Geometry and coordinate system for an axisymmetric bending of an FGM circular plate.

The components of the von Kármán strain tensor in cylindrical coordinate system for the axisymmetric case (i.e., independent of θ) are given by (see Reddy [84])

$$
\begin{aligned}
\varepsilon_{rr} &= \frac{\partial u_r}{\partial r} + \frac{1}{2}\left(\frac{\partial u_z}{\partial r}\right)^2 \\
\varepsilon_{rz} &= \frac{1}{2}\left(\frac{\partial u_r}{\partial z} + \frac{\partial u_z}{\partial r} + \frac{\partial u_z}{\partial r}\frac{\partial u_z}{\partial z}\right) \\
\varepsilon_{\theta\theta} &= \frac{u_r}{r}, \quad \varepsilon_{zz} = \frac{\partial u_z}{\partial z},
\end{aligned}
\tag{10.3.2}
$$

where (u_r, u_z) are the total displacements along the (r, z) coordinates $(u_\theta = 0)$.

10.3.2   Classical Plate Theory

10.3.2.1   Displacement and Strain Fields

In the classical plate theory (CPT), the total displacements (u_r, u_θ, u_z) along the three coordinate directions (r, θ, z), as implied by the Love–Kirchhoff hypothesis for plates, which is the same as the Bernoulli–Euler hypothesis for beams, are assumed in the form

$$
\mathbf{u} = u_r\,\hat{\mathbf{e}}_r + u_z\,\hat{\mathbf{e}}_z;
$$
$$
u_r(r, z, t) = u(r, t) - z\frac{\partial w}{\partial r}, \quad u_\theta = 0, \quad u_z(r, z, t) = w(r, t),
\tag{10.3.3}
$$

where u is the radial displacement and w is the transverse deflection of the point $(r, z = 0)$ on the midplane of the plate at time t. The displacement field in Eq. (10.3.3) is based on the Kirchhoff hypothesis that straight lines normal to the midplane before deformation remain (1) inextensible, (2) straight, and (3) normal to the midsurface after deformation.

The von Kármán strains in (10.3.2) for the classical plate theory take the form

$$
\varepsilon_{rr} = \varepsilon_{rr}^{(0)} + z\varepsilon_{rr}^{(1)}, \quad \varepsilon_{\theta\theta} = \varepsilon_{\theta\theta}^{(0)} + z\varepsilon_{\theta\theta}^{(1)},
\tag{10.3.4}
$$

where

$$
\begin{aligned}
\varepsilon_{rr}^{(0)} &= \frac{\partial u}{\partial r} + \frac{1}{2}\left(\frac{\partial w}{\partial r}\right)^2, \quad & \varepsilon_{rr}^{(1)} &= -\frac{\partial^2 w}{\partial r^2} \\
\varepsilon_{\theta\theta}^{(0)} &= \frac{u}{r}, \quad & \varepsilon_{\theta\theta}^{(1)} &= -\frac{1}{r}\frac{\partial w}{\partial r}.
\end{aligned}
\tag{10.3.5}
$$

10.3.2.2   Equations of Motion

Hamilton's principle is used to develop the weak forms and derive the equations of motion. We have

$$
0 = \int_0^T \int_b^a \left[m_0\left(\dot{u}\,\delta\dot{u} + \dot{w}\,\delta\dot{w}\right) - m_1\left(\frac{\partial\dot{w}}{\partial r}\delta\dot{u} + \dot{u}\,\frac{\partial\delta\dot{w}}{\partial r}\right) + m_2\frac{\partial\dot{w}}{\partial r}\frac{\partial\delta\dot{w}}{\partial r} \right.
$$

$$- N_{rr} \left(\frac{\partial \delta u}{\partial r} + \frac{\partial w}{\partial r} \frac{\partial \delta w}{\partial r} \right) - N_{\theta\theta} \left(\frac{\delta u}{r} \right)$$

$$+ M_{rr} \frac{\partial^2 \delta w}{\partial r^2} + \frac{1}{r} M_{\theta\theta} \frac{\partial \delta w}{\partial r} + q \delta w \bigg] r \, dr \, dt, \tag{10.3.6}$$

where $q = q(r, t)$ is the distributed transverse load, the superposed dot indicates the time derivative, and $(N_{rr}, N_{\theta\theta})$ and $(M_{rr}, M_{\theta\theta})$ are the stress resultants defined by

$$N_{rr}(r, t) = \int_{-\frac{h}{2}}^{\frac{h}{2}} \sigma_{rr} \, dz, \qquad N_{\theta\theta}(r, t) = \int_{-\frac{h}{2}}^{\frac{h}{2}} \sigma_{\theta\theta} \, dz,$$

$$M_{rr}(r, t) = \int_{-\frac{h}{2}}^{\frac{h}{2}} \sigma_{rr} z \, dz, \qquad M_{\theta\theta}(r, t) = \int_{-\frac{h}{2}}^{\frac{h}{2}} \sigma_{rr} z \, dz. \tag{10.3.7}$$

The equations of motion of the CPT are

$$\frac{1}{r} \left[\frac{\partial}{\partial r} (r N_{rr}) - N_{\theta\theta} \right] = m_0 \frac{\partial^2 u}{\partial t^2} - m_1 \frac{\partial^3 w}{\partial r \partial t^2}, \tag{10.3.8}$$

$$\frac{1}{r} \frac{\partial}{\partial r} (r V_r) + q = m_0 \frac{\partial^2 w}{\partial t^2} + \frac{1}{r} \frac{\partial}{\partial r} \left[r \left(m_1 \frac{\partial^2 u}{\partial t^2} - m_2 \frac{\partial^3 w}{\partial r \partial t^2} \right) \right], \tag{10.3.9}$$

where V_r is the effective transverse shear force acting in the rz-plane

$$V_r = Q_r + N_{rr} \frac{\partial w}{\partial r} = \frac{1}{r} \left[\frac{\partial}{\partial r} (r M_{rr}) - M_{\theta\theta} + r N_{rr} \frac{\partial w}{\partial r} \right]. \tag{10.3.10}$$

The boundary conditions involve specifying one element of each of the following pairs:

$$u \quad \text{or} \quad r N_{rr},$$

$$w \quad \text{or} \quad r \left[V_r - m_1 \frac{\partial^2 u}{\partial t^2} + m_2 \frac{\partial^3 w}{\partial r \partial t^2} \right] \equiv r \hat{V}_r, \tag{10.3.11}$$

$$\text{and} \quad -\frac{\partial w}{\partial r} \quad \text{or} \quad r M_{rr}.$$

10.3.3 First-Order Shear Deformation Theory

10.3.3.1 Displacement and Strain Fields

The first-order shear deformation plate theory (FSDT) is based on the displacement field

$$\mathbf{u} = u_r \, \hat{\mathbf{e}}_r + u_z \, \hat{\mathbf{e}}_z,$$

$$u_r(r, z, t) = u(r, t) + z \phi_r(r, t), \qquad u_z(r, z, t) = w(r, t), \tag{10.3.12}$$

where ϕ_r denotes rotation of a transverse normal in the plane $\theta = $ constant.

The nonzero von Kármán strains of the theory are

$$\varepsilon_{rr} = \varepsilon_{rr}^{(0)} + z\varepsilon_{rr}^{(1)}, \quad \varepsilon_{\theta\theta} = \varepsilon_{\theta\theta}^{(0)} + z\varepsilon_{\theta\theta}^{(1)}, \quad \varepsilon_{rz} = \varepsilon_{rz}^{(0)}, \tag{10.3.13}$$

where

$$\varepsilon_{rr}^{(0)} = \frac{\partial u}{\partial r} + \frac{1}{2}\left(\frac{\partial w}{\partial r}\right)^2, \quad \varepsilon_{rr}^{(1)} = \frac{\partial \phi_r}{\partial r},$$

$$\varepsilon_{\theta\theta}^{(0)} = \frac{u}{r}, \quad \varepsilon_{\theta\theta}^{(1)} = \frac{\phi_r}{r}, \quad 2\varepsilon_{rz}^{(0)} = \phi_r + \frac{\partial w}{\partial r}. \tag{10.3.14}$$

10.3.3.2   Equations of Motion

Hamilton's principle for the FSDT takes the form

$$0 = \int_0^T \int_b^a \left\{ m_0\left(\dot{u}\,\delta\dot{u} + \dot{w}\,\delta\dot{w}\right) + m_1\left(\dot{\phi}_r\,\delta\dot{u} + \dot{u}\,\delta\dot{\phi}_r\right) + m_2\dot{\phi}_r\,\delta\dot{\phi}_r \right.$$

$$- N_{rr}\left(\frac{\partial\delta u}{\partial r} + \frac{\partial w}{\partial r}\frac{\partial\delta w}{\partial r}\right) - N_{\theta\theta}\left(\frac{\delta u}{r}\right) - M_{rr}\frac{\partial\delta\phi_r}{\partial r}$$

$$\left. - \frac{1}{r}M_{\theta\theta}\delta\phi_r - Q_r\left(\delta\phi_r + \frac{\partial\delta w}{\partial r}\right) + q\delta w \right\} r\,dr\,dt, \tag{10.3.15}$$

where the shear force Q_r is defined by

$$Q_r = K_s \int_{-\frac{h}{2}}^{\frac{h}{2}} \sigma_{rz}\,dz, \tag{10.3.16}$$

and K_s denotes the shear correction factor.

The governing equations of motion of the FSDT are

$$\frac{1}{r}\left[\frac{\partial}{\partial r}(rN_{rr}) - N_{\theta\theta}\right] = m_0\frac{\partial^2 u}{\partial t^2} + m_1\frac{\partial^2\phi_r}{\partial t^2}, \tag{10.3.17}$$

$$\frac{1}{r}\frac{\partial}{\partial r}(rV_r) + q = m_0\frac{\partial^2 w}{\partial t^2}, \tag{10.3.18}$$

$$\frac{1}{r}\left[\frac{\partial}{\partial r}(rM_{rr}) - M_{\theta\theta}\right] - Q_r = m_1\frac{\partial^2 u}{\partial t^2} + m_2\frac{\partial^2\phi_r}{\partial t^2}, \tag{10.3.19}$$

where

$$V_r = Q_r + N_{rr}\frac{\partial w}{\partial r}. \tag{10.3.20}$$

The boundary conditions involve specifying one element of each of the following pairs:

$$u \quad \text{or} \quad rN_{rr}$$

$$w \quad \text{or} \quad rV_r \tag{10.3.21}$$

$$\phi_r \quad \text{or} \quad rM_{rr}.$$

10.3.4 Plate Constitutive Relations

For a two-constituent functionally graded linear elastic material, the stress–strain relations are

$$\left\{ \begin{array}{c} \sigma_{rr} \\ \sigma_{\theta\theta} \\ \sigma_{rz} \end{array} \right\} = \bar{E} \begin{bmatrix} 1 & \nu & 0 \\ \nu & 1 & 0 \\ 0 & 0 & \frac{1-\nu}{2} \end{bmatrix} \left\{ \begin{array}{c} \varepsilon_{rr} \\ \varepsilon_{\theta\theta} \\ 2\varepsilon_{rz} \end{array} \right\}, \quad \bar{E}(z) = \frac{E(z)}{1-\nu^2}, \quad (10.3.22)$$

where E varies with z according to Eq. (10.3.1) and ν is a constant. The stress resultants in the two theories can be expressed in terms of the displacements, as given in the next two sections.

10.3.4.1 Classical Plate Theory

The stress resultants of the CPT are related to the displacements (u, w) according to the relations

$$N_{rr} = A_{rr} \left[\frac{\partial u}{\partial r} + \frac{1}{2} \left(\frac{\partial w}{\partial r} \right)^2 + \nu \frac{u}{r} \right] - B_{rr} \left(\frac{\partial^2 w}{\partial r^2} + \frac{\nu}{r} \frac{\partial w}{\partial r} \right),$$

$$N_{\theta\theta} = A_{rr} \left[\frac{u}{r} + \nu \frac{\partial u}{\partial r} + \frac{\nu}{2} \left(\frac{\partial w}{\partial r} \right)^2 \right] - B_{rr} \left(\nu \frac{\partial^2 w}{\partial r^2} + \frac{1}{r} \frac{\partial w}{\partial r} \right),$$

$$\hspace{9cm} (10.3.23)$$

$$M_{rr} = B_{rr} \left[\frac{\partial u}{\partial r} + \frac{1}{2} \left(\frac{\partial w}{dr} \right)^2 + \nu \frac{u}{r} \right] - D_{rr} \left(\frac{\partial^2 w}{\partial r^2} + \frac{\nu}{r} \frac{\partial w}{\partial r} \right),$$

$$M_{\theta\theta} = B_{rr} \left[\frac{u}{r} + \nu \frac{\partial u}{\partial r} + \frac{\nu}{2} \left(\frac{\partial w}{\partial r} \right)^2 \right] - D_{rr} \left(\nu \frac{\partial^2 w}{\partial r^2} + \frac{1}{r} \frac{\partial w}{\partial r} \right),$$

where A_{rr}, B_{rr}, and D_{rr} are the extensional, extensional-bending, and bending stiffness coefficients

$$A_{rr} = \int_{-\frac{h}{2}}^{\frac{h}{2}} \bar{E}(z) \, dz, \quad B_{rr} = \int_{-\frac{h}{2}}^{\frac{h}{2}} \bar{E}(z) z \, dz, \quad D_{rr} = \int_{-\frac{h}{2}}^{\frac{h}{2}} \bar{E}(z) z^2 \, dz. \quad (10.3.24)$$

10.3.4.2 First-Order Plate Theory

The stress resultants in the FSDT can be expressed in terms of the generalized displacements (u, w, ϕ_r) as

$$N_{rr} = A_{rr} \left[\frac{\partial u}{\partial r} + \frac{1}{2} \left(\frac{\partial w}{\partial r} \right)^2 + \nu \frac{u}{r} \right] + B_{rr} \left(\frac{\partial \phi_r}{\partial r} + \frac{\nu}{r} \phi_r \right)$$

$$N_{\theta\theta} = A_{rr} \left[\frac{u}{r} + \nu \frac{\partial u}{\partial r} + \frac{\nu}{2} \left(\frac{\partial w}{\partial r} \right)^2 \right] + B_{rr} \left(\nu \frac{\partial \phi_r}{\partial r} + \frac{1}{r} \phi_r \right)$$

$$M_{rr} = B_{rr} \left[\frac{\partial u}{\partial r} + \frac{1}{2} \left(\frac{\partial w}{dr} \right)^2 + \nu \frac{u}{r} \right] + D_{rr} \left(\frac{\partial \phi_r}{\partial r} + \frac{\nu}{r} \phi_r \right) \qquad (10.3.25)$$

$$M_{\theta\theta} = B_{rr} \left[\frac{u}{r} + \nu \frac{\partial u}{\partial r} + \frac{\nu}{2} \left(\frac{\partial w}{\partial r} \right)^2 \right] + D_{rr} \left(\nu \frac{\partial \phi_r}{\partial r} + \frac{1}{r} \phi_r \right)$$

$$Q_r = K_s S_{rz} \left(\phi_r + \frac{\partial w}{\partial r} \right)$$

where S_{rz} is the shear stiffness coefficient defined as

$$S_{rz} = \frac{1}{2(1+\nu)} \int_{-\frac{h}{2}}^{\frac{h}{2}} E(z)\, dz, \qquad (10.3.26)$$

and K_s is the shear correction coefficient. Note that the shear modulus $G(z)$ is replaced with $E(z)$ $[G = E/2(1+\nu)]$.

10.4 A General Third-Order Plate Theory

10.4.1 Introduction

Here we develop a general third-order shear deformation plate theory (GTST) of functionally graded plates first and then specialize to the well-known plate theories (e.g., CPT and FSDT). The equations of motion are obtained using the principle of virtual displacements for the dynamic case (i.e., Hamilton's principle). Consider a plate of total thickness h and composed of FGM through the thickness. It is assumed that the material is isotropic, and the grading is assumed to be only through the thickness. The xy-plane is taken to be the undeformed midplane Ω of the plate with the z-axis positive upward from the midplane. We denote the boundary of the midplane with Γ. The plate volume is denoted as $V = \Omega \times (-h/2, h/2)$. The plate is bounded by the top surface Ω^+, bottom surface Ω^-, and the lateral surface $S = \Gamma \times (-h/2, h/2)$ (see Fig. 10.4.1).

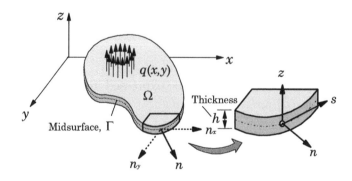

Fig. 10.4.1 Domain and various boundary segments of the domain.

When thermal effects are considered, like in the case of thermomechanical loads, the temperature distribution, which is assumed to vary only in the thickness direction, that is, $T = T(z)$, is determined by first solving a simple steady-state heat transfer equation through the thickness of the plate, with specified temperature boundary conditions at the top and bottom of the plate. The energy equation for the temperature variation through the thickness is governed by

$$\rho c_v \frac{\partial T}{\partial t} - \frac{\partial}{\partial z}\left[k(z,T)\frac{\partial T}{\partial z}\right] = 0, \quad -\frac{h}{2} \leq z \leq \frac{h}{2} \tag{10.4.1}$$

subjected to the boundary conditions

$$T(-h/2, t) = T_m(t), \quad T(h/2, t) = T_c(t), \tag{10.4.2}$$

where $k(z,T)$ and $\rho(z)$ are assumed to vary according to Eq. (10.3.1), while the specific heat c_v is assumed to be a constant.

10.4.2 Displacements and Strains

We begin with the following displacement field

$$
\begin{aligned}
u_1(x,y,z,t) &= u(x,y,t) + z\phi_x + z^2\varphi_x + z^3\psi_x, \\
u_2(x,y,z,t) &= v(x,y,t) + z\phi_y + z^2\varphi_y + z^3\psi_y, \\
u_3(x,y,z,t) &= w(x,y,t) + z\phi_z + z^2\varphi_z,
\end{aligned}
\tag{10.4.3}
$$

where (u,v,w) are the displacements along the coordinate lines of a material point on the xy-plane, that is, $u(x,y,t) = u_1(x,y,0,t)$, $v(x,y,t) = u_2(x,y,0,t)$, $w(x,y,t) = u_3(x,y,0,t)$, and

$$
\phi_x = \frac{\partial u_1}{\partial z}\bigg|_{z=0}, \quad \phi_y = \frac{\partial u_2}{\partial z}\bigg|_{z=0}, \quad \phi_z = \frac{\partial u_3}{\partial z}\bigg|_{z=0}, \quad 2\varphi_x = \frac{\partial^2 u_1}{\partial z^2}\bigg|_{z=0},
$$

$$
2\varphi_y = \frac{\partial^2 u_2}{\partial z^2}\bigg|_{z=0}, \quad 2\varphi_z = \frac{\partial^2 u_3}{\partial z^2}\bigg|_{z=0}, \quad 6\psi_x = \frac{\partial^3 u_1}{\partial z^3}\bigg|_{z=0}, \quad 6\psi_y = \frac{\partial^3 u_2}{\partial z^3}\bigg|_{z=0}.
$$

$$\tag{10.4.4}$$

The reason for expanding the in-plane displacements up to the cubic term and the transverse displacement up to the quadratic term in z is to obtain a quadratic variation of the transverse shear strains $\gamma_{xz} = 2\varepsilon_{xz}$ and $\gamma_{yz} = 2\varepsilon_{yz}$ through the plate thickness. Note that all three displacements contribute to the quadratic variation. In the most general case represented by the displacement field in Eq. (10.4.3), there are 11 generalized displacements $(u, v, w, \phi_x, \phi_y, \phi_z, \varphi_x, \varphi_y, \varphi_z, \psi_x, \psi_y)$ and, therefore, 11 differential equations will be required to determine them. Of course, the number of generalized displacement degrees of freedom per node in a displacement finite element model would be far greater than 11 because of the Hermite cubic interpolation of some of the generalized displacements.

The nonzero von Kármán type strains associated with the displacement field in Eq. (10.4.3) are given by

$$
\left\{ \begin{array}{c} \varepsilon_{xx} \\ \varepsilon_{yy} \\ \gamma_{xy} \end{array} \right\} = \sum_{i=0}^{3} z^i \left\{ \begin{array}{c} \varepsilon_{xx}^{(i)} \\ \varepsilon_{yy}^{(i)} \\ \gamma_{xy}^{(i)} \end{array} \right\}, \quad \left\{ \begin{array}{c} \varepsilon_{zz} \\ \gamma_{xz} \\ \gamma_{yz} \end{array} \right\} = \sum_{i=0}^{2} z^i \left\{ \begin{array}{c} \varepsilon_{zz}^{(i)} \\ \gamma_{xz}^{(i)} \\ \gamma_{yz}^{(i)} \end{array} \right\},
\tag{10.4.5}
$$

with

$$
\left\{ \begin{array}{c} \varepsilon_{xx}^{(0)} \\ \varepsilon_{yy}^{(0)} \\ \gamma_{xy}^{(0)} \end{array} \right\} = \left\{ \begin{array}{c} \frac{\partial u}{\partial x} + \frac{1}{2}\left(\frac{\partial w}{\partial x}\right)^2 \\ \frac{\partial v}{\partial y} + \frac{1}{2}\left(\frac{\partial w}{\partial y}\right)^2 \\ \frac{\partial u}{\partial y} + \frac{\partial v}{\partial x} + \frac{\partial w}{\partial x}\frac{\partial w}{\partial y} \end{array} \right\}, \quad \left\{ \begin{array}{c} \varepsilon_{zz}^{(0)} \\ \gamma_{xz}^{(0)} \\ \gamma_{yz}^{(0)} \end{array} \right\} = \left\{ \begin{array}{c} \phi_z \\ \phi_x + \frac{\partial w}{\partial x} \\ \phi_y + \frac{\partial w}{\partial y} \end{array} \right\},
$$

$$
\left\{ \begin{array}{c} \varepsilon_{xx}^{(1)} \\ \varepsilon_{yy}^{(1)} \\ \gamma_{xy}^{(1)} \end{array} \right\} = \left\{ \begin{array}{c} \frac{\partial \phi_x}{\partial x} \\ \frac{\partial \phi_y}{\partial y} \\ \frac{\partial \phi_x}{\partial y} + \frac{\partial \phi_y}{\partial x} \end{array} \right\}, \quad \left\{ \begin{array}{c} \varepsilon_{zz}^{(1)} \\ \gamma_{xz}^{(1)} \\ \gamma_{yz}^{(1)} \end{array} \right\} = \left\{ \begin{array}{c} 2\varphi_z \\ 2\varphi_x + \frac{\partial \phi_z}{\partial x} \\ 2\varphi_y + \frac{\partial \phi_z}{\partial y} \end{array} \right\},
\tag{10.4.6}
$$

$$
\left\{ \begin{array}{c} \varepsilon_{xx}^{(2)} \\ \varepsilon_{yy}^{(2)} \\ \gamma_{xy}^{(2)} \end{array} \right\} = \left\{ \begin{array}{c} \frac{\partial \varphi_x}{\partial x} \\ \frac{\partial \varphi_y}{\partial y} \\ \frac{\partial \varphi_x}{\partial y} + \frac{\partial \varphi_y}{\partial x} \end{array} \right\}, \quad \left\{ \begin{array}{c} \varepsilon_{zz}^{(2)} \\ \gamma_{xz}^{(2)} \\ \gamma_{yz}^{(2)} \end{array} \right\} = \left\{ \begin{array}{c} 0 \\ 3\psi_x + \frac{\partial \varphi_z}{\partial x} \\ 3\psi_y + \frac{\partial \varphi_z}{\partial y} \end{array} \right\},
$$

$$
\left\{ \begin{array}{c} \varepsilon_{xx}^{(3)} \\ \varepsilon_{yy}^{(3)} \\ \gamma_{xy}^{(3)} \end{array} \right\} = \left\{ \begin{array}{c} \frac{\partial \psi_x}{\partial x} \\ \frac{\partial \psi_y}{\partial y} \\ \frac{\partial \psi_x}{\partial y} + \frac{\partial \psi_y}{\partial x} \end{array} \right\},
\tag{10.4.7}
$$

where $(\varepsilon_{xx}^{(0)}, \varepsilon_{yy}^{(0)}, \gamma_{xy}^{(0)})$ are the *membrane strains*, $(\varepsilon_{xx}^{(1)}, \varepsilon_{yy}^{(1)}, \gamma_{xy}^{(1)})$ are the flexural (bending) strains, and $(\varepsilon_{xx}^{(2)}, \varepsilon_{yy}^{(2)}, \gamma_{xy}^{(2)})$ and $(\varepsilon_{xx}^{(3)}, \varepsilon_{yy}^{(3)}, \gamma_{xy}^{(3)})$ are higher-order strains. In writing the strain–displacement relations, we have assumed that the strains are small and rotations are moderately large.

If the transverse shear stresses, σ_{xz} and σ_{yz} are required to be zero on the top and bottom of the plate, that is, $z = \pm h/2$, as in the Reddy third-order theory [50, 113], it is necessary that γ_{xz} and γ_{yz} be zero at $z = \pm h/2$. This in turn yields

$$
2\varphi_x + \frac{\partial \phi_z}{\partial x} = 0, \quad \phi_x + \frac{\partial w}{\partial x} + \frac{h^2}{4}\left(3\psi_x + \frac{\partial \varphi_z}{\partial x}\right) = 0,
$$
$$
2\varphi_y + \frac{\partial \phi_z}{\partial y} = 0, \quad \phi_y + \frac{\partial w}{\partial y} + \frac{h^2}{4}\left(3\psi_y + \frac{\partial \varphi_z}{\partial y}\right) = 0.
\tag{10.4.8}
$$

Thus, the variables $(\varphi_x, \varphi_y, \psi_x, \psi_y)$ can be expressed in terms of $(w, \phi_x, \phi_y, \phi_z, \varphi_z)$ and thus reduce the number of generalized displacements from 11 to 7. In addition, if we set $\phi_z = \varphi_z = 0$, we obtain the displacement field of the Reddy third-order theory, which has only five variables $(u, v, w, \phi_x, \phi_y)$.

10.4.3 Equations of Motion

The equations of motion of the GTST can be derived using the principle of virtual displacements. In the derivation, we account for thermal effect with the understanding that the material properties are known functions of temperature and that the temperature change ΔT is a known function of position from the solution of Eqs (10.4.1) and (10.4.2). Thus, temperature field enters the formulation only through constitutive equations.

The principle of virtual displacements for the dynamic case requires that

$$\int_0^T (\delta \mathcal{K} - \delta \mathcal{U} - \delta \mathcal{V})\, dt = 0, \tag{10.4.9}$$

where $\delta\mathcal{K}$ is the virtual kinetic energy, $\delta\mathcal{U}$ is the virtual strain energy, and $\delta\mathcal{V}$ is the virtual work done by external forces. Each of these quantities are derived next for the general third-order theory.

The virtual kinetic energy $\delta\mathcal{K}$ is

$$\delta\mathcal{K} = \int_\Omega \int_{-\frac{h}{2}}^{\frac{h}{2}} \rho \left(\frac{\partial u_1}{\partial t} \frac{\partial \delta u_1}{\partial t} + \frac{\partial u_2}{\partial t} \frac{\partial \delta u_2}{\partial t} + \frac{\partial u_3}{\partial t} \frac{\partial \delta u_3}{\partial t} \right) dz\, dxdy$$

$$= \int_\Omega \int_{-\frac{h}{2}}^{\frac{h}{2}} \rho \Big[\left(\dot{u} + z\dot{\phi}_x + z^2 \dot{\varphi}_x + z^3 \dot{\psi}_x \right) \left(\delta\dot{u} + z\delta\dot{\phi}_x + z^2 \delta\dot{\varphi}_x + z^3 \delta\dot{\psi}_x \right)$$

$$+ \left(\dot{v} + z\dot{\phi}_y + z^2 \dot{\varphi}_y + z^3 \dot{\psi}_y \right) \left(\delta\dot{v} + z\delta\dot{\phi}_y + z^2 \delta\dot{\varphi}_y + z^3 \delta\dot{\psi}_y \right)$$

$$+ \left(\dot{w} + z\dot{\phi}_z + z^2 \dot{\varphi}_z \right) \left(\delta\dot{w} + z\delta\dot{\phi}_z + z^2 \delta\dot{\varphi}_z \right) \Big] dz\, dxdy$$

$$= \int_\Omega \Big[\left(m_0 \dot{u} + m_1 \dot{\phi}_x + m_2 \dot{\varphi}_x + m_3 \dot{\psi}_x \right) \delta\dot{u}$$

$$+ \left(m_1 \dot{u} + m_2 \dot{\phi}_x + m_3 \dot{\varphi}_x + m_4 \dot{\psi}_x \right) \delta\dot{\phi}_x$$

$$+ \left(m_2 \dot{u} + m_3 \dot{\phi}_x + m_4 \dot{\varphi}_x + m_5 \dot{\psi}_x \right) \delta\dot{\varphi}_x$$

$$+ \left(m_3 \dot{u} + m_4 \dot{\phi}_x + m_5 \dot{\varphi}_x + m_6 \dot{\psi}_x \right) \delta\dot{\psi}_x$$

$$+ \left(m_0 \dot{v} + m_1 \dot{\phi}_y + m_2 \dot{\varphi}_y + m_3 \dot{\psi}_y \right) \delta\dot{v}$$

$$+ \left(m_1 \dot{v} + m_2 \dot{\phi}_y + m_3 \dot{\varphi}_y + m_4 \dot{\psi}_y \right) \delta\dot{\phi}_y$$

$$+ \left(m_2 \dot{v} + m_3 \dot{\phi}_y + m_4 \dot{\varphi}_y + m_5 \dot{\psi}_y \right) \delta\dot{\varphi}_y$$

$$+ \left(m_3 \dot{v} + m_4 \dot{\phi}_y + m_5 \dot{\varphi}_y + m_6 \dot{\psi}_y \right) \delta\dot{\psi}_y$$

$$+ \left(m_0 \dot{w} + m_1 \dot{\phi}_z + m_2 \dot{\varphi}_z \right) \delta\dot{w} + \left(m_1 \dot{w} + m_2 \dot{\phi}_z + m_3 \dot{\varphi}_z \right) \delta\dot{\phi}_z$$

$$+ \left(m_2 \dot{w} + m_3 \dot{\phi}_z + m_4 \dot{\varphi}_z \right) \delta\dot{\varphi}_z \Big] dxdy, \tag{10.4.10}$$

where the superposed dot on a variable indicates time derivative, for example,

$\dot{u} = \partial u / \partial t$, and m_i $(i = 0, 1, 2, \ldots, 6)$ are the mass moments of inertia

$$m_i = \int_{-\frac{h}{2}}^{\frac{h}{2}} (z)^i \, \rho(z) \, dz. \tag{10.4.11}$$

The virtual strain energy is given by

$$\delta\mathcal{U} = \int_{\Omega} \left\{ \int_{-\frac{h}{2}}^{\frac{h}{2}} \left[\sigma_{xx} \left(\sum_{i=0}^{3} (z)^i \delta\varepsilon_{xx}^{(i)} \right) + \sigma_{yy} \left(\sum_{i=0}^{3} (z)^i \delta\varepsilon_{yy}^{(i)} \right) + \sigma_{xy} \left(\sum_{i=0}^{3} (z)^i \delta\gamma_{xy}^{(i)} \right) \right. \right.$$
$$\left. \left. + \sigma_{zz} \left(\sum_{i=0}^{2} (z)^i \delta\varepsilon_{zz}^{(i)} \right) + \sigma_{xz} \left(\sum_{i=0}^{2} (z)^i \delta\gamma_{xz}^{(i)} \right) + \sigma_{yz} \left(\sum_{i=0}^{2} (z)^i \delta\gamma_{yz}^{(i)} \right) \right] dz \right\} dx\,dy. \tag{10.4.12}$$

Next, we introduce thickness-integrated stress resultants

$$M_{ij}^{(k)} = \int_{-\frac{h}{2}}^{\frac{h}{2}} (z)^k \sigma_{ij} \, dz, \qquad (k = 0, 1, 2, 3). \tag{10.4.13}$$

Then the virtual strain energy can be expressed in terms of the stress resultants as

$$\delta\mathcal{U} = \int_{\Omega} \left[\sum_{i=0}^{3} \left(M_{xx}^{(i)} \delta\varepsilon_{xx}^{(i)} + M_{yy}^{(i)} \delta\varepsilon_{yy}^{(i)} + M_{xy}^{(i)} \delta\gamma_{xy}^{(i)} \right) \right.$$
$$\left. + \sum_{i=0}^{2} \left(M_{zz}^{(i)} \delta\varepsilon_{zz}^{(i)} + M_{xz}^{(i)} \delta\gamma_{xz}^{(i)} + M_{yz}^{(i)} \delta\gamma_{yz}^{(i)} \right) \right] dx\,dy. \tag{10.4.14}$$

Note that $M_{xx}^{(0)}$, $M_{yy}^{(0)}$, and $M_{xy}^{(0)}$ are the membrane forces (often denoted by N_{xx}, N_{yy}, and N_{xy}); $M_{xx}^{(1)}$, $M_{yy}^{(1)}$, and $M_{xy}^{(1)}$ are the bending moments (denoted by M_{xx}, M_{yy}, and M_{xy}); and $M_{xz}^{(0)}$ and $M_{yz}^{(0)}$ are the shear forces (denoted by Q_x and Q_y). The rest of the stress resultants are higher-order generalized forces, which are often difficult to physically interpret. Because of their similarity to the generalized physical forces identified above, they are assumed to be zero when their counterparts (i.e., generalized displacements) are not specified.

The virtual work done by external forces consists of three parts: (1) virtual work done by the body forces in $V = \Omega \times (-h/2, h/2)$, (2) virtual work done by surface tractions acting on the top and bottom surfaces of the plate Ω^+ and Ω^-, and (3) virtual work done by the surface tractions on the lateral surface $S = \Gamma \times (-h/2, h/2)$, where Ω^+ denotes the top surface of the plate, Ω is the middle surface of the plate, Ω^- is the bottom surface of the plate, and Γ is the boundary of the middle surface (see Fig. 10.4.1).

Let $(\bar{f}_x, \bar{f}_y, \bar{f}_z)$ denote the body forces (measured per unit volume), $(\bar{t}_x, \bar{t}_y, \bar{t}_z)$ denote the surface forces (measured per unit area) on S, (q_x^t, q_y^t, q_z^t) denote the forces (measured per unit area) on Ω^+, and (q_x^b, q_y^b, q_z^b) denote the forces (measured per unit area) on Ω^- in the (x, y, z) coordinate directions. Then the virtual work done by external forces is

$$\delta \mathcal{V} = -\Big[\int_V \left(\bar{f}_x \delta u_1 + \bar{f}_y \delta u_2 + \bar{f}_z \delta u_3 \right) dV + \int_S \left(\bar{t}_x \delta u_1 + \bar{t}_y \delta u_2 + \bar{t}_z \delta u_3 \right) dS$$
$$+ \int_{\Omega+} \left(q_x^t \delta u_1 + q_y^t \delta u_2 + q_z^t \delta u_3 \right) dx dy + \int_{\Omega-} \left(q_x^b \delta u_1 + q_y^b \delta u_2 + q_z^b \delta u_3 \right) dx dy \Big].$$
$$(10.4.15)$$

In view of the displacement field (10.4.3), $\delta \mathcal{V}$ can be expressed as

$$\delta \mathcal{V} = -\Big\{ \int_\Omega \Big(f_x^{(0)} \delta u + f_x^{(1)} \delta \phi_x + f_x^{(2)} \delta \varphi_x + f_x^{(3)} \delta \psi_x + f_y^{(0)} \delta v + f_y^{(1)} \delta \phi_y$$
$$+ f_y^{(2)} \delta \varphi_y + f_y^{(3)} \delta \psi_y + f_z^{(0)} \delta w + f_z^{(1)} \delta \phi_z + f_z^{(2)} \delta \varphi_z \Big) dx dy$$
$$+ \int_\Gamma \Big(t_x^{(0)} \delta u + t_x^{(1)} \delta \phi_x + t_x^{(2)} \delta \varphi_x + t_x^{(3)} \delta \psi_x + t_y^{(0)} \delta v + t_y^{(1)} \delta \phi_y + t_y^{(2)} \delta \varphi_y$$
$$+ t_y^{(3)} \delta \psi_y + t_z^{(0)} \delta w + t_z^{(1)} \delta \phi_z + t_z^{(2)} \delta \varphi_z \Big) dS$$
$$+ \int_\Omega \Big[\left(q_x^t + q_x^b \right) \delta u + \left(q_y^t + q_y^b \right) \delta v + \left(q_z^t + q_z^b \right) \delta w + \frac{h}{2} \left(q_x^t - q_x^b \right) \delta \phi_x$$
$$+ \frac{h}{2} \left(q_y^t - q_y^b \right) \delta \phi_y + \frac{h}{2} \left(q_z^t - q_z^b \right) \delta \phi_z + \frac{h^2}{4} \left(q_x^t + q_x^b \right) \delta \varphi_x + \frac{h^2}{4} \left(q_y^t + q_y^b \right) \delta \varphi_y$$
$$+ \frac{h^2}{4} \left(q_z^t + q_z^b \right) \delta \varphi_z + \frac{h^3}{8} \left(q_x^t - q_x^b \right) \delta \psi_x + \frac{h^3}{8} \left(q_y^t - q_y^b \right) \delta \psi_y \Big] dx dy \Big\},$$
$$(10.4.16)$$

where

$$f_\xi^{(i)} = \int_{-\frac{h}{2}}^{\frac{h}{2}} (z)^i \bar{f}_\xi \, dz, \quad t_\xi^{(i)} = \int_{-\frac{h}{2}}^{\frac{h}{2}} (z)^i \bar{t}_\xi \, dz. \qquad (10.4.17)$$

The equations of motion of the GTST governing FGM plates are obtained by substituting $\delta \mathcal{K}$, $\delta \mathcal{U}$, and $\delta \mathcal{V}$ from Eqs (10.4.10), (10.4.14), and (10.4.16), respectively, into Eq. (10.4.9), applying the integration by parts (i.e., the gradient theorem) to relieve all virtual generalized displacements of any differentiation, and invoking the fundamental lemma of the variational calculus. We obtain (after a lengthy algebra and manipulations) the following equations:

$$\delta u : \quad \frac{\partial M_{xx}^{(0)}}{\partial x} + \frac{\partial M_{xy}^{(0)}}{\partial y} + q_x = m_0 \ddot{u} + m_1 \ddot{\phi}_x + m_2 \ddot{\varphi}_x + m_3 \ddot{\psi}_x, \qquad (10.4.18)$$

$$\delta v : \quad \frac{\partial M_{xy}^{(0)}}{\partial x} + \frac{\partial M_{yy}^{(0)}}{\partial y} + q_y = m_0 \ddot{v} + m_1 \ddot{\phi}_y + m_2 \ddot{\varphi}_y + m_3 \ddot{\psi}_y, \qquad (10.4.19)$$

$$\delta w : \quad \frac{\partial}{\partial x}\left(\frac{\partial w}{\partial x}M_{xx}^{(0)} + \frac{\partial w}{\partial y}M_{xy}^{(0)}\right) + \frac{\partial}{\partial y}\left(\frac{\partial w}{\partial x}M_{xy}^{(0)} + \frac{\partial w}{\partial y}M_{yy}^{(0)}\right)$$

$$+ \frac{\partial M_{xz}^{(0)}}{\partial x} + \frac{\partial M_{yz}^{(0)}}{\partial y} + q_z = m_0\ddot{w} + m_1\ddot{\phi}_z + m_2\ddot{\varphi}_z, \quad (10.4.20)$$

$$\delta\phi_x : \quad \frac{\partial M_{xx}^{(1)}}{\partial x} + \frac{\partial M_{xy}^{(1)}}{\partial y} - M_{xz}^{(0)} + F_x = m_1\ddot{u} + m_2\ddot{\phi}_x + m_3\ddot{\varphi}_x + m_4\ddot{\psi}_x,$$

$$(10.4.21)$$

$$\delta\phi_y : \quad \frac{\partial M_{xy}^{(1)}}{\partial x} + \frac{\partial M_{yy}^{(1)}}{\partial y} - M_{yz}^{(0)} + F_y = m_1\ddot{v} + m_2\ddot{\phi}_y + m_3\ddot{\varphi}_y + m_4\ddot{\psi}_y,$$

$$(10.4.22)$$

$$\delta\varphi_x : \quad \frac{\partial M_{xx}^{(2)}}{\partial x} + \frac{\partial M_{xy}^{(2)}}{\partial y} - 2M_{xz}^{(1)} + G_x = m_2\ddot{u} + m_3\ddot{\phi}_x + m_4\ddot{\varphi}_x + m_5\ddot{\psi}_x,$$

$$(10.4.23)$$

$$\delta\varphi_y : \quad \frac{\partial M_{xy}^{(2)}}{\partial x} + \frac{\partial M_{yy}^{(2)}}{\partial y} - 2M_{yz}^{(1)} + G_y = m_2\ddot{v} + m_3\ddot{\phi}_y + m_4\ddot{\varphi}_y + m_5\ddot{\psi}_y,$$

$$(10.4.24)$$

$$\delta\psi_x : \quad \frac{\partial M_{xx}^{(3)}}{\partial x} + \frac{\partial M_{xy}^{(3)}}{\partial y} - 3M_{xz}^{(2)} + H_x = m_3\ddot{u} + m_4\ddot{\phi}_x + m_5\ddot{\varphi}_x + m_6\ddot{\psi}_x,$$

$$(10.4.25)$$

$$\delta\psi_y : \quad \frac{\partial M_{xy}^{(3)}}{\partial x} + \frac{\partial M_{yy}^{(3)}}{\partial y} - 3M_{yz}^{(2)} + H_y = m_3\ddot{v} + m_4\ddot{\phi}_y + m_5\ddot{\varphi}_y + m_6\ddot{\psi}_y,$$

$$(10.4.26)$$

$$\delta\phi_z : \quad \frac{\partial M_{xz}^{(1)}}{\partial x} + \frac{\partial M_{yz}^{(1)}}{\partial y} - M_{zz}^{(0)} + P = m_1\ddot{w} + m_2\ddot{\phi}_z + m_3\ddot{\varphi}_z, \quad (10.4.27)$$

$$\delta\varphi_z : \quad \frac{\partial M_{xz}^{(2)}}{\partial x} + \frac{\partial M_{yz}^{(2)}}{\partial y} - 2M_{zz}^{(1)} + Q = m_2\ddot{w} + m_3\ddot{\phi}_z + m_4\ddot{\varphi}_z, \quad (10.4.28)$$

where $q_x = f_x^{(0)} + q_x^t + q_x^b$, $q_y = f_y^{(0)} + q_y^t + q_y^b$, $q_z = f_z^{(0)} + q_z^t + q_z^b$, and

$$F_x = f_x^{(1)} + \frac{h}{2}\left(q_x^t - q_x^b\right), \quad F_y = f_y^{(1)} + \frac{h}{2}\left(q_y^t - q_y^b\right)$$

$$G_x = f_x^{(2)} + \frac{h^2}{4}\left(q_x^t + q_x^b\right), \quad G_y = f_y^{(2)} + \frac{h^2}{4}\left(q_y^t + q_y^b\right) \quad (10.4.29)$$

$$H_x = f_x^{(3)} + \frac{h^3}{8}\left(q_x^t - q_x^b\right), \quad H_y = f_y^{(3)} + \frac{h^3}{8}\left(q_y^t - q_y^b\right)$$

$$P = f_z^{(1)} + \frac{h}{2}\left(q_z^t - q_z^b\right), \quad Q = f_z^{(2)} + \frac{h^2}{4}\left(q_z^t + q_z^b\right).$$

The boundary conditions involve specifying the following generalized forces that are dual to the generalized displacement $(u, v, w, \phi_x, \phi_y, \varphi_x, \varphi_y, \psi_x, \psi_y, \phi_z, \varphi_z)$:

$$\delta u: \qquad \bar{M}_{nn}^{(0)} \equiv M_{xx}^{(0)} n_x + M_{xy}^{(0)} n_y, \tag{10.4.30}$$

$$\delta v: \qquad \bar{M}_{nt}^{(0)} \equiv M_{xy}^{(0)} n_x + M_{yy}^{(0)} n_y, \tag{10.4.31}$$

$$\delta w: \qquad \bar{M}_{nz}^{(0)} \equiv \left(\frac{\partial w}{\partial x} M_{xx}^{(0)} + \frac{\partial w}{\partial y} M_{xy}^{(0)}\right) n_x + M_{xz}^{(0)} n_x + M_{yz}^{(0)} n_y$$

$$\qquad\qquad + \left(\frac{\partial w}{\partial x} M_{xy}^{(0)} + \frac{\partial w}{\partial y} M_{yy}^{(0)}\right) n_y, \tag{10.4.32}$$

$$\delta\phi_x: \qquad \bar{M}_{nn}^{(1)} \equiv M_{xx}^{(1)} n_x + M_{xy}^{(1)} n_y, \tag{10.4.33}$$

$$\delta\phi_y: \qquad \bar{M}_{nt}^{(1)} \equiv M_{xy}^{(1)} n_x + M_{yy}^{(1)} n_y, \tag{10.4.34}$$

$$\delta\varphi_x: \qquad \bar{M}_{nn}^{(2)} \equiv M_{xx}^{(2)} n_x + M_{xy}^{(2)} n_y, \tag{10.4.35}$$

$$\delta\varphi_y: \qquad \bar{M}_{nt}^{(2)} \equiv M_{xy}^{(2)} n_x + M_{yy}^{(2)} n_y, \tag{10.4.36}$$

$$\delta\psi_x: \qquad \bar{M}_{nn}^{(3)} \equiv M_{xx}^{(3)} n_x + M_{xy}^{(3)} n_y, \tag{10.4.37}$$

$$\delta\psi_y: \qquad \bar{M}_{nt}^{(3)} \equiv M_{xy}^{(3)} n_x + M_{yy}^{(3)} n_y, \tag{10.4.38}$$

$$\delta\phi_z: \qquad \bar{M}_{nz}^{(1)} \equiv M_{xz}^{(1)} n_x + M_{yz}^{(1)} n_y, \tag{10.4.39}$$

$$\delta\varphi_z: \qquad \bar{M}_{nz}^{(2)} \equiv M_{xz}^{(2)} n_x + M_{yz}^{(2)} n_y. \tag{10.4.40}$$

10.4.4   Constitutive Relations

Here we assume that moduli E and G and density ρ [and thermal coefficient of expansion α when Eq. (10.4.1) is considered] vary according to Eq. (10.1.1), and ν is a constant. The linear constitutive relations are (see Reddy [50, 84])

$$\begin{Bmatrix} \sigma_{xx} \\ \sigma_{yy} \\ \sigma_{zz} \\ \sigma_{xy} \\ \sigma_{xz} \\ \sigma_{yz} \end{Bmatrix} = \begin{bmatrix} c_{11} & c_{12} & c_{12} & 0 & 0 & 0 \\ c_{12} & c_{11} & c_{12} & 0 & 0 & 0 \\ c_{12} & c_{12} & c_{11} & 0 & 0 & 0 \\ 0 & 0 & 0 & c_{22} & 0 & 0 \\ 0 & 0 & 0 & 0 & c_{22} & 0 \\ 0 & 0 & 0 & 0 & 0 & c_{22} \end{bmatrix} \begin{Bmatrix} \varepsilon_{xx} \\ \varepsilon_{yy} \\ \varepsilon_{zz} \\ \gamma_{xy} \\ \gamma_{xz} \\ \gamma_{yz} \end{Bmatrix} - \frac{E\alpha\Delta T}{1-2\nu} \begin{Bmatrix} 1 \\ 1 \\ 1 \\ 0 \\ 0 \\ 0 \end{Bmatrix}, \tag{10.4.41}$$

where

$$c_{11} = \frac{E(1-\nu)}{(1+\nu)(1-2\nu)}, \quad c_{12} = \frac{E\nu}{(1+\nu)(1-2\nu)}, \quad c_{22} = \frac{E}{2(1+\nu)}, \tag{10.4.42}$$

α is the coefficients of thermal expansion, and ΔT is the temperature increment from a reference temperature T_0, $\Delta T = T - T_0$.

Next, we relate the generalized forces $(M_{xx}^{(i)}, M_{yy}^{(i)}, M_{xy}^{(i)})$ to the generalized displacements $(u, v, w, \phi_x, \phi_y, \varphi_x, \varphi_y, \psi_x, \psi_y, \phi_z, \varphi_z)$. We have

$$\begin{Bmatrix} M_{xx}^{(i)} \\ M_{yy}^{(i)} \\ M_{zz}^{(i)} \end{Bmatrix} = \int_{-\frac{h}{2}}^{\frac{h}{2}} \begin{Bmatrix} \sigma_{xx} \\ \sigma_{yy} \\ \sigma_{zz} \end{Bmatrix} (z)^i dz = \sum_{k=i}^{3+i} \begin{bmatrix} A_{11}^{(k)} & A_{12}^{(k)} & A_{12}^{(k)} \\ A_{12}^{(k)} & A_{11}^{(k)} & A_{12}^{(k)} \\ A_{12}^{(k)} & A_{12}^{(k)} & A_{11}^{(k)} \end{bmatrix} \begin{Bmatrix} \varepsilon_{xx}^{(k-i)} \\ \varepsilon_{yy}^{(k-i)} \\ \varepsilon_{zz}^{(k-i)} \end{Bmatrix} - \begin{Bmatrix} X_T^{(i)} \\ Y_T^{(i)} \\ Z_T^{(i)} \end{Bmatrix},$$

$$\tag{10.4.43}$$

$$\left\{\begin{array}{c} M_{xy}^{(i)} \\ M_{xz}^{(i)} \\ M_{yz}^{(i)} \end{array}\right\} = \int_{-\frac{h}{2}}^{\frac{h}{2}} \left\{\begin{array}{c} \sigma_{xy} \\ \sigma_{xz} \\ \sigma_{yz} \end{array}\right\} (z)^i dz = \sum_{k=i}^{3+i} \left[\begin{array}{ccc} B_{11}^{(k)} & 0 & 0 \\ 0 & B_{11}^{(k)} & 0 \\ 0 & 0 & B_{11}^{(k)} \end{array}\right] \left\{\begin{array}{c} \gamma_{xy}^{(k-i)} \\ \gamma_{xz}^{(k-i)} \\ \gamma_{yz}^{(k-i)} \end{array}\right\},$$

where

$$A_{11}^{(k)} = \frac{\eta}{(1+\nu)} \int_{-\frac{h}{2}}^{\frac{h}{2}} (z)^k E(z)\, dz, \quad A_{12}^{(k)} = \frac{\zeta}{(1+\nu)} \int_{-\frac{h}{2}}^{\frac{h}{2}} (z)^k E(z)\, dz$$

$$B_{11}^{(k)} = \frac{1}{2(1+\nu)} \int_{-\frac{h}{2}}^{\frac{h}{2}} (z)^k E(z)\, dz, \quad \eta = \frac{(1-\nu)}{(1-2\nu)}, \quad \zeta = \frac{\nu}{(1-2\nu)},$$

$$(10.4.44)$$

and the generalized thermal forces $X_T^{(i)}$, $Y_T^{(i)}$, and $Z_T^{(i)}$ are defined by

$$X_T^{(i)} = Y_T^{(i)} = Z_T^{(i)} = \frac{1}{(1-2\nu)} \int_{-\frac{h}{2}}^{\frac{h}{2}} (z)^k E(z)\alpha(z)\, \Delta T\, dz. \qquad (10.4.45)$$

We note that $\varepsilon_{zz}^{(3)} = 0$, $\gamma_{xz}^{(3)} = 0$, and $\gamma_{xz}^{(3)} = 0$.

10.4.5 Specialization to Other Theories

The GTST developed herein contains all of the existing plate theories and some new theories. Some of the special cases are presented here.

10.4.5.1 A General Third-Order Plate Theory with Traction-Free Top and Bottom Surfaces

If the top and bottom surfaces of the plate are free of any tangential forces, we can invoke the conditions in Eq. (10.4.8) and eliminate φ_x, φ_y, ψ_x, and ψ_y:

$$\varphi_x = -\frac{1}{2}\frac{\partial \phi_z}{\partial x}, \quad \psi_x = -\frac{1}{3}\frac{\partial \varphi_z}{\partial x} - c_1\left(\phi_x + \frac{\partial w}{\partial x}\right),$$

$$\varphi_y = -\frac{1}{2}\frac{\partial \phi_z}{\partial y}, \quad \psi_y = -\frac{1}{3}\frac{\partial \varphi_z}{\partial y} - c_1\left(\phi_y + \frac{\partial w}{\partial y}\right). \qquad c_1 = \frac{4}{3h^2} \qquad (10.4.46)$$

Then the higher-order strains take the form

$$\left\{\begin{array}{c} \varepsilon_{xx}^{(2)} \\ \varepsilon_{yy}^{(2)} \\ \gamma_{xy}^{(2)} \end{array}\right\} = -\frac{1}{2}\left\{\begin{array}{c} \frac{\partial^2 \phi_z}{\partial x^2} \\ \frac{\partial^2 \phi_z}{\partial y^2} \\ 2\frac{\partial^2 \phi_z}{\partial x \partial y} \end{array}\right\}, \quad \left\{\begin{array}{c} \varepsilon_{xx}^{(3)} \\ \varepsilon_{yy}^{(3)} \\ \gamma_{xy}^{(3)} \end{array}\right\} = -\frac{1}{3}\left\{\begin{array}{c} \frac{\partial^2 \varphi_z}{\partial x^2} \\ \frac{\partial^2 \varphi_z}{\partial y^2} \\ 2\frac{\partial^2 \varphi_z}{\partial x \partial y} \end{array}\right\} - c_1 \left\{\begin{array}{c} \frac{\partial \phi_x}{\partial x} + \frac{\partial^2 w}{\partial x^2} \\ \frac{\partial \phi_y}{\partial y} + \frac{\partial^2 w}{\partial y^2} \\ \frac{\partial \phi_x}{\partial y} + \frac{\partial \phi_y}{\partial x} + 2\frac{\partial^2 w}{\partial x \partial y} \end{array}\right\},$$

$$(10.4.47)$$

$$\left\{\begin{array}{c} \varepsilon_{zz}^{(1)} \\ \gamma_{xz}^{(1)} \\ \gamma_{yz}^{(1)} \end{array}\right\} = \left\{\begin{array}{c} 2\varphi_z \\ 0 \\ 0 \end{array}\right\}, \quad \left\{\begin{array}{c} \varepsilon_{zz}^{(2)} \\ \gamma_{xz}^{(2)} \\ \gamma_{yz}^{(2)} \end{array}\right\} = -c_2 \left\{\begin{array}{c} 0 \\ \phi_x + \dfrac{\partial w}{\partial x} \\ \phi_y + \dfrac{\partial w}{\partial y} \end{array}\right\}, \quad c_2 = \dfrac{4}{h^2}. \qquad (10.4.48)$$

The equations of motion are

$$\delta u: \quad \frac{\partial M_{xx}^{(0)}}{\partial x} + \frac{\partial M_{xy}^{(0)}}{\partial y} + q_x = m_0 \ddot{u} + m_1 \ddot{\phi}_x - \frac{m_2}{2} \frac{\partial \ddot{\phi}_z}{\partial x}$$
$$- m_3 \left[c_1 \left(\ddot{\phi}_x + \frac{\partial \ddot{w}}{\partial x} \right) + \frac{1}{3} \frac{\partial \ddot{\varphi}_z}{\partial x} \right], \qquad (10.4.49)$$

$$\delta v: \quad \frac{\partial M_{xy}^{(0)}}{\partial x} + \frac{\partial M_{yy}^{(0)}}{\partial y} + q_y = m_0 \ddot{v} + m_1 \ddot{\phi}_y - \frac{m_2}{2} \frac{\partial \ddot{\phi}_z}{\partial y}$$
$$- m_3 \left[c_1 \left(\ddot{\phi}_y + \frac{\partial \ddot{w}}{\partial y} \right) + \frac{1}{3} \frac{\partial \ddot{\varphi}_z}{\partial y} \right], \qquad (10.4.50)$$

$$\delta w: \quad \frac{\partial}{\partial x}\left(\frac{\partial w}{\partial x} M_{xx}^{(0)} + \frac{\partial w}{\partial y} M_{xy}^{(0)} \right) + \frac{\partial}{\partial y}\left(\frac{\partial w}{\partial x} M_{xy}^{(0)} + \frac{\partial w}{\partial y} M_{yy}^{(0)} \right)$$
$$+ \frac{\partial M_{xz}^{(0)}}{\partial x} + \frac{\partial M_{yz}^{(0)}}{\partial y} + c_1 \left(\frac{\partial^2 M_{xx}^{(3)}}{\partial x^2} + 2\frac{\partial^2 M_{xy}^{(3)}}{\partial x \partial y} + \frac{\partial^2 M_{yy}^{(3)}}{\partial y^2} \right)$$
$$- c_2 \left(\frac{\partial M_{xz}^{(2)}}{\partial x} + \frac{\partial M_{yz}^{(2)}}{\partial y} \right) + q_z = m_0 \ddot{w} + m_1 \ddot{\phi}_z + m_2 \ddot{\varphi}_z$$
$$+ c_1 \left[m_3 \left(\frac{\partial \ddot{u}}{\partial x} + \frac{\partial \ddot{v}}{\partial y} \right) - c_1 m_6 \left(\frac{\partial^2 \ddot{w}}{\partial x^2} + \frac{\partial^2 \ddot{w}}{\partial y^2} \right) + \hat{m}_4 \left(\frac{\partial \ddot{\phi}_x}{\partial x} + \frac{\partial \ddot{\phi}_y}{\partial y} \right) \right.$$
$$\left. - \frac{m_5}{2} \left(\frac{\partial^2 \ddot{\phi}_z}{\partial x^2} + \frac{\partial^2 \ddot{\phi}_z}{\partial y^2} \right) - \frac{m_6}{3} \left(\frac{\partial^2 \ddot{\varphi}_z}{\partial x^2} + \frac{\partial^2 \ddot{\varphi}_z}{\partial y^2} \right) \right], \qquad (10.4.51)$$

$$\delta \phi_x: \quad \frac{\partial M_{xx}^{(1)}}{\partial x} + \frac{\partial M_{xy}^{(1)}}{\partial y} - M_{xz}^{(0)} - c_1 \left(\frac{\partial M_{xx}^{(3)}}{\partial x} + \frac{\partial M_{xy}^{(3)}}{\partial y} \right) + c_2 M_{xz}^{(2)} + F_x$$
$$= \hat{m}_1 \ddot{u} + \hat{m}_2 \ddot{\phi}_x - \frac{1}{2}\hat{m}_3 \frac{\partial \ddot{\phi}_z}{\partial x} - \frac{1}{3}\hat{m}_4 \frac{\partial \ddot{\varphi}_z}{\partial x} - c_1 \hat{m}_4 \left(\ddot{\phi}_x + \frac{\partial \ddot{w}}{\partial x} \right), \qquad (10.4.52)$$

$$\delta \phi_y: \quad \frac{\partial M_{xy}^{(1)}}{\partial x} + \frac{\partial M_{yy}^{(1)}}{\partial y} - M_{yz}^{(0)} - c_1 \left(\frac{\partial M_{xy}^{(3)}}{\partial x} + \frac{\partial M_{yy}^{(3)}}{\partial y} \right) + c_2 M_{yz}^{(2)} + F_y$$
$$= \hat{m}_1 \ddot{v} + \hat{m}_2 \ddot{\phi}_y - \frac{1}{2}\hat{m}_3 \frac{\partial \ddot{\phi}_z}{\partial y} - \frac{1}{3}\hat{m}_4 \frac{\partial \ddot{\varphi}_z}{\partial y} - c_1 \hat{m}_4 \left(\ddot{\phi}_y + \frac{\partial \ddot{w}}{\partial y} \right), \qquad (10.4.53)$$

$$\delta \phi_z: \quad \frac{1}{2}\left(\frac{\partial^2 M_{xx}^{(2)}}{\partial x^2} + 2\frac{\partial^2 M_{xy}^{(2)}}{\partial x \partial y} + \frac{\partial^2 M_{yy}^{(2)}}{\partial y^2} \right) - M_{zz}^{(0)} + P$$
$$= \frac{1}{2}\left[m_2 \left(\frac{\partial \ddot{u}}{\partial x} + \frac{\partial \ddot{v}}{\partial y} \right) + \hat{m}_3 \left(\frac{\partial \ddot{\phi}_x}{\partial x} + \frac{\partial \ddot{\phi}_y}{\partial y} \right) - \frac{m_4}{2}\left(\frac{\partial^2 \ddot{\phi}_z}{\partial x^2} + \frac{\partial^2 \ddot{\phi}_z}{\partial y^2} \right) \right]$$

$$- \frac{m_5}{2} \left[\frac{1}{3} \left(\frac{\partial^2 \ddot{\varphi}_z}{\partial x^2} + \frac{\partial^2 \ddot{\varphi}_z}{\partial y^2} \right) + c_1 \left(\frac{\partial^2 \ddot{w}}{\partial x^2} + \frac{\partial^2 \ddot{w}}{\partial y^2} \right) \right]$$

$$+ m_1 \ddot{w} + m_2 \ddot{\phi}_z + m_3 \ddot{\varphi}_z, \tag{10.4.54}$$

$$\delta\varphi_z : \quad \frac{1}{3} \left(\frac{\partial^2 M_{xx}^{(3)}}{\partial x^2} + 2 \frac{\partial^2 M_{xy}^{(3)}}{\partial x \partial y} + \frac{\partial^2 M_{yy}^{(3)}}{\partial y^2} \right) - 2M_{zz}^{(1)} + Q$$

$$= m_2 \ddot{w} + m_3 \ddot{\phi}_z + m_4 \ddot{\varphi}_z + \frac{1}{3} \left[m_3 \left(\frac{\partial \ddot{u}}{\partial x} + \frac{\partial \ddot{v}}{\partial y} \right) - c_1 m_6 \left(\frac{\partial^2 \ddot{w}}{\partial x^2} + \frac{\partial^2 \ddot{w}}{\partial y^2} \right) \right.$$

$$\left. + \hat{m}_4 \left(\frac{\partial \ddot{\phi}_x}{\partial x} + \frac{\partial \ddot{\phi}_y}{\partial y} \right) - \frac{m_5}{2} \left(\frac{\partial^2 \ddot{\phi}_z}{\partial x^2} + \frac{\partial^2 \ddot{\phi}_z}{\partial y^2} \right) - \frac{m_6}{3} \left(\frac{\partial^2 \ddot{\varphi}_z}{\partial x^2} + \frac{\partial^2 \ddot{\varphi}_z}{\partial y^2} \right) \right], \tag{10.4.55}$$

where

$$\hat{m}_i = m_i - c_1 m_{i+2}, \quad i = 1, 2, 3, 4. \tag{10.4.56}$$

The boundary conditions involve specifying the following generalized forces that are dual to the generalized displacement $(u, v, w, \phi_x, \phi_y, \phi_z, \varphi_z)$:

$$\delta u : \quad \bar{M}_{nn}^{(0)} \equiv M_{xx}^{(0)} n_x + M_{xy}^{(0)} n_y, \tag{10.4.57}$$

$$\delta v : \quad \bar{M}_{nt}^{(0)} \equiv M_{xy}^{(0)} n_x + M_{yy}^{(0)} n_y, \tag{10.4.58}$$

$$\delta w : \quad \bar{M}_{nz}^{(0)} \equiv \left(\frac{\partial w}{\partial x} M_{xx}^{(0)} + \frac{\partial w}{\partial y} M_{xy}^{(0)} \right) n_x + \left(\frac{\partial w}{\partial x} M_{xy}^{(0)} + \frac{\partial w}{\partial y} M_{yy}^{(0)} \right) n_y$$

$$+ M_{xz}^{(0)} n_x + M_{yz}^{(0)} n_y - c_2 \left(M_{xz}^{(2)} n_x + M_{yz}^{(2)} n_y \right)$$

$$+ c_1 \left[\left(\frac{\partial M_{xx}^{(3)}}{\partial x} + \frac{\partial M_{xy}^{(3)}}{\partial y} \right) n_x + \left(\frac{\partial M_{xy}^{(3)}}{\partial x} + \frac{\partial M_{yy}^{(3)}}{\partial y} \right) n_y \right]$$

$$+ c_1 \left[\left(m_3 \ddot{u} + \hat{m}_4 \ddot{\phi}_x - \frac{m_5}{2} \frac{\partial \ddot{\phi}_z}{\partial x} - \frac{m_6}{3} \frac{\partial \ddot{\varphi}_z}{\partial x} - c_1 m_6 \frac{\partial \ddot{w}}{\partial x} \right) n_x \right.$$

$$\left. + \left(m_3 \ddot{v} + \hat{m}_4 \ddot{\phi}_y - \frac{m_5}{2} \frac{\partial \ddot{\phi}_z}{\partial y} - \frac{m_6}{3} \frac{\partial \ddot{\varphi}_z}{\partial y} - c_1 m_6 \frac{\partial \ddot{w}}{\partial x} \right) n_y \right], \tag{10.4.59}$$

$$\delta\phi_x : \quad \bar{M}_{nn}^{(1)} \equiv M_{xx}^{(1)} n_x + M_{xy}^{(1)} n_y - c_1 \left(M_{xx}^{(3)} n_x + M_{xy}^{(3)} n_y \right), \tag{10.4.60}$$

$$\delta\phi_y : \quad \bar{M}_{nt}^{(1)} \equiv M_{xy}^{(1)} n_x + M_{yy}^{(1)} n_y - c_1 \left(M_{xy}^{(3)} n_x + M_{yy}^{(3)} n_y \right), \tag{10.4.61}$$

$$\delta\phi_z : \quad \left[m_2 \ddot{u} + m_3 \ddot{\phi}_x - \frac{m_4}{2} \frac{\partial \ddot{\phi}_z}{\partial x} \right.$$

$$\left. - m_5 \left(\frac{1}{3} \frac{\partial \ddot{\varphi}_z}{\partial x} - c_1 \ddot{\phi}_x - c_1 \frac{\partial \ddot{w}}{\partial x} \right) \right] n_x + \left[m_2 \ddot{v} + m_3 \ddot{\phi}_y \right.$$

$$\left. - \frac{m_4}{2} \frac{\partial \ddot{\phi}_z}{\partial y} - m_5 \left(\frac{1}{3} \frac{\partial \ddot{\varphi}_z}{\partial y} - c_1 \ddot{\phi}_y - c_1 \frac{\partial \ddot{w}}{\partial y} \right) \right] n_y, \tag{10.4.62}$$

$$\delta\varphi_z: \quad c_1\Bigg[\Bigg(m_3\ddot{u} + \hat{m}_4\ddot{\phi}_x - \frac{m_5}{2}\frac{\partial\ddot{\phi}_z}{\partial x}$$
$$- \frac{m_6}{3}\frac{\partial\ddot{\varphi}_z}{\partial x} - c_1 m_6\frac{\partial\ddot{w}}{\partial x}\Bigg)n_x + \Bigg(m_3\ddot{v} + \hat{m}_4\ddot{\phi}_y$$
$$- \frac{m_5}{2}\frac{\partial\ddot{\phi}_z}{\partial y} - \frac{m_6}{3}\frac{\partial\ddot{\varphi}_z}{\partial y} - c_1 m_6\frac{\partial\ddot{w}}{\partial y}\Bigg)n_y\Bigg]. \tag{10.4.63}$$

This third-order theory is not to be found in the literature; the closest one to this third-order theory is that due to Reddy [113].

10.4.5.2 The Reddy Third-Order Plate Theory

The Reddy third-order theory [113] is based on the displacement field in which $\phi_z = 0$ and $\varphi_z = 0$; when the top and bottom surfaces of the plate are required to be free of any tangential forces, we obtain

$$\varphi_x = 0, \quad \psi_x = -c_1\left(\phi_x + \frac{\partial w}{\partial x}\right), \quad \varphi_y = 0, \quad \psi_y = -c_1\left(\phi_y + \frac{\partial w}{\partial y}\right). \tag{10.4.64}$$

Thus, the theory is a special case of the one derived in the previous section and it deduced by setting $\phi_z = 0$, $\varphi_z = 0$, $\varphi_x = 0$, and $\varphi_y = 0$. Then the strains in Eq. (10.4.5) take the form

$$\left\{\begin{array}{c}\varepsilon_{xx}\\\varepsilon_{yy}\\\gamma_{xy}\end{array}\right\} = \left\{\begin{array}{c}\varepsilon_{xx}^{(0)}\\\varepsilon_{yy}^{(0)}\\\gamma_{xy}^{(0)}\end{array}\right\} + z\left\{\begin{array}{c}\varepsilon_{xx}^{(1)}\\\varepsilon_{yy}^{(1)}\\\gamma_{xy}^{(1)}\end{array}\right\} + z^3\left\{\begin{array}{c}\varepsilon_{xx}^{(3)}\\\varepsilon_{yy}^{(3)}\\\gamma_{xy}^{(3)}\end{array}\right\},$$

$$\left\{\begin{array}{c}\varepsilon_{zz}\\\gamma_{xz}\\\gamma_{yz}\end{array}\right\} = \left\{\begin{array}{c}\varepsilon_{zz}^{(0)}\\\gamma_{xz}^{(0)}\\\gamma_{yz}^{(0)}\end{array}\right\} + z^2\left\{\begin{array}{c}\varepsilon_{zz}^{(2)}\\\gamma_{xz}^{(2)}\\\gamma_{yz}^{(2)}\end{array}\right\}, \tag{10.4.65}$$

with

$$\left\{\begin{array}{c}\varepsilon_{xx}^{(0)}\\\varepsilon_{yy}^{(0)}\\\gamma_{xy}^{(0)}\end{array}\right\} = \left\{\begin{array}{c}\frac{\partial u}{\partial x} + \frac{1}{2}\left(\frac{\partial w}{\partial x}\right)^2\\\frac{\partial v}{\partial y} + \frac{1}{2}\left(\frac{\partial w}{\partial y}\right)^2\\\frac{\partial u}{\partial y} + \frac{\partial v}{\partial x} + \frac{\partial w}{\partial x}\frac{\partial w}{\partial y}\end{array}\right\},$$

$$\left\{\begin{array}{c}\varepsilon_{xx}^{(1)}\\\varepsilon_{yy}^{(1)}\\\gamma_{xy}^{(1)}\end{array}\right\} = \left\{\begin{array}{c}\frac{\partial\phi_x}{\partial x}\\\frac{\partial\phi_y}{\partial y}\\\frac{\partial\phi_x}{\partial y} + \frac{\partial\phi_y}{\partial x}\end{array}\right\}, \quad \left\{\begin{array}{c}\varepsilon_{zz}^{(0)}\\\gamma_{xz}^{(0)}\\\gamma_{yz}^{(0)}\end{array}\right\} = \left\{\begin{array}{c}0\\\phi_x + \frac{\partial w}{\partial x}\\\phi_y + \frac{\partial w}{\partial y}\end{array}\right\}, \tag{10.4.66}$$

$$\left\{\begin{array}{c}\varepsilon_{xx}^{(3)}\\\varepsilon_{yy}^{(3)}\\\gamma_{xy}^{(3)}\end{array}\right\} = -c_1\left\{\begin{array}{c}\frac{\partial\phi_x}{\partial x} + \frac{\partial^2 w}{\partial x^2}\\\frac{\partial\phi_y}{\partial y} + \frac{\partial^2 w}{\partial y^2}\\\frac{\partial\phi_x}{\partial y} + \frac{\partial\phi_y}{\partial x} + 2\frac{\partial^2 w}{\partial x\partial y}\end{array}\right\}, \quad \left\{\begin{array}{c}\varepsilon_{zz}^{(2)}\\\gamma_{xz}^{(2)}\\\gamma_{yz}^{(2)}\end{array}\right\} = -c_2\left\{\begin{array}{c}0\\\phi_x + \frac{\partial w}{\partial x}\\\phi_y + \frac{\partial w}{\partial y}\end{array}\right\}. \tag{10.4.67}$$

Thus the transverse normal strain ε_{zz} is identically zero. Consequently, σ_{zz} does not enter the strain energy expression and we use the plane stress-reduced constitutive relations

$$
\begin{Bmatrix} \sigma_{xx} \\ \sigma_{yy} \\ \sigma_{xy} \\ \sigma_{xz} \\ \sigma_{yz} \end{Bmatrix} = \frac{E}{1-\nu^2} \begin{bmatrix} 1 & \nu & 0 & 0 & 0 \\ \nu & 1 & 0 & 0 & 0 \\ 0 & 0 & \frac{1-\nu}{2} & 0 & 0 \\ 0 & 0 & 0 & \frac{1-\nu}{2} & 0 \\ 0 & 0 & 0 & 0 & \frac{1-\nu}{2} \end{bmatrix} \begin{Bmatrix} \varepsilon_{xx} - \alpha\Delta T \\ \varepsilon_{yy} - \alpha\Delta T \\ \gamma_{xy} \\ \gamma_{xz} \\ \gamma_{yz} \end{Bmatrix}. \tag{10.4.68}
$$

Thus, the stress resultants must be expressed in terms of the strains using the constitutive equations in Eq. (10.4.68).

The equations of motion of the Reddy third-order plate theory are

$$
\delta u: \quad \frac{\partial M_{xx}^{(0)}}{\partial x} + \frac{\partial M_{xy}^{(0)}}{\partial y} + q_x = m_0 \ddot{u} + m_1 \ddot{\phi}_x - m_3 c_1 \left(\ddot{\phi}_x + \frac{\partial \ddot{w}}{\partial x} \right), \tag{10.4.69}
$$

$$
\delta v: \quad \frac{\partial M_{xy}^{(0)}}{\partial x} + \frac{\partial M_{yy}^{(0)}}{\partial y} + q_y = m_0 \ddot{v} + m_1 \ddot{\phi}_y - m_3 c_1 \left(\ddot{\phi}_y + \frac{\partial \ddot{w}}{\partial y} \right), \tag{10.4.70}
$$

$$
\delta w: \quad \frac{\partial}{\partial x} \left(\frac{\partial w}{\partial x} M_{xx}^{(0)} + \frac{\partial w}{\partial y} M_{xy}^{(0)} \right) + \frac{\partial}{\partial y} \left(\frac{\partial w}{\partial x} M_{xy}^{(0)} + \frac{\partial w}{\partial y} M_{yy}^{(0)} \right)
$$

$$
+ \frac{\partial M_{xz}^{(0)}}{\partial x} + \frac{\partial M_{yz}^{(0)}}{\partial y} + c_1 \left(\frac{\partial^2 M_{xx}^{(3)}}{\partial x^2} + 2\frac{\partial^2 M_{xy}^{(3)}}{\partial x \partial y} + \frac{\partial^2 M_{yy}^{(3)}}{\partial y^2} \right)
$$

$$
- c_2 \left(\frac{\partial M_{xz}^{(2)}}{\partial x} + \frac{\partial M_{yz}^{(2)}}{\partial y} \right) + q_z
$$

$$
= c_1 \left[m_3 \left(\frac{\partial \ddot{u}}{\partial x} + \frac{\partial \ddot{v}}{\partial y} \right) - c_1 m_6 \left(\frac{\partial^2 \ddot{w}}{\partial x^2} + \frac{\partial^2 \ddot{w}}{\partial y^2} \right) + \hat{m}_4 \left(\frac{\partial \ddot{\phi}_x}{\partial x} + \frac{\partial \ddot{\phi}_y}{\partial y} \right) \right]
$$

$$
+ m_0 \ddot{w}, \tag{10.4.71}
$$

$$
\delta \phi_x: \quad \frac{\partial M_{xx}^{(1)}}{\partial x} + \frac{\partial M_{xy}^{(1)}}{\partial y} - M_{xz}^{(0)} - c_1 \left(\frac{\partial M_{xx}^{(3)}}{\partial x} + \frac{\partial M_{xy}^{(3)}}{\partial y} \right) + c_2 M_{xz}^{(2)} + F_x
$$

$$
= \hat{m}_1 \ddot{u} + \hat{m}_2 \ddot{\phi}_x - c_1 \hat{m}_4 \left(\ddot{\phi}_x + \frac{\partial \ddot{w}}{\partial x} \right), \tag{10.4.72}
$$

$$
\delta \phi_y: \quad \frac{\partial M_{xy}^{(1)}}{\partial x} + \frac{\partial M_{yy}^{(1)}}{\partial y} - M_{yz}^{(0)} - c_1 \left(\frac{\partial M_{xy}^{(3)}}{\partial x} + \frac{\partial M_{yy}^{(3)}}{\partial y} \right) + c_2 M_{yz}^{(2)} + F_y
$$

$$
= \hat{m}_1 \ddot{v} + \hat{m}_2 \ddot{\phi}_y - c_1 \hat{m}_4 \left(\ddot{\phi}_y + \frac{\partial \ddot{w}}{\partial y} \right). \tag{10.4.73}
$$

The generalized forces $(M_{xx}^{(i)}, M_{yy}^{(i)}, M_{xy}^{(i)})$ are related to the generalized displacements $(u, v, w, \phi_x, \phi_y)$ through Eq. (10.4.43) with

$$A_{11}^{(k)} = \frac{1}{(1-\nu^2)} \int_{-\frac{h}{2}}^{\frac{h}{2}} (z)^k E(z)\, dz,$$

$$A_{12}^{(k)} = \frac{\nu}{(1-\nu^2)} \int_{-\frac{h}{2}}^{\frac{h}{2}} (z)^k E(z)\, dz, \qquad (10.4.74)$$

$$B_{11}^{(k)} = \frac{1}{2(1+\nu)} \int_{-\frac{h}{2}}^{\frac{h}{2}} (z)^k E(z)\, dz.$$

10.4.5.3 The First-Order Plate Theory

The FSDT is based on the displacement field in which we have $\phi_z = 0$, $\varphi_z = 0$, $\varphi_x = 0$, and $\varphi_y = 0$ as well as $c_1 = 0$ and $c_2 = 0$. Then the strains in Eqs (10.4.65)–(10.4.67) simplify to

$$\left\{ \begin{array}{c} \varepsilon_{xx} \\ \varepsilon_{yy} \\ \gamma_{xy} \end{array} \right\} = \left\{ \begin{array}{c} \varepsilon_{xx}^{(0)} \\ \varepsilon_{yy}^{(0)} \\ \gamma_{xy}^{(0)} \end{array} \right\} + z \left\{ \begin{array}{c} \varepsilon_{xx}^{(1)} \\ \varepsilon_{yy}^{(1)} \\ \gamma_{xy}^{(1)} \end{array} \right\}, \quad \left\{ \begin{array}{c} \varepsilon_{zz} \\ \gamma_{xz} \\ \gamma_{yz} \end{array} \right\} = \left\{ \begin{array}{c} \varepsilon_{zz}^{(0)} \\ \gamma_{xz}^{(0)} \\ \gamma_{yz}^{(0)} \end{array} \right\} = \left\{ \begin{array}{c} 0 \\ \phi_x + \frac{\partial w}{\partial x} \\ \phi_y + \frac{\partial w}{\partial y} \end{array} \right\}$$

$$(10.4.75)$$

with

$$\left\{ \begin{array}{c} \varepsilon_{xx}^{(0)} \\ \varepsilon_{yy}^{(0)} \\ \gamma_{xy}^{(0)} \end{array} \right\} = \left\{ \begin{array}{c} \frac{\partial u}{\partial x} + \frac{1}{2}\left(\frac{\partial w}{\partial x}\right)^2 \\ \frac{\partial v}{\partial y} + \frac{1}{2}\left(\frac{\partial w}{\partial y}\right)^2 \\ \frac{\partial u}{\partial y} + \frac{\partial v}{\partial x} + \frac{\partial w}{\partial x}\frac{\partial w}{\partial y} \end{array} \right\}, \quad \left\{ \begin{array}{c} \varepsilon_{xx}^{(1)} \\ \varepsilon_{yy}^{(1)} \\ \gamma_{xy}^{(1)} \end{array} \right\} = \left\{ \begin{array}{c} \frac{\partial \phi_x}{\partial x} \\ \frac{\partial \phi_y}{\partial y} \\ \frac{\partial \phi_x}{\partial y} + \frac{\partial \phi_y}{\partial x} \end{array} \right\}. \quad (10.4.76)$$

The equations of motion are simplified to

$$\delta u : \quad \frac{\partial M_{xx}^{(0)}}{\partial x} + \frac{\partial M_{xy}^{(0)}}{\partial y} + q_x = m_0 \ddot{u} + m_1 \ddot{\phi}_x, \qquad (10.4.77)$$

$$\delta v : \quad \frac{\partial M_{xy}^{(0)}}{\partial x} + \frac{\partial M_{yy}^{(0)}}{\partial y} + q_y = m_0 \ddot{v} + m_1 \ddot{\phi}_y, \qquad (10.4.78)$$

$$\delta w : \quad \frac{\partial}{\partial x}\left(\frac{\partial w}{\partial x}M_{xx}^{(0)} + \frac{\partial w}{\partial y}M_{xy}^{(0)}\right) + \frac{\partial}{\partial y}\left(\frac{\partial w}{\partial x}M_{xy}^{(0)} + \frac{\partial w}{\partial y}M_{yy}^{(0)}\right)$$

$$+ \frac{\partial M_{xz}^{(0)}}{\partial x} + \frac{\partial M_{yz}^{(0)}}{\partial y} + q_z = m_0 \ddot{w}, \qquad (10.4.79)$$

$$\delta \phi_x : \quad \frac{\partial M_{xx}^{(1)}}{\partial x} + \frac{\partial M_{xy}^{(1)}}{\partial y} - M_{xz}^{(0)} + F_x = m_1 \ddot{u} + m_2 \ddot{\phi}_x, \qquad (10.4.80)$$

$$\delta \phi_y : \quad \frac{\partial M_{xy}^{(1)}}{\partial x} + \frac{\partial M_{yy}^{(1)}}{\partial y} - M_{yz}^{(0)} + F_y = m_1 \ddot{v} + m_2 \ddot{\phi}_y. \qquad (10.4.81)$$

The generalized forces $(M_{xx}^{(i)}, M_{yy}^{(i)}, M_{xy}^{(i)})$ in the FSDT[1] are related to the generalized displacements $(u, v, w, \phi_x, \phi_y)$ by Eq. (10.4.43) with the coefficients $A_{11}^{(k)}$, $A_{12}^{(k)}$, and $B_{11}^{(k)}$ defined by Eq. (10.4.74).

10.4.5.4   The Classical Plate Theory

The CPT is obtained by setting

$$\phi_x = -\frac{\partial w}{\partial x}, \qquad \phi_y = -\frac{\partial w}{\partial y}, \tag{10.4.82}$$

and $\varphi_x = \varphi_y = \psi_x = \psi_y = \phi_z = \varphi_z = 0$ in the displacement field (10.4.3). The equations of motion are obtained from the Reddy third-order theory by setting $c_1 = 1$ and deleting terms involving $M_{xz}^{(0)}$ and $M_{yz}^{(0)}$. We have

$$\delta u : \quad \frac{\partial M_{xx}^{(0)}}{\partial x} + \frac{\partial M_{xy}^{(0)}}{\partial y} + q_x = m_0 \ddot{u} - m_1 \frac{\partial \ddot{w}}{\partial x}, \tag{10.4.83}$$

$$\delta v : \quad \frac{\partial M_{xy}^{(0)}}{\partial x} + \frac{\partial M_{yy}^{(0)}}{\partial y} + q_y = m_0 \ddot{v} - m_1 \frac{\partial \ddot{w}}{\partial y}, \tag{10.4.84}$$

$$\delta w : \quad \frac{\partial}{\partial x}\left(\frac{\partial w}{\partial x}M_{xx}^{(0)} + \frac{\partial w}{\partial y}M_{xy}^{(0)}\right) + \frac{\partial}{\partial y}\left(\frac{\partial w}{\partial x}M_{xy}^{(0)} + \frac{\partial w}{\partial y}M_{yy}^{(0)}\right)$$

$$+ \frac{\partial^2 M_{xx}^{(1)}}{\partial x^2} + 2\frac{\partial^2 M_{xy}^{(1)}}{\partial x \partial y} + \frac{\partial^2 M_{yy}^{(1)}}{\partial y^2} + q_z$$

$$= m_0 \ddot{w} + m_1\left(\frac{\partial \ddot{u}}{\partial x} + \frac{\partial \ddot{v}}{\partial y}\right) - m_2\left(\frac{\partial^2 \ddot{w}}{\partial x^2} + \frac{\partial^2 \ddot{w}_y}{\partial y^2}\right). \tag{10.4.85}$$

10.5   Navier's Solutions

10.5.1   Preliminary Comments

In this section, we develop Navier's solutions of the linearized equations governing beams and plates for simply supported boundary conditions. The analysis of FGM beams and plates is made complicated by the fact that the governing equations are coupled in the sense that the bending solutions cannot be determined independent of the in-plane displacements. This is because of the presence of the bending–stretching stiffness coefficient, B_{xx}. Also, as before, the Navier solutions of the linearized equations can be obtained only for certain geometries and boundary conditions. Throughout this section we assume that A_{xx}, B_{xx}, D_{xx}, and S_{xz} are independent of x and y.

---

[1]The author does not like to use the phrases *Mindlin plate theory* or *Reissner-Mindlin plate theory*, and prefers to use the phrase *first-order plate theory* because Mindlin and Reissner were not the first ones to develop the theory; it goes back to Cauchy, Basset, Hencky, and others (see Reddy [50, 51] for a review of the literature).

10.5.2   Analysis of Beams

10.5.2.1   Bernoulli–Euler Beams

The linearized equations of motion of the BET are [from Eqs (10.2.30) and (10.2.31)]

$$\frac{\partial}{\partial x}\left(A_{xx}\frac{\partial u}{\partial x}+B_{xx}\frac{\partial \theta_x}{\partial x}-N_{xx}^T\right)+f_x = m_0\frac{\partial^2 u}{\partial t^2}-m_1\frac{\partial^3 w}{\partial x \partial t^2}, \tag{10.5.1}$$

$$\frac{\partial^2}{\partial x^2}\left(B_{xx}\frac{\partial u}{\partial x}+D_{xx}\frac{\partial \theta_x}{\partial x}-M_{xx}^T\right)+f_z$$

$$= m_0\frac{\partial^2 w}{\partial t^2}+m_1\frac{\partial^3 u}{\partial x \partial t^2}-m_2\frac{\partial^4 w}{\partial t^2 \partial x^2}. \tag{10.5.2}$$

Navier's solution for simply supported beams. Assuming solution of the form

$$u(x,t) = \sum_{n=1}^{\infty} U_n(t)\cos\alpha_n x, \quad w(x,t) = \sum_{n=1}^{\infty} W_n(t)\sin\alpha_n x, \tag{10.5.3}$$

and expanding the forces f_x and f_z in series as

$$f_x(x,t) = \sum_{n=1}^{\infty} F_n(t)\cos\alpha_n x, \quad f_z(x,t) = \sum_{n=1}^{\infty} Q_n(t)\sin\alpha_n x,$$

$$\tag{10.5.4}$$

$$N_{xx}^T(x,t) = \sum_{n=1}^{\infty} N_n^T(t)\sin\alpha_n x, \quad M_{xx}^T(x,t) = \sum_{n=1}^{\infty} M_n^T(t)\sin\alpha_n x,$$

where $\alpha_n = n\pi/L$, L is the length of the beam, and

$$F_n(t) = \frac{2}{L}\int_0^L f_x(x,t)\cos\alpha_n x\, dx, \quad Q_n(t) = \frac{2}{L}\int_0^L f_z(x,t)\sin\alpha_n x\, dx,$$

$$\tag{10.5.5}$$

$$N_n^T(t) = \frac{2}{L}\int_0^L N_{xx}^T(x,t)\sin\alpha_n x\, dx, \quad M_n^T(t) = \frac{2}{L}\int_0^L M_{xx}^T(x,t)\sin\alpha_n x\, dx.$$

The coefficients Q_n, for example, for some typical loads are given by

$$Q_n = \begin{cases} q_0 \ (n=1), & \text{sinusoidal load of intensity } q_0, \\ \frac{4}{n\pi}q_0, \ (n=1,3,5,\ldots), & \text{uniform load of intensity } q_0, \\ \frac{2}{L}Q_0\sin\frac{n\pi}{2} \ (n=1,2,3,\ldots), & \text{point load } Q_0 \text{ at the center.} \end{cases} \tag{10.5.6}$$

We note that the assumed solution in Eq. (10.5.3) satisfies the boundary conditions [see Eqs (10.2.9) and (10.2.10)]

$$N_{xx}(0,t)+N_{xx}^T(0,t)=0, \quad w(0,t)=0, \quad M_{xx}(0,t)+M_{xx}^T(0,t)=0,$$
$$\tag{10.5.7}$$
$$N_{xx}(L,t)+N_{xx}^T(L,t)=0, \quad w(L,t)=0, \quad M_{xx}(L,t)+M_{xx}^T(L,t)=0.$$

Substitution of the expansions from Eqs (10.5.3) and (10.5.4) into Eqs (10.5.1) and (10.5.2), we arrive at

$$-\alpha_n^2 A_{xx} U_n(t) + \alpha_n^3 B_{xx} W_n(t) - \alpha_n N_n^T(t) + F_n(t) = m_0 \ddot{U}_n(t) - m_1 \alpha_n \ddot{W}_n(t),$$
$$\alpha_n^3 B_{xx} U_n(t) - \alpha_n^4 D_{xx} W_n(t) + \alpha_n^2 M_n^T(t) + Q_n(t) \qquad (10.5.8)$$
$$= -m_1 \alpha_n \ddot{U}_n(t) + \left(m_0 + m_2 \alpha_n^2\right) \ddot{W}_n(t),$$

or in matrix form

$$\begin{bmatrix} -\alpha_n^2 A_{xx} & \alpha_n^3 B_{xx} \\ \alpha_n^3 B_{xx} & -\alpha_n^4 D_{xx} \end{bmatrix} \begin{Bmatrix} U_n \\ W_n \end{Bmatrix} + \begin{Bmatrix} F_n - \alpha_n N_n^T \\ Q_n + \alpha_n^2 M_n^T \end{Bmatrix}$$
$$= \begin{bmatrix} m_0 & -m_1 \alpha_n \\ -m_1 \alpha_n & (m_0 + m_2 \alpha_n^2) \end{bmatrix} \begin{Bmatrix} \ddot{U}_n \\ \ddot{W}_n \end{Bmatrix}. \qquad (10.5.9)$$

Static analysis. For static analysis, we set all time derivative terms to zero and obtain

$$\begin{bmatrix} \alpha_n^2 A_{xx} & -\alpha_n^3 B_{xx} \\ -\alpha_n^3 B_{xx} & \alpha_n^4 D_{xx} \end{bmatrix} \begin{Bmatrix} U_n \\ W_n \end{Bmatrix} = \begin{Bmatrix} F_n - \alpha_n N_n^T \\ Q_n + \alpha_n^2 M_n^T \end{Bmatrix}, \qquad (10.5.10)$$

whose solution is

$$U_n = \frac{1}{\alpha_n^3 \hat{D}_{xx}} \left[\left(F_n - \alpha_n N_n^T\right) D_{xx} \alpha_n + \left(Q_n + \alpha_n^2 M_n^T\right) B_{xx}\right],$$
$$\qquad (10.5.11)$$
$$W_n = \frac{1}{\alpha_n^4 \hat{D}_{xx}} \left[\left(F_n - \alpha_n N_n^T\right) B_{xx} \alpha_n + \left(Q_n + \alpha_n^2 M_n^T\right) A_{xx}\right],$$

where

$$\hat{D}_{xx} = A_{xx} D_{xx} - B_{xx}^2. \qquad (10.5.12)$$

For homogeneous isotropic beams (i.e., when $B_{xx} = 0$), the solution in Eq. (10.5.10) reduces to (with $A_{xx} = EA$ and $D_{xx} = EI$)

$$U_n = \frac{1}{\alpha_n^2 EA} \left(F_n - \alpha_n N_n^T\right); \quad W_n = \frac{1}{\alpha_n^4 EI} \left(Q_n + \alpha_n^2 M_n^T\right). \qquad (10.5.13)$$

Thus, the axial displacement $u(x)$ and transverse deflection $w(x)$ are decoupled. In particular, when the thermal resultants are zero, we obtain

$$U_n = \frac{F_n L^2}{n^2 \pi^2 EA}; \quad W_n = \frac{Q_n L^4}{n^4 \pi^4 EI}. \qquad (10.5.14)$$

Natural vibration. For natural vibration, we assume that all external forces to be zero and the solution to be periodic,

$$U_n(t) = U_n^0 e^{i\omega_n t}; \quad W_n(t) = W_n^0 e^{i\omega_n t}, \qquad (10.5.15)$$

where ω_n is the frequency of natural vibration. Substitution of Eq. (10.5.15) into Eq. (10.5.9), we obtain the following eigenvalue problem:

$$\left(\begin{bmatrix} \alpha_n^2 A_{xx} & -\alpha_n^3 B_{xx} \\ -\alpha_n^3 B_{xx} & \alpha_n^4 D_{xx} \end{bmatrix} - \omega_n^2 \begin{bmatrix} m_0 & -m_1 \alpha_n \\ -m_1 \alpha_n & (m_0 + m_2 \alpha_n^2) \end{bmatrix} \right) \left\{ \begin{matrix} U_n^0 \\ W_n^0 \end{matrix} \right\} = \left\{ \begin{matrix} 0 \\ 0 \end{matrix} \right\},$$

$$(10.5.16)$$

which can be used to determine the natural frequencies.

For the homogeneous beams, Eq. (10.5.16) reduces to (with $A_{xx} = EA$, $D_{xx} = EI$, $m_0 = \rho A$, and $m_1 = 0$, $B_{xx} = 0$, and $m_2 = \rho I$)

$$\text{Axial:} \quad \omega_n = \alpha_n \sqrt{\frac{A_{xx}}{m_0}} = \frac{n\pi}{L} \sqrt{\frac{E}{\rho}}, \tag{10.5.17}$$

$$\text{Flexural:} \quad \omega_n = \alpha_n^2 \sqrt{\frac{D_{xx}}{(m_0 + m_2 \alpha_n^2)}} = \frac{n^2 \pi^2}{L^2} \sqrt{\frac{EI}{\rho \left(A + \frac{n^2 \pi^2}{L^2} I \right)}}. \tag{10.5.18}$$

10.5.2.2 Timoshenko Beams

The linearized equations of motion of the TBT are obtained from Eqs (10.2.32)–(10.2.34)

$$\frac{\partial}{\partial x} \left(A_{xx} \frac{\partial u}{\partial x} + B_{xx} \frac{\partial \phi_x}{\partial x} - N_{xx}^T \right) + f_x = m_0 \frac{\partial^2 u}{\partial t^2} + m_1 \frac{\partial^2 \phi_x}{\partial t^2}, \tag{10.5.19}$$

$$\frac{\partial}{\partial x} \left[K_s S_{xz} \left(\phi_x + \frac{\partial w}{\partial x} \right) \right] + f_z = m_0 \frac{\partial^2 w}{\partial t^2}, \tag{10.5.20}$$

$$\frac{\partial}{\partial x} \left(B_{xx} \frac{\partial u}{\partial x} + D_{xx} \frac{\partial \phi_x}{\partial x} - M_{xx}^T \right)$$
$$- \left[K_s S_{xz} \left(\phi_x + \frac{\partial w}{\partial x} \right) \right] = m_1 \frac{\partial^2 u}{\partial t^2} + m_2 \frac{\partial^2 \phi_x}{\partial t^2}. \tag{10.5.21}$$

Navier's solution for simply supported beams. The solution is assumed to be of the form

$$u(x,t) = \sum_{n=1}^{\infty} U_n(t) \cos \alpha_n x, \quad w(x,t) = \sum_{n=1}^{\infty} W_n(t) \sin \alpha_n x,$$

$$(10.5.22)$$

$$\phi_x(x,t) = \sum_{n=1}^{\infty} X_n(t) \cos \alpha_n x,$$

and the mechanical forces (f_x, f_z) and thermal forces (N_{xx}^T, M_{xx}^T) are expanded in series as in Eqs (10.5.4) and (10.5.5). We note that the assumed solution in Eq. (10.5.22) satisfies the boundary conditions in Eq. (10.5.7).

Use of the expansions from Eqs (10.5.22) and (10.5.4) in Eqs (10.5.19)–(10.5.21) results in

$$
-\begin{bmatrix} \alpha_n^2 A_{xx} & 0 & \alpha_n^2 B_{xx} \\ 0 & \alpha_n^2 K_s S_{xz} & \alpha_n K_s S_{xz} \\ \alpha_n^2 B_{xx} & \alpha_n K_s S_{xz} & (\alpha_n^2 D_{xx} + K_s S_{xz}) \end{bmatrix} \begin{Bmatrix} U_n \\ W_n \\ X_n \end{Bmatrix} + \begin{Bmatrix} F_n - \alpha_n N_n^T \\ Q_n \\ -\alpha_n M_n^T \end{Bmatrix}
$$
$$
= \begin{bmatrix} m_0 & 0 & m_1 \\ 0 & m_0 & 0 \\ m_1 & 0 & m_2 \end{bmatrix} \begin{Bmatrix} \ddot{U}_n \\ \ddot{W}_n \\ \ddot{X}_n \end{Bmatrix}. \qquad (10.5.23)
$$

Static analysis. For static analysis, we set all time derivative terms to zero and obtain

$$
\begin{bmatrix} \alpha_n^2 A_{xx} & 0 & \alpha_n^2 B_{xx} \\ 0 & \alpha_n^2 K_s S_{xz} & \alpha_n K_s S_{xz} \\ \alpha_n^2 B_{xx} & \alpha_n K_s S_{xz} & (\alpha_n^2 D_{xx} + K_s S_{xz}) \end{bmatrix} \begin{Bmatrix} U_n \\ W_n \\ X_n \end{Bmatrix} = \begin{Bmatrix} F_n - \alpha_n N_n^T \\ Q_n \\ -\alpha_n M_n^T \end{Bmatrix}.
$$
$$
\qquad (10.5.24)
$$

For homogeneous isotropic beams (i.e., when $B_{xx} = 0$), the solution to Eq. (10.5.24) is ($A_{xx} = EA$, $D_{xx} = EI$, and $S_{xz} = GA$) is

$$
U_n = \frac{1}{\alpha_n^2 EA} \left(F_n - \alpha_n N_n^T \right), \quad X_n = -\frac{1}{\alpha_n^3 EI} \left(Q_n + \alpha_n^2 M_{xx}^T \right).
$$
$$
\qquad (10.5.25)
$$
$$
W_n = \frac{1}{\alpha_n^4 EI} \left[\left(1 + \alpha_n^2 \frac{EI}{K_s S_{xz}} \right) Q_n + \alpha_n^2 M_{xx}^T \right],
$$

When the thermal resultants are zero, we obtain

$$
U_n = \frac{F_n L^2}{n^2 \pi^2 EA}, \quad X_n = -\frac{Q_n L^3}{n^3 \pi^3 EI},
$$
$$
\qquad (10.5.26)
$$
$$
W_n = \frac{Q_n L^4}{n^4 \pi^4 EI} \left(n^2 \pi^2 \frac{EI}{K_s S_{xz} L^2} + 1 \right).
$$

Natural vibration. For natural vibration, we assume that all external forces are zero and the solution to be periodic,

$$
U_n(t) = U_n^0 e^{i\omega_n t}, \quad W_n(t) = W_n^0 e^{i\omega_n t}, \quad X_n(t) = X_n^0 e^{i\omega_n t}, \qquad (10.5.27)
$$

where ω_n is the frequency of natural vibration. Substitution of Eq. (10.5.26) into Eq. (10.5.22), we obtain the following eigenvalue problem:

$$
\left(\begin{bmatrix} \alpha_n^2 A_{xx} & 0 & \alpha_n^2 B_{xx} \\ 0 & \alpha_n^2 K_s S_{xz} & \alpha_n K_s S_{xz} \\ \alpha_n^2 B_{xx} & \alpha_n K_s S_{xz} & (\alpha_n^2 D_{xx} + K_s S_{xz}) \end{bmatrix} - \omega_n^2 \begin{bmatrix} m_0 & 0 & m_1 \\ 0 & m_0 & 0 \\ m_1 & 0 & m_2 \end{bmatrix} \right) \begin{Bmatrix} U_n^0 \\ W_n^0 \\ X_n^0 \end{Bmatrix}
$$

$$= \begin{Bmatrix} 0 \\ 0 \\ 0 \end{Bmatrix}, \tag{10.5.28}$$

which can be used to determine the natural frequencies, ω_n.

For the homogeneous beams, Eq. (10.5.28) reduces to (with $A_{xx} = EA$, $D_{xx} = EI$, $m_0 = \rho A$, and $m_1 = 0$, $B_{xx} = 0$, and $m_2 = \rho I$)

$$\text{Axial: } \omega_n = \frac{n\pi}{L}\sqrt{\frac{E}{\rho}}, \tag{10.5.29}$$

$$\text{Flexural: } \omega_n = \alpha_n^2 \sqrt{\frac{D_{xx}}{(m_0 + m_2\alpha_n^2)}} = \frac{n^2\pi^2}{L^2}\sqrt{\frac{EI}{\rho\left(A + \frac{n^2\pi^2}{L^2}I\right)}}. \tag{10.5.30}$$

10.5.2.3 Numerical Results

Bending results. We consider functionally graded beams with the following material and geometric parameters of micro-beams:

$$E_1 = 14.4 \text{ GPa}, \quad E_2 = E = 1.44 \text{ GPa}, \quad \nu = 0.38, \quad K_s = \frac{5(1+\nu)}{6+5\nu}, \tag{10.5.31}$$

$$h = 5 \times 17.6 \times 10^{-6} \text{ m}, \quad b = 2h, \quad L = 20h, \quad q_0 = 1.0 \text{ N/m}.$$

Numerical results of the normalized center deflection $\bar{w} = w(0)(E_2 I/q_0 L^4) \times 10^2$ of a simply supported FGM beams under various types of loads and for different values of the power-law index are presented in Table 10.5.1 for the two theories (BET and TBT). The effect of shear deformation on the deflection is not significant because the length-to-height ratio is 20. Figure 10.5.1 contains plots of dimensionless deflection, $\bar{w}(x) = w(x)(E_2 I/q_0 L^4)$, of FGM beams under sinusoidally distributed transverse load as a function of the dimensionless distance along the length, x/L. The difference between the BET and TBT results is small and cannot be seen in these graphs.

Table 10.5.1 Center deflections $\bar{w} \times 10^2$ of simply supported FGM beams under various types of loads and for different values of the power-law index (n)[a] $[\bar{w} = w(0)(E_2 I/q_0 L^4)]$.

n	Point load		Uniform load		Sinusoidal load	
	BET	TBT	BET	TBT	BET	TBT
0	0.2083	0.2100	0.1302	0.1310	0.1027	0.1033
1	0.4876	0.4906	0.3047	0.3062	0.2398	0.2410
5	0.9496	0.9562	0.5935	0.5968	0.4679	0.4706
10	1.0442	1.0532	0.6526	0.6571	0.5145	0.5182
100	1.6855	1.7006	1.0534	1.0610	0.8306	0.8367

[a]$n = 0$ corresponds to material 1; $n = \infty$ corresponds to material 2.

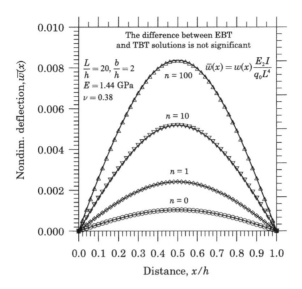

Fig. 10.5.1 Dimensionless center deflections $\bar{w}(x)$ versus the normalized distance x/L for simply supported FGM beams under sinusoidally distributed load for various values of the power-law index, n.

Free vibration results. For free vibration of homogeneous beams, the modulus, Poisson's ratio, shear correction factor, and width and length used are the same as those listed in Eq. (10.5.31). Other parameters are taken to be

$$h = 17.6 \times 10^{-6} \text{ m}, \quad \rho_1 = \rho_2 = 1.22 \times 10^3 \text{ kg/m},$$

and the moduli and densities used for the functionally graded beams are

$$E_1 = 14.4 \text{ GPa}, \quad E_2 = 1.44 \text{ GPa}, \quad \rho_1 = 12.2 \times 10^3 \text{ kg/m},$$
$$\rho_2 = 1.22 \times 10^3 \text{ kg/m}, \quad h = 17.6 \times 10^{-6} \text{ m}.$$

The rotary inertia is included in the present study. Table 10.5.2 contains the results for homogeneous ($n = 0$) and functionally graded ($n \neq 0$) beams. Figure 10.5.2 contains plots of dimensionless fundamental frequency versus the power-law index n. It is clear that, due to the nature of variation of the bending–stretching coupling coefficient B_{xx} with n, the fundamental frequency also experiences similar variation.

Table 10.5.2 First three natural frequencies $\bar{\omega}_n$ ($n = 1, 2, 3$) of homogeneous and FGM simply supported beams ($\bar{\omega}_n = \omega_n L^2 / \sqrt{\rho_2 A_0 / E_2}$).

n	$\bar{\omega}_1$ BET	$\bar{\omega}_1$ TBT	$\bar{\omega}_2$ BET	$\bar{\omega}_2$ TBT	$\bar{\omega}_3$ BET	$\bar{\omega}_3$ TBT
0	9.86	9.83	39.32	38.82	88.02	85.63
1	8.69	8.67	34.69	34.29	77.74	75.79
10	10.33	10.28	41.17	40.47	92.12	88.80

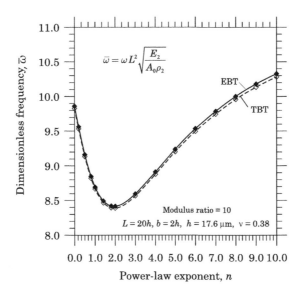

Fig. 10.5.2 The effect of the power-law index on the fundamental natural frequencies of a simply supported FGM beams.

10.5.3   Analysis of Plates

Analytical solutions for simply supported rectangular FGM plates is obtained using the Navier solution procedure. Figure 10.5.3 shows the geometry a rectangular plate, where a and b denote the in-plane dimensions along x and y coordinates and h denotes the thickness of the plate. The generalized displacements $(u, v, w, \theta_x, \theta_y, \theta_z, \phi_x, \phi_y, \phi_z, \psi_x, \psi_y)$ and external load q_z are expanded in double trigonometric series. The trigonometric functions are selected to satisfy the boundary conditions of a simply supported plate. The expansions of the dependent variables and the external load are substituted into the equation of motion, Eqs (10.4.18)–(10.4.28). For bending analysis, algebraic relations between the coefficients of the generalized displacements and load are obtained. For free vibration and buckling analysis, eigenvalue equations are derived to solve for the natural frequencies and buckling loads.

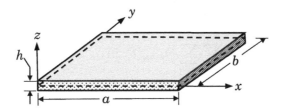

Fig. 10.5.3 The geometry and coordinate system for a rectangular plate.

10.5.3.1   Boundary Conditions

The boundary conditions of a simply supported rectangular plate for the GTST are

$$
\begin{aligned}
& u\left(x,0\right) = u\left(x,b\right) = 0, \ \ \theta_x\left(x,0\right) = \theta_x\left(x,b\right) = 0, \\
& \phi_x\left(x,0\right) = \phi_x\left(x,b\right) = 0, \ \ \psi_x\left(x,0\right) = \psi_x\left(x,b\right) = 0, \\
& v\left(0,y\right) = v\left(a,y\right) = 0, \ \ \theta_y\left(0,y\right) = \theta_y\left(a,y\right) = 0, \\
& \phi_y\left(0,y\right) = \phi_y\left(a,y\right) = 0, \ \ \psi_y\left(0,y\right) = \psi_y\left(a,y\right) = 0, \\
& w\left(x,0\right) = w\left(x,b\right) = 0, \ \ w\left(0,y\right) = w\left(a,y\right) = 0, \\
& \theta_z\left(x,0\right) = \theta_z\left(x,b\right) = 0, \ \ \theta_z\left(0,y\right) = \theta_z\left(a,y\right) = 0 \\
& \phi_z\left(x,0\right) = \phi_z\left(x,b\right) = 0, \ \ \phi_z\left(0,y\right) = \phi_z\left(a,y\right) = 0, \\
& M_{xx}^{(i)}\left(0,y\right) = M_{xx}^{(i)}\left(a,y\right) = 0, \ \ M_{yy}^{(i)}\left(x,0\right) = M_{yy}^{(i)}\left(x,b\right) = 0,
\end{aligned}
\tag{10.5.32}
$$

where $i = $0, 1, 2, 3 and $j = $0, 1, 2. The last line contains the force boundary conditions.

10.5.3.2   Expansions of Generalized Displacements

The generalized displacements are expanded in double trigonometric series that satisfy the boundary conditions in Eq. (10.5.32).

$$
u\left(x,y,z,t\right) = \sum_{m=1}^{\infty}\sum_{n=1}^{\infty} U_{mn}(t) \cos\alpha_m x \sin\beta_n y,
$$

$$
v\left(x,y,z,t\right) = \sum_{m=1}^{\infty}\sum_{n=1}^{\infty} V_{mn}(t) \sin\alpha_m x \cos\beta_n y,
$$

$$
w\left(x,y,z,t\right) = \sum_{m=1}^{\infty}\sum_{n=1}^{\infty} W_{mn}(t) \sin\alpha_m x \sin\beta_n y,
$$

$$
\theta_x\left(x,y,z,t\right) = \sum_{m=1}^{\infty}\sum_{n=1}^{\infty} \Theta_{xmn}(t) \cos\alpha_m x \sin\beta_n y,
$$

$$
\theta_y\left(x,y,z,t\right) = \sum_{m=1}^{\infty}\sum_{n=1}^{\infty} \Theta_{ymn}(t) \sin\alpha_m x \cos\beta_n y,
$$

$$
\theta_z\left(x,y,z,t\right) = \sum_{m=1}^{\infty}\sum_{n=1}^{\infty} \Theta_{zmn}(t) \sin\alpha_m x \sin\beta_n y,
\tag{10.5.33}
$$

$$
\phi_x\left(x,y,z,t\right) = \sum_{m=1}^{\infty}\sum_{n=1}^{\infty} \Phi_{xmn}(t) \cos\alpha_m x \sin\beta_n y,
$$

$$
\phi_y\left(x,y,z,t\right) = \sum_{m=1}^{\infty}\sum_{n=1}^{\infty} \Phi_{ymn}(t) \sin\alpha_m x \cos\beta_n y,
$$

$$\phi_z\left(x, y, z, t\right) = \sum_{m=1}^{\infty}\sum_{n=1}^{\infty}\Phi_{zmn}(t)\sin\alpha_m x \sin\beta_n y,$$

$$\psi_x\left(x, y, z, t\right) = \sum_{m=1}^{\infty}\sum_{n=1}^{\infty}\Psi_{xmn}(t)\cos\alpha_m x \sin\beta_n y,$$

$$\psi_y\left(x, y, z, t\right) = \sum_{m=1}^{\infty}\sum_{n=1}^{\infty}\Psi_{ymn}(t)\sin\alpha_m x \cos\beta_n y,$$

where $\alpha_m = \frac{m\pi}{a}$, $\beta_n = \frac{n\pi}{b}$. The generalized displacement coefficients (U_{mn}, V_{mn}, W_{mn}, Θ_{xmn}, Θ_{ymn}, Θ_{zmn}, Φ_{xmn}, Φ_{ymn}, Φ_{zmn}, Ψ_{xmn}, Ψ_{ymn}) are treated as time-independent variables for static bending and buckling problems.

10.5.3.3 Bending Analysis

By substituting Eq. (10.5.33) into Eqs (10.4.18)–(10.4.28) with help of Eqs (10.4.6), (10.4.7), (10.4.41), and (10.4.43) and omitting the inertia terms, we have algebraic relations for bending problems in terms of the known coefficient matrix and load vector and the unknown generalized displacement vector. The distributed transverse load is also expanded in double trigonometric series in Eq. (10.5.34):

$$q_z = \sum_{m=1}^{\infty}\sum_{n=1}^{\infty}Q_{mn}\left(t\right)\sin\alpha_m x \sin\beta_n y, \tag{10.5.34}$$

where

$$Q_{mn} = \frac{4}{ab}\int_0^a\int_0^b q_z \sin\alpha_m x \sin\beta_n y\, dxdy. \tag{10.5.35}$$

The algebraic equations are

$$\begin{bmatrix} C_{0101} & C_{0102} & C_{0103} & \cdots & C_{0110} & C_{0111} \\ C_{0201} & C_{0202} & C_{0203} & \cdots & C_{0210} & C_{0211} \\ C_{0301} & C_{0302} & C_{0303} & \cdots & C_{0310} & C_{0311} \\ \vdots & \vdots & \vdots & \ddots & \vdots & \vdots \\ C_{1001} & C_{1002} & C_{1003} & \cdots & C_{1010} & C_{1011} \\ C_{1101} & C_{1102} & C_{1103} & \cdots & C_{1110} & C_{1111} \end{bmatrix} \begin{Bmatrix} U_{mn} \\ V_{mn} \\ W_{mn} \\ \Theta_{xmn} \\ \Theta_{ymn} \\ \Theta_{zmn} \\ \Phi_{xmn} \\ \Phi_{ymn} \\ \Phi_{zmn} \\ \Psi_{xmn} \\ \Psi_{ymn} \end{Bmatrix} = \begin{Bmatrix} 0 \\ 0 \\ Q_{mn} \\ 0 \\ 0 \\ \frac{h}{2}Q_{mn} \\ 0 \\ 0 \\ \frac{h^2}{4}Q_{mn} \\ 0 \\ 0 \end{Bmatrix}, \tag{10.5.36}$$

and to distinguish $i = 1$ and $j = 11$ from $i = 11$ and $j = 1$, the subscript i and j vary 01 to 11. The known coefficients C_{ij} in Eq. (10.5.36) are

$$C_{0101} = A_{11}^{(0)}\alpha_m^2 + B_{11}^{(0)}\beta_n^2, \qquad\qquad C_{0102} = A_{12}^{(0)}\alpha_m\beta_n + B_{11}^{(0)}\alpha_m\beta_n,$$

$C_{0103} = 0,$

$C_{0104} = A_{11}^{(1)} \alpha_m^2 + B_{11}^{(1)} \beta_n^2,$

$C_{0105} = A_{12}^{(1)} \alpha_m \beta_n + B_{11}^{(1)} \alpha_m \beta_n,$

$C_{0106} = -A_{12}^{(0)} \alpha_m,$

$C_{0107} = A_{11}^{(2)} \alpha_m^2 + B_{11}^{(2)} \beta_n^2,$

$C_{0108} = A_{12}^{(2)} \alpha_m \beta_n + B_{11}^{(2)} \alpha_m \beta_n,$

$C_{0109} = -2A_{12}^{(1)} \alpha_m,$

$C_{0110} = A_{11}^{(3)} \alpha_m^2 + B_{11}^{(3)} \beta_n^2,$

$C_{0111} = A_{12}^{(3)} \alpha_m \beta_n + B_{11}^{(3)} \alpha_m \beta_n,$

$C_{0201} = A_{12}^{(0)} \alpha_m \beta_n + B_{11}^{(0)} \alpha_m \beta_n,$

$C_{0202} = A_{11}^{(0)} \beta_n^2 + B_{11}^{(0)} \alpha_m^2,$

$C_{0203} = 0,$

$C_{0204} = A_{12}^{(1)} \alpha_m \beta_n + B_{11}^{(1)} \alpha_m \beta_n,$

$C_{0205} = A_{11}^{(1)} \beta_n^2 + B_{11}^{(1)} \alpha_m^2,$

$C_{0206} = -A_{12}^{(0)} \beta_n,$

$C_{0207} = A_{12}^{(2)} \alpha_m \beta_n + B_{11}^{(2)} \alpha_m \beta_n,$

$C_{0208} = A_{11}^{(2)} \beta_n^2 + B_{11}^{(2)} \alpha_m^2,$

$C_{0209} = -2A_{12}^{(1)} \beta_n,$

$C_{0210} = A_{12}^{(3)} \alpha_m \beta_n + B_{11}^{(3)} \alpha_m \beta_n,$

$C_{0211} = A_{11}^{(3)} \beta_n^2 + B_{11}^{(3)} \alpha_m^2,$

$C_{0301} = 0,$

$C_{0302} = 0,$

$C_{0303} = B_{11}^{(0)} \left(\alpha_m^2 + \beta_n^2 \right),$

$C_{0304} = B_{11}^{(0)} \alpha_m,$

$C_{0305} = B_{11}^{(0)} \beta_n,$

$C_{0306} = B_{11}^{(1)} \left(\alpha_m^2 + \beta_n^2 \right),$

$C_{0307} = 2B_{11}^{(1)} \alpha_m,$

$C_{0308} = 2B_{11}^{(1)} \beta_n,$

$C_{0309} = B_{11}^{(2)} \left(\alpha_m^2 + \beta_n^2 \right),$

$C_{0310} = 3B_{11}^{(2)} \alpha_m,$

$C_{0311} = 3B_{11}^{(2)} \beta_n,$

$C_{0401} = A_{11}^{(1)} \alpha_m^2 + B_{11}^{(1)} \beta_n^2,$

$C_{0402} = A_{12}^{(1)} \alpha_m \beta_n + B_{11}^{(1)} \alpha_m \beta_n,$

$C_{0403} = B_{11}^{(0)} \alpha_m,$

$C_{0404} = A_{11}^{(2)} \alpha_m^2 + B_{11}^{(0)} + B_{11}^{(2)} \beta_n^2,$

$C_{0405} = A_{12}^{(2)} \alpha_m \beta_n + B_{11}^{(2)} \alpha_m \beta_n,$

$C_{0406} = -A_{12}^{(1)} \alpha_m + B_{11}^{(1)} \alpha_m,$

$C_{0407} = A_{11}^{(3)} \alpha_m^2 + 2B_{11}^{(1)} + B_{11}^{(3)} \beta_n^2,$

$C_{0408} = A_{12}^{(3)} \alpha_m \beta_n + B_{11}^{(3)} \alpha_m \beta_n,$

$C_{0409} = -2A_{12}^{(2)} \alpha_m + B_{11}^{(2)} \alpha_m$

$C_{0410} = A_{11}^{(4)} \alpha_m^2 + 3B_{11}^{(2)} + B_{11}^{(4)} \beta_n^2,$

$C_{0411} = A_{12}^{(4)} \alpha_m \beta_n + B_{11}^{(4)} \alpha_m \beta_n,$

$C_{0501} = A_{12}^{(1)} \alpha_m \beta_n + B_{11}^{(1)} \alpha_m \beta_n,$

$C_{0502} = A_{11}^{(1)} \beta_n^2 + B_{11}^{(1)} \alpha_m^2,$

$C_{0503} = B_{11}^{(0)} \beta_n,$

$C_{0504} = A_{12}^{(2)} \alpha_m \beta_n + B_{11}^{(2)} \alpha_m \beta_n,$

$C_{0505} = A_{11}^{(2)} \beta_n^2 + B_{11}^{(0)} + B_{11}^{(2)} \alpha_m^2,$

$C_{0506} = -A_{12}^{(1)} \beta_n + B_{11}^{(1)} \beta_n,$

$C_{0507} = A_{12}^{(3)} \alpha_m \beta_n + B_{11}^{(3)} \alpha_m \beta_n,$

$C_{0508} = A_{11}^{(3)} \beta_n^2 + 2B_{11}^{(1)} + B_{11}^{(3)} \alpha_m^2,$

$C_{0509} = -2A_{12}^{(2)} \beta_n + B_{11}^{(2)} \beta_n,$

$C_{0510} = A_{12}^{(4)} \alpha_m \beta_n + B_{11}^{(4)} \alpha_m \beta_n,$

$C_{0511} = A_{11}^{(4)} \beta_n^2 + 3B_{11}^{(2)} + B_{11}^{(4)} \alpha_m^2,$

$C_{0601} = -A_{12}^{(0)} \alpha_m,$

$C_{0602} = -A_{12}^{(0)} \beta_n,$

$$C_{0603} = B_{11}^{(1)} \left(\alpha_m^2 + \beta_n^2 \right),$$

$$C_{0604} = -A_{12}^{(1)} \alpha_m + B_{11}^{(1)} \alpha_m,$$

$$C_{0605} = -A_{12}^{(1)} \beta_n + B_{11}^{(1)} \beta_n,$$

$$C_{0606} = A_{11}^{(0)} + B_{11}^{(2)} \left(\alpha_m^2 + \beta_n^2 \right),$$

$$C_{0607} = -A_{12}^{(2)} \alpha_m + 2B_{11}^{(2)} \alpha_m,$$

$$C_{0608} = -A_{12}^{(2)} \beta_n + 2B_{11}^{(2)} \beta_n,$$

$$C_{0609} = 2A_{11}^{(1)} + B_{11}^{(3)} \left(\alpha_m^2 + \beta_n^2 \right),$$

$$C_{0610} = -A_{12}^{(3)} \alpha_m + 3B_{11}^{(3)} \alpha_m,$$

$$C_{0611} = -A_{12}^{(3)} \beta_n + 3B_{11}^{(3)} \beta_n,$$

$$C_{0701} = A_{11}^{(2)} \alpha_m^2 + B_{11}^{(2)} \beta_n^2,$$

$$C_{0702} = A_{12}^{(2)} \alpha_m \beta_n + B_{11}^{(2)} \alpha_m \beta_n,$$

$$C_{0703} = 2B_{11}^{(1)} \alpha_m,$$

$$C_{0704} = A_{11}^{(3)} \alpha_m^2 + 2B_{11}^{(1)} + B_{11}^{(3)} \beta_n^2,$$

$$C_{0705} = A_{12}^{(3)} \alpha_m \beta_n + B_{11}^{(3)} \alpha_m \beta_n,$$

$$C_{0706} = -A_{12}^{(2)} \alpha_m + 2B_{11}^{(2)} \alpha_m,$$

$$C_{0707} = A_{11}^{(4)} \alpha_m^2 + 4B_{11}^{(2)} + B_{11}^{(4)} \beta_n^2,$$

$$C_{0708} = A_{12}^{(4)} \alpha_m \beta_n + B_{11}^{(4)} \alpha_m \beta_n,$$

$$C_{0709} = -2A_{12}^{(3)} \alpha_m + 2B_{11}^{(3)} \alpha_m,$$

$$C_{0710} = A_{11}^{(5)} \alpha_m^2 + 6B_{11}^{(3)} + B_{11}^{(5)} \beta_n^2,$$

$$C_{0711} = A_{12}^{(5)} \alpha_m \beta_n + B_{11}^{(5)} \alpha_m \beta_n,$$

$$C_{0801} = A_{12}^{(2)} \alpha_m \beta_n + B_{11}^{(2)} \alpha_m \beta_n,$$

$$C_{0802} = A_{11}^{(2)} \beta_n^2 + B_{11}^{(2)} \alpha_m^2,$$

$$C_{0803} = 2B_{11}^{(1)} \beta_n,$$

$$C_{0804} = A_{12}^{(3)} \alpha_m \beta_n + B_{11}^{(3)} \alpha_m \beta_n,$$

$$C_{0805} = A_{11}^{(3)} \beta_n^2 + 2B_{11}^{(1)} + B_{11}^{(3)} \alpha_m^2,$$

$$C_{0806} = -A_{12}^{(2)} \beta_n + 2B_{11}^{(2)} \beta_n,$$

$$C_{0807} = A_{12}^{(4)} \alpha_m \beta_n + B_{11}^{(4)} \alpha_m \beta_n,$$

$$C_{0808} = A_{11}^{(4)} \beta_n^2 + 4B_{11}^{(2)} + B_{11}^{(4)} \alpha_m^2,$$

$$C_{0809} = -2A_{12}^{(3)} \beta_n + 2B_{11}^{(3)} \beta_n,$$

$$C_{0810} = A_{12}^{(5)} \alpha_m \beta_n + B_{11}^{(5)} \alpha_m \beta_n,$$

$$C_{0811} = A_{11}^{(5)} \beta_n^2 + 6B_{11}^{(3)} + B_{11}^{(5)} \alpha_m^2,$$

$$C_{0901} = -2A_{12}^{(1)} \alpha_m,$$

$$C_{0902} = -2A_{12}^{(1)} \beta_n,$$

$$C_{0903} = B_{11}^{(2)} \left(\alpha_m^2 + \beta_n^2 \right),$$

$$C_{0904} = -2A_{12}^{(2)} \alpha_m + B_{11}^{(2)} \alpha_m,$$

$$C_{0905} = -2A_{12}^{(2)} \beta_n + B_{11}^{(2)} \beta_n,$$

$$C_{0906} = 2A_{11}^{(1)} + B_{11}^{(3)} \left(\alpha_m^2 + \beta_n^2 \right),$$

$$C_{0907} = -2A_{12}^{(3)} \alpha_m + 2B_{11}^{(3)} \alpha_m,$$

$$C_{0908} = -2A_{12}^{(3)} \beta_n + 2B_{11}^{(3)} \beta_n,$$

$$C_{0909} = 4A_{11}^{(2)} + B_{11}^{(4)} \left(\alpha_m^2 + \beta_n^2 \right),$$

$$C_{0910} = -2A_{12}^{(4)} \alpha_m + 3B_{11}^{(4)} \alpha_m,$$

$$C_{0911} = -2A_{12}^{(4)} \beta_n + 3B_{11}^{(4)} \beta_n,$$

$$C_{1001} = A_{11}^{(3)} \alpha_m^2 + B_{11}^{(3)} \beta_n^2,$$

$$C_{1002} = A_{12}^{(3)} \alpha_m \beta_n + B_{11}^{(3)} \alpha_m \beta_n,$$

$$C_{1003} = 3B_{11}^{(2)} \alpha_m,$$

$$C_{1004} = A_{11}^{(4)} \alpha_m^2 + 3B_{11}^{(2)} + B_{11}^{(4)} \beta_n^2,$$

$$C_{1005} = A_{12}^{(4)} \alpha_m \beta_n + B_{11}^{(4)} \alpha_m \beta_n,$$

$$C_{1006} = -A_{12}^{(3)} \alpha_m + 3B_{11}^{(3)} \alpha_m,$$

$$C_{1007} = A_{11}^{(5)} \alpha_m^2 + 6B_{11}^{(3)} + B_{11}^{(5)} \beta_n^2,$$

$$C_{1008} = A_{12}^{(5)} \alpha_m \beta_n + B_{11}^{(5)} \alpha_m \beta_n,$$

$$C_{1009} = -2A_{12}^{(4)} \alpha_m + 3B_{11}^{(4)} \alpha_m,$$

$$C_{1010} = A_{11}^{(6)} \alpha_m^2 + 9B_{11}^{(4)} + B_{11}^{(6)} \beta_n^2,$$

$$C_{1011} = A_{12}^{(6)} \alpha_m \beta_n + B_{11}^{(6)} \alpha_m \beta_n,$$

$$C_{1101} = A_{12}^{(3)} \alpha_m \beta_n + B_{11}^{(3)} \alpha_m \beta_n, \qquad C_{1102} = A_{11}^{(3)} \beta_n^2 + B_{11}^{(3)} \alpha_m^2,$$

$$C_{1103} = 3 B_{11}^{(2)} \beta_n,$$

$$C_{1104} = A_{12}^{(4)} \alpha_m \beta_n + B_{11}^{(4)} \alpha_m \beta_n,$$

$$C_{1105} = A_{11}^{(4)} \beta_n^2 + 3 B_{11}^{(2)} + B_{11}^{(4)} \alpha_m^2, \qquad C_{1106} = -A_{12}^{(3)} \beta_n + 3 B_{11}^{(3)} \beta_n,$$

$$C_{1107} = A_{12}^{(5)} \alpha_m \beta_n + B_{11}^{(5)} \alpha_m \beta_n, \qquad C_{1108} = A_{11}^{(5)} \beta_n^2 + 6 B_{11}^{(3)} + B_{11}^{(5)} \alpha_m^2,$$

$$C_{1109} = -2 A_{12}^{(4)} \beta_n + 3 B_{11}^{(4)} \beta_n, \qquad C_{1110} = A_{12}^{(6)} \alpha_m \beta_n + B_{11}^{(6)} \alpha_m \beta_n,$$

$$C_{1111} = A_{11}^{(6)} \beta_n^2 + 9 B_{11}^{(4)} + B_{11}^{(6)} \alpha_m^2.$$

The coefficient for the transverse load, Q_{mn}, is obtained evaluating Eq. (10.5.35), and examples are the following:

$$Q_{mn} = \begin{cases} \dfrac{16 q_0}{mn\pi^2} & \text{for uniformly distributed load,} \\[2ex] \dfrac{4P}{ab} \sin \dfrac{m\pi}{2} \sin \dfrac{n\pi}{2} & \text{for point load at the center of the plate.} \end{cases}$$

$$(10.5.37)$$

where q_0 and P are magnitudes of the uniformly distributed load and the point load and $m = 1, 3, 5, \ldots$ and $n = 1, 3, 5, \ldots$. The unknown generalized displacement coefficients, $(U_{mn}, V_{mn}, \ldots, \Phi_{ymn})$, can be obtained from solving Eq. (10.5.36) for each m and n. The displacements $(u, v, \ldots, \psi_y)$ are evaluated by Eq. (10.5.33).

10.5.3.4   Free Vibration Analysis

For the free vibration analysis, the time-dependant variables in Eq. (10.5.33) are assumed as

$$U_{mn}(t) = U_{mn}^0 \times e^{i\omega t}, \quad \Theta_{xmn}(t) = \Theta_{xmn}^0 \times e^{i\omega t},$$

$$\Phi_{xmn}(t) = \Phi_{xmn}^0 \times e^{i\omega t}, \quad V_{mn}(t) = V_{mn}^0 \times e^{i\omega t},$$

$$\Theta_{ymn}(t) = \Theta_{ymn}^0 \times e^{i\omega t}, \quad \Phi_{ymn}(t) = \Phi_{ymn}^0 \times e^{i\omega t}, \qquad (10.5.38)$$

$$W_{mn}(t) = W_{mn}^0 \times e^{i\omega t}, \quad \Theta_{zmn}(t) = \Theta_{zmn}^0 \times e^{i\omega t},$$

$$\Phi_{zmn}(t) = \Phi_{zmn}^0 \times e^{i\omega t}, \quad \Psi_{xmn}(t) = \Psi_{xmn}^0 \times e^{i\omega t},$$

$$\Psi_{ymn}(t) = \Psi_{ymn}^0 \times e^{i\omega t},$$

where ω is the natural frequency. By substituting Eqs (10.5.38) and (10.5.33) into the equation of motion, we set up the eigenvalue problem to determine eigenfrequencies:

$$\left(\mathbf{C} - \omega_{mn}^2 \mathbf{M} \right) \mathbf{U} = \mathbf{0}. \qquad (10.5.39)$$

Here $\mathbf{C}$ is the coefficient matrix, $\mathbf{M}$ is the matrix of inertias, and $\mathbf{U}$ is the vector of generalized displacements. The matrix $\mathbf{C}$ is the same as the one in bending analysis, and the coefficients of the matrix of inertias are

$$M_{0101} = m_0, \quad M_{0102} = 0, \quad M_{0103} = 0, \quad M_{0104} = m_1, \quad M_{0105} = 0,$$
$$M_{0106} = 0, \quad M_{0107} = m_2, \quad M_{0108} = 0, \quad M_{0109} = 0, \quad M_{0110} = m_3,$$
$$M_{0111} = 0, \quad M_{0201} = 0, \quad M_{0202} = m_0, \quad M_{0203} = 0, \quad M_{0204} = 0,$$
$$M_{0205} = m_1, \quad M_{0206} = 0, \quad M_{0207} = 0, \quad M_{0208} = m_2, \quad M_{0209} = 0,$$
$$M_{0210} = 0, \quad M_{0211} = m_3, \quad M_{0301} = 0, \quad M_{0302} = 0, \quad M_{0303} = m_0,$$
$$M_{0304} = 0, \quad M_{0305} = 0, \quad M_{0306} = m_1, \quad M_{0307} = 0, \quad M_{0308} = 0,$$
$$M_{0309} = m_2, \quad M_{0310} = 0, \quad M_{0311} = 0, \quad M_{0401} = m_1, \quad M_{0402} = 0,$$
$$M_{0403} = 0, \quad M_{0404} = m_2, \quad M_{0405} = 0, \quad M_{0406} = 0, \quad M_{0407} = m_3,$$
$$M_{0408} = 0, \quad M_{0409} = 0, \quad M_{0410} = m_4, \quad M_{0411} = 0, \quad M_{0501} = 0,$$
$$M_{0502} = m_1, \quad M_{0503} = 0, \quad M_{0504} = 0, \quad M_{0505} = m_2, \quad M_{0506} = 0,$$
$$M_{0507} = 0, \quad M_{0508} = m_3, \quad M_{0509} = 0, \quad M_{0510} = 0, \quad M_{0511} = m_4,$$
$$M_{0601} = 0, \quad M_{0602} = 0, \quad M_{0603} = m_1, \quad M_{0604} = 0, \quad M_{0605} = 0,$$
$$M_{0606} = m_2, \quad M_{0607} = 0, \quad M_{0608} = 0, \quad M_{0609} = m_3, \quad M_{0610} = 0,$$
$$M_{0611} = 0, \quad M_{0701} = m_2 \quad M_{0702} = 0, \quad M_{0703} = 0, \quad M_{0704} = m_3,$$
$$M_{0705} = 0, \quad M_{0706} = 0, \quad M_{0707} = m_4, \quad M_{0708} = 0, \quad M_{0709} = 0,$$
$$M_{0710} = m_5, \quad M_{0711} = 0, \quad M_{0801} = 0, \quad M_{0802} = m_2, \quad M_{0803} = 0,$$
$$M_{0804} = 0, \quad M_{0805} = m_3, \quad M_{0806} = 0, \quad M_{0807} = 0, \quad M_{0808} = m_4,$$
$$M_{0809} = 0, \quad M_{0810} = 0, \quad M_{0811} = m_5, \quad M_{0901} = 0, \quad M_{0902} = 0,$$
$$M_{0903} = m_2, \quad M_{0904} = 0, \quad M_{0905} = 0, \quad M_{0906} = m_3, \quad M_{0907} = 0,$$
$$M_{0908} = 0, \quad M_{0909} = m_4, \quad M_{0910} = 0, \quad M_{0911} = 0, \quad M_{1001} = m_3,$$
$$M_{1002} = 0, \quad M_{1003} = 0, \quad M_{1004} = m_4, \quad M_{1005} = 0, \quad M_{1006} = 0,$$
$$M_{1007} = m_5, \quad M_{1008} = 0, \quad M_{1009} = 0, \quad M_{1010} = m_5, \quad M_{1011} = 0,$$
$$M_{1101} = 0, \quad M_{1102} = m_3, \quad M_{1103} = 0, \quad M_{1104} = 0, \quad M_{1105} = m_4,$$
$$M_{1106} = 0, \quad M_{1107} = 0, \quad M_{1108} = m_5, \quad M_{1109} = 0, \quad M_{1110} = 0,$$
$$M_{1111} = m_5, \tag{10.5.40}$$

where $m_i = \int_{-\frac{h}{2}}^{\frac{h}{2}} z^i \rho(z) \, dz$. The rule of subscripts are the same as the known coefficients matrix, $\mathbf{C}$. Small values of the plate thickness may cause the inertia matrix of GTST to be ill-conditioned because higher-order terms involving the thickness appear in the diagonal. In such cases, the governing equations, Eqs (10.4.18)–(10.4.28), are needed to be in dimensionless form.

10.5.3.5 Buckling Analysis

We assume that only in-plane forces act on each sides of simply supported plate and all other mechanical and thermal loads are zero and inertial terms are omitted for the buckling analysis. For buckling analysis, the additional terms $\hat{N}_{xx} \frac{\partial^2 w}{\partial x^2} + \hat{N}_{yy} \frac{\partial^2 w}{\partial y^2}$ are added to the right side of the equation of motion

(10.4.20) (see Reddy [50, 51]); $\hat{N}_{xx}$ and $\hat{N}_{yy}$ are the compressive loads acting on edges of a simply supported square plate and the system of equations for buckling analysis becomes

$$
\begin{bmatrix}
C_{0101} & C_{0102} & C_{0103} & \cdots & C_{0110} & C_{0111} \\
C_{0201} & C_{0202} & C_{0203} & \cdots & C_{0210} & C_{0211} \\
C_{0301} & C_{0302} & C_{0303} - N_0\left(\alpha_m^2 + k\beta_n^2\right) & \cdots & C_{0310} & C_{0311} \\
\vdots & \vdots & \vdots & \ddots & \vdots & \vdots \\
C_{1001} & C_{1002} & C_{1003} & \cdots & C_{1010} & C_{1011} \\
C_{1101} & C_{1102} & C_{1103} & \cdots & C_{1110} & C_{1111}
\end{bmatrix}
\begin{Bmatrix}
U_{mn} \\
V_{mn} \\
W_{mn} \\
\Theta_{xmn} \\
\vdots \\
\Phi_{xmn} \\
\vdots \\
\Psi_{xmn} \\
\Psi_{ymn}
\end{Bmatrix}
=
\begin{Bmatrix}
0 \\
0 \\
0 \\
0 \\
\vdots \\
0 \\
\vdots \\
0 \\
0
\end{Bmatrix},
$$

$$(10.5.41)$$

where $N_0 = -\hat{N}_{xx}$, $k = \frac{\hat{N}_{yy}}{\hat{N}_{xx}}$. To determine N_0, the static condensation of variables is applied to Eq. (10.5.41), and all unknown coefficients except W_{mn} are eliminated. The equation (10.5.42) shows an expression of generalized displacements $(U_{mn}, V_{mn}, \Theta_{xmn}, \Theta_{ymn}, \Theta_{zmn}, \Phi_{xmn}, \Phi_{ymn}, \Phi_{zmn}, \Psi_{xmn}, \Psi_{ymn})$ in terms of W_{mn}:

$$
\begin{Bmatrix}
U_{mn} \\
V_{mn} \\
\Theta_{xmn} \\
\Theta_{ymn} \\
\Theta_{zmn} \\
\Phi_{xmn} \\
\Phi_{ymn} \\
\Phi_{zmn} \\
\Psi_{xmn} \\
\Psi_{ymn}
\end{Bmatrix}
= \bar{\mathbf{C}}
\begin{Bmatrix}
C_{0103} \\
C_{0203} \\
C_{0403} \\
C_{0503} \\
C_{0603} \\
C_{0703} \\
C_{0803} \\
C_{0903} \\
C_{1003} \\
C_{1103}
\end{Bmatrix}
\{W_{mn}\},
\qquad (10.5.42)
$$

where

$$
\bar{\mathbf{C}} =
\begin{bmatrix}
C_{0101} & C_{0102} & C_{0104} & C_{0105} & \cdots & C_{0110} & C_{0111} \\
C_{0201} & C_{0202} & C_{0204} & C_{0205} & \cdots & C_{0210} & C_{0211} \\
C_{0401} & C_{0402} & C_{0404} & C_{0405} & \cdots & C_{0410} & C_{0411} \\
C_{0501} & C_{0502} & C_{0504} & C_{0505} & \cdots & C_{0510} & C_{0511} \\
\vdots & \vdots & \vdots & \vdots & \ddots & \vdots & \vdots \\
C_{1001} & C_{1002} & C_{1004} & C_{1005} & \cdots & C_{1010} & C_{1011} \\
C_{1101} & C_{1102} & C_{1104} & C_{1105} & \cdots & C_{1110} & C_{1111}
\end{bmatrix}^{-1}.
$$

After applying static condensation of the variables, we obtain the expression for the critical load N_0 as follows:

$$N_0 = \frac{1}{\alpha_m^2 + k\beta_n^2} \left(C_{0303} - \begin{Bmatrix} C_{0301} \\ C_{0302} \\ C_{0304} \\ C_{0305} \\ \vdots \\ C_{0310} \\ C_{0311} \end{Bmatrix}^T \bar{\mathbf{C}} \begin{Bmatrix} C_{0103} \\ C_{0203} \\ C_{0403} \\ C_{0503} \\ \vdots \\ C_{1003} \\ C_{1103} \end{Bmatrix} \right). \tag{10.5.43}$$

10.5.3.6 Numerical Results

Numerical examples of the analytical solution are presented here using the material properties and the dimension of the functionally graded (micro) plate from Kim and Reddy [169, 170]. The dimensions of functionally graded plate in Fig. 10.5.3 are $a = 20h$, $b = 20h$, and $h = 17.6 \times 10^{-6}$ m. The moduli and mass densities of two constituents are $E_t = 14.4$ GPa, $E_b = 1.44$ GPa, $\rho_t = 12.2 \times 10^3$ kg/m, and $\rho_b = 1.22 \times 10^3$ kg/m.

Static bending analysis
A uniformly distributed load of intensity $q_0 = 1$ N/m^2 is applied on top surface $z = \frac{h}{2}$. In this example, the number of terms in the double trigonometric series, Eqs(10.5.33) and (10.5.34), is taken to be 31. Figure 10.5.4 shows the dimensionless deflection, $\bar{w}\left(x, \frac{b}{2}, 0\right)$, versus the power-law index, n, in Eq. (10.1.1). It is clear that functionally graded plates become stiffer for smaller values of the power-law index, n, because for smaller values of n, the plate material approaches E_t, which is larger than E_b in this example.

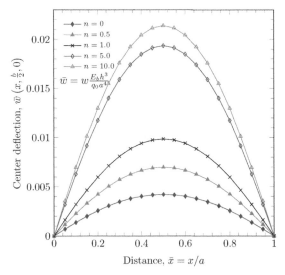

Fig. 10.5.4 Dimensionless deflection $\bar{w}\left(x, \frac{b}{2}, 0\right)$ versus distance along a FGM simply supported plate with various values of the power-law index, n.

Natural vibration analysis
Natural frequencies of simply supported square FGM plates can be determined
using Eq. (10.5.39). The fundamental frequencies are obtained when $m = 1$
and $n = 1$. Table 10.5.3 contains the natural frequencies for various power-law
index, n, and Fig. 10.5.5 shows natural frequencies of FGM plates versus the
power-law index, n.

Table 10.5.3 Dimensionless fundamental natural frequencies of simply supported plates
$\left(\bar{\omega} = \omega \sqrt{\frac{\rho_b a^4}{E_b h^2}} \right)$.

n	0	1	2	3	4	5	6	7	8	9	10
$\bar{\omega}$	6.10	5.39	5.22	5.32	5.51	5.71	5.88	6.04	6.17	6.27	6.36

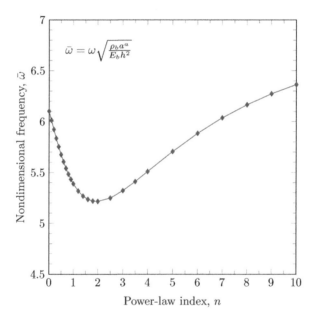

Fig. 10.5.5 Non-dimensional fundamental frequency, $\bar{\omega}$, versus power-law index, n .

Buckling analysis
The critical loads for simply supported square FGM plates are obtained using
Eq. (10.5.43). The same magnitude of compressive loads, $\hat{N}_{xx} = \hat{N}_{yy}$, are
assumed for a square plate. The minimum critical load is obtained when $m = 1$
and $n = 1$. The critical loads with various power-law indices are shown in Table
10.5.4 and Fig. 10.5.6. It is shown that we can control the critical load for the
FGM plates with predefined material variation through the plate thickness.

Table 10.5.4 The dimensionless buckling loads of simply supported FGM plates $\left(\bar{N}_0 = N_0 \frac{a^2}{E_b h^3}\right)$.

n	0	1	2	3	4	5	6	7	8	9	10
$\bar{N}_0$	18.90	8.11	5.53	4.67	4.32	4.13	4.02	3.93	3.86	3.80	3.74

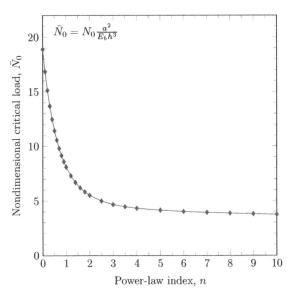

Fig. 10.5.6 Dimensionless buckling load versus the power-law index, n, for simply supported FGM plates.

10.6 Finite Element Models

10.6.1 Bending of Beams

10.6.1.1 Bernoulli–Euler Beam Theory

Here we assume linear approximation of the axial displacement u and Hermite cubic approximation of the transverse displacement w

$$u(x) \approx \sum_{j=1}^{2} u_j(t)\psi_j(x), \quad w(x) \approx \sum_{j=1}^{4} \bar{\Delta}_j(t)\varphi_j(x) \qquad (10.6.1)$$

where $\psi_j(x)$ are the linear polynomials, $\varphi_j(x)$ are the Hermite cubic polynomials, (u_1, u_2) are the nodal values of u at x_a and x_b, respectively, and $\bar{\Delta}_j$ are the nodal values $(\theta_x = -dw/dx$; see Fig. 10.6.1):

$$\bar{\Delta}_1(t) = w(x_a, t), \quad \bar{\Delta}_3(t) = w(x_b, t), \quad \bar{\Delta}_2(t) = \theta(x_a, t), \quad \bar{\Delta}_4(t) = \theta(x_b, t)$$
$$(10.6.2)$$

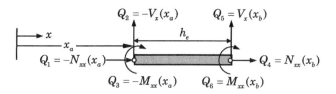

Fig. 10.6.1 Generalized forces of the finite element model; V_x for the BET is given by Eq. (10.2.10) and it is given in Eq. (10.2.20) for TBT.

Substitution of Eq. (10.6.1) for u and w and $\delta u = \psi_i$ and $\delta w = \varphi_i$ into the virtual work statements in Eq. (10.2.3), we obtain the finite element equations

$$
\begin{bmatrix} \mathbf{M}^{11} & \mathbf{M}^{12} \\ \mathbf{M}^{21} & \mathbf{M}^{22} \end{bmatrix} \begin{Bmatrix} \ddot{\mathbf{u}} \\ \ddot{\boldsymbol{\Delta}} \end{Bmatrix} + \begin{bmatrix} \mathbf{K}^{11} & \mathbf{K}^{12} \\ \mathbf{K}^{21} & \mathbf{K}^{22} \end{bmatrix} \begin{Bmatrix} \mathbf{u} \\ \boldsymbol{\Delta} \end{Bmatrix} = \begin{Bmatrix} \mathbf{F}^1 \\ \mathbf{F}^2 \end{Bmatrix} \qquad (10.6.3)
$$

where

$$
M_{ij}^{11} = \int_{x_a}^{x_b} m_0 \psi_i \psi_j \, dx, \quad M_{ij}^{22} = \int_{x_a}^{x_b} \left(m_0 \varphi_i \varphi_j + m_2 \frac{d\varphi_i}{dx} \frac{d\varphi_j}{dx} \right) dx,
$$

$$
M_{ij}^{12} = -\int_{x_a}^{x_b} m_1 \psi_i \frac{d\varphi_j}{dx} \, dx, \quad M_{ij}^{21} = -\int_{x_a}^{x_b} m_1 \frac{d\varphi_i}{dx} \psi_j \, dx,
$$

$$
K_{ij}^{11} = \int_{x_a}^{x_b} A_{xx} \frac{d\psi_i}{dx} \frac{d\psi_j}{dx} \, dx,
$$

$$
K_{ij}^{12} = -\int_{x_a}^{x_b} B_{xx} \frac{d\psi_i}{dx} \frac{d^2 \varphi_j}{dx^2} \, dx + \frac{1}{2} \int_{x_a}^{x_b} A_{xx} \frac{\partial w}{\partial x} \frac{d\psi_i}{dx} \frac{d\varphi_j}{dx} \, dx
$$

$$
K_{ij}^{21} = -\int_{x_a}^{x_b} B_{xx} \frac{d^2 \varphi_i}{dx^2} \frac{d\psi_j}{dx} \, dx + \int_{x_a}^{x_b} A_{xx} \frac{\partial w}{\partial x} \frac{d\varphi_i}{dx} \frac{d\psi_j}{dx} \, dx \qquad (10.6.4)
$$

$$
K_{ij}^{22} = \int_{x_a}^{x_b} \left[\frac{A_{xx}}{2} \left(\frac{\partial w}{\partial x} \right)^2 \frac{d\varphi_i}{dx} \frac{d\varphi_j}{dx} - B_{xx} \frac{\partial w}{\partial x} \left(\frac{1}{2} \frac{d^2 \varphi_i}{dx^2} \frac{d\varphi_j}{dx} + \frac{d\varphi_i}{dx} \frac{d^2 \varphi_j}{dx^2} \right) \right] dx
$$

$$
+ \int_{x_a}^{x_b} D_{xx} \frac{d^2 \varphi_i}{dx^2} \frac{d^2 \varphi_j}{dx^2} \, dx
$$

$$
F_i^1 = \int_{x_a}^{x_b} \left(\psi_i f_x + \frac{d\psi_i}{dx} N_{xx}^T \right) dx + \psi_i(x_a) Q_1 + \psi_i(x_b) Q_4
$$

$$
F_i^2 = \int_{x_a}^{x_b} \left(-\frac{d^2 \varphi_i}{dx^2} M_{xx}^T + \frac{d\varphi_i}{dx} \frac{\partial w}{\partial x} N_{xx}^T + \varphi_i f_z \right) dx
$$

$$
+ \varphi_i(x_a) Q_2 + \left(-\frac{d\varphi_i}{dx} \right)_{x_a} Q_3 + \varphi_i(x_b) Q_5 + \left(-\frac{d\varphi_i}{dx} \right)_{x_b} Q_6.
$$

Clearly, the stiffness matrix is not symmetric, that is, $K_{ij}^{\alpha\beta} \neq K_{ji}^{\beta\alpha}$. Further, there are two sources of the coupling between the axial displacement u and the transverse displacement w: first, the coupling is due to the extensional-bending coefficient B_{xx}, and it is independent of the von Kármán nonlinearity; second, the coupling is due to the von Kármán nonlinearity, which is independent of the coupling coefficient B_{xx}. Of course, the coefficient B_{xx} has a stronger coupling in the presence of the von Kármán nonlinearity.

10.6.1.2 Timoshenko Beam Theory

In TBT, we assume Lagrange approximation of all field variables (u, w, ϕ_x) independently:

$$u(x) \approx \sum_{j=1}^{m} u_j(t)\psi_j^{(1)}(x), \quad w(x) \approx \sum_{j=1}^{n} w_j(t)\psi_j^{(2)}(x), \quad \phi_x \approx \sum_{j=1}^{p} s_j(t)\psi_j^{(3)}(x),$$

$$(10.6.5)$$

where $\psi_j^{(\alpha)}(x)$ are the Lagrange polynomials of different order used for the three variables. Substitution of Eq. (10.6.5) for (u, w, ϕ_x) and $\delta u = \psi_i^{(1)}$, $\delta w = \psi_i^{(2)}$, and $\delta\phi_x = \psi_i^{(3)}$ into the virtual work statement in Eq. (10.2.13), we obtain the finite element equations

$$
\begin{bmatrix} \mathbf{M}^{11} & \mathbf{0} & \mathbf{M}^{13} \\ \mathbf{0} & \mathbf{M}^{22} & \mathbf{0} \\ \mathbf{M}^{31} & \mathbf{0} & \mathbf{M}^{33} \end{bmatrix} \begin{Bmatrix} \ddot{\mathbf{u}} \\ \ddot{\mathbf{w}} \\ \ddot{\mathbf{s}} \end{Bmatrix} + \begin{bmatrix} \mathbf{K}^{11} & \mathbf{K}^{12} & \mathbf{K}^{13} \\ \mathbf{K}^{21} & \mathbf{K}^{22} & \mathbf{K}^{23} \\ \mathbf{K}^{31} & \mathbf{K}^{32} & \mathbf{K}^{33} \end{bmatrix} \begin{Bmatrix} \mathbf{u} \\ \mathbf{w} \\ \mathbf{s} \end{Bmatrix} = \begin{Bmatrix} \mathbf{F}^1 \\ \mathbf{F}^2 \\ \mathbf{F}^3 \end{Bmatrix}, \quad (10.6.6)
$$

where

$$M_{ij}^{11} = \int_{x_a}^{x_b} m_0 \psi_i^{(1)} \psi_j^{(1)} \, dx, \quad M_{ij}^{22} = \int_{x_a}^{x_b} m_0 \psi_i^{(2)} \psi_j^{(2)} \, dx,$$

$$M_{ij}^{33} = \int_{x_a}^{x_b} m_2 \psi_i^{(3)} \psi_j^{(3)} \, dx, \quad M_{ij}^{13} = \int_{x_a}^{x_b} m_1 \psi_i^{(1)} \psi_j^{(3)} \, dx = M_{ji}^{31},$$

$$K_{ij}^{11} = \int_{x_a}^{x_b} A_{xx} \frac{d\psi_i^{(1)}}{dx} \frac{d\psi_j^{(1)}}{dx} \, dx, \quad K_{ij}^{12} = \frac{1}{2}\int_{x_a}^{x_b} A_{xx} \frac{\partial w}{\partial x} \frac{d\psi_i^{(1)}}{dx} \frac{d\psi_j^{(2)}}{dx} \, dx,$$

$$K_{ij}^{13} = \int_{x_a}^{x_b} B_{xx} \frac{d\psi_i^{(1)}}{dx} \frac{d\psi_j^{(3)}}{dx} \, dx, \quad K_{ij}^{21} = \int_{x_a}^{x_b} A_{xx} \frac{\partial w}{\partial x} \frac{d\psi_i^{(2)}}{dx} \frac{d\psi_j^{(1)}}{dx} \, dx,$$

$$K_{ij}^{22} = \int_{x_a}^{x_b} \left[K_s S_{xz} \frac{d^2\psi_i^{(2)}}{dx} \frac{d\psi_j^{(2)}}{dx} + \frac{A_{xx}}{2}\left(\frac{\partial w}{\partial x}\right)^2 \frac{d\psi_i^{(2)}}{dx} \frac{d\psi_j^{(2)}}{dx} \right] dx,$$

$$K_{ij}^{23} = \int_{x_a}^{x_b} \left(K_s S_{xz} \frac{d\psi_i^{(2)}}{dx} \psi_j^{(3)} + B_{xx} \frac{\partial w}{\partial x} \frac{d\psi_i^{(2)}}{dx} \frac{d\psi_j^{(3)}}{dx} \right) dx,$$

$$K_{ij}^{31} = \int_{x_a}^{x_b} B_{xx} \frac{d\psi_i^{(3)}}{dx} \frac{d\psi_j^{(1)}}{dx} \, dx,$$

$$K_{ij}^{32} = \int_{x_a}^{x_b} \left(K_s S_{xz} \psi_i^{(3)} \frac{d\psi_j^{(2)}}{dx} + \frac{B_{xx}}{2} \frac{\partial w}{\partial x} \frac{d\psi_i^{(3)}}{dx} \frac{d\psi_j^{(2)}}{dx} \right) dx,$$

$$K_{ij}^{33} = \int_{x_a}^{x_b} \left(K_s S_{xz} \psi_i^{(3)} \psi_j^{(3)} + D_{xx} \frac{d\psi_i^{(3)}}{dx} \frac{d\psi_j^{(3)}}{dx} \right) dx,$$

$$F_i^1 = \int_{x_a}^{x_b} \left(\psi_i^{(1)} f_x + \frac{d\psi_i^{(1)}}{dx} N_{xx}^T \right) dx + \psi_i^{(1)}(x_a) Q_1 + \psi_i^{(1)}(x_b) Q_4,$$

$$F_i^2 = \int_{x_a}^{x_b} \left(\psi_i^{(2)} f_z + \frac{d\psi_i^{(2)}}{dx} \frac{\partial w}{\partial x} N_{xx}^T \right) dx + \psi_i^{(2)}(x_a) Q_2 + \psi_i^{(2)}(x_b) Q_5,$$

$$F_i^3 = \int_{x_a}^{x_b} \frac{d\psi_i^{(3)}}{dx} M_{xx}^T \, dx + \psi_i^{(3)}(x_a) Q_3 + \psi_i^{(3)}(x_b) Q_6. \tag{10.6.7}$$

10.6.2   Axisymmetric Bending of Circular Plates

10.6.2.1   Classical Plate Theory

The principle of virtual displacements, Eq. (10.3.6), is equivalent to the following two statements over a typical finite element $\Omega^e = (r_a, r_b)$:

$$0 = \int_{r_a}^{r_b} \left[m_0 \ddot{u}\, \delta u - m_1 \frac{\partial \ddot{w}}{\partial r} \delta u + N_{rr} \frac{\partial \delta u}{\partial r} + N_{\theta\theta} \left(\frac{\delta u}{r} \right) \right] r \, dr$$
$$- Q_1 \delta u(r_a) - Q_4 \delta u(r_b) \tag{10.6.8}$$

$$0 = \int_{r_a}^{r_b} \left[m_0 \ddot{w}\delta w - m_1 \ddot{u} \frac{\partial \delta w}{\partial r} + m_2 \frac{\partial \ddot{w}}{\partial r} \frac{\partial \delta w}{\partial r} + N_{rr} \frac{\partial w}{\partial r} \frac{\partial \delta w}{\partial r} - M_{rr} \frac{\partial^2 \delta w}{\partial r^2} \right.$$
$$\left. - \frac{1}{r} M_{\theta\theta} \frac{\partial \delta w}{\partial r} - q \delta w \right] r \, dr$$
$$- [Q_2 \delta w(r_a) + Q_5 \delta w(r_b) + Q_3 \delta\theta(r_a) + Q_6 \delta\theta(r_b)] \tag{10.6.9}$$

where N_{rr}, $N_{\theta\theta}$, and M_{rr}, $M_{\theta\theta}$ are known in terms of u and w through Eqs (10.3.23), θ denotes the slope $\theta = -(\partial w/\partial r)$, and Q_i are the generalized forces at the nodes of the element for a circular plate (see Fig. 10.6.2):

$$Q_1 \equiv -\left[r N_{rr}\right]_{r_a}, \quad Q_2 \equiv -\left[r \hat{V}_r\right]_{r_a}, \quad Q_3 \equiv -\left[r M_{rr}\right]_{r_a},$$
$$Q_4 \equiv \left[r N_{rr}\right]_{r_b}, \quad Q_5 \equiv \left[r \hat{V}_r\right]_{r_b}, \quad Q_6 \equiv \left[r M_{rr}\right]_{r_b} \tag{10.6.10}$$

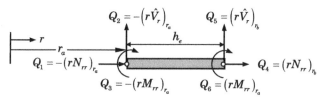

Fig. 10.6.2 Generalized forces of the finite element model; $\hat{V}_r$ for the CPT is given by Eq. (10.3.11) and it is given in Eq. (10.3.20) as $\hat{V}_r = V_r$ for FSDT.

An examination of the boundary terms in Eq. (10.2.3) shows that the Lagrange interpolation of u and Hermite interpolation of w are necessary. Let

$$u(r,t) = \sum_{j=1}^{2} \Delta_j^1(t)\psi_j(r), \quad w(r,t) = \sum_{j=1}^{4} \Delta_j^2(t)\varphi_j(r) \qquad (10.6.11)$$

where $\psi_j(r)$ $(j = 1, 2)$ are the linear polynomials, $\varphi_j(r)$ $(j = 1, 2, 3, 4)$ are the Hermite cubic polynomials, (Δ_1^1, Δ_2^1) are the nodal values of u at r_a and r_b, respectively, and Δ_j^2 are the nodal values associated with w and its derivative:

$$\Delta_1^2(t) = w(r_a, t), \quad \Delta_3^2(t) = w(r_b, t), \quad \Delta_2^2(t) = -\left.\frac{\partial w}{\partial r}\right|_{r_a, t}, \quad \Delta_4^2(t) = -\left.\frac{\partial w}{\partial r}\right|_{r_b, t}$$
$$(10.6.12)$$

Substitution of the approximations from Eq. (10.6.11) into Eqs (10.6.8) and (10.6.9), we obtain the following finite element model:

$$\begin{bmatrix} \mathbf{M}^{11} & \mathbf{M}^{12} \\ \mathbf{M}^{21} & \mathbf{M}^{22} \end{bmatrix} \begin{Bmatrix} \ddot{\Delta}^1 \\ \ddot{\Delta}^2 \end{Bmatrix} + \begin{bmatrix} \mathbf{K}^{11} & \mathbf{K}^{12} \\ \mathbf{K}^{21} & \mathbf{K}^{22} \end{bmatrix} \begin{Bmatrix} \Delta^1 \\ \Delta^2 \end{Bmatrix} = \begin{Bmatrix} \mathbf{F}^1 \\ \mathbf{F}^2 \end{Bmatrix} \qquad (10.6.13)$$

The mass and stiffness coefficients $M_{ij}^{\alpha\beta}$ and $K_{ij}^{\alpha\beta}$ and force coefficients F_i^α $(\alpha, \beta = 1, 2)$ are defined as follows:

$$M_{ij}^{11} = \int_{r_a}^{r_b} m_0 \psi_i \psi_j \, r dr, \quad M_{ij}^{12} = -\int_{r_a}^{r_b} m_1 \psi_i \frac{d\varphi_j}{dr} \, r dr,$$

$$M_{ij}^{21} = -\int_{r_a}^{r_b} m_1 \frac{d\varphi_i}{dr} \psi_j \, r dr, \quad M_{ij}^{22} = \int_{r_a}^{r_b} \left(m_0 \varphi_i \varphi_j + m_2 \frac{d\varphi_i}{dr} \frac{d\varphi_j}{dr} \right) r dr,$$

$$K_{ij}^{11} = \int_{r_a}^{r_b} \left[A_{rr} \frac{d\psi_i}{dr} \left(\frac{d\psi_j}{dr} + \frac{\nu}{r}\psi_j \right) + \frac{1}{r}\psi_i \left(\frac{1}{r}\psi_j + \nu\frac{d\psi_j}{dr} \right) \right] r dr,$$

$$K_{ij}^{12} = \int_{r_a}^{r_b} \left\{ \frac{d\psi_i}{dr} \left[A_{rr}\frac{1}{2}\frac{\partial w}{\partial r}\frac{d\varphi_j}{dr} - B_{rr}\left(\frac{d^2\varphi_j}{dr^2} + \frac{\nu}{r}\frac{d\varphi_j}{dr} \right) \right] \right.$$
$$\left. + \frac{1}{r}\psi_i \left[\frac{1}{2}A_{rr}\nu\frac{\partial w}{\partial r}\frac{d\varphi_j}{dr} - B_{rr}\left(\nu\frac{d^2\varphi_j}{dr^2} + \frac{1}{r}\frac{d\varphi_j}{dr} \right) \right] \right\} r d,$$

$$K_{ij}^{21} = \int_{r_a}^{r_b} \left[\left(A_{rr}\frac{\partial w}{\partial r}\frac{d\varphi_i}{dr} - B_{rr}\frac{d^2\varphi_i}{dr^2} \right) \left(\frac{d\psi_j}{dr} + \frac{\nu}{r}\psi_j \right) \right.$$
$$\left. - B_{rr}\frac{1}{r}\frac{d\varphi_i}{dr}\left(\frac{1}{r}\psi_j + \nu\frac{d\psi_j}{dr} \right) \right] r dr,$$

$$K_{ij}^{22} = \int_{r_a}^{r_b} \left\{ \frac{\partial w}{\partial r}\frac{d\varphi_i}{dr}\left[\frac{1}{2}A_{rr}\frac{\partial w}{\partial r}\frac{d\varphi_j}{dr} - B_{rr}\left(\frac{d^2\varphi_j}{dr^2} + \frac{\nu}{r}\frac{d\varphi_j}{dr} \right) \right] \right.$$
$$- \frac{d^2\varphi_i}{dr^2}\left[\frac{1}{2}B_{rr}\frac{\partial w}{\partial r}\frac{d\varphi_j}{dr} - D_{rr}\left(\frac{d^2\varphi_j}{dr^2} + \frac{\nu}{r}\frac{d\varphi_j}{dr} \right) \right]$$
$$\left. - \frac{1}{r}\frac{d\varphi_i}{dr}\left[\frac{1}{2}B_{rr}\nu\frac{\partial w}{\partial r}\frac{d\varphi_j}{dr} - D_{rr}\left(\nu\frac{d^2\varphi_j}{dr^2} + \frac{1}{r}\frac{d\varphi_j}{dr} \right) \right] \right\} r dr,$$

$$F_i^1 = Q_1\psi_i(r_a) + Q_4\psi_i(r_b),$$

$$F_i^2 = \int_{r_a}^{r_b} q\varphi_i\, r\, dr + Q_2\varphi_i(r_a) + Q_5\varphi_i(r_b)$$

$$+ Q_3\left[-\frac{d\varphi_i}{dx}\right]_{r_a} + Q_6\left[-\frac{d\varphi_i}{dx}\right]_{r_b}. \tag{10.6.14}$$

10.6.2.2   First-Order Shear Deformation Plate Theory

The virtual work statement in Eq. (10.3.15) is equivalent to the following three statements over a typical finite element $\Omega^e = (r_a, r_b)$:

$$0 = \int_{r_a}^{r_b}\left(m_0\ddot{u}\,\delta u + m_1\ddot{\phi}\,\delta u + N_{rr}\frac{\partial\delta u}{\partial r} + N_{\theta\theta}\frac{\delta u}{r}\right)r\,dr$$

$$- Q_1\delta u(r_a, t) - Q_4\delta u(r_b, t), \tag{10.6.15}$$

$$0 = \int_{r_a}^{r_b}\left[m_0\ddot{w}\,\delta w - q\delta w + N_{rr}\frac{\partial w}{\partial r}\frac{\partial\delta w}{\partial r} + Q_r\frac{\partial\delta w}{\partial r}\right]r\,dr$$

$$- Q_2\delta w(r_a, t) - Q_5\delta w(r_b, t), \tag{10.6.16}$$

$$0 = \int_{r_a}^{r_b}\left[m_1\ddot{u}\,\delta\phi_r + m_2\ddot{\phi}_r\,\delta\phi_r + M_{rr}\frac{\partial\delta\phi_r}{\partial r} + \frac{1}{r}M_{\theta\theta}\delta\phi_r + Q_r\delta\phi_r\right]r\,dr$$

$$- Q_3\delta\phi_r(r_a, t) - Q_6\delta\phi_r(r_b, t), \tag{10.6.17}$$

where N_{rr}, $N_{\theta\theta}$, M_{rr}, and $M_{\theta\theta}$ are known in terms of u and w through Eq. (10.3.25) and Q_i are the generalized forces at the nodes of the element for a circular plate:

$$Q_1 \equiv -[rN_{rr}]_{r_a}, \qquad Q_2 \equiv -[rV_r]_{r_a}, \qquad Q_3 \equiv -[rM_{rr}]_{r_a},$$

$$Q_4 \equiv [rN_{rr}]_{r_b}, \qquad Q_5 \equiv [rV_r]_{r_b}, \qquad Q_6 \equiv [rM_{rr}]_{r_b}. \tag{10.6.18}$$

An examination of the boundary terms in Eq. (10.6.9) shows that the Lagrange interpolations of (u, w, ϕ_r) are required. Let

$$u(r, t) = \sum_{j=1}^{m} u_j(t)\psi_j^{(1)}(r),\ \ w(r, t) = \sum_{j=1}^{n} w_j(t)\psi_j^{(2)}(r),\ \ \phi_r(r, t) = \sum_{j=1}^{n} s_j(t)\psi_j^{(3)}(r),$$

$$\tag{10.6.19}$$

where $\psi_j^{(1)}$, $\psi_j^{(2)}$, and $\psi_j^{(3)}$ are the Lagrange polynomials of different degree used for u, w, and ϕ_r, respectively. Substitution of the approximations from Eq. (10.6.19) into Eqs (10.6.15)–(10.6.17), we obtain the following finite element model:

$$\begin{bmatrix} \mathbf{M}^{11} & \mathbf{0} & \mathbf{M}^{13} \\ \mathbf{0} & \mathbf{M}^{22} & \mathbf{0} \\ \mathbf{M}^{31} & \mathbf{0} & \mathbf{M}^{33} \end{bmatrix}\begin{Bmatrix} \ddot{\mathbf{u}} \\ \ddot{\mathbf{w}} \\ \ddot{\mathbf{s}} \end{Bmatrix} + \begin{bmatrix} \mathbf{K}^{11} & \mathbf{K}^{12} & \mathbf{K}^{13} \\ \mathbf{K}^{21} & \mathbf{K}^{22} & \mathbf{K}^{23} \\ \mathbf{K}^{31} & \mathbf{K}^{32} & \mathbf{K}^{33} \end{bmatrix}\begin{Bmatrix} \mathbf{u} \\ \mathbf{w} \\ \mathbf{s} \end{Bmatrix} = \begin{Bmatrix} \mathbf{F}^1 \\ \mathbf{F}^2 \\ \mathbf{F}^3 \end{Bmatrix}. \tag{10.6.20}$$

The nonzero mass and stiffness coefficients, $M_{ij}^{\alpha\beta}$ and $K_{ij}^{\alpha\beta}$, and force coefficients F_i^{α} $(\alpha, \beta = 1, 2, 3)$ are defined as follows:

$$M_{ij}^{11} = \int_{r_a}^{r_b} m_0 \psi_i^{(1)} \psi_j^{(1)} \, r dr, \quad M_{ij}^{13} = \int_{r_a}^{r_b} m_1 \psi_i^{(1)} \psi_j^{(3)} \, r dr = M_{ij}^{31},$$

$$M_{ij}^{22} = \int_{r_a}^{r_b} m_0 \psi_i^{(2)} \psi_j^{(2)} \, r dr, \quad M_{ij}^{33} = \int_{r_a}^{r_b} m_2 \psi_i^{(3)} \psi_j^{(3)} \, r dr,$$

$$K_{ij}^{11} = \int_{r_a}^{r_b} A_{rr} \left[\frac{d\psi_i^{(1)}}{dr} \left(\frac{d\psi_j^{(1)}}{dr} + \frac{\nu}{r} \psi_j^{(1)} \right) + \frac{1}{r} \psi_i^{(1)} \left(\frac{1}{r} \psi_j^{(1)} + \nu \frac{d\psi_j^{(1)}}{dr} \right) \right] r dr,$$

$$K_{ij}^{12} = \frac{1}{2} \int_{r_a}^{r_b} A_{rr} \frac{\partial w}{\partial r} \left(\frac{d\psi_i^{(1)}}{dr} + \frac{\nu}{r} \psi_i^{(1)} \right) \frac{d\psi_j^{(2)}}{dr} r dr,$$

$$K_{ij}^{21} = \int_{r_a}^{r_b} A_{rr} \frac{\partial w}{\partial r} \frac{d\psi_i^{(2)}}{dr} \left(\frac{d\psi_j^{(1)}}{dr} + \frac{\nu}{r} \psi_j^{(1)} \right) r dr,$$

$$K_{ij}^{13} = \int_{r_a}^{r_b} B_{rr} \left[\frac{d\psi_i^{(1)}}{dr} \left(\frac{d\psi_j^{(3)}}{dr} + \frac{\nu}{r} \psi_j^{(3)} \right) + \frac{1}{r} \psi_i^{(1)} \left(\nu \frac{d\psi_j^{(3)}}{dr} + \frac{1}{r} \psi_j^{(3)} \right) \right] r dr,$$

$$K_{ij}^{22} = \int_{r_a}^{r_b} \left[\frac{1}{2} A_{rr} \left(\frac{\partial w}{\partial r} \right)^2 \frac{d\psi_i^{(2)}}{dr} \frac{d\psi_j^{(2)}}{dr} + K_s S_{rz} \frac{d\psi_i^{(2)}}{dr} \frac{d\psi_j^{(2)}}{dr} \right] r dr,$$

$$K_{ij}^{23} = \int_{r_a}^{r_b} \left[B_{rr} \frac{\partial w}{\partial r} \frac{d\psi_i^{(2)}}{dr} \left(\frac{d\psi_j^{(3)}}{dr} + \frac{\nu}{r} \psi_j^{(3)} \right) + K_s S_{rz} \frac{d\psi_i^{(2)}}{dr} \psi_j^{(3)} \right] r dr,$$

$$K_{ij}^{31} = \int_{r_a}^{r_b} B_{rr} \left[\frac{d\psi_i^{(3)}}{dr} \left(\frac{d\psi_j^{(1)}}{dr} + \frac{\nu}{r} \psi_j^{(1)} \right) + \frac{1}{r} \psi_i^{(3)} \left(\nu \frac{d\psi_j^{(1)}}{dr} + \frac{1}{r} \psi_j^{(1)} \right) \right] r dr,$$

$$K_{ij}^{32} = \int_{r_a}^{r_b} \left[\frac{1}{2} B_{rr} \frac{\partial w}{\partial r} \left(\frac{d\psi_i^{(3)}}{dr} + \frac{\nu}{r} \psi_i^{(3)} \right) \frac{d\psi_j^{(2)}}{dr} + K_s S_{rz} \psi_i^{(3)} \frac{d\psi_j^{(2)}}{dr} \right] r dr,$$

$$K_{ij}^{33} = \int_{r_a}^{r_b} \left\{ D_{rr} \left[\frac{d\psi_i^{(3)}}{dr} \left(\frac{d\psi_j^{(3)}}{dr} + \frac{\nu}{r} \psi_j^{(3)} \right) + \frac{1}{r} \psi_i^{(3)} \left(\frac{1}{r} \psi_j^{(3)} + \nu \frac{d\psi_j^{(3)}}{dr} \right) \right] \right.$$

$$\left. + K_s S_{rz} \psi_i^{(3)} \psi_j^{(3)} \right\} r dr,$$

$$F_i^1 = Q_1 \psi_i^{(1)}(r_a) + Q_4 \psi_i^{(1)}(r_b), \quad F_i^3 = Q_3 \psi_i^{(3)}(r_a) + Q_6 \psi_i^{(3)}(r_b),$$

$$F_i^2 = \int_{r_a}^{r_b} q \psi_i^{(2)} \, r dr + Q_2 \psi_i^{(2)}(r_a) + Q_5 \psi_i^{(2)}(r_b). \tag{10.6.21}$$

10.6.3   Solution of Nonlinear Equations

10.6.3.1   Time Approximation

The finite element equations in Eqs (10.6.3), (10.6.6), (10.6.13), and (10.6.20) are of the general matrix form:

$$\mathbf{M}\ddot{\boldsymbol{\Delta}} + \mathbf{K}(\boldsymbol{\Delta})\boldsymbol{\Delta} = \mathbf{F}. \tag{10.6.22}$$

The fully discretized equations are (see Reddy [46])

$$\hat{\mathbf{K}}(\boldsymbol{\Delta}_{s+1})\boldsymbol{\Delta}_{s+1} = \hat{\mathbf{F}}_{s,s+1}, \tag{10.6.23}$$

where $\boldsymbol{\Delta}_{s+1} = \boldsymbol{\Delta}(t_{s+1})$, $t_{s+1} = (s+1)\,\Delta t$, Δt being the time step, and

$$\begin{aligned}
\hat{\mathbf{K}}(\boldsymbol{\Delta}_{s+1}) &= \mathbf{K}(\boldsymbol{\Delta}_{s+1}) + a_3\,\mathbf{M}_{s+1} \\
\hat{\mathbf{F}}_{s,s+1} &= \mathbf{F}_{s+1} + \mathbf{M}_{s+1}\big(a_3\boldsymbol{\Delta}_s + a_4\dot{\boldsymbol{\Delta}}_s + a_5\ddot{\boldsymbol{\Delta}}_s\big).
\end{aligned} \tag{10.6.24}$$

At the end of each time step, the new velocity vector $\dot{\boldsymbol{\Delta}}_{s+1}$ and acceleration vector $\ddot{\boldsymbol{\Delta}}_{s+1}$ are computed using the equations

$$\begin{aligned}
\ddot{\boldsymbol{\Delta}}_{s+1} &= a_3\big(\boldsymbol{\Delta}_{s+1} - \boldsymbol{\Delta}_s\big) - a_4\,\dot{\boldsymbol{\Delta}}_s - a_5\,\ddot{\boldsymbol{\Delta}}_s \\
\dot{\boldsymbol{\Delta}}_{s+1} &= \dot{\boldsymbol{\Delta}}_s + a_2\,\ddot{\boldsymbol{\Delta}}_s + a_1\,\ddot{\boldsymbol{\Delta}}_{s+1},
\end{aligned} \tag{10.6.25}$$

and a_i are defined as

$$a_1 = \alpha\Delta t, \quad a_2 = (1-\alpha)\Delta t, \quad a_3 = \frac{2}{\gamma(\Delta t)^2},$$

$$a_4 = a_3\Delta t, \quad a_5 = \frac{1}{\gamma} - 1. \tag{10.6.26}$$

10.6.3.2   Newton's Iteration Approach

Solution of Eq. (10.6.23) by the Newton iteration method results in the following linearized equations for the incremental solution at the $(r+1)$st iteration:

$$\delta\boldsymbol{\Delta} = -(\hat{\mathbf{T}}(\boldsymbol{\Delta}_{s+1}^r))^{-1}\mathbf{R}_{s+1}^r, \tag{10.6.27}$$

where $\hat{\mathbf{T}}$ is the tangent stiffness matrix

$$\hat{\mathbf{T}}(\boldsymbol{\Delta}_{s+1}^r) \equiv \left[\frac{\partial\mathbf{R}}{\partial\boldsymbol{\Delta}}\right]_{s+1}^r, \quad \mathbf{R}_{s+1}^r = \hat{\mathbf{K}}(\boldsymbol{\Delta}_{s+1}^r)\boldsymbol{\Delta}_{s+1}^r - \hat{\mathbf{F}}_{s,s+1}. \tag{10.6.28}$$

The total solution is obtained from

$$\boldsymbol{\Delta}_{s+1}^{r+1} = \boldsymbol{\Delta}_{s+1}^r + \delta\boldsymbol{\Delta}. \tag{10.6.29}$$

Note that the tangent stiffness matrix $\hat{\mathbf{T}}^r_{s+1}$, the coefficient matrix $\hat{\mathbf{K}}^r_{s+1}$, and the residual vector $\mathbf{R}^r_{s+1}$ are evaluated using the latest known solution $\boldsymbol{\Delta}^r_{s+1}$. The right-hand side vector $\hat{\mathbf{F}}_{s,s+1}$ only depends on the known load vector (after assembly and imposition of boundary conditions) $\mathbf{F}_{s,s+1}$ and known solutions $(\boldsymbol{\Delta}_s, \dot{\boldsymbol{\Delta}}_s, \ddot{\boldsymbol{\Delta}}_s)$ from the previous time step. At the end of each time step (i.e. after nonlinear convergence is reached), the velocity and acceleration vectors are updated using the formulas in Eq. (10.6.25) (for details see Reddy [130]).

For the static case, the linearized element equations at the beginning of the rth iteration are of the form

$$\mathbf{T}^e(\boldsymbol{\Delta}^{(r)})\delta\boldsymbol{\Delta} = -\mathbf{R}^e(\boldsymbol{\Delta}^{(r)}) = (\mathbf{F}^e - \mathbf{K}^e\boldsymbol{\Delta}^e)^{(r)}, \tag{10.6.30}$$

where the tangent stiffness matrix $\mathbf{T}^e$ is calculated using the definition

$$\mathbf{T}^e \equiv \frac{\partial\mathbf{R}^e}{\partial\boldsymbol{\Delta}^e} \quad \text{or} \quad T^e_{ij} \equiv \frac{\partial R^e_i}{\partial\Delta^e_j}. \tag{10.6.31}$$

The global incremental displacement vector $\delta\mathbf{U}$ at the rth iteration is obtained by solving the assembled equations (after the imposition of the boundary conditions)

$$\delta\mathbf{U} = -[\mathbf{T}(\mathbf{U}^{(r)})]^{-1}\mathbf{R}^{(r)}, \tag{10.6.32}$$

and the total solution is computed from

$$\mathbf{U}^{(r+1)} = \mathbf{U}^{(r)} + \delta\mathbf{U}. \tag{10.6.33}$$

At the beginning of the iteration process, that is, when $r = 0$, solution $\mathbf{U}^{(0)}$ that is consistent with the problem boundary conditions must be assumed. For problems with homogeneous boundary conditions, one may take $\mathbf{U}^{(0)} = \mathbf{0}$ so that the first iteration solution is the linear solution. Using the solution from the rth iteration, we can compute the coefficient matrix $\mathbf{K}^{(r)} \equiv \mathbf{K}(\mathbf{U}^{(r)})$ and vector $\mathbf{F}^{(r-1)} \equiv \mathbf{F}(\mathbf{U}^{(r)})$. The solution at the $r+1$st iteration is determined by solving Eq. (10.6.32). Once the solution $\mathbf{U}^{(r+1)}$ is obtained, we check to see if the residual vector

$$\mathbf{R}^{(r+1)} \equiv \mathbf{K}^{(r+1)}\mathbf{U}^{(r+1)} - \mathbf{F}^{(r+1)} \tag{10.6.34}$$

is zero. The magnitude of this residual vector will be small enough if the solution has converged. In other words, we terminate the iteration if the magnitude of the residual vector, measured in a suitable norm, is less than some preselected tolerance ϵ. If the problem data are such that $\mathbf{KU}$ as well as $\mathbf{F}$ are very small, the norm of the residual vector may also be very small even when the solution $\mathbf{U}$ has not converged. Therefore, it is necessary to normalize the residual vector with respect to $\mathbf{F}$. Using the Euclidean norm, we can express the error criterion as

$$\sqrt{\frac{\mathbf{R}^{(r+1)} \cdot \mathbf{R}^{(r+1)}}{\mathbf{F}^{(r+1)} \cdot \mathbf{F}^{(r+1)}}} \leq \epsilon. \tag{10.6.35}$$

Alternatively, one may check to see if the normalized difference between solution vectors from two consecutive iterations, measured with the Euclidean norm, is less than a preselected tolerance ϵ:

$$\sqrt{\frac{\delta \mathbf{U} \cdot \delta \mathbf{U}}{\mathbf{U}^{(r+1)} \cdot \mathbf{U}^{(r+1)}}} = \sqrt{\frac{\sum_{I=1}^{N} |U_I^{(r+1)} - U_I^{(r)}|^2}{\sum_{I=1}^{N} |U_I^{(r+1)}|^2}} \leq \epsilon, \qquad (10.6.36)$$

where $\delta \mathbf{U} = \mathbf{U}^{(r+1)} - \mathbf{U}^{(r)}$. Thus, the iteration process is continued until the error criterion in Eq. (10.6.36) is satisfied. This is the error criterion used in the present study.

Acceleration of convergence for some types of nonlinearities may be achieved by using a weighted average of solutions from the last two iterations rather than the solution from the last iteration to evaluate the coefficient matrix:

$$\mathbf{U}^{(r+1)} = [\mathbf{K}(\bar{\mathbf{U}})]^{-1} \mathbf{F}(\bar{\mathbf{U}}), \quad \bar{\mathbf{U}} \equiv \rho \mathbf{U}^{(r-1)} + (1 - \rho) \mathbf{U}^{(r)}, \quad 0 \leq \rho \leq 1, \quad (10.6.37)$$

where ρ is known as the acceleration parameter. The value of ρ depends on the nature of nonlinearity and the type of problem. Often, one has to play with the value of ρ to obtain convergence.

10.6.3.3 Tangent Stiffness Coefficients for the BET

The nonlinear equations in Eq. (10.6.3) are solved using Newton's iterative method (see Reddy [130]), which involves the computation of the coefficients of the element tangent stiffness matrix $\mathbf{T}^e$ defined in Eq. (10.6.31). It is convenient to view $\mathbf{T}^e$ as one that has a structure similar to $\mathbf{K}^e$ in Eq. (10.6.3) (for the static case):

$$\begin{bmatrix} \mathbf{T}^{11} & \mathbf{T}^{12} \\ \mathbf{T}^{21} & \mathbf{T}^{22} \end{bmatrix}^{(r)} \begin{Bmatrix} \delta \boldsymbol{\Delta}^1 \\ \delta \boldsymbol{\Delta}^2 \end{Bmatrix}^{(r+1)} = - \begin{Bmatrix} \mathbf{R}^1 \\ \mathbf{R}^2 \end{Bmatrix}^{(r)}, \qquad (10.6.38)$$

where symbol $\delta \boldsymbol{\Delta}$ denotes the increment of the displacements from the rth iteration to the $(r+1)$st iteration. Also, note that $\boldsymbol{\Delta}^1 = \mathbf{u}$ and $\boldsymbol{\Delta}^2 = \bar{\boldsymbol{\Delta}}$. Then we can compute the components $\mathbf{T}^{\alpha\beta}$ from the definition (evaluated at the rth iteration)

$$T_{ij}^{\alpha\beta} = \frac{\partial R_i^\alpha}{\partial \Delta_j^\beta}, \quad \alpha, \beta = 1, 2. \qquad (10.6.39)$$

The components R_i^α of the residual vector $\mathbf{R}$ can be expressed as

$$R_i^\alpha = \sum_{\gamma=1}^{2} \sum_{p=1}^{N_\gamma} K_{ip}^{\alpha\gamma} \Delta_p^\gamma - F_i^\alpha = \sum_{p=1}^{2} K_{ip}^{\alpha 1} \Delta_p^1 + \sum_{p=1}^{4} K_{ip}^{\alpha 2} \Delta_p^2 - F_i^\alpha$$

$$= \sum_{p=1}^{2} K_{ip}^{\alpha 1} u_p + \sum_{p=1}^{4} K_{ip}^{\alpha 2} \bar{\Delta}_p - F_i^\alpha, \qquad (10.6.40)$$

where N_γ ($\gamma = 1, 2$) denotes the number of element degrees of freedom ($N_1 = 2$ and $N_2 = 4$). We have

$$
\begin{aligned}
T_{ij}^{\alpha\beta} &= \frac{\partial R_i^\alpha}{\partial \Delta_j^\beta} = \frac{\partial}{\partial \Delta_j^\beta} \left(\sum_{\gamma=1}^{2} \sum_{p=1}^{N_\gamma} K_{ip}^{\alpha\gamma} \Delta_p^\gamma - F_i^\alpha \right) \\
&= \sum_{\gamma=1}^{2} \sum_{p=1}^{N_\gamma} \left(K_{ip}^{\alpha\gamma} \frac{\partial \Delta_p^\gamma}{\partial \Delta_j^\beta} + \frac{\partial K_{ip}^{\alpha\gamma}}{\partial \Delta_j^\beta} \Delta_p^\gamma \right) - \frac{\partial F_i^\alpha}{\partial \Delta_j^\beta} \\
&= K_{ij}^{\alpha\beta} + \sum_{p=1}^{2} \frac{\partial}{\partial \Delta_j^\beta} \left(K_{ip}^{\alpha 1} \right) u_p + \sum_{p=1}^{4} \frac{\partial}{\partial \Delta_j^\beta} \left(K_{ip}^{\alpha 2} \right) \bar\Delta_p - \frac{\partial F_i^\alpha}{\partial \Delta_j^\beta}.
\end{aligned}
\tag{10.6.41}
$$

Since the coefficients $K_{ij}^{\alpha\beta}$ depend, at most, only on w and the load coefficients F_i^α are not functions of the solution, we have

$$
T_{ij}^{\alpha 1} = K_{ij}^{\alpha 1} \quad \text{for } \alpha = 1, 2.
\tag{10.6.42}
$$

The tangent stiffness coefficients that are different from their counterparts are computed, noting that $\mathbf{M}$ and $\mathbf{F}$ are not functions of the current solution, as follows:

$$
\begin{aligned}
T_{ij}^{12} &= K_{ij}^{12} + \sum_{p=1}^{4} \frac{\partial}{\partial \bar\Delta_j} \left(K_{ip}^{12} \right) \bar\Delta_p \\
&= K_{ij}^{12} + \int_{x_a}^{x_b} \frac{1}{2} A_{xx} \frac{d\psi_i}{dx} \frac{d\varphi_j}{dx} \left(\sum_{p=1}^{4} \frac{d\varphi_p}{dx} \bar\Delta_p \right) dx \\
&= K_{ij}^{12} + \int_{x_a}^{x_b} \left(\frac{1}{2} A_{xx} \frac{dw}{dx} \right) \frac{d\psi_i}{dx} \frac{d\varphi_j}{dx} \, dx = K_{ji}^{21} \\
T_{ij}^{22} &= \hat K_{ij}^{22} + \sum_{p=1}^{2} \frac{\partial}{\partial \bar\Delta_j} \left(K_{ip}^{21} \right) u_p + \sum_{p=1}^{4} \frac{\partial}{\partial \bar\Delta_j} \left(K_{ip}^{22} \right) \bar\Delta_p \\
&= K_{ij}^{22} + \sum_{p=1}^{2} \left[\int_{x_a}^{x_b} A_{xx} \frac{\partial}{\partial \bar\Delta_j} \left(\frac{dw}{dx} \right) \frac{d\varphi_i}{dx} \frac{d\psi_p}{dx} \, dx \right] u_p \\
&\quad + \sum_{p=1}^{4} \left[\int_{x_a}^{x_b} \frac{1}{2} A_{xx} \frac{\partial}{\partial \bar\Delta_j} \left(\frac{dw}{dx} \right)^2 \frac{d\varphi_i}{dx} \frac{d\varphi_p}{dx} \, dx \right] \bar\Delta_p \\
&\quad - \sum_{p=1}^{4} \left[\int_{x_a}^{x_b} B_{xx} \frac{\partial}{\partial \bar\Delta_j} \left(\frac{dw}{dx} \right) \left(\frac{1}{2} \frac{d^2\varphi_i}{dx^2} \frac{d\varphi_p}{dx} + \frac{d\varphi_i}{dx} \frac{d^2\varphi_p}{dx^2} \right) dx \right] \bar\Delta_p \\
&\quad - \int_{x_a}^{x_b} N_{xx}^T \frac{d\varphi_i}{dx} \frac{d\varphi_j}{dx} \, dx
\end{aligned}
$$

$$= K_{ij}^{22} + \int_{x_a}^{x_b} A_{xx} \frac{du}{dx} \frac{d\varphi_i}{dx} \frac{d\varphi_j}{dx} \, dx + \int_{x_a}^{x_b} A_{xx} \left(\frac{dw}{dx}\right)^2 \frac{d\varphi_i}{dx} \frac{d\varphi_j}{dx} \, dx$$

$$- \int_{x_a}^{x_b} B_{xx} \left(\frac{1}{2} \frac{dw}{dx} \frac{d^2\varphi_i}{dx^2} \frac{d\varphi_j}{dx} + \frac{d^2w}{dx^2} \frac{d\varphi_i}{dx} \frac{d\varphi_j}{dx}\right) dx$$

$$- \int_{x_a}^{x_b} N_{xx}^T \frac{d\varphi_i}{dx} \frac{d\varphi_j}{dx} \, dx. \tag{10.6.43}$$

Clearly, the tangent stiffness matrix of functionally graded BET element with the von Kármán nonlinearity is symmetric.

10.6.3.4   Tangent Stiffness Coefficients for the TBT

For the Newton's iterative procedure, we must compute the tangent stiffness matrix $\mathbf{T}^e$, which has a structure similar to $\mathbf{K}^e$ in Eq. (10.6.20) (again, for the static case):

$$\begin{bmatrix} \mathbf{T}^{11} & \mathbf{T}^{12} & \mathbf{T}^{13} \\ \mathbf{T}^{21} & \mathbf{T}^{22} & \mathbf{T}^{23} \\ \mathbf{T}^{31} & \mathbf{T}^{32} & \mathbf{T}^{33} \end{bmatrix}^{(r)} \begin{Bmatrix} \delta \mathbf{u} \\ \delta \mathbf{w} \\ \delta \mathbf{s} \end{Bmatrix}^{(r+1)} = - \begin{Bmatrix} \mathbf{R}^1 \\ \mathbf{R}^2 \\ \mathbf{R}^3 \end{Bmatrix}^{(r)}, \tag{10.6.44}$$

where

$$T_{ij}^{\alpha\beta} = \frac{\partial R_i^\alpha}{\partial \Delta_j^\beta}, \quad \alpha, \beta = 1, 2, 3; \quad \mathbf{\Delta}^1 = \mathbf{u}, \quad \mathbf{\Delta}^2 = \mathbf{w}, \quad \mathbf{\Delta}^3 = \mathbf{s} \tag{10.6.45}$$

and

$$R_i^\alpha = \sum_{\gamma=1}^{3} \sum_{k=1}^{N_\gamma} K_{ik}^{\alpha\gamma} \Delta_k^\gamma - F_i^\alpha = \sum_{k=1}^{m} K_{ik}^{\alpha1} u_k + \sum_{k=1}^{n} K_{ik}^{\alpha2} w_k + \sum_{k=1}^{p} K_{ik}^{\alpha3} s_k, \tag{10.6.46}$$

where F_i^α is assumed to be independent of the solution and $N_1 = m$, $N_2 = n$, and $N_3 = p$. Since the coefficients $K_{ij}^{\alpha\beta}$ depend, at the most, only on w and not on u and ϕ_x, we have

$$T_{ij}^{\alpha1} = K_{ij}^{\alpha1}, \quad T_{ij}^{\alpha3} = K_{ij}^{\alpha3} \quad \text{for } \alpha = 1, 2, 3. \tag{10.6.47}$$

Thus, the only tangent stiffness coefficients that need to be computed are T_{ij}^{12}, T_{ij}^{22}, and T_{ij}^{32}. We have

$$T_{ij}^{12} = K_{ij}^{12} + \sum_{k=1}^{n} \frac{\partial}{\partial w_j} \left(\hat{K}_{ik}^{12}\right) w_k$$

$$= K_{ij}^{12} + \int_{x_a}^{x_b} \frac{1}{2} A_{xx} \frac{d\psi_i^{(1)}}{dx} \frac{d\psi_j^{(2)}}{dx} \left(\sum_{k=1}^{n} \frac{d\psi_k^{(2)}}{dx} w_k\right) dx$$

$$= K_{ij}^{12} + \frac{1}{2} \int_{x_a}^{x_b} A_{xx} \frac{dw}{dx} \frac{d\psi_i^{(1)}}{dx} \frac{d\psi_j^{(2)}}{dx} \, dx = K_{ji}^{21},$$

$$T_{ij}^{22} = K_{ij}^{22} + \sum_{k=1}^{m} \frac{\partial}{\partial w_j} \left(K_{ik}^{21} \right) u_k + \sum_{k=1}^{n} \frac{\partial}{\partial w_j} \left(K_{ik}^{22} \right) w_k + \sum_{k=1}^{p} \frac{\partial}{\partial w_j} \left(K_{ik}^{23} \right) s_k$$

$$= K_{ij}^{22} + \int_{x_a}^{x_b} \left[A_{xx} \frac{du}{dx} + A_{xx} \left(\frac{dw}{dx} \right)^2 + B_{xx} \frac{d\phi_x}{dx} - N_{xx}^T \right] \frac{d\psi_i^{(2)}}{dx} \frac{d\psi_j^{(2)}}{dx} \, dx,$$

$$\hat{T}_{ij}^{32} = K_{ij}^{32} + \sum_{k=1}^{n} \frac{\partial}{\partial w_j} \left(K_{ik}^{32} \right) w_k$$

$$= K_{ij}^{32} + \frac{1}{2} \int_{x_a}^{x_b} B_{xx} \frac{dw}{dx} \frac{d\psi_i^{(3)}}{dx} \frac{d\psi_j^{(2)}}{dx} dx = K_{ji}^{23}. \tag{10.6.48}$$

Once again, we find that the tangent stiffness matrix of functionally graded TBT element with the von Kármán nonlinearity is symmetric.

10.6.3.5   Tangent Stiffness Coefficients for the CPT

The tangent stiffness coefficients for this case are given by ($T_{ij}^{11} = K_{ij}^{11}$ and $T_{ij}^{21} = K_{ij}^{21}$)

$$T_{ij}^{12} = K_{ij}^{12} + \int_{r_a}^{r_b} \frac{1}{2} A_{rr} \frac{d\bar{w}}{dr} \left(\frac{d\psi_i}{dr} \frac{d\varphi_j}{dr} + \frac{\nu}{r} \psi_i \frac{d\varphi_j}{dr} \right) r dr = T_{ji}^{21},$$

$$T_{ij}^{22} = K_{ij}^{22} + \int_{r_a}^{r_b} \left[A_{rr} \left\{ \frac{d\varphi_i}{dr} \frac{d\varphi_j}{dr} \left[\left(\frac{dw}{dr} \right)^2 + \frac{du}{dr} + \nu \frac{u}{r} \right] \right\} \right.$$

$$- B_{rr} \frac{d\varphi_i}{dr} \frac{d\varphi_j}{dr} \left(\frac{d^2 w}{dr^2} + \frac{\nu}{r} \frac{dw}{dr} \right)$$

$$\left. - \frac{1}{2} B_{rr} \frac{dw}{dr} \left(\frac{d^2 \varphi_i}{dr^2} \frac{d\varphi_j}{dr} + \frac{\nu}{r} \frac{d\varphi_i}{dr} \frac{d\varphi_j}{dr} \right) \right] r dr. \tag{10.6.49}$$

The tangent stiffness matrix is symmetric (i.e., $T_{ij}^{\alpha\beta} = T_{ji}^{\beta\alpha}$).

10.6.3.6   Tangent Stiffness Coefficients for the FSDT

The tangent stiffness coefficients (symmetric) for the FSDT are given by

$$T_{ij}^{11} = K_{ij}^{11}, \quad T_{ij}^{12} = 2K_{ij}^{12} = T_{ji}^{21}, \quad T_{ij}^{21} = K_{ij}^{21}, \quad T_{ij}^{13} = K_{ij}^{13},$$

$$T_{ij}^{31} = K_{ij}^{31} = T_{ji}^{13}, \quad T_{ij}^{23} = K_{ij}^{23}, \quad T_{ij}^{33} = K_{ij}^{33}, \quad T_{ij}^{22} = K_{ij}^{22} + \hat{K}_{ij}^{22}$$

$$\hat{K}_{ij}^{22} = \int_{r_a}^{r_b} \left\{ A_{rr} \left[\left(\frac{dw}{dr} \right)^2 + \frac{du}{dr} + \nu \frac{u}{r} \right] + B_{rr} \left(\frac{du}{dr} + \nu \frac{u}{r} \right) \right\} \frac{d\psi_i^{(2)}}{dr} \frac{d\psi_j^{(2)}}{dr} r dr,$$

$$T_{ij}^{32} = K_{ij}^{32} + \frac{1}{2} \int_{r_a}^{r_b} B_{rr} \frac{dw}{dr} \left(\frac{d\psi_i^{(3)}}{dr} + \frac{\nu}{r} \psi_i^{(3)} \right) \frac{d\psi_j^{(2)}}{dr} r dr. \tag{10.6.50}$$

10.6.4   Numerical Results for Beams and Circular Plates

10.6.4.1   Beams

The results presented herein are not representative of any real material or physical system, but they are meant to show the parametric effect of the power-law index (n) on the deflections. The axial load $f(x)$ is assumed to be zero. To eliminate shear and membrane locking (see Reddy [46, 130] for details), stiffness coefficients associated with the transverse shear and nonlinear terms are evaluated using reduced integration.

For the pinned–pinned and clamped–clamped beams considered in this study, the symmetry about $x = L/2$ is exploited to model only one-half of the beam (with $dw/dx = \phi_x = 0$ at $x = L/2$). Convergence studies using various number of elements were carried out, before settling with meshes of 8 BET elements (i.e., element with Hermite cubic approximation of w and linear approximation of u) or 4 quadratic TBT elements (i.e., element with quadratic approximation of u, w, and ϕ_x) in the half beam. One-point integration of the nonlinear terms and two-point integration of the shear terms of the stiffness matrix are used. An error tolerance of $\epsilon = 10^{-3}$ is used with a value of the acceleration parameter $\rho = 0.25$.

First, we consider a pinned–pinned (i.e., $u = w = 0$ at $x = 0$ and $x = L$) macro beam with the following geometric and material parameters (b is the width and h is the height of the beam):

$$b = 1.0 \text{ in.,} \quad L = 10 \text{ in.,} \quad h = 0.1 \text{ or } 1 \text{ in.,}$$
$$E_2 = 3 \times 10^6 \text{ psi,} \quad E_1 = 10E_2 \text{ psi,} \quad \nu = 0.3. \tag{10.6.51}$$

The nonlinear finite element analysis of the pinned–pinned beam was carried out using meshes as described before. Various values of n are considered to show the effect on the deflection parameter $\bar{w} = w/h$. Thick beams ($L/h = 10$) do not exhibit much nonlinearity (i.e., the small difference between the linear and nonlinear solutions for the loads considered herein cannot be seen in the plots), while thin beams ($L/h = 100$) exhibit significant nonlinearity, as can be seen from Fig. 10.6.3, but the effect of shear deformation is negligible and both BET and TBT predict the same results. The effect of power-law index (n) is clearly seen from the results presented in Fig. 10.6.3.

Next, we consider a clamped–clamped (i.e., fixed at $x = 0$ and $x = L$) macro beam with the following geometric and material parameters:

$$b = h = 10^{-2} \text{ m,} \quad L = 1 \text{ m,} \quad E_1 = 200 \text{ GPa,} \quad E_2 = 20 \text{ GPa,} \quad \nu = 0.25 \tag{10.6.52}$$

First, the beam is assumed to be loaded with uniformly distributed load of intensity q_0 (N/m). Because of the symmetry about $x = L/2$, only one-half of the beam is modeled using eight BET elements and four quadratic TBT elements. The boundary conditions used for the half beam are

$$\text{At } x = 0: \quad u = w = 0, \quad \frac{dw}{dx} \text{ or } \phi_x = 0,$$

$$\text{At } x = \frac{L}{2}: \quad u = 0, \quad \frac{dw}{dx} \text{ or } \phi_x = 0. \tag{10.6.53}$$

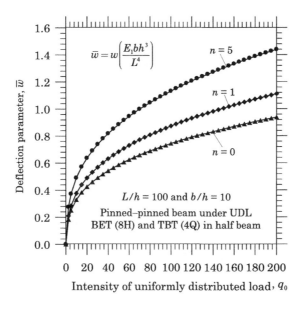

Fig. 10.6.3 Load versus dimensionless center transverse deflection $\bar{w}$ for pinned–pinned beams under uniformly distributed load (UDL).

The linear deflection for $q_0 = 2.5$ (N/m), when $n = 0$, that is, homogeneous beam made of material 1 (E_1), is $w(L/2) = 3.9062 \times 10^{-5}$ m by BET and $w(L/2) = 3.9109 \times 10^{-5}$ m by TBT (match with the exact solutions). For the same load, when $n = 1$ (i.e., FGM beam), the deflections predicted by the BET and TBT models are $w(L/2) = 9.1104 \times 10^{-5}$ m and 9.1508×10^{-5} m, respectively. As the value of n increases, the beam material approaches that of material 2 (E_2) and, therefore, the deflection increases because $E_2 < E_1$. The results predicted by the BET and TBT models are virtually the same, even for the nonlinear case, because the effect of transverse shear strain for this thin beam ($L/h = 10^2$) is negligible. Figure 10.6.4 contains plots of load versus center transverse deflection for the homogeneous and functionally graded beams for power-law index values of $n = 0, 1$, and 10.

Next, we consider the case in which the transverse load is sinusoidally distributed, $q_0 \sin \frac{\pi x}{L}$. All results were obtained with a mesh of eight quadratic TBT elements. The linear center deflection for the homogeneous ($n = 0$) TBT beam is $w(L/2) = 3.3079 \times 10^{-5}$ m when $q_0 = 2.5$ (N/m), and for functionally graded TBT beams ($n = 1$ and $n = 10$) they are $w(L/2) = 7.7400 \times 10^{-5}$ m and $w(L/2) = 1.6581 \times 10^{-4}$ m. Plots of load versus deflection for FGM beams with power-law index $n = 0, 1$, and $n = 10$ are shown in Fig. 10.6.5.

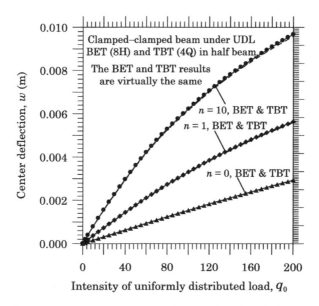

Fig. 10.6.4 Load versus center transverse deflection for clamped–clamped, functionally graded, beams under uniformly distributed load (UDL).

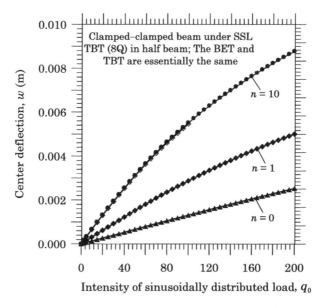

Fig. 10.6.5 Load versus center transverse deflection for clamped–clamped, functionally graded, beams under a sinusoidally distributed load (SSL).

10.6.4.2 Circular Plates

Here we consider several examples of circular plates (of radius a and thickness h) with clamped and simply supported boundary conditions to determine the parametric effects of the power-law index n and the boundary conditions on the nonlinear load-deflection behavior. The following parameters are used in obtaining the numerical results (no specific units are used) in the first three examples:

$$h = 10, \quad a = 10h, \quad \nu = 0.3, \quad E_1 = 10^6, \quad E_2 = 10^5, \quad K_s = 5/6. \quad (10.6.54)$$

Convergence studies were carried out to verify the linear and nonlinear solution of the clamped homogeneous solid circular plates under uniformly distributed transverse load by comparing with the analytical and finite element solutions from Reddy [130]. The difference between the results predicted by the two theories is found negligible for $a/h = 100$ and small for $a/h = 10$ (i.e., the shear deformation effect is small). Numerical results are presented for the power-law index of $n = 0, 1$, and 5. These values are selected only to determine the parametric effects, and they do not correspond to any specific physical system.

The first example deals with a clamped solid circular plate under uniformly distributed transverse load of intensity q_0. The boundary conditions used in the two theories are as follows:

$$\text{CPT:} \quad u = \frac{dw}{dr} = 0 \text{ at } r = 0; \quad u = w = \frac{dw}{dr} = 0 \text{ at } r = a$$
$$\text{FSDT:} \quad u = \phi_r = 0 \text{ at } r = 0; \quad u = w = \phi_r = 0 \text{ at } r = a \quad (10.6.55)$$

Eight CPT or eight quadratic FSDT elements are used, with one-point integration of the nonlinear terms and two-point integration of the shear terms of the stiffness matrix. The load parameter $P = (q_0 a^4 / E_1 h^4)$ versus the dimensionless deflection $w(0)/h$ are presented in Fig. 10.6.6 for various values of the power-law index n.

The next example deals with clamped annular plates under uniformly distributed transverse load of intensity q_0; the hole diameter is taken as $b = 2h$. The edge at $r = b$ is assumed to be free (i.e., no displacement boundary conditions are specified). Ten CPT or ten quadratic FSDT elements are used. The linear deflections, $w(b)$, of the homogeneous (with material E_1) solid and annular CPT plates at $r = b$ are 1.5725 and 1.7528 for $q_0 = 100$, respectively, which coincide with the exact solutions; the same values predicted by the FSDT elements are 1.6474 and 1.8176, respectively. For solid and annular plates with power-law index $n = 5$, the linear deflections are $w(b) = 6.4571$ and 7.2056, respectively, by the CPT; and 6.7689 and 7.4703, respectively, by the FSDT. Plots of the load parameter P versus dimensionless deflection at the inner edge, $w(b)/h$, are shown in Fig. 10.6.7 for $n = 0, 5$. For comparison, the dimensionless deflections at $r = b$ of solid FGM circular plates are also included in the figure.

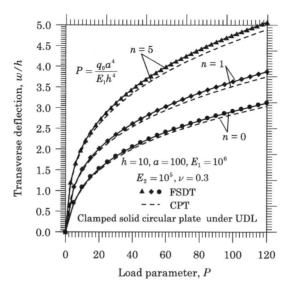

Fig. 10.6.6 Load-deflection curves for clamped solid circular plates for $n = 0, 1, 5$.

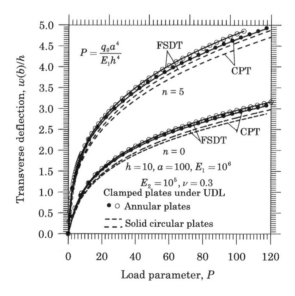

Fig. 10.6.7 Load-deflection curves for clamped annular plates for $n = 0, 1, 5$.

The effects of shear deformation as well the power-law index on load-deflection curves are clear from the figure.

The last example deals with simply supported solid circular plates. The geometric and material parameters used are the same as those in Eq. (10.6.54); the boundary conditions at $r = 0$ are the same as those listed in Eq. (10.6.55), while those at $r = a$ are taken to be $u = w = 0$ in both theories. Eight elements of CPT or eight quadratic elements of FSDT are used. It is found that both CPT and FSDT gave essentially the same results. The linear deflections $w(0)$ by the CPT and FSDT for $n = 0$ and $q_0 = 25$ are 1.7391 and 1.7586, respectively; for the same q_0 and $n = 1$, they are 3.3363 and 3.3632. Plots of the load versus deflection using the eight-element mesh of the CPT are presented in Fig. 10.6.8; the results obtained with the eight-element mesh of the FSDT are indistinguishable from those of the CPT in the plots (i.e., the effect of shear deformation in the simply supported plates is negligible even for $a/h = 10$).

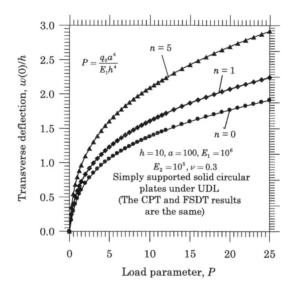

Fig. 10.6.8 Load-deflection curves for simply supported solid circular plates for $n = 0, 1, 5$.

10.7 Summary

The chapter is dedicated to theory and analysis of functionally graded beams and plates. The governing equations of functionally graded BET and TBT, CPT and FSDT circular plates, and the GTST, accounting for the von Kármán nonlinearity in all cases, are derived. The GTST of FGM plates contains 11 generalized displacements, and, as special cases, the CPT and FSDT of FGM plates can be recovered from the GTST. The Navier solutions are presented for simply supported beams and plates.

Finite element models of beams and axisymmetric circular plates are also developed, and numerical results are presented showing the influence of the power-law index and boundary conditions on the deflections. This chapter on functionally graded structures should provide the reader sufficient background to read and follow many developments in the subject area.

Problems

10.1 Verify the relations in Eq. (10.1.2) and hence Eq. (10.2.36).

10.2 Extract the two weak forms associated with the BET from Eq. (10.2.3). Express the statements in terms of the generalized displacements.

10.3 Extract the three weak forms associated with the TBT from Eq. (10.2.13). Express the statements in terms of the generalized displacements.

10.4 Extract the two weak forms associated with the classical plate theory from Eq. (10.3.6).

10.5 Extract the three weak forms associated with the first-order plate theory from Eq. (10.3.15).

10.6 Verify the equations of motion in Eqs (10.4.69)–(10.4.73) of the Reddy third-order plate theory.

10.7 Find the Navier solution for bending of a simply supported rectangular plate using the classical plate theory [use linearized form of Eqs (10.4.83)–(10.4.85)].

10.8 Find the Navier solution for bending of a simply supported rectangular plate using the first-order shear deformation plate theory [use the linearized form of Eqs (10.4.77)–(10.4.81)].

"We are what our thoughts have made us; so take care about what you think. Words are secondary. Thoughts live; they travel far."
– *Swami Vivekananda*

An Indian Hindu monk, a key figure in the introduction of the Indian philosophies of Vedanta and Yoga to the Western world.

References

1. Maxwell, J. C., "On the calculation of the equilibrium and the stiffness of frames," *Philosophical Magazine, Series 4,* **27**, 294 (1864).

2. Engesser, F., "Über statisch unbestimmte Träger bei beliebigen formän-derungsgesetz und über den satz von der kleinsten ergänzungsarbeit," *Zeitschriftdes Architektenund Ingenieur-Vereinszu Hanover,* **35**, 733–744 (1889).

3. Rayleigh, J. W. S., *The Theory of Sound,* Macmillan, London (1877). Reprinted by Dover (1945).

4. Ritz, W., "Über eine neue methode zur lösung gewisser randwertaufgeben," *Göttingener Nachsichten Mathematisch-Physikalische Klasse,* 236 (1908).

5. Galerkin, B. G., "Series solutions in rods and plates," *Vestnik Inzhenerov i Tekhnikov,* **19** (1915).

6. Hellinger, E., "Die allgemeinen Ansätze der Mechanik der Kontinua," *Enzyclopädie der Mathematischen Wissenschaften IV,* **4**, 654–655 (1914).

7. Hu, H., "On some variational principles in the theory of elasticity and the theory of plasticity," *Science Sinica,* **4**, 33–54 (1955).

8. Reissner, E., "On a variational theorem in elasticity," *Journal of Mathematical Physics,* **29**, 90–95 (1950).

9. Reissner, E., "A note on variational principles in elasticity," *International Journal of Solids and Structures,* **1**(1), 93–95 (1965).

10. Reissner, E., "On mixed variational formulations in finite elasticity," *Acta Mechanica,* **56**, 177–125 (1985).

11. Lanczos, C., *The Variational Principles of Mechanics,* 4th ed., Dover Publications, New York, NY (1986).

12. Truesdell, C. A. and Toupin, R. A., "The classical field theories," *Encyclopedia of Physics,* S. Flügge (ed.). Springer-Verlag, Berlin (1960).

13. Dugas, R., *A History of Mechanics,* Translated into English by J. R. Maddox, Dover Publications, New York (1988). First published by Éditions du Griffon, Neuchâtel, Switzerland in 1955.

14. Timoshenko, S. P., *History of Strength of Materials,* Dover, New York, NY (1983). Originally published in 1953 by McGraw-Hill, New York, NY.

15. Oravas, G. and McLean, C., "Historical development of energetical principles in elastomechanics – Part 1: From Heraclitos to Maxwell," *Applied Mechanics Reviews,* **19** (1966) 647–658

16. Oravas, G. and McLean, C., "Historical development of energetical principles in elastomechanics – Part 2: From Cotterill to Prange," *Applied Mechanics Reviews,* **19** (1966) 919–933.

17. Washizu, K., *Variational Methods in Elasticity and Plasticity,* Pergamon Press, New York, NY (1967).

Energy Principles and Variational Methods in Applied Mechanics, Third Edition. J.N. Reddy.
©2017 John Wiley & Sons Ltd. Published 2017 by John Wiley & Sons Ltd.
Companion Website: www.wiley.com/go/reddy/applied_mechanics_EPVM

18. Oden, J. T. and Reddy, J. N., *Variational Principles in Theoretical Mechanics*, 2nd ed., Springer-Verlag, Berlin (1983).

19. Argyris, J. H. and Kelsey, S., *Energy Theorems and Structural Analysis*, Butterworths, London (1954).

20. Arthurs, A. M., *Complementary Variational Principles*, Oxford University Press, London (1967).

21. Becker, M., *The Principles and Applications of Variational Methods*, M.I.T. Press, Cambridge, MA (1964).

22. Biot, M. A., *Variational Principles in Heat Transfer*, Clarendon, London (1972).

23. Charlton, T. M., *Principles of Structural Analysis*, Longmans, Harlow (1969).

24. Charlton, T. M., *Energy Principles in Theory of Structures*, Oxford University Press, Oxford, UK (1973).

25. Davies, G. A. O., *Virtual Work in Structural Analysis*, John Wiley, Chichester, UK (1982).

26. Denn, M. M. *Optimization by Variational Methods*, McGraw-Hill, New York, NY (1969).

27. Dym, C. L. and Shames, I. H., *Solid Mechanics: A Variational Approach*, McGraw-Hill, New York, NY (1973).

28. Hu, H., *Variational Principles of Theory of Elasticity with Applications*, Science Press, Beijing; Gordon and Breach Science Publishers, New York, NY (1984).

29. Finlayson, B. A., *The Method of Weighted Residuals and Variational Principles*, Academic Press, New York, NY (1972).

30. Forray, M. J., *Variational Calculus in Science and Engineering*, McGraw-Hill, New York, NY (1968).

31. Gregory, M. S., *An Introduction to Extremum Principles*, Butterworths, London (1969).

32. Hestenes, M. R., *Optimization Theory: The Finite Dimensional Case*, John Wiley, New York, NY (1975).

33. Hildebrand, F. B., *Methods of Applied Mathematics*, 2nd ed., Prentice-Hall, Englewood Cliffs, NJ (1965).

34. Kantorovitch, L. V. and Krylov, V. I., *Approximate Methods of Higher Analysis* (translated by C. D. Benster), Noordhoff, The Netherlands (1958).

35. Leipholz, H., *Direct Variational Methods and Eigenvalue Problems in Engineering*, Noordhoff, Leyden (1977).

36. Lippmann, H., *Extremum and Variational Principles in Mechanics*, Springer-Verlag, New York, NY (1972).

37. Lovelock, D., *Tensors, Differential Forms and Variational Principles*, Wiley, New York, NY (1974).

38. McGuire, W., Gallagher, R. H., and Ziemian, R. D., *Matrix Structural Analysis*, 2nd ed., John Wiley, New York, NY (2000).

39. Mikhlin, S. G., *Variational Methods in Mathematical Physics*, Pergamon (distributed by Macmillan), New York, NY (1974).

40. Mikhlin, S. G., *Mathematical Physics, An Advanced Course*, North-Holland, Amsterdam, The Netherlands (1970).

41. Mikhlin, S. G., *The Numerical Performance of Variational Methods*, Wolters-Noordhoff, Groningen, The Netherlands (1971).

42. Oden, J. T. and Ripperger, E. A., *Mechanics of Elastic Structures,* 2nd ed., Hemisphere/McGraw-Hill, New York, NY (1981).

43. Parkus, H., *Variational Principles in Thermo- and Magneto-Elasticity,* Lecture Notes, No. 58, Springer-Verlag, New York, NY (1972).

44. Petrov, I. P., *Variational Methods in Optimal Control Theory,* translated from the 1965 Russian edition, Academic Press, New York, NY (1968).

45. Prenter, P. M., *Splines and Variational Methods,* John Wiley, New York, NY (1975).

46. Reddy, J. N., *An Introduction to the Finite Element Method,* 3rd ed., McGraw-Hill, New York, NY (2006).

47. Reddy, J. N. and Rasmussen, M. L., *Advanced Engineering Analysis,* John Wiley, New York, NY (1982); reprinted by Krieger, Melbourne, FL (1990).

48. Reddy, J. N., *Energy Principles and Variational Methods in Applied Mechanics,* John Wiley, New York, NY (1984); 2nd ed. (2004).

49. Reddy, J. N., *Applied Functional Analysis and Variational Methods in Engineering,* McGraw-Hill, New York, NY (1986).

50. Reddy, J. N., *Mechanics of Laminated Composite Plates and Shells,* 2nd ed., CRC Press, Boca Raton, FL (2004).

51. Reddy, J. N., *Theory and Analysis of Elastic Plates and Shells,* 2nd ed., Taylor & Francis, Philadelphia, PA (2007).

52. Rektorys, K., *Variational Methods in Mathematics, Science and Engineering,* Reidel, Boston, MA (1977).

53. Richards, T. H., *Energy Methods in Stress Analysis,* Ellis Horwood (distributed by John Wiley & Sons), Chichester, UK (1977).

54. Rustagi, J. S., *Variational Methods in Statistics,* Academic, New York, NY (1976).

55. Schechter, R. S., *The Variational Methods in Engineering,* McGraw-Hill, New York, NY (1967).

56. Shames, I. H. and Dym, C. L., *Energy and Finite Element Methods in Structural Mechanics,* Hemisphere, Washington, DC (1985).

57. Shaw, F. S., *Virtual Displacements and Analysis of Structures,* Prentice Hall, Englewood Cliffs, NJ (1972).

58. Smith, D. K., *Variational Methods in Optimization,* Prentice-Hall, Englewood Cliffs, NJ (1974).

59. Stacey, W. M., *Variational Methods in Nuclear Reactor Physics,* Academic, New York, NY (1974).

60. Wunderlich, W. and Pilkey, W. D., *Mechanics of Structures: Variational and Computational Methods,* 2nd ed., CRC Press, Boca Raton, FL (2003).

61. Aris, R., *Vectors, Tensors, and the Basic Equations in Fluid Mechanics,* Prentice-Hall, Englewood Cliffs, NJ (1962).

62. Bedford, F. W. and Dwivedi, T. D., *Vector Calculus,* McGraw-Hill, New York, NY (1970).

63. Bellman, T. E., *Introduction to Matrix Analysis,* 2nd ed., McGraw-Hill, New York, NY (1970).

64. Borisenko, A. I. and Tarapo, I. E., *Vector and Tensor Analysis with Applications,* R. A. Silverman (Translator and Editor), Prentice-Hall, Englewood Cliffs, NJ (1968).

65. Bourne, D. E. and Kendall, P. C., *Vector Analysis and Cartesian Tensors,* Second Edition, Academic Press, New York, NY (1977).

66. Bowen, R. M. and Wang, C. C., *Introduction to Vectors and Tensors,* Plenum, New York, NY (1976).

67. Campbell, H. G., *An Introduction to Matrices, Vectors and Linear Programming,* Prentice-Hall, Englewood Cliffs, NJ (1977).

68. Chambers, L. G., *A Course in Vector Analysis,* Chapman and Hall, London, UK (1969).

69. Chisholm, J. S. R. *Vectors in Three-Dimensional Space,* Cambridge University Press, Cambridge, UK (1978).

70. Chorlton, F., *Vector and Tensor Methods,* Halsted Press, New York, NY (1964).

71. Coburn, N., *Vector and Tensor Analysis,* Mcmillan, New York, NY (1964).

72. Eisele, J. A. *Applied Matrix and Tensor Analysis,* Wiley-Interscience, New York, NY (1970).

73. Gantmacher, F. R., *The Theory of Matrices,* Chelsea, New York, NY (1959).

74. Graham, A., *Matrix Theory and Applications for Engineers and Mathematicians,* Halsted Press, New York, NY (1979).

75. Haskell, R. E., *Introduction to Vectors and Cartesian Tensors,* Prentice-Hall, Englewood Cliffs, NJ (1972).

76. Hauge, B., *An Introduction to Vector Analysis for Physicists and Engineers,* 6th ed., revised by D. Martin, Methuen, London, UK (1970).

77. Jeffreys, H., *Cartesian Tensors,* Cambridge University Press, London, UK (1965).

78. Karamcheti, K., *Vector Analysis and Cartesian Tensors,* Holden Day, San Francisco, CA (1967).

79. Lanczos, C., "Tensor Calculus," *Handbook of Physics,* London and Odishaw (Eds.), McGraw-Hill, New York, NY (1958).

80. Lichnerowicz, A., *Linear Algebra and Analysis,* Holden-Day, San Francisco, CA (1967).

81. Malvern, L. E., *Introduction to the Mechanics of a Continuous Medium,* Prentice-Hall, Englewood Cliffs, NJ (1969).

82. Moon, P. H. and Spencer, D. E., *Vectors,* Van Nostrand, Princeton, NJ (1965).

83. Phillips, H. B., *Vector Analysis,* Wiley, New York, NY (1933).

84. Reddy, J. N., *An Introduction to Continuum Mechanics,* 2nd ed., Cambridge University Press, New York, NY (2013).

85. Sokolnikoff, I.S., *Tensor Analysis,* John Wiley, New York, NY (1933).

86. Wills, A. P., *Vector Analysis with an Introduction to Tensor Analysis,* Dover, New York, NY (1953).

87. Wrede, R. C., *Introduction to Vector and Tensor Analysis,* Dover, New York, NY (1972).

88. Young, E. C., *Vector and Tensor Analysis,* Marcel Dekker, New York, NY (1978).

89. Sokolnikoff, I. S., *Mathematical Theory of Elasticity,* 2nd ed., McGraw-Hill, New York, NY; reprinted by Krieger, Melbourne, FL (1956).

90. Mushkelishvili, N. I., *Some Basic Problems of the Mathematical Theory of Elasticity,* Noordhoff, Gröningen, The Netherlands (1963).

91. Jaunzemis, W., *Continuum Mechanics,* Macmillan, New York, NY (1967).

92. Timoshenko, S. P. and Goodier, J. N., *Theory of Elasticity,* 3rd ed., McGraw-Hill, New York, NY (1970).

93. Gurtin, M. E., *An Introduction to Continuum Mechanics*, Elsevier Science & Technology, San Diego, CA (1981).

94. Hjelmstad, K. D., *Fundamentals of Structural Mechanics,* Prentice-Hall, Englewood Cliffs, NJ (1997).

95. Holzapfel, G. A., *Nonlinear Solid Mechanics,* John Wiley & Sons, New York, NY (2001).

96. Slaughter, W. S., *The Linearized Theory of Elasticity,* Birkhäser, Boston, MA (2002).

97. Sadd, M. H., *Elasticity: Theory, Applications, and Numerics,* 3rd ed., Elsevier, New York, NY (2014).

98. Fenner, R. T. and Reddy, J. N., *Mechanics of Solids and Structures,* 2nd ed., CRC Press, Boca Raton, FL (2012).

99. Castigliano, A. C., "Intorno ai sistemi elastici," Dissertazione di Laurea presentata all Commissione Esaminatrice della Reale Scuola degli Ingegneri di Torino, Vincenzo Bona, Turin, Italy (1873).

100. Crotti, F., "Esposizione del teorema di Castigliano e suo raccordo colla teoria delle lasticità," *Atti del Collegio degli Ingegneri ed Architetti in Milano*, **27**, 45 (1879); **32**, 11/12, 597 (1884).

101. Betti, E., "Teoria dell'elasticitá," *Nuovo Cimento*, Serie 2, **7-8**, 5–21; 69–97 (1872).

102. Maxwell, J. C., "On the calculation of the equilibrium and the stiffness of frames," *Philosophical Magazine, Series 4*, **27**, 294 (1864).

103. Ritz, W., "Uber eine neue Methode zur Losung gewisser Variationsprobleme der mathematischen Physik," *Journal für die reine und angewandte Mathematik,* **135**, 1–61 (1908).

104. Galerkin, B. G., "Series-solutions of some cases of equilibrium of elastic beams and plates" (in Russian), *Vestnik Inshenernov,* **1**, 897–903 (1915).

105. Galerkin, B. G., "Berechnung der frei gelagerten elliptische platte auf biegung," *Zeitschrift für Agnewandte Mathematik und Mechanik (ZAMM)*, **3**, 113–117 (1923).

106. Trefftz, E., "Zur theorie der stabilität des elastischen gleichgewichts," *Zeitschrift für Agnewandte Mathematik und Mechanik (ZAMM)*, **13**(2), 160–165 (1933).

107. Trefftz, E., "Ein gegenstück zum Ritz'schen verfahren," *Proceedings of the 2nd International Congress for Applied Mechanics,* E. Meissner (ed.), Zurich, 131–137 (1926).

108. Kirchhoff, G., *Vorlesungen über Mathematische Physik*, Vol. 1, B. G. Teubner, Leipzig, Germany (1876).

109. Basset, A. B., "On the extension and flexure of cylindrical and spherical thin elastic shells," *Philosophical Transactions of the Royal Society, Series A,* **181** (6), 433–480 (1890).

110. Hildebrand, F. B., Reissner, E., and Thomas, G. B., "Notes on the foundations of the theory of small displacements of orthotropic shells," NACA TN–1833, Washington, D.C. (1949).

111. Hencky, H., "Uber die berucksichtigung der schubverzerrung in ebenen platten," *Ingenieur Archiv,* **16**, 72–76 (1947).

112. Mindlin, R. D., "Influence of rotatory inertia and shear on flexural motions of isotropic, elastic plates," *Journal of Applied Mechanics,* **18**, 31–38 (1951).

113. Reddy, J. N., "A simple higher-order theory for laminated composite plates," *Journal of Applied Mechanics,* **51**, 745–752 (1984).

114. Reddy, J. N., "A general non-linear third-order theory of plates with moderate thickness," *International Journal of Non-Linear Mechanics,* **25**(6), 677–686 (1990).

115. Reddy, J. N., "A small strain and moderate rotation theory of laminated anisotropic plates," *Journal of Applied Mechanics,* **54**, 623–626 (1987).

116. Levinson, M., "An accurate, simple theory of the statics and dynamics of elastic plates," *Mechanics Research Communications*, **7**(6), 343–350 (1980).

117. Reissner, E., "On the theory of bending of elastic plates," *Journal of Mathematical Physics*, **23**, 184–191 (1944).

118. Reissner, E., "The effect of transverse shear deformation on the bending of elastic plates," *Journal of Applied Mechanics,* **12**, 69–77 (1945).

119. Reissner, E., "Reflections on the theory of elastic plates," *Applied Mechanics Reviews*, **38**(11), 1453–1464 (1985).

120. Kromm, A., "Verallgemeinerte Theorie der Plattenstatik," *Ingenieur-Archiv,* **21**, 266–286 (1953).

121. Kromm, A., "Über die Randquerkräfte bei gestützten Platten.," *Zeitschrift fr Angewandte Mathematik und Mechanik (ZAMM)*, **35**, 231–242 (1955).

122. Panc, V., *Theories of Elastic Plates,* Noordhoff, Leyden, The Netherlands (1975).

123. Roark, J. R. and Young, W. C., *Formulas for Stress and Strain,* McGraw-Hill, New York, NY (1975).

124. Timoshenko, S. P. and Woinowsky-Krieger, S., *Theory of Plates and Shells,* McGraw-Hill, Singapore (1970).

125. Ödman, S. T. A., "Studies of boundary value problems. Part II. Characteristic functions of rectangular plates," *Proceedings NR 24*, Swedish Cement and Concrete Research Institute, Royal Institute of Technology, Stockholm, pp. 7–62 (1955).

126. Leissa, A. W., *Vibration of Plates,* NASA SP–160, Washington, D.C. (1969).

127. Liew, K. M., Wang, C. M., Xiang, Y., and Kitipornchai, S., *Vibration of Mindlin Plates. Programming the p-Version Ritz Method,* Elsevier, Oxford, UK (1998).

128. Timoshenko, S. P. and Gere, J. M., *Theory of Elastic Stability,* 2nd ed., McGraw-Hill, New York, NY (1961).

129. Wang, C. M., Reddy, J. N., and Lee, K. H., *Shear Deformable Beams and Plates. Relationships with Classical Solutions,* Elsevier, Oxford, UK (2000).

130. Reddy, J. N., *An Introduction to Nonlinear Finite Element Analysis,* 2nd ed., Oxford University Press, Oxford, UK (2015).

131. Zienkiewicz, O. C. and Cheung, Y. K., "The finite element method for analysis of elastic isotropic and orthotropic slabs," *Proceeding of the Institute of Civil Engineers,* London, **28**, 471–488 (1964).

132. Bazeley, G. P., Cheung, Y. K., Irons, B. M., and Zienkiewicz, O. C., "Triangular elements in bending – conforming and non-conforming solutions," *Proceedings of the Conference on Matrix Methods in Structural Mechanics,* Air Force Institute of Technology, WPAFB, Dayton, Ohio, AFFDL-TR-66-80, pp. 547–576 (1965).

133. Bogner, F. K., Fox, R. L., and Schmidt, L. A., Jr., "The generation of interelement-compatible stiffness and mass matrices by the use of interpolation formulas," *Proceedings of the Conference on Matrix Methods in Structural Mechanics,* Air Force Institute of Technology, WPAFB, Dayton, Ohio, AFFDL-TR-66-80, 397–443 (1965).

134. Clough R. W. and Tocher, J. L., "Finite element stiffness matrices for analysis of plate bending," *Proceedings of the Conference on Matrix Methods in Structural Mechanics,* Air Force Institute of Technology, WPAFB, Dayton, Ohio, AFFDL-TR-66-80, pp. 515–546 (1965).

135. Herrmann, L. R. "Finite element bending analysis for plates," *Journal of Engineering Mechanics Division, ASCE* **93**(8), 13–26 (1967).

136. Fraeijis de Veubeke, B., "A conforming finite element for plate bending," *International Journal of Solids and Structures*, **4**(1), 95–108 (1968).

137. Irons, B. M., "A conforming quartic triangular element for plate bending," *International Journal for Numerical Methods in Engineering*, **1**, 29–45 (1969).

138. Melosh, R. J., "Basis of derivation of matrices for the direct stiffness method," *AIAA Journal*, **1**, 1631–1637 (1963).

139. Hrabok, M. M. and Hrudey, T. M., "A review and catalog of plate bending finite elements," *Computers & Structures* **19**(3), 479–495 (1984).

140. Barlow, J., "Optimal stress location in finite element models," *International Journal for Numerical Methods in Engineering*, **10**, 243–251 (1976).

141. Barlow, J., "More on optimal stress points – reduced integration element distortions and error estimation," *International Journal for Numerical Methods in Engineering*, **28**, 1486–1504 (1989).

142. Newmark, N. M., "A method for computation of structural dynamics," *Journal of Engineering Mechanics*, **85**, 67–94 (1959).

143. Reddy, J. N. and Tsay, C. S., "Mixed rectangular finite elements for plate bending," *Proceedings of the Oklahoma Academy of Science*, **47**, 144–148 (1977).

144. Reddy, J. N. and Tsay, C. S., "Stability and vibration of thin rectangular plates by simplified mixed finite elements," *Journal of Sound & Vibration*, **55**(2), 289–302 (1977).

145. Putcha, N. S. and Reddy, J. N., "A refined mixed shear flexible finite element for the nonlinear analysis of laminated plates, *Computers & Structures*, **22**(4), 529–538 (1986).

146. Reddy, J. N. and Sandidge, D., "Mixed finite element models for laminated composite plates," *Journal of Engineering for Industry*, **109**, 39–45 (1987).

147. Pagano, N. J., "Exact solutions for rectangular bidirectional composites and sandwich plates," *Journal of Composite Materials*, **4**(1), 20–34 (1970).

148. Anderson, B. W., "Vibration of triangular cantilever plates," *Journal of Applied Mechanics*, **18**(4), 129–134 (1954).

149. Barton, M. V., "Vibration of rectangular and skew cantilever plates," *Journal of Applied Mechanics*, **18**, 129–134 (1951).

150. Plunkett, R., "Natural frequencies of uniform and non-uniform rectangular cantilever plates," *Journal of Mechanical Engineering Science*, **5**, 146–156 (1963).

151. Carson, W. G. and Newton, R. E., "Plate buckling analysis using a fully compatible finite element," *AIAA Journal*, **7**(3), 527–529 (1969).

152. Hasselman, D. P. H. and Youngblood, G. E., "Enhanced thermal stress resistance of structural ceramics with thermal conductivity gradient," *Journal of the American Ceramic Society*, **61**(1,2), 49–53 (1978).

153. Yamanouchi, M., Koizumi, M., Hirai, T., and Shiota, I. (eds), *Proceedings of the First International Symposium on Functionally Gradient Materials*, Japan (1990).

154. Koizumi, M., "The concept of FGM," *Ceramic Transactions, Functionally Gradient Materials*, **34**, 3–10 (1993).

155. Noda, N., "Thermal stresses in materials with temperature-dependent properties," *Applied Mechanics Reviews* **44**, 383–397 (1991).

156. Noda, N. and Tsuji, T., "Steady thermal stresses in a plate of functionally gradient material with temperature-dependent properties," *Transactions of Japan Society of Mechanical Engineers Series A*, **57**, 625–631 (1991).

157. Praveen, G. N. and Reddy, J. N., "Nonlinear transient thermoelastic analysis of functionally graded ceramic-metal plates," *Journal of Solids and Structures*, **35**(33), 4457–4476 (1998).

158. Praveen, G. N., Chin, C. D., and Reddy, J. N., "Thermoelastic analysis of functionally graded ceramic-metal cylinder," *Journal of Engineering Mechanics, ASCE*, **125**(11), 1259–1266 (1999).

159. Reddy, J. N. and Chin, C. D., "Thermomechanical analysis of functionally graded cylinders and plates," *Journal of Thermal Stresses*, **26**(1), 593–626 (1998).

160. Reddy, J. N. "Analysis of functionally graded plates," *International Journal for Numerical Methods in Engineering*, **47**, 663–684 (2000).

161. Tanigawa, Y., "Some basic thermoplastic problems for nonhomogeneous structural material," *Journal of Applied Mechanics*, **48**, 377–389 (1995).

162. Vel, S. S. and Batra, R. C., "Exact solution for thermoelastic deformations of functionally graded thick rectangular plates," *AIAA Journal*, **40**, 1421–1433 (2002).

163. Shen, H.-S., "Nonlinear bending response of functionally graded plates subjected to transverse loads and in thermal environments," *International Journal of Mechanical Sciences*, **44**(3), 561–584 (2002).

164. Yang, J. and Shen, H. S., "Non-linear bending analysis of shear deformable functionally graded plates subjected to thermo-mechanical loads under various boundary conditions," *Composites, Part B*, **34**, 103–115 (2003).

165. Kitipornchai, S., Yang, J., and Liew, K. M., "Semi-analytical solution for nonlinear vibration of laminated FGM plates with geometric imperfections," *International Journal of Solids and Structures*, **41**, 305–315 (2004).

166. Aliaga, W. and Reddy, J. N., "Nonlinear thermoelastic response of functionally graded plates using the third-order plate theory," *International Journal of Computational Engineering Science*, **5**(4), 753–780 (2004).

167. Reddy, J. N., "A general nonlinear third-order theory of functionally graded plates," *International Journal of Aerospace and Lightweight Structures*, **1**(1), 1–21 (2011).

168. Reddy, J. N. and Kim, J., "A nonlinear modified couple stress-based third-order theory of functionally graded plates," *Composite Structures*, **94**(3), 1128–1143 (2012).

169. Kim, J. and Reddy, J. N., "Analytical solutions for bending, vibration, and buckling of FGM plates using a couple stress-based third-order theory, *Composite Structures*, **103**, 86–98 (2013).

170. Kim, J. and Reddy, J. N., "A general third-order theory of functionally graded plates with modified couple stress effect and the von Karman nonlinearity: theory and finite element analysis, *Acta Mechanica*, **226**(9), 2973–2998 (2015).

171. Mindlin, R. D. and Tiersten, H. F., "Effects of couple-stresses in linear elasticity," *Archive for Rational Mechanics and Analysis*, **11**, 415–448 (1962).

172. Toupin, R. A., "Elastic materials with couple stresses," *Archive for Rational Mechanics and Analysis*, **11**, 385–414 (1962).

173. Koiter, W. T., "Couple stresses in the theory of elasticity," *Proceedings of the Koninklijke Nederlandse Academie van Wetenschappen Series B: Physical Sciences*, **67**, 17–44 (1964).

174. Yang, F., Chong, A. C. M., Lam, D. C. C., and Tong, P., "Couple stress based strain gradient theory for elasticity," *International Journal of Solids and Structures*, **39**(10), 2731–2743 (2002).

175. Reddy, J. N., "Microstructure-dependent couple stress theories of functionally graded beams," *Journal of Mechanics and Physics of Solids,* **59**, 2382–2399 (2011).

176. Reddy, J. N. and Arbind, A., "Bending relationships between the modified couple stress-based functionally graded Timoshenko beams and homogeneous Bernoulli-Euler beams," *Annals of Solid and Structural Mechanics,* **3**(1), 15–26 (2012).

177. Arbind, A. and Reddy, J. N., "Nonlinear analysis of functionally graded microstructure-dependent beams," *Composite Structures,* **98**, 272–281 (2013).

178. Simsek, M. and Reddy, J. N., "Bending and vibration of functionally graded microbeams using a new higher order beam theory and the modified couple stress theory," *International Journal of Engineering Science,* **64**, 37–53 (2013).

179. Arbind, A., Reddy, J. N., and Srinivasa, A., "Modified couple stress-based third-order theory for nonlinear analysis of functionally graded beams," *Latin American Journal of Solids and Structures,* **11**(3), 459–487 (2014).

180. Komijani, M., Reddy, J. N., and Eslami, R., "Nonlinear analysis of microstructure-dependent functionally graded piezoelectric material actuators," *Journal of Mechanics and Physics of Solids,* **63**, 214–227 (2014).

181. Reddy, J. N., El-Borgi, S., and Romanoff, J., "Nonlinear analysis of functionally graded microbeams using Eringen's nonlocal differential model," *International Journal of Non-Linear Mechanics,* **67**, 308–318 (2014).

182. Reddy, J. N. and Berry, J., "Modified couple stress theory of axisymmetric bending of functionally graded circular plates," *Composites Structures,* **94**, 3664–3668 (2012).

183. Gunes, R., Apalak, M. K., and Reddy, J. N., "Experimental and numerical investigations of low velocity impact on functionally graded circular plates," *Composites, Part B,* **59**, 21–32 (2014).

184. Reddy, J. N., Romanoff, J., and Loya, J. A., "Nonlinear finite element analysis of functionally graded circular plates with modified couple stress theory," *European Journal of Mechanics-A/Solids,* **56**, 92–104 (2016).

ANSWERS TO MOST PROBLEMS

Chapter 1

1.1 $\mathbf{r} = \mathbf{A} + \mathbf{C} = \mathbf{A} + \beta\,\hat{\mathbf{e}}_B$, $\hat{\mathbf{e}}_B = \frac{\mathbf{B}}{|\mathbf{B}|}$, where β is a real number.

1.2 $(\mathbf{C} - \mathbf{A}) \times (\mathbf{B} - \mathbf{A}) \cdot (\mathbf{r} - \mathbf{A}) = 0$

1.6 (a) $\delta_{ii} = \delta_{11} + \delta_{22} + \delta_{33} = 1 + 1 + 1 = 3$ (d) $\varepsilon_{mjk}\varepsilon_{njk} = \delta_{mn}\delta_{jj} - \delta_{mj}\delta_{jn} = 2\delta_{mn}$

1.9 The vectors are linearly dependent.

1.10 The vectors are linearly independent.

1.11 (a) Linearly dependent (b) Linearly independent and spans $\Re^3$

1.12 Follows as outlined in the problem statement.

1.13 (a) $\mathbf{A} = \begin{bmatrix} \frac{1}{\sqrt{3}} & \frac{-1}{\sqrt{3}} & \frac{1}{\sqrt{3}} \\ \frac{2}{\sqrt{14}} & \frac{3}{\sqrt{14}} & \frac{1}{\sqrt{14}} \\ \frac{-4}{\sqrt{42}} & \frac{1}{\sqrt{42}} & \frac{5}{\sqrt{42}} \end{bmatrix}$ (c) $\mathbf{A} = \begin{bmatrix} \frac{\sqrt{3}}{2} & \frac{1}{2} & 0 \\ -\frac{1}{2} & \frac{\sqrt{3}}{2} & 0 \\ 0 & 0 & 1 \end{bmatrix}$

1.15 Follows from the definition $\mathbf{A} = \begin{bmatrix} \frac{1}{\sqrt{2}} & 0 & \frac{1}{\sqrt{2}} \\ \frac{1}{2} & \frac{1}{\sqrt{2}} & -\frac{1}{2} \\ -\frac{1}{2} & \frac{1}{\sqrt{2}} & \frac{1}{2} \end{bmatrix}$

1.16 The direction cosines are $(2/3, -2/3, -1/3)$.

1.17 (a) $\frac{2}{3}$ (b) $\theta = 82.34°$.

1.19 (a) $\nabla(r) = \hat{\mathbf{e}}_i x_i \left(x_j x_j \right)^{-\frac{1}{2}} = \frac{\mathbf{r}}{r}$ (c) $\nabla^2(r^n) = \frac{\partial}{\partial x_i}\left(n r^{n-2} x_i \right) = n(n+1) r^{n-2}$

(f) $\nabla \times (\mathbf{r} \times \mathbf{A}) = \varepsilon_{rst} \hat{\mathbf{e}}_i \times \hat{\mathbf{e}}_t \left(\delta_{ir} A_s + x_r \frac{\partial A_s}{\partial x_i} \right) = \varepsilon_{ist} \varepsilon_{jit} \hat{\mathbf{e}}_j A_s = -2 \hat{\mathbf{e}}_j \delta_{sj} A_s$

(h) $\nabla \times (r\mathbf{A}) = \hat{\mathbf{e}}_i \times \hat{\mathbf{e}}_k \left(\frac{x_i}{r} A_k + 0 \right) = \varepsilon_{ikt} \hat{\mathbf{e}}_t \left(\frac{x_i}{r} A_k \right) = \frac{1}{r} \varepsilon_{ikt} x_i A_k \hat{\mathbf{e}}_t = \frac{1}{r}(\mathbf{r} \times \mathbf{A})$

1.20 (b) $\nabla \cdot (\nabla \times \mathbf{A}) = \varepsilon_{jk\ell} \delta_{i\ell} \frac{\partial^2 A_k}{\partial x_i \partial x_j} = \varepsilon_{ijk} \frac{\partial^2 A_k}{\partial x_i \partial x_j} = 0$ (d) $\nabla(FG) = \hat{\mathbf{e}}_i \left(\frac{\partial F}{\partial x_i} G + F \frac{\partial G}{\partial x_i} \right)$

(f) $\nabla \times (F\mathbf{A}) = \varepsilon_{ijk} \hat{\mathbf{e}}_k \frac{\partial A_j}{\partial x_i} F - \varepsilon_{jik} \hat{\mathbf{e}}_k A_j \frac{\partial F}{\partial x_i} = F \nabla \times \mathbf{A} - \mathbf{A} \times \nabla F$

(g) $\nabla(\mathbf{A} \cdot \mathbf{B}) = \hat{\mathbf{e}}_i \left(\frac{\partial A_j}{\partial x_i} B_j + A_j \frac{\partial B_j}{\partial x_i} \right) = \nabla\mathbf{A} \cdot \mathbf{B} + \nabla\mathbf{B} \cdot \mathbf{A}$

$\mathbf{A} \times \nabla \times \mathbf{B} = \hat{\mathbf{e}}_q \varepsilon_{qip} \varepsilon_{jkp} A_i \frac{\partial B_k}{\partial x_j} = \hat{\mathbf{e}}_q (\delta_{qj} \delta_{ik} - \delta_{qk} \delta_{ij}) A_i \frac{\partial B_k}{\partial x_j} = \nabla\mathbf{B} \cdot \mathbf{A} - \mathbf{A} \cdot \nabla\mathbf{B}$

1.21 (a) $\nabla\mathbf{A} = \hat{\mathbf{e}}_r \hat{\mathbf{e}}_r \frac{\partial A_r}{\partial r} + \hat{\mathbf{e}}_r \hat{\mathbf{e}}_\theta \frac{\partial A_\theta}{\partial r} + \hat{\mathbf{e}}_\theta \hat{\mathbf{e}}_r \frac{1}{r} \left(\frac{\partial A_r}{\partial \theta} - A_\theta \right) + \hat{\mathbf{e}}_r \hat{\mathbf{e}}_z \frac{\partial A_z}{\partial r} + \hat{\mathbf{e}}_z \hat{\mathbf{e}}_r \frac{\partial A_r}{\partial z}$

$+ \hat{\mathbf{e}}_\theta \hat{\mathbf{e}}_\theta \frac{1}{r} \left(A_r + \frac{\partial A_\theta}{\partial \theta} \right) + \frac{1}{r} \hat{\mathbf{e}}_\theta \hat{\mathbf{e}}_z \frac{\partial A_z}{\partial \theta} + \hat{\mathbf{e}}_z \hat{\mathbf{e}}_\theta \frac{\partial A_\theta}{\partial z} + \hat{\mathbf{e}}_z \hat{\mathbf{e}}_z \frac{\partial A_z}{\partial z}$

(b) $\nabla\mathbf{A} = \frac{\partial A_R}{\partial R} \hat{\mathbf{e}}_R \hat{\mathbf{e}}_R + \frac{\partial A_\phi}{\partial R} \hat{\mathbf{e}}_R \hat{\mathbf{e}}_\phi + \frac{1}{R} \left(\frac{\partial A_R}{\partial \phi} - A_\phi \right) \hat{\mathbf{e}}_\phi \hat{\mathbf{e}}_R + \frac{\partial A_\theta}{\partial R} \hat{\mathbf{e}}_R \hat{\mathbf{e}}_\theta$

$+ \frac{1}{R \sin\phi} \left(\frac{\partial A_R}{\partial \theta} - A_\theta \sin\phi \right) \hat{\mathbf{e}}_\theta \hat{\mathbf{e}}_R + \frac{1}{R} \left(A_R + \frac{\partial A_\phi}{\partial \phi} \right) \hat{\mathbf{e}}_\phi \hat{\mathbf{e}}_\phi + \frac{1}{R} \frac{\partial A_\theta}{\partial \phi} \hat{\mathbf{e}}_\phi \hat{\mathbf{e}}_\theta$

$+ \frac{1}{R \sin\phi} \left(\frac{\partial A_\phi}{\partial \theta} - A_\theta \cos\phi \right) \hat{\mathbf{e}}_\theta \hat{\mathbf{e}}_\phi + \frac{1}{R \sin\phi} \left(A_R \sin\phi + A_\phi \cos\phi + \frac{\partial A_\theta}{\partial \theta} \right) \hat{\mathbf{e}}_\theta \hat{\mathbf{e}}_\theta$

1.22 Follows from the gradient theorem, Eq. (1.2.55).

1.23 $3V$, where V is the volume.

1.25 $\int_\Omega \nabla \cdot (\phi \, \nabla\psi) \, d\Omega = \oint_\Gamma \hat{\mathbf{n}} \cdot (\phi \, \nabla\psi) d\Gamma = \oint_\Gamma \phi \, \frac{\partial \psi}{\partial n} \, d\Gamma$

1.26 $\bar{\mathbf{T}} = \mathbf{A}\mathbf{T}\mathbf{A}^T = \mathbf{A} \begin{bmatrix} 1 & 0 & -1 \\ 0 & 3 & -2 \\ -1 & -2 & 0 \end{bmatrix} \mathbf{A}^T$

1.27 Let $\mathbf{A}_1 = a_{1i}\hat{\mathbf{e}}_i$, $\mathbf{A}_2 = a_{2i}\hat{\mathbf{e}}_i$, $\mathbf{A}_3 = a_{3i}\hat{\mathbf{e}}_i$, $|A| = \mathbf{A}_1 \cdot (\mathbf{A}_2 \times \mathbf{A}_3) = a_{1r}a_{2s}a_{3t}\varepsilon_{rst}$;
$\mathbf{A}_j = a_{ji}\hat{\mathbf{e}}_i$ and $\mathbf{A}_i \cdot (\mathbf{A}_j \times \mathbf{A}_k) = a_{ir}\hat{\mathbf{e}}_r \cdot (a_{js}\hat{\mathbf{e}}_s \times a_{kt}\hat{\mathbf{e}}_t) = \varepsilon_{rst}a_{ir}a_{js}a_{kt}$

$\det[(\sigma_{ij} - \lambda\delta_{ij})] = \det(\sigma_{ij}) - \frac{\lambda}{6}[\sigma_{js}\sigma_{kt}\varepsilon_{ijk}\varepsilon_{ist} + \sigma_{kt}\sigma_{ir}\varepsilon_{ijk}\varepsilon_{rjt} + \sigma_{ir}\sigma_{js}\varepsilon_{ijk}\varepsilon_{rsk}]$
$+ \lambda^2[\frac{1}{6}\varepsilon_{ijk}\varepsilon_{ijt}\sigma_{kt} + \frac{1}{6}\varepsilon_{ijk}\varepsilon_{isk}\sigma_{js} + \frac{1}{6}\varepsilon_{ijk}\varepsilon_{rjk}\sigma_{ir}] - \lambda^3\left(\frac{1}{6}\varepsilon_{ijk}\varepsilon_{ijk}\right)$
$= -\lambda^3 + \lambda^2\sigma_{ii} - \frac{\lambda}{2}(\sigma_{ii}\sigma_{kk} - \sigma_{jk}\sigma_{kj}) + \det(\sigma_{ij})$

1.28 (a) $\lambda_1 = 3$, $\lambda_2 = 2(1 + \sqrt{5})$, and $\lambda_3 = 2(1 - \sqrt{5})$; $\hat{\mathbf{A}}^{(1)} = (0, 0, 1)$,
$\hat{\mathbf{A}}^{(2)} = (-0.851, 0.526, 0)$, and $\hat{\mathbf{A}}^{(3)} = (0.526, 0.851, 0)$

(b) $\lambda_1 = 5$, $\lambda_2 = 4$, and $\lambda_3 = 1$. The eigenvector associated with eigenvalue $\lambda = 5$ is
$\hat{\mathbf{A}}^{(1)} = \pm\frac{1}{2}(1, -\sqrt{3}, 0)$

(c) $\lambda_1 = 4$, $\lambda_2 = 2$, and $\lambda_3 = 1$. The eigenvectors are $\hat{\mathbf{A}}^{(1)} = \pm\frac{1}{\sqrt{2}}(0, 1, -1)$,
$\hat{\mathbf{A}}^{(2)} = \pm\frac{1}{\sqrt{2}}(0, 1, 1)$, and $\hat{\mathbf{A}}^{(3)} = \pm(1, 0, 0)$.

(d) $\lambda_1 = 3$, $\lambda_2 = 2$, and $\lambda_3 = -1$. The eigenvectors are $\hat{\mathbf{A}}^{(1)} = \pm\frac{1}{\sqrt{2}}(1, 0, 1)$,
$\hat{\mathbf{A}}^{(2)} = \pm\frac{1}{\sqrt{3}}(-1, 1, 1)$, and $\hat{\mathbf{A}}^{(3)} = \pm\frac{1}{\sqrt{6}}(-1, -2, 1)$.

(e) $\lambda_1 = 11.824$, $\lambda_2 = 1.285$, and $\lambda_3 = -7.109$. The eigenvector associated with λ_1 is
$\hat{\mathbf{A}}^{(1)} = \pm(0.5239, 0.2462, 0.4396)$.

(f) $\lambda_1 = 3.247$, $\lambda_2 = 1.555$, and $\lambda_3 = 0.198$

1.29 (a) $I_1 = 7$, $I_2 = 4$, and $I_3 = -48$ (b) $I_1 = 10$, $I_2 = -29$, and $I_3 = 20$ (c) $I_1 = 7$, $I_2 = -14$,
and $I_3 = 8$ (d) $I_1 = 4$, $I_2 = -1$, and $I_3 = -6$ (e) $I_1 = 6$, $I_2 = 78$, and $I_3 = -108$ (f) $I_1 = 5$,
$I_2 = -6$, and $I_3 = 1$

1.30 (a) $\mathbf{t}_{\hat{n}} = 2(\hat{\mathbf{e}}_1 + 7\hat{\mathbf{e}}_2 + \hat{\mathbf{e}}_3)$, $|\mathbf{t}_{\hat{n}}| = \sqrt{204}$ ksi, $\theta = 120.89°$,
$\sigma_n = 7.33$ ksi, and $\sigma_s = 12.26$ ksi.

(b) $\mathbf{t}_{\hat{n}} = \frac{10}{3}(3\hat{\mathbf{e}}_1 + 5\hat{\mathbf{e}}_2 + 4\hat{\mathbf{e}}_3)$, $|\mathbf{t}_{\hat{n}}| = 23.57$ ksi, $\theta = 90.0°$,
$\sigma_n = 0$ ksi and $\sigma_s = 23.57$ ksi.

(c) $\mathbf{t}_{\hat{n}} = \frac{1}{3}\left[(8 + \sqrt{2})\hat{\mathbf{e}}_1 + 5 - (8 + \sqrt{2})\hat{\mathbf{e}}_2 + 4(1 + \sqrt{2})\hat{\mathbf{e}}_3\right]$.

1.32 $p(A) = [I] - 2[A] + [A][A] = \begin{bmatrix} 1 & 0 \\ 0 & 1 \end{bmatrix}$

1.33 (a) $p(\lambda) = \lambda^2 - 2\lambda + 3$ (b) $p(\lambda) = (1 - \lambda)(1 - 3\lambda + \lambda^2)$

1.34 $\phi(\lambda) = \lambda^3 - 8\lambda^2 + 18\lambda - 12 = 0$, $\mathbf{S}^{-1} = \frac{1}{12}\begin{bmatrix} 7 & -2 & 1 \\ -2 & 4 & -2 \\ 1 & -2 & 7 \end{bmatrix}$

Chapter 2

2.2 The result follows from Eq. (3) of **Problem 2.1** when Q is replaced by ρ and noting that
$\mathbf{v}_s = 0$.

2.4 This follows from Eq. (2) of **Problem 2.1** with $Q = \rho\phi(\mathbf{r}, t)$

2.6 $\nabla\left(\frac{v^2}{2}\right) - \mathbf{v} \times \nabla \times \mathbf{v} = \hat{\mathbf{e}}_i v_j \frac{\partial v_j}{\partial x_i} - \varepsilon_{rst}\varepsilon_{kit}\hat{\mathbf{e}}_k v_i \frac{\partial v_s}{\partial x_r} = \hat{\mathbf{e}}_i v_j \frac{\partial v_j}{\partial x_i} - \hat{\mathbf{e}}_k\left(v_s \frac{\partial v_s}{\partial x_k} - v_i \frac{\partial v_k}{\partial x_i}\right)$
$= v_i \frac{\partial}{\partial x_i}(v_k \hat{\mathbf{e}}_k) = \mathbf{v} \cdot \nabla\mathbf{v}$

2.7 Follows from the product rule of differentiation.

2.8 Use the following identites, which were established in **Problem 1.20**:

$$\nabla \cdot (\nabla \times \mathbf{A}) = 0. \tag{1}$$
$$(\nabla \times \mathbf{A}) \times \mathbf{A} = \mathbf{A} \cdot \nabla\mathbf{A} - \nabla\mathbf{A} \cdot \mathbf{A}. \tag{2}$$
$$\nabla \times (\mathbf{A} \times \mathbf{B}) = \mathbf{B} \cdot \nabla\mathbf{A} - \mathbf{A} \cdot \nabla\mathbf{B} + \mathbf{A}\nabla \cdot \mathbf{B} - \mathbf{B}\nabla \cdot \mathbf{A}. \tag{3}$$

2.9 Follows directly from Eq. (3) of **Problem 2.1** by replacing Q with $\rho\mathbf{v}$

2.10 Follows directly from Eq. (1) of **Problem 2.4** and the divergence theorem.

2.11 $\rho\mathbf{f} = \rho\left(\frac{\partial\mathbf{v}}{\partial t} + \mathbf{v} \cdot \nabla\mathbf{v}\right) - \nabla \cdot \boldsymbol{\sigma} = \frac{\partial}{\partial t}(\rho\mathbf{v}) + \mathbf{v}[\nabla \cdot (\rho\mathbf{v})] + \rho\mathbf{v} \cdot \nabla\mathbf{v} - \nabla \cdot \boldsymbol{\sigma} = \frac{\partial}{\partial t}(\rho\mathbf{v}) + \mathrm{div}\,(\rho\mathbf{v}\mathbf{v} - \boldsymbol{\sigma})$

2.12 Follows directly from **Problem 2.7** and **Problem 2.10**.

2.13 $\int_\Omega \rho \frac{D}{Dt}\left(e + \frac{v^2}{2}\right) d\Omega = \int_\Omega \nabla \cdot (\boldsymbol{\sigma} \cdot \mathbf{v}) \, d\Omega + \int_\Omega \rho \mathbf{f} \cdot \mathbf{v} \, d\Omega - \int_\Omega \nabla \cdot \mathbf{q} \, d\Omega$

$\rho \frac{D}{Dt}\left(e + \frac{v^2}{2}\right)$

$= \text{div}\ (\boldsymbol{\sigma} \cdot \mathbf{v}) + \rho \frac{D}{Dt}\left(\frac{v^2}{2}\right) - \mathbf{v} \cdot \text{div}\boldsymbol{\sigma} - \text{div}\ \mathbf{q}.$

2.14 $\nabla \cdot (\boldsymbol{\sigma} \cdot \mathbf{v}) = \frac{\partial}{\partial x_i}(\sigma_{ik} v_k) = \nabla \cdot \boldsymbol{\sigma} \cdot \mathbf{v} + \boldsymbol{\sigma} : (\nabla \mathbf{v})^{\mathrm{T}}$

2.15 $\sigma_n = \frac{\sigma_{11}+\sigma_{22}}{2} + \frac{\sigma_{11}-\sigma_{22}}{2}\cos 2\theta + \sigma_{12}\sin 2\theta$ and $\sigma_s = -\frac{\sigma_{11}-\sigma_{22}}{2}\sin 2\theta + \sigma_{12}\cos 2\theta$

2.16 (a) The stress field is not possible unless $c_1 = 0$ (b) The stress field is not possible
(c) The stress field is in equilibrium

2.17 $f_1 = 0$, $f_2 = 0$, and $f_3 = -4$

2.18 $\sigma_{13} = -\frac{F_0}{2I_2}\left(h^2 - x_3^2\right)$ and $\sigma_{33} = 0$ $\left(I_2 = \frac{2bh^3}{3}\right)$

2.19 $\sigma_{13} = -\frac{q_0 x_1}{2I_2}\left(h^2 - x_3^2\right)$ and $\sigma_{33} = \frac{q_0 x_3}{6I_2}\left(3h^2 - x_3^2\right) - \frac{q_0}{2b}$.

2.20 $[\sigma]_{(1,1,3)} = \begin{bmatrix} -2 & 0 & -1 \\ 0 & 16 & 0 \\ -1 & 0 & 8 \end{bmatrix}$ psi, $\left\{\begin{matrix} t_1 \\ t_2 \\ t_3 \end{matrix}\right\}_{(1,1,3)} = \frac{1}{\sqrt{3}}\left\{\begin{matrix} -3 \\ -16 \\ 7 \end{matrix}\right\}$, $t_n = \frac{20}{3}$ psi, $t_s = 9.672$ psi

2.21 $\lambda_1 = 6.856$, $\lambda_2 = -10.533$, and $\lambda_3 = -3.323$; $\hat{\mathbf{A}}^{(1)} = \pm(0.42, 0.0498, -0.905)$,
$\hat{\mathbf{A}}^{(2)} = \pm(0.257, -0.964, 0.066)$, and $\hat{\mathbf{A}}^{(3)} = \pm(0.870, 0.261, 0.418)$.

2.22 $\lambda_1 = 11.824$ (psi), $\lambda_2 = -7.109$ (psi), $\lambda_3 = 1.285$ (psi). The plane of maximum stress is
$\hat{\mathbf{n}}^{(1)} = \pm(0.730, 0.337, 0.594)$.

2.23 $e_{11} = 2(1 - X_1)X_2$, $2e_{12} = (3c_2 + c_3)X_2^2 - 2c_1$, and $e_{22} = -2c_3(1 - X_1)X_2$.

2.24 $e_{11} = 3X_1^2 X_2 + c_1\left(2c_2^3 + 3c_2^2 X_2 - X_2^3\right)$, $e_{12} = \frac{1}{2}\left[X_1^3 + 3\left(c_1 c_2^2 - 2c_1 X_2^2\right)X_1\right]$, and
$e_{22} = X_2^3 - 3X_2\left(c_2^2 + c_1 X_2^2\right) - 2c_2^3$.
The Green–Lagrange (nonlinear) strain components $E_{ij} \equiv \varepsilon_{ij}$ are
$\varepsilon_{11} = 3X_1^2 X_2 + c_1\left(2c_2^3 + 3c_2^2 X_2 - X_2^3\right) + \frac{1}{2}\left[3X_1^2 X_2 + c_1\left(2c_2^3 + 3c_2^2 X_2 - X_2^3\right)\right]^2$
$+ \frac{1}{2}\left(-3c_1 X_1 X_2^2\right)^2$,
$\varepsilon_{22} = X_2^3 - 3X_2\left(c_2^2 + c_1 X_2^2\right) - 2c_2^3 + \frac{1}{2}\left[X_1^3 + 3c_1\left(c_2^2 - X_2^2\right)X_1\right]^2$
$+ \frac{1}{2}\left[X_2^3 - 3X_2\left(c_2^2 + c_1 X_2^2\right) - 2c_2^3\right]^2$,
$\varepsilon_{12} = \frac{1}{2}\left[X_1^3 + \left(3c_1 c_2^2 - 6c_1 X_2^2\right)X_1\right] + \frac{1}{2}\left[3X_1^2 X_2 + c_1\left(2c_2^3 + 3c_2^2 X_2 - X_2^3\right)\right]$
$\times \left[X_1^3 + c_1\left(3c_2^2 - 3X_2^2\right)X_1\right] + \frac{1}{2}\left(-3c_1 X_1 X_2^2\right)\left[X_2^3 - 3X_2\left(c_2^2 + c_1 X_2^2\right) - 2c_2^3\right].$

2.25 $x_1 = X_1 + \frac{e_0}{b}X_2$, $x_2 = X_2$, and $x_3 = X_3$; $u_1 = \frac{e_0}{b}X_2$, $u_2 = 0$, and $u_3 = 0$; $2e_{12} = \frac{e_0}{b}$; $\varepsilon_{11} = 0$,
$\varepsilon_{22} = \frac{1}{2}\left(\frac{e_0}{b}\right)^2$, $\varepsilon_{12} = \frac{1}{2}\frac{e_0}{b}$.

2.26 $x_1 = X_1 + (e_0/b^2)X_2^2$, $x_2 = X_2$, and $x_3 = X_3$; $u_1 = kX_2^2$, $u_2 = 0$, and $u_3 = 0$; $2e_{12} = 2kX_2$;
$\varepsilon_{11} = 0$, $\varepsilon_{22} = 2k^2 X_2^2$, and $2\varepsilon_{12} = 2kX_2$.

2.27 $e_{11} = \frac{\partial u_1^0}{\partial x_1} + x_3\frac{\partial \phi_1}{\partial x_1}$, $e_{22} = \frac{\partial u_2^0}{\partial x_2} + x_3\frac{\partial \phi_2}{\partial x_2}$, $2e_{12} = \frac{\partial u_1^0}{\partial x_2} + \frac{\partial u_2^0}{\partial x_1} + x_3\left(\frac{\partial \phi_1}{\partial x_2} + \frac{\partial \phi_2}{\partial x_1}\right)$, $e_{33} = \frac{\partial u_3}{\partial x_3} = 0$,
$2e_{13} = \frac{\partial u_1}{\partial x_3} + \frac{\partial u_3}{\partial x_1} = \phi_1 + \frac{\partial u_3^0}{\partial x_1}$, $2e_{23} = \frac{\partial u_2}{\partial x_3} + \frac{\partial u_3}{\partial x_2} = \phi_2 + \frac{\partial u_3^0}{\partial x_2}$.

2.28 (a) $2 + 0 = 2$ (yes)   (b) $2x_3 + 0 = 2(2x_3)$ (no)

2.29 $e_{11} = 0$, $e_{12} = \frac{1}{2}\frac{e_0}{b}$, $e_{22} = \frac{1}{2}\left(\frac{e_0}{b}\right)^2$, $e'_{11}(= e'_n) = \frac{e_0}{2b}$, $e'_{12}(= e'_s) = \frac{e_0}{2b}\left(\frac{a^2 - b^2}{a^2 + b^2}\right) = 0$ (for a square block)

2.30 $\lambda_1 = 0$ and $\lambda_2 = 10 \times 10^{-4}$ (in/in). $\hat{\mathbf{A}}^{(2)} = \frac{1}{\sqrt{5}}(2, 1)$ and $\theta = 26.57°$ (clockwise).

2.31 $e_n^{AD} \approx \frac{1}{2}\left(\frac{e_0^2 + 2e_0 a}{a^2 + b^2}\right)$

2.32 $\hat{\mathbf{e}}_r$: $\quad \frac{\partial \sigma_{rr}}{\partial r} + \frac{\sigma_{rr} - \sigma_{\theta\theta}}{r} + \frac{1}{r}\frac{\partial \sigma_{r\theta}}{\partial \theta} + \frac{\partial \sigma_{rz}}{\partial z} + \rho_0 f_r = \rho_0 \frac{\partial^2 u_r}{\partial t^2}$,

$\hat{\mathbf{e}}_\theta$: $\quad \frac{\partial \sigma_{r\theta}}{\partial r} + \frac{2\sigma_{r\theta}}{r} + \frac{1}{r}\frac{\partial \sigma_{\theta\theta}}{\partial \theta} + \frac{\partial \sigma_{\theta z}}{\partial z} + \rho_0 f_\theta = \rho_0 \frac{\partial^2 u_\theta}{\partial t^2}$,

$\hat{\mathbf{e}}_z$: $\quad \frac{\partial \sigma_{rz}}{\partial r} + \frac{\sigma_{rz}}{r} + \frac{1}{r}\frac{\partial \sigma_{\theta z}}{\partial \theta} + \frac{\partial \sigma_{zz}}{\partial z} + \rho_0 f_z = \rho_0 \frac{\partial^2 u_z}{\partial t^2}$.

2.33 $\varepsilon_{rr} = \frac{\partial u_r}{\partial r}, \ \varepsilon_{\theta\theta} = \frac{u_r}{r}, \ \varepsilon_{rz} = \frac{1}{2}\left(\frac{\partial u_r}{\partial z} + \frac{\partial u_z}{\partial r}\right), \ \varepsilon_{zz} = \frac{\partial u_z}{\partial z}$

2.35 $0 = \frac{\partial^2 u_i}{\partial x_j \partial x_k}\varepsilon_{jk\ell} = \varepsilon_{jk\ell}\frac{\partial}{\partial x_k}\left(\frac{\partial u_i}{\partial x_j}\right) = \varepsilon_{jk\ell}\frac{\partial}{\partial x_k}\left(e_{ij} + \omega_{ij}\right)$. Differentiate with respect to x_m and
and multiply with ε_{imn} to obtain $0 = \varepsilon_{jk\ell}\varepsilon_{imn}e_{ij,km}$.

2.36 The body is in equilibrium for the body force components $\rho_0\, f_1 = 0$, $\rho_0\, f_2 = 0$, and $\rho_0\, f_3 = -4$.

2.37 $\frac{dT}{dx} = 0, \quad \frac{d}{dx}\left(T\frac{du}{dx}\right) + f = 0$

2.38 Begin with the relation $\sigma_{ij} = \frac{\partial U}{\partial \varepsilon_{ij}}$ to obtain the required result.

2.39 $\sigma_{ij} = 2\mu\,\varepsilon_{ij} + \lambda\,\varepsilon_{kk}\,\delta_{ij} - \alpha(2\mu + 3\lambda)\Delta T \delta_{ij}$, where $\Delta T = T - T_0$.

2.40 $\varepsilon_{ij} = \frac{1+\nu}{E}\sigma_{ij} - \frac{\nu}{E}\sigma_{kk}\,\delta_{ij} + \alpha\Delta T\,\delta_{ij}$

2.41 The relations follow directly from Eq. (4) of **Problem 2.39**.

Chapter 3

3.1 $V = (m_1 + m_2)gL_1(1 - \cos\theta_1) + m_2 gL_2(1 - \cos\theta_2)$

3.2 $V = \frac{WL}{4}\left(5\theta_1^2 + 3\theta_2^2 + \theta_3^2\right) + \frac{1}{2}\left[(k_1 + k_2)\theta_1^2 + (k_2 + k_3)\theta_2^2 + k_3\theta_3^2 - 2k_2\theta_1\theta_2 - 2k_3\theta_2\theta_3\right]$

3.3 $\Delta W^* = (u + v\theta)\Delta F_x + [v - (L + u)\theta]\,\Delta F_y$

3.4 $U = \int_0^L \frac{EA}{2}\left(\frac{du}{dx}\right)^2 dx + \frac{1}{2}k[u(L)]^2$ and $V = -\left[\int_0^L f(x)u\ dx + Pu(L)\right]$

3.5 $U = \int_0^L \frac{EI}{2}\left(\frac{d^2w}{dx^2}\right)^2 dx + \frac{1}{2}k[w(L)]^2$ and $V = -\int_0^L q_0 w\, dx$.

3.6 $U^* = \frac{1}{2EI}\left(\frac{3L}{2}M_0^2 + \frac{27L^3}{24}R_A^2 + \frac{9L^2}{4}M_0 R_A + \frac{1}{3}F_0^2 L^3 - M_0 F_0 L^2 - \frac{7}{6}R_A F_0 L^3\right)$
$\quad + \frac{f_s(1+\nu)}{EA}\left(\frac{3L}{2}R_A^2 + LF_0^2 - 2R_A F_0 L\right)$

3.7 $U = \int_0^L \frac{EI}{2}\left(\frac{d^2w}{dx^2}\right)^2 dx + \frac{k}{n+1}[w(L)]^{n+1}$ and $V = -\int_0^L q_0 w\, dx$

3.8 $U^* = \frac{1}{2EI}\left(\frac{P^2 b^3}{3} + \frac{q_0^2 b^5}{20} + \frac{Pq_0 b^4}{4}\right) + \frac{1}{2EI}\left(P + q_0 b\right)^2 \frac{a^3}{3}$
$\quad + \frac{1}{2GJ}\left(Pb + \frac{q_0 b^2}{2}\right)^2 a$

3.9 $U^* = A\sum_{i=1}^5 L_i U_{0i}^* = \frac{A}{3K^2}\left[2 \times 5 \times \left(\frac{5P}{8A}\right)^3 + 2 \times 3 \times \left(\frac{3P}{8A}\right)^3 + 4\left(\frac{P}{A}\right)^3\right]$

3.10 (a) $U = \frac{19}{25}AKv\sqrt{v}$ (b) $U^* = \left(\frac{50}{57}\right)^2\frac{P^3}{3K^2 A^2}$.

3.13 $\delta P + \delta R_A + \delta R_C = 0$

3.16 $\delta W = \int_0^L EA\frac{d\delta u}{dx}\frac{du}{dx}dx + ku(L)\delta u(L) - \left[\int_0^L f(x)\delta u\ dx + P\delta u(L)\right]$

3.17 $\delta W = \int_0^L EI\frac{d^2 w}{dx^2}\frac{d^2\delta w}{dx^2}dx + kw(L)\delta w(L) - \int_0^L q_0\delta w\, dx$

3.18 $\delta W^* = (v - \theta L\cos\theta)\,\delta F_y + (u + \theta L\sin\theta)\,\delta F_x$

3.19 $\delta W^* = \int_0^L \frac{N}{EA}\delta N\ dx - \delta P_B u(L) + \frac{1}{k}\delta F_s F_s$

3.20 $\delta W^* = \int_0^L \frac{M}{EI}\delta M\ dx - \delta F_B w(L)$

3.22 $\delta W_E^* = -(u_D\delta Q + v_D\delta P)$

3.23 $\delta W^* = L_1\left(F_x\cos\theta_1 + F_y\sin\theta_1\right)\delta\theta_1 + L_2\left(F_x\cos\theta_2 + F_y\sin\theta_2\right)\delta\theta_2$

3.24 $\delta W^* = \delta F_3\left[(v_1 - v_2)\frac{b}{a} + v_3 - v_2\right] + \delta P(v_2 - v) + \delta M\left[\frac{v_2 - v_1}{a} - \theta\right]$

3.25 $\delta U = \int_0^L \left[N_{xx}\left(\frac{d\delta u}{dx} + \frac{dw}{dx}\frac{d\delta w}{dx}\right) - M_{xx}\frac{d^2\delta w}{dx^2}\right]dx$

3.26 $0 = \int_0^L \left(N\delta\varepsilon_{xx}^{(0)} + M\delta\varepsilon_{xx}^{(1)} + V\delta\gamma_{xz}^{(0)}\right)dx$

3.28 $\delta I = \int_a^b \frac{u'}{\sqrt{1 + (u')^2}}\delta u'\ dx$

3.30 $\delta I = \int_0^1 \frac{1}{2} \sqrt{\frac{u}{1+(u')^2}} \left(\frac{2u'\delta u'}{u} - \frac{1+(u')^2}{u^2} \delta u \right) dx$

3.33 $\delta_u I = \int_\Omega \left(u_{,x}\, \delta u_{,x} + \delta u_{,x}\, v_{,x} + \delta u_{,y}\, v_{,y} + f\delta u \right) dxdy$ and

$\delta_v I = \int_\Omega \left(u_{,x}\, \delta v_{,x} + u_{,y}\, \delta v_{,y} + v_{,y}\, \delta v_{,y} + g\delta v \right) dxdy.$

3.35 $\delta_u I = \int_\Omega \left[\mu \left(\delta u_{i,j}\, u_{i,j} + u_{j,i}\, \delta u_{i,j} \right) - P\, \delta u_{i,i} \right] d\Omega - \int_\Omega f_i \delta u_i\, d\Omega - \int_{\Gamma_\sigma} \hat{t}_i \delta u_i\, d\Gamma,$

$\delta_P I = \int_\Omega \delta P\, u_{i,i}\, d\Omega$

3.37 $T = \frac{1}{\sqrt{2g}} \int_0^{x_b} \sqrt{\frac{1+(du/dx)^2}{u}}\, dx \equiv I_1(u)$

3.39 $I_4(y) = \int_a^b y(x)\, dx$

3.40 $u'''' - u'' - 1 = 0,$ in $a < x < b$, and $u' - u''' = 0,$ $u'' + u' = 0,$ at $x = a, b$

3.41 $-\frac{d}{dx} \left\{ EA \left[\frac{du}{dx} + \frac{1}{2} \left(\frac{dw}{dx} \right)^2 \right] \right\} = 0,$

$\frac{d}{dx} \left\{ EA \frac{dw}{dx} \left[\frac{du}{dx} + \frac{1}{2} \left(\frac{dw}{dx} \right)^2 \right] \right\} - \frac{d^2}{dx^2} \left(EI \frac{d^2 w}{dx^2} \right) = q$

3.43 $\mu \left(-u_{i,jj} - u_{j,ij} \right) - P_{,i} - f_i = 0$ in Ω and $\mu \left(u_{i,j} + u_{j,i} \right) n_j - Pn_i - \hat{t}_i = 0$ on Γ_σ

3.44 $D \left(\frac{\partial^4 w}{\partial x^4} + \frac{\partial^4 w}{\partial y^4} + 2 \frac{\partial^4 w}{\partial x^2 \partial y^2} \right) + kw - q = 0$

3.45 $\frac{d^2}{dr^2} \left(D_{11} r \frac{d^2 w}{dr^2} \right) - \frac{d}{dr} \left(D_{22} \frac{1}{r} \frac{dw}{dr} \right) = 0$ in $r_i < r < r_0$

3.46 $\delta w: \quad \frac{\partial \lambda_x}{\partial x} + \frac{\partial \lambda_y}{\partial y} + q = 0$

$\delta \phi_x: \quad D \frac{\partial^2 \phi_x}{\partial x^2} + \nu D \frac{\partial^2 \phi_y}{\partial x \partial y} + (1-\nu)D \frac{\partial}{\partial y} \left(\frac{\partial \phi_x}{\partial y} + \frac{\partial \phi_y}{\partial x} \right) - \lambda_x = 0$

$\delta \phi_y: \quad D \frac{\partial^2 \phi_y}{\partial y^2} + \nu D \frac{\partial^2 \phi_x}{\partial x \partial y} + (1-\nu)D \frac{\partial}{\partial x} \left(\frac{\partial \phi_y}{\partial y} + \frac{\partial \phi_x}{\partial x} \right) - \lambda_y = 0$

$\delta \lambda_x: \quad \frac{\partial w}{\partial x} + \phi_x = 0, \qquad \delta \lambda_y: \quad \frac{\partial w}{\partial y} + \phi_y = 0$

3.48 (a) $\delta u: \quad 1 + \frac{d}{dx} \left[\frac{\lambda \frac{du}{dx}}{\sqrt{1+\left(\frac{du}{dx}\right)^2}} \right] = 0, \quad \delta \lambda: \quad \int_a^b \sqrt{1+\left(\frac{du}{dx}\right)^2}\, dx - L = 0.$

(b) $1 + \gamma \frac{d}{dx} \left\{ \left[\int_a^b \sqrt{1+\left(\frac{du}{dx}\right)^2}\, dx - L \right] \frac{\frac{du}{dx}}{\sqrt{1+\left(\frac{du}{dx}\right)^2}} \right\} = 0.$

3.50 $\delta \varepsilon_{xx}: \quad EI\varepsilon_{xx} + \lambda = 0, \quad \delta \lambda: \quad \varepsilon_{xx} - \frac{d^2 w}{dx^2} = 0,$ and $\delta w: \quad -\frac{d^2 \lambda}{dx^2} - q = 0$

Chapter 4

4.1 $-\frac{d}{dx} \left(EA \frac{du}{dx} \right) - f(x) = 0, \quad 0 < x < L, \quad \left[EA \frac{du}{dx} \right]_{x=L} + ku(L) - P = 0.$

4.2 $\frac{d^2}{dx^2} \left(EI \frac{d^2 w}{dx^2} \right) - q_0 = 0, \quad 0 < x < L$

4.3 $-\frac{dN_{xx}}{dx} = f, \quad -\frac{dQ_{xz}}{dx} - \frac{d}{dx} \left(N_{xx} \frac{dw}{dx} \right) = q, \quad -\frac{dM_{xx}}{dx} + Q_{xz} = 0$

4.4 $\delta u: \quad -\frac{dN_{xx}}{dx} = f(x), \quad \delta w: \quad -\frac{d}{dx} \left(\bar{Q}_{xz} + N_{xx} \frac{dw}{dx} \right) - \frac{d^2 P_{xx}}{dx^2} = q(x),$

$\delta \phi_x: \quad -\frac{d\bar{M}_{xx}}{dx} + \bar{Q}_{xz} = 0$

4.5 $v = \left(\frac{50P}{57AE} \right)^2$

4.6 $\theta = \frac{3a}{11AEL^2} M_0, \quad \delta_B = \frac{3a}{11AEL} M_0, \quad P_B = \frac{AE}{a} \delta_B = \frac{3M_0}{11L}, \quad P_C = \frac{AE}{1.5L} \delta_C = \frac{4M_0}{11L}$

4.7 $\Delta_4 = -\frac{q_0 L^3}{48EI} + \frac{M_0 L}{4EI}, \quad F_1 = \frac{5q_0 L}{8} - \frac{3M_0}{2L}, \quad F_2 = -\frac{q_0 L^2}{8} + \frac{M_0}{2}, \quad F_3 = \frac{3q_0 L}{8} + \frac{3M_0}{2L}$

4.8 $u = 0$ and $v = \frac{2595}{32} \frac{P^2}{A^2 K^2}$

4.9 (a) $w_c = \frac{5q_0 L^4}{384EI}$

4.10 $R_B = \frac{3M_0}{2L} + \frac{3q_0 L}{8}$, $\theta_B = \frac{1}{EI}\left(\frac{M_0 L}{4} - \frac{q_0 L^3}{48}\right)$

4.11 $w_c - \frac{w_e}{2} = -\left(\frac{25F_0 L^3 + q_0 L^4}{768EI}\right)$ (one of many relations between w_c and w_e that can be obtained depending on the choice of the virtual forces) and $w_e = w(0) = \frac{81F_0 L^3}{384EI} + \frac{7q_0 L^4}{384EI}$ and $w_c = w(L/2) = \frac{28F_0 L^3}{384EI} + \frac{3q_0 L^4}{384EI}$.

4.12 $w_B = (5F_0 a^3 / 6EI)$.

4.13 $F_C = \left(\frac{L_c}{E_c A_c} + \frac{L_b}{E_b A_b}\cos^2\alpha + \frac{L_b^3}{3E_b I_b}\sin^2\alpha\right)^{-1}\left[\frac{5L_b^3}{48E_b I_b}\sin\alpha\right]F_0$

4.14 $\theta_B = \frac{170.67}{EI}$ (clockwise)

4.15 $v = \frac{1}{6EI}\left[P_v\left(2b^3 + 6ab^2\right) + 3P_h a^2 b\right] + \frac{P_v a}{EA}$ and $u = \frac{1}{6EI}\left(3P_v a^2 b + 2P_h a^3\right) + \frac{P_h b}{EA}$

4.16 $w_A = \frac{1}{EI}\left[\frac{Qa^3}{3} + \frac{(P+Q)b^3}{3}\right] + \frac{1}{GJ}\left(Qa^2 b\right)$

4.17 $u = -3\frac{P^2 L}{K^2 A^2}$ and $v = \frac{25}{\sqrt{3}}\frac{P^2 L}{K^2 A^2}$

4.18 $P_B = \frac{3M_0}{11L}$ and $P_C = \frac{4M_0}{11L}$

4.19 $w(L) = \left(\frac{q_0 L^4}{30EI} + \frac{q_0 f_s L^2}{6GA}\right)\left(1 + \frac{f_s kL}{GA} + \frac{kL^3}{3EI}\right)^{-1}$

4.20 $v = \frac{PR^3\pi}{4EI}$ and $u = \frac{PR^3}{2EI}$

4.21 $u_B = \frac{FR^3\pi}{2EI}$ and $w_B = \frac{2FR^3}{EI}$

4.22 $u_B = \frac{2pR^4}{EI}$ and $w_B = \frac{3\pi pR^4}{2EI}$

4.23 $\phi = 16(2+\nu)\frac{RT}{Ed^4}$

4.24 $w_B = \frac{PR\pi}{4EA} + \frac{PR^3\pi}{4EI} + \frac{f_s PR\pi}{4GA}$

4.25 $w_A = \frac{q_0 L^4}{8EI}\left(1 + \frac{kL^3}{3EI}\right)^{-1} = w(0)$ and $\theta_A = -\frac{q_0 L^3}{6EI}\left[\frac{3-(kL^3/EI)}{3+(8kL^3/EI)}\right]$

4.26 $w_A = \frac{1}{EI}\left(\frac{F_0 L^3}{3} + \frac{q_0 L^4}{8}\right) + \frac{f_s}{GA}\left(F_0 L + \frac{q_0 L^2}{2}\right)$ (with shear)

4.27 $w_s = -\left[\frac{F_0(L^3 - b^3)}{3EI} - \frac{F_0 b(L^2 - a^2)}{2EI} + \frac{F_0 f_s a}{GA}\right]\left(1 + \frac{kL^3}{3EI} + \frac{kLf_s}{GA}\right)^{-1}$

4.28 $v_C = 11.657\frac{PL}{EA}$ and $u_C = 2\frac{PL}{EA}$.

4.29 $D_y = 0.44P$ and $A_y = 0.56P$

4.30 $w(L) = \frac{F_0 L^3}{3EI} + f_s\frac{F_0 L}{GA}$

4.32 $w_{CA} = \frac{M_0 L^2}{16EI}$ and $\theta_{AC} = \frac{F_0 L^2}{16EI}$

4.33 $\theta_{AA} = \frac{M_0 L}{3EI}$ and $\theta_A = \frac{M_0 L}{3EI} + \frac{F_0 L^2}{16EI}$

4.35 $w_c = -\left(\frac{5F_0 L^3}{48EI} + \frac{17q_0 L^4}{384EI}\right)$

4.36 $w_A = 0.65625$ in

Chapter 5

5.1 $L = -\frac{1}{2}\left[k_1 x_1^2 + (k_2 + k_3)(x_2 - x_1)^2 + k_4(x_3 - x_2)^2\right] + Fx_3$

5.3 $\delta\dot{L}^* = -\dot{x}_s\delta F_s - \dot{x}_d\delta F_d + \dot{x}\delta F = \left(-\frac{\dot{F}}{k} - \frac{F}{\eta} + \dot{x}\right)\delta F$

5.4 $\delta\theta_1:\ -m_2 L_1 L_2\dot{\theta}_1\dot{\theta}_2\sin(\theta_1 - \theta_2) - (m_1 + m_2)gL_1\sin\theta_1$
$\quad - \frac{d}{dt}\left[(m_1 + m_2)L_1^2\dot{\theta}_1 + m_2 L_1 L_2\cos(\theta_1 - \theta_2)\dot{\theta}_2\right] = 0,$
$\delta\theta_2:\ \ m_2 L_1 L_2\dot{\theta}_1\dot{\theta}_2\sin(\theta_1 - \theta_2) - m_2 gL_2\sin\theta_2$
$\quad - m_2\frac{d}{dt}\left[L_2^2\dot{\theta}_2 + L_1 L_2\cos(\theta_1 - \theta_2)\dot{\theta}_1\right] = 0$

5.6 $0 = (m_1 + m_2)\ddot{x} - m_1\ell\left(\ddot{\theta}\sin\theta + \dot{\theta}^2\cos\theta\right) - (m_1 + m_2)g + k(x + h)$,

$\quad 0 = m_1\ell\left(\ell\ddot{\theta} - \ddot{x}\sin\theta\right) - m_1 g\ell\sin\theta$

5.7 $-(m_1 + m_2)\ddot{u}_1 + \frac{1}{\sqrt{2}}m_2\ddot{u}_2 = 0$ and $\frac{1}{\sqrt{2}}m_2 g - m_2\left(\ddot{u}_2 - \frac{1}{\sqrt{2}}\ddot{u}_1\right) = 0$

5.8 The linearized equations are $(M + m_1 + m_2)\ddot{x} + (m_1 + m_2)l_1\ddot{\theta}_1 + m_2\left(l_1\ddot{\theta}_1 + l_2\ddot{\theta}_2\right) = 0$,

$\quad - (m_1 + m_2)l_1\left(g\theta_1 + \ddot{x} + l_1\ddot{\theta}_1\right) - m_2 l_1 l_2\ddot{\theta}_2 = 0$,

$\quad - m_2 l_2\left(g\theta_2 + \ddot{x} + l_2\ddot{\theta}_2 + l_1\ddot{\theta}_1\right) = 0$.

5.9 $\ddot{x}_1 = \left(\frac{m_1 - m_2}{m_1 + m_2}\right)g$ and $\lambda = -\frac{2m_1 m_2 g}{m_1 + m_2}$

5.10 $\ddot{x}_1 = -\ddot{x}_2 = -\ddot{x}_3 = \left(\frac{m_1 - m_2 - m_3}{m_1 + m_2 + m_3}\right)g$, $\lambda_1 = -\frac{2m_1(m_2 + m_3)g}{m_1 + m_2 + m_3}$, and $\lambda_2 = \frac{2m_1 m_3 g}{m_1 + m_2 + m_3}$

5.11 $\ddot{x}_1 = \frac{g}{23}$

5.12 $m(\ddot{x} + \ell\ddot{\theta}\cos\theta - \ell\dot{\theta}^2\sin\theta) + kx = F$ and $m\left[\ell\ddot{x}\cos\theta + (\ell^2 + \Omega^2)\ddot{\theta}\right] + mg\ell\sin\theta = 2\ell F\cos\theta$.

5.14 $\delta u: -\frac{\partial N}{\partial x} - f + \frac{\partial}{\partial t}\left(m_0\frac{\partial u}{\partial t}\right) = 0$, $\quad \delta w: -\frac{\partial V}{\partial x} - \frac{\partial}{\partial x}\left(N\frac{\partial w}{\partial x}\right) - q + \frac{\partial}{\partial t}\left(m_0\frac{\partial w}{\partial t}\right) = 0$,

$\quad \delta\phi_x: -\frac{\partial M}{\partial x} + V + \frac{\partial}{\partial t}\left(m_2\frac{\partial\phi_x}{\partial t}\right) = 0$.

5.17 $\frac{\partial^2}{\partial x\partial t}\left(m_2\frac{\partial^2 w}{\partial x\partial t}\right) - \frac{\partial}{\partial t}\left(m_0\frac{\partial w}{\partial t}\right) + \frac{\partial^2 M_{xx}}{\partial x^2} - \frac{\partial^2}{\partial x^2}\left(Ic_s\frac{\partial^3 w}{\partial x^2\partial t}\right) - c\frac{\partial w}{\partial t} + q = 0$.

5.18 $\delta u: -m_0\ddot{u} + \frac{\partial N_{xx}}{\partial x} = 0$, $\quad \delta w: -m_0\ddot{w} + m_2\frac{\partial^2\ddot{w}}{\partial x^2} + \frac{\partial}{\partial x}\left(N_{xx}\frac{\partial w}{\partial x}\right) + \frac{\partial^2 M_{xx}}{\partial x^2} + q = 0$

5.19 $-\frac{\partial N_{xx}}{\partial x} - f + \frac{\partial}{\partial t}\left(m_0\frac{\partial u}{\partial t}\right) = 0$, $\quad -\frac{\partial Q_x}{\partial x} - \frac{\partial}{\partial x}\left(\frac{\partial w}{\partial x}N_{xx}\right) - q + \frac{\partial}{\partial t}\left(m_0\frac{\partial w}{\partial t}\right) = 0$,

$\quad -\frac{\partial M_{xx}}{\partial x} + Q_x + \frac{\partial}{\partial t}\left[m_2\left(c_1\frac{\partial^2 w}{\partial x\partial t} + c_2\frac{\partial\phi_x}{\partial t}\right)\right] = 0$

5.20 The linear strains are $\varepsilon_{xx} = \frac{\partial u}{\partial x} + z\frac{\partial\phi_x}{\partial x} + z^2\frac{\partial\psi_x}{\partial x}$, $2\varepsilon_{xz} = \phi_x + 2z\psi_x + \frac{\partial w}{\partial x}$.

5.22 $-\frac{1}{r}\left[\frac{\partial}{\partial r}(rN_{rr}) - N_{\theta\theta}\right] + m_0\frac{\partial^2 u}{\partial t^2} = 0$,

$\quad -\frac{1}{r}\frac{\partial}{\partial r}(rQ_r) - \frac{1}{r}\frac{\partial}{\partial r}\left(rN_{rr}\frac{\partial w}{\partial r}\right) - q + kw + m_0\frac{\partial^2 w}{\partial t^2} = 0$,

$\quad -\frac{1}{r}\left[\frac{\partial}{\partial r}(rM_{rr}) - M_{\theta\theta}\right] + Q_r + m_2\frac{\partial^2\phi_r}{\partial t^2} = 0$

5.24 $-\frac{\partial}{\partial x}\left(T_0\frac{\partial u}{\partial x}\right) - \frac{\partial}{\partial y}\left(T_0\frac{\partial u}{\partial y}\right) + \frac{\partial}{\partial t}\left(\rho\frac{\partial u}{\partial t}\right) = f(x, y, t)$

5.25 (a) $\omega_1 = \frac{9.8766}{L^2}\sqrt{\frac{EI}{\rho A}}$ (b) $\omega_1 = \frac{15.45}{L^2}\sqrt{\frac{EI}{\rho A}}$ (c) $\omega_1 = \frac{22.45}{L^2}\sqrt{\frac{EI}{\rho A}}$

Chapter 6

6.1 The set is linearly independent.

6.2 The set is linearly independent.

6.3 The set is linearly independent.

6.4 The set is linearly dependent.

6.5 The set is linearly independent.

6.6 (a) An inner product (b) Symmetric

6.7 $\|u\|_0 = \sqrt{\frac{33L^7}{35}}$

6.8 $\|u\|_1^2 = \frac{33L^7}{35} + \frac{24L^5}{5}$

6.9 $\|u\|_0^2 = \frac{8a^5}{15}\frac{8b^5}{15}$

6.10 $\|u - v\|_0 = \sqrt{\frac{29}{105}}$

6.11 Orthogonal

6.12 Not orthogonal

6.13 $a = 121/160$ and $b = -279/80$

6.14 The operator is linear

6.15 The operator is not a linear

6.16 The functional is linear

6.17 The functional is bilinear

6.18 The functional is neither bilinear nor linear

6.19 The first variation of a quadratic functional is bilinear in u and its variation δu.

6.21 $\phi_0 = h\frac{x}{L}$, $\phi_1 = x(L-x)$ and $\phi_2 = x^2(L-x)$; $\phi_0 = h\sin\frac{\pi x}{2L}$ and $\phi_n = \sin\frac{n\pi x}{L}$

6.22 $\phi_n = x^{n+1}$; $\phi_i = 1 - \cos(2i-1)\frac{\pi x}{2L}$

6.23 $\phi_1 = x(L-x)$ and $\phi_2(x) = x^2\left(\frac{3}{4}L - x\right)$ or $\phi_2(x) = x\left(\frac{3}{4}L^2 - x^2\right)$;
$\phi_i(x) = \sin\frac{(2i-1)\pi x}{L}$

6.24 $\phi_1 = x^2(L-x)$ and $\phi_2 = x^3(L-x)$; $\phi_i = 1 - \cos\frac{i\pi x}{2L}$

6.25 $\phi_1 = x(L-x)$ and $\phi_2(x) = x^2\left(\frac{3}{4}L - x\right)$ or $\phi_2(x) = x\left(\frac{3}{4}L^2 - x^2\right)$;
$\phi_i(x) = \sin\frac{(2i-1)\pi x}{L}$

6.26 $\phi_1 = (x^2 - a^2)(y^2 - a^2)$ and $\phi_2 = (x^2 + y^2)\phi_1$; $\phi_{ij} = \cos(2i-1)\frac{\pi x}{2a}\cos(2j-1)\frac{\pi y}{2a}$

6.27 (a) $\phi_1(x) = (1-x)^2$ (b) $\phi_1(x) = 3 - 4x + x^4$

6.28 $U_2(x) = \frac{f_0}{2T}x(L-x) + \frac{hx}{L}$

6.29 $W_2(x) = \frac{q_0 L^4}{24EI}\left(5\frac{x^2}{L^2} - 2\frac{x^3}{L^3}\right)$

6.30 $W_2 = \frac{q_0 L^4}{96EI}\left(4\frac{x}{L} - \frac{x^2}{L^2} - 4\frac{x^3}{L^4}\right)$

6.31 $W_2 = \frac{F_0 L^3}{64EI}\left(7\frac{x^2}{L^2} - 12\frac{x^3}{L^3} + 5\frac{x^4}{L^4}\right)$

6.32 $W_2 = \frac{q_0 L^4}{96EI}\left(\frac{x}{L} - \frac{x^2}{L^2}\right) + \frac{q_0 L^4}{96EI}\frac{(192EI - kL^3)}{(48EI + kL^3)}\left(\frac{3x}{4L} - \frac{x^3}{L^3}\right)$.

6.33 $U_2(x,y) = \frac{f_0 a^2}{T}\left[0.2922 + 0.0592\left(\frac{x^2}{a^2} + \frac{y^2}{a^2}\right)\right]\left(1 - \frac{x^2}{a^2}\right)\left(1 - \frac{y^2}{a^2}\right)$

6.34 $U_2(x) = \frac{1}{131}(186x - 75x^2 + 20x^3)$

6.36 (a) $\Delta = 12(EI)^2 L^4 + \frac{k^2 L^{12}}{25200} + \frac{11}{105}EIkL^8$, $c_1 = \frac{q_0 L^3}{12\Delta}\left(6EIL^3 + \frac{kL^7}{420}\right)$,
and $c_2 = \frac{q_0 L^3}{12\Delta}(0)$

6.37 $A_{ij} = EI\frac{ij(i+1)(j+1)}{i+j-1}(L)^{i+j-1}$ and $b_i = \frac{q_0}{i+2}(L)^{i+2}$

6.38 $\hat{N}_{cr} = 9.87\frac{EI}{L^2}$

6.40 $W_N(x,t) = \frac{4q_0}{\pi}\sum_{n=1}^{N}\frac{1}{nK_n}(1 - \cos\lambda_n t)\sin\frac{n\pi x}{L}$ for n odd

6.41 $V_2(\xi,\tau) = c_1(\tau)\xi + c_2(\tau)\xi^2$

6.43 $P_{cr} = 12\frac{EI}{L^2}$ for $N = 1, 2$, and $P_{cr} = 9.875\frac{EI}{L^2}$ for $N = 3$

6.44 For $N = 1$: $\lambda = \frac{5.8303}{a^2}$; for $N = 2$: $\lambda_1 = \frac{5.7897}{a^2}$

6.45 $U_2(x) = x(1-x)\left(\frac{71}{369} + \frac{7}{41}x\right)$

6.46 $W_1(x) = 0.8111(1-x)(2-x)$ and $U_1(x) = 0.8111(1-x)(2-x) + x$

6.47 $U_1(x,y) = \frac{3}{8}(1-x)(1-y)$; $U_2(x,y) = -\frac{8}{137}(1-x)(1-y) + \frac{95}{274}(1-x^2)(1-y^2)$

6.48 $U_2(x) = 0.46574(2x - x^2) - 0.07944(3x - x^3)$

6.49 $U_{2G}(x) = -0.08833x^2 + 0.8127x - 0.2120x^3$

6.50 $W_1(x) = \frac{5q_0 L^4}{1248}\frac{x^2}{L^2}\left(\frac{x^2}{L^2} - 4\frac{x}{L} + 6\right)$

6.51 $U_2(x,y) = (1-y)\sin \pi x - \frac{\pi}{8}\sin \pi x \sin \pi y + \frac{\pi}{24}\sin 2\pi x \sin 2\pi y$

6.52 (a) $U_1(x) = \frac{\sqrt{2}}{4}\left(3+x^2\right)$ (b) $U_1 = 2\sqrt{2} - \sqrt{3} + (\sqrt{3} - \sqrt{2})x^2$

6.53 $\lambda_1 = 4.121$ and $\lambda_2 = 25.479$; $\lambda_1 = 4.1158$ and $\lambda_2 = 24.1393$

6.54 $U_N(x,y) = \frac{16 f_0}{\pi^4}\sum_{i,j=1,3,\ldots}\frac{1}{ij(i^2+j^2)}\sin\frac{i\pi x}{a}\sin\frac{j\pi y}{b}$

6.55 $U_1 = -\frac{\sqrt{2}}{4} + \sqrt{\frac{15}{8}} + \left(\frac{5}{4}\sqrt{2} - \sqrt{\frac{15}{8}}\right)x^2$

6.56 $W_1(x) = \frac{q_0 L^4}{24 EI}\left(\frac{x^2}{L^2} - \frac{x}{L}\right)^2$

6.57 $U_2(x) = 0.4714x - 0.608x^2 + 0.2482x^3$

6.58 $U_1 = \sqrt{2} - 0.326 + 0.326x^2$

6.59 $\lambda_1 = 4.212$ and $\lambda_2 = 34.188$

6.61 $W_3(x) = \frac{q_0 L^4}{36 a_0}\left(9\frac{x^2}{L^2} - \frac{x^3}{L^3}\right)$

6.62 $U_1 = \left(-2\sqrt{2} + \sqrt{6}\right)\left(1 - x^2\right) + \sqrt{2}$

6.63 $\lambda_1 = 3.785$ and $\lambda_2 = 19.753$

6.64 $U_1 = \frac{f_0}{2\pi^2}\sin \pi x \sin \pi y$

6.65 $U_1(x,y) = e^{-\sqrt{10}y}(x - x^2)$

6.66 $U_2(x,y) = \frac{16 f_0 a^2}{27\pi^3}\left[27\left(\frac{1-\cosh \alpha_1 y}{\cosh \alpha_1 a}\right)\cos \alpha_1 x + \left(\frac{1-\cosh \alpha_3 y}{\cosh \alpha_3 a}\right)\cos \alpha_3 x\right]$

6.67 $W_1(x,t) = c_1(0)\cos \alpha t\,(1 - \cos 2\pi x)$

6.68 $\lambda_1 = \frac{\pi^2}{(2a)^2} + \frac{10}{(2b)^2}$

6.69 $\lambda_1 = \frac{\pi^2}{(2a)^2} + \frac{9.871}{(2b)^2}$

6.70 $\lambda_1 = 15\left(\frac{1}{a^2} + \frac{1}{b^2}\right)$

Chapter 7

7.1 $\frac{1}{r}\frac{d^2}{dr^2}\left(D_{11}r\frac{d^2 w}{dr^2}\right) - \frac{1}{r}\frac{d}{dr}\left(\frac{D_{22}}{r}\frac{dw}{dr}\right) - q = 0$

7.3 $D\left[\frac{1}{r}\frac{\partial}{\partial r}\left(r\frac{\partial}{\partial r}\right) + \frac{1}{r^2}\frac{\partial^2}{\partial \theta^2}\right]\left[\frac{1}{r}\frac{\partial}{\partial r}\left(r\frac{\partial w}{\partial r}\right) + \frac{1}{r^2}\frac{\partial^2 w}{\partial \theta^2}\right] + kw = q$

7.4 $w(r) = \frac{M_a a^2}{2(1+\nu)D}\left(1 - \frac{r^2}{a^2}\right)$

7.5 $w = \frac{q_0 a^4}{64 D}\left\{\frac{(5+\nu)D + \beta a}{(1+\nu)D + \beta a} - 2\left[\frac{(3+\nu)D + \beta a}{(1+\nu)D + \beta a}\right]\left(\frac{r}{a}\right)^2 + \left(\frac{r}{a}\right)^4\right\}$

$M_{rr} = \frac{q_0 a^2}{16}\left[(1+\nu)\frac{(3+\nu)D + \beta a}{(1+\nu)D + \beta a} - (3+\nu)\left(\frac{r}{a}\right)^2\right]$,

$M_{\theta\theta} = \frac{q_0 a^2}{16}\left[(1+\nu)\frac{(3+\nu)D + \beta a}{(1+\nu)D + \beta a} - (1+3\nu)\left(\frac{r}{a}\right)^2\right]$

7.7 $c_2 = -\frac{2q_1 a^2}{(1+\nu)}\left(\frac{4+\nu}{45}\right)$ and $c_4 = \frac{q_1 a^4}{(1+\nu)}\left(\frac{6+\nu}{150}\right)$

7.8 $c_2 = -\frac{12 q_0 a^2}{576}$ and $c_4 = \frac{2q_0 a^4}{576}$

7.10 $c_2 = -\frac{2q_1 a^2}{45}$ and $c_4 = \frac{3q_1 a^4}{450}$

7.11 $W_1(r) = \frac{q_1 a^4}{30(1+\nu)D}\left(1 - \frac{r^2}{a^2}\right)$

7.12 $W_1(r) = -\frac{q_1 a^4}{45 D}\left(1 - 3\frac{r^2}{a^2} + 2\frac{r^3}{a^3}\right)$

7.13 For $\nu = 0.3$, $\omega = \frac{5.5857}{a^2}\sqrt{\frac{D}{\rho h}}$

7.14 $\lambda_1 = \frac{104.39}{a^4}$

7.17 $w_c = \frac{q_0 a^4}{(1+\nu)D}\left(\frac{5+\nu}{64} - \frac{6+\nu}{150}\right)$

7.18 $w_c = \frac{43}{4800}\frac{q_0 a^4}{D}$

7.19 $w_{cb} = \frac{b^2}{16\pi D}\left(2\log\frac{b}{a} + \frac{a^2}{b^2} - 1\right)$

7.20 $w_{bc} = w(b) = \frac{a^2}{16\pi D}\left[\left(\frac{3+\nu}{1+\nu}\right)\left(1-\frac{b^2}{a^2}\right) + 2\frac{b^2}{a^2}\log\frac{b}{a}\right],$

$w_{cb} = \frac{Q_0 b^2}{16\pi D}\left[\left(\frac{3+\nu}{1+\nu}\right)\left(\frac{a^2}{b^2}-1\right) + 2\log\frac{b}{a}\right]$

7.21 $c_1 = \frac{F_0}{\pi a^2 k}\left[2 + \frac{1}{8}\left(\frac{32D(1+\nu)}{ka^4}+\frac{1}{3}\right)^{-1}\right], \quad c_2 = -\frac{Q_0}{\pi a^4 k}\left(\frac{1}{12} + \frac{8D(1+\nu)}{ka^4}\right)^{-1}$

7.23 $w(a/2, b/2) \approx 0.0028\frac{q_0 a^4}{D}$ and $M_{xx}(a/2,0) \approx 0.08 q_0 a^2$

7.24 $w(a/2,0) \approx 0.00154(q_0 a^4/D)$, $M_{xx}(a/2,0) \approx 0.0223 q_0 a^2$ and $M_{yy}(a/2,0) \approx 0.0268 q_0 a^2$

7.29 $W_1(0.5, 0.5) = \frac{16 q_0 a^4}{[16+8s^2 n^2 + 3s^4 n^4]n\pi^5 D}$

7.30 $c_{11} = \frac{8Q_0 a^4(-1)^{n+1}}{[16+8s^2 n^2+3s^4 n^4]Dn\pi^4 ab}$

7.31 $c_{11} = \frac{q_0 a^4}{4\pi^4 D(3+2s^2+3s^4)}, \quad W_1(0.5,0.5) = \frac{q_0 a^4}{\pi^4 D(3+2s^2+3s^4)}\quad (s=\frac{a}{b})$

7.32 $c_{11} = \frac{Q_0 a^4}{abD\pi^4[3+2s^2+3s^4]}, \quad s = \frac{a}{b}$

7.33 $c_{11} = (q_0 a^4)/(27.984\pi^4 D_0)$ and $w_{max} = q_0 a^4/(6.996\pi^4 D_0) = 0.0014674 q_0 a^4/D_0$

7.34 $c_1 = \frac{\sqrt{3}Q_0 a^2}{108D} = \frac{Q_0 a^2}{36\sqrt{3}D}$

7.36 $w(x,y) = \frac{q_0 a^4(x/a)}{24D(s^4+2s^2+5)}\left(1-\frac{x^2}{a^2}-\frac{y^2}{b^2}\right)^2$

7.37 $w(x,y) = \frac{q_0 a^4}{8D(1+2\nu s^2+s^4)}\left(1-\frac{x^2}{a^2}-\frac{y^2}{b^2}\right)$

7.40 $C_1 = C_2 = C_3 = 0$ and $C_4 = M^C(0)\Omega L^2$

7.41 $C_1 = \frac{3\Omega}{(1+3\Omega)L}M_{xx}^B(0)$, $C_2 = -C_1$, $C_3 = 0$, $C_4 = M_{xx}^B(0)\Omega L^2$

7.42 $w^F(x) = \frac{q_0 L^4}{24EI}\left(\frac{x^4}{L^4} - 4\frac{x^3}{L^3} + 6\frac{x^2}{L^2}\right) + \frac{q_0 L^2}{2K_s GA}\left(2\frac{x}{L} - \frac{x^2}{L^2}\right)$

Chapter 8

8.1 $U_2 = \frac{45}{59}, U_3 = \frac{35}{59}, P_1^{(1)} = \frac{35}{59}$

8.3 $U_2 = 0.2\times 10^{-3}\text{in}, \quad U_3 = -0.3333\times 10^{-3}\text{in}, \quad U_4 = -0.8667\times 10^{-3}\text{in},$
$\sigma_s = 375$ psi, $\sigma_a = -333.33$ psi, $\sigma_b = -500$ psi

8.4 $U_2 = -\frac{P}{\frac{E_c A_c}{h_c} + \frac{E_s A_s}{h_s}} = -0.005062$ in, $\sigma_s = 37.966$ ksi (comp.), $\sigma_c = 20.248$ ksi (comp.)

8.5 $U_2 = 0.4444$ in and $U_3 = .7778$ in

8.6 $U_2 = 0.02047$ in, $\sigma^{(1)} = 10,237$ psi, $\sigma^{(2)} = -38,389$ psi

8.7 $U_2 = \frac{2P}{\frac{E_s A_s}{h_1} + \frac{E_a A_a}{h_2}} = 0.303$ mm, $\sigma_a = -106$ MPa, $\sigma_s = 606$ MPa

8.8 $U_2 = u_B = -0.005$ in, $U_3 = u_C = 0.003$ in

8.9 $U_2 = u_B = 0.005005$ in, $U_3 = u_C = 0.002983$ in

8.10 $U_3 = -0.0063793$ in., $U_4 = 0.0011607$, $U_5 = -0.01032$ in.,
$Q_1^{(1)} = 1,200$ lb, $Q_2^{(1)} = -5,042$ lb-in., $Q_4^{(2)} = -5,758$ lb-in.

8.11 $U_3 = 1.1798$ mm, $U_4 = -0.01486$, $U_6 = -0.06038$; $Q_1^{(1)} = -15.895$ kN, $Q_2^{(1)} = 1.939$ kN-m,
$Q_3^{(2)} = 16.919$ kN.

8.12 $U_2 = 0.001152$, $U_4 = 0.000768$, $U_5 = -0.26266$ in., $U_6 = 0.005088$.

8.13 $U_3 = -\dfrac{q_0 L^4}{3EI(1+\frac{kL^3}{3EI})}$, $\quad U_4 = \dfrac{q_0 L^3}{6EI(1+\frac{kL^3}{3EI})}$

8.14 See the answer to **Problem 8.13** (the two problems are equivalent).

8.16 The assembled (condensed) system of equations for the unknown generalized displacements is

$$\frac{EI}{L^3}\begin{bmatrix} 24+\frac{k_1 L^3}{EI} & 0 & -12 & -6L & -\frac{k_1 L^3}{EI} & 0 \\ 0 & 8L^2 & 6L & 2L^2 & 0 & 0 \\ -12 & 6L & 12+\frac{k_2 L^3}{EI} & 6L & \times & \times \\ -6L & 2L^2 & 6L & 4L^2 & \times & \times \\ -\frac{k_1 L^3}{EI} & 0 & \times & \times & 12+\frac{k_1 L^3}{EI} & 6L \\ 0 & 0 & \times & \times & 6L & 4^2 \end{bmatrix}\begin{Bmatrix} U_1 \\ U_2 \\ U_3 \\ U_4 \\ U_5 \\ U_6 \end{Bmatrix} = -\frac{q_0 L}{12}\begin{Bmatrix} 0 \\ 0 \\ 0 \\ 0 \\ 6 \\ L \end{Bmatrix},$$

where $\times$ indicates that there is no connectivity between the degrees of freedom and zeros indicate that the stiffness coefficients are zero.

8.17 $K_{ij}^e = \int_{x_a}^{x_b} EI \dfrac{d^2\phi_i^e}{dx^2}\dfrac{d^2\phi_j^e}{dx^2}\,dx$, $\quad G_{ij}^e = \int_{x_a}^{x_b} \dfrac{d\phi_i^e}{dx}\dfrac{d\phi_j^e}{dx}\,dx$

8.18 $[G^e] = \dfrac{a}{30h}\begin{bmatrix} 36 & -3h & -36 & -3h \\ -3h & 4h^2 & 3h & -h^2 \\ -36 & 3h & 36 & 3h \\ -3h & -h^2 & 3h & 4h^2 \end{bmatrix}$

8.19 $P_{cr} = \dfrac{240 EI\lambda_2}{L^2} = 9.9439\dfrac{EI}{L^2}$

8.20 $U_3 = -\dfrac{q_0 h^4}{48EI}$, $\quad U_4 = -\dfrac{q_0 h^3}{96EI}$

8.21 $U_3 = \dfrac{q_0 h^2}{4GAK_s}$, $\quad U_4 = -\dfrac{q_0 h^3}{6(GAK_s h^2 + 4EI)} = -\dfrac{q_0 h^3}{24EI}\left[1+\left(\dfrac{h}{H}\right)^2\right]^{-1}$

8.22 $U_3 = -(1+3s^2)\dfrac{q_0 h^4}{48EI}$ and $U_4 = -\left(\dfrac{1+3s^2}{1+0.75s^2}\right)\dfrac{q_0 h^4}{96EI}$, where $s = H/h$ is the element height-to-length ratio.

8.23 $U_3 = \dfrac{q_0 h^2}{2GAK_s + kh}$ and $U_4 = 0$

8.24 $K_{ij}^e = \int_{r_a}^{r_b} D\left[\dfrac{d^2\varphi_i}{dr^2}\dfrac{d^2\varphi_j}{dr^2} + \dfrac{\nu}{r}\left(\dfrac{d\varphi_i}{dr}\dfrac{d^2\varphi_j}{dr^2} + \dfrac{d^2\varphi_i}{dr^2}\dfrac{d\varphi_j}{dr}\right) + \dfrac{1}{r^2}\dfrac{d\varphi_i}{dr}\dfrac{d\varphi_j}{dr} + k\varphi_i\varphi_j\right]r\,dr$, $\quad q_i^e = \int_{r_a}^{r_b} q\varphi_i\,r\,dr$

8.26 $K_{ij}^e = \int_{\Omega^e}\left(c_{11}\dfrac{\partial\psi_i}{\partial x}\dfrac{\partial\psi_j}{\partial x} + c_{12}\dfrac{\partial\psi_i}{\partial x}\dfrac{\partial\psi_j}{\partial y} + c_{21}\dfrac{\partial\psi_i}{\partial y}\dfrac{\partial\psi_j}{\partial x} + c_{22}\dfrac{\partial\psi_i}{\partial y}\dfrac{\partial\psi_j}{\partial y} + c_0\psi_i\psi_j\right)dxdy$, $\quad f_i^e = \int_{\Omega^e}\psi_i^e f\,dxdy$, $\quad Q_i^e = \oint_{\Gamma^e}\psi_i^e q_n^e\,ds$

Chapter 9

9.1 $-\frac{1}{k}\mathbf{v} + \nabla u = 0$, $\quad -(\nabla\cdot\mathbf{v}+\mathbf{f}) = 0$, $\quad \hat{\mathbf{n}}\cdot\mathbf{v} - q = 0$ on Γ_2, $\quad u - \hat{u} = 0$, on Γ_1

9.3 $\mu(u_{i,j}+u_{j,i})n_j - Pn_i = 0$ on Γ_1, $\quad \mu(u_{i,j}+u_{j,i})n_j - Pn_i - \hat{t}_i = 0$ on Γ_2

9.5 $\Pi_{HR}(u,\varepsilon_{xx},\mu,\lambda) = I(u) + \int_0^L \lambda\left(\varepsilon_{xx} - \dfrac{du}{dx}\right)dx + \mu(0)\left[u(0)-u_0\right]$

9.6 $I_L(u_i,\lambda) = I(u_i) + \int_V \lambda u_{i,i}\,dV$, $\lambda = -P$

9.7 Replace Π in Eq. (9.2.18) with
$$\Pi(u_i,\varepsilon_{ij}) = \int_V \{\rho\Psi(\varepsilon_{ij},T) - f_i u_i\}\,dV - \int_{\Gamma_2}\hat{t}_i u_i\,ds$$

9.9 $H(u_i,\varepsilon_{ij},\sigma_{ij},T) = W(u_i,\varepsilon_{ij},\sigma_{ij},T) + \int_\Omega\left[\dfrac{1}{2}k_{ij}\dfrac{\partial T}{\partial x_i}\dfrac{\partial T}{\partial x_j} - QT\right]dV$

9.11 The matrix coefficients are

$K_{ij}^{11} = \int_{r_a}^{r_b} A_{55}\dfrac{d\psi_i^{(1)}}{dr}\dfrac{d\psi_j^{(1)}}{dr}\,rdr$, $\quad K_{ij}^{12} = \int_{r_a}^{r_b} A_{55}\dfrac{d\psi_i^{(1)}}{dr}\psi_j^{(2)}\,rdr$,

$K_{ij}^{22} = \int_{r_b}^{r_b}\left[\left(D_{11}\dfrac{d\psi_j^{(2)}}{dr} + D_{12}\dfrac{\psi_j^{(2)}}{r}\right)\dfrac{d\psi_i^{(2)}}{dr} + \dfrac{1}{r}\left(D_{12}\dfrac{d\psi_j^{(2)}}{dr} + D_{22}\dfrac{\psi_j^{(2)}}{r}\right)\psi_i^{(2)} + A_{55}\psi_i^{(2)}\psi_j^{(2)}\right]rdr$,

$F_i^1 = \int_{r_a}^{r_b} q\psi_i^{(1)}\,rdr$, $\quad F_i^2 = 0$.

9.13 The element coefficients are ($\varphi_i^{(1)}$ and $\varphi_i^{(2)}$ are weight functions)

$K_{ij}^{11} = 0$, $\quad K_{ij}^{12} = -\int_{x_a}^{x_b} \varphi_i^{(1)} \frac{d\phi_j}{dx}\, dx$, $\quad K_{ij}^{21} = -\int_{x_a}^{x_b} \varphi_i^{(2)} \frac{d\psi_j}{dx}\, dx$, $\quad K_{ij}^{22} = \frac{1}{a}\int_{x_a}^{x_b} \varphi_i^{(2)} \phi_j\, dx$,

$F_i^1 = \int_{x_a}^{x_b} f\varphi_i^{(1)}\, dx$, $\quad F_i^2 = 0$.

In general, the weighted-residual finite element model is not symmetric

9.14 $K_{ij}^{11} = \int_{x_a}^{x_b} \frac{d\psi_i}{dx}\frac{d\psi_j}{dx}\, dx$, $\quad K_{ij}^{12} = -\int_{x_a}^{x_b} \frac{1}{a}\frac{d\psi_i}{dx}\phi_j\, dx = K_{ji}^{21}$

$K_{ij}^{22} = \int_{x_a}^{x_b}\left[\frac{1}{a^2}\phi_i\phi_j + \frac{d\phi_i}{dx}\frac{d\phi_j}{dx}\right] dx$, $\quad F_i^1 = 0$, $\quad F_i^2 = -\int_{x_a}^{x_b}\frac{d\phi_i}{dx} f(x)\, dx$

$\mathbf{K}^{11} = \frac{1}{h_e}\begin{bmatrix} 1 & -1 \\ -1 & 1 \end{bmatrix}$, $\quad \mathbf{K}^{12} = \frac{1}{2a_e}\begin{bmatrix} 1 & 1 \\ -1 & -1 \end{bmatrix}$, $\quad \mathbf{K}^{22} = \frac{h_e}{6a_e^2}\begin{bmatrix} 2 & 1 \\ 1 & 2 \end{bmatrix} + \frac{1}{h_e}\begin{bmatrix} 1 & -1 \\ -1 & 1 \end{bmatrix}$

9.15 $K_{ij}^{11} = 0$, $\quad K_{ij}^{12} = \int_{\Omega_e}\frac{\partial\psi_i}{\partial x}\psi_j\, dxdy$, $\quad K_{ij}^{13} = \int_{\Omega_e}\frac{\partial\psi_i}{\partial y}\psi_j\, dxdy$, $\quad K_{ij}^{21} = K_{ji}^{12}$,

$K_{ij}^{22} = -\int_{\Omega_e}\frac{1}{k}\psi_i\psi_j\, dxdy$, $\quad K_{ij}^{23} = 0$, $\quad K_{ij}^{31} = K_{ji}^{13}$, $\quad K_{ij}^{32} = 0$, $\quad K_{ij}^{33} = K_{ij}^{22}$,

$F_i^1 = \int_{\Gamma_e}\psi_i q_n\, ds$, $\quad F_i^2 = 0$, $\quad F_i^e = 0$

Chapter 10

10.1 Let $(0.5 + z/h) = \xi$; $\int_{-h/2}^{h/2} f(z)\, dz = \int_0^1 \xi^n (hd\xi) = \frac{h}{n+1}\left[\xi^{n+1}\right]_0^1 = \frac{h}{n+1}$

10.7 $\hat{\mathbf{C}}\mathbf{\Delta} + \hat{\mathbf{M}}\ddot{\mathbf{\Delta}} = \mathbf{F}$; $\quad F_1 = 0$, $F_2 = 0$, $F_3 = Q_{mn}$, $\Delta_1 = U_{mn}$, $\Delta_2 = V_{mn}$, $\Delta_3 = W_{mn}$

$\hat{c}_{11} = (A_{11}\alpha^2 + A_{66}\beta^2)$, $\quad \hat{c}_{12} = (A_{12} + A_{66})\alpha\beta$, $\quad \hat{c}_{13} = -B_{11}\alpha^3 - (B_{12} + 2B_{66})\alpha\beta^2$,

$\hat{c}_{22} = (A_{66}\alpha^2 + A_{22}\beta^2)$, $\quad \hat{c}_{23} = -(B_{12} + 2B_{66})\alpha^2\beta - B_{22}\beta^3$,

$\hat{c}_{33} = D_{11}\alpha^4 + 2(D_{12} + 2D_{66})\alpha^2\beta^2 + D_{22}\beta^4$, $\quad \hat{m}_{11} = \hat{m}_{22} = m_0$,

$\hat{m}_{33} = m_0 + m_2\left(\alpha^2 + \beta^2\right)$, $\quad \alpha = m\pi/a$, $\beta = n\pi/b$

10.8 $\hat{\mathbf{S}}\mathbf{\Delta} + \hat{\mathbf{M}}\ddot{\mathbf{\Delta}} = \mathbf{F}$; $\quad F_1 = 0$, $F_2 = 0$, $F_3 = Q_{mn}$, $F_4 = 0$, $F_5 = 0$,

$\Delta_1 = U_{mn}$, $\Delta_2 = V_{mn}$, $\Delta_3 = W_{mn}$, $\Delta_4 = X_{mn}$, $\Delta_5 = Y_{mn}$

$\hat{s}_{11} = (A_{11}\alpha^2 + A_{66}\beta^2)$, $\quad \hat{s}_{12} = (A_{12} + A_{66})\alpha\beta$, $\quad \hat{s}_{14} = (B_{11}\alpha^2 + B_{66}\beta^2)$,

$\hat{s}_{15} = (B_{12} + B_{66})\alpha\beta$, $\quad \hat{s}_{22} = (A_{66}\alpha^2 + A_{22}\beta^2)$, $\quad \hat{s}_{24} = \hat{s}_{15}$, $\quad \hat{s}_{25} = (B_{66}\alpha^2 + B_{22}\beta^2)$,

$\hat{s}_{33} = K(A_{55}\alpha^2 + A_{44}\beta^2)$, $\quad \hat{s}_{34} = KA_{55}\alpha$, $\quad \hat{s}_{35} = KA_{44}\beta$,

$\hat{s}_{44} = (D_{11}\alpha^2 + D_{66}\beta^2 + KA_{55})$, $\quad \hat{s}_{45} = (D_{12} + D_{66})\alpha\beta$,

$\hat{s}_{55} = (D_{66}\alpha^2 + D_{22}\beta^2 + KA_{44})$, $\quad \hat{m}_{11} = m_0$, $\quad \hat{m}_{22} = m_0$, $\quad \hat{m}_{33} = m_0$,

$\hat{m}_{44} = m_2$, $\quad \hat{m}_{55} = m_2$

Index